Perspectives on Animal Behavior

Perspectives on Animal Behavior

Third Edition

Judith Goodenough
Biology Department / University of Massachusetts, Amherst

Betty McGuire
Department of Ecology and Evolutionary Biology / Cornell University, Ithaca

Elizabeth Jakob
Psychology Department / University of Massachusetts, Amherst

John Wiley & Sons, Inc.

VICE PRESIDENT AND EXECUTIVE PUBLISHER Kaye Pace
SENIOR ACQUISITIONS EDITOR Kevin Witt
EDITORIAL PROGRAM ASSISTANT Alissa Etrheim
EDITORIAL ASSISTANT Jenna Paleski
PRODUCTION SERVICES MANAGER Dorothy Sinclair
SENIOR PRODUCTION EDITOR Janet Foxman
EXECUTIVE MARKETING MANAGER Clay Stone
CREATIVE DIRECTOR Harry Nolan
SENIOR DESIGNER Kevin Murphy
PHOTO EDITOR Sheena Goldstein
SENIOR ILLUSTRATION EDITOR Anna Melhorn
SENIOR MEDIA EDITOR Linda Muriello
PRODUCTION SERVICES Cécile Billioti de Gage/ Preparé
COVER DESIGN M77 Design
COVER PHOTO Richard Packwood/Getty Images, Inc.

This book was set in 10/12 Janson by Preparé and printed and bound by Courier/Kendallville. The cover was printed by Courier/Kendallville.

This book is printed on acid-free paper. ∞

To order books or for customer service, please call 1-800-CALL WILEY (225-5945).

Library of Congress Cataloging-in-Publication Data
Goodenough, Judith.
Perspectives on animal behavior / Judith Goodenough, Betty McGuire, Elizabeth Jakob.—3rd ed.
p. cm.
Includes bibliographical references (p.).
ISBN 978–0–470–04517–6 (cloth)
1. Animal behavior. I. McGuire, Betty. II. Jakob, Elizabeth M. III. Title.

QL751.G59 2010
591.5—dc22 2009013360

Printed in the United States of America

10 9 8 7 6 5 4 3 2 1

To my husband, Stephen Goodenough,
my daughters Aimee and Heather,
and my mother, Betty Levrat.
J. G.

In loving memory of James Patrick McGuire.
To Willy, Kate, and Owen Bemis,
and to Dora, Kevin, and Cathy McGuire.
B. M.

To Adam Porter and Margaret Dringoli, with love and gratitude.
E. J.

my husband, Stephen Goodnough,
my daughters Agnes and Heather,
and my mother, Betty Lewis.

James Patrick McQuaid

Preface

Outside the window, male ruby-throated hummingbirds newly returned from wintering in Central America are brightening this early May day in the northeastern United States. The remarkable annual migration of these tiny birds is a behavioral story in its own right, complete with a long, foodless crossing of the Gulf of Mexico to return to breeding areas they left here more than six months before. But today it is the behavior of the males that is the most immediate–they are claiming or in some cases reclaiming feeding territories through dramatic aerial displays and battles. In a week or so the females will arrive and the interactions will further intensify, as the breeding season gets under way. Watching and understanding these behaviors offers ways to engage nature in new and deeper ways, and it places our own lives in the context of the other animals that share the planet with us.

This brief foray into the easily observed lives of hummingbirds illustrates a simple point: animal behavior is engaging to watch, as evidenced by the popularity of nature shows on both small and big screens. In fact, those of us fortunate enough to offer a course in animal behavior are often told by our colleagues how lucky we are to teach a subject that people of all backgrounds find fascinating. There is something very appealing in learning about the often bizarre lives going on around us, sometimes literally under our noses.

Besides satisfying our natural curiosity, the study of animal behavior is truly interdisciplinary. To deeply understand an animal's behavior, we must look both inward to its genetic, morphological, and physiological underpinnings, outward to its environment, and back in time to its evolutionary origins. Thus, animal behavior draws together fields of study that might often seem, especially to a student early in his or her academic career, to have little to do with one another. Our hope is that readers of this book will see that the facts learned in different courses are connected after all.

The authors of this third edition share a deep interest in animal behavior, from the standpoints of teaching, writing, and research. Yet each brings to this project a very different perspective. Judith Goodenough, from the Department of Biology at the University of Massachusetts Amherst, has led the charge from the very first edition of this text. Judith has studied biological rhythms in creatures from planaria to finches to deer mice (not to mention in the green alga *Chlamydomonas*). Betty McGuire, from the Department of Ecology and Evolutionary Biology at Cornell University, who wrote eight chapters for the first edition of the book but stepped down from the second edition because of other obligations, returns to this third edition. Betty's research focuses on parental behavior, reproduction, and ecology of small mammals, and she occasionally dabbles in work with larger domestic animals such as dogs and horses. In addition to *Perspectives in Animal Behavior*, Judith and Betty have coauthored *Human Biology: Personal, Environmental, and Social Concerns* and *Biology of Humans: Concepts, Applications, and Issues*. Elizabeth Jakob, from the Department of Psychology at the University of Massachusetts Amherst, joins the team of authors. Beth studies the behavior of spiders, asking questions about their learning, perception, and interactions with conspecifics and with other species. She has carried out field projects in California, Mexico, Massachusetts, and Maine, and she also does laboratory experiments when the questions or the long Massachusetts winters demand it. She coauthored the fifth edition of *Animal Behavior: Mechanisms, Ecology, Evolution* and co-edited the laboratory manual *Learning the Skills of Research: Laboratory and Field Exercises in Animal Behavior*. Collectively, the three of us have taught courses ranging from large lectures for nonmajors to upper division courses in animal behavior, vertebrate biology, and evolution.

Our goals and basic approach remain the same as in the first two editions: to help students understand the history, mechanisms, development, function, and evolution of behavior. To this end we introduce the major approaches to the study of animal behavior and show how these diverse approaches can be integrated to provide a more complete understanding of any particular behavior. Because courses in animal behavior are offered in biology, psychology, and animal science departments, we include basic background material-for example, in genetics, neurophysiology, and endocrinology–to ensure that each chapter can stand alone and that each contains information accessible to students with different backgrounds. Within each chapter, the examples we develop provide a balance between classic and recent studies. Throughout the text we use clear and engaging writing to explain potentially complex topics such as behavioral genomics and mathematical models of behavior.

In organization, this third edition begins in much the same way as previous editions, with an introductory chapter on hypothesis testing, followed by Part 1 on approaches to the study of animal behavior. These early chapters present students with the questions asked and methods used by scientists working in subfields of animal behavior such as behavioral genetics, behavioral endocrinology, and behavioral ecology. Part 2 focuses on how behavior relates to the survival of individuals and includes chapters, for example, examining how animals find places to live, feed, and avoid being eaten. The chapters in Part 3 focus on interactions between individuals, such as those between mates, parents and offspring, competitors, or collaborators. We also describe communication in detail before moving on to exploring how it is used in conflict and cooperation.

CHANGES AND UPDATES TO THE THIRD EDITION

We added new features to this edition to promote critical thinking, active learning, and the development of vocabulary. We retained certain features from previous editions, such as the outlines at the start of each chapter and the summary at the end, to help students organize material and identify key concepts.

NEW STOP AND THINK QUESTIONS

Each chapter has at least one Stop and Think Question designed to encourage students to pause from reading the text, think about the information they have just read, and apply it to a new situation. Many of these questions focus on experimental design, methodology, or ethics.

NEW BOLD TERMS AND GLOSSARY

Key terms in each chapter are now set in bold type when first formally introduced. These terms and others of lesser importance are included in the new glossary at the end of the book.

INCREASED EMPHASIS ON ALTERNATIVE HYPOTHESES

To increase our coverage of the ways in which scientists design and carry out research in animal behavior, we added many new in-depth examples to illustrate the importance of developing and testing alternative hypotheses. These examples will help students understand how research is conducted and allow them to better design their own experiments and to evaluate the primary literature of animal behavior.

INCREASED COVERAGE OF SEVERAL KEY AREAS

In response to current events and reviewer comments, we increased our coverage of the growing interactions between the fields of animal behavior and conservation biology. In particular, we include examples that demonstrate how studies of behavior yield information critical to conservation efforts. We also address ethical issues in behavioral research, and on the flip side, we examine how behavioral research provides information essential to improving the lives of captive animals. This edition also includes more examples from the fields of human behavior and applied animal behavior.

Acknowledgments

Our editor, Kevin Witt, had confidence that three academics with very different schedules could work together to produce a current text with a single voice. We thank him for his encouragement and support. Many other dedicated people at John Wiley & Sons helped get this book into your hands. Alissa Etrheim, our editorial program assistant, helped us put this book together. She provided gentle reminders to keep us on schedule and always offered to help in any way possible. We are particularly grateful for her help with permissions. The text was improved by the skills of our copy editor, Betty Pessagno. Anna Melhorn, Senior Illustration Editor, and Sheena Goldstein, our photo editor, are largely responsible for the appearance of the text. We also thank Will Sillin for his beautiful illustrations that begin each chapter and each of the three parts of the book. Elaine Soares, our photo researcher, worked diligently to find the perfect photographs for our needs. Cécile Billioti de Gage/Preparé headed production of our book. She helped set priorities and successfully moved the project along even at the busiest times of the academic year. We are now happy to pass the project to the capable hands of Lucy Parkinson, our marketing director. Each person is a professional, a pleasure to work with, and a team player.

We thank the reviewers who provided essential feedback that helped shape this new edition:

Mitchell Baker, Queens College
Zane Barlow, University of Massachusetts Amherst
Willy Bemis, Cornell University
Renae Brodie, Mount Holyoke College
Bruce Byers, University of Massachusetts Amherst
Ethan Cloftfelter, Amherst College
Fiona Cross, University of Canterbury
Perri Eason, University of Louisville
Sarah Huber, Randolph-Macon College
Brian Kelly, University of Massachusetts Amherst
Chris Maher, University of Southern Maine
Sarah Partan, Hampshire College
Mark Petrie, University of Canterbury
Jeff Podos, University of Massachusetts Amherst
Denise Pope, Mount Holyoke College
Adam Porter, University of Massachusetts Amherst
Nancy Solomon, Miami University of Ohio
Theodore Stankowich, University of Massachusetts Amherst
Zoran Tadic, University of Zagreb
Christina Veino, University of Massachusetts Amherst
Paige Warren, University of Massachusetts Amherst
Gordon Wyse, University of Massachusetts Amherst

We are also grateful to the many authors who were generous enough to check over our portrayal of their work. Any remaining errors are, of course, ours.

FROM JUDITH GOODENOUGH

I would like to thank many of my friends and colleagues at UMass for lending their expertise to various aspects of this book. Zane Barlow's comments on genetics greatly improved the chapter. As always, Gordon Wyse provided valuable suggestions on the nervous system. Adam Porter helped smooth the writing of the most difficult sections. My e-mail friend from Croatia, Zoran Tadic, Faculty of Science, University of Zagreb, provided useful comments and a fresh perspective on many chapters. Although she did not help with writing, Margaret Ludlam helped me cope when things got tense, as they often did. She is always cheerful and calm. Most of all, I would like to thank my family who have always been my cheering squad, and who have encouraged and supported me at every stage of this revision. My husband Steve bore the brunt of my obsession. After months of my spending many more hours with the computer than with him, Steve merely quipped that he was getting rather fond of the back of my head and continued to do what was necessary to keep the household running smoothly. He's the funniest person I know, and his wit kept me sane. After 38 years of marriage, he still keeps me on my toes by introducing me as his *first* wife. My daughters, Aimee and Heather, still the joy of my life, remind me that the people you love must always come first. They'll find time to spend with me when they know I need a break. Their courage and confidence in reaching for their dreams encourages me to keep stretching toward mine—the most immediate of which is the completion of this revision. My mother, Betty Levrat, instilled in me a love of learning and was always willing to do whatever needed to be done to free time for my writing.

FROM BETTY MCGUIRE

I thank my husband, Willy Bemis, for his continuous encouragement throughout this project and willingness to read (and edit) everything I passed his way. My daughter Kate let us use some of her wonderful photographs in the third edition–at 16 she already has a keen eye and understanding of animal behavior. I thank my son, Owen, who always reminds me, by his near continuous recitation of animal facts, how fascinating animals are to all who watch them. My mother, Dora McGuire, and siblings, Kevin and Cathy McGuire, provided refuge at the Jersey shore where I could escape the deadlines and e-mail (not to mention the weather in Ithaca), and relax and reconnect for a few days. My father, James McGuire, died during the writing of this edition. He was an avid birder and a dedicated volunteer at his local animal shelter, and I know that he would have enjoyed our new text. Finally, I thank Lowell Getz, close friend and colleague of almost 25 years, for listening patiently and always being there. We will get those manuscripts done, Lowell, I promise!

FROM BETH JAKOB

I thank my husband, Adam Porter, who was incredibly supportive throughout this process, was always ready to read something and offer his frank opinion and detailed edits, and saved our credit rating by taking over paying the bills when I was too distracted to notice them. I began work on my section of the book during my sabbatical leave, and I thank the faculty, staff, and students of my host institution at the University of Canterbury, Christchurch, New Zealand for their help and friendship, Robert Jackson for his sponsorship, and the Fulbright Foundation for support. I am grateful for the patience of my graduate students and collaborators whose manuscripts and paperwork languished during the more intense periods of book preparation. Penny Jaques, Betsy Dumont, Joe Elkinton, and my other colleagues in the Organismic and Evolutionary Biology Program kept me sane and at least somewhat on track when I foolishly agreed to be Graduate Program Director when this book was underway. I am grateful to additional friends and family for forgiving my incessant conversation, deep sighs, and continuous e-mails about the book and for reminding me that there are other things beyond my computer screen, particularly Doreen Jakob, Alexis Jakob, Carolyn Truini, Jack Dringoli, Margaret Dringoli, Perri Eason, Marta Hersek, Chris Maher, Maggie Hodge, and Nancy Reynolds. Now it's finally time to get back outside!

Contents

1
Introduction

A New Caledonian crow picks up a twig, bends it into a hook with her foot, pokes it into a hole, and pulls out an insect. A male wolf spider does a rhythmic courtship dance, waving his tufted legs as he approaches an attentive female. An albatross spreads its massive wings, lifts off into the ocean breezes, and does not touch land again for two years. Two male rattlesnakes entwine themselves in a wrestling match, settling their dispute without resorting to a venomous bite. And an emperor penguin, huddled on the ice in the endless blizzards of an Antarctic winter, forgoes food for months while incubating the egg delicately balanced on his feet.

The behavior of animals is featured on endless nature shows and even the occasional big-screen blockbuster for a reason—we find fascinating these glimpses into the worlds of the other creatures on our planet. How does the world appear to them? Do they think and feel like we do? How did such strange behaviors come to exist?

At this point in human history, we have the luxury of studying animal behavior for curiosity's sake. In the past, our interest in it was based on more practical needs. In one way or another, people have been studying animal behavior for thousands of years. The most skillful hunters and fishermen are usually those who can make predictions about the behavior of their prey (Figure 1.1). It is important to know that when salmon are spawning, they will not respond to a fisherman's bait; that many rodents escape toward the dark, whereas most birds escape toward the light; and that many kinds of animals will fight, some ferociously, if they are trapped.

The study of animal behavior may have occupied the fringes of human consciousness for centuries for just such a practical reason. Later, when animals were domesticated and put to work, it was necessary to learn new things about them. Horses could be trained for riding or for pulling wagons or tools. Dogs could be trained to track prey or to protect individual humans; cats could not.

In time, the study of animal behavior took on new dimensions. The goals, as well as the techniques, changed. Animals are no longer studied simply so that we can exploit them more efficiently, although this may still be one reason for our attention. Now, however, we have become aware that increased knowledge of the behavior of specific species in their natural habitats may help us save some endangered groups from extinction. In addition, information on their normal behavior may help us ensure their welfare, not just in the wild, but also in laboratories or zoos (Blumstein et al. 2004; Sutherland et al. 1998; Swaisgood 2007). We may be interested in

FIGURE 1.1 **People have been studying animal behavior for centuries, sometimes for very practical reasons. Knowledge of the behavior of game species may make it easier to put food on the table.**

behavior as an example of a broader intellectual concern, such as evolutionary theory. Or we may be interested in studying animal behavior because it may serve as a model to help us understand human behavior. And, as we said before, sometimes we are fortunate enough to be able to study animal behavior simply because our curiosity prompts us to ask questions about some of the organisms with which we share the earth.

FOUR QUESTIONS ABOUT ANIMAL BEHAVIOR

As casual observations of animal behavior crystallized into a field of scientific study, Niko Tinbergen (1963) identified four types of questions that should be asked about behavior: What are the mechanisms that cause it? How does it develop? What is its survival value? How did it evolve? Tinbergen believed that ethology—the biological study of behavior—should "give equal attention to each of them and to their integration." Tinbergen's four questions are sometimes condensed into two categories: "how" questions, which focus on causation and development, and "why" questions, which focus on function (survival value) and evolution. It has also been suggested that consideration of the mechanisms of behavior should, at least in some cases, include both cognitive and emotional mental processes (Emery and Clayton 2005).

To better appreciate the types of questions we may ask about animal behavior, consider a familiar phenomenon: the seasonal migration of songbirds between northern and southern latitudes. As new birds appear daily at backyard feeders in early spring, we may each become curious about migration, but depending on our personal interests, each of us may ask different questions. How do they "know" it is time? How do they find their way? Such questions focus on the mechanisms that underlie the behavior. Must those making this journey for the first time learn the route from experienced travelers? Do they inherit a directional tendency from their parents? Questions such as these concern development. Why do they do it? What do they gain that outweighs the risks and demands of such a journey? These are questions about the survival value, or adaptiveness of migration. Finally, how did it all begin? Were the advancing glaciers responsible? Were the migratory paths modified during the thousands of years each species has been migrating? These questions center on the evolution of the behavior. So we see that when we ask *why* an animal behaves in a certain way, some of us may be asking about immediate causes (the machinery underlying the response) and others may be asking about the evolutionary causes.

No one type of question is better than the others. Answers to all types are necessary as we weave the fabric of our understanding. These are not competing avenues of investigation. Rather, they are complementary. Each may feed back on the others, deepening our understanding and broadening our avenues of investigation (Armstrong 1991; Halpin 1991; Stamps 1991).

ANIMAL BEHAVIOR AS AN INTERDISCIPLINARY STUDY

Marion Stamp Dawkins (1989) has drawn an analogy between Tinbergen's (1963) four aspects of investigation of behavior and the four legs of an animal. An animal that lacks one of its legs can only hobble along. Similarly, progress in the study of animal behavior is hampered by a lack of information in any one of these areas of study. This is not to imply that each investigator must ask all types of questions. Often we find that individuals are more excited by one type of question than by others. However, each investigator will be more successful in finding the answer to the question of personal interest if he or she is armed with information and techniques from all four areas of study.

THE INTERPLAY OF QUESTIONS: A CASE STUDY

To illustrate the richness of a multidimensional approach, let's consider the dwarf mongoose (*Helogale parvula*)—an appealing and active animal that has some very unusual behavior. Dwarf mongooses are the smallest African carnivores (meat eaters), weighing only 300–340 g (11–12 ounces) and only about 43 cm (16 in.) long, half of which is tail (Figure 1.2).

What makes dwarf mongooses so unusual is that they live in social groups. Mongoose packs have around 9 adults and yearlings (Creel and Waser 1994; Rood 1990), but may contain up to 40 individuals. Because we are such social creatures ourselves, sociality may not strike us as particularly unusual, but in fact it is quite rare. Most carnivores, in fact, are solitary and find each other only when it is time to mate.

Dwarf mongooses don't simply live as close neighbors, but they take social behavior to an extreme: most of the breeding is done by a dominant female (reviewed in Creel 2005). Over 70% of the pregnancies in one long-term study area were by dominant females. The top-ranking female also gets priority access to food and initiates any movements that the pack undertakes. The rest of the pack falls into a dominance hierarchy, each with their own place in the chain of command.

In spite of the fact that subordinate animals low in the hierarchy rarely get to personally breed, they help the dominant pair raise their offspring. They baby-sit, attack predators, drive away intruding mongoose groups, and warn others of a predator's approach. If a subordinate female does give birth, she will nurse the young of the top-ranking female along with her own, even though she has fewer young than the dominant female. The efforts of these helpers allow the breeders to raise more offspring than they could without help: in fact, unaided breeding pairs are almost never able to raise their offspring to independence.

FIGURE 1.2 **The dwarf mongoose lives in groups in which the members cooperate in raising the young. The dominant male and female are often the only group members that breed. Reproduction by other group members is usually suppressed. However, other high-ranking individuals are sometimes allowed to breed. The variability of reproductive suppression raises many "why" questions. The answers illustrate the interaction of physiological, behavioral, and evolutionary mechanisms.**

In Chapter 4 we will explore in detail how natural selection works, but even without more in-depth background you can easily imagine why the observation that mongooses give up some of their own chances to have offspring in favor of helping others is intriguing to animal behaviorists. This raises a host of "how" and "why" questions that involve both the mechanism and the evolution of behavior. Let's begin by examining some questions about evolution.

Evolutionary Questions About Dwarf Mongoose Behavior

When we study the evolution of behavior, we can take several different approaches. First, we can ask about the current costs and benefits of a behavior. Evolution, as we will see later in the book, is all about reproductive success, so an evolutionary approach to asking questions about mongoose behavior concerns how behavior affects the production of offspring. We can also look deeper into history, and study how a behavior first originated in mongoose ancestors.

First let's think about why dwarf mongooses might currently benefit from living in groups. If you have been lucky enough to watch dwarf mongooses in the wild or in a zoo, one of the behaviors you will notice is their constant vigilance—they stand on their tiptoes and peer alertly around. Their seeming paranoia is justified: dwarf mongooses are small and thus an appetizing prey for many other carnivores. This leads us to a hypothesis for why dwarf mongooses live in groups: to reduce the risk of predation. And in fact, researchers have found that by living in groups, dwarf mongooses benefit by each other's vigilance and by sheer safety in numbers (if a predator is going to grab a single mongoose and run off, it's better to be in a group than all alone). More than half of mongooses that venture off on their own are eaten.

Are there costs to being in a group? Certainly. As we discuss in Chapter 19, being in a group means facing competition for all sorts of resources. For dwarf mongooses, however, sharing one particularly important resource is not a problem—food. Dwarf mongooses feed primarily on arthropods—crickets, grasshoppers, termites, spiders, and scorpions—and there are plenty to go around. So, one of the biggest potential problems with being in a group isn't too important for dwarf mongooses.

So, there are plenty of advantages to being in a group for dwarf mongooses. But what about those subordinates that don't get to reproduce much, if at all, in a group? Do the costs of losing the chance to reproduce outweigh the benefits of being in a group? Let's approach this question in a way animal behaviorists often do: let's carefully outline the choices available to a subordinate mongoose. One option is to leave the original pack with a few other subordinates and together form a new pack, where it may rank higher in the hierarchy. Or it can go off and try to join an already existing pack—the chances of a new immigrant getting an opportunity to breed are fairly good, especially if the incoming mongoose manages to drive off the resident breeders (Rood 1990). However, the problem with either of these choices is that half of dispersers die in the process, generally getting eaten by predators.

Alternatively, a subordinate can stay in the pack and hope for an improvement in its social rank, either by slowly gaining status over its lifetime or by benefiting from a tragedy that befalls the dominant mongoose. In addition, subordinates do have some chance at breeding themselves. In one study, 12% of subordinate mongooses became pregnant, and DNA fingerprinting revealed that 15% of the offspring in a pack had subordinate mothers and 25% had subordinate fathers (Keane et al. 1994). The chances of subordinates getting to reproduce are especially good when extra food is around, as researchers have shown by providing food supplements. (Creel and Waser 1997). Finally, even dwarf mongooses that don't breed can help their genes pass on to the next generation by helping to raise their relatives. Thus, for any given mongoose, the current costs and benefits of its different options may vary, and indeed different individuals make different choices.

Using a different set of techniques, we can also investigate the evolutionary origins of dwarf mongoose social behavior. As we will see in many other examples later in this book, to study evolutionary origins usually involves the construction of a phylogeny, sort of a family tree of a taxonomic group. Usually we are not fortunate enough to have a good fossil record, so we have to use other techniques. Veron et al. (2004) used both gene sequences and behavioral traits of the 37 species of mongooses that are alive today in order to construct a phylogeny. Their work suggests that the ancestor of modern-day mongooses was solitary, lived in a forest, and ate vertebrates. When the environment changed during the Pliocene, some mongoose lineages switched to eating insects. Insectivory made sociality more likely because competition for food was reduced.

Proximate Questions About Dwarf Mongoose Behavior

Besides questions about current costs and benefits and the evolutionary history of mongoose social behavior, we can also ask about the mechanisms that underlie the behavior. For example, we can ask a simple question: which characteristics make an animal likely to be dominant? In one study, the dominant female was without exception the oldest in a pack. Within an age class, heavy mongooses were more likely to be dominant (Creel 2005).

We can also ask more complex and interesting questions about the interactions among mongooses in a pack. Up until now, you may have the impression that subordinate mongooses quietly give up their reproductive capabilities, but in fact reproduction is a point of contention. Dominant mongooses can reproductively suppress subordinates (Creel et al. 1992, 1995).

In general, reproductive suppression can be accomplished through either chemical or behavioral means. In the first way, chemicals released by dominant individuals, perhaps in urine or feces, suppress reproductive development or function in subordinates. In the second way, dominant individuals behave aggressively toward subordinates who attempt to breed. Sometimes both chemicals and aggressive behavior play a role.

Reproductive suppression in male and female dwarf mongooses involves different mechanisms (Creel et al. 1992). In males, reproductive suppression of subordinates is accomplished entirely through behavioral means. The dominant male attacks other males to keep them from mating with fertile females. Levels of androgens (male hormones) of subordinate and dominant males are indistinguishable. In females, suppression can occur through hormonal as well as behavioral controls. The ability to breed depends on the female's peak level of the hormone estrogen. Low-ranking subordinates have low estrogen levels compared to the breeder. However, high ranking, older subordinates have estrogen levels similar to that of the breeder, and have a better chance of breeding if they disperse. As a result, the top-ranking female must ease her behavioral suppression of the older females to keep them as helpers.

Why might both kinds of reproductive suppression evolve for females but not for males? The answer may lie in the certainty of parentage and the ultimate means of suppressing reproduction by subordinates—infanticide. The top female is likely to be able to identify her own offspring by odor cues learned at the time of birth. However, the young of a single litter can have different fathers, so a male cannot easily determine which of them are his. If resources become too scarce to support the young of a subordinate, a dominant can kill them. Because a female can recognize her own young, she can selectively kill those of subordinates. A dominant male practicing the same infanticidal policy would risk killing some of his own offspring. So, a top-ranking female will allow reproduction of subordinate females who are most likely to leave and breed elsewhere because she can veto that decision later if need be. Thus, the variation in reproductive suppression in dwarf mongooses results from an interplay between the mechanisms that control it and evolution.

The dwarf mongoose studies illustrate the extent to which proximate and ultimate questions about behavior are intertwined. In the pages that follow, we will consider aspects of the behavior of other species in light of all four of Tinbergen's questions. First, we will introduce some of the approaches to the study of animal behavior—genetics, evolution, ecology, learning, neurobiology, endocrinology, and development. We will discuss some of the basic principles and techniques used in each approach. Then, in the chapters that follow, we will consider specific types of behavior, including orientation, foraging, antipredator defense, mating, parental care, and other social behavior, and see how the various perspectives may act synergistically to broaden our understanding of animal behavior. Before we begin, however, we should take a moment to consider how scientists pose questions about animal behavior and then go about answering them.

FIGURE 1.3 **Male burrowing owls scatter mammal manure around the entrance of their nest. There are at least four alternative hypotheses for this behavior: attracting mates, signaling that the nest burrow is occupied, camouflaging the scents of an active owl nest, and attracting prey. A different set of predictions accompanies each hypothesis, and each prediction can be tested.**

HYPOTHESIS TESTING

The study of animal behavior usually begins with an observation that prompts a question, which is followed by forming **hypotheses** (logical guesses) about a possible answer. It is necessary to be able to test each hypothesis. Generally, the hypothesis leads to a prediction, which will support the hypothesis if it holds true when tested. Depending on the hypothesis, the test may involve further observations, comparisons of behavior among species, or experimental manipulation.

Different hypotheses can sometimes lead to identical predictions, and then both hypotheses are supported or refuted, depending on the outcome of the test. In this event, it is necessary to make other predictions that will allow us to reject one of the hypotheses.

There may also be alternate hypotheses to explain a particular behavior. If so, each hypothesis could lead to different predictions, and each prediction would be tested by observations and experiments.

Studies of an unusual nesting practice of burrowing owls provide an example of an observation leading to several hypotheses for the function of the behavior. Burrowing owls (*Athene cunicularis*) live and nest in desert grasslands in Canada, the western United States, and some parts of South America (Figure 1.3). The male prepares the underground nest burrow by digging and scraping out dirt or by taking over a burrow of another small mammal, such as a prairie dog or kangaroo rat. The unusual part of nest building is that the male lines the tunnel leading to the nest cup with 3 to 7 cm of mammal manure and then scatters manure around the entrance to the nest.

The observation of manure scattered around the nest entrance prompts the question, "What is the function of manure around the nest opening?" Matthew Smith and Courtney Conway (2007) developed and tested several hypotheses to answer this question.

Mate attraction The first hypothesis was that males scatter manure to attract females. Perhaps females use the manure to assess a male's quality, much like female bowerbirds use the objects in a male's bower to assess his quality. Only males scatter manure, which is consistent with the hypothesis that the manure attracts females.

Sign that burrow is occupied Suitable nest sites are limited, and resident males defend their site vigorously. Perhaps the manure signals to other males that the burrow is occupied. Nonresident males could then avoid the costs of fighting by looking elsewhere for a burrow.

Olfactory camouflage Common predators of burrowing owls include coyotes, badgers, and skunks, which are predators that use odors to locate prey. By masking the scents of an active owl nest, the manure may reduce the risk of predators detecting the owls within.

Prey attraction Burrowing owls eat what is available: mice, voles, toads, small birds, insects, spiders, and centipedes. Indeed, small arthropods make up a large part of their diet. Perhaps the manure attracts small arthropods to the owl nest, creating a home-delivery service for owl nestlings.

As you can see in Table 1.1, a different set of predictions accompanies each of these hypotheses, and each

TABLE 1.1 Hypotheses, Predictions, and Tests of the Function of Manure Scattering by Burrowing Owls

Prediction	Hypothesis 1: Mate attraction	Hypothesis 2: Burrow occupied	Hypothesis 3: Olfactory Camouflage	Hypothesis 4: Prey attraction	Test	Result	Conclusion
1. Stage of nesting cycle when manure is collected should be:	Before pair formation	Soon after arrival	Just before incubation	All stages	Estimated date of pair bond formation and dates of manure scattering	Scattering began after pair bond formation	Contradicts mate-attraction, burrow-occupied, and olfactory camouflage hypotheses
2. Presence of manure at traditional nest before owls return from migration discourages nesting by other owls	Yes/no	Yes	No	No	Scatter manure around entrance of some burrows; remove manure from some burrows of previous years	58% of nests with manure became nests; 78% of nests with removed manure became nests	Pattern in direction of burrow-occupied hypothesis but results are not statistically significant
3. Increased perception of competition increases manure scattering	Yes/no	Yes	No	No	Experimental group = presented with mount of burrowing owl and tape of its primary call; control group = presented with mount of European starling of its primary call; Measure amount of manure spread	Resident male more likely to attack an owl mount than starling mount; Increased scattering after presentation of owl mount than of European starling	Pattern in direction of prediction of burrow-occupied prediction but results are not statistically significant
4. More surviving young at nests with manure	No	No	No	Yes	Count number of surviving offspring in manure-supplemented and manure-removed nests	Manure at entrance of nest made no difference in the number of surviving young	Supports prey-attraction hypothesis Contradicts olfactory-camouflage hypothesis
5. Fewer predatory attacks in natural nests with manure	No	No	Yes	No	Select active nests and add or remove manure	Probability of predatory attacks was not altered by treatment	Contradicts olfactory-camouflage hypothesis
6. Fewer predatory attacks in artificial nests with manure	No	No	Yes	No		Probability of predatory attacks was not altered by treatment	Contradicts olfactory-camouflage hypothesis
7. Greater arthropod biomass in nests with manure	No	No	No	Yes	Compare biomass of arthropods at nest sites with and without manure and pitfall traps with and without manure	Pitfall traps with manure collected more arthropod biomass than did pitfall traps without manure	Supports prey-attraction hypothesis

prediction can be tested. For example, each hypothesis makes a different prediction about the timing of manure scattering. If the manure attracts a mate, then one would predict that manure should be present before pair bonding occurs. However, if the manure signals that the burrow is occupied, then one would predict that manure should be scattered soon after the owls return from migration, well before pair bonding. On the other hand, if manure masks the odor of the owl nest to lower the risk of predation, then one would predict that manure scattering occurred later, just before incubation. If the function of manure is to attract arthropod prey, then one would predict that scattering should increase after pair formation and be most common when nestlings are present in the nest.

The obvious test of these predictions is to find out when manure scattering takes place. The usual course of events is that single males arrive at the nesting grounds first, followed by previously mated pairs, and finally single females. Pair bonding occurs when the females arrive. The female incubates seven to ten eggs for 28 to 30 days until they hatch. She continues to sit on the brood for another week or two. During this entire time, the male brings food to the nest. For about the next six weeks, both adults feed the young until they can hunt for themselves.

Smith and Conway (2007) observed burrows every two to four days to determine the number of adults and juveniles present, when manure scattering took place, and whether any owls were killed by predators. In 87% of the 46 observed burrows, manure scattering occurred after pair bond formation, which is inconsistent with the mate-attraction and burrow-occupied hypotheses and best fits the prey-attraction hypothesis.

The burrow-occupied hypothesis predicts that a nonresident male would be less likely to enter a burrow if manure was scattered at the entrance. To test this prediction, Smith and Conway added or removed manure from burrows that had been used in the previous two years. They visited the nests twice a week to see which nests were used. Whereas 78% of the nests from which manure had been removed became nests, only 58% of the burrows with added manure became nests. This result is not statistically significant, but is in the direction that would be predicted by the hypothesis that the manure functions as an indicator that the burrow is occupied, suggesting that it is biologically meaningful.

A second prediction of the burrow-occupied hypothesis is that the amount of manure scattered would increase with perceived competition. Smith and Conway tested this prediction by presenting each resident male with a mount of a burrowing owl accompanied by a recording of its primary call and, separately, a mount of a European starling accompanied by a recording of its primary call. On three separate occasions, the calls were played for ten minutes or until the resident male attacked the mount. Before the first presentation of calls, all scattered material was removed from around the burrow entrance. After the third presentation, all scattered material was again collected and weighed. As predicted by the burrow-occupied hypothesis, resident males were more likely to attack a mount of a burrowing owl, which would be perceived as a threat, than they would a European starling mount. Furthermore, after three presentations of a burrowing owl mount, resident males scattered more manure than they did after three presentations of a European starling mount. However, once again, these results are in the right direction but are not statistically significant. Recall that the third prediction, that manure scattering will begin soon after arrival at the nesting site, was not met; manure scattering actually occurred several weeks after arrival.

The olfactory-camouflage hypothesis predicts that the amount of manure scattered around the burrow entrance will influence the risk of predation. Smith and Conway tested this prediction on active natural nests and on artificial nests. They randomly selected 26 active nest sites and added or removed manure from the nest entrances every two to four days. The artificial nests were created by baiting unoccupied burrows with chicken eggs and assigning each to one of four treatments: horse manure and signs of owls, such as feathers and pellets; coyote scat plus signs of owls; no manure or scat but signs of owls; no manure or scat or signs of owls. If the manure at the nest entrance functions to hide the scent of an active owl nest, then one would predict less predation on natural or artificial burrows with manure or scat around the entrance. This prediction was not supported; there was no difference in the probability of predation between nests with manure and those without manure.

The first prediction of the prey-attraction hypothesis is that manure attracts small arthropod species that burrowing owls typically eat. To test this prediction, Smith and Conway created 75 sampling areas, each containing two treatment sites—one with manure and a control without manure. Each sampling area also contained three pitfall traps to collect any small arthropods that were attracted to and approached the treatment site. As predicted, the average biomass (dry weight) of arthropods collected was higher at manure sites.

If more small arthropods are attracted to nests with more manure, it follows that a second prediction of the prey-attraction hypothesis is that manure increases the survival rate of nestlings by attracting arthropods to supplement the food brought to the nest by the parents. The data do not support this prediction. There was no significant difference in the number of young surviving in nests with or without manure. Furthermore, there were slightly more survivors in nests without manure, which is in the opposite direction of the prediction.

These studies point out the usefulness of testing alternate hypotheses and alternate predictions from a single hypothesis. The data do not support the hypothesis that scattered manure functions to attract a mate. Other data, though not statistically significant, suggest that manure may function as a sign that a burrow is occupied: burrows with manure were 36% less likely to become nests than burrows without manure, and the perception of competition increased manure scattering by resident males. Although the timing of manure scattering fits the olfactory-camouflage hypothesis, the failure of manure to increase survival of the young does not. The hypothesis that the function of manure scattering is to attract prey to the nest burrows is supported by both the timing of manure scattering and the increased amount of prey biomass trapped near nests with manure compared to nests without manure. However, the number of young surviving in nests was not increased by the presence of manure, which does not at first appear to be consistent with the prey-attraction hypothesis. But perhaps attracting prey to the nest allows juveniles to learn to handle prey near the safety of the burrow, or perhaps attracting prey to the nest means that the parents can make fewer foraging trips.

STOP AND THINK

Some males use materials such as feathers, grass, or dried moss to scatter at the nest entrance. Which hypothesis or hypotheses would be consistent with this observation?

STOP AND THINK

There are other possible hypotheses for the function of manure scattering at the nest entrance by male burrowing owls. Mammal manure is abundant in areas where burrowing owls nest, so perhaps manure is good nest-building material, and some gets dropped around the entrance by sloppy males. Perhaps manure serves as insulation or absorbs water to prevent flooding in the nest. How would you test these hypotheses? Can you think of others? How would you test them?

Populations of burrowing owls are declining so precipitously that they are listed as endangered in some states and as threatened in others. Learning more about their nesting behavior in general and about the function of manure scattering in particular may provide insight into whether or how the disappearance of large grazing mammals from prairies has caused the owl population to decline (Smith and Conway 2007). A study performed in Oklahoma revealed that the declining population of burrowing owls is related to the elimination of prairie dogs (*Cynomys ludovicianus*). In this region of the country, most burrowing owls made nests in abandoned prairie dog burrows. Unfortunately, prairie dogs are not welcomed guests on many ranches because they eat some of the available food and because their burrow holes may pose a risk to livestock. As prairie dogs disappear, so do their burrows. Fewer prairie dog burrows mean fewer nesting burrows for burrowing owls, so their numbers are also declining. To increase the number of burrowing owls, it would seem that the prairie dog population must also be maintained (Butts and Lewis 1982).

In the pages that follow, we will explore various perspectives on animal behavior, sometimes stopping to consider how the information might be used to serve animal welfare or conservation. We will also see many examples of hypothesis testing. Keep these general procedures in mind as you read them. Instead of passively accepting the given explanation for a particular behavior, be critical of the evidence. Try to think of alternate hypotheses for the behaviors described. Make predictions and design your own tests of those predictions.

SUMMARY

Animal behavior is studied for many reasons, both practical and intellectual. A full understanding of animal behavior requires answers to four types of questions, those about (1) immediate mechanisms, (2) development, (3) survival value, and (4) evolution. Our progress in understanding the behavior of animals will be enhanced by considering all four types of questions.

The study of animal behavior usually begins with an observation that prompts a question. The next step is to think of tentative explanations, called hypotheses, to answer that question. Each hypothesis should produce testable predictions. The tests of those predictions support or refute the hypothesis.

PART ONE

Approaches to the Study of Animal Behavior

2

History of the Study of Animal Behavior

To understand a field of study today, we must know something about its past. In this chapter, we consider the history of the study of animal behavior. Our focus is on the development of key concepts in the field.

THE BEGINNINGS

INTELLECTUAL CONTINUITY IN THE ANIMAL WORLD

It is difficult, perhaps impossible, to pinpoint the precise beginnings of the study of animal behavior. Rather than attempting this feat, we will simply consider some of the highlights in the development of the discipline. One idea, that of intellectual continuity among animals, was important in shaping some of the earliest views of animal behavior. Although the idea of intellectual continuity was summarized in 1855 by Herbert Spencer in his book *Principles of Psychology* (Spencer 1855), its roots can be traced back much further in history, to the ideas of the ancient Greek philosophers. The concept focused on continuity in mental states between "lower" and "higher" animals and was based on a picture of evolution similar to Aristotle's *scala naturae*, the great chain of being. In the *scala naturae*, the evolution of species was viewed as linear and continuous. This classification system was hierarchical, with animals ranked according to their degree of relationship, the highest point of evolution being humans (Hodos and Campbell 1969). At the bottom of the scale were creatures such as sponges; then further up the scale were insects, fish, amphibians, reptiles, birds, nonhuman mammals, and finally humans. Because evolution was seen as a linear process, with each higher species evolving from a lower one until, finally, humans emerged, it was thought that the animal mind and the human mind were simply points on a continuum.

DARWIN'S EVOLUTIONARY FRAMEWORK

A few years after the publication of Spencer's book, Charles Darwin (Figure 2.1) published his thoughts on evolution by natural selection in *On the Origin of Species* (1859). Although Darwin's focus in this book was not on animal behavior, his ideas provided a conceptual framework within which the field of animal behavior could develop. Discussed in more detail in Chapter 4, these ideas can be briefly summarized as follows:

1. Within a species, there is usually variation among individuals.
2. Some of this variation is inherited, and is passed on from mother to offspring.
3. Most of the offspring produced by animals do not survive to reproduce. Some individuals survive longer and produce more offspring than others, because of their particular inherited characteristics.

Natural selection is the differential survival and reproduction of individuals that results from genetically based variation in their behavior, morphology, physiology, and so on. Evolutionary change occurs as the heritable traits of successful individuals (i.e., those that survive and reproduce) are spread throughout the population, whereas those traits of less successful individuals are lost.

FIGURE 2.1 Charles Darwin. His ideas on evolution by natural selection provided an evolutionary framework for the study of animal behavior.

In two later books, *The Descent of Man, and Selection in Relation to Sex* (1871) and *Expression of the Emotions in Man and Animals* (1873), Darwin applied his evolutionary theory to behavior. In these volumes he recorded his careful and thorough observations on animal behavior, but his records were anecdotal and often anthropomorphic. This was not, however, sloppy science. In the tradition of his day, he believed that careful observations were useless unless they were connected to a general theory. Darwin's general theory was evolution by natural selection. Because humans evolved from other animals, he considered the minds of humans and animals to be similar in kind and to differ only in complexity. As a reflection of this belief, he described the behavior of animals by using terms that denote human emotions and feelings: Ants despaired, and dogs expressed pleasure, shame, and love. Darwin's opinion was influential, and it is not surprising that others interested in animal behavior followed his lead. Both ethologists and comparative psychologists trace the beginnings of their respective fields to the ideas of Darwin.

For about a decade after the publication of *Expression of the Emotions in Man and Animals* (1873), descriptions of animal behavior usually took the form of stories about the accomplishments of individual animals that were believed to think and experience emotions as humans do. For example, based on his subjective interpretation of what he observed, George J. Romanes, a protégé of Darwin, constructed a table of emotions that charted the evolution of the mind and listed the emotions in order of their historical or evolutionary appearance (Figure 2.2). In his books, *Animal Intelligence* (1882), *Mental Evolution in Animals* (1884), and *Mental Evolution in Man* (1889), Romanes examined the implications of Darwinian thinking about the continuity of species for the behavior of nonhuman and human animals.

In addition to Romanes, several other scientists made notable contributions to the study of animal behavior at the turn of the twentieth century. Jacques Loeb (1918) believed that all patterns of behavior were simply "forced movements" or tropisms, physiochemical reactions toward or away from stimuli. Herbert Spencer Jennings, perhaps best known for his book *Behavior of the Lower Organisms* (1906), disagreed with Loeb's ideas and instead emphasized the variability and modifiability of behavior. Of course, there were many other pioneers in the study of animal behavior, but we will move on to the turmoil that developed in the discipline during the twentieth century.

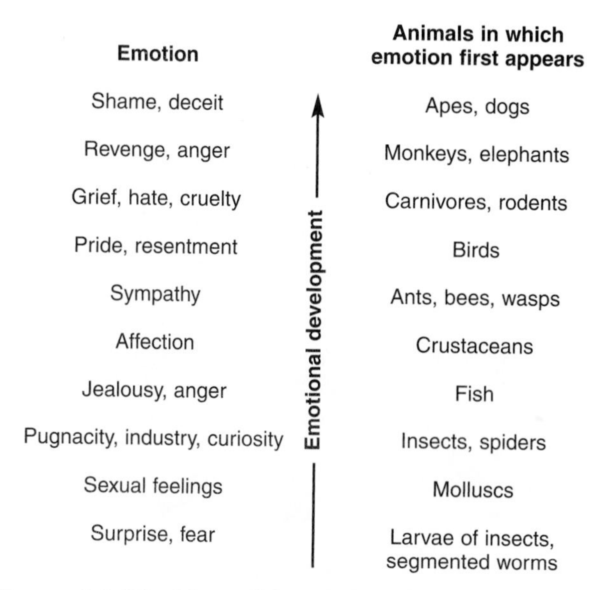

Emotion
Animals in which emotion first appears
Shame, deceit
Apes, dogs
Revenge, anger
Monkeys, elephants
Grief, hate, cruelty
Carnivores, rodents
Pride, resentment
Birds
Sympathy
Ants, bees, wasps
Affection
Crustaceans
Jealousy, anger
Fish
Pugnacity, industry, curiosity
Insects, spiders
Sexual feelings
Molluscs
Surprise, fear
Larvae of insects, segmented worms
Emotional development

FIGURE 2.2 **The ideas of George J. Romanes on the evolutionary appearance of emotions in animals. (Modified from Romanes 1889.)**

As interest in animal behavior grew, differences in opinion developed. These differences eventually led to the development of two major approaches to the study of animal behavior, ethology and comparative psychology, centered in Europe and the United States, respectively. The split that developed between the two approaches seemed at times quite severe. Indeed, one must wonder why two groups of scientists, each striving for a greater understanding of the marvels of animal behavior, could have stood worlds apart. The European ethologists and the American comparative psychologists were separated by more than the Atlantic Ocean. In fact, the basic questions they asked about animal behavior were different. (Recall from Chapter 1 that the four questions outlined by Niko Tinbergen in 1963 concerned the mechanism, function, development, and evolution of behavior.) Whereas the ethologists focused their attention on the evolution and function of behavior, the comparative psychologists concentrated on the mechanism and development of behavior. Because they asked different questions, the types of behavior they studied and even their experimental organisms differed. Whereas ethologists, by and large, studied innate behavior in birds, fish, and insects, comparative psychologists, particularly those of the behaviorist school (see discussion later in chapter), emphasized learned behavior in mammals such as the Norway rat. Furthermore, to determine the normal function of a behavior, ethologists often attempted to observe the animal in its natural habitat or in environments designed to simulate that habitat. On the other hand, comparative psychologists believed that learning was best studied in the laboratory, where variables could be controlled. Finally, ethologists were interested in differences between the species, whereas comparative psychologists searched for general "laws" of behavior.

Of course there were many exceptions to this characterization. Some ethologists made remarkable discoveries indoors in their homes and laboratories and explored the influence of experience on behavioral development (Barlow 1989), and some comparative psychologists studied a wide range of species and patterns of behavior and conducted their studies in the field (Dewsbury 1989). Indeed, within each field there were individuals interested in all four questions of animal behavior. Although traditionally ethology and comparative psychology have been portrayed as very different approaches, some accounts of the history of animal behavior have tended to downplay the differences between them (e.g., Dewsbury 1984).

CLASSICAL ETHOLOGY

THE APPROACH: EVOLUTIONARY, COMPARATIVE, DESCRIPTIVE, FIELD-ORIENTED

"Why is that animal doing that?" is perhaps the fundamental question of ethology, the approach to the study of behavior founded largely by Konrad Lorenz, Niko Tinbergen, and Karl von Frisch, European zoologists who shared the Nobel Prize in medicine and psychology in 1973 (Figure 2.3). Traditionally, **ethology** concentrated on the evolutionary basis of animal behavior. Because natural selection can act only on traits that are genetically determined, it seems a logical outcome of the ethologists' basic interest in the evolution of behavior to focus on those behavior patterns that are inherited. An emphasis on phylogeny (the evolutionary history of a species) is particularly characteristic of the work of Konrad Lorenz.

The studies of ethologists often involve comparisons among closely related species. In the words of Lorenz (1958), "Every time a biologist seeks to know why an organism looks and acts as it does, he must resort to the comparative method." Here Lorenz was referring to the method employed by comparative anatomists when they ask the same question about morphology. If comparative anatomists were to ask why a whale's flipper is structured as it is, they might compare the skeleton of the flipper with that of the forelimb of other vertebrates. They could then see that the typical vertebrate forelimb has been specialized for the aquatic life of this mammal. Similarly, if one were to wonder why a male fly of the species *Hilara sartor* spins an elaborate silken balloon to present to a female before mating (Figure 2.4), the significance of the behavior would become apparent after comparing it with the mating behavior of the other species of flies in the family Empididae. Let us consider the gift-giving behavior of *H. sartor* in more detail to illustrate the ethologist's comparative method.

Among the empidid flies are species that display almost every imaginable evolutionary step in the progression toward the balloon display. By observing the manner in which the male empidid fly approaches the female for mating, one sees that at the heart of the problem is the fact that the male may be a meal, rather than a mate, for the predacious female. In one species, *Empis trigramma*, the male approaches the female while she is eating. Because her mouth is already full, his well-timed approach increases his chances of surviving the encounter. In another species, *E. poplitea*, the male captures a prey, perhaps a fly, and gives it to the female, providing her with a meal before attempting to copulate. Males of the species *H. quadrivittata* gift-wrap the meal in a silky cocoon before offering it to the female. In another species, *H. thoracica*, the cocoon, or case, is large and elaborate, but the food inside is small and of little value. In yet another species, *H. maura*, only some males place food inside cocoons; others enclose something meaningless, such as a daisy petal. Finally, the males of *H. sartor* present the females with an empty gauze case that turns off the predatory behavior of the female, thereby allowing them to mate (Kessel 1955). Long-tailed dance fly males (*Rhamphonyia sulcata*) usually offer a female a genuine nuptial gift. Natasha LeBas and Leon

FIGURE 2.3 **Konrad Lorenz (above), Niko Tinbergen (top right), and Karl von Frisch (right), three ethologists who shared the Nobel Prize in 1973.**

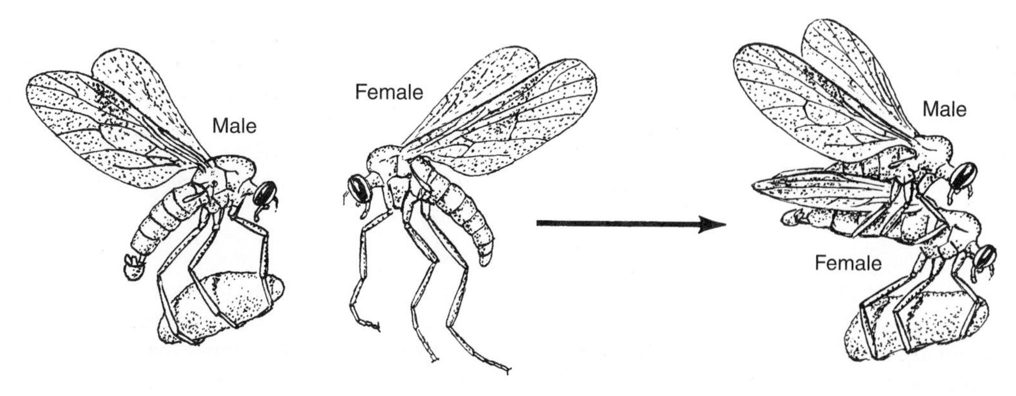

FIGURE 2.4 **Male flies of the species *Hilara sartor* present females with an empty silken balloon before mating. The evolution and function of this behavior can be understood by comparing the behavior of closely related species, that is, by using the comparative method characteristic of the ethological approach. (Drawn from descriptions in Kessel 1955.)**

Hockham (2005) altered the gift by filling it with a large prey, small prey, or a worthless nonedible token. Although females did copulate longer with males offering the largest prey, the females copulated for the same amount of time with males bearing small prey and those bearing token gifts. Thus, male cheaters can invade the population. Without a comparison of the behavior with that of other species, an observer would be hard pressed to explain why males offer silken balloons to females.

In addition to utilizing the comparative method, ethologists often work in the field rather than in the laboratory. After all, they reason, it is in the natural setting that the normal context in which the behavior is displayed is apparent. From this setting, the function of the behavior may be deduced, and knowledge of the function may allow us to understand why the behavior has been shaped to its present form by natural selection. Tinbergen and his students conducted much of their research in the field. Lorenz and his followers, on the other hand, studied captive animals, but they often attempted to simulate in captivity some characteristics of the animal's natural habitat (Barlow 1991). As we will see in Chapter 16, von Frisch's carefully designed field experiments reveal that scout honeybees communicate the distance and direction of a rich food source to recruits by "dancing" within the hive.

CLASSICAL ETHOLOGICAL CONCEPTS

The Fixed Action Pattern

At the turn of the twentieth century, Charles Otis Whitman of the University of Chicago and Oskar Heinroth of the Berlin Aquarium were pioneering the field of ethology (Lorenz 1981). Both scientists were interested in the behavior of birds, and each independently concluded that the displays of different species are often exceptionally constant. In fact, they considered these patterns of movement to be just as reliable as morphological characters in defining a particular group.

The stereotyped patterns of behavior that intrigued ethologists such as Whitman, Heinroth, and Craig were named fixed action patterns by Lorenz. By definition, a **fixed action pattern** (FAP) is a motor response that is initiated by some environmental stimulus but that can continue to completion without the influence of external stimuli. For example, Lorenz and Tinbergen (1938) showed that a female greylag goose (*Anser anser*) will retrieve an egg that has rolled just outside her nest by reaching beyond it with her bill and rolling it toward her with the underside of the bill (Figure 2.5). If the egg is experimentally removed once the rolling behavior has begun, the goose will continue the retrieval response until the now imaginary egg is safely returned to the nest. We have emphasized the fact that once initiated, FAPs continue to completion. There is little consensus, however, on their defining attributes. Other characteristics that have been used to describe them include the following: (1) the sequence of component acts of an FAP is unalterable, (2) an FAP is not learned, (3) it may be triggered under inappropriate circumstances, and (4) it is performed by all appropriate members of a species (Dewsbury 1978).

The concept of a fixed action pattern was questioned in the years following Lorenz's first introduction of the term. George Barlow (1968) suggested that, in reality, most patterns of behavior are not as stereotyped as the notion of the FAP suggests, and furthermore, most cannot easily be separated into fixed and orientation components. He suggested the alternative term *modal action pattern* (MAP). In specific cases, however, the term *fixed action pattern* may be appropriate. Finley and colleagues (1983) examined the courtship displays of mallard ducks (*Anas platyrhyncos*) and concluded that the patterns of behavior were indeed as highly stereotyped as suggested by the notion of FAP. We will continue to use the

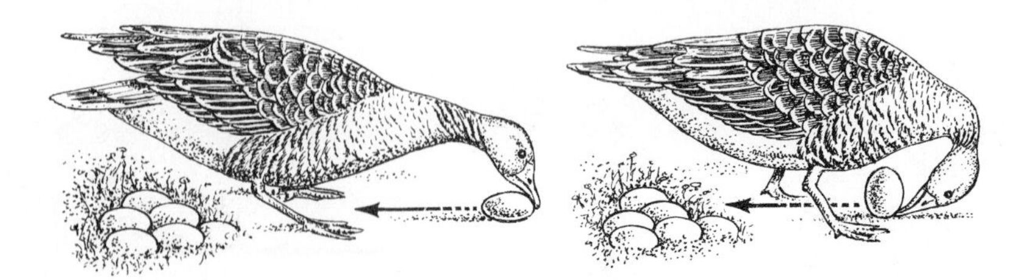

FIGURE 2.5 The egg retrieval response of the greylag goose. The chin-tucking movements used by the female as she rolls the egg back to the nest are highly stereotyped and are an example of a fixed action pattern. The side-to-side movements that correct for any deviations in the path of the egg are called the taxis component of the response. If the egg is removed, the female will continue to roll an imaginary egg back to the nest. One defining characteristic of a fixed action pattern is that it will continue to completion even in the absence of guiding stimuli. (Drawn from a photograph in Lorenz and Tinbergen 1938.)

traditional term here, keeping in mind that there is some debate over the appropriateness of its use.

Sign Stimuli and Releasers as Triggers

A fixed action pattern is obviously produced in response to something in the environment. Let's consider the nature of the stimulus that might trigger the behavior. Ethologists called such a stimulus a **sign stimulus**. If the sign stimulus is emitted by a member of the same species, it is called a social releaser or, simply, a **releaser**. Releasers are important in communication among animals, as we will see in Chapter 18. Although releasers are technically a type of sign stimulus, the terms are often used interchangeably.

Sign stimuli may be only a small part of any environmental situation. For example, a male European robin (*Erithacus rubecula*) will attack another male robin that enters its territory. Experiments have shown, however, that a tuft of red feathers is attacked as vigorously as an intruding male (Lack 1943). The attack is not stimulated by the sight of another bird but only by the sight of red feathers. Of course, in the world of male robins, red feathers usually appear on the breast of a competitor.

Sign stimuli, simple cues that may be indicative of very complex situations, get through to the animal's nervous system, where they release patterns of behavior that may consist, in large part, of fixed action patterns. For example, the attack of the male European robin may be composed of FAPs that involve pecking, clutching, and wing fluttering. The end result is that when an intruding male robin appears, it is immediately identified and effectively attacked.

Chain of Reactions

So far we have considered only relatively simple behaviors, but a great deal of complexity can be added to the behavioral repertoire by building sequences of FAPs. The final product is an intricate pattern which ethologists call a **chain of reactions**. Here, each component FAP brings the animal into the situation that evokes the next FAP.

One of the earliest analyses of a chain of reactions was conducted by N. Tinbergen (1951) on the courtship ritual of the three-spined stickleback. This complex sequence of behaviors culminates in the synchronization of gamete release, an event of obvious adaptive value in species with external fertilization. Each female behavior is triggered by the preceding male behavior, which in turn was triggered by the preceding female behavior (Figure 2.6).

A male stickleback in reproductive condition may sometimes attack a female entering his territory. If the female does not flee and instead begins to display the appropriate head-up posture in which she hangs obliquely in the water, exposing her egg-swollen abdomen, the male will begin his courtship with a zigzag dance. He repeatedly alternates a quick movement toward her with a sideways turn away. This dance releases the approach behavior of the female. Her movement induces the male to turn and swim rapidly toward the nest, an action that entices the female to follow. At the nest, he lies on his side and makes a series of rapid thrusts with his snout into the entrance while raising his dorsal spines toward his mate. This behavior is the releaser for the female to enter the nest. The presence of the female in the nest is the releaser for the male to begin to rhythmically prod the base of her tail with his snout, causing the female to release her eggs. She then swims out of the nest, making room for the male to enter and fertilize the eggs. At the completion of this ritual, the male chases the female away and continues to defend his territory against other males until another female can be enticed into the courtship routine. The male mates with three to five females and then cares for the developing eggs by guarding them from predators and fanning water over them for aeration. We see, then, that this

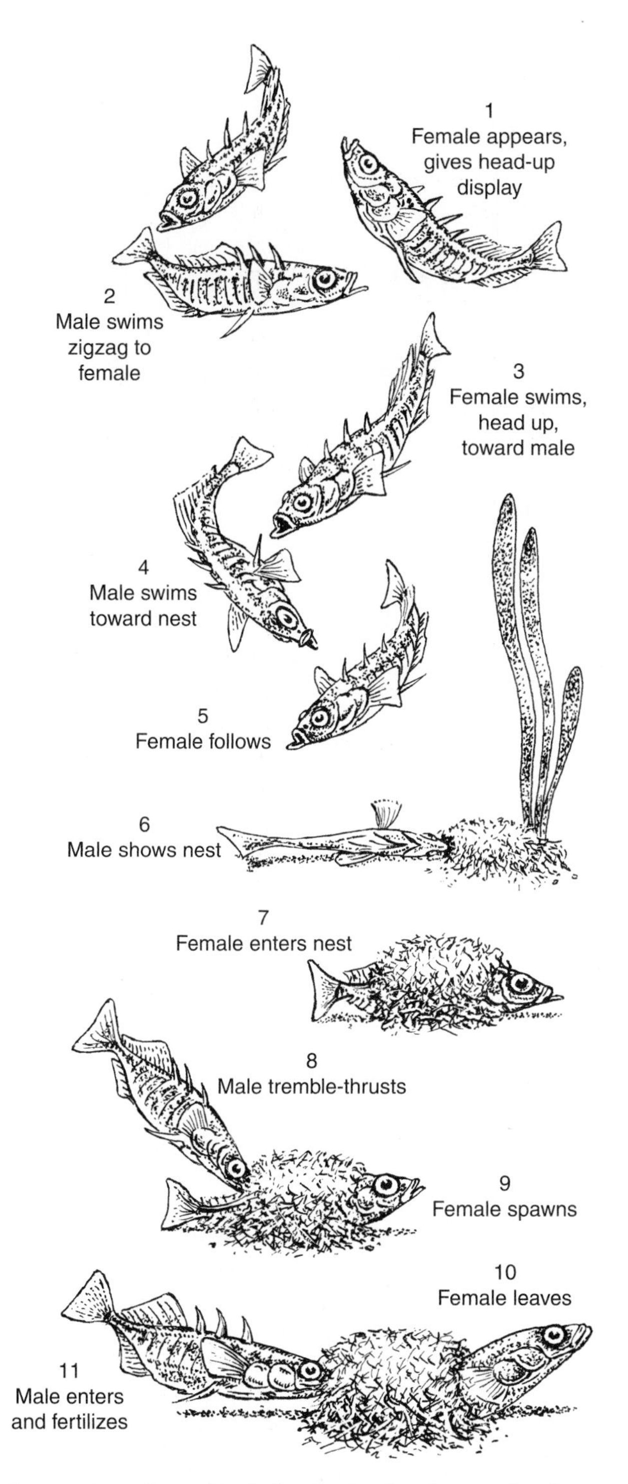
1
Female appears, gives head-up display
2
Male swims zigzag to female
3
Female swims, head up, toward male
4
Male swims toward nest
5
Female follows
6
Male shows nest
7
Female enters nest
8
Male tremble-thrusts
9
Female spawns
10
Female leaves
11
Male enters and fertilizes

FIGURE 2.6 **Courtship behavior in the three-spined stickleback. (From N. Tinbergen, 1989.)**

complex sequence is largely a chain of FAPs, each triggered by its own sign stimulus, or releaser.

The chain of reactions is not as rigid as the above description of courtship in the three-spined stickleback implies. There are actually many deviations in the precise order of the events in the ritual, and some actions must be repeated several times if one partner is less motivated than the other (Morris 1958). Such flexibility begins to make sense when the function for which the ritual evolved is considered. For the stickleback, courtship is important to time the release of the gametes, and thus males and females seem to adjust their activities so that they are physiologically ready for gamete release at the same time. Despite some flexibility, however, the component behaviors do not occur randomly. In the display, a particular behavior is more likely to be followed by certain actions than by others.

COMPARATIVE PSYCHOLOGY

THE APPROACH: PHYSIOLOGICAL, DEVELOPMENTAL, QUANTITATIVE, LABORATORY-ORIENTED

The **comparative psychologists**' emphasis on laboratory studies of observable, quantifiable patterns of behavior distinguished them from the European ethologists during the first half of the twentieth century. Recall that, at this time, many ethologists preferred to study animal behavior under natural conditions. This meant that they went into the field and observed behavior. The problem was that in the field, the unexpected is expected; one cannot control all the variables. The comparative psychologists argued that good, experimental science cannot be done under such uncontrolled conditions. The ethologists were further criticized because, although they described changes in behavior, they often neglected to quantify their results and rarely analyzed the data with statistical procedures. Given the psychologists' penchant for laboratory studies that produce quantifiable results, it is not too surprising that much of their early research focused on learning and the physiological basis of behavior. Again, however, we wish to emphasize that although learning and physiology were popular areas of study among comparative psychologists, the evolution and function of behavior were also examined by some of comparative psychology's practitioners. We will now consider some of the major conceptual developments in the field of comparative psychology.

EARLY CONCEPTS OF COMPARATIVE PSYCHOLOGY

Morgan's Canon

Recall from our previous discussion of the ideas and writings of Darwin and Romanes that the early descriptions of animal behavior were often subjective, anthropomorphic accounts. C. Lloyd Morgan helped stop the anecdotal tradition, thereby helping comparative psychology to become the objective science it is today. He argued that behavior must be explained in the simplest way that is consistent with the evidence and without the assumption that human emotions or mental abilities are involved. This idea was crystallized in Morgan's Canon (1894): "In no case may we interpret an action as the outcome of the exercise of a higher psychical faculty if it can be interpreted as the outcome of the exercise of one which stands lower in the psychological scale." In other words, when two explanations for a behavior appear equally valid, the simpler is preferred. People were urged to offer explanations of an animal's behavior without referring to the animal's presumed feelings or thought processes.

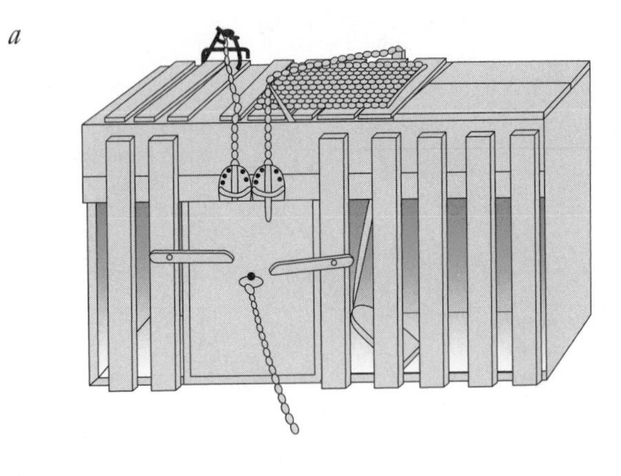

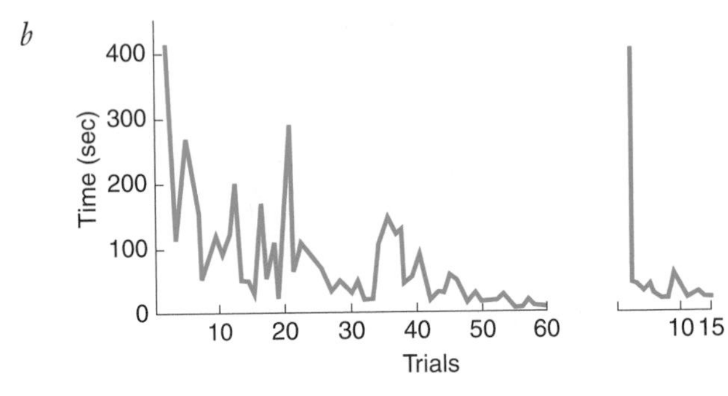

FIGURE 2.7 **(*a*) A problem box. Thorndike invented many "problem boxes" to measure the learning ability of animals. An animal would be placed inside the box and would have to learn how to operate an escape mechanism. (*b*) The time required for escape on successive trials was a measure of how quickly the animals mastered the task. (From Thorndike 1911.)**

Learning and Reinforcement

We have already mentioned that many of the early comparative psychologists focused their research efforts on learning. The early days of these studies were exciting times indeed, and some of the most important work was done by scientists in America. E. L. Thorndike (1898), for example, devised experimental techniques to study learning in the laboratory. He was a pioneer in research on what was called trial-and-error learning, now usually called operant conditioning. In operant conditioning, the animal is required to perform a behavior to receive a reward. In one series of experiments, Thorndike invented boxes that presented different problems to animals. For instance, one problem box was a crate with a trapdoor at the top through which an experimenter might drop a cat to the inside of the box. A hungry cat was left in the box until it accidentally operated a mechanism, perhaps pulling a loop or pressing a lever that opened an escape door on the side of the box, allowing it access to food that had been placed nearby. The length of time it took for each escape provides an objective, quantifiable measure of learning progress (Figure 2.7). During repeated trials, the animal became more efficient and required less time to hit the escape latch. Thorndike's studies led him to develop the Law of Effect, a cornerstone of operant conditioning. His basic notion was that responses that are rewarded, that is, followed by a "satisfying" state of affairs, will tend to be repeated (this idea was also described by C. Lloyd Morgan and other investigators of animal behavior). Thorndike began publishing studies on animal intellect and behavior in the late 1800s, and in 1911 he published a collection of his writings in a book entitled *Animal Intelligence: Experimental Studies.*

Just a few years after Thorndike introduced the idea of trial-and-error learning, Ivan Pavlov (Figure 2.8), a Russian physiologist, described the conditioned reflex. Pavlov noticed that a dog begins to salivate at the sight of food, and he reasoned that the sight of food must have come to signal the presence of food. (In science, the key observations that trigger great ideas are often quite commonplace, as in this case. It is not what you observe; it is what you make of it.) In his well-known experiment, Pavlov rang a bell immediately before feeding a dog and found that, in time, the dog came to salivate at the sound of the bell alone (Pavlov 1927). We will discuss classical conditioning in more detail in Chapter 5.

At first, comparative psychologists used operant and classical conditioning techniques to study the learning abilities of a wide variety of species. Thorndike, for example, examined learning in fish, chickens, cats, dogs, and monkeys and noted striking similarities in the learning processes of these animals. His results were therefore consistent with the idea of intellectual continuity. Thorndike concluded that although animals might differ in what they learned or in how rapidly they learned it, the process of learning must be the same in all species. In his 1911 collection of papers, he summarized his idea of intellectual continuity as follows (p. 294):

> *[Intellect's] general law is that when in a certain situation an animal acts so that pleasure results, that act is selected from all those performed and associated with that situation, so that, when that situation recurs, the act will be more likely to follow than it was before. . . . The intellectual evolution of the race consists of an increase in the number, delicacy, complexity, perma-*

FIGURE 2.8 **Ivan Pavlov described a conditioned reflex in the dog.**

> *nence and speed of formation of such associations. In man his increase reaches such a point that an apparently new type of mind results, which conceals the real continuity of the process. . . . Amongst the minds of animals that of man leads, not as a demigod from another planet, but as a king from the same race.*

Behaviorism

Another important event that steered comparative psychology toward objectivity and laboratory analysis was the birth of behaviorism, a school of psychology that restricts the study of behavior to events that can be seen—a description of the stimulus and the response it elicits. Behaviorists sought to eliminate subjectivity from their studies by concentrating their research efforts on identifying the stimuli that elicit responses and the rewards and punishments that maintain them. This was, indeed, a step toward better science. They designed experiments that would yield quantifiable data, invented equipment to measure and record responses, and developed statistical techniques that could be used to analyze behavioral data. The assumptions that an animal's mental capacity could not be measured directly, but its ability to solve a problem could, again focused attention on learning ability as a popular research subject. A learned response could be described objectively, and experiments could be conducted under the controlled conditions of the laboratory.

B. F. Skinner, one of the most famous behaviorists, devised an apparatus that was similar to Thorndike's problem box but lacked the Houdini quality. Instead of learning to operate a contrivance that provides a means of escape, a hungry animal placed in a "Skinner box" must manipulate a mechanism that provides a small food reward (Figure 2.9). A rat may learn to press a lever, and a pigeon may learn to peck at a key. Patterns of behavior that are rewarded tend to be repeated, or to increase in frequency, and so learning can be measured as the number of responses over time. Skinner believed that the control of behavior was a matter of reinforcement.

Behaviorists began to see basic principles underlying learning that were common to all species. They expected to find similarities in the learning process because at that time, the minds of all species were considered similar in kind. Thus, according to the traditional view of learning held by the followers of behaviorism, the minds of humans and animals were similar in kind and differed only in complexity. In short, there were general laws of learning that transcended all species and problems. If this was true, then it was reasonable to study the laws of acquisition, extinction, delay

FIGURE 2.9 **B. F. Skinner and his apparatus, the Skinner box. Animals placed in the box learned to operate a mechanism to obtain a food reward.**

of reinforcement, or any other aspect of the learning process in a simple and convenient animal, such as the domesticated form of the Norway rat (*Rattus norvegicus*), and the results could then be broadly applied to other species.

STOP AND THINK

Cancer is a disease that gives off odors. Assume that you are a researcher interested in knowing whether dogs can detect bladder cancer. What experiment would you design?

THE ROOTS OF PHYSIOLOGICAL PSYCHOLOGY

Although learning was a dominant focus of research during the middle of the twentieth century, it was not the only research interest of comparative psychologists. Another research topic was the physiological basis of behavior. Part of the psychological foundation of behavior is, of course, the nervous system. The comparative psychologists' interest in the neurological mechanisms of behavior can be traced back to Pierre Flourens, a protégé of Baron Cuvier, a famous scientist of nineteenth-century France who stressed the importance of laboratory research (Jaynes 1969). Flourens earned his reputation for his studies of the relationship between behavior and brain structure. For example, he did experiments in which parts of the brain were removed, such as the cerebral hemispheres from a pigeon, to look for the effect on the animal's behavior.

Karl Lashley was one comparative psychologist who maintained an interest in physiology, as well as a comparative base of study, during the years when learning by the laboratory rat dominated the field. His attempts to localize learning in the cerebral cortex resulted in the rejection of some hypotheses that were widely accepted at the time. For example, based on Pavlov's ideas, it was assumed that learning depended on the growth or strengthening of neural connections between one part of the cerebral cortex and another. To test this idea, Lashley (1950) trained rats on a variety of mazes and discrimination tests and then tried to disrupt the memory by making a cut into the cerebral cortex in a different place in each animal. After destroying varying amounts of brain tissue, he would then retest the animals to see if their behavior changed. In general, he found that when it came to complex problem solving, the entire cerebral cortex was involved, and any particular area was just as important as any other. He also experimentally addressed questions such as whether the learned response depended on a fixed pattern of muscle movements. Contrary to expectations, he found that they do not. But Lashley was not solely concerned with the neurological basis of learning in the rat. He also examined the role of the brain in emotion and in vision.

The comparative psychologist Frank Beach began his career by using brain surgery to determine the effects of lesions on the maternal behavior of the rat, but he later went on to study the effects of hormones on behavior. He analyzed the roles of nerves, hormones, and experience in the sexual behavior of fishes, amphibians, reptiles, birds, and mammals. We will discuss some of his work in the field of behavioral endocrinology in more detail in Chapter 7.

Recognizing that animal behavior is concerned with the activities of groups of animals, as well as of individuals, some comparative psychologists studied social behavior. Robert Yerkes, for example, established a research facility (later named the Yerkes Laboratory of Primate Biology) at Orange Park, Florida, to study a wide range of behavior in primates. Some researchers also began to see that although it is often easier to make measurements in the laboratory, it is not impossible to get good measurements in the field. C. R. Carpenter studied a variety of primate species, each in its natural setting: howler monkeys in Panama, spider monkeys in Central America, and gibbons in Thailand, to name a few. T. C. Schneirla used both field observation and laboratory experimentation to investigate the social behavior in army ants. In doing so, he applied the rigorous methodology of laboratory researchers to his field studies. Such pioneering studies began to help weave the two independent sciences of ethology and comparative psychology together.

SOCIOBIOLOGY AND BEHAVIORAL ECOLOGY

The field of animal behavior has grown enormously. In the 1960s and early 1970s, for example, field researchers such as John Crook (1964; 1970) and John Eisenberg and colleagues (1972) suggested that ecological context was sometimes a better correlate of social behavior than was phylogeny (remember that ethologists often focused on phylogenetic analyses of behavior). Another dramatic development was the birth of a new discipline that focused on the application of evolutionary theory to social behavior. This new discipline was called **sociobiology**. Another discipline, **behavioral ecology** examines the ways in which animals interact with their environment to learn how behavior contributes to the animal's reproductive success and survival. In either case, the key element of this approach is the idea that behavior should, on average, maximize fitness of individuals. Thus, behavioral ecologists attempt to identify the payoffs and costs that play a role in the evolution of behavior (Owens 2006).

There has been some question concerning the uniqueness of the approach of sociobiology. Some scientists, for example, question whether sociobiology is really a new discipline or simply part of contemporary ethology (e.g., Dawkins 1989). In contrast, others believe that at least early on, ethology and sociobiology could be separated in several ways (Barlow 1989). For example, whereas classical ethologists tended to derive hypotheses from detailed observations (i.e., through induction), sociobiologists tended to be more deductive, typically deriving hypotheses from larger theoretical frameworks. Whereas classical ethologists were interested in species differences, sociobiologists began to investigate individual differences, examining the costs and benefits of a particular act. Having mentioned some of the questions concerning the precise relationship between sociobiology and other fields of animal behavior, let us consider the relatively recent "history" of sociobiology.

During the late 1960s and early 1970s, most scientists were quite comfortable with the idea that natural selection acted primarily on individuals. Despite the existence of widespread agreement, however, some nagging issues that seemed inconsistent with selection at the level of the individual remained (Hinde 1982). For example, how could one explain the evolution of sterile castes in species of ants, bees, and wasps? How could the evolution of nonreproducing individuals be consistent with Darwinian selection? Similarly, how was one to explain the evolution of certain patterns of behavior, called altruistic behavior, that seemed to benefit others but were costly (with respect to survival and reproduction) to the performer? Why, for example, do some animals give alarm calls when they spot a predator, when calling may actually increase their own chances of being detected? The answer to these questions came in 1964 when W. D. Hamilton published his seminal papers, "The Genetical Evolution of Social Behaviour, I, II." Hamilton showed that evolutionary success (the contribution of genes to subsequent generations) should be measured not only by the number of surviving offspring produced by an individual but also by the effects of that individual's actions on nondescendant kin (e.g., siblings, nieces, and nephews). He coined the term *inclusive fitness* to describe an individual's collective genetic success—that is, a combination of direct fitness (own reproduction) and indirect fitness (effects on reproduction by nondescendant kin). When quantifying an individual's inclusive fitness, we count—to varying degrees, depending on how closely they are related—all the offspring, personal or of relatives, that are alive because of the actions of that individual. This concept of inclusive fitness paved the way toward an understanding of the evolution of sterile castes and altruistic acts (these topics are discussed in more detail in Chapter 19).

In the years following Hamilton's (1964) paper, many studies were conducted in which the idea of inclusive fitness was used to interpret social behavior. However, it was not until 1975, when E. O. Wilson published his landmark text, *Sociobiology*, that the true impact of sociobiological ideas was felt. The text, an engaging integration of ideas from fields such as ethology, ecology, and population biology, gained almost instant notoriety from both within and outside the scientific community. Wilson defined sociobiology as the "systematic study of the biological basis of all social behavior" and proposed that a knowledge of demography (e.g., information on population growth and age structure) and of the genetic structure of populations was essential in understanding the evolution of social behavior.

Although sociobiological ideas had been developing for several years before the publication of Wilson's book, the text crystallized many of the relevant issues and soon became the focal point for proponents and critics alike. Criticism arose from both the scientific and political arenas. First, in attempting to establish sociobiology, Wilson attacked fields such as ethology and comparative psychology and made the bold prediction that in due time sociobiology would engulf these disciplines. He specifically predicted that ethology and comparative psychology would be "cannibalized by neurophysiology and sensory physiology from one end and sociobiology and behavioral ecology from the other" (Wilson 1975). Another area of great concern, this time from the political arena, was the extension of sociobiological thinking, in the absence of sound evidence, to human social behavior (Cooper 1985). Opponents of sociobiology claimed that Wilson advocated biological determinism, the idea that the present conditions of human societies are simply the result of the biology of the human species and therefore cannot be altered. Although only the final chapter of his text was devoted specifically to humans, heated debate over the social and political implications of sociobiological theories ensued (e.g., see the collection of papers in Caplan 1978).

During the 1970s and early 1980s, research on sociobiological topics in animal behavior flourished. George Barlow (1989) suggested, "The study of animal behavior had indeed begun to stagnate by 1975, and the advent of sociobiology was just the kick in the pants the field needed to get moving again." The field of animal behavior became revitalized because sociobiology provided a framework that could be used to test hypotheses about the adaptiveness or survival value of behavior. But the "kick in the pants" was so strong that for a time almost *all* research in animal behavior was done under the banner of sociobiology (Bateson and Klopfer 1989). Of all possible questions about animal behavior, one—its function, or survival value—had come to dominate the field.

By the end of the 1980s, however, many researchers began to notice the imbalance in the study of animal behavior. It became apparent that our understanding of animal behavior would be fuller if both its immediate and evolutionary causes are considered. As Marian Stamp Dawkins (Dawkins 1989, p. 53) has said,

> *Genes operate through making bodies do things. These bodies have to develop and they need machinery (sense organs, decision centers, and means of executing action) to be able to pass their genes on to the next generation. To understand this process fully, we need a science that is not only aware of the evolutionary ebb and flow of genotypes over evolutionary time, but can look at the bridge between generations, at the bodies that grow and move and court and find food and pass their genetic cargo on through time with the frailest and most marvelous of flesh-and-blood machinery.*

MORE RECENT TRENDS

The study of animal behavior has seen some changes in the areas that are the focus of study. On the whole, however, it seems to have returned to research that considers all of Tinbergen's four questions. Michael Taborsky (2006) points out that ethology has largely regained its balance and addresses questions of mechanism as well as function. Let's consider some of the recent changes of focus in behavioral research.

FIELD STUDIES

Following the assumption that natural selection shaped behavior, we should expect that behavior observed in the field should increase the animal's chance of survival. Many of the pioneering field studies were purely descriptive, and some also included an explanation of what was described. However, today's field studies usually begin with a clearly stated hypothesis to be tested by data collection and analysis. The hypothesis generally relates either to the short-term function of the behavior or to the long-term fitness (relative number of surviving offspring) consequences of a behavior. Thus, the data collected are often the outcome or consequence of a behavior, such as the amount of food collected. Researchers might then determine the fitness consequences of natural variation in the expression of that behavior within a population (Altmann and Altmann 2003). As we will discuss in later chapters of this book, particularly in Chapters 4 and 12, today's field studies of animal behavior usually focus on the costs and benefits of a particular behavior, with the common currency being reproductive success. Natural selection is often assumed to have shaped not just an efficient but also an optimal form of behavior. For example, we would expect a starling to select the type of prey that will maximize the amount of food (energy) that can be delivered to its brood. Also, because an animal's environment includes competitors, an individual's best choice of action often depends on what other members of the population are doing. In such situations, it is often helpful to consider whether an individual's choice of action is an evolutionarily stable strategy (ESS). An ESS is a strategy that, when adopted by most members of the population, cannot be invaded by the spread of any rare alternative strategy. The concept of ESS has been applied in studies of mating systems, communication, conflict, and cooperation (Krebs and Davies 1997).

CELLULAR AND MOLECULAR BASIS OF BEHAVIOR

We are also making great strides in understanding the mechanisms of behavior, largely because of tools and techniques that were not available even a few years ago. Today, sign stimuli may be interpreted as filtering at the level of sensory receptors or as feature detection neurons that respond to specific features of a stimulus. We can identify neurons in circuits that underlie FAPs. For example, new recording techniques have made it possible to map the nervous systems of several invertebrates. In some animals, specific neurons have been linked to specific behaviors. For instance, in the grasshopper (*Omocestrus viridulus*) three different hind leg movements (FAPs) are involved in producing the courtship sound signals. By using microinjection techniques and intracellular recording, it has been shown that a specific type of brain nerve cell is responsible for each of these three FAPs. During courtship, these nerve cell types are activated in a specific sequence (Hedwig and Heinrich 1997). Fixed action patterns are now discussed in terms of neural networks, command neurons, or central pattern generators. In the chapters that follow, we will consider many other examples of how physiology, neurobiology, and molecular biology have enhanced the study of behavior.

During the last decade, some avenues of research have focused on the cellular, or even the molecular, underpinnings of animal behavior. One of the most exciting areas of research is behavioral genomics—study of the role of an organism's genetic material in behavior. An organism's genome consists of all of its DNA. One goal of genomics is to learn the sequence of all the genes in an organism's genome. Indeed, scientists have now determined the sequence of bases in the entire genomes of more than 100 organisms. This information allows scientists to zero in on the sequence of a particular gene. That sequence can then be compared to a database of known gene sequences, which contains some genes whose function is known. If the location or sequence matches that of a gene whose function is known, it may provide a clue as to the function of the gene of interest.

Genomics provides a way for researchers to consider the activity of networks of genes, instead of looking at one gene at a time. Gene sequences reveal which mRNAs to look for if the gene is active. This is often accomplished using DNA microarray analysis, which compares the activity of thousands of genes simultaneously. We will see subsets of genes become active under different conditions. Genomics also tells us which proteins to look for if the gene is active. The next goal is to determine the functions of the proteins that are produced. The biggest challenge will be to figure out how the environment and genome work together to direct the structure and behavior of an individual. As Gene Robinson (2005; p. 257) declared, the time has come to "achieve a comprehensive understanding of social life in molecular terms: how it evolved, how it is governed, and how it influences all aspects of genome structure, gene expression and organismal development, physiology and behavior."

This reductionist approach to animal behavior (trying to understand the behavior by understanding its components) has allowed many exciting discoveries, but by the early twenty-first century appeals were being made to "return to the whole organism" (e.g., Bateson 2003, 2005). According to these appeals, if you want to know why an animals behaves a certain way, you must look at the whole organism, not just neurons, genes, and molecules (Hogan 2005).

BEHAVIORAL BIOLOGY

In recent years, the term *behavioral biology* has been coined to describe behavioral research that includes more than one of Tinbergen's four questions (Taborsky 2006). The themes of returning to studying behavior at the level of the whole organism and integrated studies of the four questions are threaded through the papers celebrating the fortieth anniversary of Tinbergen's classic paper on the four questions[1]. Michael Ryan (2005) argues that an approach to animal behavior that *integrates* Tinbergen's four questions—cause, development, survival value, and evolution—is needed to provide a complete and correct understanding of behavior. More is learned by integrating the aims and methods associated with each question than by studying each question in isolation. Ryan illustrates this idea by explaining how knowledge of the evolutionary history of calls of túngara frogs (*Physalaemus pustulosus*) helped researchers understand the mechanism of male calling and female response, as well as how it develops and increases fitness. And David Sherry (2005) suggests that knowledge of the survival value or function of a behavior can assist research on the causes of the behavior. For example, knowing that the ability to sense the earth's magnetic field serves an orientation function suggests the properties that a magnetoreceptor must have.

[1]A collection of these papers can be found in *Animal Biology* 2005(4): 55.

TABLE 2.1 Disciplines in Animal Behavior

Discipline	Focus
Neuroethology	The neurological study of behavior
Behavioral endocrinology	The study of the hormonal basis of behavior
Neuroecology	The study of adaptive variation in cognition and the brain
Cognitive ecology	An approach that views cognition as an adaptive trait shaped by natural selection
Evolutionary psychology	An approach to psychology that attempts to explain human mental and psychological traits as adaptations shaped by natural selection
Behavioral genetics or genomics	The study of the influence of genetic information on behavior
Applied ethology	The study of the behavior of domestic animals or other animals kept in captivity

The integration of research focusing on Tinbergen's four questions has led to the development of new subdisciplines of animal behavior (Table 2.1). A fascinating subdiscipline is animal cognition or cognitive ethology. Animal behavior is no longer viewed only as the result of genetic programming or neural wiring or as the result of a simple stimulus–response reaction. Instead, an animal's mental capabilities are seen as a product of natural selection. The field of study began with Donald Griffin's controversial book, *The Question of Animal Awareness* (1976). Griffin later (2001) named the field *cognitive ethology*. It is an interdisciplinary area of research that brings Tinbergen's four questions to bear on the study of animals' mental experiences. Three areas of research, in particular, are making rapid progress: animal communication, seed caching and recovery, and navigation and orientation (Balda et al. 1998). You will read more about these subjects in later chapters of this book.

APPLIED ANIMAL BEHAVIOR

Professional opportunities are growing in applied animal behavior, the study of animal behavior with practical implications rather than just for the sake of accumulating knowledge. This subfield of applied animal behavior is itself divided into other disciplines. Many applied animal behaviorists focus on captive animals. Some work with companion animals, such as dogs and cats, training them and solving behavioral problems. Others study the positive effects of the human–animal bond: pets, for example, improve the mental health of many elderly. Still other researchers work with laboratory, zoo, and farm animals.

Many applied animal behaviorists work to improve the welfare of captive animals (Fraser and Weary 2005).

To understand the challenges of this discipline, consider the "Five Freedoms" for captive animals proposed by the Farm Animal Welfare Council in the United Kingdom: (1) freedom from thirst, hunger, and malnutrition, (2) freedom from discomfort due to environment, (3) freedom from pain, injury, and disease, (4) freedom to express the normal behavior of the species, and (5) freedom from fear and distress. Whether certain of these freedoms are provided is easy to determine. For instance, we can see whether the animals have ready access to fresh water and a healthy diet. It is generally possible to see that an animal is injured or diseased and to provide rapid diagnosis and treatment. It is more difficult to be sure that we are providing animals with the other freedoms. For instance, in order to ensure that animals have the freedom to express normal behavior we must know what normal behavior is. This may require new studies of captive species (or their close relatives) in their natural habitat. In order to ensure that animals are free from discomfort, pain, fear, and distress, we must understand something about the mental state of other species. Table 2.2 categorizes some of the ways that researchers assess stress in animals.

STOP AND THINK

Dust bathing is a natural behavior that chickens perform to keep their feathers in good condition and rid themselves of mites. If you were charged with designing commercial chicken cages with the welfare of chickens in mind, what experiment would you perform to determine the "importance" of the opportunity to dust bathe to chickens?

Other applied animal behaviorists work with wild animals. Among them are professionals who work in wildlife management (e.g., increasing the population of game species) and pest management. An increasingly important field is conservation behavior, in which the principles of animal behavior are used in efforts to conserve biodiversity. As human populations spill over into the habitats of animals, many populations of animals are declining or disappearing. To halt or reverse these losses, we need behavioral data about habitat preferences, migratory routes, territory size, social organization, food requirements, risk of predation, mating habits, and more. These data are crucial for designing effective nature preserves. For example, studies of tropical birds revealed the paths of their migratory routes up and down mountains, and conservationists subsequently protected corridors of land that connected preserves on the mountaintop and in the valleys. Conservation behaviorists also may breed animals in captivity for return to the wild. This requires knowledge of the communication signals used in mating. Reintroduction of a captive-born animal to the natural habitat requires training to recognize and avoid predators (see Chapter 5). Throughout the book, we will mention the work of applied animal behaviorists.

Today there is a sense of rejuvenation in the study of animal behavior, largely because many disciplines are now contributing to its study (van Staaden 1998). New techniques and interactions among disciplines allow us to ask and answer many questions about behavior that could not be addressed previously.

TABLE 2.2 Three Conceptions of Animal Welfare and Typical Measures Used to Provide Positive Evidence of Animal Welfare

Conception of animal welfare	Typical measures
Biological function	Increase in stress hormones (–) Reduction in immune competence (–) Incidence of disease and injury (–) Survival rate (+) Growth rate (+) Reproductive success (+)
Affective states	Behavioral signs of fear, pain, frustration, etc. (–) Physiological changes thought to reflect fear, pain, etc. (–) Behavioral signs of aversion or learned avoidance (–) Behavioral indicators of comfort/contentment (+) Performance of behavior (e.g., play) thought to be pleasurable (+) Behavioral signs of approach/preference (+)
Natural living	Performance of natural behavior (+) Behavioral/physiological indicators of thwarted natural behavior (–) Performance of abnormal behavior (–)

Source: D. Fraser and D. M. Weary. 2005. Applied animal behavior and animal welfare. In *The Behavior of Animals: Mechanisms, Function, and Evolution*, edited by J. J. Bolhuis and L.-A. Giraldeau, Table 15.1, p. 364. Malden, MA: Blackwell Publishing.

Study of the history of animal behavior will show us that whether our primary interest is the mechanism or the function of behavior, our efforts will be most fruitful if we keep a clear focus on *behavior* as the driving interest of research.

SUMMARY

Perhaps the most important concept in the study of animal behavior is Darwin's idea of evolution through natural selection, which provides the evolutionary framework necessary for the development of animal behavior.

In the early 1900s, the two dominant approaches to the study of animal behavior were ethology, centered in Europe, and comparative psychology, headquartered in the United States. Ethologists focused primarily on the function and evolution of behavior. Because the context in which a behavior is displayed is sometimes a clue to its function, ethologists often studied behavior under field conditions. It followed from their interest in evolution that ethologists used a comparative approach and studied primarily innate behaviors.

Early ethologists were interested in stereotyped patterns of behavior, considering them to be just as reliable as morphological characters in defining a particular group. These stereotyped behaviors were called fixed action patterns (FAPs). An FAP is triggered by a very specific stimulus. That portion of the total stimulus that releases the FAP is called the sign stimulus or releaser. Because most behaviors are not so stereotyped as implied by the notion of FAP, they have more recently been described as modal action patterns (MAPs).

In contrast to the early ethologists, comparative psychologists emphasized laboratory studies of observable, quantifiable patterns of behavior. In general, they asked questions that concerned the development or causation of behavior. Learning and the physiological bases of behavior were the focus of much of their research.

Many exciting advances were made in the study of learning. Thorndike developed the techniques for studying trial-and-error learning, and Pavlov provided the methodology for classical conditioning. Behaviorism is a school of psychology that proposes limiting the study of behavior to actions that can be observed. B. F. Skinner, a prominent behaviorist, found that patterns of behavior that are rewarded tend to be repeated or to increase in frequency, and he concluded that the control of behavior was largely a matter of reinforcement.

The physiological basis of behavior is another traditional subject investigated by comparative psychologists. Despite their emphasis on learning and physiology in the laboratory, some comparative psychologists studied the social behavior of animals in the field.

In the 1960s, a new discipline emerged in the study of animal behavior; this discipline, called sociobiology (or sometimes behavioral ecology), focused on the application of evolutionary theory to social behavior. W. D. Hamilton articulated one of its central concepts, that of inclusive fitness, in 1964. According to Hamilton, individuals behave in such a manner as to maximize their inclusive fitness (i.e., their own survival and reproduction plus that of their relatives) rather than acting simply to maximize their own fitness. Suddenly, certain issues that seemed inconsistent with selection at the level of the individual, such as the evolution of sterile castes in insects and altruistic behavior (behavior that benefits others at the expense of the performer), were explainable.

Approximately ten years after Hamilton's paper, E. O. Wilson crystallized sociobiological ideas in his landmark text, *Sociobiology*. Sociobiology and an interest in the survival value of behavior dominated the study of animal behavior for approximately a decade, but it soon became apparent that a complete understanding of behavior requires knowledge of both mechanism and function. As new technologies became available, researchers began to explore the mechanisms of behavior on a molecular or cellular level. Today, the study of animal behavior has returned to a more balanced approach that considers mechanism and function. Research is conducted in the laboratory, as well as in the field. Information gathered in this research is being applied to assist the welfare of captive animals and to study conservation biology.

3

Genetic Analysis of Behavior

Picture this: In front of you are two cages, each containing a female rat and her litter of week-old pups. Your colleague enters the room, and the door slams behind her. The pups in one cage jump in response to the sound, while those in the other cage continue grooming or sleeping. Why do the pups behave so differently when they hear loud sound? In this case, it is fair to blame the mother because, in addition to influencing the future mothering style of the female pups, the quality of maternal care affects the pups' response to stress, both now and in adulthood. The mother of the calm pups spends a great deal of time grooming and nursing her pups (Figure 3.1); the mother of the skittish pups is neglectful.

Although it might be tempting to explain these differences in behavior as being due to differences in either learning or genes, we will learn in this chapter that the relationships among genes, experience, and behavior are not that simple. For example, we will see that the experiences during the first week of life alter the activity of two genes in a nearly permanent way and influence both

FIGURE 3.1 **The quality of maternal care that rat pups receive determines how timid the pups will be and the quality of maternal care that the females will give to their own pups later in life.**

how the adult rats will respond to stress and how attentive the females will be to their own pups. Furthermore, this change in gene activity is transmitted to the next generation. So, is the behavior determined by gene activity? Yes. Is it determined by early experience? Yes. How is this possible? We will answer that question later in this chapter, after discussing some of the ways that genes can influence behavior. Surprisingly, the answer suggests that your behavior today may be influenced by your great-grandmother's lifestyle.

The relationship between genes and behavior is often difficult to decipher, but we do know that there is not a one-to-one correspondence between genes and behavior. Consider a very simple behavior: a fly extends its proboscis (a tubular structure of mouthparts) when the sense organs on its feet detect sugar. Think of all the parts of the fly that need to be in good working order for the fly to successfully perform this behavior: the physical structures (the sense organs on the feet, for example) must be functional, as must the neural circuitry to carry the information from the sense organs to the fly's brain, which must assess the information and send signals via motor neurons to the muscles of the proboscis, which must be able to contract. Each of these pieces is essential to behavior, and their structure and function are influenced by many genes. In addition, most steps are influenced by both the internal physiological state of the animal (Is it hungry? Has it already learned anything about the environment?) and the external environment.

There are two ways that genes can affect behavior. In the example of fruit fly foraging just described, genes alter behavior through their effects on development of the nervous system and physiology. We will see other examples of this relationship throughout the chapter. But we will also consider examples of behavior in which the environment, especially social interactions, trigger changes in the nervous system or physiology. These changes then alter the pattern of gene expression; that is, they turn some genes on and others off. In turn, changes in gene expression alter the nervous system and physiology, and behavior is modified (Robinson 2008). Thus, we see that the effect of genes on behavior can be dynamic. When we see animals that differ from one another in behavior, it seems almost impossible that we will be able to say precisely *how* they differ. Yet this is exactly what the thriving field of behavioral genetics is giving us the ability to do, at least for some behaviors.

BASICS OF GENE ACTION

What do genes really do, then? How do they work? As you may already know, genes can direct the synthesis of proteins. Each protein is specified by a different gene, or if the protein consists of more than one chain of amino acids, a gene specifies one of those chains. The protein may be structural and be used as a building block of the organism, or it may be regulatory. A regulatory protein may modify the activity of other genes. In other words, genes can code for specific proteins, and the proteins affect the composition and organization of the animal in ways that influence how it behaves. An animal's sensory receptors detect stimuli and send information to the nervous system, where it is further interpreted and analyzed. The nervous system may then initiate a response by effectors (muscles and glands) that results in behavior. Genes direct the development of the structure and function of receptors, nerves, muscles, and glands. Alterations in genes may change the proteins they code for, with the result that anatomy or physiology may be altered in a manner that changes behavior. We will look at a few examples of the links between genes and behavior later in this chapter.

To understand the relationship between genes and proteins, it is helpful to know a little biochemistry. Genes are made up of DNA (deoxyribonucleic acid). The structure of DNA is somewhat like a long ladder, twisted about itself like a spiral staircase. The DNA ladder is composed of two long strings of smaller molecules called nucleotides. Each nucleotide chain makes up one side of the ladder and half of each rung. A nucleotide consists of a phosphate, a nitrogenous base, and a sugar called deoxyribose. There are only four different nitrogen-containing bases in DNA: adenine (A), thymine (T), cytosine (C), and guanine (G). Although DNA has only four different nucleotides, a DNA molecule is very long and has thousands of nucleotides. When forming a rung of the ladder, adenine must pair with thymine and cytosine must pair with guanine (Figure 3.2). The specificity of these base pairs is important, not only for the accurate production of new DNA molecules, but also for conversion of the information in the gene into a protein.

The instructions for each protein are written as the sequence of bases in the DNA molecule. The first step is to transcribe the information in DNA into RNA (ribonucleic acid), specifically messenger RNA (mRNA). The DNA is unzipped for part of its length so that an mRNA molecule can be formed. This first mRNA strand is modified or edited before it leaves the nucleus. Some of the regions of the mRNA strand that do not code for a protein are then snipped out.

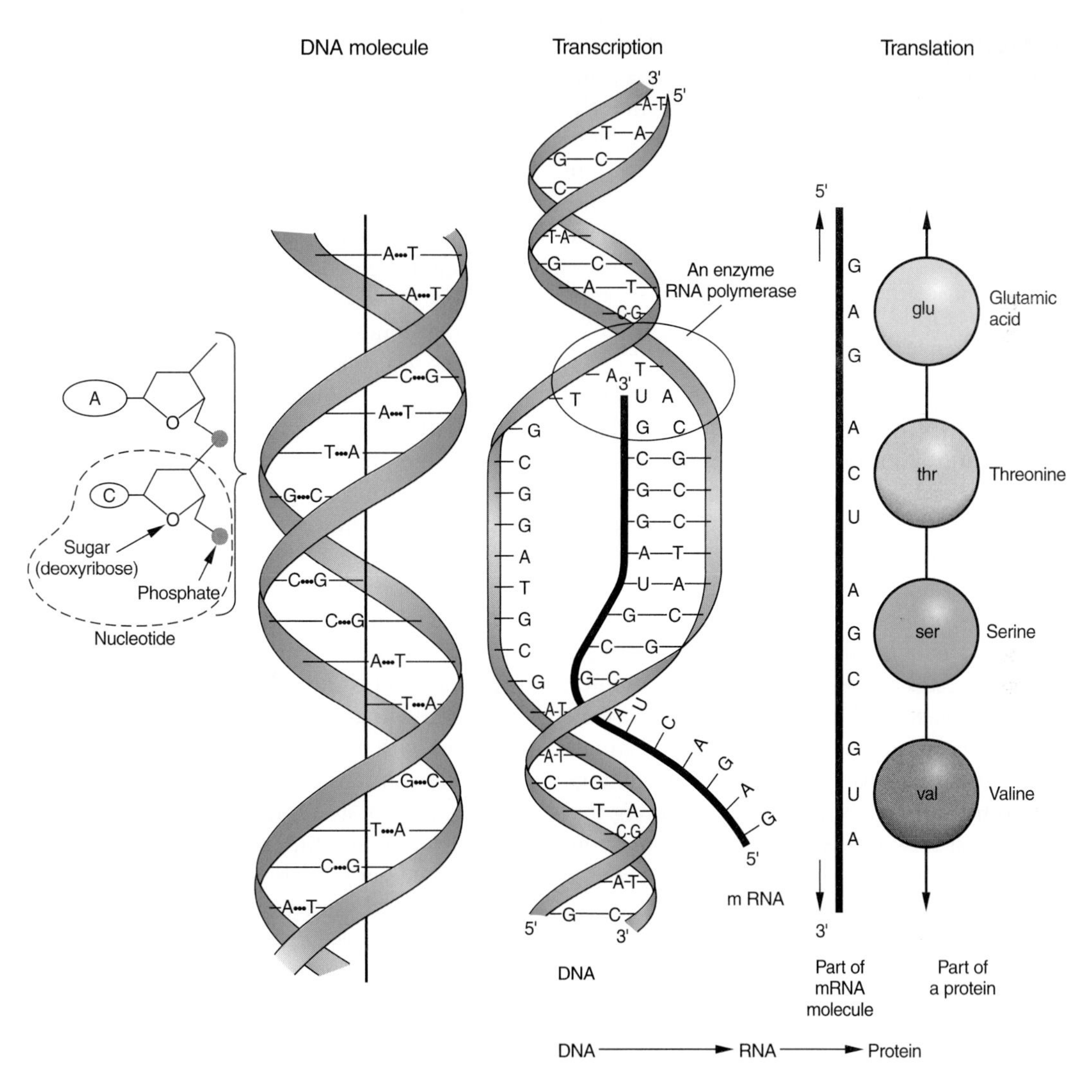

FIGURE 3.2 **A diagrammatic representation of the biochemistry of gene expression. A gene is a region of a DNA molecule that has the information needed to make a specific protein. A DNA molecule is composed of nucleotides. A nucleotide, shown on the extreme left of the figures, is a nitrogenous base, a sugar (deoxyribose), and a phosphate. In the ladderlike DNA molecule, the two uprights are composed of alternating sugar and phosphate groups, and the rungs are paired nitrogenous bases. The pairing of bases is specific: adenine with thymine and cytosine with guanine. During transcription, the synthesis of mRNA, the sequence of bases on the DNA molecule is converted to the complementary sequence of bases on the mRNA molecule. Each unit of three bases on the mRNA molecule signifies a particular amino acid. The message of messenger RNA is, therefore, its sequence of bases that determines the order and kinds of amino acids in the protein product.**

The mRNA has a structure similar to DNA except for three differences: It is single-stranded, its sugar is ribose, and the base uracil substitutes for thymine. Because adenine must pair with uracil and cytosine must pair with guanine, the sequence of bases on RNA is specified by the sequence of bases on DNA. Because the DNA molecule is so long, the four bases can be ordered in many ways. Therefore, with only four bases, DNA can encode the information needed for the synthesis of a myriad of different proteins.

The next step is to translate the order of bases on the mRNA molecule into a protein. Proteins are long chains of amino acids. Each different protein has a unique order of amino acids. A group of three bases on the mRNA molecule is translated, three bases at a time, into a protein. The order of bases on DNA specifies the sequence of bases in mRNA, which can be translated into only one array of amino acids. Thus, the information in the gene is its sequence of bases that codes for a specific protein. The transfer of information can be summarized as follows:

Sequence of bases in DNA → Sequence of bases in mRNA → Sequence of amino acids in a protein.

For most proteins, and ultimately the traits they influence, different forms exist because the underlying DNA sequences are different. For example, coat color in mice might be black or brown. In other words, this gene for coat color can be expressed in two different ways; one resulting from intense black pigment granules, and the other from chocolate brown pigment granules. These alternative forms of a gene are called **alleles**. Sometimes, as in the eye color of fruit flies, there are many possible alleles. The wild type, or most common eye color of fruit flies, is red, but other alleles of this gene can result in white or vermilion (a brilliant red tinged with orange) eyes. Furthermore, most of the animal species used in genetic studies are diploid, meaning that an individual possesses two alleles of each gene, one from each parent, and this influences how the gene is expressed. If the two alleles of a gene are identical, the individual is said to be **homozygous** for the trait. However, an individual may inherit different alleles for a gene from its mother and father. Such an individual is **heterozygous** for the gene. The genetic diversity among unrelated individuals results from the particular alleles that they possess and from whether or not they are heterozygous.

Structural genes produce proteins that become part of a structure or that have a specific function within an organism. Consider, for example, the *Shaker* gene in the fruit fly, *Drosophila melanogaster*. A particular **mutation**—that is, a specific change in the sequence of bases in the DNA—in the *Shaker* gene results in flies that shake violently under anesthesia. It turns out that the mutation changes a protein used in the formation of channels in the membranes of nerve cells that allow potassium ions to pass through. Potassium channels are critical to normal nerve cell function, and so the mutation causes a nerve cell defect and abnormal behavior (Kaplan and Trout 1969).

A recurrent theme in behavioral genetics is that behavior depends on which genes are expressed in which tissues and when. Although all the cells of an animal's body have the same genes, some of them are turned off during development. When genes are turned off, they do not produce a protein. If these genes are then turned on, however, their proteins are produced, and in some cases, they will modify a structure or function in a way that will alter behavior. Thus, an organism's behavior may *change* as specific genes are turned on or off in specific tissues. This applies even in the short term, as the animal experiences new stimuli.

The activity of specific genes is often influenced by regulatory genes, of which there are many types. Certain regulatory genes modify the activity of other genes through the production of proteins called transcription factors. Transcription factors and other products of regulatory genes increase gene expression or decrease gene expression, altering the amount of mRNA produced. Thus, gene regulation determines when and where a protein will be produced, as well as how much of that protein is produced. Mutations in regulatory genes often have a more widespread effect on the organism than do structural genes because these genes produce proteins that regulate many other genes.

Consider the mechanisms by which genes regulate courtship and mating in fruit flies. If you have ever left a ripe banana in your kitchen too long, you may know that mating is something fruit flies do well, but you may not know that the choreography is complex. The dance begins with orientation, during which the male faces the female and taps her on the abdomen with his foreleg. If she wanders away, he follows her. Next, he begins to "sing" a courtship song by fluttering a single outstretched wing. If the female does not show interest, he will repeat these actions. When the female seems receptive, he extends his proboscis and licks the female's genitalia. Next, he will try to copulate with her. If the attempt fails, he will wait a few moments before starting the ritual from the beginning (Hall 1994).

The regulatory gene *fruitless* (*fru*) affects nearly every aspect of male courtship in fruit flies and provides an example of gene regulation at several levels—sensory processing, choice of behavior, and carrying out the behavior (Dickson 2008). *Fru* is but one gene in a hierarchy of regulatory genes that influence the final behavior pattern. In such a hierarchy, the protein product of one gene regulates the activity of another gene whose protein product affects the activity of perhaps a third gene and fourth gene, and so on. Recall that a newly formed mRNA is edited or spliced before it is translated into protein. *Fru* mRNA transcripts are edited differently in males and females, under the direction of a gene higher in the hierarchy (*transformer* or *tra*) that responds to the number of X chromosomes present. When only one X chromosome is present, the male-specific versions of these fru transcription factors are produced. In turn, these fru transcription factors regulate genes in certain neurons to build the correct neural circuitry for male courtship and sex determination in fruit flies. The female-specific mRNA transcripts of *fru* do not direct the development of the same circuitry (Baker et al. 2001; Dulac 2005).

Male-specific fru proteins alter the way that the neural circuit underlying sexual behavior functions to produce male courtship behavior. This difference is most likely due to an identifiable subset of *fru*-expressing interneurons that are present only in males. In the female, these neurons are programmed to die. The presence of the male-specific fru protein prevents the death of these interneurons, which alters the way in which the neural circuit for sexual behavior functions in a male (Kimura et al. 2005).

Fru is expressed in only about 500 neurons, roughly 1.5% of the neurons in the central nervous system. However, *fru* is also expressed in nearly all of the sensory neurons involved in courtship, especially olfactory sensory neurons. Thus, as you might expect, disruptions of *fru* lead to changes in sexual behavior. Male *fru* mutants are bisexual. A group of only male *fru* mutants form courtship chains in which each male chases and courts the male in front, forming revolving circles of courting flies. Expression of male-specific *fru* causes females to display male courtship behavior instead of female behavior (Manoli et al. 2005). We see, then, that this single regulatory gene has profound influences on many aspects of mating behavior.

GOALS OF BEHAVIORAL GENETICS

When we observe an animal in nature, its actions are generally a result of many genes interacting with one another and the environment. A goal of behavioral genetics, then, is to identify the gene, or more commonly the genes, that underlie a behavior and to learn the functions of these genes. Another goal of behavior genetics is to decipher the interactions among genes and their products and between genes and the environment to understand why a particular behavior takes the form it does. In order to understand these interactions, it is often useful to quantify the heritability of a behavior.

Heritability is a statistical measure that suggests how strongly a behavior is influenced by genes. Because it is a statistical measure, we must measure the differences in the behavior of a sufficiently large number of individuals. Then, typically by comparing relatives, we can determine how much of the observed variation is due to genetic differences among the individuals and how much is caused by differences in their environments. The **heritability** of a particular trait in a specific population is the ratio of the variation caused by genetic differences to the total amount of variability in the trait in that population. Therefore, heritability can vary from 0 to 1. A value of 0.5 indicates that 50% of the variability in the population studied is due to genetic differences. The heritability of a complex trait such as behavior is rarely more than 50% (Plomin et al. 2003).

Genomics, the study of an organism's entire genome—all of its DNA—has had a major impact on behavioral genetics. The field took off during the 1990s with the beginning of gene-sequencing projects designed to discover the sequence of nucleotides in entire genomes. The goal of functional genomics is to understand the function of genes and noncoding regions of the genome. It often begins with the wealth of information created by gene-sequencing projects, but functional genomics is primarily interested in patterns of gene activity under different conditions or at different developmental stages. Comparative genomics analyzes differences in the genomes of different species. Its goal is to understand how traits have evolved, as well as how they work physiologically. The next level of study is proteomics, which strives to study the full set of proteins coded for by an organism's genes and to understand how these proteins work together to produce and modify traits. As we will see, different genes are expressed in different degrees and in different tissues at different times during development and under different environmental conditions.

METHODS OF BEHAVIORAL GENETICS

As we have seen, the observed variation in behavior among individuals results from differences in genes and in environments. Therefore, when we are interested in exploring the influence of genes on behavior, it is important to rule out environmental effects by raising the animals in the same environment. If environmental conditions are identical for all animals, then observed behavioral differences are due to genetic differences.

INBREEDING

Inbred lines, one of the tools used by behavioral geneticists, are laboratory colonies of individuals that have no, or virtually no, genetic diversity: they are homozygous for nearly all their genes. Inbred strains are usually created by mating close family members with one another for many generations and are useful in behavioral genetics because they create a population with nearly identical genes. By using inbred strains, therefore, it is possible to separate the effects of genes from those of the environment. To show the effects of genes, the behavior of members of two different inbred strains is compared in the same environment. In this case, any observed difference in behavior must be caused by a difference between the genes of the strain. Moreover, the influence of the environment on behavior can also be shown by using inbred strains. If members of the same inbred strain, individuals who are almost genetically identical, behave differently when they are raised under different conditions, then the variation must be caused by environmental effects.

Comparisons of Inbred Strains to Show the Role of Genes

The work of Ádám Miklósi and his colleagues (1997) on the antipredator behavior of paradise fish (*Macropodus opercularis*) larvae is an example of studies that compare the behavior of inbred strains. Paradise fish live in densely vegetated, shallow marshes and rice fields in Southeast Asia, along with several predator fish species. Avoiding predation is crucial to survival, and so one might suspect it has a genetic basis. Having raised strains of paradise fish that had been inbred for over 30 generations, Miklósi could investigate this possibility. He crafted model predators from plastic centrifuge tubes. Since eyes are known to be an important cue in predator recognition in many species, some models were created with black eyespots. Miklósi individually placed the larvae of two inbred strains (S and P) of paradise fish into an experimental tank and observed their responses to model predators for three minutes. There are two common antipredator responses: fleeing, in which the larva suddenly darts by slapping with its caudal fin, and backing, in which the larva swims backward with its body curved. Larvae of strain P were significantly more likely to show antipredator responses, both fleeing and backing, than were the larvae of strain S (Figure 3.3).

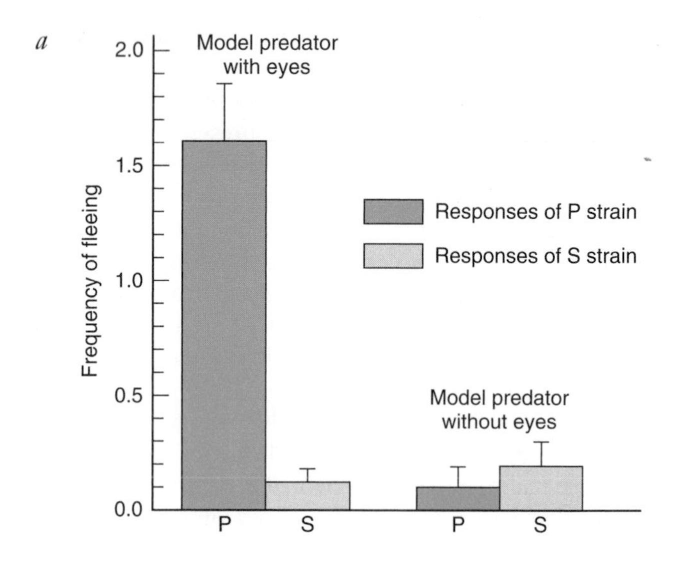

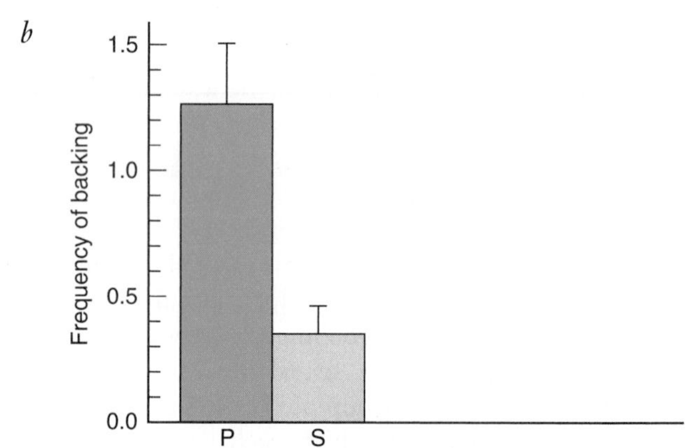

FIGURE 3.3 **The antipredator responses of larvae of two inbred strains of paradise fish when presented with model predators. Fleeing consists of darting by slapping the caudal fin. Backing involves swimming backward with a curved body in a direction perpendicular to the body axes of the larva. Larvae of strain P showed a significantly higher frequency of both fleeing and backing than did larvae of strain S, indicating genetic differences in antipredator behavior between the strains. (Data modified from Miklósi et al. 1997.)**

Comparisons of Inbred Strains to Show the Role of Environment

Because individuals of inbred strains are identical in 98% of their genes, such strains, as we mentioned, can provide a way to hold the genetic input constant while varying the environment. Thus, if differences in behavior are found, they must be due to the environment.

Some of the environmental effects on behavior take place very early in life. Eliminating the effects of early learning is tricky, but it can be done. The simplest way is by a reciprocal cross, one in which males of inbred strain A are mated with females of inbred strain B and males of strain B are mated with females of strain A. Since the individuals of each strain are homozygous for almost every gene, the hybrid offspring of both of these crosses have the same genotype, even though their mothers are from different strains. Therefore, if the behavior of the hybrid offspring of reciprocal crosses differs, it must be an effect of the parental environment. If the young are raised solely by their mothers, as they often are in the laboratory, the difference must be an effect of the maternal environment.

Theoretically, it is even possible to determine whether the maternal environment had its greatest influence on the offspring before or after birth. **Cross-fostering**, transferring the offspring shortly after birth to a mother of a different strain, is a technique for detecting maternal influences that occur after birth. If offspring that were transferred to a foster mother immediately after birth behave more like individuals of the foster mother's strain than like those of their own strain, postnatal maternal influences are implicated.

A cross-fostering experiment involving two different species of voles helped separate the influences of genes and parental environment in the expression of a particular behavior. Prairie voles, *Microtus ochrogaster*, show more parental care than do meadow voles, *M. pennsylvanicus*. For example, prairie vole females spend more time in the nest with their young, contacting their offspring by huddling over them and nursing them. Male prairie voles also show more parental care than do male meadow voles. Prairie vole males share a nest with the female and frequently groom and huddle over the young, in contrast to meadow vole males, who nest separately and rarely enter the female's nest.

To determine whether the species difference in parental care was due to genes or early experience, Betty McGuire (1988) fostered meadow vole pups with prairie vole parents. As a control, she fostered meadow vole pups to other meadow vole parents. When the foster pups became adults and had their own families, she measured the amount of parental care they gave to their second litters. The meadow vole pups that were raised by prairie voles gave more care to their own offspring than did those pups that had been fostered to other meadow vole parents. Cross-fostered females, for example, spent more time huddling over and nursing their young. The early experience of male voles influenced their parental behavior in much the same way as it affected the females' behavior. Male meadow vole pups that were fostered to parents of their own species behaved as meadow vole males usually do, in that they rarely entered the nest with the young. However, four of the eight meadow vole males raised by prairie vole parents nested with their mates and spent time in contact with the young. This cross-fostering experiment shows that the experience a vole has with its own mother can influence the way it treats its own offspring. However, not all behaviors were modified by early experience. Nonsocial behaviors such as food caching and tunnel building and overall activity level were unaffected by the species of the foster parents. Thus, using the technique of cross-fostering, it is possible to determine whether a particular behavior is influenced by the parental environment.

Of course, it would be unethical to create inbred strains in humans. However, we can take advantage of naturally occurring differences in genetic relationships among family members. Parents and offspring share 50% of their genes, and they may also share a similar environment by living together in the same household. So if children raised by strict, aggressive parents grow up to become playground bullies, is it due to genes or environment?

Adoption studies are one way to tease apart the role of genes and environment as causes for family resemblance. For example, when siblings are adopted into different families, we observe genetically related individuals who do not share the same family environment. The degree to which these siblings are similar is an indication of the contribution of genes to the family resemblance. On the other hand, when families with children of their own adopt additional children, we observe children with different genetic backgrounds living in the same environment. The degree to which the family's own children resemble the adopted children is an indication of environment in the development of that trait.

Twin studies are another favorite tool of behavioral geneticists interested in human behavior. In these studies, twins are raised in similar family environments. Researchers then compare the similarities in a particular trait between sets of identical twins (who have 100% of their genes in common) and between sets of fraternal twins (who, on average, have 50% of their genes in common). It is reasoned that similarities between identical twins that are not found in fraternal twins must be due to their genes. Perhaps the most powerful design is the adoption-twin studies. In this design, twins adopted into different families are compared to twins raised in the same family.

ARTIFICIAL SELECTION

Artificial selection is another means of demonstrating that a behavior has a genetic basis. It differs from natural selection in that an experimenter, not "nature," decides which individuals will breed and leave offspring. The rationale for artificial selection is that if the frequency of a trait in a population can be altered by choosing the appropriate breeders, it must have a genetic basis. Usually the first step is to test individuals of a genetically variable population for a particular behavior trait. Those individuals who show the desired attribute are mated with one another, and those who lack the trait are prevented from breeding. If the trait has a genetic basis, the alleles responsible for it will increase in frequency in the population because only those possessing them are producing offspring. As a result, the behavior becomes more common or exaggerated with each successive generation. If the environment is held constant, traits that change under artificial selection must have some genetic basis.

Humans often use artificial selection to create breeds of animals with traits that they consider useful. For example, even if you are not a dog lover, you have probably noticed that different breeds of dogs have different personalities. All dogs belong to the same species, but various behavior traits and hunting skills have been selected for in different breeds (strains). Terriers, for example, are fighters. Aggressiveness was selected for so that they could be used to attack small game. The beagle, on the other hand, was bred to be a scent hound.

Since beagles usually work in packs to sniff out game, they must be more tolerant with companions. Therefore, their aggressiveness was reduced by selective breeding. Shetland sheep dogs were bred for their ability to learn to herd sheep. Spaniels, used for hunting birds, are "people dogs," being more affectionate than aggressive. These behavioral differences among breeds of dogs were brought about by people who placed a premium on certain traits and then arranged matings between individual dogs that showed the desired behavior. In other words, the frequency of particular behavior patterns present in all breeds has been modified through artificial selection (Scott and Fuller 1965).

Artificial selection for nesting behavior in house mice (*Mus domesticus*) has not only demonstrated a genetic basis for the behavior but has also shed some light on how natural selection might work on the trait in wild populations. House mice usually live in fields, where they build nests of grasses and other soft plant material. In the laboratory, both male and female house mice will use cotton as nesting material, which makes it easy to quantify the size of the nest constructed. Carol Lynch (1980) noticed that some mice built larger nests than others did (Figure 3.4). These differences could be due to genetic or environmental factors, or both. Lynch suspected that there was some genetic basis in nesting behavior. To separate the genetic and environmental influences, she began to selectively mate mice, based on the size of the nest they built, and she raised all mice under the same environmental conditions. She began with a population of house mice that gathered between 13 and 18 grams of cotton over a four-day period to build their nests. Then, she selectively mated mice to create lines of high and low nest-building behavior. She mated males and females that built large nests to create high lines and males and females that built small nests to create low lines. In addition, she created control lines by randomly mating males and females from each generation. After 15 generations of artificial selection, the mice of the high nest-building line used an average of 40 grams of cotton for their nests. In contrast, the mice in the low nest-building line used an average of only 5 grams of cotton in nest construction. The mice of the control line built nests about the same size as those built by the mice in the initial population—15 grams of cotton (Figure 3.5). Selection for high and low nesting behavior was continued for over 40 generations, at which time mice from the high nest-building line collected more than 40 times the amount of cotton than did the mice of the low line.

FIGURE 3.4 **Individual differences in the size of nests built by house mice in the laboratory. The genetic basis of these differences in nest-building behavior has been demonstrated by artificial selection experiments. Whereas the mouse on the left was selected to build a small nest, the one on the right was selected to build a large nest.**

The results of this experiment confirm that nest building in house mice does indeed have a genetic basis. Thus, natural selection could have a similar influence on nest building if nest size influences fitness (the number of offspring successfully raised). House mice do, in fact, build larger nests in the north than in the south (Lynch 1992). This suggests that large nests may be a factor that helps mice in cold environments raise more offspring. In the laboratory, Lynch bred groups of mice from both lines at 22°C or at 4°C and counted the number of offspring that survived to 40 days of age. Pup survival in both lines was reduced at the lower temperature. Nonetheless, mice from the lines that built larger nests raised more pups that lived to be 40 days old at both environmental temperatures. Thus, nest building is an important component of fitness, and its genetic basis allows it to be shaped by natural selection (Bult and Lynch 1997).

It is interesting to note, however, that selection can lead to the same observed behavior—constructing large or small nests—by favoring different sets of genes. Notice in Figure 3.5 that two high lines and two low strains were created through artificial selection. Crosses between mice of different high strains (or between mice of different low strains) revealed that there were still genetic differences between strains that expressed the behavior in similar ways. This suggests that natural selection can follow different paths in different populations to produce the same adaptive behavior (Bult and Lynch 1996).

In some cases, selection influences behavior by affecting genes that code for proteins that affect the structure or function of the nervous system. For example, it is possible to selectively breed mice to be very active and to explore an open test arena or to be less active. Such selection results in specific differences in brain structure between the high- and low-activity lines: Specific regions of the hippocampus are more developed in the mice that

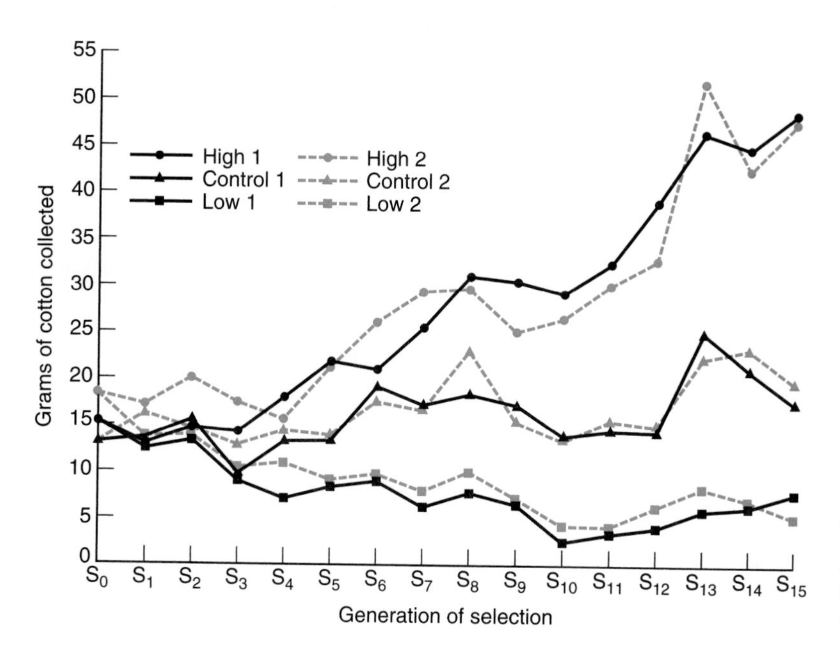

FIGURE 3.5 **Artificial selection for large or small nest size in house mice. The nests of mice in the original population consisted of 13 to 18 grams of cotton. Individuals who built the largest nests were bred with others who built large nests to create a high nest-building line. Individuals who built the smallest nests were bred with others who built small nests to create a low nest-building line. A control line was developed by randomly mating individuals of each generation. After 15 generations, the nests of the high line were an average of 8 times larger than those of the low line—40 grams and 5 grams, respectively. The nests of mice in the control line were roughly the same size as those built by mice in the initial population (15 grams). (Data from Lynch 1980.)**

are selected for high activity (Hausheer-Zarmakupi et al. 1996). As the field of behavioral genomics expands, we will be able to discover the actual genes and proteins responsible for differences such as these.

INDUCING MUTATIONS AND SCREENING FOR CHANGE IN BEHAVIOR

A mutation, a change in a gene's instructions for producing a protein, can be induced by agents that change the DNA bases. When organisms are exposed to mutagenic agents, some of them become behaviorally aberrant and can be separated from the population of normal individuals on this basis. Appropriate genetic crosses can then determine whether the behavioral change is caused by an alteration in a single gene. Even a small change may result in a difference in a specific aspect of an anatomical structure or a physiological process that mediates a behavior. Identifying the anatomical or physiological differences between mutant and normal individuals brings us closer to understanding how genes can influence behavior.

Studies on learning in the fruit fly have helped fill in a few of the missing links between genes and behavior. You have probably never met a fruit fly with remarkable intelligence, but some mutant strains are so poor at olfactory learning that they have earned the epithet *dunce.* Before considering the deficiency caused by the *dunce* mutation, olfactory learning in normal fruit flies and the way of demonstrating it should be described. Normal fruit flies can associate an odor with an unpleasant event, such as an electric shock, and learn to avoid that odor if it is encountered again. This should not be surprising because odors are important in the daily life of fruit flies for locating both food and appropriate mates. William Quinn and his colleagues (1974) demonstrated this avoidance conditioning by shocking a group of flies for 15 seconds in the presence of one odor but not in the presence of a second odor. Then they presented the odors, one at a time, to the flies without shocking them to see how many would avoid each odor. Most normal flies avoid only the odor connected with the electric shock, an association that lasts for three to six hours (Dudai et al. 1976).

Two mutants, *dunce*1 and *dunce*2, were isolated by exposing a population of fruit flies to a chemical known to cause mutations and then screening the flies based on their learning ability. The *dunce* mutations are alleles, or alternate forms, of a single gene on the X chromosome. In contrast to normal flies, *dunces* fail to learn to avoid odors associated with shock when they are taught using Quinn's experimental design (Dudai et al. 1976). Even larval *dunces* are deficient in olfactory learning (Aceves-Pina and Quinn 1979).

Why don't *dunce* fruit flies learn as well as normal flies? The first guess—that the sensory system was defective so that the *dunce* flies could not detect either the odors or the shock—was incorrect. Experiments revealed that the mutants are able to detect both (Dudai et al. 1976). In spite of this ability, mutant flies are unable to remember the association between the shock and an odor.

Apparently, the *dunce* mutants have a problem with the early stages of memory formation. If they are shocked in the presence of an odor without subsequent exposure to a second odor, they do in fact associate the odor with the aversive stimulus and avoid it if tested immediately after training, but the association fades quickly. The association between the shock and an odor is short-lived, and the experience with a control odor during the interval between the training and the test of learning seems to eliminate associations that may have formed. For these two reasons, we conclude that an early stage of memory formation is defective in the *dunce* mutants (Dudai 1979).

What does the *dunce* gene do? It codes for a form of an enzyme called cyclic AMP (adenosine monophosphate) phosphodiesterase. This enzyme is important because it breaks down cyclic AMP (cAMP), which is a mediator of many biochemical processes in different types of cells. Thus, we see that the *dunce* mutation causes a reduced level of cAMP phosphodiesterase and, therefore, an increased level of cAMP. In addition, the mutation impairs an early stage of memory formation in olfactory learning. This should lead you to suspect that the enzyme or cAMP might play a role in the underlying physiological process involved in this type of memory formation. This idea has been tested by inhibiting cAMP phosphodiesterase in normal flies and testing the olfactory learning abilities. When treated in this way, normal flies learn no better than *dunces* do (Byers et al. 1981). Thus, the behavior of *dunce* flies suggests that cAMP has a role in learning and memory.

Another single-gene mutation in fruit flies, *rutabaga*, also causes poor learning and memory. This mutation has filled in some of the details of the connection between cAMP and memory formation. The level of cAMP within a cell actually depends on two enzymes. As we've seen, cAMP phosphodiesterase breaks down cAMP, lowering its concentration. In contrast, another enzyme, called adenylyl cyclase, raises cAMP levels by causing the formation of cAMP from ATP (adenosine triphosphate) (Figure 3.6). *Rutabaga* mutants have a defective form of adenylyl cyclase (Levin et al. 1992). As a result, *rutabaga*'s adenylyl cyclase is not activated by the stimulus involved in learning in the way it would be in a normal fly. Here, the learning stimulus doesn't cause cAMP levels to rise. Thus, *dunce* and *rutabaga* have opposite effects on the level of cAMP within a cell.

As is often the case in science, the more we learn, the more questions we have. In this case, the *dunce* and *rutabaga* mutants prompt us to wonder how cAMP is involved in memory formation. Figure 3.6 summarizes a pathway involving cAMP that is thought to be important in part of a regulatory pathway for memory formation. Notice that cAMP binds to another enzyme, protein kinase A (PKA), and activates it. All protein kinases work by adding a phosphate group to another molecule, activating or inactivating the other molecule. In this case, PKA activates the *CREB* gene, which codes for the protein CREB (*c*AMP *r*esponse *b*inding protein)

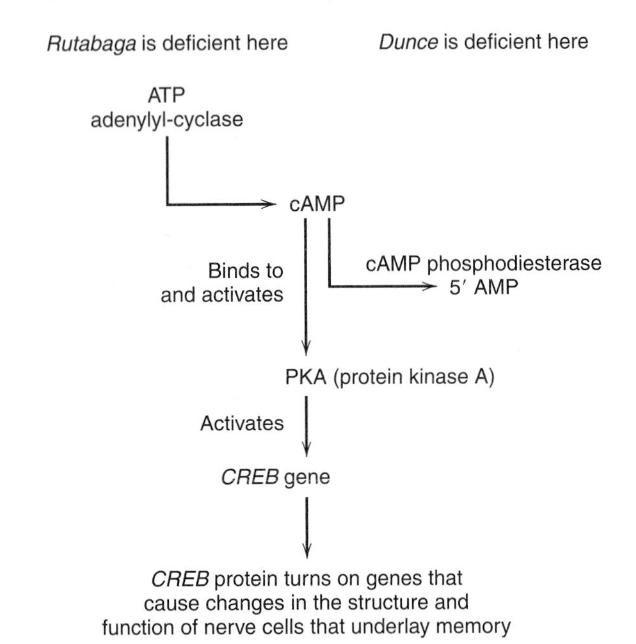

FIGURE 3.6 **A summary of the molecular events that accompany memory formation in fruit flies. Single-gene mutations that result in poor olfactory learning were helpful in uncovering many of the details of memory formation. The *dunce* mutant has a defective form of cAMP phosphodiesterase, an enzyme that breaks down cAMP. *Rutabaga* has a defective form of adenylyl cyclase, an enzyme that forms cAMP from ATP. Thus, both mutations affect the levels of cAMP within nerve cells. The cAMP binds to another enzyme, PKA (protein kinase A), and activates it. Active PKA then turns on the *CREB* gene, whose protein regulates the activity of other genes so that new connections can be made among nerve cells. These connections are responsible, in part, for long-term memory.**

that, in turn, activates other genes. It is not cAMP itselt that produces the behavior, but rather cAMP is part of the regulatory pathway that affects the behavior. These other genes affected by the regulatory hierarcy then control the growth of connections between brain cells, changes in the nervous system that are responsible for memory (Davis et al. 1995).

FINDING NATURAL VARIANTS AND LOOKING FOR GENETIC DIFFERENCES

Most of us would not find foraging fruit fly larvae of great interest, unless the foraging was taking place in a fruit bowl on our kitchen table. But it is fascinating to many behavioral geneticists, because it helps us to understand much more complicated behaviors. The interest began when Marla Sokolowski (1980) noticed two forms of feeding behavior in natural populations of larval fruit flies (*D. melanogaster*). "Rover" larvae move around continually on their food and often leave their food source to look for another. In contrast, "sitter" larvae travel only short distances and tend to remain on a food source. In fact, when these larvae were brought into the laboratory and allowed to feed on yeast paste in a petri dish, the distances traveled by rovers were nearly four times longer than those of sitters. When food is not present, all flies move rapidly, as rovers do. Thus, the difference in foraging styles is not because sitters are energy-deficient or sick (Pereira and Sokolowski 1993; Sokolowski et al. 1997).

HYBRIDIZATION

Sokolowski immediately suspected that a behavior with two distinct forms would be genetically controlled. To investigate the genetic basis of these rover and sitter foraging strategies, Sokolowski performed a series of hybridization experiments using adult rovers and adult sitters in which she mated rovers to sitters, and she compared these "hybrid" offspring to pure rovers and pure sitters.

Sokolowski began by crossing adult rovers with other rovers and adult sitters with other sitters, creating parental strains of sitters and rovers. The responses of male larvae resulting from these crosses are shown in Figure 3.7*a*. Notice that the frequency distribution of path lengths of rovers is easily distinguished from that of sitters. Next, Sokolowski crossed adults of sitter larvae with adults of rover larvae. The path lengths of the first generation (F_1) offspring of these crosses are shown in Figure 3.7*b*. Nearly all the larvae were rovers. When the F_1 adults were crossed with one another, the resulting second-generation (F_2) larvae consisted of both rovers and sitters in a ratio of three rovers to one sitter (Figure 3.7*c*). This ratio of offspring is what would be expected if the variation in foraging strategies in a natural population could be explained by variation in a single gene, dubbed *foraging* (*for*), with two alleles (alternative forms). The two alleles of *for* are for^R (rover) and for^S (*sitter*), and for^R is dominant to for^S. Thus, rover flies are for^Rfor^R or for^Rfor^S and sitters are for^Sfor^S (de Belle and Sokolowski 1987).

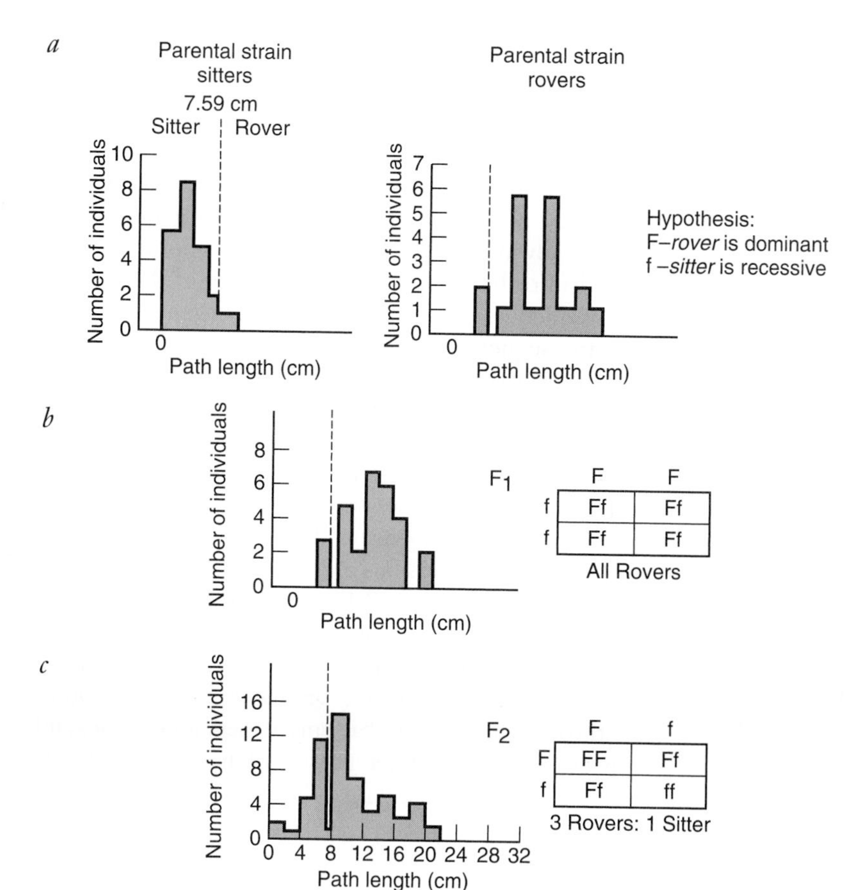

FIGURE 3.7 **The results of mating experiments on fruit flies showing "rover" or "sitter" foraging strategies as larvae. Rover larvae forage over significantly longer distances than do sitter larvae. (*a*) The parental strains were created by crossing rovers with rovers or sitters with sitters. (*b*) The frequency distributions show foraging path distances of male (F1) larvae of crosses between adults of the parental rover strain and adults of the parental sitter strain. Almost all the resulting larvae were rovers. (*c*) The frequency distributions show foraging path distances of male (F2) larvae resulting from crosses between F1 flies. The results are consistent with the hypothesis that the trait is controlled by a single gene, called *foraging (for)*, and that the *rover* allele is dominant to the *sitter* allele. Punnett squares of the crosses are shown on the right. (Data modified from de Belle and Sokolowski 1987.)**

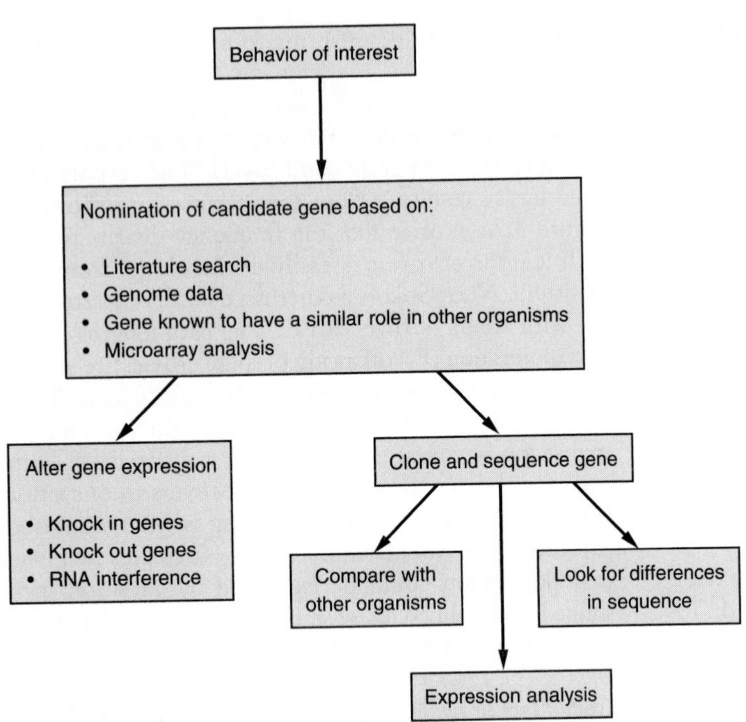

FIGURE 3.8 **The candidate gene approach for identifying genes underlying a behavior of interest nominates genes based on their location or their role in other organisms.**

THE *FORAGING* GENE AS AN EXAMPLE OF BEHAVIORAL GENETICS IN ACTION

We will continue with the story of the quest to understand the role of genes and environment in fruit fly foraging. This story will illustrate some basic principles of behavioral genetics and some of the techniques used to look at the role of genes in behavior.

CANDIDATE GENES

A series of breeding experiments showed that *for* is located on *Drosophila* chromosome 2 (de Belle et al. 1989) close to that of a gene for an enzyme PKG (cyclic GMP[1]-dependent kinase), which was already known to be important in signaling pathways within the cell (reviewed in Barinaga 1994). Next, the gene was cloned (many identical copies of it were created) and was shown to be identical to a known *Drosophila gene* (*dg2*) that produces PKG. Indeed, the heads of adult rover flies had significantly more of the enzyme PKG than did the heads of adult sitter flies. The function of *for* and the difference in the level of PKG in rover and sitter flies are consistent with the idea that PKG causes the difference in behavior, and is therefore part of the regulatory hierarchy affecting foraging behavior of these flies (Osborne et al. 1997).

Over the course of evolution, the DNA sequences that influence a particular behavior tend to be conserved between species. Thus, once researchers know the genetic basis of a trait in one species, they have a good hint about the genetic basis of the same trait in other species. For example, after the links between the *for* gene, PKG, and fruit fly foraging behavior were identified, researchers began to suspect that the *for* gene might play a role in food-related behaviors in other organisms. In other words, *for* became a candidate gene for foraging behavior. The nomination of a **candidate gene** may be based on a search of the literature for genes known to be involved in producing a similar behavior in another organism or by comparing the sequence of the gene to the sequences of genes in other organisms using genome-sequencing data (Figure 3.8).

In this case, a search of databases of gene sequences revealed genes with similar sequences and function as the fruit fly *for* gene in three other organisms: the honeybee (*Apis mellifera*), red harvester ants (*Pogonomyrmex barbatus*), and the nematode *Caenorhabditis elegans* (Figure 3.9). Slight differences in the sequence of the gene have arisen during evolution, but the versions of *for* in all these species affect the level of PKG (Table 3.1). The genes in these organisms are orthologues, meaning that they descended from the same ancestral gene and have the same function. As we will see, however, *for* activity and *for*-PKG affects foraging through different mechanisms in these four species and may be regulated in different ways (Tan and Tang 2006).

Honeybees

Fruit flies forage to satisfy their own hunger. In contrast, honeybee workers forage in order to bring food back to their colony, and their own hunger is not lessened by

[1]Guanosine monophosphate.

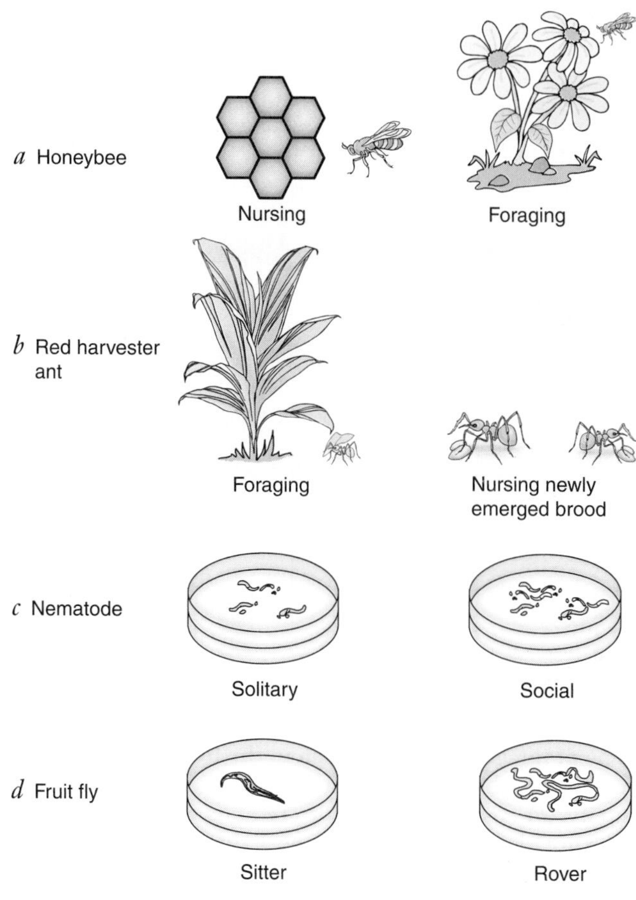

FIGURE 3.9 **When the foraging gene is expressed, PKG levels increase, changing food-related behavior in honeybees, red harvester ants, nematodes (*C. elegans*), and fruit flies. (Modified from Tan and Tang 2006.)**

foraging. Nonetheless, the same *for* gene regulates foraging in both organisms, albeit in very different ways.

Unlike fly larvae, whose main job in life is to eat, the duties assigned to a honeybee worker depend largely on her age. For the first few days after a honeybee worker emerges from her brood cell, her responsibilities are janitorial—she cleans the cells. Next, she begins taking on the duties of a "nurse," which include caring for and feeding the brood. When she is about 2 or 3 weeks old, she ventures out of the hive and begins to collect pollen or nectar in her new role as a forager.

The age-related switch from nurse to forager is associated with an increase in the activity of the *for* gene in the brains of honeybees. The level of *for* mRNA indicates *for* gene activity, that is, how much the gene is being expressed. The greater the *for* gene activity, the greater the *for* mRNA level. Recall that *for* mRNA encodes PKG. As you can see in Figure 3.10, levels of *for* mRNA in the brain are higher in foragers than in nurses (Ben-Shahar et al. 2002).

Is this conclusive evidence that *for* gene activity underlies the switch from nurse to forager? No. Age is another variable that differs between nurses and foragers. To rule out age as the cause of the change in colony responsibilities, Yehuda Ben-Shahar and his colleagues (2002) manipulated colonies so that all members were initially one day old. The absence of foragers caused some members of the colony to switch to foraging as much as two weeks sooner than usual. The nurse bees and the forager bees were the same age, but the brains of foragers were still higher in *for* mRNA. Thus, the switch in behavior from nurse to forager is due to the *for* gene activity and is not an age-related progression.

Harvester Ants

Still another variation on exactly how *for* affects behavior comes from red harvester ants. Like bees, red harvester ants live in large colonies with 10,000 to 12,000 workers. Some workers perform tasks inside the colony, such as caring for newly emerged ants, while others forage outside the colony. The number of workers changes with environmental conditions. As we saw in honeybees,

TABLE 3.1 **Orthologues of the *Foraging* Gene**

Gene	Organism	Behavior	Inheritance
for	Fruit fly (*Drosophila melanogaster*)	Natural variation in an individual fly's foraging behavior; rovers have longer foraging paths than sitters	Allelic variation: rover (for^R) is dominant to for^S (Sokolowski et al. 1997)
egl-4	A nematode worm (*Caenorhabditis elegans*)	Solitary to social feeding	Mutation in *egl-4* causes solitary feeding; expression of *egl-4* causes worms to aggregate in social feeding (Fugiwara et al. 2002)
amfor	Honey bee (*Apis mellifera*)	Change in caste from nurse (within-hive duties) to forager (out-of-hive duties)	Gene expression of *amfor* is lowest in nurses and highest in foragers (Ben-Shahar et al. 2002)

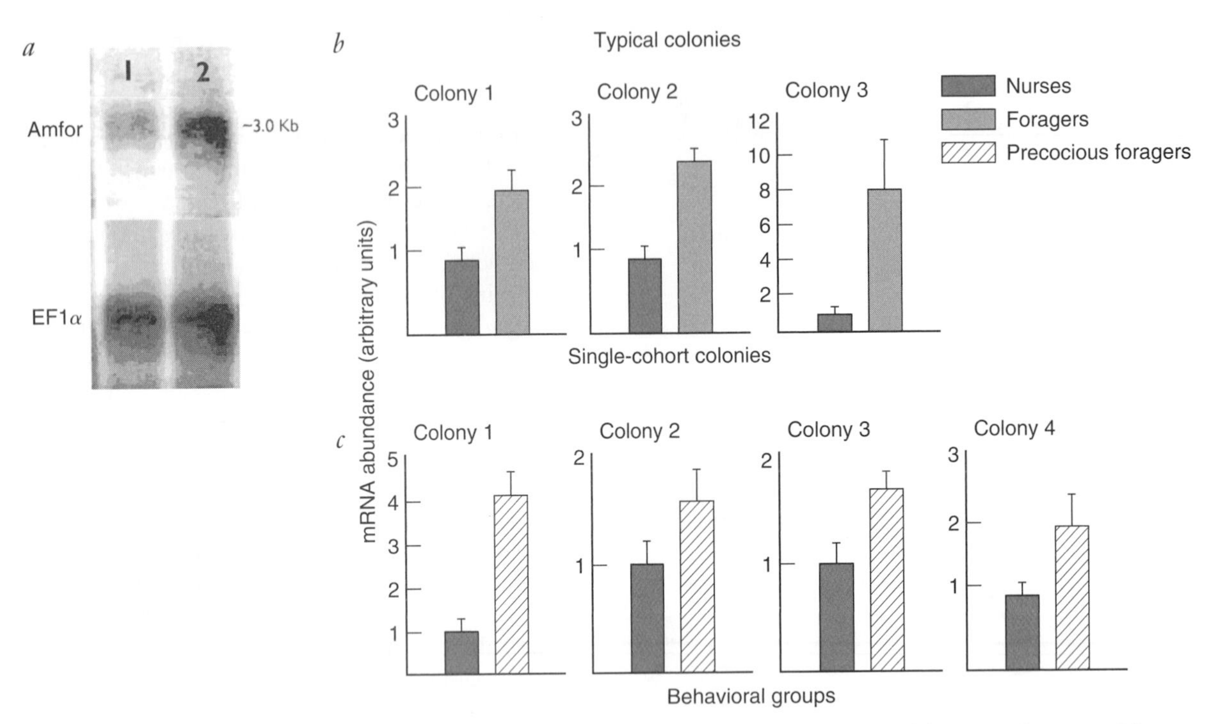

FIGURE 3.10 The *foraging* gene becomes active as nurse honeybees leave the hive and become foragers. The level of *for* mRNA, which encodes PKG, is higher in the brains of forager honeybees than in the brains of nurse honeybees. (*a*) Northern blots of the brains of nurse and forager honeybees. The darkness of the band reflects the level of *for* mRNA and, therefore, the degree of gene activity. (*b*) Histograms of arbitrary units of *for* mRNA in the brains of nurses and foragers indicate greater *for* gene expression in foragers. (*c*) The activity of the *for* gene in foragers is greater than that in nurse bees of the same age. (From Ben-Shahar et al. 2002.)

the harvester ant version of *for* varies in expression as workers switch from within-colony duties to foraging. However, in harvester ants the relationship between *for* expression and behavior is the opposite of the relationship in honeybees: *for* expression is greater in the brains of workers within the colony than it is in the brains of foragers (Ingram et al. 2005).

Nematodes

Similar to the fruit fly, the nematode (roundworm) *Caenorhabditis elegans* has two forms of foraging: roaming and dwelling. A roamer travels long distances without stopping. In contrast, a dweller travels short distances and makes frequent stops. As in the fruit fly, differences in the alleles of the nematode *for* gene result in roaming or dwelling during foraging (Fujiwara et al. 2002). However, unlike the fruit fly where roving increases with *for* gene activity, in *C. elegans* roaming decreases with *for* gene activity.

LINKING A PROTEIN TO A TRAIT

So far, we have presented very strong circumstantial evidence that foraging behavior is associated with the protein PKG. How can we confirm this? A particularly elegant technique to confirm whether a particular candidate gene underlies a behavior is to alter the gene's expression. This can be done by increasing gene activity by adding copies (knocking in) of the gene or by decreasing gene activity by disabling (knocking out) genes. Knocking in genes increases the amount of that gene's protein product, causing a greater effect on behavior. Knocking out a gene eliminates the product of the disabled gene. If the probability that the behavior of interest is displayed is changed by altering the activity of the gene, it suggests that the gene is involved in producing that behavior (Fitzpatrick et al. 2005).

The technique of knocking in was used to confirm the mechanisms by which the *for* gene alters the feeding strategy. By adding four extra copies of the *for* gene to the nuclei of eggs from a sitter fruit fly, Sokolowski's team showed conclusively that the *for* gene affects the rover vs. sitter behavior through its effects on PKG activity. Recall that sitter flies normally have lower PKG activity than do rover flies. When extra copies of *for* were added to sitter fly egg nuclei, the resulting flies had PKG levels similar to those of rover flies and their foraging behavior was similar to that of rovers (Osborne et al. 1997). In nature, then, mutations in *for* reduce the amount of PKG produced, causing the sitter foraging behavior.

The technique of knocking out genes has confirmed that *for* may act differently depending on the organism it is in. When the nematode *for* gene is

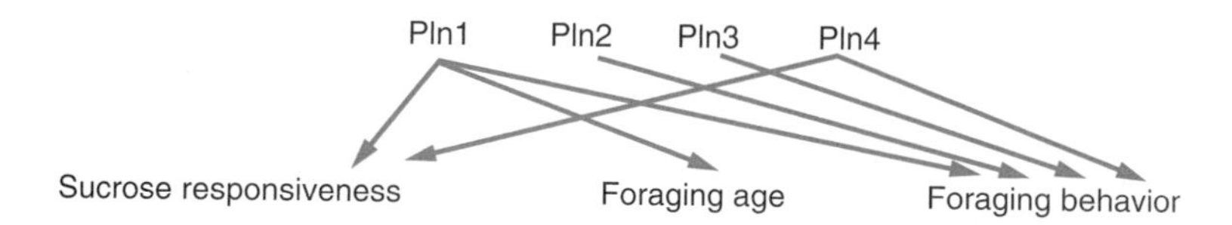

FIGURE 3.11 Interactions between the "pollen" QTLs (*pln* 1-4), the genes involved in foraging and sucrose responsiveness in honeybees. (Modified from Hunt et al. 2007.)

knocked out, there is an increase in roaming behavior (Fujiwara et al. 2002). Note that this is the reverse of what happens in the fruit fly, where roving increases with PKG activity.

LOCATING ALL THE GENES ASSOCIATED WITH A TRAIT

The expression of most behaviors does not fall into discrete categories, such as rover or sitter. Instead, the expression varies continuously, similar to the way that height varies throughout a population. The reason for continuous variation is that many genes play a role in shaping behavior. We describe a trait (behavior in this case) as a **quantitative trait**[2] when many genes are involved. A region of DNA associated with a particular quantitative trait is called a quantitative trait locus (QTL). Thus, there are many quantitative trait loci (QTLs) associated with any complex behavior. QTL analysis tells us whether the trait is determined by a few genes, each having a large effect, or many genes, each having a small effect. Thus a goal of QTL research is to identify the many genes underlying a trait and the extent to which each of them alters the trait (Plomin et al. 2003). Furthermore, once a QTL has been identified, that region of DNA can be sequenced and compared to a database of DNA containing genes whose functions are known. This information can help identify candidate genes, that is, genes that might be involved in the trait in question.

We have seen that a honeybee worker's change from within-nest duties, such as caring for the brood, to foraging is caused by an increase in the activity of the *foraging* gene, which causes an increase in PKG, an important cell-signaling molecule. Once a worker has become a forager, she has many decisions to make. The first decision concerns the type of food. Bees forage for pollen, nectar, or both. Many factors influence this choice, including genetic makeup, brood status, colony resources, and availability of food resources (Rüppell et al. 2004). Strains of bees were bred and selected on the basis of the amount of pollen the workers stored in the colony. The workers in colonies that store large amounts of pollen tend to specialize more in pollen (rather than nectar) collection and carry heavier loads of pollen than do bees from colonies that do not hoard pollen. Those traits might be expected, but other behavioral traits associated with pollen-hoarding were not so easy to predict: workers mature to foragers at a younger age, they are more active after emerging, they are better at learning various tasks, and they are more responsive to sucrose (Rüppell et al. 2006).

There are four genes known to underlie this collection of behaviors associated with foraging. These "pollen" QTLs are named *pln-1*, *pln-2*, *pln-3*, and *pln-4*. *Pln-1* and *pln-2*, which are associated with the size of the pollen loads collected by workers, are responsible for 59% of the variation in the amount of pollen stored in honeybee colonies (Rüppell et al. 2004). *Pln-2* and *pln-3* influence the bee's ability to determine the concentration of sugar in the nectar collected. *Pln-4* is either the honeybee *for* or a gene located close to it. Interactions of these genes influence the age at the onset of foraging and sucrose responsiveness, which affect the choice of food source (Figure 3.11). Thus, each of these genes has a very specific role to play in creating the behavior of pollen collection.

MICROARRAY ANALYSIS

Another innovative technique that allows us to simultaneously investigate the effect of many genes on a behavior is microarray analysis. As you now know, a number of genes may be present in an animal, but not all of them are expressed (active) during a particular behavior. **Microarray analysis** enables us to create a gene expression profile by monitoring the expression of hundreds or even thousands of genes at once. A DNA microarray consists of thousands of DNA sequences stamped onto a solid surface, such as a glass slide. Molecular tags are used to identify mRNA produced by each of the genes. The greater the mRNA production, the more active the gene. In this way, it is possible to compare gene activity in different tissues or in the same tissue at different times. The genes that are active *only* during a particular behavior may play a role in producing that behavior (Hofmann 2003).

Charles Whitfield and his colleagues (2003) looked at patterns of gene expression of about 5500 genes in nurse bees (5 to 9 days old) and forager bees (28 to 32 days old) from a typical colony. Nurses and foragers showed significant differences in the expression of 2200 (39%) of the 5500 genes tested (Figure 3.12). Although the expression levels of many genes differ, the magnitude of the difference is not great. Thus, modest

[2]Quantitative traits are also called polygenic traits.

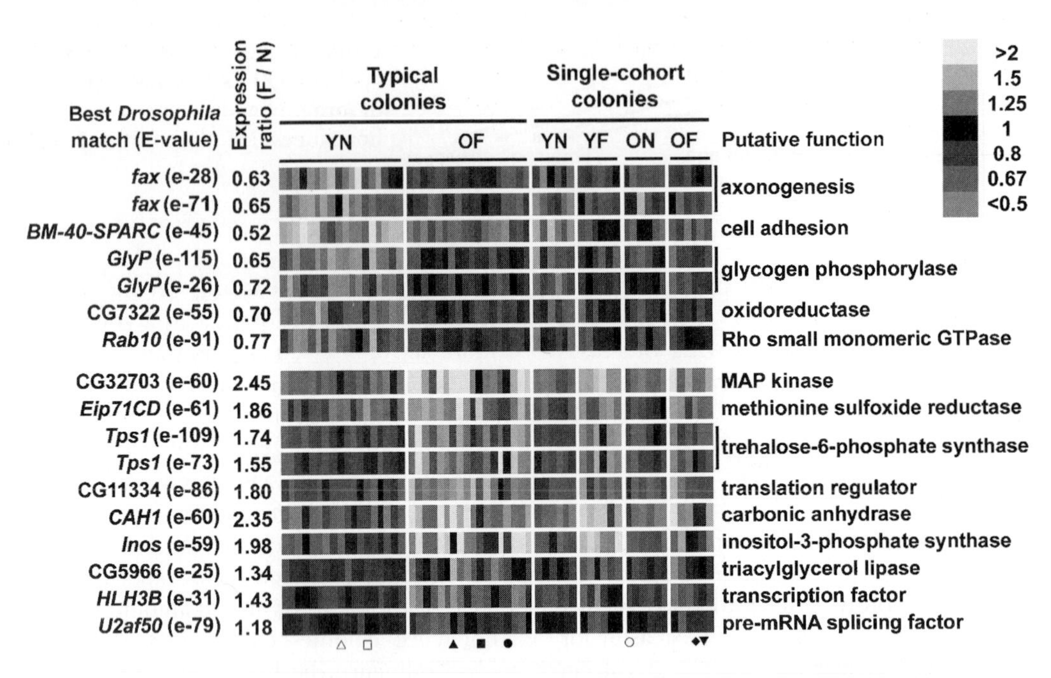

FIGURE 3.12 The gene expression profile in the brain of a honeybee nurse differs from that of the forager brain (left), even if the nurse and forager are the same age (right). Each bar shows the activity of a particular gene, named on the left. The intensity of activity is indicated by color. (From Whitfield et al. 2003.)

changes in gene activity are correlated with large behavioral differences.

DNA microanalysis was also used to confirm the earlier study that looked only at *for* gene activity—the pattern of gene activity is correlated with the behavioral chores performed by nurses or foragers, not with their age. Recall that in colonies created so that they are initially composed of only young bees, some workers will mature into foragers as much as two weeks earlier than usual and others will remain nurses for much longer than usual. Thus, there are precocious foragers and overage nurses. The gene expression patterns in the brains of age-matched nurses and precocious foragers were compared, as were the gene expression patterns of age-matched foragers and overage nurses. Those four groups were analyzed along with nurses and foragers from typical colonies. The gene expression patterns reflect the bee's behavior, not its age. Furthermore, the team was able to predict with 92% to 95% accuracy whether an individual bee was a nurse or a forager by its pattern of gene expression alone (Whitfield et al. 2003).

IMPORTANT PRINCIPLES OF BEHAVIORAL GENETICS

The *for* gene has been useful to illustrate the techniques used to discover the relationship between genes and behavior. Now, as we turn to broader issues of behavioral genetics, we will include both the *for* gene and genes that influence other behaviors, including social behaviors.

ONE GENE USUALLY AFFECTS SEVERAL TRAITS

PKG is an important cell-signaling molecule in the regulatory hierarchy of many traits, so it should not be surprising that the *for* gene that codes for it influences behaviors other than locomotion during foraging. For example, *for* influences how fruit flies respond to sugar. When stimulated with a drop of sucrose, a fruit fly extends its proboscis to feed. Rovers do this more readily than do sitters. Furthermore, rover flies continue to respond to repeated sucrose stimulation for a longer time than sitter flies do. In nature, adult rovers move further away from the food source after feeding than adult sitters. Perhaps rovers and sitters differ in their evaluation of the food source. Rovers may continue searching for food because they remain responsive to food; sitters may stay put because they lose responsiveness to food. As a result, rover foraging behavior may have an advantage in places where food sources are scattered (Scheiner et al. 2004).

Rovers and sitters also differ in the loss of responsiveness to repeated electrical stimulation of the brain. When a stimulus occurs repeatedly without consequence, animals usually lose responsiveness to that stimulus. This loss of responsiveness, called habituation, is discussed in Chapter 5. In fruit flies, electrical brain stimulation activates the giant fiber jump-and-flight escape response. When fruit fly brains are electrically stimulated repeatedly, sitter fruit flies habituate more rapidly than do rovers. Thus, as we saw with responsiveness to sugar, rovers remain responsive to electrical stimulation longer than sitters do (Engel and Hoy 2000). Similar patterns of responsiveness to stimuli affecting different sensory modalities indicate that the *for* gene affects behavior at the level of the brain, not at the sensory level (Scheiner et al. 2004).

In honeybees, the *for* gene is regulated by vitellogenin, the protein product of the *vitellogenin* gene, which has multiple effects on social organization. Vitellogenin is responsible for the onset of foraging because it inhibits juvenile hormone (JH), a hormone that stimulates the transition to forager. The level of vitellogenin naturally declines with age, removing the inhibition of JH and allowing the transition to forager. In addition, high levels of vitellogenin earlier in life prime foragers to specialize in collecting pollen. Vitellogenin also slows aging by protecting against free radicals.

Mindy Nelson and her co-workers (2007) used a technique called gene knockdown to demonstrate that the honeybee *vitellogenin* gene regulates social organization through multiple (pleiotropic) effects on the onset of foraging and foraging specialization on nectar or pollen (Figure 3.13*a*). In this case, gene activity was knocked down using RNA interference (RNAi), which is double-stranded RNA that is complementary to the mRNA from the gene of interest. When introduced into a cell, the RNAi binds to a protein to form a complex that destroys the particular mRNA of interest. In

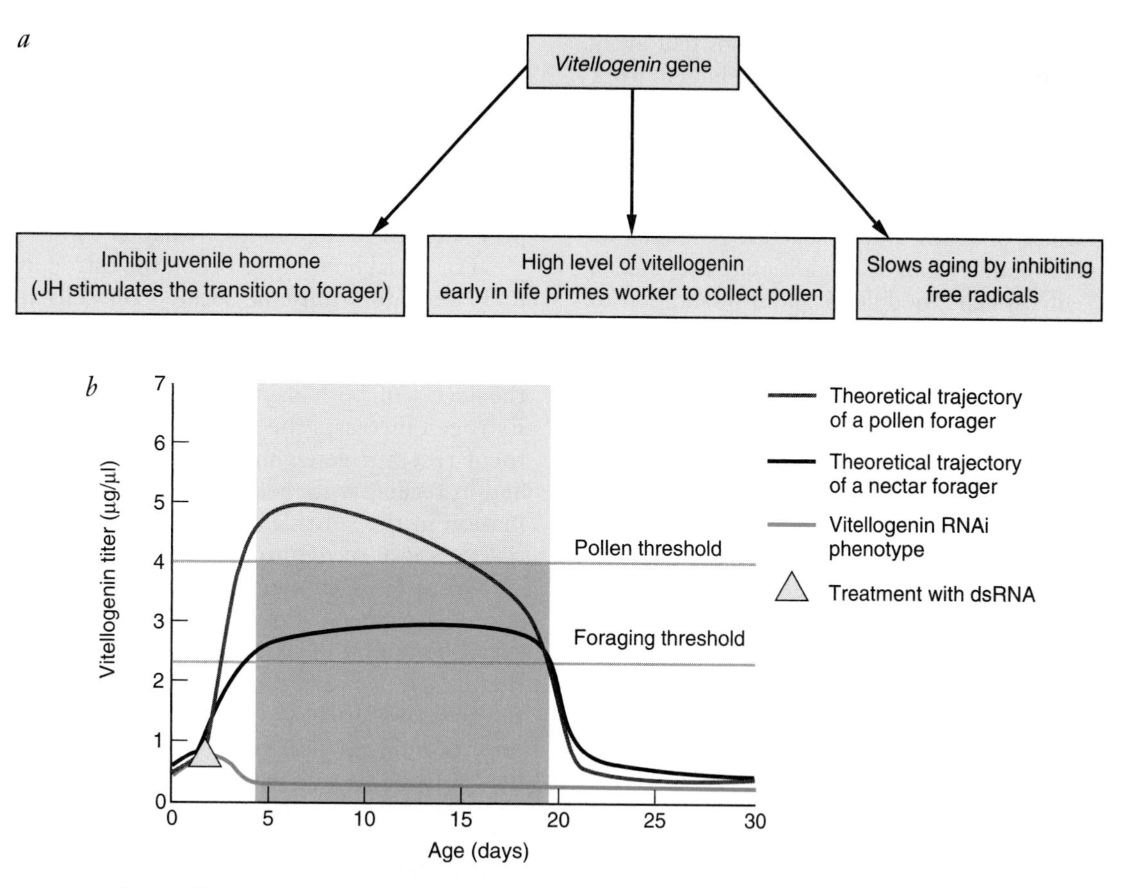

FIGURE 3.13 The vitellogenin gene has several effects. (*a*) The vitellogenin gene affects the timing of the transition to foraging, foraging specialization, and aging. (*b*) A single gene, *vitellogenin*, coordinates the transition from nurse to forager and primes the forgers to collect either nectar or pollen. The activity of the *vitellogenin* gene was lowered using RNA interference, which resulted in very low vitellogenin production. If vitellogenin levels are above the foraging threshold, the change from nurse to forager is suppressed. The level of vitellogenin decreases with age, and when it drops below the foraging threshold, the likelihood of the onset of forging increases. The level of vitellogenin before it drops below the foraging threshold influences whether the forager will specialize in collecting pollen or nectar. If the vitellogenin level was high, the forager specialized in pollen collection. When vitellogenin production was knocked down using RNAi, the workers began foraging at an earlier age than usual and collected pollen. (From Nelson et al. 2007.)

this experiment, RNAi lowered the mRNA produced by the *vitellogenin* gene and lowered vitellogenin levels. Worker honeybees whose vitellogenin levels were lowered by RNAi switched from nurse to forager at an earlier age, preferred nectar as a food source, and lived shorter lives (Figure 3.13*b*). Many genes, like *vitellogenin*, affect many traits. This has implications for the evolution of behavior, as a variety of traits may share an underlying genetic basis and may thus not evolve independently from one another. We'll return to this point in the next chapter.

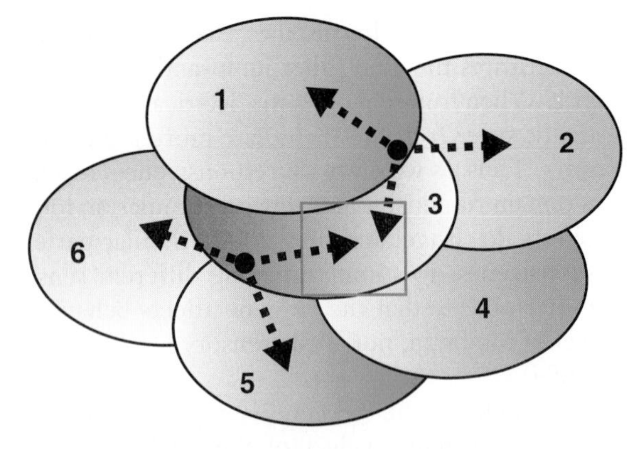

FIGURE 3.14 **Hypothetical diagram of overlapping genetic modules. Each circle represents a genetic module that controls a specific behavior. Each genetic module consists of interacting networks of genes. The arrows represent mRNA transcripts that can affect certain genetic modules. In this scheme, gene module 1 affects gene modules 2 and 3 in addition to itself. Gene module 3 affects gene modules 5 and 6 in addition to itself. As a result, gene module 1 can affect gene module 5 through its effects on gene module 3. (From Amholt 2004.)**

GENES WORK IN INTERACTING NETWORKS

While genes such as *for* have strong influence on behavior, they do not act alone. Microarray analysis has taught us that genes work in functional networks and that those networks interact to form genetic modules, genes that work together closely, that are responsible for a certain behavior. The genetic modules have regions that overlap in some of their genes, allowing the modules to work together and to regulate one another. Figure 3.14 shows a hypothetical diagram of overlapping genetic modules. Each circle represents a genetic module that controls a specific behavior. Each genetic module consists of interacting networks of genes. In this scheme, gene module 1 affects gene modules 2 and 3 in addition to itself. Gene module 3 affects gene modules 5 and 6 in addition to itself. As a result, gene module 1 can affect gene module 5 through its effects on gene module 3. Thus, a change in a single gene can cause a change that can have a ripple effect that causes changes in interacting gene modules and have effects on many related traits or behaviors (Amholt 2004). This property allows many behaviors and parts of an animal's physiology to be coregulated in important ways.

We see the importance of each gene in a network in the interactions among four genes that affect the formation of social bonds in rodents. Social bonding requires that an individual be motivated to approach and engage another individual and that the two animals are able to identify one another through remembered social cues. Then, under appropriate conditions a social bond can form (Lim and Young 2006).

Social recognition between female mice depends on interactions among the protein products of four genes: estrogen receptor α, estrogen receptor β, oxytocin, and an oxytocin receptor. Estrogen and oxytocin are hormones, which are chemical messengers that are released into the bloodstream and carried to cells throughout the body. (Hormones and behavior are discussed in more detail in Chapter 7.) Some hormones, including oxytocin, are produced by neurons. For a cell to respond to a hormone, the cell must have a receptor for that hormone. If it does, the hormone binds to the receptor, and then the cell responds the way that type of cell is programmed to respond (see Chapter 7).

For social bonding to occur, the estrogen and oxytocin receptors must be located on cells in specific regions of the brain. The estrogen receptors are important because estrogen plays a role in regulating the levels of both oxytocin and oxytocin receptors. Estrogen increases the level of expression of the oxytocin receptor genes in the hypothalamus. Oxytocin and its receptors are needed to process olfactory information in the amygdala of the brain, which is essential for social recognition.

Elena Choleris and her co-workers (2004) demonstrated that interactions of all four genes are needed for social bonding to occur. They created knockout mice that lacked one of the four components. If even one of the genes was disabled, the mutant mice couldn't distinguish a familiar mouse from a stranger. Based on this evidence, Choleris and her colleagues have proposed that a small network involving these four genes underlies the olfactory basis of social recognition (Figure 3.15). In this model, olfactory cues are detected by the usual sensory receptors, and the information is sent to the amygdala, a reward region of the brain. Estrogens from the ovaries bind to estrogen receptors β in a region of the hypothalamus (known as the paraventricular nucleus). Estrogen is a steroid hormone that often acts by binding to a receptor, entering a cell's nucleus, and turning on certain genes. In this case, estrogen turns on the oxytocin gene. The oxytocin reaches the amygdala through

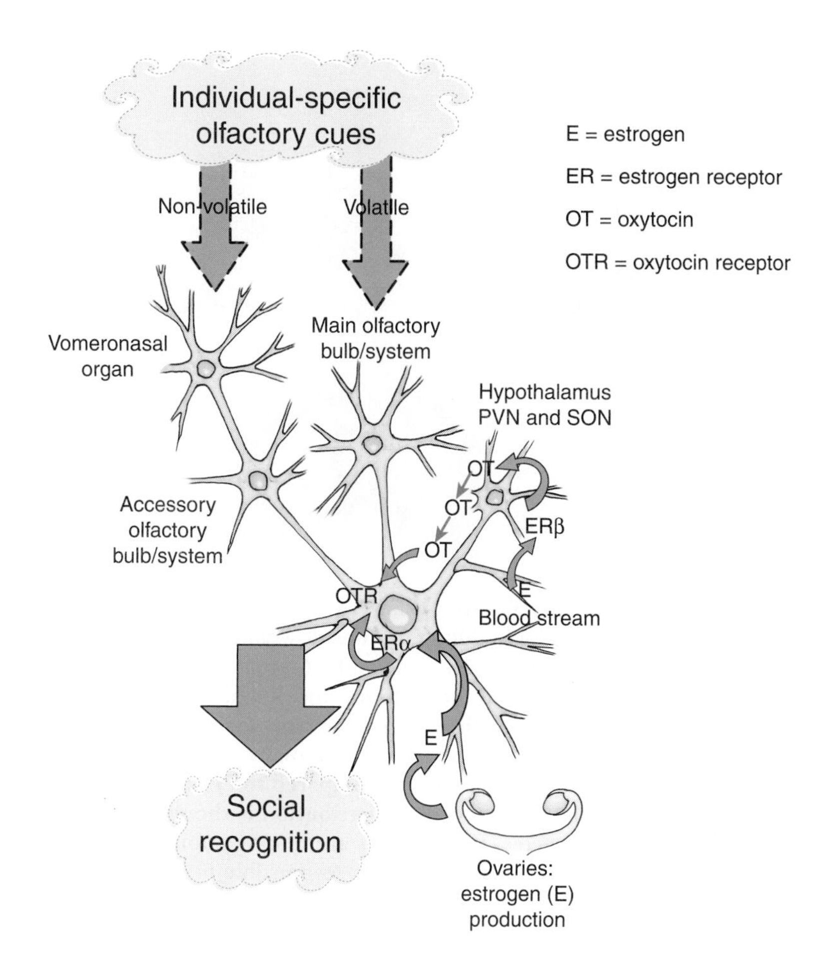

FIGURE 3.15 **A diagram of a four-gene micronet underlying social recognition. Estrogen produced by the ovaries binds to estrogen receptor α in the amygdala of the brain, regulating the expression of the oxytocin receptor gene. Estrogens also bind to estrogen receptor β in the hypothalamus of the brain, which regulates the expression of oxytocin. The binding of oxytocin to its receptors underlies social recognition. (From Choleris et al. 2004.)**

axons from the hypothalamus. In the amygdala, oxytocin binds to estrogen receptor α and turns on oxytocin receptor genes. We see, then, that any disruption to the gene interactions in this module causes the entire system to fail.

BEHAVIORAL VARIATION AND GENES

As we have seen, genetically controlled behavior is not necessarily fixed and stereotyped. Instead, it can be dynamic and responsive to the environment. What roles might genes play in such variation in behavior?

Behavioral variability might be caused by differences in the alleles present in an individual. Recall that the foraging style of fruit flies depends on which alleles of *for* are present. If for^R is present, the larva or fly is a rover, but if both alleles are for^S it would be a sitter.

Alternatively, behavioral variation might be caused by differences in gene regulation. Microarray analysis shows us the pattern of gene activity in a given tissue under certain conditions. If the conditions change, so does the pattern of gene activity. New genes will be turned on, and other genes will be turned off. A gene's level of expression can also be modified. Regulatory regions of DNA control gene activity and are, therefore, responsible for the pattern of gene activity.

Gene regulation might vary over time, as the honeybee *for* gene does. Recall that the transition from nurse to forager occurs as the activity of *for* increases, thereby raising the level of PKG (Ben-Shahar et al. 2002).

The behavior of other organisms may vary because of differences in the tissues in which the gene is expressed. We see this effect in social bonding of voles. Prairie voles are monogamous, meaning that they remain with a female after mating. In contrast, a male meadow vole is nonmonogamous; he loves them and leaves them. Why?

Male prairie voles respond to the hormone vasopressin differently than male meadow voles because of differences in the distribution of vasopressin receptors (V1aR) in the brains of the two species (Figure 3.16). As with all hormones, vasopressin must bind to a receptor to bring about its effects. Only cells with vasopressin receptors can respond to vasopressin. Male prairie voles have many more vasopressin receptors than do male meadow voles, and their receptors are concentrated in the "pleasure center," a reward system of the brain. In contrast, male meadow voles have few vasopressin receptors in the reward system of the brain (Young et al. 1999).

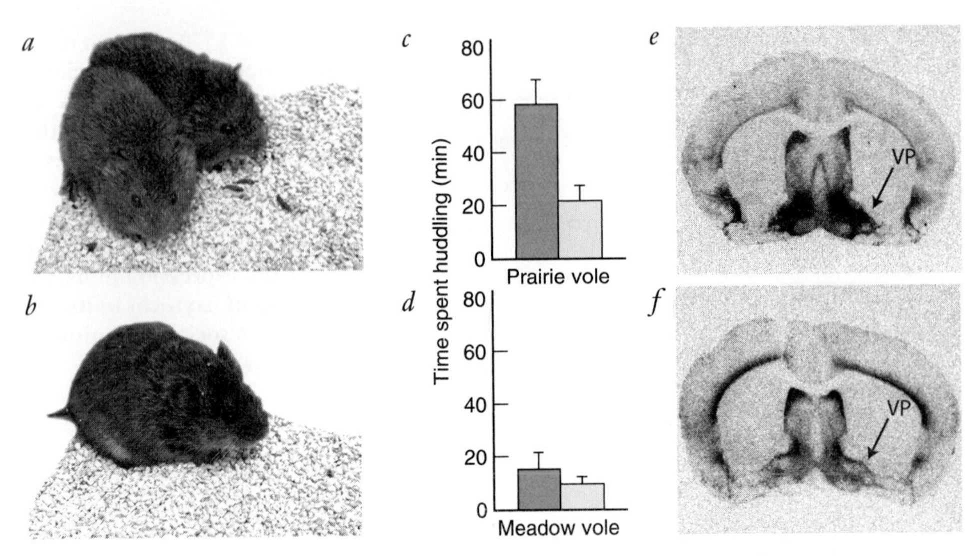

FIGURE 3.16 **Social bond formation and vasopressin receptor distribution. Whereas a (*a*) prairie vole male forms a pair bond with his partner after mating, as shown by huddling, (*b*) meadow vole males remain solitary. Histograms showing the amount of time a (*c*) male prairie vole and a (*d*) male meadow vole spend huddling with a female after they have mated with her. Monogamous male prairie voles spend time huddling with a female after he has mated with her. Promiscuous male meadow voles spend little time with a female after mating. (*e*) Male prairie voles have many vasopressin receptors (VP) in the reward centers of the brain (ventral pallidum), as compared to (*f*) the number of vasopressin receptors in meadow voles. The difference in receptor distribution is thought to be responsible for the difference in pair bonding in these species. Prairie voles are monogamous; meadow voles are not. A promoter found in prairie voles but not meadow voles is responsible for the differences in receptor distribution. (From Lim et al. 2004.)**

After mating, equal amounts of vasopressin surge in the brains of males of both species. In the prairie vole, the vasopressin binds to its receptors and stimulates the reward system. Afterward, the male prairie vole associates the odor of his female mate with the feeling of reward caused by vasopressin, resulting in a preference for the female mate over a strange female. Male meadow voles have fewer vasopressin receptors in the brain's reward system, so the vasopressin released during mating does not stimulate the reward system, and no pair bond forms.

Miranda Lim and her colleagues (Lim et al. 2004) hypothesized that genetic variation in the vasopressin receptor gene underlies the differences in the distribution of vasopressin receptors in the brains of monogamous and polygamous voles. They tested the hypothesis by comparing the DNA sequences of the vasopressin receptor gene in several vole species, and found that the coding sequences are 99% identical. Thus the protein produced by the gene is virtually the same in every vole species tested.

Monogamous and polygamous voles do differ in the promoter region of the vasopressin receptor gene. The promoter is a region of DNA needed to turn on the gene. Monogamous species, such as the prairie vole, have an expanded region of repetitive DNA in the promoter region. The polygamous meadow vole lacks this region in the promoter.

Lim and her colleagues then demonstrated that this promoter is likely to be the cause for the difference in the number of and distribution of receptors, as well as the difference in partner preference in these species. They injected the promoter attached to a harmless viral carrier into a reward center (ventral pallium) of meadow vole males and measured the length of time each male spent huddling in close contact with the female after mating. Compared with untreated meadow vole males, males who received the promoter spend significantly more time huddling with the female after mating. Similarly, if that promoter sequence is inserted into male mice, which are not closely related to voles and are never monogamous, the mice show an increased preference for a partner with whom they have mated, similar to the effect on monogamous prairie vole.

STOP AND THINK

Before identifying the vasopressin receptor distribution as the cause of the difference in social bonding in monogamous prairie voles and in nonmonogamous montane voles, researchers injected vasopressin into the brains of both prairie voles and montane voles (Young et al. 1999). Why was this step necessary? What outcome led the researchers to conclude that the distribution of receptors was the critical difference between the responses of these species?

Based on the important roles for vasopressin, oxytocin, their receptors, and estrogen receptors in social bonding in rodents, researchers suspect that these hormones play a role in human social behavior, including autism (Lim et al. 2005). Autism spectrum disorders are developmental disorders characterized by poor communication and social skills, usually accompanied by repetitive, stereotyped behavior patterns. One in every 166 children in the United States is diagnosed with an autism spectrum disorder. One way to explore this hypothesis that alterations in the oxytocin or vasopressin signaling pathways are related to autism is to look for differences in levels of these hormones among autistic children or age-matched nonautistic children. In fact, oxytocin and vasopressin levels are lower in autistic children (Lim and Young 2006). One study has shown that oxytocin administration boosts some social skills in autistic patients (Bartz and Hollander 2006). In addition, some studies, taken together, suggest that variation in the vasopressin receptor promoter gene in humans is associated with autism (Insel 2006).

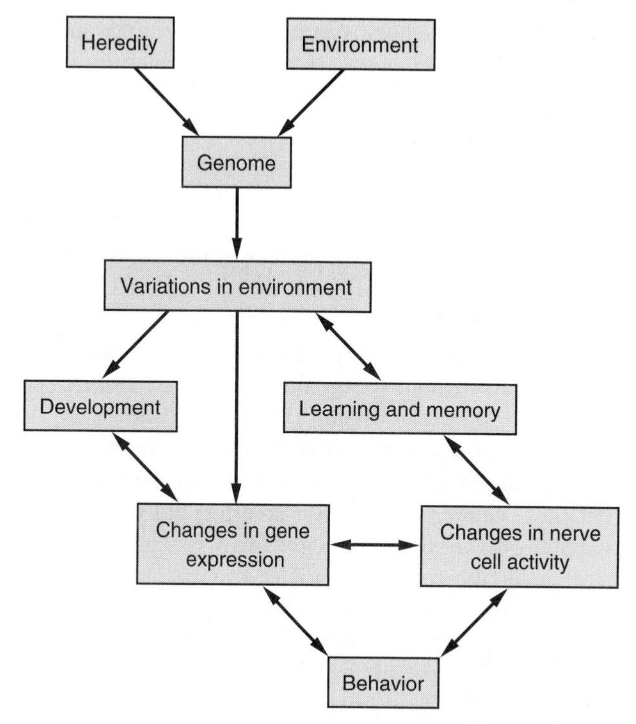

FIGURE 3.17 **Genes and the environment interact to produce behavior.**

ENVIRONMENTAL REGULATION OF GENE EXPRESSION

Earlier in this chapter, we presented the idea of heritability, which is the part of the variation in a trait that can be attributed to genetics. The remainder of the variability is attributed to environmental influences. The implication is that the effects of genes and the effects of the environment can be neatly divided into distinct parcels.

We now know that it is not nature or nurture, but instead nature *and* nurture. The observed behavior is the product of genes and environment acting on the genome. A change in the pattern of gene expression is often the first quantifiable sign of those interactions of genes and environment affecting the behavior (Figure 3.17; Robinson 2004).

As we will see in Chapter 6, a nerve cell can respond to appropriate stimuli within milliseconds by generating an action potential. Thus, the immediate response to a stimulus is neural. However, that neural response often initiates a broader response by activating immediate early genes. These genes become active within minutes to hours after a nerve cell is stimulated and code for proteins that regulate the activity of other genes. These other genes then produce proteins that are important in the behavioral response, including turning on a gene for a specific hormone or its receptor or affecting the growth of nerve cells and nerve cell activity. We will consider dominance relationships in cichlid fish and song learning in male songbirds to better understand how the environment can turn on immediate early genes, altering gene expression in the brain to produce behavior adapted to the environment.

DOMINANCE RELATIONSHIPS IN CICHLID FISH

There are two types of males in the cichlid fish *Astatotilapia*[3] *burtoni*. Dominant males are brightly colored, aggressively defend territories, and have greater reproductive success than subordinate males. Subordinates are nonterritorial, have camouflage coloration, and are less likely to be chosen as mates. The physical environment of the cichlid's natural habitat fluctuates, causing frequent changes in dominance relationships. As a male rises in social status, he becomes sexually mature and his growth rate slows.

Interacting genes simultaneously affect physiology, behavior, and social status in these fish. Gonadotropin-releasing hormone (GnRH), a small protein encoded by the *GnRH* gene, plays an important role in vertebrate reproductive physiology. Dominant males have larger GnRH-releasing neurons in the preoptic area of the hypothalamus of the brain than do subordinate males. The larger size of these neurons is a result of increased activity of the *GnRH* gene.

In *A. burtoni*, the *GnRH* gene is regulated by social stimuli resulting from dominance or submission. As a

[3] *Astatotilapia* was formerly known as *Haplochromis*.

nonterritorial male begins to win aggressive encounters, his *GnRH* gene increases activity, as does the number of GnRH receptors in the pituitary gland. As GnRH, the protein product of the *GnRH* gene, increases, he moves up in social rank and acquires a territory. GnRH triggers the release of hormones that, in turn, trigger the release of sex hormones (gonadotropins) from the pituitary gland, which leads to development of the testes and the production of sperm. At the same time, GnRH brings about changes in coloration. His gray body becomes blue or yellow and a black bar appears across his face. The colorful transformation signals to both male and female fish that he is now dominant (Hofmann 2006).

Body size is an important factor in a male's ability to attain dominance, and so it should not be surprising that these characteristics are also affected by social status. Subordinate males typically grow faster than dominant males. Once dominance is attained, growth rate slows and the male devotes more of his energy to reproduction. Social stimuli associated with dominance bring about the changes in growth rate by increasing the expression of a gene encoding the hormone somatostatin. Somatostatin *inhibits* the release of growth hormone (GH). Thus, once a male has established dominance and becomes reproductively active (because of GnRH), his somatostatin level increases, causing a decrease in GH. Lower levels of GH lead to slower growth. If he then loses challenges by subordinate males and falls in dominance, becoming a nonterritorial male, somatostatin levels decrease and GH levels increase, causing his growth rate to increase again. The change in growth rate may reflect energy trade-offs between growth and reproduction. A subordinate male invests in growth; a dominant male invests in reproduction (Hofmann 2006).

Opportunity to rise in social status may occur rapidly, as when a dominant male is plucked from the water by a predator. The immediate early gene, *erg-1*, orchestrates a subordinate male's response to this opportunity to rise in social status. Although changes in a male's aggressiveness and fertility may take a week or so to occur, within minutes of the opportunity to rise in social status he becomes brilliantly colored and begins to make threatening displays and chases other males. Sabrina Burmeister and her colleagues (2005) observed a group of four females, a dominant male, and a subordinate male. Adjacent to the observation tank were tanks containing large communities of fish. The fish in the observation tank could see the other fish but could not interact with them. The researchers used infrared night-vision goggles to remove the resident dominant male from the observation tank 1 hour before the lights were turned on. Because cichlids respond primarily to visual cues, they do very little in the dark. However, when the lights came on, the remaining male perceived an opportunity to rise in social status and within minutes his coloration and behavior changed (Figure 3.18). The changes in coloration and behavior were accompanied by an increase in the activity of the *erg-1* gene in the brain region rich in GnRH-releasing neurons. *Erg-1* codes for proteins that regulate the activity of other genes. Among these genes is *GnRH*. As we have seen, the protein product of *GnRH* acts in the pituitary gland to increase the levels of sex hormones. Other genes activated by *erg-1* produce proteins that are thought to be important in the growth and activity of nerve cells, which underlie behavioral changes.

SONG LEARNING IN MALE SONGBIRDS

We see an example of interactions among genes, and the internal and external environment when a male zebra finch (*Taeniopygia guttata*) or canary (*Serinus canaria*) is first exposed to the song of its own species. Young males learn to sing their species' song by imitating the songs of adult males. It is important that they learn the correct song because, like other songbirds, these birds sing to defend territories and to attract mates.

But what changes in the nervous system underlie a male songbird's learning his species' song? Claudio Mello, David Clayton, and their colleagues have uncovered some answers to this question. They suspected that changes, if they occurred, would be found in the bird's forebrain because this region is important in auditory processing. They further hypothesized that if gene activity changed, it would most likely be the activity of one of the so-called immediate-early genes, which become active within minutes after a nerve cell is stimulated and code for proteins that regulate the activity of other genes. In this example, the other genes then produce proteins that are thought to be important in the growth of nerve cells and to affect nerve cell activity, which, in turn, are involved in forming long-term memories. *Zenk*[4] is one of these genes. If *zenk* were turned on by exposure to the song, the levels of zenk mRNA would be expected to rise. So, Mello and his colleagues (1992) played a 45-minute tape recording to young male zebra finches or canaries. The recording was that of the song of its own species, one of another species, or simply bursts of sound. They then measured the level of zenk mRNA in the birds' forebrains. It turns out that *zenk* activity is increased greatly when males hear the song of their own species and much less so in response to the song of another species. Exposure to simple bursts of sound did not increase *zenk* activity above that found in birds without any auditory stimu-

[4]*Zenk* is a homologue of *erg-1*.

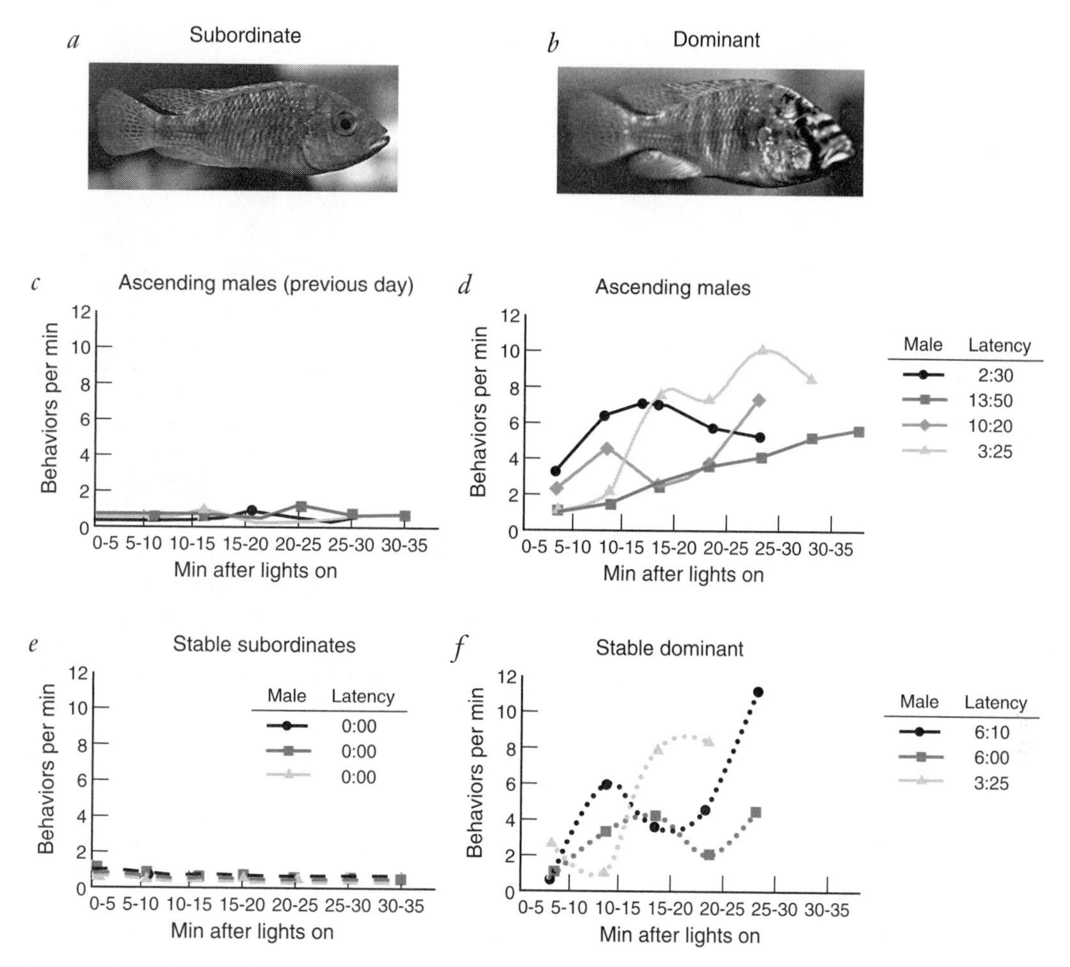

FIGURE 3.18 The behavioral response to social opportunity. (*a*) Subordinate male appearance. (*b*) Dominant male appearance. (*c*) The day before the opportunity to rise in social status, subordinate males do not display dominant behaviors, such as chasing, threats, courtship solicitations, and spawning site entries. (*d*) Within minutes of the perception that the dominant male was gone, a subordinate male displays behavior associated with dominance. (*e*) These changes do not occur in a subordinate male if the dominat male remains in the tank. (*f*) The behavioral changes in a male rising in rank are similar to the behaviors displayed by a stable dominant male. (From Burmeister et al. 2005.)

lation (Figure 3.19). The results of this experiment are consistent with the idea that *zenk* is activated when a male songbird hears the song of its species.

One function of bird song is to defend territories, and it would be a waste of energy to continuously defend borders that are not contested. It should not be surprising, then, that male songbirds can discriminate among the songs of other singers. Such discrimination requires a male to learn the characteristics of songs of specific males, its neighbors, for instance. Mello and his colleagues (1995) hypothesized that *zenk* activity underlies the *formation* of long-term memories. Therefore, they predicted that *zenk* activity would increase in response to the songs of unfamiliar males but would not increase in response to the songs of familiar males. To test this hypothesis, they measured the levels of zenk mRNA in the auditory forebrain regions of male zebra finches following repeated exposure to the song of the same zebra finch. As expected, zenk mRNA increased during the first 30 minutes. However, in spite of continued stimulation by the same song, zenk mRNA fell to baseline levels. Furthermore, when the same male's song was played after a full day of silence, there was no increase in *zenk* activity. Nonetheless, *zenk* activity did increase following exposure to the species song of another, unfamiliar zebra finch individual (Figure 3.20).

Similar changes in *zenk* activity have been shown in freely ranging song sparrows (*Melospiza melodia*). When the species' song was played through a loudspeaker within a male's territory, he approached the speaker, searched for the intruder, and began actively singing in defense of his territory. *Zenk* activity in

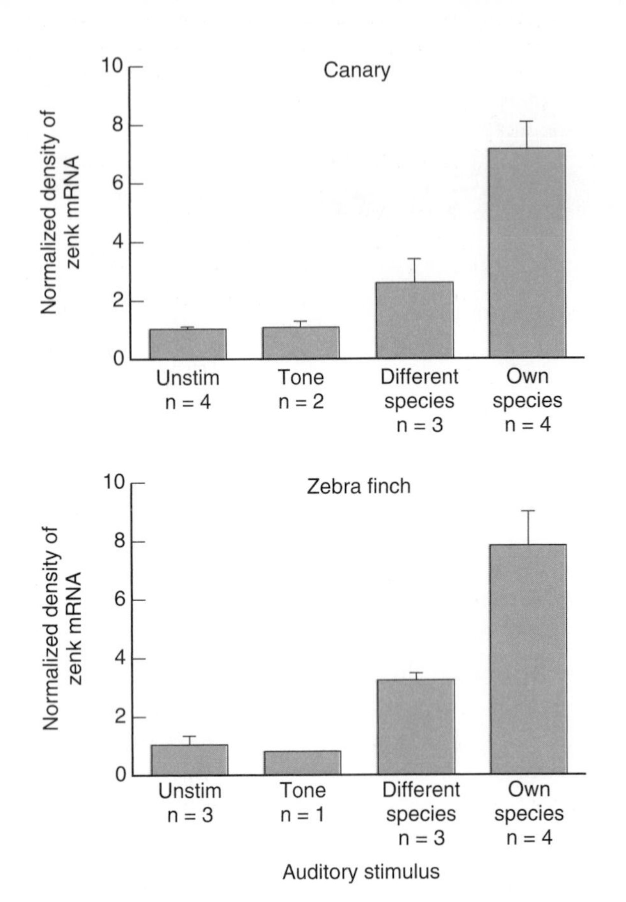

FIGURE 3.19 ***Zenk*** **gene activity in the forebrains of male songbirds following exposure to songs of their own species, songs of another species, bursts of tones that are not song, or no auditory stimulation. The density of zenk mRNA, which reflects gene activity, increases following exposure to birdsong. The increase following exposure to the song of the listener's own species is significantly greater than that following the song of another species.** ***Zenk*** **activity is not a simple response to noise because it is not greater following nonsong tones than when the bird is not stimulated by any sound. (Data from Mello et al. 1992.)**

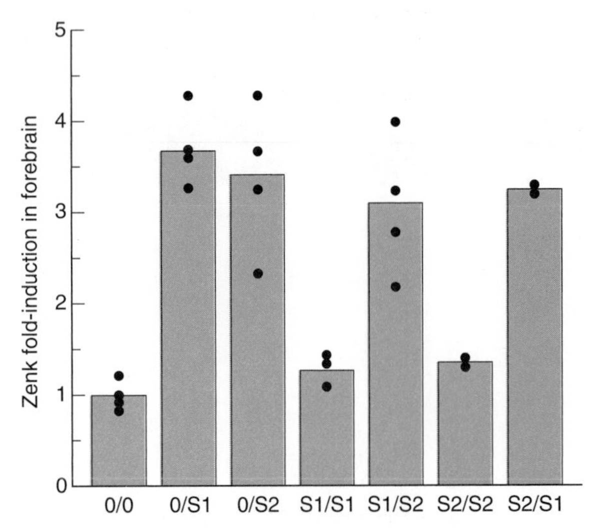

FIGURE 3.20 ***Zenk*** **activity in auditory regions of the forebrain of a male zebra finch drops following repeated exposure to the song of one male, but it can be reactivated by a new song. The birds were exposed to repeated song stimulus for 2.5 hours, which was immediately followed by a second "test" stimulus lasting 30 minutes. The levels of zenk mRNA in the birds' forebrains were then measured. Each dot represents the amount of zenk mRNA in a single bird. The bars represent the mean zenk mRNA level for the experimental group. These results are consistent with the hypothesis that** ***zenk*** **activity underlies the formation of long-term memories. We would not expect** ***zenk*** **to be activated after the memories have been formed. The 0 represents silence. S1 and S2 indicate specific songs of two different individuals from another aviary. (Data from Mello et al. 1995.)**

several brain auditory structures was higher in the males whose territories had been challenged by a loudspeaker intruder than in unstimulated controls. Thus, natural behaviors in the field affect gene activity in much the same way they do in the laboratory (Jarvis et al. 1997).

We see, then, that the relationship between genes and behavior need not be static. Instead, the environment can continuously alter gene activity and, therefore, modify an organism's behavior.

THE IMPORTANCE OF GENETIC BACKGROUND TO BEHAVIORAL GENETICS

The expression of a gene of interest may depend on which alleles of all the other genes in the genome (the genetic background) are present. In other words, different genetic backgrounds may respond differently to the same mutation. The fruit fly rover versus sitter behaviors again provide an example. Recall that the rover genotype is either $for^R for^R$ or $for^R for^S$. The sitter genotype, on the other hand, is $for^S for^S$. We have seen that as cGMP increases, so do PKG levels and rover foraging behavior. Rovers have higher levels of PKG than do sitters. Although it seemed logical to predict that a mutation in a gene other than *for* that decreased cGMP would also decrease PKG and locomotion during foraging, it didn't happen that way. Instead, both rover and sitter larvae with mutations in this gene had increased PKG and increased locomotor activity.

Craig Riedl and his colleagues (2005) used microarray analysis to compare gene activity in rover and in sitter flies after this mutation decreasing cGMP was introduced. Some genes responded in the same manner in both rovers and sitters. For example, this mutation increased the activity of 37 genes and reduced the gene activity of 14 other genes in both rovers and sitters.

However, in a subset of nine commonly affected genes the changes in gene activity were in the opposite direction and depended on whether the fly was a rover or a sitter. Thus, the mutation had a different effect on the pattern of gene expression depending on whether the fly's genotype was rover or sitter.

NETWORKS OF GENES ARE RESPONSIVE TO THE ENVIRONMENT

Complex behaviors are produced by genes interacting with the environment. For those genes, some of the "environmental" factors are hormones within the same organism. We will consider the ways in which hormones affect behavior in Chapter 7, so we will not cover that here. When you read Chapter 7, keep in mind that an important way that certain hormones affect behavior is by altering gene expression patterns.

Rearing Conditions Affect Gene Networks

Males in the same population of Atlantic salmon (*Salmo salar*) have alternative reproductive life histories. Some males reach sexual maturity in the first three years of life and become sneakers; that is, they steal matings. Sneaker males are ten times smaller than migratory males and are able to reproduce without leaving freshwater. The early sexual maturity of sneaker males is associated with larger gonads, smaller body size, changes in feeding and hormone levels, increased response to female pheromones, and a male body scent (Aubin-Horth et al. 2005a). Other males migrate out to sea, where they mature, and return years later as large fish to breed in freshwater (Aubin-Horth et al. 2005b). Thus, within the same population, males of the same age will either be immature, nonreproductive males that will migrate out to sea before breeding or reproductive sneaker males.

A different subset of genes is expressed in the brains of sneaker males compared with those of immature males. Microarray analysis of 2917 genes revealed that 15% (432) of the genes differed in activity in the brains of sneaker males compared with immature males (Figure 3.21). Most of these genes have only small differences in activity but caused large differences in morphology and behavior. Whereas the genes that showed increased activity in sneaker males were associated with reproduction, the genes showing

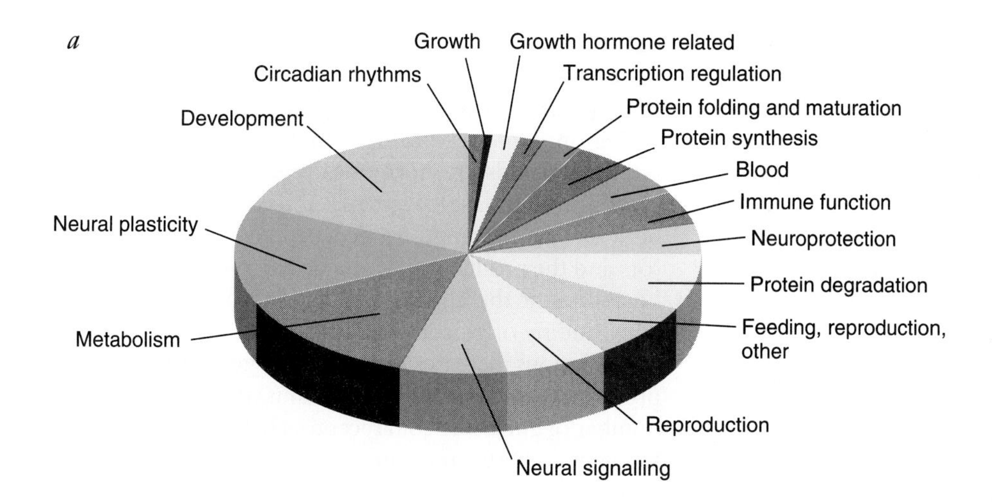

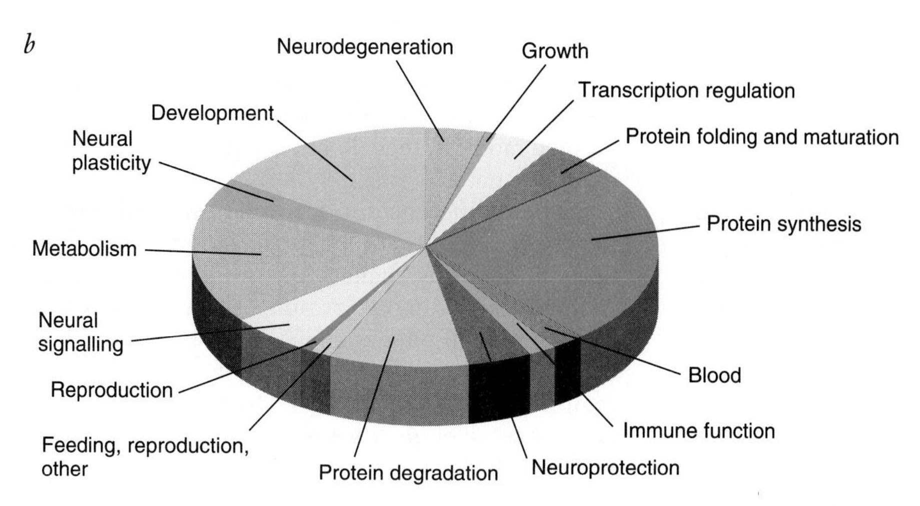

FIGURE 3.21 **Proportion of genes with increased activity in (*a*) sneaker male and (*b*) immature male Atlantic salmon. Genes showing higher activity in sneaker males tend to be associated with reproduction. Immature males had higher activity in genes associated with body growth. (From Aubin-Horth et al. 2005b.)**

increased activity in immature males were associated with growth (Aubin-Horth et al. 2005b).

Reproductive strategy is not the only factor affecting gene expression. Environmental cues, such as population density, food availability, or temperature and lighting also influence gene activity. Nadia Aubin-Horth and her colleagues (2005a) compared gene expression patterns in the brains of sneaker males to the gene expression patterns in the brains of immature males as in the previous experiment. This time, however, she also considered rearing conditions. Gene expression patterns of sneaker and immature males from a wild population were compared with those from laboratory-raised salmon. The fish that were raised in the lab were originally obtained from the same population as the wild-caught fish. Reproductive strategy and rearing environment interact in interesting ways. The activity of one subset of genes differs between wild-caught fish and laboratory-raised fish whose ancestors came from the same wild population, regardless of their reproductive strategy. Presumably these genes are related to survival in different environments.

There is also a subset of genes that is more active in sneaker males than in immature males, regardless of the rearing environment. As we saw in the previous experiment, the genes that are more active in sneaker males are associated with reproduction, and the genes that are more active in immature males are associated with growth. Interestingly, more than half the genes that varied in activity between sneakers and immature males varied in only one environment. In other words, the activity of a subset of the genes in males with the same reproductive strategy varied depending on which environment they experienced. So, in fish raised in different environments, different regulatory networks may lead to the same reproductive tactic.

EPIGENETICS AND BEHAVIORAL GENETICS

Now that we are familiar with some of the ways that genes influence behavior and some of the ways that genes and environment interact, let's return to the story that opened this chapter: the maternal care in rats influencing stress responses and the maternal behavior of rats in future generations. The effects of maternal care on later behavior of the pups occurs through epigenetics, a means of gene regulation that is only recently becoming understood as a mechanism that plays a role in shaping behavior. **Epigenetics** involves a stable alteration in gene expression *without* changes in DNA sequence. In other words, it regulates how genes are expressed without changing the proteins they encode. We will consider the roles that two epigenetic processes—DNA methylation and histone acetylation—play in behavior. These processes alter gene expression by affecting how tightly packaged the DNA molecule is. DNA is packaged with proteins, mostly histones to form chromatin, which forms chromosomes. DNA methylation (adding a methyl group to the cytosine bases in DNA) turns off the activity of a gene by bringing in proteins that act to compact DNA into a tighter form, reducing access by regulatory proteins that promote transcription. On the other hand, histone acetylation makes the DNA less tightly coiled and gene expression easier. At one time, researchers believed that DNA methylation and acetylation occurred during development and permanently altered gene expression in certain tissues. We now know, however, that these processes can be affected by the environment and that the pattern of DNA methylation is dynamic and changes over time.

One of the first demonstrations of epigenetic control of the activity of a gene being nearly permanently turned off by early life experiences come from Norway rats (*Rattus norwegicus*). As described in the chapter opening, some rat mothers nurture their young more than other mothers do. Nurturing mothers extensively lick, groom, and nurse their young. Their pups tolerate stress better than pups of less attentive mothers. When the female pups are old enough to reproduce, they too become nurturing mothers.

Through what physiological mechanism does the quality of maternal care affect adult stress responses and maternal care? The frequent contact with pups increases the expression of a gene for a glucocorticoid receptor in a region of the brain called the hippocampus. Glucocorticoids are hormones released by the adrenal glands in response to stress. A feedback relationship between the binding of glucocorticoids to their receptors and the production of glucocorticoids maintains the hormones at the appropriate level to cope with stress. The binding of glucocorticoids to their receptors in the hippocampus increases the activity of genes that dampen production of glucocorticoids. Thus, the increase in the number of glucocorticoid receptors helps the well-cared-for pups to better regulate their response to stress, and so they are less fearful. In contrast, pups that did not receive appropriate maternal care during the first week of life do not produce as many glucocorticoid receptors in the hippocampus, so they are anxious and fearful in response to stress. Furthermore, when the female pups become mothers, they do not adequately lick or groom their own pups.

These differences in maternal care and response to stress are due to changes in gene regulation brought about by changes in chromosome structure. At birth the promoter region of the glucocorticoid receptors of all pups is demethylated, and at the end of the first day of life methylation of the promoter region had begun in all pups. However, by the end of the first week of life, the promoter of a glucocorticoid receptor in the pups of neglectful mothers was more methylated than the same gene in pups of nurturing mothers. Licking and grooming increase the level of serotonin, a neurotransmitter used in communication between certain neurons. In turn, serotonin results in the removal of methyl groups and the addition of acetyl groups to histones. Both changes make DNA less tightly coiled and more easily expressed. The resulting increased expression of the glucocorticoid receptor increases the number of glucocorticoid receptors in the hippocampus and lessens the pups' responses to stress (Weaver et al. 2006).

Recall that the quality of maternal care also influences how nurturing a female pup will be toward her own young when she becomes an adult. The effect on future mothering styles is brought about by epigenetic changes in another gene; a particular estrogen receptor (ER α) in the preoptic area of the hypothalamus becomes methylated in female offspring of mothers who show little maternal care. As we saw earlier in this chapter, this estrogen receptor is necessary for the regulation of oxytocin receptor binding in the same region of the brain. Oxytocin is a hormone crucial for maternal care. Thus, poor maternal care results in methylation of genes for receptors of glucocorticoid hormone, a hormone that mediates stress responses, and genes for ER α, which is needed for proper response to oxytocin. In this way, mother–pup interactions cause stable genetic changes that influence the behavior of adult offspring (Champagne et al. 2006).

Although DNA methylation patterns are considered to be stable, some studies suggest that methylation can be reversed in adulthood. Injection of trichostatin A, a chemical that causes demethylation, into the brains of pups who had received poor maternal care made the pups less fearful and better mothers. The effects of good maternal care can also be reversed. Methionine, which is a chemical found in the diet, can alter DNA methylation by donating a methyl group. When methionine is infused into the brains of the adult offspring who had received either good or poor maternal care, the degree of DNA methylation increased. The adult offspring who had received good maternal care now became anxious in stressful situations and the females showed little maternal care (Weaver et al. 2006).

STOP AND THINK

Experiments were performed in which the pups of mothers who offered poor maternal care were cross-fostered to mothers who offered good maternal care. Pups from good mothers were cross-fostered to mothers who offered poor maternal care. As adults, the pups raised by good mothers handled stress well, and the females were good mothers. The pups raised by poor mothers were fearful as adults, and the females were poor mothers (Francis et al. 1999). Why were these experiments necessary? What can you conclude from them?

DNA methylation patterns can be affected by environmental factors, cause disease, be transmitted through generations, and, potentially, influence evolution. Consider, for example, the effects that exposure to the fungicide vinclozolin has on rats. Vinclozolin, which is still used today on some agricultural crops, is an environmental endocrine disrupter. It interferes with hormone signaling of androgens (male sex hormones), progestins, glucocorticoids, and mineralcorticoids. The DNA methylation pattern differs between rats that were exposed to vinclozolin and rats that were not. The differences in DNA methylation patterns are believed to be the reason that brief exposure of a rat embryo to vinclozolin increases the likelihood of diseases that do not begin until adulthood. If male embryos are exposed to vinclozolin during the time that testes are forming, the adult male has an increased risk of abnormal testes and low fertility. The increased risk of adult-onset diseases and reduced fertility are passed through four generations, even if the offspring have never been in contact with vinclozolin (Anway et al. 2006). Vinclozolin causes these effects by changing the methylation pattern of 15 genes, many of which are regulatory regions of DNA, reducing the expression of these genes (Chang et al. 2006). The acquired patterns of DNA methylation and chromatin condensation are then passed from generation to generation.

Perhaps the most interesting aspect of vinclozolin exposure is that it influences mate choice for at least three generations. As we will see in Chapter 4, evolution occurs when individuals with one set of alleles enjoy more reproductive success than individuals with other alleles. Therefore, it is in an individual's best interest to choose a mate with "good genes." Because vinclozolin increases the risk of adult diseases and lowers male fertility, it seems reasonable that a female should avoid mating with a male who had been exposed to vinclozolin. Regardless of whether a female has been exposed to vinclozolin herself, when given equal access to both a male with a history of exposure and a male that has had no exposure, she shows a significant preference for the unexposed male. In fact, she prefers an unexposed male to the great-grandson of an exposed male! This female choice could affect evolution (Crews et al. 2007).

Epigenetic changes in DNA may also underlie learning. Courtney Miller and David Sweatt (2007) conditioned rats to be fearful of a location by giving them an electric shock when they were in the training chamber. If the rats froze when they were placed in the chamber at a later time, the researchers concluded that the rats had formed fearful memories of the chamber. Miller and Sweatt demonstrated that the pattern of DNA methylation in the hippocampus of the brain changes when memories are formed by treating the rats with drugs that prevent methylation between their training and testing. Specifically, as fearful memories form, rapid methylation (silencing) of a memory-suppressing gene (*protein phosphase 1, PP1*) and demethylation of a memory-promoting gene *reelin* take place.

One take-home message is that we should look more closely at the role of epigenetics and disease. Epigenetics is believed to play a role in human behavioral disorders, such as autism spectrum disorders, Rett syndrome (a developmental disorder that affects the nervous system), and fragile-X syndrome (an inherited form of mental impairment; Schanen 2006).

A second take-home message is that DNA is sensitive to the environment, so what we eat and the chemicals we are exposed to may influence our health by affecting our gene expression patterns. We are beginning to realize that the lifestyle of a person's ancestors, and not just the individual's behavior, can influence health. Or, put another way, your lifestyle can influence the health of your great-grandchildren. Maternal nutrition during pregnancy causes epigenetic changes in gene activity in the fetus that increases susceptibility to obesity, type-2 diabetes, heart disease, and cancer (Martin-Gronert and Ozanne 2006).

COMPLEX RELATIONSHIPS AMONG GENES

We have seen that complex interactions among genes can affect behavior. The environment and social interactions can also alter the pattern of gene expression, bringing about changes in behavior. As you can see in Figure 3.22, these relationships may occur rapidly by effects on the nervous system. The relationships can also change behavior on a slower time scale by affecting brain development or altering the genome. As we will see in Chapter 4, evolution can also modify the genome (Robinson et al. 2008).

A BROADER PERSPECTIVE

Technical revolutions have vastly changed the field of behavioral genetics in just a few short years. As never before, we are able to follow the thread that ties together genes, cells, the entire organism and its behaviors, all the way to the adaptation of the organism in its natural environment. A new interdisciplinary field called evolutionary and ecological functional genomics seeks to understand the processes that are biologically important to both adaptation to the environment and evolutionary fitness. It brings together people with divergent interests—molecular, cellular, organismal, and ecological—to investigate how the mechanisms that underlie a behavior increase function in a natural environment to increase evolutionary fitness (Fitzpatrick et al. 2005).

Behavioral ecologists are interested in adaptive behavior, which results from natural selection acting on variations in phenotypes (traits) that have a genetic basis. In nature, individuals of the same species may have slight differences, perhaps only a single base pair, in their genes. These differences (polymorphisms) may produce subtle differences in phenotype on which natural selection can act. Sequencing genes identifies these polymorphisms. Microarray analysis would be useful in identifying genes whose activity is correlated with important ecological conditions (Feder and Mitchell-Olds 2003). In this chapter, we have considered several examples of ecological genomics: social recognition in rodents (Choleris et al. 2004), dominance relations in cichlid fish (Hofmann 2006; Trainor and Hofmann 2006), monogamy vs. polygamy in voles (Lim et al. 2004), and alternative reproductive strategies in salmon (Aubin-Horth et al. 2005b).

Our knowledge of genome sequences has expanded well beyond lab animals and now includes a wide variety of taxa. Even if we have only the gene sequences of lab animals, such as the house mouse or Norway rat, the similarity in the sequences often allows us to transfer what we know about their DNA sequences to wild populations of rodents, such as the deer mouse. These genome sequences may help identify the specific gene changes that are responsible for adaptation to a specific environment and to speciation (Storz and Hoekstra 2007). Polymorphisms occur in a natural population. These small changes in gene structure may result in subtle changes in the way in which genes interact, giving rise to differences in the trait (behavior). The genetic polymorphisms that result in changes in the trait are what natural selection "selects" during evolution. Techniques have been developed to use microarray analysis to look at gene expression in native populations in the field to gain an understanding of the genetic basis of an organism's response to environmental change (Travers et al. 2007). For example, there are about a dozen species of finches on the Galapagos Islands, each differing in beak shape and size. Different beak shapes adapt different species to eating different types of food. DNA microanalysis has been used to discover that natural selection worked on a calcium-signaling molecule, calmodulin, to change beak shapes in these finches (Abzhanov et al. 2006).

Evolutionary functional genetics focuses on the interface of genomics and evolution. In this chapter, we have seen that evolution can occur by selection of new structural alleles or by altering gene regulation. In the examples of food-related behavior in fruit flies, honeybees, harvester ants, and nematodes, we have seen that a relationship between a signaling molecule (PKG) and a type of behavior (obtaining food) can be conserved in evolution, but molecular pathways between the two may

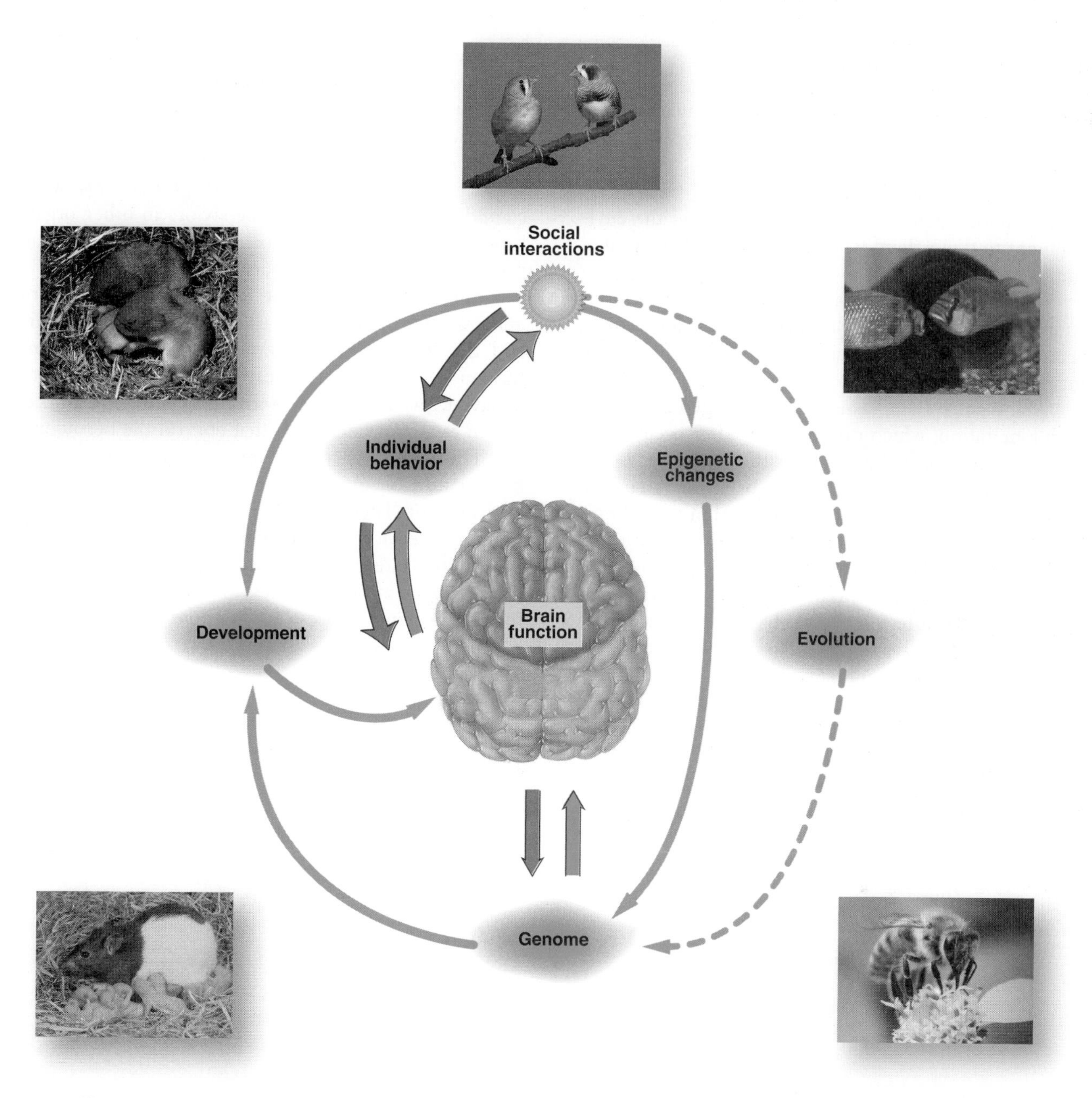

FIGURE 3.22 **Complex relationships among genes, the nervous system, and behavior occur over different time scales. These relationships can vary over physiological time by affecting brain activity. These relationships can also work over developmental time, that is, via effects on brain development and modification of the pattern of gene activity. The genome and behavior can also be altered over evolutionary time. (Modified from Robinson et al. 2008.)**

differ. Recall that high PKG increased foraging activity in fruit flies and honeybees but decreased foraging activity in harvester ants. The fact that increased levels of PKG have the opposite effect on behavior in the fruit fly and honeybee as they do in the harvester ant and nematode could suggest that the association evolved twice (Fitzpatrick and Sokolowski 2004).

Some researchers apply the techniques of evolutionary and ecological functional genomics specifically to social behavior, that is, behavior that involves interactions among members of the same species. This growing field, called sociogenomics, endeavors to understand how genes influence social behavior. Sociogenomics goes beyond simply identifying genes. It seeks to understand the functions of those genes, and how they affect the mechanisms that bring about behavior, primarily those of the nervous and endocrine systems (Robinson 1999). Indeed, most of the examples we discussed in this chapter could be considered examples of sociogenomics at work.

SUMMARY

Genes are made up of DNA (deoxyribonucleic acid). Behavior depends on which genes are being expressed, when, and in which tissues.

A goal of behavior genetics is to decipher the interactions among genes and between genes and the environment to understand why a particular behavior takes the form it does.

The classical methods of experimentally demonstrating the genetic basis of behavior include inbreeding, artificial selection, and hybridization. Inbreeding is the mating of close relatives.

Many animal studies on behavioral genetics begin with a mutation, which may be induced or occur naturally. Then a researcher identifies the gene and discovers how the gene influences a behavior. Artificial selection is a different breeding regimen in which individuals showing a desired behavior are bred with one another. If the frequency of the trait in the population increases when the appropriate breeders are mated, then the behavior must have a genetic basis. Hybridization is another breeding system used to demonstrate that a given behavior has a genetic basis. Individuals that display the behavior in distinct but different ways are mated with one another, and the behavior of the hybrid offspring is observed.

The *foraging (for)* gene provides a good example of the techniques used by behavioral geneticists. *For* was first identified in fruit flies, in which it exists naturally in two alleles for^R and for^S. For^R is dominant and results in larvae and adult flies called rovers because they travel long distances while foraging. For^S is recessive and results in flies described as sitters because they travel only short distances during foraging. PKG, an important cell-signaling molecule, is *for*'s protein product. *For* was a candidate gene for foraging behaviors in other organisms. *For* underlies foraging behaviors in honeybees, harvester ants, and the nematode.

Among the important principles of behavioral genetics are that one gene usually affects several traits and that genes work in interacting networks. Thus, though we often use the shorthand that a gene is "for a behavior," what we mean is that such genes are embedded in regulatory networks that control the behavior.

The activity of immediate-early genes is often the first quantifiable indication of genes interacting with the environment. These genes become active within minutes to hours after nerve cell activity. Their protein products help orchestrate the behavioral response to stimuli.

Changes in the tightness of the coiling of DNA influence gene expression without altering DNA sequence. We see this in the effect of the quality of maternal care on the stress responses and future maternal behavior of pups.

Evolutionary and ecological evolution seeks to understand how natural selection works to select the genes that are responsible for ecologically adaptive behavior. Sociogenomics is the evolutionary and ecological genomics of social behavior.

4

Natural Selection and Behavior

If a meddlesome biologist scooped up kittiwake gull chicks from neighboring nests and swapped them when their parents were away, the consequences would not be too severe. The returning parents would adopt and feed the new chicks, as if they were following the rule that any chick in the nest should be treated as offspring. However, playing the same trick on herring gulls would cause greater commotion: these gulls recognize even very young chicks as their own and refuse to care for the chicks of others.

These observations make sense in light of the ecology of the two species. Even without an experimenter's intervention, herring gulls are likely to encounter chicks from other nests: they nest on the ground in large colonies, and neighboring chicks often wander (Figure 4.1). Kittiwake gulls, on the other hand, nest on steep cliffs, where the chances of the wrong chick ending up in the nest are minimal (Figure 4.2). Herring gulls and kittiwakes differ in other traits as well: herring gulls have wider and shallower nests, whereas the deeper nests of kittiwakes are less likely to allow eggs to roll off cliffs. Predators such as foxes can move freely through a herring gull colony but can't reach the tiny ledges where the kittiwake nests are perched. Thus, herring gull parents would do well to make the nest less obvious to predators. In fact, this seems to be the case: herring gulls remove the tell-tale signs of eggshells and droppings from the nest area, thus reducing the chance that predators will detect the nests. Kittiwakes, in contrast, are less tidy, leaving egg shells and droppings in the vicinity of their nest (Cullen 1957a, b).

FIGURE 4.1 The nesting grounds of the herring gull. Nests are close together and accessible to predators.

FIGURE 4.2 Nesting area of the kittiwake. The shift in nesting sites from the ground to cliffside has rendered the kittiwake nests inaccessible to predators and is correlated with many behavioral changes.

The behaviors of these species are therefore well suited to their respective environments. But by what process does this match come about? How does it happen that even closely related species such as these can differ so much? The answer, of course, lies in evolution. As the scientist Theodosius Dobzhansky famously put it, "Nothing in biology makes sense except in the light of evolution."

We define **evolution** as a change in the frequencies of different alleles in a population of organisms over generations. An allele is an alternative form of a gene (Chapter 3). A population is an interbreeding group of organisms of the same species. Although other types of evolutionary forces have an impact on animal behavior, natural selection is arguably the most important and is the reason why species so often appear to be well suited for their environments. We'll begin by examining natural selection in detail, and we'll save our discussion of other evolutionary forces for later in the chapter.

NATURAL SELECTION

Charles Darwin was already convinced that species were not fixed, but could change over time, when he had a flash of insight about the process by which this change occurs. He spent several more decades immersed in research, deep thought, and extensive writing before he published his first paper on the subject of the causes of evolutionary change. During those years, Darwin contemplated many esoteric topics, including pigeons. Pigeon breeding was a common pastime in the 1800s. Darwin was interested in the great variety of pigeon breeds. Some had fancy feathers around their heads, others had flashy colors, and still others had strange tumbling behavior during flight. Although no one yet knew about the existence of genes, pigeon fanciers understood that they could obtain pigeons with the traits they wanted by mating males and females that had these traits. Often, the offspring had even more extreme traits than did either of their parents. Slowly, over generations, extraordinarily bizarre pigeons could be bred through this process of **artificial selection**. Other breeders had long made similar modifications in other species of domesticated animals, selecting for chickens that reliably laid eggs, cows that produced a great deal of milk, and dogs that were good at hunting or herding.

Darwin noticed a parallel between the process that was happening in pigeon lofts and what might be going on in nature. Reluctant to publish his ideas before they were fully fleshed out, he painstakingly assembled his evidence over the decades. Finally, pushed by the fact that another naturalist, Alfred Russell Wallace, had converged on the same ideas, Darwin published *On the Origin of Species by Natural Selection or The Preservation of Favored Races in the Struggle for Life*. This quote sums up the essence of Darwin's view (1859, Chapter 1):

> *As many more individuals of a species are born than can possibly survive and, as there is a frequently recurring struggle for existence, it follows that any being, if it vary however slightly in any manner profitable to itself, under the complex and sometimes varying conditions of life, will have a better chance of surviving, and thus be naturally selected. From the strong principle of inheritance, any selected variety will tend to propagate its new and modified form.*

Since 1859, Darwin's ideas have been expanded and built upon in an explosion of both empirical and theoretical research. However, the core concepts of natural selection remain the same. Let's put them into simpler language.

Observation 1: Individuals in a population vary: they differ in appearance, behavior, physiology, or some other part of their phenotype.

Observation 2: Some of these variable traits are genetically based. They have been inherited from parents and can be passed on to offspring. As we have seen in Chapter 3, many behavioral traits are influenced by the genotype.

Observation 3: Among the inherited traits are a few that improve an individual's chances of leaving more offspring than other individuals can. These traits may be important to an animal's survival (e.g., how well can it avoid a predator or tolerate a cold winter?), its reproduction (e.g., how well can it attract mates?), or both. Notice that Darwin's quote above from *The Origin of Species* focuses on competition, but in fact, many different traits may determine an animal's reproductive success.

Conclusion: Because offspring are likely to inherit their parents' beneficial traits, these traits become more common in the population relative to the traits borne by less successful reproducers. This is evolution by natural selection.

Whenever we find the simple prerequisites of heritable variation along with differential survival and reproduction of some phenotypes, we face a logically inevitable conclusion: evolution must occur. This type of evolutionary change is called natural selection because nature "selects" those traits that enhance reproductive success. For example, male bighorn sheep that are victorious in head-to-head butting contests with other males generally leave more offspring than do the losers (Figure 4.3, Geist 1971). Winning males are likely to have a suite of characteristics: perhaps they are larger, or their horns are stronger, or they are better strategists during fights. Any of these traits with a heritable component will be passed on to their offspring. Because the winning males leave more offspring than losing males, the composition of the

FIGURE 4.3 **Two bighorn rams clashing heads. The winner will have priority in mating and will be more likely to leave offspring.**

population will change in the next generation: more individuals will have alleles that code for the "winning" traits rather than the "losing" traits. These traits are then said to be under selection. Traits that have evolved because they allow individuals to survive and reproduce better than their competitors are called **adaptations**. (We will restrict the term *adaptation* to those traits that have a genetic basis. This definition excludes learned behaviors, although the *capacity* to learn may be an adaptation. We'll return to this point in Chapter 5.) The word "adaptation" also refers to the process of change over evolutionary time that occurs through natural selection. Note that biologists say "adaptation" but not "adaption."

COMMON MISUNDERSTANDINGS ABOUT NATURAL SELECTION

Although the logic of natural selection is straightforward, nonscientists often share some common misunderstandings about the concept, which are sometimes reinforced by poorly written articles in the popular press. This problem is exacerbated by the terminology of evolution, which uses words that have other, more commonly used meanings. For example, by now you should see that the word "selection" is not used in quite the same way in the phrase "natural selection" as in the phrase "artificial selection." In artificial selection, the selective force is imposed by humans' particular goals. In contrast, natural selection is not capable of long-range or even short-range planning. Although it might be beneficial for humans to have wheels built into their feet, natural selection cannot see us through generations of humans with rudimentary and useless axles in our arches, or other necessary precursors of those useful wheels.

Another troublesome phrase is "survival of the fittest." First, "survival" is only one of many traits that natural selection might act upon: an animal must not only survive, but must compete for resources, find a home and a mate, and ultimately produce offspring that carry its genes. Any one of these abilities, plus many more, might be improved through natural selection. The second part of the phrase, "the fittest," is also misleading: it evokes the picture of the most physically fit, strongest, and most aggressive individuals dominating all others in order to pass their genes along. However, in an evolutionary sense, **fitness** has a more subtle meaning: it is the reproductive success of an allele or an individual relative to other alleles or individuals in the same population. Sometimes the biggest and most muscular individuals do have the highest fitness in an evolutionary sense, but sometimes other traits can be more important to evolutionary success. For example, sometimes small sneaky males mate with more females than do large dominant males. We can identify two components of fitness. The most commonly measured is **direct fitness**, the number of surviving offspring an individual produces. An individual can also increase the number of its alleles that survive in the next generation by helping relatives who share its alleles (for example, by helping a sister raise her offspring). Fitness gained by helping relatives is called **indirect fitness**. Direct and indirect fitness are together called **inclusive fitness**. We'll discuss inclusive fitness in detail in Chapter 17, but for the present, we'll focus on direct fitness. We are generally most interested in relative fitness, or the average fitness of a gene or individual compared with the rest of the population.

A related phrase that should be avoided is "for the good of the species." A common misconception is that traits evolve in order to help a *species* survive. However, natural selection cannot act with the future of the species in mind. If a trait increases an animal's fitness relative to that of other animals, the trait will increase in the population, even if it means trouble for the species in the long term. For instance, if a genetic mutation arises that increases the number of individuals' offspring, natural selection will likely cause the mutation to become much more frequent. The population may grow quickly and outstrip its available food resources, and then crash. Thus, a trait that is favorable to the individual may increase in frequency, even though it is *not* favorable for the population or species. We will explore this concept in more detail in later chapters.

For many people, "evolution" conjures up images only of **macroevolution**, or large-scale changes over geological time, such as birds evolving from reptilian ancestors. This is indeed an example of evolutionary change, but it is not the whole story. Remember our definition: evolution is a change in allele frequencies in a population of organisms over generations. Thus, the concept of evolution also encompasses small changes that happen over only a few generations. For example, the Colorado potato beetle damages crops. When farmers spray pesticides on their fields, most beetles die, but a lucky few happen to have alleles that allow them to survive the poison. Soon the population is primarily comprised of the offspring of these fortunate beetles, all of which have genes that confer resistance to the pesticide. This is an example of **microevolution**, an evolutionary change within species.

Finally, remember that it is populations, not individuals, that evolve. Animals may develop over the course of their lives, but evolutionary change only happens in populations from one generation to the next.

GENETIC VARIATION

VARIATION IS COMMON

Variation across individuals is the rule rather than the exception. All house flies may look much the same to you, but if you examined them under a microscope you would see many differences: their wings are of slightly different lengths, the color is different, the hairs on their abdomens vary. In most cases, not even the offspring of the same parents are identical, as is easily demonstrated by a glance at your own parents or your cat's kittens (Figure 4.4). Sometimes the differences between individuals are obvious, such as in size or color pattern; in other cases, they are harder to detect, such as differences in metabolic rate or in mating behaviors that are exhibited very briefly at particular times of the year. Sometimes variants fall into distinct classes. In other cases, variation is continuous, changing gradually from one extreme to the other. A common pattern is the familiar bell-shaped normal distribution, with most individuals falling about midway between the extremes (Figure 4.5*a*). Regardless of the exact form it takes, it is clear that variation is a common feature of populations.

Variation in the traits of organisms is called phenotypic variation. As described in Chapter 3, phenotypic variation arises from two sources: the underlying genetics and the environment. For evolution, the key component of variation is genetic variation. Only traits that are at least partially based on genes can evolve. Even if there is differential survival and reproduction, if all individuals are genetically identical for a particular trait, that trait cannot evolve by natural selection.

It is also important to remember, however, that although traits cannot evolve unless they have a genetic basis, evolution does not act directly on the genotype (the genetic makeup), but rather on the phenotype (the observable traits). Selection therefore cannot act on genetic differences if they have no effect on the phenotype. For example, recessive alleles in heterozygous individuals are not expressed in the phenotype, so natural selection cannot act on these traits when they occur in heterozygous individuals: the number of offspring that a heterozygous individual has is not affected by the presence of that recessive allele. Natural selection can only "see" recessive traits when they are in homozygous individuals. Thus, to fully understand how selection acts on a trait, we need to understand the relationship between the genotype and the phenotype. In Chapter 3, we discussed some of the continually evolving methods that behavioral geneticists use to link phenotype to genotype.

THE RAW MATERIAL OF GENETIC VARIATION

Natural selection does not create genetic variability, but only acts on the variability present in a population. So where does genetic variability come from? The two main sources are mutation and recombination.

Mutation

Mutation is a change in the DNA sequence of an organism. Of concern to evolutionary biologists are mutations that occur in sperm and eggs, or the tissue that gives rise to them, and thus can be passed on to offspring. (In contrast, somatic mutations, such as most cancers, cannot be passed onto offspring.) Mutations come in many forms. Some mutations affect only a small part of the genotype:

FIGURE 4.4 **Variation in offspring of the same two parents. During the meiotic cell division that forms gametes, the alleles of each parent are shuffled and recombined. Thus, sexual reproduction would scramble any "perfect" combination of genes.**

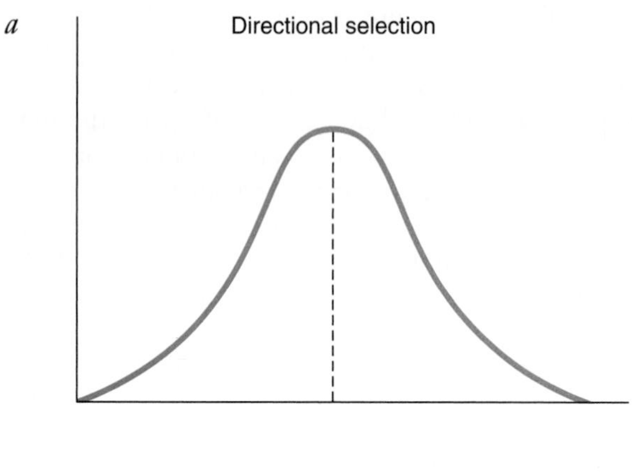

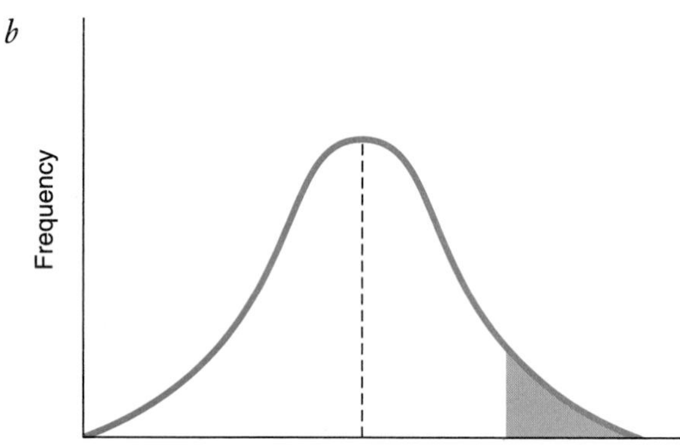

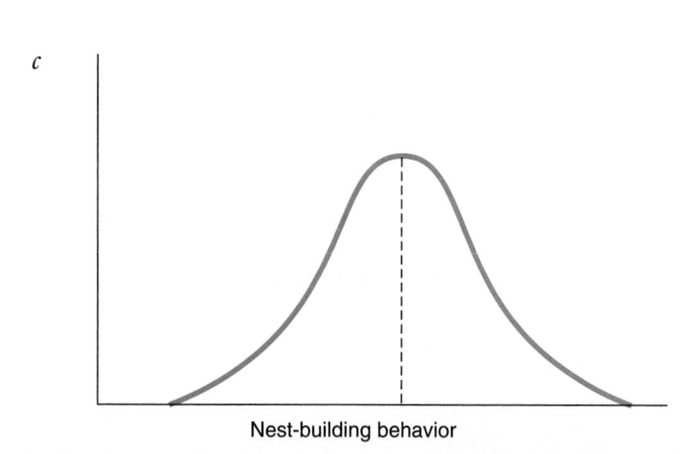

FIGURE 4.5 **Directional selection. *(a)* The familiar bell curve, illustrating a fictional data set of nest-building behavior in mice. The dotted line indicates the population average. Most individuals build medium-sized nests, with some building very large nests and others very small nests. *(b)* If the weather grows colder, only the mice with the largest and warmest nests will successfully reproduce. These lucky mice are at the right-hand side of the bell curve. *(c)* In the next generation, the population is shifted to the right. Note that the average is larger than in *(a)*.**

perhaps one base pair of nucleotides is substituted for another, or base pairs are inserted or deleted. However, in many cases even these small changes have a dramatic effect on the function of a gene, regardless of whether the affected gene's job is to make proteins (a structural gene) or to turn other genes on or off (a regulatory gene). Other mutations are larger: genes may be duplicated or deleted, and entire pieces of chromosomes can move from one chromosome to another or be reversed in their orientation.

Any type of mutation, if its owner survives, increases genetic variation in the population. However, variation produced by mutation is likely to be disadvantageous. After all, a random change in any finely tuned machine, whether it is a living body or a laptop computer, is unlikely to be an improvement. Nonetheless, in spite of the fact that only a tiny percentage of mutations are beneficial, mutation is the ultimate source of variation.

Recombination

In organisms that undergo sexual reproduction, another source of genetic diversity is the recombination of alleles that occurs during meiosis, the type of cell division that results in the formation of gametes (eggs or sperm). During meiosis, a diploid parent (with a pair of homologous chromosomes) produces haploid gametes. A gamete thus contains only half of the genetic material of the parent that produced it, so a parent that produces an offspring by sexual reproduction shares only half its alleles. Chromosomes containing different copies of an allele can show up in different gametes in any combination. Thus, two sisters share half of each parent's genes but not necessarily the same half. Furthermore, at the start of meiosis, when homologous chromosomes are lined up, pieces of chromosomes containing alleles for the same gene are swapped in a process called crossing-over. Individual chromosomes in the gametes are therefore unlikely to be identical to those in either parent, further increasing the differences among siblings. Thus, even if natural selection were somehow able to produce an individual with a perfect combination of alleles, the offspring of a sexually-reproducing individual would not have that same combination.

VARIATION AND THE RESPONSE TO NATURAL SELECTION

The genetic variation provided by mutation and recombination provides raw material on which natural selection can work. Remember that evolution happens at the population level, not the individual level, so let's think explicitly about how populations change when they undergo natural selection.

Return to the normal distribution in Figure 4.5*a*. What might we expect to happen to this distribution over subsequent generations? Under stable environmental conditions, the animals with traits at the center of the distribution would be expected to do best: they might be most successful in the current environment, whereas those at the extreme ends of the distribution are less well suited to current conditions. Under conditions of *stabilizing selection* such as this one, where the optimum phenotype is near the population's average, the mean (or average) phenotype in the population is unlikely to shift much, if at all, from one generation to the next, although the distribution may become narrower. In other cases, the environment may change and the optimum phenotype may shift over time. In this case, those at one extreme would come to be favored (Figure 4.5*b*), and the curve that represents the population's phenotype would shift in that direction (Figure 4.5*c*). This phenomenon, called *directional selection*, can be weak—resulting in a curve that shifts just a little from generation to generation—or strong—resulting in a big change from one generation to the next. Recall the example of microevolution in a field of insect pests sprayed with pesticide. Pesticide application is an extremely strong directional selection pressure, and generally only a handful of individuals are resistant. Their offspring quickly make up the bulk of the population, and thus the change in the genetic makeup of the population can be enormous in just a few generations.

We have already seen another way to visualize the effects of selection. Refer to Figure 3.5, which illustrates the change over time in nest-building behavior in house mice. The top lines are groups that underwent directional selection for large nests, and you can see that nest-building behavior increased over time. The low lines show the response of lines undergoing directional selection for small nests. The middle lines were from a control group. You can see that in the low lines there was an initial decrease in nest size, but then the response to selection flattened out around the tenth generation. We commonly see this pattern: as directional selection proceeds, eventually we "run out" of genetic variation upon which natural selection can act. So, when directional selection is applied consistently, we expect to see the population respond by becoming more adapted, eventually leading to a decrease in genetic variation. Mutation can restore this variation, but on a longer time scale.

THE MAINTENANCE OF VARIATION

Since natural selection favors those individuals with certain traits, why hasn't it eliminated from the population those individuals that bear other traits? Shouldn't natural selection weed out the less optimal animals and keep only the best? Given that mutations are rare, why do we still see variation in most populations? Now we return to the topic we set aside earlier: natural selection is not the only evolutionary force that changes the allele frequencies in populations.

GENE FLOW AND GENETIC DRIFT

Imagine a group of populations, all physically near each other but in slightly different ecological conditions and thus under different selection pressures. If these populations were completely separated from one another, then one would expect them to diverge over time, with local adaptation to the conditions of each region. But what would happen if the populations are not completely isolated and individuals moved between them? Genes from the populations would mix in a process called **gene flow**. Gene flow makes populations more similar to one another. Depending on its strength, gene flow can slow or even halt the effect of local adaptation.

An excellent example of the effect of gene flow on behavior is provided by Susan Riechert and her colleagues. They documented genetically based differences in territorial behavior among populations of the funnel-web-building spider (*Agelenopsis aperta*) living in different habitats. These spiders compete for sites in which to build their webs, and they defend these sites from conspecifics (Figure 4.6; Riechert 1986). This species occupies a wide variety of habitats in its range from northern Wyoming to southern Mexico. Some spiders live in relatively lush riparian vegetation along the rivers and lakes of Arizona where insect prey are abundant but predators are also common. Other spiders live in desert grassland, a much harsher environment. In this environment, insect prey are few, and the scorching sun makes it impossible to forage during much of the day. There are fewer good places in which to build a web in the grassland than in the riparian habitat.

FIGURE 4.6 **A funnel-web spider at the entrance to the funnel that extends from its web. There are differences in the expression of territorial behavior in populations of this spider that live in different ecological conditions.**

Spiders living in these two habitats are strikingly different in behavior. Grassland spiders are much more aggressive and will not allow other spiders near their webs (Figure 4.7). Furthermore, the intensity of territorial disputes between grassland spiders is greater than those between riparian spiders. Threat displays of grassland spiders are more likely to escalate into battles, and the fighting more often results in physical injury or death (Riechert 1979, 1981, 1982). Here, in this unforgiving environment, web sites are extremely valuable, so it is worth taking the risk of engaging in a dangerous fight. Grassland spiders are also very aggressive toward prey and attack prey that land in their web much more quickly than do riparian spiders (Hedrick and Riechert 1989). In contrast, in a riparian habitat, web sites are easier to find and therefore not so valuable. Risking injury is not as likely to be worthwhile. Prey are abundant, so missing the occasional insect by being a bit slow to attack is not a big deal. Riparian spiders have other problems, however: birds and other predators also prefer the riparian habitat. When researchers mimic a predation attempt by disturbing their webs, riparian spiders are very cautious, hiding for a long time compared to desert spiders before venturing from the safety of their funnel (Riechert and Hedrick 1990).

These behavioral differences are not simply a response to environmental conditions, such as food availability, nor are they learned from territorial disputes or other experiences. Rather, they are traceable to genetic differences between the desert and riparian populations (Maynard Smith and Riechert 1984; Riechert and Maynard Smith 1989). Spiders were collected from an arid grassland environment in New Mexico and from a riparian environment in Arizona. Purebred lines were established by allowing individuals from a particular habitat to mate with one another. After the spiderlings emerged from the eggs, each was raised separately and given all the prey they could eat. When they were mature, females from each population line were placed in experimental enclosures where they could build webs. Just as in field populations, the average distance between laboratory-raised females from riparian populations was less than that between females from grassland populations. Even under lush conditions, grassland spiders still maintain a large web—an indication that the behaviors responsible for territory size are genetically, rather than environmentally, determined. In another experiment, spiders were collected from either grassland or riparian habitats and transplanted to the other (Riechert and Hall 2000). Phenotypes that were inappropriate to the new habitats were selected against. That is, they did not survive as well as phenotypes native to the habitat.

These populations of spiders thus seem well adapted to their respective environments. However, spiders of one particular riparian population had characteristics that do not fit this pattern. Instead of showing the mellow behavior typical of populations in riparian areas, this population showed significantly more variability in behavioral traits, including the highly aggressive territorial behavior typical of desert populations. Why was this? It turns out that this riparian population was not isolated from desert populations but was constantly faced with an influx of immigrants. When researchers installed a drift fence to prevent individuals from moving from one population to another, they cut off gene flow. The

FIGURE 4.7 Variation in ecological conditions in different regions of the range of the funnel-web spider. The desert grassland of New Mexico (top) has few suitable web sites and low prey abundance. In contrast, most of the woodlands near rivers in Arizona (bottom) have many suitable web sites and prey is plentiful.

population then evolved over several generations to become less aggressive and more cautious, and thus more adapted to its local environment (Riechert 1993a). Thus, gene flow, by making populations genetically more similar to one another, can keep populations from being as well adapted to their local environment as they might be.

Genetic drift is another evolutionary process that may influence behavior. Genetic drift is the change in allele frequencies in a population due purely to chance events. For example, imagine that the only individuals in a population that carry a rare allele happen to die before they breed, not because the allele itself is unfavorable, but just because those individuals were unlucky. The rare allele would then be gone from the population. Allele frequencies in populations drift up or down over generations. A particular allele might even drift to fixation, so that it is carried in every member of the population. As you may have deduced, genetic drift is more important as population size gets smaller.

Populations that have gone through a bottleneck—a sharp reduction in population size analogous to the narrowing of a bottle at its neck—may show long-lasting evidence of genetic drift. Populations sometimes experience bottlenecks because of natural events, but bottlenecks are especially likely to occur in rarer animals of conservation concern. For example, a bottleneck in cheetah population size may account for the increase in the frequency of deformed sperm in male cheetahs (O'Brien 1994). Other populations for which drift is important include any stock of animals built up from a small number of individuals, such as laboratory animals, domesticated livestock, fish in hatcheries, or insects reared for biological control. In practice, it can be very hard to be certain when genetic drift has acted: demonstrating with confidence that allele frequencies have changed by chance means that all other explanations need to be examined and eliminated. For example, island populations of bumblebees vary in how they search for and handle flowers. Chittka et al. (2004) examined a number of possible adaptive explanations for this variation, such as differences in the array of flower species on the different islands, before concluding that drift was the most likely explanation for the pattern. We can become more confident of the importance of drift in a particular population if we have historical information about population size, but unless the underlying genetic variation does not affect the phenotype in any way, it is challenging to be confident that the changes could have been caused by some selective factor that was not studied.

CORRELATED TRAITS

It would be a mistake to think of any organism as an assortment of traits that evolve independently of one another. Traits may be correlated with one another for a number of reasons. One gene, such as a regulatory gene, may affect several traits (a phenomenon known as *pleiotropy*). Similarly, genes are sometimes tightly linked when they are physically close together on the same chromosome. One gene may be dragged along for a time when there is selection on its linked partner, at least until recombination and selection allow the correlation to be broken. Finally, two traits may share an underlying morphological and physiological basis that may make it difficult to uncouple them. Whatever the reason, when traits are tightly correlated, even negative traits might be maintained in the population if the net effect on the genotype is a positive one. For example, an individual's behavior is often consistent across different environments (Sih 2004): an animal that is bold and thus very successful in seeking out mates may also be inappropriately bold when investigating that strange noise in the underbrush. Boldness in both situations may be influenced by the same underlying genotype, levels of hormones, and the like, and thus behaving optimally in every situation might be impossible.

Let's look at an example of correlated traits. One of the most famous studies is on the evolution of beak size and shape in Darwin's finches in the Galapagos Islands in Ecuador. Charles Darwin was first to posit why many finch species—all quite similar in appearance—nonetheless differed in striking ways, particularly in their beaks. He suggested that the species shared a common ancestor, but that over time they diverged and specialized on different food resources. Some species, for example, have crushing beaks useful for seeds, and others have beaks useful for poking into flowers. So far, this is simply a wonderful illustration of the effect of natural selection on the traits of beak shape and size.

However, finches use their beaks for more than just feeding: male birds also sing, and how they sing is influenced by their beak shape. Females base their choice of mates on song (Podos and Nowicki 2004). Thus, selective forces may not act on just the single behavioral pattern of feeding, but there may be a ripple effect on singing behavior and mating behavior. Correlated traits such as these do not evolve independently from one another.

CHANGING ENVIRONMENTAL CONDITIONS

When we examine a population of animals today, we must remember that we are looking only at a single snapshot in time. A population may appear to be poorly adapted to current conditions because there is an evolutionary time lag between selection and its effects. Fluctuation of selection pressures from generation to generation means that natural selection must play catch-up, so today's traits may reflect past evolutionary pressures, not current ones. Thus, paradoxically, natural selection may actually be the reason that a trait we see today does not appear to be adaptive. Experiments are needed to sort this out.

Humans are the cause of a great deal of rapid environmental change. Not all of the change is detrimental to animals. In some cases, animals actually benefit from close proximity to humans. For example, in northern Massachusetts, opossums (*Didelphis virginiana*) are at the northernmost edge of their range; their naked tails and ears make them vulnerable to the cold. Opossums that survive a harsh winter are those that live near humans and can take advantage of shelter and food, such as leftover corn in the fields inadvertently provided by humans (Kanda 2005; Kanda et al. 2006; Figure 4.8). One might predict that over time there will be selection for opossums that are less fearful of humans.

Global warming is also causing new selection pressures on species (Walther et al. 2002). For example, newts and frogs both breed in ponds in the spring. In Britain, newts (*Triturus* spp.) have responded to warming temperatures by entering ponds earlier than they used to, but frogs (*Rana temporaria*) still reproduce at the same time. This means that frog eggs and tadpoles are now exposed to more newt predators than before (Beebe 1995). Humans cause changing selection pressures in other ways. To take just a few examples, habitat is lost to development, pollutants and fertilizers change water chemistry, traditional migratory stopovers disappear, and light pollution (caused by use of artificial lights throughout the night) can interfere with animal navigation. These environmental changes are occurring so rapidly that they are just a blink of an eye in evolutionary time, and many populations cannot evolve fast enough to keep up.

The environment of a species includes not only abiotic factors, such as climate and weather, but also biotic factors, such as other animals. For example, until recently the island nation of New Zealand had only two species of bats but no other mammals. Birds there have not faced mammalian predators in their evolutionary past, so they do not have the antipredator skills needed to cope with them. Many New Zealand birds will alight fearlessly near dangerous animals. Some, such as a parrot called the kakapo, have even lost their ability to fly (Figure 4.9). Now that cats, rats, stoats, weasels, and other predators have become established, many bird populations are in dramatic decline. Some researchers have tried to instill fear of predators into native birds by using aversive conditioning techniques. For example, McLean and his colleagues (1999) presented dead stuffed cats and ferrets to New Zealand robin chicks (*Petroica australis*) while playing robin alarm calls and distress calls. Robins learned to associate cats and ferrets with danger and reduced their tendency to approach them. Techniques such as these are obviously extremely time-intensive but may be useful as a last-ditch measure to save severely threatened populations. Most preservation efforts revolve instead around intensive trapping of the predators, but even that approach is only feasible on small islands or fenced-in reserves. Antipredator behavior may sometimes evolve in response to new predators, but some species are likely to be lost forever.

Adaptations that evolve in one species may change the selection pressures on other species, which in turn may change the selection pressures on the first (Van Valen 1973). For example, insectivorous bats use sonar to locate flying moths. In response, some moth species have evolved the ability to detect the ultrasonic signals emitted by the bat and to undertake evasive action with a fast erratic flight. Bats are then under even greater pressure to detect

FIGURE 4.8 North American oppossums can benefit from proximity to human habitation.

FIGURE 4.9 **A flightless New Zealand parrot called the kakapo. This species is vulnerable to introduced predators, such as stoats (a kind of weasel).**

and follow moths. This sort of coevolution is known as an evolutionary arms race, analogous to the mutual back-and-forth escalation of weaponry between the United States and Russia in the decades after World War II. Species engaged in arms races are also said to be behaving like the Red Queen in Lewis Carroll's *Alice in Wonderland*, who had to run as fast as possible just to stay in place.

FREQUENCY-DEPENDENT SELECTION

Sometimes variation is maintained in a population because different genotypes are favored at different times. One such mechanism is frequency-dependent selection, in which an allele has a greater selective advantage when it is rare than when it is common in the population. As a result, the frequency of any given allele fluctuates: it increases until it is common and then decreases once the alternative allele is favored. There are many examples of frequency-dependent selection (Ayala and Campbell 1974). We will consider two types—frequency-dependent predation and frequency-dependent reproduction.

Frequency-Dependent Predation

It is easy to see how frequency-dependent predation could maintain variation in a population of prey. Although predators usually have a varied diet, they often attack one prey type more often than is expected by chance. For example, when the members of a prey species differ in some characteristic, such as color, a predator might concentrate on the most common form and ignore the less common forms. The more common individuals are preferentially attacked until their numbers, and thus their alleles, decline in frequency. Meanwhile, the rarer form survives and reproduces, and its relative frequency increases. Then the predator switches to the new most common form of prey, which eventually then decreases in number, and the cycle begins again.

Evidence from many species shows that predators do indeed choose the most common type of prey, especially if the prey density is low (reviewed in Allen 1988). Alan Bond and Alan Kamil (1998) found experimental support for the hypothesis that frequency-dependent predation can maintain genetic diversity in a population. In these ingenious experiments, blue jays (*Cyanocitta cristata*) "preyed on" virtual moths presented on touchpad computer screens. Photographs of the dark form of a moth (*Catocala relicta*) commonly eaten by blue jays were scanned to create digital images that were then modified to create four different forms of moths. A fifth form was generated in a similar manner but with a different moth species (*C. retecta*). The virtual moths were presented on a computer screen against one of five different backgrounds that altered the difficulty of detecting the moths. The blue jays preyed on the moths by pecking at the screen. The "dead" moth was removed from the display, and the jay received a food reward. The relative numbers of moth forms that escaped detection determined the subsequent abundance of each prey type. The blue jays preyed on the most common form of moth and switched to alternative forms when that form became less common. This was true as long as the virtual moths were not too cryptic. Thus, frequency-dependent predation can maintain variation in the appearance of prey. In nature, the maintenance of this prey polymorphism (literally, "many forms") would also maintain the genetic variation underlying it. As we discussed earlier, selection acts on the phenotype, but evolution happens through changes in allele frequencies. We'll see more examples of frequency-dependent predation in Chapter 13.

Frequency-Dependent Reproduction

In this type of mating, sometimes called the rare-male effect, a male with a rare phenotype enjoys more than his expected share of matings. The alleles of the rare phenotype increase in the population until they become common and are no longer favored. The allele frequencies of different phenotypes can thus seesaw back and forth over time. The rare-male effect has been demonstrated in guppies (*Poecilia reticulata*), a fish species in which male coloration is extremely variable even within a single population. Females were allowed to examine males through a glass partition. They were then given a choice between mating with a male of a familiar phenotype or a male of a novel phenotype. Females chose males with novel color patterns—rare males—over males with a color pattern with which they were familiar (Hughes et al. 1999). Over time, frequency-dependent mating can maintain a variety of male phenotypes in the population, in the same way that frequency-dependent predation maintained variation in the types of moths in the example above.

An example that combines these types of frequency dependence comes from Texas field crickets, *Gryllus texensis*. Male crickets chirp by rubbing together a special "file" and "scraper" on their wings. This calling behavior attracts females. Males vary tremendously in the time they devote to calling every night—some rarely or never call, and others call for more than ten hours in a night. Why would there be this much variation? Unfortunately for the males, calling also attracts parasitoid flies, which lay their eggs on the crickets. The fly larvae burrow into the males, killing them within about a week. Flies are especially active early in the evening and are most common in the fall. Thus, when flies are common, the calling males are soon parasitized, and the males calling less end up with more mates over their (longer) lives. Of course, when the flies are rare, the calling males have the obvious advantage (Bertram 2002).

NEGATIVE-ASSORTATIVE MATING

Negative-assortative mating also preserves genetic variation in a population. This, in essence, means that opposites attract. More generally, the term describes the situation in which individuals tend to choose mates with a different phenotype than their own. It differs from rare-male advantage, where females of all phenotypes prefer the unusual males. Here, females of different phenotypes have different preferences. Obviously, if the phenotypic difference has a genetic basis, genetic variability will be enhanced. Admittedly, such assortment is not common in nature.

Negative-assortative mating maintains the tan-striped and white-striped morphs of white-throated sparrows (*Zonotrichia albicollis*) in approximately equal numbers in a population. Both female morphs prefer tan-striped males, which are better parents because they spend more time feeding the chicks. However, the white-striped females outcompete the tan-striped females for access to the tan-striped males. Tan-striped females then pair with the leftover white-striped males. As a result, 93 to 98% of the population mates with an individual of the opposite morph (Houtman and Falls 1994).

In some species, negative-assortative mating is a mechanism that prevents inbreeding. For example, in some strains of mice, individuals can determine whether others share certain of their alleles by the smell of urine. They then choose mates that have different odors from themselves (reviewed in Penn and Potts 1999). Patrick Bateson (1983) has suggested that some species are able to recognize kin (individuals sharing many alleles) and then choose mates that differ from kin. Bateson's ideas are discussed further in Chapter 8.

EVOLUTIONARILY STABLE STRATEGIES: FITNESS AND THE BEHAVIOR OF OTHERS

Sometimes the success of a strategy depends on what other individuals are doing. Recall the rules of the childhood game of rock–paper–scissors: rock breaks scissors; scissors cut paper; paper covers rock. Any one of the moves could win or lose depending on the actions of the other players. Play scissors and you win *if* your opponent plays paper. But if you play scissors too often, your opponent will catch on and defeat you by playing rock. The optimal behavior is dependent on frequency: whether you win depends on the frequency of the strategies played by others. If you don't have any knowledge of their plans, the best way to have a chance of breaking even is to play all three strategies—rock, paper, and scissors—in random order with equal frequency (Maynard Smith 1976).

Just as rock, paper, and scissors may be considered alternate strategies in a game, an animal's behavior may be described as a strategy (Maynard Smith 1976, 1982). Our use of the word "strategy" does not imply that the animal consciously plans the best course of action for maximizing reproductive success. "Strategies" are simply the set of behaviors available to an animal, and "winning" means that the individual's fitness increases more than its competitor's does (i.e., it leaves more offspring).

The optimal strategy for an individual to follow when the rewards (called payoffs) depend on what others are doing is called an **evolutionarily stable strategy** (ESS). By definition, an ESS is a strategy that, when adopted by nearly all members of a population, cannot be beaten by a different strategy: no other strategy confers more fitness benefits. For example, remember that a herring gull will not take care of a neighboring chick that wanders into its nest. This is an ESS because there is no alternative behavior that will yield greater reproductive success: the alternative strategy of caring for other birds' chicks would mean that herring gull parents would waste time and energy caring for offspring that are not their own. As a result, an ESS is unbeatable and uncheatable in the long run.

An ESS may be "pure" and consist of a single strategy, as in the example above, or it may be "mixed," consisting of several strategies in a stable equilibrium. Consider, for instance, a hypothetical population of fish-catching birds. There are two strategies for getting dinner—catch your own fish or steal one from another bird. Natural selection is assumed to favor the strategy that maximizes benefit. Since the thief minimizes its costs and gets full benefits from the efforts of others, thievery is favored initially. However, as the proportion of bandits in the population increases, so does the likelihood of encountering either another robber or a bird that has already had its fish stolen. Then, honesty becomes the best policy. When hard-working birds become common, thievery once again becomes profitable (Dawkins 1980). As the relative frequencies of alternative strategies fluctuate, they reach some ratio at which both strategies result in equal reproductive success. That particular mix of strategies will be an ESS.

Mixed ESSs can arise in two ways. First, different genotypes could be responsible for producing each strategy. In this case, each individual of the population would always adopt the same strategy, but individuals would differ on which strategy they adopted. At equilibrium, there would be a particular frequency of each type of individual. Second, each individual could vary its strategy and play each one with a certain frequency. Using the rock–paper–scissors example again to illustrate this difference, we see that stability would result if one-third of the population played rock, one-third played scissors, and one-third played paper. Alternatively, each member of the population could play rock, paper, and scissors with equal frequency.

Now let's examine a few biological examples of evolutionarily stable strategies.

The Nesting Strategies of Digger Wasps The nesting behavior of female digger wasps (*Sphex ichneumoneus*) is a clear illustration of a mixed ESS in which two strategies coexist (Brockmann et al. 1979). A female lays her eggs in underground nests that consist of a burrow with one or more side tunnels, each ending in a brood chamber (Figure 4.10). She lays a single egg in a brood chamber after provisioning it with one to six katydids, a process that can take as long as ten days.

Here is the choice female wasps face: to dig or not to dig (Figure 4.11). On some occasions a female will dig her own nest, while on others she will enter an existing burrow. Digging has an obvious cost in time and energy since it takes a female an average of 100 minutes to dig

FIGURE 4.10 **A female digger wasp at the entrance of the burrow.**

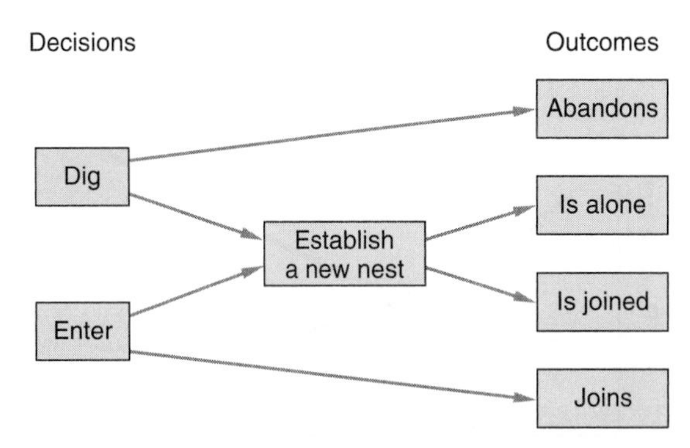

FIGURE 4.11 **A female digger wasp's alternative nesting strategies and their outcomes. The two available strategies are to "dig" and to "enter" a burrow dug by someone else. There are three possible outcomes of a decision to dig: the female may remain alone and retain exclusive use of the burrow; she may be joined by another female; or she may have to abandon the nest because of some catastrophe. If a female decides to enter an existing burrow, she may be alone or she may be joining another female. (From Brockmann et al. 1979.)**

a burrow. Furthermore, the investment is not risk-free. There is no guarantee that she will not be joined by another female, and if she is, she may lose her investment. In addition, temporary catastrophes can take place—for instance, an invasion by ants or a centipede—that may force a female to abandon her burrow. Once the intruders have gone, the abandoned nest is quite suitable for nesting again. A female that finds an abandoned nest reaps the benefits without incurring the costs. So, it might seem that entering an existing burrow would be the favored strategy. Indeed, it is—*if* the burrow is actually abandoned. Unfortunately, there is no way to determine whether the nest is abandoned or whether the resident is just out hunting. A female who is provisioning a nest is gone most of the time; it may be hours or days before she learns that another female is occupying the nest. Eventually the two females will meet, and when they do, they fight, sometimes to the death, and winner takes all. If the intruder wins, she continues to provision the nest and lays her eggs on the jointly provided supply of katydids.

Whether it is best to dig or to enter depends on what the other members of the population are doing. The strategy of entering an existing burrow is most successful when it is rare. As entering becomes more common, there are fewer diggers. As a result, the chances of entering an occupied nest increase, along with costly fights, and eventually digging becomes a better strategy.

For one digger wasp population studied in New Hampshire, 41% of the decisions made by wasps were to "enter" and 59% were to "dig" (Brockmann et al. 1979; Figure 4.12). Is that an evolutionary stable mix of strategies? If so, the reproductive success of females who

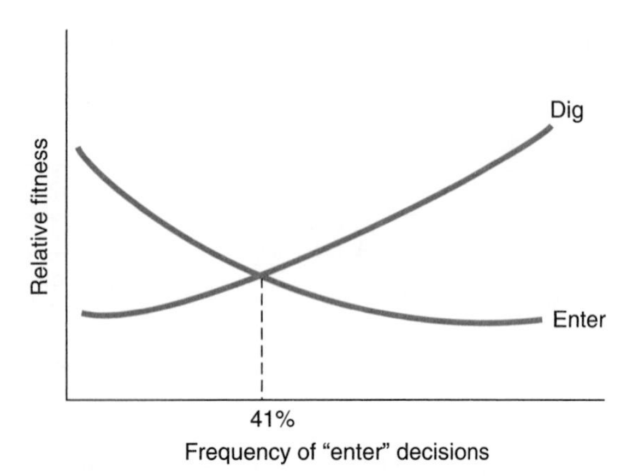

FIGURE 4.12 **The digger wasp's strategies to "dig" and to "enter" are a mixed ESS: neither can win out over the other under all conditions. The success of either strategy depends on that adopted by other members of the population. For a New Hampshire population of digger wasps, the strategies have equal fitness when they exist in the ratio of 41% "enter" and 59% "dig."**

enter must equal that of females who dig. If the payoffs of the strategies are not equal, females choosing the more successful strategy will become more prevalent in the population. To determine the reproductive success of these strategies, Jane Brockmann and her colleagues (1979) spent over 1500 hours observing 410 burrows. The reproductive success of dig versus enter decisions was calculated. In the study population, the researchers found no significant difference in the reproductive success of individuals who adopted the digging versus entering strategies. Whereas the reproductive success of individuals who decided to dig a nest was an average of 0.96 egg laid per 100 hours, the reproductive success of those who decided to enter an existing nest was 0.84 egg per 100 hours. Although the average for nondiggers was slightly less than that for diggers, the range of variation in reproductive success between the two groups overlapped broadly. Therefore, the nesting strategies of female digger wasps comprise a mixed ESS.

Reproductive Strategies of Male Lizards The changes in fitness among alternative male reproductive strategies of the side-blotched lizard (*Uta stansburiana*) provide an example in which the strategies of a mixed ESS cycle from one generation to the next. This small iguanid lizard lives in the inner coast range of California. Male lizards come in three throat colors: orange, yellow, or blue. Each of these three genetically determined color morphs also displays a different reproductive strategy. Males with orange throats are very aggressive and defend large territories, within which live several females. However, a population of only orange-throated males is not evolutionarily stable because yellow-throated males can steal their mates. Yellow males don't defend territories. Instead, these "sneaker" males mate covertly with females on the territories of orange-throated males. They get away with this because orange-throated males cannot defend all their females. However, a population of yellow-throated males is also not evolutionary stable because this strategy can be invaded by males with blue throats. Blue-throated males defend territories that are only big enough to hold a single female, so they can successfully defend her against sneaky yellow-throated males. However, when the yellow-throated sneaker males are rare, it once again pays to defend large territories with several females, and the reproductive success of orange-throated males exceeds that of blue-throated males. Thus, orange-throated males can successfully invade a population of blue-throated males, completing the dynamic cycle. In this mixed ESS, then, yellow beats orange, blue beats yellow, and orange beats blue. Notice the similarity between changes in morph frequency and the rock–paper–scissors game. The predominant color morph in a natural population was observed to fluctuate in the manner predicted by these frequency-dependent changes in fitness: blue was predominant in 1991, orange in 1992, yellow in 1993-1994, and blue again in 1995 (Sinervo and Lively 1996).

TESTING HYPOTHESES ABOUT NATURAL SELECTION AND ADAPTATION

Questions about the evolution and adaptive significance of behavior have been central to ethology since its beginning. Recall that two of Tinbergen's (1963) four questions were: What is a trait's function (survival value)? How did the trait evolve? These are questions about, respectively, the current adaptive value of a trait and its evolutionary history. These questions still drive the research of many ethologists, behavioral ecologists, and sociobiologists.

We began the chapter by observing that many animal traits appear to be well suited to their environments, and we attributed this match to natural selection. Natural selection is a powerful force, but we must keep in mind that the notion that any particular trait is an adaptation is a hypothesis that must be tested. Stephen J. Gould and Richard Lewontin (1979) called behaviorists (and others) to task for too often accepting the adaptive hypothesis without proof. Indeed, they ridiculed the adaptationist approach, which they claim breaks an individual into separate traits and assumes that each of those traits is adaptive. Gould and Lewontin named this practice a Panglossian paradigm, after Dr. Pangloss in Voltaire's satire *Candide* who made the obviously absurd assertion that "Things cannot be other than they are. . . . Everything is made for the best purpose. Our noses were made to carry spectacles, so we have spectacles. Legs were clearly intended for breeches, and we wear them."

Although Gould and Lewontin suggested that a Panglossian philosophy is inherent in the thinking of all adaptationists, Ernst Mayr (1983) argued that it is not. Mayr asserted that adaptationists claim only that natural selection produces the best genotypes possible given the many constraints placed on a population, not that all traits are perfect. As we have seen, a trait may not be optimal for several reasons. First, natural selection must act on the total phenotype of the individual, which is usually a mixed bag of traits—some good, some bad—and many traits (whether good or bad) are influenced by the same sets of genes so that perfection is elusive. Second, natural selection can act only on the available alternatives. Those alternatives will depend on the constraints imposed by the population's evolutionary history and each individual's present conditions—ecological, anatomical, and physiological. The "perfect" mutation or allele combination may not have yet arisen, if it can at all. Finally, natural selection works in a given environment, and as we have seen, conditions may vary from place to place or change over time. In other words, Mayr argued that Gould and Lewontin were setting up a straw man (an argument that is particularly weak and thus easily countered). Mayr suggested that scientists who study adaptation are (generally) well aware of other forces that can cause traits to be suboptimal, and that adaptationists do not actually hold the views that Gould and Lewontin ascribed to them. Nevertheless, the cautionary tale has had an important influence on evolutionary biologists, including behaviorists, who now pay much more attention to testing adaptive hypotheses instead of simply assuming that a trait has a current adaptive function.

So if we cannot just assume that a given trait is an adaptation, how do we decide whether or not it is? Let's look in more detail at funnel-web spiders. Earlier, we saw that the expression of territoriality in these spiders is well suited to the existing ecological conditions. We hypothesized that natural selection was responsible for the appropriateness of the behavior and, therefore, that territoriality is an adaptation. For any characteristic to be an adaptation, individuals bearing the trait must leave more offspring than those lacking it. Data support the hypothesis that territory quality influences reproductive success. In a lava bed in central New Mexico, spiders that defend high-quality web sites, those with the best ambient temperatures and prey abundance, have 13 times the reproductive potential of their neighbors in poor-quality areas (Riechert and Tracy 1975). These data support our hypothesis that territoriality is adaptive.

When we ask questions about the adaptiveness of behavior, we are necessarily asking about its value for enhancing reproductive success. The aim in answering such a question is to understand why those animals that behave in a certain way survive and reproduce better than those who behave *in some other way*. In our consideration of territoriality in funnel-web spiders, the alternative forms of behavior were easy to identify. The fitness of spiders that defend territories in areas that are hot and have few prey could be compared to the fitness of others that build webs in areas with more favorable thermal conditions and features that attract prey. But the alternatives are not always this easily identifiable because the losers of the competition may be long gone. Nonetheless, if we are to demonstrate adaptiveness, we must always identify the alternatives from which natural selection had to choose (Dawkins 1986).

Testing whether particular traits are adaptive has stimulated interesting research that might have been neglected if one readily assumed the nonadaptiveness of traits. Consider, for instance, Niko Tinbergen's observation of a seemingly nonadaptive behavior of black-headed gulls (*Larus ridibundus*). Tinbergen observed that the gull parent does not immediately remove the broken eggshells from the nest. Tinbergen and his colleagues had already demonstrated that the presence of eggshells in the nest attracted predators such as herring gulls and carrion crows. But here was the black-headed gull, sitting for hours among the conspicuous shell fragments. Tinbergen first thought that this delay must be dangerous and only explainable as a pleiotropic and nonadaptive effect associated with the removal behavior. He warned, however, that leaping to such conclusions is, in essence, a refusal to investigate.

Tinbergen then observed the black-headed gull colonies more carefully, looking for evidence of the adaptiveness of the delay. He noted that chicks were commonly eaten by neighboring gulls and that cannibalistic neighbors took many more chicks that were newly hatched and still wet than chicks that had dried and become fluffy. Whereas a gull could swallow a wet chick within a few seconds, it took about ten minutes to down a dry chick. One might imagine the difference as similar to trying to swallow a peeled grape as opposed to a cotton ball. Nest-robbing gulls were observed to snatch the wet young within a fraction of a second if the parent was distracted by a predator. In fact, one chick was gulped down while its parent was carrying off some eggshells. So, Tinbergen deduced that although removing the shells reduced predation by other species of birds, delaying removal until the chicks were dry decreased the likelihood of the chicks' being cannibalized by neighboring gulls while their parents were away on their chores (Tinbergen et al. 1962).

In this case, Tinbergen tested his hypothesis by simple observation, but other research on the adaptiveness of behavior incorporates several different approaches—experiments, comparative studies, monitoring natural selection in the field, and mathematical modeling. Each of these approaches to testing hypotheses about adaptation involves determining the reproductive success of different forms of the same trait. The observational approach compares the observed forms to other (sometimes hypothetical) forms. The comparative approach compares the behavior of the same or related species in different environments. The experimental approach compares different forms of the behavior, and usually conditions are manipulated and the behavior is observed. Monitoring selection in the field involves documenting changes in the frequency of behaviors over time. Mathematical models are used to understand the logic of how complex suites of variables influence behaviors and to compare the potential success of different strategies under a range of conditions that is difficult to replicate in the field or laboratory.

THE EXPERIMENTAL APPROACH

We opened the chapter by describing some differences between cliff-nesting kittiwakes and ground-nesting gulls that are correlated with the degree of the risk of predation. Cullen (1957a, b) found that kittiwakes, which have low predation rates, leave eggshell pieces in the nest, but ground-nesting gulls, which have high predation rates, generally remove broken eggshells. This correlation prompted Tinbergen to hypothesize that the survival value of eggshell removal was to reduce predation on the young (Figure 4.13). However, several other hypotheses are possible: the sharp edges on shells might injure chicks or interfere with brooding; an empty shell might slip over an unhatched egg, encasing the chick in an impenetrable double layer of shell; or the egg remains might serve as a breeding ground for parasites, bacteria, or fungi.

FIGURE 4.13 **A bittern removing pieces of eggshell from its nest. This behavior is typical of many bird species that rely heavily on nest concealment to reduce predation of the young. The white inner surface of the eggshell may make the nest more noticeable to a predator.**

Tinbergen and his colleagues (1962) tested the hypothesis that eggshell removal reduced predation on chicks. The study began with observation. It was noted that the eggs, chicks, and nest were camouflaged and might be difficult for a predator to spot. However, the bright white inner surface of a piece of eggshell might catch a predator's eye and reveal the nest site. So the researchers began by painting some black-headed gull eggs white to test the idea that white eggs might be more vulnerable to predators than the naturally camouflaged eggs. Of 68 naturally colored gull eggs, only 13 were taken by predators. However, 43 of the 69 white eggs were taken. The difference in predation rates lent credence to the idea that the white inner surface of eggshell pieces might endanger nearby eggs or chicks. Because all black-headed gulls remove eggshells, Tinbergen could not compare survival rates in natural nests with and without eggshell pieces. Instead, he created artificial variation to observe the effects of natural selection. He made his own gull nests with eggs and placed white pieces of shell at various distances from some of the nests. The broken eggshell bits did attract predators. Furthermore, the risk of predation decreased as eggshell pieces were placed at increasing distances from the nest. Note, however, that this is a test of only one of the possible hypotheses listed above, and other hypotheses may also stand up to testing: these hypotheses are not mutually exclusive.

Another example of testing multiple competing hypotheses is the study of the peculiar behavior of snake scent application (Clucas et al. 2008). California ground squirrels (*Spermophilus beecheyi*) and rock squirrels (*S. variegatus*) chew the shed skins of rattlesnakes, one of their major predators, and then lick their fur. Clucas and her colleagues considered three hypotheses for this behavior: it may serve as a defense against ectoparasites (e.g., fleas, ticks, or mites); it may distract conspecifics during aggressive interaction; or it may deter predators. Each hypothesis leads to different predictions about which squirrels should engage in this behavior more frequently. Because juveniles have more fleas than adults, the antiparasite hypothesis predicts that juveniles should apply snake scent more frequently. Males engage in aggressive interactions more often than do juveniles and adult females, so if distracting conspecifics is the primary function, males should be the most likely to apply snake scent. Finally, juveniles and adult females are most vulnerable to predators, so the antipredator hypothesis predicts that they will be the ones most likely to apply scent. Juveniles and adult females were indeed most likely to apply scent, so an antipredator function seems most likely and can now be further tested using controlled experiments.

THE COMPARATIVE APPROACH

The **comparative approach** to the investigation of adaptation involves taking into account the evolutionary relationships among a set of study species. For example, one might study individuals of the same or related species that inhabit different kinds of environments. These individuals will have inherited some common genes because they have a common ancestor. But if they have come to live in different ecological situations, they now experience different selection pressures and thus may have diverged in their traits. Similarly, the converse might also be true—unrelated species that have come to inhabit the same environment, and thus experience the same selection pressures, may display similar traits.

We have already seen the comparative approach in action in the herring and kittiwake gull example: these are species that descended from a recent, common ground-nesting ancestor but now live in very different ecological situations and have behavioral differences to match their ecological circumstances. However, a comparison of only kittiwake and herring gulls is quite limited: these two species essentially provide a single observation of different environmental effects on the behavior of close relatives. The presence of a correlation between the behavior and the environmental conditions, however logical, might still be just a coincidence. Larger sample sizes are needed in order to rule out random chance as an explanation. (If you flip a coin once and get "heads," does that mean the coin always lands on "heads"?)

How can the comparative method be used to its best effect? If an animal behaviorist is lucky enough to work on a taxonomic group that has been well studied by systematists, she may be fortunate enough to have a phylogeny available. A phylogeny is an "evolutionary tree" that shows the historical relationships among a group of organisms, in particular the order in which different subgroups branched off from one another. The reconstruction of a phylogeny is generally done by comparing the living members of a group (and fossils, where possible) and grouping together those that share relatively newer (derived) traits that differ from the traits of their immediate ancestors. Historically, morphological traits were used to construct phylogenies. Now, however, DNA sequences are often used. (The methods of phylogenetic reconstruction are worthy of several books on their own, so we will leave aside the details.)

Once a phylogeny is established, behavioral traits can be mapped onto it. If the phylogeny is definitive and detailed enough, it may allow us to determine the order in which behavioral and morphological traits evolved. For example, males of some species in the swordtail fish genus *Xiphophorus* have long tailfin extensions called swords, but males in other species do not. Females prefer males with swords. Strangely, even females in species in which the males have no swords prefer males that have plastic swords artificially attached. A phylogenetic reconstruction of the swordtail genus suggested that the female's preference for swords evolved before the sword itself: females were predisposed to be attracted to swords even before males evolved them (Basolo 1990, 1995a, b). This sort of study requires a robust phylogeny, which is often difficult to obtain. Phylogenies are hypotheses, too, and can change depending on the data on which they are based. For example, a swordtail phylogeny deduced from different data had a different shape and suggested a different evolutionary order for these behavioral traits (Meyer 1997), but even further evidence supports the initial interpretation (Meyer et al. 2006). As more and more data are used to reconstruct a phylogeny, its shape becomes more stable, along with the behavioral deductions we can make from it.

Besides determining the relative order in which phylogenetic traits evolve, we can also use phylogenetic information to examine the relationship between behavior and various ecological variables (such as the risk of predation in the gull example). As we have mentioned, data from a single pair of species limits the confidence we can place in the conclusions that we draw. Ideally, it would be best to have multiple species found in different environments: for example, the Galapagos swallow-tailed gull, which chooses nest sites with characteristics intermediate between those of kittiwakes and herring gulls, also shows behavioral patterns that are intermediate (Hailman 1967). Better still is to include not just multiple species, but those with multiple evolutionary origins. If a behavior has evolved a number of times and is significantly correlated with a particular ecological context, we can be more confident that a common evolutionary explanation exists. Examine the phylogenetic trees in Figure 4.14 to better understand the role of phylogeny in drawing strong conclusions.

Limitations of the Comparative Approach

Although the comparative method can be helpful in the study of adaptation, it must be applied carefully (Clutton-Brock and Harvey 1979, 1984; Gittleman 1989). Just as in other evolutionary interpretations, we should consider, test, and rule out alternative hypotheses. Sometimes this can be done by listing the competing hypotheses and developing predictions for each. The confirmation of the predictions should lend more weight to some hypotheses than to others.

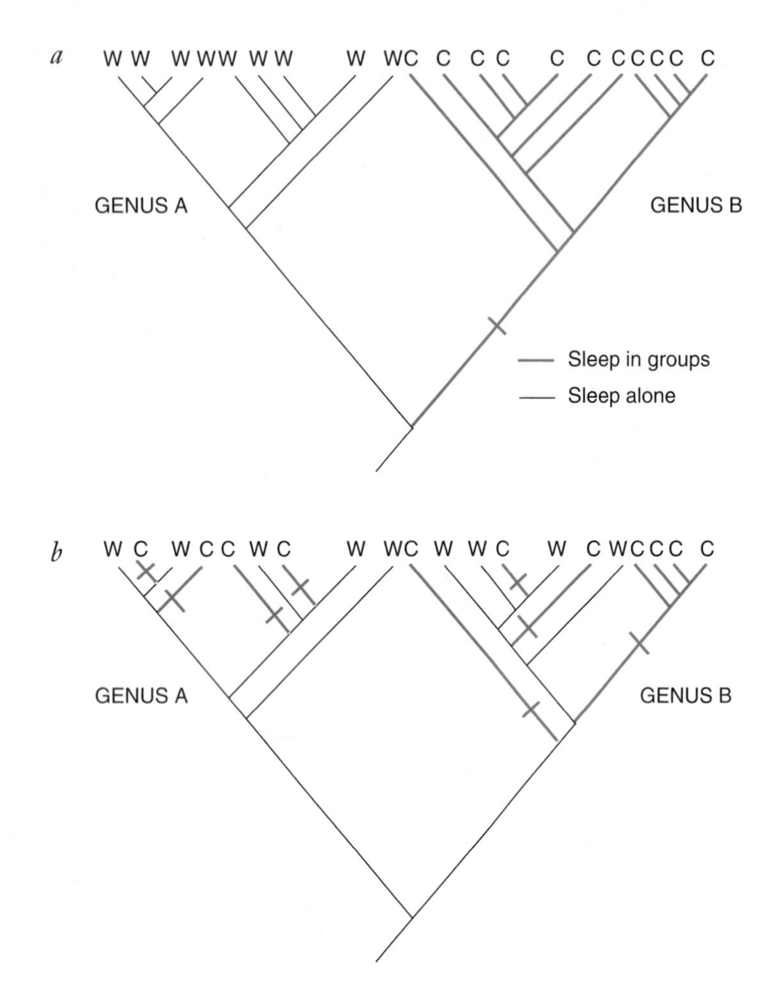

FIGURE 4.14 A hypothetical example illustrating the effect of wise choice of species on our ability to test hypotheses about the influence of the environment on traits. Imagine that we are interested in testing the hypothesis that sleeping in large groups (rather than alone) is associated with cold temperatures. Pictured are two hypothetical phylogenies that illustrate the possible relationships among 20 species. The modern-day species are at the "tips" of the phylogeny at the top of each tree, and their ancestors are nearer the bottom. Branch points indicate places where lineages diverged. In both phylogenies, we have 20 species. In each, 10 species sleep in large groups (indicated by a colored line), and all of these are found in cold regions (indicated by C). The remaining 10 species sleep alone (indicated by a black line) and are found in warm regions (W). From other evidence, we know that the common ancestor (at the bottom) slept alone and lived in a cold region. If the phylogeny is as illustrated in *(a)*, we essentially have a sample size of one for our test: sleeping in groups evolved only once. All the species that sleep in groups are related, and all the species that sleep alone are closely related. Thus, it is hard to draw any conclusions about the evolutionary origin of sleeping in groups. In *(b)*, sleeping in groups evolved eight times, so our sample size for our test of the correlation between sleeping in groups and temperature is eight. Here we can be more confident that temperature and sleeping in groups are related. (Modified from Rosenheim 1993.)

We should also keep in mind that correlations between traits and ecological variables, such as those shown in Figure 4.14, identified by the comparative approach, do not prove there is a common cause. Consider the difficulties in determining whether the diet was a cause or an effect of sociality in John Crook's (1964) comparative study of over 90 species of weaver birds (*Ploceinae*). He noticed that species living in the forest generally eat insects and forage alone. In contrast, species inhabiting the savannah eat seeds and feed in flocks (Figure 4.15). Crook identified correlations between the degree of sociality and two factors—diet and predation. But which is cause and which is effect? Seeds often have a patchy distribution, and groups of foragers are more likely to find a patch that can feed them all. One might infer, therefore, that diet is the cause of flocking in weaver birds. However, living in groups is also a good defense against predators, and after animals began living in groups, seeds may have been the only food source that could supply enough food for an entire flock. So, was sociality a cause or an effect of diet? The correlation does not answer the question. As with any correlational study, it is also important to remember that a third, unmeasured variable may be the underlying connection between the variables under study.

In some cases, the *lack* of a pattern between traits and ecological variables helps to eliminate a hypothesis. For example, thermoregulation is one hypothesis explaining why birds often roost in groups at night. By huddling together as they sleep, they may conserve body warmth. This hypothesis predicts that species that spend more time in cold areas and species that have lower body masses would be especially likely to need this thermoregulatory boost. A phylogenetically based study that examined species from many distantly related groups of birds did not find this pattern, so the thermoregulatory hypothesis is not supported (Beauchamp 1999). As with any study where we accept the null hypothesis, considerations of statistical power are key: in this case, the sample size is large enough to provide confidence that we would have been able to detect a biologically meaningful pattern if one had existed.

FIGURE 4.15 **Weaver birds. John Crook compared the diet and degree of sociality of weaver birds that inhabit different environments. He observed that solitary species, such as *Ploceus nelicouri* (shown on the left; nest not in proportion), are usually insectivores that defend large territories in the forest. In contrast, social species, such as *P. phillipinus* (shown on the right), eat seeds and live in the open savannah.**

MONITORING SELECTION IN THE FIELD

It's relatively straightforward to measure evolution over generations in the laboratory (although even that can be extremely time consuming), but it is much, much more difficult to do so in the field. We've already mentioned the variation in beak size and shape across Darwin's finch species on the Galapagos Islands. Beaks of these closely related species range from robust to slender, according to the kind of food they eat. In most such cases of adaptive radiation (the evolution of an ancestral species into many different ecological niches), we must rely on comparative analyses in order to trace the pattern that evolution has taken. In finches, we can see natural selection in action. In an intensive 30-year-long field study, Rosemary and Peter Grant and their colleagues documented changes in beak size in medium ground finches (*Geospiza fortis*) in response to the environment (Boag and Grant 1981; Price et al. 1984). During periods of drought, medium ground finches that had deeper beaks were better able to eat the hard seeds available for food. These birds produced more offspring. Because beak size has a genetic component, the offspring also had deeper beaks and the population mean shifted. During rainy years, many more small seeds were available; birds with smaller bills had the advantage, shifting the population mean back. The finch work is beautifully summarized in Jonathan Weiner's (1995) book *The Beak of the Finch: A Story of Evolution in Our Time*, winner of a Pulitzer Prize.

MODELING THE COSTS AND BENEFITS OF TRAITS

Most actions have costs and benefits. When considering the best course of action, people often create lists of the pros and cons of each strategy. Lists are helpful because they identify factors that should be considered in the decision. But the decision may still be difficult because it may require integration of concerns along different dimensions. For instance, if you were deciding whether to get your own apartment or live with your parents, you might have to weigh factors that represent two dimensions—money (the difference in rent) against freedom (gained by being on your own). These different factors might seem to be apples and oranges, and so they might be difficult to compare.

A similar example for animals is the decision of whether to stay in a safe place where there is not much to eat or whether to go out to forage in a place where there is abundant food but where dangerous predators might lurk. How do we decide which of these strategies is optimal under current conditions? As the number of influences on a behavior increases, it can be increasingly difficult to identify the best solution.

Animal behaviorists often use optimality modeling to weigh the pros and cons or, to use the proper jargon, the costs and benefits, of each available strategy. A model is a mathematical expression of the costs and benefits of each strategy. First, all costs and benefits are translated into common units that represent a measure of fitness. This common measure of fitness is called "currency" (a holdover from economic theory, where modeling approaches now used in animal behavior were first developed), and it allows different strategies to be compared. The optimal strategy is the behavioral alternative that maximizes the difference between the costs and benefits. In economic terms, the alternative that maximizes the difference between costs and benefits is the one that yields the greatest profit. In evolutionary terms, this is the choice that maximizes fitness. If a behavior is at least in part genetically based, the successful alternative is the one that would contribute most to the next generation. We'll explore some detailed examples of mathematical models in later chapters.

Measuring costs and benefits in terms of fitness is extremely difficult to do empirically. It is generally impossible, especially for long-lived species, to follow individuals in the field over the course of their lives and to measure their reproductive success. Instead, researchers often measure a behavior that they assume correlates with fitness. For example, the rate at which animals acquire calories has been used as a way to compare different foraging strategies. The assumption is that the faster the rate of energy intake, the higher the fitness. Of course, this assumption is very much an oversimplification: surely an animal that is intent on eating as fast as possible may miss an approaching predator. Therefore, in constructing a model, it is important to think carefully about what an animal faces during the time period under consideration—is it really free to focus on foraging, or should it take into account predation risk as well? If the latter is true, then predation risk should be included as a variable in the model. Once all the variables are identified and included, the model can be used to identify the optimal behaviors under different conditions, such as high prey abundance and low predation risk. These predictions can then be tested using experiments in which conditions are manipulated.

This may sound like complex, higher-level thinking that is above and beyond the capabilities of most animals. Surely a bumblebee does not calculate the rate of nectar production by different flower species, the density of the flowers, and the energy it uses in flying from plant to plant before choosing where to forage. In using language like "decision rules," behaviorists are not implying that animals make conscious decisions or "think things through" to find the optimal course of action. Instead, we assume that natural selection has shaped behavior over generations so that the animal responds appropriately under a particular set of circumstances. What may appear to be very complex behavior may come down to following a fairly simple strategy.

Models can give us insight into how well simple behavioral rules can generate complex-looking behavior. For example, in one model, researchers defined a habitat grid and added simulated animals. The animals were allowed to move around the grid as if they were playing a video game. Given just a few rules to follow, such as "Fight other animals that you encounter" and "Reduce the probability of returning to places where you have had a fight," the animals settled into a pattern that looks remarkably like territoriality: each stayed in its own area and defended it against other animals (Stamps and Krishnan 2001). This sort of result does not prove, of course, that these *are* the rules that animals follow—other rules may generate the same result—but it gives us a logically coherent hypothesis to test against data.

SUMMARY

Animals frequently appear to match their environments very well, and natural selection is the process that creates this match. Natural selection occurs when there is phenotypic variation in a population, the variation has at least some genetic basis, and some of these inherited traits improve their owners' chance of leaving more offspring. When these conditions are present, the allele frequencies in a population change over generations, and evolution by natural selection occurs.

In spite of the power of natural selection to shape animals and their behavior, we cannot assume that organisms are perfectly adapted to their environment. Animals may migrate into a population from nearby areas, bringing their genes with them. In small populations, genetic drift (changes in allele frequencies by chance alone) becomes increasingly important. Phenotypic traits are often correlated with one another, so selection on one trait might drag along another trait. Selection pressures also change over time, and evolution lags behind: the genetic makeup of today's populations results from selection on previous generations.

Selection pressure on a particular genotype may depend on its frequency in the population: for example, predators may preferentially feed on the most common prey, driving down its numbers, and then switch to an alternative prey whose numbers are on the rise. The success of a particular genotype may also depend on the behavior of other members of the population. An evolutionarily stable strategy, or ESS, is one that cannot be beaten by another. An ESS might be a single strategy or a combination of strategies, each played with a particular frequency.

How are hypotheses about the adaptive nature of behavior tested? Several approaches are available to us. We can conduct experiments to measure the present-day costs and benefits of particular behaviors. This is probably the most common approach taken by modern-day behavioral ecologists. If we know something about the phylogenetic relationships among a group of species, we can use the comparative method to tease apart the evolutionary history of a particular behavioral trait. In rare cases, we can monitor populations in the field and actually see natural selection at work. Finally, we can use mathematical techniques to model the costs and benefits of behavior and compare the value of different strategies.

5

Learning and Cognition

Most people have presuppositions about the mental lives of animals. We tend to discount the abilities of some species, especially those very different from us, but we anthropomorphize our pets and other primates and assume that they think like we do. For example, imagine watching a chimpanzee break off a twig, strip its leaves off, carry it to a termite mound, and insert the stick into the hole on top. It then draws out the stick, now covered with swarming termites, and licks the termites off. Does it seem to you that the chimp planned its actions? When it first broke off the twig, did it understand that it would result in a snack? Now imagine watching an ant lion, a little larval insect, dig a pit in the sand. The ant lion waits in the bottom of the pit. Finally, an ant slips over the edge and begins to slide toward the ant lion's waiting jaws. The ant tries to scramble back out, and the ant lion hurls sand upward. The ant is knocked back into the pit, and the ant lion feeds. What do you think the ant lion understands about its own behavior? Did it anticipate catching ants when it dug its pit? Would your opinion change if you knew that ant lions can learn to associate a human-made vibratory cue with the presentation of food (Guillette et al. 2009)? Most of us would attribute more cognitive abilities to the chimp than the insect, but are the underlying mechanisms really different? How might we critically examine these questions? In this chapter, we will discuss how we know what animals know: how they learn, why species differ, how we can test their ability to solve challenging cognitive problems, and even how they view themselves and others.

DEFINITION OF LEARNING

Learning has proved so tricky to define that one textbook on the topic begins by defining it as "a term devised to embarrass learning psychologists, who tie themselves in knots trying to define it" (Lieberman 1993). For example, we could say that learning is "a process through which experience changes an individual's behavior." This certainly encompasses what we usually think of as learning, but it also includes phenomena that we would *not* call learning. For instance, an athlete might run the last mile of a marathon at a slower speed than she ran the first mile. This is the result of the experience of running, but it isn't learning. Similarly, a man entering a dark movie theater from the sunlit street may inadvertently step on a discarded candy box on the floor because the photoreceptors in his eyes have not yet adapted to the dim light. In a few minutes, his eyes adjust and he steps over the trash. Thus, sensory adaptation can also lead to a change in behavior, but we wouldn't want to call this learning either.

Let's expand our definition: learning is a change in our capacity for behavior as a result of experience, excluding the effects of fatigue, sensory adaptation, or maturation of the nervous system (Hinde 1970). "Experience" includes exposure to particular combinations of environmental stimuli, as well as practicing a behavior.

Complicating matters further is that behavioral changes that result from learning are not always expressed immediately. For example, a student may not demonstrate that she has learned course material until the day of the exam. In addition, the learned behavior may not be consistently expressed every time the opportunity presents itself—sometimes the student may be able to articulate what she has learned, and sometimes she may not. Thus, the change in behavior that results from learning is perhaps more accurately described as a change in the *probability* that a certain behavior will occur.

This discussion should make it clear that we can't necessarily know whether an animal has learned something just by seeing a change in its behavior. To be absolutely sure that learning has occurred, we must manipulate the experiences of different groups of animals and then compare their performance on the same test (Shettleworth 1998). Because of this need for carefully controlled experiments, the vast majority of work on learning has been done in laboratory settings. However, as we will see, we can still address evolutionarily important questions, and we can even do some field studies. The modern study of learning incorporates both proximate and ultimate questions, all the way from the cellular level to the phylogenetic level, where we take into account evolutionary relationships among species.

Researchers classify patterns of learning into different categories, but keep in mind that the relationships among them may be more complex than they first appear—they may overlap, and the distinctions between them may not be clear-cut. In fact, researchers do not all agree on the nature of the categories or how many there should be. We'll postpone discussion of several types of learning until later chapters. For example, imprinting and song learning, which generally occur early in life, are discussed in the chapter on the development of behavior (Chapter 8).

TYPES OF LEARNING

HABITUATION

We usually think of learning as resulting in the expansion of an individual's behavioral repertoire—perhaps learning a new skill or a new association. However, in habituation, the animal learns *not* to respond to a particular stimulus because the stimulus has proven to be harmless. A bird must learn not to fly away every time the wind rustles the leaves. Habituation can be defined more precisely as the waning of a response after repeated presentation of a stimulus. Once habituation occurs, its effects are long lasting. Habituation is everywhere, from unicellular protozoans to humans (Wyers et al. 1973). It is generally considered to be the simplest form of learning.

A classic example that illustrates the essential characteristics of habituation is the clamworm, *Nereis pelagica*. This marine polychaete lives in underwater tube-shaped burrows it constructs out of mud. It filters tiny bits of food from the water. When it feeds, it partially emerges from its tube. However, it withdraws quickly back into the safety of the tube when it senses sudden stimuli such as a shadow that could signal the approach of a predator.

Clark (1960) kept clamworms in shallow pans of water in the laboratory. When he passed a shadow over them, they withdrew into their tubes. The second time he presented the shadow, slightly fewer worms responded. The third presentation elicited even fewer withdrawals. As shown in Figure 5.1, subsequent presentations resulted in a continued decline in escape responses. The clamworms had habituated, and the effects of habituation lasted for several hours. The clamworms' decline in responsiveness was not because the sense organs became adapted to the stimulus, because sensory adaptation occurs much more quickly than this. Nor was the decline due to muscle fatigue, because habituated worms still withdrew in response to prodding. The clamworms had learned to stop responding to the shadow.

A characteristic of habituation is that it is specific to a particular stimulus. For instance, young turkeys, chickens, and pheasants innately show antipredator behaviors, such as crouching and giving alarm calls, at

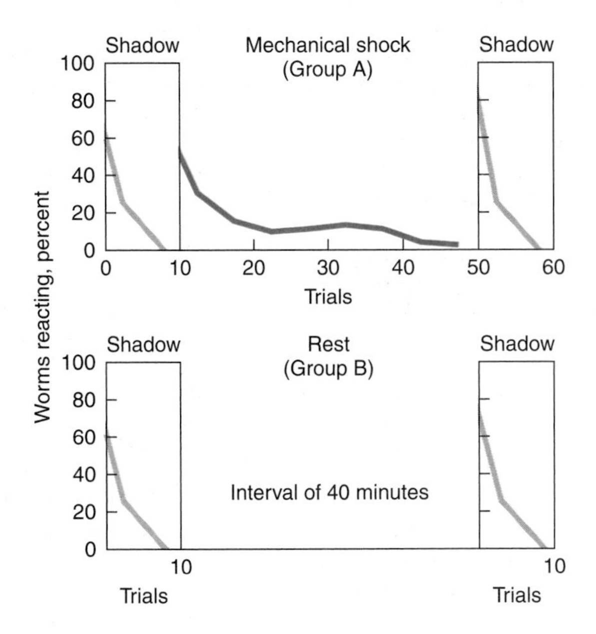

FIGURE 5.1 **Habituation of the withdrawal response to a shadow by the clamworm, *Nereis*. In habituation, the simplest form of learning, the animal learns *not* to respond to frequently encountered stimuli that are not associated with reward or punishment. Both groups of worms habituated to the shadow during trials 1–10. A mechanical shock was then administered to group A at 1-minute intervals. The worms initially responded to the new stimulus, showing that the loss of the response to the shadow was not a result of fatigue. Group B rested for 40 minutes while group A received the mechanical shocks. Both groups were equally responsive to the shadow during the last 10 stimulus presentations. Thus, habituation was specific to a stimulus. (Modified from R. B. Clark 1960.)**

the sight of objects moving overhead. The chicks initially respond to a great variety of objects, only a few of which are dangerous. By the time they are adults, the birds respond only to the image of a predator such as a hawk flying overhead. Schleidt (1961a, b) tested whether habituation could underlie the development of the specificity of these responses. He showed that models of various shapes—a gooselike silhouette, and even a circle and a square—all effectively elicited alarm calls from young turkey chicks during the initial two days of testing. When any one of these models was presented frequently, it elicited fewer and fewer alarm calls. Chicks still called in response to stimuli they encountered only occasionally.

The Adaptive Value of Habituation

Obviously, it is important for a clamworm to withdraw to the safety of its burrow when a shadow is that of a predator. However, a recurring shadow that is not followed by an attack is more likely caused by something harmless, perhaps a patch of algae that is repeatedly blocking the sun as it undulates with the waves. In this case, responding to the shadow every time it appears would mean that the clamworm loses opportunities to feed. Unnecessary responses also waste energy. Habituation, like other kinds of learning, thus focuses attention and energy on the important aspects of the environment (Leibrecht and Askew 1980).

Habituation has been documented in nearly every species that has been tested, but it may vary in its details in ways that make sense given the habitat of the species. For instance, consider two closely related species of crabs. On one hand, *Chasmagnathus* is a semiterrestrial crab that lives on the mudflats amid patches of cord grass along the coast of South America. On the other hand, *Pachygrapsus* inhabits the rocky intertidal zone. Crabs of both species begin to run when a shadow passes overhead, and both habituate to repeated presentation of shadows. However, habituation lasts much longer in *Chasmagnathus*. This makes sense because the windblown grass of the habitat of *Chasmagnathus* casts many harmless moving shadows, but shadows on the bare rocks of the habitat of *Pachygrapsus* are more likely to signal the arrival of a predator (Tomsic et al. 1993).

Habituation may also occur in the context of interactions within species. For instance, animals that defend territories encounter their next-door neighbors time and again. Over time, many species reduce their aggressive responses directed toward these familiar neighbors: there is little point to fighting day after day over a boundary that has already been settled. However, unfamiliar intruders will still provoke a territory holder to attack. This phenomenon can be nicely demonstrated in species with acoustic communication by playing back recorded calls. A number of bird species (e.g., Falls 1982) as well as bullfrogs (Davis 1987) respond aggressively to a playback of a stranger's call but not to that of a familiar call. One hypothesis is that habituation is the mechanism that mediates this process: perhaps frogs stop responding aggressively to a call when they hear it repeatedly (Peeke 1984). Bee and Gerhardt (2001) created a "new neighbor" bullfrog by synthesizing a new call and playing it back from a previously unoccupied territory. Initially, this mysterious new voice elicited quite a response: male bullfrogs called back and charged at the speaker. After repeated presentations, their aggressiveness declined. Because this decline carried over between nights and was specific to particular characteristics of the calls, it meets the criteria of "relatively permanent" and "specific to a stimulus."

STOP AND THINK

Can you think of other hypotheses, besides habituation, explaining why animals might stop responding to neighbors' calls?

Habituation as an Experimental Tool

Habituation is a very useful tool for the study of cognitive processes in animals, including humans. In a typical protocol, a subject is habituated to a stimulus, and then a new stimulus is presented. If the subject's response changes, the experimenter knows it can detect the difference between the two stimuli. This is especially useful in the study of cognition in infants, where it's not always easy to determine what the infant perceives.

CLASSICAL CONDITIONING

In **associative learning**, some sort of a mental connection is formed between representations of two stimuli (Shettleworth 1998). The first type of associative learning we will examine is classical conditioning. For many people, the phrases "classical conditioning" and "Pavlov's dogs" are intertwined. Pavlov first laid out the principles of classical conditioning in 1927. Pavlov was a Russian physiologist whose main interest was not learning but digestion. He wondered why a dog salivates at the *anticipation* of food rather than just its presence. He hypothesized that the animal had made a connection between the sight or smell of food and the food itself, and he became interested in exactly how dogs made these associations.

To measure saliva, Pavlov made a small opening, or fistula, in the dog's cheek so that the saliva would drain into a funnel outside the dog's body. The hungry dog was harnessed into position on a stand and then presented with various stimuli. As expected, the dog salivated when powdered food was blown into its mouth. In contrast, it did not salivate when it heard the sound of a bell. Then Pavlov began pairing these two stimuli: immediately before food powder was presented, the bell sounded. Pavlov presented these paired stimuli repeatedly at intervals over several days. After 30 presentations, the dog salivated in response to the bell alone. As trials continued, the dog salivated more profusely and responded more quickly to the bell (Pavlov 1927).

Let's phrase these results in more general terms. To begin, an animal has a particular inborn response to a certain stimulus. This is called the **unconditioned stimulus** (US) because the animal did not have to learn the response to it. In Pavlov's study, the US is food. The response to the US is called the **unconditioned response** (UR). In Pavlov's study, the UR is salivation. A second stimulus is paired repeatedly with the US until eventually it, too, is able to elicit the response. At this point, the new stimulus is called the **conditioned stimulus** (CS) because the animal's response has become conditional upon its presentation; here, the CS is the bell. The response to the conditioned stimulus is called the **conditioned response** (CR). The conditioned response may differ slightly from the unconditioned response. In Pavlov's study, the CR is salivation, like the UR. The new connection between the US and CS is called a conditioned reflex.

Over thousands of controlled studies, researchers have found some remarkably consistent characteristics of classical conditioning. These general features hold across a wide range of stimuli, as well as across species. Next, we'll describe three of the most important characteristics of classical conditioning.

First, the order of the presentation of the US and CS is important. Conditioning is most effective when the CS (such as a tone) precedes the US (such as food). The CS serves as a signal that the US will appear; a cue is of little value if it occurs after the fact. Also, the two stimuli must occur fairly close together if an association between them is to be made. Thus, if you want your new love to associate you with something nice, stand on the doorstep with flowers hidden behind your back. Right after the object of your affection sees your smiling face, present the flowers.

A second characteristic arises from the fact that useful signals are reliable: they predict that a particular event or stimulus will follow. A signal is useless if it merely indicates that any one of a dozen events may follow. Therefore, it should not be surprising that for classical conditioning to be most effective, the CS must precede the US more often than it does other stimuli (Rescorla 1988a, b).

Finally, after an association between a CS and US is formed, it can be lost again. If the CS is no longer reliable because it is presented frequently without being followed by the US (for example, if a tone is given time and again with no food), the subject stops responding to the tone. The loss of the conditioned response is called **extinction**. Thus, it is important to remember to continue to bring flowers to your love! Of course, in a changing environment, it is fortunate that learned responses can be extinguished.

The Adaptive Value of Classical Conditioning

Pavlov (1927) suggested, and other researchers agree, that learning through classical conditioning is likely to provide fitness advantages to wild animals. However, most studies of classical conditioning over the decades have focused on determining the rules under which it functions. Most effort has been channeled toward characterizing the process of conditioning, such as the most effective interval between the conditioned and unconditioned stimuli and the time course of extinction. Relatively few studies have addressed the potential value of classical conditioning in the everyday life of an animal. We'll discuss three of them here. Our first example comes from Karen Hollis (1984, 1999) on territorial and reproductive behaviors in blue gouramis (*Trichogaster trichopterus*), fish that inhabit shallow pools and streams in Africa and Southeast Asia.

A male blue gourami defends its territory with an aggressive display: it swims rapidly toward the intruding fish, with all fins erect. If the intruder does not respond with a submissive posture or retreat, the contest escalates into a heated battle that can result in serious injury. The males bite each other and flip their tails to beat water against the opponent's sensitive lateral line organ. The lateral line is a row of receptors running in a line down the side of fish that detect movement and vibration in the water. Dangerous fights such as these gourami fights are most likely to evolve when the value of the resource is great. Success is crucial for male gouramis because females rarely mate with a male without a territory.

If a male were to learn the signals that indicate the approach of a rival—perhaps visual, chemical, or mechanical cues—he might be better prepared for battle and gain a competitive edge. Hollis selected pairs of male fish with similar body size and aggression levels, and placed them on opposite sides of a divider in an aquarium. For one member of each pair, a 10-second light (the CS) preceded a 15-second viewing of a rival (the US). As a control, the CS and US were also shown to the other member of the pair, but their presentations were not paired: they occurred randomly with respect to one another. During the test trials, the light signal was given, and then the barrier that separated the fish was removed, allowing them to interact. The males that had been classically conditioned to associate the light with the imminent appearance of a rival were superior in territorial defense. They approached the territorial border with their fins already erect. During the ensuing fights, they delivered significantly more tailbeats and bites than did their competitors (Hollis 1984). This response may have been mediated through hormones: in conditioned males, the presentation of the light led to an increase in androgens, male sex hormones known to heighten aggressiveness in many species of vertebrates (Hollis 1990; see Chapter 7).

The conditioned male gains a long-term benefit in addition to his immediate competitive edge: the experience of winning increases the probability of winning again in battles with new opponents. Thus, conditioned males not only win the first battle but are likely to keep winning. In contrast, fish that lose their first battle are likely to lose later battles as well. In one experiment, all fish that lost the first battle also lost the second one (Hollis et al. 1995). Many species, both vertebrates and invertebrates, show a similar "winner effect."

Male blue gouramis that successfully defend a territory are more likely to attract females, but excessive aggressiveness could actually harm mating success. A territorial male is likely to attack all visitors to his territory, even females. If a male is conditioned with a light signal to expect the arrival of a female, he is less likely to attack her when she appears (Hollis et al. 1997). As can be seen in Figure 5.2, conditioned males bit females fewer times than did unconditioned males. These conditioned males also spent more time building a nest. The shift in behavior from aggressive to reproductive activities paid off in reproductive success. Conditioned males spawned

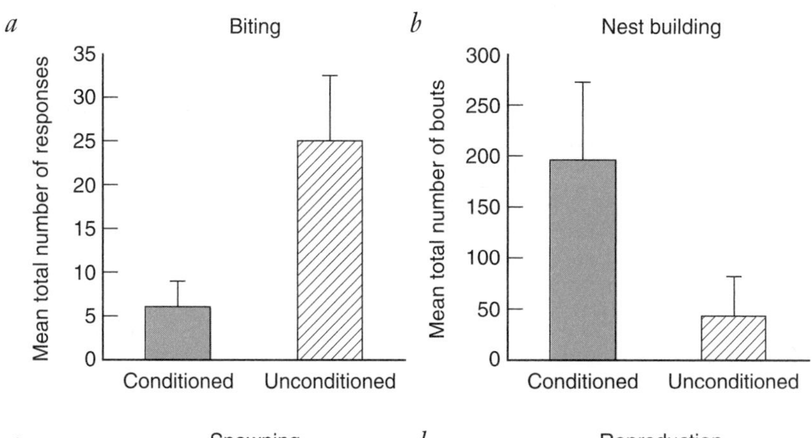

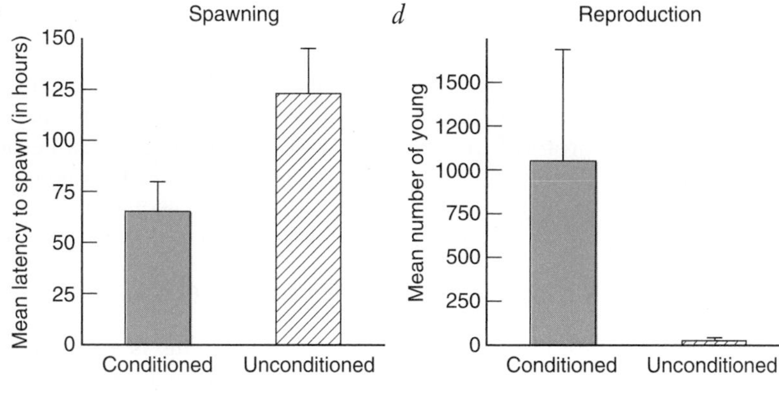

FIGURE 5.2 **Male blue gourami fish learned, through classical conditioning, that a light signaled the imminent appearance of a receptive female. Conditioned males had greater reproductive success than unconditioned males. Conditioned males (*a*) bit approaching females less frequently, (*b*) spent more time building nests, (*c*) were quicker to spawn, and (*d*) produced more young than did unconditioned males. (Data from Hollis et al. 1997.)**

more quickly and produced more fry than did unconditioned males.

In nature, of course, flashing lights are unlikely cues. However, many natural signals are likely to be available to be learned (Hollis 1999). For example, territorial invaders might be seen, heard, or scented by a territorial holder. The shape of a gravid belly might reliably indicate a willing female's approach.

Let's look at another taxon in which classical conditioning functions in an evolutionarily relevant context. Male field crickets (*Gryllus bimaculatus*) mate by transferring a sperm in a packet called a spermatophore. Because spermatophores are costly to produce, males would do best not to transfer larger spermatophores than necessary. However, when females mate more than once, the sperm from different mates compete inside her body to fertilize her eggs. Thus, if a male faces another male in a competition for a female's attentions, he will increase his chances of fathering more offspring by transferring a larger spermatophore and thus more sperm to the female (Mallard and Barnard 2003). Interestingly, males can learn to associate environmental cues with the presence of male competitors (Lyons and Barnard 2006). The experimental design is illustrated in Figure 5.3. Males were placed in one side of a terrarium along with topographical cues: either two or four Lego bricks. The other side of the terrarium, visible through a clear wall, either held a potential competitor or was empty. Each subject male had four opportunities to mate during training. For any particular subject, a particular quantity of bricks (two or four) was always associated with the presence of a competitor, while the other quantity of bricks never was. Thus, males had a chance to learn that a particular number of bricks signaled the presence of a competitor. After training, the males were allowed to mate next to each arrangement of bricks, with no competitors present. Males produced larger spermatophores in the environment that, for them, had been associated with a competitor. Odor cues (peppermint- and vanilla-scented oils from a cosmetics store) could also serve as signals. When tested with scents that had been associated with the presence of a competitor, males again produced larger spermatophores.

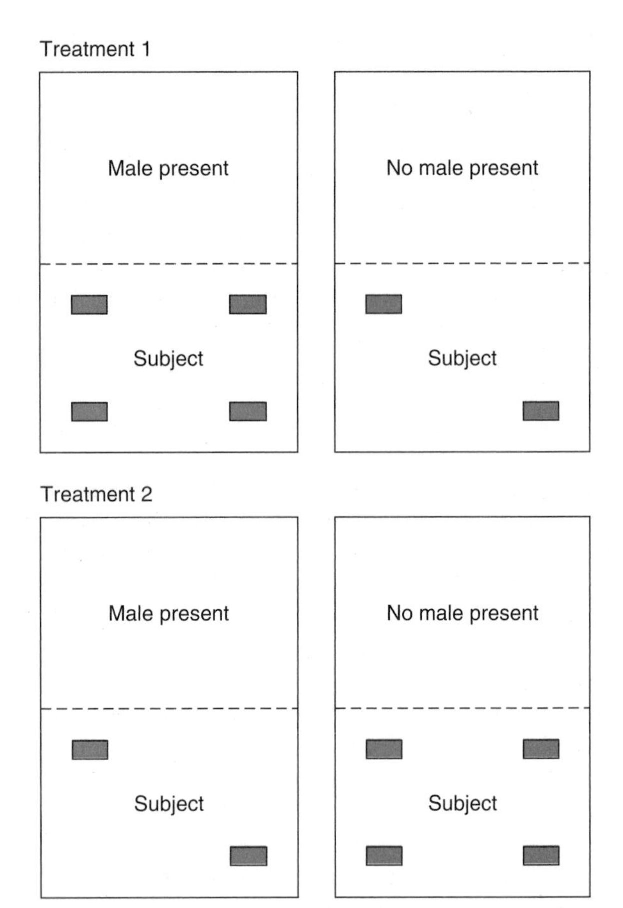

FIGURE 5.3 Field crickets learned about spatial cues that signaled the presence of a competitor. Males were placed on one side of a terrarium and either two or four Lego bricks. For males in Treatment 1, four bricks were always associated with the presence of a male competitor on the other side of the partition, and two bricks were never associated with a competitor. For males in Treatment 2, the situation was reversed. After training, subjects were allowed to mate next to each arrangement of bricks, with no competitors present. Males produced larger sperm packets in the environment associated with a competitor. (From Lyons and Barnard 2006.)

Our final example of classical conditioning in an ecological context concerns feeding behaviors. A wonderful model system is the honeybee. A foraging bee has a lot to learn. Flowers bloom and fade, so the best places to forage are constantly changing. Individual flowers can vary in the amount and quality of nectar and pollen they produce, as well as whether they've recently been visited by another forager. Flower species also vary in their shape, and bees must access each flower shape differently. Location, color, shape, pattern, texture, and odor of flowers are all characteristics that bees learn about. For example, bees can be rapidly conditioned to respond to odor. Carefully controlled odor cues are presented to bees that are held in tubes (Takeda 1961). When the antennae of a bee are touched with a sucrose solution (the US), the bee extends its proboscis to lick it (the UR). When an odor is presented just before the sucrose solution is presented, the bee rapidly forms an association between the odor (the CS) and the sucrose, and begins to extend its proboscis to the odor alone (the CR). This easy-to-use protocol has enabled researchers to test numerous hypotheses about learning, including how bees generalize from one stimulus to others (Menzel 2007).

OPERANT CONDITIONING

Another form of associative learning is operant (or instrumental) conditioning. The formal study of this topic was begun by Thorndike, who invented a "puzzle box" with a door that could be opened with a latch on the inside. We already described this box in Chapter 2, but will review it briefly here. Thorndike would place a hungry cat in the box with a tempting bowl of food outside the box. The cat would leap around in an attempt to get to the food. Eventually, the cat would accidentally hit the lever in the correct way, the door would open, and the cat would get to eat. Thorndike would then scoop up the cat and pop it back in the box again. Over successive trials, a typical cat would get faster and faster at performing the correct behavior to release the latch. This type of learning is called operant conditioning to emphasize that the animal operates on the environment to produce consequences. It is also called trial-and-error learning.

B. F. Skinner later invented the Skinner box, an apparatus that was even easier to use than Thorndike's puzzle box, and is still used today. A hungry animal is placed in the Skinner box, where it must learn to manipulate a mechanism (such as pressing a lever or pecking a key) in order to get a food reward (Figure 5.4). Data collection (number of bar or key presses) is generally automated and very rapid.

FIGURE 5.4 **A rat in a Skinner box. The hungry animal explores the box and eventually presses the bar. This automatically results in the delivery of a small food pellet that the rat quickly consumes. The food reward increases the probability that the rat will press the bar again.**

A stimulus, such as a bit of food, that changes the probability that an animal will repeat its behavior is called a **reinforcer**. In the experiments described so far, positive reinforcers were used. A positive reinforcer is one that increases the probability that a behavior will be repeated, such as food offered to a hungry rat or a drink to a thirsty one. The definition of a negative reinforcer may seem counterintuitive: it increases the probability of a response once it is removed. If an unpleasant or painful stimulus stops when an animal performs a certain act, it is likely to repeat that action. For example, a rat will learn to push a bar to turn off a bright electric light for 60 seconds (Keller 1941). Negative reinforcement is thus different from punishment, which is an aversive stimulus that results in a decrease in a response.

In operant conditioning, as in classical conditioning, the timing of events is critical. When the animal spontaneously performs a behavior, reinforcement must follow closely. In a sense, a cause-and-effect relationship develops between the performance of the act and the delivery of the reinforcer. When reinforcement is withheld, the response rate will gradually decline and become extinguished, just as the strength of the conditioned reflex decreases when the CS is presented many times without the US.

Shaping

Operant conditioning can be used to teach animals to perform novel and sometimes complex acts. Hollywood animal trainers rely on a method called **shaping**, which has parallels to the gradual way in which a sculptor molds a lump of clay (Skinner 1953). At first, the trainer reinforces any gross approximation of the desired act but then requires better and better performances to get a reward. For example, to train a dolphin to jump from the water through a hoop, first reward it for approaching the hoop. When it learns to approach, reward it only when it swims through the hoop. Then raise the hoop on successive trials until it is clear of the water, and offer your dolphin a fish only when it makes the leap.

Shaping works on people as well as other animals. When a writer, Amy Sutherland, was researching animal training techniques for a book, it struck her that many of the techniques might be useful in her marriage. "After all," she writes, "you don't get a sea lion to balance a ball on the end of its nose by nagging." She quietly began to use shaping techniques to train her husband not to throw laundry on the floor and to change some of his other annoying habits. Eventually, she couldn't resist explaining what she was doing—it worked very well!— and he began using the same techniques on her. Sutherland described her experiences in a humorous piece in the New York Times ("What Shamu Taught Me About a Happy Marriage"). Apparently the piece resonated with readers: it was the paper's most emailed article in 2006.

Reinforcement Schedules

In real life, reward seldom follows every performance of an act. Instead, the reward is usually intermittent. For example, a honeybee will find nectar rewards in a flower only if the flower hasn't been recently visited by another bee. The frequency with which rewards are offered is called the reinforcement schedule. Partial reinforcement schedules vary either the ratio of nonreinforced to reinforced response or the time period between successive reinforcements. Alternatively, rewards may be doled out in no particular pattern (Ferster and Skinner 1957).

Each reinforcement schedule has predictable effects on the rate of response and on how long the animal will continue responding when it is no longer rewarded. We will highlight just a few examples from an extensive body of work. A **continuous reinforcement schedule**, in which each occurrence of the behavior is rewarded, is best during the initial training to establish and shape a response. A **fixed ratio schedule**, one in which the animal must respond a set number of times before reinforcement is given, usually results in very high response rates because the individual has control over how quickly it will be rewarded. The faster it responds, the sooner it completes the number of responses required to receive the reward. A fixed ratio reinforcement schedule is similar to piecework in factories, in which the employee gets paid when a certain number of items are completed. Employers like the system because of the very high production rate it generates. In a **variable ratio schedule**, the number of responses required for reinforcement varies randomly. This also generates very high response rates because the individual is rewarded for fast responses. The variability means that there aren't detectable patterns of reinforcement, so the subject is unable to discern immediately when reinforcement has stopped. Thus, the response tends to persist even if the reward is withheld for a while. This is exactly the behavior that casino owners want to encourage in their customers, so slot machines are programmed with a variable ratio schedule.

LATENT LEARNING

Sometimes animals seem to learn without any obvious immediate reward. For instance, an animal can learn important characteristics of its environment during unrewarded explorations and then use this information later. Even though the knowledge is not put to immediate use (i.e., it is latent), it may later prove to be lifesaving.

The value of latent learning seems intuitively obvious. Several studies have shown that familiarity with the terrain improves survival (Metzgar 1967). Pairs of white-footed mice (*Peromyscus leucopus*) were released into a room with a screech owl (*Otus asio*). One of the pair previously had the opportunity to explore the room for a few days. The other mouse had no experience in the room. On 13 of the 17 trials, the owl caught one of the mice. Only two of the captured mice were from the group that was familiar with the room, suggesting that their knowledge of the environment helped them evade the predator.

Even ants seem to be able to gather information for later use. *Temnothorax albipennis* ants build nests in flat rock crevices. When their nests are damaged, they have to move to a new area. Like apartment hunters, ants evaluate prospective nest sites based on a range of criteria: floor area, headroom, entrance size, darkness, hygiene, and the proximity of hostile neighbors (reviewed in Franks et al. 2007). All this evaluation takes time, and if their nest is destroyed, the ants must find a new home very quickly. Researchers tested whether ants keep track of the local housing options even before they need to move. Ants turn out to be quite content with nest sites made of cardboard sandwiched between glass slides, so it is easy to design laboratory experiments, as shown in Figure 5.5. Researchers placed a new nest site, Alternative #1, near the ants' current nest for a week so that the ants could become familiar with it. Next, the researchers introduced a second nest of exactly the same quality, Alternative #2. They then immediately destroyed the ants' current nest and forced them to move. Any difference in whether ants favored Alternative #1 or #2 would suggest that they had learned something about Alternative #1 during the week of reconnaissance.

In the first experiment, the ants' initial nest was of high quality, but both Alternatives #1 and #2 were of low quality. Of 30 colonies tested, only two chose Alternative #1, 23 chose Alternative #2, and 5 were split in their

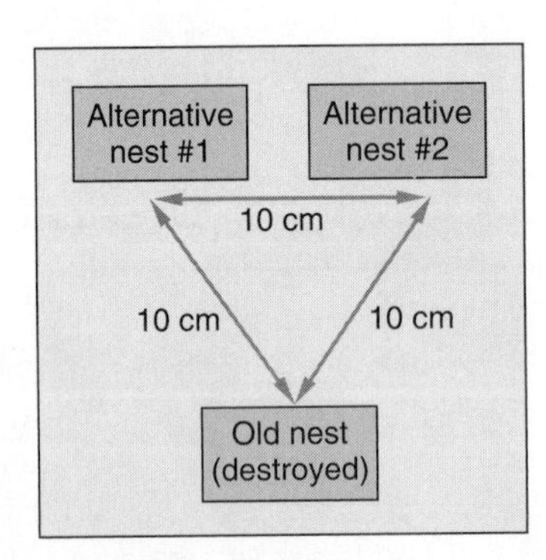

FIGURE 5.5 The experimental design for a study of latent learning in ants. Ants were living in the nest site pictured at the bottom of the figure. A new nest (Alternative nest #1) was added to their cage and left there for a week. Then, a second nest (Alternative nest #2) was added, and the original nest was immediately destroyed. When the two alternative nests were of identical low quality, ants preferred to settle in nest #2. When the two alternative nests were of identical high quality, ants had no preference. The locations of the two alternative nests were randomized. (Modified from Franks et al. 2007.)

choice, colonizing both nests. Thus, ants discarded a familiar, but low-quality, alternative in favor of an identical nest. This decision might seem illogical—why choose one poor nest over an identical one?—but the authors argue that it makes sense for ants to ignore a familiar but unattractive option in favor of exploring for a better one, even if they ultimately settle for something that is also unattractive. The experiment was repeated, but this time with both alternatives of the same high quality as the old nests. Here, the ants had no preference: 23 of 24 colonies were split between the choices, suggesting that they indeed are assessing site quality when they explore, and they can use this information later at an appropriate time.

SOCIAL LEARNING

Learning from others is a fundamental part of human learning (Bandura 1962; Meltzoff 1988), but it is not part of every animal's behavioral repertoire. Clearly, social species have much greater opportunity for social learning than do solitary species.

The term **social learning** encompasses a broad range of phenomena, some of which suggest a higher level of cognitive skill on the part of the animal than do others. In some cases, animals inadvertently provide information to other animals. In other cases, individuals actively share information through specific signals. Researchers who study social learning distinguish several categories.

In **stimulus enhancement**, an animal may be attracted to a particular *object* because a conspecific is near it or is interacting with it. Similarly, in **local enhancement**, an animal may be attracted to a particular *location* because a conspecific is there. Thus, in both types of enhancement, information is not being actively communicated by one animal to another (Galef 1988; Giraldeau 1997; Marler 1996). As an example of stimulus enhancement, rats can learn dietary preferences from other rats by smelling their breath. In one experiment, a "demonstrator" rat ate food flavored with cocoa or cinnamon. The demonstrator was then anesthetized and placed 2 inches away from the wire cage of an awake "observer" rat. Although the demonstrator slept through the demonstration, the observer later showed a preference for the food the demonstrator had eaten (Galef 1990a).

In nature, local and stimulus enhancement occur frequently in the context of foraging. For instance, when bumblebees first visit a new flower species, they are more likely to land on flowers that are already occupied by other bees (Worden and Papaj 2005). Once they learn about nectar availability and quantity, they decide for themselves rather than following conspecifics (reviewed in Leadbeater and Chittka 2007). Many other animals also use conspecifics as cues to good foraging patches.

Observational conditioning is a type of classical conditioning that occurs in social situations. For instance, some animals can learn to avoid dangerous situations by watching conspecifics. For example, rhesus monkeys learn to fear and avoid snakes by watching other monkeys show their fear (Mineka and Cook 1988). Interestingly, monkeys did *not* show a fear of flowers after watching other monkeys respond fearfully to them (Cook and Mineka 1990), suggesting that fear of snakes results from a combination of experience and a predisposition to learn this particular association. Similarly, fathead minnows do not innately show fear of one of their natural predators, the northern pike. However, minnows learn to show fear responses to pike odor when they are paired with minnows that have had experience with pike, but not when paired with inexperienced minnows. After learning to recognize the pike as a predator, the minnows have a better chance of surviving future encounters, and they are able to transmit the information to naive fathead minnows (Mathis et al. 1996). Note that in observational learning, we do not need to assume that observers understand anything about the mental state of the animals they are learning from.

In **goal-directed emulation**, an observer seems to learn from observation what goal is to be achieved but does not copy precisely what the demonstrator does. For example, chimpanzees and children both watched an adult human demonstrator retrieve artificial fruit from a clear plastic box by untwisting bolts. Chimps showed goal-directed emulation: they directed their attention at the correct part of the box but did not imitate the action of the demonstrator. Children, in contrast, imitated the actions of the observer exactly, even extraneous motions (Whiten et al. 1999).

That brings us to the evidence for **imitation**, where an observer copies exactly what a demonstrator does. Imitation is quite rare and difficult to document, especially without knowing the detailed history of what the animal has previously learned or seen. One method for studying imitation is the two-action test. The subject is presented with a task that has two equally easy solutions. If subjects are more likely to choose the solution that they have just seen demonstrated, it is taken as evidence of imitation. For example, budgerigars (pet-store parakeets, *Melopsittacus undulatus*) were trained as demonstrators. Each learned one of a series of techniques for removing the cover from a food dish: using their feet, pecking with their bills, or pulling with their bills. Observer budgies watched a demonstrator open a dish. When presented with a similar dish, the observer used the same technique it had just witnessed (Dawson and Foss 1965). Subsequent efforts by others to replicate this test either failed to do so or produced only transient effects. One reason for these conflicting results might have been variation in the performance of the demonstrators. Mottley and Heyes (2003) controlled for this variation by letting budgies watch videos of demonstrators rather than live ones. Budgies were more likely to remove a stopper from a food box using the method they had seen demonstrated in the videos than the alternative method.

The Adaptive Value of Social Learning

The potential adaptive value of social learning is clear. It saves some of the time and energy that might be wasted as an individual learned the business of survival by trial and error. Although each member of a population may have the capacity to learn appropriate responses for themselves, it is often more efficient and less dangerous to learn about the world from others (Galef 1976).

Let's examine some potential benefits of social learning about food. Rats eat a wide range of food types. However, not all potential foods are safe or nutritious. Rats not only learn about food by smelling each other's breath, as described above, but by observation. When a rat observes another eating a novel food, the observer is more likely to try it than if it observed another rat eating a familiar food. This is a safe way to add breadth to the diet (Galef 1993). As a result, groups of rats will learn to select a nutritionally balanced diet more quickly than do rats that are housed alone (Galef and Wright 1995). Other species may learn routes to food from conspecifics. Guppies (*Poecilia reticulata*), for example, quickly learn a safe route by shoaling, or swimming in large groups (Laland and Williams 1997).

Animals may also learn from other species. We see this in different populations of Zenaida doves in Barbados, which live only a few hundred meters apart. Group-foraging doves learn more quickly from other doves, but territorial doves learn more readily from Carib grackles, the species they most often feed with in mixed flocks (Carlier and Lefebvre 1997).

Traditions

Many socially learned behaviors are transient and disappear quickly. Others, called **traditions**, spread through a group and are stable over time. For instance, a larcenous tradition began in England around 1921 when a bird species called the blue tit (*Parus caeruleus*) learned to break into milk bottles to steal the cream, which, in the days before homogenization, floated to the top. This technique spread throughout Great Britain as other birds acquired the habit (Fisher and Hinde 1949) (Figure 5.6).

Primate groups show a great deal of behavioral variation, suggesting the importance of traditional behaviors. For example, chimpanzee groups vary markedly in their behaviors: 39 behavior patterns, including tool use, grooming, and courtship, occur frequently in some communities but are absent in others (Whiten et al. 1999). An interesting tradition is the food-washing habit that spread within a group of snow monkeys. As the story goes, a young female snow monkey of Japan, named Imo, developed new techniques for the treatment of sweet potatoes and wheat, food provided by the researchers who study the social behavior of the snow monkeys. First, Imo discovered that washing the sweet potato in the sea not only cleaned it but also enhanced the flavor by lightly salting it (Figure 5.7). One of Imo's playmates observed her and followed suit. Then Imo's mother caught on. And so the tradition spread, usually from youngsters to mothers and siblings. When the youngsters became mothers, their offspring imitated the behavior as if food had always been cleaned in this way.

FIGURE 5.6 **The tradition among birds of opening milk bottles to steal sips of cream spread rapidly from one area in England. This trick may have been spread by social learning.**

FIGURE 5.7 **The tradition of washing sweet potatoes in the sea was begun by a young Japanese snow monkey, and it spread rapidly to other members of the troop.**

Several years later Imo started a new custom. The researchers spread wheat on the sand, from which the snow monkeys had to painstakingly pick each grain. One day Imo tossed a handful of sand and wheat into the sea. The sand sank but the wheat floated so that it could be scooped up from the surface. This ploy was also picked up by most monkeys in the troop during the next few years (Kawai 1965; Kawamura 1959; Lefebvre 1995).

However, we must be careful. The division between individual learning (learning through one's own experience) and social learning is not always clear-cut: both may occur simultaneously and can be difficult to distinguish. For example, sweet potato washing may occur through stimulus enhancement. A monkey may pick up a dropped potato that has been washed, like the taste, and then be primed to learn to wash potatoes on its own (de Waal 2001). In addition, differential reinforcement may maintain the behavior. The monkeys' only source of sweet potatoes is the caretaker. Since the food washing interests researchers and amuses tourists, the caretakers give more sweet potatoes to those members of the troop that were known to wash them than to those that did not (Galef 1990b). Although the habit clearly spread throughout the population, we cannot be sure of the mechanism.

SPECIES DIFFERENCES IN LEARNING: COMPARATIVE STUDIES

For decades, the dominant view in the study of learning was that it is a general process that occurs in essentially the same way across mammal species. This view is certainly not groundless: many essential characteristics of learning, such as the most effective order of presentation of conditioned versus unconditioned stimuli, are indeed similar, whether studied in rats or humans. However, in recent years, researchers have been intrigued not just by similarities across species in how and what they learn, but also by their differences. We've already mentioned several studies that document differences across species that seem to correlate with the ecological conditions they face. In this section, we will more explicitly consider the evidence that differences across species are rooted in natural selection.

THE ABILITY TO LEARN AS A HERITABLE TRAIT

As we have seen, in order for natural selection to act, the trait in question must be at least partly heritable. What, exactly, about a learned behavior is inherited? A jumping spider (*Phidippus princeps*) can learn that red and black milkweed bugs (*Oncopeltus fasciatus*) are not good to eat: the first time a spider sees one, it leaps on it, but by the eighth trial, the spider ignores a bug that crawls right past it (Skow and Jakob 2006). However, the knowledge that milkweed bugs taste nasty is not passed onto to the spider's offspring. Its offspring must learn this for themselves. Learned knowledge is not genetically heritable, although, in some species, offspring can learn from watching their parents. What *is* heritable, and thus subject to natural selection, is the *capacity* to learn.

The heritability of the ability to learn has been experimentally demonstrated in several species, including that standby of behavioral genetics, *Drosophila melanogaster*. Mery and Kawecki (2002) carried out an artificial selection experiment like those experiments described in Chapter 3. They gave the fruit flies a choice of two places to oviposit: media flavored with pineapple versus that flavored with orange juice. One of these media also contained a quinine solution. Quinine tastes bitter to humans, and it also deters flies. After experience with this arrangement, flies were then offered a choice between orange and pineapple media that had no quinine. Flies that had learned the association between quinine and a particular flavor avoided that flavor and laid their eggs on the neutral flavor. These eggs were collected and reared up to adulthood on an unflavored cornmeal mix. Thus, only fruit flies that learned to avoid the flavor associated with quinine contributed their alleles to the next generation. Each generation, flies were tested on the same learning task. After 15 generations, flies from these selected lines were able to learn the task faster and remember it longer than were flies from control lines. These abilities were not confined to the original task: they could also learn about novel flavors (apple and tomato). Thus, the experimenters were able to select for an increased ability to learn to identify odors of fruits suitable for egg laying.

EVOLUTION AND THE VARIATION IN LEARNING ACROSS SPECIES

Learning allows an animal to adjust its behavior to new situations, even those to which its ancestors were never exposed. Our anthropocentric view is that the ability to learn is undeniably a positive trait, and it may seem counterintuitive that the ability to learn may not always be advantageous (Shettleworth 1998). Learning has its costs. First, it takes time to learn: a spider that is born with an innate aversion to eating red and black bugs is saved the time and trouble of repeated mistakes. Second, the ability to learn requires the dedication of neurons to the task. Because neurons cannot be infinitely reduced in size, there is just so much space available for different functions in a brain of a given size. If, like a spider, your brain were smaller than the size of a pinhead, perhaps it would be better to devote your neuronal space to something else, such as large olfactory centers that might allow you to detect and interpret the chemical scents left by prey. Finally, learning seems to have an "operating cost"—it takes energy to collect, process, and store information. *Drosophila* lines forced to use their ability to learn had fewer offspring than ones that were not (Mery and Kawecki 2004), implying that natural selection should act against flies that learn "too well."

Because the ability to learn can be heritable and has costs and benefits, we predict that we should see differences in learning ability across species (Kamil and Mauldin 1988; Kamil and Yoerg 1982). The environment and evolutionary history of a species should influence the degree to which a particular type of learning will increase their fitness. We have already seen that animals cannot learn all tasks with the same ease: there appear to be biological constraints on learning, and members of a particular species may be prepared to learn certain things and not others (Chapter 2). Now we will focus on the evidence for differences among species in their ability to learn different tasks.

The most complete example of species-specific differences in learning ability comes to us from three related species of birds: Clark's nutcrackers (*Nucifraga columbiana*), pinyon jays (*Gymnorhinus cyanocephalus*), and scrub jays (*Aphelocoma coerulescens*) (Figure 5.8). These birds are among the species that cache (store) seeds: they collect pine seeds in autumn and dig small holes in which to hide them so that the seeds will not be stolen by other animals. The birds recover and eat the seeds throughout the winter and spring when food is scarce.

a

b

c

FIGURE 5.8 **Seed-caching birds. (*a*) Clark's nutcrackers, (*b*) pinyon jays, and (*c*) scrub jays are birds that hide seeds in holes in the ground during the autumn and return to find the seeds during the winter and spring, when food is less plentiful.**

These three species differ in their ecology and in the extent to which they rely on seeds (see review in Gibson and Kamil 2005). Clark's nutcrackers are the champion seed-storers (Balda 1980; Balda and Kamil 1998). This species lives at high elevations in coniferous forests in western North America, where winters are harsh and long. During a three-week period in the fall, they may cache as many as 33,000 pine seeds in several thousand separate locations as far as 22 km from the harvesting site. Throughout the winter, nutcrackers survive almost entirely on these stored seeds. The second species, the pinyon jay, lives in pine woodlands at lower elevations. This species relies less on caching than do nutcrackers but still caches an impressive 20,000 seeds as far as 11 km away from the collecting site (Balda 1980; Balda and Kamil 1998). About 70 to 90% of the pinyon jays' winter diet consists of cached seeds. Scrub jays, the final species in our trio, store "only" about 6000 seeds a year, and these comprise less than 60% of the winter diet (Balda 1980; Vander Wall and Balda 1977).

These birds routinely find individual beakfuls of seeds, months after they've hidden them, in a landscape that may be transformed by snow. How do they do this? It's not that they simply smell the seeds or sense them in some other way: they actually remember the exact locations (Balda 1980; Vander Wall 1982). This is one of the most impressive examples of spatial memory on record, far surpassing human abilities.

The ecological differences among these three species led investigators to a prediction: species that rely more heavily on caching to survive the winter have a better spatial memory. To test this prediction, Balda and Kamil (1989) devised an elegant experiment that mimics the process of caching but under controlled conditions. Birds were permitted to store seeds in sand-filled holes in the floor of an indoor aviary. The floor had 90 holes, any of which could be open and filled with sand suitable for burying seeds, or blocked with a wooden plug. This arrangement allowed the experimenters to vary the position and number of the holes available for caching. Each bird's ability to recover caches was tested in two conditions. In one, only 15 holes were open; in the other, all 90 holes were available. After a bird had placed eight caches, it was removed from the room. One week later, when it was returned to the aviary, all 90 holes were open, and the bird's task was to probe in the subset of the holes where it had cached seeds. The accuracy of recovery was measured as the proportion of holes probed that had contained their seeds. All three species performed better than expected by chance alone. However, nutcrackers and pinyon jays, the species that depend most heavily on finding their stored seeds to survive the winter, did significantly better than the scrub jays in both experimental conditions (Figure 5.9). Species differences were small when only 15 holes were available for caching but much larger when all 90 holes were available.

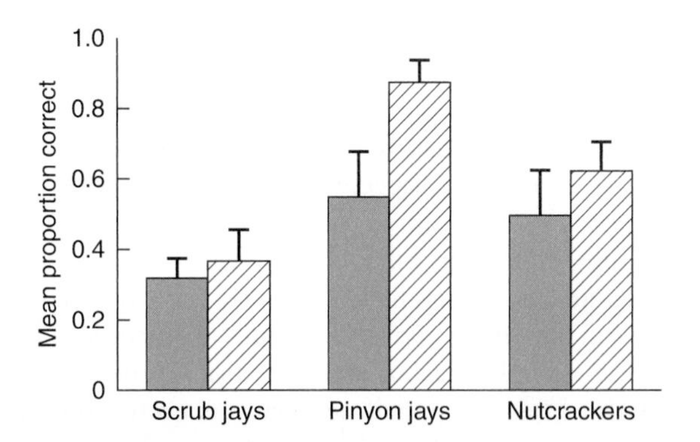

FIGURE 5.9 **Histograms that show the accuracy with which scrub jays, pinyon jays, and Clark's nutcrackers find their caches. Each bird first hid seeds in sand-filled holes in an indoor aviary. The aviary floor had 90 holes, each of which could be filled with sand or plugged. In one experimental condition, indicated with the solid bars, 15 holes were available for caching. In the other, indicated by striped bars, all 90 holes were available. After caching, the birds were removed from the room for a week. When they were returned, all 90 holes had been filled with sand. To recover the seeds, the birds would probe the sand with their beaks. Accuracy was measured as the proportion of holes probed that contained seeds. Clark's nutcrackers and pinyon jays, the species most dependent on cached seeds for winter survival, were significantly better than scrub jays at recovering their caches. (From Balda and Kamil 1989.)**

These data make sense, then, in light of the ecological differences between the species. But is this test enough? Perhaps there is something about these particular experimental conditions that make the test difficult for scrub jays. Perhaps, for example, they are not as motivated as the other species to cache and recover seeds in the aviary. This problem gets at the heart of one of the difficulties of interpreting comparative data from different species on learning: are differences the result of the way the species respond to the test conditions, or do they reflect true cognitive differences?

One way to resolve this difficulty is to test species in multiple experimental situations that present the same type of task but in different ways (Kamil and Mauldin 1988). In this case, we need other tests of spatial memory that do not require birds to cache seeds. In a second test, birds were given the chance to collect seeds from holes in an aviary, much like the first test (Kamil et al. 1994). However, this task differed because they were not allowed to cache seeds themselves but were required to learn the locations of seeds cached by the experimenter. Birds were trained in a room with four open holes, each with seeds. After training, they were given access to the room with those same four holes open but with no seeds in them. In addition, an extra four holes were also open, and these did have seeds. Thus, the task was to learn to remember, and then bypass, the holes where they had cached seeds earlier. Again, Clark's nutcrackers and pinyon jays were better at this task than were scrub jays. In contrast to the previous test, when the task got harder such that the birds had to remember the locations for a longer time, the species differences decreased. Differences between the tests reiterate the value of multiple kinds of tests in unraveling species differences in learning ability: even small differences in experimental design can have meaningful effects on the outcome. This, of course, complicates our interpretation of the results.

A third test of spatial memory was quite different from the first two. Olson et al. (1995) used an experimental design with the descriptive name of a "delayed operant nonmatching-to-sample procedure" to test nutcrackers, pinyon jays, scrub jays, along with a fourth species, Mexican jays. Mexican jays also cache food, and they live at a higher elevation than scrub jays. Birds were first trained to peck at a monitor at an illuminated circle in a particular location. During the testing session, two circles were illuminated: one in the location the bird had been trained with and one in a new location. Birds were rewarded for pecking at the key in the new location. By increasing the delay between the training sessions and the testing sessions, the researchers could determine how long the birds could remember the location of the key that was rewarded during training. Nutcrackers performed better than other species (Figure 5.10*a*).

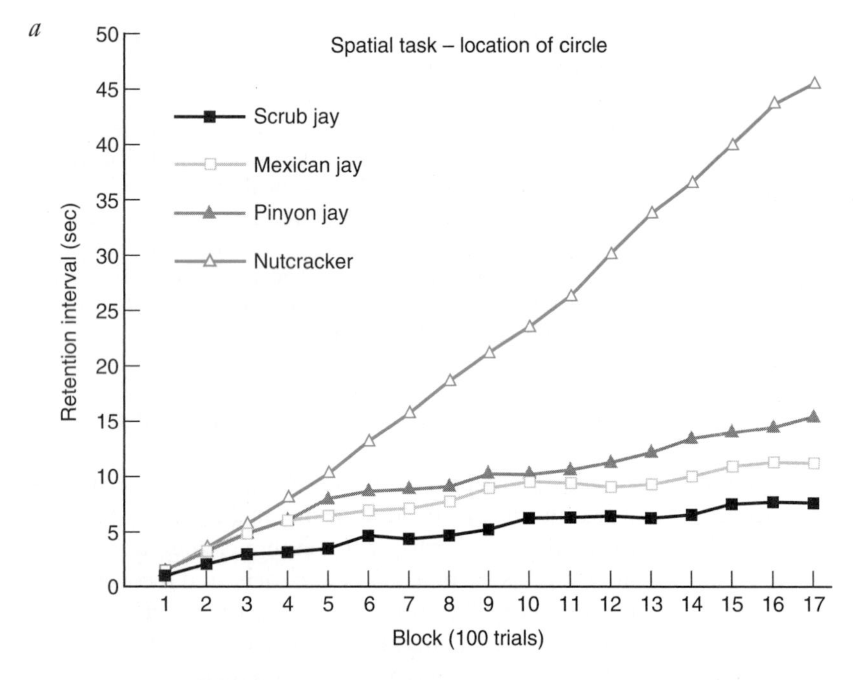

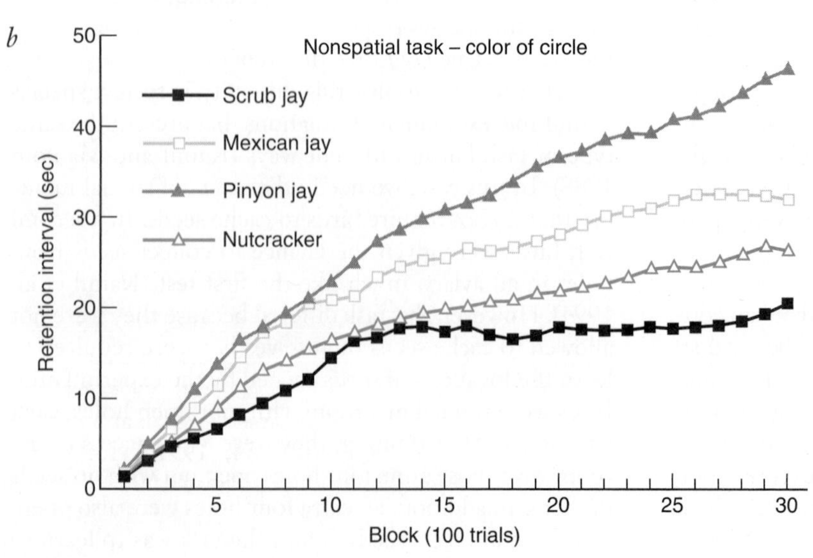

FIGURE 5.10 Comparison of learning abilities among food-storing and nonstoring corvids on spatial and nonspatial tasks. Shown here is the length of time each type of bird could remember a particular task. (*a*) Clark's nutcracker, the species most highly dependent on stored food, was an obvious champion on a spatial task, remembering the location of a circle displayed on a screen. (*b*) However, species differences in memory on a nonspatial task, remembering the color of a circle, were not related to the species' dependence on stored food. Thus, the outstanding performance of Clark's nutcracker on spatial memory tasks is not simply because these birds adapt to life in captivity better than other species or because nutcrackers are generally more "intelligent." These observations add strength to the hypothesis that learning abilities are shaped by natural selection. (Data from Olson et al. 1995.)

These experiments show that differences in spatial memory among Clark's nutcrackers, pinyon jays, and scrub jays are not limited to the recovery of cached seeds. These results increase our confidence in our conclusion that spatial memory is an adaptive specialization. However, we still can't dismiss the alternative hypotheses that the species differences arise because the food-caching species happen to be better suited to life in the laboratory environment, or possibly are more motivated to complete the tasks. The likelihood of these alternatives can be examined by testing species on a nonspatial task. If the differences between species persist even in a nonspatial task, then we should suspect that they are driven by something besides an adaptive specialization for spatial learning. Olson et al. (1995) gave birds a nonspatial task, a nonmatching-to-sample procedure based on color, rather than location. In contrast to the spatial test, there were no differences among species (Figure 5.10b). This again supports the idea that spatial learning ability, rather than learning in general, is predicted by the ecology of the species.

Yet another hypothesis could explain species differences that we see in various spatial tests. Perhaps these species differ not in their spatial learning ability, but in their ability to perceive appropriate details of the environment (Macphail and Bolhuis 2001). Gibson and Kamil (2005) tested the perceptual ability of these species by requiring them to discriminate the distance between two landmarks on a computer screen. All three species were able to make fine discriminations, but nutcrackers were no better—indeed, they were slightly worse—than the other species. Thus, this experiment provides no evidence that differences among these species in the learning tasks were due to how they perceive spatial information. However, proving that perceptual differences are not driving differences among species—in essence, proving a null hypothesis—is incredibly difficult, if not impossible.

This correlation between spatial memory and food caching is consistent with our hypothesis that evolution may shape learning ability, but it is also possible that the observed differences in spatial memory are simply chance differences among this admittedly small group of species that have been studied. The case would be stronger if the same patterns were found among other groups of related species. Food-storing behavior has also evolved among some species of the Paridae (titmice and chickadees), a family of birds that is phylogenetically distinct from the Corvidae (nutcrackers and jays). Parids store seeds and insects in hundreds of widely scattered sites for shorter time periods than do the corvids—hours to weeks as opposed to months. In several tests, food-storing species perform better than non–food-storing species (Giraldeau 1997; Krebs et al. 1996; Shettleworth 1995), although the evidence is not as consistent as that in corvids (reviewed in Shettleworth 1995 and Pravosudov 2007).

Far fewer comparative studies of spatial learning have been done in caching mammals, but a similar relationship between ecology and spatial skills has been documented in several species. For instance, the Great Basin kangaroo rat (*Dipodomys microps*) eats leaves of the saltbush. This is a common and abundant food, and presents no navigational challenges. In contrast, the Merriam's kangaroo rat (*D. merriami*) prefers seeds and stores them in scattered locations, much as nutcrackers do. As predicted from their ecological differences, the Merriam's kangaroo rat performed better on a spatial test than the Great Basin rat (Barkley and Jacobs 2007).

Earlier in this section we suggested that learning ability depends on the number of neurons devoted to the task. In birds and mammals, spatial learning is known to be at least partly based in the region of the brain called the hippocampus. Does the size of the hippocampus vary with the amount of food hoarding? Results from earlier studies were mixed, with some finding a relationship (e.g., Basil et al. 1996; Krebs et al. 1989; Sherry et al. 1989) and others not (e.g., Brodin and Lundbord 2003). A recent large comparative study supports a relationship between hippocampal size and spatial learning. This study included 55 bird species and controlled carefully for phylogenetic relationships. Closely related species may have more similar cognitive abilities than more distantly related species, which can bias the results of a comparative study unless phylogeny is accounted for. Both the relative volume of the hippocampus and the total volume of the brain were positively correlated with the amount of food hoarding (Garamszegi and Eens 2004). However, Provosudov and de Kort (2006) found that scrub jay brains are much larger than previously reported, and argued that methodological issues cloud the interpretation of the existing data on hippocampal size. Clearly, the issue is not yet settled, and research continues.

OTHER EVIDENCE OF COGNITIVE ABILITIES IN ANIMALS

The study of the mental processes of animals goes beyond the study of learning. Some scientists have wondered whether animals have mental experiences—thoughts and feelings, for instance (Bateson and Klopfer 1991; Griffin 1981, 1982, 1984, 1991; Hoage and Goldman 1986; Mellgren 1983; Ristau 1991). But how could we ever *know* whether other animals think or whether they are self-aware? What constitutes good evidence for complex cognitive abilities? In this section, we'll discuss experiments that address the idea of animal understanding.

TOOL USE

A sea otter, floating on its back, uses a rock to break a clam shell on its belly. A vulture drops a rock on an ostrich egg, which cracks open. A chimpanzee strips the leaves off a stick and uses it to pull juicy termites out of a mound. All of these are examples of tool use—the use of an object in order to obtain a goal. Once considered to be a hallmark trait that separated humans from other animals, tool use is now known in many species.

Tool use *seems* to demonstrate a high level of cognition. In order to use a tool, must not an animal understand how it works? Animals using a tool to solve a problem can often appear to be thinking it through. Let's begin by considering the famous example of Köhler's (1927) chimpanzees, particularly one named Sultan. Sultan first learned to use a stick as a tool to extend his reach and rake in a banana on the ground outside his cage. Having mastered this, he was given two sticks that when put together end to end were just long enough to reach the fruit. Sultan tried unsuccessfully to reach the reward with each of the sticks. He even managed to prod one stick with the tip of the other until it touched the banana, but since the sticks were not joined, he could not retrieve the fruit. For over an hour, Sultan persistently tried, and failed, to get the banana. Finally, he seemed to give up and began to play with the sticks (Figure 5.11). As he was playing, he happened to hold one in each hand so the ends were pointed toward one another. At this point he fitted one end of the stick into the other, thus lengthening the tool. Immediately, he ran to the bars of his cage and began to rake in the banana. As he was drawing the banana toward him, the two sticks separated. That Sultan quickly recovered the sticks and rejoined them was evidence to Köhler that the chimp understood that fitting two bamboo poles together was an effective way to increase his reach far enough to obtain the fruit. Because of the suddenness of Sultan's solution, Köhler called his behavior **insight**. He documented other similar situations where a flash of understanding seemed to occur, such as when chimps stacked boxes and climbed on top in order to knock down a hanging banana with a stick.

FIGURE 5.11 **Sultan playing with sticks. After getting experience with the sticks, Sultan fit them together end to end to reach bananas.**

Perhaps the chimps were able to see new relationships among events and were able to consider the problem as a whole. Perhaps they even formed a mental representation of the problem and then mentally worked through solutions to it. However, we must be careful here. All the details of the chimps' prior experience were not known. Perhaps chimps that moved boxes and then climbed on them to reach a banana had previously learned two separate behaviors—moving boxes toward targets and climbing on an object to reach another object. Pigeons can chain together similar learned tasks. Pigeons that were trained both to push a box in a particular direction and to climb on the box to get a reward were able to put both behaviors together. Epstein et al. (1984) concluded that seemingly insightful behavior might be built from specific stimulus–response relationships learned through operant conditioning.

These studies emphasize the need to control for an animal's prior experience in order to understand exactly what it knows when it manipulates objects. Let's consider another example of apparently insightful behavior in a bird, this time in the common raven (*Corvus corax*). Ravens are known to pull up ice fishing lines to steal fish, as shown in the illustration at the start of this chapter. Bernd Heinrich (1995) presented hand-reared ravens with meat suspended from string, a problem he knew they had never encountered before. To reach the suspended food, a bird had to pull up a loop of string, step on the loop to hold it in place, and then reach down and pull up another loop. The bird had to repeat this cycle six to eight times to obtain the food. At least ten species of birds can be taught by operant conditioning to pull up food dangling on a string if the distance between the food and the perch is gradually lengthened. However, a few of the ravens in Heinrich's study solved the problem immediately without any indication of going through a learning process. In fact, one bird went through the entire sequence of 30 steps and obtained the food the first time it approached the string, even though no other bird in the group had previously shown the behavior. Ravens can correctly solve more complicated versions of this task. When given two strings, one with a rock and one with a piece of meat, that are crossed over, ravens can pull on the correct string on their first trial.

In nature, ravens don't normally pull on one object to obtain another one. When they pull on food, such as the entrails of a dead animal, they eat the food while pulling. Thus, Heinrich argues that it is unlikely that this complicated behavior was learned, was genetically programmed, or occurred by chance. The ravens apparently have the ability to find insightful solutions to new problems, using string as a tool.

Insight alone may not always be enough. A recent extension of the research on ravens demonstrated that familiarity with a simpler task may be required to succeed on a new task. Here, the string was looped up and through the cage, then down again. The birds had to pull the string *down* in order to raise the meat. Ravens that were familiar with the pull-up task could quickly do the pull-down task, but naive birds could not (Heinrich and Bugnyar 2005).

On the South Pacific island of New Caledonia, native crows are especially adept at making and using tools. In the wild, crows craft tools of several different varieties out of twigs and leaves (Hunt 1996; Hunt and Gray 2004; Hunt et al. 2006). For example, to make a hook, they snap off a twig, strip off the leaves, and then use their bill to sculpt the end of the twig. The crows then poke the hooked twigs into holes in order to extract insects and other small prey (Figure 5.12). They can even make hooks out of unfamiliar material. In the lab, crows were given a puzzle consisting of a clear vertical tube with a small bucket in it. The bucket contained food and had a handle that could be reached from above, but only with the aid of a tool. One crow was given a straight piece of wire, a substance she had no experience with. She quickly bent it into a hook and retrieved the bucket (Weir et al. 2002).

New Caledonian crows also use "metatools": they can use one tool on another. In this task, the crows needed to use a short stick in order to retrieve a longer stick from a "toolbox." They could then use the longer stick to retrieve a piece of meat. Six of seven crows correctly tried to extract the long tool with the short tool, and four crows successfully solved the problem on their first attempt. In a follow-up experiment, the positions of the tools were reversed—the small stick was in the "toolbox," and the long tool was given directly to the crows. In this setup, the crows did not need to use the "metatool" approach—they simply needed to pick up the long tool and extract the food. All six crows that were tested briefly attempted the unnecessary step of using the long tool to access the short tool in the "toolbox" but quickly rectified their mistake and began going straight to the food with the long tool (Taylor et al. 2007). The authors argue that the crows may well have understood the more general causal relationship that tools can be used to access out-of-reach objects, even other tools.

As an interesting note, scientists have recently been able to attach tiny video cameras to wild New Caledonian crows, enabling us to see them use tools in nature when no experimenters are nearby. These videos have revealed that crows appear to keep particularly good tools for future use (Rutz et al. 2007).

FIGURE 5.12 **A New Caledonian crow using a stick as a tool.**

These experiments on ravens and crows seem to reinforce the idea that animals might fully understand tool-using tasks, but let's look at a case where this is clearly not so (see Shettleworth 1998 for a more detailed review of this literature). Capuchin monkeys were shown a Plexiglas tube mounted horizontally on a stand with a peanut in the middle. The tube was too narrow for the monkeys to reach in with their arms. When they were provided with sticks, monkeys could quickly learn to poke a stick into the tube to push the peanut out (Visalberghi and Trinca 1989). But then an interesting twist was introduced into the design: a trap was placed into the tube (Figure 5.13). If a monkey inserted the stick in the wrong end of the tube, it would push the reward into the trap. Of four monkeys tested, only a single monkey consistently inserted the stick in the correct end of the tube, and only after 90 trials (Visalberghi and Limongelli 1994). Further tests demonstrated that even this individual didn't fully grasp the task. The researchers suspected that she might be following the rule, learned by trial and error, of "Push the stick in the side of the tube furthest from the treat." Indeed, when the tube was rotated so that the trap was on top, the monkey followed this rule, even though it was no longer necessary. Chimpanzees, in contrast, showed more understanding of the task (Limongelli et al. 1995). Human children under three years of age behaved more like capuchins than chimpanzees (Visalberghi and Limongelli 1996).

Tools are, of course, of great importance in human evolution, and perhaps we can gain some insight into our own past by examining the behavior of our close relatives. As we have already seen, chimpanzees are accomplished tool users. In the wild, they use sticks to forage for termites and rocks as a hammer and anvil to pound open nuts. Recently, chimpanzees have been observed fashioning spears out of sticks, trimming branches off and using their incisors to sharpen the end. The chimps then jabbed the spears into hollow trees. One was seen extracting a bushbaby (a small primate) from a hole in a tree after jabbing with the spear (Pruetz and Bertolani 2007). It is likely that foraging and hunting were the first contexts in which our ancestors used tools.

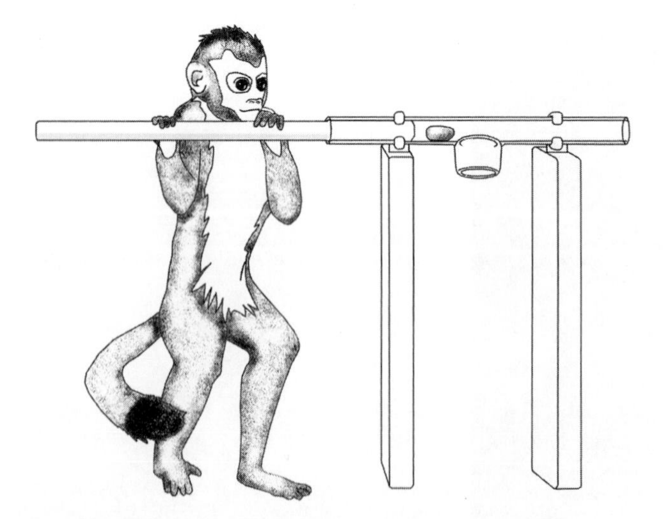

FIGURE 5.13 **The tube task but with a trap added. The subject must push a treat out of the tube with a stick. Because of the trap in the center of the tube, the stick must be inserted in a particular end of the tube. Here, a capuchin monkey is about to make an error.**

As we see, the investigation of tool use can be tricky. The experimenter can easily miss, or misinterpret, a moment of sudden insight. In addition, small changes in the task can greatly alter the outcome. Carefully controlled experiments are the key, with attention to potential alternative interpretations.

DETOURS

Detouring is the ability to identify an alternative route to a reward when the direct route is blocked. Although animals can often improve on detour tests with experience, here we are most interested in how they respond the first time they are confronted with a test. Nearly everyone who has taken an unruly dog for a walk has had a demonstration that not all animals understand how to solve detour problems. A dog that has wrapped its leash around the legs of its owner while trying to get to a squirrel does not comprehend that sometimes the best route from point A to point B is not necessarily the most direct. In controlled experiments, dogs were clearly motivated to reach a toy or food on the other side of a wire-mesh fence, but they often tried digging under the fence rather than detouring around it (Pongrácz et al. 2001). In contrast, if you have watched tree squirrels, you know that other species are very good at detour problems. Squirrels seem to immediately "see" that to get from tree to tree, they must choose the branch that reaches between the trees. Another comparative study of detour ability shows differences even among more closely related species. Quail and herring gulls were easily able to solve a detour task in which they were required to walk around a barrier, but canaries could not (Zucca et al. 2005). The difference might be because canaries don't walk much in the wild—when they face a similar sort of detour task in daily life, they can solve it by flying.

How exactly do animals solve detour tasks? One species that excels at detours might come as a surprise: jumping spiders in the genus *Portia*. Like other jumping spiders, *Portia* has large anterior eyes specialized for acute vision (Figure 5.14). However, whereas most jumping spiders attack insect prey by stalking and tackling it, *Portia* prefers to hunt other spiders. To do it, *Portia*

FIGURE 5.14 **A jumping spider of the genus *Portia*, showing its large specialized anterior eyes.**

climbs right into the webs of other spiders and lures them in with a series of vibratory signals similar to those given by struggling prey. When the web owners get close, *Portia* grabs them (Jackson and Wilcox 1993a). *Portia* can spot spider webs from some distance away, but often, in order to reach them in their complicated three-dimensional environments, it must perform a detour (Jackson and Wilcox 1993b). In the lab, it can, before setting out, choose correctly between detours that lead to prey versus those that do not (Tarsitano and Jackson 1997), and detour routes that are complete versus those that have gaps (Tarsitano and Andrew 1999) (Figure 5.15). *Portia* solves a detour problem by looking at the lure, then slowly scanning along the horizontal features of the potential route. If the route ends, the spider turns back to look at the lure again and then begins once more. This example demonstrates how fairly simple behavioral rules can give rise to behaviors that appear to be quite complex.

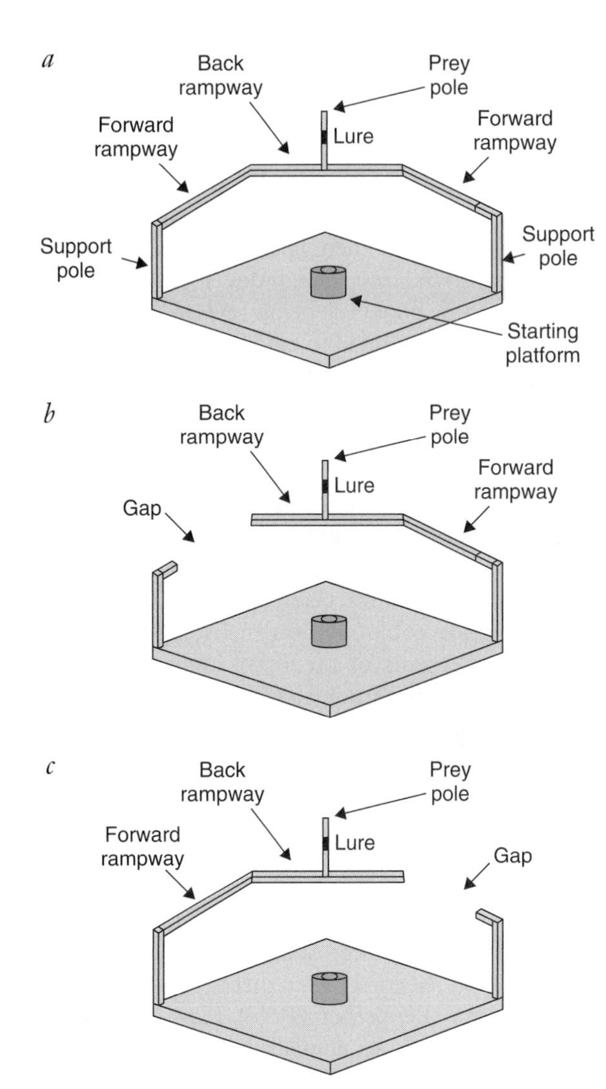

FIGURE 5.15 **A detour test for a jumping spider, *Portia*. A lure is placed on the prey pole. *Portia* must leave the starting platform and climb the supports to reach the prey. In (*a*), both support poles lead to the prey, but in (*b*) and (*c*) only one side of the route is complete, and the other side has a gap. Spiders can correctly choose the complete path at the start of their route. (From Tarsitano and Andrew 1999.)**

UNDERSTANDING NUMBERS AND OTHER ABSTRACT CONCEPTS

It seems quite valuable for animals to have a sense of quantity: Is there more fruit on this tree or that tree? Are there more ducks in the pond to my left or to my right? The ability to discriminate these sorts of relative numbers is quite common and does not seem to require difficult mental gymnastics.

What is more difficult is the ability to **count**—to assign a tag such as "1, 2, 3," to individual quantities. An animal that can count can apply tags to different kinds of objects. The ability to count things, no matter what they are, demonstrates some understanding of the abstract concept of numbers.

Research into animal counting had an inauspicious beginning (described in Shettleworth 1998). Clever Hans was a horse in the early 1900s who would answer arithmetic questions extremely accurately by tapping with his hoof. He caused quite a sensation, until a young psychologist tested him when he could not see any people who knew the answer. It turned out that Clever Hans really was clever, but at reading very subtle signals of humans rather than at math. As Hans tapped his way toward the correct answer, the body posture of the humans in the room would subtly change, and Hans would stop.

Clever Hans unwittingly taught researchers quite a bit about the importance of designing careful experiments to test the ability to count. We now have good evidence that several nonhuman animals can count. One that demonstrated this talent vocally was an African gray parrot (*Psittacus erithacus*) named Alex (Figure 5.16), who, sadly, died as this book was being revised. We all know that parrots can be trained to talk, but most of us would guess that they are mimicking their trainers. This was certainly not true of Alex. Irene Pepperberg (2000) detailed more than 20 years of research on Alex in *The Alex Studies*. He learned labels (names) for over 35 different objects. By combining labels, he could identify, request, refuse, or comment on more than 100 different objects. Furthermore, he used language to show that he understood certain abstract concepts. One such concept was quantity. He could say how many items were in a group for collections of up to six items, even if the objects were scattered around a tray (Pepperberg 1987a). Even more remarkable was Alex's ability to accurately count specific items in what is called a confounded number set, which are items that vary in more than one characteristic. For example, a set might consist of two types of objects, say balls and keys, that appear in two colors, red or blue. When presented with all these objects mixed together on a tray, Alex could say the number of items of a specific type and color, such as the number of blue keys. He responded correctly to these types of questions 83% of the time (Pepperberg 1994). He could even add up the total from two sequentially presented collections (Pepperberg 2006). Alex also may have had a limited understanding of the concept of zero. He spontaneously began to say "none" in response to the absence of objects on the tray (Pepperberg and Gordon 2005). However, in a follow-up experiment, when asked explicitly how many objects were underneath an empty cup, he either refused to answer or said "one" (Pepperberg 2006). Alex starred in many nature and science television shows, and it is well worth watching him in action.

Another impressive demonstration of counting ability comes from Sheba, a chimpanzee. Using a touch screen, Sheba can indicate the Arabic numeral that describes a group of objects. She can also add numbers: if three small groups of objects are put in three separate places around the room, she can visit them in turn and then correctly choose the numeral that represents the sum. Finally, if the three groups of objects are replaced with numeral cards, she can still choose the numeral that represents the correct total (Boysen and Berntson 1989).

Counting is one abstract concept, but there are others. Alex the parrot understood other abstract concepts—the concepts of same and different. He demonstrated this ability in experiments in which he was shown two objects

FIGURE 5.16 **Alex, an African gray parrot who learned several concepts. Alex knew the concept of same/different, an idea once thought to characterize only humans and their closest primate relatives.**

at a time. The objects would differ in one of three qualities: color, shape, or material. He might be shown a yellow, rawhide pentagon and a gray, wooden pentagon or a green, wooden triangle and a blue, wooden triangle. Then Alex would be asked, "What's same?" or "What's different?" A correct answer to the first question is to name the category of the similar shared characteristic. When he saw the first of the previous examples, Alex would have to answer "shape," not "pentagon." In the second example, a correct response to "What's different?" would be "color," not "green." When shown objects he had seen before, Alex correctly identified the characteristic that was the same or different 76% of the time. He was also shown pairs of objects that he had never seen before, and 85% of the time he correctly identified the characteristic that was the same or different (Pepperberg 1987b).

Alex's abilities were particularly impressive given that his brain was the size of a walnut. However, as Srinivasan and Zhang (2003) argue, brain size does not reliably indicate the ability of animals to do tasks such as concept learning. Even honeybees can learn to distinguish between same and different. For example, they can learn that if they see a particular pattern when they enter a testing apparatus, they must then choose the matching pattern (a delayed matching-to-sample procedure). Once they learn this task, they can immediately correctly perform a similar task with colors, without further training. Thus, they appear to learn the concept of matching (reviewed in Srinivasan and Zhang 2003).

Some very interesting experiments on pigeons (*Columba livia*) show that they are able to form concepts such as "tree" or "water" or "human." The typical protocol is a training session with a series of slides, generally a variety of photographs. Pigeons are rewarded when they peck at a slide that has an example of a particular category, such as a person. For instance, they might see 40 photographs with people pictured from different angles, in partial view, and so on, as well as 40 photographs with no people. After they have learned to classify the photos correctly, they are then given new slides they have never seen before, and they are asked to classify them. Pigeons can recognize water, for instance, in various forms—a droplet, a river, a lake (Herrnstein et al. 1976; Mallot and Siddall 1972; Siegel and Honig 1970). They can even learn to distinguish paintings by Monet from those of Picasso (Watanabe et al. 1995).

SELF-RECOGNITION AND PERSPECTIVE TAKING

How do animals perceive themselves? Do they see their bodies as unique entities, separate from the rest of the world? Can they adopt the point of view of other animals? On the surface, these questions seem to be untestable—but research on a variety of fronts has shed some light on them.

The "mark test" was devised nearly 40 years ago (Gallup 1970). It is still in use, and its interpretation is still controversial. Here, the subject is given a mirror and is given time to adapt to it. Initially, animals often treat the mirror as a conspecific, making threats to it or greeting it. After some time, some species, such as chimps, begin to use the mirror to groom otherwise out-of-sight areas, pick their teeth, and the like. At this point, the animal is given general anesthesia, and a harmless, odorless dye is applied to some area of the face. After the subject recovers from anesthesia, its behavior is observed without the mirror for a baseline sample. Then the animal is shown the mirror. If the subject recognizes itself, it should see that it now has an odd new mark, and direct touches and grooming toward the marked area. The best-controlled versions of this experiment compare these touches to the marked area with touches toward predefined control areas on the head that have no mark (Povinelli et al. 1997; Shettleworth 1998). Species that have "passed" the mark test include chimpanzees (*Pan troglodytes*) (Povinelli et al. 1997), dolphins (*Tursiops truncatus*) (which of course have no way to touch marks but do turn their bodies to inspect marks in the mirror; Reiss and Marino 2001), and Asian elephants (*Elephas maximus*) (which touch marks with their trunks; Plotnik et al. 2006). Even in these species, often only some of the individuals tested are successful.

But what does this mean? Certainly this helps to demonstrate that an animal has self-perception and a knowledge of its physical body, but does it mean the animal has a concept of self in the same way we do? A different approach to the question of self versus nonself is to ask whether animals can take the perspective of other individuals, and to understand what others know and do not know. To clarify this idea, let's look at an example. Hare and Tomasello (2000) set up a competitive situation over two pieces of food between a subordinate and a dominant chimpanzee. If a dominant chimp sees a subordinate eating food, it will take the food away. In the experiment, the dominant chimp could only see one piece of food because the other was hidden by a barrier. The subordinate chimp could see both pieces of food and could also see the dominant chimp. The question was whether the subordinate chimp was aware of which piece of food the dominant could see—could the subordinate understand the perspective of the dominant? In fact, this seemed to be the case: when given a choice, subordinates selected the piece that was not visible to the dominant individual. Capuchins, in contrast, did not show evidence of perspective taking (Hare et al. 2003). There is a growing literature on "seeing and knowing" and the attribution of knowledge and mental states, and many species have been tested with a range of clever experiments.

A compelling idea about the evolution of the "theory of mind" is that it is driven by social complexity: the social environment creates new selection pressures for the evolution of "social intelligence." For example, the ability to learn and keep track of relationships among other individuals may well be evolutionarily advantageous. We'll return to the topic of sociality in Chapters 18 and 19.

SUMMARY

Types of Learning

Learning is a change in behavior as a result of experience, excluding changes as a result of maturation of the nervous system, fatigue, or sensory adaptation. Learning is traditionally divided into categories:

Habituation. The animal learns *not* to respond to a specific stimulus because it has been encountered frequently without important consequences. Habituation is adaptive because it conserves energy and leaves more time for other important activities.

Classical Conditioning. Classical conditioning is a type of associative learning. The animal learns to give a response normally elicited by one stimulus (the unconditioned stimulus, or US) to a new stimulus (the conditioned stimulus, or CS) because the two are repeatedly paired. Conditioning is most effective if the CS reliably precedes the US. If the CS is presented many times without the US, the response to the new stimulus will be gradually lost. This is called extinction.

Operant Conditioning. This is another type of associative learning. Here, the frequency of a behavior is increased because it is reinforced. Novel behaviors can be introduced into the repertoire through shaping. During shaping, the reward is made contingent upon closer and closer approximation to the desired action. Sometimes not every response is reinforced. The frequency with which the reward is given is called the reinforcement schedule.

Latent Learning. Latent learning occurs without any obvious reinforcement, and is not obvious until sometime later in life. The information gained through exploration is an example.

Social Learning. The animal learns from others. Types of social learning include stimulus and local enhancement (where animals are attracted to an object or location by conspecifics), observational conditioning (a type of classical conditioning that occurs in social contexts), goal-directed emulation (where an animal learns the goal of a task by watching another animal but not exactly how to perform it), and imitation (where an animal copies another's action). Traditions spread through a group and are stable over time.

Species Differences in Learning: Comparative Studies

The ability to learn has a genetic basis, as artificial selection experiments show. Thus, differences in learning ability may be due to natural selection. Comparative studies of learning have addressed the question of whether learning is adaptively specialized across different species. The most complete case study is that of food-storing birds, where evidence suggests that species that rely heavily on being able to find stored seeds have better spatial memory than species that do not.

Other Evidence of Cognitive Abilities in Animals

Animals demonstrate cognitive skills in other tasks besides learning. A common theme is that the design of experiments is crucial: subtle differences can produce profoundly different outcomes.

Tool Use. Tools are objects that an animal uses to reach a goal. In some cases, animals seem to understand *how* a tool works; in other cases, their understanding is more limited.

Detour Behavior. An animal takes an indirect route to a goal. Species differences in the ability to detour are well documented.

Understanding Numbers and Other Abstract Concepts. Many species understand relative numbers, but only a few have been shown to be able to "tag" particular quantities in the way that we do. Many animals understand other concepts, such as "same" vs. "different," and can classify objects into various categories.

Self-Recognition and Perspective Taking. Do animals perceive themselves as separate from others? The "mark test," whereby animals are given a mark without their knowledge and then allowed to examine themselves in a mirror, provides evidence that some animals can recognize themselves. Experiments on perspective-taking suggest that some animals can understand that other individuals do not have the same knowledge that they do.

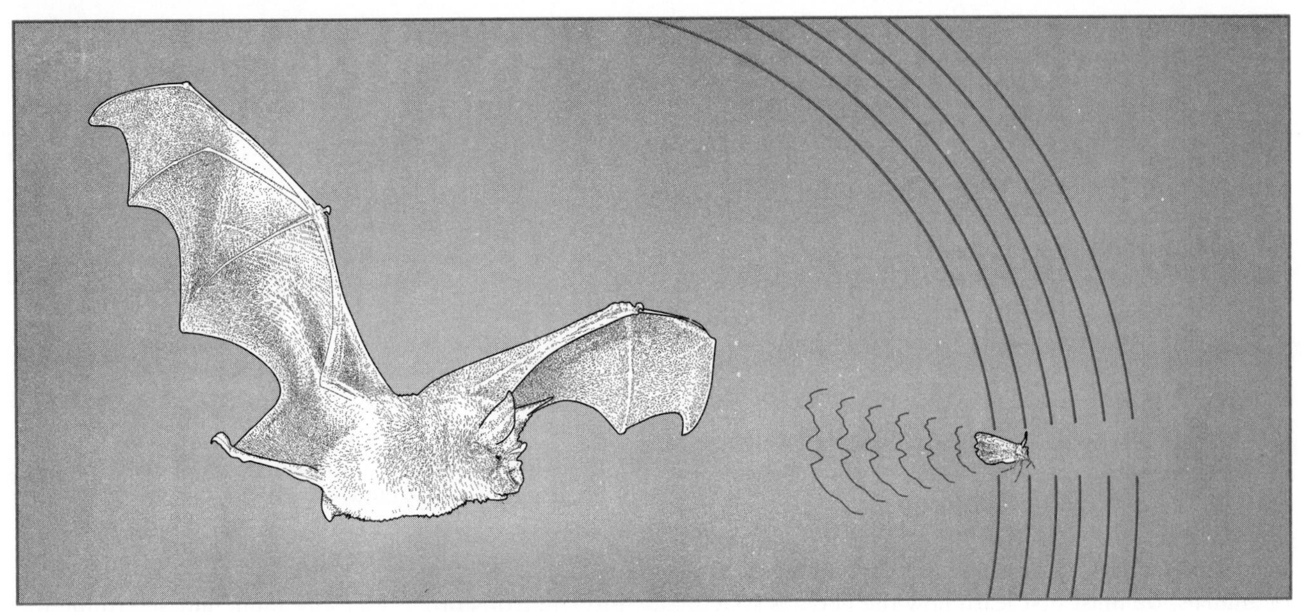

6

Physiological Analysis—Nerve Cells and Behavior

Just as the house cat raises its paw to strike, the cockroach (*Periplaneta americana*) dashes across the floor and disappears into a tiny crevice. If we were to film this sequence and then replay it in slow motion, we would see that the cat's paw was still several centimeters away from the cockroach when the intended victim turned its body away from the cat and ran. Indeed, if you have tried to step on a cockroach, you may have been unpleasantly surprised at its speed, 70 to 80 cm per second, and its ability to change direction rapidly enough to avoid your well-aimed foot. Such a quick response depends on the nervous system; hormones, the subject of the next chapter, could not trigger such a speedy response. Although the cockroach escape response may not impress you, it has inspired some researchers' efforts to build antimissile technology and crash devices for automobiles. Our goals in this chapter will not be that lofty. We will consider the escape response of the cockroach as an introduction to the types of neurons and the concept that neurons are organized into networks.

Adaptive behavior depends on interactions among the components of the nervous system, the body, and the environment. Sensory receptors must detect critical stimuli, and sensory input must be filtered to extract the most biologically relevant information. Based on this information, the nervous system must then produce adaptive responses. For instance, a moth's simple nervous system must process information from a hunting bat's calls to avoid predation. The nervous system of a barn owl processes auditory information so that the sounds of a scurrying mouse can be precisely located and a direct strike executed in complete darkness. The brain responds to incoming stimuli in a dynamic fashion, which involves changes in connections between neurons, the growth of new neurons, and interactions between brain regions. The movements involved in an animal's response are also sculpted by interactions among nerve cells. One of the most interesting questions in the field of motor control is how rhythmic motor patterns, such as locust flight, are generated by groups of neurons called central pattern generators.

These are just a few of the issues we will address in this chapter. Although we will focus on mechanism here, keep in mind that mechanism cannot be considered apart from evolution. Nervous systems are the product of evolution, and they in turn affect the direction of evolution.

Before we can understand the behavioral responses of animals, we must first learn how the nervous system is put together. We will present a series of examples demonstrating that nervous systems have evolved so that animals can quickly (1) detect pertinent events in their environment, (2) choose appropriate responses to such events, and (3) coordinate the parts of their bodies necessary to execute the responses.

CONCEPTS FROM CELLULAR NEUROBIOLOGY

TYPES OF NEURONS AND THEIR JOBS

How did the cockroach in the opening scenario detect the predator in time to take evasive action? Kenneth Roeder, and later Jeffrey Camhi and his colleagues, studied the escape response of cockroaches and discovered that these unlovable house guests respond to gusts of air that are created by even the slightest movements of their enemies (Camhi 1984, 1988; Camhi et al. 1978). Cockroaches, it turns out, have numerous hairlike receptors that are sensitive to wind, and these receptors are located on two posterior appendages called cerci (singular, cercus; Figure 6.1). When these wind-sensitive receptors are stimulated, they alert the nervous system of the cockroach, and within a matter of milliseconds, the cockroach turns away from the direction of the wind and starts to run.

The escape of the cockroach is orchestrated by the interactions among nerve cells, which are also called **neurons**. Neurons can be classified into three groups based on their function. Neurons that carry signals from a receptor organ at the periphery toward the central nervous system (in vertebrates, the brain and spinal cord, and in invertebrates, the brain and nerve cord) are called **sensory** or **afferent neurons**. Those that carry signals away from the central nervous system to muscles and glands are called **efferent** or **motor neurons**. Interneurons, found within the central nervous system, connect neurons to each other. Interactions among

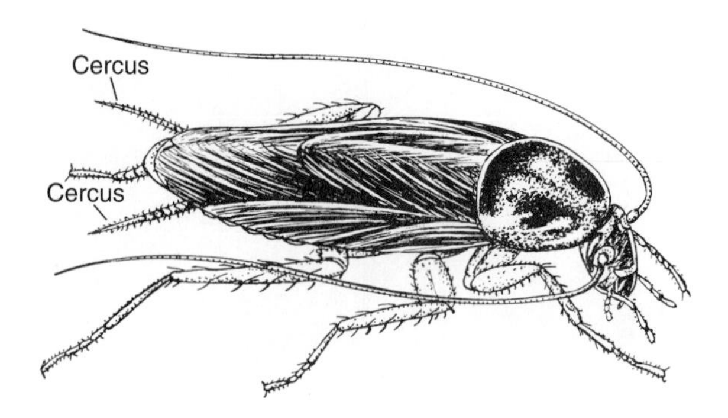

FIGURE 6.1 **A cockroach, showing the two cerci, each with approximately 200 wind-sensitive hairs.**

interneurons process sensory input and determine the motor response.

Look at these three types of neurons in the cockroach as we consider the role each plays in the escape response (Figure 6.2). At the base of each of the wind-sensitive hairs on the cercus is a single sensory neuron that relays pertinent information from the external environment into the central nervous system. In the central nervous system, the sensory neuron makes contact with an interneuron; in this case, the interneuron is described as a giant because of its exceptionally large diameter. This giant interneuron ascends the nerve cord to the head. Before reaching the head, however, the giant interneuron makes contact with an interneuron in the thoracic area, which in turn connects with motor neurons that relay messages to the hind leg muscles (Schaefer et al. 1994). (An advantage of studying the neural basis of behavior in an invertebrate animal such as the cockroach is that it is sometimes possible to identify the individual neurons involved in a specific behavior, particularly a pattern of behavior associated with escape. Because escape requires fast action, the neurons involved in escape responses are often large in diameter to permit the rapid conduction of messages. The result is that these large neurons are somewhat easier to identify than their smaller counterparts are.)

How does the cockroach determine the direction of a wind gust, so that it can run away from cats rather than straight into them? Most of the segments on each cercus have a row of sensory hairs that can be deflected slightly by wind or touch. Each wind-sensitive hair responds differently to a gust of wind from a particular direction. Thus, the pattern of output from the sensory hairs encodes information about the direction of the wind. This information is sent to the giant interneurons. Seven giant interneurons run along each side of the cockroach's ventral nerve cord (Levi and Camhi 1995). The pattern of output from the sensory hairs caused by a wind gust will stimulate each giant interneuron differently. The firing rates are summed to determine the direction of the gust of wind, and the cockroach will turn

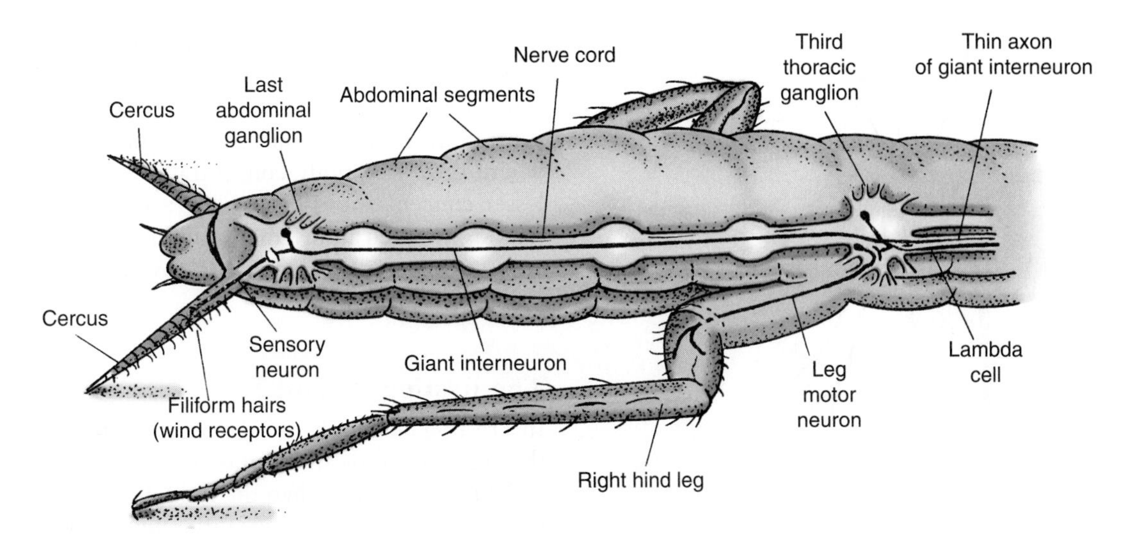

FIGURE 6.2 **Some of the cells in the underlying neural circuitry of the cockroach escape response. Note that the sensory neuron, also called the wind-receptor neuron, that is leaving the cercus makes contact with a giant interneuron in the central nervous system, which in turn makes contact with another interneuron that synapses with a motor neuron in the leg. (Modified from Camhi 1980 with new information from Ritzmann 1986.)**

and run in the opposite direction (Levi and Camhi 2000). If the cockroach were to run directly away from the threat each time it escaped, the predator could predict the escape direction and capture the roach. Instead, the cockroach escapes in one of a preferred set of paths away from the threat—usually about 90°, 120°, 150°, or 180° away from the threat. This unpredictability in the escape path keeps the predator guessing (Domenici et al 2008).

Although there is no such thing as a typical neuron, it is possible to identify characteristics common to some neurons. We will use a motor neuron (Figure 6.3), in this case from a mammal, as our example. The nucleus of a motor neuron is contained in the cell body (soma), from which small-diameter processes (**neurites**) typically extend. In the traditional view, information enters a neuron via a collection of branching neurites and then travels down a single, long neurite to be passed on to other neurons. The neurites that receive the information are called **dendrites**; the single, long, cable-like neurite that transmits the information to other neurons is called an **axon**. In most vertebrates, some axons have a fatty wrapping called the **myelin sheath**. In our example of the motor neuron, the axon ends on a muscle or a gland (an effector), which brings about the animal's behavioral response. Although the terms *dendrite* and *axon* are well established in the literature, it is now recognized that the flow of information through a neuron is often not so neatly divided into separate receiving and transmitting processes. We will continue to use the terms, keeping in mind that in many cases the specific direction of the informational flow has not actually been demonstrated.

The myelin sheath is formed by the plasma membranes of glial cells, which are supporting cells in the nervous system that become wrapped around the axon many times. Since a single glial cell encloses only a small region, about 1 mm, of an axon, the myelin sheath is not continuous. The regions along an axon between adjacent glial cells are exposed to the extracellular environment. This arrangement is important to the speed at which the nerve cell conducts messages. The message "jumps" successively from one exposed region to the next, increasing the rate of transmission as much as 100 times. For this reason, axons that conduct signals over long distances are usually myelinated.

THE MESSAGE OF A NEURON

Let us now delve more deeply into the details of the how and why of ion movements that are responsible for a neuron's message, called an action potential. An action potential is an electrochemical signal caused by electrically charged atoms, called ions, moving across the membrane. Ions can cross the membrane of a nerve cell by means of either the sodium-potassium pump or ion channels. The pump uses cellular energy to move three sodium ions (Na^+) outward while transporting two potassium ions (K^+) inward. An ion channel, on the other hand, is a small pore that extends through the membrane of a nerve cell. There are different types of channels, each type forming a specific passageway for only one or a few kinds of ions. Whereas some are passive ion channels that are always open, others are active ion channels (also called gated channels) that open in response to a specific triggering signal. Triggering signals may include the presence of chemicals (neurotransmitters) in the space between the membranes of neurons, changes in the

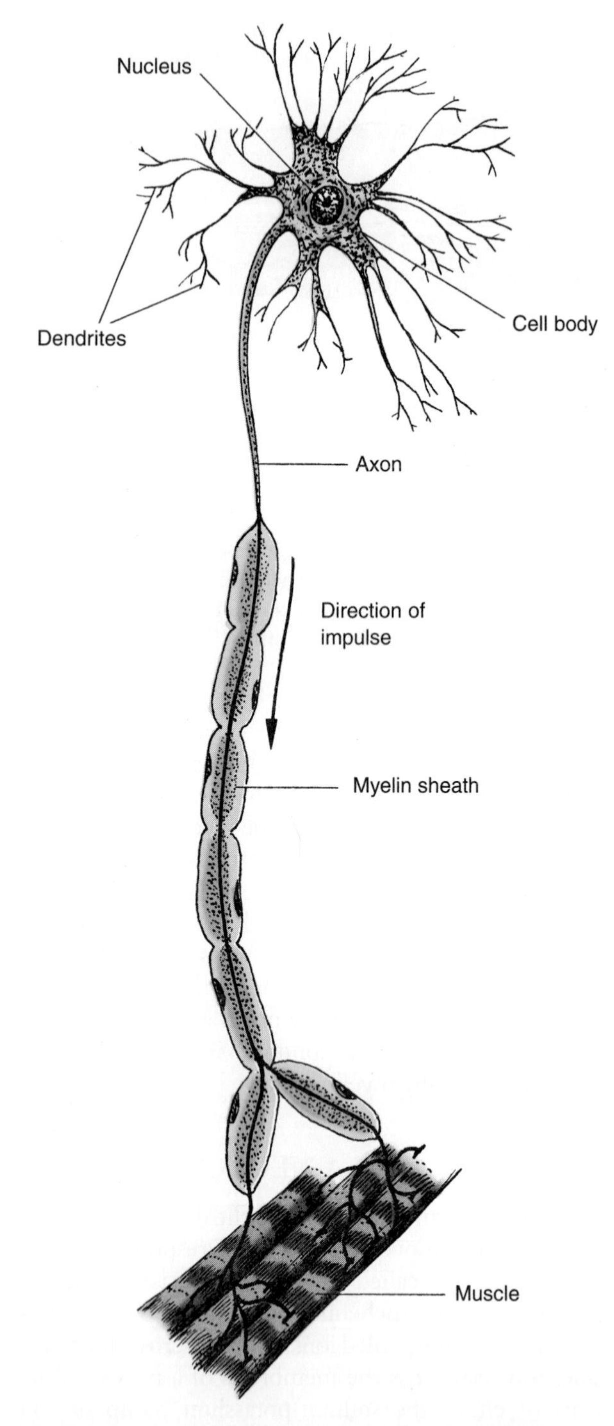

FIGURE 6.3 **A motor neuron. The soma or cell body maintains the cell. The dendrites are extensions specialized for receiving input from other cells. The axon is specialized to conduct the message away from the cell and to release a chemical that will communicate with another cell. In vertebrates, some axons are covered by a fatty myelin sheath.**

charge difference across the membrane (the membrane potential), changes in the concentration of intracellular calcium ions (Ca^{++}), or any combination of these factors (see discussion of synaptic transmission later in this chapter). Once an ion channel is open, the ion it admits may move across the membrane in response to either a concentration gradient (ions tend to move from an area where they are highly concentrated toward an area of lesser concentration) or an electrical gradient (because ions are charged atoms, they tend to move away from an area with a similar charge and toward an area of the opposite charge).

The Resting Potential

In a resting nerve cell, one that is not relaying a message, the area just inside the membrane is about 60 millivolts (mV) more negative than the fluid immediately outside the membrane. This charge difference across the membrane is called the resting potential of the neuron. A membrane in this resting state is described as polarized (Figure 6.4).

The resting potential results from the unequal distribution of certain ions across the membrane. The concentration of Na^+ is much greater outside the neuron than within. The concentration of K^+ shows just the opposite pattern, and is greater inside the cell than outside. Certain large, negatively charged proteins are held within the neuron either because the membrane is impermeable to them or because they are bound to intracellular structures. These proteins are primarily responsible for the negative charge within the neuron.

Most of the active ion channels in the membrane of a resting neuron are closed, but passive channels are, of course, open. Because most of the passive channels are specific for K^+, the membrane is much more permeable to K^+ than it is to other ions. Drawn by the negative charge within, positively charged K^+ will enter the neuron and accumulate there. At some point, when there is roughly 20 to 30 times more K^+ inside than outside, the concentration gradient counteracts the electrical gradient. When the two forces—an electrical gradient that draws K^+ inward and a concentration gradient that pushes K^+ outward—are equally balanced and there is no further net movement of K^+; the cell has reached its resting potential.

Why is Na^+ more concentrated outside the neuron? Although Na^+, like K^+, is attracted by the negative charge inside the neuron, the membrane is relatively impermeable to it, and so only a few Na^+ can leak through. Furthermore, the sodium-potassium pump actively removes Na^+ from within the cell, transporting it outward against electrical and concentration gradients.

The Action Potential

The **action potential** (nerve impulse) is an electrochemical event that lasts about 1 millisecond. The action potential consists of a wave of depolarization followed by

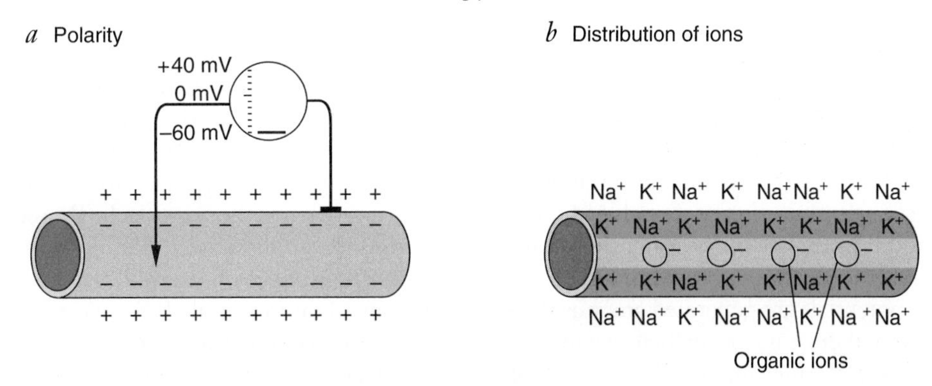

FIGURE 6.4 **The resting potential. (*a*) In the resting state, the inside of an axon is more negative than the outside. (*b*) This charge difference is caused by the unequal distribution of ions inside and outside the cell. There are more sodium ions outside and more potassium ions inside. In addition, there are large, negatively charged proteins (organic ions) held inside the cell, giving the interior an overall negative charge.**

repolarization that spreads along the axon. The depolarization, or loss of the negative charge within, is caused by the inward movement of Na^+. However, the repolarization, or restoration of the negative charge within the neuron, is caused by K^+ leaving the cell (Figure 6.5).

Let's see how depolarization and repolarization occur. The membrane becomes slightly depolarized when some of the active Na^+ channels open and Na^+ enters the cell, drawn by both electrical and concentration gradients. The positive charge on Na^+ slightly offsets the negative charge inside the cell, and the membrane becomes slightly depolarized. If the depolarization is great enough, that is, if threshold is reached, voltage-sensitive sodium channels open and Na^+ ions rush to the interior of the cell. At roughly the peak of the depolarization, about 0.5 millisecond after the voltage-sensitive sodium gates open, they close and cannot reopen again for a few milliseconds. Almost simultaneously, voltage-sensitive potassium channels open, greatly increasing the membrane's permeability to K^+. Potassium ions then leave the cell, driven by the temporary positive charge within and by the concentration gradient. The exodus of K^+ restores the negative charge to the inner boundary of the membrane. In fact, enough potassium ions may leave to temporarily make the cell's interior even more negatively charged than usual, a condition called hyperpolarization. Notice that although the original resting potential is eventually restored, the distribution of ions is different. This situation is corrected by the sodium-potassium pump, which moves K^+ back in and Na^+ back out of the cell.

This depolarization and repolarization of the neuronal membrane spreads rapidly along the axon, generated at each spot in the same manner in which it was started. The local depolarization at one point of the membrane opens the voltage-sensitive sodium channels in the neighboring region of membrane, thereby triggering its depolarization. The net result is that a wave of excitation travels down the axon.

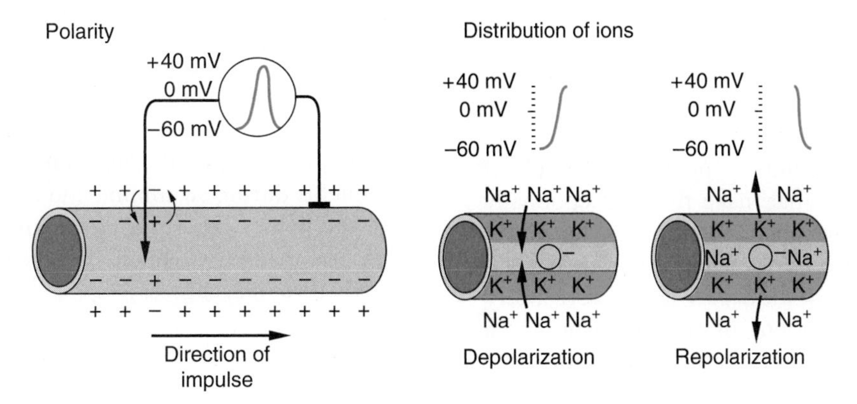

FIGURE 6.5 **The action potential of a neuron. Depolarization is caused by sodium ions entering the cell, and repolarization is caused by potassium ions exiting the cell.**

Immediately after an action potential, the neuron cannot be stimulated to fire again for 0.5 to 2 milliseconds because the sodium channels cannot be reopened right away. At the start of this absolute refractory period, no amount of stimulation can generate an action potential. During the latter part of this interval, the relative refractory period, stimulation must be greater than the usual threshold value to generate an action potential. Although the refractory period is brief, it is biologically significant because it determines the maximum rate of firing.

Because action potentials are generated anew as they travel down the neuron, there is no loss in their magnitude: an action potential has the same amplitude at its point of initiation as at every other point along the neuron. We therefore describe the action potential as an all-or-none phenomenon: it either occurs or it doesn't, and doesn't occur halfway. This characteristic makes action potentials well-suited to be the nervous system's long-distance signal.

If an action potential is always the same, how can differences in the intensity of stimuli be sensed? The intensity is encoded in the firing rate of the neuron and by the number of neurons responding. For example, if we place our hand on a hot stove rather than on a warm one, the firing rates of neurons in our hand that respond to heat or pain may be increased. Also, touching a very hot stove will activate more neurons than will touching a warm stovetop because the thresholds of neurons that register heat vary—more neurons reach their thresholds and fire at higher temperatures.

IONS, MEMBRANE PERMEABILITY, AND BEHAVIOR

Although ions and their movements through the channels of nerve cell membranes may seem, at best, to be only remotely related to an animal's behavior, we will see that this is not the case. Let's consider how ions and changes in membrane permeability relate directly to what we see an animal doing. Here we describe an example of how changes in the membrane permeability caused by mutation in the *Shaker* gene of the fruit fly (*Drosophila melanogaster*) produce atypical behavior.

When fruit flies are anesthetized with ether, one occasionally sees a mutant fly that shakes its legs, wings, or abdomen. Among the mutations that result in shaking under ether anesthesia are *Shaker*, *hyperkinetic*, and *ether-a-go-go*. All this shaking is apparently a result of neurons with mutations that make them exceptionally excitable. More is known about what makes the *Shaker* mutants so jittery, so we will concentrate on them.

It was first shown that the *Shaker* larvae were jittery because an excessive amount of neurotransmitter, a chemical released by a neuron that allows communication with another neuron or a muscle cell (discussed shortly), was released at the junction between a motor neuron and a muscle cell, causing extreme muscle contractions. Furthermore, the muscle contractions were uncoordinated because the release of the transmitter from different neurons was asynchronous (Jan et al. 1977). Recall that a neuron's message is an electrochemical signal consisting of a wave of depolarization caused by sodium ions (Na^+) entering the nerve cell followed by a wave of repolarization caused by potassium ions (K^+) leaving the nerve cell. By recording from the neurons of adult *Shaker* flies, it was shown that the mutant neurons do not repolarize as quickly as normal neurons (Tanouye et al. 1981). As a result, an excess of calcium ions enters the neuron and causes the release of neurotransmitter to continue longer than is typical.

Why doesn't the cell repolarize normally? Apparently, the *Shaker* gene codes for a protein that forms part of the potassium channels involved in repolarization; a mutation at the *Shaker* locus results in certain potassium channels not being formed, and this disrupts the process of repolarization (Kaplan and Trout 1969; Molina et al. 1997). That the behavioral defect, that is, shaking, results from the absence of K^+ channels was elegantly shown by experiments in which a functional *Shaker* gene was inserted into mutant flies. This experimental procedure resulted in a normal flow of potassium across the membranes of nerve cells and an end to the jittery behavior caused by the mutations (Zagotta et al. 1989).

BEHAVIORAL CHANGE AND SYNAPTIC TRANSMISSION

Synapses are important structures because they are decision and integration points within the nervous system. The molecular events that occur at synapses determine whether the message of one neuron will generate an action potential in the next neuron. Typically, a neuron receives input through thousands of synapses. That input is integrated in ways that make complex behaviors possible. Changes in the functioning of synapses or in the number of synaptic connections often explain why behavior can change because of experience or maturation. As we will see shortly, the gill-withdrawal reflex of the sea hare *Aplysia* can be modified through various forms of learning, each involving changes in how neurons communicate with one another at synapses. We will also learn how the modification of synaptic function caused by chemicals called neuromodulators changes leg movements during swimming and courtship in the blue crab. But before exploring the cellular mechanisms for behavioral change, we should become more familiar with the structure of synapses.

THE STRUCTURE OF THE SYNAPSE

The gap between neurons is called a synapse, and at a specific **synapse** information is typically transferred in

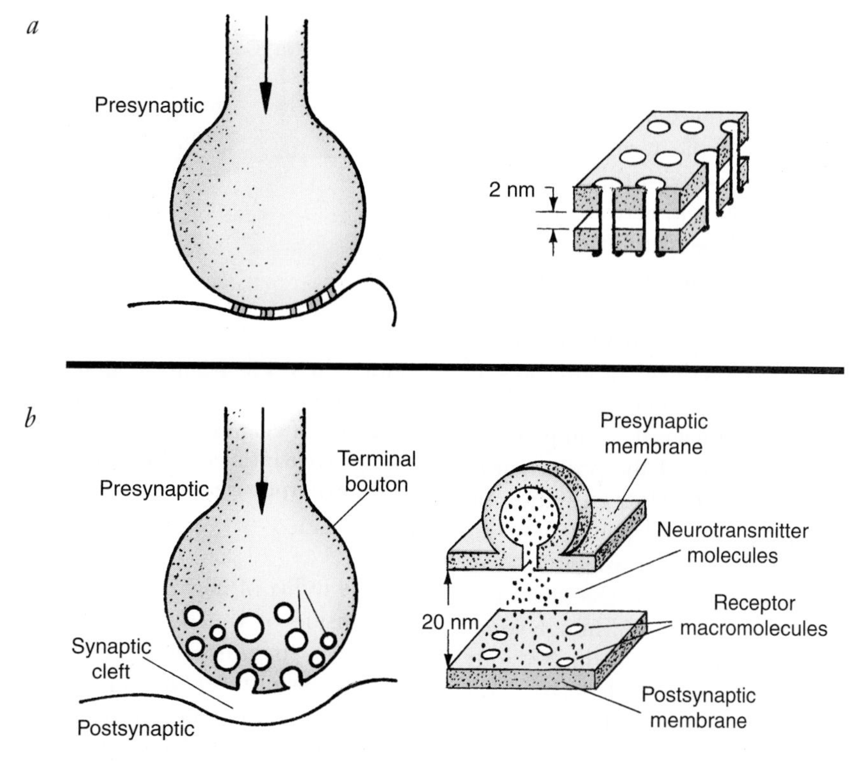

FIGURE 6.6 **The gap between neurons is called a synapse. There are two types of synapses: (*a*) electrical and (*b*) chemical. In an electrical synapse, the distance between neurons is very small, and ions can flow directly from one neuron to the next. In a chemical synapse, however, the distance between neurons is greater, and one neuron affects the activity of the other by releasing chemicals into the gap between them. The original distribution of ions is later restored by the sodium-potassium pump.**

one direction, from the presynaptic neuron to the postsynaptic neuron. (The descriptor *presynaptic* or *postsynaptic* refers to the direction of information flow at a specific synapse.) There are two major categories of synapses: electrical and chemical (Figure 6.6).

Electrical Synapses

We will discuss electrical synapses first because they are much less common and the behavioral examples that follow pertain to chemical synapses. Electrical synapses are known for their speed of transmission. Whereas a signal may cross an electrical synapse in about 0.1 millisecond, durations about 0.5 or 1 millisecond are typical at chemical synapses. Not surprisingly, then, electrical synapses are often part of the neural circuitry that underlies patterns of behavior when sheer speed is essential, such as the escape responses exhibited by animals confronted with a predator.

In an electrical synapse, the gap between the neurons is small, only about 2 nanometers (1 nm = 10^{-9} m), and is bridged by tiny connecting tubes that allow ions to flow directly from one neuron to the other. When an action potential arrives at the axon terminal of the presynaptic neuron, Na^+ enters this terminal, causing a potential difference between the inside of this cell and the postsynaptic cell. Because of the difference, positively charged ions, mostly K^+, move from the presynaptic cell through the tiny tubular connections into the postsynaptic neuron. These newly arriving ions may sufficiently depolarize the postsynaptic cell to induce an action potential.

Chemical Synapses

Chemical synapses are characterized by a larger space between the membranes of the two neurons (typically 20–30 nm) than is found in electrical synapses. Rather than information being transmitted from one neuron to the next via direct electrical connections, it is transmitted across a chemical synapse in the form of a chemical called a **neurotransmitter**. There are several steps in this process, and these steps account for the slower speed of transmission at a chemical synapse than at an electrical one. First, the action potential travels down the axon to small swellings called terminal boutons at the end of the axon. Second, at the terminal boutons, the action potential causes the neuron to release a neurotransmitter from small storage sacs called synaptic vesicles.

Third, the release of neurotransmitter occurs because the action potential opens voltage-sensitive calcium ion channels. Fourth, the calcium ions that flood to the inside initiate events that cause the synaptic vesicles to move toward the membrane of the terminal bouton. Fifth, once there, the vesicles fuse with the membrane and dump their contents into the gap between the cells (the synaptic cleft). Finally, the neurotransmitter then diffuses the short distance across the cleft and binds to receptors on either another neuron or a muscle cell.

When the neurotransmitter binds to a receptor on the postsynaptic cell, it either directly opens ion gates or indirectly affects ion gates through biochemical mechanisms. In either case, the neurotransmitter will increase the permeability of the membrane of the postsynaptic cell to specific ions. This results in either excitation or inhibition of the postsynaptic neuron, depending on the particular ions involved. In the case of excitation, the neurotransmitter causes the opening of channels that allow both Na^+ and K^+ to pass through. Although some K^+ ions move out, they are outnumbered by the Na^+ ions moving in. This causes a slight temporary depolarization of the membrane of the postsynaptic cell. This depolarization is called an excitatory postsynaptic potential, or EPSP (Figure 6.7). If this depolarization reaches a certain point, the threshold, an action potential is generated in the postsynaptic cell.

On the other hand, a neurotransmitter may act in an inhibitory fashion, making it less likely that an action potential will be generated in the postsynaptic neuron. When the neurotransmitter binds to the receptors in an inhibitory synapse, either K^+ channels or K^+/Cl^- channels open. The exit of positively charged potassium ions or the influx of negatively charged chloride ions makes the inside of the postsynaptic membrane even more negative than usual. In other words, an inhibitory neurotransmitter momentarily hyperpolarizes the membrane. The hyperpolarization is called an inhibitory postsynaptic potential, or IPSP (Figure 6.8).

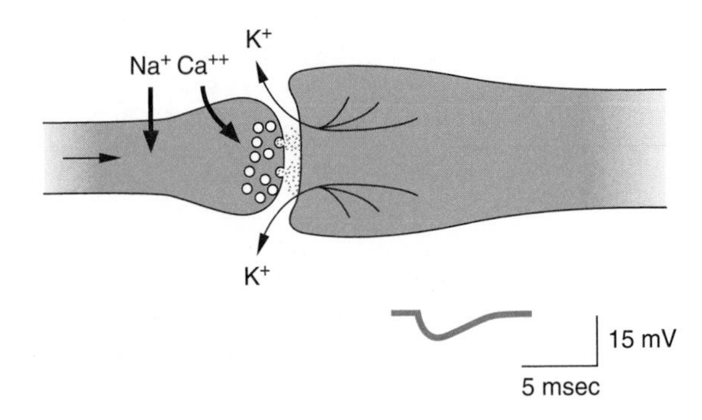

FIGURE 6.8 The events at an inhibitory synapse. The neurotransmitter binds to a receptor and causes the opening of channels that allow potassium ions to leave the postsynaptic cell. Thus it becomes hyperpolarized. This hyperpolarization is called an IPSP.

As long as the neurotransmitter remains in the synapse, it will continue to excite or inhibit the postsynaptic cell. However, the effect of a neurotransmitter can be halted in a variety of ways. In some cases, the neurotransmitter is broken down by an enzyme, and its component parts are absorbed for resynthesis. In other instances, however, the molecules of the neurotransmitter are released intact after acting on the postsynaptic cell, absorbed by the presynaptic cell, and repackaged for subsequent release.

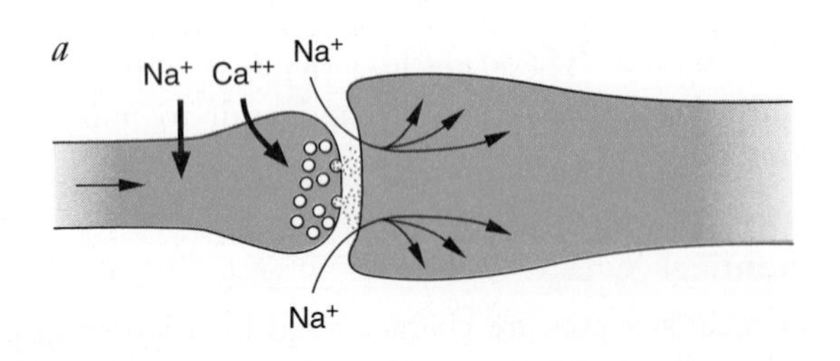

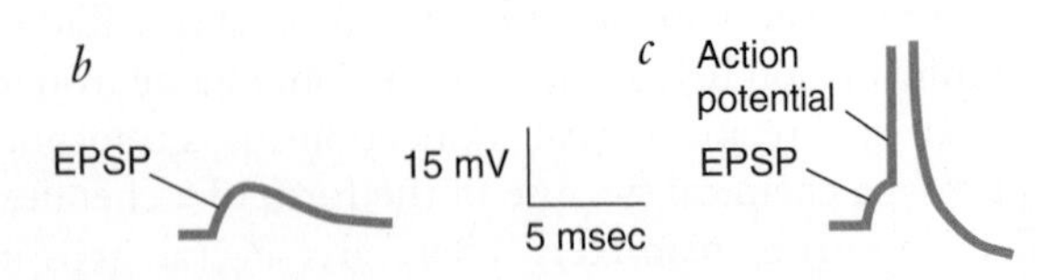

FIGURE 6.7 The events at an excitatory synapse. (*a*) The neurotransmitter binds to a receptor and causes the opening of channels that allow sodium ions to enter the cell, thereby slightly depolarizing the postsynaptic cell. (*b*) This slight depolarization is called an EPSP. (*c*) If the depolarization reaches the threshold, an action potential is generated.

INTEGRATION

Most neurons receive input from many other neurons. In fact, a given neuron may communicate with hundreds or thousands of other cells. The slight depolarization (EPSPs) and hyperpolarization (IPSPs) that result from input from all the synapses are summed on the postsynaptic membrane, either because one neuron sends a repeated signal or because many neurons send messages to one postsynaptic cell. In other words, the EPSPs and IPSPs combine with one another as they arrive at the cell body. If these interacting changes in membrane potential combine to produce a large enough depolarization, voltage-sensitive Na^+ gates are opened and an action potential is triggered.

Neuromodulators are chemicals that cause voltage changes that occur over a slower time course than those caused by "traditional" neurotransmitters—seconds, minutes, hours, and perhaps even days. The fast changes are brought about by traditional neurotransmitters—those chemicals, previously discussed, that open ion gates, causing EPSPs or IPSPs. In contrast, neuromodulators alter neuronal activity slowly, by biochemical means.

The effects of neuromodulators appear to be mediated by substances within the postsynaptic neuron called second messengers (Breedlove et al. 2007). These second messengers (e.g., calcium and the cyclic nucleotides cAMP and cGMP couple the membrane receptors of the postsynaptic cell with the movements of ions through one or more enzymatic steps. Neuromodulators may, for example, upon reaching the receptor on the postsynaptic neuron, trigger the formation of the second messenger cAMP within the neuron, which in turn activates an enzyme that changes the shape of proteins in certain ion channels. Once the ion channels have been altered in this manner, the permeability of the membrane to specific ions is also changed, thereby affecting the activity of the neuron. It is the relatively slow pace of the enzymatic activities that produces the typically prolonged effects of neuromodulators.

Functionally, neuromodulators appear to be intermediate to classic neurotransmitters and hormones. Whereas neurotransmitters are released at specific synaptic clefts and hormones are broadcast throughout the body via the bloodstream, neuromodulators are released in the general vicinity of their target tissue. It is, however, difficult to establish the precise point at which a neurotransmitter becomes a neuromodulator, and a neuromodulator a hormone. In fact, the same chemical may have different functions in different places. Some chemicals, dopamine, for instance, act as neurotransmitters at some synapses and as neuromodulators at others. Similarly, a chemical may act locally in the nervous system as a neurotransmitter, whereas in other places in the body it is released into the bloodstream and acts at a distant site as a hormone does (see Chapter 7 for a discussion of hormones). Despite some fuzziness in definition, there is no question that neuromodulators, through their actions on neurons, glands, and muscles, can produce profound effects on behavior.

Consider the behavior of the male blue crab (*Callinectes sapidus*) as an example of neuromodulation of rhythmic movements of the swimming legs. Just before a female blue crab matures, she releases a pheromone, a chemical used to communicate with other members of the species, in her urine. When a male blue crab senses the pheromone, he begins his courtship display. He spreads his claws apart in front, extends his walking legs, and raises his swimming legs in the rear. He then waves his paddle-shaped swimming legs from side to side above the carapace (Kamio et al. 2008). Besides courtship, two other distinct stereotyped behaviors, sideway swimming and backward swimming, involve the rhythmic movement of swimming legs. In each behavior, the swimming legs are waved in a slightly different way and the crab assumes a different posture. However, because these three behaviors are so similar, it is likely that they share common neural elements (Wood 1995a).

Neuromodulators, combined with the proper olfactory stimulation, affect whether the crab will perform the courtship display instead of the two swimming behaviors. When the neuromodulators proctolin, dopamine, octopamine, serotonin, and norepinephrine are separately injected into blue crabs, each drug produces a unique posture or combination of limb movements. The postures and limb movements are the same as those observed in freely moving, untreated crabs. Injection of dopamine produces the posture of the courtship display, in which the male stands high on the tips of his walking legs; injection of proctolin produces the rhythmic leg movements characteristic of courtship that serve to fan chemicals in his urine (pheromones) toward a female to attract her (Wood 1995b).

Electrical stimulation of specific neurons under different conditions has identified interneurons in the esophageal connectives (neural connections between the brain and ventral nerve cord) that trigger rhythmic waving of the swimming legs. Some of these interneurons trigger rhythmic leg waving when sex pheromone is applied to the antennule of the crab. Under these conditions, the leg waving is not distinctly characteristic of any of the three rhythmic behavior patterns. However, when proctolin is applied while these interneurons are being stimulated, the motor output changes to the rhythmic waving of the courtship display (Wood 1995b). The natural source of the proctolin that initiates courtship leg movements is thought to be a cluster of nerve cells in the subesophageal ganglia (Wood et al. 1996).

SPECIALIZATIONS FOR PERCEPTION OF BIOLOGICALLY RELEVANT STIMULI—SENSORY PROCESSING

In spite of the wide variety of stimuli bombarding an animal from its environment, it is able to detect only a limited range, and of those that it detects, it may ignore all but a few key stimuli. The job of an animal's sensory system is not to transmit all available information but rather to be selective and provide only information that is vital to the animal's lifestyle or, more to the point, information that influences its reproductive success.

In some cases, the receptors themselves are "tuned" to detect biologically relevant stimuli. Consider, for example, the relationship between the sensitivity of photoreceptors and the dominant wavelengths of light in the habitat of certain teleost fish. The wavelengths of light that actually reach the eye of a fish will depend on many factors. The color of the water is one such factor, and it varies among habitats. When the sensitivity of the eyes of fish from different habitats is compared, we see that the sensitivity to different wavelengths (colors) of light has been adjusted by natural selection so that peak sensitivity occurs at the most common wavelength of light in a given habitat (Bowmaker et al. 1994; Lythgoe et al. 1994). We will see more examples of sensory tuning when we explore the evolution of communication in Chapter 17.

Besides detecting stimuli critical to survival, animals must be able to pick out that stimulus from a background of "noise" in the same sensory modality. For example, recall that the cockroach runs when exposed to a gust of wind from a predator. Yet, it ignores non-threatening or irrelevant sources of wind, such as the wind that it creates itself while walking. How is such selectivity possible? In the laboratory, cockroaches run when exposed to wind with a peak velocity of only 12 mm per second, the approximate velocity of wind created by the lunge of a predator such as a toad. Cockroaches, however, do not respond to the 100-mm-per-second wind that they create by normal walking. How is it that cockroaches manage not to respond to the relatively strong wind created by walking and yet run when exposed to much softer wind signals? It turns out that it is not the velocity of the wind that is the critical factor but the acceleration (rate of change of wind speed), and a strike by a predator delivers wind with greater acceleration than the wind produced by the stepping legs of a cockroach. In fact, when cockroaches were tested with wind puffs that had the same peak velocity but differed in acceleration, they ran more frequently when exposed to winds of higher acceleration (Plummer and Camhi 1981). Winds with low acceleration typically produced no response. Thus, cockroaches appear to pay particular attention to the acceleration of the wind stimulus, and this allows them to ignore irrelevant wind signals and to focus on important information in their environment.

PROCESSING OF SENSORY INFORMATION FOR SOUND LOCALIZATION

We will consider the mechanisms of sound localization as an example of stimulus processing of biologically important stimuli. It is often important for an organism to locate the source of sound. For example, a potential mate may be producing the sound. A male mosquito finds a female by the sound of her beating wings. It would do a male little good to know that a female was present and be unable to locate her. Similarly, many predators determine their prey's position by localizing sounds generated by the prey. Locating the source of a sound has obvious importance to prey animals as well—the crunching sound of brush under a leopard's foot has fixed its position for many a wary baboon.

What properties of sound enable its source to be located? Actually, part of the answer is remarkably simple: Sounds can be localized by how loud they are, that is, by their intensity. A simple rule might be that sound seems louder when the receptor is closer to the sound. If only one ear is involved in locating the sound source, however, the rule may not hold (Camhi 1984). Let's say that the left ear hears a soft sound; was the sound soft because it was produced by a weak source on the left side or by a strong source on the right side? To eliminate such confusion, both ears must be used in the sound localization process—this is called binaural comparison. Some animals use binaural comparison of sound intensity to locate the source of sound.

Timing is also important in locating the source of a sound. Two differences in timing could be of potential use, and both rely on binaural comparison. The first occurs at the onset of sound—sound begins and ends sooner in the ear that is closest to the source. The second difference in timing occurs during an ongoing sound. During a continuing sound, there are differences in the phase (the point in the wave of compression or rarefaction) of the sound wave reaching each ear. The extent of the phase difference will depend both on the wavelength of the sound and on the distance between the ears. When the wavelength of the sound is twice the width of the head, the peak of a sound wave arrives at one ear and the trough arrives at the other. Under these conditions, the sound is easily localized. In contrast, when the wavelength of the sound equals the head width, the phase of the sound wave is the same in each ear, and the sound is difficult to localize (Figure 6.9).

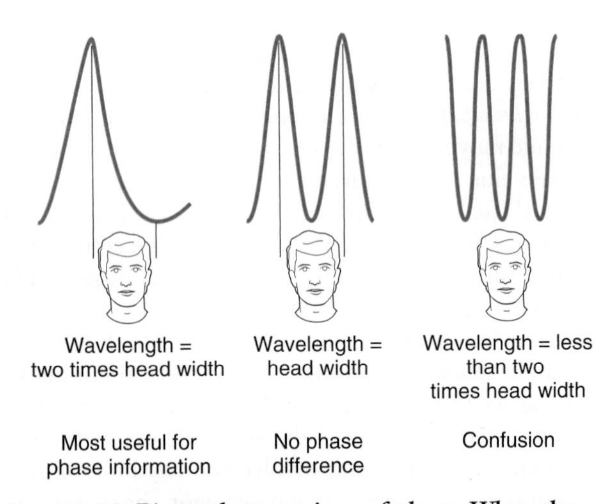

FIGURE 6.9 Binaural comparison of phase. When the sound is prolonged, differences in the phase of the sound wave at each ear may indicate the direction of the source. The usefulness of this cue depends on the wavelength and the distance between the ears.

PREDATORS AND PREY: THE NEUROETHOLOGY OF LIFE-AND-DEATH STRUGGLES

Now let's consider how nervous systems gather and process information about the source of sounds to produce adaptive behaviors—escape behavior by prey and prey localization by a predator. We will first consider how sound information is processed by the relatively simple nervous system of a night-flying (noctuid) moth, allowing it to escape from an echolocating bat. Then we will consider how a barn owl obtains and processes sound information to locate its prey.

Escape Responses of Noctuid Moths

Noctuid moths are a favorite prey of certain bats. Indeed, moths typically make up more than half of a bat's diet. The bats capture their prey on the wing, locating these flying insects by echolocation—that is, by emitting high-frequency sounds that bounce back to the bat from any structure in the environment. Here we will focus on how moths escape predation.

Kenneth Roeder (1967) has provided a fascinating account of how the relatively simple auditory apparatus of the moth is used to detect an approaching bat and how the moth then takes evasive action. When the bat's ultrasonic echolocation pulses are soft, indicating that the bat is still at a distance, the moth turns and flies directly away. However, loud ultrasonic pulses mean that the bat is very close, and emergency actions are needed—erratic, unpredictable looping and wingfolding to produce a free fall. Moths that hear a bat's approach and take evasive action are about 40% less likely to be eaten.

Roeder found that these moths have two ears, one on either side of the thorax (the insect's midsection), and that each ear has only two auditory receptor cells (Roeder and Treat 1957). The receptors are tuned to the frequencies of the echolocation calls of species of bats living in their vicinity, which is generally between 20 and 50 kHz. One receptor, called the A1 cell, is about ten times more sensitive than the other cell, the A2 cell. The A1 cell begins to respond when the sound is soft, indicating that the bat is still at a distance. The sensitivity of this cell is important because it will determine how much time the moth will have to take appropriate evasive action. The A2 cell responds only to loud sounds (Pérez and Coro 1984; Roeder and Payne 1966), as would come from a nearby bat.

The moth responds to bat sounds long before the bat can detect the moth (Roeder and Payne 1966). North American moths can detect a hunting big brown bat (*Eptescicus fuscus*) from a distance of nearly 100 ft, whereas the bat must be within about 15 ft to detect a moth-size target (Fenton 1992). The A1 cell, then, warns the moth that there is a hunting bat in the vicinity, in much the same way that your car's radar detector alerts you of a police radar trap.

How does the moth's nervous system analyze the available information and direct effective evasive maneuvers? The sensitive A1 cell responds to the sounds of a distant bat, and its input reveals the direction and distance of the bat (Figure 6.10). If the bat, for example, is on the left side, the left A1 cell is exposed to louder sounds because the A1 cell on the right is somewhat shielded by the moth's body. Therefore, the left receptor fires sooner and more frequently upon receiving each sound of the bat. When the bat is directly behind or in front of the moth, both neurons will fire simultaneously. A slight turn of the moth's body will then result in differences in the right and left receptors, which will reveal whether the bat is approaching from the front or rear. What about its altitude? If the bat is above the moth, the bat's sounds are louder during the upward beat of the moth's wings when the moth's ears are uncovered than when the moth's wings are down, covering the ears and muffling the bat's cries. However, if the bat is beneath the moth, the bat's echolocation cries will reach the moth's ears unimpeded regardless of the position of the moth's wings. Therefore, the moth's wingbeats will have no effect on the pattern of neural firing. The moth, then, is able to decode the incoming data, so that it detects both the presence and precise location of the bat.

How is this information processed to produce an appropriate escape pattern? If the bat is passing some distance away, the A1 cell begins to fire. Its firing rate will increase as the bat gets closer and its cries become louder. Up to a certain firing rate of the A1 cell, the distance between predator and prey is too great for the bat to detect the moth. Therefore, the most adaptive response of the moth would be to turn and fly directly away, thus decreasing the likelihood of detection by increasing the intervening distance and by exposing less surface area to the bat. This escape pattern results when the moth turns its body until the A1 firing from each ear is equalized. When the bat changes direction, so does the moth (Roeder and Treat 1961).

Bats fly faster than moths, though, and if the bat gets too close, then the moth's evasive maneuver switches to an erratic flight pattern. The moth's wings begin to beat in either peculiar, irregular patterns or not at all. The insect itself probably has no way of knowing where it is going as it begins a series of loops, rolls, and dives. But it is also very difficult for the bat to pilot a course to intercept the moth. If the moth crashes into the ground, so much the better. It is safe here because the earth will mask its echoes.

How does the moth determine whether the bat is gaining on it? One clue is that the sound of an approaching bat grows louder. Recall that the A2 cell is less sensitive than the A1 cell and doesn't begin to fire

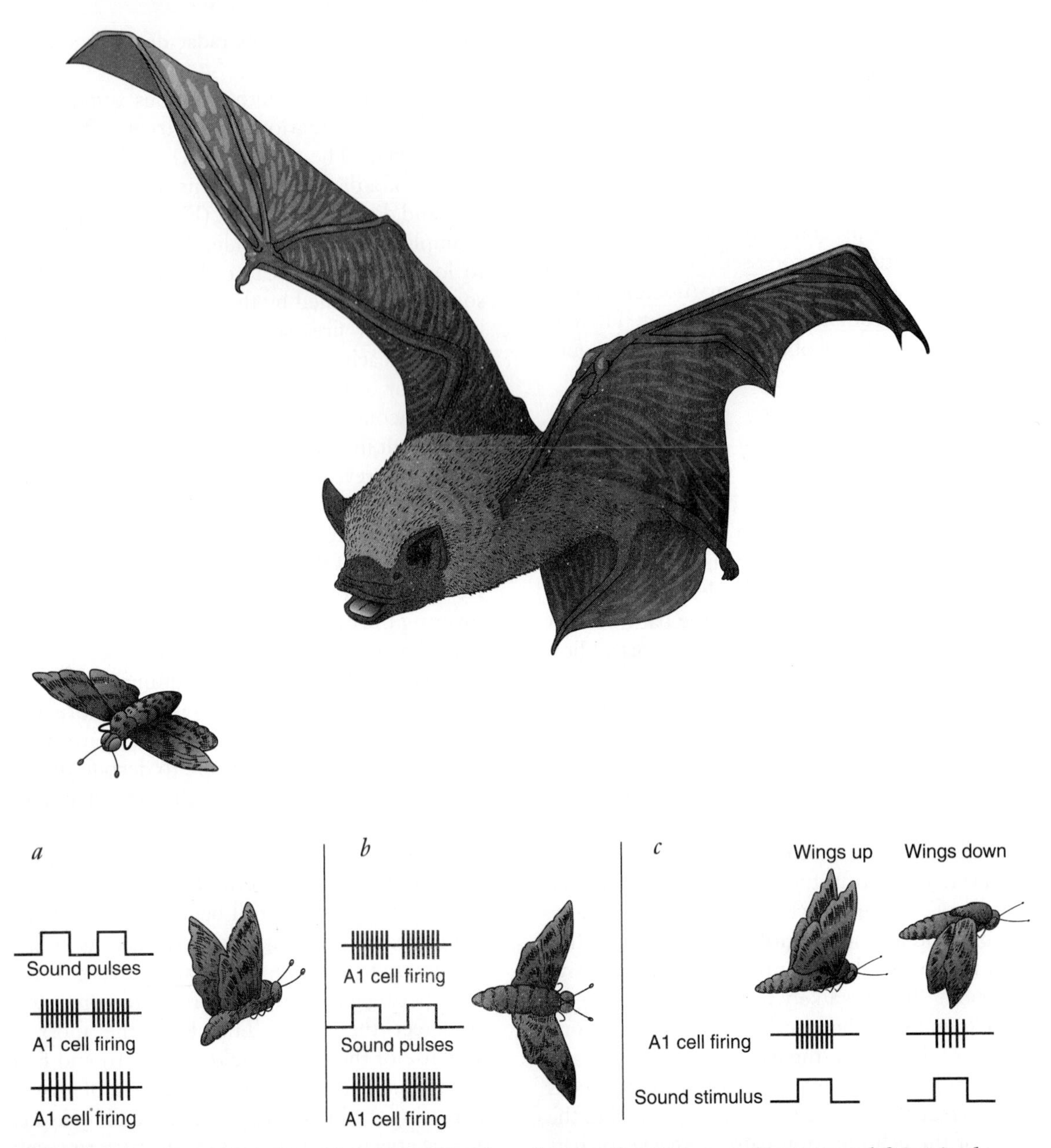

FIGURE 6.10 The relationship of sound pulses from a hunting bat and an auditory neural firing in the hunted moth. (*a*) When a hunting bat, emitting its high-pitched sounds, approaches a noctuid moth from the side, the receptors on that side fire slightly sooner and more rapidly than those on the shielded side. (*b*) When the bat is behind the moth, the moth's receptors on both sides fire with a similar rapid pattern. (*c*) When the bat is above the moth, the moth's auditory receptors fire when its wings are up but not when its wings cover the receptors on the down stroke. (Redrawn from Alcock 2001.)

until the bat is close by. Based on these differences in threshold, Roeder suggested that the A1 cell functions as an "early warning" cell and the A2 cell as an "emergency" neuron that switches the moth's evasive response to an erratic flight pattern. As reasonable as the hypothesis seems, it is not consistent with the data. One would predict, for instance, that if the activity of the A2 cell was the switch that changes the evasive response from flying directly away to erratic flight, then a moth with only one type of A cell would not switch to erratic flight when the intensity of the bat's call increased. Notodontid moths have a single type of A cell, but they display both types of evasive behavior (Surlykke 1984). Thus in noctuid moths, which have two auditory cells, the A2 cell does begin firing when the bat is nearby, but this activity may not be responsible for the change to erratic flight.

Another clue to the bat's proximity is provided by the type of echolocation sounds the bat produces because these change during the hunt. While the bat is searching for prey, its pulses are relatively long (about 10 ms) and are repeated slowly (about 10 per second). When prey has been detected, the bat switches to the approach phase of the hunt. The sound pulses get shorter (about 5 ms) and are repeated more rapidly (about 20 per second). In the final approach, which begins when the bat is within a meter of its prey, the bat begins a feeding buzz, consisting of short pulses (0.5 to 2 ms) repeated rapidly (100 to 200 per second) (Boyan and Miller 1991).

The response of the A1 cell can follow the bat's call rates at all phases of the hunt up to about 150 ms before the bat would capture the moth (Fullard et al. 2003). Since the call rate changes as the bat gets closer to its prey, the output of the A1 cell provides information about the distance of the bat. The A1 cell sends this information directly to two interneurons, called 501 and 504. These interneurons respond differently to the same input from A1. The differences in interneuron responses somehow encode information about the distance of the bat and direct the appropriate escape response (Boyan and Fullard 1986; Boyan and Miller 1991).

Prey Localization by Barn Owls

Silently and suddenly, a barn owl (*Tyto alba*) sweeps from the sky to strike its prey with astonishing accuracy (Figure 6.11). How does it find its prey? Although in nature the barn owl's keen night vision is important in locating prey, the sounds of a scurrying mouse are sufficient for the owl to strike with deadly precision. Laboratory tests have revealed that birds such as the barn owl are able to locate the source of sounds within 1° or 2° in both the horizontal and vertical planes (1° is approximately the width of your little finger held at arm's length). Because of its astounding ability to detect and

FIGURE 6.11 A hunting barn owl. A barn owl can locate its prey by using sound cues alone.

locate the source of sound, this nocturnal predator can pinpoint its prey by the rustlings the prey makes, and it can precisely determine not only the prey's location along the ground but also its own angle of elevation above the prey.

How do we know that the hunting owl uses the prey's sound? For one thing, we know that barn owls can catch a mouse in a completely darkened room (Payne 1962). In experiments, a barn owl was able to capture a skittering leaf pulled along the floor by a string in a dark room (indicating that sight and smell are not involved), and if unable to see, it will leap into the middle of an expensive loudspeaker from which mouse sounds emanate.

To locate its prey by using sound cues, the barn owl must place the source of the sound on a horizontal plane from left to right (i.e., its azimuth), as well as on a vertical plane (i.e., its elevation). We now know that a barn owl uses different cues for locating sound cues in horizontal and vertical planes.

The owl uses time differences in the arrival of sound in each ear to place it on a horizontal plane and differences in intensity between the two ears to determine the elevation of the sound source (reviewed in Konishi 2003). Masakazu (Mark) Konishi (1993a, b) learned this by playing sound in a barn owl's ear through small earphones. An owl turns its entire head to face the direction from which it perceives the sound source, because its eyes are fixed in their sockets. When the sound in one ear preceded that in the other, the owl turned its head in the direction of the leading ear. The longer the time difference, the further the owl turned its head.

The intensity differences in the two ears vary with the elevation of the sound source largely because of the arrangement of the ear canals and facial feathers (von Campenhausen and Wagner 2006). The two ear canals that channel the sound toward the inner ears are, oddly enough, situated asymmetrically, with the right one higher than the left. Because of this difference in ear placement, each ear responds differently to a sound at a given elevation. This helps the owl determine its own elevation above the sound source, information critical to an aerial predator. Also, the face of the barn owl is composed of rows of densely packed feathers, called the facial ruff, that act as a focusing apparatus for sound (Figure 6.12). Troughs in the facial ruff, like a hand cupped behind the ear, both amplify the sound and make the ear more sensitive to sound from certain directions.

The facial ruff assists the owl in localizing sounds by creating differences in intensity of the sound in both ears. Loudness is a cue to localizing the sound in both the horizontal and the vertical dimensions. Sound is generally louder in the ear closer to the source. Because of the structure of the facial ruff, the left ear collects

FIGURE 6.12 The barn owl, a night hunter, has a facial disc of densely packed feathers that may gather sounds and aid in detecting their source.

low-frequency sounds primarily from the left side, and the right ear collects low-frequency sounds from the right side. A comparison of the intensity of low-frequency sounds in each ear helps the owl determine from which side of the head the sound originates. However, the facial ruff channels high-frequency sound to each ear differently, depending on the elevation of the sound source. As a result, the right ear is more sensitive to high-frequency sounds that originate above the head, and the left ear is more sensitive to high-frequency sounds from below the head. The owl compares the loudness of high-frequency sounds in each ear to determine its position above or below the sound source. As a sound source moves upward from below the bird to a position above the owl's head, the high-frequency sounds would first be loudest in the left ear and then gradually become louder in the right ear (Knudsen 1981).

Information on the timing and loudness of sounds in each ear is then sent to the owl's central nervous system over the auditory nerve in a pattern of nerve impulses. The information is first sent to the cochlear nuclei. Each side of the brain (cerebral hemisphere) has two cochlear nuclei: the magnocellular nucleus and the angular nucleus. Every axon in the auditory nerve sends a branch to both of these nuclei. Whereas the branch of the auditory nerve that goes to the magnocellular nuclei conveys timing information, the branch to the angular nuclei transmits intensity information. Thus, the timing data that place the sound on a horizontal plane are processed separately from the intensity data that place the sound on a vertical plane. These different features of sound are processed in parallel along nearly independent pathways to higher processing stations, where a map of auditory space is eventually formed (Konishi 1993b).

The map of auditory space is formed in the external nucleus of the inferior colliculus of the midbrain. Within the inferior colliculus are certain neurons that respond selectively to specific degrees of binaural differences in sound (Figure 6.13). For example, one neuron may respond maximally to differences that correspond to a sound originating 30° to the right of the owl. The sound would arrive a certain amount sooner and be a certain degree louder in the right ear than in

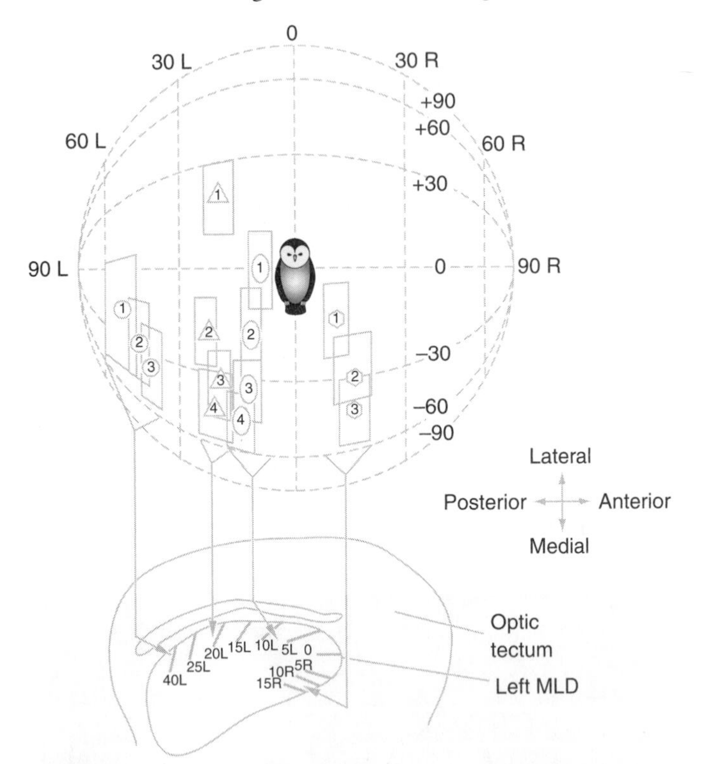

FIGURE 6.13 Auditory neurons in the midbrain area (the inferior colliculus) of a barn owl. The top figure shows a hemisphere of space in front of the owl's head. Neurons in the inferior colliculus respond to sounds that originate at different points. The numbered rectangles indicate 14 areas to which specific inferior colliculus neurons are tuned. The lower figure indicates the manner in which the auditory space of the owl is represented in the inferior colliculus. A horizontal section of the inferior colliculus is shown with bars, indicating the position of specific neurons. The point in space to which that neuron responds is indicated. Notice that the neurons in the inferior colliculus are spatially organized. (From Knudsen and Konishi 1978.)

FIGURE 6.14 **A young owl wears goggles that shift its visual field about 10° to the left. When reaching toward a visual target, the bird first misses the target by about 10° to the left but gradually adjusts. The owl's auditory spatial map is calibrated to its visual spatial map, so once it adjusts to the goggles, sounds from a particular location are also excited by visual stimuli presented 10° to the left of the target.**

the left. Those exact differences in timing and loudness stimulate that particular neuron in the inferior colliculus. The degree of binaural difference varies with the location of the sound, and the binaural difference that stimulates cells of the inferior colliculus varies from neuron to neuron (Knudsen 1982). The resulting auditory space map is then transmitted to the optic tectum (an area of the brain involved in localizing and orienting to visual information).

Auditory maps can be formed without visual input, but the precision of the map is increased by visual experience. Indeed, when the information from the auditory map conflicts with visual input, owls trust their vision over their hearing. Normally, the auditory space map in the inferior colliculus and the visual space map in the optic tectum are aligned. When the two maps are misaligned by blocking one ear, the owl initially mislocates sound in the direction of the open ear. After many weeks experience with an earplug, a young owl learns new associations between auditory and visual cues and orients itself correctly. The auditory and visual space maps also become misaligned when an owl wears goggles that shift the visual field by 10°. A young owl gradually adjusts the location of sound localization to match its distorted visual map (reviewed in Knudsen 2002) (Figure 6.14).

PROCESSING IN THE CENTRAL NERVOUS SYSTEM

The brain is a dynamic structure that changes with experience. Brain regions operate in networks. For example, there is a social behavior network consisting of six brain regions. The pattern of activity among the brain regions in this network determines social behavior.

BRAIN CHANGES UNDERLYING BEHAVIORAL CHANGE

When behavior changes because of the animal's experience, we usually say that the animal has learned something. However, we cannot be sure that learning has occurred unless we can elicit a memory. There are several stages of memory formation. **Short-term memories** can be retrieved for minutes, hours, or perhaps a day. Short-term memories may then be strengthened to form **intermediate-term memories**, which may last a day or so. We will discuss habituation and sensitization in the sea slug *Aplysia* as examples of intermediate-term memory formation. **Long-term memories** last weeks, months, or years.

What are memories made of? A philosopher might have a poetic answer, but from our point of view, memories are made from synaptic modifications—changes in the strength or number of synapses. Different molecular mechanisms are responsible for the formation of memories of different durations. Nonetheless, this statement holds true for all memory formation: Following an experience, some synapses will get stronger, and others will get weaker. The pattern of the changes in synapses throughout the nervous system underlies memories (Beer 1999). Possible changes that leap to mind generally include functional changes, such as an increase or decrease in the amount of neurotransmitter released at a synapse and changes in the number of receptors on the postsynaptic membrane that alter the responsiveness of the postsynaptic cell. Structural changes in neurons could alter the number of synapses that a particular neuron makes. The growth of new neurons may also underlie learning and memory (Bruel-Jungerman et al. 2007).

Intermediate-Term Learning in *Aplysia*

Let's consider the molecular events that occur at synapses when behavior can change with experience with a single stimulus (nonassociative learning), as occurs when the sea hare *Aplysia* learns. While *Aplysia* moves across the ocean bottom eating seaweed, its siphon is extended, and its gills, the respiratory organs, are spread out on the dorsal side. The gills are partly covered by a protective sheet called the mantle shelf, which terminates in the siphon, a fleshy spout through which *Aplysia* can squirt out excess seawater and wastes. When the siphon is touched, the gills, the siphon, and the mantle shelf withdraw into the mantle cavity. This defensive response, called the gill-withdrawal reflex, protects the gills from predators (Figure 6.15).

The gill-withdrawal response can be modified by experience, that is, through learning. One form of

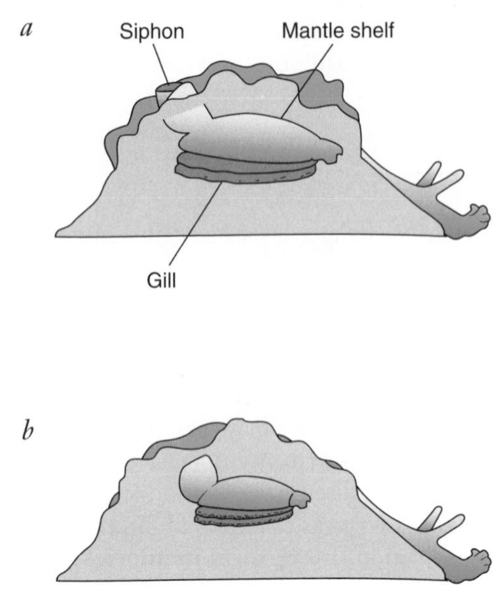

FIGURE 6.15 **The gill-withdrawal reflex in the sea hare, *Aplysia*. The gills, mantle shelf, and siphon are drawn here as if the animal were transparent. Normally, the gills are spread out and are only partially protected by the mantle shelf. (*a*) The siphon, through which water is drawn in over the gills and excess water is expelled, is extended so that just the tip is visible when the animal is seen from the side. (*b*) If the siphon is touched, the gills, mantle shelf, and siphon are withdrawn into the mantle cavity. The gill-withdrawal reflex can be modified by learning.**

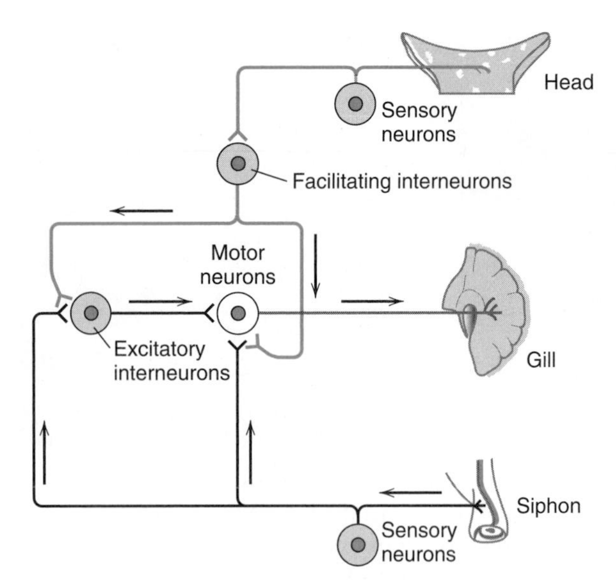

FIGURE 6.16 **Neural circuitry for habituation and sensitization of the gill-withdrawal reflex in *Aplysia*.**

learning, called **habituation**, occurs when an animal learns not to respond to a repeated stimulus that proves to be harmless (also discussed in Chapter 5). In other words, *Aplysia* learns to ignore an irrelevant stimulus. Habituation is adaptive because it saves energy. We can demonstrate habituation by disturbing an *Aplysia*'s siphon many times, by touch or by a brief jet of seawater, for instance. After 15 such stimuli administered ten minutes apart, the reflexive responses are only half of their initial value. In **sensitization**, a second form of learning in *Aplysia*, the withdrawal reflex becomes stronger when a stimulus that elicits withdrawal is preceded by a strong, noxious stimulus, such as an electric shock, administered almost anywhere on *Aplysia*'s surface. Depending on the number and strength of the noxious stimuli, sensitization can last seconds to days (Carew et al. 1972; Kandel 1979; Pinsker et al. 1970).

Through years of intensive research, Eric Kandel and other investigators have determined the neural circuits and many of the molecular mechanisms that underlie these forms of learning in *Aplysia* (Kandel 1976; Kandel 1979a, b). *Aplysia* is an ideal organism for a neurobiologist because it has just 20,000 neurons and the neurons are large—10 to 50 times larger than neurons in a mammalian brain. As a result, neurobiologists know many of these nerve cells by name. A diagram of the neural circuitry for the gill-withdrawal reflex is shown in Figure 6.16. Although there are actually 24 sensory neurons serving the siphon skin that terminate on 6 motor neurons to the muscle for gill retraction, this simplified diagram shows only one of each type of neuron. You can see that the sensory neuron from the siphon skin synapses directly on a motor neuron for gill withdrawal.

Beginning in the late 1960s, neurobiologists began to work out the details of the neural mechanism of habituation. One hypothesis was that the sensory neuron was giving a weaker response to repeated stimuli. This idea was shown to be incorrect by inserting a microelectrode into a sensory neuron to measure its electrical responses to stimulation. After repeated stimulation, the sensory neuron still responds normally. However, it fails to excite the motor neuron as it initially did. Could habituation be due to a decrease in the motor neuron's responsiveness? No. If a motor neuron is repeatedly stimulated directly with an electrical current, it remains fully responsive. Another guess was muscle fatigue, but this, too, was ruled out. Even after habituation, direct stimulation of the motor neuron causes full contraction of the gill muscle (Kupfermann et al. 1970).

It turns out that habituation occurs because the sensory neuron releases less neurotransmitter because of repeated stimulation (Figure 6.17). This, in turn, results in fewer action potentials in the motor neuron for gill withdrawal. Recall that many molecules of neurotransmitter are released from a synaptic vesicle. This packet of many molecules of neurotransmitter is called a quantum. The change in the EPSP caused by the release of neurotransmitter from one synaptic vesicle is called a quantum of response. During habituation, the EPSP decreases with repeated stimulation of the sensory neuron, and it does so in integral multiples of a quantum of

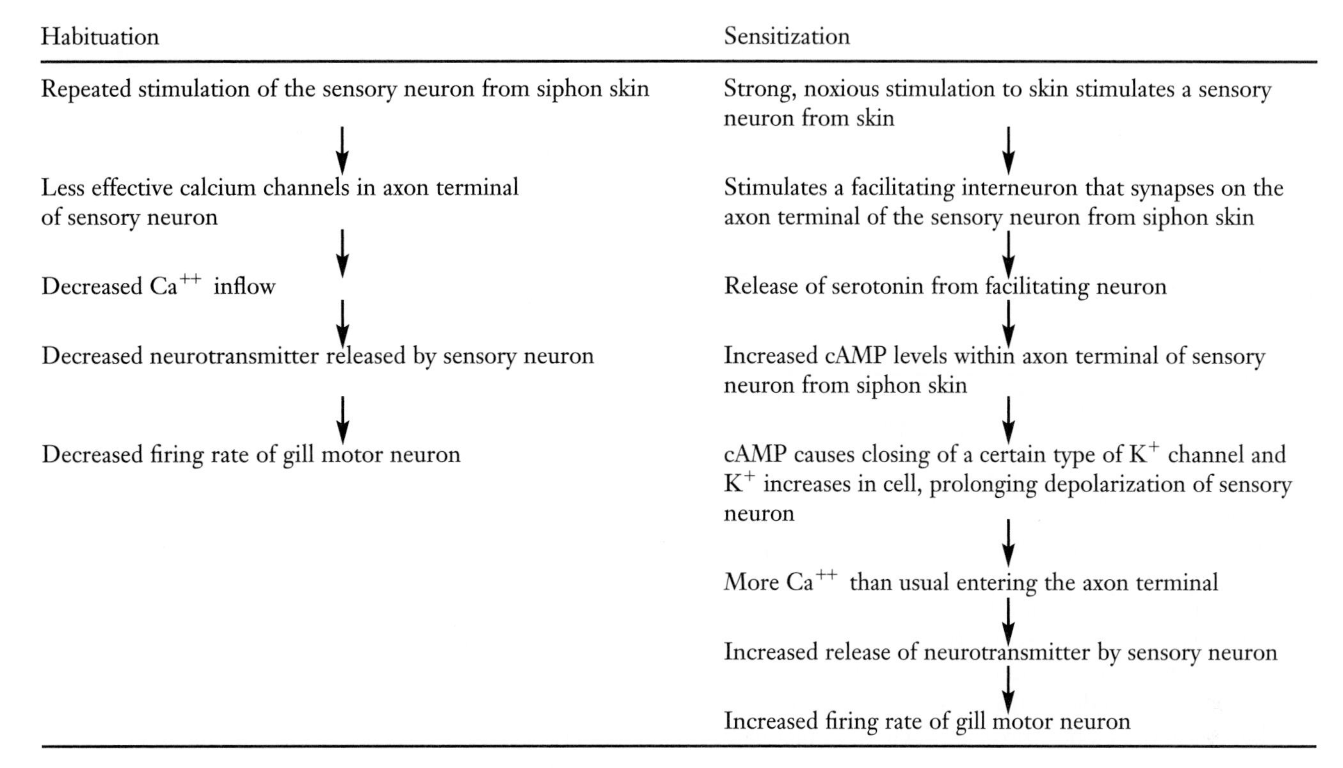

Habituation
Sensitization
Repeated stimulation of the sensory neuron from siphon skin
Less effective calcium channels in axon terminal of sensory neuron
Decreased Ca^{++} inflow
Decreased neurotransmitter released by sensory neuron
Decreased firing rate of gill motor neuron
Strong, noxious stimulation to skin stimulates a sensory neuron from skin
Stimulates a facilitating interneuron that synapses on the axon terminal of the sensory neuron from siphon skin
Release of serotonin from facilitating neuron
Increased cAMP levels within axon terminal of sensory neuron from siphon skin
cAMP causes closing of a certain type of K^+ channel and K^+ increases in cell, prolonging depolarization of sensory neuron
More Ca^{++} than usual entering the axon terminal
Increased release of neurotransmitter by sensory neuron
Increased firing rate of gill motor neuron

FIGURE 6.17 **Changes in synaptic functioning that accompany habituation and sensitization.**

response (Castellucci and Kandel 1974). In other words, during habituation, successive action potentials cause increasingly fewer synaptic vesicles to fuse with the membrane and release their contents into the synaptic cleft. The reason is that calcium channels become less effective because of repeated stimulation, and so they allow less Ca^{++} into the axon terminal (Byrne 1987; Hochner 1986). Calcium ions are needed for the synaptic vesicles to fuse with the presynaptic membrane.

Sensitization also involves changes in the functioning of synapses, but in this case, the amount of neurotransmitter released by the sensory neuron from the siphon onto the gill motor neuron is *increased*, thereby increasing the motor neuron's rate of firing. Sensitization requires a facilitating interneuron. The process begins when strong stimulation to the body surface of *Aplysia* stimulates a sensory neuron, which in turn stimulates a facilitating interneuron. These interneurons release serotonin onto the axon terminal of the sensory neuron from the siphon skin. Serotonin increases intracellular concentration of a second messenger, in this case cyclic adenosine monophosphate (cAMP), which causes the closing of a certain type of K^+ channel of the cell membrane[1]. This keeps K^+ inside the cell and keeps the sensory neuron depolarized longer than normal, allowing additional Ca^{++} to enter the cell. The elevated Ca^{++} levels cause more neurotransmitter to be released (Kandel and Schwartz 1982).

Long-Term Memory Formation

We look in the fridge and notice we need milk. The next afternoon, we stop at the store to pick up a carton. How is this memory stored? We now know quite a bit about the molecular basis of memory. Memory involves changes in synaptic connections, which involves four processes: long-term potentiation, long-term depression, synaptic remodeling, and neurogenesis (Bruel-Jungerman et al. 2007). These mechanisms alter the strength of existing synapses, add new synapses, or remove old synapses. We will consider each mechanism, in turn.

Long-Term Potentiation Repeated stimuli, as might occur when an animal is learning a task, result in **long-term potentiation**, or **LTP**, which strengthens the connections between the adjacent neurons. LTP occurs at many synapses, possibly every excitatory synapse, in the mammalian brain. It is the molecular mechanism that underlies the acquisition and storage of memories (Malenka and Bear 2004). In the laboratory, we can simulate the events that occur at a synapse during learning by electrically stimulating a presynaptic neuron. If the neuron is stimulated once a second, the response of the postsynaptic cell to each stimulus remains the same. However, if the neuron is stimulated rapidly and

[1]At some synapses, such as this one, the neurotransmitter opens or closes ion channels through indirect biochemical mechanisms. These indirect mechanisms are similar to those used by neuromodulators.

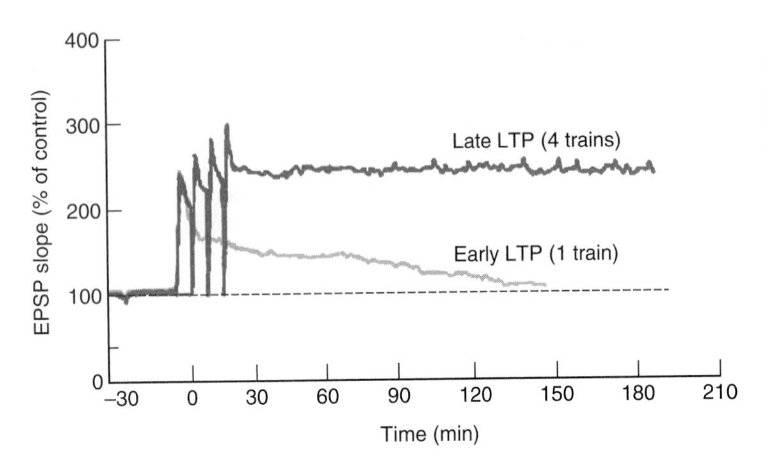

FIGURE 6.18 **A rapid train of stimuli enhances the responsiveness of the postsynaptic cell. This phenomenon, called long-term potentiation, is induced by learning and underlies memory formation. (From Kandel 2001.)**

repeatedly for a second or two, the response of the postsynaptic neuron is elevated and remains elevated (potentiated) for minutes or hours (Figure 6.18).

Here is how LTP works (Figure 6.19). LTP is believed to involve two proteins embedded in the membrane of nerve cells, called AMPA receptors[2] and NMDA[3] receptors. These proteins open molecular gates that allow ions to pass in and out of the nerve cell, similar to the sodium and potassium channels we discussed earlier. In the early stage of LTP, while an animal is learning to associate a stimulus with a response, the stimulus causes the presynaptic neuron to release a neurotransmitter called glutamate into the synapse. Glutamate binds to both AMPA and NMDA receptors, but at this point NMDA receptors are blocked by magnesium ions and cannot respond to glutamate. When glutamate binds to AMPA receptors, sodium channels open. Sodium ions enter the nerve cell, and the postsynaptic cell becomes depolarized. If depolarization reaches threshold, an action potential is generated in the postsynaptic cell. Because of the depolarization, the magnesium ions blocking the NMDA channel are driven out. The unblocked NMDA receptors can now respond to the glutamate, and calcium ions enter along with additional sodium ions. The calcium ions activate several proteins within the postsynaptic neuron. Some of these activated proteins further increase the sensitivity of existing AMPA receptors, and others cause additional AMPA receptors to move to the postsynaptic membrane. Thus, the cell becomes even more responsive to glutamate. As we will see shortly, during the late phase of LTP, another protein in the postsynaptic cell, CREB, is activated and turns on certain genes, resulting in long-lasting structural changes in the connections between neurons.

Does LTP underlie long-term memory formation? They do have common features. Both can be induced within seconds and last for days or weeks. Furthermore, long-term memory formation and the later stages of LTP require protein synthesis. However, these features could be coincidental.

Jonathan Whitlock and colleagues (2006) tested the hypothesis that learning induces LTP, which underlies long-term memory formation. They trained rats to associate a painful stimulus with a location. Rats prefer dark areas to light areas but will avoid a dark area where they have received a foot shock. Whitlock and colleagues reasoned that if LTP underlies learning, then the responsiveness of neurons in a region of the hippocampus (CA1) known to be essential for spatial learning would be increased by training. They measured the responsiveness of neurons in the area of the hippocampus before and after training. We would not expect that *all* the synapses in this brain region would be altered by training. Whitlock's team used a biochemical marker to identify the synapses that had been altered by learning. When they looked at these synapses specifically, they found enhanced responsiveness, which indicates LTP. The altered synapses occurred in about 25% of the neurons in the CA1 area.

Long-Term Depression When we learn something new, the pattern of synaptic excitability of many neurons changes. Some synapses become stronger, and others become weaker. **Long-term depression** is a mechanism that weakens the effectiveness of a synapse, decreasing the magnitude of a response by the postsynaptic cell. It occurs after a slow train of stimuli (in contrast to the rapid train of stimuli that causes LTP) has activated the presynaptic cell. Thus, long-term depression may also play a role in memory formation.

Synaptic Remodeling Long-term memory requires structural changes in synapses. Several kinds of structural changes in the brain are reviewed by Julie Markham and William Greenough (2004). One change is in dendritic spines—short extensions on dendrites that form half of a synapse. Changes in the size, shape,

[2]A glutamate receptor that also binds AMPA (α-amino-3-hydroxy-5-methyl-4-isoxazolepropionic acid).
[3]A glutamate receptor that also binds NMDA (N-methyl-D-aspartate).

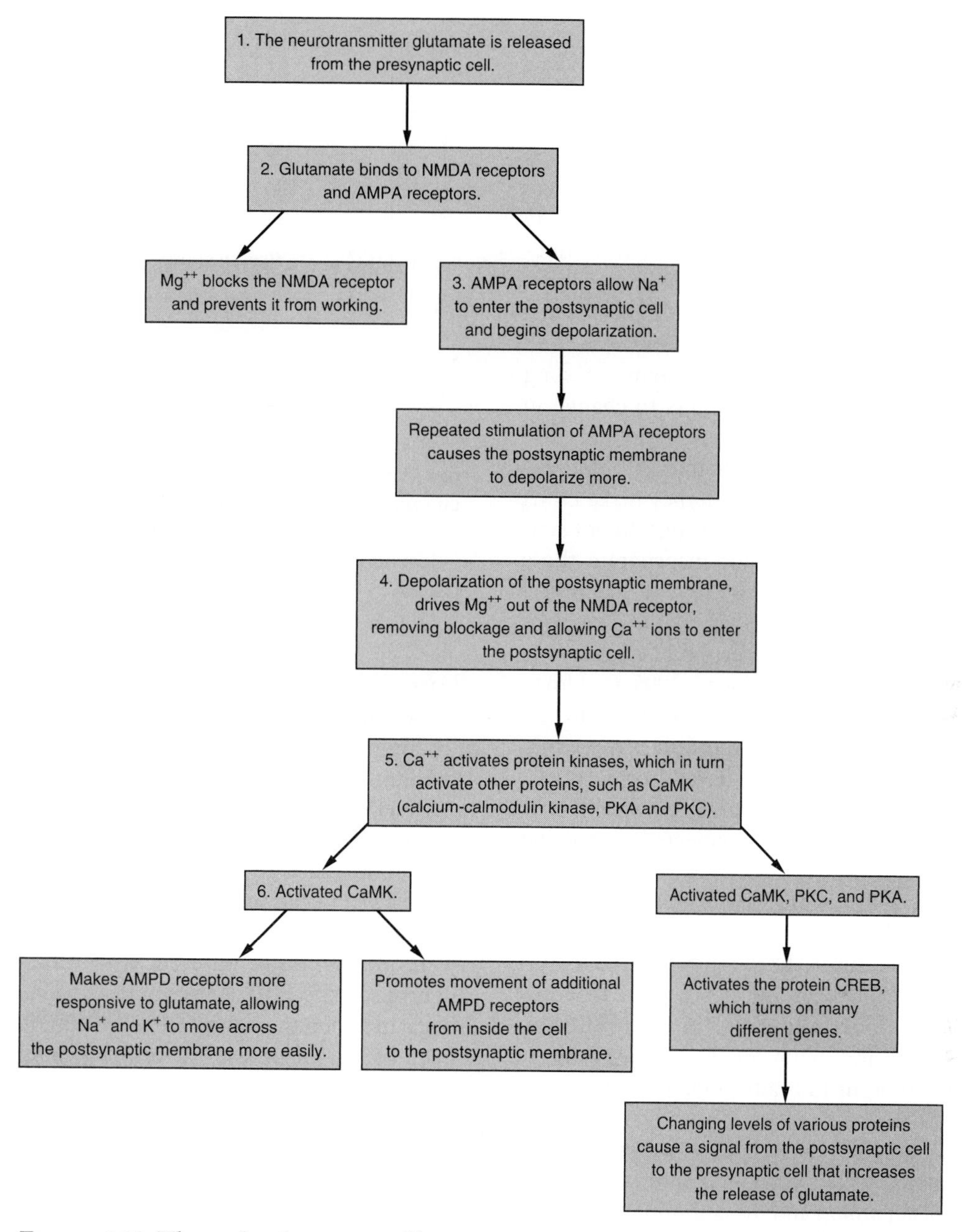

FIGURE 6.19 **The molecular events of long-term potentiation. These changes lead to a long-lasting increase in the responsiveness of the synapse, making communication between the neurons easier.**

and number of dendritic spines enhance communication between neurons, thereby reinforcing a particular neural pathway. Other structural changes involve glial cells, which produce growth factors that guide axons to form synapses in the correct destinations and influence communication at synapses by removing neurotransmitters from the synapse.

Structural changes such as these require protein synthesis. Recall that during long-term potentiation, calcium ions enter the postsynaptic cell and activate certain proteins. CREB is among these. One of the functions of CREB is to turn on sets of genes that synthesize additional proteins. These proteins then bring about long-lasting structural changes in the synapses.

Neurogenesis At one time, we believed that the growth of new neurons was limited to development and that no new neurons formed in adulthood. This belief implied that adult learning and memory were due only to changes in synapses, as described above. However, recent evidence shows that the birth of new neurons, or **neurogenesis**, can occur even in older animals.

Song Learning in Birds Fernando Nottebohm was one of the first to suggest that brain changes that occur with behavioral changes might include the birth of new neurons, along with the death of old neurons. Nottebohm studied the neurobiology of bird songs (2005). As we will see in Chapter 8, discrete but interconnected centers in the songbird's brain, collectively called the song system, make the learning and production of song possible. Nottebohm (1981) discovered that the volume of two song centers, the HVC and the RA (robust nucleus of the arcopallium), increases and decreases seasonally, along with the amount of song learning. What causes these song centers to change in size? Researchers can test whether new neurons are added to an area by adding marker chemicals. New neurons incorporate these chemicals when they make DNA in preparation for cell division; old neurons do not synthesize DNA and therefore do not incorporate these markers. As it turns out, the increase in the size of the RA is caused by an increase in both the size of neurons and the space between them, but in the HVC it is due to the formation of new neurons (Sherry 2005; Goldman and Nottebohm 1983). The neurons are born in the lateral ventricles of the forebrain and differentiate into specific types of neurons as they migrate to the HVC. Most of the new neurons die within a few weeks. The survivors are those that become connected to existing neural networks, and they replace cells that died. The number of neurons dying in the HVC peaks in late summer, and the rate at which new neurons are added to the HVC peaks in the autumn. Through the fall and winter, the rate of neuron birth and death must be equal because the total number of neurons in the HVC does not increase until spring. Thus, neurogenesis regularly replaces neurons in a bird's brain (Sherry 2005; Strand and Deviche 2007).

How does neurogenesis relate to song learning? Most of the time, song learning by young male zebra finches (*Taeniopygia guttata*) and the addition of new neurons to the HVC are correlated. When reared with adult males, young zebra finches learn their song by imitating the songs of their adult companions from age 30 through 65 days, eventually perfecting their stereotyped song between days 65 and 90. As you can see in the control group of Figure 6.20, neurogenesis is highest as song learning begins, declines as the song is learned, and is lowest after 60 days, around the time that the song stabilizes (Wilbrecht et al. 2002).

We see, then, that a correlation exists between the timing of the additional new neurons to the HVC and song learning. But which event causes the other? Do the new neurons allow song learning to take place, or does song learning cause the addition of new neurons to the HVC? Researchers have explored the question of which comes first by manipulating the opportunity for birds to

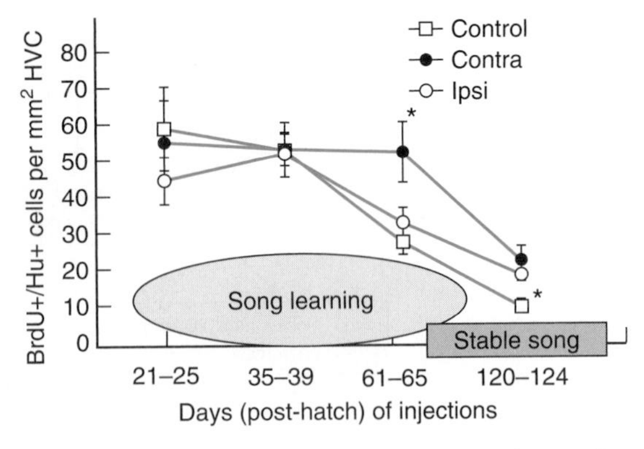

FIGURE 6.20 **The addition of new neurons to one of the song control centers (HVC) of male zebra finches changes during the course of song learning. It is high as song learning begins, and it slowly declines as the song is learned. It is lowest after a stable song has developed. (From Wilbrecht et al. 2002.)**

learn their songs. They predicted that if song learning causes the addition of new neurons to the HVC, then limiting the opportunity for song learning will reduce the number of neurons added.

Let's consider how the opportunity for song learning has been manipulated in male zebra finches. One can deafen the males toward the beginning of song learning (26 days) and can prevent them from singing by cutting the nerves to the syrinx (vocal organ of birds). There were no significant differences in the number of new neurons in the HVC between hearing and deaf birds or birds in which the nerves to both sides of the syrinx had been cut. Compared with controls, there were twice as many new neurons in the HVC of birds in which the nerve on one side of the syrinx had been cut (Figure 6.20). The total number of neurons in the HVC of all adult birds was not affected by any procedures. Thus, song learning doesn't alter the total number of cells in the adult HVC, but it may affect the addition of new neurons during a limited amount of time and special circumstances, such as having the nerve to the syrinx cut on only one side (Wilbrecht et al. 2002).

A second approach to manipulating the opportunity for song learning is to extend the sensitive period (a developmental stage when key experiences alter behavior more than at other ages) during which a young male imitates and perfects his song to match the songs of adult males. The sensitive period is extended when young males are raised alone without an opportunity to see or hear adult males. Males raised in isolation improvise a song that, even after day 80, they can modify to match the song of an adult male if one becomes available. If learning causes the addition of neurons to the HVC, then we predict that extending the sensitive period

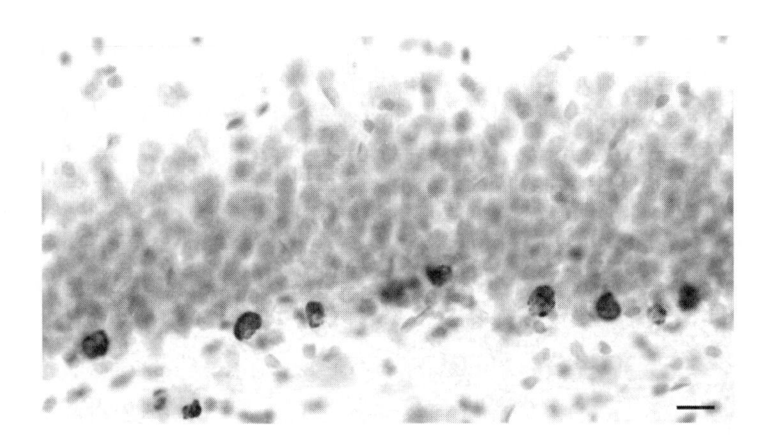

FIGURE 6.21 **A section through the dentate gyrus of the hippocampus showing new neurons. The new neurons are darker in color because they are labeled with BrdU, which is taken into cells that are synthesizing DNA in preparation for cell division.**

would increase the number of new neurons in the HVC. This is indeed the case. Isolated males added about 1.6 times more neurons than control males added during posthatching age 65 to 150 days (Wilbrecht et al. 2006).

The alternative hypothesis—that new neurons are needed for song learning—appears to be correct. The current hypothesis is that seasonal changes in photoperiod cause changes in blood testosterone levels in male songbirds. As the days get longer during the spring, testosterone levels increase. Testosterone then causes an increase in singing, which increases the production of brain-derived neurotrophic factor, a protein that enhances the survival of new neurons in the HVC. In this way, older, underused neurons will be replaced by new, heavily used neurons (Nottebohm 2002). This hypothesis is supported by data from male house finches (*Carpodacus mexianus*) (Strand and Deviche 2007) and male Gambel's white-crowned sparrows (*Zonotrichia leucophrys*) (Brenowitz et al. 2007).

Spatial Learning in Mammals As in birds, we also see the appearance of new neurons and the death of old ones. However, in mammals neurogenesis occurs only in a region of the hippocampus called the dentate gyrus and the olfactory bulb (Figure 6.21).

What is the role of neurogenesis in the hippocampus? Because the hippocampus is especially important in spatial learning, we might expect spatial learning tasks to be more affected by blocking neurogenesis than learning tasks that are independent of the hippocampus. Tracey Shors and colleagues (2001) investigated the effect of a reduction in neuronal cell division on memory formation in rats. They treated rats with saline or with a chemical that suppresses cell division (methylazoxymethanol acetate; MAM) and found that the production of new neurons is important in memory formation that requires the hippocampus (Figure 6.22).

David Dupret and his colleagues (2007) concluded that cell death is as important as the birth of new neurons during spatial learning. A common test of spatial learning in mice and rats is the water maze. In this test, the animal is placed in a pool of opaque liquid. It must swim until it finds a platform on which it can rest. The platform is hidden beneath the liquid's surface, and so it is not visible to the animal. The animals must learn the location of the platform using visual cues placed around

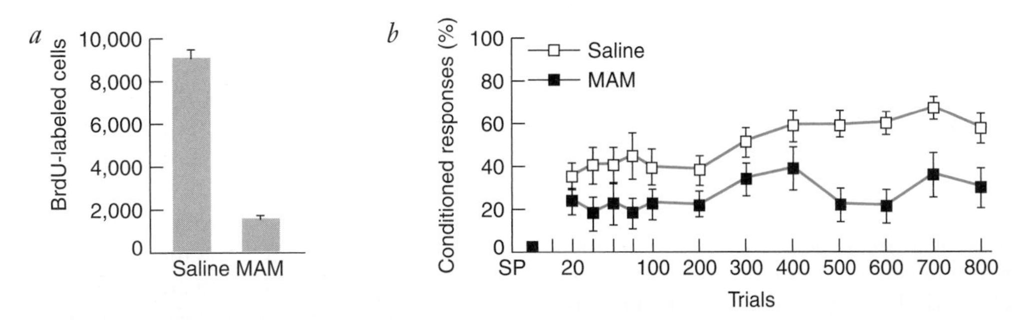

FIGURE 6.22 **Adult neurogenesis is involved in conditioning that depends on the hippocampus. Before training, the animals were treated with saline or with a chemical that suppresses cell division (MAM). New neurons were labeled by administering a chemical that is incorporated into DNA during DNA synthesis. (*a*) The animals treated with MAM produce fewer new neurons and did not learn the task. (*b*) The control animals (treated with saline) continued to produce new neurons, and their performance on the task improved. (From Shors et al. 2001.)**

the water tank. The hidden platform water maze is a form of spatial learning, and it depends on the hippocampus.

When cell death was inhibited by treating the animals before training with a chemical, spatial learning was impaired. The rats could learn the position of the hidden platform within a training session but could not remember it the next day.

Dupret suggests that spatial learning causes a cascade of events similar to the formation of functional neural networks during development. Thousands of new neurons are formed in the hippocampus each day. During the early phase of learning in the hidden-platform water maze, when the animal's performance is improving quickly, the survival of neurons that are about one week old is increased (Figure 6.23; Dupret et al. 2007). After the animal has begun to master the task, additional new neurons—those that have not yet made connections with other neurons—die. Cell death is necessary for the survival of the earlier born neurons and for an increase in neurogenesis, which increases the pool of new neurons.

Changes in the number of new neurons in the hippocampus are also associated with spatial learning during food storing and retrieval in birds. For example,

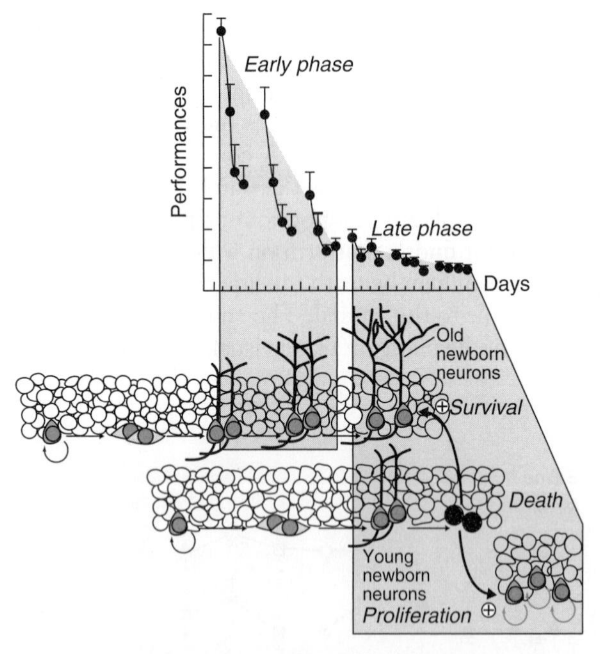

FIGURE 6.23 **Spatial learning depends on the production of new neurons, the selective survival of new neurons, and selective cell death. During the early phase of learning in the hidden-platform water maze, new neurons begin to make synaptic connections with existing neurons. Learning increases the survival of these neurons. In the late phase of learning, slightly younger neurons die by apoptosis. This cell death enhances the survival of the new neurons that have made synaptic connections and the rate of neurogenesis. (From Dupret et al. 2007.)**

hand-reared marsh tits (*Parus palustris*) that were allowed to store and retrieve sunflower seeds had more new neurons in the ventricular zone border of the hippocampus than did age-matched control birds that were not permitted to store food. The birth of new neurons in the hippocampus of black-capped chickadees (*Parus atricapillus*) increases beginning in February and March and peaks in October (Barnea and Nottebohm 1994). It is also at this time of year that food storing is thought to peak, but see Pravosudov (2006) who suggests that there may be an additional peak of food storing in the spring that is not accompanied by an increase in new neurons in the hippocampus.

Does spatial learning increase neurogenesis and alter the anatomy of the hippocampus in humans as well? Licensed London taxi drivers provided a pool of subjects on which to explore this possibility because they must complete two years of training, colloquially known as "being on The Knowledge," to learn to navigate thousands of streets in London. Eleanor Maguire and her colleagues (2000) hypothesized that the anatomy of the hippocampus would be altered by years of navigational training. They compared structural MRI brain scans of 16 healthy, right-handed licensed taxi drivers with scans of 50 right-handed males of similar age but without experience driving taxis. The only structural differences in the brains of men in these two groups were in the regions of the hippocampus. Compared with control males, the posterior hippocampus of taxi drivers was larger. In contrast, the anterior region of the hippocampus was smaller in taxi drivers than in control males. Furthermore, the volume of the posterior hippocampus increased, and the volume of the anterior hippocampus decreased with the amount of time spent as a taxi cab driver (Figure 6.24). These results would be predicted if the posterior hippocampus stores a "mental map" of the city and its volume can expand with the amount of information encoded in this mental map.

SOCIAL BEHAVIOR NETWORK

Now that we have considered changes in the brain that underlie learning, we will explore interactions among different regions of the brain involved in social behavior. Sarah Newman (1999) originally suggested that mammalian social behavior is controlled by six regions (nodes) of the brain: (1) the medial extended amygdala, including the medial bed nucleus and the nucleus of the stria terminalis (BSTm), (2) the lateral septum (LS), (3) the preoptic area (POA), (4) the anterior hypothalamus (AH), (5) the ventromedial hypothalamus (VMH), and (6) the midbrain. These regions of the brain are thought to make up a social behavior network for several reasons. Each of these regions plays a role in multiple social behaviors, including aggression, sexual behavior, social

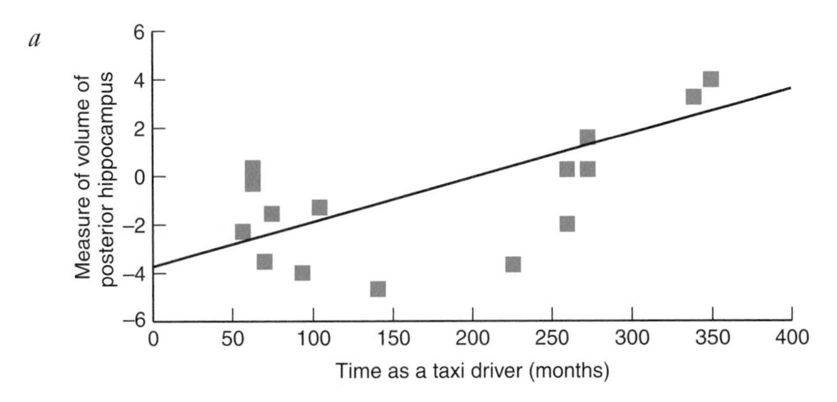

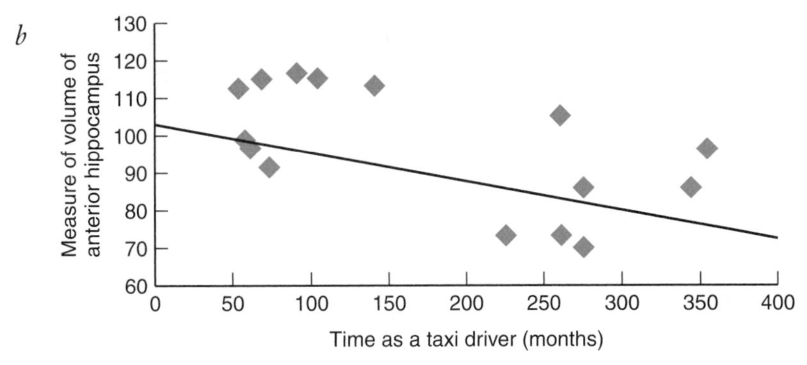

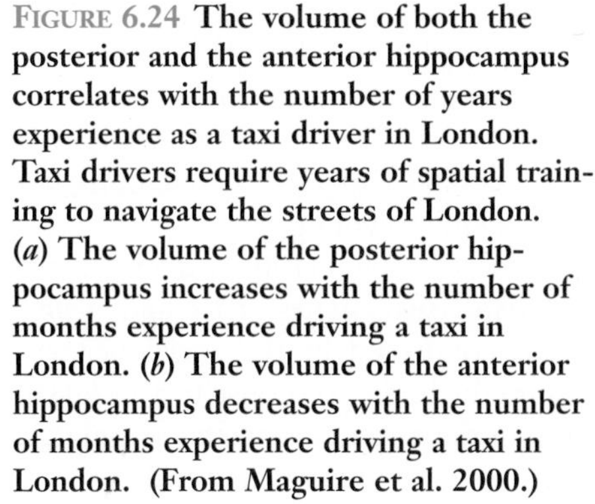

FIGURE 6.24 **The volume of both the posterior and the anterior hippocampus correlates with the number of years experience as a taxi driver in London. Taxi drivers require years of spatial training to navigate the streets of London. (*a*) The volume of the posterior hippocampus increases with the number of months experience driving a taxi in London. (*b*) The volume of the anterior hippocampus decreases with the number of months experience driving a taxi in London. (From Maguire et al. 2000.)**

recognition, affiliation, bonding, parental behavior, and response to stress. We have looked at examples of involvement of these areas in social behavior in Chapter 3, and we will consider other examples in Chapter 7. Interactions between these regions are possible because they are bidirectionally connected. Furthermore, each of these regions contains sex steroid receptors, which are needed for sexual differentiation and coordination of social behavior. Each brain region is responsive to a variety of stimuli. The observed social behavior is not the result of independent responses of brain regions but rather the *pattern* of response across *all* regions (Figure 6.25) (Goodson 2005).

James Goodson proposes that this social behavior network can be found in the brains of all vertebrates; natural selection produces diversity in social behavior by acting on the responses of each region to steroid hormones and neuropeptides. Goodson's work on the midshipman fish (*Porichthys notatus*) demonstrates that this social behavior network was present in the earliest vertebrates. In addition, many of the functional, structural, and neuroendocrine responses of the social behavior network of fish and birds are similar to those of mammals (Goodson 2005; Goodson et al. 2005).

Goodson and his colleagues hypothesized that the pattern of activity across the brain regions forming the social behavior network would vary in species of birds that differ in the typical size of their social group. They identified four species of birds that differ in their degree of sociality. The subjects included four species of estrildid finches, two of which are monogamous, show biparental care, and live in similar habitats. The finches included the zebra finch (*Taeniopygia guttata*) and the spice finch (*Lonchura punctulata*), which live in colonies of from 90 to 300. Two species of waxbills were also

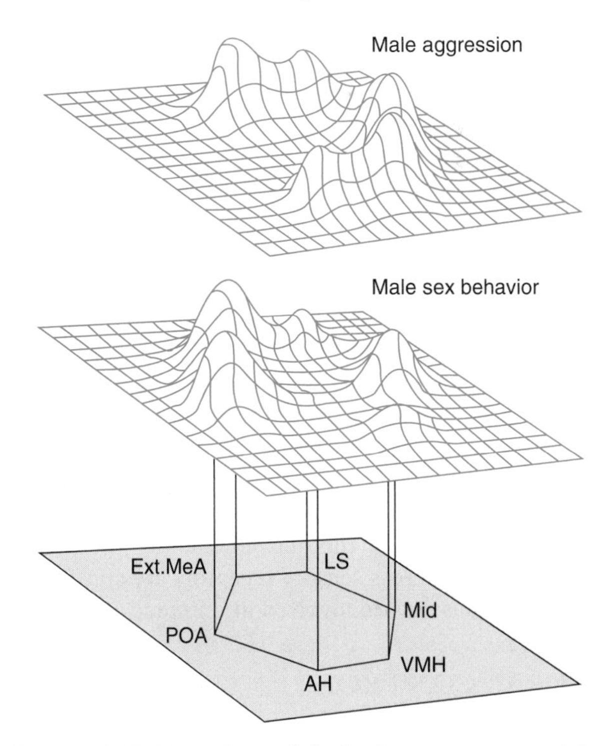

FIGURE 6.25 **Six regions of the brain compose a social behavior network that is active in many types of social behavior. The pattern of activity of these regions varies with the type of behavior. (From Newman 1999.)**

subjects. The waxbills included the moderately gregarious Angolan blue waxbill (*Uraeginthus angolensis*), which lives in groups of from 8 to 40 individuals, and the territorial violet-eared waxbill (*Uraeginthus granatina*), which lives in small groups containing both parents and their dependent young.

Goodson's team (2005) examined the activity of the immediate early genes, *c-fos* and *zenk*, in the brain regions connected in the social behavior network to measure the brain's response to a social stimulus—exposure to a same-sex conspecific. (Immediate early genes signal changes in neural activity.) Indeed, the activity of two brain regions does differ among species of birds that differ in their degree of sociality. The medial extended amygdala functions in social arousal and avoidance. The activity of this region of the brain (as measured by the activity of immediate-early genes) increased from the colonial, gregarious species of finches to the moderately gregarious Angolan blue waxbill to the territorial violet-eared waxbill. There were also changes in activity in brain regions involved in social stress and dominance-related behaviors. We might predict that exposure to a same-sex conspecific bird would be more stressful to a territorial bird than to a colonial bird. This prediction held true when activities in the brain regions involved in stress and dominance were compared. The immediate-early gene responses were clearly higher in territorial species than in colonial species. As you can see in Figure 6.26, the pattern of activity across the brain regions is different in a gregarious species and in a territorial species of waxbills.

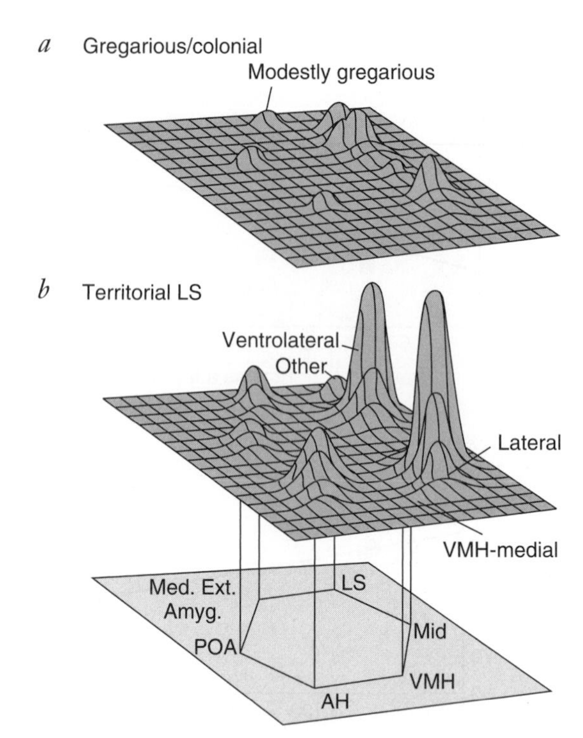

FIGURE 6.26 **The pattern of activity across the brain regions is different in a gregarious species and a territorial species of waxbills. At the base of the figure are each of the six brain regions making up the social behavior network. (*a*) The pattern of activity across the brain regions in a gregarious species, the Angolan blue waxbill (group size 8–40). (*b*) The pattern of activity across the same brain regions in a territorial species, the violet-eared waxbill. (Goodson et al. 2005.)**

RESPONDING—MOTOR SYSTEMS

Animals receive information about their environment via their sensory systems and then respond by means of their motor systems. Two principal components of motor systems are motor organs (typically muscles) and the neural circuits that control them. Like sensory receptors, muscles are often described as biological transducers. There are, however, some important distinctions between the two. Whereas sensory receptors transduce environmental energy such as light into the electrochemical signals of the nervous system, muscles convert the signals of the nervous system into the movements of the body. Furthermore, whereas sensory receptors are concerned with input, muscles are involved in output.

NEURAL CONTROL IN MOTOR SYSTEMS

A particular movement, or behavior, is produced by muscles, whose activity is controlled by motor neurons. Recall that motor neurons, in turn, usually receive their information from interneurons. Thus, each movement is ultimately controlled by the activity of a neural circuit. There are three major ways in which neural circuits control and coordinate movement: (1) the sensory reflex, (2) the central pattern generator, and (3) motor command.

In sensory reflexes, sensory neurons initiate activity in motor neurons, sometimes through direct synaptic connections but more typically through connections with a small number of interneurons. We have already considered the gill-withdrawal reflex in *Aplysia*, so here we will consider the ways in which a central pattern generator and motor command control the locust's flight.

LOCUST FLIGHT

Locusts, those species of "short-horned" grasshoppers found in the family Acrididae, exhibit legendary mass migrations. In fact, accounts of locust plagues date back to the Book of Exodus, written in about 1500 B.C. (Williams 1965). Swarms of locusts have also been recorded in recent times. Although representatives of the family are found throughout the world, the migratory species are found in the tropics or subtropics, typically in the drier regions of these areas. As evidence of the amazing flight behavior of locusts, consider this account

FIGURE 6.27 **A desert locust.**

written by the entomologist C. B. Williams (1965, pp. 77–78) while working in East Africa:

> *A most spectacular flight occurred on 29th January, 1929, at Amani, in the Usambra Hills in north-eastern Tanganyika. . . . We received a telephone warning shortly after breakfast that an immense swarm of locusts was passing in our direction over an estate about six miles to the north. . . . An hour or so later the first outfliers began to appear—gigantic grasshoppers about six inches across the wings, and of a deep purple-brown. Minute by minute the numbers increased, like a brown mist over the tops of the trees. When they settled they changed the colour of the forest; by the weight of their numbers they broke branches of trees up to three inches in diameter; the noise of their slipping up and down on the corrugated iron roofs of the houses made conversation difficult. . . . The swarm was over a mile wide, over a hundred feet deep, and passed for nine hours at a speed of about six miles per hour.*

The species of locusts described in Williams's account was the desert locust, *Schistocerca gregaria* (Figure 6.27). Although this species has been used in some laboratory studies of flight behavior, the migratory locust (*Locusta migratoria*) is more commonly studied. Let us now look at the neural control of flight behavior in these remarkable insects.

The Motor Pattern of Locust Flight

Locusts have two pairs of wings: the forewings, located on the second thoracic segment, and the hind wings, located on the third thoracic segment. The wings of free-flying locusts move up and down, in a rhythmic manner, about 20 times per second. Because locusts maintained in the laboratory and tethered to a holder exhibit close to normal flight, it is possible to obtain a detailed analysis of the motor pattern (Figure 6.28). During flight, the two pairs of wings do not move in a precisely synchronous manner, but rather the hind wings lead the forewings. The entire cycle of movement lasts about 50 milliseconds.

Two sets of muscles act on each wing. One set, the elevators, raises the wing; the other set, the depressors, lowers it. Because these two sets of muscles have opposite effects on the wing, they are called antagonists (muscles with the same effects on a specific structure are called synergists).

Patterns of activity in the flight muscles of a tethered locust can be recorded by inserting tiny wire electrodes into the muscles (see Figure 6.28*a*). The signals (a recording of muscle action potentials) from the elevator and depressor muscles of the flying insect are then displayed on an oscilloscope; this type of recording is called a myogram (Figure 6.28*b*). Myograms have shown that the depressors are activated when the wings are up, and the elevators are activated when the wings are down (Figure 6.28*c*). The relative timing of the activation of these muscles is critical, and we will now examine how such timing may be controlled by the central nervous system.

The Neural Basis of Locust Flight

There are two general hypotheses for the neural basis of rhythmic behavior such as locust flight. The **peripheral-control hypothesis** is that each movement stimulates sensory receptors, which in turn trigger the next movement in the sequence. The second movement stimulates other sensory receptors that trigger the first component. Thus, sensory feedback is necessary for this hypothesis. An alternative hypothesis is that a central pattern generator controls locust flight. A **central pattern generator** is a neuron or network of neurons that is capable of generating patterned activity in motor neurons, even when all sensory input has been removed from the system (Carew 2000).

Donald Wilson's paper in 1961 was pivotal in our understanding of rhythmic behavior (reviewed in Edwards 2006). Wilson reasoned that if the peripheral-control hypothesis were correct, then elimination of sensory feedback would eliminate the pattern of activity in the motor neurons that raise and lower the wings. In contrast, the central pattern generator hypothesis predicts that the elimination of sensory feedback would not prevent the pattern of activity in the motor neurons controlling the wing movements. Locusts have mechanoreceptors on their wings that send information on wing position, via sensory neurons, to the central nervous system. This control system can be eliminated by sectioning the sensory nerve or by injecting the chemical phentolamine, which blocks the activation of mechanoreceptors without affecting the central nervous system (Ramirez and Pearson 1990). The removal of

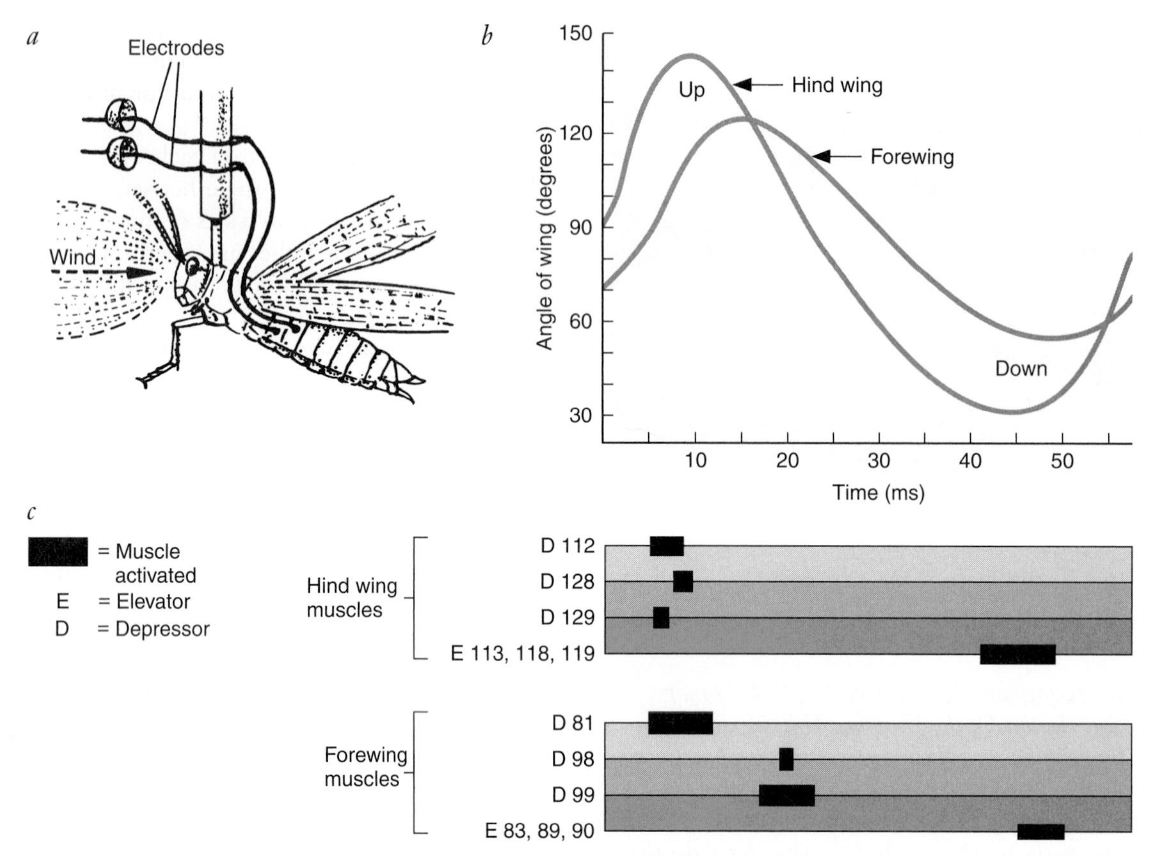

FIGURE 6.28 Analysis of the flight pattern of locusts. (*a*) If wind is directed at the head, a tethered locust can be induced to fly. The activity of flight muscles of the second and third thoracic segments can then be measured by inserting small wire electrodes in them. (*b*) The pattern of wing movements during flight in a tethered locust. (*c*) The pattern of activity in individual flight muscles as recorded with microelectrodes. (*a:* Modified from Horsmannet al. 1983. *b* and *c:* Modified from Young 1989. Original data from Wilson 1961 and Wilson and Weis-Fogh 1962.)

sensory input by either technique is called deafferentation. (Recall that sensory neurons are also called afferent neurons.) Although some changes in the flight pattern are observed after the surgery (e.g., the frequency of wing beats drops from 20 to about 10 per second), the basic pattern of wing movements is still present. The removal of sensory feedback from the wings can be taken one step further than simply sectioning the sensory nerves. The wings and thoracic muscles can be completely removed. Even after such extensive dissection, a pattern of neural activity similar to that seen during flight occurs in the motor nerves that emerge from the thoracic ganglia. Taken together, these results suggest that the basic flight pattern of the locust is not dependent on sensory feedback but is generated in the central nervous system. In other words, a neuronal circuit in the central nervous system generates the alternating bursts of activity in motor neurons that innervate the elevator and depressor muscles. We see, then, that a central pattern generator controls locust flight.

Although the basic rhythm of flight is generated by a central pattern generator in the thoracic ganglia, information gathered by sense organs located on the wings enables the locust to adjust its flight pattern on a cycle-by-cycle basis. Sensory information thus imparts some flexibility to the system of motor control, allowing the animal to respond to air turbulence and other environmental uncertainties. The system of motor command plays a particularly important role in that the interneurons that descend from the brain initiate (and modulate) activity in the central pattern generator. These interneurons appear to play a role in correctional steering, when the locust compensates for deviations in the flight course. The combination of central and peripheral control observed in locust flight probably characterizes most patterns of rhythmic behavior in animals (Delcomyn 1980).

SUMMARY

The basic unit of the nervous system is the neuron, a cell that usually has three parts: a cell body, or soma, which maintains the cell; dendrites, which receive information and conduct it toward the soma; and an axon, which conducts the nerve impulse away from the soma. There are three functional classifications of neurons. A *sensory neuron* is specialized to detect stimuli or to receive information from sensory receptors and conduct it to the central nervous system. *Interneurons*, located in the central nervous system, link one neuron to another. A *motor neuron* carries the information from the central nervous system to a muscle or gland.

When a neuron is resting, that is, not conducting an impulse, it is more negative inside than outside. Whereas potassium ions (K^+) are more concentrated inside the neuron's membrane than outside, sodium ions (Na^+) are more concentrated outside the membrane.

The message of a neuron is called an action potential or nerve impulse. The action potential is a wave of depolarization followed rapidly by repolarization that travels to the end of the axon with no loss in strength. Should a stimulus open ion channels that will allow sodium ions to cross the membrane, sodium ions will enter the neuron and depolarize the membrane. Potassium ions will then move out of the neuron, repolarizing the membrane.

Information travels from one neuron to the next by crossing the gap, or synapse, between them. The two types of synapses are electrical and chemical. Electrical synapses involve such tight connections between the neurons that the impulse spreads directly from cell to cell. Several steps are involved in transferring information across a chemical synapse: (1) The action potential travels down the axon to small swellings called terminal boutons. (2) Here, the action potential causes a neurotransmitter to be released from storage sacs (synaptic vesicles) into the gap between cells (synaptic cleft). (3) The neurotransmitter diffuses across the gap and binds to a special receptor. (4) When this occurs, ion channels open, changing the charge difference across the membrane. (5) If the synapse is excitatory, the postsynaptic cell is slightly depolarized. However, if the synapse is inhibitory, the postsynaptic cell is slightly hyperpolarized, and no new action potential will be generated.

The membrane potential of the postsynaptic cell can also be changed by neuromodulators. These substances work more slowly than neurotransmitters and bring about changes by biochemical means. Neuromodulators can have profound effects on behavior, as demonstrated by the effects of the neuromodulators on rhythmic leg movements in the blue crab.

Animals receive information from the environment at their sense organs. Sense organs, such as eyes and ears, are selective, containing specific receptor cells that respond to a particular form of environmental energy. Noctuid moths have only a single receptor cell (A1 cell) that is involved in detecting a hunting bat and evoking an escape response. Barn owls have an auditory space map that allows them to accurately locate sounds made by their prey in complete darkness.

When the siphon of *Aplysia* is touched, the animal withdraws its gills, mantle shelf, and siphon into the mantle cavity. This is called the gill-withdrawal reflex. This reflex shows habituation, that is, a decrease in responsiveness because of repeated stimulation. It also shows sensitization, in which a strong stimulus anywhere on *Aplysia*'s surface will cause an exaggerated gill-withdrawal reflex from a light touch on the siphon.

Short-term memory involves changes in the strength of synaptic connections. In *Aplysia*, habituation occurs because the sensory neuron releases less neurotransmitter when it is repeatedly stimulated. As a result, the gill motor neuron is less likely to fire. Sensitization occurs when a strong stimulus causes a sensory neuron to activate a facilitating interneuron to release serotonin on its synaptic ending near the axon terminal of the sensory neuron from the siphon skin. Serotonin then causes biochemical changes that lead the sensory neuron to release more neurotransmitter than usual. This increases the likelihood that the gill motor neuron will fire.

Long-term memory involves long-lasting changes in synapses. These changes occur through four mechanisms: long-term potentiation, long-term depression, synaptic remodeling, and neurogenesis. Neurogenesis in a song center (HVC) in the brain underlies song learning in birds. Neurogenesis in the hippocampus underlies spatial learning.

Despite the diverse types of stimuli that bombard an animal from its environment, it is able to detect only a limited range, and of those, it may ignore all but a few key stimuli.

Animals use their motor systems to respond to information picked up by their sensory systems. Central pattern generators can activate motor neurons in the absence of all sensory feedback.

7

Physiological Analysis of Behavior—The Endocrine System

After nightfall, from late spring through summer, along the western coast of North America, one can hear the male plainfin midshipman fish (*Porichthys notatus*) "singing" to attract females to his nest. The song, which is more of a droning hum than a melodic lovesong, is produced when the male contracts a pair of vocal muscles against the swim bladder (a buoyancy-regulating organ). The nests are built under rocky shelters in the intertidal zone. Although a female mates with only one male each breeding season, a male may mate with five or six females. He then guards 1000 to 1200 eggs.

There are actually two types of male plainfin midshipman fish, and they display different reproductive strategies. The males we've described, those that build nests and sing to court females, are type I males. In contrast, type II males don't build nests or hum to attract females. Instead, they either sneak into the type I's nest and spawn or they lie outside the entrance and deposit sperm there while fanning water toward the nest's opening. The sperm of the type II male will then compete with that of the type I male, who did all the work (Bass et al. 1997). Type II males are sneaky cuckolders (males that steal matings from other males) whose success depends on avoiding detection by territorial males.

There are other differences between type I and type II males. Type I males are larger and take longer to reach sexual maturity. Although type II males become sexually mature sooner, they can't attract mates. The vocal muscles of type I males used to produce the droning hum that is so attractive to females are much more developed than those of a type II male. The brain of a type I male is also specialized to allow him to hum his courtship song. The size of the motor neurons to the vocal muscles and the brain center controlling those muscles are also larger in a type I male. Surprisingly, however, the ratio of testis to body weight of a type II male is nine times that of a type I. Because of his bulging gonads, a type II male resembles a gravid female (Figure 7.1). The coloration of a type II male is also more similar to that of a female than a type I male. These similarities to females probably make it easier for type II males to lurk around the nests of type I males.

Sneaky cuckoldry by type II males seems to be a "fixed" behavioral strategy because these males will not hold territories or court females even when both options are offered to them (Brantley and Bass 1994). The morphology of type II males—small bodies, vocal muscles, and underlying neural circuitry—seems to prevent behavioral plasticity, the ability to switch to an alternative behavior. In contrast, field and laboratory observations of type I males show that these males will cuckold other type I males when given the chance (Lee and Bass 2004). The morphology of type I males apparently does

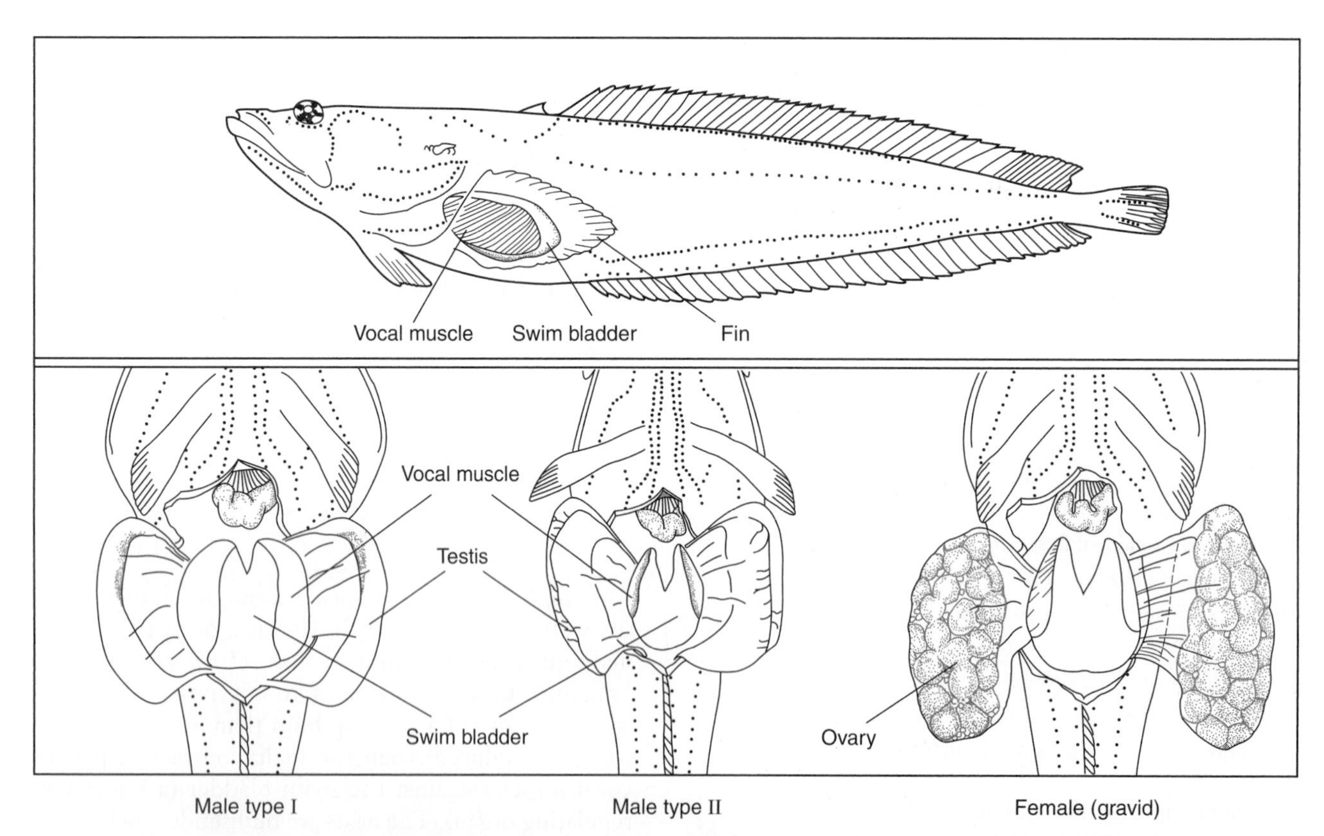

FIGURE 7.1 **In plainfin midshipman fish, there are two types of males, each with a distinct mating strategy. Type I males build nests, sing to attract females, and guard nests. To sing, a type I male contracts his vocal muscles against the swim bladder to create the droning hum of his courtship song (top). A type II male waits for a type I male to attract a female and then sneaks into the nest and spawns or releases sperm just outside the nest and fans water into the nest opening. The vocal muscles of a type I male (bottom left) are much more developed than those of a type II male (bottom center) or a female (bottom right). The ratio of vocal muscle to body weight is six times greater in a type I male than in a type II. Type I males are larger, but the ratio of gonads to body weight is nine times smaller than that of a type II male. The large testes make a type II male look like a gravid female. (Modified from Bass 1996.)**

not preclude behavioral plasticity. Indeed, type I cuckolders seem to take advantage of their large size and aggressively resist ejection by resident type I males. Thus, whereas type II males engage in sneaky cuckoldry, type I males exhibit aggressive cuckoldry when not humming and defending a nest.

Hormonal differences are responsible for the physical and behavioral differences among type I males, type II males, and females. What are hormones, and how do they produce such dramatic effects on the nervous system, muscles, and behavior? What role do hormones play in the development and display of behavioral differences between the sexes and between individuals of the same sex? Do hormonal effects on behavior vary as a function of the physical or social environment? Can behavior, in turn, influence the levels of hormones? It is to these issues that we now turn.

THE ENDOCRINE SYSTEM

ENDOCRINE GLANDS AND HORMONES

We begin with a definition. Hormones are chemical substances secreted in one part of the body that cause changes in other parts of the body. Hormones are secreted either by endocrine glands or by neurons. Endocrine glands, unlike exocrine glands (e.g., sweat, salivary, or scent glands), which have specialized ducts for secretion of products, lack ducts and secrete their products into the spaces between cells, from which the hormones diffuse into the bloodstream. Once in the blood, hormones travel along the vast network of vessels to virtually every part of the body. Hormones secreted by nerve cells are called neurohormones or neurosecretions. These are produced in the cell body of the nerve cell, travel along the axon, and are released at the axon tip. Functioning as chemical messengers in an elaborate system of internal communication, hormones and neurohormones exert their effects at the cellular level by altering metabolic activity or by inducing growth and differentiation. Changes at the cellular level can eventually influence behavior.

In this chapter we will focus on how chemical signals function *within* an individual to influence its behavior, morphology, and physiology. We will see that a given hormone can have diverse effects within an individual. It is also worth mentioning, however, that some chemical signals that function in communication within individuals also function in communication *between* individuals. This is the case for some chemical signals in arthropods, a large group of invertebrates that includes insects, crustaceans, spiders, and scorpions. Molting hormones in arthropods, known as ecdysteroids, help orchestrate the periodic shedding of the exoskeleton in these animals. One of these molting hormones in crustaceans, 20-hydroxyecdysone (fortunately abbreviated 20HE), also functions as a feeding deterrent between individual shore crabs (*Carcinus maenas*) (Hayden et al. 2007). Mating in shore crabs is a delicate affair that coincides with molting in females. Prior to a female molting, a male will guard her from rival males and predators, cradling her beneath his abdomen for a few days. Once a female molts, copulation soon follows. A soft-shelled female is extremely vulnerable to predation by fish, octopus, and other crustaceans, including members of her own species. It turns out that 20HE released by the female at this time effectively deters her mate from eating her. The feeding deterrent effect of 20HE is sex-specific, deterring *C. maenas* males but not females. Here, then, 20HE is acting not as a hormone (as it does when regulating molting within the individual) but as a pheromone or chemical substance that functions in communication between individuals of the same species (see Chapter 16 for a further discussion of pheromones). It is not yet known whether the 20HE released by newly molted *C. maenas* females deters feeding in other species of predatory crustaceans. It is known, however, that 20HE released by *Pycnogonum litorale*, another marine arthropod, deters feeding by *C. maenas*, one of its predators (Hoffmann et al. 2006; Tomaschko 1994). Thus, the chemical signal 20HE functions within individual arthropods (as molting hormone), between individuals of the same species (as a sex-specific feeding deterrent during mating interactions in *C. maenas*), and between individuals of different species (as a feeding deterrent in *P. litorale* against predatory crustaceans such as *C. maenas*). Having hopefully impressed you with the varied roles chemical signals play in communication, we now turn our attention exclusively to how such signals function within individuals.

HORMONAL VERSUS NEURAL COMMUNICATION

The endocrine system of an animal is closely associated with its nervous system. As mentioned, some hormones, in fact, are made by nerve cells. In addition, neurons and hormones often work together to control a single process. For example, in some interactions neurons respond to hormones, whereas in others endocrine glands receive information and directions from the brain. Nervous and endocrine systems are so closely associated that they are often discussed as a single system, the neuroendocrine system. Despite this close association, neural and hormonal modes of information transfer have different purposes within the body, and each system is essential in its own right.

In comparing communication through nervous or endocrine pathways, we should first briefly review how neurons transfer information. In the nervous system, information is transmitted along distinct pathways (chains of neurons) at speeds of up to 100 meters per second. After the impulse arrives at its destination in the body, neural information is transmitted via a series of electrical events that culminate in the release of neurotransmitter near its target, such as an effector tissue—a muscle, for example. Neurotransmitters are rapidly destroyed after they are secreted. As a result, information delivered by the nervous system usually produces a response that is rapid in onset, short in duration, and highly localized.

In contrast to neural communication, hormonal transfer of information tends to occur in a more leisurely, persistent manner. Typically, hormones are secreted slowly and remain in the bloodstream for some time. Rather than traveling to a precise location, these chemical messengers contact virtually all cells in the body, although only some cells are able to respond to the particular hormonal stimulus. The cells that respond, called target cells, have receptor molecules that recognize and bind to specific hormones. This binding activates the receptor and initiates the cell's response to the hormone, perhaps by turning on certain genes or by altering the cell's secretory activity or the properties of its plasma membrane. The precise nature of the response depends on the type of target tissue because different types of cells are specialized to perform specific functions in the body. Only cells with receptors for a particular hormone can respond to it. The concentration of receptors for a particular hormone determines the cell's sensitivity to it. In short, transfer of information by the endocrine system often occurs more slowly than that of the nervous system, and it usually produces effects that are more general and long lasting.

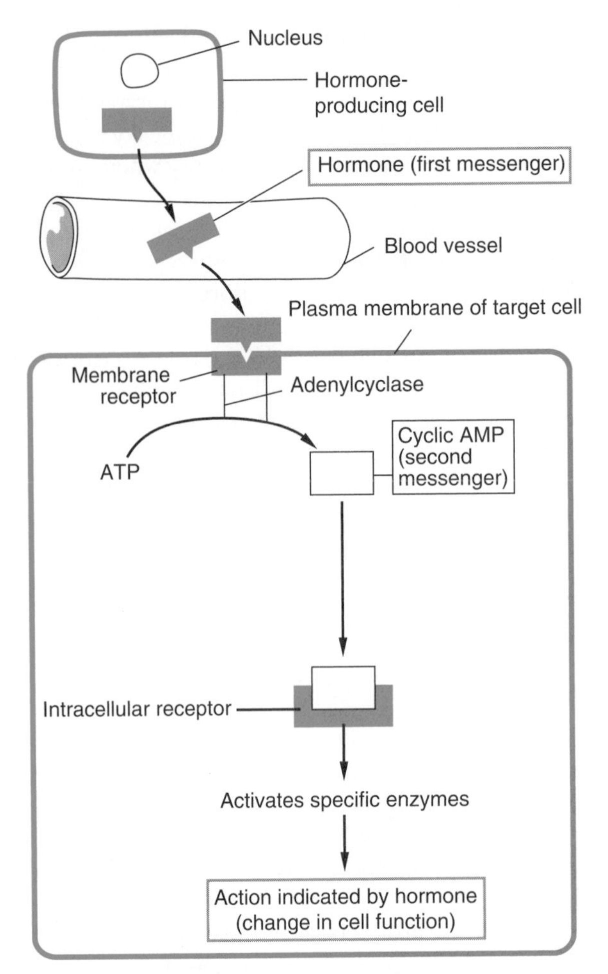

FIGURE 7.2 **Mechanism of action of some peptide hormones.**

TYPES OF HORMONES AND THEIR MODES OF ACTION

Animals produce two major classes of hormones: peptides and steroids. Although differing in structure and mode of action, both types cause changes within the cell that eventually influence behavior.

Peptide Hormones and Amino Acid Derivatives

Peptide hormones are amino acid chains, ranging in size from about 3 to 300 amino acids. These hormones, along with amino acid derivatives, are water-soluble and usually affect cells by binding to receptor molecules on the cell surface (Figure 7.2). Through a complex sequence of molecular interactions, often including the use of a secondary messenger (such as cyclic adenosine monophosphate, or cAMP), peptide hormones create short-term changes in cell membrane properties and long-term changes in protein function, often by activating enzymes. Examples of peptide hormones are luteinizing hormone (LH) and follicle-stimulating hormone (FSH) produced by the anterior pituitary gland.

Steroid Hormones

Steroid hormones are a group of closely related hormones secreted primarily by the gonads and adrenal glands in vertebrates. The four major classes of steroids include progestogens, androgens, estrogens, and corticosteroids. The first three classes are secreted primarily by the gonads and are often referred to as the sex steroids. All steroid hormones are chemically derived from cholesterol and hence are highly fat-soluble. As a result of their solubility in lipids, steroid hormones move easily through the lipid boundaries of cells and into the cell interior, or cytoplasm (Figure 7.3). Once inside a cell, steroids combine with receptor molecules in the nucleus and, in some cases, in the cytoplasm. If binding occurs in the cytoplasm, then the steroid-receptor complex moves to the nucleus. In the nucleus, the complex attaches to DNA and affects subsequent gene expression and protein synthesis, a process that may take several hours to days and one that produces relatively long-lasting effects on behavior.

Because we will be discussing the role of steroid hormones in reproductive behavior, two points about the sex steroids should be emphasized here. First, hormonal output is not rigidly determined by sex. Females generally have small amounts of "male hormones" such as testosterone, and males typically have low levels of "female hormones" including estrogens. A second point to emphasize is that although different hormones produce different effects, the sex steroids are chemically very similar (Figure 7.4). Some hormones lie along the pathway of synthesis of other hormones. Testosterone, for example, is an intermediate step in the synthesis of estradiol. As a result of the common structure of these two steroid hormones, some behavior patterns may be activated by injections of either testosterone or estradiol. In other words, a certain degree of substitutability is associated with steroid hormones.

We have presented the traditional dichotomy of peptide versus steroid hormones and their different mechanisms of action, but things are never so simple. For example, we describe steroids as binding to receptors inside the cell, and typically modifying gene expression and protein synthesis over several hours or days to produce relatively long-lasting effects on behavior; steroid effects such as these are described as genomic.

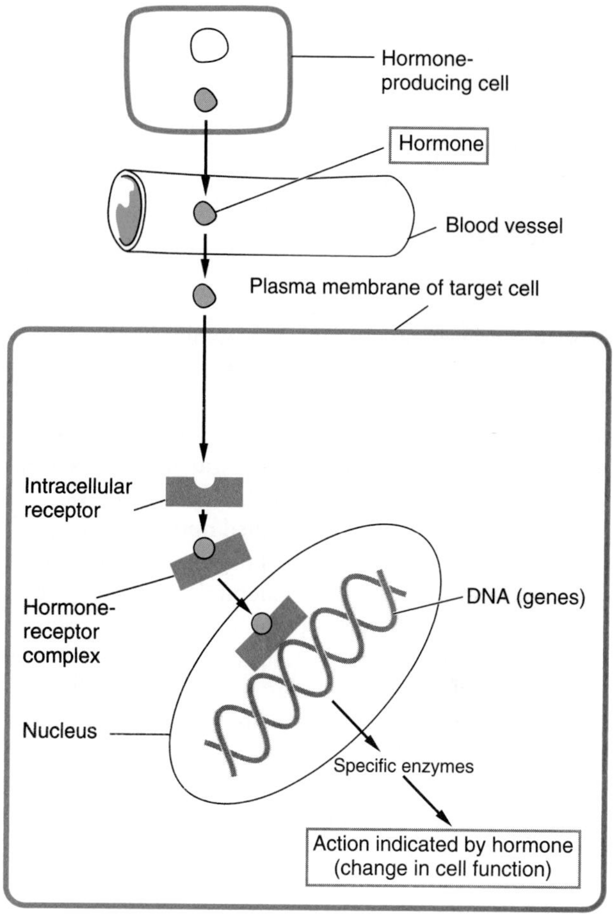

FIGURE 7.3 Mechanism of action of steroid hormones in which gene expression is altered.

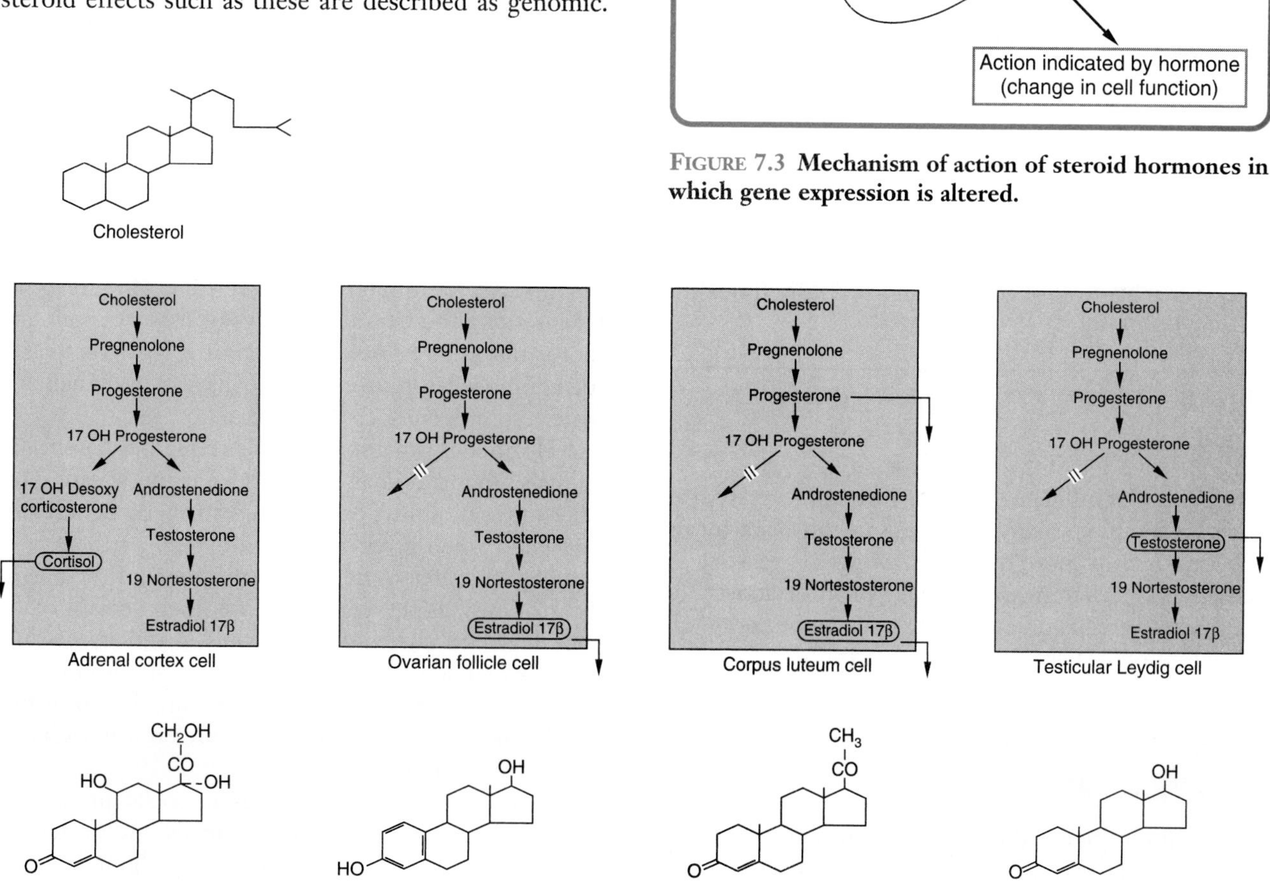

FIGURE 7.4 Biochemical pathways by which steroid hormones are synthesized. Note that many of the steroid hormones are chemically very similar. (Modified from Daly and Wilson 1983; Tepperman 1980.)

In recent years, steroids have been found to work in additional ways. It is now known that steroids, acting on the brain through diverse pathways that include interacting with membrane receptors and altering cell-signaling pathways, can produce rapid behavioral responses, on the order of seconds or a few minutes (reviewed by Moore and Evans 1999; Remage-Healey and Bass 2006). Steroid effects that occur rapidly after steroid application are described as nongenomic because the short latency indicates that modifications to gene expression and protein synthesis are probably not involved.

Finally, we have also described steroids as being produced primarily by the gonads and adrenal glands of vertebrates. It is important for us to mention that steroid hormones are also produced by the nervous system (Compagnone and Mellon 2000; Plassart-Schiess and Baulieu 2001). These steroids, called **neurosteroids**, differ from those produced by the gonads and adrenal glands (sometimes called peripheral steroids) in at least one major way. Whereas a peripheral steroid travels in the bloodstream and acts on target cells some distance from its gland of origin, a neurosteroid acts in the nervous system on either nearby cells or the same cell that produced it. In terms of mechanism of action, some neurosteroids act in the classic way we described for peripheral steroids (binding to receptors inside the target cell and affecting gene expression); others influence receptors on the membranes of neurons; and still others act on microtubules within neurons (microtubules are components of the cytoskeleton involved in transport within cells and cell division and movement). Thus, we now recognize that steroid hormones are produced by more locations within the vertebrate body and have more diverse mechanisms of action than previously thought.

HOW HORMONES INFLUENCE BEHAVIOR

Hormones can influence behavior through several pathways. Generally speaking, hormones modify behavior by affecting one or more of the following: (1) sensory or perceptual mechanisms, (2) development or activity of the central nervous system, and (3) muscles important in the execution of behavior.

EFFECTS ON SENSATION AND PERCEPTION

Hormones influence the ability to detect certain stimuli, as well as the responses to and preferences for particular stimuli. In some species, mate choice is at least partially based on hormone-mediated differences in the ability to detect stimuli. In domestic pigs (*Sus scrofa*), for instance, females are able to detect lower quantities of the boar pheromone, 16-androsterone, than are males (Dorries et al. 1995). Females are attracted to this chemical and assume a sexually receptive posture in response to it. Males are not attracted by the boar pheromone. However, if a male is castrated to remove the source of testosterone before the age of five months and is given the female hormone estradiol as an adult, he shows the usual female responses to a boar (Adkins-Regan et al. 1989).

Sometimes hormones mediate behavioral responses to particular stimuli. As an example, we will consider the effects of vasotocin, a peptide hormone found in nonmammalian vertebrates that is very similar to the hormone vasopressin in mammals. Vasotocin affects how male roughskin newts (*Taricha granulosa*) react to visual, chemical, and tactile stimuli from females (Rose and Moore 2002). When identifying prospective mates, male newts first rely on visual cues such as size, shape, and color, and then switch to olfactory cues for close-up confirmation of species, gender, and reproductive state. If all checks out, then the male clasps the gravid female for several hours in a posture known as amplexus (Figure 7.5). During this time, the female becomes sexually receptive and will pick up the spermatophore (package of sperm) that the male subsequently deposits. Vasotocin has been shown to influence the early stages of mate choice in males—identification of potential mates based on visual and chemical stimuli—as well as the clasping response of amplexus. Males injected with vasotocin spend more time in the vicinity of female visual and olfactory cues than do control males injected with saline, and they show enhanced responsiveness to tactile stimuli from females during amplexus. Although the precise location(s) at which vasotocin exerts its effects along the neural pathways of sensory input and motor output are not yet known, it is clear that vasotocin influences reproductive behavior by enhancing male responsiveness to female stimuli.

Hormones can also cause a change in preference in animals. Young animals that receive parental care typically interact almost exclusively with parents and siblings, and when given a choice of social partner, most youngsters prefer to be around members of their family. However, as the young mature, their social preference tends to switch to nonfamily members (as any parent of an adolescent knows), especially those of the opposite sex. Sex steroids have been implicated in this developmental change in social preference (Adkins-Regan and Leung 2006).

Hormone-mediated changes in social preference are not limited to maturing animals. Adult female meadow voles, *Microtus pennsylvanicus*, exhibit an adaptive seasonal change in odor and social preference. During the winter, female meadow voles nest communally with other females and, at this time, prefer female odors to male odors. However, in the spring and sum-

FIGURE 7.5 **A male roughskin newt (on top) in amplexus with a female. The hormone vasotocin enhances male responsiveness to visual, chemical, and tactile stimuli from the female.**

mer, when they defend territories against other females and mate with males, they prefer the scent of a male to that of a female. Thus, changing odor preferences help female voles choose their company so they can successfully raise as many offspring as possible. The reversal in odor preference is caused by changes in the amount of estrogen the female produces. Estrogen levels fluctuate in response to seasonal changes in the length of daylight, with higher levels associated with longer days (Ferkin and Zucker 1991).

EFFECTS ON DEVELOPMENT AND ACTIVITY OF THE CENTRAL NERVOUS SYSTEM

Circulating hormones can affect behavior by influencing the central nervous system. In fact, hormones have been found to influence several characteristics of different regions of the brain, including (1) the volume of brain tissue, (2) the number of cells in brain tissue, (3) the size of cell bodies of neurons, (4) the extent of dendritic branching of neurons, (5) the percentage of neurons sensitive to particular hormones, and (6) the survival of neurons.

An example in which hormones influence the central nervous system concerns the development of singing behavior in birds. In the zebra finch (*Taeniopygia guttata*), sex differences in the brain nuclei that control song are established around the time of hatching. Early exposure to steroid hormones regulates the size of song nuclei, the size and number of neurons within these brain areas, the extent of dendritic branching, and the number of androgen receptors (reviewed in Wade and Arnold 2004). Thus in the zebra finch, sex differences in adult singing behavior (males sing and females do not) are linked to differences that are established in the brains of males and females as a result of the hormonal milieu (environment) soon after hatching. The steroid hormones involved in the early masculinization of the zebra finch brain (and indeed in the general organization of the developing brain in this species) appear to be neurosteroids rather than gonadal steroids (London and Schlinger 2007; Wade and Arnold 1996). (See Chapter 8 for a detailed discussion of bird song.)

EFFECTS ON MUSCLES

Hormones can influence behavior by affecting muscles and motor neurons. Consider, for example, two cases of sexually dimorphic patterns of behavior—calling behavior in frogs and copulatory movements in rodents—that illustrate hormonal influences on muscles.

Our first example concerns the calling behavior of the South African clawed frog, *Xenopus laevis* (reviewed in Kelley 1996). Clawed frogs occur in sub-Saharan Africa, where they inhabit shallow, and often murky, bodies of water. Males of this species emit six different calls, the most common of which is the advertisement call (Tobias et al. 2004). This metallic-sounding call consists of alternating fast and slow trills; during fast trills, there may be a progressive increase in volume. The advertisement call allows females to find males in their typically soupy locations. Sexually receptive females approach calling males and produce a rapping call, which stimulates the males to answer, and a duet ensues (Tobias et al. 1998). Receptive females permit males to clasp them around the waist for several hours while their eggs are released and fertilized. Whereas receptive females rap, females that are not sexually ready tick. The ticking call consists of slow, monotonous clicks with no change in intensity; the rapping call is similar but has a somewhat shorter interval between clicks. The advertisement call of males and the ticking call of unreceptive females have been most completely studied and are shown in Figure 7.6.

How does it come about that male clawed frogs produce advertisement calls of rapid trills, whereas unreceptive females produce the slow ticking call? Darcy Kelley and co-workers demonstrated that characteristics of the muscles and neuromuscular junctions of the larynx are responsible for sex differences in the rate at which calls are produced (in males the muscles of the larynx contract and relax 71 times per second and only 6 times per second in females). Adult males have 8 times as many muscle fibers in their larynx as do females. Also, male muscle cells are of the fast-twitch, fatigue-resistant type, whereas most muscle cells in the larynx of females are slow-twitch and fatigue-prone.

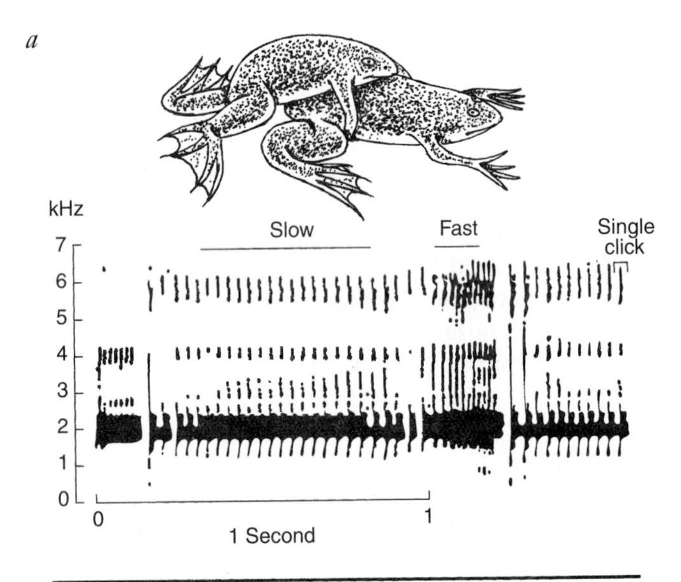

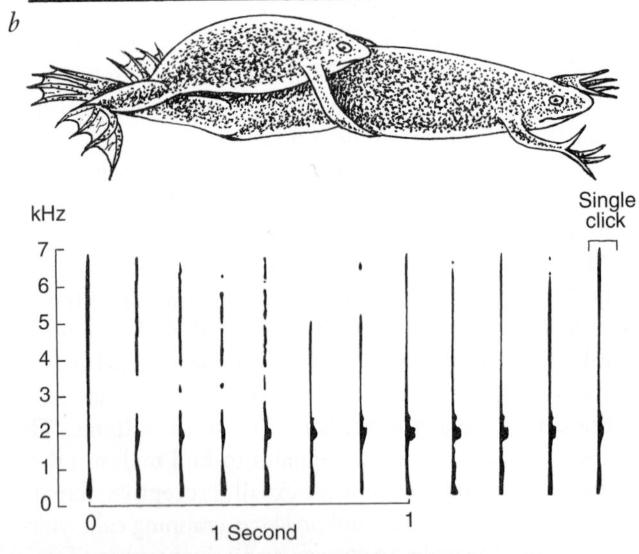

FIGURE 7.6 **Differences in the calling behavior of male and female clawed frogs result from the effects of hormones on the muscles of the larynx. (*a*) Male clawed frog clasping a sexually receptive female that responded to his mating call (shown below pair). The male's call consists of slow and fast trills. (*b*) Male clasping a sexually unreceptive female that responds to his clasp by emitting the ticking call (shown below pair). (From Kelley and Gorlick 1990.)**

What causes these differences in laryngeal muscles? At the time of metamorphosis (the change from tadpole to frog), the number of muscle fibers in the larynx of males and females is identical to the number in the larynx of adult females. Apparently, as males mature and their levels of androgens rise, new fibers are added. In addition to increasing the number of muscle fibers, androgens also influence the type of fiber, promoting expression of the fast-twitch cells. Hormones other than androgens, including prolactin and secretions from the thyroid gland, also appear to play a role in masculinizing the larynx. In short, sex differences in the calling behavior of *X. laevis* can be traced, in part, to hormone-induced changes in the muscles of the larynx.

Our second example of sex differences in behavior resulting from hormonal influences on muscles concerns the perineal muscles involved in mating (the perineum is the area between the urogenital and anal openings). The levator ani/bulbocavernosus muscles control copulatory reflexes in male Norway rats (Hart 1980). Although these muscles are present in the perineum of both sexes at birth, they are completely absent in adult females. Breedlove and Arnold (1983) have shown that the levator ani/bulbocavernosus muscles shrink and fold inward in females unless supplied with androgen. The lack of androgen in females during the perinatal period (the time surrounding birth) also results in the death of the motor neurons that supply these muscles. Thus, sex differences in the copulatory movements of adult rats result, in part, from early hormonal influences on the growth and maintenance of the specific muscles and motor neurons involved in mating.

Most mammals studied to date are similar to Norway rats in having sexually dimorphic perineal muscles and motor neurons; that is, the muscles and motor neurons that innervate them are absent or reduced in females. An interesting exception to this pattern has emerged in the case of the naked mole rat (*Heterocephalus glaber*). We begin by providing some necessary background on this unique species. Naked mole rats are cooperatively breeding rodents that live in large colonies (Jarvis 1981). Each colony contains a queen (the only breeding female), one or a few breeding males, and numerous nonreproductive subordinates, almost all of which will never breed (a subordinate can attain breeding status only when a breeder dies or upon dispersing from the colony and encountering a potential mate). Male and female subordinates are remarkably similar in their behavior and body size, and even in their external genitalia, although internal reproductive organs are normally sexually differentiated (Jarvis 1981; Lacey and

Sherman 1991). Consistent with these observations of few sex differences in subordinate naked mole rats are data showing that the perineal muscles and their motor neurons are also sexually monomorphic (i.e., the same in males and females) (Peroulakis et al. 2002). It seems that sexual differentiation in behavior and reproductive structures is limited in subordinate naked mole rats, and this makes sense given the similar nonbreeding roles played by low-ranking males and females in colony life. Among breeders, the levator ani muscle is actually larger in the queen than in breeding males, while other perineal muscles are similar in size (Seney et al. 2006). The results for size comparisons of the levator ani muscles of breeders and nonbreeders are shown in Figure 7.7. In species of mammals studied to date, the perineal muscles control reflexes of the penis; their role, if any, in females with reduced muscles and motor neurons is unknown. The suggestion for naked mole rats is that the enlarged levator ani muscle helps the queen deliver the enormous number of offspring that she will produce over the course of her lifetime (a queen can produce litters every 80 or so days and can have up to 28 young in a litter!). Although much work remains to be done on the endocrinology of naked mole rats and other cooperatively breeding mammals, the current data demonstrate that hormonal influences on perineal muscles reflect reproductive life history.

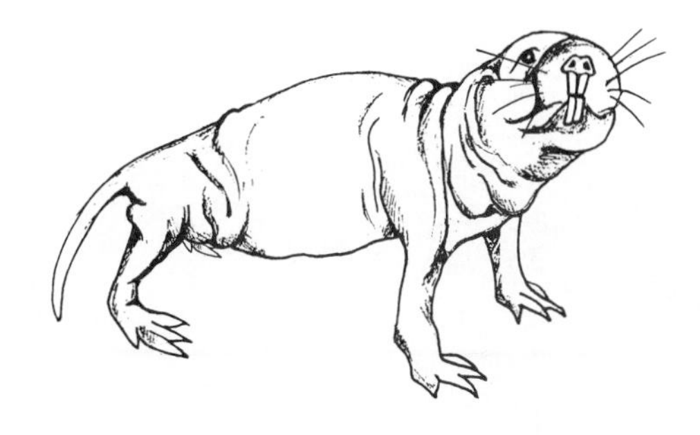

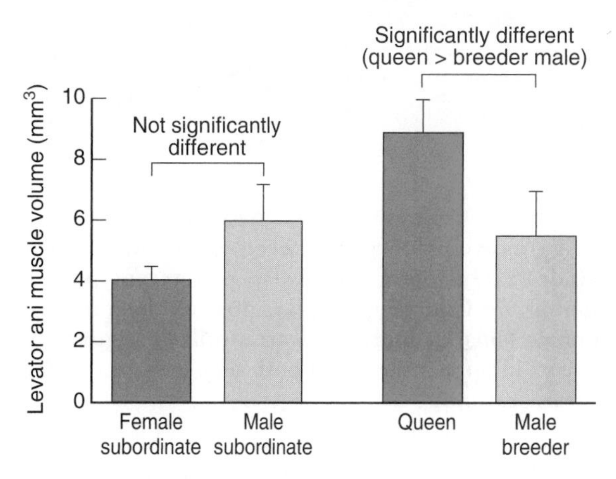

FIGURE 7.7 **Muscles of the perineum are sexually dimorphic in most mammals, being larger in males than in females as a result of the early effects of androgens. In naked mole rats, however, perineal muscles do not differ in size between subordinate males and females, perhaps reflecting their similar nonbreeding roles in colony life (results for one perineal muscle, the levator ani, are shown here). The levator ani muscle of the queen is actually larger than that of breeding males. (From Seney et al. 2006.)**

METHODS OF STUDYING HORMONE–BEHAVIOR RELATIONSHIPS

Several techniques are available for the study of hormonal influences on behavior. Here we examine two general approaches to questions about behavioral endocrinology. The first approach might be called interventional because the experimenter manipulates the hormones of the animal. This often involves the removal of a gland, followed by hormone replacement therapy. In the second approach, researchers look for changes in behavior that parallel fluctuations in hormone levels. Studies that use this second approach are called correlational studies.

INTERVENTIONAL STUDIES

Fairly conclusive evidence of the function of a hormone can be gained by removing its source, that is, the endocrine gland, and recording the subsequent effects. The hormone is then replaced by implanting a new gland or by administering the hormone. If the effects of gland removal are reversed by replacing the hormone, we conclude that the hormone was responsible for the changes.

David Crews (1974, 1979a) used gland removal (castration) and hormone (androgen) replacement therapy in the study of hormonal control of sexual and aggressive behavior in lizards. Among his favorite subjects is *Anolis carolinensis*, the green anole. This small iguanid lizard inhabits the southeastern United States and displays a social system in which males fiercely defend their territories against male intruders. The territory of a single male often encompasses the home ranges of two or three females. As you might expect from these living arrangements, male anoles have an interesting repertoire of aggressive and sexual behaviors (Figure 7.8).

Both agonistic and sexual displays of male *A. carolinensis* share a species-typical bobbing movement, which is made even more dramatic by extension of the red throat fan, or dewlap. When confronted by a male intruder, a resident male anole immediately begins to

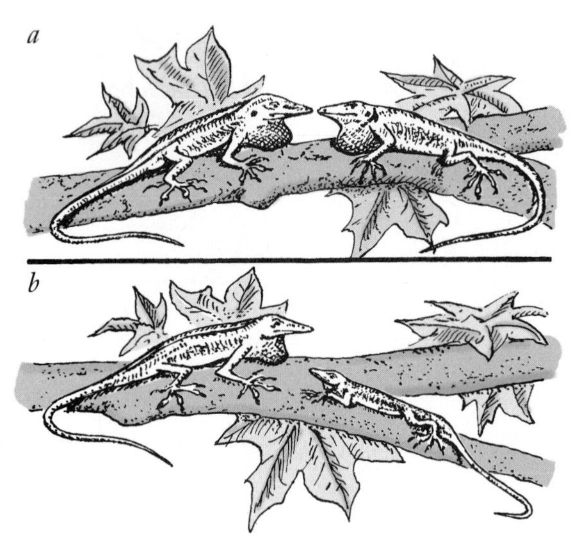

FIGURE 7.8 **Displays of the male green anole. (*a*) Aggressive posturing between two males often includes extension of the dewlap and stereotyped bobbing. (*b*) Courtship displays directed by a male to a female (smaller individual) are similar to aggressive displays in appearance, and both displays are mediated by testosterone.**

display—usually by compressing his body and adjusting his posture in such a way as to present the intruder with a lateral view of his impressive physique. As if this were not enough, the resident then lowers his hyoid apparatus (a structure in the back of the throat that is responsible for movements of the tongue) and exhibits a highly stereotyped bobbing pattern. The display ends at this point if the intruder rapidly nods his head, thereby acknowledging his subservient position. However, if the intruder fails to display the submissive posture, the display of the resident male escalates to ever-increasing frequencies. In the heat of confrontation, the two combatants acquire a crest along the back and neck and a black spot behind each eye. A wrestling match ensues as the resident and intruder circle, with locked jaws, in an attempt to dislodge the other from the prized perch. Engaging in aggressive behavior is apparently quite rewarding to males, for they prefer to spend time at the sites of their previous aggressive encounters (Farrell and Wilczynski 2006). The courtship behavior of male green anoles is very similar to their aggressive behavior. Typically, however, the body is not laterally compressed, and the bobbing dewlap display is less stereotyped in courtship. In effect, each male has his own version of how best to attract females.

Once Crews had documented the display repertoire of the feisty *A. carolinensis*, he set out to examine hormonal control of male aggressive and sexual behavior through castration and androgen replacement therapy. Removal of the testes led to a sharp decline in sexual behavior, but administration of testosterone implants reinstated this behavior to precastration levels (Figure 7.9) (Crews 1974; Crews et al. 1978). Thus, Crews and his co-workers concluded that testosterone regulates courtship and copulation in the male green anole.

The relationship between testosterone and aggressive behavior was not so simple. If a male was castrated and returned to his home cage, he continued to be aggressive toward intruders. However, if the male was castrated and placed in a new cage, his aggressive behavior declined in a manner similar to that noted for sexual behavior. Thus, unlike sexual behavior, aggressive behavior appears to be only partially dependent on gonadal hormones and subject to influence by social factors such as residence status.

Although simple in concept, interventional studies have become quite sophisticated as a result of major advances in techniques for manipulating hormone levels. For example, cannulation techniques now allow administration of minute amounts of hormone to specific brain regions. Other advances utilize techniques

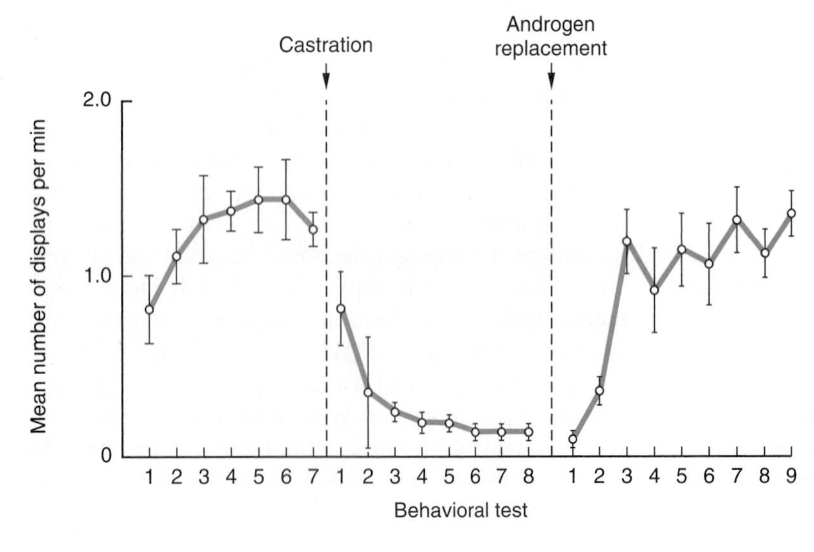

FIGURE 7.9 **Effect of castration and testosterone replacement therapy on the courtship behavior of the male green anole. (Modified from Crews 1979b.)**

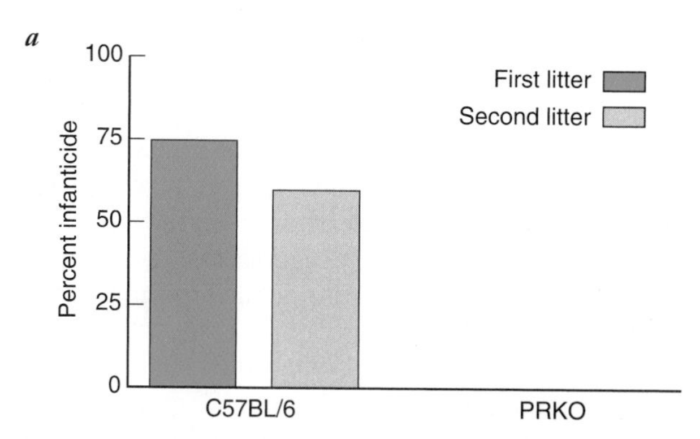

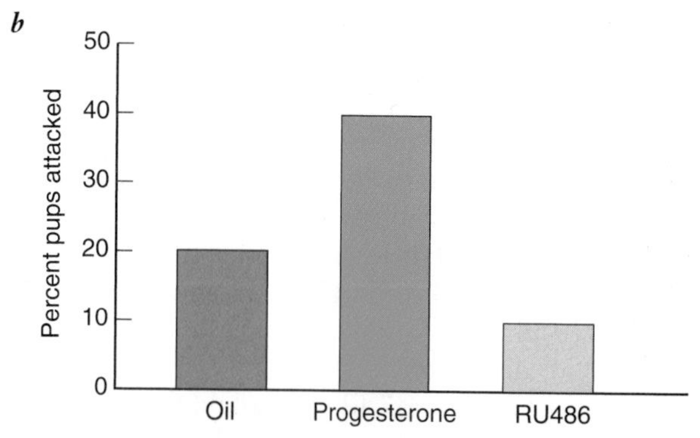

FIGURE 7.10 **Several different interventions demonstrate that progesterone regulates aggression shown by male mice toward infants. (*a*) Control males from the C57BL/6 laboratory strain kill offspring in their first litters about 74% of the time; instances of infanticide decline somewhat with second litters but are still high (58%). PRKO males are insensitive to progesterone because they lack progesterone receptors; these males show no infanticide toward young in either first or second litters. (*b*) Treatment with progesterone increases attacks by males on young relative to control males receiving oil, and treatment with RU486, an antihormone that blocks progesterone receptors, decreases attacks. Males used in the experiments shown in (*b*) had never mated, so were tested with pups from other pairs. (From Schneider et al. 2003.)**

whereby hormones are labeled with radioactivity and their paths traced through the body. The discovery of **antihormones**, drugs that can temporarily and reversibly suppress the actions of specific hormones, has also aided investigation of hormonal influences on behavior.

Genetic "knockout" mice also offer new opportunities to manipulate hormone levels to study the relationship between hormones and behavior. A knockout mouse is one in which a specific gene is targeted and inactivated to eliminate the gene product. In this case, the gene product may be a hormone or a receptor for a hormone. For example, there are progesterone receptor knockout (PRKO) mice; these mice do not respond to progesterone because they lack the appropriate receptors. Whereas males of most strains of laboratory mice behave aggressively toward infants and often kill them, PRKO males are not aggressive to infants (Figure 7.10*a*) (Schneider et al. 2003). Like PRKO males, male mice whose progesterone receptors have been temporarily blocked by administration of the antihormone RU486 rarely attack pups. Finally, male mice that have received progesterone implants (silastic implants filled with progesterone and sesame oil) are more aggressive toward infants than control mice receiving implants filled only with sesame oil (Figure 7.10*b*). Taken together, these findings indicate that progesterone mediates the aggression shown by adult male mice toward infants. Aside from demonstrating several techniques used in research in behavioral endocrinology, this study is interesting for at least two more reasons. First, progesterone has traditionally been viewed as a hormone that influences female behavior and physiology, and here we have an effect of progesterone on male behavior. Second, different hormones seem to mediate different forms of aggression in adult male mice; it has long been known that testosterone mediates aggression directed at other adult males, and now we see that progesterone mediates aggression directed at infants.

CORRELATIONAL STUDIES

We can also study hormonal influences on behavior with correlational studies. In using this approach, researchers look for changes in behavior that parallel fluctuations in hormone levels. Correlational studies are useful, but they are not as conclusive as experimental work because there is no evidence of causation. Consider, for example, a correlational study that revealed the relationship between the level of testosterone and aggressive behavior in a songbird.

John Wingfield has examined the behavioral endocrinology of birds under natural conditions (for a review, see Wingfield and Moore 1987). In one study of song sparrows, *Melospiza melodia* (Figure 7.11), Wingfield

FIGURE 7.11 **Male song sparrow.**

(1984) captured males in mist nets or traps baited with seed, collected a small blood sample from the wing vein, and marked each individual with a unique combination of leg bands. Birds were released at the site of capture and seemed relatively unperturbed by the sampling procedure; in fact, some individuals sang within 15 to 30 minutes of release. A given male was sampled from five to ten times during a single breeding season, and each sample was analyzed for testosterone. Wingfield also observed their behavior during this period.

As shown in Figure 7.12, Wingfield found a close correlation between peak levels of male territorial and aggressive behavior and maximum levels of testosterone. A male song sparrow defends his territory most intensely during its initial establishment and when his mate is laying the first clutch. At the time of egg laying, females are sexually receptive and males aggressively guard them from other would-be suitors. Testosterone reaches peak levels during the initial period of territory establishment and during the laying of eggs for the first brood. Interestingly, not only does testosterone increase aggressive behavior in song sparrows at these times, but the reverse is also true; that is, aggressive interactions can increase plasma levels of testosterone. So, we see that a hormone may influence a particular behavior, and that behavior, in turn, may influence levels of the hormone. It is also interesting that testosterone does not peak during the period when the female is sexually receptive and laying the second clutch. Aggressiveness is correlated with testosterone levels. A male guards his mate with less enthusiasm during the second laying period than during the first period. This pattern may be related to the fact that while a female is laying the second clutch, a male is often responsible for feeding fledglings from their first clutch. Wingfield (1984) speculates that high levels of testosterone and the resulting heightened levels of male aggression would interfere with paternal behavior. Field studies of the song sparrow demonstrate that circulating levels of testosterone wax and wane in parallel with changing patterns of male territorial aggression. This correlational evidence strongly suggests that testosterone mediates aggressive behavior during the breeding season in this species. Finally, we have the **challenge hypothesis** (Wingfield et al. 1990), which states that levels of hormones, such as testosterone, that regulate aggression and dominance are influenced by the social environment, and that their levels rise during times of social challenge or instability (e.g., during the initial period of territory establishment in the case of song sparrows). This hypothesis is now being tested in other species of vertebrates (Hirschenhauser and Oliveira 2006) and also in insects (Trumbo 2007). The relationship between testosterone and aggressive behavior is also discussed in Chapter 18.

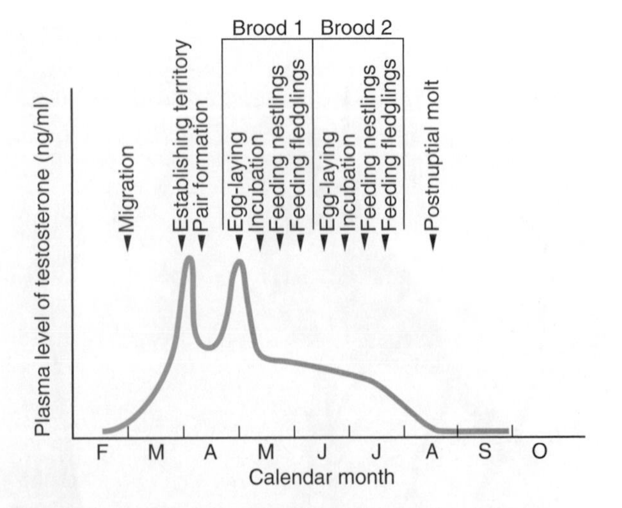

FIGURE 7.12 Changes in circulating levels of testosterone in free-living male song sparrows as a function of the stage of the breeding cycle. (From Wingfield 1984.)

In male song sparrows in sedentary (nonmigratory) populations, territorial aggression is not limited to the breeding season. Indeed, such males also exhibit territorial aggression during the nonbreeding season, after they have completed molting in late summer. Does testosterone regulate aggression during the nonbreeding season when testes have regressed? Apparently not, because levels of testosterone are undetectable in the plasma of nonbreeding males and do not increase following aggressive interactions. Instead, the culprit appears to be estradiol. It seems that dehydroepiandrosterone (DHEA), a precursor that can be converted into active sex steroids in appropriate tissues, is metabolized in the brain of males to form estradiol, which fuels aggression during the nonbreeding period (Wingfield and Soma 2002). DHEA may originate in the adrenal glands or the regressed testes. It is also possible that estradiol forms directly in the brain from cholesterol (recall that the nervous system can also make steroid hormones). Why might a different hormone mediate territorial aggression during breeding and nonbreeding periods? Several studies have shown that high levels of testosterone can be costly to birds in terms of increased metabolic rate and decreased body mass, fat stores, and immune function (reviewed in Soma 2006). The energetic costs of testosterone would be most critical during the nonbreeding season when birds experience lower temperatures and reduced food supplies. Researchers therefore suggest that the mechanism of DHEA-estradiol mediated aggression may have evolved so that males could avoid the costs of high testosterone during the nonbreeding season.

In recent years, it has become possible to monitor hormone levels through the analysis of urine and feces rather than blood. These less invasive procedures are often used in field studies and when repeated sampling is necessary (Whitten et al. 1998).

STOP AND THINK

Corticosterone is a steroid hormone released from the adrenal glands in response to stress. If you were charged with measuring levels of corticosterone in free-living squirrels, how would you proceed? Specifically, would you collect blood, urine, or feces? What would you consider when making your choice? Would your choice differ if you were measuring a different hormone?

ORGANIZATIONAL AND ACTIVATIONAL EFFECTS OF HORMONES

The modes by which steroid hormones influence behavior may be classified as organizational and activational (Phoenix et al. 1959). We first define the dichotomy, and then we provide two examples of these different effects, the first focusing on sex differences in the mating behavior of rats and the second on individual differences in aggressive behavior of male tree lizards. We conclude this section by presenting some questions raised by researchers about the usefulness of the dichotomy.

DEFINING THE DICHOTOMY

In **organizational effects**, steroids organize neural pathways responsible for certain patterns of behavior. Organizational effects occur early in life, usually just before or after birth, and tend to be permanent. This permanence implies structural changes in the brain or other long-term cellular changes, such as in the responsiveness of neurons to steroid hormones (Arnold and Breedlove 1985). Steroid hormones can also have organizational effects on nonneural systems. For example, testosterone during the late embryonic period organizes development of the anal fin and its skeletal support structures in western mosquitofish, a species in which the male's modified anal fin is used to fertilize eggs inside the female (Rosa-Molinar et al. 1996). These changes to the male fin and its skeletal supports occur early in life, are permanent, and form the basis for sex differences in mating behavior.

Steroid hormones also affect behavior by activating neural systems responsible for mediating specific patterns of behavior. In contrast to organizational effects, **activational effects** usually occur in adulthood and tend to be transient, lasting only as long as the hormone is present at relatively high levels. In keeping with their impermanence, activational effects are thought to involve subtle changes in previously established connections (such as slight changes in neurotransmitter production or release along established pathways) rather than gross reorganization of neural pathways. At this point, we will consider the organizational and activational effects of steroid hormones as they relate to the development and display of sexual behavior in the Norway rat.

SEX DIFFERENCES IN THE BEHAVIOR OF NORWAY RATS

Not surprisingly, adult male and female rats differ in their sexual behavior (Figure 7.13). Whereas social investigation, mounting, intromission, and ejaculation typify mating in males, behavioral patterns associated with solicitation and acceptance characterize the sexual behavior of females. The lordosis posture, for example, is a copulatory position that female rats assume when grasped on the flanks by an interested male. The intensity of the lordosis response varies across the ovulatory cycle, being most pronounced when mature eggs are ready to be fertilized. The sexual behavior of female rats also includes a variety of solicitation behaviors, such as ear wiggling and a hopping and darting gait, that typically precede display of the lordosis posture (Beach 1976). Although mounting is almost always associated with males and lordosis with females, these behavior patterns occasionally occur in the other sex. Every once in a while, females will mount other females and, similarly, males will occasionally accept mounts from their cagemates. However, by and large, males display mounting and females assume the lordosis posture. These differences in patterns of

FIGURE 7.13 **Male and female Norway rats differ in their sexual behavior. Whereas mounting is characteristic of males, the acceptance posture, called lordosis, is characteristic of females. These sex differences in adult behavior are established through the action of steroid hormones around the time of birth.**

adult copulatory behavior are due to differences in the brains of male and female rats, differences that were induced by the irreversible actions of androgens in late fetal and early neonatal life. Let's now consider in some detail the organizational effects of gonadal steroids on sexual behavior.

It is clear that testosterone in the bloodstream of neonatal rats produces organizational effects. During perinatal life, male and female rats have the potential to develop neural control mechanisms for both masculine and feminine sexual behavior. Certain neurons in the brains of both males and females have the capacity to bind sex hormones. During a brief period, starting about two weeks after conception and extending until approximately four to five days after birth, however, testosterone secreted by the testes of developing males is bound to receptors in the target neurons (testosterone also binds to target muscles at this time). Once there, testosterone initiates the production of enzymes that will switch development onto the "male track." The neonatal testosterone causes males to (1) develop the capacity to express masculine sexual behavior and (2) lose the capacity to express feminine copulatory behavior.

Experiments involving castration and hormone replacement techniques have demonstrated the organizational effects of early secretion of testosterone. Removal of the testes in a rat soon after birth results in an adult with a reduced capacity to display masculine patterns of sexual behavior and an enhanced capacity to display feminine patterns. These males are capable of high levels of female solicitation and lordosis as adults. However, if removal of the testes is followed by an experimental injection of testosterone before five days of age and the proper male hormones are administered in adulthood, the rat will display normal male sexual behavior. Normal female fetuses produce low levels of testosterone, so the male developmental pattern is not initiated. A single injection of testosterone into a female rat soon after birth, however, produces irreversible effects on her adult sexual behavior. The testosterone-treated female shows fewer feminine and more masculine patterns of copulatory behavior than does a normal female. Thus, the development of a "male" brain requires the presence of testosterone around the time of birth. In the absence of testosterone, a "female" brain develops. The effects of perinatal testosterone secretion on adult sexual behavior are organizational in that they occur early in life and involve permanent structural changes in the brain. Figure 7.14 summarizes sexual differentiation in the brain and behavior of the young rat.

Before moving to the activational effects of sex steroids, we should mention that masculinization of the brain and behavior is somewhat more complex than just described. In laboratory rats, testosterone appears to be only an intermediate chemical in the process, and it is estradiol, a hormone usually associated with females, that actually directs development along the masculine track. Testosterone enters neurons in specific regions of the brain and is converted intracellularly to estradiol, which in turn causes masculinization. Look again at Figure 7.4 to see that steroid hormones are chemically very similar and that testosterone lies along the pathway of synthesis of estradiol. For the record, androgens, acting through androgen receptors, still have a role in copulatory behavior; recall that androgens masculinize the muscles of the penis (levator ani/bulbocavernosus) and their associated motor neurons.

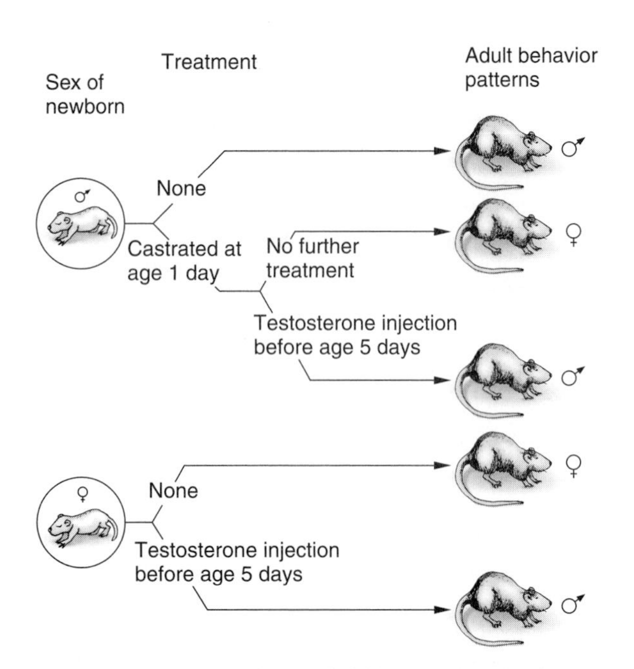

FIGURE 7.14 **Pattern of sexual differentiation in the brain and behavior of the Norway rat.**

The main question that arises from estradiol's role in the masculinization process is this: Why doesn't estradiol have the same effect in young female rats? To begin with, the levels of estradiol in young females are very low. In addition, during this critical period of brain development, an estrogen-binding protein, called alpha-fetoprotein, is produced in the livers of the fetuses. This protein, found in the cerebrospinal fluid of newborn males and females, persists in ever-decreasing amounts during the first three weeks of life. During this time, alpha-fetoprotein prevents estradiol from reaching target neurons in the brain. In female rats, then, alpha-fetoprotein binds any circulating estradiol and thereby prevents it from initiating the male pattern (McEwen 1976). Alpha-fetoprotein does not, however, bind testosterone. Thus, in male rats, testosterone produced by the testes can reach the brain, be converted to estradiol, and result in sexual differentiation.

In adulthood, steroid hormones produce activational effects on sexual behavior in male and female rats. Female rats with high blood levels of estrogen and progesterone display feminine sexual behaviors in the presence of a sexually active male, but these patterns rarely occur when levels of these ovarian hormones are low. In fact, an adult female whose ovaries have been removed will not copulate unless she receives injections of estrogen and progesterone. Similarly, removal of the testes in an adult male eventually eliminates copulatory behavior, unless he is given injections of testosterone. In these cases, the effects of steroid hormones on sexual behavior are described as activational because estrogen and progesterone in females and testosterone in males presumably exert their effects by activating existing neural pathways. High levels of the gonadal steroids activate specific patterns of sexual behavior. Thus, in contrast to permanent changes in sexual behavior caused by administration of testosterone during the neonatal period, only a transient activational effect on copulatory behavior is produced by sex steroids in adulthood.

One final point will help to distinguish organizational and activational effects of steroid hormones on sexual behavior. Males and females that have had their reproductive organs removed in adulthood generally cannot be induced to behave like members of the opposite sex. For example, a female rat whose ovaries have been removed in adulthood cannot, through injections of testosterone, be induced to show mounting behavior. By adulthood, the nervous systems of adult males and females have already differentiated (i.e., the organizational effects of early steroid secretion have long since occurred), and the mature brains are not capable of responding to hormonal signals of the opposite sex.

INDIVIDUAL DIFFERENCES IN THE BEHAVIOR OF MALE TREE LIZARDS

We have focused on how organizational and activational effects of steroid hormones explain differences between the sexes, using sexual behavior of the Norway rat as our example. More recently, the organizational/activational dichotomy has been used to understand differences between individuals of the same sex. Because individual differences are most pronounced in species with naturally occurring alternative male phenotypes, these species have been the focus of this line of research (Crews and Moore 2005). In the chapter opener we described the alternative male phenotypes of plainfin midshipman fish; now we consider those of the tree lizard.

The tree lizard (*Urosaurus ornatus*) has alternative male phenotypes that differ in aggressive behavior and color of the dewlap. Aggressive males have orange-blue dewlaps, whereas nonaggressive males have orange dewlaps (Moore et al. 1998). Males are one phenotype or the other and remain so for life; such alternative phenotypes are said to be *developmentally fixed*. The two types of males have similar hormone profiles in adulthood, so activational effects are not indicated. It has been found that differences between the male phenotypes are organized by steroid hormones during the neonatal period. Further work with tree lizards has shown that the nonaggressive males are flexible in their territorial behavior, switching between sedentary and nomadic behavior in response to environmental conditions (Knapp et al. 2003); these alternative phenotypes are thus said to be *developmentally plastic*. When conditions become stressful, the hormone corticosterone rises and ultimately (through its effects on testosterone) reduces site fidelity, triggering nomadic behavior in nonaggressive males. This switch in territorial behavior is prompted by changes in hormone levels in adulthood and thus reflects activational effects of steroid hormones. Males of the aggressive phenotype do not show plasticity in their territorial behavior; they are always territorial. The findings for tree lizards support the **relative plasticity hypothesis** proposed by Michael Moore concerning the actions of steroid hormones and alternative male phenotypes. Moore (1991) hypothesized that alternative phenotypes that are developmentally fixed rely on organizational effects of steroid hormones, whereas alternative phenotypes that are developmentally plastic rely on activational effects.

QUESTIONING THE DICHOTOMY

Although many researchers employ the distinction of organizational and activational effects of steroid hormones on behavior, Arnold and Breedlove (1985) questioned the usefulness of this dichotomy. They reviewed experimental findings from the previous decade and concluded that the organizational–activational distinction was too restrictive. How would one classify, for example, effects produced by steroid hormones that were both organizational and activational in nature, such as the production of permanent effects in adulthood? Also, while acknowledging the wealth of behavioral evidence supporting the organizational–activational dichotomy, their attempts to uncover biochemical, anatomical, or physiological evidence of two fundamentally different ways in which steroid hormones act on the nervous system were unsuccessful. In their opinion, failure to find specific cellular processes uniquely associated with each type of effect further blurs the organizational–activational distinction. Although it is important to keep such concerns in mind when discussing steroid influences, we believe that the traditional distinction of organizational and activational effects is still useful in categorizing hormonal effects on behavior.

THE DYNAMIC RELATIONSHIP BETWEEN HORMONES AND BEHAVIOR

The interaction between hormones and behavior is a dynamic one. As an illustration, we will consider the reciprocal relationship between hormones and behavior, and also show how hormones can rapidly and adaptively suppress a behavior.

A RECIPROCAL RELATIONSHIP

Whereas hormones can activate specific forms of behavior, behavioral stimuli can, in turn, induce rapid changes in the levels of those hormones. We already have mentioned that aggressive interactions in male song sparrows cause levels of testosterone to rise. Sexual stimuli have also been shown to trigger rapid increases in androgen levels. The marine toad (*Bufo marinus*), a native of Central and South America, is an explosive breeder and the first amphibian species in which it was shown that sexual behavior could affect hormonal state.

Orchinik, Licht, and Crews (1988) studied two populations of marine toads in Hawaii, where the species breeds year-round, with bursts of mating activity following heavy rainfall. During these breeding explosions, males typically compete to clasp the limited number of females, and mating involves prolonged amplexus. When male toads were allowed to clasp stimulus females for zero, one, two, or three hours, concentrations of androgens (testosterone and a form of testosterone called 5-alpha-dihydrotestosterone, or 5-DHT) increased with the number of hours spent in amplexus (Figure 7.15). In addition, in field-sampled males, androgen concentrations were higher in amplexing males than in unpaired, "bachelor" males. The apparent rise in androgens during amplexus suggests that mating behavior induced the hormonal response rather than vice versa. Although a similar relationship between amplexus and androgen level has been found in several other species of frogs and toads, the pattern is not found in all species examined to date (Moore et al. 2005).

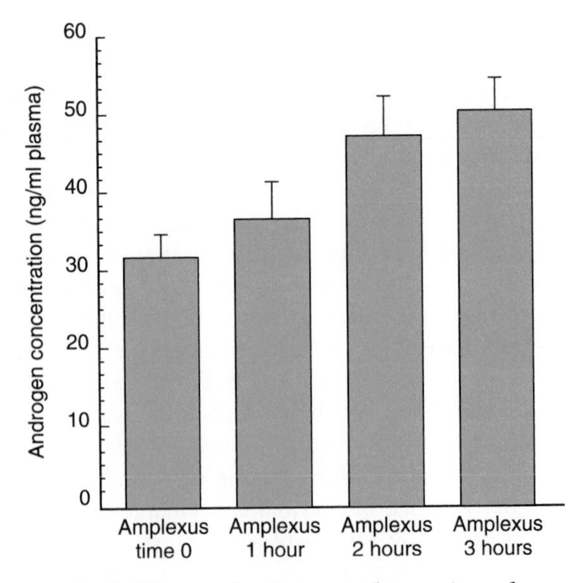

FIGURE 7.15 Changes in plasma androgen in male marine toads as a function of the time spent in amplexus. The rise in androgens during amplexus suggests that mating behavior induced the hormonal response rather than vice versa. (From Orchinik et al. 1988.)

HORMONAL SUPPRESSION OF BEHAVIOR

Hormones can also rapidly and selectively suppress a behavior, when such suppression is appropriate. We return to the roughskin newt (*Taricha granulosa*), whose mate choice behavior we described earlier in the chapter. Androgens and vasotocin mediate amplexus in roughskin newts, and undisturbed males will clasp a gravid female for several hours while she becomes sexually receptive. But what happens when a male in amplexus detects a predator nearby? Does he continue with amplexus and "hope for the best," or does he terminate the behavior and seek a safe hiding place? Researchers have found that the hormone corticosterone rapidly suppresses amplectic-clasping behavior in male roughskin newts. This was discovered by administering corticosterone to amplexing males and also by exposing males to stressful conditions known to prompt corticosterone secretion (Moore and Miller 1984). In addition to its observable suppressive effects on clasping behavior, corticosterone also reduces the activity of certain neurons that are typically active when tactile stimuli trigger amplexus (Rose and Moore 1999). The actions of corticosterone on neurons and clasping behavior occur within minutes of its administration; these rapid responses indicate that corticosterone is acting via a receptor in the membrane of neurons rather than by binding to an intracellular receptor and altering gene expression and protein synthesis (Moore and Evans 1999; Orchinik et al. 1991). We see, then, that the dynamic interactions between hormones, behavior, and the nervous system allow the behavioral state of an animal to shift rapidly and adaptively (if not always conveniently!) to changing environmental circumstances.

INTERACTIONS BETWEEN HORMONES, BEHAVIOR, AND THE ENVIRONMENT

Our example of the roughskin newt shows that hormones provide a mechanism through which an animal can adjust its behavior so that it is appropriate for the situation at hand. Next, we take a look at how hormones, behavior, and the environment interact to generate adaptive behavior in the long term. With respect to the role of the environment, we first consider the physical environment and then the social environment.

ADJUSTING TO THE HARSHNESS AND PREDICTABILITY OF THE PHYSICAL ENVIRONMENT

The habitats of different species vary in the number of mating opportunities they provide. As a result, the association between gonadal hormones and sexual behavior varies among species in ways that allow animals to produce the greatest number of surviving offspring.

David Crews (1984, 1987) compared patterns of reproduction in a wide variety of vertebrates and found numerous exceptions to the "rule" of hormone dependence of mating behavior that we observed in the Norway rat. In his survey, Crews considered relationships among the following three components of the reproductive process: (1) production of gametes, (2) secretion of sex steroids by the gonads, and (3) timing of mating behavior. Amid the diversity of reproductive tactics of vertebrates, the following three general patterns of reproduction emerged: associated, dissociated, and constant (Figure 7.16). As we describe these three general patterns, keep in mind that even these three categories do not cover all of the reproductive patterns exhibited by vertebrates. Indeed, as data on additional species accumulate, it is clear that some species fall between these three categories and that the diversity of reproductive patterns is better described as a continuum rather than as consisting of several discrete categories (Woolley et al. 2004).

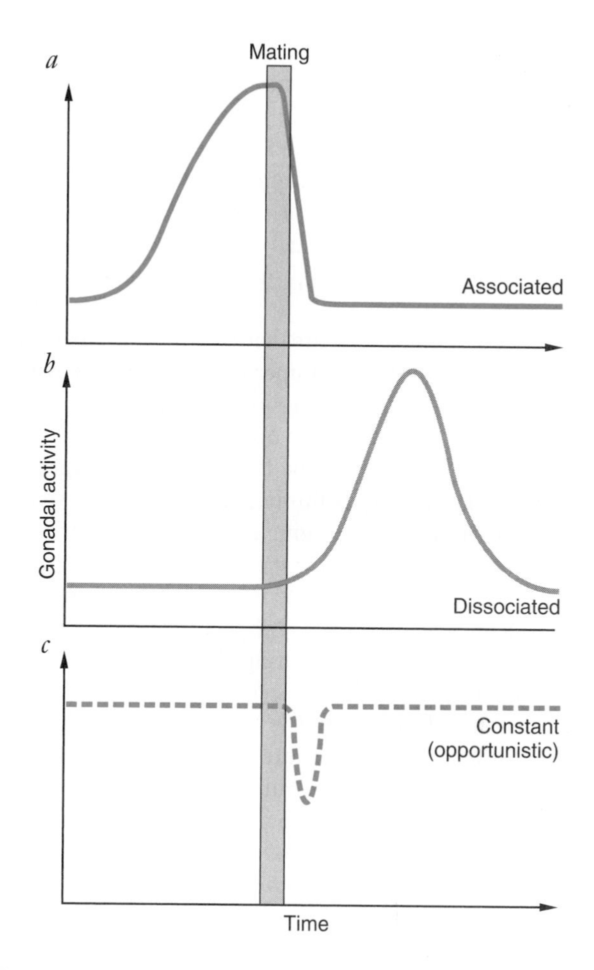

FIGURE 7.16 **Vertebrates exhibit three general patterns of reproduction. (*a*) In species exhibiting the associated reproductive pattern, mating occurs at the time of maximum gonadal activity as measured by the maturation of gametes and peak levels of sex steroids. (*b*) In species exhibiting the dissociated pattern, mating occurs at a time of minimal gonadal activity. (*c*) In species exhibiting the constant (also called the opportunistic) reproductive pattern, gonadal activity is maintained at or near maximum levels at all times. (From Crews 1987.)**

Some animals, such as the Norway rat, exhibit a close temporal association between gonadal activity and mating; specifically, gonadal growth and an increase in circulating levels of sex steroids activate mating behavior. This pattern of gonadal activity in relation to mating has been termed an **associated reproductive pattern** (Figure 7.16*a*) and has been found in most vertebrates studied to date (Crews and Moore 2005).

Some species, however, exhibit a dramatically different pattern of reproduction in which mating behavior is completely uncoupled from gamete maturation and secretion of sex steroids. In species that exhibit the **dissociated reproductive pattern** (Figure 7.16*b*), gonadal activity occurs only after all breeding activity for the current season has ceased, and gametes are thus produced and stored for the next breeding season. Gonadal hormones may not play any role in the activation of sexual behavior in species that display the dissociated pattern.

Typically, species with a dissociated reproductive pattern inhabit harsh environments in which there is a predictable, but narrow, window of opportunity to breed, and a specific physical or behavioral cue triggers mating behavior. Consider, for example, the red-sided garter snake (*Thamnophis sirtalis parietalis*), a species that ranges farther north than any other reptile in the Western Hemisphere. The window of mating opportunity for this snake is from one to four weeks, and courtship behavior of adult male garter snakes is activated by an increase in ambient temperature following winter dormancy, rather than by a surge in testicular hormones (reviewed by Crews 1983; Crews and Moore 2005).

Garter snakes in western Canada emerge in early spring from subterranean limestone caverns, where they have hibernated in aggregations of up to 10,000 individuals. Males emerge first, en masse, and congregate at the den opening. Females emerge singly or in small groups over the next one to three weeks and mate with males that are hanging out at the entrance. Because of this timing difference in the emergence of males and females, males greatly outnumber females at the den opening, sometimes 500 to 1 (Figure 7.17). In view of these odds, it is not unusual for a writhing mass of snakes, called a "mating ball," to form, in which over 100 males attempt to mate with a single female. Against all odds, females usually mate with only one male and immediately disperse to summer feeding grounds, where they give birth to live young in August. Males, on the other hand, remain at the den opening and move to feeding grounds only after all the females have emerged.

Testicular activity is minimal in male garter snakes during the period of emerging and mating. In fact, it is five to ten weeks later, after the males have left the den site and will no longer court females, that the testes grow and androgen levels increase. Sperm produced during this time is stored for use during the next spring.

Male red-sided garter snakes use environmental cues instead of circulating levels of sex hormones to determine the appropriate season for mating. Numerous experiments, utilizing castration and replacement techniques and destruction of either the temperature-sensing areas of the brain or the pineal gland, have revealed that, rather than relying on surges of sex steroids, the neural mechanisms that activate sexual behavior in male garter snakes are triggered by a shift in temperature (Crews et al. 1988; Krohmer and Crews 1987). More specifically, it is the increase in temperature in the spring that follows a long period of cold temperatures and dormancy. This is not to say that courtship behavior is completely independent of hormonal control or that sex steroids do not play an organizational role in the development of sexual behavior, but rather that mating does not occur at the time of maximum gonadal activity.

Like the male, the female red-sided garter snake mates when her gonads are small, gametes immature, and circulating levels of sex steroids low. In the case of the female, however, changes in sexual attractivity and receptivity are mediated by physiological changes that occur as a consequence of mating. Thus, although both male and female red-sided garter snakes display a dissociated pattern of reproduction, they differ in the type of stimulus that triggers breeding behavior. Whereas a change in ambient temperature triggers courtship behavior in males, stimuli associated with mating appear to activate physiological changes in females.

The third type of reproductive tactic, described by Crews (1987) as the **constant or opportunistic reproductive pattern** (Figure 7.16*c*), is characteristic of species that inhabit harsh environments, such as certain deserts, where suitable breeding conditions occur suddenly and unpredictably. In the case of desert-dwelling animals, reproduction is often initiated by rainfall. While waiting for suitable circumstances in which to breed, these species maintain large gonads, mature gametes, and high circulating levels of sex steroids for prolonged periods of time.

In Chapter 6 and earlier in this chapter we introduced you to the singing behavior and underlying changes in the brain of male zebra finches, *Taeniopygia guttata*. Here we focus on the reproductive behavior and physiology of male and female zebra finches, with an emphasis on life history. Zebra finches live in the deserts of Australia, where rainfall occurs rarely and unpredictably. Through droughts that can last for years, males

FIGURE 7.17 **Male red-sided garter snakes wait at den openings for emerging females. Dense mating aggregations form as females emerge singly or in small groups. The activation of sexual behavior in this species is independent of sex steroids.**

and females maintain their reproductive systems in a constant state of readiness. No matter how long the wait, each sex is poised, prepared for the opportunity to breed (Serventy 1971). The connection between reproduction and rainfall is based on food; the rains produce flushes of grass seeds that the adults feed to their young.

Courtship among males and females in desert populations begins shortly after the rain starts to fall; copulation occurs within hours, and nest building can begin as early as the next day. To maintain this accelerated pace, both males and females carry material to the nest. It is interesting that in more humid areas of their range, where the reproductive needs are not so immediate, males and females exhibit the division of labor characteristic of finches—that is, the male alone carries grass to the nest, and the female waits at the nest for each delivery and arranges the new material as it arrives (Immelmann 1963). Consistent with these early observations of behavioral differences in zebra finches living in climatically different environments are more recent data showing that degree of breeding readiness—as measured by size of testes in males and ovarian follicles in females—also varies with habitat predictability. Zebra finches living in arid rangelands of central Australia, a habitat with highly unpredictable rainfall patterns, maintain higher levels of reproductive readiness than do those living in a seasonal, more predictable habitat in southeastern Australia (Perfito et al. 2007). This comparison of populations of the same species occupying habitats that differ in predictability and harshness provides strong evidence of the close tie between environmental conditions and the reproductive behavior and physiology of a species.

ADJUSTING TO ONLOOKERS IN THE SOCIAL ENVIRONMENT

Like the physical environment, the social environment of an animal can influence hormone–behavior relationships. We will consider how the behavior and hormone levels of individuals can change when conspecifics are watching.

Male Siamese fighting fish (*Betta splendens*) are spectacularly colorful creatures with long flowing fins and a propensity for building nests at the water surface. These nests, made of mucus-covered bubbles blown by the males, attract females and serve as home for eggs and newly hatched fry. But don't let the beauty and home-making tendencies of male Siamese fighting fish fool you, for as the common name of this fish suggests, they are also known for their pugnacity. During aggressive contests, males flare their gills, beat their tails, and bite one another. Under natural conditions, males defend territories centered on their nest, and such territories may be closely spaced in the environment.

Given the apparent importance of the nest and neighbors in the natural history of Siamese fighting fish, Teresa Dzieweczynski and colleagues hypothesized that social environment (in this case, presence or absence of an audience) and territory status (here, presence or absence of a nest) would influence aggressive behavior and hormone levels in males of this species (Dzieweczynski et al. 2005, 2006). The experimental setup consisted of three abutting tanks that were separated by opaque partitions prior to testing; the two males to be scored for aggressive behavior were each placed in a tank (Figure 7.18). For the audience conditions, either a male or a female was confined in a small transparent container in the third tank, or the container in the third tank was left empty. Once all fish were in their respective tanks, they were given 24 hours to adjust to their new surroundings. At the time of testing the next day, first the opaque partition that separated the two test fish from the audience fish was removed and test fish were given five minutes to view either the audience or the empty container in the third tank. Then, the opaque partition between the two test fish was removed and their behavior scored for 20 minutes. In a subset of the males, the authors also measured levels of 11-ketotestosterone (11 KT), a major androgen in fish known to mediate aggressive behavior. This hormone was extracted from water into which each test fish was moved and housed for two hours after the behavioral test was completed. (This noninvasive method of hormone collection is another example of the efforts being made to measure hormones without stressing or killing the test subjects.)

Aggressive behavior and levels of 11KT were influenced by male territory status and audience. When neither test male had a nest, they were less aggressive

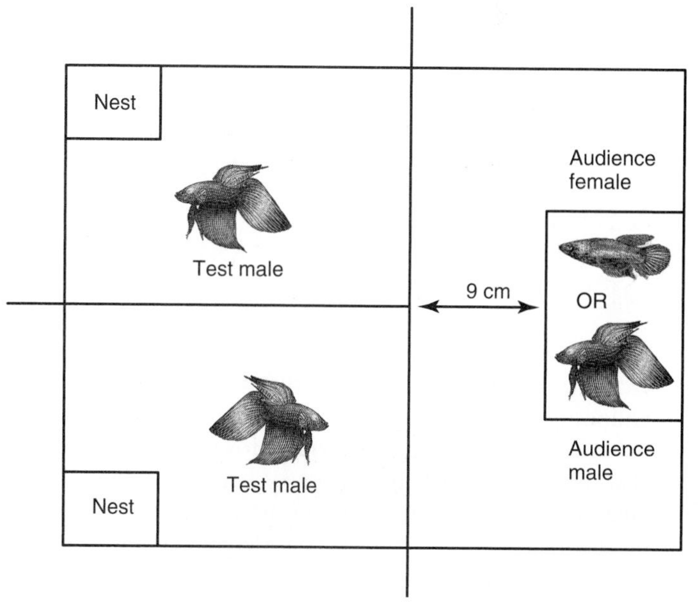

FIGURE 7.18 **Experimental setup for testing the response of male Siamese fighting fish to presence and type (male or female) of audience. (Modified from Dzieweczynski et al. 2005.)**

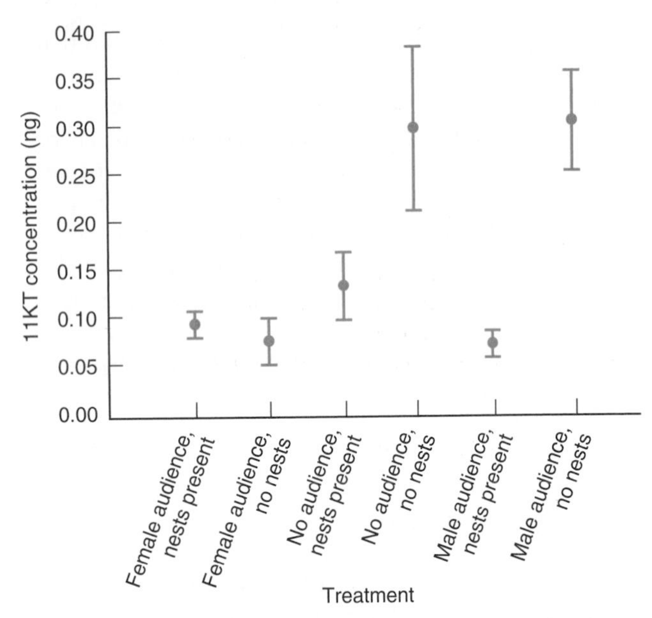

FIGURE 7.19 **In male Siamese fighting fish, levels of 11-ketotestosterone (11KT) are influenced by presence of a nest and audience. Generally, levels of 11KT were lower in males with nests and when a female audience was present. A similar pattern was found for aggressive behavior. (From Dzieweczynski et al. 2006.)**

when a female audience or no audience was present than when a male audience was present. In addition, in the presence of a male audience, males were less aggressive when both had nests as compared to when neither had a nest. The results for hormone levels generally paralleled those for behavior; levels of 11KT were lower in test males with nests as compared to those without nests and lower with a female audience (Figure 7.19). We see, then, that in Siamese fighting fish, levels of aggressive behavior during male–male encounters and the mediating androgen 11KT are influenced by nesting status of the male and who, if anyone, is watching.

A DETAILED LOOK AT THE HORMONAL BASIS OF SELECTED BEHAVIORS

As you have undoubtedly noticed while reading this chapter, much of the work in the field of behavioral endocrinology focuses on the hormonal bases of sexual behavior and aggressive behavior. However, hormones also influence many other types of behavior. Here we take an in-depth look at the hormonal bases of three behaviors that we have not yet considered: helping at the nest, scent-marking, and migrating.

HELPING AT THE NEST

Helpers—nonbreeding animals that assist the breeding pair in rearing their young—have assorted parental duties, including providing food and protecting the young. Helping has been reported in over 200 species of birds and more than 100 species of mammals.

One of the favorite species to study is the Florida scrub jay (*Aphelocoma coerulescens*), a bird that lives in south-central Florida, usually in the dry oak scrub. Although a group can range from two to six members, it usually has three. As we will discuss in Chapter 19, the helpers are offspring of the breeding pair from a previous year who remain on the territory and help raise their siblings. Because the siblings share some of the helpers' genes, the helpers manage to get some of their genes into the population, even though the lack of a territory prevents them from breeding. In this way, the helpers make the best of a bad situation. Thus, we can see the evolutionary causes of helping. But what about the physiological basis of helping behavior?

An obvious first question to ask is whether helpers help because they are physiologically incapable of breeding themselves. Ronald Mumme and colleagues have tested this hypothesis in a population of Florida scrub jays at the Archbold Biological Station. Their data suggest that there are no physiological reasons that cause helpers to delay reproduction (Schoech et al. 1996; Schoech 1998). Helpers do produce the hormones important to reproduction. In both males and females, breeders and nonbreeders have the same levels of luteinizing hormone (LH), the hormone from the anterior pituitary gland that stimulates the growth and development of the ovaries and testes at the start of each breeding season. Although male breeders have somewhat higher levels of testosterone than do male helpers, the seasonal pattern of testosterone production is essentially the same. In females, breeders and helpers have the same level of estradiol, their primary sex steroid. However, the seasonal pattern of estradiol production is different in female breeders and female helpers. Nevertheless, these observations reveal that the testes and ovaries of helpers are functional, at least for hormone production.

Another hypothesis is that helpers might be physiologically capable of breeding, but that they delay reproduction because they are unable to gather enough food in order to breed successfully. This, too, does not appear to be the case. Helpers are indeed slightly smaller than breeders, but this probably is not because they are less successful foragers, but because they are younger. When the size difference is taken into account, the weights of male helpers and breeders are equivalent. Female helpers are apparently able to forage as well as female breeders during the winter months because they weigh about the same at the start of the breeding season. If the helpers delay breeding because they cannot gather enough food, we would expect that supplying supplemental food to the population would allow more helpers than usual to switch to being breeders. When this was tried, however, the additional food did not increase the number of helpers that became breeders. Failure of food supplementation to prompt helpers to become breeders is even more striking, given recent data indicating that food supplementation has positive effects on reproduction in scrub jays: breeders in supplemented groups of jays initiated clutches earlier than did those in control groups that were not supplemented (Schoech et al. 2004).

Another hypothesis that might explain why helpers delay reproduction is stress: perhaps the presence of the dominant breeding pair stimulates the helper's adrenal gland to produce the stress hormone corticosterone, which is known to suppress the production of the hormones needed for reproduction. This doesn't seem to occur. Helpers and breeders have equivalent amounts of corticosterone throughout the breeding season.

Thus, it appears that helpers are physiologically capable of reproduction, but simply lack the opportunity to do so. The next proximate question about helping behavior concerns the role of hormones in initiating parental behaviors, such as feeding the nestlings. Parental behaviors may be caused by the pituitary hormone prolactin in both helpers and breeders. Prolactin production is stimulated by cues from the nest, eggs, and nestlings. For this reason, prolactin levels increase throughout the breeding season. Birds that spend the most time caring for the eggs and young produce the most prolactin. In general, females produce more prolactin than males, and breeders of either sex produce more than helpers. Prolactin levels are lower in helpers because breeders won't allow them near the nest until the young have hatched. But both breeders and helpers feed the young. There is a direct relationship between a helper's level of prolactin and the feeding score (a measure of how much a bird fed the nestlings) (Figure 7.20). Notice that some helpers didn't actually help; they have feeding scores of zero. The prolactin levels of the helpers that *did* help are much higher than those of birds that did not help. It is interesting that there is no correlation between the prolactin levels of breeders and their feeding score. Prolactin is known to affect many aspects of parental behavior besides feeding the young. A relationship between prolactin and parental behavior among breeders may not be seen in these data because parental behaviors other than feeding were not measured (Schoech et al. 1996).

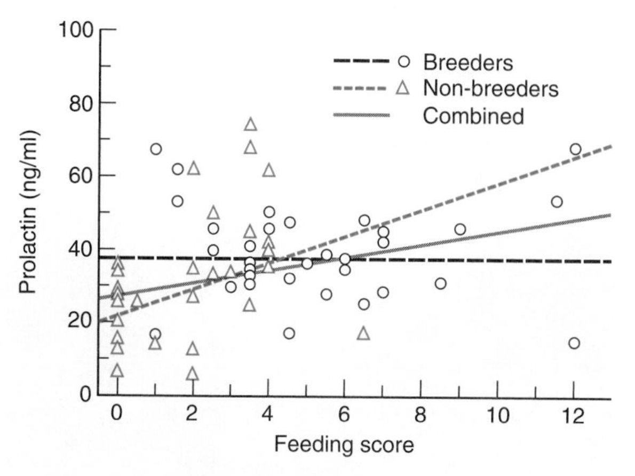

FIGURE 7.20 In Florida scrub jays, there is a direct relationship between how much a non-breeder feeds nestlings and its prolactin levels. (Data from Schoech et al. 1996.)

SCENT-MARKING

Scent-marking is the act of strategically placing a chemical mark in the environment; many mammals apply urine, feces, or secretions from special scent glands. (Recall that scent glands are exocrine glands and, as such, differ from endocrine glands in having ducts that release their products to the body surface. Scent glands are found in many locations on the body and, depending on the species, can occur in such locations as between the digits, on the legs, chest, or belly, on the head, or in the anal canal. Many mammals have scent glands at multiple locations.) Scent marks likely convey information about individual identification, age, and reproductive state, and function to establish and maintain territories and breeding relationships (see Chapter 16 for more information on scent-marking). Our next example concerns a behavior that probably everyone has observed—urine-marking by domestic dogs. Urine-marking in dogs is distinguished from simple elimination by the fact that urine is directed at a specific object in the environment, such as a tuft of long grass or a fire hydrant.

The urinary behavior of domestic dogs (*Canis lupus*; yes, dogs now have the same scientific name as wolves and no longer go by *Canis familiaris*) is sexually dimorphic. Adult males urinate more frequently than do adult females and are more likely to direct their urine at objects in the environment (i.e., to urine-mark). Even urinary posture is sexually dimorphic; whereas males lift a leg to urinate, females typically squat. Sex differences in urinary posture are organized by sex steroids (testosterone) around the time of birth (Beach 1974; Ranson and Beach 1985). Testosterone is not, however, required to activate the leg-lifting posture in adulthood. As any owner of a castrated dog can attest to, even though the source of testosterone has been removed, a neutered male still lifts his leg to urinate. We see from this example that whereas some sexually dimorphic patterns of behavior, such as sexual behavior in Norway rats, are organized and activated by sex steroids, others are simply organized by these hormones.

As we have said several times in this chapter, things are often more complicated than they first appear, and further research often reveals new details about particular behaviors and their hormonal control. Recent studies with female Jack Russell terriers have shown that marking with urine is more common than previously described for female dogs (about 60% of urinations by females are directed at objects in the environment) and that not all female dogs squat (Wirant and McGuire 2004). Indeed, there is considerable variation among females in the postures used while urinating, with some females displaying the traditional squatting posture, while others use squat-raises (one leg is raised while in a squatting posture) and even handstands (Figure 7.21). Furthermore, the propensity to mark with urine varies across the estrous cycle, being most common just before and during estrus. These data suggest that female dogs mark with urine to convey information about their reproductive state (Wirant et al. 2007). Whether these patterns of urinary behavior in female Jack Russell terriers will generalize to other breeds of dogs or mixed breeds remains to be seen. The precise physiological bases for changes across the estrous cycle and for variation among females in urinary behavior also remain to be determined.

MIGRATING

We have seen that hormones influence behavior in diverse ways, sometimes altering a preference for certain stimuli or perhaps influencing the nervous system or muscles. We have also seen that the relationship between hormones and behavior is dynamic, allowing animals to adjust to their physical and social environments. We end the chapter by considering hormone-induced changes in the behavior, physiology, and morphology of the Atlantic salmon (*Salmo salar*), a species whose complex and fascinating life history centers on making repeated movements between freshwater and saltwater. We begin with a brief summary of the species' life cycle.

Adult Atlantic salmon leave the ocean and return to their natal stream to spawn. Spawning typically occurs in the fall, and the adults bury thousands of large, fertilized eggs in a gravel depression called a redd. In the spring, the eggs hatch into yolk-sac larvae (alevins) that remain in the nest, living off their yolk for about six weeks. Eventually, when the yolk runs out, young salmon emerge from the nest as fry and begin to feed independently. As time passes and the salmon reach about 5 cm in total length, they develop vertical marks on the sides of their bodies. Salmon at this stage are known as parr, and the marks, called parr marks, function as camouflage. When parr reach about 15 cm in length and environmental conditions are right, they undergo a process

a

b

c

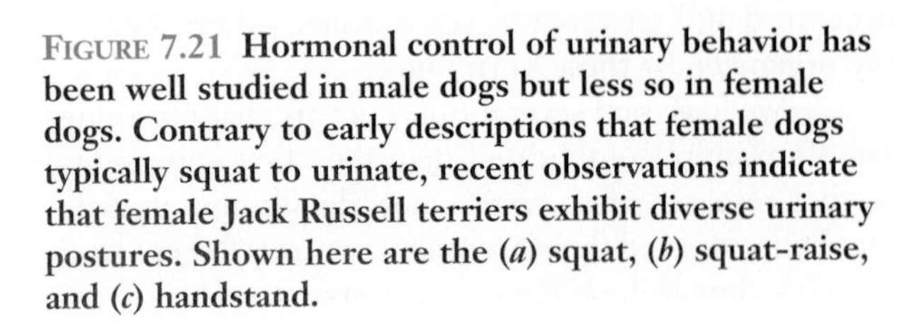

FIGURE 7.21 **Hormonal control of urinary behavior has been well studied in male dogs but less so in female dogs. Contrary to early descriptions that female dogs typically squat to urinate, recent observations indicate that female Jack Russell terriers exhibit diverse urinary postures. Shown here are the (*a*) squat, (*b*) squat-raise, and (*c*) handstand.**

called smoltification after which they are known as smolts. Smoltification is a critical developmental process that prepares young salmon for migrating to the ocean and living there for one or more years before returning to their natal stream as adults to spawn and begin the cycle anew. Because Atlantic salmon are iteroparous (reproducing repeatedly rather than dying after spawning as is characteristic of Pacific salmon), they may make multiple forays between freshwater and saltwater over the course of their lives. Our focus will be on the hormone-mediated changes associated with the first trip from their natal stream to the sea.

Dramatic changes in physiology, morphology, and behavior occur during the developmental change from parr to smolt. The osmoregulatory system of parr is set up for life in freshwater and must change for life in saltwater, so we see changes in the cells of the gills, gut, and kidneys in preparation for seaward migration and life. Also during smoltification, the parr marks that functioned as camouflage for young bottom-dwelling salmon in the shallow stream environment are replaced by the reflective silver color so characteristic of fishes evading predators in pelagic environments. Even the way young salmon position themselves in the current changes, and this undoubtedly involves adjustments in their nervous and sensory systems. Parr face into the current (a behavior called positive rheotaxis), and this position allows them to see food coming their way from upstream. Smolts, on the other hand, face downstream (negative rheotaxis), an orientation necessary for their impending migration to the sea. We also see changes in social behavior. Whereas parr aggressively defend individual feeding territories, smolts exhibit decreased territorial and agonistic behavior and eventually form schools.

What environmental factors initiate and which hormones mediate the changes associated with becoming a smolt and preparing for life at sea? Although we do not have all of the answers yet, we are beginning to understand key aspects of this transformation (Figure 7.22). Environmental factors, such as high water flow and increasing photoperiod and temperature, appear to

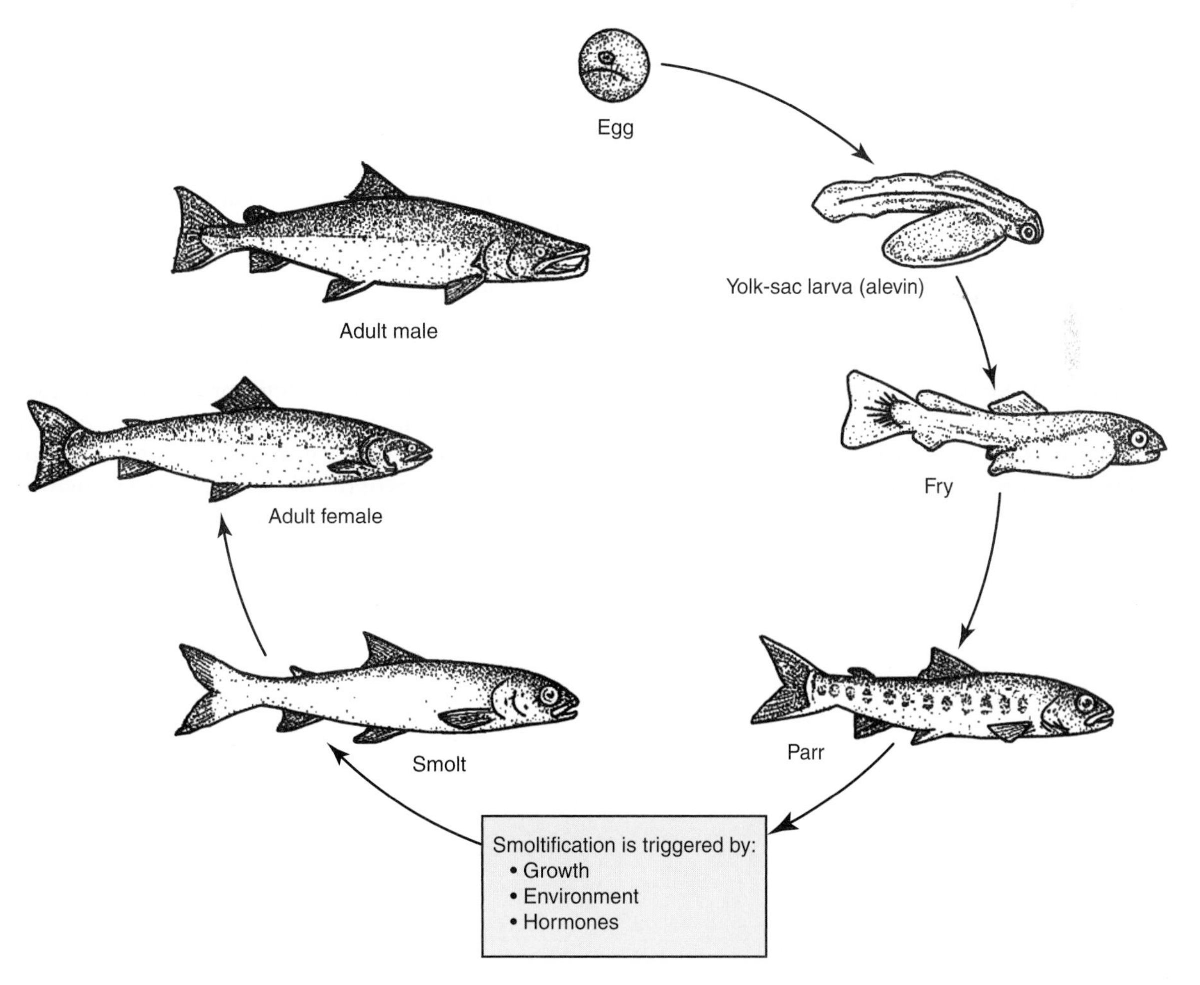

FIGURE 7.22 **Environmental cues (high water and increasing temperature and day length) prompt the developmental change from parr to smolt in Atlantic salmon that have reached a critical body length. Several hormones mediate the necessary changes in physiology, morphology, and behavior associated with the move from freshwater to saltwater.**

prompt smoltification in fish of sufficient size (McCormick et al. 1998). Several hormones have also been identified as important in mediating the necessary physiological and morphological changes. Whereas prolactin promotes osmoregulation in freshwater and decreases during smoltification, growth hormone increases and works antagonistically to increase tolerance for saltwater (Sakamoto and McCormick 2006). Cortisol appears to interact with both hormones to promote acclimation to a particular environment (McCormick and Bradshaw 2006). Thyroid hormones are responsible for the replacement of parr marks by silver coloration (Hoar 1988). The hormones involved in many of the behavioral changes of smoltification have yet to be identified.

This example shows the intricate interactions of the developing organism and its external environment, and how hormones help to orchestrate adjustments in behavior, physiology, and morphology in different environments. Development is the focus of our next chapter.

SUMMARY

Animals have two closely associated, yet different, systems of internal communication: the nervous system and the endocrine system. Typically, transfer of information occurs more slowly by the endocrine system than by the nervous system, and the effects produced are more general and long lasting. Whereas neural information is transmitted via a series of electrical events, communication by the endocrine system occurs through hormones, chemical substances that are secreted by either endocrine glands or neurons.

Hormones and neurohormones produce changes at the cellular level that ultimately influence behavior. Peptide hormones are water-soluble amino acid chains that bind to receptors at the cell surface, which activates a cascade of chemical reactions within the cell. In contrast, steroid hormones are derived from cholesterol, and because they are fat-soluble, they can move through the cell boundaries and bind to receptors inside the cell. The hormone-receptor complex then enters the nucleus and turns on certain genes. Recent evidence indicates that steroid hormones also interact with membrane receptors. This pathway produces more rapid changes in behavior than is possible in the pathway involving gene expression.

The mechanisms by which hormones influence behavior include alterations in (1) sensation or perception, (2) development and activity of the central nervous system, or (3) muscles responsible for the execution of behavior.

Traditionally, the effects of steroid hormones on behavior have been divided into organizational and activational effects. Organizational effects occur early in life and tend to be permanent. In contrast, activational effects occur in adulthood and tend to be transient, lasting only as long as the hormone is present in relatively high concentrations. In activational effects, steroids serve only to activate existing neural pathways responsible for a specific behavior rather than to organize neural pathways. Though still useful, the traditional distinction between organizational and activational effects has been questioned because of the lack of biochemical, anatomical, and physiological evidence for two fundamentally different ways in which steroids produce their effects.

Hormonal effects on behavior can be studied by the removal of the gland and hormone replacement (so-called interventional studies) or by correlational studies. In the latter, researchers look for changes in behavior that parallel fluctuations in hormone levels.

The interaction between hormones and behavior is a dynamic one. In some instances, hormones initiate changes in behavior; in others, behavior causes changes in the levels of circulating hormones. Hormones can also rapidly suppress a behavior when appropriate. The interactions between hormones and behavior are sensitive to aspects of the physical and social environment. Behaviors mediated by hormones include aggression, courtship, mating, caring for young, scent-marking, and migrating.

8
The Development of Behavior

A mallard duckling (*Anas platyrhynchos*), nestled inside its egg, hears the muffled voice of its mother overhead. After spending approximately four weeks in the egg, the youngster pecks through the shell and frees itself (Figure 8.1). It then spends one more day beneath its mother. Should a predator, sometime during the first day posthatching, wander into the area around the nest, the duckling, in concert with its eight or so siblings, responds rapidly to the alarm call of its mother by "freezing"—crouching low and ceasing all movement and vocalization (Miller 1980). If the duckling goes unnoticed by the predator, the very next morning it responds promptly to yet another of its mother's calls, this time the assembly call, by following her from the nest through the long grass to a nearby pond (Miller and Gottlieb 1978). Here, the duckling will string along behind its mother and siblings for some time to come. As the weeks pass and the young bird continues to associate with family members, it learns the characteristics of an appropriate mate (Schutz 1965). This information, though obtained early in life, will not come in handy until the first breeding season. Indeed, the duckling learns, soon after hatching, many of the things it needs to survive and reproduce. We see, then, that the changes that occur during behavioral development may contribute to fitness immediately (as in the duckling's freeze response to its mother's alarm call) and in adulthood (as in mate preference).

FIGURE 8.1 **Newly hatched mallard ducklings. Experiences prior to and soon after hatching will profoundly influence each duckling's behavioral development.**

Several questions arise from this brief description of the early behavioral development of a mallard duck. How does the genetic makeup of the duckling interact with its internal and external environment to produce such behavior? How do the nervous and endocrine systems influence behavioral development? What role do visual, auditory, or social stimuli play in the development of freezing, following, and sexual behaviors? Are experiences prior to hatching important to the development of posthatching behavior? What happens when a behavior, such as the following response, ceases to be a part of the individual's behavioral repertoire? If we look beyond the single duckling in our example and consider the species as a whole, does behavioral development always proceed in a predictable and reliable fashion? In this chapter we address these questions about the behavioral development (change in behavior over time) of mallard ducks and other animals. Recall that Niko Tinbergen (1963) considered questions about development to be one of the four main types of questions that should be asked about behavior.

INFLUENCES ON BEHAVIORAL DEVELOPMENT

Patterns of behavior come and go throughout development. A behavior may appear in an animal's repertoire, only to disappear or change shortly thereafter. Here we consider some of the factors that influence the development of behavior. Keep in mind, however, that these factors are not mutually exclusive and that they likely interact with one another throughout development. We will not focus on the role of genes in behavioral development because this topic is covered in Chapter 3.

DEVELOPMENT OF THE NERVOUS SYSTEM

Behavior is primarily controlled by the nervous system, and so is intimately linked with the development of this system. This is especially obvious early in an animal's life when development of the nervous system is rapid and dramatic. Consider, for example, the neural and behavioral development of embryonic Atlantic salmon, *Salmo salar* (Abu-Gideiri 1966; Huntingford 1986). Stages in the development of this fish are depicted in Figure 8.2.

The first movements of the embryo are seen in the feeble twitches of the heart, soon followed by movements in the dorsal musculature. Interestingly, these movements begin before the nervous system has formed and are thus myogenic, or muscular, in origin. The impulse begins in the muscle itself. Approximately halfway through embryonic life, the major motor systems appear in the spinal cord. A short time later, the motor neurons make contact with anterior muscles, giving the embryo the ability to flex its body. Soon, with the development of neural connections at different points and on both sides of the body, the embryo displays the first undulating movements associated with swimming. Development of the sensory system of the trunk and its connection to the skin occurs a short time later; after this, the embryo can move in response to tactile stimulation. Finally, the neural circuits that underlie both fin and jaw movements are complete, allowing independent and coordinated movement of these structures. Neural and behavioral development continues (in fact, the young salmon has not yet even hatched). We can see from this example that development of key parts of the nervous system underlies the appearance of new patterns of behavior.

When a behavior disappears from an animal's repertoire, does this mean that the underlying neural circuits have also disappeared? In some cases, the neural circuits are dismantled or permanently altered in concert with cessation of the behavior. As an example of this situation, we will consider changes that occur in the tobacco hornworm (*Manduca sexta*) during complete metamorphosis, when the insect changes from caterpillar to pupa to moth. Much of this work has been carried out in the laboratory of Janis Weeks (Hazelett and Weeks 2005; Weeks et al. 1989; Weeks 2003).

Complete metamorphosis entails dramatic changes, not only in the animal's morphology, but also in its behavioral repertoire (during incomplete metamorphosis, as occurs in insects such as grasshoppers, juveniles look much like adults). During the remarkable transformation of complete metamorphosis, the nervous system of the tobacco hornworm must sequentially control three very different stages: the larva or caterpillar, the pupa, and the adult moth (Figure 8.3). The animal is

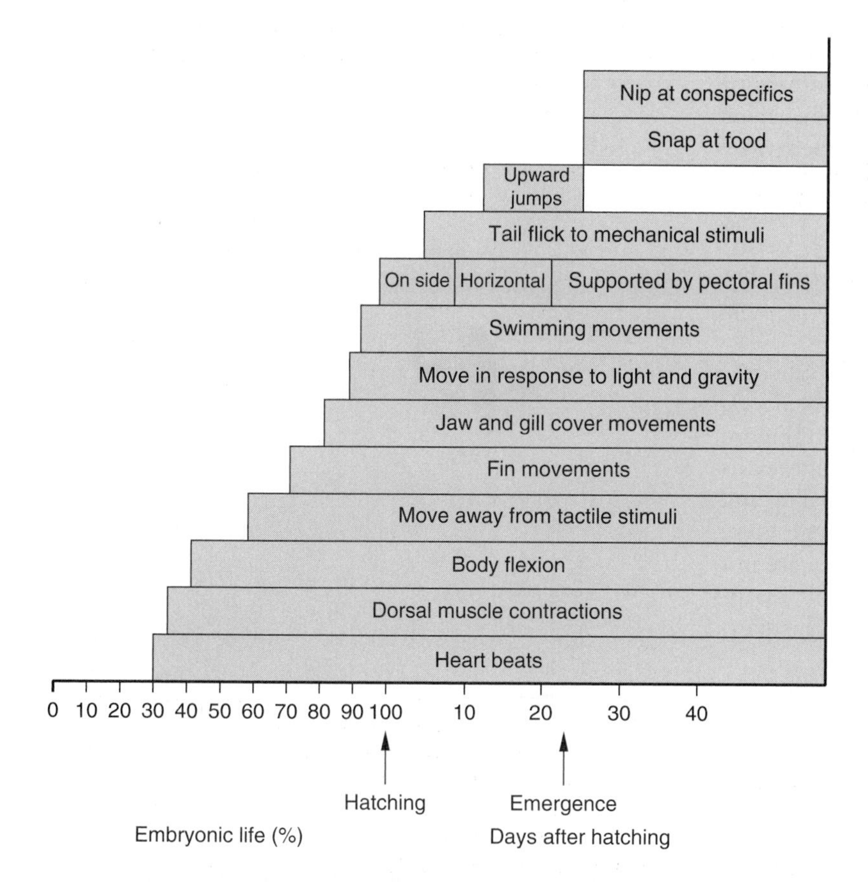

FIGURE 8.2 **Behavioral development in the Atlantic salmon. Patterns of behavior emerge in parallel with the development of neural structures necessary for their performance. (Modified from Huntingford 1986; drawn from the data of Abu-Gideiri 1966; Dill 1977.)**

a

b

c

FIGURE 8.3 **Metamorphosis in the tobacco hornworm, *Manduca sexta*: (*a*) larva or caterpillar, (*b*) pupa, and (*c*) adult.**

transformed from a crawling, eating machine to a flying, reproducing machine. Although some patterns of behavior are exhibited in all three stages (e.g., behavior associated with shedding of the cuticle), many behaviors are restricted to a single stage (e.g., crawling in the larva and flight in the adult). Much of the neural circuitry controlling stage-specific patterns of behavior is assembled and dismantled during development. Let's see what happens when crawling behavior is lost at the larval–pupal transformation.

Caterpillars of the tobacco hornworm have abdominal prolegs, stumplike appendages that are not developmentally related to the legs of the adult moth (Figure 8.4*a*). The prolegs act in simple withdrawal reflexes, as well as in more complex behaviors such as crawling and helping the animal grasp the substrate. Although these behaviors are important to the caterpillar, they are not to the pupa; and the proleg behaviors gradually disappear during the larval–pupal transformation. The question is, then, what causes their disappearance?

While we often associate the formation of new neurons with behavioral development, sometimes the death of old neurons is key to changes in behavior over time. Most proleg movements are accomplished by retractor muscles (Figure 8.4*b*) that are innervated by motor neurons with densely branching arbors, or dendrites. During the larval–pupal transformation, substantial regression of the dendrites of many of the motor neurons occurs (Figure 8.4*c*). These motor neurons die, and their associated proleg muscles degenerate and become nonfunctional. Specifically, the proleg muscles and motor neurons in abdominal segments 1, 4, 5, and 6 degenerate; we will focus on what causes the death of the motor neurons and muscles in these segments.

The demise of the proleg neuromuscular system, and hence proleg behaviors, is prompted by a peak in ecdysteroid hormones secreted by the prothoracic gland at the back of the head. This peak occurs just before the transition to the pupal stage (Figure 8.4*d*). At this time, high levels of ecdysteroids trigger regression of the dendrites of the motor neurons that innervate the proleg muscles. As a result, the motor neurons are removed from behavioral circuits, and proleg behaviors are lost in the pupa.

Does the underlying neural circuitry always disappear when a behavior is lost from an animal's repertoire? Not always. Chickens (*Gallus gallus domesticus*) typically hatch over a 45- to 90-minute period at the end of incubation. During hatching, the chick escapes from the confines of its shell through a highly stereotyped series of movements, rotating its upper body and thrusting its head and legs. Because these patterns of behavior associated with hatching later disappear from the repertoire of chickens, Anne Bekoff and Julie Kauer (1984) became interested in the fate of the neural circuitry underlying, in particular, the leg movements of hatching. Their

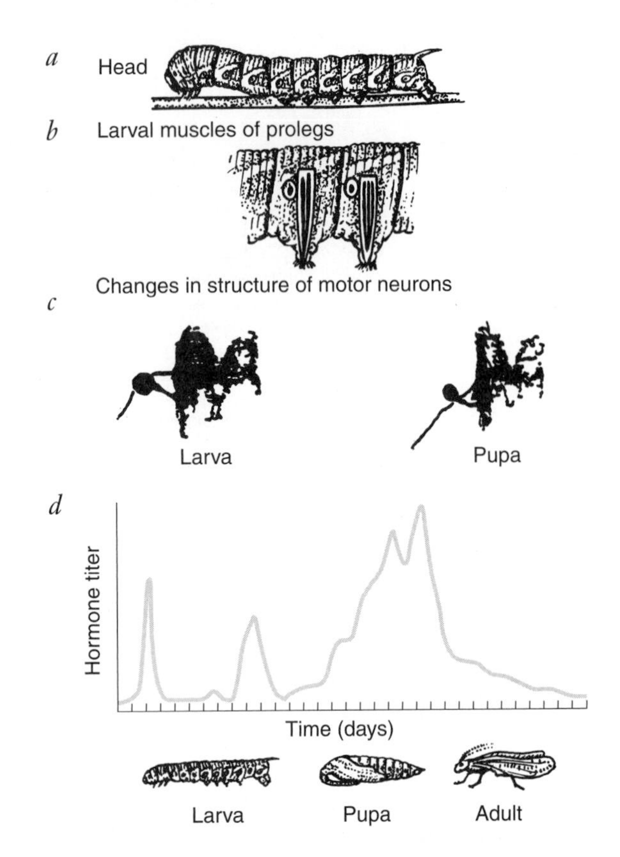

FIGURE 8.4 **During metamorphosis in the tobacco hornworm, modifications to the nervous system (induced by changing levels of hormones) eliminate patterns of behavior. (*a*) The abdominal prolegs of the caterpillar are involved in withdrawal reflexes, crawling, and grasping the substrate. These behaviors are important to the larva but disappear from the animal's repertoire once it reaches the pupal stage. (*b*) Cutaway view showing the retractor muscles of the prolegs. (*c*) The dendrites of the motor neurons that innervate the retractor muscles regress during the larval–pupal transformation. (*d*) The peak in ecdysteroid hormones just before transformation to the pupa triggers dendritic regression. (Modified from Weeks et al. 1989; data from Bollenbacher et al. 1981; Weeks and Truman 1984.)**

method was an interesting one. Rather than asking the age-old question, "Which came first, the chicken or the egg?" they asked, "What happens when you put a chicken back into an egg?" Bekoff and Kauer placed posthatching chicks up to 61 days of age (chicks at this age have molted their fuzzy down, are fully feathered, and basically resemble small chickens) in artificial glass eggs and recorded their behavior and muscle movements. Each chick was gently folded into the hatching position and placed into a ventilated glass egg of the appropriate size. Within two minutes of being placed in the artificial eggs, chicks of all ages began to produce a behavior that qualitatively and quantitatively resembled that of normal hatching. Rather than being dismantled or permanently altered after hatching, the neural circuitry for the leg movements of hatching clearly remains functional in older chickens. Additional studies suggest that a basic neural circuit for leg movements is built early in embryonic development and that it is modified to produce movements associated with hatching and later walking (Bekoff 1992). Thus, we see that in some cases the disappearance of a behavior is not associated with the complete dismantling of its underlying neural circuitry; instead, the circuitry is modified to serve other patterns of behavior.

DEVELOPMENT OF NONNEURAL STRUCTURES

Sometimes behavioral change is driven by morphological changes that are not neural. Obviously, a behavior can't be performed unless the animal has developed the morphological structures necessary for its performance. Consider the changes in feeding behavior that occur in the paddlefish *Polyodon spathula*. This fish, from the Mississippi and Ohio River drainages of North America, has a bizarre, paddle-shaped snout, which can be almost half the length of the body. Although larval paddlefish feed by chasing and selectively plucking individual zooplankton from the water, adult paddlefish are indiscriminate filter feeders, dropping their lower jaw and consuming all material strained from the water as they plow through their environment (Figure 8.5). Changes in the feeding behavior of paddlefish parallel the development of gill rakers (Rosen and Hales 1981). These bony structures, comb-like in appearance, project from the gill arches into the oral cavity and strain food particles from the water. Absent from larval paddlefish, gill rakers begin to develop when young fish are about 100 mm long (about 4 in.). Gill rakers are well-developed when the young paddlers reach 300 mm in length, and it is at this stage that feeding behavior takes on the fully adult pattern of indiscriminate filter feeding. Here, then, we have an example of how changes in behavior are coordinated with the development of specific morphological structures.

FIGURE 8.5 A paddlefish exhibiting the adult feeding pattern, filter feeding. Rather than indiscriminate filter feeding, larval paddlefish selectively pick zooplankton out of the water column. The eventual development of the adult mode of feeding parallels the development of gill rakers in the young fish.

HORMONAL MILIEU

We have seen that hormones prompt regression in parts of the neural circuitry of the tobacco hornworm caterpillar, ultimately resulting in loss of crawling behaviors in the pupa. Now we'll consider two additional examples of how developing animals are influenced by the hormonal milieu (environment) they experience early in life. Both examples relate to the effects of exposure to androgens, such as testosterone, during the prenatal period; the first example concerns house mice and the second, black-headed gulls.

Like peas in a pod, house mouse fetuses (*Mus musculus*) line the uterine horns of their mother (Figure 8.6). Each fetus has its own personal placenta (vascular connection to the mother) and floats within a fluid-filled compartment called the amniotic sac. Even before birth, the endocrine glands of these tiny individuals are producing hormones that may permanently alter not only

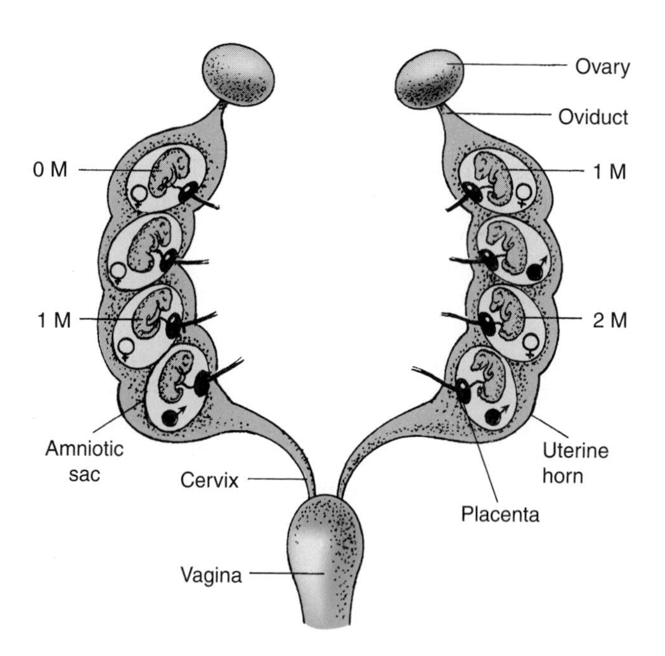

FIGURE 8.6 Mouse fetuses line the uterine horns of a pregnant female. Because the fetuses are in such close quarters, hormones from one fetus can influence behavioral development of contiguous fetuses. Female fetuses can occupy the following three positions relative to male fetuses: 2M, between two males; 1M, next to one male; 0M, not next to a male. The 2M females differ substantially from 0M females in their adult behavior and physiology as a result of proximity to male fetuses in the intrauterine environment. (Modified from McLaren and Michie 1960.)

their own but also their neighbor's morphology, physiology, and behavior. In many fascinating experiments, Frederick vom Saal, John Vandenbergh, and others have shown that development of mouse fetuses can be modified by exposure to hormones secreted by contiguous littermates (studies reviewed in Ryan and Vandenbergh 2002). As shown in Figure 8.6, fetuses can occupy three intrauterine positions relative to siblings of the opposite sex. Females can be positioned between two male fetuses (2M females), next to one male (1M females), or not next to a male (0M females). But what does intrauterine position have to do with adult patterns of behavior? It turns out that by day 17 of gestation, levels of testosterone are three times higher in the blood of male fetuses than in the blood of female fetuses. Even more intriguing is the finding that on this same day, 2M female fetuses (i.e., those females nestled between two male littermates) have significantly higher concentrations of testosterone in their blood and amniotic fluid than do female fetuses not next to males (i.e., 0M females). Apparently, hormones pass through the amniotic fluid and possibly via uterine blood vessels to contiguous littermates. As a result of prenatal exposure to testosterone, adult 2M females display traits that distinguish them from 0M females. Specifically, 2M females (1) are less attractive to males, (2) are more aggressive to female intruders, (3) mark a novel environment at a higher rate, (4) maintain larger home ranges, (5) experience their first estrus at a later age, and (6) produce fewer viable litters. These differences in physiology and adult behavior exist despite the fact that after birth, testosterone levels do not differ between the two groups of female mice. Thus, behavioral differences in adulthood can be traced to differential exposure to hormones in the intrauterine environment. Position in the uterus also affects the behavioral development of male mice, but males are somewhat less sensitive than females to intrauterine effects.

Prenatal exposure to androgens also influences the growth, physiology, and behavioral development of young birds (Groothuis et al. 2005). While developing within its egg, a bird is exposed to androgens (testosterone, dihydrotestosterone, and androstenedione) deposited into the yolk by its mother while the egg formed in her ovaries. The precise mechanism by which the mother deposits androgens is not yet known. In black-headed gulls (*Larus ridibundus*), yolk levels of testosterone and androstenedione increase with each egg laid. Because first-laid eggs have the lowest androgen levels, they are used in experimental manipulations. Androgens can be injected into the yolk of first-laid eggs without exceeding the levels typically found in later-laid eggs, thus keeping experimental hormone levels within the normal limits experienced by the species. Injections are usually given at the start of incubation, with sesame oil serving as the vehicle (substance in which the hormone is dissolved). Thus, comparisons can be made between experimental birds (those hatched from first-laid eggs injected with androgens and sesame oil) and control birds (those hatched from first-laid eggs injected only with sesame oil). Such comparisons for black-headed gulls have shown that experimental birds exhibit the following effects during the prefledging period (period before leaving the nest): (1) earlier hatching, (2) enhanced growth, (3) enhanced begging for food, and (4) suppressed immune function. Maintaining immune function is energetically expensive; the latter effect is therefore consistent with the overall pattern of enhanced androgen exposure resulting in chicks that direct energy to growth and competitiveness rather than to immune function. Less is known about the long-term effects of enhanced yolk androgens, but the data so far indicate that experimental birds show more frequent courtship and threat displays, win more aggressive encounters, and have lower adult survival when compared to control birds.

The study of avian maternal hormones is relatively new, and much remains to be learned about the effects of differing levels of androgens on developing young and the possible adaptive value of androgen deposition. Early indications are that deposition of androgens may be a way for mothers to adjust the developmental trajectories of their offspring to current environmental conditions. For example, black-headed gulls begin to incubate their eggs when the first egg is laid rather than waiting until the clutch of three eggs is complete. As a result, the chicks hatch asynchronously, with the first-laid egg hatching first, the second-laid egg hatching next, and the last-laid egg hatching last. Chicks that hatch last are smaller and not as strong as their older siblings. Perhaps, then, the greater deposition of androgens in last-laid eggs helps to lessen the effects of being the last to hatch by enhancing the growth and competitiveness of these chicks. Avian mothers also adjust androgen deposition from one clutch to the next. There is some evidence that female black-headed gulls exposed to intense competition deposit higher levels of androgens in their eggs. It is possible, then, that these mothers are preparing their offspring for the highly competitive environment they may face once they are free from the confines of their shell (Groothuis et al. 2005).

STOP AND THINK

In this section we describe the effects on behavioral development of intrauterine position in rodents and maternal hormones in bird eggs. What might be some advantages and disadvantages of each model system for exploring hormone-mediated maternal effects on developing offspring? Also, for those of you who have already read Chapter 7, are the effects of intrauterine position in rodents and maternal hormones in bird eggs organizational or activational? Explain your answer.

PHYSICAL CHARACTERISTICS OF THE ENVIRONMENT

Imagine a situation in which the mate choice of an adult animal is shaped by the environmental temperature it experienced during embryonic development. This is precisely the case for the leopard gecko (*Eublepharis macularius*), a lizard with temperature-dependent sex determination. In this species (as well as in several other species of lizards, many turtles, and all crocodilians), whether an individual is a male or a female is determined not by sex chromosomes, but by the temperature experienced during a window of time about midway through its incubation period. More specifically, in leopard geckos, females are typically produced at low and high incubation temperatures (26° and 34°C, respectively), and different sex ratios are produced at intermediate temperatures (for example, mostly females at 30°C and mostly males at 32.5°C). Incubation temperature, however, determines more than an individual's gender—it also influences the individual's adult aggressive and sexual behavior in comparison to others of its sex. Here we will focus on the behavior of mate choice as studied by Oliver Putz and David Crews (2006). When males reared at 30°C were placed in a Y-maze and given a choice between spending time near a female reared at 30°C and one reared at 34°C, they preferred the female from a 34°C incubation temperature (Figure 8.7*a*). In contrast, when males reared at 32.5°C were placed in the Y-maze and given a choice between spending time near a female reared at 30°C and one reared at 34°C, they preferred the female from a 30°C incubation temperature (Figure 8.7*b*). The authors suggest that incubation temperature influences brain development and leads to different perceptions of individuals of the opposite sex. Here, then, we have an example of a physical characteristic of the *prenatal* environment influencing the development of adult behavior.

Physical characteristics of the *postnatal* environment also influence development of the brain and behavior. This can be seen in the results of numerous studies comparing neural and behavioral development in captive animals reared either in enriched laboratory environments or in standard laboratory housing conditions. Enriched environments typically mean larger cages, a more complex and variable physical environment with nesting material, foraging devices, toys, hiding places, and the opportunity for voluntary exercise, as well as more complex social groups (we will consider the specific effects of social isolation later in the chapter). In rodents, environmental enrichment results in several structural changes in the brain, which include increased numbers of neurons, synapses, and dendritic branches, and increased brain weight and size (van Praag et al. 2000; Würbel 2001). Rodents from enriched environments also exhibit enhanced learning and memory. In captive nonhuman primates, individuals housed in enriched environments exhibit a more balanced repertoire of natural behaviors and display these behaviors at species-typical frequencies and intensities. Primates from enriched environments also exhibit fewer abnormal behaviors such as self-injurious behaviors (e.g., self-biting and head-banging) and stereotypic behaviors (repetitive behavior, such as pacing and flipping, with no apparent biological function).

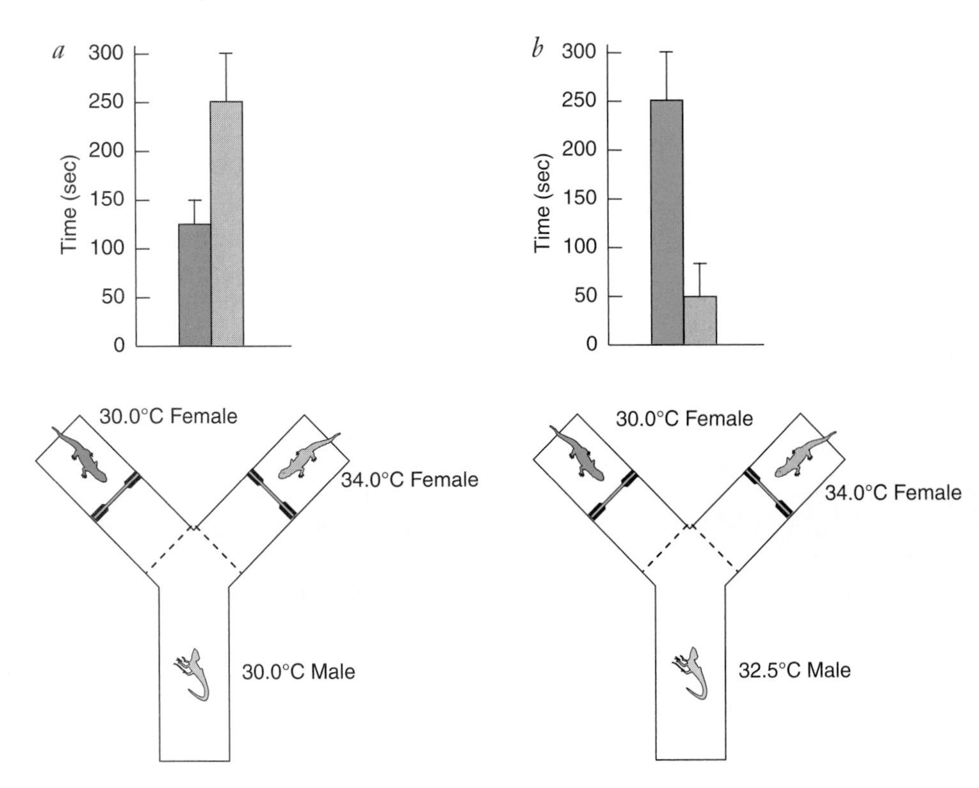

FIGURE 8.7 **Characteristics of the physical environment influence behavioral development in leopard geckos. Shown here are the effects of incubation temperature on the mate preferences of adult males, as measured by the time that they spend near individual females in a Y-maze. (*a*) Males from a 30°C incubation temperature prefer females from a 34°C incubation temperature. (*b*) In contrast, males from a 32.5°C incubation temperature prefer females from a 30°C incubation temperature. (From Crews and Groothuis 2005.)**

These abnormal behaviors appear to develop in response to unavoidable stress or fear and to the inactivity, boredom, and frustration experienced in environments lacking physical and social complexity (Honess and Marin 2006; Lutz and Novak 2005). Indeed, environmental enrichment is now advocated as the preferred means to treat diverse zoo animals that have developed stereotypic behaviors in captivity (Mason et al. 2007).

EXPERIENCE THROUGH PLAY

Play is thought to be important in the normal development of behavior in many mammals, as well as in some birds and even a few reptiles. Play behavior also has been described in octopuses, thus extending its occurrence to invertebrates (Kuba et al. 2006; Mather and Anderson 1999). We know play when we do it and when we see other animals doing it (Figure 8.8). During a visit to the zoo, we even recognize play in species we have never seen before. The attribute that pervades all play, and by which we most commonly identify it, is our subjective judgment of a lack of purposefulness.

Although play is easy to spot, it is difficult to define, mainly because no specific behavior pattern or series of activities exclusively characterizes it. Play borrows pieces of other behavior patterns, usually incomplete sequences and often in an exaggerated form. It consists of elements drawn from other, functionally different behavior patterns juxtaposed in new sequences. Some species may rapidly alternate prey-catching movements and aggressive behavior, while others mix components from hunting and sexual behavior. The movements of play are often repeated more often than during nonplay interactions, and play may be slightly modified from another behavior. When a dog is aggressive, it bares its teeth and growls. Its hair stands on end, adding to its ferocious appearance. However, during a play fight, the growl is not accompanied by the raising of hair. Play is also defined by whether the behavior is spontaneous and voluntary and exhibited by healthy individuals (Burghardt 2005).

FIGURE 8.8 **Lion cubs playing. Play is thought to be vital to the development of behavior in mammals.**

Play is also difficult to define because there are several types. First is social play, which includes play fighting or play chasing, as well as sexual play. We have all been amused by the friendly tussles of kittens and puppies as they chase, wrestle, and pounce on one another. Sexual play includes playful mounting by gazelles and precocious courtship in some turtles. A second form is locomotor play—that is, exercise. Foals kick up their heels and gallop. Young primates, including human children, may swing and roll and slide. Polar bear cubs climb ashore only to leap back into the water. Object play is the third form. In this form of play, objects are manipulated. When first presented with a novel object, a young animal typically explores it by touching, sniffing, or viewing it from different angles. After the initial sensory investigation, the object may become a toy (Fagan 1981). Sometimes a young animal will flit among the types of play in rapid succession, and the predominant form of play may change as the animal matures (Burghardt 1998).

But why do animals play? In other words, what function does this frolicking serve? What role does play have in the development of behavior? The hypotheses for the long-term significance of play can be grouped into three categories (Thompson 1996):

1. Physical: training for strength, endurance, and muscular coordination, particularly the skills relating to intraspecific fighting and prey capture.
2. Social: practice of social skills such as grooming and sexual behavior; establishment and maintenance of social bonds. Play also helps develop an animal's ability to read and send signals to communicate with other members of its species.
3. Cognitive: learning specific skills or improving overall perceptual abilities.

As we explore these hypotheses further, we will see that some examples of play fit into more than one category. One of the physical benefits of play may be that it helps form connections between neurons in the brain, especially in the cerebellum, a brain region important in motor coordination and memory of motor patterns. There is a limited time period during which synapses are being formed between neurons in the cerebellum. During this time, experience affects the number and pattern of the synapses. In house mice, Norway rats, and domestic cats, locomotor play coincides with the period when the cerebellar architecture is being shaped. In other words, locomotor play begins just when experience can modify the connections within the cerebellum and ends when the cerebellar structure is set (Byers and Walker 1995).

Another physical benefit of play is that it affords animals the opportunity to practice skills that will be essential to later survival (Caro 1988). Hunting games, for instance, may help perfect the movements of catching prey such as stalking and shaking. Some examples of these actions are familiar. Kittens stalk leaves, and puppies often "shake the life" from toys as they would a prey animal. During object play, fledgling American kestrels (*Falco sparverius*) prefer objects that resemble their natural prey (Negro et al. 1996). Play may have some immediate benefits for cheetah (*Acinonyx jubatus*) cubs by honing their predatory skills. Cubs that playfully crouched and stalked littermates also crouched and stalked prey more often than less playful companions. And when the mother released live prey for her cubs to catch, those that showed the highest rates of object play and contact social play were more likely to be successful (Caro 1995).

The play fighting of young animals may serve as practice for the battles of adults that establish dominance hierarchies and defend territories. In the fury of a play fight, no serious biting and no threat behavior take place. In some species, larger, older, and more dominant animals seem to handicap themselves in tussles with weaker playmates. Strength and skill are often matched to those of the partner. Some animals seem to practice territory defense as well. Young deer and goats vie for possession of an area in a game reminiscent of King of the Mountain.

It has also been suggested, however, that rather than being training for serious adult fighting, play fighting develops beneficial cognitive and social skills. During play fights among squirrel monkeys (*Saimiri sciureus*), the young males may reverse dominant and submissive roles (Biben 1998). Without experience in a dominant role, young monkeys may grow up to be overly submissive, and without experience in the submissive role, they may grow up to be bullies. Play fighting may also help a juvenile learn to read the intentions of others. Is the opponent bluffing? How motivated is this opponent? These social and cognitive skills may, in fact, prove to be more important than physical skills.

Although we typically associate behavioral development with youth, changes in behavior continue throughout adulthood. After all, the nervous and endocrine systems of animals change with age, as do the conditions individuals experience in their physical and social environments. Though less playful than juveniles, some adults do play. For example, spontaneous play is common among adult dogs (Bauer and Smuts 2007) and their ancestors, wolves (Mech 1970). Does play behavior differ between adults and young? And if engaging in play helps young animals to hone their physical, social, and cognitive skills for use in adulthood, then what function might play have *in* adulthood? Let's take a look at stone handling, a type of solitary object play exhibited by some adult, juvenile, and infant Japanese macaques (*Macaca fuscata*). The behavior consists of gathering, manipulating, and scattering stones. Stone handling has been observed in some provisioned troops; it has never been observed in nonprovisioned troops. Stone handling presumably occurs in provisioned troops because troop members have substantial free time and don't have to search for food. The behavior is only acquired by young individuals, but once acquired, it is practiced into adulthood and old age. The performance of stone handling in older individuals makes the behavior different from many other forms of play in primates, which disappear from the behavioral repertoire with age.

Charmalie Nahallage and Michael Huffman (2007) conducted daily observations of a provisioned troop of Japanese macaques known to exhibit stone handling. The researchers indicate that young Japanese macaques engage in frequent, short-duration bouts of stone handling, often accompanied by locomotion and energetic movements. In contrast, adults stone-handle less frequently and for longer durations. Adults also exhibit more complex manipulative patterns and stone-handle when stationary, often in a favorite location. Based on these and other differences, the authors hypothesize that the functions of stone handling may differ for young and adult macaques. Nahallage and Huffman propose that stone handling functions in the development of motor and perceptual skills in young macaques, and in the maintenance and regeneration of neural pathways in adults. They further suggest that stone handling may slow the deterioration of cognitive function associated with aging in this long-lived species. Testing their hypotheses will require additional behavioral observations, cognitive testing, and neuroanatomical data.

THE CONCEPT OF SENSITIVE PERIODS

We see, then, that factors such as the experience of play, characteristics of the physical environment, or changes in the animal's nervous system or hormonal state can cause behavioral change during development. Now we will see that often these factors have their effects on development during particular windows of time, called sensitive periods.

CHANGING TERMINOLOGY—FROM CRITICAL PERIODS TO SENSITIVE PERIODS

Early on, windows of opportunity for learning were called *critical periods*. Konrad Lorenz (1935) borrowed this term from embryology, where it was used to describe times in early development that were characterized by rapid changes in organization. During these brief, well-defined periods, an experimental interruption of the normal sequence of events produced profound and irreversible effects on the developing embryo. Thus, as first used by Lorenz, a critical period was a phase of susceptibility to environmental stimuli that was brief, well defined, and within which exposure to certain stimuli produced irreversible effects on subsequent behavior.

Recently, terms such as *sensitive period*, *sensitive phase*, *susceptible period*, and *optimal period* have been used in place of critical period. The newer terms indeed reflect certain modifications in the definition of this period in light of more recent research (reviewed in Michel and Tyler 2005). In fact, many of Lorenz's (1935) basic precepts have now been modified. Specifically, we now know that such periods (1) are fairly extended, (2) are not sharply defined but gradual in their onset and termination, (3) differ in duration among species, individuals, and functional systems, and (4) depend on the nature and intensity of environmental stimuli both before and during the sensitive period. Moreover, most phenomena based on sensitive periods are not irreversible. Instead, patterns of behavior developed during sensitive periods can usually be altered or suppressed under certain conditions, especially those associated with high levels of stress. Deprivation (e.g., rearing an animal in isolation or in darkness) can reverse or destroy a pattern of behavior established during a sensitive period. It is important, however, not to overemphasize the reversibility of patterns of behavior established during sensitive periods. Conditions such as rearing in complete isolation or total darkness are unlikely to be encountered by most animals outside the laboratory environment. Furthermore, even in the laboratory, behaviors established during sensitive periods are usually more resistant to change than those learned at other times (Immelmann and Suomi 1981). We will use the term *sensitive period* because it is now commonly found in the literature. Our working definition of **sensitive period** is a time during development when certain experiences have a greater influence on the characteristics of an individual than at other stages.

TIMING OF SENSITIVE PERIODS

In most animals, sensitive periods occur early in life. Why is this so? We usually assume that this is the time when animals have the greatest opportunity to gain knowledge from parents and close relatives, knowledge that is particularly important in species recognition. Later, they might not interact with them so intimately, and in some cases they will, in fact, be exposed to intense stimuli from other species (Immelmann and Suomi 1981). For example, in some species of birds, the young remain in the nest for only a few weeks after hatching and then leave to join mixed-species flocks. It would not be surprising if these young learn to recognize conspecifics during a brief sensitive period before leaving the nest. Otherwise, choosing an appropriate mate could later be a confusing exercise indeed because birds that waited too long to learn the defining qualities of their species might very well learn the plumage and song characteristics of another species.

Some animals have little or no contact with their parents or other close relatives after birth or hatching. One might wonder, then, would sensitive periods occur early in development in these species? And is early learning limited to acquiring knowledge about appropriate social partners, or do animals also learn characteristics of appropriate places to live or breed? Consider the case of Pacific salmon in the genus *Oncorhynchus*. Adult salmon spawn in freshwater, usually streams, and depending on the species and population, they may or may not die after spawning. When the eggs hatch and the fry eventually emerge from their gravel nests, they typically pass through several developmental stages in their home streams, the last of which, called the smolt stage, prepares them for migrating thousands of kilometers downstream to enter oceanic feeding grounds. After a time at sea, virtually all surviving adults return to their natal stream to spawn. It is a remarkable feat of navigation, and when they reach the freshwater inlets, they unerringly swim up the appropriate tributaries, making all the correct decisions at every fork until they reach the very stream where they were spawned—and they seem to do it by smell. Apparently, before their migration to the sea and during a sensitive period, juvenile salmon learn the odor of water at the site where they were spawned. The water at the natal spawning site has a unique chemical composition known as the "home stream olfactory bouquet," or HSOB. The sensitive period for learning the HSOB seems to coincide with smoltification, the developmental transformation of young salmon from parr (freshwater residents) to smolts ready for seaward migration and life (see Chapter 7) (Carruth et al. 2002). It is also possible that the sensitive period begins somewhat earlier (Dittman and Quinn 1996). In any case, upon returning to their natal stream, adult salmon are stimulated to swim upstream by the familiar odor. Why is it important that salmon learn the precise location of their natal stream? The answer is that each population is finely adapted to its home water, so much so that salmon experimentally introduced into other streams show a higher mortality rate than locally adapted individuals (Quinn and Dittman 1990). The period of early learning thus ensures that it is the odor of the fish's own spawning place that is remembered (for a more detailed discussion of salmon homing, see Chapter 10). We see, then, that during sensitive periods, animals may learn the appropriate cues, not only of conspecifics, but also of the local physical environment.

STOP AND THINK

A common strategy for restoring or enhancing populations of anadromous salmonids (i.e., those that return from the sea to breed in freshwater) is to artificially rear young in hatcheries and then release them into streams with the expectation that they would eventually migrate to the sea. Given what you know about olfactory imprinting in salmon, how might the release be orchestrated and its timing planned to result in good return rates to freshwater spawning grounds?

Onset of Sensitivity

It is clear that young animals have a heightened sensitivity to certain environmental stimuli such as the physical appearance of a parent or the smell of a home stream. One might ask, then, what causes this increased sensitivity to certain cues? One suggestion is that the onset of sensitivity may be due both to endogenous (internal) and exogenous (external) factors (Bateson 1979). Increases in sensitivity generally begin as soon as the relevant motor and sensory capacities of the young animal are developed. Changes in the internal state, such as fluctuations in hormone levels, may also influence sensitivity. Then, endogenous factors interact with environmental variables to produce the start of the sensitive period. For example, although the visual component of filial imprinting in birds (the response of newly hatched young to follow their mother; see later discussion) begins once hatchlings are able to perceive and process optical stimuli, experience with light appears to interact with the internal conditions to initiate this particular sensitive period (e.g., Bateson 1976). Here, then, we have endogenous factors (ability of the nervous system to process optical stimuli) interacting with environmental factors (exposure to light) to produce the start of the sensitive period.

Decline in Sensitivity

Several explanations have been proposed for the termination of sensitive periods (reviewed in Johnson 2005); we consider two of them here. One explanation assumes that the decline in sensitivity is under endogenous control, perhaps influenced by a maturational timetable. According to this idea, some physiological factor, intrinsic to the animal, ends the period of receptivity to external stimulation once a specific stage of maturation has been reached. Another explanation is based on the idea that learning is a self-terminating process. In essence, learning causes neurobiological changes that reduce plasticity and ultimately terminate the sensitive period. An example of this idea is the proposition that learning a first language causes neurobiological changes in humans that effectively bring the sensitive period for language learning to a close. This might explain, for example, why learning a second language later in life is often more difficult than learning the first language.

MULTIPLE SENSITIVE PERIODS

Individuals typically experience multiple sensitive periods during their development (Bischof 2007). We will see that a young male songbird experiences sensitive periods during song learning. The same young bird will experience sensitive periods associated with sexual imprinting (learning the characteristics of an appropriate mate; see later discussion). Another example of multiple sensitive periods within individuals concerns visual development in humans (Lewis and Maurer 2005). Researchers comparing visual development in visually normal children with that of children deprived of early visual experience because they were either born with cataracts or developed them later have discovered that there are different sensitive periods for different aspects of vision (e.g., visual acuity, peripheral vision, motion detection), and even within each of these aspects there may be more than one sensitive period.

SOME EXAMPLES OF SENSITIVE PERIODS IN BEHAVIORAL DEVELOPMENT

Now that we have discussed the definition and timing of sensitive periods, and the fact that a given individual—be it a songbird or a human—will experience multiple sensitive periods, let's consider some examples of behavioral development that depend, to varying degrees, on specific experiences during a window of time.

Filial Imprinting

Anyone who has ever watched chicks in a farmyard or ducklings and goslings on a pond knows that the young generally follow their mother wherever she goes (Figure 8.9). How does such following behavior develop? Konrad Lorenz (1935), working with newly hatched goslings, was the first to systematically study this behavior. In one experiment, he divided a clutch of eggs laid by a greylag goose (*Anser anser*) into two groups. One group was hatched by the mother, and as expected, these goslings trailed behind her. The second group was hatched in an incubator. The first moving object these goslings encountered was Lorenz, and they responded to him as they normally would to their mother. Lorenz marked the goslings so that he could determine in which group they belonged and placed them all under a box. When the box was lifted, liberating all the goslings simultaneously, they streamed toward their respective "parents," normally reared goslings toward their mother and incubator-reared ones toward Lorenz. The goslings

FIGURE 8.9 Young Canada geese following their mother. The following response results from filial imprinting, the process by which young precocial birds learn the characteristics of their mother and then preferentially follow her.

had developed a preference for characteristics associated with their "mother" and expressed this preference through their following behavior. The attachment was unfailing, and from that point on Lorenz had goslings following in his footsteps (Figure 8.10).

Because social attachment evidenced by following seemed to be immediate and irreversible, Lorenz named the process *Pragung*, which means, "stamping." The English translation is "imprinting." Used in this context, the term suggests that during the first encounter with a moving object, its image is somehow permanently stamped on the nervous system of the young animal.

We now know that at least two distinct processes are involved in the development by young birds such as chicks, ducklings, and goslings of a preference for following their mother (studies reviewed by Bolhuis and Honey 1998; Hogan and Bolhuis 2005). In one process, a predisposition to approach objects with the general characteristics of conspecifics emerges in the young bird, even without previous exposure to a conspecific. For example, chicks without any previous exposure to an adult conspecific or a red box preferentially approach a stuffed conspecific when given a choice between it and the red box. (This is not to say that experience is unimportant in the development of the predisposition for conspecific characteristics. It turns out that other nonspecific experiences, such as being handled or placed in a running wheel, can induce the predisposition as long as these experiences occur during the sensitive period.) In the second process, called **filial imprinting**, the young bird learns, through exposure to its mother, her particular characteristics and then preferentially follows her. The biological function of filial imprinting is probably to allow young birds to recognize close relatives and thereby distinguish their parents from other adults that might attack them (Bateson 1990). The two processes—development of the predisposition and filial imprinting—seem to be localized in different regions of the brain. We will focus our discussion on the development of the following response in mallard ducklings.

FIGURE 8.10 Goslings following their "mother," Konrad Lorenz. Lorenz was one of the first scientists to study imprinting experimentally.

Mallard ducklings, like most young in the orders Anseriformes (ducks, geese, and swans) and Galliformes (chickens, turkeys, and quail), are *precocial*, that is, quite capable of moving about and feeding on their own just a short time after hatching. Filial imprinting is usually studied in species with precocial young. The following response is nonexistent—or much less evident—in species such as the songbirds discussed later, whose young are *altricial*, that is, virtually helpless and incapable of feeding on their own or following their parents for the first few weeks after hatching. We begin with a brief description of reproduction and early development in mallards.

Upon finding a suitable nesting site, typically a shallow crevice in the ground, the mallard hen begins to lay her eggs, at the rate of one egg per day (Miller and Gottlieb 1978). The average clutch size is eight to ten, and after the last egg is laid the hen begins incubation, a process that lasts approximately 26 days. Encouraged by the warmth of the mother's body, the embryo inside each egg begins to develop. Two to three days before hatching, each embryo moves its head into the air space within its egg and begins to vocalize; these vocalizations are called contentment calls. About 24 hours later, the embryos pip the outer shell and then take another day to break through the rest of the shell and hatch. Most of the ducklings in a clutch will hatch within an interval of ten hours. The hen broods her young for a day and then leaves the nest and emits calls to encourage the ducklings to follow. Although she vocalizes during incubation and brooding, the frequency increases dramatically at the time of the nest exodus. Prompted by their mother's assembly calls, the young leave the nest and follow her to a nearby pond or lake, where they will paddle behind her.

Observations such as these stimulate many questions. What characteristics of the mother form the basis for the ducklings' attachment? Do the ducklings imprint on the mother's call, physical appearance, or some combination of the two? What role might siblings play in the development of the following response? Finally, is there a sensitive period during which exposure to certain cues must occur for the normal development of filial behavior?

Many of the answers to these questions came from the laboratory of Gilbert Gottlieb, a pioneer in behavioral development who passed away in 2006. For several decades, Gottlieb and co-workers examined the development of the following response in Peking ducks, a domestic form of the mallard (despite their domestication, Peking ducks are quite similar to their wild counterparts in their behavior). Here we consider some of the work of Gottlieb and his colleagues.

In one experiment with ducklings that had never had contact with the mother, Gottlieb (1978) examined (1) the relative importance of the hen's auditory and visual cues in the development of the following response and (2) whether the parental call of a duckling's own species would be more effective than that of other species in inducing and maintaining attachment behavior. As part of the study, 224 eggs were hatched in an incubator; thus, the ducklings never came in contact with their mothers. The ducklings were divided into four groups and tested for their following response. In all four groups, the ducklings were tested with a stuffed replica of a Peking hen as it moved about a circular runway. Individuals in one of the groups, however, were tested with a silent hen, whereas individuals in the other three groups were tested with hens that emitted assembly calls through a speaker concealed on their undersides. Of the ducklings that heard assembly calls, one group was exposed to mallard calls (Peking calls), one group to wood duck calls, and one group to domestic chicken calls. Each duckling was given a 20-minute test to determine whether it would follow the stuffed model around the circular arena.

The results are presented in Figure 8.11. As you can see, the auditory stimulus of the maternal call is important in filial imprinting: all conditions with calls were much more effective than no call at all. Furthermore, even though the ducklings had never before heard the maternal call of their species, they responded selectively to it. The maternal call of the mallard was far more effective than that of the wood duck or chicken in inducing following by the ducklings.

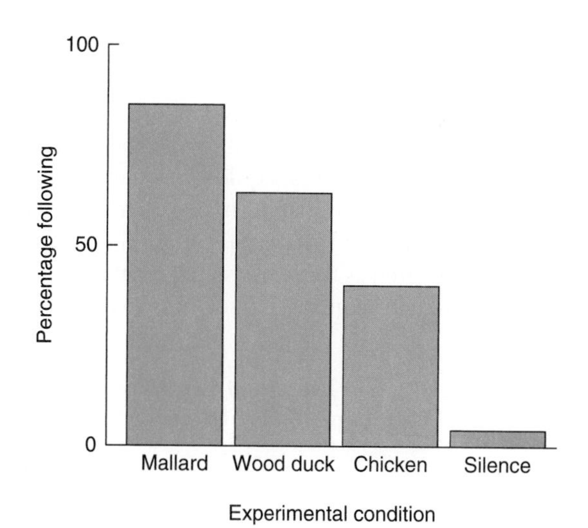

FIGURE 8.11 A stuffed Peking hen emitting mallard calls was more effective in eliciting the following response in incubator-reared Peking ducklings than a hen emitting either wood duck or chicken calls (Peking ducks are a domestic form of the mallard duck). All hens with calls were more effective than a silent hen. Thus, auditory cues from the mother are important in controlling the early behavior of ducklings, and ducklings respond selectively to the call of their own species without previous exposure to it. (Drawn from the data of Gottlieb 1965.)

In a second experiment, incubator-reared ducklings were given a choice of following either a stuffed Peking hen that was emitting mallard calls or a stuffed Peking hen that was emitting chicken calls. When placed in this simultaneous choice situation approximately one day after hatching, the majority of ducklings (76%) followed the model that was emitting the mallard call.

Taken together, these experiments demonstrate that auditory stimuli from the mother are an important influence on the behavior of newly hatched ducklings. Furthermore, ducklings respond selectively to the maternal assembly call of their own species without any previous exposure to it (remember that all ducklings were reared in incubators and therefore had no contact with hens). This preferential response by the ducklings to the assembly call of their species is an example of the predisposition described earlier. Recall that young precocial birds may develop a preference for stimuli from conspecifics without prior exposure to the particular stimuli.

These findings do not, however, suggest that experience is unimportant in the development of a preference for the assembly call. Experience, as it turns out, is critical, but it occurs prenatally (before hatching). For Peking ducklings to exhibit a preference for the mallard hen's call, they must hear their own contentment calls or those of their siblings before hatching (Gottlieb 1978). If ducklings are reared in isolation (and therefore are not exposed to the calls of their siblings) and made mute just before they begin to vocalize within the egg (and therefore are not exposed to their own calls), they no longer display their highly selective response to the maternal call of their species. When these ducklings without normal embryonic auditory experience are placed in a test apparatus equidistant between two speakers, one speaker emitting a mallard's maternal call and the other a chicken's maternal call, they choose the latter almost as often as they choose the former 48 hours after hatching. In contrast, ducklings with normal embryonic auditory experience always choose the mallard's call over the chicken's (Table 8.1). Thus, auditory experience before hatching is important to the development of the following response in mallard ducklings because it induces a predisposition to approach the mallard assembly call.

Is there a sensitive period during which exposure to contentment calls must occur for ducklings to exhibit a preference for the call of the mallard hen? If ducklings heard contentment calls after hatching, would they display the normal preference for mallard calls? Gottlieb (1985) set out to answer these questions. He began by raising ducklings in isolation and making them mute before they began to vocalize within the egg (again, these manipulations ensure that the ducklings are not exposed to their own calls or the calls of their siblings). One group was exposed to contentment calls during the embryonic period (approximately 24 hours prior to hatching), and another group was exposed to contentment calls during the postnatal period (approximately 24 hours after hatching). Ducklings in each group were then tested for their preference for the mallard hen's call. This time, however, they were given the choice of approaching a speaker that emitted normal calls of a mallard hen or a speaker that emitted artificially slowed calls of a mallard hen. As shown in Table 8.2, ducklings must be exposed to contentment calls before hatching if they are to show a preference for the normal call; exposure after hatching is ineffective in producing the preference for the normal call of their species. Here, then, we have an example of a sensitive period occurring during embryonic development.

What role, if any, might visual stimuli play in the development of the following response? Several experiments have shown that two conditions must be simultaneously met if visual imprinting on a mallard hen is to occur in the ducklings. The ducklings must be reared with other ducklings and allowed to actively follow a mallard hen, or a model, if they are to prefer the appear-

TABLE 8.1 The Effects of Embryonic Auditory Experience on Call Preferences of Peking Ducklings 48 Hours after Hatching

		Preference		
	N	Mallard call	Chicken call	Both
Vocal-communal	24	24	0	0
Mute-isolated				
First experiment	22	12	9	1
Replication	21	14	6	1
Total	43	26	15	2

Ducklings raised in the vocal-communal group could hear themselves and the calls of siblings prior to hatching; mute-isolated ducklings had no such auditory experience. *N* = number of ducklings that responded to calls.

Source: Data from Gottlieb (1978).

TABLE 8.2 Effects on Preferences of Mute-Isolated Ducklings of Embryonic Versus Postnatal Exposure to Contentment Calls

		Preference		
Time of exposure to contentment calls	N	Normal mallard	Slowed mallard	Both
Embryonic	32	21	8	3
Postnatal	37	14	20	3

Source: Data from Gottlieb (1985).

ance of a hen of their own species to that of a hen of another species at later testing (e.g., Dyer et al. 1989; Lickliter and Gottlieb 1988). Ducklings that are reared in isolation and given passive exposure to a stuffed mallard hen (i.e., housed with a stationary model) do not develop a preference for the mallard hen. These results suggest that under natural conditions, ducklings learn the visual characteristics of their mother after leaving the nest and following her around.

We see, then, that both auditory and visual stimuli are important in development of the following response. Auditory stimulation from the mother appears to be largely responsible for prompting the ducklings to leave the nest and for influencing their earliest following behavior. However, the hen's appearance becomes important after the nest exodus.

The importance in filial imprinting of auditory cues, visual cues, and active following was well summarized by Lorenz (1952, pp. 42–43). In an experiment with Peking ducklings, he found himself in a rather embarrassing position for one destined to become a Nobel Laureate. In his words,

> *The freshly hatched ducklings have an inborn reaction to the call-note, but not to the optical picture of the mother. Anything that emits the right quack note will be considered as mother, whether it is a fat white Peking duck or a still fatter man. However, the substituted object must not exceed a certain height. At the beginning of these experiments, I had sat myself down in the grass amongst the ducklings and, in order to make them follow me, had dragged myself, sitting, away from them. So it came about, on a certain Whit-Sunday, that, in company with my ducklings, I was wandering about, squatting and quacking, in a May-green meadow at the upper part of our garden. I was congratulating myself on the obedience and exactitude with which my ducklings came waddling after me, when I suddenly looked up and saw the garden fence framed by a row of dead-white faces: a group of tourists was standing at the fence and staring horrified in my direction. Forgivable! For all they could see was a big man with a beard dragging himself, crouching, round the meadow, in figures of eight, glancing constantly over his shoulder and quacking—but the ducklings, the all-revealing and all-explaining ducklings were hidden in the tall spring grass from the view of the astonished crowd!*

To summarize, the experiments of Gilbert Gottlieb and co-workers have demonstrated that the following response in Peking ducklings results from a complex interaction of auditory, visual, and social stimuli provided by the hen and siblings. Several important generalizations about early behavioral development have arisen from their work. First, we can no longer think of experience as occurring only after birth or hatching; embryonic experience can also influence behavior. In the case of Peking ducklings, listening to the contentment calls of siblings before hatching is critical to development of their preference for the maternal call of their own species after hatching. Second, the experiments with ducklings demonstrate that a variety of stimuli may be involved in developing a single pattern of behavior and that the relative importance of different stimuli may change as the young animal matures. Although the following behavior of ducklings soon after hatching is influenced largely by auditory cues from their mother, only a few days later visual stimuli (in combination with auditory stimuli) become important in the following response. The relative priorities of different cues match the timing for development of the auditory and visual systems; in ducklings, as in all birds, the auditory system develops before the visual system (Gottlieb 1968). These results emphasize the close interaction between physical maturation and experience in early behavioral development. Finally, the study of filial imprinting in Peking ducklings illustrates that we must be open-minded when trying to sort out just which experiences affect a given behavior. Who would have thought that listening to siblings before hatching or interacting with siblings after hatching would be critical in the development of the ducklings' attachment to their mother? We now know that such nonobvious experiential factors are indeed essential to the development of following behavior in this species.

Sexual Imprinting

We have seen that early experience influences a duckling's attachment to its mother, an attachment that the young bird demonstrates by trailing behind her in the days and weeks following the exodus from the nest. Early experience also has important consequences for the development of mate preferences in birds. In many species, experience with parents and siblings early in life influences sexual preferences in adulthood. The learning process in this case is called **sexual imprinting**. Typically, sexual preferences develop after filial preferences, although the sensitive periods may overlap to some degree (Bateson 1979). Whereas filial imprinting is indicated by the following response of young birds, sexual imprinting is typically shown in the preferences of sexually mature birds for individuals of the opposite sex. It is important to note that unlike filial imprinting, sexual imprinting occurs in both altricial and precocial birds. Our first example concerns the altricial zebra finch, and our second, the precocial Japanese quail.

One early and dramatic demonstration of the importance of early experience to subsequent mate preference came from cross-fostering experiments with finches. Klaus Immelmann (1969) placed eggs of zebra finches (*Taeniopygia guttata*) in clutches belonging to Bengalese finches (*Lonchura striata*). The Bengalese foster parents raised the entire brood until the young could feed themselves. From then on, young zebra finch males were reared in isolation until sexually mature. When they were later given a choice between a zebra finch female and a Bengalese finch female, they courted Bengalese females almost exclusively.

A second study demonstrated that brief contact with foster parents early in life could exert a more powerful, longer-lasting influence on mate preference than long-term social contact in adulthood. Cross-fostered zebra finch males were again separated from their foster Bengalese parents, but this time they were provided with a conspecific female and nesting supplies. Most of these males eventually mated with conspecific females and successfully produced young. When they were tested several months or years later, however, the males still displayed a preference for Bengalese females (Immelmann 1972).

More recent experiments have shown that sexual imprinting is a two-stage process (Bischof 1994, 2003). The first stage is the acquisition stage. This stage begins around ten days after hatching (when the visual system of a zebra finch has matured sufficiently to detect structured visual information) and ends at about 40 to 60 days of age. It is during this time that a young male forms a social bond to its parents. Because of this bond, the male prefers to socialize with members of its parents' species, and this social preference guides his first courtship attempts. The second stage is the consolidation stage (sometimes called the stabilization stage). This stage occurs when the male actually courts a female for the first time. During courtship, the social preference for the parental species becomes linked to sexual behavior and is consolidated or stabilized. The precise timing of the consolidation stage requires further research, but it appears to begin when males become sexually mature around 70 days of age and to end at about 100 to 150 days of age. Once the consolidation stage ends, no new preferences can be established.

The acquisition and consolidation phases are different processes, and each can be modified by different factors. The strength of the social preference for the parental species that develops during the acquisition phase is influenced by the amount of food that the young male is given by his parents. However, consolidation, the linking of the social preference for the parental species to sexual behavior, is most affected by the degree to which the male is aroused at the time of courtship (Oetting et al. 1995).

At first glance, it might seem curious that an individual has to learn to identify an appropriate mate. Wouldn't it be a safer evolutionary strategy to have a mating preference that cannot be modified by early social experience? Apparently not. For now, however, we can only hypothesize about the importance of early learning in choosing mates.

One idea, put forth by Patrick Bateson (1983), provides an interesting explanation for the functional significance of sexual imprinting. Bateson suggested that animals learn to identify and selectively respond to kin. Armed with information on what their relatives look like, individuals then choose mates similar but not identical to their family members. Given that both extreme inbreeding and outbreeding may have costs (see Chapter 14), sexual imprinting provides information that allows animals to strike a balance between the two. Bateson used evidence from studies of quail to support his argument.

Japanese quail (*Coturnix japonica*) prefer to mate with individuals that are similar to, yet slightly different from, members of their immediate family (Bateson 1982). In one study, chicks were reared with siblings for the first 30 days after hatching and then socially isolated until they became sexually mature. At 60 days of age, males and females were tested for mate preference in an apparatus that permitted viewing of several other Japanese quail (Figure 8.12*a*). The birds that were viewed were of the opposite sex of the test animals and belonged to one of the following five groups: (1) familiar sibling, (2) novel (unfamiliar) sibling, (3) novel first cousin, (4) novel third cousin, or (5) novel unrelated individual. Similarity in plumage between Japanese quail is considered to be proportional to genetic relatedness, and thus test animals could presumably judge genetic distance on the basis of plumage characteristics. As we see in Figure 8.12*b*, both

a

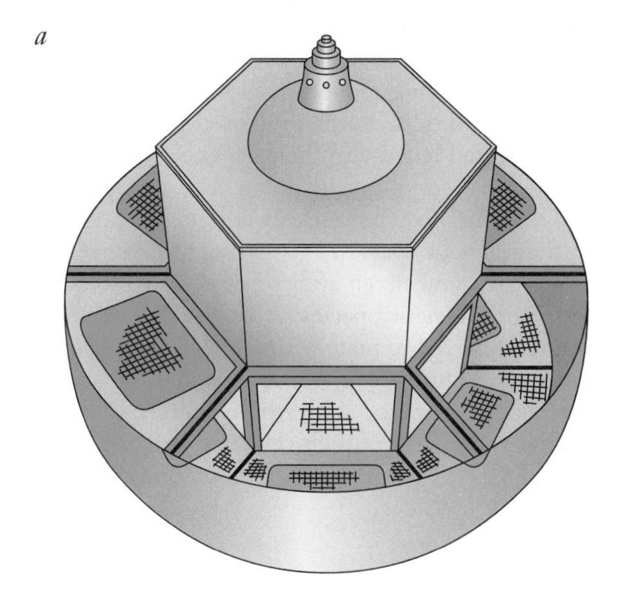

b

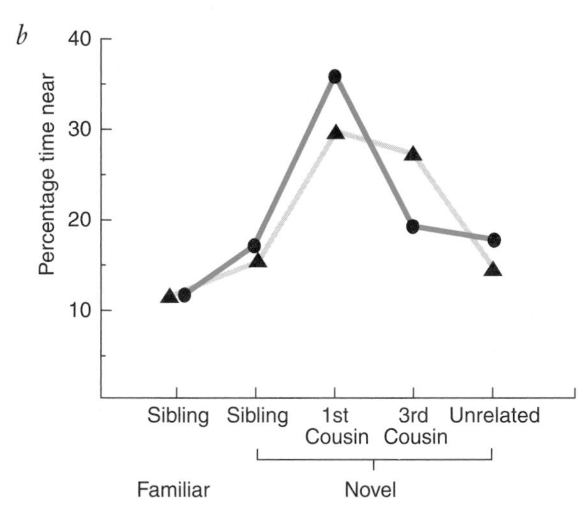

FIGURE 8.12 **Japanese quail prefer to spend time near individuals that are similar to, yet slightly different from, members of their immediate family. (*a*) Apparatus used by Patrick Bateson to test preferences of adult quail. (*b*) Both male (shown as triangles) and female quail (shown as circles) prefer first cousins, possibly striking an optimal balance between inbreeding and outbreeding. (*a*: From Bateson 1982. *b*: Modified from Bateson 1982.)**

males and females preferred to spend time near first cousins. The sexual preferences displayed by quail reflect a choice slightly displaced from the familiar characteristics of immediate family members: siblings are too familiar, novel unrelated animals are too different, but first cousins are the perfect mix of familiarity and novelty.

The results from studies of Japanese quail certainly seem to support Bateson's contention that through sexual imprinting some young animals learn the characteristics of their close relatives and then, in adulthood, choose a mate similar but not identical to their family members. A note of caution, however, is in order. As Bateson (1983) himself points out, we must be careful in generalizing results from laboratory studies to animals in their natural environment. Clearly, laboratory conditions of rearing and testing animals are vastly different from the natural conditions under which animals live and choose mates. We should also be aware that our choice of measures, such as the number of approaches an animal makes toward another, might not reflect mate preference or actual mating in nature. In the best of all possible worlds, we would compare results from the controlled environment of the laboratory with field observations on the impact of early experience on subsequent mate choice. In the field, however, it is often difficult to know the precise genetic relationships of animals and to track animals from birth to adulthood as would be necessary to chronicle their early experiences and subsequent mating behavior.

Although many factors make the study of sexual imprinting and later mate choice in free-living animals difficult, Tore Slagsvold and colleagues have performed cross-fostering experiments with wild birds and then followed the birds' subsequent survival and mating success (Slagsvold 2004; Slagsvold et al. 2002). In one experiment, they cross-fostered whole broods of great tits (*Parus major*) to nests of blue tits (*Cyanistes caeruleus*) and found that the young experienced low mating success because they strongly preferred their host species. Thus, sexual imprinting had occurred in this field experiment. However, when the researchers performed the reciprocal cross, that is, cross-fostered blue tits to great tits, most young mated with a conspecific, indicating that they had not sexually imprinted on their host species. Similarly, when pied flycatchers (*Ficedula hypoleuca*) were cross-fostered to nests of blue tits and great tits, there was no evidence that the young had sexually imprinted on either host species. The cross-fostered flycatchers did not differ from control birds (flycatchers from unmanipulated nests) in their mating success or breeding success (e.g., clutch size, number of young hatched and fledged). In addition, cross-fostered male flycatchers were similar to control male flycatchers in their response to intruders presented in a cage close to their nest box. The cross-fostered males reacted aggressively when the cage contained a conspecific male, displayed at their nest when it contained a conspecific female, and ignored the cage when it contained a female of their host species. Taken together, these experiments with free-living birds suggest that species vary in their sensitivity to sexual imprinting. Although the causes of this variation are not yet known, Slagsvold and co-workers hypothesize that several factors may influence the sensitivity of a species to sexual imprinting. They predict decreased sensitivity to sexual imprinting in species that are solitary rather than social and polygynous rather than monogamous. The researchers also predict that species will be less sensitive to sexual imprinting when cross-fostered to a host species that is either socially dominant or distantly related. The study of additional species in the field is necessary to test these hypotheses.

So far, we have limited our discussion of sexual imprinting to birds. Is there evidence of a similar process in mammals? Keith Kendrick and colleagues reciprocally cross-fostered offspring of domestic sheep (*Ovis aries*) and goats (*Capra hircus*) at birth (Kendrick et al. 1998, 2001). The young grew up with their heterospecific mothers (designated the maternal species), but also had social contact with their genetic species throughout development (e.g., lambs raised by goat mothers were allowed social contact with other sheep). When tested at one year of age, cross-fostered males exhibited strong social and mating preferences for females of their maternal species, spending about 89% of their time with these females. In contrast, normally reared male sheep and goats preferred females of their genetic species, spending 96% of their time with them. For the next three years, the cross-fostered males were housed exclusively with their genetic species, and mating preferences were assessed each year. Even after three years of sole exposure to their genetic species, cross-fostered males still strongly preferred to mate with females of their maternal species (Figure 8.13*a*). Cross-fostered females preferred social contact with their maternal species, spending about 69% of their time with females of their

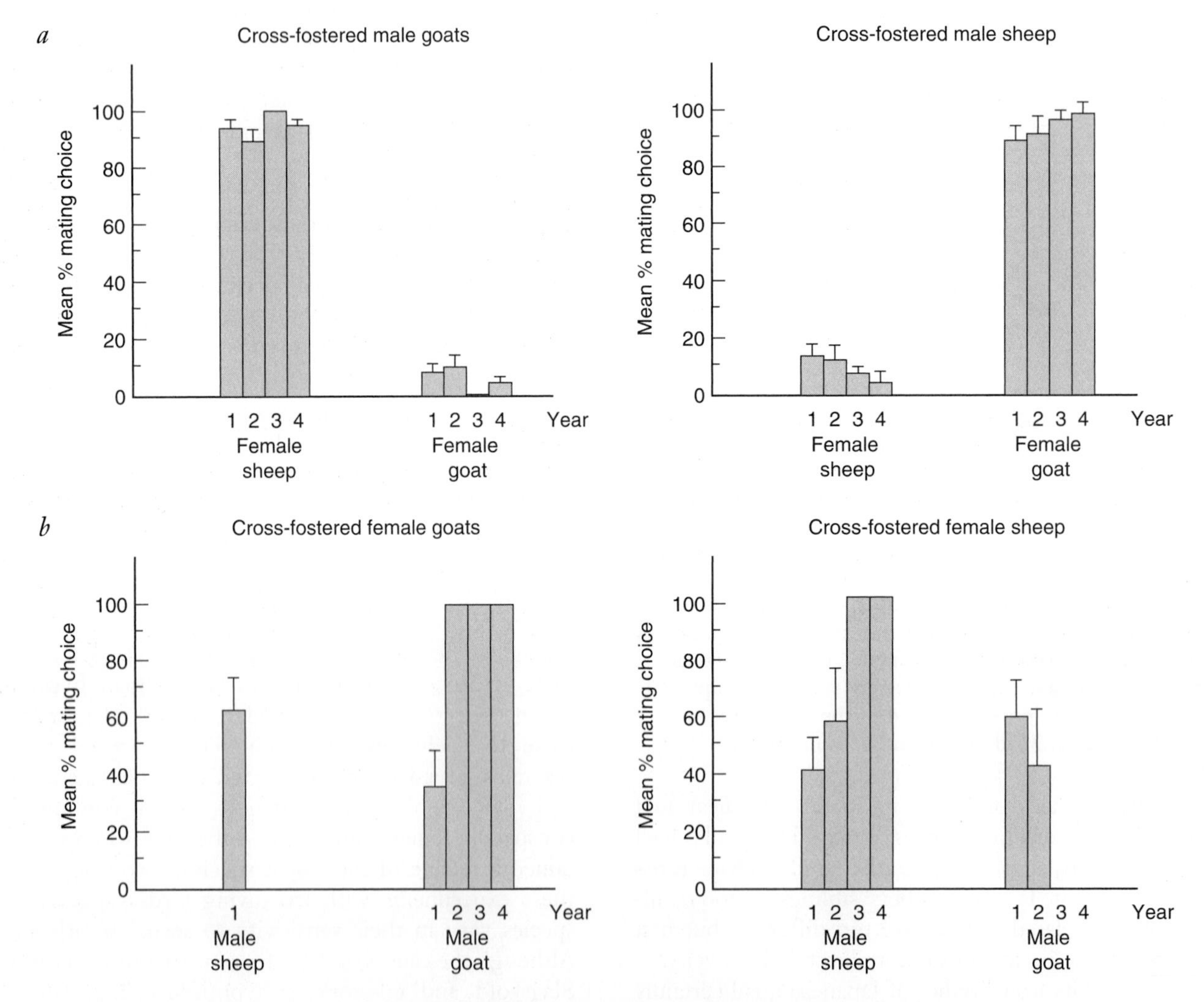

FIGURE 8.13 Mate preferences in adult goats and sheep are influenced by exposure to their mother early in life. (*a*) Male mating preferences. Cross-fostered male goats and sheep prefer to mate with females of their maternal species, and this preference is maintained even after several years of sole exposure to their genetic species. (*b*) Female mating preferences. Cross-fostered female goats and sheep exhibit weaker mating preferences for the maternal species than do males, and female preference can be reversed over time by exposure to their genetic species. (Modified from Kendrick et al. 2001.)

maternal species, but their preferences were weaker than those shown by cross-fostered males. Cross-fostered females also exhibited weaker mating preferences for males of the maternal species, and such preferences were reversible within one to two years of housing with their genetic species (Figure 8.13*b*). Thus, sheep and goats appear to learn the characteristics of an appropriate mate during early exposure to their mother, but males are more strongly influenced than females by such exposure.

What features of the mother (genetic or foster) might young goats and sheep attend to? Further experiments demonstrated that patterns of preference identical to those found when sheep and goats were tested with live stimulus animals could be obtained by showing test animals pictures of sheep and goat faces. These data suggest that features of the face may be critical in the formation of social and mating preferences in the two species. Interestingly, another report from this research group has shown that sheep have a remarkable ability to remember the faces of other sheep; perhaps to some people's surprise, sheep can remember 50 different sheep faces for two or more years (Kendrick et al. 2007)! Also of interest are findings showing the importance of parental facial features in the development of mate preferences in our own species (Bereczkei et al. 2004).

Maternal Attachment

Another imprinting-like process that occurs in mammals is maternal attachment. In some species of ungulates, a lasting bond between the mother and young is established rapidly after birth and results in the mother directing her care exclusively toward her own offspring. Domestic sheep and goats have been extensively studied in this regard, so we will return to them for our discussion of the development of maternal attachment. Like many ungulates, sheep and goats live in large social groups in which females tend to give birth synchronously. Young of both species are precocial, capable of wandering away from their mother soon after birth and mixing with other members of the group. Thus, it is important that a bond between mother and offspring be established early. Furthermore, young sheep and goats will initially approach any mother to nurse, so it also pays for a mother to be able to tell her own young from others. Mothers do not provide care for one another's offspring. Indeed, rejection of the young of other mothers is a no-nonsense affair that can be quite violent, involving head-butts, high-pitched vocalizations, and withdrawal from the unrelated youngster attempting to suckle (Figure 8.14).

Although the formation of mother–young relationships in sheep and goats depends on mutual recognition and bonding between the mother and her offspring, we focus our discussion on the development of the mother's attachment to her offspring. Formation of this attachment entails two processes, each of which requires maternal contact with the offspring during sensitive periods in the immediate hours after parturition. The sensitive periods for maternal attachment in sheep and goats thus serve as an example of sensitive periods occurring in adult animals.

FIGURE 8.14 **A mother goat rejecting an unfamiliar offspring. Mother goats develop an attachment to their own young shortly after birth and butt away unfamiliar young.**

The development of maternal attachment in sheep and goats consists of two interrelated processes: (1) the activation of maternal responsiveness (i.e., the display of maternal behavior when in the presence of young) and (2) the establishment of maternal selectivity (i.e., directing maternal behavior only toward those young with which the mother has bonded) (studies reviewed by Poindron et al. 2007). The existence of two processes was initially suggested by findings showing that right after parturition mothers will accept any young; in other words, there is a period of time during which mothers are maternally responsive but not selective. Here is a description of the two processes in more detail, beginning with maternal responsiveness. Undisturbed female sheep and goats display maternal behavior within minutes of giving birth. However, if their offspring is removed at birth before they have had any contact with it, then maternal responsiveness fades within hours; these females fail to show maternal behavior when reunited with their offspring 4 to 12 hours later. Even a few hours of contact with their offspring, or a few minutes in some cases, are sufficient to maintain maternal responsiveness. Thus, the sensitive period for maternal responsiveness seems to extend from parturition to no more than 12 hours postpartum in most mothers. Experiments involving separation and subsequent testing for acceptance of own or unrelated offspring suggest that the sensitive period for development of selectivity extends from parturition to perhaps 1 or 2 hours postpartum; thus, this sensitive period appears to lie within the sensitive period for maternal responsiveness. Only 30 minutes of contact with their offspring during the sensitive period appears sufficient for most mothers to display selective maternal behavior.

During the sensitive periods, what sensory cues from offspring are most important in the development of maternal responsiveness and selectivity? In both processes, the olfactory system of the mother plays a prominent role, and olfactory cues from amniotic fluid seem particularly important. However, the two processes differ in their ability to compensate for loss of olfaction induced through experimental procedures such as cutting the main olfactory nerves, destroying the olfactory bulbs, or blocking olfaction at the nostrils. Whereas all of these experimental treatments performed individually on mothers prevent the development of maternal selectivity, they do not interfere with maternal responsiveness. In the absence of olfactory cues, mothers appear to compensate and rely on other sensory cues from neonates to display maternal behavior at parturition and beyond. However, they do not selectively respond to their own offspring.

Brood Care

Until now we have been considering the effects of experience during specific windows of time on the development of social behavior in birds and mammals. We can now ask, what evidence have we of critical periods in behavioral development in other groups of animals? Our next example comes from the insects.

The life of an ant can often be quite complicated. In many species, individuals live in complex societies in which labor is divided among colony members, and their roles can change dramatically and repeatedly over the course of a few weeks. Given their intricate social system and frequent changes in job status, it is not surprising that ants have become popular subjects for researchers interested in behavioral development. Of particular interest is the Neotropical ant *Ectatomma tuberculatum*; in this species, an individual's job changes with age.

Annette Champalbert and Jean-Paul Lachaud (1990) examined the role of early social experience in the development of behavior of *E. tuberculatum* workers. The two researchers were interested in what effects a ten-day period of social isolation would have on behavioral development in workers of different ages. The researchers began their task on a coffee plantation in southern Mexico, where they collected four ant colonies from the bases of different coffee trees, each colony containing 200 to 300 individuals. Each colony was placed into an artificial nest made of plaster that had several interconnected chambers and a foraging area. A glass pane permitted the ants to be observed. A few hours after their emergence from cocoons, 15 workers from each of the four colonies were individually labeled with a small numbered tag glued to their thorax and then reintroduced into their respective colonies. These were the control workers. Other ants, the experimental workers, were labeled in the same manner and then isolated in a glass tube equipped with food and water. In one colony the isolation period began at emergence; in the second, third, and fourth colonies, the period of isolation began when workers were two, four, or eight days old, respectively. Like the controls, there were 15 workers in each of the four isolation groups. Champalbert and Lachaud recorded the behavior of control and experimental ants for 45 days after emergence.

The question was, what effects, if any, would social isolation have on the behavioral development of workers in the experimental groups? Also, would the effects of isolation depend on when they were isolated? During the first week after emergence, the control workers spent most of their time feeding or being groomed by other colony members. By the second week, however, work began, and the ants started to specialize in nursing activities, first caring for larvae, then cocoons, and finally eggs. Sometime during the third week, the workers changed their job status again and began to explore the nest and engage in domestic tasks. Finally, about a month after emergence, the control workers focused on activities related to their new careers as either guards at the nest or foragers outside the nest. Thus, over the course of only a few weeks, *E. tuberculatum* workers typically underwent several changes in behavioral specialization, and in a very specific sequence.

So, how did the behavior of ants in the experimental groups compare with that of controls? What were the effects of isolation immediately at emergence, or two, four, or eight days later? It is interesting that the workers isolated at two days after emergence showed the most abnormal behavioral development. They differed from controls in the order of appearance of their activities and in the level of performance of their tasks, being particularly lax in brood care (Figure 8.15). In contrast, the behavioral development of workers isolated either at emergence or four or eight days later was more similar to that of the controls. In the case of workers isolated at emergence, behavioral development was simply delayed. The reintroduction of these workers into society seemed to correspond to a second emergence; although 11 days old at the time of reintroduction, these workers behaved like newly emerged ants. Soon after reintroduction into the colony, however, the various specializations in activities appeared progressively in the same sequence as that of the control workers. Only minor abnormalities were noted in the workers isolated at four days, and the behavioral development of those isolated at eight days was almost identical to that of the controls. In summary, the extent of behavioral abnormalities observed in *E. tuberculatum* workers as a result of a ten-day period of social isolation depended on the age of the workers at the time of isolation, with the most

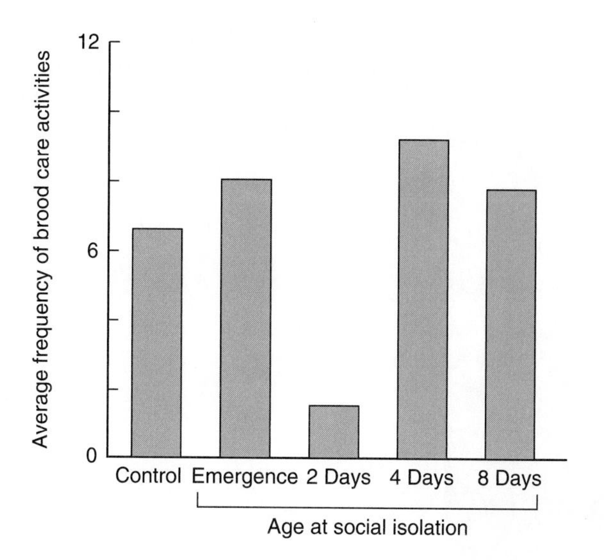

FIGURE 8.15 **A sensitive period for the development of brood care behavior in worker ants. Here, a ten-day period of social isolation beginning two days after emergence produces more severe disruptions in the performance of brood care activities by *Ectatomma tuberculatum* workers than do periods of isolation beginning either at emergence or four or eight days later. (Drawn from the data of Champalbert and Lachaud 1990.)**

serious abnormalities occurring in ants isolated two days after emergence.

Champalbert and Lachaud (1990) concluded from their observations of *E. tuberculatum* worker ants that there is a sensitive period during the first four days after emergence in which exposure to social stimuli affects the establishment of behavior, especially that related to brood care. One might ask, then, why would workers isolated at emergence (and therefore removed from social stimuli during the four-day critical period) display fairly normal behavioral development? The authors suggest that when workers are isolated immediately after emergence, their development is simply put on hold. The reintroduction of these workers into the colony after their period of social isolation mimics an emergence. During the four days following reintroduction, these workers receive the social stimuli important in the development of brood care activities. In other words, it is not the physiologically defined first four days of life that is important for the development of brood care behavior but rather the first four days of social contact with colony members. These four days may follow either natural emergence or the artificial second emergence produced by the reintroduction of workers into the colony after isolation. Here, again, we see an example of flexibility associated with sensitive periods in behavioral development; in ants isolated at emergence, the lack of exposure to appropriate social stimuli delays closure of the sensitive period until a few days after reintroduction to the colony. In view of the poor performance of workers isolated at two days, it also appears that the entire sensitive period must occur in an uninterrupted fashion to ensure normal behavioral development. Because these workers received some portion of the necessary social stimulation during their first two days following emergence, their developmental system could not be put on hold until reintroduction into the colony.

So far we have described the factors that influence behavioral development and the fact that many of these factors exert their influence during sensitive periods. Now we will describe the development of singing behavior in birds, an example that illustrates the combined influences of genes, experience, and neural and hormonal control during particular sensitive periods.

PULLING IT ALL TOGETHER—THE DEVELOPMENT OF BIRD SONG

All of us have probably, at one time or another, been struck by the beautiful and often complex songs produced by the birds around us. Consider, for example, the rich and melodious soliloquies of the canary (*Serinus canaria*) or the odd mix of rattles, whistles, and clicks woven throughout the songs of the European starling (*Sturnus vulgaris*), a talented bird, indeed, but one that is largely unappreciated by North American birdwatchers because of its status as an introduced species that aggressively preempts native cavity-nesting species from choice nest sites. Canaries and European starlings are members of the order Passeriformes. Some species within this order learn to sing (those within the oscine suborder, the so-called songbirds), while other species do not (those within the suboscine suborder, but there may be some exceptions; Kroodsma 2005; Saranathan et al. 2007). Of the 22 remaining orders of birds, song learning is found in parrots (order Psittaciformes) and hummingbirds (order Apodiformes). We will focus on the species that learn their song, and our examples will come from the order Passeriformes. Nevertheless, it is important to realize that even if species do not *learn* to sing, they may still produce fairly complex vocalizations. Let's take a look at how bird song develops.

Song development has been examined from the standpoint of evolution and ecology, as well as from the more mechanistic approach of assessing genetic, neural, hormonal, and social influences on behavior. Together, these approaches have revealed the continuous interplay between the developing bird and its internal and external environment. As we will see, the factors that influence the development of song range from interactions between cells to those between individuals. We begin by describing the development of singing behavior in the zebra

FIGURE 8.16 **Male zebra finches sing, and females are attracted to their song. Here a female zebra finch (left) perches near a male (right).**

finch, an extremely social species native to Australia (Figure 8.16). Zebra finches do very well in captivity (in fact, many of you have probably seen them in pet stores and heard the "beeps" in their song), so they are well-suited to behavioral studies in the laboratory. Along the way we will also discuss other species that differ from zebra finches to emphasize the variation across species.

GENETIC, HORMONAL, AND NEURAL CONTROL OF SONG

In zebra finches, as in many songbirds, only males sing and females are attracted to their song (e.g., Holveck and Riebel 2007). Males of many bird species use song not only to attract females, but also to advertise the boundaries of their territories. Zebra finches, however, breed in colonies, and males seem quite tolerant of other birds except in the immediate vicinity of their nests, so song in this species seems to function largely in mate attraction. We'll look first at what is known about the neural substrates of song.

Researchers have identified several areas of the brain involved in song. These areas are called song nuclei (a nucleus is a collection of neurons that is anatomically distinct), and they are linked to one another in neural pathways (Figure 8.17*a*). Two major pathways have been identified. The posterior pathway (shown by white arrows in Figure 8.17*a*) seems to be involved in the production of song, and the anterior pathway (shown by black arrows in Figure 8.17*a*) in the acquisition of song. Still other areas of the brain function in the perception (hearing) of song and in the storage of memories of song heard when young (Bolhuis and Eda-Fujiwara 2003; Gobes and Bolhuis 2007). The overall size of song nuclei in the two major pathways is sometimes linked to the size of the song repertoire (number of songs sung by a bird). For example, only male zebra finches sing, and song nuclei are larger in males than in females of this species (Figure 8.17*b*) (Nottebohm and Arnold 1976). In zebra finches, sex differences in neuroanatomy are also evident at the cellular level; male song nuclei have more and larger neurons than do the same regions in females (Gurney 1981).

The apparent association in zebra finches between sex differences in singing behavior and sex differences in the size of song nuclei in the brain is not found in all species of songbird. As an example of a disassociation between singing behavior and the volume of song nuclei, we turn to one of several tropical species of birds in which both sexes sing. Male and female bush shrikes (*Laniarius funebris*) sing duets; they sing equally complex songs and have similar song repertoire sizes. Based on this information, we might predict that the song nuclei would be the same size in male and female bush shrikes. In contrast to our prediction, however, males still have larger song nuclei (HVC and RA) than females and have more neurons in these areas (Gahr et al. 1998). Examples such as this one suggest that the relationship between singing behavior and morphology of the song system in the brain may be species-specific or at least more complicated than the zebra finch example suggests (Bolhuis and Gahr 2006).

How do sex differences in song nuclei arise? The chromosomal difference between the sexes is thought to initiate the dramatic differences in male and female brain structure. In birds, males have two Z chromosomes, and females have one Z and one W chromosome. The differential action of genes encoded on the sex chromosomes is thought to prompt sexual differentiation of the song system in zebra finches. However, much remains

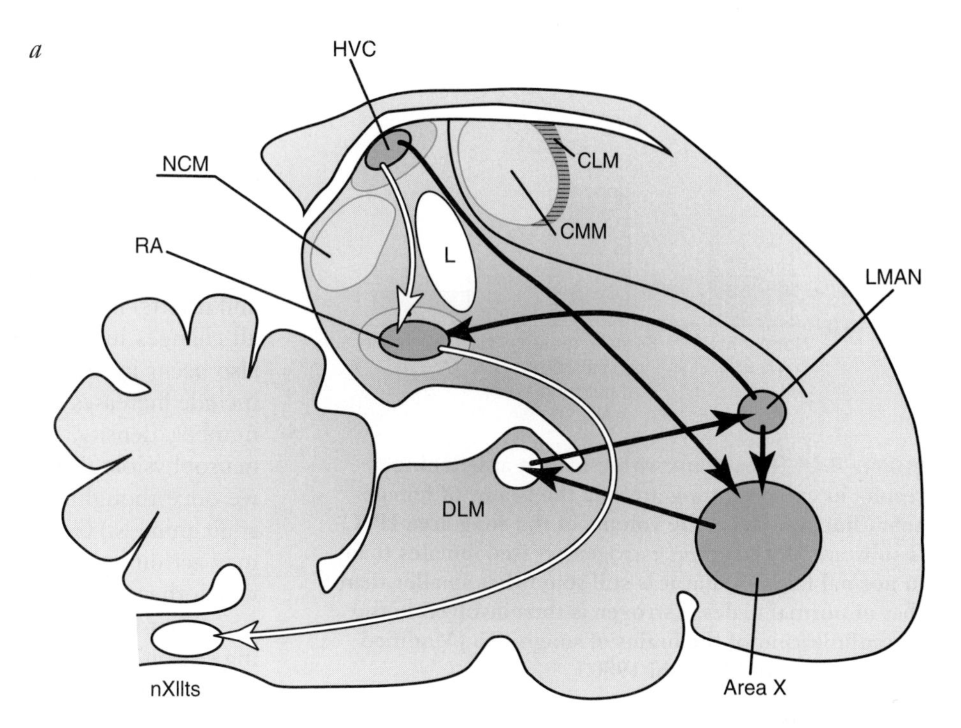

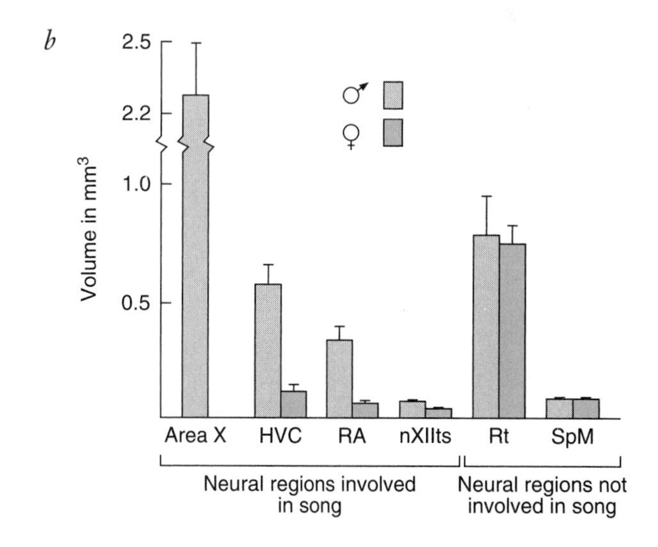

FIGURE 8.17 **(*a*) The songbird brain showing the nuclei involved in song. Although all structures are found on both sides of the brain, those on only one side are shown for simplicity. White arrows show the posterior pathway thought to be involved in song production; black arrows show the anterior pathway thought to be involved in song acquisition. For completeness, areas involved in auditory processing and perception are also shown (field L, NCM, CLM, and CMM). Abbreviations: CLM, caudal lateral mesopallium; CMM, caudal medial mesopallium; DLM, nucleus dorsolateralis anterior, pars medialis; HVC, a letter-based name; field L; LMAN, lateral magnocellular nucleus of the anterior nidopallium; NCM, caudal medial nidopallium; nXIIts, tracheosyringeal portion of the nucleus hypoglossus; RA, robust nucleus of the arcopallium. The nomenclature of the avian brain has been revised (Reiner et al. 2004; see Reiner, Perkel, Mello, and Jarvis 2004 for specific reference to songbirds), so don't be surprised when the names used in this figure differ somewhat from those in earlier publications. (*b*) Sexual dimorphism in the brains of zebra finches. Note that four of the brain regions involved in song (Area X, HVC, RA, and nXIIts) are substantially larger in males than in females. No such difference is found in two regions (Rt and SpM) that are not involved in song. (*a*: Modified from Bolhuis 2008. *b*: Nottebohm and Arnold 1976.)**

to be learned about which genes are expressed at which times in development. Hormones also play a role in sexual differentiation of the song system. Estrogen, in particular, seems critical. A remarkable group of experiments with zebra finches illustrates the powerful effects of estrogen on parts of the neural system that control singing behavior (Gurney and Konishi 1980). If pellets containing estrogen are implanted in newly hatched male zebra finches, they have no effect (i.e., in adulthood the brains and songs of estrogen-treated males are similar to those of untreated males). In contrast, treatment of nestling females with estrogen results in enlarged song areas in the brain. These areas are larger than those of untreated females, though they are not as large as corresponding areas in the normal male brain (Figure 8.18). Thus, estrogen appears to masculinize the brain. What is the source of the estrogen that masculinizes the brain? Most evidence points to the brain itself. (Recall from Chapter 7 that steroid hormones are produced by the gonads, adrenal glands, *and* the brain.) One hypothesis for sexual differentiation of the zebra finch brain suggests that genes on the sex chromosomes cause differences in the brain's synthesis of estrogen, and the higher levels of estrogen in male brain tissue masculinize the song system (Wade and Arnold 2004).

Hormones, we will see, have both organizational and activational effects on singing behavior. Indeed, experiments have revealed that the effects of hormones

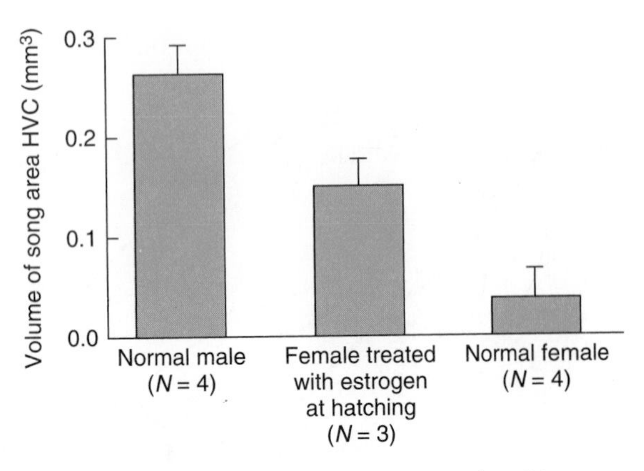

FIGURE 8.18 Treatment with estrogen at hatching results in enlarged song areas in the brains of female zebra finches. Here, the volume of the song area HVC is substantially larger in estrogen-treated females than in normal females, but it is still somewhat smaller than that of normal males. Estrogen is thus involved in the masculinization of the brains of songbirds. (Modified from Gurney and Konishi 1980.)

on the brain and singing behavior of zebra finches do not end around the time of hatching. Although the typical male has a masculinized brain, elevated levels of testosterone in the bloodstream are necessary to stimulate singing in adulthood. Similarly, those females that receive estrogen implants soon after hatching, and thus develop enlarged song areas, do not sing in adulthood unless testosterone is administered (Gurney and Konishi 1980). Females that do not receive estrogen implants at hatching, but do receive testosterone in adulthood, do not sing. Apparently, early exposure to estrogen establishes a sensitivity to testosterone in the brains of experimental females (and normal males), and exposure to testosterone in adulthood stimulates song. In other words, whereas estrogen early in life *organizes* the development of a male song system, a high circulating level of testosterone in adulthood appears necessary to *activate* singing behavior. Finally, the brains of birds are not always sensitive to the organizing effects of estrogen. There appears to be a sensitive period, around the time of hatching, in which the brain is particularly sensitive to hormonal influences. During this window of time, estrogen exerts its powerful effects on the developing nervous system. After that, the neural pathways that control song cannot be switched to the male track. (See Chapter 7 for a further discussion of the organizational and activational effects of steroid hormones.)

For many species of songbirds, the hormonal activation of song in free-living adult birds occurs on a seasonal basis. Song occurs most often or only during the breeding season when levels of testosterone and its metabolites (the products of the breakdown of testosterone, estrogen, and 5α-dihydrotestosterone) are higher than during the nonbreeding season (higher levels of steroid hormones occur during the breeding season because increasing day length induces increases in the size and activity of the gonads). Increased levels of testosterone and its metabolites prompt seasonal changes in the structure and physiology of song nuclei, which in turn promote singing behavior (Brenowitz 2004; Brenowitz et al. 2007). Indeed, during the breeding season, the volumes of song nuclei (HVC, RA, X, and nXIIts) may increase by up to 200%! Beyond overall changes in volume, cellular changes in song nuclei also occur in seasonally breeding birds. These changes include increases during the breeding season in neuron number, density, metabolic capacity, and spontaneous neurophysiological activity. (Recall from Chapter 6 that we once thought that new neurons could not form in adult animals.) Gonadal testosterone and its metabolites may act directly on the song nucleus called the HVC, which then influences other song nuclei.

What happens with the approach of the nonbreeding season? As the hours of daylight shorten and levels of testosterone and its metabolites decline, the song system in the brain regresses. Regression of song nuclei can occur quite rapidly, revealing the amazing plasticity in structure (and function) of the avian brain. For example, when adult male Gambel's white-crowned sparrows (*Zonotrichia leucophrys gambeli*) were maintained in the laboratory under conditions consistent with the breeding season (long day length and implants containing testosterone) and then castrated and shifted to short day lengths (testosterone implants were removed at the time of castration), the volume of the HVC regressed 22% within 12 hours of the withdrawal of testosterone (Thompson et al. 2007). With regression of the song system, singing declines or ceases altogether. In species in which singing declines but does not stop completely, the structure of song in nonbreeding birds may become more variable than song during the breeding season. Regression of the song system during the nonbreeding season has been interpreted as a way to reduce the energetic costs of maintaining the system during the fall and winter when many birds face significant energetic challenges, including those associated with migration, low temperatures, and low food availability.

We have seen that genetic information, the hormonal milieu, and development of the nervous system all influence singing behavior. We can now ask what role learning plays in the development of song.

ROLE OF LEARNING IN SONG DEVELOPMENT

Whereas zebra finches have frequently been used to study the physiological bases for song development, white-crowned sparrows (*Zonotrichia leucophrys*), songbirds from North America, have been a favorite subject for those curious about the role of learning (Figure 8.19).

FIGURE 8.19 **White-crowned sparrows, such as the male shown here, have often been used in studies of learning and song development.**

The importance of learning to song development in adult male white-crowned sparrows was demonstrated in isolation experiments conducted by Peter Marler and his colleagues (Marler 1970; Marler and Tamura 1964). Under natural conditions, a young male sparrow hears the songs of his father and other adult males around him during his first summer and autumn. During these first few months of life, the young male produces only subsong, a highly variable, rambling series of sounds with none of the syllables typical of the full song of adult males. As the male's first breeding season approaches, his song begins to contain elements recognizable as white-crowned sparrow syllables. This vocalization, which occurs in late winter or early spring of the male's first year, is called plastic song. The male continues to refine his song and eventually his song crystallizes. Thus, at the start of his first breeding season, he begins to produce the full song characteristic of adult male white-crowned sparrows. The songs of the male are copies (imitations) of those he heard many months before.

Young male white-crowned sparrows reared in isolation develop abnormal song (Figure 8.20). However, if isolated males hear tapes of white-crown song during the period from 10 to 50 days of age, they will develop their normal species song. Young white-crowned sparrows that hear taped white-crown songs either before 10 days of age or after 50 days do not copy it. Thus, under the conditions of social isolation and "tape tutoring," the sensitive period for song learning occurs between 10 and 50 days posthatching. There is some flexibility, however, in the timing and duration of this sensitive period: the sensitive period can be extended beyond 50 days when white-crowned males are exposed to live white-crowned tutors rather than taped songs (Baptista and Petrinovich 1984). Thus, we see that environmental conditions can influence the sensitive period for song learning (we return to this topic shortly). The important point now, however, is that, for song to develop normally, male white-crowned sparrows must be exposed early in life to the songs of adult males of their species.

In addition to hearing adult conspecific song, young male white-crowned sparrows must also hear themselves sing if they are to produce normal song. A series of classic experiments revealed the critical role of auditory feedback in song development (Konishi 1965). Young male white-crowned sparrows that are exposed to the songs of adult male conspecifics early in life and then deafened before the onset of subsong produce a rambling and variable song. Apparently, to develop full song, a sparrow must be able to hear his own voice and compare his vocal output to songs that he memorized months previously. Further studies revealed that isolated white-crowned males with intact hearing produce songs that are somewhat more normal in structure than those produced by males deafened prior to subsong (Refer again to Figure 8.20.) Thus, if a white-crowned sparrow can hear himself sing, he can produce a song with a few of the normal species-specific qualities. Overall, however, the song of an isolated male is still

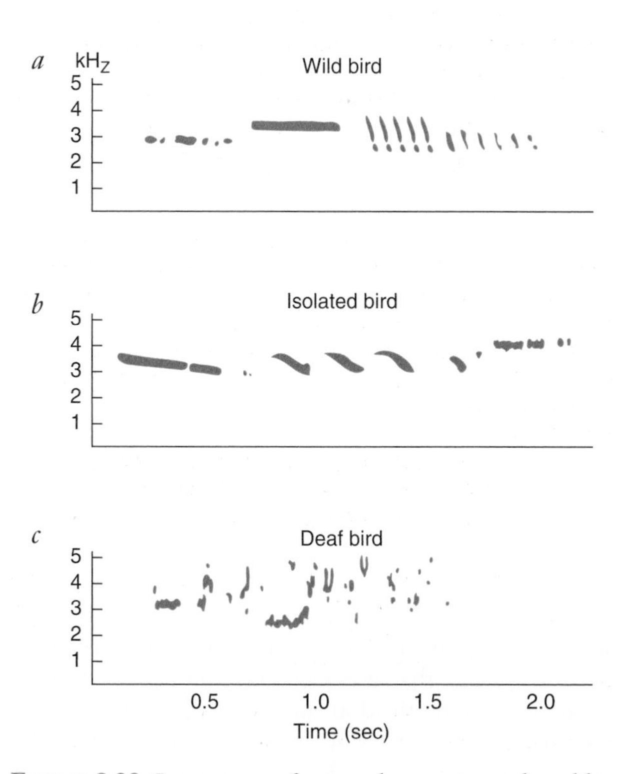

FIGURE 8.20 **Sonograms of songs that were produced by white-crowned sparrows (*a*) reared under natural conditions, (*b*) reared in social isolation, and (*c*) deafened at an early age. (Modified from Konishi 1965.)**

quite abnormal. Finally, although auditory feedback is important for song development, it is not essential for the maintenance of song in male white-crowned sparrows. Males deafened after their songs have crystallized continue to produce relatively normal songs for at least one year after deafening (Konishi 1965). There are, however, species in which auditory feedback appears essential to the maintenance of normal song in adulthood. For example, deafening of adult males results in the deterioration of song in zebra finches (Nordeen and Nordeen 1992), Bengalese finches (Woolley and Rubel 1997), and canaries (Nottebohm et al. 1976). Species differences in the importance of auditory feedback to maintenance of adult song may relate to species differences in song complexity or the ability to modify songs in adulthood (Brainard and Doupe 2000). It is also possible that the conflicting results relate more to the age at which males were deafened, with auditory feedback being less important to song maintenance in older males (Konishi 2004; Lombardino and Nottebohm 2000).

The results of these and other laboratory studies suggest that song development in sparrows consists of two phases—a sensory phase, during which songs are learned and stored in memory, and a sensorimotor stage, when singing a learned song actually begins. In the sensory phase, sounds that are heard during the first few months of life are stored in memory for months without rehearsal. During this time, young males produce only subsong, which does not involve retrieval or rehearsal of previously learned material. In the species of sparrows studied to date, the sensorimotor phase begins at about seven or eight months of age. At this time, birds retrieve a learned song from memory and rehearse it, constantly matching their sounds to the sounds they memorized months earlier. Over the next few months song patterns crystallize, and males begin to produce full song at the start of their first breeding season. The final adult song will remain virtually unchanged for the rest of the sparrow's life. The phases of song development are outlined in Table 8.3.

TABLE 8.3 Phases of Song Development in Sparrows

1. Sensory phase
   a. Acquisition—song learned
   b. Storage—song retained in memory
2. Sensorimotor phase
   a. Retrieval and production—motor rehearsal of learned song
   b. Motor stabilization—song crystallized into final adult song

Source: Marler (1987).

As young males move from the plastic phase of song to the crystallized phase, they sometimes drop songs from their plastic song repertoire. For example, in the months leading up to their first breeding season, young male song sparrows (*Melospiza melodia*) drop one to three songs from their plastic song repertoire, eventually ending up with a crystallized song repertoire of 8 to 11 songs. The young males do not drop songs at random. Indeed, the discarded songs match the songs of fewer neighboring males or match them less well than the songs that are retained. Sharing songs may facilitate communication among neighboring males. For example, shared songs are a reliable signal of a young male being a "local" and perhaps less likely to trespass or to attempt to take over another male's territory (Beecher and Brenowitz 2005). By facilitating communication, song sharing may allow neighboring males to avoid territorial battles that are costly in time and energy. In support of this idea, there is evidence that sharing songs is beneficial to male song sparrows: degree of song sharing is positively correlated with male survival (Wilson et al. 2000) and territory tenure in this species (Beecher et al. 2000). Interestingly, a comparison of plastic and crystallized versions of retained songs revealed that some of the retained songs are modified to make them less similar to the songs of neighbors, perhaps promoting individual recognition. Thus, song learning in young male song sparrows seems to involve a delicate balance between two opposing forces—the tendency to copy and keep the songs of neighboring males and the need for each male to make at least some of his songs individually distinctive (Nordby et al. 2007).

So far, we have confined our discussion of song learning to the well-studied sparrows. However, a great deal of variation exists across species. This variation is especially evident when it comes to sensitive periods in song learning, our next topic.

SENSITIVE PERIODS IN SONG LEARNING

We learned from experiments on male white-crowned sparrows raised in isolation and tutored with tape recordings of white-crown songs that the sensitive period for song learning extends from 10 to 50 days of age. We also learned that the sensitive period could be extended by exposing young males to live conspecific tutors instead of taped song. These results suggest some degree of flexibility or variability in sensitive periods and indicate that sensitive periods can be influenced by environmental conditions. It has been found that there is considerable variation both among and within species of songbirds in the time course for song learning. Here, we consider some of that variation and what it might mean.

The length and timing of the sensitive period for song learning vary greatly across species. Canaries, for example, have been called lifelong learners (or an open-ended species) as a result of their ability to continually revise their songs throughout adulthood. In contrast, white-crowned sparrows and zebra finches have a more restricted period of learning that occurs during the first weeks or months of life; these species are called age-limited learners (or closed-ended species). The discovery of some degree of song plasticity in adulthood in some age-limited learners has prompted the suggestion that these two developmental patterns—lifelong learners and age-limited learners—are not dichotomous (i.e., they are not the only two possible options), but instead points on a continuum (Brenowitz 2004).

Species differences in length and timing of sensitive periods for song learning often reflect species differences in life history. For example, we have described how young male song sparrows learn to sing by listening to and memorizing the songs of adult males that they hear during the first few months of life. These young males then prune their plastic song repertoires to produce full adult song by their first breeding season, when they are not quite one year old. Now consider the odd life history of brown-headed cowbirds (*Molothrus ater*). These birds are brood parasites. Females of this species lay their eggs in the nests of other bird species, thereby relinquishing all parental duties to foster parents (Figure 8.21). If young male cowbirds relied on learning characteristics of their songs by listening to male birds in the vicinity of the nest during their first few weeks of life, they would most likely sing the song of their foster parents. But field and laboratory observations indicate that this is not the case at all; when male cowbirds become sexually mature, they sing the cowbird song. This raises an interesting question: when does the sensitive period for song learning occur in brood parasites such as cowbirds? After fledging from the nests of their foster parents in late summer, young brown-headed cowbirds join flocks containing other cowbirds—is this when they learn their song? But this too can be a difficult time for song learning because by this time, many adults have already left breeding grounds, and those males that remain have ceased singing. Given these details of the cowbird life history, it should come as no surprise that song learning is delayed in brown-headed cowbirds relative to song sparrows: male cowbirds do not perfect their song until their second winter. This delay gives them sufficient time to hear as yearlings the songs of adult male conspecifics (O'Loghlen and Rothstein 1993).

FIGURE 8.21 Brown-headed cowbirds lay their eggs in the nests of other species, leaving their offspring in the care of the host parents. Here, a host parent, a yellow warbler, feeds a cowbird nestling.

Sensitive periods for song learning also vary within a species. Individual differences in the timing of song learning are commonly reported in both field and laboratory studies, and for some well-studied species, the variation can be understood in the context of their life history. In the marsh wren, *Cistothorus palustris*, the time course for vocal learning is influenced, in part, by a parameter of the physical environment—photoperiod (Kroodsma and Pickert 1980). In the northeastern United States, young wrens hatch from mid-June to late August. Males that hatch in June experience long day lengths, whereas those that hatch in August experience relatively shorter day lengths. Does exposure to different photoperiods after hatching affect the time course for song learning in this species? Laboratory experiments revealed that males raised under day lengths that simulate hatching late in the breeding season (August) learned songs the following spring, whereas males raised under conditions that simulate early hatching (June) did not. Thus, the sensitive period for late-hatched males appears to extend beyond that for early-hatched males. A flexible sensitive period determined, in part, by environmental conditions may make sense for some songbirds, especially those species in which late-hatched young tend to disperse further than early-hatched young. Such flexibility permits the learning of song the following spring in the area in which the late hatched males have settled (recall that singing the songs of neighbors is a successful strategy for many songbirds).

Thus, we see that variation in the length and timing of sensitive periods for song learning both among and within species is perhaps best viewed in the context of life history. Furthermore, such variation probably serves to maximize the chances that each male develops species-appropriate singing behavior.

OWN-SPECIES BIAS IN SONG LEARNING

When given a choice, most young male birds learn their songs from members of their own species. For example, male white-crowned sparrows copied only the songs of their own species when tutored with tapes of the song sparrow and white-crowned sparrow (Marler 1970). This preference is called own-species bias, and it also influences the speed and accuracy with which young males learn. Typically, birds learn the songs of their own species more rapidly and accurately than those of a different species, although this tendency varies across species and with method of song presentation. In light of evidence that own-species bias characterizes song learning, researchers are now investigating the neural basis for this bias. These research efforts often involve recording responses from neurons in the auditory area known as field L (refer, again, to Figure 8.17*a*) while exposing anesthetized male birds to a variety of sounds, including conspecific song (e.g., Grace et al. 2003).

When a young male is exposed to tape recordings of conspecifics, live tutors of another species can sometimes override the bias for learning conspecific songs. When young white-crowned sparrows are presented with visible, singing song sparrow tutors and can hear the songs of white-crowned sparrows in the background only, they learn the song of the song sparrow (Baptista and Petrinovich 1986). That young male sparrows learn the song of the adult that they are allowed to interact with, even if this adult belongs to another species, emphasizes the importance of social interactions in the song-learning process. Next we examine how social factors influence the development of song.

SOCIAL FACTORS AND SONG DEVELOPMENT

As described previously, young cowbirds are reared by parents of different species, and they begin to associate with conspecifics after fledging, when they join cowbird flocks. Several studies have revealed that interactions with conspecific males and females in flocks on breeding and wintering grounds influence song development in cowbirds.

Interest in the role of social interactions in the development of song in cowbirds grew from the finding that the songs of male cowbirds reared in isolation from male conspecifics were substantially more appealing to female cowbirds than the songs of males reared in groups, as measured by the willingness of females to display a copulatory posture (Figure 8.22) (West et al. 1981). Why are the songs of isolates more potent than those of normal males? The answer comes from observations of what occurred when group-reared males and isolation-reared males were individually introduced into an aviary with an established colony of cowbirds. Group-reared males kept a low profile in their new environment; they did not sing in the presence of resident males and were not attacked. In contrast, the isolates repeatedly sang their high-potency song in the presence of resident males and females, and the resident males responded by attacking and sometimes killing the energetic songsters.

a

b

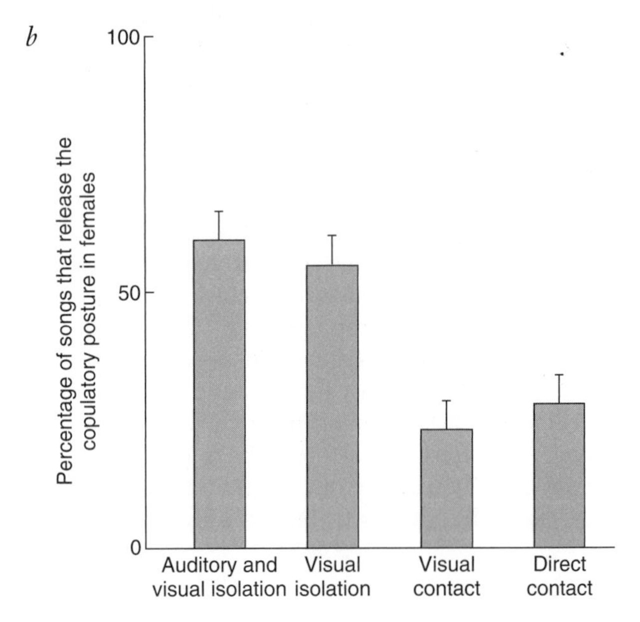

FIGURE 8.22 **Female cowbirds prefer the songs of males reared in isolation to those of males reared socially. (*a*) Females display their preference for songs by assuming the copulatory posture. (*b*) Males without visual or direct contact with dominant males produce more potent songs than do males that have visual or direct contact. In cowbirds, then, young males learn to modify their songs to avoid attack by adult males. (Drawn from the data of West et al. 1981.)**

Apparently, males living in mixed-sex groups, such as the flocks that occur in nature, learn to suppress their song and thereby avoid attacks from dominant males. Subsequent observations of young males maintained in flocks with and without adult males reveal that adult males influence not only the potency of song eventually produced by young males, but also the rate at which song develops: specifically, presence of adult males slows the rate of song development in young male cowbirds (White et al. 2002).

What role might female cowbirds play in shaping the songs of males? To examine this question, King and West (1983) used two different subspecies of cowbirds, one found in the area of North Carolina (*M. ater ater*) and the other around Texas (*M. ater obscurus*). We will refer to these cowbirds as subspecies A and subspecies O, respectively. Male cowbirds of subspecies A were hand-reared in acoustic isolation until they were 50 days old. At this time they were housed for an entire year with (1) individuals of another species (canaries or starlings), (2) adult female cowbirds of subspecies A, or (3) adult female cowbirds of subspecies O. At the end of this period, the researchers examined the songs of the males. The three groups of males of subspecies A had developed remarkably different songs (Figure 8.23). Males that were housed with canaries or starlings had diverse repertoires, singing a mix of songs from the two cowbird subspecies, along with imitations of the songs of other species in their cage. Whereas males housed with cowbird females of subspecies A sang all A songs, those housed with cowbird females of subspecies O sang predominantly O songs. Male cowbirds thus develop song repertoires that are biased toward the preferences of their female companions. More recently, it has been shown that female cowbirds also influence the rate of song development in young males: males housed with females from their own population progress from subsong to plastic song to stereotyped (full) song more rapidly than males housed with females from a different population (Smith et al. 2000).

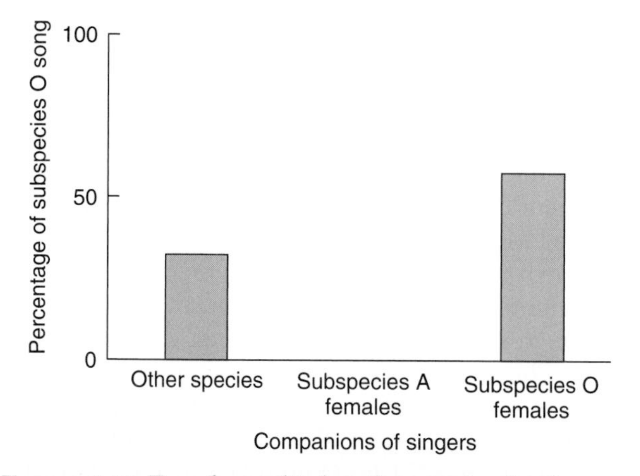

FIGURE 8.23 **Female cowbirds influence the development of song in males. When males of subspecies A are housed with either subspecies A or subspecies O females, they mold their song to the preferences of their female companion. Note that subspecies A males housed with females of their own subspecies do not sing any subspecies O song, but those housed with subspecies O females sing subspecies O songs predominantly. (Drawn from the data of King and West 1983.)**

Because female cowbirds do not sing, their influence on male song development cannot be through song. Instead, they may encourage the male's singing by a simple display called the wing stroke, in which one or both wings are moved rapidly away from the body (West and King 1988). These wing strokes are not doled out indiscriminately: approximately 94% of the males' songs elicit no visible change in the behavior of female cowbirds, and on average, a single wing stroke occurs for every 100 songs. It seems, then, that male cowbirds adjust their songs to the whims of their audience, molding their song structure to avoid conflict with dominant males and to stimulate available females. Just how they achieve this delicate balance in natural flocks is an interesting question.

Social factors are also important in shaping the songs of other songbird species. As mentioned previously, young male songbirds often drop a few songs from their plastic song repertoire along the way to reaching their final song repertoire. This so-called selective attrition phase is thought to occur at the start of the male's first breeding season when he seeks to establish a territory and engages in countersinging interactions with neighboring males (Nelson and Marler 1994). During these interactions, the young male attempts to match the songs of his neighbors. Over time, he drops those songs that are the poorest matches. These direct interactions with other males at the start of the breeding season therefore shape song development. More recent evidence indicates that some young songbirds choose the songs to keep in their final repertoire not only based on the singing interactions in which they directly participate, but also by eavesdropping on the singing interactions of other males. Indeed, at least under laboratory conditions, young male song sparrows seem to learn more by eavesdropping (indirect social interactions) than by direct social interactions, possibly because overheard interactions are less threatening (Beecher et al. 2007; Burt et al. 2007). We see, then, that song development can be influenced by direct and indirect social factors.

A DIVERSITY OF SONG LEARNING STRATEGIES

As studies of different species of songbirds have accumulated over the years, it has become clear that songbirds exhibit diverse song learning strategies (reviewed by Beecher and Brenowitz 2005). We have seen, for example, that the period within which young males learn song ranges from a relatively brief sensitive period early in life to learning throughout a lifetime. We described how social isolation produces abnormal song in white-crowned sparrows; however, while this is true for many other species of songbirds, it is not true for all species (Kroodsma et al. 1997; Leitner et al. 2002). Diversity also exists in the number of songs that birds learn. While the song repertoires of most species are small (less than five songs) to moderate (about ten songs), some repertoires are truly astronomical in size (over 2000 songs). There also is variation in how closely young males copy tutor songs. While males in some species faithfully copy (imitate) tutor songs, males in other species modify tutor songs or invent species-typical songs that bear no resemblance to tutor song. Song learning strategies among songbirds clearly differ along several dimensions. One challenge today is to understand the diversity of song learning programs in the context of songbird phylogeny with the aim of determining whether certain programs have evolved within certain lineages (Beecher and Brenowitz 2005).

STOP AND THINK

Spencer et al. (2007) raise the issue of how best to raise young songbirds brought to wildlife rehabilitation hospitals whose aim is to ultimately return the birds to their natural habitats. We have discussed the effects of rearing conditions on the development of song and the importance of song to subsequent survival and breeding success (e.g., song functions in territory maintenance and mate attraction). If you were charged with developing guidelines for housing and tutoring young songbirds brought to your rehabilitation facility, what factors would you consider in making your recommendations?

DEVELOPMENTAL HOMEOSTASIS

Throughout this chapter we have seen that experience plays an important role in behavioral development. Having said this, we might also note that the manipulations used to demonstrate the sensitivity of the developing animal to external influence are often quite severe. Such manipulations typically involve rearing in isolation or rearing by another species; neither condition is likely to be encountered by most animals under natural conditions. Indeed, despite the fact that in their natural environments, individuals within a species often develop under a diverse array of physical and social conditions, the vast majority of adults display species-typical, normal behavior. Under field conditions, after all, most white-crowned sparrows sing white-crown songs, most mallard ducklings follow their mother, and most worker ants perform their brood care responsibilities in an admirable fashion. Why do such complex patterns of behavior develop so reliably when there seems to be so much room for error?

The developmental processes appear to be capable of buffering themselves against potentially harmful influences to produce functional adults. This buffering capacity is called **developmental homeostasis.** A buffer is something that dampens drastic changes. We can look around us and see that despite diverse experiences, behavioral development in most individuals proceeds in a very predictable and reliable manner. In effect, the developmental process demonstrates a certain stability and resilience in the face of a host of constantly changing variables, including, as we have seen, many of those we introduce experimentally. Our final example, meant to illustrate the resilience of the developmental process and the efforts being made today to improve the lives of captive nonhuman primates, concerns the recovery of chimpanzees from years of profound social isolation.

REHABILITATION OF CHIMPANZEES AFTER LONG-TERM ISOLATION

Michaela Reimers and colleagues (2007) studied 13 male chimpanzees (*Pan troglodytes*) that had been separated as infants from their mothers and conspecifics and imported from Africa to a laboratory in Europe between 1976 and 1986 for use in hepatitis and HIV research. When being used in biomedical research, the chimpanzees were individually housed in indoor cages at the laboratory. Although all 13 males experienced early maternal separation, spatial confinement, and lack of control over their daily lives, they differed in the timing at which social deprivation began and the total duration of social isolation. Six males arrived at the research laboratory when they were between one and two years of age, and did not experience peer housing before placement into individual cages. These males were categorized as early deprived (ED). The remaining seven males arrived at the laboratory when they were between three and four years of age, and then spent one year in a peer group before being separated into individual cages. These males were categorized as late deprived (LD). At the time of the study by Reimers and colleagues, ED males were about 22 years old and had spent, on average, 21 years in isolation, and LD males were about 19 years old and had spent, on average, 17 years in isolation. The study consisted of four phases: laboratory (three months); transport to the new facility (seven days, which included the three-hour trip and subsequent adjustment period); habituation to the new facility (six weeks); and resocialization (one year). The new facility contained a spacious indoor enclosure with tree trunks and other wooden structures. During the habituation phase, males were individually and repeatedly introduced to the indoor enclosure. During the resocialization phase, males were introduced into the enclosure with a small number of other males for a few hours; the precise length of a session was determined by the well-being and progress of the participants, and thus was effectively controlled by them. Socialization sessions were interspersed with resting periods in individual cages, and the composition of the social groups changed over days. Eventually, ten of the males became members of one all-male group, and the remaining three males became members of mixed-sex groups. Reimers and colleagues monitored levels of glucocorticoids (stress hormones) in fecal samples across the four phases of the study, conducted two standardized novelty tests, and observed social interactions in the all-male group during the first year of group living. The first novelty test was conducted during the laboratory phase, and consisted of scoring the response of each chimp to a novel plush toy placed in the food drawer of its individual cage. The second novelty test was conducted during the habituation phase and consisted of scoring the response of each chimp when first introduced to the new and spacious indoor enclosure.

The timing and duration of social deprivation influenced the behavioral and hormonal responses of the chimpanzees. Chimpanzees in the ED group took longer than those in the LD group to take the novel toy out of the food drawer and to enter the novel environment (Figure 8.24*a*). In addition, whereas chimpanzees in the ED group explored the new toy and environment in a cautious manner, those in the LD group exhibited exploratory behavior that was active and bold. ED and LD chimpanzees also differed in their social behavior, with ED males initiating and receiving fewer social interactions than LD males. When considering all of the study chimpanzees, stress hormone levels increased during the transport phase, declined during the habituation phase, and declined still further during the resocialization phase (Figure 8.24*b*). Overall, levels of stress hormones were higher in ED males than in LD males, and the patterns of change across phases also differed between the two groups. Glucocorticoid levels in ED males remained fairly high during habituation and then declined significantly during resocialization to below habituation levels. In contrast, glucocorticoid levels in LD males were already lower during habituation and then declined still further during resocialization to levels below those for the laboratory phase. Taken together, the behavioral and hormonal data indicate that chimpanzees separated from their mothers and conspecifics at a younger age and kept in isolation for more years are less explorative, less social, and more susceptible to stress than chimpanzees that experienced somewhat less severe deprivation. In addition, the patterns of change in levels of stress hormones across the study suggest that chimps are better able to cope with social stressors (as

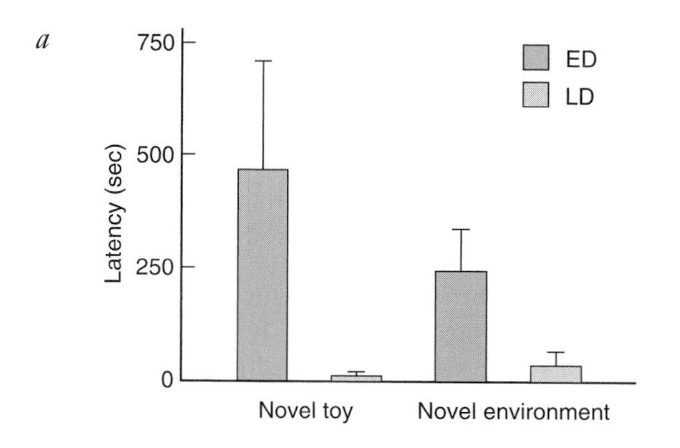

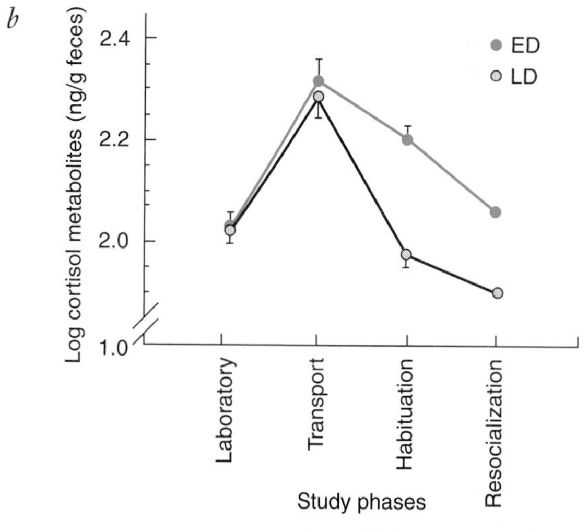

FIGURE 8.24 **Timing and duration of social isolation influence the behavioral and hormonal profiles of male chimpanzees previously used in biomedical research. (*a*) Chimpanzees in the early deprivation group (separated from mother and conspecifics at 1 to 2 years of age; ED) take longer then those in the late deprivation group (separated from mothers and conspecifics at 3 to 4 years of age; LD) to explore a novel toy and a novel environment. (*b*) Levels of stress hormones (as measured by metabolites of the hormone cortisol in feces) also differ between ED and LD groups, although both groups showed a decreased stress response to resocialization. Thus, chimpanzees isolated for many years can be rehabilitated to live less stressful, social lives, thereby illustrating the resilience of developmental processes. (From Reimers et al. 2007.)**

during the resocialization phase) than environmental stressors (as during the habituation phase)(refer again to Figure 8.24*b*); the authors suggest that this may reflect the chimps' perceived greater control over social situations than environmental situations. The main point for our discussion of developmental homeostasis, however, is that chimpanzees that have experienced profound social isolation, in some cases lasting two decades, can recover and live less stressful lives with conspecifics. Here, then, we have an example of careful and therapeutic rehabilitation revealing the resilience of developmental processes in our closest relatives.

SUMMARY

Patterns of behavior appear, disappear, and alter in form as animals develop. There are several causes of behavioral change, and one of the most significant is the development of the nervous system. During embryonic life, patterns of behavior often emerge in parallel with development of the sensory and neural structures necessary for their performance. Sometimes neuronal growth, death, or alterations in structure can be linked to developmental changes in a particular behavior. Behavioral change may also come about through the development of specific morphological structures. In developing paddlefish, the change in feeding behavior from picking select items out of the water column to indiscriminate filter feeding parallels the development of gill rakers. Changes in hormonal state can also trigger behavioral change, as in intrauterine effects in mice and the deposition by avian mothers of varying amounts of hormones in their eggs. Finally, experience also affects the development of behavior. For example, play is an experience that is vital to behavioral development in many species. It is expressed in several ways: social play; locomotor play (exercise); and object manipulation. Hypotheses for the function of play include the following: it improves physical condition, it is important in developing social skills and bonds, and it helps develop and maintain cognitive skills. Neural, hormonal, morphological, and experiential causes of behavioral change often interact during the continuous interplay between the developing animal and its internal and external environment.

Genetic factors also enter into the complex interaction between the developing animal and its environment. The interaction between genetic and environmental factors is perhaps best illustrated by the development of sexually dimorphic patterns of behavior, such as singing in birds. In general, only male birds sing, and in some species this is reflected in their neuroanatomy. The developmental basis for sex differences in the brains and vocal behavior of songbirds is the chromosomal difference between the sexes. This genetic difference dictates patterns of secretion of steroid hormones that, in males, lead to the growth and differentiation of the regions of the brain involved in song. In addition to neural and hormonal influences, experience plays an important role in the development of singing behavior. If they are to produce normal songs in adulthood, young males of many species must listen to the songs of adult conspecifics and must be able to hear themselves sing. Social interactions, whether direct (e.g., countersinging) or indirect (e.g., eavesdropping), also influence the development of song.

In male songbirds, the pulse of steroid hormones that affects the developing nervous system and the experience of hearing conspecific songs are not always capable of influencing the development of singing behavior. Indeed, to be effective, the change in hormone level and the experience of listening to conspecifics often must occur during somewhat restricted periods of time, when the young bird is particularly sensitive to such influences. These periods of enhanced sensitivity to environmental stimuli are called sensitive periods, and they characterize diverse behavioral phenomena, including filial imprinting (the response of young precocial birds to follow their mother), sexual imprinting (learning the characteristics of an appropriate mate), and maternal attachment (the development of a bond between a mother and her offspring such that she selectively cares for her own young). There are tremendous differences among species in the timing of sensitive periods for the development of specific behaviors, and even within a species there is flexibility in the time course of sensitive periods.

We have said that experience, especially that occurring during a sensitive period in the young animal's life, can profoundly influence behavioral development. Under normal circumstances, however, animals encounter a wide variety of physical and social conditions, and most develop normally. The ability of development to proceed in a normal fashion in the face of environmental perturbation is called developmental homeostasis. The successful rehabilitation of research chimpanzees subjected to long-term social isolation speaks to the resilience of behavioral development.

PART TWO

Survival

9

Biological Clocks

Imagine for a moment that you have a pet hamster, a friendly fellow who quietly shares your bedroom. However, late one Thursday night you are studying for a test but finding it hard to concentrate because a hamster is most active during night hours, and his running wheel is squeaking. So, you place the hamster and his cage in your closet and continue to study. The next morning, you take the test and then leave for the weekend. On Monday night you are back, and you notice a squeaking in your closet. You have forgotten about the hamster. He has had plenty of food and water, but he has been in the dark for three days. As you retrieve him, you notice that he begins to run on his wheel at about the same time as he normally did.

How could he know what time it was? We attribute the ability to measure time without any obvious environmental cues to an internal, living clock. When any hamster is experimentally sequestered in the constant darkness and temperature of the laboratory so that each turn of its running wheel can be recorded automatically for months or even years, a record similar to the one shown in Figure 9.1 usually results. Notice that the hamster woke up almost exactly 12 minutes later each day during the entire study. Its bouts of activity alternate with rest with such regularity that it is often described as an activity rhythm. The ability to measure time is common not just in hamsters but also in most animals. In fact, biological clocks

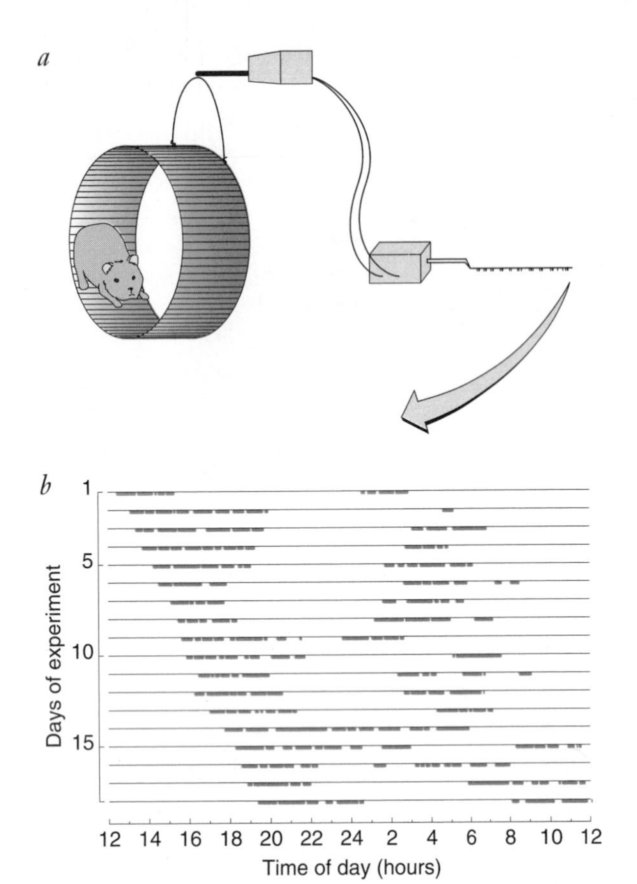

FIGURE 9.1 **(*a*) A hamster in a running wheel equipped to record each turn of the wheel shows periodic bouts of activity that alternate with rest. (*b*) In constant dim light, the cycle length of this activity rhythm is slightly longer than 24 hours. For each rotation of the running wheel, a vertical line is automatically made on a chart at the time of day when the activity occurred. In this record, the bouts of activity were so intense that the vertical lines appear to have fused, forming dark horizontal bands. Notice that although the animal had no light or temperature cycle as a cue to the time, it would awaken about 12 minutes later each day.**

have been found in every eukaryotic organism tested, as well as in cyanobacteria (Paranjpe and Sharma 2005; Woelfle and Johnson 2006).

The rhythmical nature makes sense in the light of evolutionary principles. Life evolved under cyclical conditions, and the ecological conditions under which animals find themselves differ tremendously at different times of the cycle. As we discuss in more detail later in the chapter, it is adaptive to *predict* upcoming changes in a cycle—for example, that darkness is coming or that winter is nearly upon us—rather than just respond to these events as they occur.

Every living thing is subjected to the regularly varying environmental conditions on earth orchestrated by the relative movements of the earth, moon, and sun. As the earth spins on its axis, life is exposed to rhythmic variations in light intensity, temperature, relative humidity, barometric pressure, geomagnetism, cosmic radiation, and the electrostatic field. The earth also rotates relative to the moon once every lunar day (24.8 hours). The moon's gravitational pull draws the water on the earth's surface toward it, causing it to "pile up" and thus resulting in high tide. These tidal cycles cause dramatic changes in the environment of intertidal organisms—flooding followed by desiccation when exposed to air. The relative positions of the earth, moon, and sun result in the fortnightly (biweekly) alternation between spring and neap tides, as will be explained shortly. The moon revolves about the earth once every lunar month (29.5 days), generating changes in the intensity of nocturnal illumination and causing fluctuations in the earth's magnetic field. Finally, the earth, tilted on its axis, circles the sun, causing the progression of the seasons, with its sometimes dramatic alterations in photoperiod and temperature.

Although the environmental modifications may be extreme, they are generally predictable. Often it is advantageous to gear an activity to occur at a specific time relative to some rhythmic aspect of the environment. So, biological clocks may have evolved as adaptations to these environmental cycles. Biological clocks also provide a mechanism to synchronize various internal processes with other internal processes. Thus, internal synchronization may be another adaptive value (Sharma 2003).

In this chapter, we will see examples of the various approaches to the study of behavior discussed in previous chapters: Rhythms are so pervasive that they have piqued the interest of scientists studying their adaptive value and evolution, genetic underpinnings, the hormonal control, and the neural control. We are now beginning to understand the connections among these levels of analyses. First, we will describe some rhythmic behaviors and then the properties of the clock that drives them.

DEFINING PROPERTIES OF CLOCK-CONTROLLED RHYTHMS

Like any good clock, biological clocks measure time at the same rate under nearly all conditions, and they have mechanisms that reset them as needed to keep them synchronized with environmental cycles.

PERSISTENCE IN CONSTANT CONDITIONS

A defining property of clock-controlled rhythms is that cycles continue in the absence of environmental cues such as light-dark and temperature cycles. This means

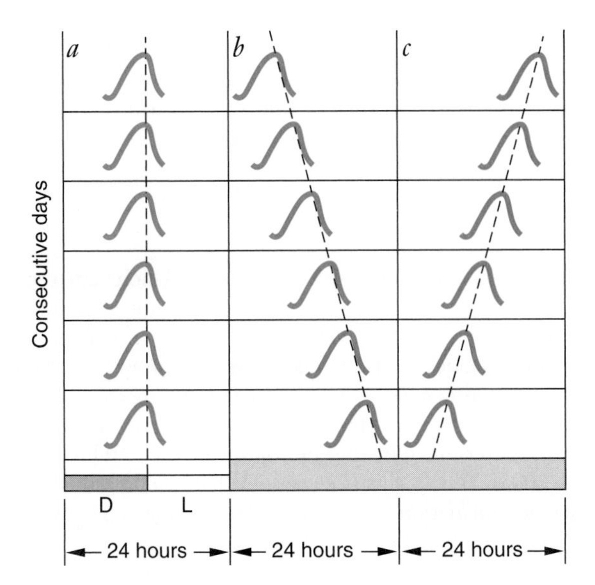

FIGURE 9.2 **Diagram illustrating a biological rhythm in the entrained and free-running state. (*a*) The clock is entrained to the light–dark cycle indicated by the bars at the bottom of the column. Entrainment is the establishment of a stable phase relationship between the rhythm and the light–dark cycle, thus ensuring that the activities programmed by the clock occur at the appropriate times. When an organism is placed in constant conditions, the period length of its rhythms is seldom exactly 24 hours. Depending on the organism, the light intensity, and the temperature, the period length may be (*b*) slightly longer than 24 hours or (*c*) slightly shorter than 24 hours. The adjective *circadian* is used to describe this change in period. (From Brown et al. 1970.)**

that the external day-night cycles in light or temperature are not causing the rhythms. Instead, we attribute the ability to keep time without external cues to an internal, (endogenous) biological clock.

However, in the constancy of the laboratory, the **period** (the interval between two identical points in the cycle) of the rhythm rarely is exactly what it was in nature; that is, it becomes slightly longer or shorter. This change in the period is described with the prefix *circa*. So, a daily rhythm, one that is 24 hours in nature, is described as being **circadian**—*circa*, "about;" *diem*, "a day" (Figure 9.2). A lunar day (tidal) rhythm is described as being **circalunidian**; a monthly rhythm, **circamonthly**; and an annual rhythm, **circannual**. In other words, a laboratory hamster kept in constant conditions may begin to run a little later every night. If it starts to run 10 minutes later in each cycle, after two weeks its activity will be about 2.5 hours out of phase with the actual daily cycle.

When an animal is kept in constant conditions in the laboratory, the period length of its rhythms generally deviates from that observed in nature. We assume that the period length of the rhythm in constant conditions is a reflection of the rate at which the clock is running. Sometimes this point is emphasized by describing the circadian period length in constant conditions as **free-running**, implying that it is no longer manipulated by environmental cycles.

If any clock is to be useful, it must be precise, and the biological clock is no exception. When an animal is cloistered in unvarying conditions and the free-running period length of its activity rhythm is determined on successive days of several months, the measurements are usually found to be extremely consistent. For some animals the precision is astounding. For example, the biological clock of the flying squirrel measures a day to within minutes without external time cues (Mistlberger and Rusak 2005).

ENTRAINMENT BY ENVIRONMENTAL CYCLES

By our definition, biological rhythms are generally not exactly the same length as the natural cycles—a circadian rhythm, for example, has a period of approximately one day ("circa" means "about"), which we see when animals are held under constant conditions. To keep from getting wildly out of synchrony with the natural cycle, biological clocks need to be reset, or **entrained**, to the cycle. Let's focus on some examples of the environmental cycles that reset circadian clocks.

For circadian rhythms, the most powerful phase-setting agent is a light–dark cycle (Johnson et al. 2003). In the laboratory, the biological clock can be reset at will by manipulating the light–dark cycle. If a hamster is kept in a light–dark cycle with 12 hours of light alternating with 12 hours of darkness such that the light is turned off at 6 P.M., its activity begins shortly after 6 each evening. If the light–dark cycle is be changed abruptly so that darkness begins at midnight, over the next few days the hamster's clock would be gradually reset so that at the end of about five days, activity would begin shortly after midnight real time. If, after several weeks of this lighting regime, the hamster is returned to constant conditions, its activity rhythm would have a period that approximates 24 hours, and, more important, the rhythm would initially be in phase with the second light–dark cycle.

Phase resetting occurs because a cue, a change from dark to light for instance, affects the clock differently depending on when in the clock's cycle it occurs. Although the rhythms are separate from the clock itself, we assume that they indicate what time the clock is signaling. If an organism is kept in constant dark, its rhythms will free-run, and we refer to points in the cycle as circadian time. For example, a hamster is active at night. So, when the hamster is kept in constant darkness, we refer to the time when it begins activity as early circadian nighttime. If a brief light pulse interrupts the darkness during early circadian night, it causes a phase delay. In other words, it resets the clock so that the hamster will become active later than expected the next day. But if a brief light pulse interrupts darkness during the late circadian night, it causes a phase advance; that is, the animal becomes active sooner than expected in the next cycle. In most animals, a brief light pulse occurring during circadian daytime has little or no effect. In nature, the clock is reset by light at dawn and dusk each day so that it keeps accurate time and is set to local time.

Although entrainment is important for all organisms because it adjusts biological rhythms to prevailing environmental cycles, it is especially important for animals living in temperate regions. Here, the length of daylight within each 24-hour day varies throughout the year. In the northern temperate zone, the interval of daylight gets progressively shorter each day from the summer solstice to the winter solstice. Then, the interval of daylight gets progressively longer each day until the summer solstice. Therefore, if it were advantageous for an animal to become active at dawn or at twilight, the onset of activity would have to change each day to stay appropriately synchronized with environmental light–dark cycle.

Consider the activity of a flying squirrel (*Glaucomys volans*). Flying squirrels nest in deep holes in trees where they cannot see whether it is light or dark outside. Flying squirrels are nocturnal, and each has a circadian clock that serves as an alarm clock to awaken them at twilight, which allows the squirrel to synchronize its activity to a specific light intensity shortly after sunset, regardless of the time of year (DeCoursey 2004a).

When awakened by its circadian clock, the squirrel goes to the den opening and briefly samples the prevailing light. If it is still light outside, the squirrel returns to its nest to sleep a bit longer and its circadian clock is reset so that activity begins slightly later the next night. Patricia DeCoursey (1986) demonstrated that light sampling behavior resets the circadian clock. She captured flying squirrels and kept them in a light-proof nest box connected by a tube to an open area with food and water. The outside area had a standard light–dark cycle. On most days, the squirrels did not see light: they slept in their light-proof nest box, leaving only after the lights went off and returning before the lights came on. The period of the activity rhythm of a flying squirrel is shorter than 24 hours in constant conditions. Therefore, it would awaken a few minutes earlier each evening. After a few nights, it encountered light when it reached the exit of the nest box. Upon seeing the light, it immediately returned to its dark nest box and rested before making a second attempt to exit. This brief exposure to light reset the squirrel's circadian clock; the next night the squirrel woke up 30 minutes later, and when it left the nest box, it encountered darkness once again. This pattern was repeated every four to five days. In this way, the squirrel's activity rhythm entrained to the light–dark cycle with only a few minutes of light exposure each week. In nature, this pattern of phase adjustment also causes the squirrel's onset of activity to follow sundown as it changes through the seasons.

STOP AND THINK

A stable phase relationship between a 24-hour environmental cycle and a rhythmic activity is not enough to demonstrate that the environmental cycle is entraining the biological clock. Instead, it is necessary to show that when the environmental cycle is eliminated and the animal is in constant conditions, the clock begins to free-run with a phase determined by the environmental cycle (Johnson et al. 2003). Why is this necessary? What could be an alternative explanation for a stable phase relationship between an environmental cycle and a rhythmic activity?

TEMPERATURE COMPENSATION

The biological clock remains accurate in spite of large changes in environmental temperature. This is somewhat surprising if one assumes that the timing mechanism is rooted in the cell's biochemistry. As a rule, chemical reactions double or triple in rate for each 10°C change in temperature. However, the effect of an equal temperature rise on the rate at which the clock runs is usually minor, rarely as large as 20%. This insensitivity to the effects of temperature suggests that the clock somehow compensates for them (reviewed in Mistlberger and Rusak 2005). It should be apparent that if the clock were as sensitive as most other chemical reactions are to temperature changes, it would function as a thermometer, indicating the ambient temperature by its rate of running, rather than as a timepiece.

Consider some possible consequences if the biological clock were affected by changes in temperature. First, the biological clocks of some animals would run at different rates at different times of day. This would obviously affect poikilothermic animals, but it could also be a problem for certain mammals. Consider, for example, the cave-dwelling, insectivorous bats of the temperate zone. Many species roost in cool, deep caves during the daytime. They allow their body temperature to drop to that of the cave while they are resting, which conserves metabolic energy. However, when they leave the cave to forage at night, their body temperature rises to 37°C. If the biological clock were not temperature compensated, how could it remain accurate when faced with such drastic changes in body temperature (DeCoursey 2004a)? The phase relationship between two oscillators (in this case, the biological clock and the LD cycle) varies with the period length of the oscillations. Thus, the second consequence of the clock running at different rates due to daily or seasonal changes in environmental temperature would be a change in the phase relationship between the rhythmic activity and the light–dark cycle, possibly causing the activity to occur at an inappropriate time of day. For example, activity might begin earlier or later than it usually does. The change in the phase relationship would occur because the free-running period length is an important factor determining the phase relationship between the rhythmic activity and the light–dark cycle (Gunawan and Doyle 2007).

RHYTHMIC BEHAVIOR

Rhythmicity in behavior and physiology is so common that it must be considered by anyone studying animal behavior. An animal is not perpetually the same. Rather, its behavior may fluctuate so that it is appropriate to the time of day or the state of the tides or the phase of the moon or the season of the year. We will begin by describing a variety of biological rhythms, each of which is synchronized with a geophysical cycle in nature. For a biological rhythm to be attributed to a biological clock, the rhythm must be shown to continue in constant conditions but with a slight change in period length.

DAILY RHYTHMS

The predominant geophysical cycle is the daily light–dark cycle, caused by the rotation of the earth relative to the sun. The activity of most animals is synchronized with the daily light–dark cycle, with most animals restricting their activity to a specific portion of the day. Nocturnal animals, including the familiar hamsters, cockroaches, bats, mice, and rats, are busiest at night. Diurnal animals, such as most songbirds and humans, are active during the day. Crepuscular animals are active primarily at dawn and dusk.

LUNAR DAY RHYTHMS

As the moon passes over the surface of the earth, its gravitational field draws up a bulge in the ocean waters. One bulge occurs beneath the moon and another on the opposite side of the earth. These bulges sweep across the seas as the earth rotates beneath the moon, thus causing high tides when they reach the shoreline. Since there are two "heaps" of water, there are usually two high tides each lunar day, one every 12.4 hours. The tides may cause some rather dramatic changes in the environment, particularly for organisms living on the seashore.

The activity of the fiddler crab, *Uca pugnax*, a resident of the intertidal zone, is synchronized with the tidal changes. Fiddler crabs can be seen scurrying along the marsh during low tide in search of food and mates. Before the sea floods the area, the crabs return to their burrows to wait out the inundation. When a fiddler crab is removed from the beach and sequestered in the laboratory, away from tidal changes, its behavior remains rhythmic. Periods of activity alternate with quiescence every 12.4 hours, the usual interval between high tides (reviewed by Palmer 1995, 2000; Figure 9.3). (Palmer [1995; 2000] presents evidence that rhythms synchronized with tides are lunar day rhythms with two peaks each lunar day.)

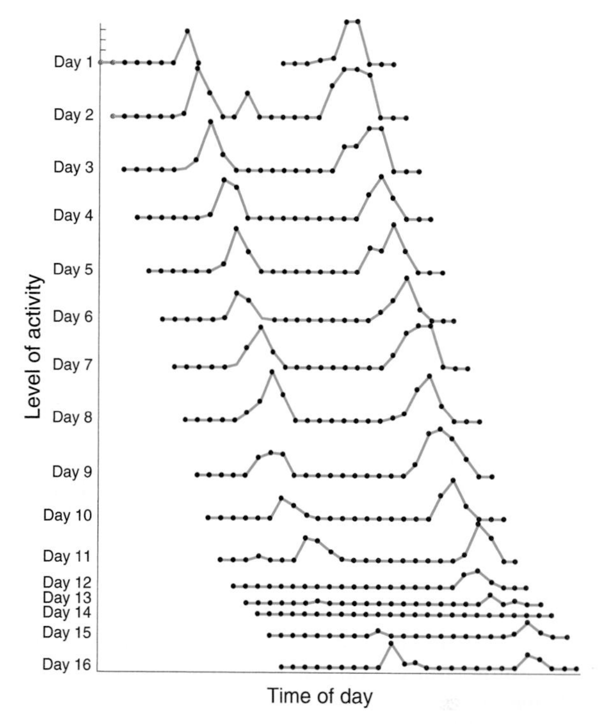

FIGURE 9.3 **An activity rhythm of a fiddler crab (*Uca pugnax*). Although the crab was maintained in constant darkness and temperature (20°C), the animal was active at approximately the times of low tide at its home beach. (From Palmer 1970.)**

SEMILUNAR RHYTHMS

The height of the tides is also influenced by the gravitational field of the sun. In fact, the highest tides are caused when the gravitational fields of the moon and the sun are operating together. At new moon, when the moon appears as a dark disk, and full moon, the earth, the moon, and the sun are in line, causing the gravitational fields of the sun and the moon to augment each other (Figure 9.4). Thus, the earth experiences the highest high tides and lowest low tides at new and full moons. These periods of greatest tidal exchange are referred to as the spring tides. At the quarters of the moon, the gravitational fields of the moon and the sun are at right angles to each other. Because their pulls are now antagonistic, the tidal exchange is smaller than at other times of the month. These periods of lowest high tides and highest low tides are called the neap tides. Some organisms possess a biological clock that allows them to

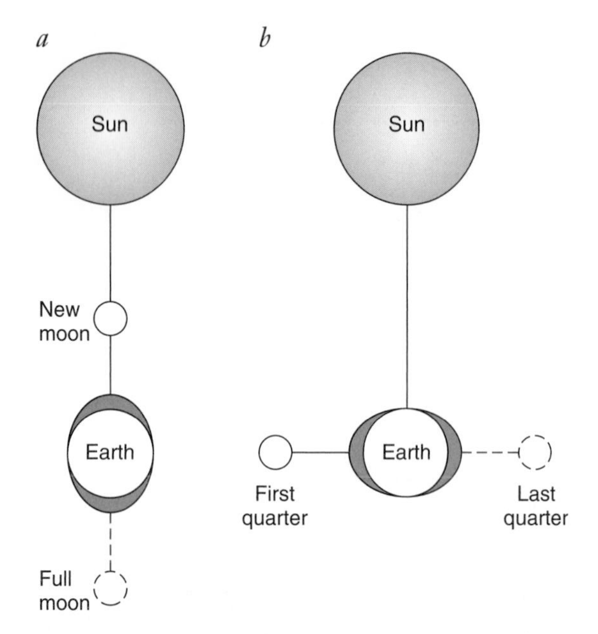

FIGURE 9.4 **The effect of the relative positions of the earth, moon, and sun on the amplitude of tidal exchange. (*a*) At the times of new and full moons, the gravitational fields of the moon and the sun assist each other, causing the spring tides. (*b*) During the first and last quarters of the moon, the gravitational fields of the moon and the sun are perpendicular to each other. This results in the smallest tidal exchange, the neap tides.**

predict the times of spring tides or neap tides and gear their activities to these regular changes.

An example of a fortnightly rhythm is seen in the tiny chironomid midge, *Clunio marinus*. In *Clunio*, the end of development, the emergence of adults from their pupal cases, is programmed to coincide with tidal changes. These insects live at the lowest extreme of the intertidal zone so that they are exposed to the air for only a few hours once every two weeks, during each spring low tide. When the tide recedes, the males are first to break free of their puparia, the cases in which they developed. Each one locates a female and assists in her emergence. They have little time to waste, for their habitat will soon be submerged again, so copulation follows quickly. Then the winged male carries his mate to where she will lay her eggs. All these activities must be precisely timed so that they occur during the short, two-hour period during which the habitat is exposed.

Dietrich Neumann (1976) found that if a population is brought into the laboratory and maintained in a light–dark cycle in which 12 hours of light alternate with 12 hours of darkness, emergence from the puparium is random. If, however, one simulates the light of the full moon by leaving a dim light on for four consecutive nights, the emergence of adults from the puparia becomes synchronized. Under these conditions, just as in nature, emergence occurs at approximately fortnightly intervals for about two months.

MONTHLY RHYTHMS

The interval from full moon to full moon, a synodic lunar month (29.5 days), corresponds to the length of time it takes the moon to revolve once around the earth. Some organisms have a clock that allows them to program their activities to occur at specific times during this cycle.

The ant lion (*Myrmeleon obscurus*) shows a monthly rhythm in the size of the pit it builds. A lazy hunter, it builds a steep-sided conical pit in the sand and then lies in ambush at the base, with all but its immense mandibles covered with sand, waiting for some small arthropod, such as an ant, to slide into the pit toward its outstretched jaws. The ant lion then sucks out the prey's body fluids (Figure 9.5). Like a werewolf, the ant lion changes its

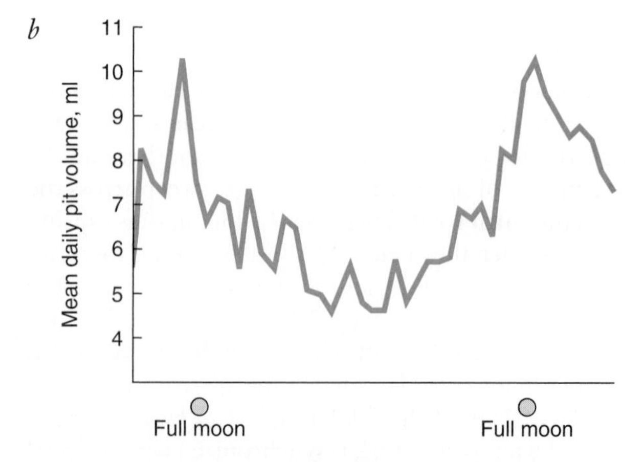

FIGURE 9.5 **A monthly rhythm in the pit size of the predatory ant lion. (*a*) The ant lion waits at the bottom of a self-constructed pit with only its pincers exposed. When a small arthropod, such as this ant, slips into the pit, the ant lion sucks the prey's body juices. (*b*) Monthly rhythms in the pit size of 50 ant lions maintained in constant conditions in the laboratory. Each of the predators was fed one ant a day. Larger pits were constructed at full moon than at new moon, even though the ant lions did not experience changes in the phase of the moon. (From Youthed and Moran 1969.)**

behavior at full moon. It constructs larger pits at the time of full moon than at new moon. Careful daily measurements of the size of the pits of ant lions cloistered in constant conditions in the laboratory have revealed that this is a clock-controlled rhythm and not a simple response to some aspect of the environment such as the amount of moonlight (Youthed and Moran 1969).

ANNUAL RHYTHMS

The seasonal changes in the environment can be dramatic, especially in the temperate zone. As the days shorten and the temperature drops, plants and animals prepare themselves for severe and frigid weather. Some species avoid the cold and limited food of winter by migrating. An annual biological clock is important in timing migration. We see this, for example, in the activity of garden warblers, *Sylvia borin*. The bird's activity can be monitored by using microswitches mounted beneath its perch. The bird whose activity is shown in Figure 9.6 was maintained in the laboratory at a constant temperature and with an unvarying length of day (12 hours of light alternating with 12 hours of darkness), so that it was deprived of the most obvious cues for the onset of winter or of spring. Notice that during the summer and winter months, its activity was limited to the daylight hours. However, during the autumn and spring, when it would be migrating in nature, the caged bird also became somewhat active at night. This nocturnal activity, called *Zugunruhe*, or migratory restlessness, is a cage-adapted form of migratory activity.

This timing function of the annual clock is particularly important for birds that winter close to the equator, where there are few cues to the changing season. At the equator, the photoperiod is constant throughout the year, just as it is in the laboratory, and rainfall and food abundance are too variable from year to year to serve as reliable cues signaling the appropriate time to begin migrating.

An annual clock also physiologically readies birds for migration and reproduction. The bird gets fatter (indicated by body mass) during the winter, which helps provide fuel for the spring migration; it molts during the winter; and its testes enlarge for summer reproductive activity. These cycles are free-running for many years in constant conditions, and the length of the cycle is generally slightly longer or shorter than a year (Gwinner 1996). (The mechanisms of orientation during migration are discussed in Chapter 10. The costs and benefits of migration are discussed in Chapter 11.)

One must be cautious in describing a behavior or physiological process that fluctuates annually as one that is controlled by an annual clock. Seasonal changes in behavior may be controlled not by an annual clock, but by a response to the changing photoperiod, the shortening of days during the winter months and the increasing

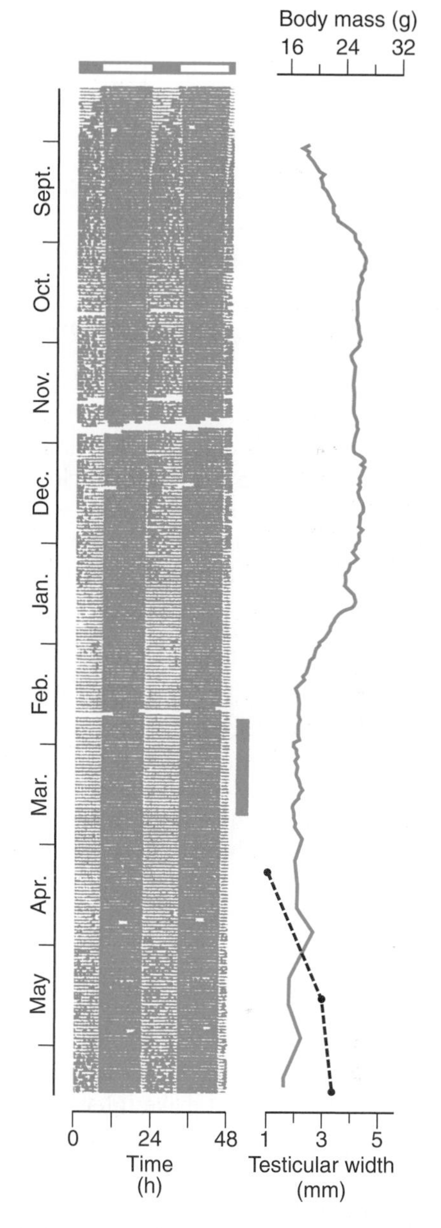

FIGURE 9.6 Annual cycles in migratory restlessness, body weight, testis size, and molting in a garden warbler held in a constant light–dark cycle (12 hours of light alternating with 12 hours of darkness) and at a constant temperature. Activity was measured with a microswitch mounted under the perch. Successive days are mounted underneath each other. The original record (0–24) is repeated on the right (24–48). Most of the bird's perch-hopping activity occurred during the day. When birds in nature are migrating, in the autumn and spring, the caged bird showed increased activity (migratory restlessness) during the night. The body weight changes throughout the year such that the bird fattens during the winter. These energy stores will increase the chances of successful spring migration. The testes enlarge during the spring in preparation for summer breeding. The molt (indicated by the vertical colored bar) occurred in late February to March. (Data from Gwinner 1996.)

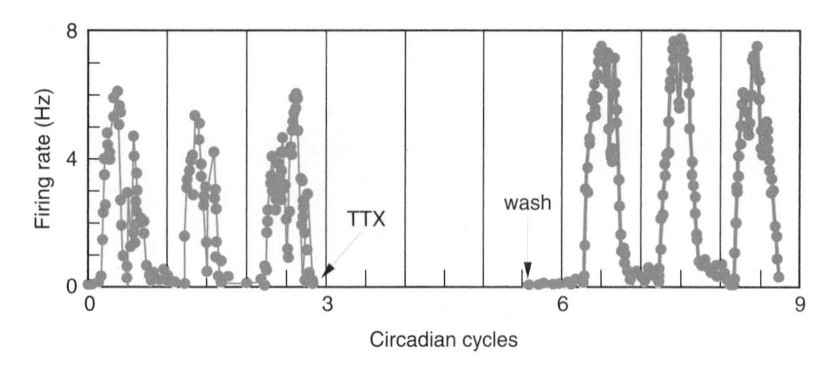

FIGURE 9.7 **The rhythm in spontaneous electrical activity in isolated neurons from rat SCN. Sodium-dependent action potentials are blocked for 2.5 cycles by tetrodotoxin (TTX). When the inhibitor is washed out, the rhythm in spontaneous nerve firing returns with a phase predicted by the initial cycles. This demonstrates that the biological clock is separate from the rhythmic process it drives. (From Welsh et al. 1995.)**

daylight of the spring and summer. To measure a change in day length requires only a circadian clock. Unlike a response governed by photoperiod alone, a rhythm that is controlled by an annual clock will continue to be rhythmic even in the absence of changing day length.

STOP AND THINK

When one wants to determine whether a daily, tidal, or lunar rhythm is controlled by an endogenous clock, the organism is placed in constant light or constant darkness. When one wants to determine whether an annual rhythm is controlled by an annual clock, the animal is kept in a constant photoperiod. Why are the procedures different?

THE CLOCK VERSUS THE HANDS OF THE CLOCK

When we study biological rhythms, we actually look at the rhythmic processes and make inferences about the clock itself. However, it is important to remember that the biological clock is separate from the processes it drives. Perhaps an analogy to a more familiar timepiece will emphasize this important point. Think of a clock with hands. If you tear the hands from its face, the internal gears will continue to run undaunted. And so it is with biological clocks. You can alter the behavior that is controlled by a clock without necessarily stopping the clock mechanism itself.

David Welsh and his colleagues (1995) performed the biological equivalent of tearing the hands from the clock. The suprachiasmatic nuclei (SCN) in the brain of a mammal are a "master" biological clock that drives rhythms in other processes, as we will discuss in detail later. Welsh removed neurons from the SCN of a newborn rat and grew them in tissue culture. The spontaneous rate of firing of these single neurons varies regularly during each day, even in tissue culture, and so we can assume that the rhythm is driven by an internal cellular clock. This nerve firing was completely stopped by the addition of tetrodotoxin, a chemical that prevents action potentials that require sodium (Na^+) ions. However, 2.5 days later, when the tetrodotoxin was washed out of the cells, the rhythm reappeared with a phase predicted by the initial cycles (Figure 9.7). This suggests that although nerve firing had been halted, the clock was running accurately the entire time. Therefore, like the hands of a clock, the rhythmic process—nerve firing in this case—is separate from the clock mechanism. Processes are made to be rhythmic because they are coupled to and driven by a biological clock.

ADVANTAGES OF CLOCK-CONTROLLED BEHAVIOR

We have seen that many behavioral rhythms match the prominent geophysical cycles—a day, a lunar day, a lunar month, and a year. The geophysical cycles generate rhythmic changes in environmental conditions. One might wonder, then, why biological clocks exist at all. If the clocks cause changes that are correlated with environmental cues, why not just respond to the cues themselves?

ANTICIPATION OF ENVIRONMENTAL CHANGE

One reason for timing an event with a biological clock rather than responding directly to periodic environmental fluctuations is that it lets an animal anticipate the change and allow adequate time for behavioral preparation. For example, in nature, adult fruit flies (*Drosophila*) emerge from their pupal cases during a short interval around dawn. At this time, the atmosphere is cool and moist, allowing the flies an opportunity to expand their wings with a minimal loss of water through the still permeable cuticle. This procedure takes several hours to complete. However, the relative humidity drops rapidly after the sun rises. If the flies waited until there was a change in light intensity, temperature, or relative humidity before beginning the preparations for emergence, they would emerge later in the day, when the water loss to the arid air could prevent the wings from expanding properly.

SYNCHRONIZATION OF A BEHAVIOR WITH AN EVENT THAT CANNOT BE SENSED DIRECTLY

Another advantage of the clock's control of a behavior is that it allows the behavior to be synchronized with a factor in the environment that the animal cannot sense directly. An example is the timing of bee flights to patches of flowers that the bees have learned are open only during restricted times of the day (Figure 9.8). The flowers visited for nectar may be far away from the hive, and so the bee could not use vision or olfaction to determine whether the flowers were open. Their time sense was experimentally demonstrated during the early part of this century by individually marking them and offering them sugar water at a feeding station during a restricted time each day, between 10 A.M. and noon. After six to eight days of this training, most of the bees frequented the feeding station only during the learned hours. The real test, however, was on subsequent days, when no food was present at the feeding station. As seen in Figure 9.8*c*, the greatest number of bees returned to the empty feeding station only at the time at which food had been previously available (Beling 1929). In subsequent tests, it was found that the bees' time sense is astonishingly accurate. Bees can be trained to go to nine different feeding stations at nine different times of the day. They are able to distinguish points in time separated by as few as 20 minutes (Koltermann 1971). The adaptiveness of such abilities for bees is clear. Flowers have a rhythm in nectar secretion, producing more at some times of the day than at others. The biological clock allows bees to time their visits to flowers so that they arrive when the flower is secreting nectar. This means that the bees can gather the maximum amount of food with the minimum effort.

a

b

c

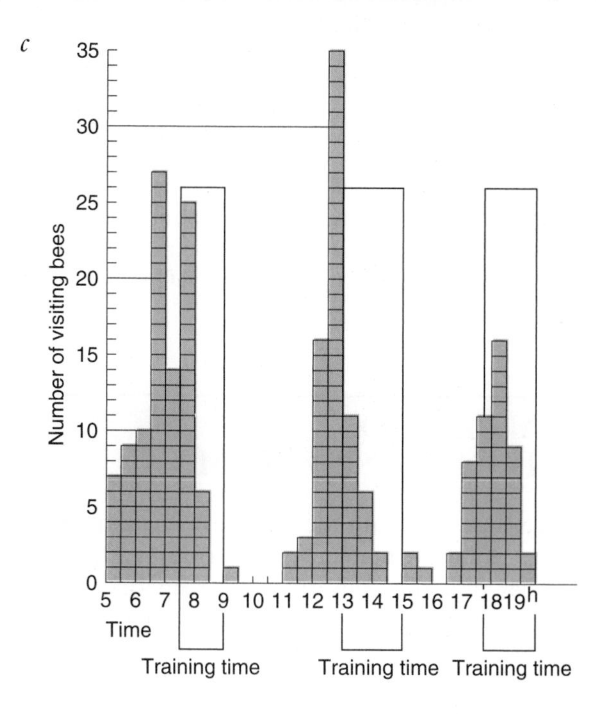

FIGURE 9.8 **The time sense in bees. (*a*) Honeybees can use their biological clock to time their visits to distant patches of flowers so that they arrive when the flowers are open and nectar is available. (*b*) Bees were marked for individual recognition and trained to come to a feeding dish only at the specific times at which food was made available. (*c*) After six days of training, the feeding dishes were left empty and the number of bees arriving throughout the day was recorded. The bees arrived at the feeding station only when food had been previously present. (Modified from Beling 1929.)**

CONTINUOUS MEASUREMENT OF TIME

Sometimes an animal may consult its clock to determine what time it is. As we have seen, this information is necessary to anticipate periodic environmental changes or to synchronize behavior with other events. However, at other times a clock is consulted to measure an interval of time. This, then, is a third benefit of a biological clock.

The ability to measure the passage of time continuously is crucial to an animal's time-compensated orientation. For example, a worker honeybee (*Apis* spp.) indicates the direction to a nectar source to recruit bees through a dance that tells them of the proper flight bearing relative to the sun. Because the sun is a moving reference point, the honeybee must know not only the time of day when it discovered the nectar but also how much time has passed since then. The biological clock provides this information. The use of the sun as a compass will be explored in more detail in Chapter 10.

ADAPTIVENESS OF BIOLOGICAL CLOCKS

Keeping these advantages of the clock in mind, we may wonder whether there is any evidence that a biological clock actually does increase fitness. Surprisingly few people have addressed this question. There is evidence that the clock enhances fitness in cyanobacteria (Woelfle and Johnson 2006), but that may be of marginal interest to those interested in animal behavior. However, Patricia DeCoursey and her colleagues have gathered evidence that the clock is indeed adaptive for antelope ground squirrels (*Ammospermophilus leucurus*). In their study, these researchers destroyed the SCN of some animals and compared their survival rate outdoors to that of intact animals. Because the SCN is the master biological clock in mammals (as we will see shortly), this procedure allowed them to compare the survival of animals with and without clocks. In their initial study, 12 intact control animals and 10 SCN-lesioned squirrels were monitored in a desert enclosure. Their activity was continuously monitored in several ways, including a motion detector and a video camera. All the ground squirrels were primarily active during daylight.

However, an important difference in the behavior of the two groups is that the SCN-lesioned animals were more likely to be active on the ground surface of the enclosure during the nighttime than were intact animals. Whereas the amount of activity occurring during the night in SCN-lesioned animals ranged from 16% to 52.1%, nighttime activity represented no more than 1.3% of the activity of intact animals. Nine of the 12 control animals were active only in the day. This difference in nighttime activity had unfortunate consequences for the SCN-lesioned animals. One night, when seven control animals and five SCN-lesioned animals had been introduced to the enclosure, a feral cat treated the enclosure as a kitty-convenience store. The videotape recorded the cat picking off ground squirrels that were active that night. As a result, the cat killed 60% of the SCN-lesioned animals but only 29% of the intact controls (DeCoursey et al. 1997). Thus, it seems that an important function of the clock for these ground squirrels may be to reduce activity at dangerous times, such as nighttime.

DeCoursey then asked whether a biological clock enhances survival in eastern chipmunks, *Tamias striatus*. Animals were captured in the wild and taken to the laboratory for surgery. The SCN was destroyed in 30 animals, and 24 others were given sham lesions; that is, they were anesthetized and their skulls were opened, but they were not lesioned. The sham-lesioned animals served as surgical controls because they experienced the same surgical procedures except for lesioning and they suffered the consequences of removal from their habitat, such as the possible takeover of their dens by other animals, but did not undergo lesioning. Both groups of squirrels were equipped with radio telemetry collars. The survival of these two groups of animals was compared with that of 20 intact controls (animals that did not undergo surgery or removal from their habitat). Although none of the chipmunks was active outside of their burrows after dark, some chipmunks were active within the den. SCN-lesioned chipmunks were much more active than chipmunks in either of the control groups. The primary cause of mortality was predation, probably by a weasel. There were significant differences in mortality rates between treatment groups and both control groups. The largest differences in mortality were seen on days 15–80 of the study. The chipmunks in the SCN-lesioned group died at the rate of 0.798% per day. In contrast, those in the sham-operated group died at the rate of 0.169% per day, and those in the intact control group at the rate of 0.276% per day. The authors hypothesize that the movements of the SCN-lesioned chipmunks within their dens at night alerted a predator to their locations (DeCoursey et al. 2000).

ORGANIZATION OF CIRCADIAN SYSTEMS

Single cells may contain the necessary equipment for biological timing. Unicellular organisms have biological clocks, and the cells that make up tissues and organs often have their own independent clocks. Thus, a complex nervous system or endocrine system is not an essential component of the biological clock.

MULTIPLE CLOCKS

There is no such thing as "the" biological clock. Instead, there are clocks scattered throughout an animal's body (Reppert and Weaver 2002). Fruit flies (*Drosophila*) have a multitude of independent clocks located throughout their bodies, and these clocks respond to changes in light–dark cycles without any help from the head (Plautz et al. 1997a, b). This was shown by using an interesting technique that caused cells with clocks to glow. The technique has since proved to be a valuable tool; it has advanced the study of rhythms because it allows researchers to observe the molecular activity of important clock genes in a single, living, intact animal. Before this, clock gene activity had to be studied by synchronizing the clocks of members of a large population of fruit flies with a light–dark or temperature cycle and then periodically selecting a group of flies from the population, grinding them up, and testing for gene activity.

Many researchers investigating the fruit fly's clock focus their efforts on the *period* (*per*) gene, which is an integral part of the clock mechanism. To monitor the clock's activity, the research groups headed by Jeffrey Hall and by Steve Kay genetically engineered fruit flies to contain the firefly luciferase gene. Luciferase is an enzyme that acts on luciferin to produce light, allowing the firefly to glow. Whenever the *per* gene was turned on, the luciferase gene was also switched on. Because the fruit flies' diets were laced with luciferin, the fruit flies glowed whenever luciferase was present. Thus, whenever the fly glowed, it meant that the *per* gene was turned on. Special cameras and video equipment measured the glow, and computers traced and recorded the glow pattern.

These glow rhythms that indicate *per* gene activity will synchronize with light–dark cycles and will continue in constant darkness with a free-running period length. Not only do intact flies glow rhythmically, but so will cultures of head, thorax, or abdomen. Furthermore, separate cultures of body parts exposed to the same light–dark cycle will glow in unison, showing that each piece of cultured tissue has its own independent clocks and that these clocks have their own photoreceptors. Moreover, this raises the possibility that the insect's brain is not required as a master clock to synchronize rhythms throughout the body.

The glow rhythm of a fly kept in constant darkness gradually decreases in amplitude because the clocks in different cells run at slightly different rates without a light–dark cycle; thus, the independent clocks gradually become asynchronous. It is interesting to note that the head is the only body part in which the clocks remain synchronized in the prolonged absence of light. However, when exposed to a new light–dark cycle, the clocks throughout the fly entrain within one cycle and the glow becomes rhythmic again. In nature, asynchrony among peripheral clocks is not a problem: Fruit flies usually have an environmental light cycle that is able to synchronize their many independent clocks because each has its own photoreceptor, as we will see shortly (Figure 9.9).

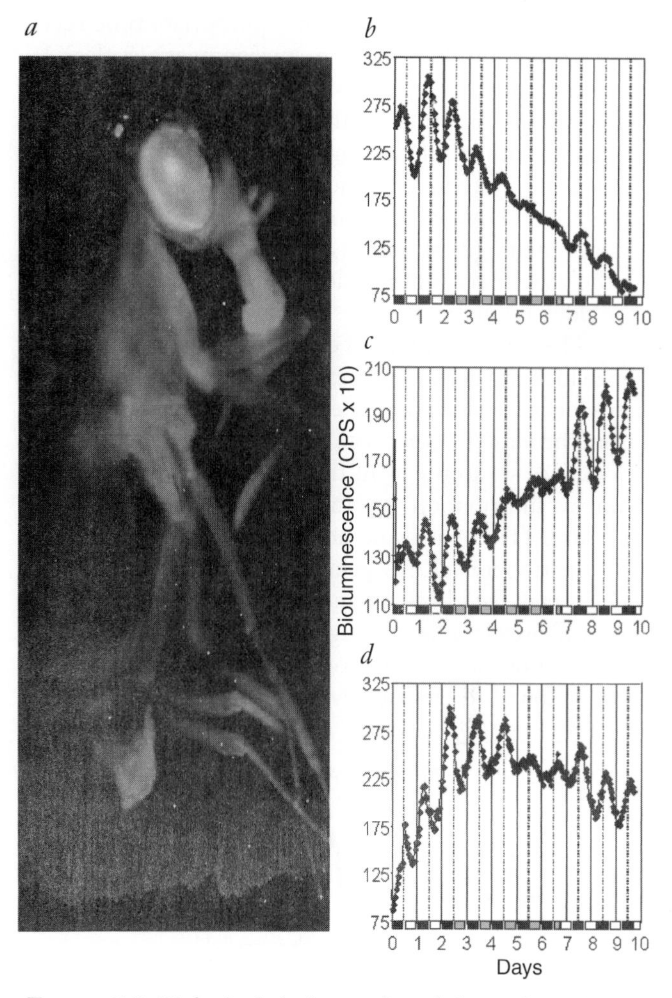

FIGURE 9.9 **Biological clocks are found throughout fruit flies, not just in the brain. The *period* (*per*) gene is thought be an integral part of the clock's mechanism. To measure *per* activity, the firefly gene for luciferase was linked to the *per* promoter, which turns on the gene. Luciferase is the enzyme that causes a firefly to glow in the presence of luciferin. As a result, (*a*) the fruit fly glowed with an eerie green color whenever the per gene was turned on. Computers measured and recorded the pattern of glow. The glow rhythm persisted in constant darkness for several cycles and could be synchronized with light–dark cycles. When parts of the fly were cultured separately, the cultured segments continued to glow rhythmically and could still be set by light–dark cycles. Rhythmic glow can be seen in separately cultured (*b*) heads, (*c*) thoraxes, and (*d*) abdomens. Thus, these peripheral clocks do not require input from the brain. (From Plautz et al. 1997.)**

COORDINATION OF CIRCADIAN TIMING

Most rhythmic animals have a multitude of independent peripheral clocks in cells throughout the body. However, information about environmental cycles may not reach each clock directly. How, then, are an individual's many clocks synchronized so that all the rhythmic processes occur at the appropriate time relative to one another and the environment's cycles? It appears that there is at least one "master" clock in the brain that is entrained by the light–dark cycle and that regulates other clocks through the nervous and/or endocrine system. Therefore, we can consider four questions: (1) What photoreceptors are responsible for entrainment? (2) Where is the master clock? (3) What is the genetic basis of the clock? (4) How does the master clock regulate the other clocks in the body?

The general scheme of circadian organization is that one clock or several interacting clocks function as master clocks to synchronize peripheral clocks. The output from the master clock(s) can be neural or hormonal. The clocks are set to the right time because photoreceptors convey information on the light–dark cycle to the clock(s). (There may be additional entraining input.) The peripheral clocks generate the rhythmic output, which may feed back on and affect the master clock(s). We will use circadian timing in mammals as an example. Information on circadian organization in other species can be found in DeCoursey (2004b).

Photoreceptors for Entrainment

The eyes contain the photoreceptors for light entrainment in mammals. However, the photoreceptors for entrainment are in a different part of the retina than those involved in vision. Instead, the photoreceptors for entrainment are ganglion cells in the retina that contain the photopigment melanopsin. (The photopigment cryptochrome may also play a role in entrainment.) The information about the lighting conditions reaches the clock through the retinohypothalamic tract (RHT), a bundle of nerve fibers connecting the retina with the hypothalamus (Reppert and Weaver 2002).

Master Clock

In mammals, the circadian system is arranged as a hierarchy of clocks with the SCN as the master biological clock (Reppert and Weaver 2002) (Figure 9.10). What is the evidence that the SCN is the master clock? The activity of the SCN remains rhythmic in tissue culture, confirming that it is an independent clock. Indeed, when cultured SCN neurons are separated, the spontaneous

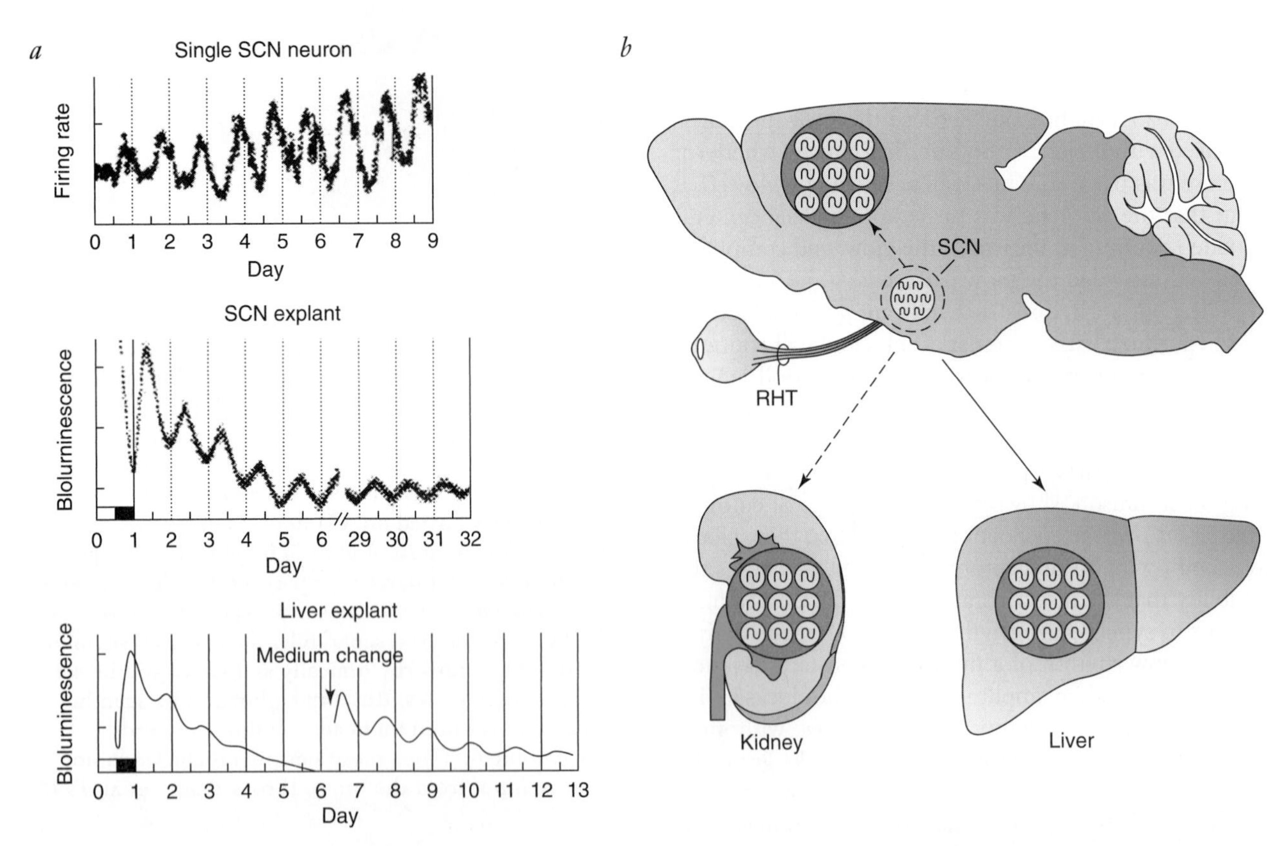

FIGURE 9.10 **The master biological clock of mammals is the suprachiasmatic nucleus (SCN) in the brain. (*a*) The firing rate of a single neuron from the SCN continues to fire rhythmically is tissue culture. Neurons from the SCN that are modified to glow when a clock gene turns on also remain rhythmic in tissue. There are also circadian oscillators in other regions of the brain and in other tissues throughout the body. Here, the rhythm in liver tissue in culture dampened after a few days but was restored by changing the medium. (*b*) Photoreceptors in the mammalian circadian system reach the SCN via the retinohypothalmic tract (RHT). The SCN synchronizes the oscillators throughout the body. (From Reppert and Weaver 2002.)**

electrical firing of individual neurons is rhythmic in constant conditions, each of them with a slightly different period (Welsh et al. 1995). In another experiment, the SCN was isolated from neural input by a knife cut that created an island of hypothalamic tissue. Following this treatment, the neural activity in the hypothalamic island remained rhythmic in constant darkness, but activity in other brain regions was continuous (Figure 9.11). This strongly suggests that the SCN is a self-sustaining oscillator that instills rhythmicity in other brain regions through neural connections (Inouye and Kawamura 1979).

The SCN was finally established as the primary clock in mammals by transplantation studies. When an SCN from a conspecific is transplanted into the brains of rats or hamsters that have been made arrhythmic by destroying their own SCN, their activity becomes rhythmic once again. Importantly, the period length of the restored activity rhythm matches that of the transplanted SCN rather than the period length previously displayed by the recipient. This result is what one would expect if the SCN were the clock that was providing timing information and not just a component needed to make the host's clock function (discussed in Mistlberger and Rusak 2005).

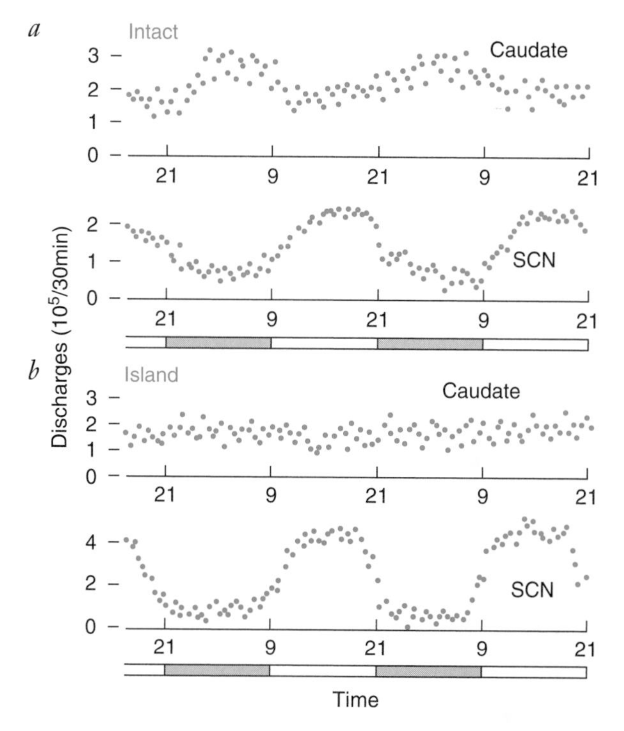

FIGURE 9.11 **Nerve activity in specific brain regions (the caudate and the suprachiasmatic nuclei) of a rat before and after isolating an "island" of brain tissue containing the suprachiasmatic nuclei. (*a*) The activity in a normal, intact rat is rhythmic in both regions of the brain. (*b*) After a region of the hypothalamus that contains the suprachiasmatic nuclei was isolated as an island, the nerve activity was rhythmic within the SCN but not outside the island in the caudate region. This supports the idea that the suprachiasmatic nuclei are self-sustaining oscillators. (Courtesy of S. T. Inouye.)**

Genetic Basis of Mammalian Circadian Timing

What are the molecular gears that make the clock tick? Rhythmic gene activity seems to be involved in the clock mechanism. The products of one gene or set of genes activate or inhibit the activity of other genes, which in turn affect the activity of the first genes. This creates a self-regulated feedback loop of gene activity that measures an approximately 24-hour interval.

We will begin discussing the genetic basis of the circadian cycle with two proteins—Clock and Bmal 1—that bind together, forming a complex that enters the nucleus (Figure 9.12). The Clock/Bmal 1 complex turns on the activity of both the *period* (*per*) and the *cryptochrome* (*cry*) genes. The protein products of these genes (Per and Cry) bind together along with the protein product of the *tau* gene (casein kinase I epsilon) to form a complex. The Per/Cry/Tau complex suppresses the action of the Clock/Bmal 1 complex, resulting in less activity of *per* and *cry*. With less Per and Cry being produced and the degradation of Per, Cry, and Tau, the level of the Per/Cry/Tau complex declines. Now, with less inhibition of their activity, *per* and *cry* are turned on again. This cycle takes about 24 hours to complete.

Peripheral Clocks

Although the SCN may be the master biological clock, other circadian clocks tick throughout the body, keeping their own internal time (Figure 9.13). These so-called peripheral oscillators have been demonstrated with a technique described earlier—using the *luciferase* gene to indicate the activity of genes involved in the clock mechanism. We now know that rhythms in bioluminescence persist for more than 20 days in cultures of cells from the SCN, liver, and lung (Yoo et al. 2004). Indeed, robust bioluminescence rhythms persisted in individual fibroblasts that were maintained in tissue culture for decades. Fibroblasts are "generic" cells found throughout the body, so if fibroblasts have a personal clock perhaps all cells do (Welsh et al. 2004).

Clock Output

The role of the SCN is to entrain the peripheral circadian oscillators so that they are correctly set to environmental time. The phase relationship between rhythmic output of the SCN and rhythmic clock genes in peripheral tissues varies. Peak clock gene expression occurs at distinct times of the day and varies in different tissues, suggesting that the clock is not directly causing the rhythmic output of peripheral tissues (Herzog and Tosini 2001).

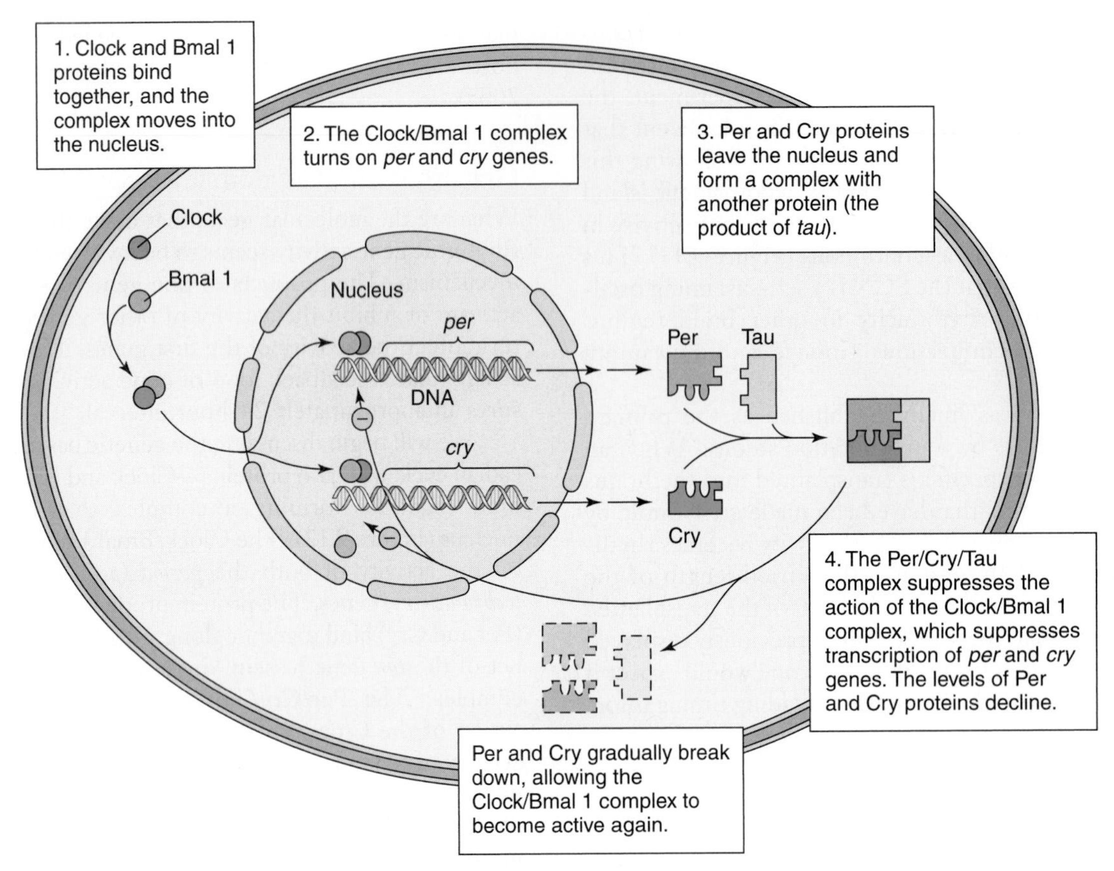
1. Clock and Bmal 1 proteins bind together, and the complex moves into the nucleus.
2. The Clock/Bmal 1 complex turns on *per* and *cry* genes.
3. Per and Cry proteins leave the nucleus and form a complex with another protein (the product of *tau*).
Clock
Bmal 1
Nucleus
per
DNA
cry
Per
Tau
Cry
4. The Per/Cry/Tau complex suppresses the action of the Clock/Bmal 1 complex, which suppresses transcription of *per* and *cry* genes. The levels of Per and Cry proteins decline.
Per and Cry gradually break down, allowing the Clock/Bmal 1 complex to become active again.

FIGURE 9.12 **The genetic basis of circadian timing in mammals consists of two feedback loops in gene activity. This is a simplified diagram. There are actually two *per* genes and two *cry* genes. The protein product of *tau* (Tau in diagram) is casein kinase 1 epsilon.**

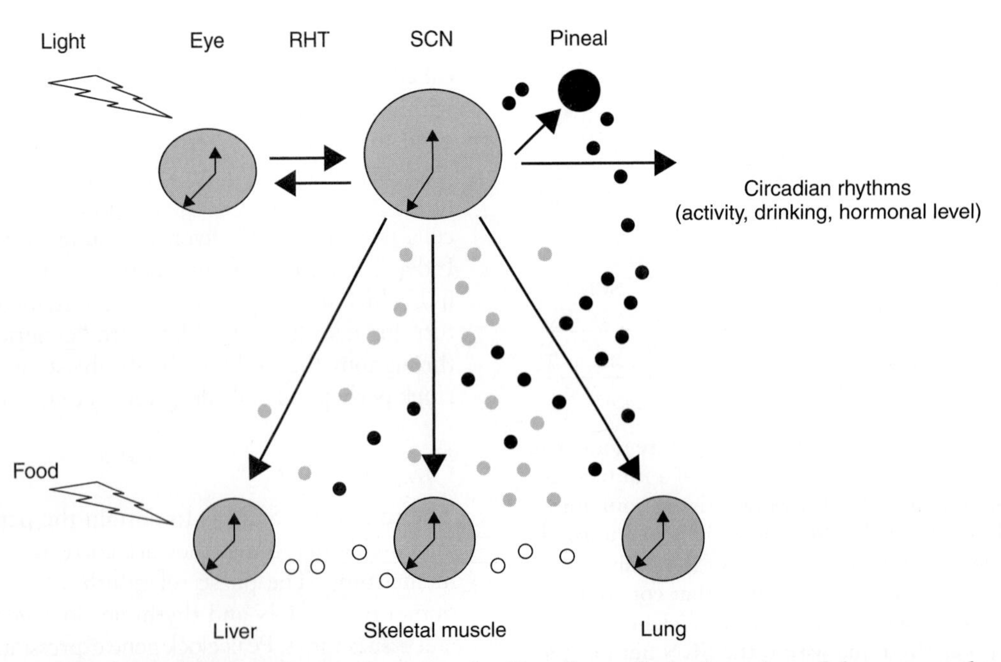
Light
Eye
RHT
SCN
Pineal
Circadian rhythms
(activity, drinking, hormonal level)
Food
Liver
Skeletal muscle
Lung

FIGURE 9.13 **A model of circadian organization in mammals. Ganglion cells in the retina of the eye are the photoreceptors for the circadian system. Light information travels to the SCN, which is the master clock. Neural output (indicated by arrows) and hormonal output (indicated by dots) from the SCN entrain peripheral circadian clocks located throughout the body. (From Herzog and Tosini 2001.)**

Some signals from the SCN are sent over neural pathways, and others are hormonal. Most of the neural connections from the SCN are to other regions of the hypothalamus and therefore to the autonomic nervous system, which influences the level of many hormones. The SCN has at least two neural output pathways that affect rhythms. One of these is a pathway to the preoptic nucleus of the hypothalamus, and this seems to control the rhythm in ovulation but does not affect the activity rhythm. The second neural pathway leads to the paraventricular nucleus in the hypothalamus, an important brain center that integrates neuroendocrine and autonomic functions. This neural pathway then leads to the pineal, which rhythmically produces melatonin. Both reproductive responses to the length of day, which depend on melatonin from the pineal, and rhythms in hormones generally require neural connections, but some may also be influenced by small molecules that diffuse to their target (Tousson and Meissl 2004).

Activity rhythms, on the other hand, are caused by the SCN's release of chemical signals to other parts of the brain without the help of neural connections. Rae Silver and her colleagues (1996) demonstrated this through transplant experiments similar to those described earlier, but with one important difference. The donor SCN tissue was enclosed within a capsule that allowed nutrients and diffusible molecules to flow between the host and graft tissue but did not allow neural processes to grow (Figure 9.14*a*). As in previous transplant experiments, both SCN of the host were destroyed prior to transplant, making the animal arrhythmic. The transplants were made between hamsters whose clocks ran at different rates because some carried the *tau* mutation, which alters the period length observed in constant conditions. The encapsulated grafts restored the activity rhythm with the period length characteristic of the donor SCN (Figure 9.14*b*).

In mammals, the rhythmic production of two hormones—melatonin from the pineal gland and glucocorticoids (cortisol) from the adrenal glands—are thought to be important in entraining peripheral oscillators (Herzog and Tosini 2001). Neural connections

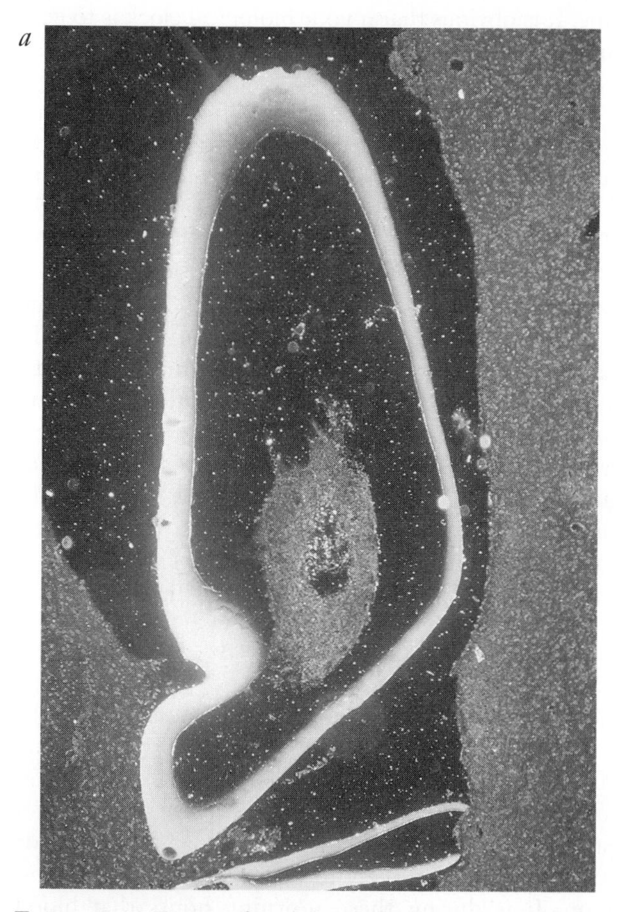

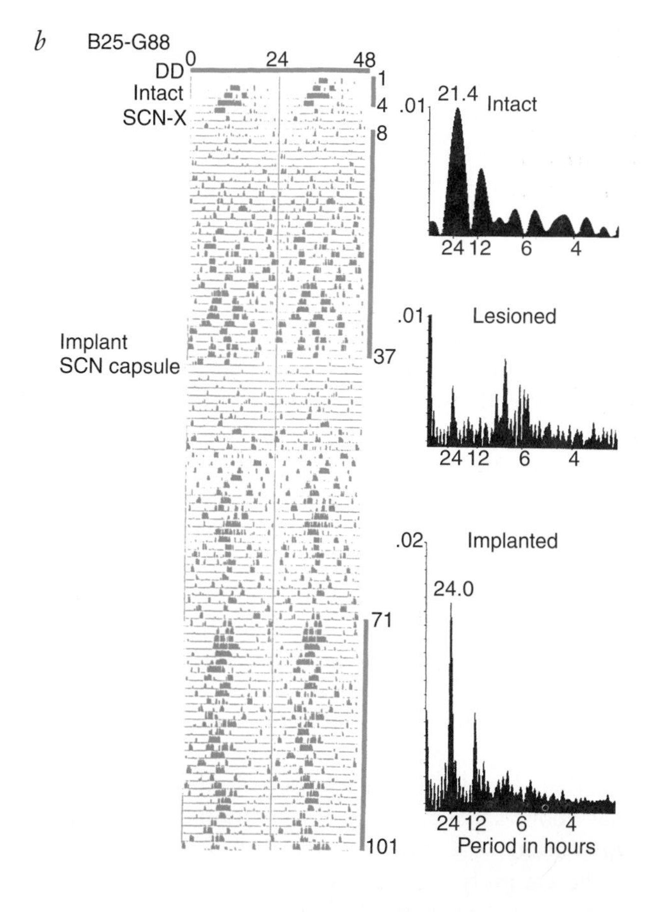

FIGURE 9.14 **Encapsulated donor SCN can reinstill rhythmic activity in SCN-lesioned hamsters. (*a*) The capsule, shown here in white, surrounds the donor SCN. Although the capsule allowed nutrients and small molecules to diffuse between the host and graft tissue, it prevented the formation of neural connections. (*b*) The activity rhythm of the intact host hamster (wild-type/tau) had a period length of 21.4 hours in constant darkness. After the SCN was destroyed, the animal was arrhythmic. Implantation of a capsule containing wild-type SCN (period length of 24 hours) restored the activity rhythm with a period characteristic of the implant. (From Silver et al. 1996.)**

from the SCN cause the pineal to produce more melatonin at night. Melatonin then amplifies the body temperature rhythm, facilitates sleep, and controls photoperiodic responses. Neural connections from the SCN to the anterior hypothalamus set into action a cascade of hormonal events that result in the rhythmic production of glucocorticoids by the adrenal gland. Glucocorticoids are steroid hormones that control many physiological functions (DeCoursey 2004b).

STOP AND THINK

Light cannot reset the clock in the SCN of people who are totally blind. Consequently, their sleep–wakefulness rhythms drift out of phase with the day–night cycle. As a result, they are often sleepy during the day or wide awake at night. In one experiment, blind people were able to set their clocks by taking a dose (10 mg) of melatonin at bedtime (Sack et al. 2000). Why do you think this is possible?

What is the molecular basis of circadian clock output? The clock controls the pattern of gene activity in cells, and that pattern is different in different types of tissue. As you may recall from Chapter 3, a gene encodes the information needed to make a protein. When a gene is active, that protein is produced. The protein either forms part of a structure in the cell or plays a functional role, such as turning other genes on or off or altering metabolic processes in the cells. Thus, if the pattern of gene activity is different in different types of tissue, the response of different tissues also differs. However, since the clock is rhythmically controlling the pattern of gene activity, the response of the cell is also rhythmic.

DNA microarray analysis, a technique that can reveal the activity of thousands of genes at a time (see Chapter 3), shows us the pattern of gene expression. Of the genes examined in rodents, 0.5 to 9% of those in the SCN, pineal gland, heart, liver, and kidney are clock-controlled. The pattern of gene expression differs in different types of tissue. Only about 10% of the clock-controlled genes are similar in two or more tissues, and these are usually genes involved directly in the mechanism of the clock. The remainder of the active clock-controlled genes play roles in a variety of cellular pathways, including cell-signaling pathways, regulation of the cell cycle, and protein metabolism (Duffield 2003). Jonathan Arnold and his colleagues (2008) discovered that about 25% of the genes in the mold *Neurospora crassa* are clock-controlled. Many of those are involved in the assembly of ribosomes (the structures on which proteins are constructed). Many of these proteins are enzymes that control key metabolic processes that are important in a specific cell type. Thus, the clock controls when and where specific biochemical reactions will occur, and those reactions are specific to a particular type of tissue (Hastings et al. 2003).

HUMAN IMPLICATIONS OF CIRCADIAN RHYTHMS

JET LAG

Perhaps the most familiar way that circadian clocks affect humans is jet lag, a syndrome of effects that frequently includes a decrease in mental alertness and an increase in gastric distress. Jet lag is caused by a disruption of circadian timing. We have seen that one function of the biological clock is to time certain activities so that they occur at the best point of some predictable cycle in the environment. To be useful, then, they must be set to local time. Light resets the master clock in the SCN, which then resets peripheral clocks so that each peaks at the most adaptive time relative to other clocks (Yamazaki et al. 2000).

Traveling across time zones is a problem for humans because of the speed with which we can do so. If you were to fly from Cape Cod, Massachusetts, across several time zones to Big Sur, California, the first thing that you would want to do is to set your watch to the local time. It is obvious that if your biological clock is to gear your activities to the appropriate time of day in the new locale, it too must be reset.

As you disembark from a plane after traveling across time zones, your biological clocks are still set to the local time of your home. The clocks will gradually adjust to the day–night cycle in the new locale (Figure 9.15). However, this shift cannot occur immediately; it may take several days. The length of time required for the biological clock to be reset to the new local time increases with the number of time zones traversed. To make matters worse, not all your body functions adjust at the same rate, so the normal phase relationship among physiological processes is upset. Therefore, for a few days after longitudinal travel, your body time is out of phase with local time, and your rhythms may be peaking at inappropriate times relative to one another. During this time, you often suffer psychological and physiological disturbances associated with jet lag.

HUMAN HEALTH

Nearly every physiological process in humans is rhythmic, with each process peaking at the appropriate time of day. It should not be surprising, then, that certain acute medical conditions are more likely to occur at a particular time of day. Heart attacks and strokes, for example, are most likely to occur between 6 A.M. and noon. It is during these morning hours that blood pressure rises, platelets become stickier and more likely to form blood clots, and the mechanism that breaks down blood clots is least active. On the other hand, asthma attacks are most likely to occur at night when the level of epinephrine, a hormone that causes the air

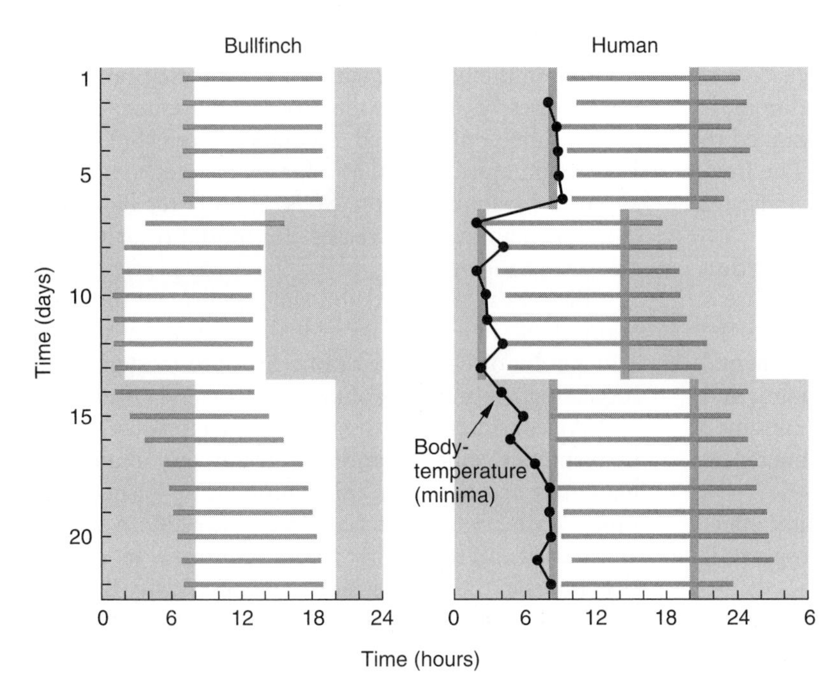

FIGURE 9.15 The resetting of the biological clock by light–dark cycles. Immediately following a trip across time zones, the biological clock is still set to home time rather than that of the vacation locale. It may take several days for the clock to be reset so that it has the proper phase relationship with the new light–dark cycle. In addition, various rhythms (clocks) may rephase at different rates. During the interval of readjustment, the individual suffers from jet lag and may not feel well. (From Aschoff 1967.)

tubules to dilate, and cortisol, a hormone that suppresses the immune system, are low (Waterhouse and DeCoursey 2004).

Because of the current epidemics of obesity and sleep loss, it is interesting to note the relationships between circadian rhythms, sleep, and energy metabolism. In one of the earliest studies showing this relationship, four healthy males were allowed to sleep only four hours each night for six consecutive nights. Following breakfast on the last day of sleep restriction, the males had a high blood level of glucose and reduced sensitivity to insulin. The extent of the changes in glucose and insulin was comparable to changes that accompany aging or diabetes. A second study looked at the level of the hormones leptin, which decreases appetite, and ghrelin, which increases appetite. Following two consecutive nights with only four hours of sleep each night, the subjects' leptin levels were down and ghelin levels were up. The males also reported that they had a heartier appetite than usual. A much larger study, the Wisconsin Sleep Cohort Study, included over 1000 subjects. The subjects who reported they spent little time sleeping each night had lower leptin levels and higher ghrelin levels than did those who reported sleeping longer (Laposky et al. 2008).

Recall that the *clock* gene is an important part of the circadian mechanism in mammals. Within six weeks of birth, homozygous *clock* mutant mice eat more than normal mice and gain more weight. The mutant mice are also active much longer than normal mice, and they eat frequently throughout their active period. In contrast, normal mice eat at the beginning and end of their active period. As the mutant clock mice aged, they developed high cholesterol, high triglycerides, high blood sugar, low insulin, and bloated fat cells (Turek et al. 2005).

Paolo Sassone-Corsi and his colleagues (Asher et al. 2008) have found the molecular link between the circadian clock and metabolism, opening new possibilities for treating diabetes and obesity. The activity of a gene called *Sirt1* is clock controlled, and it is modulated by how many nutrients a cell is consuming. *Sirt1* responds to the energy state of a cell and transmits that information on the energy state to the clock by binding to the Clock/Bmal 1 complex (see earlier discussion). Thus, these findings help explain why lack of sleep can increase hunger and lead to obesity and diabetes.

STOP AND THINK

If you were traveling from Tampa, Florida, to San Diego, California, to compete in an important athletic event, what steps could you take before you left to minimize jet lag? If you could choose the time of the event, would you choose morning or afternoon? Why?

SUMMARY

Life evolved in a cyclic environment caused by the relative movements of the earth, sun, and moon. Often it is advantageous to gear an activity to occur at a specific time relative to some rhythmic aspect of the environment. Thus clocks evolved as adaptations to these environmental cycles. Clock-controlled biological rhythms have three defining properties: persistence in constant conditions, entrainment to environmental cycles, and temperature compensation. In the constancy of the laboratory, the period length of biological rhythms may deviate slightly from the one displayed in nature. For this reason, periods are described with the prefix *circa*, meaning "about," and are called circadian, circalunidian, or circannual. The period length in constant conditions is described as the free-running period and is assumed to reflect the rate at which the clock is running. The free-running period is generally kept constant, which indicates that the biological clock is very accurate.

Although the period length of a biological rhythm is "circa" in the constancy of the laboratory, in nature it matches that of the geophysical cycle exactly because the clock is entrained to (locked onto) an environmental cycle. Entrainment adjusts both the period length and the phase of the rhythm. Daily rhythms can be entrained to light–dark cycles and, in some species, to temperature cycles.

Environmental temperature has only a slight effect on the rate at which the clock runs. This property is called temperature compensation.

There are many examples of rhythmic processes in animals that match the basic geophysical periods: a day (24 hours), the tides (12.4 hours), a lunar day (24.8 hours), a fortnight (14 days), a lunar month (29.5 days), and a year (365 days). Many of these processes remain rhythmic when the individual is isolated from the obvious environmental cycles that might be thought to provide time cues. For instance, many daily rhythms persist when the individual is kept in the laboratory without light–dark or temperature cycles. Therefore, we say that the rhythms are caused by an internal biological clock.

The biological clock is separate from the rhythms it drives. Processes become rhythmic when they are coupled to the biological clock.

There are several reasons why it may be advantageous to have a biological clock to measure time rather than responding directly to environmental changes: (1) anticipation of the environmental changes with enough time to prepare for the behavior, (2) synchronization of the behavior with some event that cannot be sensed directly, and (3) continuous measurement of time so that time-compensated orientation is possible. There is now some evidence that a functional clock enhances survival. Lesioning of the SCN, the master biological clock in mammals, increases the mortality of certain free-living rodents.

Biological clocks exist in single cells, and there are many clocks in a single individual. It seems that there is a hierarchy of clocks, with one or more master clocks regulating the activities of other, peripheral clocks.

We considered the organization of circadian timing in mammals. The photoreceptors for entrainment in mammals are the eyes. The information reaches the suprachiasmatic nucleus via nerves (the retinohypothalamic tract). The master clock control for certain rhythms is in the SCN, which regulates activity through nerves and chemicals. The genetic basis of circadian clocks involves self-regulated feedback loops in which the products of one or more genes affect the activity of other genes. The master circadian clock is the SCN, which exerts control over peripheral oscillators through nerves and hormones.

The genetic basis of mammalian clocks involves feedback loops. The activity of *Bmal 1* and *Clock* is counterbalanced by *per* and *cry* activity.

Many mammalian tissues have their own circadian clocks. These peripheral clocks persist in cell culture. Within the body, the peripheral clocks are entrained by neural and humoral output of the SCN. In these ways, the SCN controls the pattern of gene expression in tissues.

The human circadian clock is related to health in several ways: jet lag, onset of acute medical conditions, and energy metabolism.

10

Mechanisms of Orientation and Navigation

Many of us have been moved by a crisp autumn day, enveloped in the reds, yellows, and browns of the season and watching formations of ducks or geese fly against a steely sky. We might have noticed that if it is early in the day, the flocks may be heading almost due south. If it is nearing dusk or if fields of grain are nearby, they may be temporarily diverted to resting or feeding areas. But when they resume their flight, they will head southward again.

In the following spring, we may stand beside a swift-moving river in the Pacific Northwest and watch salmon below a dam or a fish ladder. As they lie in deeper pools, resting before the next powerful drive that will carry them one step nearer the spawning ground, they all face one way—upstream.

Both the birds and the fish are responding to a complex and changing environment by positioning themselves correctly in it and by moving from one part of it to another. Although the feats of migration are astounding, they are no more crucial to survival than are mundane daily activities such as seeking a suitable habitat, looking for food and returning home again, searching for a mate, or identifying offspring. These actions also depend on the proper orientation to key aspects of the environment. Indeed, an animal's life depends on oriented movements both within and between habitats.

In this chapter we will explore some of the *mechanisms* by which animals orient themselves in space. (The costs and benefits of dispersal, habitat selection, and migration are covered in Chapter 11.)

LEVELS OF NAVIGATIONAL ABILITY

Many animals often travel between home and a goal, but they do not all accomplish this feat in the same manner. We group animal strategies for finding their way into three levels of ability (Bingman and Cheng 2005; Ronacher 2008).

PILOTING

One level is **piloting**, the ability to find a goal by referring to familiar landmarks. The animal may search either randomly or systematically for the relevant landmarks.

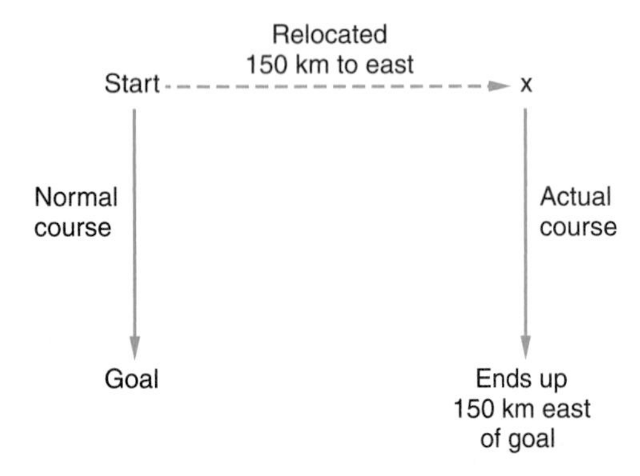

FIGURE 10.1 **Experimental relocation of an animal that is using compass orientation causes it to miss the goal by the amount of its displacement.**

Although we usually think of landmarks as visual, the guidepost may be in any sensory modality. As we will see shortly, magnetic cues guide sea turtles during their oceanic travels, and olfactory cues guide salmon during their upstream migration.

COMPASS ORIENTATION

A second level, called **compass orientation**, is the ability to head in a geographical direction without the use of landmarks. The sun, the stars, and even the earth's magnetic field may be used as compasses by many different species. One way to demonstrate that an animal is using compass orientation is to move it to a distant location and determine whether it continues in the same direction or compensates for the displacement. If it does not compensate for the relocation, compass orientation is indicated (Figure 10.1). When immature birds of certain migratory species, such as European starlings, were displaced experimentally, they flew in the same direction as the parent group that had not been moved, and they flew for the same distance (Perdeck 1967). In other words, they migrated in a path parallel to their original migratory direction. However, because they had been experimentally displaced before beginning their migration, they did not reach their normal destination. In some cases, this meant that they ended up in ecologically unsatisfactory places (Figure 10.2).

Uses for Compass Orientation

Compass orientation can be used in different ways—in both short-distance and long-distance navigation.

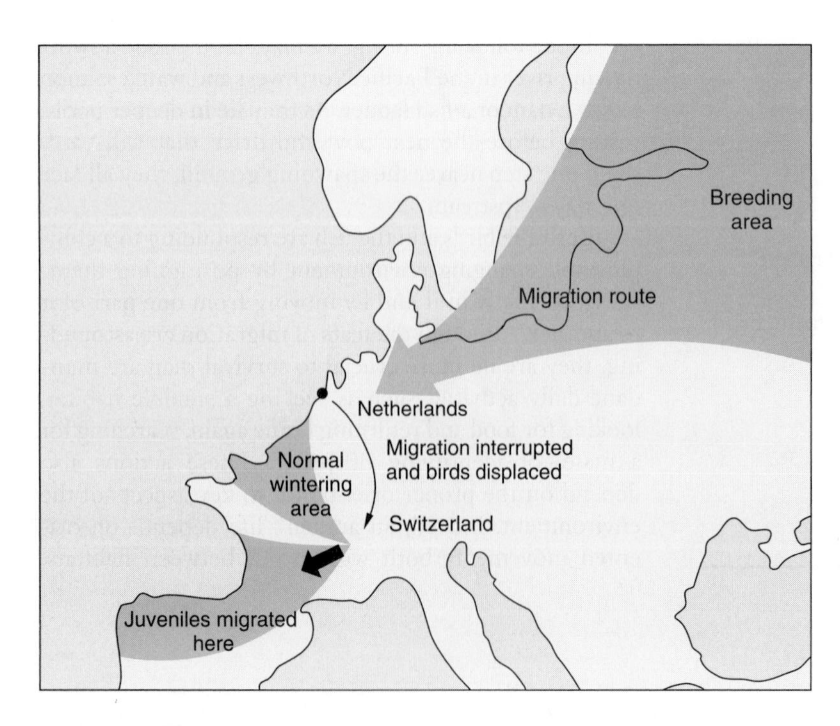

FIGURE 10.2 **Immature starlings captured in the Netherlands and released in Switzerland did not compensate for the relocation during their autumn migration. Instead, they traveled southwest, their normal migratory direction, and ended up in incorrect wintering areas. (Modified from Perdeck 1958.)**

Migratory Direction of Juvenile Birds Most first time migrant birds reach their destination without knowing where that goal is located. They are guided by an inherited program that tells the juveniles in which direction to fly and how long to fly. This innate program is sometimes called **vector navigation** (Berthold 2001; Bingman et al. 2006).

What observations have supported the idea of vector navigation? Individual birds held in the laboratory flutter in the direction in which they would be flying if they were free. When their cousins in nature have completed their migratory journey, the captive birds also cease their directional activity. Furthermore, many species, particularly those that fly from Central Europe to Africa, change compass bearing during their flight. Garden warblers (*Sylvia borin*) and blackcaps (*S. atricapilla*) held in the laboratory change the direction in which they flutter in their cages at the time that free-flying members of their population change direction (Gwinner and Wiltschko 1978; Helbig et al. 1989). Cross-breeding studies have also shown the inheritance of migratory direction. Andreas Helbig (1991) cross-bred members of two populations of blackcaps that had very different migratory directions. The orientation of the offspring was intermediate between those of the parents. Indeed, migratory direction is inherited by the additive effects of a number of genes (Berthold 2001).

Path Integration Besides their use in long-distance navigation, compasses can be used to improve in another type of navigation, called **path integration** or **dead reckoning**. In path integration, the animal integrates information on the sequence of direction and distance traveled during each leg of the outward journey (Figure 10.3). Then, knowing its location relative to home, the animal can head directly there, using its compass(es). A compass may also be used to determine the direction traveled on each leg of the outward journey, or the direction may be estimated from the twists and turns taken, sounds, smells, or even the earth's magnetic field. Information from the outward journey is used to calculate the homeward direction (vector). (Thus, some authors consider path integration to be a type of vector navigation [Ronacher 2008].) The estimates of distance and direction are often adjusted for any displacement due to current or wind. Once close to home, landmarks may be used to pinpoint the exact location of home.

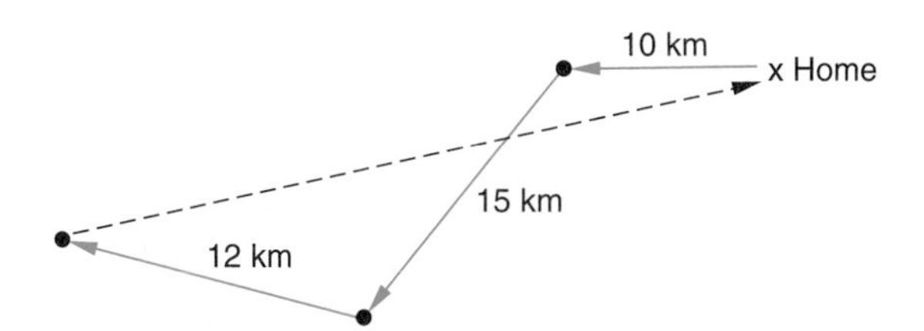

FIGURE 10.3 **Navigation by path integration. This involves determining one's position by using the direction and distance of each successive leg of the outward trip. A compass can then be used to steer a course directly toward home.**

Many types of animals use path integration to find their way around. Consider, for example, the desert ant (*Cataglyphis bicolor*). During its foraging forays, this insect wanders far from its nest over almost featureless terrain. After prey is located, sometimes 100 meters away from the nest, roughly the distance of a football field, the ant turns and heads directly toward home. It appears that the ant knows its position relative to its nest by taking into account each turn and the distance traveled on each leg of its outward trip. If a researcher captures an ant as it is leaving a feeding station headed for home and relocates the ant to a distant site, the ant's path is in a direction that would have led it home if it had not been experimentally moved (Wehner and Srinivasan 1981).

How does a desert ant determine the direction and distance of its outward route? The direction is determined using the pattern of polarization of skylight. Ants determine their direction by using the pattern of skylight polarization, which is caused by the sun's position (discussed shortly) (Müller and Wehner 2007). Desert ants determine the distance they travel using a mechanism that integrates the number of strides required to reach the goal with stride length. Matthias Wittlinger and colleagues (2007) demonstrated this internal pedometer in a very clever way. As we all know, a person with longer legs requires fewer steps to reach a goal than does a person with short legs. Therefore, the researchers predicted that manipulating the length of ant's legs would cause the ants to misestimate the distance to the nest. The researchers collected ants at an experimental feeder and manipulated the length of the ants' legs. They lengthened the legs of some ants by attaching pig's bristles to the ant's legs, creating stilts. They shortened the legs of other ants by partial amputation. The ants walking on stilts overestimated the distance to the nest, whereas the ants with stubby legs underestimated the distance. An added complication to this means of calculating the distance traveled from home is that stride length varies with rate of travel. Thus, as remarkable as this stride counting might seem, the actual mechanism of distance determination also includes an estimation of stride length. Once at home, cues from inside the nest reset the path integrator to zero, so that it can be set again by the next outward journey (Knaden and Wehner 2006).

Map and Compass A compass may also be used with a map to calculate a homeward path. Imagine yourself abandoned in an unfamiliar place with only a compass to guide your homeward journey. Before you could head home, you would also need a map so that you could know where you were relative to home. Only then could you use your compass and orient yourself correctly.

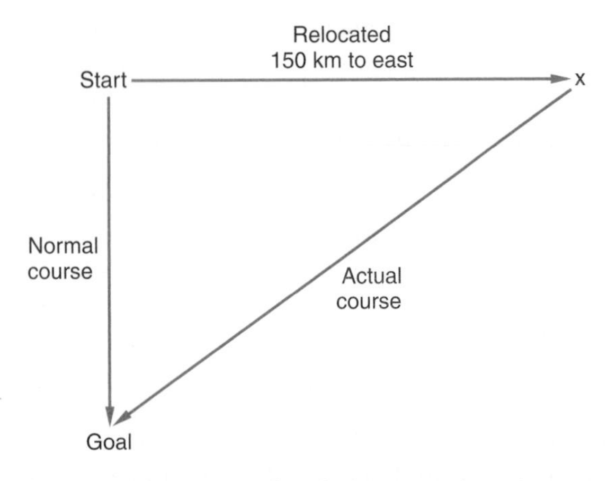

FIGURE 10.4 **An animal that finds its way by using true navigation can compensate for experimental relocation and travel toward the goal. This implies that the animal cannot directly sense its goal and that it is not using familiar landmarks to direct its journey.**

TRUE NAVIGATION

A third level of orientation, sometimes called true navigation[1] is the ability to maintain or establish reference to a goal, regardless of its location, without the use of landmarks (Bingman and Cheng 2005; Ronacher 2008). Generally, this implies that the animal cannot directly sense its goal and that if it is displaced while en route, it compensates by changing direction, thereby heading once again toward the goal (Figure 10.4).

Only a few species, most notably the homing pigeon (*Columba livia*), have been shown to have true navigational ability. Certain other groups of birds, including oceanic seabirds and swallows, are also known to home with great accuracy (Able 1980; Emlen 1975), as do sea turtles (Lohmann and Lohmann 2006). Interestingly, an invertebrate, the spiny lobster (*Panulirus argus*), also seems to have true navigation abilities (Boles and Lohmann 2003).

MULTIPLICITY OF ORIENTATION CUES

The feats of migration are indeed astounding—an arctic tern circumnavigating the globe, a monarch butterfly fluttering thousands of miles to winter in Mexico, a salmon returning to the stream in which it hatched after years in the open sea. How do they do it? There is no simple answer. Different species may use different mechanisms, and any given species usually has several navigational mechanisms available. Indeed, common themes in orientation systems are the use of multiple cues, a hierarchy of systems, and transfer of information among various systems (Berthold 2001; Bingman and Cheng 2005; Walcott 2005). When one mechanism becomes temporarily inoperative, a backup is used. Furthermore, a navigational system may involve more than one sensory system. These interactions can be quite complex, but we will simplify matters by considering each sensory mechanism separately.

VISUAL CUES

Visual mechanisms of orientation include the use of visual landmarks and celestial cues such as the sun, stars, and polarized light.

LANDMARKS

A **landmark** is an easily recognizable cue along a route that can be quickly stored in memory to guide a later journey. Although landmarks can be based on any sensory modality, we most commonly think of visual landmarks. Indeed, landmark recognition is perhaps the most obvious way that vision may be used for orientation or navigation. Humans use landmarks frequently when giving directions: "turn left before the bank" or "make a right just after the gas station." Because the use of landmarks is so familiar to us, it is probably not too surprising to learn that many animals also use them to find their way.

Demonstrating Landmark Use

There are various ways to show that landmarks play a role in orientation. One way is to move the landmark and see whether this alters the orientation of the animal. In a classic study, Niko Tinbergen demonstrated that the digger wasp, *Philanthus triangulum*, relies on landmarks to relocate its nest after a foraging flight. While a female wasp was inside the nest, a ring of 20 pine cones was placed around the opening. When she left the nest, she flew around the area, apparently noting local landmarks, and then flew off in search of prey. During her absence, the ring of pine cones was moved a short distance (1 foot) away. On each of 13 observed trips, the returning wasp searched the middle of the pine cone ring for the nest opening. However, she did not find it until the pine cones were returned to their original position (Tinbergen and Kruyt 1938).

Animals can also be prevented from using landmarks by clouding their vision. Consider, for example, the ingenious way that Klaus Schmidt-Koenig and Hans Schlichte (1972) demonstrated that homing pigeons do

[1]*True navigation* is an unfortunate term since it carries with it the implication that other means of finding one's was from place to place are not real methods of navigating. This is certainly not true. Nevertheless, we will use the term simply to distinguish this method of maintaining a course from the others.

FIGURE 10.5 Homing pigeons that are wearing frosted contact lenses are unable to use landmarks for navigation. However, these pigeons head home just as accurately as those with normal vision do. Therefore, although pigeons may use landmarks if they are available, they do not require them to home.

FIGURE 10.6 The desert ant uses a remembered sequence of landmark images to find its way home in a familiar area.

not require landmarks to return to the vicinity of their home loft: they created frosted contact lenses for the pigeons (Figure 10.5). Through these lenses, pigeons could only vaguely see nearby objects and distant ones not at all. Nonetheless, the flight paths of these pigeons were oriented toward home just as accurately as those of control pigeons. Thus, the pigeons cannot be depending on familiar landmarks to guide their journey home. Note that this does not mean that they do not use landmarks when they are available, just that they can determine the homeward direction without them. Also, although pigeons with frosted lenses get to the general area of their home loft, they often cannot find the loft itself. Landmarks, then, may be important in pinpointing the exact loft location but are not necessary for determining the direction of home.

Models of Landmark Use

Knowing that an animal uses landmarks to find its way does not tell us *how* those landmarks are used. Do other animals use landmarks as humans do, as part of a mental map of the area? Perhaps some species do, but others might use landmarks in different ways. A simple model of landmark use is that the animal stores the image of a group of landmarks in its memory, almost like a photograph of the scene. Then it moves about the environment until its view of nearby objects matches the remembered "snapshot" (Emery and Clayton 2005). Rüdiger Wehner (1981) suggested that a whole series of memory snapshots might be filed in the order in which they are encountered. He added that invertebrates might be able to use landmarks by comparing the successive images of surrounding objects with a series of memory snapshots of the landmarks along a familiar route.

One animal that appears to use memory snapshots of landmarks is the desert ant (Figure 10.6). As previously mentioned, desert ants are able to plot a course back to the nest by path integration; that is, they integrate the directions and distances traveled on all legs of the journey away from the nest to plot a direct course back. However, they also use landmarks, especially when they have almost reached the nest on their return from the foraging site (Åkesson and Wehner 2002). Once the ants are close to the nest entrance, they use a systematic search to find the opening of the burrow. The search strategy varies with the species of desert ant and the number of natural landmarks in their native habitat (Narendra et al. 2008). Desert ants tend to follow familiar routes. In fact, if landmarks are available, desert ants often use landmarks instead of path integration. If the most direct path is an unfamiliar route, it could lead over rocks or be blocked by scrub, and so landmarks are favored. Nonetheless, if the ant comes across a clearing, it can use path integration to take the most direct course home (Collett et al. 1998).

SUN COMPASS

Many animals use the sun as a celestial compass. In other words, these animals can determine compass direction from the position of the sun. Because of the earth's rotation, the sun appears to move through the sky at an average rate of 15° per hour. The sun rises in the east and moves across the sky to set on the western horizon. The specific course that the sun appears to take varies with the latitude of the observer and the season of the year, but it is predictable (Figure 10.7). Therefore, if the sun's path and the time of day are known, the sun can be used as a compass.

Knowledge of one compass bearing is all that is necessary for orientation in any direction. Consider this simplified example. Suppose you decided to camp in the woods a short distance north of your home. As you headed for your campsite at 9 A.M., the sun would be in the east, so you would keep the sun on your right to travel north. However, during your homeward trek the next morning, you would keep the sun on your left to travel south.

The use of the sun for orientation is complicated by its apparent motion through the sky. The sun appears to move at an average rate of 15° an hour. Therefore, an animal heading straight for its goal and navigating by keeping a constant angle between its path and the sun would, after one hour, be following a path that would be off by 15°. Some species take only short trips, so errors due to the sun's apparent motion are inconsequential. These species do not adjust their course with the sun's. But if the sun is to be used as an orientation cue for a prolonged period, the animal must compensate for the sun's movement. To do so, it must be able to measure the passage of time and correctly adjust its angle with the position of the sun. At 9 A.M. an animal wishing to travel south might keep the sun at an angle of 45° to its left. By 3 P.M., however, the sun will have moved approximately 90° at an average rate of 15° an hour. To maintain the same southward bearing,

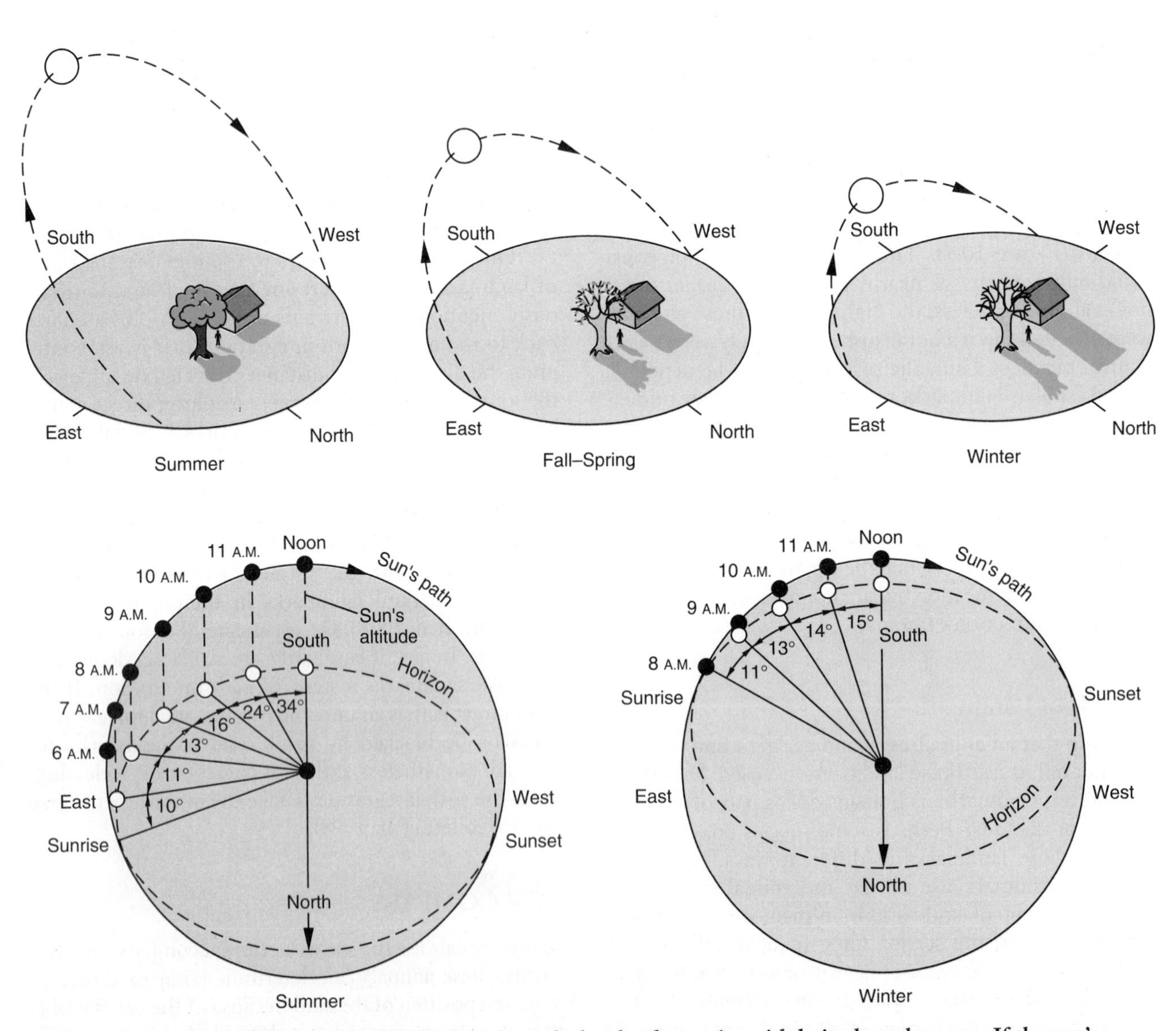

FIGURE 10.7 The sun follows a predictable path through the sky that varies with latitude and season. If the sun's course and the time of day are known, the sun's bearing (azimuth) provides a compass bearing. The sun appears to move across the sky at an average rate of 15° an hour. Therefore, if the sun is to be used as a compass for a long time, the animal must compensate for its movement.

the animal must now assume a 45° angle, with the sun on its right. Time is measured by using a biological clock (discussed in Chapter 9; time-compensated orientation of bee dances is discussed in Chapter 16).

The first work on sun compass orientation was done on birds and bees in the laboratories of Gustav Kramer (1950) and Karl von Frisch (1950), respectively. Although these two investigative groups worked at the same time, neither knew of the other's work. Nevertheless, they often used similar experimental designs to reveal the details of sun compass orientation. We will take a closer look at the experiments of Gustav Kramer here, but if you want to compare these studies to those of von Frisch, consult von Frisch's (1967) fascinating book, *The Dance Language and Orientation of Bees* or the discussion of bee dances in Chapter 16.

Gustav Kramer (1949) began his studies by trapping migrant birds and caring for them in cages. He then noticed that they became restless during their normal migration season. Furthermore, most of their activity took place on the side of the cage corresponding to the direction in which the birds would be flying if they were free to migrate. This activity has been aptly named migratory restlessness. In noting these tendencies, Kramer set the stage for a series of experiments that would yield valuable evidence in the quest for the navigational mechanisms of birds.

The indication that birds migrating during the day use the sun as a navigational cue was that the orientation (directionality) of migratory restlessness was lost when the sun was blocked from view. Kramer (1951) set up outdoor experiments with caged starlings, *Sturnus vulgaris* (Figure 10.8), which are daytime migrators, and found that they oriented in the normal migratory direction unless the sky was overcast, in which case they lost their directional ability and moved about randomly. When the sun reappeared, they oriented correctly again, suggesting that they were using the sun as a compass. Then Kramer devised experiments in which the sun was blocked from view and a mirror was used to change the *apparent* position of the sun. The birds reoriented according to the direction of the new "sun."

Because migration occurs during limited periods in the fall and spring, experiments using migratory restlessness to study orientation mechanisms are limited to two brief intervals a year. To eliminate this problem, Kramer (1951) devised an orientation cage in which there were 12 identical food boxes encircling a central birdcage (Figure 10.9). Kramer and his students trained birds to expect food in a box that lay in a particular compass direction. This ring of food boxes could be rotated so that a bird trained to get food in a given compass direction would not always be going to the same food box. This eliminated the possibility that the bird might learn to recognize the food dish by some characteristic,

FIGURE 10.8 **Starlings are daytime migrators and were the subject of Gustav Kramer's pioneering work on bird navigation.**

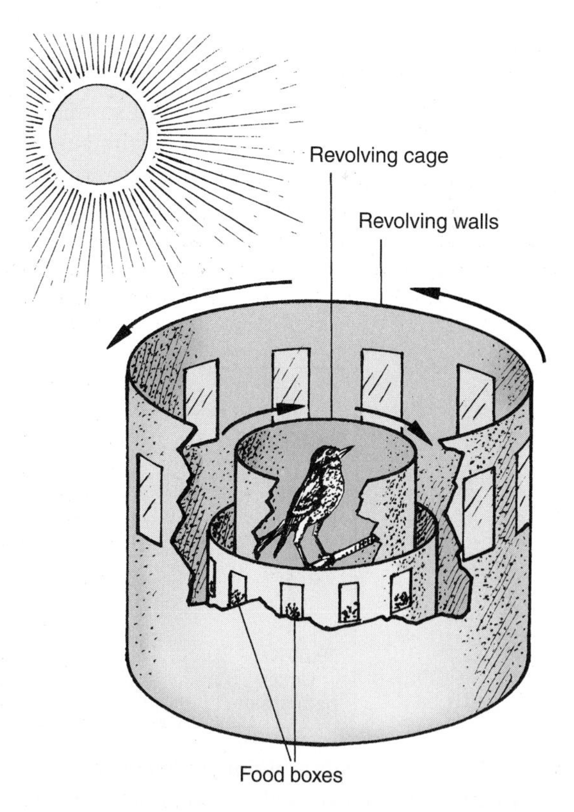

FIGURE 10.9 **Kramer's orientation cage. The bird can see the sky through the glass roof but is prevented from seeing the surrounding landscape. It is trained to look for food in a food box that is placed in a particular compass direction.**

such as a dent. As long as the birds could see the sun, they would approach the proper food box. However, on overcast days the birds were often disoriented, as would be expected if they were using a sun compass.

The results of experiments with birds in Kramer's orientation cages not only confirm those on migratory restlessness (Kramer 1951), but also indicate that the birds compensate for the movement of the sun. Actually, the idea of time-compensated sun compass orientation began when Kramer noticed that the birds in his orientation cages were able to orient in the proper direction even as the sun moved across the sky. When the real sun was replaced with a stationary light source, the birds continually adjusted their orientation with the stationary sun as though it were moving. The orientation with the artificial sun changed at a rate of about 15° an hour, just as it would to maintain a constant compass bearing using the real sun.

The birds are able to compensate for the sun's apparent movement; therefore, they must possess some sort of independent timing mechanism. As we saw in Chapter 9, the biological clock that allows birds to compensate for the movement of the sun can be reset by artificially altering the light–dark regime. Initially, the birds are placed in an artificial light–dark cycle that corresponds to the natural lighting conditions outside; the lights are on from 6 A.M. to 6 P.M. The light period is then shifted so that it begins earlier or later than the actual time of dawn. For example, if the animal is exposed to a light–dark cycle that is shifted so that the lights come on at noon instead of 6 A.M., the animal's biological clock is gradually reset. In this case, the animal's body time would be set six hours later than real time. Therefore, if the biological clock is used to compensate for the movement of the sun, orientation should be off by the amount that the sun had moved during that interval. In this example, orientation should be shifted 90° (6 x 15°) clockwise, for example, west instead of south (Figure 10.10).

STOP AND THINK

How would orientation change if the light–dark cycle was changed so that the lights came on at midnight instead of 6 A.M.?

One of Kramer's students, Klaus Hoffmann (1954), was the first to use the clock-shift experiment to demonstrate the involvement of the biological clock in sun compass orientation. After resetting the internal clock of starlings by keeping them in an artificial light–dark cycle for several days, the birds' orientation was shifted by the predicted amount.

Using experiments similar to the classic studies described above, we have confirmed that a time-compensated sun compass exists in a wide variety of organisms (Åkesson and Hedenström 2007, Bingman 2005; Rozhok 2008). We also know more details about time-compensated sun orientation. For example, even with limited exposure to the sun (experience with a partial arc), many animals develop a sun compass that can be used all day (discussed in Rozhok 2008). Furthermore, the apparent movement of the sun through the sky varies with the time of day; it appears to move faster at noon than at sunrise or sunset. The internal clock of birds compensates for daily variation in the rate of the sun's apparent movement (Wiltschcko et al. 2000). Importantly, the compasses used by animals—sun, the stars, and the earth's magnetic field—interact in some interesting ways, as we will see shortly.

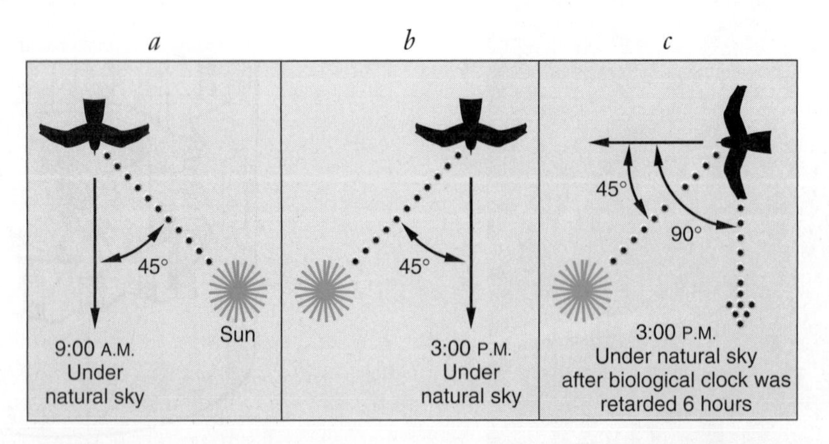

FIGURE 10.10 **A clock-shift experiment demonstrates time-compensated sun compass orientation. (*a*) The flight path of a bird flying south at 9 A.M. might be at an angle of 45° to the right of the sun. (*b*) By 3 P.M., the sun would have moved roughly 90°, so to continue flying in the same direction, the bird's flight path might be at an angle of 45° to the left of the sun. (*c*) If the bird's biological clock were delayed by six hours and the bird's orientation tested at 3 P.M. (when the bird's body time was 9 A.M.), it would orient to the west. The flight path of the bird would be determined by the bird's biological clock. The flight path would, therefore, be appropriate for 9 A.M., and orientation would be shifted by 90° clockwise. (From Palmer 1966.)**

STAR COMPASS

Many species of bird migrants travel at night. Even if they set their bearings by the position of the setting sun, how do they steer their course throughout the night? One important cue is the stars. This was first demonstrated by Franz and Eleonore Sauer (Sauer 1957, 1961; Sauer and Sauer 1960). Using several species of sylviid warblers, they performed a series of experiments aimed at discovering just which objects in the nighttime sky the birds use as cues. The Sauers kept their caged warblers inside a planetarium so that the nighttime sky could be controlled. They first lined up the planetarium sky with the sky outside and found that the birds oriented themselves in the proper migratory direction for that time of year. Then the lights were turned out, and the star pattern of the sky was rotated. The birds continued to orient according to the new direction of the planetarium sky. When the dome was diffusely lit, the birds were disoriented and moved about randomly. In some experiments, even though the moon and planets were not projected, the birds oriented correctly, apparently taking their bearings from the stars.

We know the most about the mechanism of star compass orientation in the indigo bunting (*Passerina cyanea*). Our knowledge has been gained primarily through Stephen Emlen's systematic planetarium studies. These indicate that the indigo bunting relies on the region of the sky within 35° of Polaris (Figure 10.11). Since Polaris is the pole star, it shows little apparent movement and, therefore, provides the most stationary reference point in the northern sky. The other constellations rotate around this point (Figure 10.12). The stars nearer Polaris move through smaller arcs than do those farther away, closer to the celestial equator. The birds learn that the center of rotation of the stars is in the north, information that is used to guide their migration either northward or southward. The major constellations in this region are the Big Dipper, the Little Dipper, Draco, Cepheus, and Cassiopeia. Experiments have shown that it is not necessary for all these constellations to be visible at once. If one constellation is blocked by cloud cover, the bird simply relies on an alternative constellation (Emlen 1967a, b).

a

b

FIGURE 10.11 (*a*) Star compass orientation was explored by exposing nocturnal migrants, indigo buntings, to a planetarium sky. During the normal time of migration, caged birds will flutter in the proper migratory direction if the stars are visible. (*b*) In some studies, a bird's feet were inked, thus creating a record of its activity on the sides of a funnel-shaped cage.

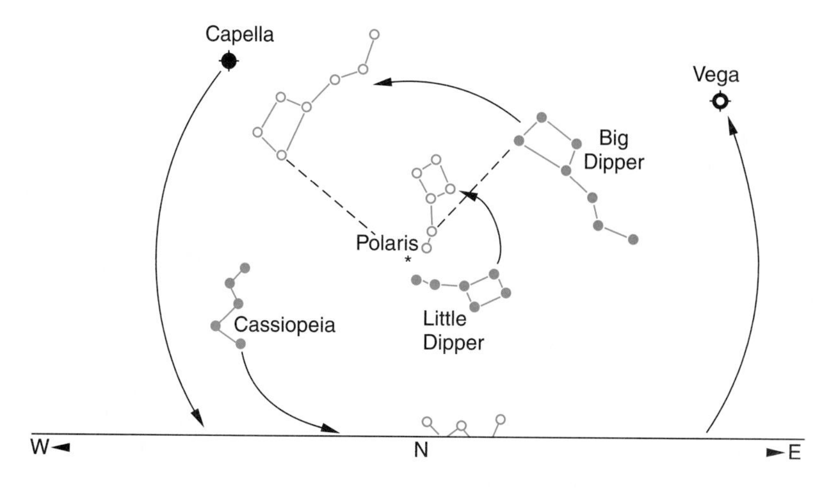

FIGURE 10.12 The stars rotate around Polaris, the North Star. The center of rotation of the stars tells birds which way is north. The positions of stars in the northern sky during the spring are shown here. The closed circles indicate star positions during the early evening, and the open circles indicate the positions of the same stars six hours later.

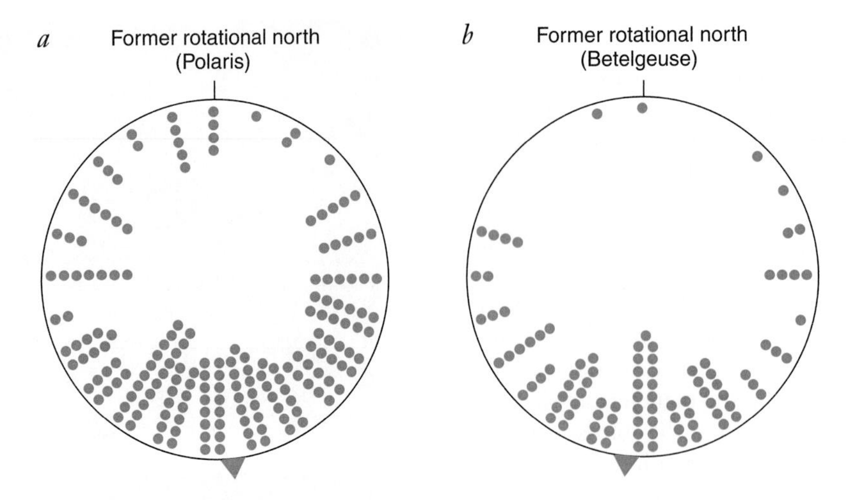

FIGURE 10.13 **The orientation of indigo buntings to a stationary planetarium sky after exposure to different celestial rotations. During their first summer, indigo buntings learn that the center of celestial rotation is north. This was demonstrated by exposing a group of young birds to a planetarium sky that rotated (*a*) around Polaris (the North Star) or (*b*) around Betelgeuse. During their first autumn, when they would be migrating south, they were exposed to a stationary planetarium sky. Each dot is the mean direction of activity for a single test. The arrow on the periphery of the circle is the overall mean direction of activity. Each group oriented away from the star that had been the center of rotation. (Modified from data of Able and Bingman 1987; Emlen 1970.)**

Young birds learn that the center of rotation of stars is north. The axis of rotation then gives directional meaning to the configuration of constellations. Once their star compass has been set in this way, the birds do not need to see the constellations rotate. Simply viewing certain constellations is sufficient for orientation. This was first demonstrated by exposing groups of young indigo buntings to normal star patterns in a planetarium sky. One group saw a normal pattern of rotation, one that rotated around Polaris. The other group viewed the normal pattern of stars, but instead of rotating around Polaris, these stars rotated around Betelgeuse, a bright star closer to the equator. When the birds came into a migratory condition, their orientation was tested under a stationary sky. Although each group was headed in a different geographic direction, both groups were well oriented in the appropriate migratory direction relative to the center of rotation they had experienced, either Betelgeuse or Polaris (Figure 10.13). In other words, in the autumn, when the birds would be heading south for the winter, those that had experienced Betelgeuse as the center of rotation interpreted the position of that star as north and headed away from it (Emlen 1969, 1970, 1972).

The development of the star compass has been studied in only a few species other than the indigo bunting. Garden warblers (Wiltschko 1982; Wiltschko et al. 1987) and pied flycatchers, *Ficedula hypoleuca* (Bingman 1984) also learn that the center of celestial rotation indicates north.

POLARIZED LIGHT AND ORIENTATION

One of the puzzling facets of sun compass orientation is that many animals continue to orient correctly even when their view of most of the sky is blocked. How is this possible? For at least some of these animals, another celestial orientation cue is available in patches of blue sky—polarized light. Before considering how animals orient to polarized light, let's examine the nature of polarized light and how the pattern of skylight polarization depends on the position of the sun.

The Nature of Polarized Light

Light consists of many electromagnetic waves, all vibrating perpendicularly to the direction of propagation (Figure 10.14). As a crude analogy, think of a rope held loosely between two people as a light beam. The rope itself would define the direction of propagation of the light beam. If one person repeatedly flicked his or her wrist, the rope would begin to wave or oscillate. These oscillations would also be perpendicular to the length of the rope, but they could be vertical, horizontal, or any angle in between, depending on how she flicked her wrist. The same is true of light waves. Most light con-

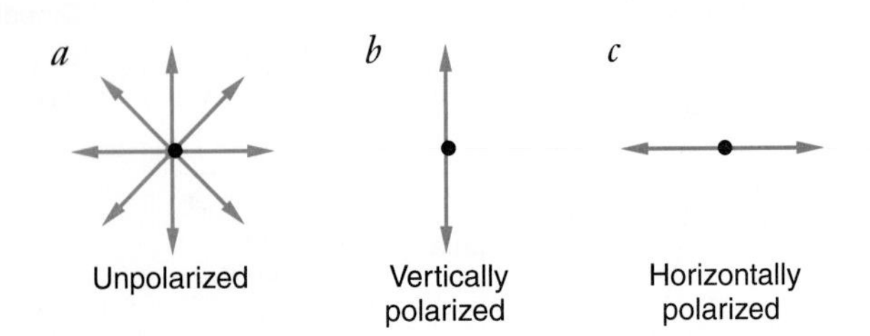

FIGURE 10.14 **Unpolarized and polarized light. The arrows show the planes of vibration of a light beam that is coming straight out of the page.**

sists of a great many waves that are vibrating in all possible planes perpendicular to the direction in which the wave is traveling. Such light is described as unpolarized. In fully polarized light, however, all waves vibrate in only one plane. Our rope light beam, for instance, would become vertically polarized if the person's wrist were flicked only up and down. In this case, the rope might oscillate vertically in the spaces between the boards of a picket fence.

As sunlight passes through the atmosphere, it becomes polarized by air molecules and particles in the air, but the degree and direction of polarization in a given region of the sky depend on the position of the sun. In other words, there is a pattern of polarized light in the sky that is directly related to the sun's position (Figure 10.15). One aspect of this pattern is the degree of polarization. To picture the pattern of polarization, think of the sky as a celestial sphere with the sun at one pole and an "antisun" at the other. The light at the poles is unpolarized, but it becomes gradually more strongly polarized with increasing distance from the poles. Thus, between the sun and the antisun, there is a band where the light in the sky is more highly polarized than in other regions. This region is described as the band of maximum polarization. But there is more to the pattern than this: the direction of the plane of polarization (called the e-vector) also varies according to the position of the sun. The plane of polarization of sunlight is always perpendicular to the direction in which the light beam is traveling. If you were to draw imaginary lines of latitude on the celestial sphere so that they formed concentric circles around the sun and antisun, these lines would indicate the plane of polarization at any point in the sky. Since the entire pattern of polarization of light in the sky is determined by the sun's position, the pattern moves westward as the sun moves through the sky (Waterman 1989).

Uses of Polarized Light in Orientation

Polarized light reflected from shiny surfaces, such as water or a moist substrate, is used by some aquatic insects to detect suitable habitat. Indeed, polarized light may actually attract them (Schwind 1991). For the backswimmer, *Notonecta glauca* (Figure 10.16), not only is the horizontally polarized light that is reflected from the surface of a pond a beacon that helps the insect, as it flies overhead, locate a new body of water during dispersal, but it also triggers a plunge reaction that brings the insect closer to a new home (Schwind 1983).

The plane of polarization of the light in the sky is used as an orientation cue in two possible ways. First, polarized light is used as an axis for orientation. In other words, an animal might move at some angle with respect to the plane of polarization. Many animals use polarized light in this way. Salamanders living near a shoreline, for instance, can use the plane of polarization to direct their

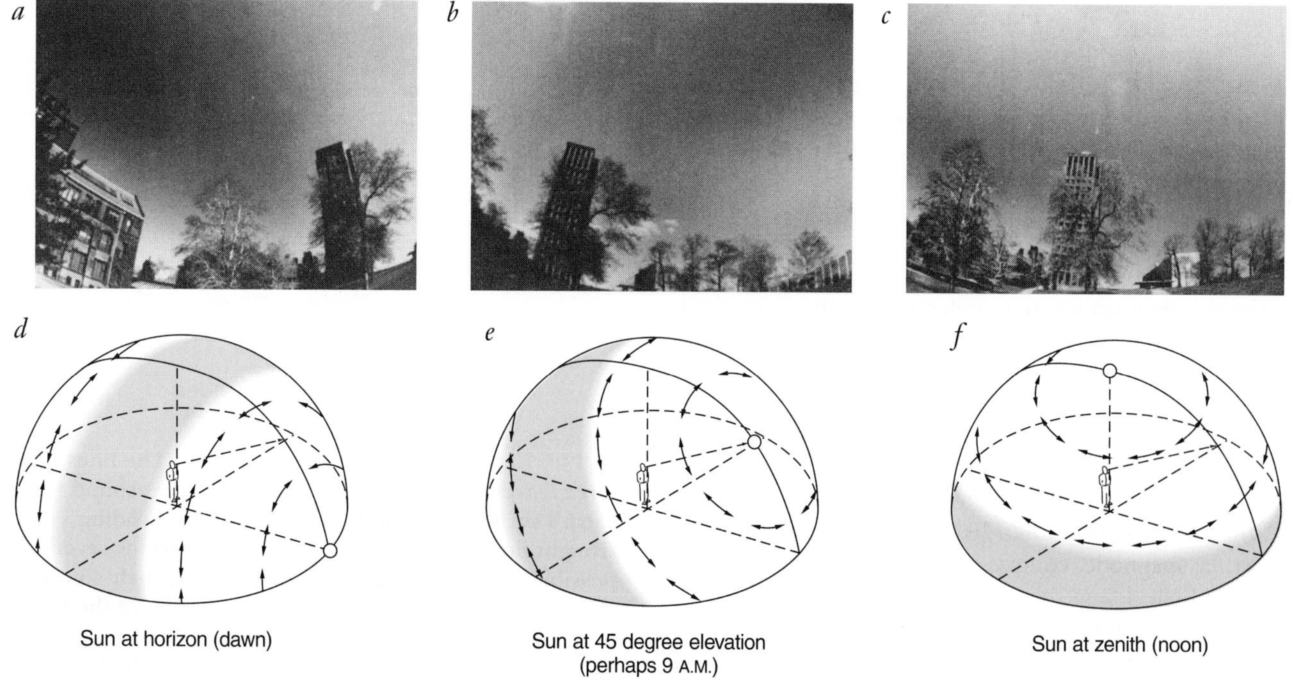

FIGURE 10.15 **The sky viewed through a polarizing filter to show the pattern of skylight polarization at (*a*) 9 A.M., (*b*) noon, and (*c*) 3 P.M. The darker region of the sky is the band of maximum polarization. The diagrams below show the pattern of polarization (*d*) with the sun on the horizon, (*e*) at 45° elevation, and (*f*) at zenith. The arrows indicate the direction of the plane of polarization. The small circle denotes the position of the sun. The pattern of polarization depends on the position of the sun. The blue sky provides an orientation cue for animals that can perceive the plane of polarization.**

FIGURE 10.16 **Many aquatic insects, such as this backswimmer, use polarized light reflected from water or a moist surface to locate an appropriate habitat. A backswimmer spends almost its entire life underwater. These insects are commonly seen in ponds, suspended beneath the water surface, as this one is. Adults can fly, however, and may disperse to a new pond before laying the second batch of eggs of the season.**

movements toward land or water (Adler 1976). Second, the pattern of polarization of sunlight might be used to determine the sun's position when it is blocked from view. The polarization of light in the sky could also provide an orientation cue at dawn and dusk, when the sun is below the horizon. Many birds that migrate at night set their bearings at sunset. Apparently, the pattern of skylight polarization at sunset (Able 1982) and at sunrise (Moore 1986) assists the orientation of birds migrating at these times because some experiments have shown that the birds' directional tendencies are altered when the plane of polarized light to which they are exposed is experimentally shifted by rotating polarizing filters. Indeed, when a bird is setting its bearings for the night, polarized light is a more important orientation cue than the sun's position along the horizon at dusk or the geomagnetic field (Able 1993; Able and Able 1996).

MAGNETIC CUES

Many organisms, ranging from bacteria to certain vertebrates, orient their activities relative to the earth's magnetic field. These activities include direction finding and navigation over long and short distances—the long-distance migrations of birds (reviewed in Wiltschko and Wiltschko 2005) to the nightly foraging forays of spiny lobsters (Lohmann et al. 2007). Magnetic sense may also help an organism locate a preferred direction, as when bacteria swim downward, toward the muddy bottom they call home (Blakemore and Frankel 1981). The earth's magnetic field may also orient nest building, as in the Ansell's mole rat, a rodent that lives underground (Marhold et al. 1997), or roosting place of bats (Wang et al. 2007). Indeed, Wolfgang and Roswitha Wiltschko (Wiltschko and Wiltschko 2007) suggest that, in birds at least, a magnetic compass evolved in nonmigratory species first. These species probably used the magnetic compass for optimizing paths to and from various goals, such as nest sites, feeding sites, and drinking sites. Later, when some species began to migrate, the migrants use the magnetic compass to orient during migration.

The ability to use the earth's magnetic field as a compass has its advantages. It can be used in places where visual cues are limited or absent, such as a roosting cave, underground tunnel, or the depths of an ocean. And, unlike celestial cues, it is constant year round, night and day.

CUES FROM THE EARTH'S MAGNETIC FIELD

To picture the geomagnetic field around the earth, imagine an immense bar magnet through the earth's core from north to south. However, this bar magnet is tilted slightly from the geographic north-south axis, and the magnetic poles are shifted slightly from the geographic, or rotational, poles (Figure 10.17). The difference

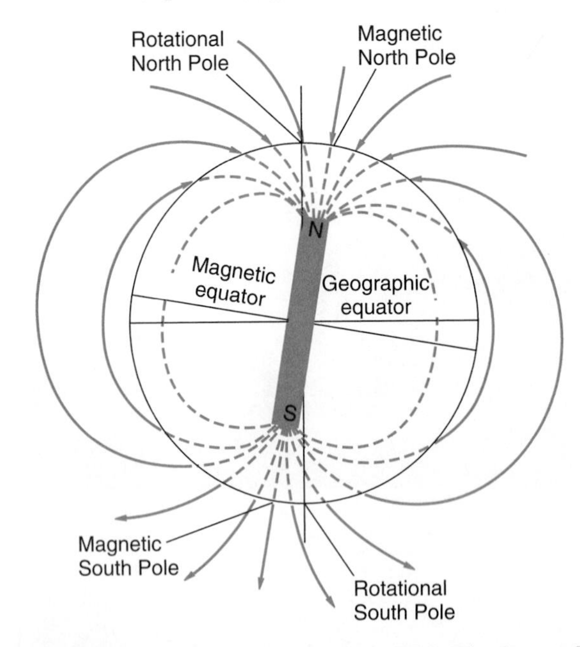

FIGURE 10.17 **The earth's magnetic field. The lines of force leave magnetic south vertically; curve around the earth's surface; and enter magnetic north, heading straight down. The geomagnetic field provides several possible cues for navigation: polarity, the north-south axis of the lines of force, and the inclination of the lines of force. The magnetic compass of most animals appears to be an inclination compass. They determine the north-south axis from the orientation of the lines of force but assign direction to this by the inclination of the force lines. In the northern hemisphere, north is the direction in which the force lines dip toward the earth.**

between the magnetic pole and the geographic pole is called the declination of the earth's magnetic field. Because the declination is small in most places, usually less than 20°, magnetic north is usually a reasonably good indicator of geographic north. (On maps and nautical charts used for precise navigation, both geographic north and magnetic north are indicated, so that a navigator or backpacker can adjust her compass readings for declination.) The declination is, of course, greatest near the poles.

Several aspects of the earth's magnetic field vary in a predictable manner and could, therefore, provide directional cues. One aspect is polarity. The magnetic north pole is called the positive pole, and the magnetic south, the negative pole. The second aspect is the angle of the lines of force with respect to earth's surface. These leave the magnetic south pole vertically; curve around the surface of the earth; become level with the surface at the magnetic equator; and reenter the magnetic north pole, going straight down. The angle of inclination, or dip, of the magnetic field is the angle that the line of force makes with the horizon. The angle of inclination is steepest (vertical) near the poles and near zero (horizontal) near the equator. The third aspect that varies predictably is the intensity (or strength) of the geomagnetic field. It is greatest at the poles and least at the equator.

Thus, we see that the polarity, inclination, and intensity of the earth's magnetic field vary systematically with latitude, providing three potential orientation cues. Which of these are used? Our own experience with compasses immediately brings polarity to mind. When the needle on a compass points north, it is responding to the polarity of the earth's field. Indeed, some species of animals seem to respond to polarity (Table 10.1). This list includes invertebrates, the spiny lobster, for instance, as well as vertebrates, including some fish and birds; the mole rat, a rodent that lives underground (Wiltschko and Wiltschko 2006); and a bat (Wang et al. 2007). We know that an animal responds to polarity when its orientation changes in response to an experimental shift in the direction of magnetic north.

Other animals, including most birds and sea turtles, appear to use the magnetic field inclination. Instead of north or south, they distinguish between "poleward," where the lines of force are steepest, and "equatorward," where the lines of force are parallel to the earth's surface. Although the horizontal component of the earth's field (the direction of magnetic north), which runs between magnetic north and magnetic south, indicates to the animal the north-south axis, the vertical component of the earth's magnetic field (the inclination of the field) is the cue that tells the animal whether it is going toward the pole or toward the equator (Wiltschko and Wiltschko 2006).

We can determine whether an animal is using the polarity or the angle of inclination of the pole by separately altering the horizontal and the vertical components of the experimental magnetic field and observing the effect of the animal's orientation. If an animal uses a polarity compass, it will shift its orientation when the horizontal component of the field is shifted. In contrast, an animal using an inclination compass will shift its orientation when the vertical component of the experimental field is altered.

Ansell's mole rats (*Cryptomys anselli*) orient using the polarity of the magnetic field. These small rodents normally live in darkness in subterranean colonies. When housed in circular arenas in captivity, they reliably and spontaneously build their nests in the southeastern region of the arena. Researchers placed mole rats of the

TABLE 10.1 Animals Demonstrated to Use a Magnetic Compass

Systematic group				Type of compass
Molluscs				
Snails	1 order	1 family	1 species	???
Arthropods				
Crustaceans	3 orders	3 families	5 species	Polarity compass
Insects	6 orders	7 families	9 species	Polarity compass
Vertebrata				
Cartilageous fish	1 order	1 family	1 species	???
Bony fish	2 orders	2 families	4 species	Polarity compass
Amphibians	1 order	2 families	2 species	Inclination compass
Reptilians	1 order	2 families	2 species	Inclination compass
Birds	4 orders	12 families	21 species	Inclination compass
Mammals	2 orders	2 families	3 species	Polarity compass

same family group into a circular test arena. Within hours, the animals gathered nesting materials and built a nest in the southeast sector of the arena. Then researchers used a Helmholtz coil, a device that generates a magnetic field when an electric current runs through it, to alter the magnetic field experienced by the mole rats. The magnetic field experienced by the birds can be altered by reversing the direction of current flow through the coil. When researchers reversed the horizontal component (the polarity) of the magnetic field, the mole rats began to build nests in the northwest sector of the arena. However, when researchers inverted the vertical component (the angle of inclination) of the magnetic field, the mole rats continued nesting in the southeast sector (Figure 10.18*a*) (Marthold et al. 1997).

In contrast, birds use the inclination angle of the earth's magnetic field for orientation. For example, in the laboratory the migratory restlessness of European robins remains oriented in the proper direction even when the birds have no visual cues. When the magnetic world that the birds experienced was reversed by switching the polarity of an experimental field, there was no effect on their orientation. However, the birds reoriented if the inclination in the experimental field was altered (Figure 10.18*b*). It is interesting that these birds were not able to orient according to magnetic field lines that were horizontal to the earth's surface. Horizontal field lines occur around the equator. A bird could determine the north-south axis in a horizontal field, but without the inclination it would not know which direction is north or south (Wiltschko and Wiltschko 1972).

The results of an experiment on free-flying homing pigeons are also consistent with the idea that a bird's magnetic compass is based on the inclination of the magnetic lines of force. Small Helmholtz coil hats were fitted onto the heads of homing pigeons (Figure 10.19*a*). A Helmholtz coil is a device that generates a magnetic field when an electric current runs through it. The magnetic field experienced by the birds can be altered by reversing the direction of current flow through the coil. On cloudy days, when the pigeons relied on magnetic cues rather than their sun compass, they oriented as if they considered north to be the direction in which the magnetic lines of force dip into the earth. Those birds that experienced the greatest dip in the magnetic field in the north, as it is in the nor-

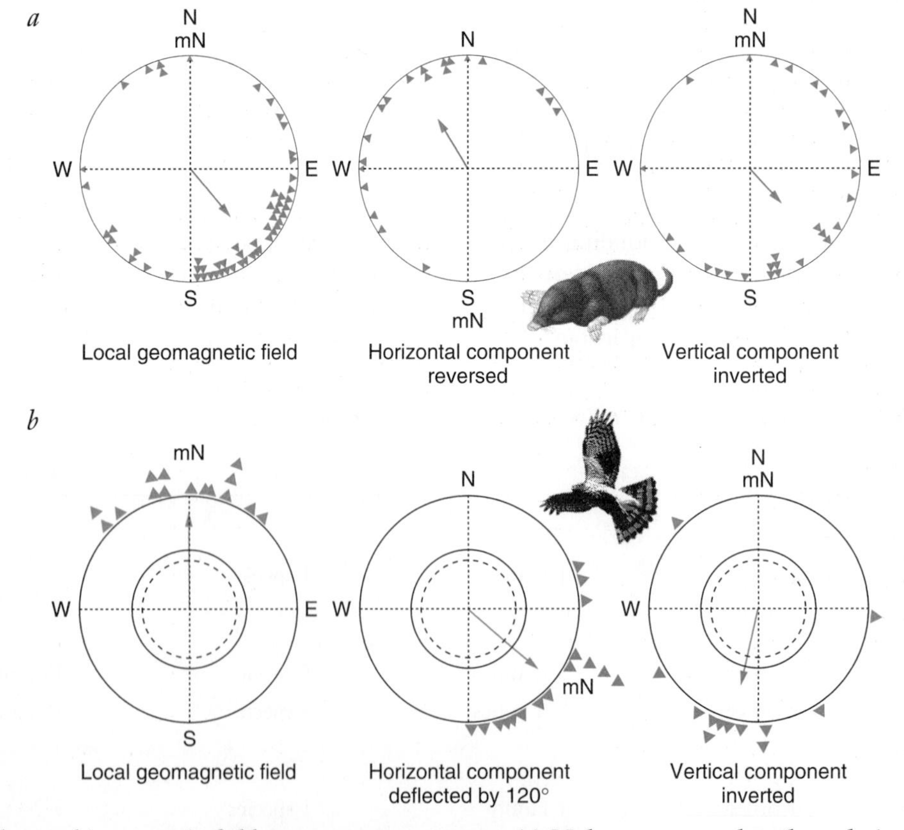

FIGURE 10.18 **The earth's magnetic field can serve as a compass. (*a*) Mole rats respond to the polarity (horizontal component) of the ambient magnetic field. They build their nests in the southeast portion of a circular arena. If the magnetic field is experimentally reversed, mole rats build their nests in the northwest portion of the arena. However, if the vertical component of the ambient magnetic field is reversed, mole rats do not change their orientation. (*b*) Birds use the inclination of the lines of force (vertical component of the earth's magnetic field) as a compass. The lines of force are steepest at the poles and horizontal at the equator. Birds reverse their orientation when the inclination of the magnetic field is reversed, but they do not alter their orientation if the polarity of the magnetic field is changed. (From Wiltschko and Wiltschko 2005.)**

a

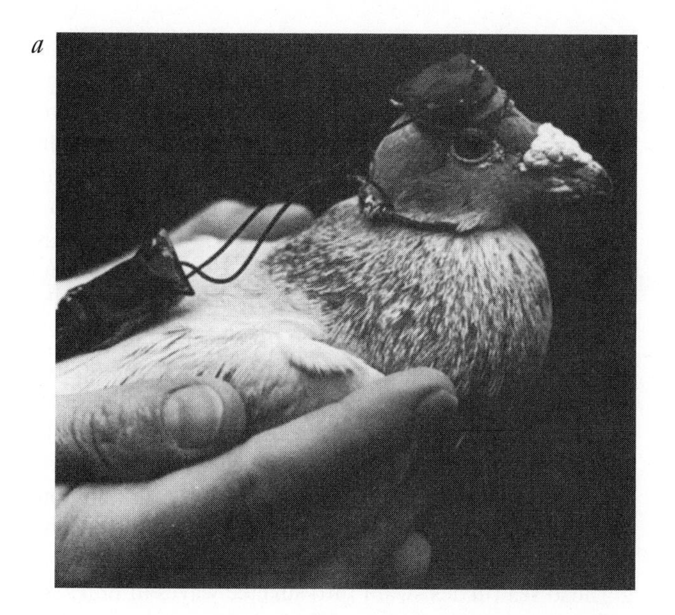

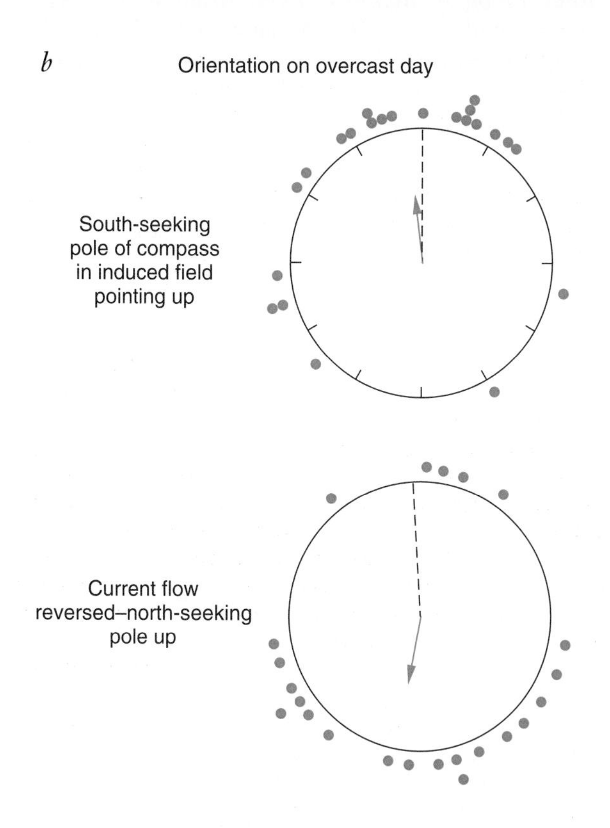

FIGURE 10.19 **(*a*) A pigeon with a Helmholtz coil, a device that generates a magnetic field, on its head. (*b*) The magnetic field experienced by the pigeon can be altered by changing the direction in which the electric current runs through the coil. On overcast days, when the birds could not use the sun as a compass, the magnetic field influenced their orientation. They oriented as if they interpreted north as the direction in which the magnetic lines of force dip toward the earth. Each dot indicates the direction in which a bird vanished from sight after being released. The arrow in the center indicates the mean vanishing bearing. (Modified from data of Walcott and Green 1974.)**

mal geomagnetic field, headed home. In contrast, the birds that experienced the greatest dip in the magnetic field in the south were misdirected by the reversed magnetic information and headed directly away from home (Figure 10.19*b*) (Visalberghi and Alleva 1979; Walcott and Green 1974).

There are also some indications that several species respond to the small differences in the intensity of the geomagnetic field. Among these animals are bees (Kirschvink et al. 1997; Walker and Bitterman 1989), homing pigeons (Dennis et al. 2007; Keeton et al. 1974; Kowalski et al. 1988), sea turtles (Lohmann and Lohmann 1996a), and the American alligator (Rodda 1984). If changes in magnetic intensity can be sensed, the gradual increase in strength between the equator and the poles could also serve as a crude compass.

DIRECTIONAL INFORMATION FROM THE EARTH'S MAGNETIC FIELD: A MAGNETIC COMPASS

If we keep in mind that orientation is essential to the survival of migrating or homing animals, it should not come as a surprise that orientation is affected by the interaction of many cues, as well as many variables, including experience, species differences, and amount of stored energy. We will separate some of these interacting variables to try to understand just how animals remain oriented when faced with the real problems of navigating. Many animals can obtain directional information from the earth's magnetic field; that is, the earth's magnetic field can serve as a **magnetic compass**.

The Magnetic Compass and Bird Navigation

As we have seen, birds use the earth's magnetic field as a compass. They determine whether they are headed toward the pole or the equator by the angle of inclination of the magnetic lines of force.

Inherited Migratory Program Migratory birds inherit a program that tells them to travel in a certain geographical direction, based on magnetic cues, for a certain amount of time. Because the magnetic compass of birds is an inclination compass, migrants from either the northern or the southern hemisphere might use the same migratory program—fly toward the equator (where the lines of force are more horizontal) in the fall and toward the pole (where the lines of force are more vertical) in the spring (Wiltschko and Wiltschko 1996).

Some birds, however, cross the equator during migration and then keep going. We might wonder, then, how a bird from northern regions that crosses the equator can continue to fly south in the southern hemisphere. To continue flying in the same geographical direction when the equator is crossed, the birds must reverse their migratory direction with respect to the inclination compass: they must now fly "poleward" instead of "equatorward." Experience with the horizontal magnetic field around the equator is the switch that causes the birds to begin flying "poleward" (Wiltschko and Wiltschko 1996).

The sensitivity of the magnetic compass of birds corresponds to the strength of the earth's magnetic field. A bird generally does not respond to magnetic fields that are much stronger or weaker than that which is typical in the area where it has been living. In fact, the range of intensities to which a bird may respond on a given day is usually narrower than those that it might experience during migration. However, it seems that the range of sensitivity may be adjusted by exposure to a field of a new strength for a period of time. Thus, responsiveness may be fine-tuned during migration (Wiltschko 1978; Wiltschko and Wiltschko 1999).

The Magnetic Compass of Sea Turtles

Some sea turtles travel tens of thousands of kilometers during their lifetimes, a feat that can require continuous swimming for periods of several weeks, with no land in sight. As a loggerhead sea turtle, *Caretta caretta* (Figure 10.20), makes its way across the featureless Atlantic Ocean from the coast of Florida (perhaps to the Sargasso Sea and back), it is guided by the earth's magnetic field (Lohmann and Lohmann 1992). The hatchlings swim toward *magnetic* northeast in the normal geomagnetic field and continue to do so when the field is experimentally reversed (Figure 10.21) (Lohmann 1991). And, similar to a bird's magnetic compass, that of the sea turtle is based on the inclination of the magnetic lines of force (Light et al. 1993). Indeed, the magnetic compass of sea turtles has many of the characteristics of the avian magnetic compass.

FIGURE 10.20 **A hatchling loggerhead sea turtle. These turtles may use the earth's magnetic field to guide their travels through the open ocean.**

A sea turtle begins its journey immediately after hatching. It uses local cues to head toward the ocean. When sea turtle hatchlings first enter the ocean, they simply swim into the waves to maintain an offshore heading. Near the shore, the waves come directly toward land, so swimming into the waves takes the turtles out to sea. The course that is initiated by swimming into the waves is later transferred to the magnetic compass.

In the open ocean, waves can no longer serve as a navigational cue because they can come from any direction. Here, sea turtles maintain the same angle with the magnetic field that they assumed while swimming into the waves. In this way, they stay on course. Simultaneous experience with both cues seems to be important. This was revealed in an experiment in which hatchling loggerhead sea turtles swam into surface waves in tanks for either 15 or 30 minutes. Their orientation was then tested in still water and in a magnetic field. Only those hatchlings with 30 minutes of experience swimming into waves in a magnetic field were able to maintain their orientation in still water (Goff et al. 1998).

POSITIONAL INFORMATION FROM THE EARTH'S MAGNETIC FIELD: A MAGNETIC MAP?

As we have seen, true navigation requires not only a compass but also a map. The map is necessary to know one's position relative to the goal, and then a compass is needed to guide the journey in a homeward direction. Kenneth and Catherine Lohmann (2006; Lohmann et al. 2007) caution that the magnetic maps of animals have not been fully characterized and may function in a very different way than human maps do. Investigation of magnetic maps has been hampered because there is no standard definition of the term *map* among researchers. For some researchers, a map requires a mental image—an internal spatial representation—of the region, but that view is increasingly giving way to a broader view of a map. For example, by the Lohmanns' definition, an animal has a **magnetic map** if it can obtain positional information from the earth's magnetic field, that is, if the animal can use the earth's magnetic field to determine its position relative to a target or goal. In this construct of a magnetic map, the map may be inherited or learned and specific or very general. We will use the Lohmanns' definition of a magnetic map in this text.

What features of the earth's magnetic field could provide positional information? As we have seen, the angle of inclination varies predictably with latitude, so an animal that could detect this feature could determine whether its position is north or south of the goal. If an animal could detect the intensity of the total magnetic field, the horizontal component of the field and/or the vertical component of the field, it could determine its

a

b

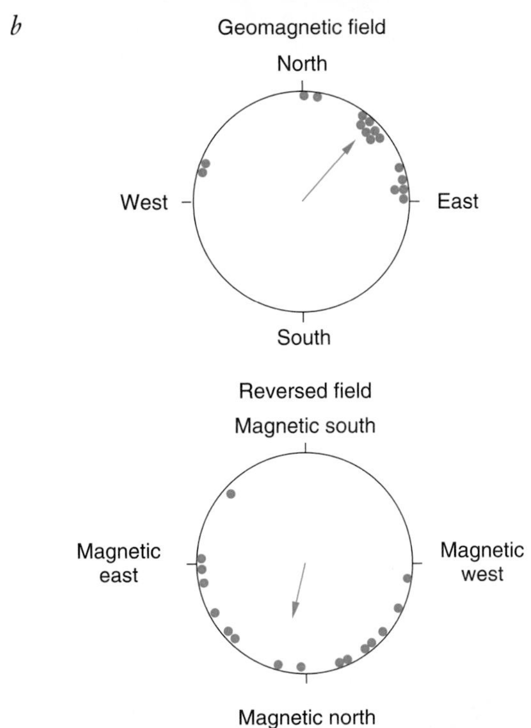

FIGURE 10.21 **A demonstration of the ability of loggerhead sea turtle hatchlings to orient to magnetic fields. (*a*) A sea turtle is harnessed in a small tank so that its swimming direction can be determined. A coil that can alter the magnetic field experienced by the turtles surrounds the tank. (*b*) When exposed to the earth's magnetic field, the turtles orient toward magnetic northeast. When the field is reversed, the hatchlings still orient to magnetic northeast, even though this is in the opposite geographic direction. (From Lohmann 1991.)**

position relative to the goal. Declination (the difference between geographic north and magnetic north) also varies in a regular pattern and could potentially be used as a clue to position. We will see that animals can use cues from the earth's magnetic field to navigate, but the cues used may differ among animals or as an animal ages (Lohmann and Lohmann 2006; Lohmann et al. 2007).

Magnetic Signposts

The magnetic "maps" of some animals may consist of inherited responses to magnetic landmarks, or signposts, that trigger changes in direction. We see such magnetic triggers along the migratory pathways of certain birds, for instance, the pied flycatcher. The Central European population of pied flycatchers first flies southwest to Iberia and then southeast. The change in migratory direction allows the birds to avoid the Alps, Mediterranean Sea, and the central Sahara (Figure 10.22).The birds have an inherited program that causes them to change migratory direction when they experience a magnetic field characteristic of key geographical locations at the appropriate time. Flycatchers held in captivity will flutter their wings and head in the correct migratory direction when they are exposed to a magnetic field characteristic of Frankfurt, Germany, where their free-flying comrades begin their migration. If captive flycatchers are then exposed to the magnetic field characteristic of Iberia, where the migrating flycatchers change direction, the captive flycatchers shift the direction of their fluttering to southeast. Captive flycatchers who continue to experience the same magnetic field throughout the migratory time period or who experience the magnetic field characteristic of the end point do not appropriately shift direction. Thus, the local magnetic field of Iberia acts as a signpost telling the migrating birds to shift flight direction slightly to the left (Beck and Wiltschcko 1988; Wiltschko and Wiltschko 2005).

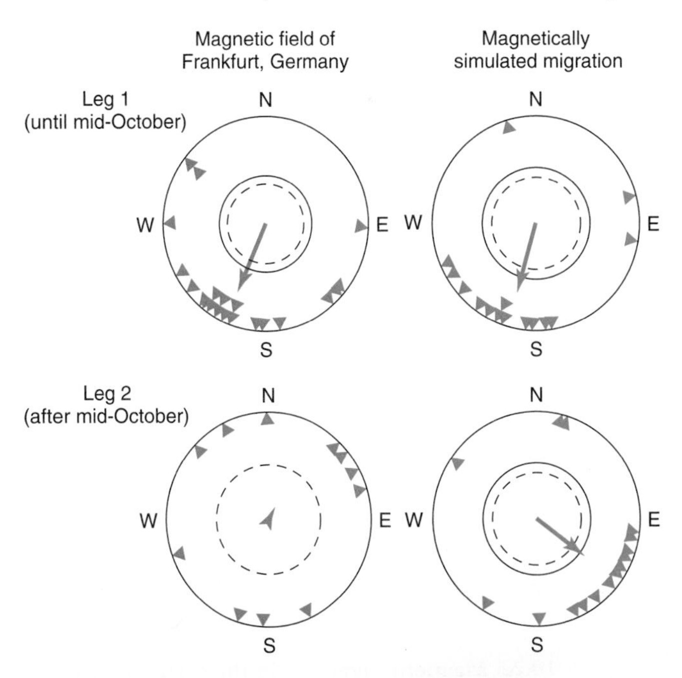

FIGURE 10.22 **Orientation of young pied flycatchers held in captivity during their first migration and exposed to magnetic fields typical of those along the route. Only the birds exposed to the correct magnetically simulated journey oriented properly. Each triangle represents the direction in which a bird oriented. The arrow indicates the mean direction of all birds. (Modified from data of Beck and Wiltschko 1988.)**

Magnetic signposts also trigger changes in swimming direction during the open-sea navigation of sea turtles. When loggerhead hatchlings are exposed to a magnetic field typical of northern Florida, they swim east-southeast using the earth's magnetic field as a compass. This heading will bring the sea turtles to the Gulf Stream, which will lead them to the North Atlantic gyre, a circular current that flows clockwise around the Sargasso Sea. Young loggerheads remain in the warm, rich water of this gyre for five to ten years.

These inherited orientation responses to magnetic fields help to keep the young loggerheads from straying out of the gyre (reviewed in Lohmann et al. 2008). This was demonstrated by recording the preferred swimming direction of hatchling loggerheads that had never been in the ocean. The turtles were exposed to magnetic fields characteristic of three widely separated regions along the migratory route of the North Atlantic gyre. The young loggerheads oriented to each field by swimming in a direction that would keep them in the favorable waters of the gyre if they had been migrating (Figure 10.23).

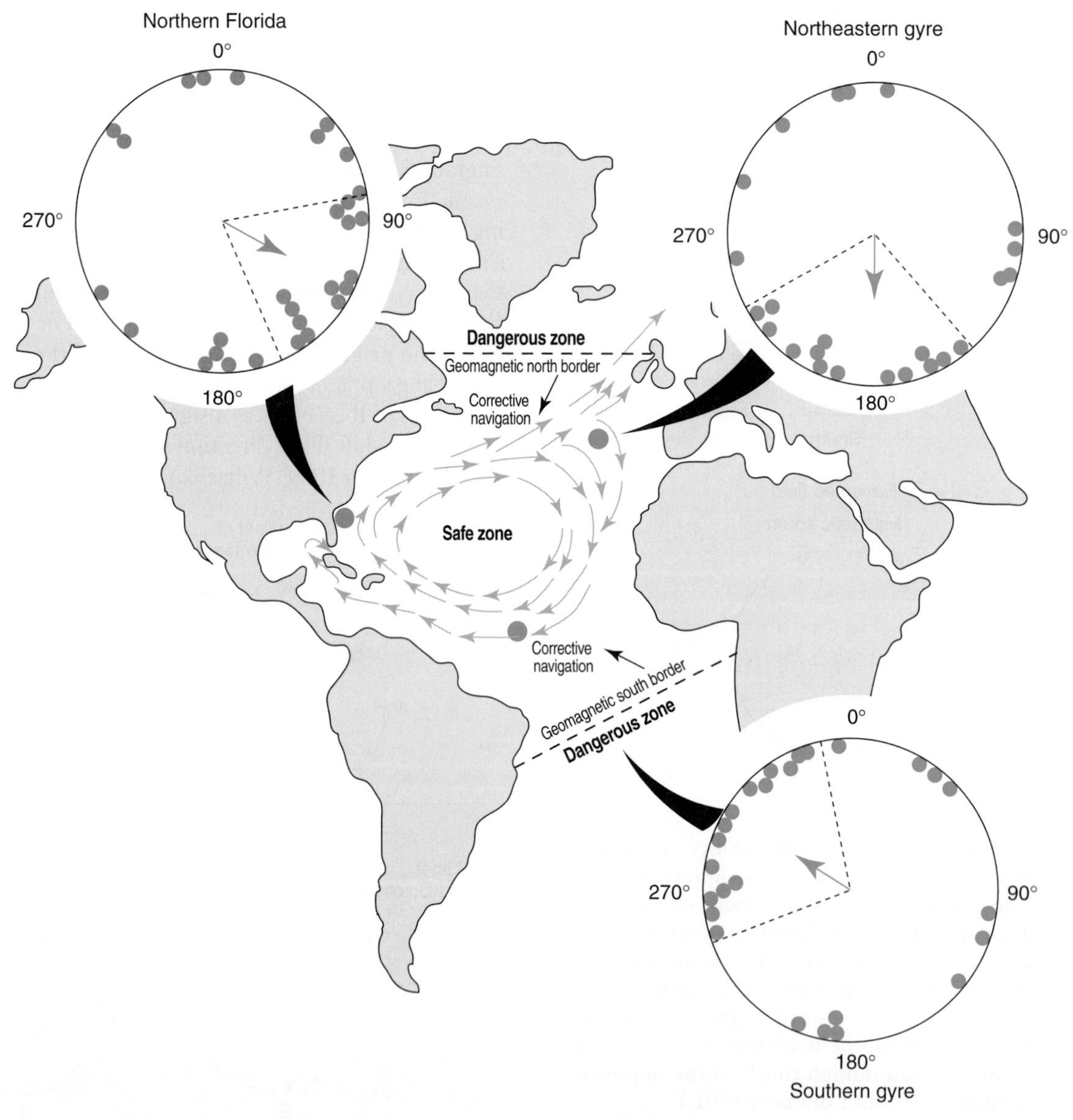

FIGURE 10.23 **Magnetic signposts in the earth's magnetic field may direct juvenile sea turtles in the proper direction to remain within the North Atlantic gyre, a circular current in the Sargasso Sea. The arrows in the ocean indicate the direction of the major currents of the gyre. Juvenile sea turtles normally swim within the gyre for several years. In the laboratory, juvenile sea turtles exposed to magnetic fields characteristic of three locations along the migratory route preferred to swim in the direction that would keep them swimming within the gyre if they had been migrating. The arrows leading to each circle show the location of the magnetic field to which the turtles were exposed. Each dot indicates the direction in which a harnessed juvenile sea turtle swam. The arrow in the center indicates the mean swimming bearing. (Modified from Lohmann et al. 2001.)**

Thus, hatchling loggerheads are programmed to swim in a particular direction when they encounter magnetic fields found in critical regions of the gyre—places where leaving the gyre would lead the juveniles to unfavorable waters. Regional differences in earth's magnetic field serve as navigational beacons that guide the open-sea migration of young loggerheads, without the turtles having a conception of their geographic position or their position relative to a goal (Lohmann et al. 2001).

Position Relative to Goal

Certain animals may use an aspect or aspects of earth's magnetic field as a map to locate their position relative to a goal. We do know some animals can detect both the inclination and the intensity of earth's magnetic field. Both of these features vary across the earth's surface, and they vary in different directions. Thus, animals could use either of these features to "know" the direction to the goal.

Some of the magnetic effects on pigeon homing seem to be more than interference with the magnetic compass and, therefore, may support the idea of a magnetic map. One example is the disorientation of pigeons released in magnetic anomalies, places where the earth's magnetic field is extremely irregular. Pigeons relying on the predictable changes in the geomagnetic field would become confused in areas where the field is abnormal. Some magnetic anomalies disorient pigeons even under sunny skies, when presumably they would be using the sun as a compass (Frei 1982; Frei and Wagner 1976; Wagner 1976; Walcott 1978). A perfect compass (the sun) cannot help if the map is messed up. This suggests that the geomagnetic field may be more than just a compass. As you can see in Figure 10.24, some birds released at magnetic anomalies appear to follow the magnetic topography, usually preferring the magnetic valleys, where the lower field strength is closer to home values.

In a more recent study, Todd Dennis and his colleagues (2007) equipped homing pigeons with GPS-based tracking devices and tracked their flight paths near places with magnetic anomalies. Regardless of the direction to home, the pigeons flew either parallel or perpendicular to the local lines with similar intensity of the geomagnetic field. The alignment of flight paths with magnetic intensity lines is interpreted as an indication that the pigeons can detect and respond to spatial variability of the geomagnetic field.

As a sea turtle matures, it learns the geomagnetic topography of specific areas and uses that information as at least part of the map it uses to locate an isolated target, such as a nesting beach (Lohmann and Lohmann 1996a, b). After spending several years swimming in the North Atlantic gyre, juvenile loggerhead turtles and green turtles (*Chelonia mydas*) that hatched along the eastern coast of the United States move toward the coastline to feeding sites. Certain sea turtles migrate along the east coast between summer feeding grounds in temperate regions and winter feeding grounds in the south. These juvenile turtles migrate to the same specific feeding locations each autumn and spring (Avens and Lohmann 2004). Every few years, adult sea turtles of nearly all species migrate from their feeding locations to nesting areas and back again. Adults of many populations return to nest on the same beaches where they hatched (reviewed in Lohmann et al. 2008).

How do sea turtles migrate with such precision? The earth's magnetic field provides a global positioning system that tells them their position relative to a goal. Kenneth Lohmann and colleagues (2004) demonstrated that juvenile and adult sea turtles use the geomagnetic field as a navigational map—a more complex use than hatchlings. The researchers captured juvenile green turtles from their feeding grounds located at about the midpoint of the eastern coast of Florida. The swimming

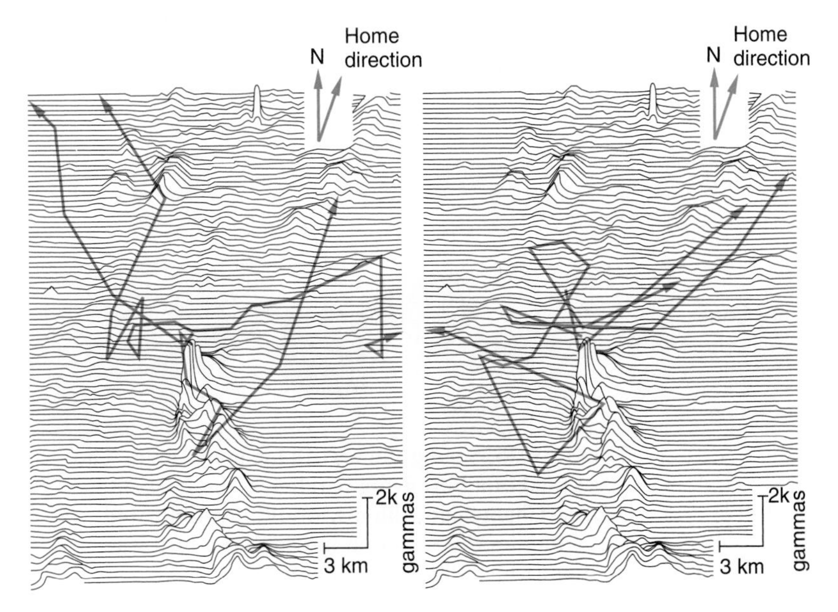

FIGURE 10.24 **The flight paths of pigeons in magnetic anomalies. In some places the geomagnetic field is highly irregular. Pigeons released in these areas may be completely disoriented, even on sunny days. The paths of these pigeons seem to follow the magnetic valleys, where the field strength is closer to the value at the home loft. (From Gould 1980.)**

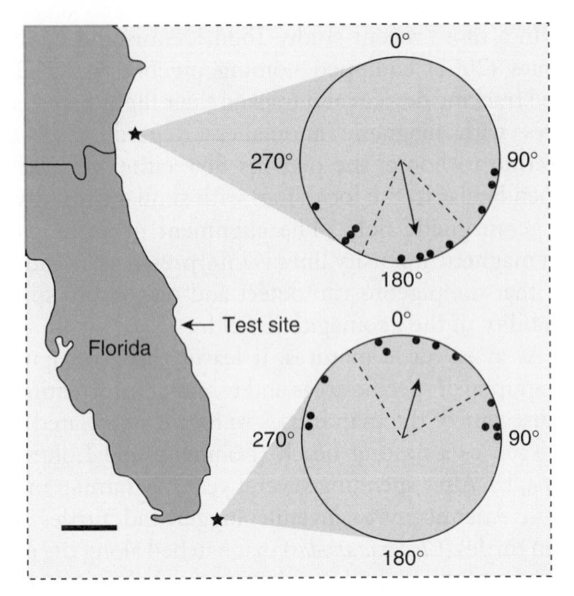

FIGURE 10.25 **As sea turtles mature, they use the earth's magnetic field to determine their location relative to home. Sea turtles return to the same feeding grounds every year. Researchers captured sea turtles at their feeding ground along the east coast of Florida. The preferred direction of swimming of each turtle (indicated by a black dot in the circle) was determined as previously described. The turtles were exposed to a magnetic field similar to the field that exists north of the site or to the field that exists south of the site. The sites are indicated by stars. The turtles swam in a direction that would return them to their feeding grounds (the test site) if they actually had been displaced. (From Lohmann et al. 2004.)**

direction of tethered turtles was monitored as in previous experiments. Turtles were then exposed to either a geomagnetic field that would be found 337 km north of the test site or a magnetic field that would be found 337 km south of the test site. Turtles exposed to a northern magnetic field swam approximately southward; those exposed to a southern magnetic field swam northward (Figure 10.25). The magnetic field may tell the turtle whether it is north or south of its goal. The turtle might then move in the appropriate direction until it encounters other cues, perhaps chemical, that identify the feeding grounds (reviewed in Lohmann and Lohmann 2006).

MAGNETORECEPTION

Humans do not sense magnetic fields—at least not consciously. We might wonder, then, how animals sense the earth's magnetic field. There are at least two types of magnetoreceptors. One type involves specialized photoreceptors and is light dependent. Thus, certain animals may "see" the earth's magnetic field. The basic idea of this light-dependent model of magnetoreception is that photoreceptor molecules absorb light better under certain magnetic conditions. Thus, the amount of light absorption also provides information about the local magnetic field. The second hypothesis involves magnetite, a magnetic mineral found in many animals that orient to the geomagnetic field. In this model, the magnetite responds to the earth's magnetic field. This response could then affect other sensory receptors, perhaps mechanoreceptors, open ion channels, or act on the cell physically.

Light-Dependent Magnetoreception

Because birds are the best-studied group, we will tell their story. We must add, however, that similar observations of a relationship between photoreception and magnetoreception have been discovered in other animals (Rozhok 2008).

What initial observations suggest that photoreception and magnetoreception are linked in birds? First, the magnetoreceptor is located in the eye, specifically the right eye. Second, birds cannot remain oriented to a magnetic field in darkness. Not only is light required, but it must be light of specific wavelengths. Birds usually require blue light to remain oriented to a magnetic field but may be able to orient in red light if they are given time to adjust (Wiltschko and Wiltschko 2006).

Cryptochrome, a photopigment involved in magnetoreception, stimulates the photoreceptors differently depending on the orientation of the magnetic field. Thus, it seems that migratory birds sense the magnetic field as a visual pattern (Figure 10.26) (Ritz et al. 2000). Unlike some photopigments, which change shape when they absorb light, cryptochrome uses photons to transfer electrons forming radical pairs (pairs or triplets of spinning electrons). The radical pairs lead to further reactions in a cascading pathway, and magnetic fields alter the functioning of radical pairs. Cryptochromes absorb blue-green light—the wavelengths important for magnetic orientation. In migratory birds, cryptochromes are pro-

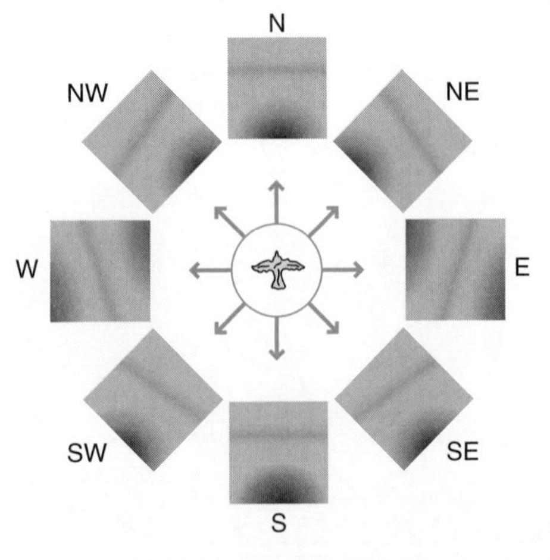

FIGURE 10.26 **Seeing the earth's magnetic field. The visual field of a bird flying parallel to the horizon in Urbana-Champaign, Illinois, would be modulated depending on the direction in which the bird was looking. (From Ritz et al. 2000.)**

duced (the genes for cryptochromes are active) at night, when many migrants are flying. Nonmigratory birds produce cryptochromes only during the day. The difference in the times of cryptochrome production suggests that all birds may need magnetic information during the day, but only night-flying migrants also need it at night. Notably, cryptochromes are found in the ganglion cells of a migratory garden warbler's retina and in large displaced ganglion cells, which project to brain areas where magnetically sensitive neurons have been reported and these areas show high levels of neuronal activity during magnetic orientation (Mouritsen et al. 2004).

The cryptochrome-containing cells of the retina connect to neurons in a brain region called Cluster N, where neurons are especially active when night-flying migratory birds are orienting to a magnetic field. The retina and Cluster N are connected via pathways through the thalamus, a brain region important for vision. Dominik Heyers and his colleagues (2007) demonstrated this connection using special dyes that can be traced as they travel along nerve fibers. They injected one type of tracer dye into the cryptochrome-containing cells of the retina and another type in the neurons in Cluster N. The tracers met in the thalamus, which supports the hypothesis that birds use their visual system to sense magnetic fields.

Magnetite

Many organisms known to have the ability to sense geomagnetic fields, including honeybees, trout, salmon, birds, and sea turtles, have deposits of magnetic material, magnetite, which often forms chains or clumps. In vertebrates, these deposits are commonly found in the head or skull. The magnetite crystals can twist into alignment with the earth's magnetic field if permitted to do so. Such movement might stimulate a stretch receptor.

If the magnetite deposits function as magnetoreceptors in larger organisms, the information they provide would have to be transmitted to the nervous system. Therefore, associations between magnetite and the nervous system are of particular interest. So far, the closest we have come to identifying the actual magnetoreceptor cells is in the rainbow trout (*Oncorhynchus mykiss*). Michael Walker and his colleagues first confirmed that the ophthalmic branch of the trigeminal nerve (a cranial nerve that carries sensory information from the front of the head) contains fibers that respond to magnetic fields. Then they used a special dye to trace these fibers both to the brain and to cells in the olfactory epithelium in the nose of the trout. These cells, the candidate magnetoreceptor cells, contain small amounts of a material thought to be magnetite (Walker et al. 1997).

In birds, magnetite deposits are found in the area of the upper beak. Interestingly, branches of the bobolink's trigeminal nerve appear to innervate the region in which magnetite deposits are found. These branches respond to earth-strength changes in the direction of the magnetic field (Semm and Beason 1990).

A popular way to demonstrate a role of magnetite in magnetoreception is to remagnetize the magnetite using a brief, strong magnetic pulse to the region of the animal where magnetite is located. If a strong magnetic pulse alters orientation, the conclusion is that magnetite is a part of the magnetoreceptor. In this way, researchers have demonstrated that the polarity compass of bats is based on magnetite (Holland et al. 2008).

In addition to their light-dependent inclination compass, birds have a magnetite receptor. Experiments on Australian silvereyes (*Zosterops l. lateralis*) provide an example (Holland et al. 2008). When adult silvereyes that were flying northward from Tasmania during their autumn migration were exposed to a strong magnetic pulse, their orientation was shifted clockwise by about 90° toward the east. Similar results were obtained when adult silvereyes were exposed to a strong magnetic pulse during the spring migration. These observations support the idea that a magnetite-based receptor plays a role in orientation, but they don't indicate whether it is involved in the compass sense or the map sense. However, when juvenile silvereyes are exposed to a magnetic pulse shortly after fledging, before they begin to migrate, the pulse had little effect on their orientation. The juveniles continued to orient in their normal autumnal migratory direction. Unlike adult migrants, which have established a navigational map during previous migrations, the juveniles rely on an innate migratory program that heads them in the appropriate compass direction for their first migration. Because a magnetic pulse disrupts orientation in adults but not in juveniles, it is thought that the earth's magnetic field is part of the navigational map of adults (reviewed in Wiltschko et al. 2005; Wiltschko and Wiltschko 2006, 2007).

Two Magnetoreceptor Systems

Recent studies aimed at exploring the physiological basis for magnetoreception support the idea that animals might have more than one type of magnetic sensitivity. As we have seen, there are two proposed mechanisms for magnetoreception, one light-dependent and the other based on magnetite. Table 10.2 presents mechanisms and their proposed functions.

Certain species seem to have both types of magnetoreception systems, each serving a different purpose. For example, the eastern red-spotted newt (*Notophthalmus viridescens*) uses a magnetic compass based on the inclination of the magnetic lines of force when orienting toward the shore. We know this because their orientation was shifted by about 180° when the vertical component of the magnetic field was inverted. These newts are also able to home, that is, to return to the point of origin after being moved to an unfamiliar location. During homing, the newt's orientation is unaffected by an inversion of the vertical component of the magnetic field (Phillips 1986), but is shifted

TABLE 10.2

	Photopigment-dependent magnetoreceptor	Magnetite magnetoreceptor
Feature of geomagnetic field detected	Inclination or polarity	Intensity
Tasks in which it is used in birds	Compass (direction finding)	Map (position, signpost, or trigger)
Site of reception	Retina of right eye	Upper beak and/or ethmoid region
Nerve	Optic nerve	Ophthalmic branch of trigeminal nerve
Brain structures involved	Nucleus of the basal optic root (nBOR); optic tectum	Trigeminal ganglion

by a change in polarity (Phillips 1987). Thus, these initial observations suggest that, in the newt at least, the mechanism(s) for magnetoreception involved in homing differs from the one involved in shoreward compass orientation.

The magnetic compass used by the eastern newt when orienting toward the shore is light-dependent (Phillips and Borland 1992). The orientation of newts during homing is also affected by exposure to different wavelengths of light. However, the effects of long wavelengths on homing are different from those on shoreward orientation. Furthermore, light-dependent processes are not expected to respond to the polarity of a magnetic field, and we know that a newt's homing ability is sensitive to polarity changes. This again suggests two magnetoreception mechanisms in newts (Phillips and Borland 1994).

Migratory birds may also have two mechanisms of magnetoreception that serve different functions. The light-dependent mechanism serves as a magnetic compass (Ritz et al. 2009; Rodgers and Hore 2009). Because a magnetite-based mechanism is theoretically capable of detecting minute variations in the earth's magnetic field, it may be part of the magnetic "map" receptor. To use the geomagnetic field as a map, an animal might merely compare the local intensity of the field with that at the goal. A receptor system used in a map sense, then, would not have to respond to the direction of the field, but it would be expected to respond to slight variations, less than 0.1%, in the intensity of the magnetic field experienced. The amount of magnetic material typically found in pigeons' skulls could comprise a receptor that would provide enough sensitivity to small differences in magnetic field to fit the bill. A comparison of the effects of a strong magnetic pulse on the orientation of juvenile and adult Australian silvereyes supports the idea that a magnetite-based receptor system is part of a "map." It is commonly believed that whereas adult migrants have established a navigational map, juveniles have not. As we have seen, the orientation of adult silvereyes is shifted by a magnetic pulse, presumably because their navigational map was affected. In contrast, the juvenile silvereyes remained oriented in the appropriate migratory direction after a magnetic pulse. The magnetic pulse may not affect the orientation of juveniles because they have not yet formed a magnetic map. Instead, their orientation was based on an innate migratory program. They use their magnetic compass, which is based on a light-dependent magnetoreception process, to head in the appropriate direction according to their inherited migratory program (reviewed in Wiltschko et al. 2005; Wiltschko and Wiltschko 2006, 2007).

CHEMICAL CUES

In this section, we will focus on the use of olfactory cues for orientation during homing. We will discover that salmon are guided to the stream where they hatched by chemical landmarks, and we will examine the more recent suggestion that pigeons also use olfactory cues when homing.

OLFACTION AND SALMON HOMING

One of the most remarkable stories in the annals of animal behavior concerns the travels of the salmon. Salmon hatch in the cold, clear freshwater of rivers or lakes and then descend from the streams that flow from those areas and swim to sea, fanning out in all directions. Once they reach the ocean, depending on the species, they may spend one to five years there until they reach their breeding condition. Now large, glistening, beautifully colored creatures, they head from their feeding grounds back through the trackless sea to the very river from which they came. When they reach the river, they swim upstream, turning up the correct tributary until they reach the very one where they spent their youth.

Wild salmon return to the specific location of the natal stream in which they were born with remarkable precision. Thomas Quinn and his colleagues (2006) demonstrated this site fidelity by using temperature changes during incubation of prehatch sockeye salmon embryos to cause banding patterns on the ear bones of the fish. These banding patterns marked fish for later identification. The researchers chose a pond associated with Hansen Creek in southwestern Alaska as the site where the embryos would emerge and buried the

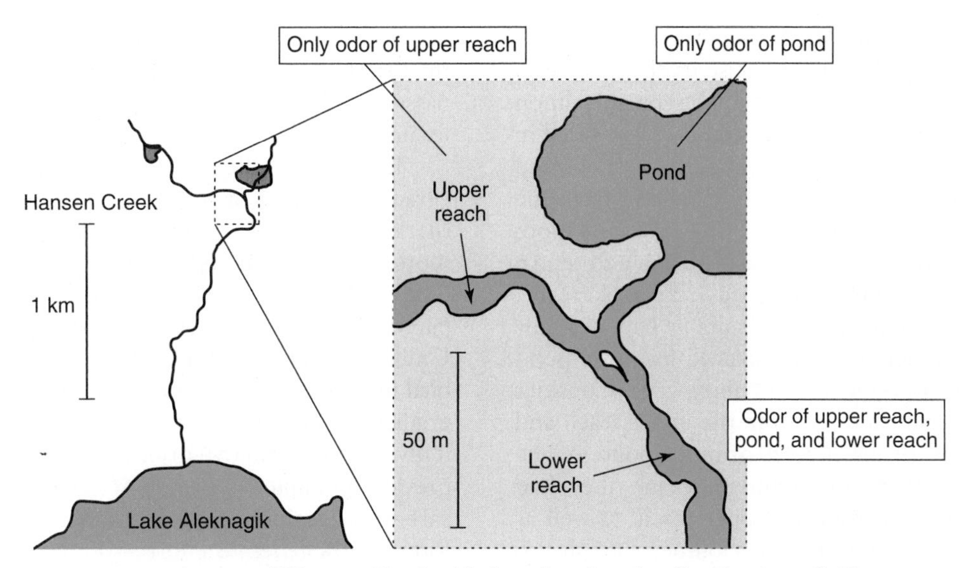

FIGURE 10.27 **A map of Hansen Creek, Alaska, showing the distribution of olfactory cues in different regions of the creek area.**

embryos at the bottom of the pond (see the map in Figure 10.27). The embryos emerged, migrated to the sea along with unmarked fish from the Hansen Creek area, and then migrated back to the creek. Sockeye salmon die after spawning. The carcasses of salmon along the creek and in the pond were examined for banding patterns on the ear bones. Of the 324 salmon carcasses in the pond, 12 were marked, but none of the 138 carcasses found in the creek were marked. Thus, the marked salmon returned to the site of their incubation—the pond associated with Hansen Creek.

Although navigation in the open seas appears to depend on the integration of several sensory cues, including magnetism (Lohmann et al. 2008), sun compass, polarized light, and perhaps even odors, navigation up the rivers is based primarily on olfactory cues (reviewed in Dittman and Quinn 1996). According to the olfactory hypothesis of salmon homing, young salmon learn the odors of the home stream before they begin their downstream migration (Hasler and Wisby 1951). The odor of the home stream is most likely the particular mixture of amino acids in the water (Shoji et al. 2000; Shoji et al. 2003; Yamamoto and Ueda 2007). After spending time at sea, the salmon return to the coast and use olfactory cues to locate the mouth of the river in which they hatched. During their upstream migration, the salmon follow a chemical trail back to the tributary where they hatched. When they come to a fork in the river, they may swim back and forth across the two branches. If they mistakenly swim up the wrong branch and lose the scent of the home stream, they retreat downstream until the scent is encountered again. Then, they usually take the correct route. Researchers have hypothesized such nondirect homing (choosing the wrong tributary and returning to the fork to choose another) for many years, but it has only recently been verified. Radio-tagged spring-summer Chinook salmon (*Oncorhynchus tshawyscha*) tracked in the Columbia River system that chose the wrong branch of the river returned to the fork and swam up another branch (Keefer et al. 2008).

Sensory deprivation experiments have demonstrated the importance of olfaction in salmon homing. Blinding the fish had no effect, but plugging their nasal cavities impaired their ability to home correctly. Coho salmon (*Oncorhynchus kisutch*) were trapped shortly after they had made their choice of forks in a Y-shaped stream. The nasal cavities of half of those caught in each branch were plugged. The other half were untreated. All the fish were then released downstream from the fork and allowed to repeat their upstream migration. Whereas 89% of the control fish returned to the branch where they were originally captured, only 60% of the fish with nose plugs made the correct choice (Wisby and Hasler 1954). In another study, a fish with its nose plugged swam with others of its kind to the opening of its home pond. However, unable to smell the special characteristics of its home waters, it did not enter the pond (Cooper et al. 1976).

Olfactory cues, not qualities of the habitat, guide salmon to their birthplace. These conclusions are consistent with a study done on sockeye salmon in Hansen Creek in Alaska. As you can see in Figure 10.27, Hansen Creek has an upper and a lower reach (a reach is the region of a river or creek between two bends). It is also fed by water from a pond (the pond in which marked fish hatched in the study illustrating site fidelity discussed earlier). During the spawning season, salmon were collected and tagged from both reaches of Hansen Creek and from the pond. The olfactory cues available at these sites differed. The upper creek had only the odor of the upper creek, and the pond had only the odor of the pond. The pond was a better quality area because its characteristics made predation on salmon less likely. The control salmon were released at their capture site, and they remained in that immediate area. The experimental fish were released at a site other than their capture site. Salmon from the pond that were released in the lower reach, where more olfactory cues were available, were more likely to return to the pond than were salmon released in the upper reach. But recall that the pond is a more suitable habitat. How do we know whether the fish displaced from the pond returned because of odor cues or habitat cues? Consider the behavior of fish captured from the upper reach and released either in the lower reach or in the pond. As the displaced upper-reach fish swam upstream, they had olfactory information from the upper reach, as well as from the pond. During the upstream journey, the fish also assessed habitat-quality cues. Most of the homing salmon bypassed the habitat- quality cues from the pond and followed olfactory cues to the upper reach. Most upper-reach fish displaced to the pond stayed in the pond; they did not have olfactory cues to guide them back to the upper reach (Stewart et al. 2004).

STOP AND THINK

What would you have concluded if fish from the upper reach that were released in the pond had stayed in the pond?

OLFACTION AND PIGEON HOMING

No one denies that olfactory cues are of paramount importance during the upstream migration of salmon, but the role olfaction plays in pigeon homing has been controversial (Wallraff 2004, 2005). Let's look at the evidence.

Models of Avian Olfactory Navigation

Two models for olfactory navigation have been suggested. According to Floriani Papi's "mosaic" model, pigeons form a mosaic map of environmental odors within a radius of 70 to 100 km of their home loft. Some of this map would take shape as the young birds experienced odors at specific locations during exercise and training flights. More distant features of the map would be filled in as wind carried faraway odors to the loft. One odor might be brought by wind from the north and another by wind from the east. The bird would associate each odor with the direction of the wind carrying it. When the wind shifted direction, the odors that arrived first would be closer than those that took longer to arrive (Papi et al. 1972). For instance, a hypothetical pigeon might learn that the sea is to the west, an evergreen forest is south, a large city is north, and a garbage dump is east. If the bird in this example smelled pine needles at its release site, it would assume that it was in the forest south of its loft and would use one of its compasses, perhaps the sun or the earth's magnetic field, to fly north.

Hans G. Wallraff (1980, 1981) has suggested a "gradient" model of olfactory navigation that assumes that there are stable gradients in the intensity of one or more environmental odors. Then, wherever it was, the bird would determine the strength of the odor and compare it to the remembered intensity at the home loft. Unlike the mosaic model, which requires only that the bird make qualitative discriminations among odors, the gradient model demands that the bird make both qualitative and quantitative discriminations. Reconsider the previous example. The smell of the ocean might form an east-west gradient, and the fragrance of the evergreen forest might generate a north-south gradient. If the bird in the previous example smelled the air at a release site and determined that the scent of the sea was stronger but the smell of the forest was weaker than at the home loft, it would determine that its current position was northwest of home.

Tests of the Models

These models of olfactory navigation have stimulated intensive research, and it is becoming clear that odors are important in the navigation of homing pigeons. Let us see how different researchers have approached the question.

Distorting the Olfactory Map A method of testing olfactory hypotheses is to manipulate olfactory information to distort the bird's olfactory map. This has been done by deflecting the natural winds to make it seem that odors are coming from another direction. The deflector lofts used in these experiments typically have wooden baffles that shift wind flow in a predictable manner (Figure 10.28). For instance, wind from the south might be deflected so that it seemed to come from the east. A pigeon in this loft would form an olfactory map that was shifted counterclockwise by 45°. When it was released south of its loft, we would expect it to interpret the local odors as being east of its loft and fly west to get home.

Deflector loft experiments have shown consistent shifts in the orientation of homing pigeons (Baldaccini et al. 1975; Kiepenheuer 1978; Waldvogel et al. 1978). However, there are reasons to believe that the shift in orientation observed in pigeons from deflector lofts might be due to something other than a distorted olfactory map. We would expect pigeons that were temporarily prevented from smelling at the time of their release to be unable to read their olfactory map and to orient randomly. But this is not the case: the orientation of smell-blind (anosmic) pigeons from deflector lofts is

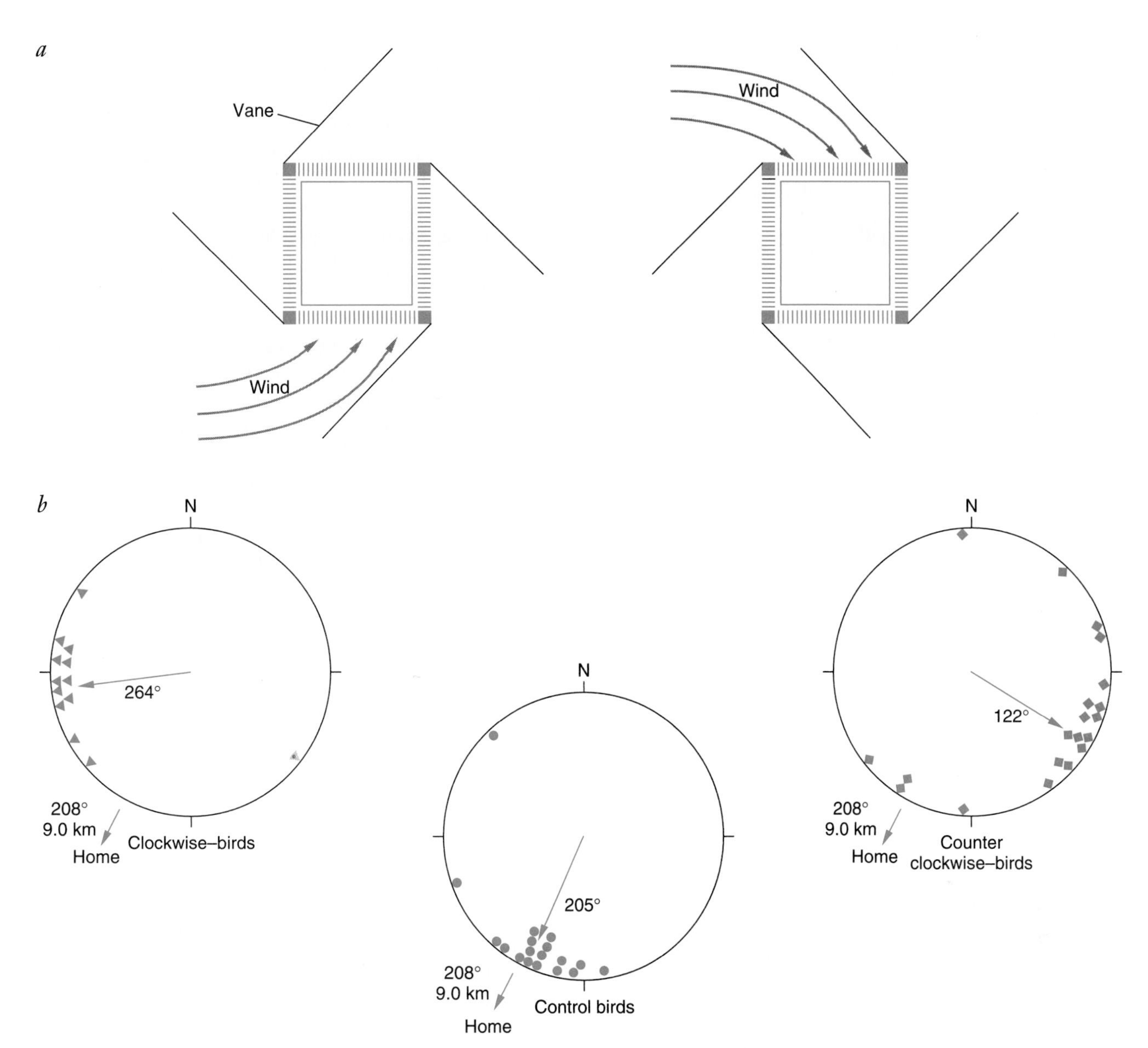

FIGURE 10.28 **Deflector lofts shift the orientation of pigeons. (*a*) Deflector lofts have baffles that shift the apparent direction of the wind by 90°. Pigeons living in deflector lofts should form shifted olfactory maps. (*b*) The vanishing directions of these pigeons are shifted by about 90°. The dots at the periphery of the circle denote the direction in which the pigeon flew out of sight. The arrow within the circle indicates the mean bearing of all birds. Although a shift in orientation is reported in all deflector loft experiments, it may be due to the deflection of light rather than a shift in the olfactory map. (Data from Baldaccini et al. 1975.)**

still shifted (Kiepenheuer 1979). Accordingly, it has been concluded that the baffles in these lofts also deflect sunlight and that the consistent shift in pigeon orientation is caused by an alteration in the sun compass (Phillips et al. 2006).

Manipulating Olfactory Information Although the interpretation of olfactory deprivation and deflector loft experiments is quite controversial, the experiments in which olfactory information predictably alters the orientation of pigeons remain as unshaken support for an olfactory hypothesis. For example, the orientation of pigeons was influenced by their experience with an unnatural odor, benzaldehyde (Figure 10.29). Pigeons were kept in lofts where they were fully exposed to the wind. The experimental birds were exposed to an air current coming from a specific direction and carrying the odor of benzaldehyde in addition to the natural breezes. We would expect these pigeons to incorporate this information into their olfactory maps. The control birds were exposed to only the natural winds, so they would not have an area with the odor of benzaldehyde

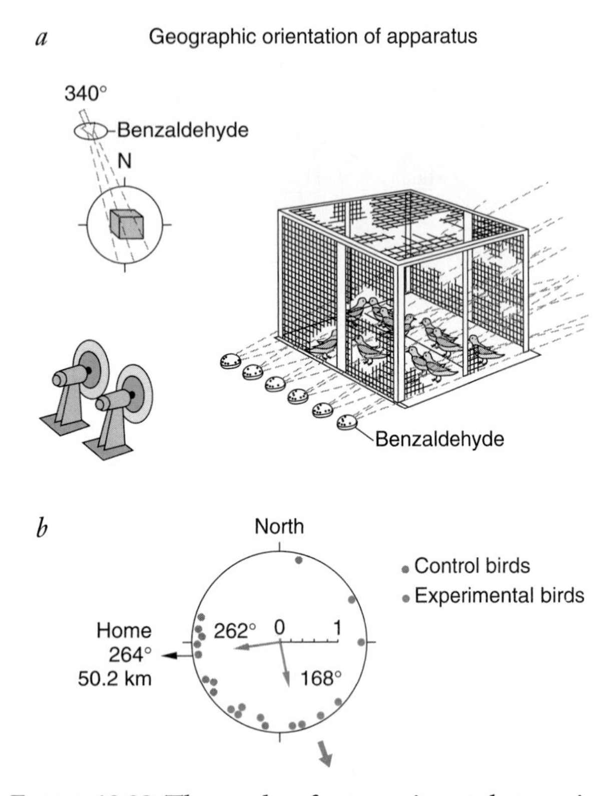

FIGURE 10.29 **The results of an experiment that manipulated a pigeon's olfactory information. (*a*) The experimental pigeons were kept in a loft that was exposed to natural odors, as well as to a breeze carrying the odor of benzaldehyde from a source northwest of the loft. Control birds were exposed to only natural odors. While they were transported to the release site, all birds were exposed to the odor of benzaldehyde. (*b*) The orientation of the experimental birds, but not the control birds, was altered by exposure to benzaldehyde. The initial orientation of control birds was homeward. However, the initial orientation of experimental birds was toward the southeast, as would be expected if they had interpreted the odor of benzaldehyde as an indication that the release site was northwest of the loft. The experimental birds oriented as if they formed an olfactory map containing an area with the odor of benzaldehyde. (Data from Ioalé et al. 1990.)**

in their olfactory map. All the birds were exposed to benzaldehyde while they were transported to the release site and at that site. The experimental birds took off in a direction opposite to that from which they had experienced benzaldehyde at the loft. In other words, they oriented as if they used an olfactory map that contained an area scented with benzaldehyde. If the release site did not smell of benzaldehyde, the experimental birds were homeward oriented. The control birds were not confused by the smell of benzaldehyde at the release site and flew home. Since benzaldehyde was not part of their olfactory map, they did not associate it with a particular direction. They used other cues to guide them home (Ioalé et al. 1990).

Depriving Birds of Their Sense of Smell Another approach in testing olfactory hypotheses is to deprive the pigeon of its sense of smell and observe the effect on its orientation and homing success. These anosmic pigeons are less accurate in their initial orientation, and fewer return home from an unfamiliar, but not from a familiar, release site. Regardless of its effect on orientation, olfactory deprivation always delays the bird's departure from the release site (Able 1996). These results are consistent with the idea that olfaction plays an important role in pigeon homing.

Besides its effect on the pigeon's sense of smell, perhaps olfactory deprivation affects another behavior, one not primarily controlled by olfaction, and this other behavior alters homing performance. Suppose the procedures that impair the sense of smell also affect the pigeons' motivation or their ability to process information. Though possible, the evidence does not support these possibilities. Anosmic pigeons home as well as control pigeons when they are released from familiar sites. Thus, the procedures do not seem to affect the birds' motivation to return home. Furthermore, pigeons whose sense of smell is temporarily blocked by an application of zinc sulfate to the olfactory epithelium have problems in returning home from unfamiliar locations, but they perform as well as controls in a spatial memory task that does not involve homing (Budzynski et al. 1998).

Could it be that some other sense, say sensitivity to magnetism, is blocked along with olfaction? The discovery of magnetite deposits in the beaks of homing pigeons, which are thought to be magnetoreceptors, makes this an intriguing possibility (Tian et al. 2007). Recall that information from the magnetite magnetoreceptors travels to the brain over the trigeminal nerve. Information about odors travels to the brain over the olfactory nerve. To evaluate the relative importance of magnetic and olfactory information, Anna Gagliardo and her colleagues (2006) severed the trigeminal nerves of one group of pigeons to deprive the pigeons of magnetic information and severed the olfactory nerves of another group of pigeons to deprive the pigeons of olfactory information. A control group of pigeons underwent sham surgery, in which the pigeons underwent similar surgical procedures as the experimental birds but the nerves were not severed. None of the pigeons had experience outside of its loft. The pigeons were released more than 50 km from home. As you can see in Figure 10.30, the initial orientation of the released pigeons of the sham-operated control group and the group that had the trigeminal nerve severed was in the general direction of home. In contrast, the initial orientation of pigeons with severed olfactory nerves was in the opposite direction. Furthermore, the number of pigeons that returned home within 24 hours (23 out of 24) was the same in the sham-operated control group and the experimental group without input from their

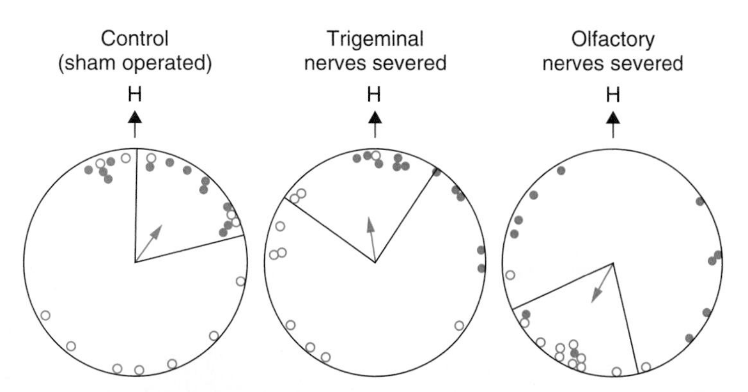

FIGURE 10.30 **Information from olfactory receptors is necessary for homing pigeons to return from unfamiliar locations, but input from magnetoreceptors is neither necessary nor sufficient for homing ability. The dots within the circles represent the vanishing direction of each pigeon in each. The arrow within the circle indicates the mean vector of the group's vanishing direction. The arrow outside the circle pointing to H indicates the home direction. Sham-operated pigeons and pigeons with the trigeminal nerve cut headed in the direction of home, but pigeons with the olfactory nerve cut headed away from home. (Gagliardo et al. 2006.)**

magnetite receptors. However, only 4 of the 24 pigeons lacking olfactory information made it home. These results are consistent with the hypothesis that olfactory cues are more important than magnetic cues in a homing pigeon's navigational map.

ELECTRICAL CUES AND ELECTROLOCATION

Electrical cues have a variety of potential uses for those organisms that can sense them. As we will see in Chapter 12, certain predators use the electrical cues given off by living organisms to detect their prey. In addition, electrical fields generated by nonliving sources, such as the motion of great ocean currents, waves and tides, and rivers, could provide cues for navigation. Although there is currently no evidence that migrating fish such as salmon, shad, herring, or tuna are electroreceptive, there is some evidence that electrical features of the ocean floor may help guide the movements of bottom-feeding species such as the dogfish shark (Waterman 1989).

Although most living organisms generate weak electrical fields in water, only a few species have electric organs that generate pulses, creating electrical fields that can be used in communication (discussed in Chapter 16) and orientation (reviewed in Caputi and Budell 2006). The electric organs of weak electric fish (mormyriforms and gymnotidforms), located near their tail, for instance, generate a continuous stream of brief electrical pulses. The result is an electrical field around the fish in which the head acts as the positive pole and the tail as the negative pole. Nearby objects distort the field, and the distortions are detected by numerous electroreceptors in the lateral lines along the sides of the fish. A weakly electric fish generally keeps its body rigid, a posture that simplifies the analysis of the electrical signals.

These fish examine their surroundings by using their electrical sense. Since they live in muddy water, where vision is limited, and since they are active at night, electrolocation is quite useful. Objects whose electrical conductivity differs from that of water disturb this electrical field. An object with greater conductivity than that of water—another animal, for instance—directs current toward itself. Objects that are less conductive than water, such as a rock jutting into its path, deflect the current (Figure 10.31). Thus, the fish can distinguish between living and nonliving objects in its environment.

The distortions in the electrical field create an electrical image of objects that can tell the fish a great deal about its environment. The distortion varies according to the location of the object relative to the fish, so the location of the image on its skin tells the fish where in relation to its own body the object is located. If the distortion is greatest on the right, the object is located on the right. An object near the fish's head creates the greatest distortion near the head (Caputi et al. 1998). The degree to which the electrical field is distorted by an object (the amplitude of the image) is greater in the center of the image than at the periphery. The fish often performs a series of movements close to the object under investigation. These actions might provide sensory input that helps the fish determine the object's size or shape (von der Emde 1999).

Electric fish can even measure the distance of most objects accurately, regardless of the object's size, shape, or material of which it is composed. In contrast to a visual image, the size (width) of an electrical image

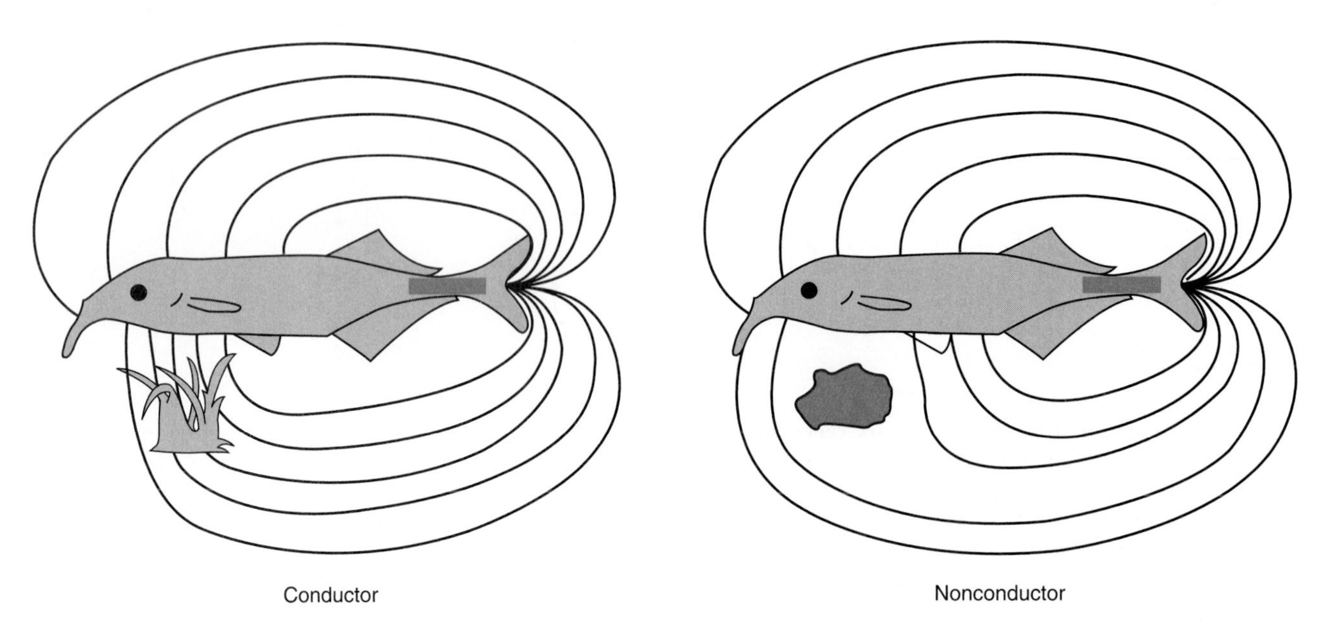

FIGURE 10.31 **Electroreception. The electrical field generated by this fish is distorted by nearby objects. A good conductor, such as another living organism, draws the lines of force together. A nonconductor, such as a rock, spreads them out. Using electroreceptors distributed over its body surface, the fish senses the changes in the electrical field to "picture" its environment. (From von der Emde 1999.)**

increases with distance. In addition, the amplitude differences between the center and the edges of an electrical image become smaller with the increasing distance of the object (Figure 10.32). The fish uses both of these features—size and amplitude—together to determine the distance of an object. A large, nearby object might cast the same-sized image as a smaller, distant object, but the more distant object would have smaller amplitude differences between the central and outer areas of the image. The electrical images of a 2-cm cube of metal or plastic presented at different distances and measured along the midline of an electric fish are shown in Figure 10.33 (von der Emde 1999).

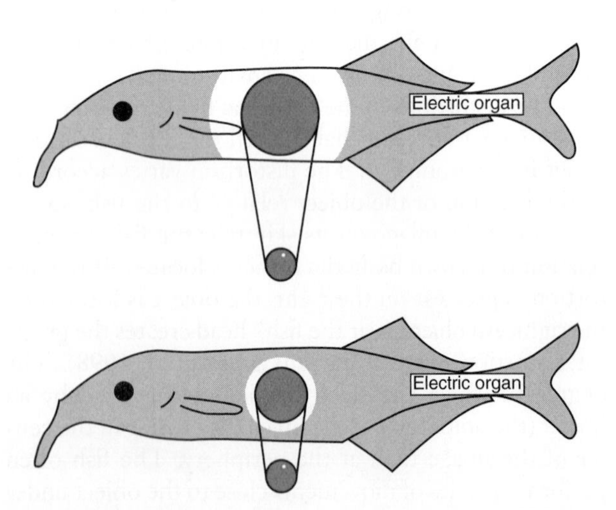

FIGURE 10.32 **The electrical image of a metal sphere at different distances from the fish. The size (width) of the electrical image increases with distance. The amplitude differences in the degree of distortion of the electrical field between the center and the periphery of the electrical image decrease with increasing distance. (From von der Emde 1999.)**

SUMMARY

Navigational strategies can be grouped into three levels. One level of orientation, called piloting, is the ability to locate a goal by referring to landmarks. A second level is compass orientation, in which an animal orients in a particular compass direction without referring to landmarks. This is the type of navigation used by most bird migrants. A young bird migrant uses its compasses for vector navigation, an inherited program that tells the bird to fly in a given direction for a certain length of time. Some animals use a compass in path integration: they memorize the sequence of direction and distance on the outward journey to determine their location relative to home, and then they use a compass to travel directly home. A third level of navigational skill describes an animal's ability to locate the goal without the use of landmarks, even if it is released in an unfamiliar location. True navigation requires a map to determine location and a compass to guide the journey.

Animals have access to and use many different cues for orientation and navigation. The sensory modality of the primary cue varies among species, and many species have a hierarchy of cues. Although the interactions among cues can be complex, we have considered each sensory basis separately.

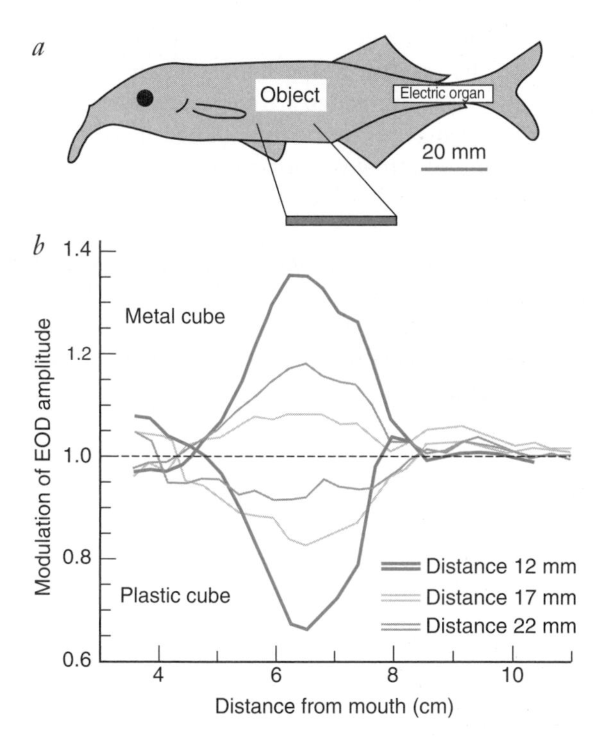

FIGURE 10.33 **(*a*) A weakly electric fish, *Gnathonemus petersii*, with a 2-cm cube positioned for electrical image measurement. (*b*) The electrical images of a metal or plastic cube at three distances from the fish's surface, measured at the midline. The electrical image of the metal cube is shown as a peak, and the image of the plastic cube is shown as a trough because metal (a conductor) pulls the lines of force together, and plastic (a nonconductor) spreads them out. Regardless of the composition of the cube, the width of the electrical image increases with increasing distance. The difference in amplitude between the center and the periphery of the image gets smaller with increasing distance. The fish uses the ratio of two features of the image—size and the amplitude differences between the core and the rim—to determine the distance of an object. (From von der Emde 1999.)**

Visual cues include landmarks; the sun, stars, or moon; and the pattern of skylight polarization. Methods of demonstrating that an animal uses landmarks in navigation include moving the landmark to see whether the animal reorients or becomes disoriented and impairing the animal's vision so that it cannot use landmarks. Some species use landmarks by matching the objects viewed with the remembered image of the array of landmarks. When landmarks are used in this way, the animal must always follow the familiar path. The sun may be used as a point of reference by assuming some angle relative to it during the journey and then reversing the angle to get home. Alternatively, since the sun follows a predictable path through the sky, if the time of day is known, the sun's position provides a compass bearing. If the sun is used as an orientation cue over a long interval of time, the animal must compensate for the sun's movement. Animals must learn to use the sun as a compass. The point of sunset is also an orientation cue that some nocturnal migrants use to select their flight direction, which is then maintained throughout the night by using other cues.

The stars provide an orientation cue for some nocturnal avian migrants. Birds such as the indigo bunting learn that the center of celestial rotation is north. This gives directional meaning to the constellations in the circumpolar area. Since the spatial relationship among these constellations is constant, if one is blocked by cloud cover, the birds can use the others to determine the direction of north.

Sunlight becomes polarized as it passes through the atmosphere. The pattern of polarization of light in the sky varies with the position of the sun. Polarized light may provide an axis for orientation, or it may allow animals to locate the sun from a patch of blue sky even when their view of the sun is blocked.

The earth's magnetic field provides several cues that could be used for orientation: polarity, inclination, and intensity. Some animals use a polarity compass, but most animals use an inclination compass, which distinguishes between equatorward (where the magnetic lines of force are horizontal) and poleward (where the lines of force dip toward the earth's surface).

Birds and sea turtles use an inclination magnetic compass for directional information. Birds use their compass to follow an inherited migratory program based on magnetic cues. Hatchling sea turtles use a magnetic compass while migrating across the Atlantic Ocean. They calibrate their magnetic compass relative to the direction of the surface waves that they experienced as they initially swam offshore.

A magnetic map provides information from earth's magnetic field that an animal can use to determine its position relative to a goal or target. Migrating birds and sea turtles may have a general magnetic map consisting of an inherited program of changes in direction of travel in response to magnetic signposts (magnetic fields characteristic of specific locations). Homing pigeons and sea turtles develop a more detailed magnetic map with experience living in a region. For example, with experience swimming in specific regions, sea turtles form a magnetic map based on the learned topography of the geomagnetic field. This information helps a turtle navigate to a specific target area.

There are (at least) two types of magnetoreceptors. One is light dependent. In birds, the magnetoreceptor, located in the right eye, contains the photopigment cryptochrome. Cryptochrome absorbs light differently depending on the orientation of the magnetic field. Information from these receptors connects to a part of the brain called Cluster N, which connects to a region of the brain that analyzes visual information. A second type of magnetoreceptor contains deposits of a magnetic material called magnetite. The crystals of magnetite twist in alignment of the magnetic field. This twisting could stimulate a stretch receptor. Salamanders and birds are among the animals that have two magnetoreceptor mechanisms, each serving a different function.

Chemical cues are also used for orientation. Salmon are guided to their natal stream by chemical cues. Young salmon learn (imprint on) the characteristic odors of their natal stream and then follow the odor trail back to that place.

Homing pigeons may also rely on olfactory navigation. Although the results of deflector lofts are consistent, they may not be due to a shifted olfactory map. However, the results of experiments in which olfactory information is manipulated are consistent with an olfactory basis for pigeons' navigation. The role of olfaction in their homing remains controversial.

Some aquatic species can detect electrical fields. These could be of use in navigation. A few species have electric organs that can generate electrical fields, which can be used in communication and navigation. The weak electric fish generate a stream of electrical pulses and then sense objects by the disturbance created in this symmetrical field.

11

The Ecology and Evolution of Spatial Distribution

Sturgeons and their ancestors have spawned in rivers for millions of years. Today, however, nearly all of the 25 or so living species of sturgeons are endangered, including the Chinese sturgeon (*Acipenser sinensis*). This magnificent fish can reach 4 or 5 m in length, weigh more than 550 kg (1000 lb), and live for a century (Figure 11.1). Like many other species of sturgeons, *A. sinensis* is anadromous, spending most of its life in the sea but returning to freshwater to breed (Bemis and Kynard 1997). In better times, Chinese sturgeon spawned at sites far upstream in the Yangtze River (Yang et al. 2006). From there, larvae would drift or swim downstream, eventually reaching the sea at Shanghai as juveniles. The young sturgeon spend years at sea and finally begin their first upstream spawning migration about ten years later. The Yangtze River is the third longest river in the world; in addition to being lengthy, it has powerful currents, so it took about 18 months for the sturgeon to migrate 3000 km to their traditional spawning grounds. All this changed in the 1980s, however, with construction of the Gezhouba Dam on the Yangtze, a project that blocked Chinese sturgeon halfway along their migration route to the traditional spawning grounds.

Chinese sturgeon responded by breeding below the Gezhouba Dam, in a much smaller stretch of the Yangtze. The number of spawning adults dropped from 2000 before construction of the dam to a few hundred, reflecting disruption by the dam, water pollution, overfishing, and heavy boat traffic (boat propellers kill several sturgeon every year). Efforts are under way to spawn Chinese sturgeon in captivity for release in the Yangtze. In addition, because characteristics of preferred spawning habitats are known (e.g., rocky substrate and moderate water velocity), some researchers propose constructing artificial spawning grounds. It is still unclear whether these efforts will save the now endangered Chinese sturgeon.

FIGURE 11.1 **The endangered Chinese sturgeon. In the past, these fish migrated 3000 km up the Yangtze River to their traditional spawning grounds. A dam now blocks their migratory route, forcing them to spawn in lower reaches of the river. Here, researchers are shown measuring a Chinese sturgeon. They will also attach an acoustic tag to the fish, which will allow them to track its movements after release.**

The story of *Acipenser sinensis* raises several questions. Why do animals undergo long-distance migrations? What are the costs and benefits of this behavior? And if we know what features endangered animals seek in a breeding habitat, then how can we best use this knowledge to conserve them? These are some of the questions that we will consider in this chapter as we describe the movements that animals make when searching for places to live and breed.

REMAINING AT HOME VERSUS LEAVING

Some animals are born in one place and then move to another location, where they breed, never to return to their birthplace. This behavior is called natal dispersal, and it has come to be defined in many ways. Here, we will use Clobert et al.'s (2001, p. xvii) definition of **natal dispersal** as "the movement between the natal area or social group and the area or social group where breeding first takes place." Dispersal can also occur after reproduction. This second type of dispersal, called **breeding dispersal** (or sometimes postbreeding dispersal), is defined by Clobert et al. (2001, p. xvii) as "the movement between two successive breeding areas or social groups." For example, some female rodents, upon weaning their litters at one location, leave the nest and litter and move to a new site where they will give birth to their next litter. In this chapter, we will focus on natal, rather than breeding, dispersal.

Natal dispersal can be contrasted with **natal philopatry**, in which offspring remain at their natal area and share the home range or territory with their parents (Waser and Jones 1983).

COSTS AND BENEFITS OF NATAL PHILOPATRY

What determines whether a juvenile should remain in the area of its birth or disperse? There is no simple answer to that question. Indeed, there are probably multiple influences on dispersal and philopatry (Boinski et al. 2005; Dobson and Jones 1985; Solomon 2003). Furthermore, such factors differ in their importance between species, sexes, and individuals. Here we will consider some of those factors as they relate to the costs and benefits of natal philopatry and dispersal.

One potential cost of remaining in the birth area is inbreeding—mating between relatives. Extreme inbreeding involves mating between parents and offspring or between siblings. Inbreeding is costly because it reduces variation among offspring and increases the risk of producing offspring that are homozygous for harmful or lethal recessive alleles. All organisms probably carry some harmful alleles that are recessive and therefore not expressed. Close relatives are more likely to have inherited the same versions of these alleles than are other members of the population. If close relatives mate, then their offspring are at increased risk of inheriting a copy of the deleterious allele from each parent. These offspring will then display the harmful trait (Shields 1982).

It can be difficult to determine the fitness costs of inbreeding because the frequency of inbreeding is usually low. However, a long-term study of the Mexican jay (*Aphelocoma ultramarina*) did find fitness costs associated with inbreeding. The brood sizes of inbred pairs were smaller than those of outbred pairs, which suggests hatching failure. Furthermore, compared with outbred nestlings, significantly fewer of the inbred nestlings survived to the next year (Brown and Brown 1998).

A second potential cost of staying at home is reproductive suppression. Adult breeders may suppress the reproductive development of philopatric young through chemical means (e.g., pheromones) or behavioral methods (e.g., aggression). Although suppressed young may not themselves reproduce, they may still be successful (from the standpoint of having their genes represented in future generations) by helping their parents rear subsequent broods or litters. We will have more to say about group living and helping by older offspring in Chapter 19.

A third possible cost of philopatry is increased competition. If conditions are crowded at home, then young that remain in their birth area may have to compete with relatives for food, nest sites, or mates. Sometimes the competition among relatives for breeding opportunities within the natal group is so intense that battles become deadly (reviewed by Griffin and West 2002). Thus, limited access to critical resources could result in lower survival and/or reproductive success than could be achieved by dispersing (Shields 1987). In other words, sometimes the grass really is greener on the other side of the street, so juveniles are better off leaving home than competing with relatives for limited resources.

What benefits might be associated with philopatry? Populations of animals can become adapted to local conditions. Combinations of genes that work well together under these local conditions might be favored by selection. An animal that disperses from its birth site and settles elsewhere, in an area with different ecological conditions, may not be as well adapted to its new home.

Another benefit of remaining near the birthplace is familiarity with the local physical and social setting. Such familiarity may enable philopatric young to be efficient not only at finding and controlling food but also in escaping from predators. Familiarity with family and neighbors is also likely to reduce the levels of aggression and stress associated with social interactions. In short, philopatric young may live longer and leave more offspring because of the relatively low risks and energy use associated with living in familiar surroundings (Shields 1982).

Sometimes offspring remain at home due to constraints, such as a shortage of potential mates or lack of suitable territories in which they might settle. Constraints on dispersal are discussed in more detail in Chapter 19.

COSTS AND BENEFITS OF NATAL DISPERSAL

Dispersing also has costs and benefits. Dispersers often face high energy costs and increased predation risk (Metzgar 1967; Nunes and Holekamp 1996). This is particularly true for small mammals that inhabit underground burrows (a lifestyle described as fossorial) that provide a fairly constant physical environment and safety from some types of predators (for example, hawks and owls). Long-distance above-ground dispersal by individuals of fossorial or semifossorial species exposes them to potentially harsh weather conditions and a suite of predators unable to enter their burrow systems. These costs may be compounded by a lack of familiarity with the terrain and high levels of aggression from residents of the population in which they are attempting to establish a new home. Male prairie voles released at unfamiliar locations move greater distances and take longer to find refuge than those released at familiar locations (Jacquot and Solomon 1997). In addition, nonresident woodland voles (Back et al. 2002) and Norway rats (McGuire et al. 2006) face harsh treatment by residents when attempting to settle in an area. Indeed, all in all, estimates of dispersal-induced mortality in animals can exceed 50% (e.g., Daniels and Walters 2000; Johnson and Gaines 1990). At the opposite extreme, dispersal costs appear to be negligible in some species (Gillis and Krebs 2000).

Dispersal also appears to have several benefits. One potential benefit is that dispersers avoid competition with kin for critical resources in their natal area (e.g., Perrin and Lehmann 2001). Some evidence to support this benefit comes from wolf spiders (*Pardosa monticola*), a species in which mothers help their offspring disperse. Dries Bonte and colleagues (2007) studied natal dispersal in wolf spiders in the laboratory and in coastal dune grasslands of Belgium. Before describing their findings, we should mention a few things about the natural history of wolf spiders. Wolf spiders do not live in webs; instead, they employ a combination of active foraging for prey and sit and wait strategies, depending on energy demands. Young hatch from an egg sac that the female carries, and then climb onto her abdomen where they live off their own stored energy reserves (i.e., they do not feed at this stage). Observations of newly independent spiderlings suggest that they display limited mobility with respect to ballooning (using strands of silk to travel by wind) and moving on the ground. Given their limited mobility, Bonte et al. (2007) predicted that spiderlings would face strong kin competition if they all simultaneously left their mother's abdomen to occupy the same general location (brood size, they discovered, can be as large as 72 spiderlings). It turns out that spiderlings do not simultaneously disperse from mom's abdomen; instead, broodmates dismount gradually over the course of about 185 hours following hatching (Figure 11.2*a*). Equally interesting is the observation that females carrying spiderlings are much more mobile than either females carrying egg sacs or females without egg sacs or spiderlings (Figure 11.2*b*). Bonte et al. (2007) suggest that through their increased movement at this particular time, wolf spider mothers actively spread their offspring throughout the environment. According to the

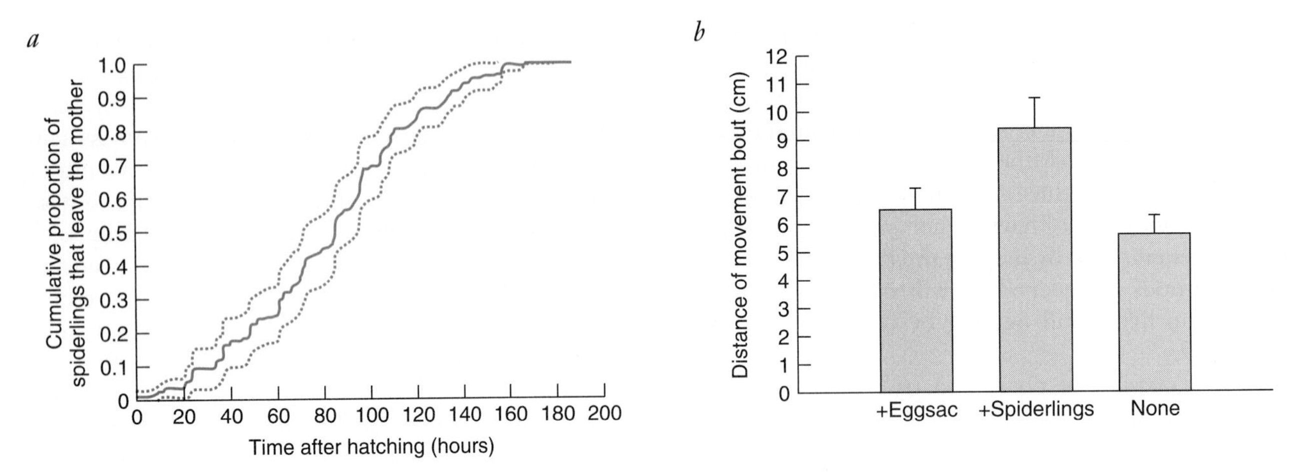

FIGURE 11.2 **Patterns of natal dispersal and maternal movement in wolf spiders ensure that siblings do not compete with one another. (*a*) Spiderlings dismount from their mother's abdomen gradually over a period of about 185 hours after hatching, rather than leaving all at once. (*b*) Female spiders carrying spiderlings move greater distances than females carrying egg sacs or females without egg sacs or spiderlings (labeled as "None" in the graph). (Modified from Bonte et al. 2007.)**

Bonte study, this behavior, combined with the asynchronous departures of individual young from their abdomen, promotes the avoidance of competition with kin for resources.

In other species, avoiding competition with kin cannot explain patterns of natal dispersal. For example, if competition for resources in the natal area is an important factor leading to dispersal, then we might predict that within a species, dispersal would increase with increases in litter size. However, in brown bears (Zedrosser et al. 2007) and prairie voles (McGuire et al. 1993), the likelihood of natal dispersal is not associated with litter size. In fact, natal dispersal is more common from small than large groups of prairie voles. This finding is the precise opposite of what we would predict if avoidance of competition for resources at home was an important function of natal dispersal in this species. Competition for mates also does not explain natal dispersal in prairie voles, for dispersal occurs at similar levels from natal groups having high levels of mate competition (number of competitors > number of potential mates) and low levels of mate competition (number of competitors < number of potential mates).

Avoidance of inbreeding is another potential benefit of natal dispersal. Evidence that some young animals leave home to avoid breeding with close relatives comes in several forms. For example, some evidence involves observations that young disperse in response to the presence of an opposite-sex parent (e.g., Gundersen and Andreassen 1998; Le Galliard et al. 2003; Wolff 1992). Other evidence comes in the form of all members of one sex dispersing no matter what the ecological or social conditions. In Belding's ground squirrels, for example, all males leave home (and very few females do) regardless of the levels of competition for mates or resources; avoidance of inbreeding remains a possible explanation for this pattern (Holekamp and Sherman 1989). Similarly, observations of brown bears in two different study sites in Sweden revealed that nearly all males leave home (94% of those studied), regardless of ecological factors such as population density and sex ratio. In contrast, only 41% of female brown bears dispersed from their natal site (Zedrosser et al. 2007). Predominant (or more distant) dispersal by one sex is predicted if individuals disperse to avoid inbreeding. For the most effective inbreeding avoidance, we would also expect dispersers to leave home before becoming reproductive, and many do (Dobson 1982).

There has been considerable debate over the role of inbreeding in natal dispersal, with some scientists arguing that it plays a significant role and others describing its role as irrelevant (reviewed in Perrin and Goudet 2001). Indeed, some authors have argued that animals should rely on mate choice rather than a risky behavior like dispersal to avoid inbreeding (Moore and Ali 1984). Young female mammals, for example, may avoid selecting their brothers as mates, and in response, their brothers disperse from the natal site to find females willing to mate with them. According to this argument, individuals do not disperse from home to avoid breeding with relatives; instead, natal dispersal is the *result* of females not choosing male relatives as mates.

Spotted hyenas (*Crocuta crocuta*) seem to be a species in which female mate choice drives natal dispersal by males. In this species, dispersal by females from the natal clan is very rare; in contrast, most, but not all, males leave their natal clan. Thus, predominant dispersal by one sex occurs in spotted hyenas, but does this reflect female mate choice or inbreeding avoidance? In an effort to sort out the causes of natal dispersal in male spotted hyenas, Höner and colleagues (2007) studied eight clans in northern Tanzania from April 1996 to April 2006. Using behavioral observations, demographic data, and genetic analyses of paternity, the researchers examined patterns of female mate choice and tested four hypotheses for natal dispersal by males. Specifically, they examined whether males disperse in order to avoid (1) competition with other males for mates, (2) breeding with close female relatives, or (3) competition for food resources. The fourth hypothesis tested was that males disperse in response to patterns of mate choice by females. To test these four hypotheses, Höner et al. (2007) examined which of the following four variables (keyed to each hypothesis) predicted the clan in which males began their reproductive career: (1) intensity of male–male competition in the clan (estimated by the number of reproductively active natal males and immigrant males), (2) number of unrelated adult females in the clan (defined as those with a coefficient of relatedness < 0.5), (3) number of main prey animals in the clan territory per adult or yearling hyena at the time of clan selection, or (4) number of young females most likely to breed with males. Clan selection by males was independent of the first three variables (intensity of male–male competition, number of unrelated females, and per capita number of available prey). These findings suggest that competition for mates, inbreeding avoidance, and competition for food do not cause natal dispersal by males. What about patterns of mate choice by females? Using paternity analyses of litters, Höner et al. (2007) found that females adhered to the following general rule to avoid mating with close relatives: "avoid mating with males that were members of your clan when you were born and select as mates those males that arrived in your clan (through birth or immigration) after your birth." This rule would reduce the chances that females would mate with their father and older brothers. The paternity analyses showed that nearly 90% of 134 litters examined were sired by males that were born into or immigrated into the female's clan after she was born (Figure 11.3*a*). Paternity analyses also revealed another aspect of female choice: young females were more apt than older females to select males with shorter residency times in their clan (refer again to Figure 11.3*a*). Given these patterns of female mate preferences, the researchers predicted that males would begin their reproductive career in the clan (natal or otherwise) that had the greatest number of young females. This was, in fact, what they found. After making short-term forays into the territories of nearby clans (presumably to evaluate dispersal options), males select as their place to begin reproduction the clan with the highest number of young females (remember, these are the females most likely to choose them as mates). Importantly, such a choice results in long-term fitness benefits for the males (Figure 11.3*b*). We see from this example that patterns of mate choice can be key to understanding patterns of natal dispersal. We will have more to say about mate choice in Chapter 14.

We have focused on the ultimate causes of natal dispersal, such as avoidance of inbreeding and competition with kin for resources or mates. There are also proximate causes of natal dispersal (reviewed in Lawson Handley and Perrin 2007; Nunes 2007). Factors suggested to trigger natal dispersal include attainment of sufficient body

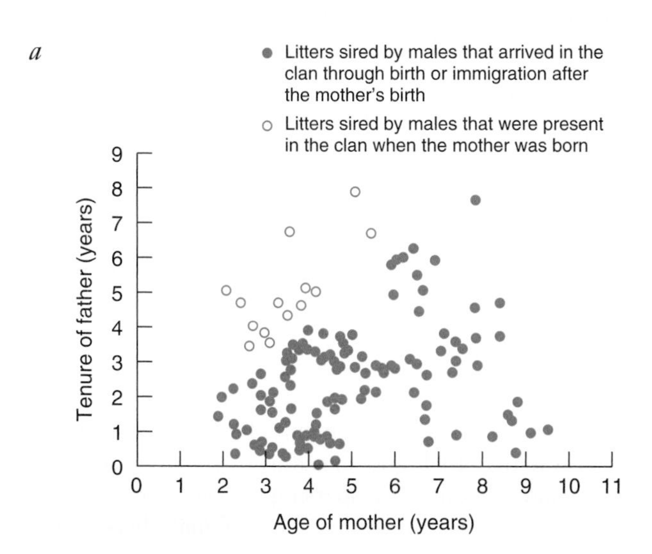

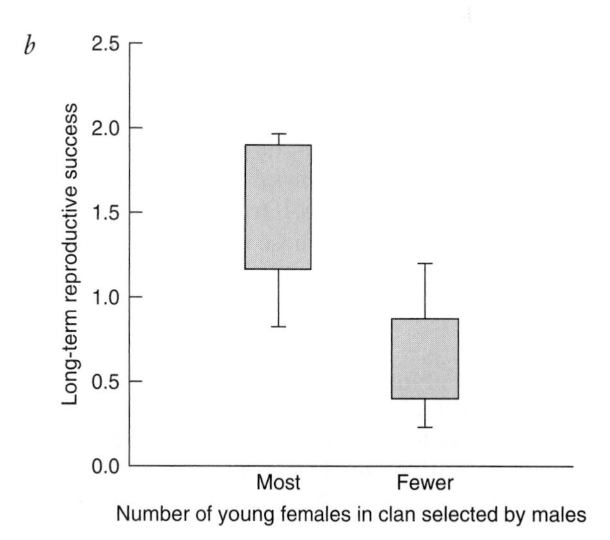

FIGURE 11.3 Natal dispersal by male spotted hyenas is driven by patterns of female mate choice. (*a*) Shown here are data from 134 litters (each circle represents a litter). Most litters are sired by males that arrive (through birth or immigration) in the mother's clan after her birth. In addition, young females are more apt than older females to select males with shorter residency times in their clan. (*b*) Males that begin their reproductive careers in clans with the highest number of young females have higher long-term reproductive success than males that select clans with fewer young females. (From Höner et al. 2007.)

size or fat reserves, aggression from other group members, shortage of food at the natal site, attraction to opposite sex individuals in other groups, and weakening of social bonds with members of the natal group. There is evidence from some species that androgens organize dispersal behavior (see Chapter 7 for a discussion of the organizational effects of steroid hormones on behavior).

Natal dispersal has also been linked to individual differences in personality traits such as "boldness" or exploratory behavior. In great tits, for example, postfledging movement distances of individual females in a wild population correlate with their scores on a laboratory test of exploratory behavior. More specifically, females that score high on the laboratory test (a score equals the number of movements made during a two-minute test in a novel environment) move further during postfledging movements than females with low exploratory test scores (Dingemanse et al. 2003). The positive correlation between exploratory behavior and distance dispersed does not characterize all species, however. Radiotracking of Siberian flying squirrels before and after dispersal indicates that long-distance dispersers explore less than short-distance dispersers (Selonen and Hanski 2006).

SEX BIASES IN NATAL DISPERSAL

As we have seen, males and females of a particular species often differ in whether or not they disperse from their birthplace. Even more striking is the observation that the direction of the sex bias in natal dispersal differs between birds and mammals (Table 11.1). In the majority of bird species that have been studied, females are more likely to disperse than males. In mammals, however, just the reverse is true—males are more likely to disperse than females.

We can ask what drives sex biases in natal dispersal, and why does the bias differ in direction between birds and mammals? Several hypotheses have been proposed to explain sex-biased dispersal (reviewed by Lawson Handley and Perrin 2007). Those that we will consider suggest that sex differences in natal dispersal reflect the following factors, sometimes in combination: (1) inbreeding avoidance; (2) local resource competition; (3) local mate competition; and (4) cooperative behavior among kin. Mammals have been best studied with respect to natal dispersal (Lawson Handley and Perrin 2007), and some explanations focus on them. The first hypothesis we discuss, however, was developed specifically to explain the different dispersal patterns in mammals and birds.

TABLE 11.1 Number of Species of Mammals and Birds in Which Natal Dispersal Is Male-Biased, Is Female-Biased, or in Which Offspring of Both Sexes Disperse

	Predominant dispersing sex		
	Male	Female	Both
Mammals	45	5	15
Birds	3	21	16

Source: Data from Greenwood (1980).

Local Resource Competition and Inbreeding

One hypothesis is that a sex bias in dispersal evolved as a way to avoid the genetic costs of inbreeding while enjoying the benefits of familiarity with local physical and social conditions. A sex bias in dispersal seems to be the perfect compromise: extreme inbreeding is prevented because members of one sex disperse, and individuals of the other sex experience the benefits of philopatry (Greenwood 1980).

Although this hypothesis explains why sex biases in dispersal tendencies occur, it does not explain which sex leaves home and why the direction of the bias differs in birds and mammals. Two explanations have been suggested for the direction of the sex bias.

The first idea is that the sex most involved in territory acquisition and defense should stay home because it benefits most from familiarity with the natal territory (Greenwood 1980). Most birds are socially monogamous (i.e., they live in male-female pairs), and males usually compete for territories that attract females, rather than competing for females directly. This is called a resource-defense mating system. Under conditions such as these, familiarity with a particular area might be more important to males than to females; thus males should be philopatric. Female birds might disperse to avoid the genetic costs of close inbreeding and to choose territories with the best resources.

In contrast, most mammals exhibit mate-defense polygyny; that is, a single male defends a group of females. In this mating system, males directly compete with one another for females rather than territories with resources. Young or subordinate males, unable to compete successfully for access to females, may disperse to increase their chances of mating. Also, female mammals often live in matrilineal social groups (groups of mothers, daughters, and granddaughters) in which the benefits of living with kin may be quite high. Because of this social system, females benefit most by staying home and males may benefit by dispersing to avoid the genetic costs of extreme inbreeding. Thus, female-biased dispersal in birds seems to be linked to resource-defense mating systems, and male-biased dispersal in mammals seems to be linked to mate-defense mating systems (reviewed in Clarke et al. 1997).

What about the exceptions, for instance, mammals that display resource-defense (rather than mate defense) polygyny? What is the direction of sex bias in natal dispersal in these species? In some mammals with resource-defense mating strategies, natal dispersal is female biased, as would be predicted by Greenwood (1980). This is the case, for example, for sac-winged bats (Nagy et al. 2007). However, European roe deer also defend resources rather than mates and dispersal is not female biased in this species, illustrating that a resource-defense mating system does not necessarily lead to female-biased natal dispersal in mammals (Coulon et al. 2006).

The second idea to explain the direction of the sex bias is that the sex that gets first choice of breeding sites is the one that remains in the natal area; the other sex disperses. This model was first developed to explain sex-biased dispersal patterns in mammals (Clutton-Brock 1989), but it was later extended to birds (Wolff and Plissner 1998). In either case, the model assumes that philopatry is more desirable than dispersal. According to this model, mating systems affect the dispersal patterns of mammals indirectly by influencing whether the father will be present when his daughters are old enough to breed. If he is not, females have first choice of the breeding site, and they choose to stay at home. However, if the father is still around when his daughters reach sexual maturity, he has first choice of the breeding site, and so females disperse to avoid inbreeding.

Because female mammals nurse their young, a role for males in early parental care is necessarily limited, and so in most species males have little involvement in offspring care. This allows males to avoid long-term pair bonds, and they are free to wander over large areas. When competition over mates is intense, as among elephant seals or red deer, a male's opportunity to breed may be limited. As a result, he is likely to be gone before his daughters are old enough to reproduce. Thus, daughters don't have to disperse to avoid inbreeding. However, in those species in which a male's reproductive life span is long and he is present when his daughters are old enough to breed, as in chimpanzees, the females usually disperse (Clutton-Brock 1989; Wolff 1994).

Local Mate Competition

Another hypothesis is that differences between males and females in levels of competition for mates might be involved in sex differences in the dispersal tendency in mammals (Dobson 1982). Because most mammals are polygynous, competition for mates would be more intense among males than among females, and thus dispersal should be more common in males. Furthermore, in species with monogamous mating systems, levels of competition for mates would be more equal between the sexes, and males and females should disperse in similar proportions.

TABLE 11.2 Number of Species of Mammals in Which Natal Dispersal Is Male-Biased, Is Female-Biased, or in Which Offspring of Both Sexes Disperse, as a Function of Type of Mating System

	Predominant dispersing sex		
Mating system	Male	Female	Both
Monogamous	0	1	11
Polygynous or promiscuous	46	2	9

Source: Data from Dobson (1982).

When dispersal data for species of mammals with different mating systems were examined, they revealed remarkable agreement with this hypothesis (Table 11.2). However, avoidance of inbreeding is probably a significant influence on natal dispersal in many species because reduction in competition for mates can't explain all sex differences in natal dispersal. (For instance, it doesn't explain why females are more likely than males to disperse in monogamous birds.)

Cooperative Behavior Among Kin

When discussing the benefits of natal dispersal, we mentioned that leaving home allows young animals to avoid competing with kin (for resources or mates) and to avoid breeding with kin. However, if kin exhibit cooperative behavior, then it might be beneficial to stay at home. Furthermore, if cooperation benefits one sex more than the other, then we would expect the cooperative sex to be philopatric. We would expect the other sex to disperse to avoid inbreeding because inbreeding costs for this sex are not counterbalanced by advantages gained through kin cooperation (Perrin and Goudet 2001). In this way, cooperative behavior can contribute to sex biases in natal dispersal.

One prediction that arises from the kin cooperation hypothesis is that the magnitude of sex-biased dispersal should increase with increases in social complexity. Consistent with this prediction is the observation that particularly dramatic sex differences in natal dispersal often characterize highly social polygynous mammals (Pusey 1987; Smale et al. 1997). Other evidence comes from polygynous ground-dwelling sciurids, a group of rodents that includes the ground squirrels (*Spermophilus* spp.), marmots (*Marmota* spp.), and prairie dogs (*Cynomys* spp.). This group is particularly well suited for testing the relationship between degree of sex bias in natal dispersal and social complexity because social structures range from solitary to large social groups with

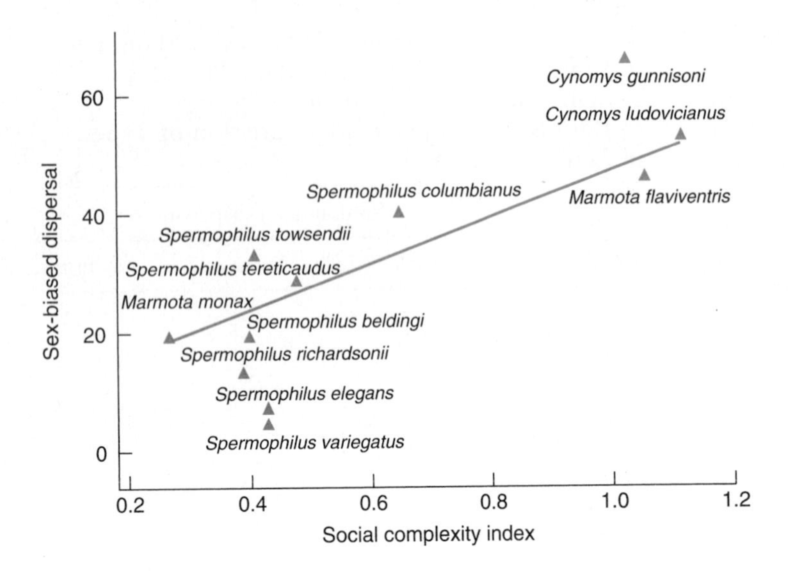

FIGURE 11.4 In polygynous ground-dwelling sciurids (members of the squirrel family), the degree of sex bias in natal dispersal (calculated as the difference between male and female dispersal rates) increases with social complexity. Shown here are data from two species of marmots (*Marmota* spp.), two species of prairie dogs (*Cynomys* spp.), and seven species of ground squirrels (*Spermophilus* spp.). (From Devillard et al. 2004.)

individuals from several generations. Sébastien Devillard and colleagues (2004) reviewed the literature on sciurids and extracted for 11 species natal dispersal rates (number of animals that leave their natal site or social group before their first reproduction/total number of potential dispersers in the population). They used an index of social complexity previously developed for ground-dwelling sciurids. The index, thought to better describe degree of social complexity than group size alone, considers the types of social interactions possible within groups (e.g., the extent of interaction between individuals of different age and sex classes) (Blumstein and Armitage 1997). When Devillard et al. (2004) analyzed data from the literature, they found that male-biased natal dispersal characterized all 11 species. In addition, plots of sex bias in natal dispersal (calculated as the difference between male and female dispersal rates) versus index of social complexity revealed that male-biased dispersal increased with social complexity (Figure 11.4). However, the increased bias resulted solely from increases in male dispersal and not from expected concomitant decreases in female dispersal. So, in ground-dwelling sciurids, data for males, but not females, support Perrin and Goudet's (2001) model regarding the relationship between sex-biased dispersal and social complexity.

Sex biases in natal dispersal are likely caused by several interacting factors, including mating system, inbreeding avoidance, kin competition, and kin cooperation. Although mammals have been most studied in this regard, studies with birds are increasing, and sex biases in dispersal have also been documented in other groups of vertebrates, including fishes and amphibians. Increased study of natal dispersal in these groups, as well as in invertebrates, will permit testing the generality of the hypotheses put forth largely from mammalian data (Lawson Handley and Perrin 2007).

NATAL DISPERSAL AND CONSERVATION BIOLOGY

Before moving to the topic of habitat selection, we should mention that natal dispersal has important consequences for the genetic structure of populations. Small populations cut off from other populations may lose genetic diversity through genetic drift, but immigration brings in new genes. Not surprisingly, then, natal dispersal has important implications for the conservation of populations and species. Habitat destruction by humans has fragmented populations of many species and has made inhospitable the areas between habitat fragments. Such changes hinder dispersal. As part of our efforts to conserve threatened populations, it may be critically important to preserve dispersal corridors (narrow areas that connect other, larger areas of habitat) that promote gene flow. Knowledge of dispersal patterns will also help to predict range expansions of introduced predators and pests (Macdonald and Johnson 2001). We see, then, that understanding a behavior such as natal dispersal can have important conservation implications.

STOP AND THINK

In the field, natal dispersal and philopatry are often measured by mark and recapture methods (e.g., catch an animal, mark it for individual identification, release it, and then subsequently recapture it at the same or a different location) (Figure 11.5). Sometimes genetic methods are used in combination with mark-recapture approaches to examine how dispersal translates into gene flow. Given what you know about natal dispersal, what are some potential problems in using mark-recapture methods to study dispersal in natural populations? For example, how easy would it be to know whether a particular animal dispersed from home or died? How might you design a study to minimize this issue?

a

b

FIGURE 11.5 **Mark-recapture methods are often used to monitor natural populations of small mammals. Repeated live-trapping provides data on behavior (e.g., dispersal and social organization) and demography (e.g., changes in population density). (*a*) A student at the University of Illinois checking a live trap. (*b*) Another student handling a prairie vole removed from a live trap at the same study site.**

HABITAT SELECTION

Animals that disperse from their natal site or breeding site must eventually select a new location in which to settle. This process of habitat selection can be divided into three phases: (1) search (animal searches for a new habitat), (2) settlement (animal arrives in a new habitat and begins to establish a home range or territory), and (3) residency (animal lives in the new habitat) (Stamps 2001). Generally speaking, the phases of search and settlement are costly times for dispersers, while residency is when the benefits of habitat selection accrue. We will focus largely on the search phase, and we begin with the question, "What features of the habitat might dispersers use to judge habitat quality?"

INDICATORS OF HABITAT QUALITY

Animals seem to have clear habitat preferences. For example, the forest buffalo (*Syncerus caffer nanus*) of western and central Africa prefers grassy clearings and open stands of forest with large trees and open canopy to other types of forest (e.g., mixed, riparian, and forest dominated by a single tree species) (Melletti et al. 2007). This preference is particularly striking given that clearings are quite rare, representing only 1% of habitats in some areas. Because forest buffalo are social creatures, they may prefer clearings, particularly as resting spots, because members of the herd can easily maintain visual and physical contact. Let's take a look at what general features animals might evaluate when selecting a habitat.

Indicators of habitat quality include the presence of resources (e.g., food, nest, and rest sites), conspecifics, and heterospecifics (members of another species) (Stamps 2001). As you might imagine, dispersers may lack the time and energy needed to complete in-depth assessments of critical resources in the habitats they encounter in their search for a new home, so how do they evaluate real estate? Evidence suggests that some dispersers rely on "quick and dirty" cues to assess the relative quality of prospective settlement sites. For example, young lizards (*Anolis aeneus*) that leave their natal site to search for feeding territories spend only about six hours evaluating a particular location. Given their varied diet of arthropods, this time period is probably too short to permit a detailed assessment of prey availability. So, rather than assessing arthropod availability at each site, the lizards seem to assess habitat characteristics such as light intensity and amount of leaf litter. These characteristics correlate with prey availability and lend themselves to more rapid evaluation than the painstaking task of evaluating the local availability of several different prey species (Stamps 1994).

What about conspecifics? Should dispersers evaluate their presence or absence in the vicinity of a prospective home? Absence of conspecifics from a particular site might be a good thing. After all, if an individual settles into an unoccupied site, then it would avoid all that nasty intraspecific competition for resources. There are two main ideas concerning how fitness might change with number of conspecifics in an area. One model, called the ideal free distribution, predicts that individual fitness will decline as the number of conspecifics in a patch increases (Fretwell 1972). Another idea suggests that individual fitness increases with the number of conspecifics at low to moderate densities, and then declines from moderate to high densities (Allee 1951). According to this "Allee effect," having a few neighbors is beneficial, especially when it comes to defense against predators and access to potential mates.

Presence of conspecifics in an area could also serve as a source of information about habitat quality (Stamps 1988). Juvenile lizards (*A. aenus*), for example, seem to use conspecific presence as an indirect cue to habitat quality. Juveniles were allowed to view two sites of equivalent quality; a territory owner was present on one site but not on the other. After ten days, the territory owner was removed and the juveniles were allowed to select between the two sites. The young lizards preferred the previously occupied site to the equivalent, unoccupied site (Stamps 1987). To summarize, then, when it comes to habitat selection, there are two general explanations for the phenomenon of conspecific attraction: (1) Allee effects in which individuals benefit from the presence of conspecifics after arriving at a site (e.g., through enhanced detection of predators or access to mates), and (2) individuals' use of conspecifics as an indicator of habitat quality.

Beyond the mere presence of conspecifics, might individuals looking for a place to settle evaluate certain *characteristics* of resident conspecifics? The answer appears to be yes. There is good evidence, for example, that breeding birds monitor the reproductive success of conspecifics in their local area and then use this information to decide where to nest during future breeding efforts (Doligez et al. 2002; Parejo et al. 2007). Information concerning local conspecifics is described as "public information" in order to differentiate it from "personal information." Personal information is the focal bird's own breeding success at a particular site, which also affects whether it will remain at that site or leave.

The presence of heterospecifics in a particular habitat can have costs and benefits for an animal considering whether to settle there. For example, interactions between species that share mutual resources could be negative due to interspecific competition. Interactions between heterospecific individuals can also be beneficial, however, such as when birds form mixed species flocks and experience the benefits of enhanced food acquisition and antipredator behavior. Thus, moving in where there are neighbors of other species may be beneficial under certain circumstances. This idea has been formalized as the heterospecific attraction hypothesis, which states that individuals choose habitat patches based on the presence of established residents of another species (Mönkkönen et al. 1999). This model predicts that individuals searching for a new home will display the strongest attraction to heterospecifics when the benefits of social aggregation outweigh the costs of competition and when the costs of independent sampling of habitats (i.e., evaluating habitat quality on one's own rather than using the presence of heterospecifics as a cue) are high.

An interesting test of the heterospecific attraction hypothesis concerns the attraction of migrant birds of one species to resident birds of other species. There are several possible explanations for why migrants might be attracted to heterospecific residents. First, residents could indicate high-quality habitat. Second, migrants may experience food or safety benefits from grouping with heterospecific residents (as described for mixed-species flocks). Third, using presence of heterospecific residents as an indicator of habitat quality may be a faster and more accurate method of habitat assessment than independent sampling upon arrival at potential breeding sites (residents, after all, have all year to assess the quality of various habitat patches as compared to migrants who have a much shorter time period).

Jukka T. Forsman and colleagues (2002) examined whether the presence of resident titmice (*Parus major*, *P. montanus*, and *Cyanistes caeruleus*) influences the settlement and fitness of migrant pied flycatchers (*Ficedula hypoleuca*). Titmice and pied flycatchers overlap in their use of resources; all species nest in cavities and forage for arboreal arthropods using hovering and foliage gleaning. Forsman et al. (2002) conducted their experiments on two spatial scales: landscape and nest site. In the landscape scale experiment, they removed all titmice from some forest patches (to achieve a density of zero) and added titmice to other forest patches (to achieve a density of about five pairs per 10 ha); these density manipulations were performed prior to the arrival of flycatchers in the spring, and the achieved densities were maintained throughout the breeding season. In the nest-site scale experiment, flycatchers were allowed to choose between a nest box close to an active titmouse nest (distance of 25 m) or farther away from an active titmouse nest (distance of 100 m). In both experiments, the researchers recorded the arrival dates of flycatchers and several measures of fitness for flycatcher broods produced at the different sites. At both landscape and nest-site scales, flycatchers were attracted to the vicinity of titmice. Flycatchers tended to arrive earlier on forest patches where titmice numbers were increased than on patches where titmice had been removed, and they preferred nest boxes placed near an active titmouse nest to those located farther away. With regard to the fitness effects of habitat selection, brood sizes of flycatchers were larger in patches where numbers of titmice had been experimentally increased than in patches from which titmice had been removed (landscape scale experiment), and flycatchers breeding closer to titmice had larger nestlings than those breeding farther away (nest-site scale experiment) (Figure 11.6). These results support the heterospecific attraction hypothesis. (Note also that the results are just the opposite of what we would expect if interspecific competition were the defining interaction between titmice and flycatchers.)

We see that individuals searching for a home may evaluate multiple and diverse characteristics of potential settlement sites, ranging from physical features of the

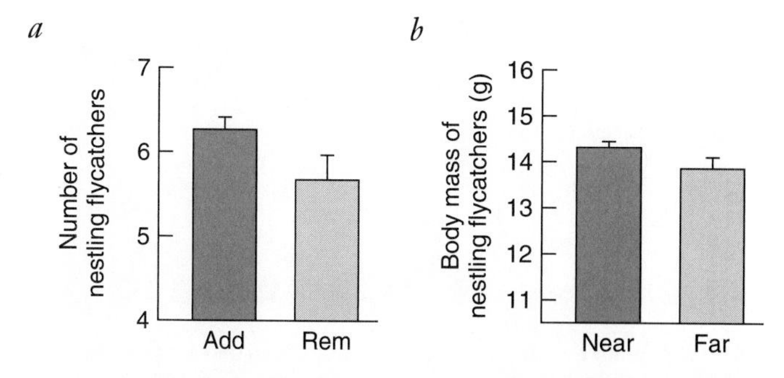

FIGURE 11.6 **In support of the heterospecific attraction hypothesis, migrant pied flycatchers are attracted to sites with resident titmice, and this attraction results in higher fitness for the flycatchers. (*a*) Numbers of flycatcher nestlings 13 days post hatching were greater in patches where numbers of titmice were experimentally increased (Add) than in patches from which titmice had been removed (Rem). (*b*) Flycatchers breeding near titmice had larger nestlings than those breeding farther away.**

habitat to details of the social scene. We next consider the strategies used by animals when searching for a new home.

SEARCH TACTICS

Animals use two general search strategies during habitat selection: comparison and sequential search (Figure 11.7) (reviewed in Stamps and Davis 2006). The first tactic, comparison, entails visiting several areas, revisiting some, and then choosing the area judged to be the highest quality. This tactic is often described as a best-of-N strategy because the animal visits N habitats and then selects the best one. In contrast, animals using sequential search tactics arrive at a location, decide whether to accept or reject it as a place to live, and in the event they reject it, continue their search. Animals using sequential search tactics do not return to areas they have already visited (except by chance), and they often travel long distances in relatively straight lines before establishing residence in an area. Decisions by dispersers using sequential search are influenced by factors such as the total time available for the search, quality of available habitats, and how often high-quality habitats are encountered. In addition, dispersers using sequential search tactics may have an acceptance threshold, and this threshold may decline as the search continues (i.e., an animal may be highly selective at the start of the search and less selective over time). Comparative and sequential search tactics do not sort out by taxon; indeed, both tactics can occur within a single population of a species (e.g., Byrom and Krebs 1999).

EFFECTS OF NATAL EXPERIENCE

Experience in the natal environment appears to be an important influence on habitat selection. As an example, let's consider the host preferences exhibited by the parasitoid wasp (*Pachycrepoideus vindemia*). A parasitoid is an organism whose offspring develop on or within a host, eventually killing the host; in our discussion of habitat selection, then, the host of a parasitoid is the equivalent of a habitat. In other words, just as other animals select habitats within which to reproduce, female parasitoids select hosts on (or within) which their offspring will develop. Young *P. vindemia* develop on the pupae of several different fly species. R. J. Morris and M. D. E. Fellowes (2002) conducted laboratory experiments with female *P. vindemia* to determine the effects of natal host and experience on subsequent host preference. In terms of natal host, female wasps were allowed to develop on either fruit flies or houseflies. Following development on their host, the experience of females was manipulated by placing females in vials (for 24 hours) either alone, with two fruit fly pupae, or with two housefly pupae. Thus, there were three different groups regarding experience: (1) no opportunity to attack either fruit flies or houseflies; (2) opportunity

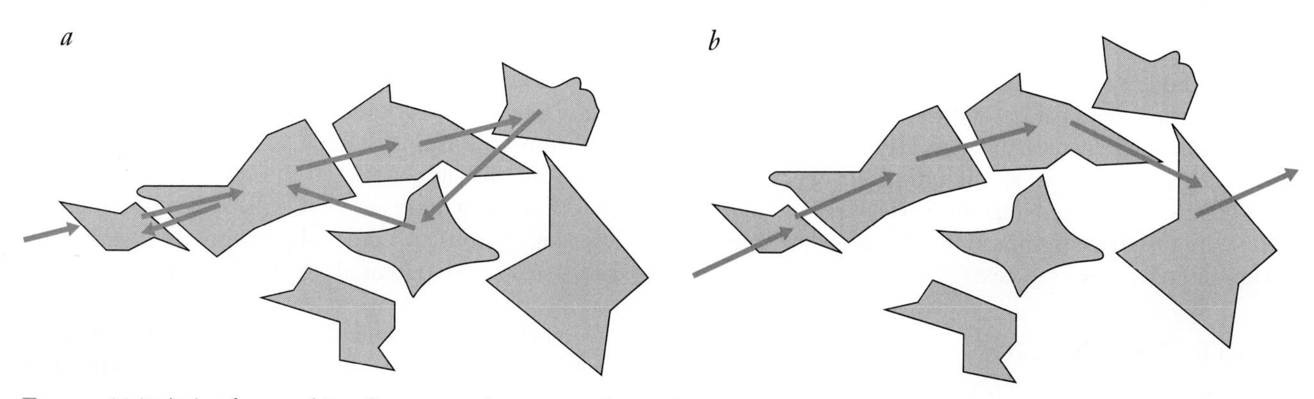

FIGURE 11.7 **Animals searching for a new place to settle employ two general strategies called comparison and sequential search. (*a*) In comparison (best of N) strategies, an animal visits several areas, revisits some, and then chooses the area judged to be the highest quality. (*b*) During sequential search, an animal does not return to sites except by chance, and tends to travel long distances in relatively straight lines.**

to attack fruit flies; and (3) opportunity to attack houseflies. At the end of their 24-hour experience, females were tested for host preference by placing them in a Petri dish for two hours with ten fruit fly pupae and ten housefly pupae. To assess host preference, the researchers monitored the number of emerged parisitoids, emerged flies, and dead pupae in each dish. The results indicate that host preference in *P. vindemia* is influenced by an interaction between natal host and subsequent experience. Female wasps reared on fruit flies preferred fruit flies as hosts, but only when allowed to gain experience attacking fruit flies, which they did preferentially. Similarly, female wasps reared on houseflies preferred houseflies as hosts, but only when allowed to gain experience attacking houseflies, which they, too, did preferentially. The authors suggest that cues from the natal host prime females to respond to these cues should they be encountered again, and when reencountered, the cues are learned by females, establishing host preference. We see, then, that the natal environment (here, the natal host) can influence selection of a site for reproduction.

The general phenomenon whereby an animal's experience in its natal habitat induces a preference for a postdispersal habitat with similar qualities is called natal habitat preference induction (NHPI) (Davis and Stamps 2004). Recall that the search phase of habitat selection is thought to be costly in terms of time and energy, not to mention the risk of predation associated with dispersing through unfamiliar terrain. Thus, it is possible that NHPI helps a disperser to more quickly and efficiently recognize a suitable habitat, thereby minimizing the costs of the search phase. Also, it seems reasonable to expect that a habitat similar to the one in which the disperser grew up would be of sufficient quality to settle in because the dispersing individual itself has survived to leave home. Another explanation for NHPI is that dispersers will have greater fitness if they settle in a habitat similar to their natal habitat because their particular phenotype has been shaped by this type of habitat. For example, while living at home, individuals may develop specific methods for finding and capturing prey, and these methods might work best in postdispersal habitats similar to their natal habitat and less well in habitats different from their natal habitat. NHPI has been documented in diverse taxa, including insects, fishes, amphibians, birds, and mammals (Davis and Stamps 2004) and may have significant implications for conservation biology, as we see next.

HABITAT SELECTION AND CONSERVATION BIOLOGY

Conservation efforts using translocation (moving animals from one part of their natural range to another) and captive-release programs (breeding animals in captivity

FIGURE 11.8 **Many translocated animals travel long distances from the site of release. In cougars, this is especially true for adult males.**

and then releasing them to the wild) often fail because animals move rapidly away from the site of release, often traveling long distances (Figure 11.8) (reviewed in Stamps and Swaisgood 2007). Long-distance travel is associated with high mortality, and those individuals that escape death and eventually settle may exhibit declines in condition and depressed reproduction. In addition, rapid, long-distance movements from the release site make postrelease provisioning and monitoring extremely difficult, further compromising conservation efforts. Indeed, translocations and captive-release programs tend to have low success rates (11–75%).

Rapid, long-distance movement from a release site suggests that the animal rejects the habitat at the release site and is searching for suitable habitat in which to settle (Stamps and Swaisgood 2007). An understanding of habitat selection, in general, and natal habitat preference induction, in particular, may inspire modifications to these programs that could help animals find the release site a more acceptable place to settle. For example, if animals developing in captivity were provided with conspicuous stimuli and cues (for example, odors or sounds) similar to those at the intended release site, then perhaps they would be more inclined to stay in the vicinity of the release site. In translocation efforts, placing stimuli and cues from the original habitat at the release site might reduce the disparity between the two habitats, making the release site a more acceptable place to settle. If taken, these suggestions and others by Stamps and Swaisgood (2007) have the potential to increase the success of translocation and captive-release programs.

Another area in which habitat selection by animals is relevant to conservation biology concerns ecological traps. An ecological trap is a low-quality habitat that animals prefer over a high-quality habitat (Battin 2004). One of the best examples concerns Cooper's hawks (*Accipiter cooperii*) living in Tucson, Arizona (Boal and

Mannan 1999). Cooper's hawks seem to select the city as a home because of its plentiful nest sites and prey; indeed, urban hawks nest earlier and have larger clutches than hawks nesting in less urban areas outside Tucson. However, nestling mortality is substantially higher in Tucson (> 50%) than in areas surrounding the city (< 5%), making Tucson an ecological trap for Cooper's hawks. The very pigeons and doves that attract Cooper's hawks to the city because of their abundance (these prey species comprise 84% of the diet of city hawks) carry trichomoniasis; this disease is the primary cause of nestling mortality in the city-dwelling Cooper's hawks. Thus, the cues apparently used by Cooper's hawks to select a habitat (availability of nest sites and prey) lead them to select an inferior habitat in which disease causes dramatically reduced survival of their offspring. Ecological models suggest that populations living in an ecological trap move toward extinction, especially if the initial population size is small (Battin 2004). The Cooper's hawks in Tucson are not disappearing, however. Instead, the number of hawks in the city is at least stable, probably because birds continue to move into the city from the surrounding areas. Unfortunately, we know little about the capacity of (and time required for) animals to evolve new habitat preferences or to adapt to new environmental conditions. This lack of knowledge makes it difficult for us to predict whether species can escape from ecological traps.

STOP AND THINK

Many examples of ecological traps are associated with human activities. However, ecological traps can occur without human influence. Can you think of a situation without human involvement that would qualify as an ecological trap?

MIGRATION

In some species, spatial distribution varies over time, often with the seasons. Indeed, there may be dramatic mass movements of animals, and some of these are migrations. The term *migration* has been defined in many ways (reviewed in Dingle and Drake 2007). The definition that we will use compares migration to other movements made by animals that take them beyond their current home range for a significant period of time (e.g., natal and breeding dispersal). During dispersal, individuals cease moving and settle into a new home range once a suitable location has been found. In contrast, **migration** is movement away from the home range that does not stop upon encountering the first suitable location. Instead, migrating animals continue to move until they eventually become responsive to the presence of resources, such as nest sites and food, and then they stop (Dingle and Drake 2007). From this definition, we see that within a particular species, migratory movements occur over greater distances than dispersal movements.

In some species, migration involves the movement of animals away from an area and the subsequent return to that area. In these cases, animals usually migrate between breeding areas and overwintering, or feeding, areas (Figure 11.9). In other species, however, migration is a one-way affair (i.e., it is not a round trip); many migratory insects, for example, permanently abandon their site of origin. In general, round-trip migrations are associated with long-lived species (e.g., vertebrates) and one-way migrations with short-lived species (e.g., insects) (Dingle and Drake 2007). There is also variation in the distance moved during migration. For example, a salamander may travel less than a kilometer from its woodland home to the pond where it breeds. In other

FIGURE 11.9 **Migrating caribou marching through Alaska in July. During the spring, many caribou breed in the tundra. Beginning in July, they migrate south, where food will be available through the winter.**

species, the distances are truly astounding. Northern elephant seals migrate twice a year—once to breed and again mostly to eat—from beaches in Southern California and Baja California, Mexico, to northern feeding grounds in the Aleutian Islands. Thus they migrate about 8000 km each year, and that is just the horizontal distance. They make frequent, deep dives that can add another 3000 km of vertical distance to their journey (Tennesen 1999). The arctic tern migrates even further, about 20,000 km one way, between its southern and its northern breeding area (Baker 1980). There are also long-distance migrants among the insects, perhaps best represented by desert locusts and monarch butterflies, with one-way migration distances of about 5000 km (Waloff 1959) and 3600 km (Brower 1996), respectively.

To further complicate matters (or to make things more interesting we should say!), migration can take several forms (Berthold 2001; Dingle and Drake 2007). Migration can be *obligate* (an individual always migrates) or *facultative* (an individual migrates if local conditions deteriorate but stays put if conditions remain suitable). There is also *differential* migration, in which the migratory patterns of individuals within a population differ by age class or gender. In some small passerines (perching birds), individuals migrate in their first year but remain on breeding grounds in subsequent years (the species may even be described as resident in the breeding area), while in others, females tend to be more migratory than males. Sometimes both age and gender influence who migrates; in blue tits, for example, there is a predominance of juveniles and females among migrants. These age and gender differences are usually attributed to juveniles and females being less able to compete for food on the breeding grounds when it becomes scarce during the nonbreeding period (Berthold 2001).

Why should an animal bother to travel hundreds or thousands of kilometers to one location only to return to its starting point half a year later? This question probably has many answers, and no single one could apply to the diverse species that migrate. However, the simplicity of one explanation hides its profundity: those animals that migrate do so because they produce more offspring this way. Although the actual costs and benefits of migration vary among species, the benefits must result in the production of more offspring than would be possible if the individual stayed put. To explore this question further, we will consider some of the possible costs and benefits that might accompany migration.

COSTS OF MIGRATION

Migration takes a tremendous toll. Only half of the songbirds that leave the coast of Massachusetts each year ever return, and less than half of the waterfowl in North America that migrate south each fall return to their breeding grounds (Fisher 1979). The mortality rate of black-throated blue warblers is at least 15 times higher during spring and autumn migration than during periods when individuals are not migrating (Sillett and Holmes 2002).

Energy

One reason for the enormous losses associated with migration is that traveling such long distances requires a great deal of energy. For instance, a bird uses about six to eight times more energy when flying than when resting. Imagine yourself running 4-minute miles continuously for 80 hours. This would require roughly the same amount of energy per kilogram of body mass as a blackpoll's nonstop transoceanic flight of 105 to 155 hours (Williams and Williams 1978). It is a good thing, however, that the blackpoll is not walking. It turns out that the energy costs of traveling a certain distance vary with the mode of locomotion. If each vertebrate animal in our comparison obtained the same amount of energy from each gram of fat, then using 1 gram, a mammal could walk 15 km, a bird could fly 54 km, and a fish could swim 154 km (Aidley 1981). However, even if it is easier for birds to travel great distances than it is for you, migration is still metabolically demanding.

Natural selection favors behaviors that reduce the risk of starvation during migration. One way to do this is to store fat before the journey begins. Gram for gram, fat provides more than twice the energy of carbohydrate or protein. Thus it is an extremely important energy source during migration. It should not be surprising, then, that migratory animals as diverse as insects, fish, birds, and mammals put on fat reserves prior to migration (Berthold 2001). Indeed, the body mass of birds that migrate over long distances may more than double before migrating (Klaassen 1996). (Increasing body mass to avoid starvation, however, must be balanced against the potential negative effects of a heavy fuel load on flight performance.)

Even with premigratory fat deposition, some animals do not have sufficient fat stores to complete their journey without stopping. Certain small birds reduce their risk of starvation by refueling along the way. Rather than making a nonstop trip, their migration consists of alternating periods of flight and stopover. These birds tend to spend much more time in stopover than in flight, making time spent in stopover the principal determinant of the total duration of their migration (Alerstam 2003; Hedenström and Alerstam 1997). Dividing migration into short episodes also allows birds to store smaller energy reserves that do not compromise flight performance. How do they know where to stop for food along the migratory route? Some evidence for insectivorous migrants moving between the Neotropical (Central and South America) and Nearctic regions (North America) suggests that they use landscape features, such as the amount of hardwood forest, when deciding where to take a refueling break (Buler et al. 2007). Amount of hardwood forest, it has been found, is positively correlated with arthropod abundance in the understory, something that should be of interest to hungry insect-eating birds.

Not all migratory birds, however, make extended stopovers. Osprey (*Pandion haliaetus*) often exhibit fly-and-forage migration, a strategy that combines foraging with covering migration distance (Strandberg and Alerstam 2007). Individual ospreys deviate somewhat from their migratory route to forage for fish in nearby bodies of water but tend to move on in less than 12 hours (Figure 11.10). This strategy may be used by other birds that fly extensively while foraging, such as falcons and seabirds.

Risk of Predation

Many weary migrants fall to predators. For example, songbirds are "fast food" for the Eleanora's falcon. The songbirds, worn out by their flight across the Mediterranean, land in the nesting area of this falcon, where they become easy prey for the predatory birds in the midst of feeding newly hatched young (Walter 1979). Risk of predation is present not only at the end of a long journey, but also during the journey itself. Some migratory species experience heavy predation because their predators follow their seasonal movements. Lions, cheetahs, and hyenas often track the movements of African ungulates (Schaller 1972); wolves follow North American caribou (Sinclair 1983); and water pythons migrate seasonally to exploit their migratory prey, the dusky rat (Madsen and Shine 1996). Songbirds leaving North America in the autumn for southern overwintering grounds face some 5 million raptors (birds of prey) that are also making the trip and feeding along the way. And, of course, there are nonmigratory predators at stopovers. We see, then, that the predator landscape for migrants has two components: one related to nonmigratory predators located along the route and at the end points, and a second consisting of migratory predators. It has been suggested that for birds the timing and routing of migration have been shaped, at least in part, by predation risk (Ydenberg et al. 2007).

Risk of Inclement Weather

Migration generally occurs during the spring and fall, times of notoriously unstable weather, which can drastically raise the cost of migration. Severe rainstorms and snowstorms kill millions of migrating

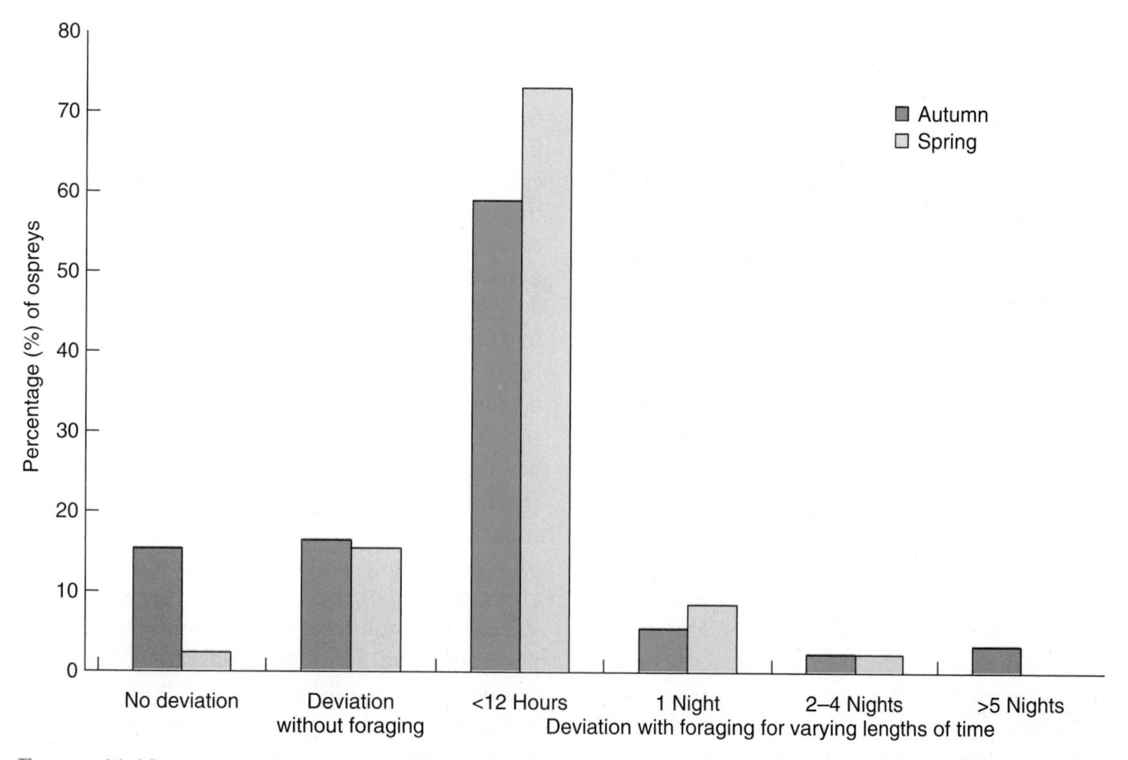

FIGURE 11.10 Osprey often exhibit fly-and-forage migration, a strategy that combines foraging with covering migration distance. Ospreys employing this strategy deviate from their migratory route to forage for fish in nearby bodies of water, but typically stay for relatively short periods of time (<12 hours). In contrast, many other migratory birds alternate periods of flight with relatively lengthy stopovers. (Modified from Strandberg and Alerstam 2007.)

monarch butterflies. It is not uncommon to see thousands of dead or dying monarchs on the shores of Lake Ontario or Lake Erie following a severe storm (Urquhart 1987). In birds, the most devastating mortality occurs when land species encounter storms over water, where they cannot take shelter (reviewed in Newton 2007). Many of these birds are probably lost without a trace, some wash up on shore, and others die upon reaching shore. A flock of Lapland longspurs migrating through Minnesota encountered a sudden snowstorm one night. The next morning, 750,000 of these small birds were found dead on the ice of two lakes, each lake about 1 square mile in size; carcasses were reported from a much larger area, however, leading to a mortality estimate of 1.5 million individuals from this one storm (Roberts 1907a,b). Mortality in migratory birds also results from unseasonably cold temperatures soon after arriving at breeding areas or before departing from such areas in late summer or autumn (reviewed in Newton 2007).

FIGURE 11.11 **Wind turbines that produce electricity, such as this one at the Shoals Marine Laboratory on Appledore Island, Maine, generate little or no pollution and do not expel greenhouse gases. They can, however, pose problems for bats and nocturnally active birds, especially when operated as part of large scale wind farms.**

Obstacles

The cost of migration is high in many cases because large areas of inhospitable terrain must be crossed (e.g., deserts in some cases and water for land birds). And then there is the matter of obstacles. Birds often crash into tall structures such as lighthouses, skyscrapers, and TV towers. In a single night, seven towers in Illinois felled 3200 birds (Fisher 1979). In addition, in the push to develop alternative and renewable sources of energy, wind-powered turbines that generate electricity have been erected in many locations, sometimes on a small scale (e.g., a single turbine) (Figure 11.11) but also on the much larger scale of wind farms. Although wind turbines generate little or no pollution and do not expel greenhouse gases, they can pose problems for nocturnally active birds and bats (Kunz et al. 2007; Kuvlesky et al. 2007). These problems can be categorized as direct or indirect. Direct problems refer to fatalities resulting from night-flying birds and bats colliding with wind turbine rotors and monopoles. Indirect problems result from the alteration of the landscape associated with the development of wind farms (e.g., the construction of roads, buildings, and electrical transmission lines). Monitoring bird and bat fatalities at wind facilities can be challenging owing to issues of searcher efficiency (most searches are done by humans, but in some cases trained dogs are used) and removal of carcasses by scavengers before they can be counted. Nevertheless, the data available suggest that migrating passerines make up a significant portion of the fatalities, as do tree-dwelling migratory species of bats.

In addition to such dangers, territorial animals, such as birds, must relinquish the rights to a hard-won territory each year and compete vigorously to become reestablished the following year. So what rewards could possibly override such disadvantages?

BENEFITS OF MIGRATION

Energy Profit

We can intuitively understand the advantages of moving from approaching arctic winters to the sunny tropics. Even a cursory familiarity with the elements of nature could also convince us of the advantages of simply moving from a mountaintop to a valley every winter. In each case, the animals are trading a less hospitable habitat for a more hospitable one.

The severe weather during the northern winters has favored migration. Each fall, millions of monarch butterflies migrate southward from the central and eastern regions of Canada and the United States to fir forests in central Mexico—sites with particular characteristics that enable the survival of the butterflies (Figure 11.12*a*). These forests are found about 3000 m (nearly 2 mi) above sea level on the southwest slopes of a very small area of mountaintops. An important characteristic of these forests is that their temperature is cool but not freezing. Exposure to freezing temperatures would kill the monarchs. However, warmer temperatures would unnecessarily elevate their metabolic rates and waste energy reserves (Calvert and Brower 1986).

Another characteristic of these overwintering sites that enhances the survival of monarchs is the tall trees, primarily oyamel firs. Besides providing branches on which the butterflies can roost in large numbers, the

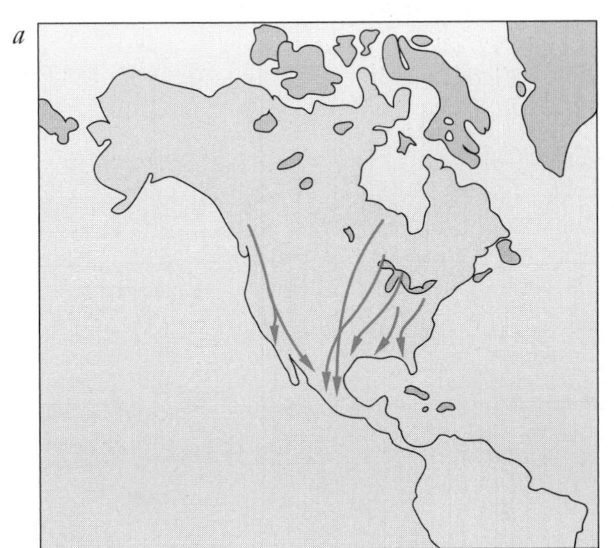

FIGURE 11.12 **The migration of monarch butterflies is truly phenomenal. (*a*) The monarchs travel in large groups from northeastern North America to overwinter in Mexico or from central California to the coast. (*b*) Monarch butterflies roost on trees in their overwintering sites in the mountains west of Mexico City.**

trees form a thick, protective canopy over the butterflies (Figure 11.12*b*). The canopy serves as an umbrella, shielding the butterflies from rain, snow, or hail. This increases survival because a dry monarch can withstand colder temperatures better than can one with water on its surface (Figure 11.13). If monarch butterflies are wet, 50% of the population will freeze at –4.2°C. If they are dry, however, the temperature can dip to –7.7°C before 50% of the population freezes. The canopy also serves as a blanket that keeps the butterflies warm. Openings in the forest canopy increase radiational cooling, which can lower body temperatures to as much as 4°C below the ambient air temperature. The body temperature of monarchs under a dense canopy is approximately the same as the air temperature. However, body temperature drops in proportion to the degree of exposure, increasing the chances that the butterflies will freeze to death (Anderson and Brower 1996).

Seasonal changes in climate also affect food supply. In some species, migration is an adaptation that permits the exploitation of temporary or moving resources. The larvae of monarch butterflies feed only on milkweed. In regions of the eastern United States, however, the milkweed plants grow only during the spring and summer months (Urquhart 1987). Certain species of insectivorous bats may migrate in response to the size of the insect supply. Mexican free-tailed bats, for instance, leave the southwestern United States as the harsh winter climate causes the insect supply to dwindle. They migrate to regions of Mexico where insects are available throughout the winter (Fenton 1983).

As winter approaches, increasing the animals' energy needs, the food supply drops and forces any resident species into more severe competition for such commodities. So, in spite of the energy required for migration, it may result in an overall energetic savings. For example, a study of the dickcissel revealed that despite the energy costs of migration, this bird enjoys an energetic advantage from both its southward autumn migration and its northward spring migration. Studies of the junco, white-throated sparrow, and American tree sparrow show that by avoiding the temperature stresses of northern winters, these species compensate for at least some of the energy spent on migration (reviewed in Dingle 1980).

The question arises, then, if there is so much food in the warmer winter habitats, why do species migrate from such areas? Why do they return to their summer homes at all?

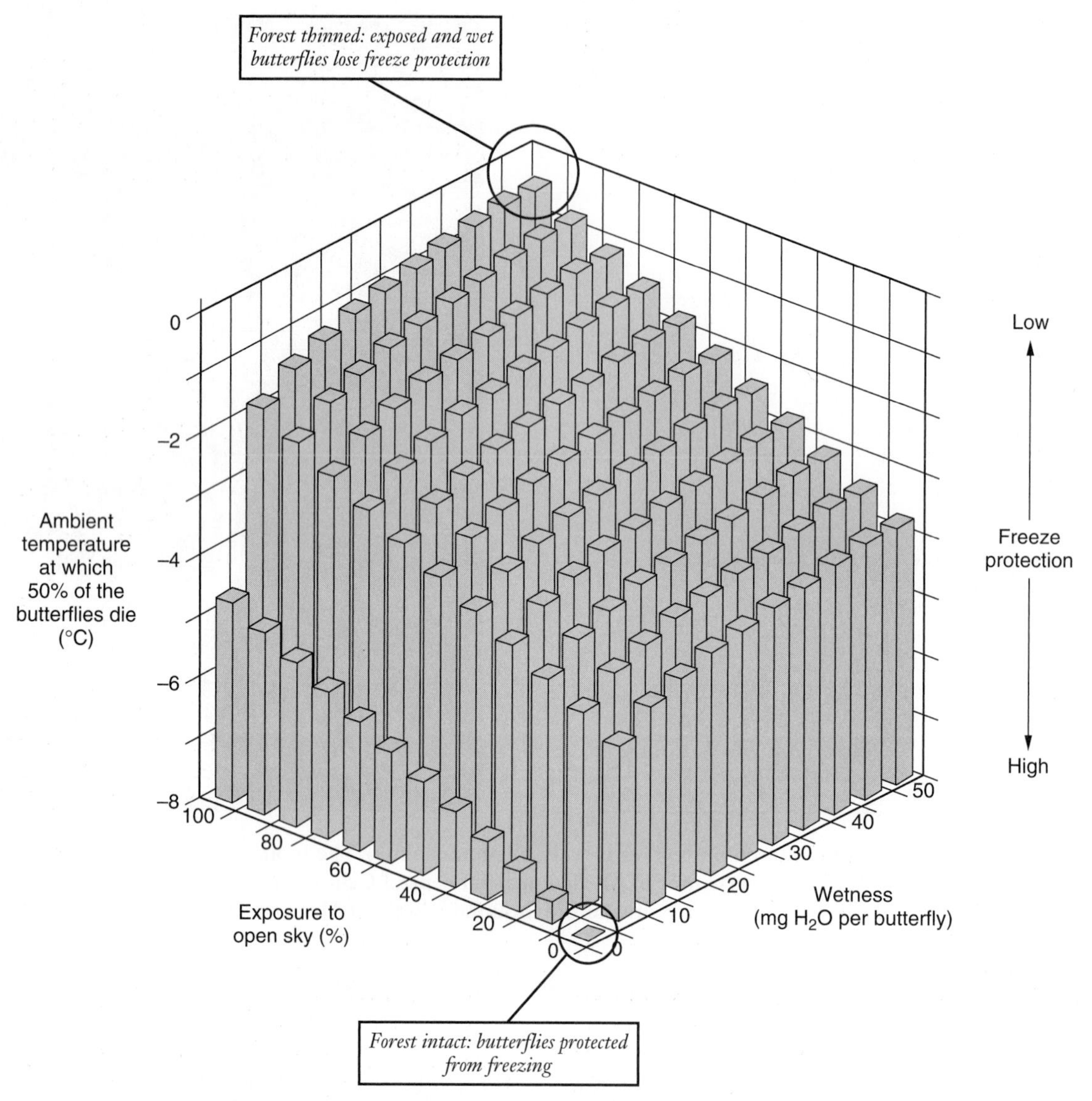

FIGURE 11.13 **The intact forest canopy of the monarch butterfly's overwintering site normally serves as both an umbrella and a blanket, which keeps the monarchs from freezing. The vertical axis indicates the ambient air temperature at which 50% of a butterfly sample will freeze to death. A wet butterfly freezes at a warmer ambient air temperature than a dry one. Because of radiational cooling, exposure to open sky also allows a butterfly to freeze at a higher ambient temperature than if it were protected by the forest canopy. When the forest is thinned because local farmers need the wood, monarchs are vulnerable to freezing, especially on clear nights after wet winter storms. Thus, thinning of the forests in the monarch's overwintering area in Mexico is endangering the monarch population. (Data from Anderson and Brower 1996.)**

Reproductive Benefits

One answer might be that there are important advantages in rearing broods in the summer habitats. For example, days in the far north are long, and the birds' working day can be extended—they can bring more food to their offspring in a given period of time and perhaps rear the brood faster. Another result of long days is that more food is available for offspring and more young can be raised (Figure 11.14). Although factors other than food availability may also play a role, generally the farther north from the tropics a species breeds, the larger is its brood (Welty 1962).

Some species migrate to areas that provide the necessary conditions for breeding or that offer some protection from predators. Gray and humpback whales, for example, breed in coastal bays and lagoons that provide the warmer temperatures needed for calving and help protect the calves from predation. The need for protected rookery sites may prompt seal, sea lion, and walrus migrations, as they come ashore on their traditional beaches after months at sea. Sea turtles also regularly migrate thousands of kilometers between feeding grounds and breeding areas, usually the same beaches on which they hatched. For exam-

FIGURE 11.14 **The arctic tern, a champion of migration, moves between its breeding grounds in the Arctic and its winter home in the Antarctic. By breeding in northern regions, it can take advantage of long days to gather more food for its young.**

ple, female green turtles (Figure 11.15) feed in the warm marine pastures off the coast of Brazil and then swim roughly 1800 km to the sandy shores of Ascension Island, where they lay about 100 eggs, each the size of a golf ball (Lohmann 1992). The beaches where sea turtles lay their eggs are on isolated stretches of continental shores or small remote islands (Lohmann and Lohmann 1996b). Because of their isolation, these beaches might have fewer predators.

FIGURE 11.15 **Some green turtles migrate from their feeding ground off the coast of Brazil to Ascension Island to breed on sheltered beaches, where it is safe from predators.**

Reduction in Competition

Another advantage in returning to the temperate zone is that of escaping the high level of competition that exists in a warmer, more densely populated area. The annual flush of life in the temperate zones provides a predictable supply of food that can be exploited readily by certain species without competition from the large number of nonmigrants that inhabit the tropics (Lack 1968).

Reduction in Predation and Parasitism

A third advantage in returning to temperate zones to breed lies in escaping predation. If predators are unable to follow herds of migratory ungulates, for instance, each individual's chance of survival is enhanced. Thus, escape from predation has been suggested as the reason that the number of migratory ungulates is so much greater than the number of nonmigratory ungulates (Fryxell et al. 1988).

In the far north, breeding periods are very short because of the weather cycles. This can be an advantage to nesting birds, which are in danger from predators. The short season results in a great number of birds nesting simultaneously, thereby reducing the likelihood of any single individual being taken by a predator. Also, since there is no extended period of food availability for predators, their numbers are kept low. By leaving certain geographical areas each year, migratory species deprive many parasites and microorganisms of permanent hosts to which they can closely adapt. Long, harsh winters in the frozen north further reduce the number of these threats.

MIGRATION AND CONSERVATION BIOLOGY

Long-distance migrations are one of the most spectacular of biological phenomena; sadly, they are becoming increasingly rare events. For example, in the Greater Yellowstone Ecosystem, a 19-million-acre temperate ecosystem that spans parts of Idaho, Wyoming, and Montana, many of the historic and current migration routes of ungulates have already been lost. More specifically, all 14 migration routes used by bison have been lost, as well as 78% of the routes used by pronghorn and 58% of those used by elk. The causes for these losses include increases in the human population and associated losses of habitat; in many cases, fences, highways, and housing subdivisions block migration routes, and petroleum development threatens wintering areas. A possible landscape-level solution to this problem lies in creating a protected network of national wildlife migration corridors; in the Yellowstone region, this could protect long-distance migrations by all three ungulate species (Berger 2004).

SUMMARY

Natal dispersal involves permanent movement away from the natal area or social group to where breeding first occurs. In natal philopatry, offspring remain at their birthplace. Dispersal can also occur after reproduction; the movement between two successive breeding areas or social groups is called breeding dispersal.

Costs and benefits are associated with philopatry and natal dispersal. Potential costs of philopatry include those of inbreeding, reproductive suppression, and competition with relatives for mates or resources. On the other hand, a certain level of inbreeding may actually be beneficial due to the maintenance of gene complexes that are particularly well suited to local conditions. Another advantage of staying home is familiarity with the local physical and social setting—animals can find food and escape predators more easily in a familiar area. Dispersers, in contrast, are thought to face high energy costs and risks of predation as a result of increased movement and lack of familiarity with the physical and social environment.

The direction of sex differences in natal dispersal differs between birds and mammals. In most birds, females are more likely to disperse than males. In most mammals, however, males are more likely to disperse than females. At least three hypotheses have been suggested to explain patterns of dispersal in birds and mammals: (1) inbreeding avoidance/local resource competition; (2) local mate competition; and (3) cooperative behavior among kin.

Animals that disperse from their natal site or breeding site must eventually select a new location in which to settle. This process of habitat selection can be divided into search, settlement, and residency. Animals searching for a new place to settle employ comparison (best of N) strategies and sequential search strategies, and likely evaluate resources, and presence of conspecifics and heterospecifics when assessing the quality of a site. There is evidence that natal experience can influence habitat selection.

Migration occurs over greater distances than dispersal, and also has costs and benefits. Among the costs are increased energy expenditure, predation, and exposure to severe weather. A major advantage concerns attaining a favorable net energy balance. In spite of the energetic cost of migration, an animal may experience a net energy profit. For example, North American songbirds that migrate south each fall gain by escaping the metabolically draining harsh winter temperatures and by avoiding the increase in competition that accompanies a reduction in winter food supplies. Migratory individuals may also gain by a reduction in predation or parasitism.

12

Foraging Behavior

On a college campus on a warm spring day, a gray squirrel peers down from a tree and makes a chittering sound. A group of students sitting under the tree finish takeout coffees and snacks and get to their feet, brushing crumbs from their clothes. As they walk away, the squirrel dashes down the tree and seizes a grape-sized piece of muffin from the ground. Without pausing, it turns and runs back up the tree, where it sits on a branch and finishes off the muffin in less than a minute.

Across campus, another squirrel is sniffing the ground near a picnic table on a grassy lawn. There is a line of maple trees about 10 m away. This squirrel also finds a small piece of food, but instead of racing back to the tree, he sits next to the table and eats it. He continues searching, and finds a half a cookie. Lifting his head high to hold it somewhat awkwardly off the ground, he runs to the nearest tree and climbs up. Not a moment too soon—a jogger, with her dog ranging in front of her, appears at the edge of the lawn and soon heads past the picnic table.

If you attend college on a North American campus, you've probably seen something similar to what we just described. What you may not realize is that you've witnessed a series of decisions about **foraging**—finding, processing, and eating food—that might surprise you in their complexity. Squirrels base their foraging decisions on a food item's size, its energetic value, how easy it is to carry, and its distance from cover—and how these variables interact (Lima et al. 1985, Lima and Valone 1986). In this chapter, we will explore how animals make decisions such as these.

OBTAINING FOOD

Animals must consume either plants or other animals in order to live. Animals can neither capture energy directly from the sun, as plants do, nor get carbon or nitrogen (components of biologically important molecules) directly from the environment. Instead, animals must acquire these essentials by feeding on other organisms. Although we are all intimately familiar with foraging, animals have a surprising range of adaptations for acquiring food. We'll explore the diversity of foraging techniques in this section.

SUSPENSION FEEDING

Many aquatic species feed by removing small suspended food particles from the surrounding water. Species vary in the techniques they use. A few species sieve water and strain out the food. Others trap particles on sticky surfaces of mucus. For example, the annelid worm *Chaetopterus* sits in a U-shaped burrow, where it creates a mucous net. As it pumps water through its burrow, food particles are trapped in the mucus, which the worm periodically rolls into a ball and swallows (Figure 12.1).

You might think that suspension feeding would only be a feasible way to make a living for small animals, but the largest animal species on earth feeds this way. Blue whales (*Balaenoptera musculus*), a species of baleen whale, can often weigh over 100 tons, yet survive wholly on tiny shrimp-like creatures called krill. Instead of teeth, baleen whales have rows of fringed plates made of keratin (similar to human fingernails). These whales take in enormous mouthfuls of water, then expel the water through the plates, leaving behind the krill to be swallowed. It has been estimated that blue whales can filter out up to four tons of krill per day.

Suspension feeders are often sessile (stationary) and must take the food that comes their way, but others (like the whales) can move about in order to select good foraging areas. Some species can even adjust their filtration pattern to choose particular types of particles at different times. Among these selective species are dabbling ducks, a group that includes northern shovelers and the

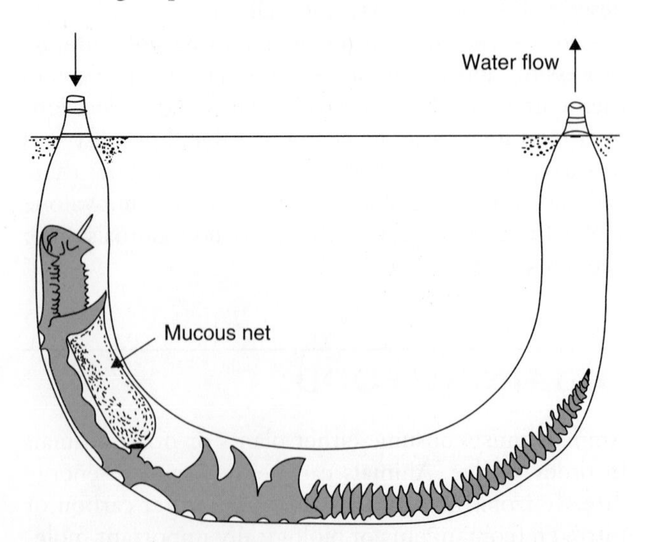

FIGURE 12.1 ***Chaetopterus*, an annelid worm that lives in U-shaped tunnels in the sand, filter feeding with a mucous net. (From Brusca and Brusca 1990.)**

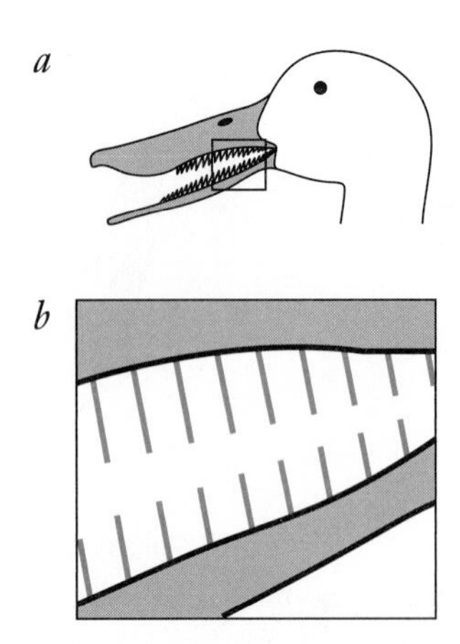

FIGURE 12.2 **The filtering apparatus of dabbling ducks. (*a*) Duck beak, with area of closeup in (*b*). From Gurd (2007).**

familiar mallards. Dabbling ducks strain food from water through filters that fringe the edge of their bills. They can adjust this filtering apparatus by changing the position of their bill while feeding (Figure 12.2; Gurd 2007). When they feed with their bill in a more closed position, they can filter out small particles. When they feed with their bill in a more open position, only larger particles are retained. However, because closing the bill during feeding helps force more water out of the bill, foraging in a more open position reduces the rate at which water, and therefore food, is pumped through the bill. Reducing the foraging rate may be advantageous to ducks foraging in muddy wetlands. Gurd (2007) found that when ducks were given water that had small, unwanted debris particles in it, as well as larger food particles, they only consumed the food, but when they were given water with only food in it, they filtered the food faster. These results show that the ducks found it better to accept a lower foraging rate, but avoid the detritus, than to consume both food and detritus at a higher rate.

OMNIVORY

We will spend most of our time in this chapter treating **herbivory** (plant eating) and **carnivory** (meat eating) separately, for this simplifies our discussion a bit. However, keep in mind that many animals exhibit **omnivory** and eat both plants and animals. Animals may be omnivorous for a variety of reasons, including the limited availability of a preferred food, a need for nutritional variety, and to minimize exposure to risks associated with a particular food type (e.g., predators or toxins that may have cumulative effects) (reviewed in Singer and Bernays 2003).

HERBIVORY

Plants have a variety of parts—roots, leaves, stems, fruits, and flowers—each of which may be consumed by different species of animals in countless interesting ways. To understand the diversity of behaviors associated with herbivory, we need to understand the plant's perspective: a plant may well suffer a loss of fitness if an animal eats its roots or leaves, but it benefits if its pollen is transferred to another plant by a pollinator, or if the seeds in its fruit are carried off to a new germination site. Thus, the challenges facing an herbivore vary depending on whether the plant is marshalling a defense that must be overcome or encouraging an animal's attentions. Let's look at some examples.

Spiny cacti, poisonous hemlock, thorny rosebushes, and prickly thistles are all familiar examples of defended plants. A more unusual and particularly unpleasant defense is used by plants in the milkweed family, which get their common name because their leaves and stems, when cut, ooze a white sap. However, only in appearance is the sap milky—it is sticky and thick, and, in many species, it has chemical irritants that make it even more noxious to touch or taste. Nevertheless, many insects are able to feed on milkweed. Milkweed beetles (*Tetraopes tetrophthalmus*) overcome this defense in an innovative way: by biting repeatedly into the midrib of the milkweed leaf. The sap leaks out through the bite holes (Figure 12.3) before it reaches the tips of the leaves, where the beetle can then feed without gumming up its mouthparts (Dussourd 1999). This isn't good for the plant because midrib cutting and feeding reduce the plant's ability to photosynthesize (Delaney and Higley 2006). Perhaps beetles are, for the moment, winning this particular arms race. (See Chapter 13 for a look at how milkweed and monarch butterflies interact.)

FIGURE 12.3 **Vein-cutting behavior of milkweed beetles (*Tetraopes tetrophthalmus*). After a beetle cuts the midrib of the leave, the sticky noxious sap oozes out. The beetle can then feed at the tips of the leaves.**

While roots, stems, and leaves are often defended, fruits and flowers have evolved to attract animals. Most ripe fruit is red (think of apples and raspberries) or black (think blackberries). Is this because these colors have evolved to attract frugivores (fruit eaters) that will disperse the seeds? Birds, one of the main frugivores, have four types of retinal cones that give them excellent color vision. Foraging birds *do* prefer red and black fruits—but not because they are attracted to the color itself. Instead, red and black fruits contrast well with the typical green background of foliage, and it is this contrast that attracts birds (Schmidt et al. 2004).

Flowers have also evolved visual cues attractive to pollinators. As you probably know, many plants rely on animals for sexual reproduction. A bee, hummingbird, bat, or other pollinator that visits a flower is dusted with pollen. As it moves from flower to flower, it transfers the pollen and flowers are fertilized. Many plants have evolved means of enticing pollinators to visit, such as by offering nectar rewards. Some flowers have markings called nectar guides that act like runway lights, directing pollinators to the part of the flower where the nectar is stored. Even flowers that appear white to us may have nectar guides: bees and many other insects can see in the ultraviolet, and many flowers have markings that are visible only under UV. Other flowers offer specialized rewards. Flowers that are pollinated by carrion flies (which feed on dead flesh) are not recommended for household decoration—their putrid odor has evolved to attract their pollinators, not humans (e.g., Burgess et al. 2004). An especially clear example of a coadaptation between plants and pollinators comes from a species of hummingbird (*Elampis jugularis*) that feeds on two species of *Heliconia*. Female birds have a long, curved bill that matches the long, curved flowers of their preferred flower species, *H. bihai*. The short, straight bill of the male matches their preferred flower, *H. caribaea* (Temeles et al. 2000).

A different approach to herbivory is taken by species that cultivate some or all of the food they need. Agriculture has evolved independently in three insect orders: ants, termites, and ambrosia beetles. Leaf cutter ants cut fresh leaves and carry the pieces back to the nest under the ground (Figure 12.4). There they encourage the growth of a special fungus on the leaves. This fungus, whose existence is unknown outside the ant nests, is the

FIGURE 12.4 **Leaf cutter ants are transporting leaves to fertilize their fungus gardens. These ants cultivate a special fungus that serves as their primary source for food.**

primary food source for these ants (Weber 1972). The ants prepare their fungus gardens by licking the leaves on both sides to remove the waxy layer covering the leaf and reduce the population of microorganisms that might compete with the desired fungus. The leaf fragments are then chewed to a pulp, placed in the fungus garden, and inoculated with hyphae of fungus. This preparation makes the leaves a richer source of nourishment for the fungus. The ants are attentive farmers. When they tidy their gardens, they collect debris and compress it into a pocket inside their mouth. This pocket functions as a sterilization area that kills the spores of a garden parasite. Sterilization is caused by bacteria that live inside the pocket and produce antibiotics (Little et al. 2006). Fungus gardens in abandoned nests degenerate quickly as they become overrun with other microbes (Quinlan and Cherrett 1977).

Other species also modify their own food supplies. Limpets, which are related to snails, leave behind mucous trails as they crawl along on underwater rocks and feed on microalgae. In several species, the mucus not only traps microalgae but stimulates its growth. Limpets then revisit their old slime trails to harvest the crop (Connor 1986). Gorillas are herbivorous and will rip down large plants, resulting in a surge in the growth of young, fast-growing vegetation, which is encouraged by the new light (Watts 1987).

STOP AND THINK

Which is likely to be stronger, selection on the predator to be better at capturing prey, or selection on the prey to be better at escaping the predator?

CARNIVORY

Whereas plants sometimes benefit from herbivores, carnivores must capture food that never benefits from being eaten. Here, we often see evidence of an arms race: prey species have evolved myriad defenses against predation, and predators have evolved to overcome those defenses.

In this section, we will examine the strategies a successful predator might use, and we will devote Chapter 13 to the perspective of the prey. To quote Jerry Seinfeld, when we watch a nature show about antelope, we root "Run, antelope, run! Use your speed, get away!"—but the very next week, watching a show about lions, we cheer "Get the antelope, eat him, bite his head! Trap him, don't let him use his speed!" In this section, prepare to cheer for the predators.

Pursuit

The classic nature-show scene like the one we have just described is what generally comes to mind when we consider predation. It's no wonder: no one can forget the sight of a cheetah in full pursuit of a gazelle. Adapted for speed, with long, slender legs and a flexible spine that allows for an even longer stride, cheetahs can reach a top speed of 70 mph (113 km/h) (Caro 1994).

Other species that chase their prey include several seabirds. Northern gannets (*Sula bassana*) feed by spectacular plunge dives: they plummet down into the ocean to grab unsuspecting fish. By attaching data-logging units to the gannets that measured depth, researchers were able to characterize the dives in detail. Gannets perform both long, deep (more than 8 m) U-shaped dives, during which they use their wings to propel themselves, as well as short, shallower V-shaped dives (Garthe et al. 2000).

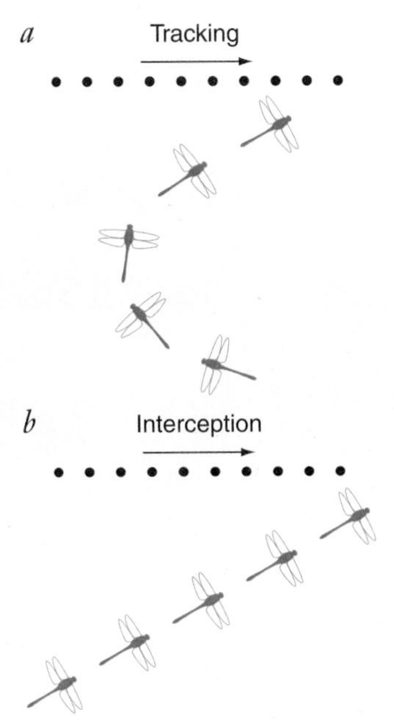

FIGURE 12.5 **Different methods of pursuing a target. (*a*) Tracking behavior, where the predator steers to keep the moving prey, represented by dots, in front. (*b*) Interception, where the predator aims in front of the target. From Olberg et al. (2000).**

Many insects also pursue prey. You might imagine that it could be very tricky for one insect to capture another flying at high speed. Insects use several alternative mechanisms to accomplish this feat (Collett and Land 1978; reviewed in Olberg et al. 2000). First, predators can steer so as to keep the moving image directly in front of them. This results in tracking behavior (Figure 12.5*a*), and if the predator is faster than the prey, the prey will lose the race. A number of insects (e.g., houseflies and beetles) use tracking to pursue objects, be they mates or food. Other insects use the strategy of interception. Instead of aiming at the target's current position, the pursuer intercepts the prey's flight path by aiming at a place in front of it (Figure 12.5*b*). This is the strategy used by dragonflies. Dragonflies have excellent vision and typically perch on vegetation until a prey happens by; then they pursue it and intercept it. Dragonflies are extremely effective predators. In one study, they captured 97% of the prey they pursued (Olberg et al. 2000).

How do dragonflies, or, for that matter, other animals, calculate the necessary course to intercept a prey? It seems like a tricky problem from physics class, but it can be solved by a simple rule. A dragonfly will intercept its prey if the dragonfly moves so that the position of the target on the retina remains at a single point (Olberg et al. 2000). Both fielders catching baseballs (McBeath et al. 1995) and dogs catching frisbees (Shaffer et al. 2004) use a similar strategy. Fielders and dogs, unlike dragonflies, are confined to a two-dimensional field, so they can't restrict the targets on the retina to a single point. Instead, the fielder or dog runs so that the target appears to rise in a straight line. This strategy results in a curved path to the target.

A common strategy of animals that engage in pursuit is to hunt in groups. We'll investigate this behavior in more detail in Chapter 19.

Stealth

Full-out pursuit is exhausting and dangerous. Even predators built for speed like cheetahs do not always engage in a long chase, but will approach their prey slowly until quite close, using a hunting method called stalk-and-rush. They also flush prey concealed in vegetation and pursue them only over short distances (reviewed in Caro 1994). To facilitate surprise attacks, predators often choose hunting sites that provide good cover. Hopcraft et al. (2005) tested two alternative hypotheses: do lions spend more time hunting in habitat where there is good cover or in habitat with a higher density of prey? In the field, lions followed large groups of prey (e.g., migrating herds), but on a smaller scale, they more frequently chose to hunt where there was cover, even when fewer prey were present.

In some cases the predator's ability to draw near is enhanced by camouflage, rendering the predator more difficult to detect. Often this takes the form of markings, colors, or even behaviors that make the predator blend into the background. Jumping spiders of the genus *Portia* use environmental disturbances as a smokescreen to camouflage their approach to the prey. *Portia* preys on other spiders, and often does so by climbing right into their webs. The movement of *Portia* shakes the web slightly, and this can alert the prey spider to the approaching danger. In the field, researchers noticed that *Portia* seemed to move more when the wind blew the web, and to pause when the air was still. In the lab, Wilcox et al. (1996) used a fan to create a breeze to blow on the web. By turning it on and off, they demonstrated that *Portia* does indeed use environmental disturbance to mask its approach, a behavior called smokescreening (Wilcox et al. 1996).

Aggressive Mimicry

In **aggressive mimicry**, a predator gets close to its prey because it mimics a signal that is not avoided by the prey and may even be attractive to it. Some predators have specialized structures that are used as lures. The alligator snapping turtle lies on the muddy bottom of streams and lakes and holds its mouth wide open. Its tongue has a wormlike outgrowth that, when wiggled enticingly, attracts fish that the turtle then snaps up. The deep-sea angler fish lives at extreme ocean depths, where virtually total darkness prevails. Nonetheless, the female has a long, fleshy appendage attached to the top of the head that is luminous and acts as a lure, bringing curious fish within reach of its toothy jaws (Wickler 1972).

a

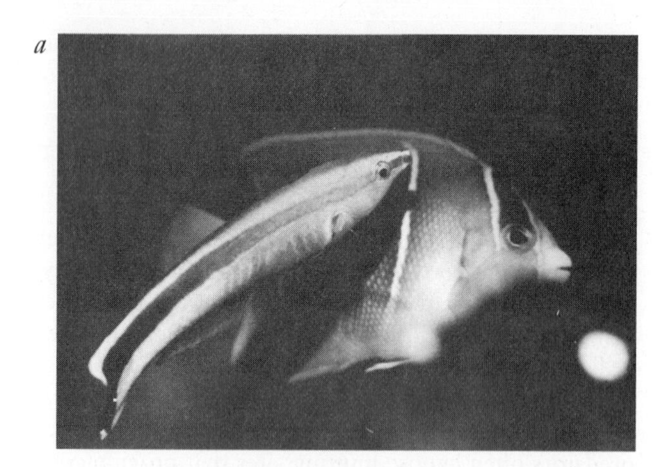

b

FIGURE 12.6 **(*a*) A cleaner wrasse is removing parasites from another fish. The service of cleaner fish is beneficial because it removes external parasites, diseased tissue, fungi, and bacteria. The cleaner wrasse has distinctive markings and behavior patterns that advertise its services. (*b*) *Aspidontus* is disguised as a cleaner wrasse and mimics its behavior. As a result, fish in need of the cleaning service allow the phony cleaner to approach. *Aspidontus* then takes a bite out of the would-be customer.**

Some predatory species draw within striking distance of their unsuspecting prey by mimicking beneficial species. For example, some blennid fish look and act like beneficial cleaner fish to lure their prey. The cleaning wrasse (*Labroides dimidiatus*) removes parasites, diseased tissue, fungi, and bacteria from other fish (Figure 12.6*a*), even swimming into their mouths. Indeed, parasite-laden fish line up at coral reef "cleaning stations" much like cars at a car wash to avail themselves of the services of the cleaning wrasse. Some cleaning stations may have hundreds of patrons each day. The cleaning wrasses generally advertise their services with distinctive swimming motions performed above the cleaning station. Customers show their willingness to be cleaned by postures that are characteristic of the species. A phony cleaner, a blennid fish (*Aspidontus taeniatus*), looks and behaves like the cleaning wrasse (Figure 12.6*b*), but when it is invited to approach, it takes a bite out of gills or other soft parts of the customer (Wickler 1972).

Many predators attract their prey by sending signals that mimic the mate of the prey species. One of the most famous examples is found in fireflies. Not flies at all, these beetles have bioluminescent organs on the ventral side of their abdomens. In northeastern North America, a number of firefly species live in the same habitat. At night, males fly about and flash in species-specific patterns. A receptive female, waiting in the vegetation below, flashes back to a conspecific male. The male flies down to land near her, whereupon they signal back and forth in a "lengthy courtship dialogue" (Woods et al. 2007). However, this does not always work out as a male might wish: sometimes a *Photuris* female signals back to a *Photinus* male. When the *Photinus* male lands to find his mate, the *Photuris* female eats him (Lloyd 1975, 1986). *Photuris* females get more than just calories from this deception. *Photinus* fireflies of both sexes produce defensive compounds. *Photuris* females can only get these compounds by preying on *Photinus* (Eisner et al. 1978, 1997). The severity of this selection pressure is demonstrated by *Lucidota atra*, another firefly that also has defensive chemicals and that is also eagerly consumed by *Photuris*. In contrast to *Photinus*, *L. atra* is diurnal, with only vestigial bioluminescent organs, suggesting that it may have evolved to be day-active in order to escape predation by *Photuris* (Gronquist et al. 2006).

We will discuss more examples of deception in predation and in other contexts in Chapters 13 and 19.

Traps

Some predators are able to trap their prey: they manipulate objects or alter their environment in such a way as to capture, or at least restrain, prey. For example, humpback whales (*Megaptera novaeangliae*) build bubble "nets" to trap, or more accurately to corral, their prey. Beginning about 15 m deep, the whale blows bubbles by forcing bursts of air from its blowhole while swimming in an upward spiral. The bubbles form a cylindrical net that concentrates krill and small fish (Figure 12.7). The whale then swims upward through the center of the bubble net with an open mouth and devours the prey in a single gulp (Earle 1979). Researchers created artificial bubble nets so they could observe the response of prey more closely. Herring were very reluctant to swim through bubble nets, although they were more willing to do so in order to reach a larger group of conspecifics on the other side (Sharpe and Dill 1997).

Some of the most familiar trap builders are spiders. Spider webs come in a variety of forms. The familiar

FIGURE 12.7 The humpback whale traps krill in cylindrical bubble "nets" that it constructs by forcing air out of its blowhole. In this photo, the bubble net is seen as a ring of bubbles on the surface of the water. Using a bubble net, a whale can corral the fast-swimming krill and consume them in a single gulp.

"Charlotte's web"-style orb web incorporates as many as seven different kinds of silk to make the radii, the center hubs, and the sticky spiral (Foelix 1996). Sheet weavers make flat sheets that can be quite plain—just a flat or slightly curved sheet—or much more elaborate. For example, you can probably make a good guess as to what a bowl-and-doily spider web (*Frontinella communis*) looks like: a flat sheet, the "doily," with a bowl-shaped structure sitting over the top of it, all surrounded by support strands. The spider hangs upside down under the web, grabbing insects through the silk that have hit the support strands and tumbled into the bowl above the spider. Other spiders have very reduced webs. *Deinopsis*, for example, stretches a small net of silk between the tips of its feet and waits patiently with its arms outspread. When an insect happens past, it claps its legs together and traps its victim. The prize for the most minimalist web must go to the bolas spider, which uses only a silk line with a drop of glue at the end. It swings the line around with its leg, smacks a passing insect with the drop of glue, and reels it in (Figure 12.8) (reviewed in Yeargan 1994). Some species of bolas spiders lure moths to them by emitting female sex pheromones (Stowe et al. 1987).

FIGURE 12.8 A bolas spider, which catches insects with a drop of glue swung on the end of a silk line.

How spider webs are designed for catching prey has been the subject of a great deal of research—it's generally easy to measure webs and to quantify prey capture success in the field, so it is often possible to quantify the relationship between web architecture and fitness. For example, Venner and Casas (2005) observed the orb webs of *Zygiella x-notata* and found that most of the prey the spiders caught were quite small (less than 2 mm). Large prey (more than 10 mm long) were quite rare and are captured only about every 20 days. However, these large prey are crucial to fitness: spiders cannot survive and reproduce eggs without catching at least some large prey. Venner and Casas suggest that webs are designed to take advantage of these rare events.

ADAPTATIONS FOR DETECTING PREY

Sensory Specializations

One of the recurring themes in this book is that we must continually be aware of the sensory systems of our study organisms. As we look around the animal kingdom, we find many sensory specializations for prey detection that are poorly developed or lacking in humans, and thus not immediately obvious to researchers. Here, we'll make that point by looking at a few examples.

The snakes in two large families, the Crotalidae, or pit vipers (e.g., rattlesnake, water moccasin, and copperhead), and the Boidae (e.g., boa constrictor, python, and anaconda), use their prey's body heat to help guide their hunt. They have special receptors that are so sensitive to infrared radiation (heat) that these snakes can locate their warm-blooded prey even in the darkness of night (Figure 12.9). It's been estimated that the sensory endings must be able to respond to

a

b

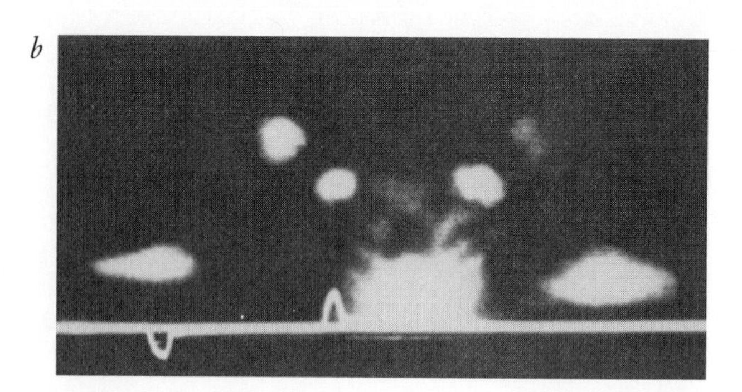

FIGURE 12.9 **(*a*) The heat sensors of the black-tailed rattlesnake, located in pit organs between each eye and nostril (the dark circular structures in this photo), allow it to detect its warm-blooded prey in complete darkness. Snakes of the families Crotalidae and Boidae have infrared heat sensors that help them find their prey. (*b*) A rattlesnake's "picture" of a mouse facing toward it.**

contrasts in temperature of 0.001°C (Bakken and Krochmal 2007). In spite of the ability of snakes to strike precisely at their prey (Figure 12.10), the image on the pit organ is apparently quite poorly focused, which suggests that snakes must select environments to hunt that provide good thermal contrast (Bakken and Krochmal 2007).

In the quiet of the Mojave Desert, running insects transmit vibrations through the sand. Predatory sand scorpions (*Paruoctonus mesaensis*) can detect these disturbances from up to 30 cm away. Scorpions sense vibrations by structures called slit sensilla located on each of their eight legs. Having so many legs is a convenience: by comparing the timing of the arrival of a vibration at each leg, the scorpion can precisely determine the direction of nearby vibrational sources, and unerringly and swiftly attack (Brownell 1984; Brownell and van Hemmen 2001).

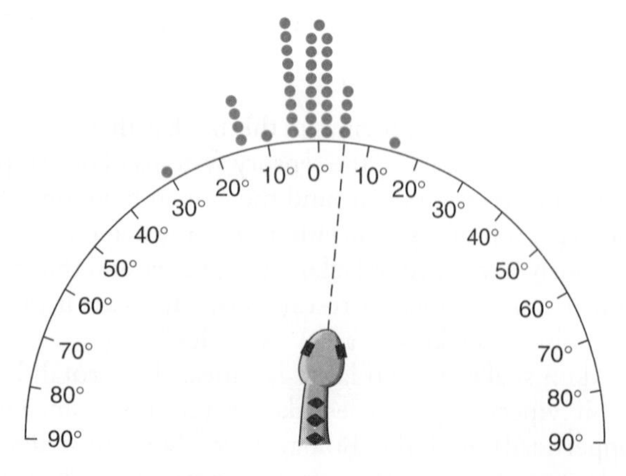

FIGURE 12.10 **A rattlesnake's heat sensors can direct an amazingly accurate strike. Each circle indicates the angular error of a strike at a warm soldering iron by a blindfolded rattlesnake. The average error of the strikes was less than 5°. (Redrawn from Newman and Hartline 1982.)**

The star-nosed mole (*Condylura cristata*) finds prey by using its sense of touch. Critical to that sense is its unusual star-shaped nose, consisting of 22 fleshy, mobile appendages. Aside from its snout, this mole looks like most moles. It is about 6 inches (15 cm) long and has powerful digging claws and poorly developed eyes. Rarely coming to the surface, the star-nosed mole lives in the extensive tunnel system it digs in the wetlands. It searches these tunnels for earthworms and other small prey by touching the walls with its nose as it moves forward. Each second, the nose makes an average of 13 touches. The movements are controlled by muscles that are attached to the skull and then connected to the base of each appendage. Each appendage is sheathed with touch receptors, called Eimer's organs, that communicate with the somatosensory area of the cortex of the mole's brain. One particular pair of appendages is used for detailed investigation, analogous to the foveal area of the human eye that is tightly packed with sensory receptors (Catania and Kaas 1996; Catania and Remple 2004). The mole's Eimar organs are so sensitive and fast-moving that it's been estimated that a mole can examine 300 m of tunnel floor every day (Catania and Remple 2004). High-speed video recording has revealed that star-nosed moles can also detect odors underwater, a feat previously thought impossible. Moles accomplish this by exhaling air bubbles onto objects and then sniffing the bubbles back up (Catania 2006).

Procellariiform seabirds, including the petrels and albatrosses, seem to sniff out seafood patches scattered over vast expanses of open ocean (Nevitt 1999a). These are sometimes called tube-nose seabirds because of their long tubular nostrils, used to excrete excess salt as well as to smell. What can they be smelling as they soar over the featureless ocean, hundreds of miles from land? Like the humpback whales discussed earlier, tube-nose seabirds feed on krill. As predators eat krill, the macerated krill bodies release volatile compounds that seabirds can smell. Krill themselves feed on phytoplankton, which releases an odorous gas, dimethyl sulfide (DMS). DMS can also lead the seabirds to krill.

Gabrielle Nevitt and her colleagues conducted experiments out at sea demonstrating that the odors of macerated krill and of DMS attract seabirds. They created vegetable-oil slicks, some of which were perfumed with krill extract; plain vegetable-oil slicks served as controls. Observers who did not know which of the slicks were krill-scented counted the number of seabirds attracted to the slicks and measured how quickly they appeared. Cape petrels, southern giant petrels, and black-browed albatrosses showed up within minutes. Five times more of these birds appeared at the krill-scented slicks than at the control slicks. The odor did not attract storm petrels or Antarctic fulmars. Equal numbers of these birds showed up at control and krill-scented slicks (Nevitt 1999b). Similar experiments compared the attractiveness of DMS-scented slicks with plain vegetable-oil slicks. A number of bird species, including prions, white-chinned petrels, and two species of storm petrels showed up at DMS-scented slicks twice as often as at control slicks. But three species of albatrosses and Cape petrels were just as attracted to the control slicks as to the scented ones (Nevitt et al. 1995). Thus, it seems that the odors of krill or DMS can provide an immediate and direct way to assess the potential productivity of an area but that different species of seabirds use these cues in different ways.

Recently, researchers have begun investigating the development of the seabirds' ability to follow their noses. Even young blue petrel chicks, one to six days before they fledge, prefer the scent of DMS over a control odor in a choice test. Thus, chicks do not have to learn the scent of DMS through their own foraging experience, but instead embark on their first foraging trip ready to follow the odor to food (Bonadonna et al. 2006).

Other remarkable sensory specializations for detecting prey are found in those most infamous of predators, sharks. Many senses are involved in the shark's astounding ability to detect and track down prey. Sharks can hear prey from a distance of over nine football fields and will turn and swim toward it. When the shark is within a few hundred meters of its prey, its nose can help direct its search. As the shark nears its prey, its lateral line organs detect small disturbances in the water caused by the swimming motion of prey, and so the shark knows the location of the prey even if it cannot see it. Within close range, the shark can see its prey. However, an even more surprising sense is perhaps the most impressive. The story of how this bizarre sense came to be understood is described by Fields (2007) and illustrates how tricky it can be to figure out a behavior that is beyond the normal sensory experience of humans.

The first hint of something unusual came in 1678, when the Italian anatomist Stefano Lorenzini described strange pores speckling the head of sharks near their mouths. By peeling back the skin, he found that the pores led to gel-filled tubes inside the sharks' heads. What could these be? By the late nineteenth century, using improved microscopy, scientists could trace nerves leading from these tubes to the brain, suggesting a sensory function. Other biologists during the 1900s described that the firing rate of these nerves varied with changes in pressure, touch, temperature, and salinity. One scientist, R. W. Murray, happened to switch on a magnetic field near a shark neuron preparation and noticed that the nerve's firing rate skyrocketed. As Fields (2007) describes: "Astonishingly, Murray determined that the organs could respond to fields as weak as one millionth of a volt applied across a centimeter of seawater. This effect is equivalent to the intensity of the voltage gradient that would be produced in the sea by connecting up a 1.5-volt AA battery with one pole dipped in the Long Island Sound and the other pole in the waters off Jacksonville, Fla. Theoretically, a shark swimming between these points could easily tell when the battery was switched on or off."

What could this electromagnetic sense be for? Neural and muscular activity in animals generates electromagnetic signals: their pumping hearts and brain waves, for example, produce short, weak pulses. However, these are not the signals that the sharks' sense organs are tuned to detect. They detect only slow-changing fields, such as those generated by batteries. This is exactly the sort of signal produced by a fish's body in seawater: the salt solution in the fish's body differs from that of seawater.

By the 1970s, Adrianus Kalmijn (1966, 1971) was the first to demonstrate that sharks can use this sense to locate prey by electrical fields. He found that captive sharks (*Scyliorhinus canicula*) would attack electrodes buried in the sand of an aquarium. In the ocean, Fields and his team used a fiberglass (nonmetallic) boat. They lowered pairs of electrodes into the water but activated only one. Observers were not told which was active. Then the biologists dumped ground-up fish into the water and waited for the sharks to come. Time and again sharks attacked the "live" electrode, demonstrating that in natural conditions, sharks respond to electrical cues.

The use of bioelectric cues is not restricted to sharks but is also found in rays, skates, lungfish (Figure 12.11), and some larval amphibians. An example with unexpectedly similar talents is the platypus (*Ornithorhynchus anatinus*), an oddball egg-laying Australian mammal. The platypus has both electroreceptors and mechanoreceptors on its ducklike bill that help in locating prey (Figure 12.12) (Griffiths 1988; Scheich et al. 1986). (Electroreception was lost with the move to land and reevolved in monotremes, the group of mammals that includes the

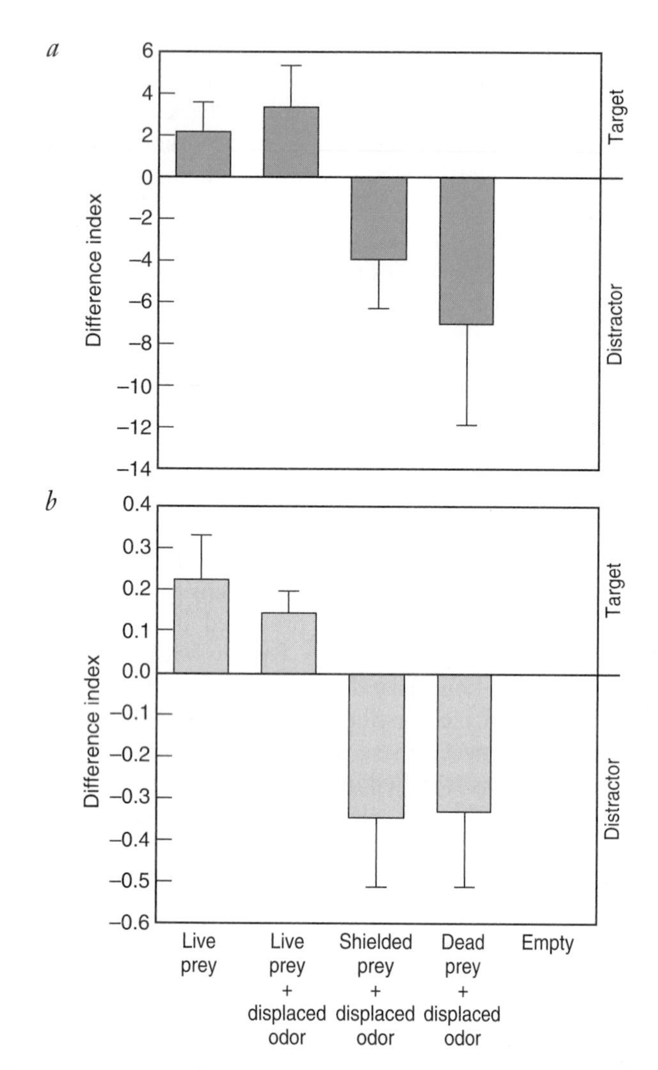

FIGURE 12.11 **The results of experiments demonstrating that Australian lungfish can locate their prey by using the electrical field generated by a living organism in seawater. The results are shown as the difference between the amount of foraging activity above the target, the hidden chamber where the prey was located, and the distractor, the region of the aquarium to which the prey's odor was displaced. A positive score indicates that the fish were foraging above the target; a negative score indicates that the fish were foraging over the distractor. The difference indices are shown for (*a*) the foraging intensity and (*b*) the foraging accuracy of each treatment. Although the lungfish can and do use chemical cues to locate prey, bioelectric cues are more important. (From Watt et al. 1999.)**

platypus.) The platypus hunts at night in murky streams. It swims slowly, patrolling for food with its eyes, ears, and nostrils closed. As it swims, it makes intermittent head sweeps, scanning the area for electrical signals. When it detects an electrical stimulus, it swims two to three times faster and makes larger, regular head sweeps at an average rate of two sweeps per second. The platypus is most sensitive to electrical stimuli that originate slightly below it and almost completely to its side (80° lateral from the tip of the bill). A platypus can even detect a battery lying on the bottom of the creek. A hunting platypus will often dive to the bottom of the body of water and dig in the silt with its paws and bill. As it probes, it stirs up small animals, including crayfish, insect larvae, and freshwater shrimp, which dart away. As the prey swim off, they generate electrical fields sensed by the platypus's electroreceptors, while the mechanical waves in the water are sensed by mechanoreceptors on the platypus's bill. Together these sensory organs give the platypus a complete, three-dimensional fix on the position of its underwater prey (Manger and Pettigrew 1995; Pettigrew et al. 1998).

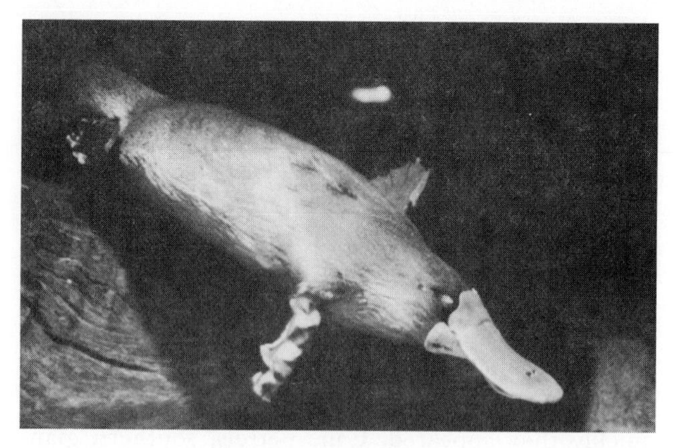

FIGURE 12.12 **The duck-billed platypus is among the predators that can locate prey by detecting their electric fields.**

The lungfish and the platypus use electrical cues given off by animals to locate prey, but other species locate prey by generating their own electrical field and sensing disturbances in that field created by the prey. The black ghost knifefish (*Apteronotus albifrons*) is a nocturnal predator that uses its electrical sense to feed on small prey such as the insect larvae and small crustaceans in the freshwater rivers of South America. This fish has two types of electroreceptor organs distributed over its entire body. One type, the tuberous electroreceptor organ, is specialized for detecting alterations in the electrical field that the fish generates itself and uses for orientation (electrolocation is discussed further in Chapter 10). In contrast, the ampullary organs are specialized to detect electrical fields generated by external sources, such as prey animals. These receptors, in coordination with the mechanosensory lateral line system, help the fish hunt for prey at night or in muddy water. An agile swimmer, the black ghost kitefish can swim backward as easily as forward and hover in place. It can even swim upside down or horizontally while making sweeping searches for prey. By changing its position, velocity, and orientation, it can influence the pattern of incoming electrical signals, making it a formidable predator.

Search Image Formation

So far we've been talking about adaptations of the sensory organs that enable animals to detect prey. Now let's consider a more subtle change that happens as an animal forages. Luuk Tinbergen (1960) was watching birds bringing insects to their chicks in the Dutch pinewoods when he became aware of an interesting pattern. As the season progressed, the relative abundance of different insect species varied—some species became more plentiful, while others hit a population peak and then faded away. One might expect that birds would capture the different species in proportion to their abundance. Instead, when a species was in low numbers, it was taken less frequently than would be expected. As it became more common, it was taken more frequently than expected. What could be going on?

Perhaps the underlying process is similar to what we commonly experience when we are searching for an object. As you walk along a pond's edge, you may find that frogs seem to materialize in front of you and leap into the water with an "Eep!" However, if you take time to scan the area for a while, you can begin to pick out froggy shapes, and soon you will see many frogs where previously there had appeared to be none. The heightened ability to detect a target is called forming a **search image**. Perhaps this is what drove the pattern Tinbergen saw: after the birds got experience with a particular species of prey, they began to focus their attention on it, and they ignored other prey that did not fit their search image. Since Tinbergen's observation, this phenomenon of "overselection" has been found under controlled conditions in the laboratory (e.g., Bond 1983; Langley 1996; Langley et al. 1996). In a typical experiment, birds are given a chance to forage for food, which can be manipulated to either stand out from or blend in with the background. Overselection is more likely to be exhibited when prey are camouflaged and difficult to pick out against the background—obvious prey are generally eaten in the expected proportions.

An assumption of the search image hypothesis is that a search image is specific. If animals are really forming a search image, we would expect that they improve in their ability to detect only a particular kind of prey, not all camouflaged prey. Pietrewicz and Kamil (1979, 1981) used an elegant approach to simulate predator–prey interactions between blue jays (*Cyanocitta cristata*) and one of their normal prey types, underwing moths (*Catocala* spp.). Hungry blue jays were given a slide show (Figure 12.13). Some slides contained a moth; some did not. If a moth was present and the bird saw it, it could peck at the slide and receive a mealworm reward. After a short interval, the next trial began. When no moth was present, the blue jay could abandon its search of that area and look elsewhere by pecking at a key that would then advance the projector to the next slide. The bird could

a

b

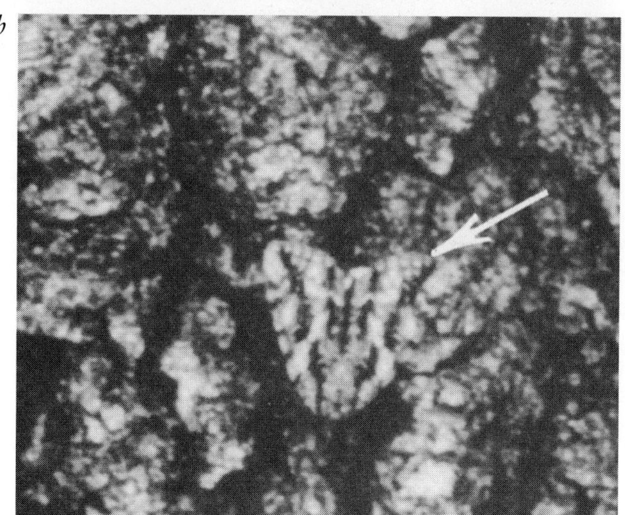

c

FIGURE 12.13 **(*a*) A blue jay is "foraging" for cryptic prey. The blue jay was shown slides, some of which contained a cryptic moth. Among the cryptic moths shown to the blue jays on slides were (*b*) *Catocala retecta* and (*c*) *C. relicta*. If the bird was shown a slide that contained a moth and pecked the appropriate key after spotting it, the bird received a mealworm reward.**

therefore make two types of mistakes. One was a false alarm—incorrectly responding as if a moth were present. The other was a miss, in which the bird failed to see the cryptic moth and pecked at the advance key. Either of these mistakes resulted in a delay before the hungry predator could search for prey in the next slide. This experimental design, though it seems quite artificial, has the advantage over experiments using free-living birds because it allows the experimenters to control the order in which the birds encounter prey.

The search image hypothesis predicts that recent experience with a prey type would allow the birds to learn its key characteristics and prime them to look for others of that type. Encounters with one species of camouflaged moth would help jays find that species more accurately and quickly, but would not improve their ability to find a second camouflaged species. The jays were shown a series of slides, half of which contained a moth. They were tested under three conditions: (1) the moth-containing slides showed *C. retecta* only, (2) the moth-containing slides showed *C. relicta* only, and (3) the two moth species were shown in random order. The results, shown in Figure 12.14, are exactly what would be predicted by the search image hypothesis. The jays' ability to detect one prey type improved with consecutive encounters. However, prey detection did not improve when the two species were encountered in random order. Thus, we see that experience with one type of cryptic prey improved the predator's ability to find that type of prey but not other kinds of cryptic prey.

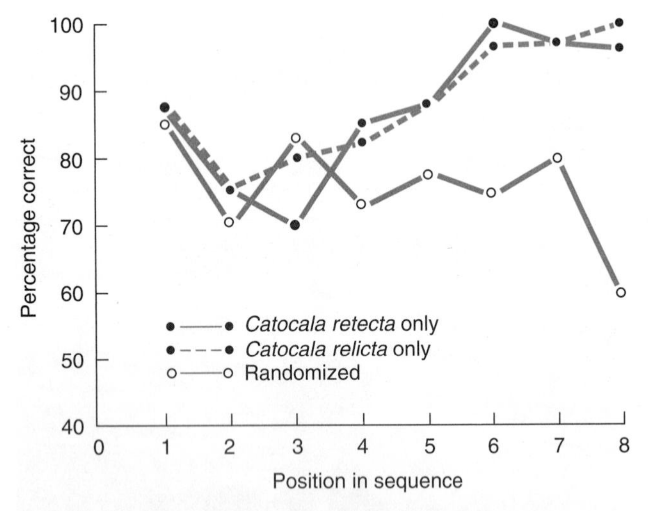

FIGURE 12.14 **The percentage of correct responses by blue jays when shown moth slides in sequences of the same species (runs) or a random sequence of both species. After experience with one cryptic species of moth, blue jays became better able to detect that species. However, if the species were shown in random order, the jays' performances did not improve. This is consistent with the search image hypothesis. (Data from Pietrewicz and Kamil 1981.)**

What exactly comprises a search image? If you are looking for a book with a green cover and white lettering among the papers, tape dispenser, tissue box, and empty coffee cups on your desk, are you trying to compare a mental image of the book against each of the objects on your desk? Or, instead, are you looking for a particular feature (e.g., green items)? An interesting experiment suggests that, at least sometimes, animals are looking for particular features. Photographs of wheat and beans were digitally manipulated so that either color, shape, or both were changed. After experience with a run of normal stimuli, pigeons were given one of these manipulated stimuli. Pigeons attended to both color and shape of beans, but only to color of wheat—and this is the feature that best allowed them to discriminate the grain from the background (Langley 1996).

The study of search image formation has new practical implications. The detection of hard-to-find objects is exactly the problem faced by airport personnel looking for weapons on X-ray screens or in bag searches. One particular type of airport employee has been shown to use search images: sniffer dogs trained to detect explosives in luggage. When dogs were exposed to a high percentage of TNT containers relative to other explosives and were thus given the opportunity to form an olfactory search image, they were better able to detect TNT (Gazit et al. 2005). (Be reassured—the dogs' ability to detect explosives was very high even without the chance to develop a search image.) The sniffer dogs also illustrate that search images need not be visual.

OPTIMAL FORAGING

Are there any rules to make foraging as efficient as possible? In this part of the chapter, we shift away from the details of specific examples and expand to a broader perspective. We will use optimality theory to understand the decisions that foraging animals make. In Chapter 4, we introduced the technique of **optimality modeling**. Animals have different behavioral options (strategies) available to them, and we use models to weigh the costs and benefits of each strategy. A model is a mathematical expression of all the costs and benefits of each strategy. These costs and benefits are measured by a common **currency** that represents some measure of fitness. In other words, animals that make the best choice are favored by natural selection. Remember that we do not need to assume that the animal is able to work out complicated solutions in its head—natural selection can, in a sense, do the hard work of giving animals the appropriate abilities for solving the optimality problem.

Foraging has been a favorite subject for testing optimality theory because it is relatively easy to fit into a modeling framework. First, we can often break down foraging into a series of decisions and then focus on one type of decision at a time. Examples of some of these decisions are what to eat, where to look for food, how long to search one area before moving on, and what sort of path to take through an area.

Second, often we can identify a logical currency, or common measure, by which to compare these decisions. For example, for many species it is reasonable to assume that it is useful to maximize the rate of energy gain over a particular time period. Increased food intake increases survival and fecundity (number of offspring) in many species. (To take just two of many examples, both spiders and songbirds produce more offspring when given more food; Frey-Roos et al. 1995; Sherman 1994.) Thus, we are often reasonably comfortable in using short-term measures of foraging success as an indicator of long-term fitness. Energy consumption (caloric intake) has the additional advantage that it can often be measured directly. Thus, energy-based models are a reasonable place to start. However, we should always think critically about, and try to test, whether our study system meets the assumption that the foraging behavior we are interested in has evolutionarily meaningful consequences.

Finally, we can often identify limitations, or **constraints**, on foraging behavior. For example, an animal's ability to gather food may be constrained by its gut capacity, its ability to detect food, or the presence of predators in the environment.

In the next section, we will describe two models that illustrate different types of foraging decisions: diet selection and movement from one patch of food to another. These are both simple models with only a handful of variables and some restrictive assumptions. We will then explore ways to improve their realism.

DIET SELECTION: A SIMPLE MODEL

Foraging animals often encounter many items that are possible to eat. The question is whether they should include all these different types of food in their diet, or whether they should instead focus on just some of them and ignore the others. You can imagine that, in real life, all sorts of considerations might go into this decision—how hard it is to gather particular items, how rare they are, how tasty, their nutritional value, if they are dangerous to catch, and so on. All of these are variables that describe some aspect of each food item, and each might well be important for a given species. However, it would be very difficult to account for all of them at once.

Instead, we are going to begin by stripping this problem down to its bare bones and focusing on only a few variables. In general, models oversimplify nature. As a rule of thumb, simpler models have more general conclusions that apply to a wider range of examples, but more detailed models tend to make more precise and accurate predictions for a given situation. This example is a very simple model. It's best to read this sample model slowly and try to follow each step. (If you are one of the many readers who tends to skip over equations, this is a good chance to practice translating math into English.)

Here is the scenario that we are modeling. A forager is searching for food. Two kinds of food are available in the environment, and the forager only finds one piece of food at a time. Thus, when the forager encounters a piece of food, it has a decision to make: it could eat it, or it could ignore it and keep looking (and perhaps find a better piece of food). In modeling jargon, two strategies are available to the forager.

Next, in order to decide which strategy is best, we must compare them using a common currency. In this model, we will assume that the best decision that a forager can make is the one that produces the fastest rate of energy gain. Thus, our currency is the rate of energy gain.

Next, we must identify the constraints in our model. One constraint is the amount of time it takes to process food: a nut, for example, must be shelled, a crab must have its carapace removed, but a juicy caterpillar need only be swallowed. We will call this processing time the **handling time**. We will also assume that an animal cannot eat food and look for food at the same time.

It also takes time to find food, and different types of food may be easier or more difficult to find. For example, moths on trees may be difficult to discover, and other prey items may be relatively rare. We call this the **search time**. These are simple ideas, and it is easy to describe them using variables:

E_1 = the amount of energy gained by eating prey type 1 (in calories)

E_2 = the amount of energy gained by eating prey type 2 (in calories)

For handling food:

h_1 = the time it takes to eat prey type 1 (in seconds)

h_2 = the time it takes to eat prey type 2 (in seconds)

For searching:

S_1 = the amount of time it takes to find prey type 1 (in seconds)

S_2 = the amount of time it takes to find prey type 2 (in seconds)

To see what the next step in constructing our model should be, consider the currency. We want to measure the rate of energy gain from each prey type. A good clue can be found in the units: rate of energy gain is measured in calories/second.

Once a forager has found a prey, what is the rate of energy gain? We call this the profitability of each prey.

$$\text{profitability of prey type 1} = \frac{E_1}{h_1} \text{ in units of } \frac{\text{calories}}{\text{second}}$$

$$\text{profitability of prey type 2} = \frac{E_2}{h_1} \text{ in units of } \frac{\text{calories}}{\text{second}}$$

To make this model easier to discuss, let's define prey type 1 as the type with the highest profitability. Now we are ready to make our first prediction based on this model. Imagine a forager has just found prey type 1. Should it eat it? (Would you? Think about this for a moment before you continue reading.)

In our model, a forager should always eat prey type 1. It has the highest profitability, so the forager can never do better than to eat it. This is the first prediction from our model.

Now consider a more challenging question. Imagine a forager has just found prey type 2. Should it eat it, or should it ignore it and keep looking?

To answer this question, compare the following rates of energy gain. The first is the rate of energy gained by eating prey type 2. The second is the rate of energy gained by ignoring prey type 2 and continuing to search. (Take a minute to write out what these would be in symbols before you look ahead.)

The profitability of prey type 2, once a forager has found it, is E_2/h_2 The rate of gain of finding and eating prey type 1 is a little different because the forager must find it first, and then it can eat it. Thus, the two rates we must compare are:

$$\frac{E_2}{h_2} > \frac{E_1}{S_1 + h_1}$$

The left side of the equation says "Now that I have found prey type 2, what is the energy per second that I will gain if I eat it?" The right side of the equation says "What is the energy I will gain per second if I look for and eat prey type 1 instead?" When the rate on the left side of the equation is greater than the right side, the forager should eat prey type 2. This is the second prediction of the model: the predator should switch instantaneously between including prey type 2 in the diet or not, depending on which side of the equation gives the higher rate of energy gain.

Take a minute to examine this equation a bit further. Which of the variables is missing? You will notice that S_2 is not in the equation. It may seem counterintuitive, but the third prediction of the model is that the forager should not take into account the search time for prey type 2 when making this choice. Another way to say this is that the number of the less profitable prey in the habitat (type 2) should not influence the choice. If the equation above says that the forager should not eat prey type 2, then the forager should *always* ignore it, even if prey type 2 is piled up to the forager's chin!

Tests of the Diet Model

How does one test the optimal diet model? Let's begin with a few specific examples. Observations that are qualitatively consistent with these predictions were found in a field study of redshanks (*Tringa totanus*), shorebirds that feed on worms (Figure 12.15). Redshanks feed on both large and small worms. Large worms provide more energy than small ones. In some locations, large worms are common (search time, S_1, is short) and in other locations, they are rare (S_1 is long). At sites where large worms were common, the birds were indeed more selective and ignored the small worms (Goss-Custard et al. 1998).

In field studies such as this one, the rate at which the subject encounters different prey types is largely out of the experimenter's control. To better control for encounter rate, John Krebs and his co-workers (1977) designed an apparatus that would allow them to control the abundance of two different food items presented to great tits (*Parus major*). The birds were allowed to pick mealworms off of a conveyor belt that ran past their cage. This allowed Krebs to control both the type of prey encountered and the abundance of the different prey. The prey items were small and large pieces of mealworms. The energy value of large mealworms was greater than that of small ones, but the handling time was equal for the two prey types. As predicted by the model, birds always ate the large mealworms. As the large mealworms were made more plentiful, the birds became selective and generally ignored the small ones. However, the shift between including and excluding small worms in the diet was not sudden, as predicted by the model, but gradual. A number of explanations have been proposed for what is termed *partial preferences* for high-ranking prey, including discrimination errors (confusion between large and small prey) and the time required to learn the values of *S*, the rates of encounter of different prey types (Krebs and McCleery1984). Even when Krebs's experiment was repeated with modifications, including giving the birds time to estimate *S*, birds still never excluded the smaller prey from their diet (Berec et al. 2003).

So, how successful is optimal diet theory? Sih and Christensen (2001) reviewed 134 studies from 1986 to 1995 and scored each on the degree of fit of the data to predictions. They found a great deal of variation in the performance success of the model. Interestingly, the model did a good job in explaining the diets of foragers that consume immobile or essentially immobile prey (such as leaves, seeds, nectar, mealworms, and clams) but did not do well when prey were active. This makes sense: the optimal diet theory, as originally formulated, does

FIGURE 12.15 **The redshank is a shorebird that feeds on worms. Large worms are more profitable than small ones. When redshanks forage in areas where large worms are abundant, they are more selective and eat more large worms than they do in areas where large worms are rare.**

not take into account the behavior of prey. Two types of prey may be of equal abundance, but if one is better at hiding or is more likely to escape successfully, then the predator's diet will reflect that.

DECIDING WHEN TO LEAVE A PATCH: THE MARGINAL VALUE THEOREM

As an animal forages within a particular location, food in that patch may become more difficult to obtain. Imagine picking blueberries along a roadside. There are patches where blueberries are common, separated by areas without berry bushes. As you pick berries from a particular patch, your success rate will change over time. The longer you stay, the slower your bucket will fill, as the berries get more and more rare. Similarly, prey may become increasingly rare as a predator hunts because they take evasive action. Regardless of the cause, at some point it will become advantageous for the forager to move to a new patch, where food will be easier to find. Here again we can model the decision of the forager: when should it stay in a patch, and when should it go? This problem was modeled by Charnov (1976) in the **marginal value theorem**.

This model is exceptionally easy to present as a graph. The curved line in Figure 12.16 is the gain curve. It is the cumulative amount of energy that a forager has gained as it stays in a patch. To return to our berry-picking example, you can imagine that the *y*-axis is the total number of berries that you have accumulated in your bucket. The curve flattens and nears an asymptote as the berries run out. So, when should you give up and move to another patch?

STOP AND THINK

It's a bad year for berries. Each patch of bushes has fewer total berries, but the travel time between bushes remains the same. Should you stay in each patch for a longer, shorter, or the same amount of time than in a normal year?

To answer this question, you need to know more information: how far away is the next patch? If it's a few steps away, you might be more likely to leave and move to a better patch as soon as berries start to become difficult to find. However, if it's a 15-minute walk, you might be more likely to strip every last berry off the bushes. We can add the travel time to the next patch as a point on the left-hand side of the *x*-axis. It represents the time it will take before you can start getting energy from the next patch. By drawing a tangent from this point to the gain curve, and dropping a line straight down, we can determine the best time to leave your current patch. This is called the marginal value. If an animal leaves a patch before this time, it will be traveling when it could still be profitably foraging, but if it stays too long, it will waste time searching for food in a depleted area.

Tests of the Marginal Value Theorem

The marginal value theorem has been tested in a wide range of organisms, both vertebrates and invertebrates. Many studies mimic the example above almost exactly: the experiment measures the duration that foragers spend in either natural or artificially constructed patches. The marginal value theorem can also be applied to slightly different questions, such as predicting how much food an animal should carry in a single load back to a central place, such as a nest or burrow. For example, eastern chipmunks carry food stuffed in their cheeks. It gets harder and harder to add more food as their cheek pouches get full, resulting in a gain curve just like that in Figure 12.16. The amount of time that chipmunks spent at a seed tray (patch time) increased with the distance between the seed tray and the burrow (Giraldeau and Kramer 1982).

Reviews of the performance of the marginal value theorem (Stephens and Krebs 1986, Nonacs 2001) show that there is good qualitative support for it. In general, animals prefer rich patches to poor ones, patch residence times correlate with patch quality, and increased travel time leads to longer time in patches. However, foragers

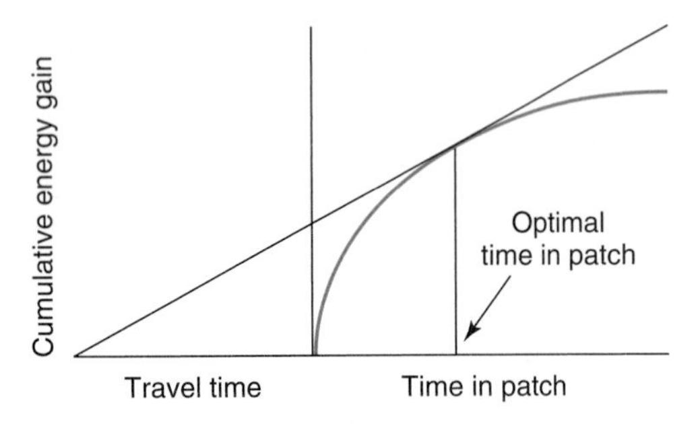

FIGURE 12.16 **The marginal value theorem. The curved line is the gain curve. To the right side of the *x*-axis, the time that an animal stays in the patch increases. To the left side, travel time increases. A line drawn from the travel time and tangent to the gain curve indicates the optimal time to spend in a patch. A shorter travel time means a shorter optimal time in the patch.**

consistently stay longer than predicted (Nonacs 2001), suggesting that this simple version of the patch model might be *too* simple. Better results are obtained if, for example, the nutritional state of the forager is considered (Nonacs 2001). We'll return to the role of nutritional state later in this chapter.

ADDING COMPLEXITY AND REALISM

These two simple models give us a starting point for thinking about foraging, but they do not make accurate predictions for every case. Now let's look at some added variations that can help us better predict the foraging strategies of animals.

Energy Alone Is not Enough

So far we have assumed that the only thing a forager has to consider is maximizing its energy intake, but we should not be surprised that animals often have specific nutritional requirements. The commonly touted nutritional guideline for humans, "Eat a variety of foods," may also apply to other animals. For instance, nestlings of the European bee-eater (*Merops apiaster*) convert food to body weight more efficiently if they are fed a mixture of bees and dragonflies than if they eat only bees or only dragonflies (Krebs and Avery 1984).

A classic example of nutritional constraints comes from the moose (*Alces alces*) that live on Isle Royale in Lake Superior. They must obtain enough energy for the growth and maintenance of their huge bodies (Figure 12.17), but they also have a minimum daily requirement for sodium. The leaves of deciduous trees on the shore contain more calories than aquatic plants, so to maximize energy intake the moose should eat only land plants. However, land plants have a low sodium content. In contrast, low-calorie aquatic plants

FIGURE 12.17 **Moose must obtain enough energy for growth and maintenance, and the leaves of deciduous trees contain more energy than aquatic plants. However, moose must also obtain a minimal amount of sodium. So, although sodium-containing aquatic plants provide less energy, moose must eat some low-calorie plants. This is an example of how nutrient requirements can influence optimal foraging.**

have a higher concentration of sodium. A moose balances these needs by eating a mixture of plants so that energy intake is as great as it can be while sufficient sodium is still obtained. When sodium needs are added to the model, moose behavior is more accurately predicted (Belovsky 1978). Another forager that attends to nutrients is the wild stripe-tailed hummingbird. Hummingbirds spent more time at feeders that had vitamin tablets dissolved in the sugar solution than at control feeders with only sugar water (Carroll and Moore 1993).

Insects also regulate their nutrient intake. For instance, when locusts (*Schistocerca gregaria*) that have been kept on a diet low in protein or carbohydrate are given a choice of food, they select the food that redresses their nutritional deficiencies (reviewed in Simpson et al. 2004). Similarly, tent caterpillars (*Malacosoma disstria*) can select nutritionally balanced food over unbalanced food. However, there is an odd twist to this story: these caterpillars are extremely gregarious, and they steadfastly follow the silk lines of their nestmates. If the first caterpillar chooses the wrong food type, the rest follow the trail and become "trapped," unable to reverse the suboptimal choice (Dussutour et al. 2007).

Incomplete Information

You may have noticed something very unrealistic about the models we have discussed: they assume a great deal of knowledge on the part of the forager. In the diet model, for example, our calculations assume that the ani-

FIGURE 12.18 **Chipmunks generally forage on seeds beneath deciduous trees. Thus, their food is found in patches of fluctuating abundance. As the quality of their current feeding location declines, they spend more time sampling the food abundance at other locations.**

mal knows the search time for each of the two types of prey. This is not necessarily true: a forager may be able to learn that with experience, but it takes some time. Similarly, with the marginal value theorem we assume that the forager knows average travel time between patches, and what the gain curve of other patches looks like. What if you don't know where all the blueberry bushes are? To gain this knowledge, the animal must be familiar with the area and must perhaps even periodically sample other patches.

Some foragers, in fact, do monitor their environment in this way. For instance, chipmunks feed on seeds from deciduous trees, found in patches of fluctuating abundance (Figure 12.18). The supply of seeds below a particular tree may be here today and gone tomorrow, varying with the schedule of ripening, amount of wind, and activities of other animals. As a result, chipmunks must decide how often to check the seed supply at other trees to determine whether it might be beneficial to switch foraging locations. Of course, time spent checking other trees is time lost feeding, so there is an optimal amount of sampling. In one experiment, the chipmunks fed from artificial patches—trays of sunflower seeds—and the value of a given patch was manipulated by varying the number of seeds on the tray. The chipmunks spent more time sampling the food density at other locations as the quality of the patch being exploited decreased (Kramer and Weary 1991).

Although chipmunks seem to successfully monitor their environments, other animals may not, or may not yet have had time to accurately sample. In a few species, animals are able to assess their environment by watching conspecifics; we will return to this issue in Chapter 19. When evaluating the predictions of a model, we must carefully assess whether the animal has all the information at its disposal that the model assumes it does.

Rules of Thumb

We know that animals do not actually make the complex calculations that models do, any more than you solve equations before selecting a snack from the refrigerator. Animals, however, may be able to get close enough to an optimal behavior by following an approximation called a **rule of thumb**. Let's look at an example.

Northwestern crows (*Corvus caurinus*) search along the waterline during low tide for whelks, which are large snails. A crow will pick up a whelk, carry it over the rocks, fly almost vertically upward, and then drop the whelk. If the whelk smashes open, the crow eats the meat. If not, the bird retrieves the whelk, flies upward, and drops it again, repeating the procedure until it does break.

Reto Zach wondered whether the crows were selective about which whelks they dropped. He also wondered whether there were differences in the ease of handling whelks of different sizes. One might expect that flying upward would be the most costly part of eating whelks in terms of energy. Then Zach asked whether crows adjust their cost according to the expected caloric value of the meal. How many times would a crow continue to drop a whelk that was difficult to crack open?

Zach (1978) found that the crows preferentially prey on large whelks. He demonstrated this finding in several ways. First, he collected broken pieces of whelk shells, and by comparing the size of the base of the shells to those of living whelks, he estimated the size of the whelks that had been eaten. The whelks selected by the crows appeared to be among the largest and heaviest on the beach. Larger whelks provide more energy, so this observation was consistent with the hypothesis that the crows could distinguish profitable prey. However, Zach was cautious in his interpretation. Perhaps, he reasoned, the pieces of smaller shells were more easily washed out to sea by wave action, leaving the larger pieces overrepresented in this sample. So, Zach offered each of three pairs of crows equal numbers of small, medium, and large whelks and, at hourly intervals, recorded the number of each size taken. Crows selected large whelks (Table 12.1). Even after the crows had eaten most of the large whelks, they continued to ignore smaller ones. Was this because large whelks are more palatable? No. By removing whelks from their shells and presenting equal numbers of each size class to crows, Zach demonstrated that all size classes were equally palatable.

Why, then, do crows accept only the largest and heaviest whelks on the beach? One reason seemed simple enough—larger whelks have a higher caloric value. However, Zach wondered whether size affects the ease

TABLE 12.1 Numbers of Small, Medium, and Large Whelks Laid Out and Cumulative Numbers of Whelks Taken Over the Subsequent 5 Hours*

	Size of whelk		
	Small	Medium	Large
Total laid out	75	75	75
Taken after 1 hour	0	2	28
2 hours	0	3	56
3 hours	0	4	65
4 hours	0	4	68
5 hours	0	6	71

*Results from three pairs of crows were homogeneous (replicated goodness-of-fit test) and therefore combined. Right from the start whelks were taken nonrandomly ($p < .005$; single classification goodness–of-fit test).
(From Zach 1978.)

of breaking the shell as well. To answer this question, he collected whelks of different sizes and dropped them onto rocks from different heights. He found that large whelks were more likely to break than the medium and small whelks; the large ones required fewer drops from any given height.

We see in Figure 12.19 the total height to which a crow would have to fly, on average, to break whelks of the three different sizes at different heights. Notice that the probability that a whelk will break increases with the height of the drop. The lower the dropping height, the more drops are required to break it. The total vertical height needed to break the shell can be determined by multiplying the number of drops by the height of the drop. There is a height—slightly more than 5 m—at which the total height required to break a whelk is minimized. Zach (1979) found that the crows dropped whelks from a mean height of 5.23 m (±0.07 m).

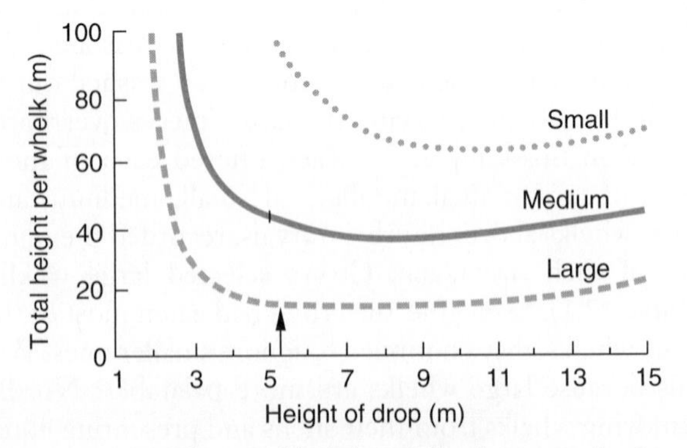

FIGURE 12.19 The total height of the drop required to break the shells of different-sized whelks. Northwestern crows choose large whelks and drop them from a height of about 5 m (indicated by the arrow). This minimizes the energy used in obtaining food and yields an energy profit. The crows would use more energy than they obtain by feeding on small or medium whelks. (From Zach 1979.)

Given this information, Zach (1979) was able to calculate a crow's expected energy profit from whelks of different sizes. He found that crows are likely to gain 2.04 kilocalories from a large whelk, but they use 0.55 kilocalories to obtain it. Thus, the net energy gain for a large whelk is 1.49 kilocalories per whelk. Medium and small whelks, on the other hand, require more energy to handle because they are harder to break. They also contain fewer calories. Zach calculated that medium whelks actually *cost* energy to eat: the net energy gain for a medium whelk is –0.30 kilocalories. Obviously, small whelks, which are even harder to break and contain fewer calories, would be even more costly to eat. Crows thus attempt to feed upon only profitable whelks, which are large whelks.

Crows make similar decisions when feeding on littleneck clams (*Tapes philippinarum*). Again, the largest clams are the most profitable, and again, crows choose the largest clams to drop (Richardson and Verbeek 1986).

So far, the crows' behavior is just as predicted by the models. Let's add another twist. If crows are given a choice between clams and whelks, what should they do? Clams, per unit of weight, offer more calories than whelks, so even a slightly smaller clam should be preferred to a larger whelk. However, crows did not always pick the most profitable prey. Instead, they followed the rule of thumb, "Take the heaviest prey item." This rule led them, on some occasions, to pick a hefty whelk over a not-quite-so-heavy clam, and thus not get as many calories as they might have. However, these losses were not huge, and the rule of thumb provides a reasonably good approximation of optimal behavior (O'Brien et al. 2005).

We often see rules of thumb such of these. It is very often impossible for an animal to gather the data it would need to follow an optimal strategy, as behavioral, physiological and time constraints limit its ability to assess its environment (Stephens and Krebs 1986). One may

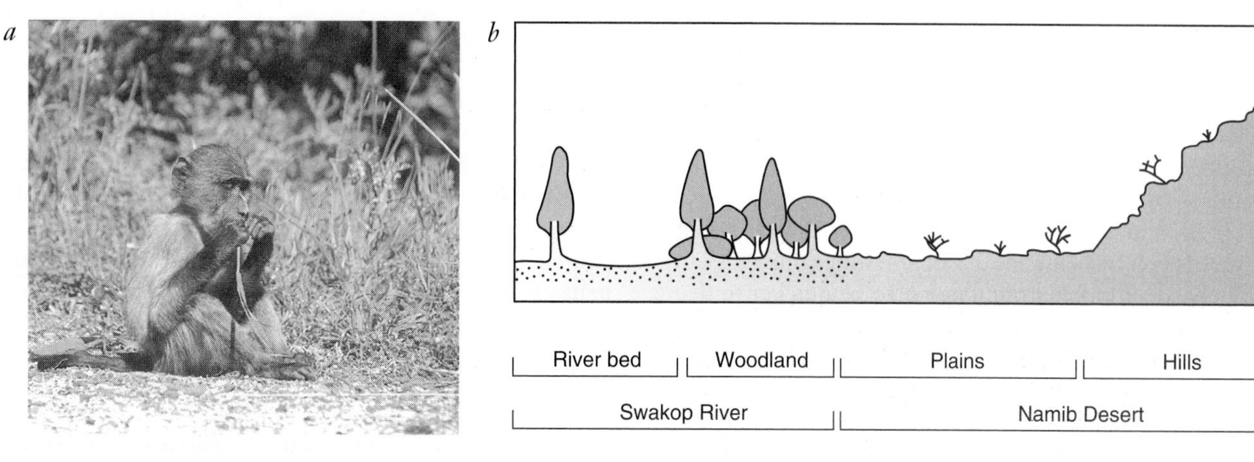

FIGURE 12.20 **(*a*) Chacma baboons make a trade-off between predation risk and energy gain. They feed on leaves, flowers, fruits, and pods. They sleep in trees or on cliffs, but they spend most of the day on the ground, where they are exposed to predators. The primary predators are leopards and lions. (*b*) The four habitats available to the baboons in the study are riverbed, woodland, plains, and hills. Both the risk of predation and the abundance of food differ in each habitat. The baboons do not spend much time foraging in the woodland, where food is most abundant. Instead, they feed mostly in the riverbed habitat. Although less food is available, this habitat is safer from predators. (Diagram from Cowlishaw 1997.)**

approach this problem by comparing the performance of these rules of thumb against the performance of an optimal strategy, as we did for the crows, or by modifying the model to include more realistic constraints. Either approach gives us insight into exactly what determines animal decisions.

Avoiding Predators

Animals foraging for food are potential meals themselves and would do well to take that into account. Numerous studies confirm that animals do indeed assess predation risk (Brown and Kotler 2004; Lima and Dill 1990).

How Foragers Minimize Predation Risk Changes in foraging behavior in the face of predation risk are manifested in a number of ways, including: (1) avoiding dangerous places, (2) avoiding dangerous times of day, (3) increasing vigilance when foraging, and (4) selecting portable foods. Let's look at each of these in turn.

Species from a range of taxa forage in less profitable but safer sites, rather than in more profitable but more dangerous areas. To take an example, the foraging behavior of a desert population of chacma baboons, *Papio cynocephalus ursinus* (Figure 12.20*a*) was studied by Cowlishaw (1997). The site was Tsaobis Leopard Park, Namibia, located in southwest Africa. The park's environment is rugged, with mountains and ravines, as well as gravel and alluvial plains. Four different habitats can be identified in the reserve—riverbed, woodland, plains, and hills—each differing in food availability and predation risk (Figure 12.20*b;* Table 12.2). These baboons have a simple diet. During the time span of this study (late winter), 92 to 97% of the baboons' feeding time was spent gathering leaves, flowers, fruits, and pods from only five plant species. Over 90% of the food energy was found in the woodland and almost all of the rest in the riverbed. The most common predators of baboons are leopards and lions, which hunt by stalk-and-rush methods. They have better luck capturing prey from shorter ambush distances and are, therefore, more successful in habitats with cover that hides their approach. Cowlishaw estimated the predation risk in each habitat. The woodland had the highest estimated risk of predation, the riverbed and the plains had a modest risk, and the hills had the lowest risk. The woodland, unfortunately for the baboons, had the most food. When foraging, the baboons spent more time in the safer, riverbed area, even though much less food could be found there. Other activities, such as resting and grooming, were usually conducted in the hills, where the predation risk was lowest.

The threat of predation can also affect when an animal chooses to forage. Scorpions, fearsome predators

TABLE 12.2 **Relative Food Abundance and Predation Risk in Four Habitats Available to Baboons in Tsaobis Leopard Park**

	Food abundance	Predation risk
Riverbed	Modest	Modest
Woodland	High	High
Plains	Negligible	Modest
Hills	Negligible	Low

themselves, can also be victims. One species forages less on moonlit nights when they are likely to be more visible to predators such as owls (Skutelsky 1996). Because scorpions detect their own prey by its vibrations, as described above, their own foraging success is not compromised by this cautious behavior.

If an animal is foraging in a dangerous place, it can reduce its risk by being more vigilant: for example, it can lift its head and look around more, or be quicker to flee if it hears a noise. Vigilance may come with an energetic cost, as it is often difficult or impossible to be vigilant while foraging. For example, when wolves were reintroduced into Yellowstone National Park, greater elk (*Cervus elaphe*) spent more time being vigilant and less time feeding (Laundre et al. 2001).

Perhaps less obviously, predation can also affect diet choice. Remember the example that opened the chapter. Under risky conditions, foraging gray squirrels will sometimes reject items with a higher profitability (energy gained per second of handling time), in favor of less profitable items that are easier to carry to a safe place (Lima and Valone 1986).

Quantifying How an Animal Perceives Risk An interesting tack that many researchers have taken is to "ask" animals exactly how risky they perceive a particular site to be. For example, gerbils (*Gerbillus a. allenbyi*) were given a choice between foraging for seeds in a safe plot or a "risky" plot. "Risky" plots were either exposed to the simulated light of a full moon or were visited by a trained owl flying overhead. As you might expect, gerbils preferred the safe plot over the risky plot when both held the same number of seeds. However, gerbils were willing to accept increased risk when the price was right: if enough seeds were added to the risky plot, gerbils would forage there (Abramsky et al. 2002).

The Cascading Effect of Risk Avoidance The following experiment hammers home the wide-reaching ecological consequences of antipredator behavior by foragers. Beckerman et al. (1997) studied an ecosystem of spiders, their grasshopper prey, and plants that the grasshoppers fed on. Grasshoppers decrease their movements when in the presence of spiders, presumably to avoid detection. The question was whether this antipredator behavior affected the amount of plant damage. The researchers devised a clever way to distinguish between the effect of the spiders consuming the grasshoppers and the effect of the antipredator behavior adopted by the grasshoppers in response to the spiders. The research team created "safe" spiders by gluing their mouthparts together. These spiders could move around normally but could not kill prey. Grasshoppers responded to glued spiders in the same way as they did to normal spiders—by decreasing their movement. The researchers set up various types of enclosures with plants, grasshoppers, and glued or normal spiders. The presence of spiders—glued or unglued—reduced the amount of plant damage by grasshoppers. The spiders' mere presence changed the behavior of the grasshoppers enough to have a significant, measurable impact on plants. Thus, this example illustrates how antipredator behavior can have a cascade of consequences for an ecological community.

The Presence of Competitors

American crows (*Corvus brachyrhynchos*) drop walnuts to break them open in the same way that northwestern crows drop whelks (Figure 12.21). Recall that the upward flight to drop the food item is energetically costly. Cristol and Switzer (1999) have shown that American crows adjust the height from which they drop a walnut according to the circumstances. English walnuts break more easily than black walnuts, and the crows drop the English walnuts from lower heights than black ones. The crows adjust the height of the drop to account for substrate hardness. American crows often feed in large flocks, and so there is always the threat that a dropped nut will be stolen by another bird. The crows also adjust the height to minimize the chances of theft. The risk of theft was determined by using an index that combined the num-

FIGURE 12.21 American crows adjust the manner in which they handle food so that energy gain is maximized. The crows eat both English and black walnuts, which they break open by flying up to 30 m upward and dropping them onto the ground. The energetic cost is the upward flight to drop the walnut. American crows adjust the drop height to account for the hardness of both the walnut and the ground. Since they feed in flocks, there is a risk that a dropped walnut will be stolen by another bird. The crows lower the drop height when the threat that the walnut would be stolen exceeds a certain threshold.

ber and proximity of conspecifics. When the risk exceeded a certain threshold, the crows lowered the drop height so that they could recover the nut before it was snatched by another bird.

You can see that some crows do all the hard work of flying up, whereas others get something for nothing and grab the nut that someone else has discovered and dropped. This sort of pilfering is not at all unusual, and a whole class of models has been created to explore it. These are called producer/scrounger models, where a producer is the animal that makes the resource available and the scrounger is the one that steals it.

The Role of Internal State

The simplest foraging models assume that all animals behave the same way. For example, the diet selection model took into account only energy, search time, and handling time. However, individual foragers vary. Some might be more experienced foragers than others, in better condition and thus able to move more quickly, closer to reproduction, or hungrier. These variations across individuals may mean that the optimal decisions of foragers differ. For example, the colonial spider (*Metepeira incrassata*) lives in large colonies with a shared frame web. Each spider puts up an individual orb within the frame. The spider's position within the colony influences foraging success and the risk of predation or parasitism. Individuals on the periphery of the colony are more successful at capturing prey, getting 24 to 42% more flying insects than individuals in the colony's core. Unfortunately, because predators (wasps and birds) approach from the edges of the colony, peripheral spiders are also more likely to be eaten. Thus, as we have seen in other species, there is a trade-off between foraging success and predation risk. This trade-off is especially pronounced for large females because predators prefer them. Furthermore, if the eggs are left unguarded at the periphery, there is an increased chance of egg-mass parasitism. As a result, most individuals hatch and begin life in the central regions of the colony. Younger, smaller spiders may take a chance and build their orbs on the edges of the colony, where they can obtain more food and grow faster, thereby increasing the odds of reaching sexual maturity. But as the spiderlings mature, the balance of risks changes and safety becomes more important than foraging success. So, the larger spiders that have reached sexual maturity prefer the core positions (Rayor and Uetz 1990).

Any situation where each individual's traits influence the decisions that are optimal requires a much more sophisticated model than we have presented so far. One way to approach this problem is with dynamic state-variable models (Clark and Mangel 2000). Every individual is described by a set of variables, each of which symbolizes one of its changeable attributes (its state). So, for example, a modeler might represent hunger with one state variable, and size with another. Thus, even given the same set of ecological circumstances, different individuals may make different decisions. The decisions an animal makes on one day influence its state, which in turn influences the decisions it makes the next day, making the system change dynamically. These models thus capture more of the complexity of real-life foraging but are more computationally complex.

Risk Sensitivity: Response to Variability

In the models we've considered so far, we've been assuming that foragers are maximizing the long-term rate of energy gain. However, foragers may also respond to the variability in food availability. That is, an individual may have to choose between a site that reliably supplies a moderate amount of food and one that fluctuates between a rich and poor food supply. The individual that chooses a variable site could get lucky and find plenty of food quite easily, but there is always the risk that food will be scarce. In the jargon of foraging theory, the term *risk* in these models refers to variability in food abundance. Some animals are gamblers and choose the variable site. They are called risk-prone. Others, those who are risk-averse, tend to choose reliable sites where they are more or less guaranteed of finding at least some food (Stephens and Krebs 1986).

Risk sensitivity has been documented in a variety of taxa, including insects, fish, birds, and mammals (Kacelnik and Bateson 1996). In a typical experiment, an animal is offered a choice between a constant option that always offers the same amount of food, and a variable option that offers the same *average* amount of food as the constant option, but with variability. The animal is given the chance to learn about both these options before making its choice.

One interesting pattern in many studies is that an animal's hunger level often determines whether it is risk-prone or risk-averse (Bateson 2002; Kacelnik and Bateson 1996). Why might this be so? The reasoning is as follows: an animal that fails to find a certain minimal amount of energy each day will die of starvation. If enough food can be found at the site that provides a stable food supply, there is no benefit in gambling on finding sufficient food at the variable location. However, if the stable site does not provide enough food to prevent starvation, the only chance for survival is to forage in the location where the food supply is variable and hope for the best. In general, because animals that are full are less likely to starve, they should be risk-averse, whereas hungry animals should be risk-prone (Stephens and Charnov 1982).

THE UTILITY OF MODELS

Now that we have seen that few animals behave exactly as predicted by our simplest models, it's worth asking whether it is useful to develop them at all. The perspective of models, as we described in Chapter 4, is that modeling provides a chance for us to clarify our assumptions about our study animal, such as how its behavior relates to evolutionary fitness and what information it has about its environment. Models are really just a way to formally state a hypothesis about a behavior. Based on that hypothesis, we can generate testable and often quantitative predictions. If we can successfully predict animal behavior, we are more confident—though rarely positive—that we understand it.

SUMMARY

There are many ways in which an animal can obtain food. Many aquatic animals filter food from the surrounding water. Some animals are omnivores, eating both plants and animals. Herbivores forage on plants. Many behaviors have evolved to overcome plant defenses. Others foragers are drawn in by plants, which benefit by having their seeds dispersed or flowers pollinated. Some animals work as farmers and cultivate their own food food. The leaf cutter ant, for instance, maintains fungus gardens.

Carnivores are often engaged in arms races with their prey. Some carnivores rely primarily on speed to capture their prey, whereas others use stealth to approach without detection. Some lure their prey with specialized structures or behaviors that mimic the courtship displays of the prey species. Other predators use traps.

Many predators have sensory specializations that improve their ability to detect their prey. For example, some species of snakes detect the body heat that emanates from their warm-blooded victims. The sand scorpion, on the other hand, is exceedingly sensitive to vibrations in the sand that are created by its prey. Star-nosed moles have bizarre nasal appendages that are highly sensitive to touch. Petrels and albatrosses have an extraordinary sense of smell. From a human perspective, perhaps the most unusual sensory specialization is the ability of some animals, including sharks, to detect the electrical fields that living organisms generate when they are in seawater.

Another adaptation for finding cryptic prey is the use of search images. Here, a predator learns the key features of particular prey and focuses on them while hunting.

According to optimality theory, natural selection favors the behavioral alternative whose benefits outweigh its costs by the greatest amount. Several simple models form the basis for making predictions about optimal foraging. The first model is the diet selection model, which assumes that foragers should attempt to maximize their rate of energy gain. The profitability of a food item is defined as the energy it provides divided by the time it takes to find, capture, prepare, and digest that item. The model predicts that when the most profitable items are abundant and search time for them is short, less profitable items will be eliminated from the diet. In general, tests of the model have shown that it does a good job when predicting the behavior of foragers that consume immobile or essentially immobile prey.

A second decision that animals must make when they are foraging is whether or not to leave a patch of food and move onto the next patch. Again, we assume that animals are maximizing their rate of energy gain. The marginal value theorem, a graphical model, uses the energy gain within a patch and travel time to the next patch to predict when an animal should leave the patch. When travel time increases, animals should stay longer in their patch. The predictions of the marginal value theorem are generally supported in empirical tests, but foragers consistently stay longer than predicted.

There are numerous reasons why these simple models may not be adequate to predict behavior. Animals may choose less profitable food items because of specific nutritional requirements. They may not know everything that the model assumes they know; they may have to spend time sampling their environment. They may be using "rules of thumb" to approximate the optimal behavior. They may be taking the possible presence of predators or competitors into consideration.

Our basic models assume that all individuals behave the same. However, this may not be the case. Individual characteristics of foragers, such as hunger and age, may mean that different behaviors are optimal for different individuals. More complex models are needed to deal with this situation.

Some animals are sensitive not only to rate of energy gain, but also to variability in the availability of food. Risk-sensitive animals have preferences when given a choice between a feeding location that supplies a moderate, but constant, supply of food and one that fluctuates between a rich and a poor food supply. Often, animals that are hungry tend to be risk-prone, choosing the variable site, whereas satiated animals tend to be risk-averse, choosing the constant site.

The simplest models often cannot predict exactly how animals forage. However, models force us to clarify our assumptions, generate testable predictions about behavior, and increase our confidence that we understand the factors that influence behavior.

13

Antipredator Behavior

What could be better to eat than a butterfly? Soft and juicy, with no teeth to nip you or claws to scratch you, butterflies are preyed on by birds, mammals, and spiders alike. In light of the constant threat of predation, butterflies have developed an impressive array of devices to outsmart their enemies, and their protective strategies appear to work, at least some of the time. Here we consider the antipredator strategies employed by the monarch butterfly (*Danaus plexippus*).

Like many animals, monarch butterflies use a combination of color pattern and behavior to avoid being eaten. Their boldly patterned, orange, black, and white wings warn potential predators that they taste bad, their unpalatability being due to their assimilation of noxious chemicals from food plants. In particular, monarch larvae feed on milkweed plants (Asclepiadaceae) and incorporate toxins, called cardiac glycosides, into their own tissues (Brower et al. 1968). Predatory birds that eat one of these insects, even as adults, have severe vomiting and tend to avoid butterflies of similar appearance in the future. However, one might wonder what good it is to be filled with toxins if the individual must be eaten before the poisons will work. The advantage is that many predators release, unharmed, prey that are brightly colored and bad tasting (some predators may even have an innate aversion to bright colors). Poisons, stolen from plants, may thus deter some predators. However, not all milkweed plants contain the same amount of cardiac glycosides. Butterflies reared on plants that do not contain substantial amounts of the poison are quite palatable, although they may still be avoided by predators who have had experience with more noxious members of the species.

No defense system works all the time. Even for the monarch butterfly, the effectiveness of protective strategies varies with the season, species of predator, and context of the predator–prey encounter. For example, in the late summer and autumn, monarch butterflies from eastern North America migrate to the mountains of central Mexico (Brower 1996). The months spent in Mexico, however, are far from a winter vacation (Figure 13.1). Birds of two species, the black-backed oriole and the black-headed grosbeak, have penetrated the monarch's chemical defense system; these two species eat an estimated 4,550 to 34,300 butterflies per day in some overwintering colonies (Brower and Calvert 1985). The oriole selectively strips off relatively palatable portions of the butterflies' bodies (e.g., the thoracic muscle and abdominal contents), and the grosbeak appears insensitive to the cardiac glycosides (Fink and Brower 1981). However, all is not lost for the monarch. Facing the prospects of an avian feeding frenzy each winter, the butterflies reinforce their antipredator system by converging in enormous numbers at their overwintering colonies; some of these colonies have tens of millions of individuals (Brower et al. 2004; Calvert et al. 1979). By forming dense aggregations, it is likely that the monarchs dramatically dilute the predation risk to any one individual. Also, because predation is most intense at the periphery of the colony, central positions are highly sought after and are quickly assumed by the first individuals to arrive. In the life of a monarch butterfly, it does not pay to be fashionably late in arriving at the overwintering site. Forming dense aggregations, of course, can have benefits in addition to avoiding being eaten (see Chapter 19), and monarchs may gain additional benefits from overwintering in large groups.

Our example of the monarch butterfly illustrates a general point about antipredator behavior: animals typically have multiple devices to avoid being eaten. Monarchs also illustrate a second point that we will discuss next: the colors that function in antipredator devices often have other functions as well. In monarch butterflies, the orange color that deters predation when paired with black and white in bold patterns also functions in reproduction. Males with deeper orange on their wings mate more often than those with a lighter shade of orange (Davis et al. 2007). In monarchs, then, the intense orange color works in concert in two different behavioral contexts—it helps to deter predators and it enhances male mating success. It is not hard to imagine, however, a scenario in which a color or pattern that functions as an antipredator device might not be ideal for mating success. For example, a species that relies on having dull colors that blend with the environment might face problems when trying to attract a mate, a time when bright, conspicuous colors are usually best.

Because predation is such a pervasive theme in the pageant of life, we can ask how animals cope with its constant threat. Which devices aid in escaping detection by a predator, and which come into play once a prey animal has been detected and capture seems all too imminent? Does membership in a group always confer antipredator privileges? And what compromises do animals reach when avoiding being eaten conflicts with other critical behaviors such as feeding and reproducing? We begin with camouflage.

FIGURE 13.1 Avian predators penetrate the chemical defense system of monarch butterflies at their overwintering sites in Mexico.

CAMOUFLAGE

Animals have several forms of camouflage (Endler 1981, 2006). Some forms, such as disruptive coloration, countershading, transparency, and coloration matching the visual background, are thought to help prey avoid detection by visually hunting predators. Another type, termed masquerade, relies on prey animals appearing to be inedible to predators. Thus, while such prey are detected as distinct from their background, their uncanny resemblance to a leaf, twig, or bird dropping makes them of little interest to a hungry predator searching for another animal to eat. Although camouflage comes in several forms, each with its own unique twist, its general message is simple: "I am not here." In the sections that follow, we will define and provide examples of each type of camouflage and illustrate how researchers test hypotheses concerning the disguises of animals. Most forms of animal camouflage were described more than a century ago. Surprisingly, direct experimental evidence to support their roles as antipredator devices is often quite limited. The good news, however, is that studies of camouflage are now flourishing and yielding incredible insights into the interactions between predators and prey, and the evolution of prey coloration.

COLORATION MATCHING THE VISUAL BACKGROUND

The coloration of some animals resembles their background and appears to reduce the risk of detection by visually hunting predators. It is sometimes called cryptic coloration, and the strategy is called background matching. We will avoid using the term *cryptic coloration* because it means different things to different people; some people use it in a narrow sense to refer only to coloration matching the visual background, whereas others use it in a more general sense to mean any coloration aimed at avoiding detection. In its more general usage, cryptic coloration would include other types of coloration, such as disruptive coloration (see below). So, we will use coloration matching the visual background because it clearly describes the phenomenon.

Probably all of us, at one time or another, have marveled at the ability of certain animals to blend with the background on which they are resting. While walking through the forest, we might struggle to see a moth against the bark of a tree or a grouse in its nest on the forest floor (Figure 13.2). We might even describe their coloration as "incredible camouflage" to our hiking partner. But how do researchers study this phenomenon under controlled experimental conditions? Can the degree to which animals match the visual background be quantified? One way to assess such coloration stems from a particular definition of it. John Endler (1978) defined it as concealment that results from an animal's resemblance to a random sample of the visual background. Thus, in some studies, degree of background matching is assessed by quantifying the similarity between prey coloration and its background, and often the researcher chooses the relevant aspects to compare (perhaps the density or distribution of elements in the color pattern). Another option is to let predators "tell" us about the effectiveness of different prey colorations. For example, by monitoring the search times of predators presented with different prey, researchers can gauge the effectiveness of the prey's camouflage; the longer it takes the predator to find the prey, the closer the match to the visual background. We will consider a study that used the second method.

Sami Merilaita and Johan Lind (2005) used artificial prey and an artificial background, but real avian predators, in an experiment designed to test the hypothesis that background matching is maximized when coloration visually matches a random sample of the background.

FIGURE 13.2 Animals, such as this grouse, that blend with their background are often described as cryptic. Technically, the grouse displays coloration matching the visual background.

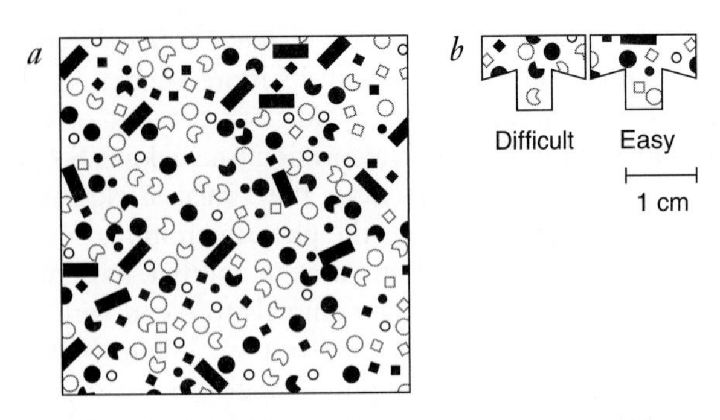

FIGURE 13.3 **Testing the effectiveness of camouflage. (*a*) The background used to test whether background matching is maximized when prey coloration matches a random sample of the background. (*b*) Artificial prey were classified as either difficult to detect or easy to detect. (Modified from Merilaita and Lind 2005.)**

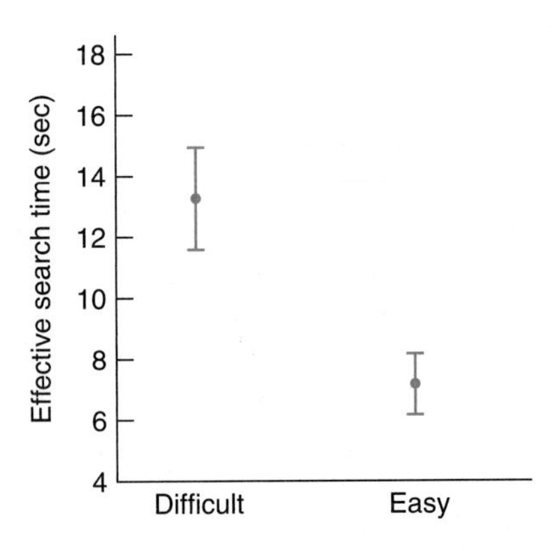

FIGURE 13.4 **Search times of great tits were longer for the prey classified as difficult to detect than for prey classified as easy to detect. These results indicate that different samples of a background may provide different degrees of camouflage through background matching. (Modified from Merilaita and Lind 2005.)**

The predators were great tits (*Parus major*) that were captured in mist nets and individually housed indoors at the research station in Sweden where the work was carried out. Merilaita and Lind designed paper T-shaped prey that the birds could learn to recognize. They also designed a background of sufficient complexity to make detection of the prey relatively difficult (Figure 13.3*a*). The paper background was mounted on cardboard. Prey patterns were random samples from the black and white background pattern and were of two types—those judged difficult to detect and those easy to detect (Figure 13.3*b*). Ease of detection for the two prey types was judged by the researchers and by a small number of birds used in a pilot experiment, but not in the "real" experiment described next. In the "real" experiment, a bird was moved to a new cage and given about an hour to adjust to its new surroundings before testing began. At testing, each prey item was presented singly to the bird; the paper prey item had been lightly glued to the background to cover a hole that contained a food treat, a peanut chip (the birds had been previously trained to associate the paper prey with food). The researchers measured with a stopwatch the time that a bird spent on the background searching for the prey. Merilaita and Lind found that search time was significantly longer for the prey expected to be more difficult to detect than for the prey expected to be easier to detect (Figure 13.4). These results indicate that all samples of a background do not provide equally camouflaged prey coloration. The authors suggest that prey coloration matching a random visual sample of the background may only maximize background matching on very simple backgrounds. This study speaks to the benefits of designing experimental conditions that mimic, to some extent, the conditions faced by prey animals in their natural habitats (i.e., real predators and complex backgrounds).

Up until now we have focused on the physical appearance of prey. However, there is more to background matching than simply color and pattern. Many animals appear to select "correct" backgrounds, and once there they exhibit behavior that maximizes their camouflage. The California yellow-legged frog (*Rana muscosa*) inhabits swift-flowing streams in the woodlands of southern California. The light gray granite boulders that line the streams seem conspicuous resting spots for the yellow-brown frog. Below the water, however, these same boulders are covered by a yellow-brown layer of algae. At a moment's notice, *R. muscosa* leaps into the water and lies motionless against a background to which it is perfectly matched (Norris and Lowe 1964).

Is the combination of matching and selecting the appropriate background adaptive? If it is, then prey should experience less predation when sitting on the substrates that they tend to select as resting spots than they do when sitting on other surfaces. Blair Feltmate and D. Dudley Williams (1989) tested this idea by using rainbow trout (*Oncorhynchus mykiss*) as predators and stonefly nymphs (*Paragnetina media*) as prey. Background color preferences of stoneflies, stream insects that are dark brown to black in color, were first tested by placing each of 24 nymphs into its own aquarium along with one dark brown and one light gray commercial tile on the bottom of the aquarium. Nymphs were left to settle for 24 hours, and then at 1400 hours (2 P.M.) the researchers recorded whether the nymphs rested on the dark brown or light gray substrate. The experiment was repeated with recordings of nymphal position at 0200, 0600, 0800, and 2100 hours to test whether selection of substrate varied as a function of time of day (lights in the laboratory were on timers and were off from 1900 to 0700 hours). Thus, independent replicates of the experiment were run at five different times of the day, three in the dark (2100, 0200, and 0600 hours) and two in the light (0800 and 1400 hours). The results, depicted in Figure 13.5, demonstrate that stoneflies selected the dark brown substrate rather than the light gray one at 0800, 1400, and 2100 hours; no selection was observed at 0200 or 0600 hours. Although stonefly nymphs selected the dark over the light substrate, this selection ceased approximately two hours after the lights in the laboratory went off and resumed within one hour of their being turned on.

In the next experiment, Feltmate and Williams (1989) examined whether stoneflies resting on the light substrate were more vulnerable to predation by rainbow trout. As before, each stonefly was introduced into its own aquarium. This time, however, the tank contained either light or dark tiles (not both, as in the first experiment). A trout was released into each tank after the nymphs had two hours to adjust to their new surroundings. Twenty-four hours after releasing the nymphs, the authors recorded the number of stoneflies consumed in tanks containing either the light or dark substrate. The consumption of nymphs by trout was lower in tanks that contained the dark substrate (3 of 24 nymphs eaten) than in tanks that contained the light substrate (19 of 24 nymphs eaten). These data suggest that the choice of dark resting spots by stoneflies has been favored by natural selection, at least in part, because it reduces the risk of being found and eaten by visually hunting fish. The breakdown in substrate color selection during the hours of darkness links visual predation to the distribution of nymphs. After all, animals need to be cryptic only when they are most vulnerable to predation by visual hunters (Endler 1978). Note, however, that we do not know how the visual abilities of stoneflies change under dark conditions and what role this might play in the observed breakdown of substrate selection. The choice of substrate by stoneflies may also conceal them from their own prey, as has been shown for other aquatic insects (Moum and Baker 1990).

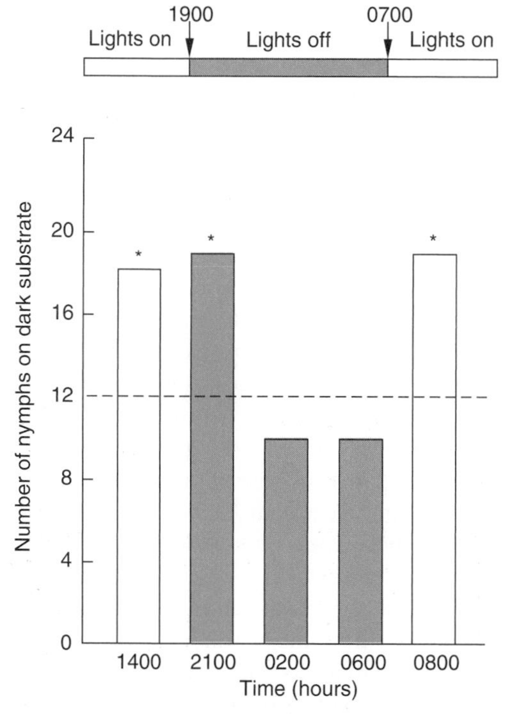

FIGURE 13.5 **Substrate selection in stonefly nymphs at various times of the day (white bars represent data when lights are on in the laboratory and colored bars when lights are off) when given the option of resting on dark brown or light gray tiles. During the lights-on period (0800 and 1400 hours) and shortly after the lights go off (2100 hours), a larger number of nymphs were observed on dark tiles, a background on which they were camouflaged, than on light tiles. Substrate selection was not apparent during the two remaining dark observations (0200 and 0600 hours). The dashed line represents expected results if no selection occurred; the asterisks indicate selection for dark brown tiles. (From Feltmate and Williams 1989.)**

Usually, animals that employ background matching are camouflaged in some habitats but not in others, and thus their occurrence is often restricted to those particular areas where they are best concealed. One way some species get around this restriction is by changing color as they change backgrounds. The cuttlefish (*Sepia officinalis*), which is not a fish at all but a cephalopod mollusk related to such creatures as squid, octopus, and nautilus, is perhaps the true master of color change, known for its swift and dramatic changes in body color and pattern (Messenger 2001). (Chances are that you have actually seen part of the white, internal shell support of this animal hanging in a birdcage to help parakeets and other birds keep their beaks sharp. Called cuttlebone, it is not bone at all.) When resting on the bottom, *Sepia* adjusts its color to that of the substrate at hand. Indeed, within a matter of seconds of settling on a sandy bottom, the dorsal color can change from gray (Figure 13.6*a*) to sandy brown, the latter pattern rendering it virtually invisible to predators (Figure 13.6*b*). These changes in body pattern are considered examples of coloration matching the visual background. On certain backgrounds, however, the cuttlefish displays color patterns with bold contrasting elements (Figure 13.6*c-d*). Is this another form of camouflage, and if so, how could it make the cuttlefish less likely to be detected by visually hunting predators? It turns out that the bold elements of the cuttlefish are part of disruptive coloration, the form of camouflage to which we turn next.

STOP AND THINK

Martin Stevens (2007) strongly suggests that when studying the effectiveness of prey camouflage (as well as other protective markings) efforts be made to consider the visual and cognitive abilities of the prey's predators. Early studies often relied on assessment by humans. Why might it be unwise to rely solely on human assessment to determine the effectiveness of a particular example of camouflage? What would you want to know about a predator to gauge the effectiveness of its prey's camouflage?

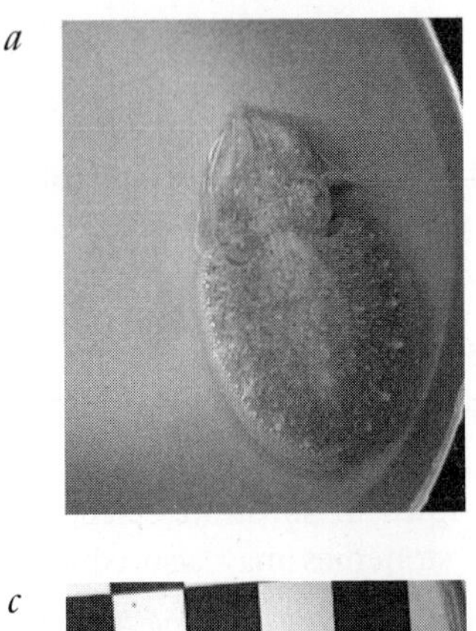

FIGURE 13.6 **Many animals that employ background matching are restricted to portions of their habitat in which they are well concealed. The cuttlefish gets around this restriction by changing its color to match the particular background on which it rests. Shown here is a cuttlefish resting on (*a*) a uniformly gray artificial background; (*b*) a natural sand background; (*c*) an artificial checkerboard background; and (*d*) a natural background of dark and light rocks. The coloration of the cuttlefish can match the background as in (*b*) or be disruptive as in (*c*) and (*d*).**

DISRUPTIVE COLORATION

Many animals avoid being seen by matching their background, but sometimes such matching is not enough because visually hunting predators may recognize prey by their body outline. Some animals seem to break up their body outline by developing bizarre projections; other species appear to do so with bold contrasting markings. These bold patches on prey are thought to function in preventing, or at least delaying, visual recognition of the prey by a predator. How do the patches achieve this? Bold patches may catch the eye of a predator, thereby drawing attention away from the outline of the prey's body; in addition, patches at the periphery of the prey's body may break up the continuity of the body outline (Cott 1940; Merilaita 1998). Coloration designed to prevent perception of a prey animal's form is called **disruptive coloration.** Although disruptive coloration is often described as a widespread device in animal camouflage, direct experimental evidence for it is quite limited. Here we describe a study in which scientists conducted direct field tests for disruptive coloration, using birds as predators and artificial moths as prey.

Innes Cuthill and colleagues (2005) conducted a field study designed to test the following two hypotheses about disruptive coloration: (1) color patterns at the periphery of an animal should provide better concealment than those placed randomly (tested in experiment 1), and (2) colors of high contrast should provide better concealment than those of low contrast (tested in experiment 2). The researchers constructed artificial moth-like prey in which the paper wings were shaped like triangles and the edible body was a dead mealworm (killed by freezing the night before and then thawed). They then pinned the artificial prey to oak trees in a forest and monitored "survival" of the prey 2, 4, 6, and 24 hours later. Predation by birds could be distinguished from that by slugs and spiders. Whereas birds simply took most or all of the mealworm, slugs left telltale slime trails and spiders sucked out the fluids of the mealworm, leaving the empty exoskeleton. In both experiments, numerous replicates were run in different areas of the forest over a six-month period.

In experiment 1, the markings of the prey (printed patterns on the paper) either overlapped the edges of the wings (the "Edge" treatment) or were located toward the inside of the prey, away from the edges (the "Inside" treatment), or the prey was a single color (Figure 13.7*a*). There were two different examples in the Inside treatment, Inside 1 (the same markings used in Edge were moved inward so that they did not overlap the edges) and Inside 2 (other randomly selected markings were placed inside the prey; this example was included because movement of the pattern elements from the periphery to the inside for Inside 1 created straight lines that could increase conspicuousness). In these three bicolored treatments, the markings were black on a dark brown background designed to match the ridge patterns of mature oak trees. Importantly, because these three prey treatments had patterns derived from photographs of tree bark, they were expected to be equally camouflaged from the standpoint of background matching. Thus, only disruptive coloration would predict that the treatment Edge would survive better than the two Inside treatments. The researchers also

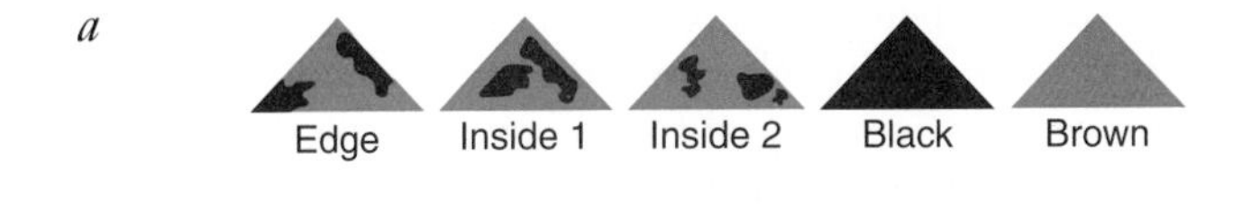

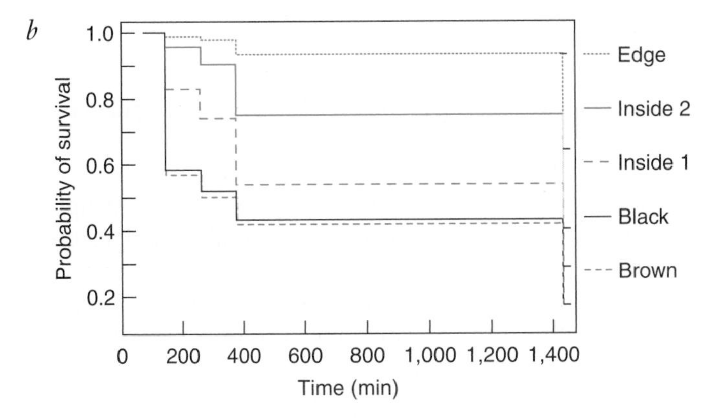

FIGURE 13.7 **Disruptive coloration. (*a*) Artificial moth-like prey designed to test whether color patterns at the periphery provide better concealment than either patterns at the interior or a single color. (*b*) Survival over time of artificial moth-like prey pinned to oak trees in a forest. Patterns at the edge of the body increase survival of artificial prey. (From Cuthill et al. 2005.)**

include prey with wings that were monochrome brown or monochrome black; according to background matching, the prey in these two treatments ("Black" or "Brown") should be less camouflaged than those in the three treatments with bicolored prey (Edge, Inside 1, and Inside 2). The results support disruptive coloration as an effective form of camouflage (Figure 13.7*b*). As you can see, prey in the Edge treatment survived better than those in the two Inside treatments, and prey in the three bicolored treatments, in turn, survived better than those in the two monochrome treatments.

In experiment 2, artificial moths were designed to test whether high-contrast colors provide better concealment than low-contrast colors. Cuthill and colleagues used edge, inside, and monochrome patterns as in experiment 1, but this time included high- and low-contrast prey in each type. Thus, the following six treatments were run: (1) Edge, high contrast; (2) Edge, low contrast; (3) Inside, high contrast; (4) Inside, low contrast; (5) Average color of the high-contrast color pair; and (6) Average color of the low-contrast color pair. As predicted only by the hypothesis of disruptive coloration, prey in the Edge, high-contrast treatment survived best (Figure 13.8). Taken together, the results from experiments 1 and 2 indicate that disruptive coloration is an effective camouflage device against birds, above and beyond that of background matching.

What, then, is the relationship between disruptive coloration and coloration matching the visual background, and how might these forms of camouflage interact with other aspects of the lives of prey animals? John Endler (2006) suggests that there may be a three-way relationship (or trade-off) between the two types of coloration and habitat specialization. Species that rely purely on background matching may successfully evade detection by predators as long as they are living in a particular habitat; such species may thus be habitat specialists. Disruptive coloration, in comparison, works on a greater variety of visual backgrounds and may be a

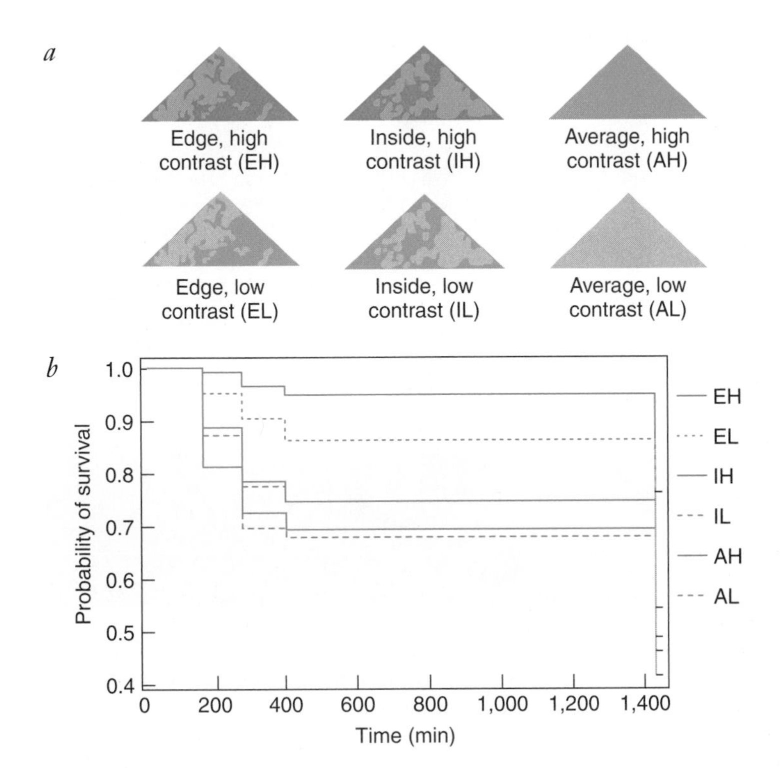

FIGURE 13.8 **Disruptive coloration. (*a*) Artificial moth-like prey designed to test whether high-contrast color patterns provide better concealment than low-contrast patterns. (*b*) Survival over time of the artificial moth-like prey pinned to oak trees in a forest. High-contrast disruptive patterns increase survival of artificial prey. (From Cuthill et al. 2005.)**

strategy employed by habitat generalists. The existence of a trade-off between the different forms of camouflage and degree of habitat specialization warrants further investigation (and we urge you to do so!).

COUNTERSHADING

Many animals have dark backs and light bellies, a pattern called **countershading**. Over a century ago, the painter and naturalist A. H. Thayer (1896) suggested that countershading makes animals difficult to detect because it allows them to obscure their own shadows. As the argument goes, because light normally comes from above, the ventral surface of the body is typically in shadow and predators could cue in on darkened bellies (Figure 13.9). Thayer suggested that by being darker dorsally and paler ventrally, animals could obscure the ventral shadow. This mechanism by which countershading is proposed to achieve camouflage is called **self-shadow concealment** (Kiltie 1988; Ruxton et al. 2004). An alternative mechanism through which countershading could make animals difficult to detect is one that we have already discussed, background matching. Background matching might be particularly common in aquatic animals because they are likely to be viewed from above and from below by predators. With light coming from above, a light belly would help an animal match the bright background when viewed from below by a predator. Similarly, a dark back would help an animal match the background of dark, deep waters when viewed from above by a predator.

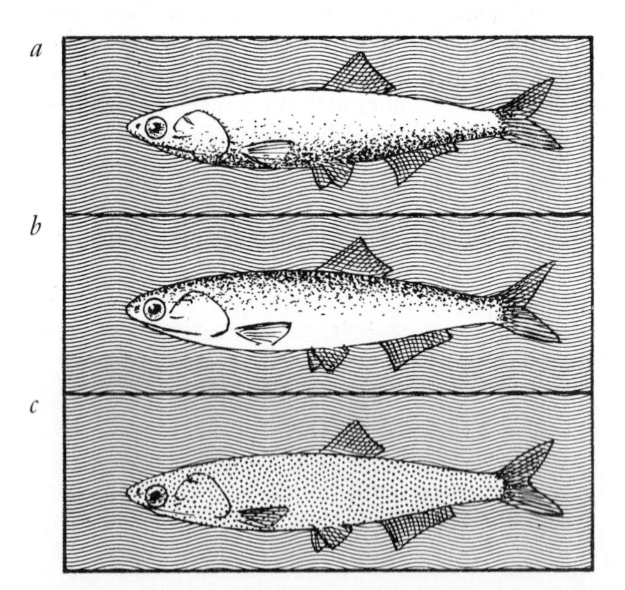

FIGURE 13.9 **The effects of countershading on conspicuousness. (*a*) Light normally comes from above, and under these circumstances fish that are uniformly colored have a conspicuous ventral outline. (*b*) A countershaded fish is darker dorsally than ventrally (as shown here, illuminated from all sides). (*c*) Thus, its body outline is obscured when light comes only from above. This mechanism by which countershading is proposed to achieve camouflage is called self-shadow concealment. (After Cott 1940.)**

Although many animals display countershading, direct evidence that this pattern achieves camouflage through self-shadow concealment is embarrassingly meager (Ruxton et al. 2004). Indeed, the alternative mechanism of background matching cannot be ruled out in most purported cases. Furthermore, for many examples of countershading, we cannot even discount the possibility that the combination of dark backs and light bellies is completely unrelated to camouflage and instead functions in thermoregulation or protection from ultraviolet radiation (Kiltie 1988). We will consider a study that examined the possible functions of countershading in the naked mole rat, a small mammal with a big reputation for its highly social lifestyle. The study is important because it examines multiple potential functions of the countershading pattern.

Naked mole rats (*Heterocephalus glaber*) are fossorial, spending most of their time in extensive underground burrow systems in the dry areas of East Africa. They are also eusocial, living in large colonies that contain a single breeding female (the queen), a few breeding males, and numerous nonbreeding workers that care for the queen's offspring, maintain the burrow system, and feed and defend the colony. (Eusociality is discussed in more detail in Chapter 19.) Stanton Braude and colleagues (2001) have studied the behavior of naked mole rats in captive colonies at the University of Michigan; they have also monitored the demography and behavior of wild colonies in East Africa for many years. Their early, anecdotal observations on the color of field and laboratory animals suggested that most colony members display countershading, having a darker dorsal coloration

FIGURE 13.10 **Although most members of naked mole rat colonies exhibit countershading (darker backs than bellies), the queen does not.**

(purple-brown-gray) than ventral coloration (pink). They also noticed, however, that queens were interesting (and all pink) exceptions (Figure 13.10). A few other colony members also lacked the dark dorsum. On the basis of these early observations, Braude and co-workers decided to look more systematically at color in naked mole rats. As part of their long-term field study, they began to meticulously record the occurrence of live-trapped naked mole rats with extreme pink dorsal coloration. They also began to carefully quantify the color of known age animals in their laboratory colonies. They discovered that most naked mole rats are indeed countershaded; exceptions to the countershaded pattern included newborns, queens, most breeding males, and very old individuals, all of whom are uniformly pink. In naked mole rats, countershading begins to develop a few weeks after birth and is fully developed by 3 months of age; interestingly, it begins to disappear at about 7 years of age.

What is the function of countershading in naked mole rats, especially given that these animals spend most of their time in dark underground burrow systems? Can the distribution of countershading among colony members or the timing of its development and its loss tell us anything about its function(s)? Braude et al. (2001) used the field and laboratory data they had collected to test five hypotheses for the existence of countershading in naked mole rats. The first four hypotheses are not mutually exclusive; they state that countershading (1) protects individuals from ultraviolet light, (2) facilitates thermoregulation, (3) protects against abrasion because the pigment melanin may strengthen skin, and (4) provides camouflage for individuals dispersing above ground. These hypotheses can be contrasted with the fifth and final hypothesis that countershading has no current function in naked mole rats and is simply a remnant of adaptive countershading in their surface-dwelling ancestors. We will not review all of the evidence for or against each hypothesis here; we will simply say that Braude and co-workers concluded that their data were most consistent with the camouflage hypothesis. Known dispersers at their field site were, on average, 2.3 years of age, with the oldest recorded disperser being 5 years old. Thus, the age of dispersal coincides with the time when naked mole rats have well-developed countershading. Other observations at their field site indicate that naked mole rats tend to disperse on the surface at night rather than during the day, and often on moonlit nights. Thus, a darkened dorsum could provide protection against nocturnal avian predators (and would not be needed at night for protection against ultraviolet radiation). Although this study does not distinguish between self-shadow concealment and background matching as possible mechanisms through which countershading might achieve its camouflage function, it is a good example of keeping an open mind when searching for the functions of a particular color pattern and testing alternative hypotheses.

STOP AND THINK

The study by Braude et al. (2001) is nonmanipulative and thus represents an indirect test of the hypothesis that countershading functions in camouflage. How would you design a more direct test of the hypothesis that countershading contributes to camouflage in naked mole rats? What experimental manipulations would you propose? Would your manipulations allow you to distinguish between self-shadow concealment and background matching as the mechanism by which camouflage is achieved?

TRANSPARENCY

Some animals are camouflaged simply by being transparent. Although no animal is completely transparent, organisms such as cnidarians (e.g., hydroids and jellyfish), ctenophores (e.g., comb jellies), and the pelagic (open ocean), larval stages of many fish achieve near transparency by such means as high water content of tissues, small size, and reduced number of light-absorbing molecules or pigments (McFall-Ngai 1990).

Frequently neglected in discussions of camouflage mechanisms, transparency is probably the dominant form of camouflage in aquatic environments, particularly in pelagic habitats where organisms have no surfaces to match or places to hide (Johnsen 2001). Transparency is extremely rare in terrestrial habitats for at least two reasons. The first reason concerns refractive indexes (the angle at which light bends when passing from one medium into another) of water and air. Let's first consider the situation for aquatic organisms, and then we'll look at that of terrestrial ones. Because animals' bodies are largely water, when light travels from the surrounding water into the tissues of an aquatic animal, the angle of light is virtually unchanged; in the absence of light-scattering or light-absorbing elements, the animal appears to be transparent (light, then, is basically passing from water into water). In contrast, in a terrestrial environment, light must pass from air into the water-filled tissues of an animal. The difference in the refractive indexes of air and the terrestrial animal's tissues creates an obvious body outline, greatly diminishing transparency. The second reason transparency is rarely used as a camouflaging mechanism by terrestrial animals has to do with the deleterious effects of ultraviolet radiation on land. Animals on land need protective pigments, making transparency difficult, if not impossible. In aquatic habitats, much of the ultraviolet radiation is filtered out within a few meters of the water's surface, and thus animals living beyond this distance are not subject to the same radiation damage as terrestrial organisms.

MASQUERADE

As mentioned at the beginning of our discussion of camouflage, masquerade differs from the other forms because the prey may be detected but deemed inedible by predators searching for animals to eat (Endler 1981, 2006). Leaf resemblance is a particularly common disguise. Among insects that resemble leaves, we find green or brown coloration, leaf-like patterns of venation on their bodies, and flattened shapes. These morphological specializations are often accompanied by behavioral ones, ranging from remaining still during daylight hours to swaying like a leaf in the wind.

Leaf resemblance also occurs in some small vertebrates, including amphibians that normally inhabit the leaf litter of the forest floor (Figure 13.11*a*). Resemblance to dead leaves, however, is not restricted to terrestrial species. Indeed, *Tetranematichthys wallacei*, a nocturnal catfish that inhabits small Amazonian streams, displays a remarkable resemblance to dead leaves, and its disguise includes aspects of its physical appearance and its behavior (Sazima et al. 2006). The body and fins of *T. wallacei* are the color of dead leaves, and the body is laterally compressed or flattened, giving it a leaf-like shape. A nocturnal forager, the catfish spends daylight hours lying on its side amid dead leaves on the stream bottom (Figure 13.11*b*). If disturbed from its resting place by a potential predator (or a curious scientist), the catfish, with no discernible movement of its fins, drifts slowly downstream like a waterlogged leaf.

a

b

FIGURE 13.11 Some prey masquerade as objects that appear inedible to predators searching for another animal to eat. Here, (*a*) frogs from Malaysia and (*b*) a catfish from Amazonian streams resemble dead leaves.

OTHER FUNCTIONS OF COLOR

Evasion of predators is not the only function of color pattern in animals. Color affects heat balance and thus plays a role in thermoregulation. Color and pattern are also important in many aspects of communication, including mate recognition, courtship, male–male competition, and territorial defense.

As mentioned at the start of this chapter, the various functions of animal color and pattern may act in concert or in opposition. Let us consider a case in which they act in opposition. If color and pattern are adjusted for thermoregulation, how can animals communicate effectively with mates and competitors and at the same time be inconspicuous to visually hunting predators? Although some animals that employ background matching have evolved alternative means of exchanging information (e.g., relying on auditory or olfactory signals to communicate with conspecifics), many still rely on visual cues. As we will see, the color pattern displayed by a particular animal may be a compromise between factors that favor camouflage and those that favor conspicuousness.

John Endler's (1978) work with wild populations of guppies (*Poecilia reticulata*) in northeastern Venezuela and Trinidad provides an excellent example of how color patterns may represent a balance between mate acquisition and camouflage. Whereas a female's choice of mate and competition among males favors brighter colors and more visible patterns in guppies, selection by diurnal visual predators (at least six species of fish and one freshwater prawn) favors less colorful and less conspicuous patterns. It is interesting that as predation risk increases across communities, the colors and patterns of guppies become less obvious because of (1) shifts to less conspicuous colors, (2) reductions in the number of spots, (3) reductions in the size of spots, and (4) slight reductions in the diversity of colors and patterns (Figure 13.12). In areas in which guppies encounter low-predation pressure, however, the balance shifts toward attracting mates, and colors and patterns become more conspicuous.

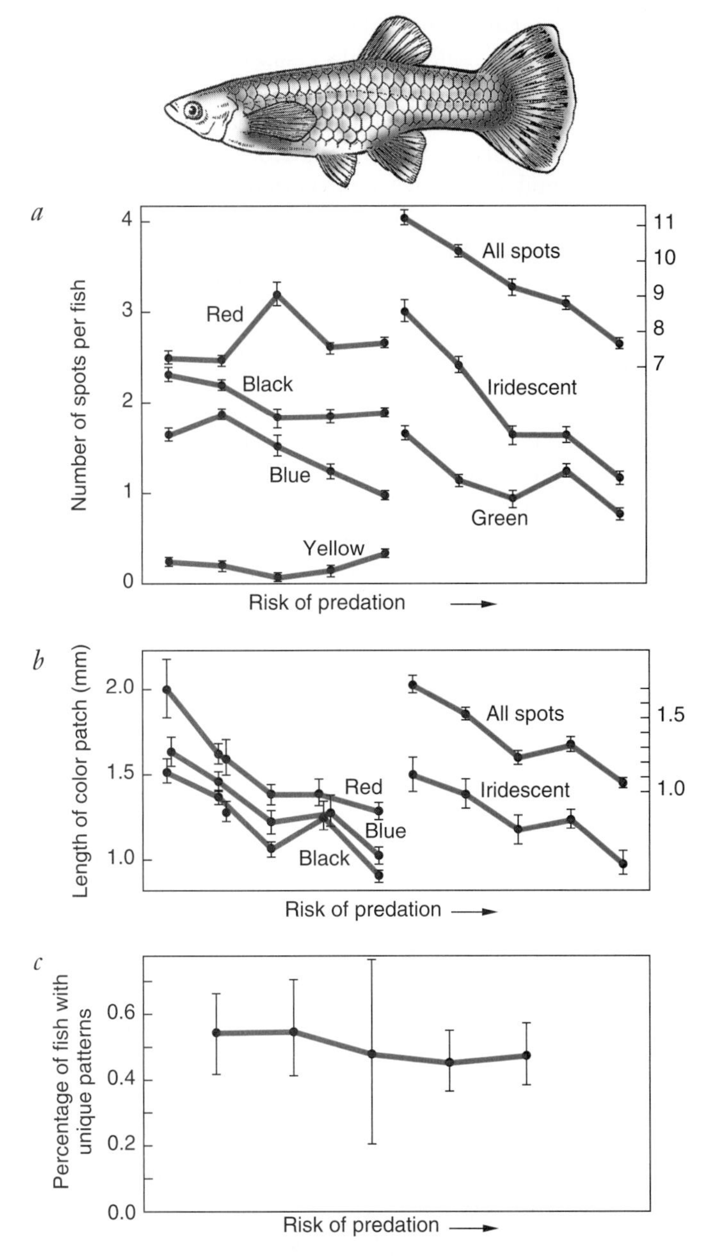

FIGURE 13.12 **An animal's color is often a compromise solution to the problem of selection for conspicuousness in courtship displays and selection for inconspicuousness to visually hunting predators. Changes in the color and pattern of guppies as a function of predation pressure reflect the fine balance between these two selective forces. As predation risk increases, the (*a*) number of spots, (*b*) length of color patches, and (*c*) diversity of patterns decrease. (Modified from Endler 1978.)**

POLYMORPHISM

Like most things, background matching is not foolproof. Although individuals may blend with their background, predators in a given area may develop a search image for that particular species and systematically search out and consume remaining individuals (a search image is the heightened ability to detect a prey species; see Chapter 12). If individuals of the prey species are widely spaced, however, predators will rarely encounter them and will soon forget the search image. Indeed, individuals of many species that employ background matching occur at widely spaced locations throughout their environment.

Other species get around the problem of search images by occurring in several different shapes and/or color forms, that is, by exhibiting **polymorphism**. We will consider an example of color polymorphism that concerns fox squirrels (*Sciurus niger*) in the eastern United States. Fox squirrels have been described as the most variable in color of all mammals in North America (e.g., Cahalane 1961). Color varies both among and within populations. Dorsal coloration may range from gray or tan to black, and coloration on the head and ear region is often distinctive (Figure 13.13). Even within a single litter, both melanistic (black) and nonmelanistic young can be found. Intrigued by the variation in coat color of fox squirrels, Richard Kiltie (1989) examined close to 2000 museum specimens of this species. He determined the percentage of dorsal black for each skin and compiled information on the occurrence of wildfires in the eastern United States. Taken together, his data on coat color and fires show that the incidence of melanistic individuals is correlated with the frequency of wildfires over the total range of the fox squirrel. Both wildfires and melanistic squirrels are more common in the southeastern United States (Figure 13.14).

In fox squirrels, the melanistic polymorphism in coat colors may thus be maintained by the periodic blackening of the ground and lower portions of tree trunks by wildfires. One would imagine that dark squirrels are less conspicuous to hawks than are light or variably colored individuals against a blackened background. However, the advantage does not remain with the black squirrels for

FIGURE 13.13 **Some color morphs of the fox squirrel. Note the variation in the percentages of dorsal black and the pattern around the head and ear region. (After Kiltie 1989.)**

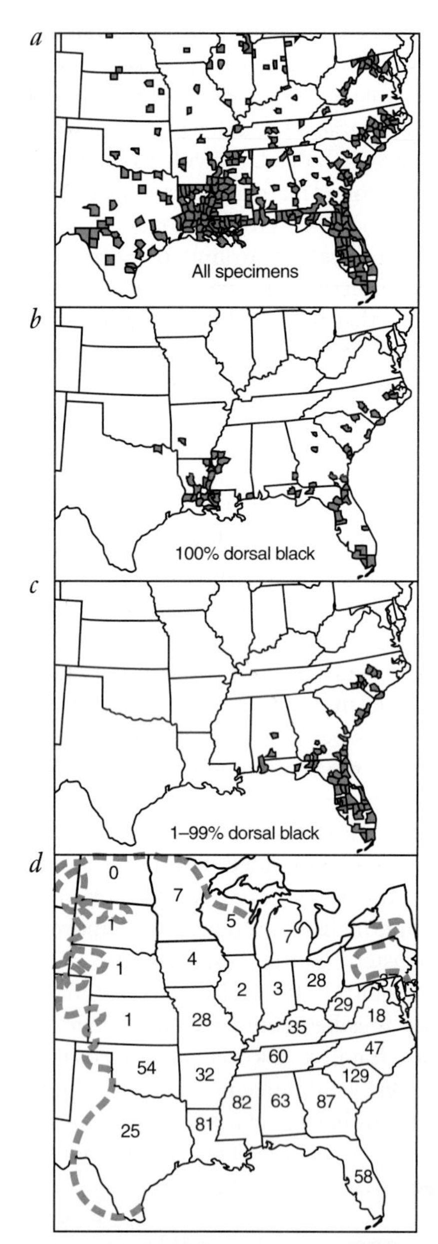

FIGURE 13.14 **In the fox squirrel, the incidence of melanism is correlated with the frequency of wildfires. (*a*) Most of the counties in the eastern United States from which Kiltie (1989) examined museum specimens of fox squirrels. (*b*) Counties from which specimens with 100% dorsal black were noted. (*c*) Counties from which specimens with intermediate levels of dorsal black (1–99%) were recorded. (*d*) Average number of wildfires per state in protected forestlands during the years 1978–1982; values have been normalized to take into account the area of land under wildfire surveillance (dashed lines depict limits of the fox squirrel's range). Note that melanistic fox squirrels occur primarily (*b*) in the southern portions of the species range and that individuals with intermediate levels of dark coloration are limited to (*c*) the eastern Gulf and Atlantic coastal plains. These areas in the southeastern United States are also the areas in which wildfires are most common (*d*). (From Kiltie 1989.)**

long. As rainfall and new plant growth convert a charred area into a less uniformly black substrate, fox squirrels with variable amounts and patterns of black dorsal coloration would be more difficult to see than uniformly black individuals against the patches of light and dark underground. Finally, when the period of regrowth of the pine and oak forest is almost complete, the advantage may shift to squirrels that are uniformly light in coloration. Thus, variable coat color in fox squirrels may result from the alternating superiority of light and dark individuals in matching the background of an environment that periodically burns and regenerates.

In some cases, polymorphic species do not have coloration that matches the visual background and rely solely on their diverse appearance to evade detection by predators. Whether camouflaged or not, by being different, individuals of prey species can occur at higher densities without suffering increased mortality from predators searching for individuals with a specific appearance. Some species that occur at very high densities exhibit extreme polymorphism, making it almost impossible to find two individuals that look alike (Figure 13.15).

Gairdner Moment (1962) described the phenomenon in which members of a population look as little like one another as possible. In such populations the probability of an individual's having a certain appearance is inversely related to the number of other individuals in the population that have that appearance. If one morph in a polymorphic population is much more common than another morph, predators are likely to develop a search image for the more common, rather than the rare,

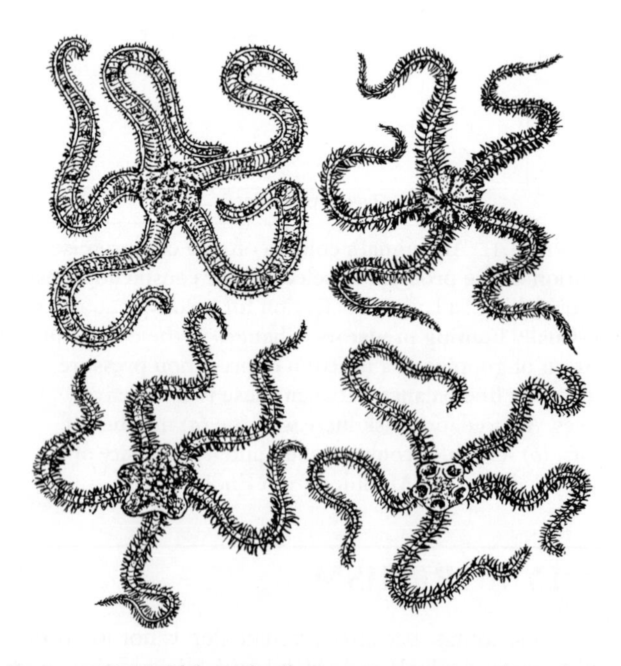

FIGURE 13.15 **Although all the same species, these four brittle stars are dramatically different in appearance, thereby inhibiting the formation of search images in predators. (Drawn from photograph in Moment 1962.)**

morph. The end result is that predators take more of the common form relative to its frequency in the population. Thus, for example, when two morphs are equally camouflaged and are exposed to predators that use search images when hunting, the rare morph will have a selective advantage over the common morph. We described this form of frequency-dependent selection in Chapter 4. This form of selection has been called apostatic selection (Clarke 1969). Its strength varies as a function of factors such as density, palatability, and conspicuousness of prey (Allen 1988). Furthermore, Jeremy Greenwood (1984) indicates that predators need not hunt by search image to cause apostatic selection in prey. Some predators, for example, may simply have an aversion to prey that are rare or unfamiliar to them.

What experimental evidence do we have that being different pays off? Croze (1970), working on a sandy peninsula in England, placed 27 painted mussel shells with pieces of meat under them on the ground and exposed them to predation by carrion crows (*Corvus corone*). In some of the 14 trials, the shells were monomorphic (i.e., all the same color), whereas in others they were trimorphic (9 red, 9 yellow, and 9 black). The results, summarized in Table 13.1, show that the crows took fewer of the trimorphic than the monomorphic prey. The percentage of survival for each of the three morphs in a trimorphic population was two to three times higher than in monomorphic populations. Thus, a morph had a twofold to threefold selective advantage when occurring as part of a trimorphic population. Croze's results demonstrate that when prey populations occur at the same density, individuals in polymorphic populations experience less predation than those in monomorphic populations.

TABLE 13.1 Percent Survival of Painted Mussel Shells in Either Monomorphic or Trimorphic Populations When Exposed to Predation by Carrion Crows

	Type of population	
Shell color	Monomorphic	Trimorphic
Yellow	10	31
Black	12	40
Red	19	45

Source: Data from Croze (1970).

WARNING COLORATION

Many animals that have dangerous or unpleasant attributes appear to advertise this fact with bright colors and contrasting patterns. Bold markings, typically in black, white, red, or yellow, warn the predator of the prey's noxious qualities, and through this warning discourage an attack. For example, the dramatic black and white markings of spotted and striped skunks may serve, in part, to warn predators of the foul-smelling repellent that may, upon further harassment, be released from the skunks' anal scent glands (Figure 13.16). Many insects, such as the social wasps, have a boldly patterned yellow and black body thought to warn of their painful sting. The phenomenon by which a conspicuous appearance (often coloration) serves to advertise dangerous or unpleasant attributes is called **aposematism**; sometimes, it is simply called **warning coloration**. Although it is not difficult to find apparent examples of aposematism in nature, by now you know that we are also interested in direct tests of presumed antipredator devices. Here we consider some of the direct evidence for warning coloration in dendrobatid frogs.

Frogs within the family Dendrobatidae are best known for their bright coloration and toxic skin secretions, although there is substantial variation among and within species in both of these characteristics. Some species are red, yellow, blue, or some combination of these colors, and the colors may contrast with black markings. Perhaps the most notorious of the toxic species is *Phyllobates terribilis*, a single individual of which has enough toxin in its skin to kill about 20,000 mice or, in more familiar currency, 100 humans. The Choco Indians of western Colombia make deadly weapons by simply wiping their blowgun darts across the back of one of these metallic yellow frogs; so lethal is the poison in the frog's skin that a dart poisoned in this manner can remain deadly for more than a year. The bright coloration of dendrobatid frogs has been widely viewed as an example of aposematic coloration, a warning to potential

FIGURE 13.16 **The bold black and white patterns of skunks are thought to be aposematic, warning potential predators of the foul liquid that may be sprayed from glands beneath the tail.**

predators of the poisons in their skin. What evidence do we have that their coloration serves an aposematic function? Kyle Summers and Mark Clough (2001) reasoned that if the coloration of dendrobatid frogs is indeed aposematic, then we would predict that the more toxic species will have brighter, more extensive coloration than the less toxic or nontoxic species. They set out to test the hypothesis that color will evolve in tandem with toxicity in dendrobatid frogs using a phylogenetically controlled comparative analysis. (This type of analysis controls for associations between toxicity and coloration caused by shared ancestry. See Chapter 4 for discussion of the comparative approach). Summers and Clough obtained toxicity data for 21 species from the literature, and using information on toxin diversity, amount, and lethality, determined a total toxicity score for each species. They assessed the brightness and extent of coloration of each of the 21 species using two methods. In the first method, human observers were presented with color photographs of each species and asked to rank them. In the second, color photographs of each species were scanned into a computer where a program measured the brightness of colors and the proportion of the frog covered by each color. The same measurements were made on scanned color photographs of leaf litter to produce an overall measure of contrast of each species of frog against a leaf litter background. Summers and Clough found a significant association between toxicity and coloration (whether assessed by humans or the computer): the more toxic species were the most colorful. This finding is consistent with the hypothesis that bright coloration in dendrobatids serves an aposematic function.

Additional data on the color and toxicity of dendrobatid frogs have been collected in the years since the study by Summers and Clough (2001). What have these data shown? Results from one study conducted by Catherine Darst, Molly Cummings, and David Cannatella (2006) indicate that degree of toxicity and brightness can be decoupled in some dendrobatid frogs. As part of a larger study, these researchers evaluated the conspicuousness and toxicity of three closely related species within the genus *Epipedobates*. Their methods differed in several ways from those of Summers and Clough (2001). For example, they assessed toxicity by monitoring the recovery times of mice injected with minute quantities of extracts from the skin of the three frog species (the quantities were chosen so as not to kill the mice but to cause mild irritation). Despite methodological differences, we might still expect to see a pattern similar to that found by Summers and Clough (2001), that is, that toxicity and conspicuousness are positively associated in the three frog species studied. Instead, Darst and colleagues found that the most toxic species (*E. parvulus*) was not the most conspicuous, and the most conspicuous species (*E. bilinguis*) was only moderately toxic.

How can we reconcile the divergent findings of the two studies? Although without further studies we can't be sure of the precise role of methodological differences in yielding the different findings, there is at least one other possibility. A recently developed theoretical model of the costs and benefits of defensive traits suggests that the relationship between conspicuousness and toxicity will vary under different ecological conditions, being positively correlated under some conditions and even negatively correlated under others (Speed and Ruxton 2007). A negative relationship between conspicuousness and toxicity might occur if the costs of conspicuousness increase to the point that prey are better off decreasing their investment in bright colors and increasing their investment in toxins to compensate. The studies of aposematism in dendrobatid frogs thus highlight the diversity that can exist even within a particular form of defense.

Animals with warning coloration often enhance their conspicuousness behaviorally. Many are active during the daytime, and individuals of some species form dense, obvious aggregations. Although rare forms in aposematic animals are typically selected against (predators will not be as familiar with the rare form as they are with the common form and may attack), they are at less of a disadvantage when they occur in clusters (Greenwood et al. 1989). Thus, dense aggregations of aposematic prey not only emphasize the warning but also function as areas in which rare forms may arise and survive.

The response of predators to aposematic coloration may be learned or innate. In the first case, predators sample some of the prey, discover their unpleasantness, and learn to avoid animals of similar appearance when searching for subsequent meals. For example, garter snakes (*Thamnophis radix*) develop a much stronger avoidance of conspicuously colored noxious prey than of nonaposematic prey, even though olfaction plays an important role in detection and ingestion of prey. Two different types of prey, earthworms and fish, were offered to garter snakes on forceps that were aposematically colored (yellow and black) or nonaposematically colored (green). After they had consumed fish that was presented on forceps of either type of coloration, snakes in the experimental group were injected with lithium chloride to induce illness. Immediately following the induced illness, the garter snakes avoided all fish, regardless of the color of the forceps. However, the garter snakes that had been offered fish on aposematically colored forceps had a much longer-lasting aversion to fish than those who had been offered fish on green forceps (Figure 13.17). Seven days after the induced illness, only one of five snakes in the aposematic treatment group ingested any fish during the 120-second test interval. In contrast, all five of the snakes in the nonaposematic treatment group eventually attacked fish. Thus we see that predators learn to avoid unpalatable prey more readily if the prey are conspicuously colored (Terrick et al. 1995).

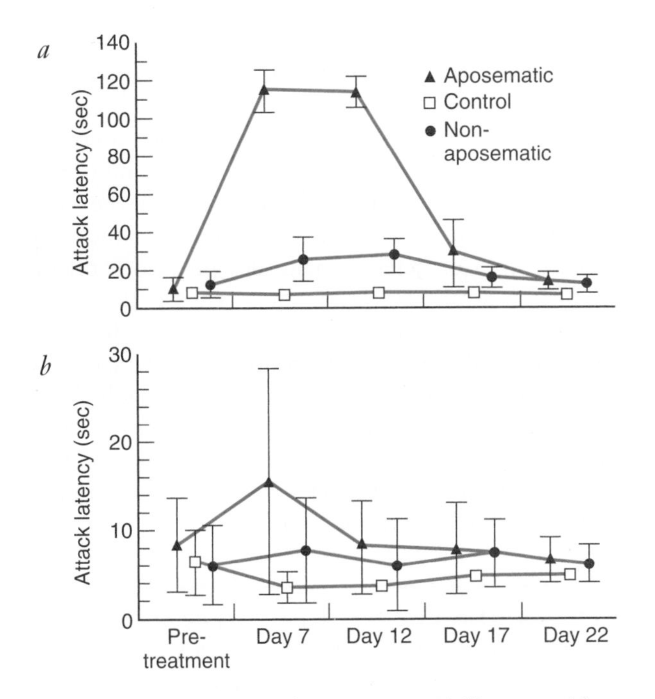

FIGURE 13.17 **Predators learn more quickly to avoid distasteful prey that are conspicuous. Here, garter snakes were first offered pieces of fish on aposematic (yellow and black) or on nonaposematic (green) forceps. The snakes were then made ill by an injection of lithium chloride. The post-treatment attack latencies to (*a*) pieces of fish or (*b*) earthworms indicate that snakes in the aposematic treatment group had a stronger aversion to fish than did snakes in the nonaposematic treatment group. (Data from Terrick et al. 1995.)**

Sometimes two warningly colored species look alike. Apparently, two noxious species can benefit from a shared pattern because predators consume fewer of each species in the process of learning to avoid all animals of that general appearance. This phenomenon is called **Müllerian mimicry**.

Although some predators learn through memorable experiences to avoid aposematic prey, others display innate avoidance. An innate response to warning coloration might be favored over a learned response when the secondary defense of the prey has the potential of being fatal to the predator. Learning at the moment of death is of little value.

Sometimes, like advice, warning coloration is ignored. A predator that is starving might tackle a noxious prey that it would normally pass up during better times. Wolves will attack skunks and porcupines when other prey is scarce. In addition, some predators are specialists and are able to eat certain aposematic animals, or the least noxious parts of them, as we saw for the black-headed grosbeak and black-backed oriole that were preying on the unpalatable monarch butterfly. However, as long as an antipredator device confers a net advantage in survival and reproduction, it will continue in the population.

BATESIAN MIMICRY

Batesian mimicry is named after the nineteenth-century English naturalist Henry Walter Bates, and it refers to a palatable species that has adopted the warning characteristics of a noxious or harmful species. The harmless species is called the mimic and the noxious one, the model. By resembling a noxious species, the mimic gains protection from predators. The evolution of mimicry has been the center of lively discussion (e.g., Holmgren and Enquist 1999; Joron and Mallet 1998; Mappes and Alatalo 1997).

The degree of protection experienced by the mimic varies as a function of numerous factors. The ratio of models to mimic is important, for example. The mimic does better when it is rare and therefore less likely to be detected by the predator than the noxious model (Turner 1977). And the more distasteful the model, the better the mimic fares. These predictions have been experimentally tested, using great tits (*Parus major*) as the predators. The birds were offered model and mimic prey one at a time, as prey would be encountered in nature. Both prey types survived better when there were fewer mimics. The models survived significantly better the more distasteful they were, and the degree of unpalatability also affected the survival of the mimics (Lindstrom et al. 1997). The memory of predators, availability of alternate prey, and whether mimics and models are encountered simultaneously or separately may also play a role (Speed and Turner 1999). As a result, a mimic may gain most if its habits and daily activity overlap those of its model species.

This increase in benefits was demonstrated by using naive birds, brown-eared bulbuls (*Hypsipetes amaurotis pryeri*), that were trained to take food from two feeders in captivity. The model prey was *Pachliopta aristolochiae*, a butterfly that sequesters alkaloids as a larva and is, therefore, distasteful. A noxious model butterfly was placed at one of the feeders. After an unpleasant experience eating the model, the bird took less of the palatable food from the feeder. This suggests that the bird associated the unpleasantness not just with the model but also with the place where it was experienced. The birds were then offered female swallowtail butterflies (*Papilio polytes*), which come in mimic and nonmimic forms. The birds avoided the mimetic forms of swallowtails (Uesugi 1996).

Although in some instances the resemblance between model and mimic seems almost exact, the likeness usually does not have to be perfect because predators appear to generalize conspicuous features of noxious prey. In some cases, poor mimics may exist because they exploit some constraint in the predator's visual or learning mechanisms. An example is provided by hoverflies, which mimic certain wasps. Studies have shown that pigeons rank hoverflies according to their similarity to the wasp model. To human eyes, the two most common types of hoverflies show the least resemblance to wasps, yet the pigeons rank them as being very similar to wasps. It is thought that these wasps have some key feature that is used by pigeons in pattern recognition (Dittrich et al. 1993). This is a good reminder that when studying the defenses of prey animals, it is important to consider the cognitive and perceptual abilities of their predators.

Numerous mimetic resemblances are recounted in the literature, but only a few studies demonstrate that the purported mimics actually gain protection from their natural enemies. T. E. Reimchen (1989) first described a system of Batesian mimicry involving the juvenile stage of a snail (the mimic) and the tubes of a polychaete worm (the model) and then provided evidence that the resemblance actually conferred some degree of protection to the young snails. The snail, *Littorina mariae*, lives in the intertidal zone of the North Atlantic. The shells of some juveniles have a conspicuous white spiral, and the shells of others are yellow or brown. When adult, the snails are either yellow or brown, and the white spiral possessed by some is visible only as a white apex on the shell. Egg masses of the snail are deposited directly on algal fronds, and once the juveniles hatch they disperse on the fronds. Snails with white-spiral coloration were observed only in habitats where the polychaete *Spirobis* was present. In these habitats, white-spiral phase juveniles are virtually indistinguishable from the tubes of *Spirobis* that are cemented to the fronds (Figure 13.18).

Reimchen collected the intertidal fish *Blennius pholis*, an important predator on juvenile snails, and conducted predation experiments in aquaria in the laboratory. Although the polychaete tubes are not noxious to the fish, they represent a substantial investment in time and energy because they are difficult to remove from the substrate, and once removed, they may prove to be unoccupied. In the experiments, blennies were housed alone in an aquarium and were presented with juvenile snails on either an algal frond with polychaete tubes or an algal frond without tubes. At each presentation, three juvenile snails (one white-spiral, one yellow, and one brown) were randomly positioned on the frond and the frond was lowered to the bottom of the tank. Once blennies detected a snail, they plucked it off the frond and swallowed the shell whole. Reimchen recorded the first snail taken at each trial. Overall, white-spiral snails suffered the lowest number of attacks, and attacks were less common on fronds with polychaete tubes (9.4%) than on fronds devoid of tubes (22.9%). Thus, in this unusual system of snail-polychaete mimicry, resemblance to the model does appear to confer a protective advantage to the mimic.

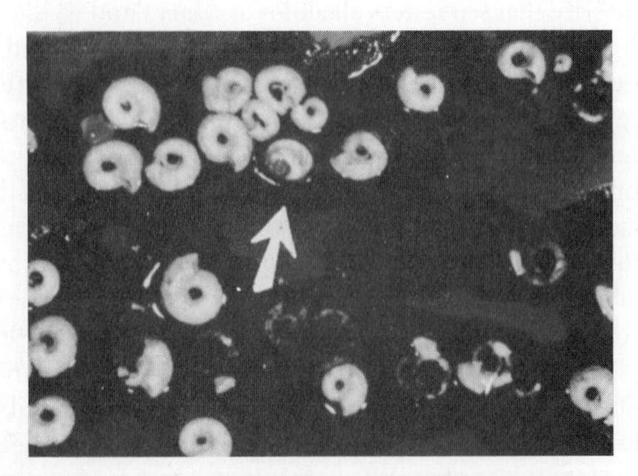

FIGURE 13.18 **The white phase juvenile of a snail (shown by the arrow) is a Batesian mimic of the tubes of certain polychaetes. By resembling the tubes, young snails may gain protection from fish that are searching for food on algal fronds.**

DIVERTING COLORATION, STRUCTURES, AND BEHAVIOR

Many animals have evolved colors, structures, and patterns of behavior that seem to divert a predator's attention, while they, and in some cases their offspring, escape with little or no damage. Whereas many of the antipredator devices discussed so far help prey to avoid an encounter with a predator, distraction devices come into play once a prey animal has been discovered or when discovery seems all too imminent. We next consider some of the devices of prey animals that operate during an encounter with a predator.

FALSE HEADS

Many predators direct their initial attack at the head of the prey. Some prey species have taken advantage of this tendency by evolving false heads that are located at their posterior end, a safe distance from their true heads. Lycaenid butterflies (Lepidoptera: Lycaenidae) display patterns of color, structure, and behavior that are consistent with deflecting predator attacks toward a false head (e.g., Robbins 1981). Individuals of the species *Thecla togarna*, for example, have a false head, complete with dummy antennae, at the tips of their hind wings (Figure 13.19). These butterflies enhance the structural illusion of a head at their hind end by performing two rather convincing behavioral displays. First, upon landing, the butterfly jerks its hind wings, thereby moving the dummy antennae up and down while keeping the true antennae motionless. The second ploy occurs at the instant of landing, when the butterfly quickly turns so that its false head points in the direction of previous flight. An approaching predator is thus confronted with a prey that flutters off in the direction opposite to that expected. Experimental tests have demonstrated that markings associated with false heads misdirect the

FIGURE 13.19 **The false head of a butterfly. Note the pattern of markings that tends to focus attention on the posterior end of the butterfly and the dummy antennae and eyes at the tips of the hind wings. Markings and structure combine with behavior (e.g., movement of dummy, rather than true, antennae) to divert the attention of a predator away from the true head. (Redrawn from Wickler 1968.)**

attacks of avian predators and, in particular, increase the possibility of escape if the prey is caught to begin with (Wourms and Wasserman 1985).

AUTOTOMY

Rather than simply diverting a predator's attack toward a nonvital portion of the anatomy, some prey actually hand over a "disposable" body part to their attacker, almost as a consolation prize. **Autotomy**, the ability to break off a body part when attacked, has evolved as a defense mechanism in both vertebrates and invertebrates. Tail autotomy in lizards, for example, is commonly reported, as well as in some salamanders, a few snakes, and even some rodents. A more dramatic autotomy, however, is seen in sea cucumbers (members of the phylum Echinodermata), which, upon being attacked, forcefully expel their visceral organs (guts, in the vernacular) through a rupture in the cloacal region or body wall. The predator may then begin to feed on the sea cucumber's offering as it makes its slow escape. In most autotomy cases, the disposable body part is subsequently regenerated. As an example of the phenomenon, we will focus on tail autotomy in lizards.

Tail autotomy benefits the lizard in two ways: first, it allows the lizard to break away from its attacker, and, second, if the detached tail continues to thrash and writhe, the attacker is distracted as the lizard runs away (Arnold 1988). Although the vigor and duration of postautotomy tail movement varies among species, in some lizards the tail may thrash for as long as five minutes. The effectiveness of tail autotomy is underscored by the presence of tails in the stomachs of predators, as well as the occurrence of tailless lizards and lizards with regenerated tails in natural populations.

Direct experimental evidence for the importance of tail autotomy as an antipredator device comes from a laboratory study by Benjamin Dial and Lloyd Fitzpatrick (1983). These researchers tested the effectiveness of tail autotomy and postautotomy tail movements in permitting the escape of lizards from mammalian and snake predators. In the first study, staged encounters were conducted between a feral cat and two species of lizards, *Scincella lateralis* (a species with vigorous postautotomy tail thrashing) and *Anolis carolinensis* (a species with less vigorous thrashing). Dial and Fitzpatrick recorded the cat's reaction to lizards of both species under two conditions: (1) thrashing tail trials—lizards and their autotomized tails were placed in front of the cat immediately after autotomy, and (2) exhausted tail trials—tails were allowed to thrash to exhaustion and then lizards and their autotomized tails were placed in front of the cat. In both types of trials, autotomy was induced by the experimenters. They gripped the lizards' tails at the caudal fracture plane with forceps (in many species of lizards, tail breakage takes place at preformed areas of weakness). The results, summarized in Table 13.2, show that the dramatic postautotomy tail thrashing of *S. lateralis* is an effective escape tactic, whereas the more subdued tail movement of *A. carolinensis* is not. Note that in all of the thrashing-tail trials with *S. lateralis*, the cat attacked the tail, rather than the lizard, and in all cases the lizard escaped. In 100% of the exhausted-tail trials with this species, however, the cat attacked and captured the lizards. The results for *A. carolinensis* were quite different: the cat attacked the lizards and ignored the tails in all trials.

TABLE 13.2 **Responses of a Feral Cat to Simultaneous Presentation of Autotomized Tails (Either Thrashing or Exhausted) and Live Tailless Bodies of Two Species of Lizards**

	Number of responses		
Tail	Attack to tail	Attack to body	Escape of lizard
		Anolis carolinensis	
Exhausted	0	8	3
Thrashing	0	6	1
		Scincella lateralis	
Exhausted	0	6	0
Thrashing	7	0	7

Source: Data from Dial and Fitzpatrick (1983).

In the second experiment, Dial and Fitzpatrick (1983) examined whether postautotomy tail movement influenced the predator's handling time. The authors staged encounters between *S. lateralis* and the snake *Lampropeltis triangulum*, again using autotomized tails that were either thrashing or exhausted. On average, the snakes required 37 seconds longer to handle thrashing tails than exhausted tails, providing the tailless lizard with more time to escape. Thus, for the lizard *S. lateralis*, postautotomy tail movement supplements the simple mechanism of breaking away from the predator's grasp and, depending on the type of predator, may either attract the predator's attention (as in the case of the cat) or increase the time required to handle the autotomized tail (as in the case of the snake). Either way, postautotomy tail movement enhances the opportunity for the lizard to escape.

Until this point we have focused on the benefits of tail autotomy without mentioning potential costs. Depending on the species of lizard, tail loss may lead to reductions in speed, balance, swimming, climbing, or mating ability, and when the tail is used as a display, it may even lead to declines in social status (Cooper et al. 2004; Fox and Rostker 1982; Langkilde et al. 2005). Furthermore, regeneration of the tail must certainly entail costs in energy and materials. Many lizards, after all, have substantial fat deposits in their tails that are also lost with the tail. Finally, unlike many other antipredator devices, once used, autotomy cannot be employed again, at least until the tail regenerates.

FEIGNING INJURY OR DEATH

In ground-nesting birds such as the killdeer (*Charadrius vociferus*), a parent may feign injury in an elaborate effort to divert the attention of an approaching predator away from its nest and young, particularly soon after hatching, when offspring are most vulnerable (Brunton 1990). Upon spying a predator, an adult may suddenly begin dragging its wing, as it flutters away from its nest. The predator follows, and as it closes in the killdeer suddenly recovers and flies away, giving a loud call. If all goes as planned, the predator will continue to wander off.

Some animals rely not on diverting the attention of a predator but on causing the predator to lose interest. Because some predators kill only when their prey is moving, an animal that feigns death may fail to release the predator's killing behavior, and with any luck the predator will lose interest and move along in search of a livelier victim. Perhaps the most familiar feigner of death is the opossum, *Didelphis virginiana* (Figure 13.20). Hence the phrase "playing possum" has come to be synonymous with "playing dead." Although their performance is less well publicized than that of opossums, juvenile caimans (*Caiman crocodilus*) react aggressively toward humans when approached on land but feign death when handled in water (Gorzula 1978). The response of an individual to a particular predator may thus vary as a function of context.

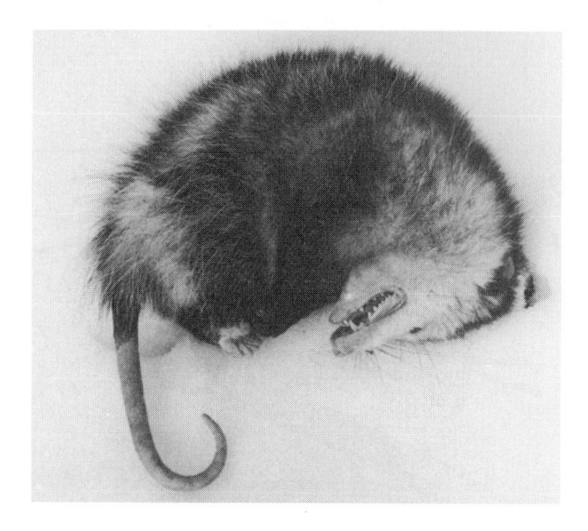

FIGURE 13.20 **An opossum playing dead.**

Hognose snakes (*Heterodon platirhinos*) have a complex repertoire of antipredator mechanisms, and feigning death is one option. These fairly large nonvenomous or slightly venomous snakes occur in sandy habitats in the eastern United States. When first disturbed, the hognose opts for bluffing—it flattens and expands the front third of its body and head, forming a hood and causing it to look larger. It then curls into an exaggerated S-coil and hisses, occasionally making false strikes at its tormentor. When further provoked, however, it drops the bluff and begins to writhe violently and to defecate. Then it rolls over, belly up, with its mouth open and tongue lolling. If the predator loses interest in the now immobile "corpse" and moves away, the snake slowly rights itself and crawls off. Natural encounters between predators and prey are rarely witnessed events, so most research on death feigning has been conducted under laboratory conditions with staged predator–prey encounters. We will consider two such studies, but keep in mind that there is a general need for field observations of death feigning during natural predator–prey encounters (Gregory et al. 2007).

The complete repertoire of antipredator mechanisms occurs in young hognose snakes, and Gordon Burghardt and Harry Greene (1988) have shown that newborn snakes are capable of making very subtle assessments of the degree of threat posed by a particular predator. The researchers conducted two experiments in which they monitored the recovery from feigning death (i.e., crawling away) of newly hatched snakes under various conditions. In experiment 1, the recovery of snakes was monitored in the presence or absence of a stuffed screech owl (*Otus asio*) mounted on a tripod 1 meter from the belly-up snake. In experiment 2, the snake recovered (1) in the presence of a human who was staring at the snake from a distance of 1 m, (2) in the presence of the same person in the same location but whose eyes were averted, and (3) in a control condition in which no human being was visible. Both the presence of the owl (experiment 1) and the direct human gaze (experiment

2) resulted in longer recovery times than the respective control conditions (Figure 13.21). When the human being averted his or her eyes, the recovery time was intermediate. Thus, young snakes are capable of using rather subtle cues from predators to make adjustments in their antipredator behavior.

From our discussion so far, you can see that the antipredator behavior typically described as death feigning eventually involves immobility elicited by the presence (or grasp) of a predator. Some observers have noted, however, that the rigid postures assumed by some supposedly death-feigning animals barely resemble those of dead animals. Could immobility in these cases serve an antipredator function other than mimicking death? The answer appears to be yes. Atsushi Honma and colleagues (2006) conducted predation experiments in which pygmy grasshoppers (*Criotettix japonicus*) were exposed to four different potential predators found in the grasshopper's habitat: a frog, a bird, a praying mantis, and a wolf spider. The results of the predation experiments are presented in Table 13.3. When grasped by a frog, a grasshopper immediately assumes a rigid T-shape (formed by firmly bending its hind legs downward), which it maintains for several minutes, even after release by the frogs, most of which could not swallow the immobile grasshopper. During encounters with other potential predators, however, the grasshopper never becomes immobile. Instead, it exhibits different behaviors, such as struggling to free itself from the grasp of a praying mantis, hopping away if the initial attack by the bird was unsuccessful, or swaying back and forth in an apparent attempt to threaten an approaching spider. In a second experiment, Honma et al. (2006) tied the hind legs of grasshoppers with fishing line so they could not assume the T-shape, and presented them to frogs (the life of a grasshopper destined for use in predation experiments can be harsh indeed!). Videotape analysis revealed that the frogs easily swallowed the bound grasshoppers; each frog could readily adjust the position of the grasshopper in its mouth. On the basis of their results, Honma and co-workers suggest that the rigid posture assumed by pygmy grasshoppers is a specialized adaptation for avoiding frog predation. They further propose that the immobile posture deters predation not by mimicking death, but by enlarging the functional body size of the grasshopper, making it difficult to swallow. Praying mantises are obviously quite effective predators of grasshoppers (refer again to Table 13.3), so why don't grasshoppers assume the rigid posture when captured by them? According to the authors, the size-enhancing rigid posture of grasshoppers would be ineffective against praying mantises because these predators gnaw on captured prey rather than swallowing it whole. Findings such as these remind us that when evaluating the defensive behavior of a prey animal, it is always good to know something about the foraging mode of its predators. The results with pygmy grasshoppers also suggest a need to revise the way in which we view examples of predator-induced immobility in prey: in some cases, prey immobility may function to mimic death, while in others it may serve to enhance body size, a strategy that also seems to intimidate predators, as we discuss next.

TABLE 13.3 Results from Predation Experiments Involving Pygmy Grasshoppers (as Prey) and Four Different Predators

Predator	Number of attacks	Death feigning occurred	Predation successful
Frog	20	17	4
Bird	10	0	2
Mantis	9	0	8
Spider	20	0	0

Source: Data from Honma et al. (2006).

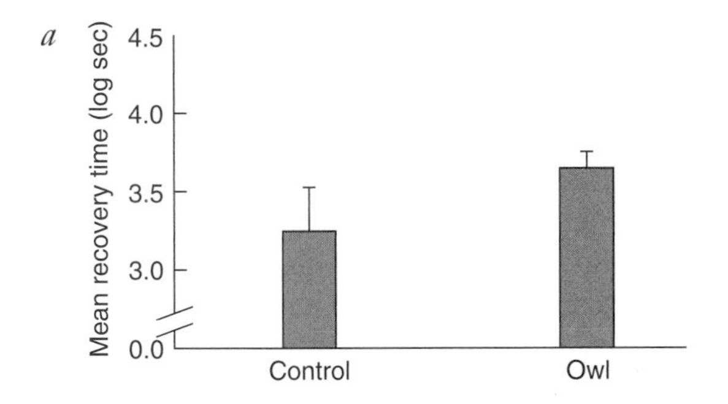

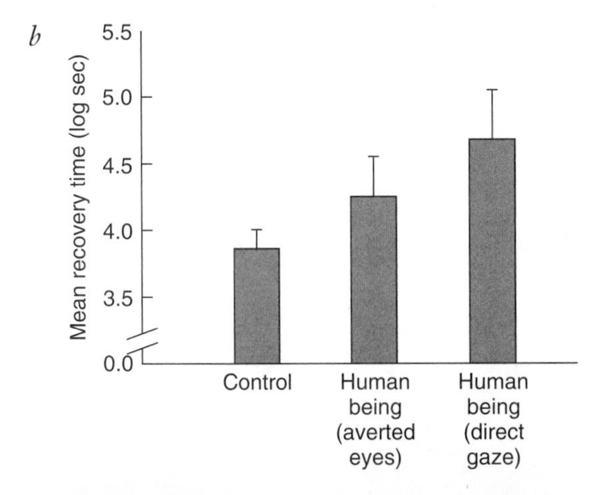

FIGURE 13.21 **Mean time to recovery from feigning death in neonatal hognose snakes exposed to various recovery conditions. (*a*) In experiment 1, snakes recovered in the presence of a stuffed owl or in the absence of a stuffed owl (control condition). (*b*) In experiment 2, snakes recovered in the presence of a human being (with eyes staring at the snake or with eyes averted) or in the absence of a human being (control condition). Because the recovery times were skewed, the data were transformed into logarithms of seconds. (From Burghardt and Greene 1988.)**

INTIMIDATION AND FIGHTING BACK

Prey animals have many ways of communicating "I am formidable" to a predator. Presumably, when a predator encounters a large, threatening, well-armed prey, it will continue on its way, searching for a less challenging meal. Here are some ways that prey can intimidate predators and even fight back under certain circumstances.

ENHANCEMENT OF BODY SIZE AND DISPLAY OF WEAPONRY

When dealing with potential predators, some animals employ the size-maximization principle. A cat hunches its back and erects its fur in the presence of a dog. Some toads and fishes inflate themselves when disturbed. In each case, the animal increases its size and appears more formidable or unswallowable. Several displays of intimidation through an increase in size are shown in Figure 13.22, but threat maximization need not always be visual. Loud calls, hisses, or growls may also cause a predator to look elsewhere for its next meal.

Some animals display their weapons when confronting a predator. Ungulates often display their horns and paw at the ground, perhaps to draw attention to their dangerous hooves. Porcupines erect their spines and cats display their teeth. All these postures are probably meant to intimidate a predator.

FIGURE 13.22 **Intimidation displays in several species of animals. The displays make the animal appear larger and more formidable to predators. The animals shown here are (counterclockwise) the frilled lizard, cat, short-eared owl, and porcupine fish. (Modified from Johnsgard 1967.)**

EYESPOTS

Some animals have spots that resemble eyes in the oddest of places (Figure 13.23). What possible function could such markings have? It turns out that eyespots can serve at least two defensive functions (Owen 1980). First, if the spots are relatively small, then they may serve as targets to misdirect a predator in a manner similar to that described for false heads. Such eyespots are typically located on nonvital portions of the body, and thus prey can often escape a predatory encounter with less than fatal damage. Second, if the spots are large, few in number, and suddenly flashed, they may startle or frighten a predator; it is this function of eyespots that will be our focus here.

What evidence is there to support the claim that by flashing eyespots a harmless prey animal can increase its chances of surviving an encounter with a potential predator? We will consider the case of the peacock butterfly (*Inachis io*). At rest, peacock butterflies are dark-colored leaf mimics. The margins of their wings are irregular, and they keep their wings closed so that only the dark ventral side of each is displayed. When disturbed, however, the butterfly transforms into a very different creature, suddenly flicking open its wings to expose bright colors and four large eyespots, and at the same time emitting a hissing sound. (The sound is produced when the butterfly rubs the veins on the base of its forewings against those on the top of its hindwings.) The butterfly repeatedly performs this sequence of wing flicking and hissing, all the while appearing to track changes in predator position. The suddenly flashed eyespots and the hissing sounds certainly seem like devices to intimidate potential predators, but what experimental evidence do we have to support this function for the two traits?

Adrian Vallin and colleagues (2005), armed with a black permanent marker and a pair of scissors, set out to test whether the eyespot display and hissing sound of the harmless peacock butterfly increase its chances of surviving an encounter with an avian predator, the blue tit (*Cyanistes caeruleus;* previously *Parus caeruleus*). The

FIGURE 13.23 **A toad directs its backside toward an attacker, revealing a pair of eyespots.**

researchers presented wild caught blue tits with butterflies from one of the following six treatments (Figure 13.24): (1) butterflies whose eyespots had been colored over with a black marker (experimental group to determine the effects of eyespots); (2) butterflies whose eyespots were intact but whose wings on the dorsal side, close to the body, had been colored over with a black marker (control group for eyespot manipulation); (3) butterflies whose sound production abilities had been stopped through removal of a small part of the forewings involved in producing the hissing sound (experimental group to determine the effects of sound production); (4) butterflies whose sound production capabilities were intact but who had a small part of the base of their hindwings removed (control group for the sound production manipulation); (5) butterflies whose eyespots had been colored over and whose sound production had been stopped (experimental group to determine the combined effects of eyespots and sound production); and (6) butterflies whose wing bases had been colored with a black marker and who had a small part of the base of their hindwings removed (control group for eyespot and

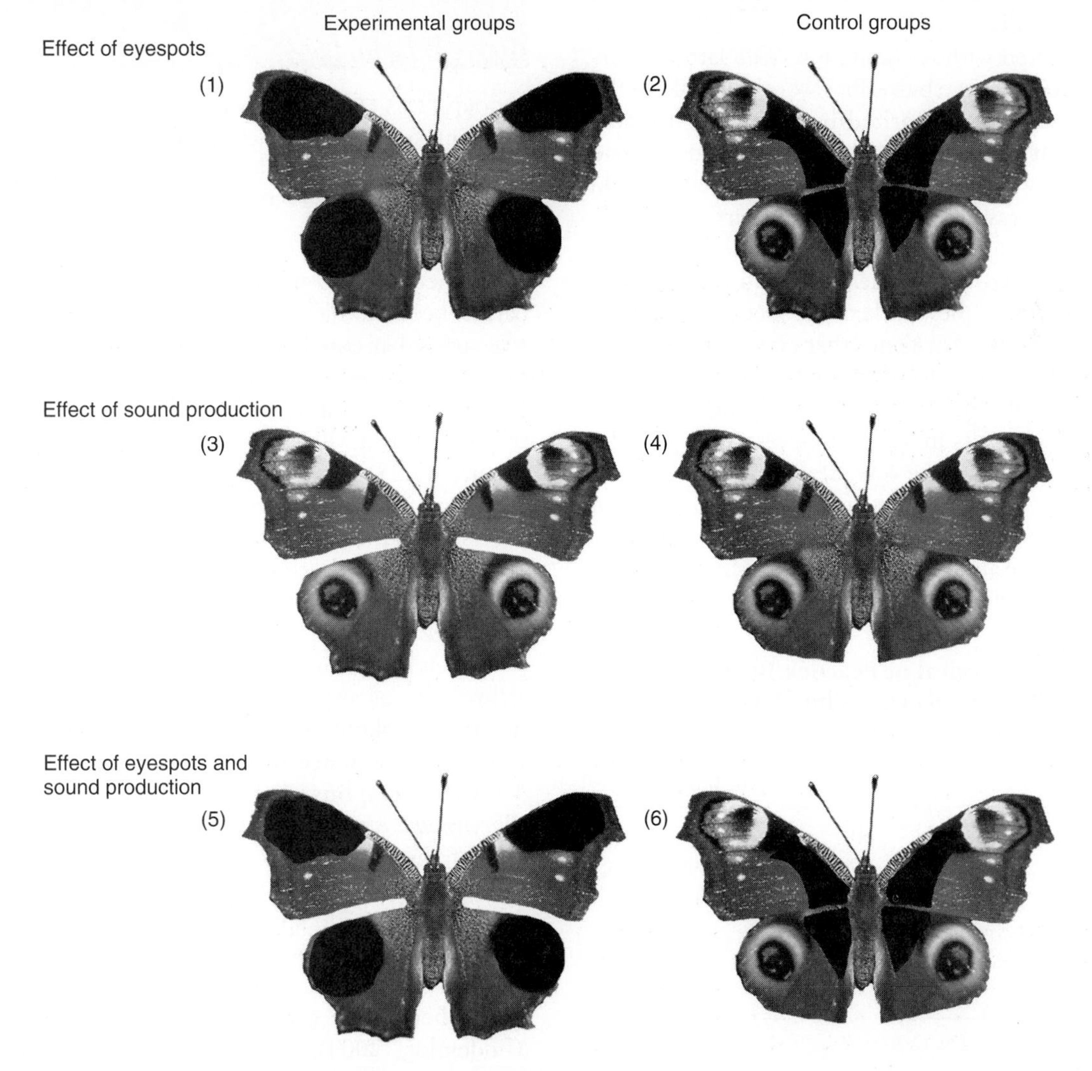

FIGURE 13.24 Six different treatment groups of peacock butterflies used to test the effectiveness of eyespots and sound in deterring an avian predator. Upper row: (1) butterflies whose eyespots had been colored over with a black marker (experimental group to determine the effects of eyespots) and (2) butterflies whose eyespots were intact but whose wings on the dorsal side, close to the body, had been colored over with a black marker (control group for eyespot manipulation). Middle row: (3) butterflies whose sound production abilities had been stopped through removal of a small part of the forewings involved in producing the hissing sound (experimental group to determine the effects of sound production) and (4) butterflies whose sound production capabilities were intact but who had a small part of the base of their hindwings removed (control group for the sound production manipulation). Bottom row: (5) butterflies whose eyespots had been colored over and whose sound production had been stopped (experimental group to determine the combined effects of eyespots and sound production) and (6) butterflies whose wing bases had been colored with a black marker and who had a small part of the base of their hindwings removed (control group for eyespot and sound production manipulations). (Modified from Vallin et al. 2005.)

sound production manipulations). During predation trials, a butterfly from one of the six groups was placed on a willow log in the middle of a small room, and a blue tit was released into the room. Each trial lasted 30 minutes, and during this time Vallin and co-workers directly observed and videotaped the bird–butterfly interactions. They conducted from eight to ten trials in each of the six treatments. No individual bird or butterfly was used more than once, and birds were released at their site of capture after use in a trial. All birds were banded prior to release, so their individual identity would be known to the researchers, and they would not be used again. All butterflies were discovered by the birds, and all that were seized were eaten without hesitation. The latter observations suggest that the butterflies were palatable to the birds. The results, summarized by treatment in Table 13.4, show that butterflies survived better when they had eyespots than when the eyespots were colored over. In contrast, there was no difference in the survival of butterflies in the sound and no sound treatments. Finally, butterflies survived better when they had eyespots and sound than when they lacked eyespots and sound, although there was no evidence that eyespots and sound were any better than eyespots alone (compare "eyespots and sound" with the "no sound" treatment). Taken together, these results indicate that eyespot displays by a harmless butterfly can be very effective in deterring an avian predator. Although sound production did not seem critical to the peacock butterfly surviving its encounters with blue tits, the authors suggest that the hissing sound might prove effective in deterring other predators.

TABLE 13.4 Survival of Peacock Butterflies during Predation Trials with Blue Tits

Treatment	Number of trials	% Survival of butterflies
No eyespots	10	50.0
Eyespots	9	100.0
No sound	8	100.0
Sound	8	87.5
No eyespots or sound	10	20.0
Eyespots and sound	9	100.0

Source: Data from Vallin et al. (2005).

CHEMICAL REPELLENTS

A wide variety of insects can discharge noxious chemicals when they are captured. Some of these chemicals are powerful toxins or irritants, and in some species they can be shot with considerable accuracy in several directions. The assassin bug (*Platymeris rhadamantus*) reacts to a disturbance by spitting copious amounts of fluid in the direction of the attacker. The saliva is rich in enzymes and causes intense local pain when it comes in contact with membranes of the eyes or nose.

FIGURE 13.25 **The bombardier beetle ejects a hot, irritating spray at its attackers. This beetle, tethered to a wire fastened to its back with wax, responds with excellent aim to a forceps pinch on its left foreleg.**

Other masters of chemical warfare are the bombardier beetles, which deter predators by emitting a defensive spray that contains substances stored in two glands that open at the tip of the abdomen (Dean et al. 1990; Eisner 1958). Because the tip of the abdomen acts as a revolving turret, the spray can be aimed in all directions (Figure 13.25). The chemical reactants from the two glands are mixed just before they are discharged, producing a sudden increase in temperature of the mixture. The hot spray is ejected, accompanied by audible pops, in quick pulses. The effect has been likened to that of the German V-1 "buzz" bomb of World War II (Dean et al. 1990).

Chemical deterrents are by no means limited to arthropods, as anyone who has had the misfortune of surprising a skunk or who owns a dog that has enjoyed the same experience must surely know. Although the defensive response of the Texas horned lizard (*Phrynosoma cornutum*) is perhaps less well known than that of the skunk, it is certainly no less spectacular. When disturbed by canid predators such as kit foxes and coyotes, this small, spine-covered lizard of the southwestern United States typically spatters its attacker with a stream of blood ejected from the sinus surrounding its eyes (Sherbrooke and Mason 2005; Sherbrooke and Middendorf 2004). When squirted, coyotes and kit foxes respond with exaggerated mouth movements and lateral head shaking. Blood squirting in *P. cornutum* is rarely directed at predators other than canids. At the turn of the twentieth century, Charles Holder (1901) examined blood squirting behavior and suggested, on the basis of trials in which his fox terrier posed as a predator, that the ejected blood contained noxious components. Apparently, contact between the ejected blood and nasal membranes of the dog was particularly irritating, and only a single encounter was required to produce "a wholesome dread" in the lizard's canine tormentor. More recent data suggest that the squirted blood primarily affects oral receptors (Sherbrooke and Mason 2005) and that the source of the defensive compounds in the lizard's blood may be its diet of ants (Sherbrooke and Middendorf 2004).

PRONOUNCEMENT OF VIGILANCE

Prey that scan their surroundings for potential predators are said to be vigilant. You might be surprised to learn that having detected a predator, some prey actually approach it rather than head as quickly as possible in the opposite direction. Other prey may leave, but not before performing a display seemingly directed at the predator. What could be the function(s) of these prey reactions? One function might be in communicating to the predator that it is dealing with an alert and aware prey. Sit-and-wait predators might leave their ambush sites once prey communicate to them that they have been spotted. An excellent example concerns the interactions between prey and timber rattlesnakes (*Crotalus horridus*), predators that may spend hours or even days at an ambush site waiting for the appropriate moment to strike. Rulon Clark (2005) captured timber rattlesnakes at a nature preserve in New York, surgically implanted radiotransmitters in their body cavities, and returned the snakes to their site of capture. He then located individual snakes on a daily basis and trained videocameras on those that were either basking (snake in the sunlight, with loose coils) or hunting (snake with tight coils and head oriented in the direction of a runway used by small mammals). Using this combination of radiotracking and videotaping, Clark recorded natural encounters between the snakes and their prey, which included chipmunks, gray squirrels, and a wood thrush. In 12 instances, individual prey were recorded harassing a snake; six of the snakes were basking and six hunting. During harassment displays, prey made repeated approaches to the snake.

FIGURE 13.26 **Stotting by a Thomson's gazelle.**

TABLE 13.5 **Hypotheses to Explain the Function of Stotting**

Benefits to the individual
Signaling to the predator
1. Pursuit invitation
2. Predator detection
3. Pursuit deterrence
4. Prey is healthy
5. Startle
6. Confusion effect
Signaling to conspecifics
7. Social cohesion
8. Attract mother's attention
Signaling not involved
9. Antiambush behavior[a]
10. Play
Benefits to other individuals
11. Warn conspecifics

[a]Height gained through stotting allows fleeing individuals a better view of their surroundings.

Source: After Caro (1986a).

The displays had visual components (for example, tail-flagging by the rodents) and may have had auditory components, but this could not be determined because the videocameras were not equipped with microphones. Hunting snakes that received the displays were 4.3 times more likely to abandon their site after harassment than hunting snakes that were not harassed. Also, hunting snakes that were harassed tended to move greater distances before resuming foraging than those that were not harassed and were simply changing ambush sites. Snakes that were basking showed no increased tendency to abandon their site following harassment. These data provide evidence from natural encounters between predators and their prey to support the notion that prey may benefit by communicating to an actively hunting sit-and-wait predator that it has been spotted. In this case, the specific benefit to the prey is that the predator leaves the general area following harassment.

Stalking predators may also abandon the hunt once they receive the signal that they have been detected. Stotting, a stiff-legged bounding display performed by pronghorn, and many species of deer and antelope (Figure 13.26), appears to be just such a signal. The stotting display has attracted the attention of many investigators, and arrival at pronouncement of vigilance (i.e., announcing to a predator that it has been detected) as a plausible function has involved testing predictions from several hypotheses. At least 11 hypotheses have been proposed for the function of stotting (Caro 1986a; Table 13.5). Although not mutually exclusive, the hypotheses range from the interpretation of stotting as a signal given by a hunted animal to either a predator or a conspecific to the interpretation that stotting has no signal value at

TABLE 13.6 Outcome of 31 Cheetah Hunts Involving Thomson's Gazelles That Did or Did Not Stott

	Chase occurred			
	Chase successful	Chase unsuccessful	Hunt abandoned	Total
Gazelle stotts	0	2	5	7
Gazelle does not stott	5	7	12	24

Source: Modified from Caro (1986b).

all and is simply a form of play or, alternatively, a means to get a better view of the surrounding vegetation (the antiambush hypothesis). In the first true effort to distinguish among the hypotheses, Tim Caro (1986b) recorded the response of Thomson's gazelles (*Gazella thomsoni*) to naturally occurring predators, usually cheetahs (*Acinonyx jubatus*), in the Serengeti National Park of Tanzania. He analyzed prey behavior, cheetah behavior, and the outcome of hunts and found that cheetahs were more likely to abandon hunts when their prey stotted than when they did not (Table 13.6). These results, combined with other data that refuted many of the remaining hypotheses, suggested that stotting typically functioned to inform the predator that it had been detected. Two other functions for stotting were supported by Caro's observations. First, mothers may stott to distract a predator from their fawn, a function much like the broken wing displays described for killdeer. Second, fawns appear to stott to inform their mother that they have been disturbed at their hiding place.

A subsequent study suggests that the context of the cheetah–gazelle encounter and age of the performing gazelle are not the only factors that influence the function of stotting. The type of predator is another consideration. When hunted by coursing predators that rely on stamina to outrun their prey, gazelles appear to use stotting as an honest signal of their ability to outrun predators (FitzGibbon and Fanshawe 1988). Coursing predators such as African wild dogs (*Lycaon pictus*) concentrate their chases on those individuals within a group that stott at lower rates. Thus they appear to use information conveyed in stotting to select their prey. In the study by FitzGibbon and Fanshawe, the mean rate of stotting by gazelles that were chased was 1.64 stotts per second, and for those not chased, 1.86 stotts per second. By signaling their ability to escape at the start of a hunt, those gazelles with high stamina and/or running speeds may not have to prove their physical prowess by outrunning wild dogs in long, exhausting, and potentially dangerous chases.

If the function of stotting varies with the species of predator, then we should not be surprised if future studies reveal that the function varies with the species of prey as well. Indeed, although Caro (1986b) discarded the antiambush hypothesis for Thomson's gazelles (recall that this hypothesis states that stotting is not a signal but simply a way to gain a better view of the surroundings), Stankowich and Coss (2007a) found some empirical support for this hypothesis for Columbian black-tailed deer (*Odocoileus hemionus columbianus*). When approached by a single human observer, black-tailed deer stotted more often in taller vegetation, a finding consistent with one prediction of the antiambush hypothesis. Stankowich and Coss predict, however, that stotting in black-tailed deer likely has multiple functions, as it does in Thomson's gazelles. Finally, although often performed in the presence of predators, stotting also occurs during intraspecific encounters in many species, and we can only guess what its function is under these circumstances.

GROUP DEFENSE

Until now we have focused almost exclusively on strategies employed by individual animals to avoid being eaten. Some animals, however, are social, and membership in a group makes accessible several antipredator tactics that are not available to solitary individuals. We next consider some examples of how social animals cope with predators. Although we discuss each tactic separately, some may interact for any given species. Finally, as you consider these antipredator mechanisms, keep in mind that group living has many advantages, including those totally unrelated to protection from predators (see Chapter 19).

ALARM SIGNALS

When a predator approaches a group of prey, one or more individuals within the aggregation may give a signal that alerts other members of the group to the predator's presence. Alarm signals may be visual, auditory, or chemical, and they often inspire retreat by prey to a safe location. In some cases, the alarm may aid the signaler or its relatives. In other instances, an alarm appears to benefit all those exposed to the signal, including in some cases individuals of different species and even different taxonomic groups. Eurasian red squirrels, for example, respond with flight or increased vigilance to the alarm calls of Eurasian jays (Randler 2006). The proposed selective advantages of signaling alarm are covered in more detail in Chapter 19. We will focus our discussion here on the chemical alarm system of an amphibian, the western toad.

Injured tadpoles of the western toad, *Bufo boreas*, produce an alarm substance, and there is experimental evidence that it functions as an effective antipredator device. Individuals of this species live in the ponds and lakes of western North America, where the tadpoles form dense aggregations. Diana Hews (1988) first documented the response of toad tadpoles to release of the

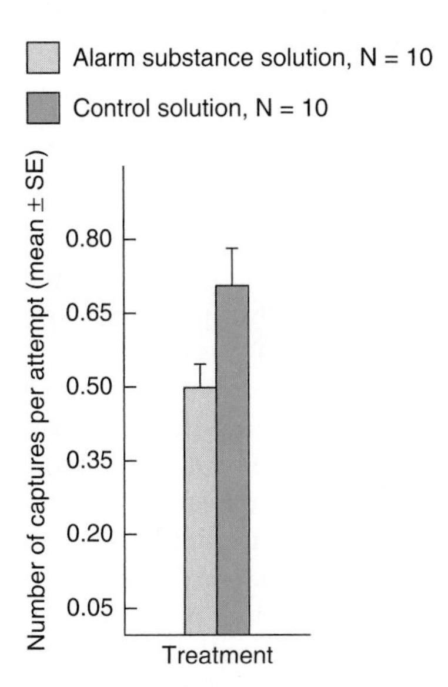

FIGURE 13.27 **Effects of alarm substance of toad tadpoles on the attack success rate of dragonfly naiads. Tadpoles exposed to the alarm substance were less vulnerable to predation by naiads than those exposed to the control substance, water. (Modified from Hews 1988.)**

alarm substance and then tested whether tadpoles alerted by the substance had higher survival rates than those not exposed. Two natural predators of western toad tadpoles, giant water bugs and dragonfly naiads, were used in the experiments.

When tested in aquaria, western toad tadpoles avoided the side of the tank that contained a giant water bug that was feeding on a conspecific tadpole (in a visually isolated but interconnected container) and increased their activity. Tadpoles did not avoid the predator's side of the tank when the water bug was feeding on a tadpole of another species, though they did show a slight, short-term increase in activity. Regarding the latter results for activity, Hews suggests that there might be a small and nonspecific activity response to substances released from injured tadpoles of any species, but that the response to an injured conspecific is much stronger and longer lasting. It is important to note that toad tadpoles alerted by the conspecific alarm substance were less vulnerable to predation. Dragonfly naiads had fewer captures per attack in tests with tadpoles exposed to the toad extract containing the alarm substance than with tadpoles exposed to the control extract, water (Figure 13.27).

IMPROVED DETECTION

Early detection of a predator can often translate into escape for prey, and groups are typically superior to lone animals in their ability to spot predators. Increases in the number of group members (and hence the number of eyes, ears, noses, etc.) often result in increases in the immediacy with which approaching predators are detected, and the escape response of a vigilant individual can alert others to the approaching danger.

The benefits of increased predator-detecting ability accrue to members of many different kinds of groups. Florida scrub jays (*Aphelocoma coerulescens*) form single-species groupings, usually consisting of two to eight family members. Because these birds live in small, permanent groups of stable composition, it is possible for individuals to coordinate their vigilance in a highly structured sentinel system (McGowan and Woolfenden 1989). At any given time, only one family member typically sits on an exposed perch and continually scans the surroundings for predators. If a predator is spotted, the sentinel sounds the alarm, and family members respond by either mobbing a ground predator or fleeing, or monitoring the movements of an aerial attacker. Periodic exchanges among family members occur to relieve the sentinel bird of its duties. Sentinel systems have also been reported for some mammals that live in family-based social groups, including dwarf mongooses (Rasa 1986) and meerkats (Moran 1984). The benefit of improved detection, however, is not reserved solely for family-based social groups with relatively permanent membership. Even members of temporary groupings, such as Iberian green frogs that fortuitously come together at favorable foraging locations, appear to benefit from improved detection of predators (Martín et al. 2006). Finally, members of mixed-species groups also benefit from improved detection, providing that they are on the lookout for the same species of predators and that they communicate detection to other group members.

DILUTION EFFECT

Individuals in groups are safer not only because of their enhanced ability to detect predators but also because each individual has a smaller chance of becoming the next victim. Called the **dilution effect**, this advantage for grouped prey operates if predators encounter single individuals or small groups as often as large groups and if there is a limit to the number of prey killed per encounter. As group size increases, the dilution effect becomes more effective, and improved vigilance appears to provide relatively less benefit (Dehn 1990).

Although this notion of safety in numbers has intuitive appeal, in some cases, predators aggregate in areas where prey are abundant. As a result, some grouped prey may actually suffer higher predation rates. In an examination of the balance between the forces of the dilution effect and the aggregating response of predators, Turchin and Kareiva (1989) studied grouping in aphids (*Aphis varians*). These small insects form dense clusters on the flowerheads of fireweed, and it is here that they are preyed on by ladybird beetles, typically *Hippodamia convergens*. In one experiment, the researchers quantified per capita population growth rates (a measure of individual survivorship) for aphids living singly and for those living in colonies of over 1000 individuals and found that individual aphids benefited by forming groups. Grouping was only advantageous, however, in the

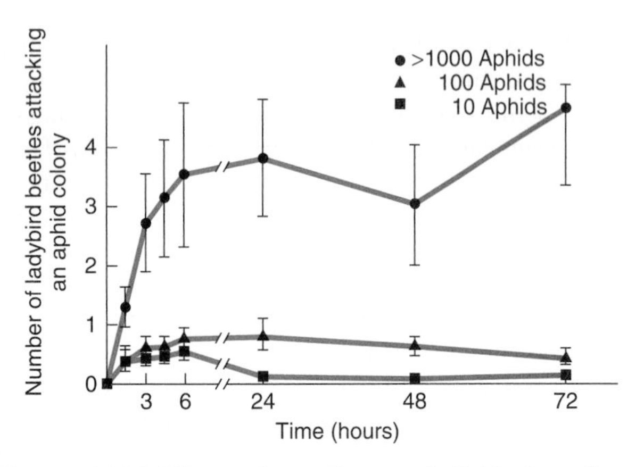

FIGURE 13.28 **The tendency for prey individuals to form large groups and thereby dilute their chances of becoming the next victim may be countered by the tendency by their predators to aggregate where prey are most common. Here, predatory ladybird beetles gather in larger numbers at aphid colonies that contain the most individuals. (Modified from Turchin and Kareiva 1989.)**

presence of predators; when ladybird beetles were excluded from fireweed plants, the individual survivorship of aphids did not increase with the colony's size.

The next question, then, is how do ladybird beetles respond to the grouping of their prey? Turchin and Kareiva found that beetles exhibited a strong aggregation response: more than four times as many beetles were found at aphid colonies of over 1000 individuals than at small and medium-sized colonies (Figure 13.28). In addition to gathering at large colonies, ladybird beetles also increased their feeding rate as the aphids' density increased. On average, beetles consumed 0.9 aphids per ten minutes in colonies of 10 individuals and 2.4 aphids per ten minutes in colonies of 1000 individuals. Thus, the group size of aphids appears to affect the per capita growth rate of the aphid colony, the number of predators attracted to the colony, and the rate at which predators feed. Given all these factors, does grouping reduce predation risk for aphids? Apparently so. When the researchers calculated the instantaneous risk of predation to an individual aphid in a ten-minute period, they obtained values of 0.05 for colonies of 10 and 0.008 for colonies of 1000 or more. Thus, in the aphid-ladybird system, the dilution effect still occurs despite the strong tendency of predators to aggregate at large colonies of prey. Turchin and Kareiva are quick to point out, however, that predators are not the only enemies of aphids. Parasitoids and pathogens may increase rapidly in large groups of aphids and may profoundly affect mortality, perhaps even eliminating the antipredator advantages of the dilution effect.

SELFISH HERD

In most groups, as we saw with those of overwintering monarch butterflies at the start of this chapter, centrally located animals appear to be safer than those at the edges. By obtaining a central position, animals can decrease their chances of being attacked and increase the probability that one of their more peripheral colleagues will be eaten instead. This antipredator mechanism, often referred to as the **selfish herd** (Hamilton 1971), emphasizes that although a given group appears to consist of members that coordinate their escape efforts, it is actually composed of selfish individuals, each trying to position as many others as possible between itself and the predator. The selfish herd hypothesis differs from the dilution effect and improved detection because it considers the spatial arrangement of individuals within a group (papers reviewed by Beauchamp 2007).

According to the selfish herd hypothesis, individuals at the center of a group are safer than those at the periphery when attacked by a predator from outside the group. Thus, during an attack, we might expect members of a group to seek the safer, more central positions, and this would lead to a reduction in nearest neighbor distances. Here we consider one study that examined the way in which individuals exposed to an alarm signal place themselves in a group as compared to individuals who were habituated to the signal. Schreckstoff is an alarm chemical produced by some species of fish when physically attacked. Jens Krause (1993) habituated 14 dace (*Leuciscus leuciscus*) to the odor of Schreckstoff, so that they no longer reacted to its presence. A lone minnow (*Phoxinus phoxinus*), which was still responsive to Schreckstoff, was added to the group. Before Schreckstoff was added to the water, the minnow randomly intermingled in the shoal of dace. In repeated tests, when Schreckstoff was added, however, the minnow moved closer to the other fish and positioned itself so that it was surrounded by them. Only the alarmed fish chose a central location in the group, thus supporting predictions of the selfish herd hypothesis.

One might ask, then, are central locations within the group *always* the best? The answer is no. In fact, a study on the antipredator advantages of schooling in fish suggests that the center is sometimes the most dangerous place to be. When in the company of a predatory seabass (*Centropristis striata*), silversides (*Menidia menidia*) at the center of a school suffered the most attacks (Parrish 1989). Rather than assaulting the margins, seabass swim toward the center of the school, split the school into two groups, and then strike at the tail end of one of the groups, where individuals that were in the center now find themselves. The relative safety of a location within a group thus depends on the predator's method of attack. Because schools of fish undoubtedly cope with a number of predators, each possibly using a different attack strategy, the relative advantage of central versus peripheral locations may change. In addition, factors such as foraging efficiency (those in the front see the food first) and the energetics of locomotion (fish in the front of a school may experience more "drag" than those at the back) probably also influence optimal positions within the school.

CONFUSION EFFECT

Predators that direct their attacks at a single animal in a group may hesitate or become confused when confronted with several potential meals at once. No matter how brief, any delay in the attack will operate in favor of the prey. The **confusion effect** describes the situation in which predators are less successful in attacking prey because they are unable to single out and attack individual prey (Krause and Ruxton 2002; Miller 1922). Few experiments have investigated whether specific predators are susceptible to the confusion effect, but the limited data available suggest that tactile predators (including nematodes and the larvae of certain insects) are particularly susceptible, as are visual predators with highly agile prey (Jeschke and Tollrian 2007). Our example concerns visual predators.

The confusion effect is thought to be one of the primary antipredator advantages of schooling in fish. When the fish in the school scatter, it makes it difficult for visual predators to focus on a single one. If you've ever tried to use a little net to catch a single member of a school of fish in an aquarium, this may seem especially believable to you. Neill and Cullen (1974) examined the effects of the size of the school on the hunting success of two cephalopod predators (squid, *Loligo vulgaris*, and cuttlefish, *S. officinalis*) and two fish predators (pike, *Esox lucius*, and perch, *Perca fluviatilis*). Whereas squid, cuttlefish, and pike are ambush predators, perch typically chase their intended victims. In most cases, predators were tested with fish of their natural prey species in schools of 1, 6, and 20 individuals. For all four predators, attack success per encounter decreased as the size of the school of prey increased (Figure 13.29). In the three ambush predators, the increased size of the prey's school appeared to produce hesitation and behavior characteristic of conflict (such as alternating between approach and avoidance). Perch, however, switched targets more frequently as the size of the school increased and, with each switch, reverted to an earlier stage of the hunting sequence. Under natural conditions, predators of fish may achieve hunting success by restricting their attacks to individuals that have either strayed from the school or have a conspicuous appearance. In both cases, the predator can concentrate on the odd target.

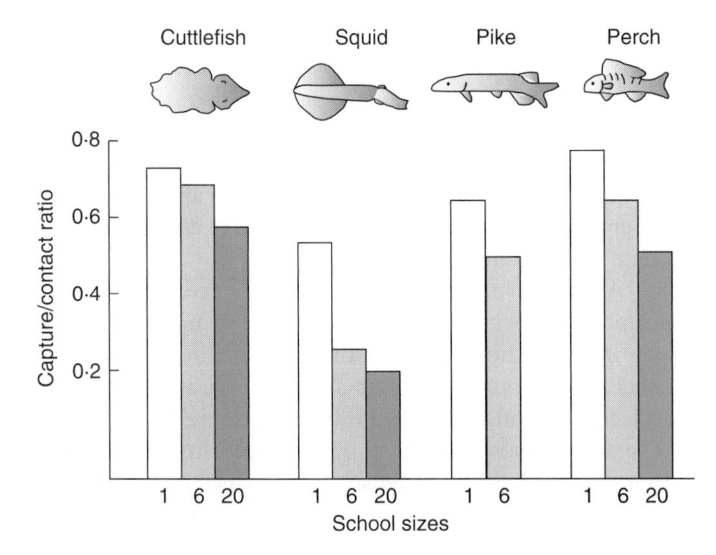

FIGURE 13.29 **The confusion effect. As the size of the school of prey increases, four predators have reduced hunting success because of either hesitation and conflict behavior (in the case of the cuttlefish, squid, and pike) or frequent switching of targets (perch). (Modified from Neill and Cullen 1974.)**

MOBBING

Sometimes prey attack predators. Approaching, gathering around, and harassing one's enemies is called **mobbing** (Curio 1978). This antipredator strategy typically involves visual and vocal displays, as well as swoops, runs, and direct hits on the predator. Mobbing is usually initiated by a single individual, and then conspecifics, or members of another species, join in the fracas. The possible functions of mobbing include, but are not limited to, (1) confusing the predator; (2) discouraging the predator either through harassment or by the announcement that it has been spotted early in its hunting sequence; (3) alerting others, particularly relatives, of the danger; and (4) providing an opportunity for others, again particularly relatives, to learn to recognize and fear the object that is being mobbed (Curio 1978). Most evidence suggests that mobbing is not an act performed by a cooperative group of individuals that is attempting to protect the group as a whole, but rather it is the selfish act of individuals that are attempting to protect only those that will benefit them directly, that is, themselves and their mates, offspring, and relatives (Ostreiher 2003; Shields 1984; Tamura 1989).

Is mobbing costly? Mobbing certainly takes time and energy away from other activities. In addition, when in hot pursuit of a predator, mobbers appear to have a greater chance of being preyed on than nonmobbers, although there is some disagreement over whether mobbing actually entails a deadly risk to those that participate (Curio and Regelmann 1986; Hennessy 1986). One factor influencing the degree of risk might be the size of the mob. We can imagine, for example, that as more individuals join the mob, the risk to any one of them could be reduced through dilution or confusion effects. But does risk arise only from the predator being mobbed? At least for pied flycatchers (*Ficedula hypoleuca*), the answer appears to be no. Tatjana Krama and Indrikis Krams (2005) played tapes of flycatcher mobbing calls or blank tapes from the top of nest boxes that contained parts of recently abandoned flycatcher nests and a quail egg. The tapes were played during daylight hours when flycatchers give mobbing calls in response to predators in the vicinity of their nest. The researchers were interested in whether martens (*Martes martes*), weasel-like predators that significantly decrease the reproductive success of pied flycatchers through nocturnal raids on nests, might eavesdrop on mobbing calls while resting nearby during the day. Krama and Krams found that more nest boxes with mobbing calls were depredated by martens (identified by their tracks) than nest boxes with blank tapes (13/56 as compared to 2/56, respectively). These data suggest that beyond the potential risk from the predator being mobbed, prey also face risks from other predators that may be eavesdropping on their mobbing displays.

In this chapter we have described the many ways in which prey species respond to predators. We will end our coverage with a discussion of what happens to these patterns of behavior when prey species lose their predators; loss of predators can occur naturally (say, when the prey expands its range into an area where the predators do not occur) or unnaturally (for example, if the prey species is brought into captivity). When predators are lost, do antipredator behaviors remain in the repertoire of the prey species, or do the behaviors eventually disappear? And if they do disappear, what factors influence the speed with which they depart?

MAINTENANCE OF ANTIPREDATOR BEHAVIOR

If an animal freezes or flees when it detects a predator, that means it's not doing something else: for instance, it's not foraging, looking for mates, or resting. So, responding willy-nilly to everything in the environment as if it were dangerous may well decrease an animal's fitness.

If antipredator behavior is costly, then we expect that it should be lost when it is no longer needed. One case in which animals may suddenly find themselves free of predators is on islands. We've already described how birds on the island nation of New Zealand often exhibit remarkable naiveté when it comes to potentially dangerous mammals, which are not naturally found there (Chapter 4). Marine iguanas (*Amblyrhynchus cristatus*) on the Galapagos Islands have also been free of predators for the last 5 to 15 million years, until feral cats and dogs were introduced about 150 years ago. Although iguanas become more wary following experience with a perceived threat of predation, they still are not cautious enough to escape from predators (Rodl et al. 2007). Marsupials (kangaroos, wallabies, and their relatives) also lose some (but not all) of their antipredator behaviors after they are isolated on islands (Blumstein and Daniel 2005).

Relaxed selection for antipredator behavior is not confined to islands. Predators that were once historically important might become locally extinct. Ted Stankowich hid behind cover, then popped up life-sized photographs of large, predatory cats to elicit alarm responses from Columbian black-tailed deer (*Odocoileus hemionus columbianus*). Deer responded more strongly to the model of their current predator, a puma, than to a model bearing a spotted pattern similar to the jaguar that roamed the area up until about 600,000 years ago (Stankowich and Coss 2007b).

Populations do not lose antipredator behaviors at the same rate once a particular predator disappears. The time course of the loss of antipredator behavior depends on, among other things, the costliness of fleeing from the predator (Blumstein et al. 2006), and whether all or just some predators are lost (Blumstein 2006).

SUMMARY

Some antipredator mechanisms decrease the probability of an encounter with a potential predator. The prey may go undetected if it blends with its background (coloration matching the visual background) or if the outline of its body is broken up by bold contrasting patterns (disruptive coloration). Other prey occur in several shapes and colors (polymorphism), presumably to prevent the formation of search images by predators. Alternatively, the prey may be detected by a predator and either be recognized as inedible (warning coloration) or go unrecognized as a potentially tasty meal (masquerade and Batesian mimicry). Although many of these defenses involve colors and patterns, the behavior of a prey animal is critical to the success of these mechanisms.

Other defenses operate during an encounter with a predator and increase an animal's chances of surviving the encounter. Amid the many options available, an individual may divert the predator's attention, inform the predator that it has been spotted early in the hunt, or turn the tables and fight back.

Membership in a group has antipredator advantages. Generally, group-living prey are better than solitary prey at detecting, confusing, and discouraging predators. In addition, during any given attack, an individual in a large group has a lower probability of being the one selected by the predator (dilution effect) and may use other group members as a shield between itself and the enemy (selfish herd).

Although for convenience we have discussed antipredator behavior as distinct defensive mechanisms, our intent is not to imply that a given individual or species is characterized by a single protective strategy. Indeed, most animals face a multitude of predators with diverse methods for detecting and capturing prey, and thus having several defensive tactics is crucial. The behavior and color patterns of animals must be interpreted in the context of several selective forces; after all, animals must not only avoid being eaten but also must feed and reproduce. Finally, antipredator behaviors have costs, as is nicely illustrated by the loss of such behaviors from the repertoire of a prey species when its predators disappear.

PART THREE

Interactions Between Individuals

14

Reproductive Behavior

Male fiddler crabs of the genus *Uca* are spectacular creatures (Figure 14.1). Unlike females that have two small claws for feeding, males have one small feeding claw and one supersized claw (often called the major claw) that can make up nearly half of their total body mass. When feeding, fiddler crabs use the feeding claw(s) to scoop mud or sand into the mouth where specialized mouthparts remove algae and detritus from the sediment. To partially compensate for having only one feeding claw, males feed at faster rates and for longer periods than do females, but still females—with their two feeding claws—gain energy more quickly (Weissburg 1992, 1993). Males incur other energetic costs associated with their greatly enlarged claw. In sand fiddler crabs (*Uca pugilator*), for example, males with their major claw have higher metabolic rates than males without their major claw (Allen and Levinton 2007). In this same study, when males were matched for body size and exercised on a

FIGURE 14.1 **A male fiddler crab has one enormous claw and one small feeding claw (a female, not shown, has two small feeding claws). The greatly enlarged claw is expensive to produce and maintain, so why do males have such a structure?**

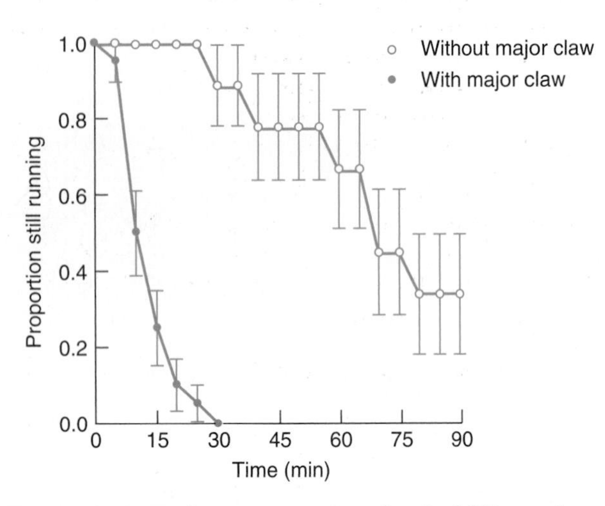

FIGURE 14.2 **Endurance capacity of male fiddler crabs with their major claw and without their major claw when run on a treadmill. Males without their major claw exhibit greater endurance than those with their major claw, indicating that the greatly enlarged claw is costly. (Modified from Allen and Levinton 2007.)**

treadmill, those with their major claw had lower endurance capacities than those without their major claw (Figure 14.2). Finally, observations of males that lose their major claw and then regenerate it indicate that significant production costs are associated with the claw (Backwell et al. 2000; Hopkins 1982).

This costly claw presents a puzzle. What is the function (or functions) of the enormous claw of male fiddler crabs? Given all the documented costs, how did the claw evolve? How can we reconcile the production of such a bizarre structure with natural selection?

SEXUAL SELECTION: HISTORICAL AND THEORETICAL BACKGROUND

Darwin (1871) was the first to suggest that spectacular structures (such as the enormous claw of male fiddler crabs) could arise and be maintained through the process of sexual selection. According to Darwin, sexual selection can occur through two mechanisms: male competition for access to mates and female choice of mates. In the first, called **intrasexual selection**, individuals of one sex, usually males, gain a competitive edge by fighting with each other, and winners could claim the spoils of victory—females. Intense fighting and competition for mates could lead to selection for increased size and elaborate weapons. Male fiddler crabs, it turns out, use their major claw in battles with other males to secure breeding burrows in preferred locations. Sexual selection can also involve selective mate choice, in which individuals of the sex in demand, usually females, choose mates with certain preferred characteristics; this form of sexual selection is called **intersexual selection**. Thus, males not only fight with each other for access to females but also compete to attract females through the elaboration of structures or behavior patterns. Male fiddler crabs also use their major claw to attract females, holding it at certain angles and executing precision movements. In fiddler crabs then, the claw has dual functions, male–male combat and mate attraction. In other animals, a spectacular structure or behavior may function in both capacities, or only in one.

Our focus in this chapter will be on reproductive behavior, with a specific emphasis on sexual selection. Although Darwin (1871) envisioned sexual selection occurring before copulation, we now know that sexual selection operates after copulation as well (see discussions of sperm competition and cryptic female choice).

EXPLANATIONS FOR SEX DIFFERENCES IN REPRODUCTIVE BEHAVIOR

Throughout the animal kingdom, males typically compete for females and females actively choose their mates. Why? Some people think that these differences in mating strategies between the sexes are related to differences in investment in gamete production and parental care.

At the core of the gamete argument are differences in the size and number of male and female gametes. More specifically, whereas females produce a small number of large, energetically expensive eggs, males produce millions of small, relatively inexpensive sperm. A. J. Bateman (1948), working with fruit flies, was the first to suggest that as a result of differences in the number of gametes produced by the sexes, females (or more correctly, their eggs) become a limited resource for which males compete. A female, he reasoned, is likely to maximize her reproductive success by finding the best-quality male to fertilize the limited number of eggs she produces. Whereas a male's reproductive success is predicted to increase with the number of mates he acquires, a female's reproductive success is not expected to increase by mating with more than one male. Also, among his fruit flies, some males appeared quite unsuccessful in their attempts to acquire mates, while others garnered a disproportionate share of matings. Females, on the other hand, were usually similar in their number of matings, probably because they could quickly obtain sufficient sperm to fertilize their limited supply of eggs. In short, Bateman (1948) found that variation in reproductive success was greater among males than among females. Taken together, Bateman's experiments with fruit flies led him to suggest that the sex difference in gametes may help to explain the "undiscriminating eagerness" of males and the "discriminating passivity" of females.

Later, Robert Trivers (1972) suggested that, in addition to differential investment by the sexes in gametes, there was differential investment in offspring, and this too was responsible for competition and mate choice. In Trivers's view, the sex that provides more parental investment for offspring (usually the female) becomes a limiting resource for which the sex that invests less (usually the male) competes. As a result, males compete for access to females and females have the luxury of choosing among available suitors.

Within-species variation in the courtship roles of some insects supports the idea that the degree of parental investment is important in determining which sex competes for access to individuals of the other sex. For example, courtship roles are far from fixed in certain species of katydids (relatives of the grasshoppers). Indeed, in one species (as yet not formally classified by scientists), whether males compete for females or females compete for males varies with the relative importance of male parental investment. In katydids, male parental investment consists of a nuptial meal. At the time of mating, a male katydid transfers his spermatophore (a packet of sperm and fluids) to the female. Following separation of the couple, the female bends around and eats part of the spermatophore that is still attached to her abdomen (Figure 14.3). This protein-rich meal is important in successful reproduction because both the number and fitness of her offspring are enhanced by the male's gift. The relative importance of the male's gift, however, varies with food availability.

FIGURE 14.3 **A female katydid is eating part of the spermatophore deposited by her mate. This protein-rich gift constitutes the male katydid's parental investment. When food is scarce, the gift has relatively greater value, and rather than females choosing mates, males choose females.**

The gift is especially important to females when food is scarce, and it is during these times that females compete for males. In contrast, when food is plentiful, the relative value of the gift declines, and males compete for females (Gwynne and Simmons 1990). We see, then, that a change in courtship roles within a species of katydid results from variation in the relative importance of male parental investment.

Trivers's theory can also be tested by examining the select group of nontraditional species—those in which males invest more than females in the care of offspring. If Trivers were correct, then we would predict that females of these species would compete for males, who in turn would be quite discriminating in their choice of mates. As we will see in the following chapter on parental care and mating systems, Trivers's theory is often, but not always, supported by these exceptions to the rule. In many species with sex-role reversals, males are choosy and females actively compete for mates, as predicted by Trivers's theory. In others, however, females also are choosy (even though they do not provide parental care), and males still seem to compete with one another for access to mates.

Findings that are inconsistent with Trivers's predictions have caused some scientists to challenge the notion that patterns of parental investment are prime determinants of the nature and strength of sexual selection (Gowaty 2003). Others have suggested that while patterns of parental investment may explain which sex competes for mates in most animals (i.e., the sex that invests less in parental care competes more intensely), it may not explain patterns of reproductive competition in *all* animals. Meerkats (*Suricata suricatta*), for example, are cooperatively breeding African mammals in which a single dominant female within a group monopolizes reproduction. Despite displaying the typical mammalian pattern of greater parental investment by females than males (females, after all, are responsible for gestation and lactation), competition for reproductive opportunities is more intense among females than among males (Clutton-Brock et al. 2006). Interestingly, female meerkats that achieve breeder status show changes in their behavior and morphology that are either not shown by males or shown to a lesser extent, including increases in body mass, levels of testosterone, and aggression. Here, then, is an example of a species in which patterns of parental investment do not allow us to correctly predict patterns of intrasexual competition for breeding opportunities.

Clearly, there is more variation in the reproductive roles of males and females than previously imagined or acknowledged. It should not be too surprising, then, that with the discovery of diverse mating strategies also came questions about the ideas originally put forth by Bateman (1948).

REVISITING THE IDEAS OF BATEMAN

In recent years, some of Bateman's ideas and their near uncritical acceptance and use by scientists have been challenged (Tang-Martinez and Ryder 2005). The challenges include methodological problems with Bateman's (1948) experiments, subsequent simplification and selective use of his data by other scientists, and evidence collected in the years following publication of his paper that contradicts his ideas on male mating costs and female reproductive success and behavior. Our focus here will be on some of this new evidence.

Male Mating Costs and Selectivity

Some researchers have questioned the long-held notion that males incur trivial costs when producing gametes (Dewsbury 1982a; Simmons 1988). They are quick to point out that although, gamete for gamete, sperm are vastly smaller and cheaper to produce than eggs, sperm are probably never passed along one at a time. Instead, millions of sperm are transferred in groups along with accessory fluids in ejaculates (unpackaged fluids) or spermatophores (packaged fluids). Although the cost of producing a single gamete may be minuscule, the costs of producing sperm groups and accessory fluids may limit the reproductive potentials of males. In field crickets (*Gryllus bimaculatus*), for example, the costs of spermatophore production (calculated by determining the percentage of the donor male's body weight made up by his spermatophore) are relatively greater for small males than for larger ones. Small males appear to cope with these higher costs not by producing smaller spermatophores but by increasing the refractory period between matings (time from when the male attaches a spermatophore to a female until the onset of the next attempt at courtship, when another spermatophore is ready). Thus, for small male field crickets, the costs of spermatophore production appear to limit their number of matings (Simmons 1988).

Spermatophores also appear costly to males of other invertebrate species. In some butterflies, for example, the costs are reflected in the decreased size and increased water content of successive spermatophores (Ferkau and Fischer 2006). In one species of stalk-eyed fly, the high costs of spermatophore production, particularly those related to secretions from accessory glands, reduce life span in males with high mating rates (Pomiankowski et al. 2005). Finally, in some male crayfish, ejaculate size decreases with successive fertilizations (Rubolini et al. 2007).

Males of some vertebrate species also appear to experience costs related to sperm production. Although most of us might assume that remaining immobile and building up sperm supplies would be less energetically expensive than engaging in active reproductive behaviors such as searching for and courting potential mates, we would be wrong when it comes to the snake *Vipera berus*. In this species, sperm production and overt reproductive behavior occur sequentially (rather than simultaneously) and are separated in time by an obvious indicator, sloughing of the skin. Importantly, males do not eat during either period, so the rate at which they lose body mass can serve as an estimate of the energetic costs of each period. The rate of mass loss while immobile and producing sperm is at least as great as that during the subsequent period of active reproductive behaviors (searching for females, competing with other males, courtship, and copulation), and is sometimes greater (Olsson et al. 1997). There is also evidence from other vertebrate species that males cannot produce unlimited numbers of sperm and may, in fact, face sperm depletion. In male deer mice, sperm counts decrease in successive ejaculates, and counts remain depressed for at least one day (Dewsbury and Sawrey 1984). Dominant male Soay sheep mate frequently and as a result ejaculate insufficient sperm for fertilization (Preston et al. 2001).

Data such as these for invertebrates and vertebrates, collected in the years following the publication of Bateman's (1948) paper, indicate that there can be significant costs associated with producing sperm and secretions from accessory glands, and sperm supplies may be limited. Given these findings, we should not be surprised to learn that males are somewhat prudent in their allocation of ejaculates (Wedell et al. 2002), not to mention selective in their choice of mates. Indeed, male selectivity has been reported for many species (e.g., fruit flies, Gowaty et al. 2003; great snipe, Saether et al. 2001; house mice, Drickamer et al. 2003; and savannah baboons, Alberts et al. 2006). We will continue to evaluate Bateman's ideas in light of new evidence that has emerged in recent years and will return to mate choice later in the chapter.

Female Behavior and Reproductive Success

As mentioned previously, Bateman (1948) described females as somewhat passive participants in courtship and mating. However, more recent data indicate that females are active participants in reproductive activities, often soliciting copulations from one male or several different males. Even though glass laboratory vials might not seem like a place that would inspire courtship, female fruit flies (*Drosophila spp.*) often approach and pursue males when placed together in such vials, and some show interest in particular males that is equal to or greater than the interest that these males show in them (Gowaty et al. 2003). Also, *Drosophila* females may signal their enthusiasm for a male and willingness to mate with him by becoming immobile, but this in no way should be interpreted as "passive" behavior on the

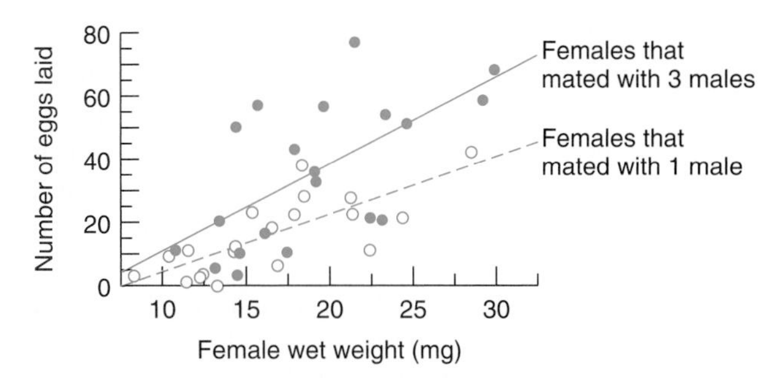

FIGURE 14.4 **Fecundity of female fireflies increases with number of mates, when controlled for female body weight. Here, females that mated with three males produced more eggs than females that mated with one male. (Modified from Rooney and Lewis 2002.)**

part of females. Bonobos (*Pan paniscus*), sometimes described as "our other closest relatives" because they are in the same genus as chimpanzees (*Pan troglodytes*), exhibit interesting social and sexual dynamics, to say the least. Female bonobos are dominant to males and mate promiscuously (and publicly) with males from within their community and with males from other communities (Parish and de Waal 2000). Thus, more recent observations from species as diverse as fruit flies and bonobos indicate that females are anything but passive in mating interactions.

Bateman's (1948) prediction that female reproductive success should not increase with number of mates has also been challenged. We first consider an example in which female reproductive success increases with number of mates. In fireflies (*Photinus ignitus*), females that mate with three males have higher fecundity (number of eggs) than those that mate with a single male (Figure 14.4) (Lewis et al. 2004; Rooney and Lewis 2002). Why should this be the case? During mating, males transfer a spermatophore to the female. Sperm at the anterior end of the spermatophore are released into the female's spermatheca (sperm storage organ) to be used later. The remainder of the spermatophore passes to the spermatophore-digesting organ where it will break down over a few days, freeing nutrients for use by the female. (Thus, unlike the katydids discussed earlier, females can get nutrients from the spermatophore simply by absorption rather than eating it.) *Photinus ignitus* females use the protein from the digested spermatophore to fuel development of their eggs (specifically the production of yolk). Because these fireflies feed as larvae but not as adults, the nutrients gained from the spermatophore supplement the female's larval energy reserves. Thus, we see that *P. ignitus* females benefit from additional matings because such matings supplement their declining larval energy reserves and lead to greater egg production.

Finally, although Bateman (1948) found lower variance in reproductive success among female than male fruit flies, there is now evidence that variance in the reproductive success of females is not always lower than that of males. For example, variance in reproductive success is greater in females than in males for red sea urchins (Levitan 2005) and meadow voles (Sheridan and Tamarin 1988). It has also been suggested that we might expect greater variance in female than male reproductive success in cooperatively breeding mammals in which social groups contain several nonbreeding adults and females experience greater reproductive suppression than males (Hauber and Lacey 2005). These examples show that there are exceptions to the rule that variance in female reproductive success is lower than variance in male reproductive success, and of course the interesting question for such exceptions is "Why?"

With this introduction to some of the historical and theoretical background for sexual selection, and recent challenges to some of these ideas, we will consider intrasexual selection, and then intersexual selection, in more detail.

INTRASEXUAL SELECTION—COMPETITION FOR MATES

Intrasexual selection has led males to evolve a broad spectrum of attributes related to intense competition for mates. Some mechanisms operate prior to copulation, whereas others have their effect once mating has occurred. We will now consider a few of the many tactics males use to gain a competitive edge in the mating game.

ADAPTATIONS THAT HELP A MALE SECURE COPULATIONS

Dominance Behavior

Males in some species may secure copulations by dominating other males and thereby excluding competitors from females. This behavior, of course, places a certain premium on greater male strength and more effective weaponry. Because the female makes no such investments, males tend to diverge from females in both appearance and aggressiveness.

In many birds and mammals, males are larger than females, presumably because large body size improves the fighting ability and hence the reproductive success of males. It can be expected that the greatest levels of **sexual dimorphism** (difference in the appearance and behavior of the sexes) will be found in species in which the competition between males is most intense.

This is, in fact, often the case, but sexual dimorphism can result from other factors as well (Isaac 2005). In many organisms, the effect of female body size on female reproductive success is equally important in determining which sex is more "built to last" than the other. Also, although ecological factors usually affect males and females in a similar manner, sexual dimorphism in body size sometimes results from sex differences in predation pressures or resource use. Indeed, some sexually dimorphic traits may be caused by divergence in food use. Some of the best examples of this concern feeding structures rather than overall body size. Males and females of certain species of mosquito have different mouthparts, and this dimorphism is related to the nectar diet of males and the blood diet of females (Proctor et al. 1996). Similarly, sexual differences in bill lengths and bill curvatures of the purple-throated carib, a species of hummingbird from the West Indies, also are caused by a divergence in food use; males and females sip nectar from flowers of different lengths and curvatures on different plant species (Temeles et al. 2000). Thus sex differences in appearance are sometimes caused by sexual selection and other times by ecological factors.

How would you determine whether a difference in body size between the sexes is caused by sexual selection or ecological factors? As an example of how you might go about making such a determination, we'll consider the case of the cichlid fish *Lamprologus callipterus*, a species endemic to Lake Tanganyika, East Africa. Males of this species are a whopping 12 times heavier than the females with which they spawn. To understand the factors that influence the extreme sexual size dimorphism in *L. callipterus*, we must first know something about the species' life history. Territorial males guard nests of empty snail shells; they inherit shells from previous nest owners, collect some unclaimed shells on their own, and steal shells from the nests of neighbors. At first glance, this male obsession with empty snail shells might seem strange, but it is perfectly understandable in light of the observation that female *L. callipterus* will only breed in empty shells. Females arrive on the territory of a male, enter an empty shell, spawn, and then care for their brood within the shell for about two weeks. In an effort to determine the influence of sexual selection and ecological constraints on the extreme sexual size dimorphism in *L. callipterus*, Dolores Schütz and Michael Taborsky (2005) directly observed the behavior of free-living males and females in Lake Tanganyika and conducted experiments in aquaria in the laboratory. The researchers tested the following four nonexclusive hypotheses concerning sexual size dimorphism (SSD) in *L. callipterus*: (1) SSD results from intrasexual selection (male–male competition), (2) SSD results from intersexual selection (female choice), (3) SSD results from an ecological constraint on male body size (the ability to transport shells depends on large body size), and (4) SSD results from an ecological constraint on female body size (the ability to enter and breed in shells depends on small body size). The specific predictions associated with each hypothesis, general testing methods, and the results are shown in Table 14.1. As you can see from the table, Schütz and Taborsky found strong support for hypotheses 3 and 4. Thus, it appears that the ecological constraint of shell size influences the body sizes of males and females in a divergent manner, favoring males large enough to carry shells and females small enough to fit into shells. The findings also suggest that intrasexual selection and intersexual selection have relatively minor and indirect effects on the evolution and maintenance of sexual size dimorphism in *L. callipterus*. For example, field observations revealed that large males do not have larger nests than small males, and laboratory data indicated that large and small males do not differ in the speed with which they obtain nests. However, large males tend to hold nests longer and in the end sire more offspring. We see from this study that carefully conducted field observations and laboratory experiments can tease apart the influences on body size of sexual selection and ecological factors, and that sometimes several factors can influence body size, and in different ways.

In species in which males form dominance relationships, high-ranking males typically engage in more sexual activity than low-ranking males and experience greater reproductive success (Dewsbury 1982b; Ellis 1995). In their attempts to monopolize sexual access to females, dominant males sometimes interfere directly with the copulations of subordinate males. Daniel Estep and colleagues (1988) examined interference by dominant males and the consequent inhibition of sexual behavior among subordinate males of the stumptail macaque (*Macaca arctoides*). Long-term observations of a captive colony of this Old World primate revealed the existence of a stable, linear dominance hierarchy among adult males. Dominant males monopolize access to females by inhibiting subordinate males through aggressive threats and disruption of mating attempts. Within the dominance hierarchy, copulation frequencies were rank-related; the alpha (highest-ranking) male had almost as many copulations as the other 17 adult males combined (Table 14.2). Within this group, the three lowest-ranking males failed to mate even once during the 34-week study. If low-ranking males mated at all, they usually did so surreptitiously, out of view of dominant males. When tested in the absence of dominant males, low-ranking males copulated far more frequently than they did in their presence. (Dominance relationships are discussed further in Chapter 18.)

TABLE 14.1 Testing Whether Sexual Size Dimorphism in a Cichlid Fish Results from Sexual Selection or Ecological Constraints

Hypothesis	Prediction	Testing condition	Prediction supported?
(1) Intrasexual selection	(1a) Because of enhanced competitive abilities, large males have larger nests than small males	Field observations	No
	(1b) Large males acquire nests more rapidly and hold onto nests longer than do small males	Laboratory experiment in which two different sized males directly competed for nests and females	No/Yes: both large and small males acquire nests rapidly, but large males are more aggressive, hold nests longer, and as a result sire more offspring
(2) Intersexual selection	(2a) Females prefer larger or heavier males	Field observations Laboratory experiment in which females were allowed to choose between two males that differed in standard length and condition (weight/standard length^3)	No No
	(2b) Females choose males with larger nests	Field observations Laboratory experiment in which females were allowed to choose between two males that differed in the number of shells in their nest (5 shells versus 15 shells)	Yes No
(3) Ecological constraints of male body size	(3a) Males must reach a minimum size to carry shells	Laboratory experiment in which males were presented with shells of different sizes and observed to see whether they could carry the shells to their nest	Yes
	(3b) Larger males carry shells more efficiently	Same experiment as directly above	Yes
(4) Ecological constraints of female body size	(4a) Females prefer large shells for spawning	Laboratory experiment in which 3–4 females were placed in a tank with a territorial male and 10 shells of different sizes	Yes
	(4b) Female reproductive success increases with relative shell size	Same experiment as directly above	Yes
	(4c) Females adjust their growth to available shell sizes	Laboratory experiment in which two groups of 6 small, wild-caught females were housed with a territorial male and either 10 large shells or 10 small shells	Yes

Modified from Schütz and Taborsky (2005).

TABLE 14.2 Patterns of Copulation in Relation to Dominance in a Captive Colony of Stumptail Macaques

Male dominance rank	Type of copulation	Total copulations[a]
Alpha male	Visible	424
	Surreptitious	—
14 Other adult males[b]	Visible	69
	Surreptitious	393

[a] Per 1050 hours of observation.

[b] Of 18 males in the colony, 3 failed to secure a copulation.

Source: Estep et al. (1988).

STOP AND THINK

Spritzer et al. (2006) studied the relationship between social dominance of male meadow voles (*Microtus pennsylvanicus*) and reproductive success. They first tested males for dominance in neutral arenas in the laboratory and then released them into field enclosures. Each of eight enclosures received four males (two with the highest dominance ranks and two with the lowest ranks from the original groups of eight males) along with four females, each of which was confined to a wooden nest box. The nest boxes had one-way doors so that males could enter and mate, but females could not leave; the boxes were checked twice daily to release males, and females were checked weekly for pregnancy. Pregnant females were brought into the laboratory to give birth and to check the paternity of their offspring. Females removed to give birth in captivity were replaced with new females; males that disappeared from enclosures were replaced with males of known dominance rank from their original group of eight. Over the course of several weeks, Spritzer and colleagues monitored voles in the enclosures through live trapping. Their data from trapping and paternity analyses revealed that meadow vole males with high-dominance rank had larger home ranges but sired significantly fewer litters than males with low-dominance rank. How might you explain these results? What questions would you want to ask the researchers about the details of their study methods and results?

Alternative Reproductive Strategies

In keeping with the old adage "out of sight, out of mind," subordinate stumptail macaques cope with dominant males by copulating on the sly. This question arises, then: how do males of other species deal with large, dominant males? Males of many species use alternative reproductive strategies. We begin by describing the breeding antics of bluegill sunfish, a species with external fertilization (fertilization occurs in the external environment and not the female reproductive tract).

Male bluegill sunfish (*Lepomis macrochirus*) display a discrete life history polymorphism and are either parentals or cuckholders (Gross 1982; Neff et al. 2003; Stoltz and Neff 2006). Parental males become sexually mature around 7 years of age. During the breeding season, these males compete for sites at which to build their nest, guard females that enter their nest to spawn, and then provide sole care to the developing young. Cuckholders, on the other hand, are precocial males, maturing at 2 years of age to begin their life of stealing fertilizations from parental males. Cuckholders do not build and defend nests, nor do they provide parental care. When small (2 to 3 years old), they are termed "sneakers" because they hide in vegetation at the edge of a parental male's nest and then dart into the nest to release sperm at the time the parental male is spawning with a female. Although parental males actively defend their nests against sneakers, they are not always successful in keeping the small males out. Once cuckholders reach a substantial body size at about 4 to 5 years of age, they become female mimics (sometimes called "satellites"), taking on the coloration and behavior of mature females. Female mimics succeed in entering the nests of parental males, who presumably are deceived by their feminine ways and attire. Once inside the nest, the female mimic positions himself between a mature female and the parental male, all the time engaging in behavior typical of an egg-bearing female (e.g., slow dipping movements and exaggerated rubbing of the side of the male). At the time of spawning, both the parental male and the female mimic release sperm.

Alternative male mating tactics have also been documented in species of amphibians in which males employ loud (and energetically expensive) calls to attract females. Although all males may be sexually mature and capable of calling, some remain silent and associate closely with a calling male, ready to intercept females attracted to the calls of the other male. The silent males are typically called satellite males. We'll take a look at callers and satellites in the natterjack toad.

On spring evenings, male natterjack toads (*Bufo calamita*) gather at the edges of temporary ponds and call

FIGURE 14.5 A male natterjack toad calls for females at the edge of a pond. Calling males compete with other males, called satellite males, which crouch silently nearby and intercept females on their way to the pond.

loudly to attract females. Callers typically adopt a head-up posture and energetically belt out their call (Figure 14.5). Satellites, on the other hand, keep a low profile and remain stationary and silent in a crouched position next to a calling male. If a female is attracted to the caller, a satellite male will make every effort to intercept and clasp her. In natterjack toads, body size and call intensity are highly correlated with mating success; in general, small males with weak calls are at a reproductive disadvantage. Not surprisingly, the consistency with which males in a population adopt the calling or satellite tactic is correlated with their body length: small males tend to be satellites; large males tend to be callers; and intermediate males are switch hitters, alternating opportunistically between the two strategies (Arak 1988).

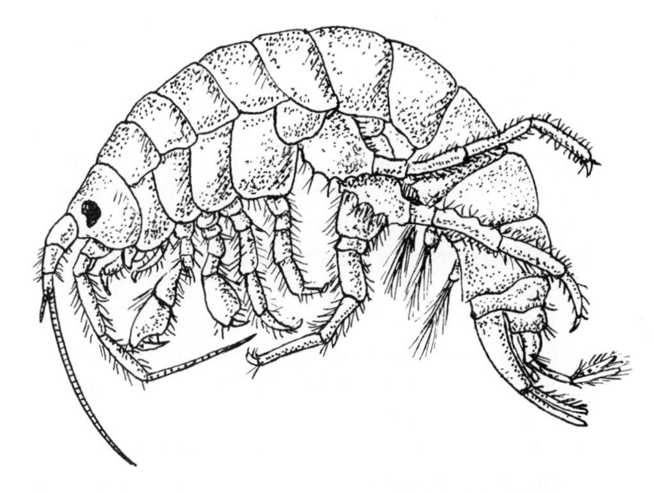

FIGURE 14.6 **Male amphipods guard females during the precopulatory period, just before the female molts. By maintaining constant contact with a premolt female, the male ensures that his sperm rather than the sperm of a rival are used in fertilization.**

ADAPTATIONS THAT FAVOR THE USE OF A MALE'S SPERM

Mate Guarding

In addition to the option of adopting alternative reproductive strategies, males display a host of adaptations that increase the probability that their sperm, and not the sperm of a competitor, will fertilize the eggs of a particular female. An obvious strategy is to be present when eggs can be fertilized and prevent other males from gaining access to the female during this time. Depending on whether the female stores sperm between copulation and fertilization, mate guarding can occur before or after copulation, or both.

In some species, including many crustaceans, mating is often restricted to a short period of time after the female molts, and females don't store sperm. Consequently, the first male present fertilizes all the eggs. In such species, we generally find precopulatory mate guarding (Birkhead and Parker 1997). The tiny amphipod *Gammarus lawrencianus* (Figure 14.6), for example, lives in estuaries along the coast of North America. Typically, a free-swimming male grabs a passing female before she molts and draws her to his ventral surface. Once the male has a firm hold on the female, he moves her in such a way that the long axis of her body is at right angles to his. In this posture the male palpates the female with various appendages, and after a minute or so, rotates her into the precopula position in which her body is now underneath and parallel to his. It is during the period of palpation and inspection that mate-guarding decisions are made. One factor in the decision is whether the female has already mated. Males are more likely to guard females with an empty brood pouch than those with a brood pouch full of juveniles or recently fertilized eggs (Dunham 1986). A second factor is the length of time the male will have to guard the female before mating. Although females early in their reproductive cycle are also typically rejected (presumably because of the greater time investment in guarding), males that have gone a long time without female social contact will accept them (Dunham and Hurshman 1990). Once in the precopula position, the two swim as a single unit until the female molts and sperm transfer is complete. Thus, the male has ensured that his sperm, rather than the sperm of a competitor, are used in fertilization.

In some cases, a male may protect his reproductive interests by guarding his mate after copulation as well. If a female were to mate with more than one male, the first male to copulate with her would presumably have a reduced chance of fertilizing her eggs, as a result of sperm competition with the second male (see later discussion). Male blue crabs (*Callinectes sapidus*) guard females before and after mating. A male guards a female before she molts so that he will be present when her eggs can be fertilized. For the record, a male blue crab's version of guarding his mate is to carry her beneath him, with his sternum against her dorsal carapace; once she has completed molting, he flips her over such that her abdomen is against his and copulation can begin. But about 12% of the females mate again within a few days of the final molt. A female that mates more than once stores all the sperm from both males. Thus, sperm from both males have equal access to her eggs. So, to protect his interests, a male blue crab also guards his mate after copulation. If many other males are present, a male guards a female for a longer time after copulation than he does if few males are present (Jivoff 1997).

Male *Gammarus lawrencianus* and *Callinectes sapidus* guard females by holding onto them. In other animals, however, guarding males stay close to females but do not maintain direct physical contact with them. Especially in species in which males employ the "stay close by" method, how can we distinguish mate guarding from some other behavior involving close association, such as when individuals stay close to one another to detect or deter predators? Pamela Willis and Lawrence Dill (2007) followed Dall's porpoises (*Phocoenoides dalli*) in the coastal waters off southwestern Canada to determine whether mate guarding occurs in this species. Female Dall's porpoises give birth to a single offspring each summer and shortly thereafter enter estrus. Most females with newborn calves are accompanied by a male, but does this association represent mate guarding by the male? Willis and Dill observed 18 male–female pairs during the weeks following parturition and compared their behavior while at the surface with that of 24 associations involving two males (hereafter called male–male pairs). The researchers made numerous predictions, including the following: if the association between males and females does indeed represent mate guarding, then when compared to members of male–male pairs, members of male–female pairs will (1) maintain shorter distances between one another, (2) remain in proximity to one another for longer durations, (3) be more likely to surface in synchrony, and (4) be less likely to join other individuals or groups. Perhaps most importantly, Willis and Dill predicted that if males were guarding females, then such males would respond more aggressively than males traveling with another male to approaches by adult males. Their results provide strong behavioral evidence that male Dall's porpoises guard females during the period following parturition when females are in estrus. As you can see in Figure 14.7, members of male–female pairs stayed together longer, maintained shorter distances between one another, and were more likely to surface at the same time than were members of male–male pairs. Male–female pairs were also less likely than male–male pairs to join other individuals. Finally, also as predicted, when paired with a female, males were more likely to respond aggressively to approaches by other adult males.

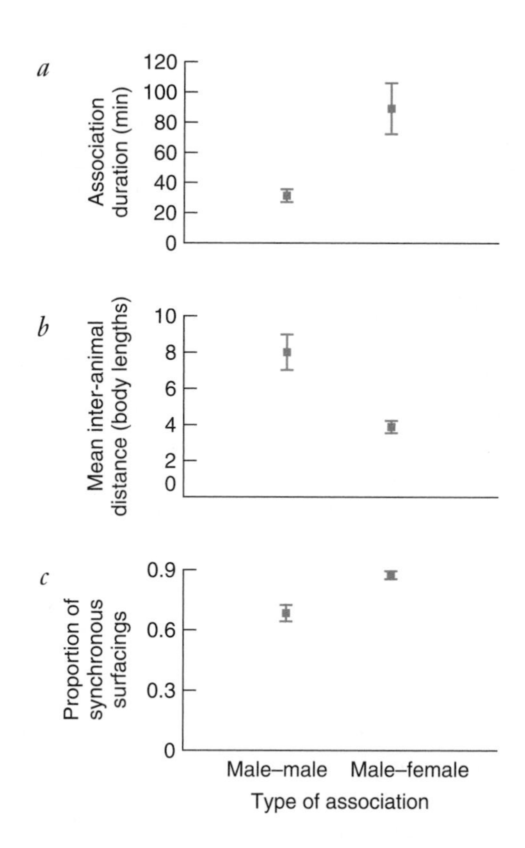

FIGURE 14.7 **Behavioral evidence that male–female associations (pairs) of Dall's porpoises during the period of female receptivity represent mate guarding. Here, (*a*) the duration of such associations, (*b*) the distance between members of the associations, and (*c*) the proportion of synchronous surfacings by association members are compared with those of male–male associations. (Modified from Willis and Dill 2007.)**

Although all research has its challenges, working with cetaceans (whales, porpoises, and dolphins) is extremely difficult given their aquatic existence, wide dispersal, high mobility, and typically large body size. Because of these challenges, behavioral observations of wild individuals are often limited to surface sightings, and genetic analysis of parentage to determine the effectiveness of mate guarding seems extremely difficult, if not logistically impossible, in natural populations. So, we will turn to a smaller, more terrestrial creature in an effort to answer the question of whether mate guarding is effective. More specifically, we will examine a study that addressed the question, "Does mate guarding reduce extra-pair fertilizations?"

Jan Komdeur et al. (2007) examined the effectiveness of mate guarding in reducing extra-pair fertilizations in the Seychelles warbler (*Acrocephalus sechellensis*). Males of this species closely follow their mate during the five or so days prior to egg laying, the time when the female is most receptive. Such guarding seems to make sense given that each breeding season a pair typically produces a single clutch with a single egg, and males invest heavily in the care of the long-dependent youngster. With restricted opportunities to reproduce and substantial parental investment, it presumably pays for a male to keep a close eye on his mate to ensure that the nestling he will be feeding for the next four months is his own. Previous studies on the effectiveness of mate guarding by male birds have typically removed the male from a pair and then monitored the occurrence of extra-pair copulations or extra-pair fertilizations involving the experimentally "widowed" female. One problem with this approach is that the female may dramatically change her behavior in response to her mate's disappearance, perhaps stepping up efforts to pair with a new male or males. So, instead of removing the male from a pair, Komdeur and colleagues left males with their mates but induced them to stop mate guarding. The researchers took advantage of the fact that a male Seychelles warbler abruptly stops guarding his mate once he spots an egg in their nest. The researchers placed a model egg in the nest of 20 experimental pairs during the fertile period of the female, but before egg laying, and thereby induced males in these pairs to stop mate guarding (the cessation of mate guarding was confirmed by behavioral observations). These nests and adults were monitored along with another 20 control nests without model eggs. Of the 20 control nests, 9 eggs hatched, 1 failed to hatch, and predators took 10. Of the 20 experimental nests, 8 eggs hatched, 2 failed to hatch, and predators took the remaining 10 eggs. Parentage analysis revealed that of those eggs that hatched in the two groups, 44% of control nestlings were not sired by the female's social mate (male with which she shared a territory and nest) as compared to 75% of nestlings in the experimental group in which males were induced to stop mate guarding. Thus, in the Seychelles warbler, male mate guarding reduces the risk of cuckoldry. We'll have more to say about the generally high level of extra-pair fertilizations in birds in Chapter 15.

Mechanisms to Displace or Inactivate Rival Sperm

Although physical struggles among males for females are often conspicuous, perhaps the fiercest clashes related to mating occur more quietly—in the female reproductive tract. If the female has mated with more than one male, their sperm must compete for the opportunity to fertilize her eggs; this phenomenon is called **sperm competition**. For example, the quality of a male's sperm—measured, perhaps, by their motility or the number of viable sperm per ejaculate—may determine the proportion of young that he sires. Another factor that may come into play is his position in the line of suitors to mate with a specific female. In some species, the advantage goes to the first male to mate with the female, in others the last male, and in still others mating order does not seem to be an important factor in determining patterns of paternity. Geoff A. Parker (1970, 1984) pioneered research in the area of sperm competition and described the phenomenon as a "push-pull" relationship between two evolutionary forces—one that acts on males to displace previous ejaculates left by rivals, and one that acts on early males to prevent such displacement.

In some cases, the interference is rather crude—the male simply removes rival sperm. For example, some male damselflies use their penis not only to transfer sperm but also to remove sperm previously deposited by competitors (Waage 1979). Backward-pointing hairs on the horns of the damselfly's penis appear to aid in scooping out clumps of entangled sperm left by earlier rivals (Figure 14.8). Some male crustaceans employ equally subtle tactics. Rather than scooping out their rival's sperm, male spider crabs push the ejaculates of earlier males to the top of the female's sperm storage receptacle; seal them off in this new location with a gel that hardens; and then place their own sperm near the female's oviduct, the prime location for fertilization to occur (Diesel 1990).

Males of other species stimulate the female into ejecting the sperm of another male. During copulation,

a

b

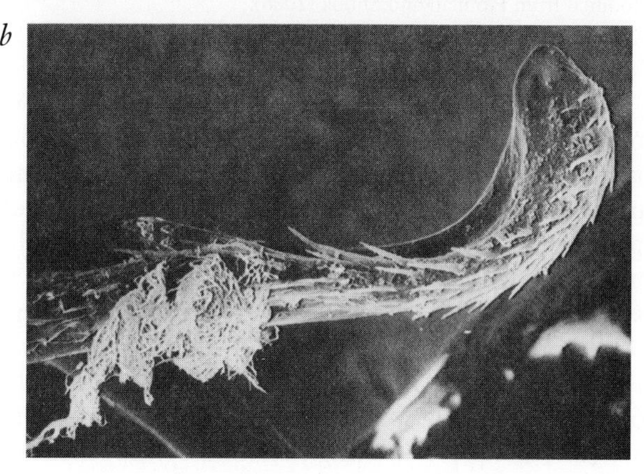

FIGURE 14.8 **Male damselflies remove the sperm left by rivals. (*a*) A copulating pair of damselflies. (*b*) The penis of a male damselfly serves not only to transfer sperm to the female but also to remove sperm previously left by competitors; backward-pointing hairs on the horns of the penis remove clumps of rival sperm.**

a male of another species of damselfly stimulates the female sensory system that controls egg laying and fertilization, causing her to release sperm from her sperm storage structure. This allows him to gain access to rival sperm that would otherwise be unreachable (Cordoba-Aguilar 1999). A similar strategy is used by male dunnocks, which are small songbirds. If the male sees his mate near another male, he pecks at the female's cloaca, causing her to eject sperm-containing fluid (Davies 1983).

Another phenomenon that is often linked to sperm competition is sperm heteromorphism (the simultaneous production by a single male of at least two types of sperm in the same ejaculate) (Holman and Snook 2006). In most cases of heteromorphism, there are two sperm morphs, one that can fertilize eggs and one that cannot. The two morphs are called eusperm and parasperm, respectively. Sperm heteromorphism occurs in diverse animals, including many species of mollusks, insects, and fishes (see papers reviewed by Hayakawa 2007). But what is the function of infertile sperm, and what is their connection with sperm competition?

TABLE 14.3 **Hypotheses for the Evolution of Parasperm (Infertile Sperm)**

Hypothesis	Description
Nonadaptive	Parasperm have no adaptive function
Sperm competition	Parasperm help eusperm compete to fertilize eggs
Offensive	Parasperm displace or kill rival sperm
Defensive	Parasperm prevent rival sperm from reaching the eggs or inhibit remating by the female
Provisioning	Parasperm, upon degeneration, provide nutrition to the female, her eggs, or eusperm
Facilitation	Parasperm facilitate transport or readiness of eusperm
Sacrificial sperm	Parasperm dilute the effects of spermicides, thereby increasing survival of eusperm
Cryptic female choice	Parasperm increase the chances that eusperm will be preferred during cryptic female choice

Modified from Holman and Snook (2006).

There are several hypotheses for the evolution of infertile sperm (Table 14.3). First, it is possible that parasperm have no adaptive function. Most consider this unlikely given the large number of parasperm produced (parasperm may make up half of an ejaculate) and the presumed costs of their production. Alternatively, parasperm could have one or more adaptive functions. For example, parasperm may play a role in sperm competition. Such a role could be offensive (for example, killing or displacing sperm from another male) or defensive (for example, inhibiting remating by the female or blocking sperm from another male). Other potential adaptive functions of parasperm include their ability to facilitate the transport or readiness of eusperm or to provide (upon their degeneration in the female reproductive tract) nutrients to the female, her eggs, or the eusperm. Holman and Snook (2006) provide two additional adaptive hypotheses: (1) parasperm may perform a sacrificial function by diluting the effects of spermicides produced by females and thereby promoting survival of eusperm (the authors define spermicides as anything that kills, disables, or dumps sperm), or (2) parasperm could increase the chances that eusperm will be selected during cryptic female choice (choice by a female who has mated with more than one male regarding which sperm will fertilize her eggs). In terms of the cryptic female choice hypothesis, it is possible, for example, that females assess males by the number or quality of parasperm they produce. We will discuss cryptic female choice in more detail later in the chapter.

At present, the functions of parasperm remain unknown for most species. Notable exceptions include some species of Lepidoptera (moths and butterflies) in which parasperm facilitate readiness of eusperm for participation in fertilization (Sahara and Takemura 2003). Parasperm also appear to function in sperm competition in Lepidoptera. The functional equivalent of "cheap filler" parasperm may delay further matings by females. More specifically, distention of a female's sperm storage organ by parasperm could signal a successful insemination and thereby reduce the female's receptivity to further matings (Cook and Wedell 1999; Silberglied et al. 1984). Finally, in some fish that employ external fertilization, there is evidence that parasperm function in at least two ways (Hayakawa 2007). Parasperm increase the distance that semen travels at spawning, thereby increasing the number of eusperm reaching an egg mass. Then, upon arrival at the egg mass, parasperm form defensive lumps at the outskirts of the mass that prevent sperm from other males from reaching the eggs.

Mechanisms to Avoid Sperm Displacement

How can males ensure that their sperm are not displaced by those of rival males? We'll consider the case of Rocky Mountain bighorn sheep (*Ovis canadensis canadensis*), a species in which females usually mate with several males during a single period of estrus. Sperm competition in this species is intense because subordinate males (called coursing rams) are remarkably successful in forcing copulations on ewes guarded by tending rams (Hogg 1988; Hogg and Forbes 1997). Tending and coursing rams copulate with estrous ewes at extremely high rates. Frequent copulations by individual male bighorns presumably function to increase the proportion of their own sperm in the female reproductive tract. Tending rams copulate at especially high rates immediately after successful copulation by a coursing ram. Such "retaliatory" copulations by dominant males (Figure 14.9) may serve a function akin to that of the damselfly penis—mechanical displacement of rival sperm. Alternatively, retaliation may be advantageous to the dominant male because sperm from the last mating are more likely to fertilize eggs.

Some males mate longer than is necessary for the release of sperm. Males may use prolonged mating to avoid or reduce competition with the sperm of other males. Two hypotheses have been suggested to explain the function of prolonged mating. The first, called the extended mate-guarding hypothesis, suggests that prolonged mating reduces the chances that the female will be inseminated by another male. The second, the ejaculate transfer hypothesis, suggests that prolonged mat-

FIGURE 14.9 **Competition for mates is intense among males of Rocky Mountain bighorn sheep. Although dominance and access to mates are established by fierce physical clashes, male competition continues in the reproductive tracts of females. Dominant males copulate at high rates immediately after successful copulation by a subordinate male. Such retaliatory copulations may displace the sperm of the subordinate male.**

ing results in more of the male's sperm being transferred to the female. Christopher Linn and colleagues (2007) tested these two alternative hypotheses in *Nephila clavipes*, an orb-weaving spider with a penchant for copulation. In the two days following a female's molt to adulthood, the largest male on her web copulates with her several hundred times, with individual copulations lasting up to 14 minutes. Smaller, peripheral males on the web hang around the preoccupied couple, but rarely are successful in gaining copulations. At the end of two days, the mating comes to an end and the female eventually abandons the web. To ascertain the advantage of prolonged copulation in this species, Linn et al. (2007) allowed members of male–female dyads to mate for one of the following three durations: (1) two natural copulations, each lasting about 14 minutes, (2) 2 hours, during which numerous copulations occurred, or (3) 48 hours (the natural mating period following the female's molt). In each of the three treatments, some females were used to quantify the number of sperm that had been transferred to them by the male; sperm storage organs (spermathecae) were removed from females, and the sperm within each organ were counted. The remaining females in each treatment were presented with a second male four days after the molt to determine their likelihood of mating again (termed remating by the authors). The data show that the longer a female was allowed to copulate with the first male, the lesser the likelihood of her mating with a second male (percent of encounters in which a female mated with a second male: 93% in the two-copulation treatment; 90% in the 2-hour treatment; and 28% in the 48-hour treatment). These data on likelihood of remating support the mate guarding hypothesis. The duration that the female and first male were allowed to mate did not, however, influence the number of sperm found in the female's spermathecae. These data on sperm counts thus fail to support the ejaculate transfer hypothesis. In summary then, prolonged copulation in *Nephila clavipes* does not result in the transfer of more sperm, but it does function to reduce the likelihood of the female mating with another male.

Repellents and Copulatory Plugs

In an effort to reduce the likelihood of future matings by competitors, males of many taxa apply a repellent odor to their mates. In the case of a neotropical butterfly, a male transfers an "antiaphrodisiac" pheromone to his mate during copulation, which makes her repulsive to future suitors (Gilbert 1988).

In other species, a male may deposit a copulatory plug made of thick, viscous secretions in the female's reproductive tract. Copulatory plugs occur in many vertebrates, including snakes, lizards, marsupials, rodents, bats, and primates. Several functions have been suggested for copulatory plugs, among them: (1) "enforcing chastity," with the plug acting as a barrier to subsequent inseminations (Martan and Shepherd 1976; Voss 1979); (2) ensuring the retention of sperm in the female reproductive tract (Dewsbury 1988); (3) aiding the transport of sperm within the female reproductive tract (Carballada and Esponda 1992); (4) providing for the gradual release of sperm as the plug disintegrates (Asdell 1946); and (5) providing a means by which a male can scent-mark a female's body and convey information regarding his identity and dominance status (Moreira et al. 2006). For this last proposed function, the information conveyed in the plug might be used by the male himself (for example, to identify females with which he has already mated), by rival males, or by the female (for example, during cryptic female choice). Unfortunately, for the vast majority of species the precise function of copulatory plugs remains a mystery (Ramm et al. 2005; Reeder 2003). Nevertheless, researchers continue to make intriguing observations—for example, males and females of some species remove copulatory plugs—that may someday shed light on copulatory plug function(s).

Placing plugs in the female reproductive tract is not confined to male vertebrates. Indeed, some male spiders intent on protecting their paternity employ quite dramatic techniques to plug the reproductive tracts of females with which they have mated. A male spider, by the way, copulates by inserting his pedipalps (paired extremities posterior to the fangs that serve as copulatory organs) into the genital opening(s) of a female (in some groups of spiders, the genital openings of females are paired, while in others they are not). In *Argiope aurantia*, within seconds of inserting his second pedipalp into the female's genital opening, the male becomes unresponsive and dies shortly thereafter, becoming, in essence, a whole-body mating plug (Foellmer and

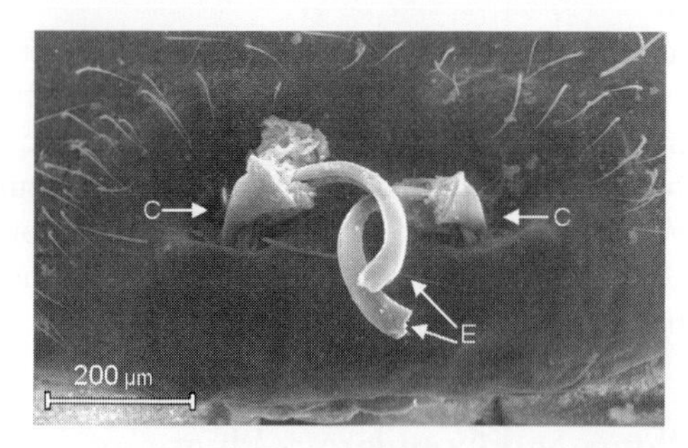

FIGURE 14.10 **Some male spiders leave parts of their copulatory organs in the genital tracts of females with which they have mated. Such parts appear to obstruct matings by subsequent males. This image from a scanning electron microscope shows parts of a male's pedipalps (indicated by the letters E, for embolus, and C, for conductor) that were left in each of the two genital openings of a female.**

Fairbairn 2003). In a slightly less dramatic manner, copulating males of the golden orb web spider *Nephila fenestrata* protect their paternity by leaving parts of their pedipalps in the genital openings of females (Figure 14.10). These pedipalp fragments obstruct copulations by subsequent suitors (Fromhage and Schneider 2006).

So far in our discussion of sperm competition, we have focused on interactions between males. Do females benefit by mating with more than one male? Multiple matings may, in fact, be to a female's advantage. Mating with several males may (1) increase the probability of fertilization, (2) increase the genetic diversity of offspring, (3) result in the accumulation of material benefits if males provide nutritional gifts at copulation, or (4) ensure that a female's sons are good at the game of sperm competition, if the trait is heritable. These and other suggestions are reviewed in more detail by Møller and Birkhead (1989).

SEXUAL INTERFERENCE: DECREASING THE REPRODUCTIVE SUCCESS OF RIVAL MALES

As we have seen, mating success is often measured by how well one advances one's own reproductive efforts and how effectively one interferes with a competitor's efforts. We will now concentrate on the latter. Any behavior that reduces a rival's fitness by decreasing his mating success is called **sexual interference** (Arnold 1976).

Some of the most effective animals in sexual interference are male newts. Adrianne Massey (1988) studied sexual interactions in field populations of red-spotted newts (*Notophthalmus viridescens*) and noted three tactics of sexual interference: (1) spermatophore transfer interference, (2) pseudofemale behavior, and (3) amplexus interference. These are all methods through which male newts decrease the reproductive success of competing males.

The first method, spermatophore transfer interference, occurs when a rival male inserts himself between a female and the courting male that has just dismounted. The rival male not only induces spermatophore deposition by the first male but also slips his own spermatophore into position so that the female picks that up rather than the spermatophore of the first male. This form of sexual interference reduces the first male's supply of spermatophores for future inseminations and prevents him from inseminating the female (as well as permitting the intruding male to inseminate her). The second method, pseudofemale behavior, also causes courting males to waste spermatophores, but this time in the context of male–male pairings (Figure 14.11). Male red-spotted newts often clasp other males, although such pairings are usually brief because clasped males give a head-down display that elicits release. In some cases,

a

b

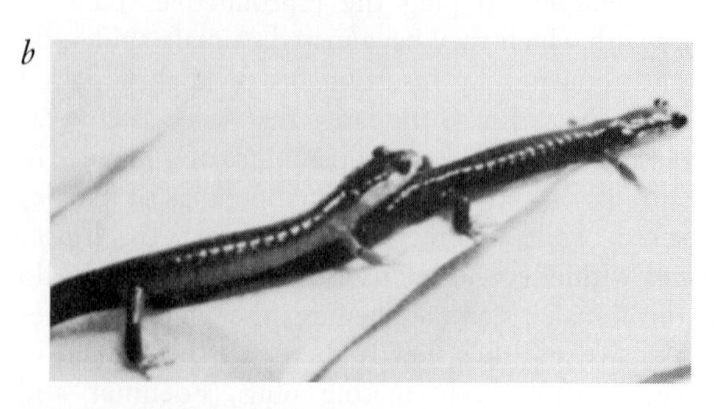

FIGURE 14.11 **There are few differences in heterosexual and homosexual courtship in newts. (*a*) A receptive female has straddled the tail of a male to stimulate the male to deposit a spermatophore. (*b*) A male mimics female behavior to cause the courting male to deposit a spermatophore. Pseudofemale behavior is one form of sexual interference used by male newts to decrease the reproductive success of competitors.**

FIGURE 14.12 **Although infanticide may serve several functions, it is a dramatic example of sexual interference. Here, a lion has killed the cub of a rival male.**

however, the clasped male does not signal his maleness to the clasping male. Furthermore, once the clasping male dismounts, the clasped male may nudge him, in the manner of females, and get him to uselessly deposit his spermatophore. Males of the red-spotted newt occasionally engage in a third method of interference, amplexus interference. Here, an intruder simply inspects a pair in amplexus at close range, usually just before the mating male dismounts to deposit a spermatophore. The presence of a voyeur leads the amplexing male to pause. When he resumes his mating behavior, he usually picks it up at an earlier stage. The interference, then, causes the mating male to waste time and energy without increasing his probability of fertilizing the eggs.

Other species exhibit more dramatic forms of sexual interference, such as infanticide, the killing of a competitor's offspring (Figure 14.12). Infanticide has been reported for diverse invertebrate and vertebrate taxa (Ebensperger and Blumstein 2007; Hausfater and Hrdy 1984; Trumbo 2006; Van Schaik and Janson 2000). In some mammals, sexually selected infanticide occurs when one or more males from outside the social group usurp the resident male. In Hanuman langurs, the primate species in which infanticidal male intruders were first reported, infanticide occurs when individuals from all-male bands invade harems (Sugiyama 1965). With her offspring gone, the female quickly returns to estrus; in Yukimaru Sugiyama's (1984) words, infants are attacked because they are "little more than an obstacle to activation of the mother's receptivity."

Although most primatologists accept the hypothesis that male infanticide in Hanuman langurs is sexually selected, acceptance is not universal (Brown 1996; Dixson 1998). Nonetheless, supporting evidence is slowly accumulating. One criticism was that the acts of killing were not actually observed and were only assumed to have been committed by the males that took over the groups. However, many instances have now been observed (Sommer 1993). Other critics argue that infanticide is a result of crowded living conditions, but it turns out that male infanticide is equally common in low-density and high-density populations (Dixson 1998; Newton 1986). Furthermore, two predictions of the sexual selection hypothesis for male infanticide—that the infanticidal males will kill only infants that are unrelated to them and that infanticide will increase a male's chance of siring the next infants—have now been supported by DNA analysis (Borries et al. 1999).

We have focused on infanticide by males and the hypothesis that the killing of infants is adaptive because intruding males gain more rapid access to mates. There are, however, alternative hypotheses for infanticidal behavior (Table 14.4), and predictions from these hypotheses should also be tested when searching to explain the killing of conspecific young.

TABLE 14.4 **Hypotheses for Infanticide by Males**

Hypothesis	Predictions
Nonadaptive	Infanticide is associated with severe disturbance to the physical environment (e.g., habitat reduction) Infanticide is associated with severe disturbance to the social environment (e.g., overcrowding) Performance and context of infanticide should resemble that of an accident
Acquisition of mates	Infanticidal males kill only offspring they do not sire Killing infants shortens the interbirth interval of victimized females Infanticidal males mate with the female whose offspring they killed
Acquisition of food	Males consume infants they have killed Infanticide and eating of young will increase when food supply is limited Performance and context of infanticide should resemble that of a predatory attack
Acquisition of space	Infanticide will increase when territories or nest sites are limited Infanticidal males take over the nests or territories used by the individual whose offspring they killed

Based on table and text of Ebensperger and Blumstein (2007).

Females of many species also commit infanticide; explanations for female infanticide include all of those listed in Table 14.4, with the exception of the access to mates hypothesis.

INTERSEXUAL SELECTION—MATE CHOICE

Intersexual selection occurs when one sex is in the position of choosing individuals of the other sex as mates. Although there are exceptions, females often do the choosing while males compete among themselves to be chosen. We will consider this general case, realizing that sometimes males choose and females compete. First, we consider some criteria on which females may base their choice of mate.

CRITERIA BY WHICH FEMALES CHOOSE MATES

Characteristics used by females to select a mate should affect female fitness, be assessable, and vary among males (Searcy 1979). Given these criteria, females are thought to choose mates on the basis of their ability to provide sufficient sperm, useful resources, parental care, and genes that are good or compatible. Sufficient sperm, useful resources (e.g., food or nest sites), and parental care are considered direct material benefits; these benefits are obtained by the female doing the choosing. Good or compatible genes, on the other hand, are considered indirect because they benefit the female's offspring in the next generation (Andersson 1994). Although some females may evaluate potential mates on only one characteristic, others may base their choice on multiple criteria. Here we consider some of the characteristics that females might evaluate when searching for "Mr. Right."

Ability to Provide Sufficient Sperm

Earlier we stated that the sperm supplies of males are no longer viewed as limitless and that successive matings may deplete a male's sperm supplies. But do females in the market for a mate consider the number of sperm a particular male has available? In the case of female stone crabs (*Hapalogaster dentata*), females appear to do just that. Taku Sato and Seiji Goshima (2007) collected stone crabs off the coast of Japan and transported them to a marine station where they conducted several experiments in aquaria filled with seawater. In the first three experiments, they determined that female stone crabs prefer large males to small males and that females use chemical cues, rather than visual ones, when making their selection. In the fourth experiment they examined the influence of sperm limitation. Thirty large males were each allowed to mate with four small females, one female on each of four successive days; these males were considered "depleted males." (The researchers had males mate with four females because previous data had shown that the number of ejaculated sperm found in females decreases with a male's successive matings, and this is especially true for the male's fifth mate [Sato and Goshima 2006].) Another 30 large males were categorized as "unmated males" because they were not given any prior access to females. In mate-choice experiments conducted one to three days after the depleted males' fourth day of mating (female stone crabs only mate right after molting, so this window of time was necessary to ensure the availability of newly molted females), the researchers placed a depleted male and an unmated male each into an opaque plastic pipe in the test aquarium (Figure 14.13*a*). Each plastic pipe had six very narrow openings to allow the exchange of seawater (and chemical cues) between the inside and outside of each pipe; the openings in the pipes were narrow enough, however, to prevent a crab on the outside from seeing the individual within the pipe and vice versa (these pipes had been developed for use in the earlier experiments to eliminate visual cues and test for the importance of chemical cues in female choice). Fifteen minutes after the males were introduced into the test aquarium, a large female was introduced, and she too was given time to acclimate to her surroundings. Sato and Goshima used large females for the testing because large females spawn more eggs than small females, thus it would be particularly challenging for males to fertilize all the eggs of a large female. Once the female acclimation period had ended, the researchers recorded the total time out of the 40-minute test that the female spent in the circular area surrounding each pipe (refer, again, to Figure 14.13*a*). Thirty such choice tests, each with different crabs, were conducted. On average, females spent more time near unmated males than near depleted males (Figure 14.13*b*).

Female stone crabs prefer unmated males and appear to use chemical cues to assess male ability to transfer sufficient sperm. The authors suggest that such a preference makes sense in this species because males experience sperm depletion with successive matings (there are other species of crabs in which males can replenish sperm between matings), and females typically mate with one male during a reproductive season, so their choice is critical. Finally, males guard newly molted females by holding onto them, so the use of chemical cues by females allows them to assess at a distance the ability of males to provide sufficient sperm and to only approach a male that passes the test. If females assessed males by visual cues, then this presumably would require closer inspection by the female and increase the chances that she would be grabbed and guarded, and prevented from exercising mate choice.

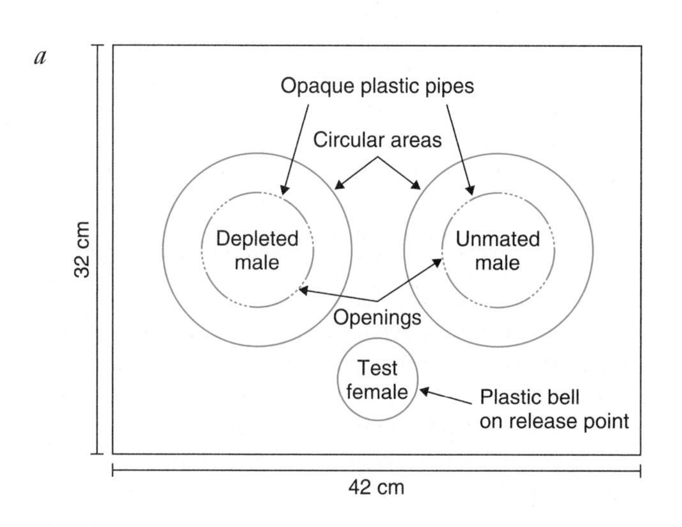

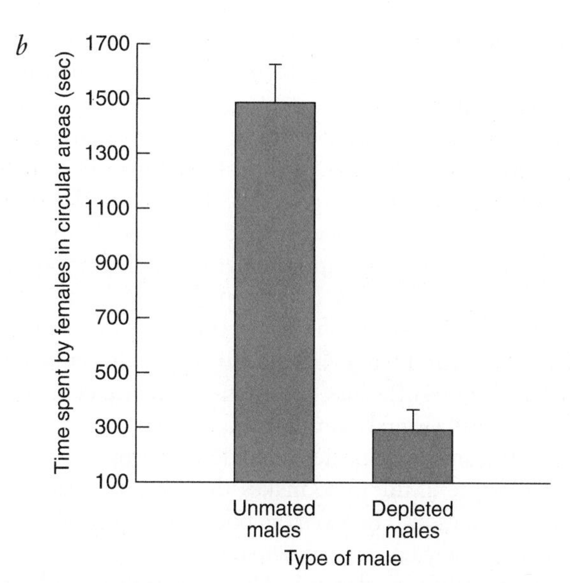

FIGURE 14.13 **Female stone crabs evaluate whether males have sufficient sperm. (*a*) The experimental aquarium in which female stone crabs could choose between spending time near a depleted male (a male that had mated with four other females) or a male that had not mated. Each male was placed in an opaque plastic pipe with narrow openings that allowed the passage of chemical cues. Females were initially placed under a plastic bell and allowed to acclimate to testing conditions. Female preference was assessed by the amount of time they spent in the circular area surrounding each pipe. (Modified from Sato and Goshima 2007.) (*b*) Female stone crabs prefer unmated males to depleted males. (Drawn from data in Sato and Goshima 2007.)**

Ability to Provide Useful Resources

Females of some species may base their choice of a mate on the quality of resources provided by males. By so doing, they could receive either immediate gains from gifts presented during the courtship period or more long-term benefits from access to valuable resources, such as food or nest sites, that are controlled by males. Females that exchange mating for material goods could place themselves at an advantage. The increased commodities could obviously enhance reproductive output by enabling the female to live longer by being well fed or having access to a protected nest site. Not only would she have a competitive advantage over females without mates, but also high-quality resources would probably improve the survivorship and competitive ability of her offspring. Here we consider a few of the diverse material goods that males may provide to females and upon which females may evaluate potential mates.

Nuptial Gifts and Cannibalism The males of many birds and some species of insects offer nutrition or other valuable substances to the female during courtship. These nuptial gifts may take various forms, including prey, glandular secretions, and the spermatophore. Indeed, a male may even offer parts or all of his body for the female to feed on (Andersson 1994). The functions of the gifts generally fall into two categories, which are not mutually exclusive. First, they may increase the male's chances of mating by making him more attractive to the female, by making it easier to copulate, or by maximizing the amount of sperm transferred. Second, they may serve as paternal investment by increasing the number or fitness of his offspring (Vahed 1998).

In some species, for example, the katydid mentioned previously, the spermatophore that the male presents as a nuptial gift provides nutrients for the female or the eggs. Depending on her own nutritional state, a female can either put a large portion of the male's contribution into the eggs or use the materials from the spermatophore for herself, increasing the chances that she will live long enough to breed again (Gwynne and Brown 1994).

The benefit obtained from the nuptial gift isn't always energy. There are many other valuable substances. Larvae of the arctiid moth *Utetheisa ornatrix* sequester alkaloids from the plants on which they feed. These alkaloids make them quite distasteful to certain predatory spiders. Males transmit some of these alkaloids to the female with their spermatophores. The female bestows some of the alkaloids on her eggs, giving them protection from predators (Eisner and Meinwald 1995). The female is also protected by this gift. Almost immediately, she becomes unacceptable prey to these spiders (Gonzalez et al. 1999). In other moths, limited micronutrients are passed to the female or her eggs in the nuptial gift (Smedley and Eisner 1996).

Males of some insect species present a gift of prey that the female feeds on during copulation, buying them time to transfer sperm. Male hangingflies (Bittacidae) usually offer a large prey item, one that will take at least 20 minutes to consume. It takes the male about 20 minutes to completely transfer enough sperm and fluids to make the female unreceptive to other males and begin laying her eggs (Thornhill 1976). In other insects, the male's nuptial gift is a part or all of his body. During copulation, a female sagebrush cricket feeds on the male's fleshy hind wings and the hemolymph that oozes from the wounds. This courtship feeding keeps the female mounted while the male transfers his sperm. Males whose hind wings have been surgically removed transfer significantly less sperm than do intact males (Figure 14.14) (Eggert and Sakaluk 1994).

Finally, in some species of spiders, scorpions, mantids, and diptera, the male makes the ultimate sacrifice during copulation. He gives his body to the cannibalistic female. Oddly enough, under some circumstances this ultimate sacrifice is adaptive. The male redback spider (*Latrodectus hasselti*) stores his sperm in a tightly coiled structure on his head, called a palp. To transfer sperm, he scrapes the palp across the female's genital opening, which causes the coil to unspring. The open coil is then inserted into the genital opening and sperm is transferred. A few seconds later the male does a somersault, placing his abdomen directly above the female's jaws. About 65% of the time, the female eats him while he's in the somersault position. His chances of being consumed are much greater if the female is hungry. The males gain two paternity advantages from this suicidal behavior. One advantage is that a cannibalized male fertilizes about twice as many eggs as a male who survives copulation because he copulates about twice as long. Another advantage is that the female is about 17 times less likely to mate again after consuming the first mate (Andrade 1996).

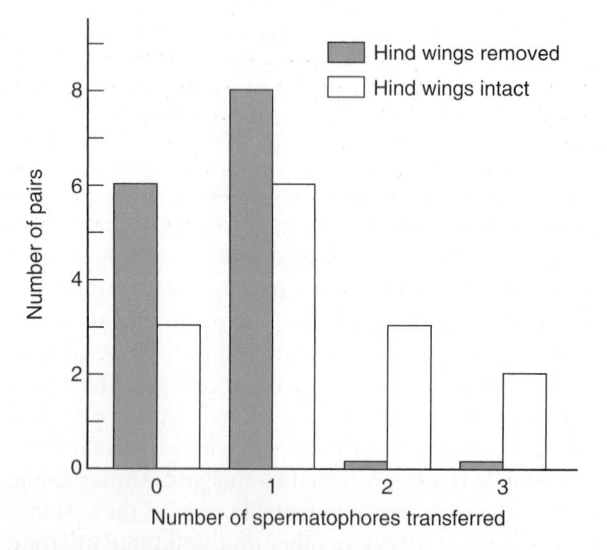

FIGURE 14.14 **Male sagebrush crickets with intact hind wings transfer significantly more sperm packets to females in a 12-hour period than do males whose hind wings have been surgically removed. A female sagebrush cricket feeds on the male's fleshy hind wings during copulation. This helps to keep her mounted while the male transfers his sperm. (Modified from Eggert and Sakaluk 1994.)**

Territory In many species, females selecting mates appear to assess the quality of territories held by males. For example, female house wrens (*Troglodytes aedon*) select a mate based, at least in part, on the number of nest sites in his territory. House wrens are secondary cavity nesters (i.e., they are not capable of excavating their own nest cavities, so they instead use nest holes made by other species, such as woodpeckers) that will readily use nest boxes. By experimentally adding from one to three nest boxes to male territories already containing a single nest box, Kevin Eckerle and Charles Thompson (2006) found that time to pairing for males was associated with the number of nest boxes on their territory: males with the most nest boxes were selected first by females, leading to shorter times to pairing for these males. Extra nest sites might decrease the chances that the female's own nest would be found by predators or provide alternative sites should re-nesting be necessary. In other species, females might evaluate the amount or quality of food on a male's territory.

Ability to Provide Parental Care Females may also assess a male's parental abilities. This seems even more challenging than estimating the value of material resources, but it appears to occur nonetheless. Indeed, evidence from diverse species indicates that females may use physical or behavioral features of males to predict parental quality. In some species of birds, for example, females may judge male parental ability on the basis of the quality of nutritional gifts provided during the period of courtship (Figure 14.15). A large number of high-quality gifts may signal a male's superb foraging skill and willingness to feed his mate and offspring during incubation and posthatching stages. Some females may use the success of previous nesting attempts to judge the parental ability of males (Coulson 1966). We will focus on how certain female fish use the speed of water currents to judge the parental abilities of potential mates.

Gobies of the *Rhinogobius brunneus* species complex inhabit streams with steep gradients in Japan. Male gobies build nests, provide sole care to the eggs, and engage in aggressive displays (Figure 14.16*a*). Although males and females occur in pools, they only court in currents. At the start of courtship, a female remains stationary on the stream bottom; the male, on the other hand, rises off the bottom and begins to dance. If not carried away by the swift-flowing water, the male approaches the female and shakes his head from side to side. If the female bends her body in response, then he tries repeatedly to lead her to his nest, all the while spreading his fins and vibrating his body. At this stage, many females reject males, and

FIGURE 14.15 **A male European crossbill passes regurgitated seeds to a female. Female birds may judge the parental ability of a male by the quality and quantity of gifts provided during courtship feeding.**

a

b

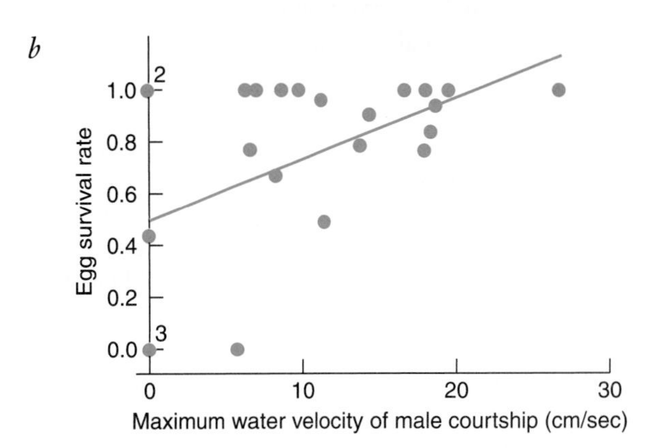

FIGURE 14.16 **Female gobies choose males that can court in fast currents because such males are in superior physical condition and will not eat their offspring when guarding them. (*a*) A male goby in an aggressive posture. (*b*) Shown here is the relationship between maximum current speed at which a male can court and the survival of his eggs. (Modified from Takahashi and Kohda 2004.)**

rejected males leave the swift-flowing water and return to the pool. If the female does not reject the male and instead follows him to his nest, she then enters the nest and deposits her eggs in a single layer on the ceiling and the male fertilizes them. Following her departure, the male arranges the substrate to close off the nest, effectively walling himself in with the kids. He remains in the nest, fanning the eggs for about two weeks. While confined, his eggs are the only food available, and some males cannibalize some or all of their eggs.

On what basis does a female goby accept or reject a male? Daisuke Takahashi and Masanori Kohda (2004) conducted mate-choice experiments in the laboratory and discovered that female gobies favor males that perform their courtship dance in the fastest water currents. But why? It turns out that only males in exceptional physical condition can dance in fast currents and not be swept away. In addition, the maximum water velocity at which a male can dance is positively correlated with the survival of his eggs (Figure 14.16*b*). Males unable to court in fast currents are in less than exceptional physical condition and more likely to consume their eggs during the two-week guarding period. So, female gobies appear to evaluate the velocity of the current in which a male displays, and such information indicates his condition and ability to abstain from eating his offspring. Thus, courting in fast current is an "honest signal" of a male's ability to offer paternal care.

In some species, males do not provide direct material benefits to their mates or parental care to their offspring. How might females evaluate potential mates in these species? There is evidence that females evaluate male traits that signal genetic quality.

Ability to Provide Good Genes

Barring direct examination of a male's genotype, how could a female evaluate variation in genetic quality among suitors? Females might judge genetic quality by examining a male's general physical well-being, capacity to dominate rival males, or capacity for prolonged survival. We'll first describe how a showy secondary sex characteristic can be a reliable, honest indicator of a male's health and good genes. (Honest signals are discussed further in Chapter 17.)

William D. Hamilton and Marlene Zuk (1982) were the first to propose that the elaborate ornaments of males represent reliable signals of health and nutritional status. They examined plumage coloration in North American birds and suggested that only males in top physical condition would be able to maintain bright, showy plumage. Because bird species vary in their susceptibility to parasitic infection, Hamilton and Zuk predicted that the degree of male brightness would be correlated with the risk of attack by parasites. Accordingly, if males of a species vary substantially in their parasite load, it would behoove a female to choose a male that honestly signals his good health, for the offspring of this male may experience increased viability if they inherit his resistance to parasites. Because the brightness of plumage is closely tied to a male's general health, it is a reliable signal that cannot be faked by a parasite-laden male.

Based on this hypothesis, Hamilton and Zuk reasoned that in species in which the risk of infection by parasites is minimal, information regarding parasite load has little value to females in the market for a mate, and thus females should not display a preference for males with showy features. Hamilton and Zuk therefore predicted that in species with low risk of parasite infection, males would not be brightly colored.

In testing their hypothesis, Hamilton and Zuk surveyed the literature on avian parasites and determined the risk of infection for each bird species. They then ranked each species from 1 (very dull) to 6 (very striking) on a plumage showiness scale. In support of their ideas, there was a significant association between showiness and the risk of parasitic infection: those species with the highest risk of infection from blood parasites had the showiest males.

In the years since Hamilton and Zuk (1982) published their findings, the influence of parasites on mate choice has been examined in many species of birds (e.g., Borgia 1986; Møller 1991; Spencer et al. 2005), as well as in a host of other animals, including amphibians (Hausfater et al. 1990); fish (Barbosa and Magurran 2006); and humans (Low 1990). Some studies support the idea that females choose mates by using cues that will lead to more viable offspring; others do not. Taken together, however, the overall pattern that emerges is that showy males have fewer parasites and are better able than dull males to mount a stronger immune response against a wide variety of parasites. These observations support the idea that females choose showy males to obtain good, parasite-resistance genes for their offspring (Møller et al. 1999).

The extent of asymmetry, called fluctuating asymmetry, in otherwise bilaterally symmetrical traits is another potential indicator of overall genetic quality. Because the growth and development of both sides of the body are controlled by the same set of genes, we would expect the right and left sides to be identical. This is not always the case, however. Environmental insults or genetic defects can cause one side of the body to develop in a slightly different manner from the other, causing the traits to be asymmetrical. Females in many species prefer males with symmetrical traits, presumably because symmetry signals a healthy condition and good genes (Møller and Pomiankowski 1993; Møller and Swaddle 1997).

Consider the sailfin molly *Poecilia latipinna*, a small tropical fish commonly found in brackish coastal marshes. It is so named because the males have a large dorsal fin that resembles a sail. The fin is displayed to the female during courtship. Males are brightly colored and may have vertical bars on the sides of their bodies. The number of bars may differ between sides. In nature males differ in the presence of bars, number of bars, and symmetry in the number of bars. In laboratory choice tests, female sailfin mollies prefer males with vertical bars to those without bars. Furthermore, when bars are present, females prefer symmetrical males—those with the same number of bars on each side of the body (Schluter et al. 1998).

Studies on barn swallows suggest a way in which trait symmetry might reflect good genes. Female barn swallows (*Hirundo rustica*) prefer to mate with males with long tails (Møller 1988). Mite infestations stunt the growth of tail feathers, and so tail length advertises whether a male has been previously infected with mites. Mite infestations also increase asymmetry in tail length. Besides preferring males with longer tails, female swallows prefer males with tails that are symmetrical in length (Møller 1990). So, by choosing males with long, symmetrical tails, females are choosing males with genes for parasite resistance (Møller 1992).

We mentioned at the start of this section that females might also evaluate male genetic quality by assessing a male's physical condition and ability to dominate other males. Our next two examples focus on male physical condition and dominance as indicators of genetic quality. The first of the two examples involves elaborate male ornaments; the second example does not.

Male satin bowerbirds (*Ptilonorhynchus violaceus*) build unique stick structures called bowers in which they display to females in attempts to secure copulations (Figure 14.17). Males decorate their bowers with artifi-

FIGURE 14.17 **Male satin bowerbirds decorate their bowers with natural and artificial materials. The number and rarity of decorations on a bower are important determinants of mating success. Female bowerbirds may judge the genetic quality of a male by the attractiveness of his bower.**

FIGURE 14.18 **Pronghorn females may choose males based on good genes, as assessed by the male's ability to attract and defend females from other males during the breeding season.**

cial materials (e.g., bottle caps), snail shells, feathers, and flowers. They display strong preferences for inflorescences of certain colors; blue and purple flowers are relatively rare in the environment of satin bowerbirds and are preferred over yellow and white flowers, whereas orange, red, and pink inflorescences are completely unacceptable (Borgia et al. 1987). Males also masticate leaves and then paint the resulting mixture on the walls of their bower. Bowers and their decorations appear to have no intrinsic value to either sex outside the context of sexual display and probably serve primarily as indicators of male quality. Indeed, the number and type of decorations on a bower are important determinants of a male's mating success, as are characteristics of his behavioral display (Coleman et al. 2004; Patricelli et al. 2003). Not surprisingly, males go to great lengths to steal rare decorations from the bowers of competitors (Wojcieszek et al. 2007). Females may favor males who exhibit exotic decorations because the ability to accumulate and hold these decorations indicates that a male is in top physical condition. Because the number of inflorescences on bowers is correlated with age, females could use the number of flowers, and specifically the number of rare decorations, to assess the experience of a male. Thus, although female bowerbirds cannot directly examine the genotype of potential mates, they may evaluate a male's genetic quality based, at least in part, on the attractiveness of his bower.

Female choice based on good genes need not always involve elaborate ornaments, whether worn by the male himself or meticulously arranged by him on his display arena. Our next example of female choice for good genes concerns pronghorn (*Antilocapra americana*), a species of ungulate endemic to the Nearctic region (Figure 14.18). Pronghorn are the fastest terrestrial mammals in the New World, reaching speeds near those of cheetahs, the record holders of the Old World. Endurance and vigor are important for male and female pronghorn when escaping predators; these traits are also evaluated by females when selecting a mate. Like male satin bowerbirds, male pronghorn do not monopolize resources critical to females and do not provide paternal care, yet females still actively choose their mate. All this suggests that females might be basing their choice on good genes. John Byers and colleagues have been studying a population of pronghorn in Montana for over two decades. Their observations indicate that females copulate once per estrous period and that most use an energetically expensive sampling strategy to select a mate (Byers et al. 1994, 2006). For the two weeks prior to estrus, females move independently and make short-duration visits to widely spaced harem-holding males (harems are one or more females defended by a single male). Although harem-holding males make every effort to prevent visiting females from leaving their site, females always leave a male that fails to maintain a sufficient zone of tranquility (an area free of other males trying to gain access to the females) around his harem. After visiting numerous males, females make their decision, and return and mate with the chosen male, leaving him after about a day. Most females in the population converge on a small subset of harem-holding males, and these males achieve the majority of the copulations. What is it about these select males that leads to their choice by most females?

Male pronghorn mating success is not correlated with body size, horn size, age, or rate of display (during displays, males scent-mark, parallel-walk, and emit snort-wheezes). However, a male's mating success is strongly correlated with his ability to attract and retain females, as measured by haremdays, the sum of a male's mean harem size for each day across all days of the breeding period (Figure 14.19) (Byers et al. 1994). Number of haremdays, in essence, is a measure of a male's vigor, especially as it pertains to his ability to maintain a large zone of tranquility for his harem. But is male vigor a good predictor of offspring quality? Apparently so. Offspring of vigorous males are more likely to survive to weaning and beyond than offspring of less vigorous males (Byers and Waits 2006). Thus, in pronghorn, females evaluate male vigor, and male vigor indicates good genes.

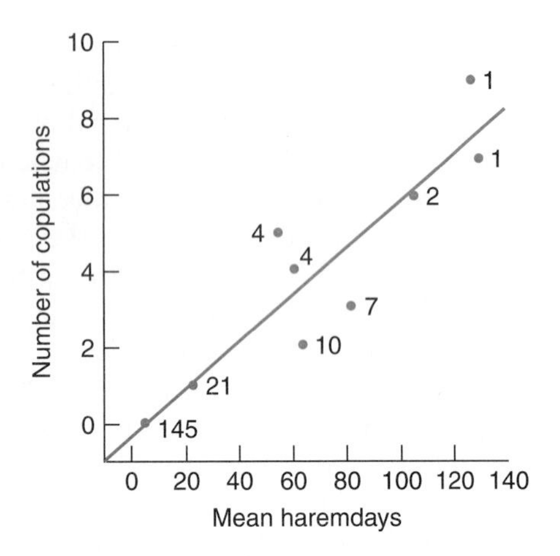

FIGURE 14.19 **Number of copulations per breeding season in relation to haremdays, defined as the sum of a male's mean harem size for each day across all days of the breeding season. Number beside each point indicates sample size. (Modified from Byers et al. 1994.)**

Ability to Provide Compatible Genes

Sometimes the fitness consequences of mating are not simply a function of the male's genetic quality; instead, the fitness consequences may depend on the "fit" between the genes of the female and those of the male (Zeh and Zeh 1996; Mays and Hill 2004). In other words, females may evaluate the ability of males to provide genes compatible with their own. Sometimes the most compatible genes are those that are dissimilar to the female's genes. We'll consider an example involving genes of the major histocompatibility complex (MHC). The MHC is a large chromosomal region that varies tremendously among individuals. It is important in the immune responses that protect animals from disease-causing organisms. (The role of the MHC in kin discrimination is discussed in Chapter 19.) We will focus on the role of the MHC in mate choice in house mice, the species in which the phenomenon was first described by Yamazaki et al. (1976). Although the initial discovery concerned male mice preferring to mate with females carrying dissimilar MHC genes, some subsequent studies have tested the preferences of females, and we will focus on these investigations.

Several laboratory studies have shown that female house mice prefer males that differ from themselves in the makeup of the MHC region (studies reviewed in Penn and Potts 1999; Solomon and Keane 2007). Also, studies in seminatural enclosures show that female mice leave their territories to engage in extra-pair matings with males that differ from them in the MHC region (Carroll and Potts 2007). Mice use odor cues to determine the degree of similarity of the potential mate's MHC region to their own. Peptides, bound to MHC molecules, are the source of the odors, and they are detected by the mouse's vomeronasal organ (Leinders-Zufall et al. 2004). (The vomeronasal organ is an accessory olfactory organ found in the anterior roof of the mouth of many mammals and some other terrestrial vertebrates.)

We still aren't sure *why* MHC-dependent mating preferences exist, but two hypotheses have been suggested. The first is that they evaluate the degree of genetic relatedness of potential mates to avoid inbreeding. Inbreeding caused by mating with close relatives increases homozygosity and hence the risk of producing offspring that are homozygous for deleterious or lethal recessive alleles. On the other hand, extreme outbreeding may cause the breakup of successful parental complexes of genes. Given the potential costs associated with mating with either very close relatives or complete strangers, females may strike a balance between extreme inbreeding and outbreeding when choosing a mate. The second hypothesis is that MHC-dependent mating preferences allow a female to increase her offspring's resistance to disease. Recall that the genes of the MHC region are important in protecting against disease-causing organisms. By choosing a male whose MHC alleles differ from hers, she increases the variability of the MHC region in her offspring. This may make them resistant to a wider variety of disease-causing organisms.

There are additional questions to be answered about the MHC and mate preferences. For example, preference for an MHC-dissimilar mate is reported in some, but not all, studies of mate choice in house mice (reviewed in Solomon and Keane 2007). How can we explain the different findings of the different studies? Also, it remains to be determined just how MHC-based odors compete with other genetically based odors known to be important in individual and kin recognition in house mice (for example, those associated with major urinary proteins) (Carroll and Potts 2007; Cheetham et al. 2007).

STOP AND THINK

Several methodological problems are commonly encountered when studying female choice. Aside from the fact that mate choice may be very subtle, it is often masked or confounded by extraneous effects, such as motivation of the female at the time of assessment. If a female does not display a preference for a male during a choice test, is it because she has rejected him as a potential mate, or is it simply because she is not receptive? Also, as we have seen, female choice in the laboratory is often measured by orientation toward or time spent near a particular male; yet in some cases, these behavioral responses may not relate to actual mating inclination. And then there is the problem of determining the precise roles of male competition and female choice within a single mating system. Although females may appear to choose large males over small males, is it really mate choice or is it because large males effectively exclude small males from courtship activities? What factors would you take into consideration when designing a test of female choice? What would you want to know about your study species before designing such a test? How would you separate the effects of female choice and male–male competition?

ORIGIN AND MAINTENANCE OF MATE-CHOICE PREFERENCES

It isn't too difficult to see why a female should be choosy or why she would choose the male that offers her the most in material benefits, such as nuptial gifts, a good territory, or high-quality parental care. In these instances, a female's preference evolves because it increases her fitness. But how do female mate preferences originate when there are no direct material benefits, that is, when the female's preference affects the fitness of her offspring rather than her own fitness? We'll consider two explanations for the origin of preferences that do not involve direct material benefits: indicator and runaway mechanisms. Some people prefer not to make a distinction between these two explanations (Kokko et al. 2003), but we find it useful to discuss them separately. We end with a third explanation for the origin of female preferences called sensory bias.

Indicator Mechanisms

Indicator mechanisms are often called good genes models. These models assume that a particular trait in males indicates viability and that both the trait and viability have a genetic basis. If a female preference (also genetically based) for the male trait should arise, then these females mate with males carrying genes for the trait and enhanced viability. In this way, genes for the male trait, high viability, and the female preference become associated. High viability is favored by natural selection; because the genes for the male trait and female preference are associated with viability, they too are favored.

An example of a good genes model is the handicap principle proposed by Amotz Zahavi (1975). According to this principle, females prefer a male with a trait that reduces his chances of survival but announces his superior genetic quality precisely because he has managed to survive despite his "handicap." In short, male secondary sexual characteristics act as honest signals, indicating high fitness, and females choose males with the greatest handicaps because their superior genes may help produce viable offspring.

The hypothesis that females choose males with good genes predicts that the mating preferences of females should increase the viability of their offspring. Some supporting evidence comes from studies that show a relationship between exaggerated male traits and offspring viability. One of the best known examples of an exaggerated male trait is the ostentatious tail of the peacock. Peafowl (*Pavo cristatus*) are also an example of a species in which there is good evidence that the females (known as peahens) get genetic benefits that increase survival of their offspring through mate-choice preferences. Marion Petrie has studied a free-ranging population of peafowl in Whipsnade Park in England. During the early spring, adult males gather to court females on a communal breeding ground called a lek. An adult peacock is a thing of beauty (Figure 14.20). His chief glory is a train of long, beautifully marked feathers, each tipped with an iridescent "eye." A courting male lifts his tail, which lies under the train. This elevates the train and spreads it out like a fan. He then struts around on his small territory on the lek, vibrating his tail rapidly. In turn, the tail vibrations cause the plumes in his train to rattle audibly. He attempts to copulate with a "hoot-dash," in which he begins to lower his train and rushes toward the female while giving the "hoot" call. If his copulation attempt is successful, the female will squat in front of him, allowing him to mount her.

FIGURE 14.20 **A peacock displays his train. Peahens choose males with the most elaborate trains. Elaborate trains appear to indicate good genes; thus, females that select highly ornamented males increase the viability of their offspring.**

Peahens are choosy, preferring males with elaborate trains. When it comes to peacock mating success, the "eyes" have it; that is, mating success is significantly correlated with the number of eyespots in the male's train. On one lek, which consisted of ten courting males, the most successful male copulated 12 times, but the least successful males never did. A female never accepted the first suitor she saw. On average, she visited about three males before copulating. In 10 out of 11 observed courtship displays that ended with successful copulation, the female chose the male with the highest number of eyespots in his train of those she visited (Petrie et al. 1991).

When 20 eyespots were experimentally cut out of the trains of some peacocks, the male's mating success was significantly less than his success during the previous year. The attractiveness of control males, whose tails were left intact after being captured and handled, remained the same (Petrie and Halliday 1994).

Besides boosting a peacock's mating success, the extent of sexual ornamentation appears to be correlated with survivorship. In the spring of 1990, of 33 displaying males in the same population, 22 copulated successfully at least once and 11 were never successful. During the following winter, two foxes managed to enter the park, and they killed five peacocks. Four of the birds that were killed were among those that had been unsuccessful in gaining any copulations in the previous season. The fifth bird had copulated only twice. The males that were killed also had shorter trains with fewer eyespots than surviving males had. This observation is consistent with the hypothesis that only the healthiest males can develop long, elaborate trains. Thus females are choosing males with good genes (Petrie 1992).

It also seems that the offspring of females who choose mates with elaborate trains benefit from their father's good genes. Petrie (1994) demonstrated this by pairing males with females chosen at random. In each large cage, she placed one male and four females. The mated males varied in attractiveness to females, as measured by the mean area of eyespots in the train. All the offspring were raised under common conditions. When they were 84 days old, the offspring were weighed. The offspring of males with more elaborate trains weighed more than those with less showy fathers. After two years, more of the offspring of the highly ornamented males were still alive than were those of less attractive males. Because the matings were arranged by Petrie, the differences in offspring viability cannot be due to differences in the quality of females. By ruling out maternal effects, Petrie has provided strong evidence that a female can enhance the survival chances of her offspring through mate-choice preferences. Here, then, is an example in which females choose males on the basis of a trait that indicates male genetic quality, and this choice, in turn, results in more viable sons and daughters.

Runaway Mechanisms

Runaway mechanisms are also called Fisher or Fisherian mechanisms, after R. A. Fisher (1930), who first described the process. The basic idea is that the male trait is correlated with the female preference for that trait. It begins when females evolve a preference for a particular male characteristic. A female that mates with a male that has the attractive characteristic will have sons with the trait, providing that the attractive character of the male is inherited, and she will have daughters who show a preference for that trait. The attractive sons will acquire more mates with a preference for that characteristic than other males and thus will leave more progeny. In this way, runaway selection can produce increasingly exaggerated male traits and a stronger female preference for them. So, whereas indicator mechanisms require that females acquire good genes that increase the viability of their sons and daughters, a runaway selection mechanism requires that females acquire genes that make their sons particularly attractive.

Lande (1981) and Kirkpatrick (1982) developed mathematical models of Fisher's (1930) ideas for the evolution of female choice. These models demonstrate that runaway selection can indeed result in mate choice for characteristics that are arbitrary or even disadvantageous to the health and survival of individuals, providing that females prefer to mate with males that possess them. Runaway selection would favor ever more exaggerated male characteristics and females that find them attractive. So when does it all stop? The process will be stabilized only when natural selection balances sexual selection (Fisher 1930). In other words, when the male trait becomes too energetically costly to produce or when it makes males less likely to escape from predators, selection will no longer favor further exaggeration of the trait.

Experimental support for the runaway selection hypothesis comes from sandflies of the *Lutzomyia longipalpis* species complex. At night, male sandflies aggregate on the back of a vertebrate host, perhaps a chicken, where they defend small (radius of 2 cm), mobile territories by bumping one another with their abdomens. Females visit the host and evaluate several males but copulate with just one. Mating success among males is highly skewed, with some males copulating with many females and others unsuccessful, reflecting the considerable agreement displayed by females in their choice of mate. Therésa M. Jones and colleagues (1998) conducted laboratory experiments designed to evaluate the fitness consequences of mate choice for female sandflies and their offspring. Female sandflies were allowed to choose among several males. Females and their offspring were then monitored for survival and fecundity. The results indicated that females do not gain direct benefits by mating with attractive males (here, attractiveness is defined as the number of females with which a male mates) because females that mated with attractive males did not themselves have higher survival or fecundity. But what about indirect benefits, that is, benefits to their offspring? Jones and colleagues found no evidence that a female's choice of mate influenced the viability of her offspring; the survival of sons and daughters, and the fecun-

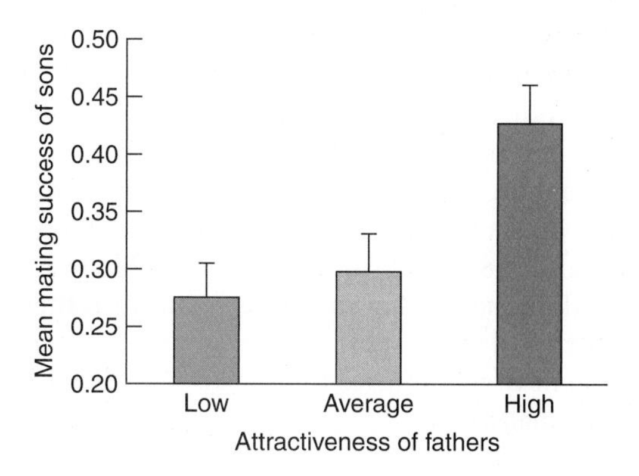

FIGURE 14.21 **In sandflies, the mating success of sons increases with the attractiveness of their fathers. These data support runaway selection models for the origin of female preferences.**

dity of daughters, did not increase with attractiveness of their father. These findings thus do not support good genes models. However, the mating success of sons increased with the attractiveness of their father (Figure 14.21). Thus, female sandflies that select an attractive male appear to benefit by having attractive sons; these findings support runaway selection.

Sensory Bias

According to the sensory bias model, female preferences for certain traits in males could evolve because male traits stimulate an existing bias in the female's sensory system. The original bias might relate to feeding or avoiding predators. For example, females might have a sensory bias to help them find food of a particular color; males can then exploit this bias by using the same color to attract females during courtship. Consider how such a scenario might work. By chance, a male has a mutation that produces a trait that exploits a sensory bias of females. This male, though he provides neither the resources nor the good genes that other males provide, is successful in convincing females to mate with him. Over time, males with such a trait would become more common in the population. Eventually, however, there would be selection for females to resist mating with such males because they receive no benefits from selecting them. In response to increasing female resistance, selection would then favor males with mutations that further exaggerate the trait. The cycle of increasing female resistance to mating with males that possess the trait and male exaggeration of the trait leads to expensive male traits of no use to females but needed by males to obtain matings. This specific explanation for the evolution of extreme traits in males that exploit a sensory bias of females and provide no benefits to females is called the **chase-away model** (Holland and Rice 1998). It emphasizes conflict between males and females, a topic that we consider in more detail at the end of this chapter.

Recall that male satin bowerbirds decorate their bowers with natural and artificial objects, and that they are especially picky about the color of their decorations. Recall also that a male satin bowerbird prefers blue and that the color of his decorations is an important determinant of his mating success. J. R. Madden and K. Tanner (2003) investigated a possible example of sensory bias in bowerbirds. They found that when given a choice of grapes of different colors (commercial food dyes were used to make brown, yellow, red, green, or blue grapes), female regent bowerbirds and female satin bowerbirds (to a somewhat lesser extent) preferred to eat blue grapes and males of both species preferred to decorate with blue grapes. These data are thus consistent with the sensory bias model—choice of bower decoration could have evolved to exploit a sensory bias in females that was originally related to foraging. See Chapter 17 for additional examples of sensory bias.

CRYPTIC FEMALE CHOICE

The observation that "It ain't over 'til it's over" may ring true for female mate choice. Females of certain species appear able to choose the sperm that will fertilize their eggs after copulating with several males. This is called cryptic female choice because it is a hidden, internal decision made after copulation (Eberhard 1996). In the sperm wars, this choice is the female equivalent of male sperm competition.

So far, evidence for cryptic female choice is somewhat limited because it is difficult to distinguish from sperm competition (Andersson and Simmons 2006). Nevertheless, cryptic female choice has been demonstrated in several species, most notably insects. The mechanisms by which females choose sperm after copulation are diverse and include (but are not limited to) the following: premature termination of copulation; premature removal of spermatophores; failure to store sperm that have been transferred; removal of stored sperm; and differential use of stored sperm in fertilization. Females of a given species may use one or more mechanisms. We will focus on the red flour beetle (*Tribolium castaneum*), a species in which females mate with two or more males in quick succession.

In a somewhat grisly experiment, Tatyana Fedina (2007) investigated whether female flour beetles control the quantity of sperm transferred to them per spermatophore. In this species, sperm are not packaged into spermatophores before copulation (as in insects like crickets and katydids), but during copulation. She manipulated male phenotypic quality (body condition) by using fed males (each male was housed in a vial that

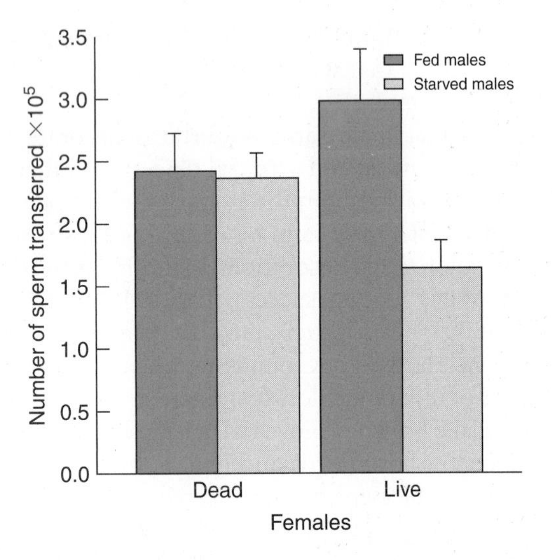

FIGURE 14.22 **Female flour beetles exhibit cryptic female choice. In this experiment, fed males transferred significantly more sperm than did starved males, but only when mating with live females. Thus, live female flour beetles control the number of sperm transferred to them by males, favoring males in good physical condition.**

contained wheat flour for seven days) and starved males (each male was housed in a vial without wheat flour for seven days). She also manipulated whether females could control the amount of sperm transferred to them during copulation by using live females (which can exert control over copulations) or dead females (which cannot exert control) (yes, male flour beetles will mate with recently killed females). The experimental design thus contained the following treatments, with 10 to 12 male–female pairs in each: (1) live females and fed males, (2) live females and starved males, (3) dead females and fed males, (4) dead females and starved males. The design allowed female influence to be distinguished from male influence on sperm transfer. Immediately after members of male-female pairs finished copulating, Fedina dissected out the reproductive tract of the female and estimated the number of sperm present. As you can see in Figure 14.22, fed males transferred significantly more sperm than starved males, but only when mating with live females. Thus, when female flour beetles have the ability to do so (i.e., when alive), they control the number of sperm transferred to them by males, favoring males in good physical condition. Although the precise physiological mechanism by which females exert such control remains to be determined, Fedina (2007) suggests that females may contract or relax certain muscles in their reproductive tract (bursal muscles), thereby controlling the flow of sperm.

Controlling the amount of sperm transferred per spermatophore is not the only mechanism of cryptic female choice in flour beetles. Females also selectively prevent the transfer of spermatophores at copulation, expel spermatophores soon after mating, and may control the amount of sperm stored in the spermatheca (Bloch Qazi et al. 1998; Fedina and Lewis 2004, 2006). Indeed, female flour beetles appear able to make rather fine adjustments during sperm transfer and storage that allow them to manipulate the sperm representation of different males.

SEXUAL CONFLICT

Parker (1979, 2006) defines **sexual conflict** as a conflict between the evolutionary interests of males and females. It sometimes is generated by sexual selection. For example, males compete with one another to mate with females. That competition may result in the evolution of reproductive behaviors and adaptations that are harmful to females. For example, male fruit flies produce accessory gland proteins that promote a male's success in sperm competition, but decrease the longevity and reproductive success of frequently mating females (Chapman et al. 1995, 2000). Sexual conflict takes several forms, but the two main ones concern mating/fertilization and parental investment. These situations involve interactions between males and females, during which each individual's fitness depends on its own strategy as well as that of its partner. Our focus here will be on sexual conflict during mating/fertilization.

As mentioned, sexual conflict can lead to traits evolving in males that are beneficial to them but damaging to females. Females, in turn, may evolve a counter-adaptation that reduces or overcomes the harmful effects of male sexual behavior. Coevolution between males and females that is propelled by sexual conflict is called sexually antagonistic coevolution. We will briefly describe two examples of this phenomenon—one concerns bedbugs and the other, seed beetles.

Rather than mating in the traditional manner, male bedbugs (Heteroptera: Cimicidae) engage in traumatic insemination whereby they stab their intromittent organ through the abdomen of a female, and inject sperm and accessory gland fluids directly into her body cavity (Morrow and Arnqvist 2003; Stutt and Siva-Jothy 2001). This form of male sexual behavior might have evolved because it allowed males to inseminate females that had already mated (in the usual way) and were resistant to further inseminations. While beneficial to males, traumatic insemination is harmful to females; females with high mating rates have shorter life spans.

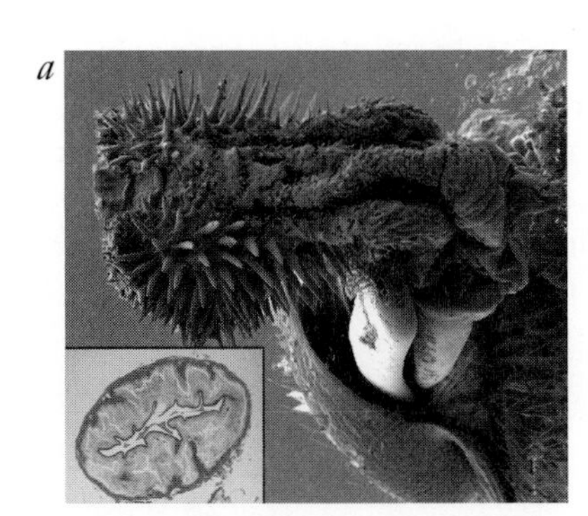

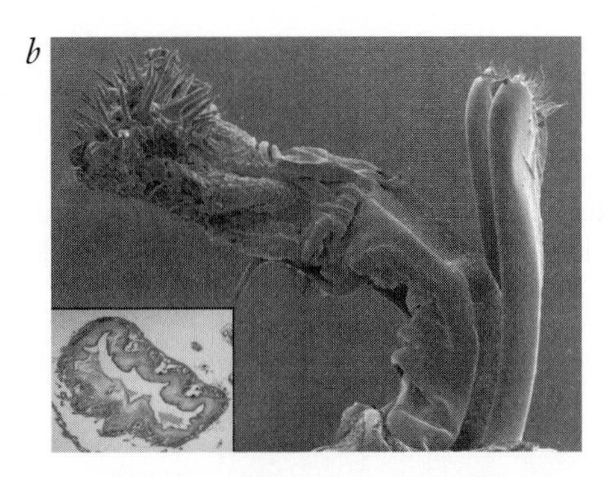

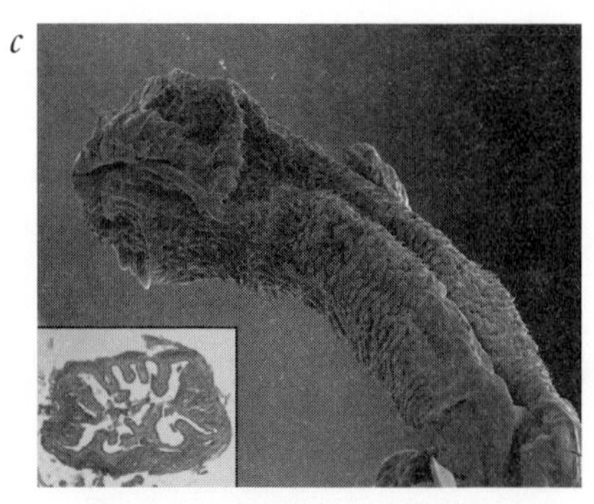

FIGURE 14.23 **Sometimes sexual conflict leads to sexually antagonistic coevolution, in which an adaptation in one sex is met with a counteradaptation in the other sex. Shown here are scanning electron micrographs of male genitalia and cross sections of female copulatory ducts (insets) for three species of seed beetles. Note that the degree of robustness of female copulatory ducts parallels the degree of spininess of male genitalia: both are greatest in (*a*), intermediate in (*b*), and least in (*c*). Comparative data like these suggest that in seed beetles, the male adaptation of spiny genitalia has been met with a female counteradaptation, robust copulatory ducts.**

Female bedbugs, however, have responded with a counteradaptation. Just beneath their cuticle, at the exact spot on their abdomen where the male typically inserts his intromittent organ, the females have a spermalege. This special organ somehow reduces the trauma associated with the piercing of the female's abdomen. Indeed, experimental piercing of the female's abdomen outside the spermalege causes dramatic decreases in female life span *and* lifetime egg production (Morrow and Arnqvist 2003). Generally, males should benefit by harming their mates as little as possible, and so they typically inseminate females at the site of the spermalege. In the case of bedbugs, then, male competition may have led to the evolution of traumatic insemination by males, which in turn led to the evolution of the spermalege in females. This provides us with an example of how sexual selection can generate sexual conflict, which in turn can lead to sexually antagonistic coevolution.

Our second example of sexually antagonistic coevolution concerns seed beetles (Coleoptera: Bruchidae). Males of some species have genitalia armed with elaborate spines that help to anchor them during copulation. These spines, however, are harmful to females, penetrating their copulatory duct during mating and leaving scars where the damaged tissue heals (Rönn et al. 2007). The degree of harm experienced by females increases with the spininess of male genitalia. Evidence for sexually antagonistic coevolution comes from comparative data showing that the male adaptation of spiny genitalia appears to have been met by a female counteradaptation, robust copulatory ducts (Figure 14.23). We'll have much more to say about the battles between the sexes in Chapter 15 when we describe sexual conflict in relation to parental investment.

SUMMARY

Sexual selection results from (1) competition within one sex for mates (intrasexual selection) and (2) preferences exhibited by one sex for certain traits in the opposite sex (intersexual selection). Early observations regarding sexual selection suggested that because females invest more in gametes and parental care they are more selective than males in their choice of mates and are usually a limited resource for which males compete. There have been recent challenges to these ideas. Nevertheless, competition among males for access to females appears responsible for the evolution in males of a broad spectrum of physical and behavioral attributes that enhance fighting prowess. Weapons, large size, and physical strength have evolved in males of many species. In addition, males devote much time and energy to ensuring that their sperm and not the sperm of a competitor are used to fertilize a female's eggs. In some species this may take the form of dominance behavior, mate guarding, or alternative reproductive strategies; in others, repellents or copulatory plugs are used. Because reproductive success is measured in relative terms, males may also enhance their position by employing tactics of sexual interference that decrease the reproductive success of other males.

Females may base their choice of mate on the male's ability to provide material benefits such as sufficient sperm, nuptial gifts, food or nest sites, or parental care. Females may also consider whether the male has good or compatible genes. In some species females evaluate males on a single criterion, while in others females consider multiple criteria. Explanations for the origin and maintenance of female mate-choice preferences when the male does not provide material benefits include indicator mechanisms (good genes) and runaway selection. A third explanation, sensory bias, suggests that female preferences for certain traits in males could evolve because male traits stimulate an existing bias in the female's sensory system. Female choice that occurs after copulation is described as cryptic because it is a hidden, internal choice.

Sexual selection may lead to sexual conflict, defined as a conflict between the evolutionary interests of males and females. Such conflict is often associated with mating and fertilization, and can lead to sexually antagonistic coevolution in which an adaptation in one sex is met with a counteradaptation in the other sex.

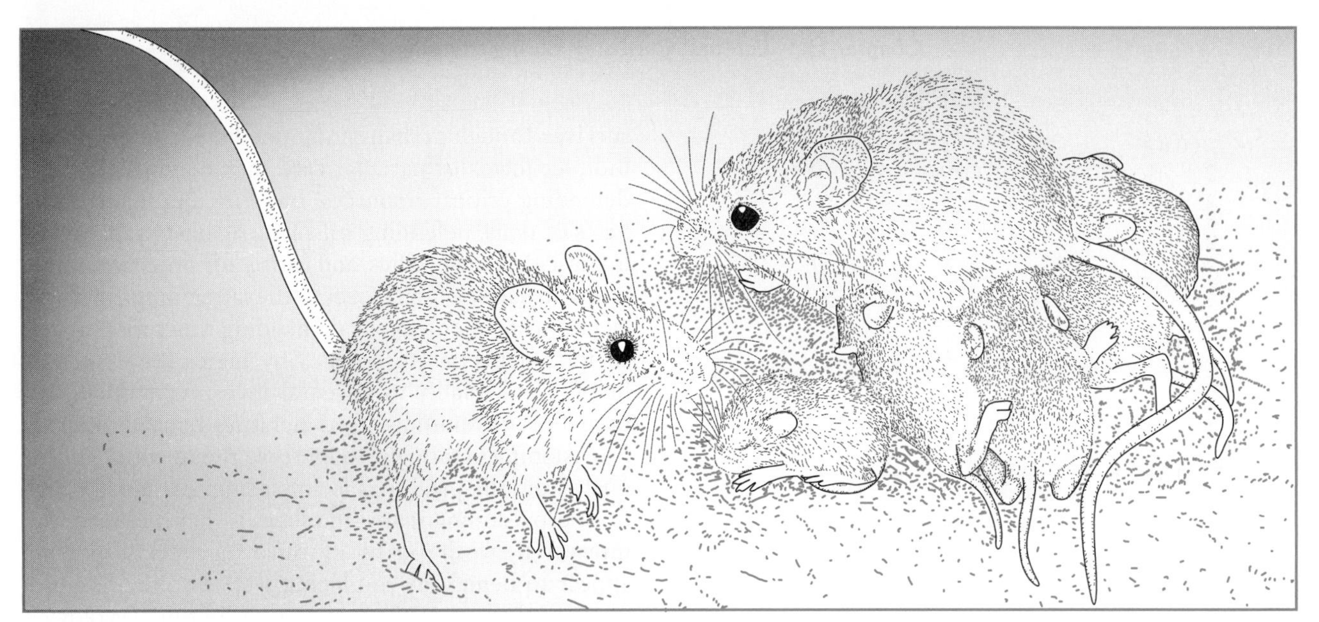

15

Parental Care and Mating Systems

In several species of mammals, including humans, mothers that produce sons incur greater costs than those that produce daughters. The costs experienced by mammalian mothers that bear sons vary with species, but include such things as higher parasite loads, delay in the next reproductive effort, reduced likelihood of future reproduction, and reduced longevity. The costs, however, do not stop there. Indeed, a study of human pedigree data gleaned from population registers containing records of births, marriages, and deaths from areas of preindustrial Finland (roughly from the early 1700s through the early 1800s) reveals an intergenerational cost to bearing sons. Specifically, after controlling for ecological conditions (different locations) and social class, Ian Rickard and colleagues (2007) found that offspring born after elder brothers had similar survival but lower lifetime reproductive success than offspring born after elder sisters (Figure 15.1). The reduced lifetime reproductive success was due largely to lower lifetime fecundity (number of offspring produced); lower survival rates of offspring played a lesser role. The authors suggest that mothers who produced sons experienced a substantial reproductive cost, which made them less able to invest in their next child. Their lower investment in the next child was then manifested in the lower lifetime reproductive success of that child.

Reproduction certainly seems to be a complicated affair, with both short- and long-term ramifications for family members. In this chapter, we focus on the costs and benefits associated with reproduction. We begin with the parent–offspring relationship and examine conflicts among family members and how parents allocate resources to their offspring. We also examine the diversity of animal caregivers, including mothers, fathers, and nonbiological parents. Our second focus is on the mating relationship and the evolution of different mating systems.

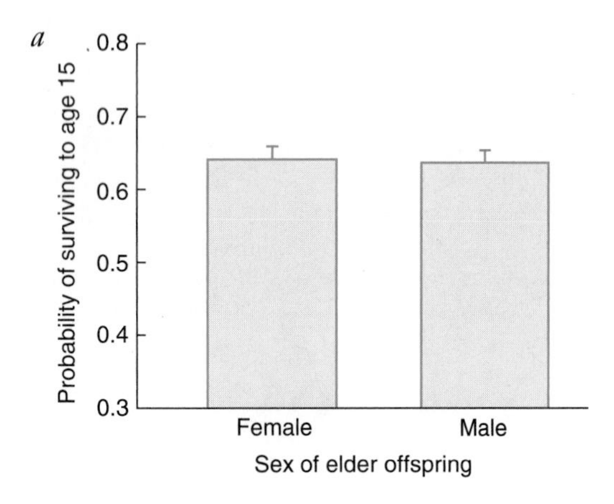

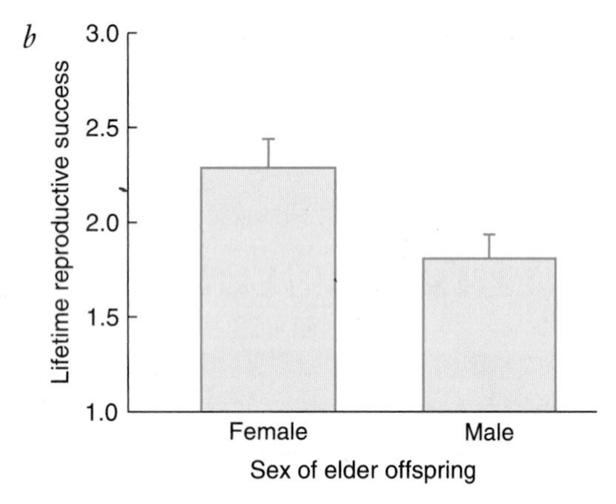

FIGURE 15.1 **In a preindustrial human population in Finland, producing sons does not affect the survival of subsequent offspring but does affect the lifetime reproductive success of subsequent offspring. (*a*) The probability of surviving to 15 years of age in relation to the sex of the elder offspring. (*b*) Lifetime reproductive success, defined as number of children raised to 15 years of age, in relation to the sex of the elder offspring. (From Rickard et al. 2007.)**

PARENTAL CARE

Trivers (1972) defined **parental investment** as any investment by parents in an offspring that increases the survival of that offspring while decreasing the ability of the parent to invest in other offspring. Parental behaviors are characterized as either direct or indirect (Kleiman and Malcolm 1981). Direct parental care includes behaviors that have an immediate physical impact on offspring and their survival. In mammals, for example, direct parental care includes behaviors such as nursing (and feeding), grooming, transporting, and huddling with young. Indirect parental care includes behaviors that parents may perform while away from the young, so these behaviors may not involve direct physical contact with offspring. Nevertheless, behaviors classified as indirect parental care still affect offspring survival, though perhaps not immediately. In mammals, indirect forms of parental care include acquiring and defending critical resources, building and maintaining nests or dens, defending offspring against predators or infanticidal conspecifics, and caring for pregnant or lactating females. With respect to the latter, pregnancy and lactation are energetically demanding times for females, so food deliveries to females by mates are especially helpful and qualify as parental investment. Male owl monkeys, for example, feed lactating females and this provisioning is thought to increase the quantity and/or quality of milk produced for offspring. An added benefit of provisioning in owl monkeys is that it reduces the interval between births of offspring. Thus, parents appear to benefit by producing well-fed offspring and more of them over the course of a lifetime (Wolovich et al. 2008).

As a rule, patterns of parental investment should maximize an individual adult's lifetime reproductive success and not necessarily each reproductive event. Thus, all parents must make two important decisions. First, they must decide how much of their own resources to devote to reproduction instead of to their own growth and survival. Second, they must decide how to allocate the available resources among their offspring (Clutton-Brock and Godfray 1991). It is easy to see how these decisions can lead to conflicts of interest between parents and offspring and among siblings. We first consider the nature of these conflicts and then look at some of the specific factors that influence parental decisions regarding resource provision.

CONFLICTS AMONG FAMILY MEMBERS OVER PARENTAL INVESTMENT

Evolutionary conflicts over parental investment include sexual conflict, intrabrood conflict, and interbrood conflict (Mock and Parker 1997; Trivers 1974). Recall from Chapter 14 that sexual conflict is conflict between the evolutionary interests of males and females (Parker 1979, 2006). This conflict takes several forms, but the two main ones concern mating (discussed in Chapter 14) and parental investment (our focus here). Both of these situations involve interactions between males and females during which each individual's fitness depends on its own strategy as well as the strategy of its partner. For parental care, conflict emerges because the costs of providing care are paid separately by each parent, whereas the benefits accrue to both parents regardless of which one provides the care (Wedell et al. 2006). As a result of this arrangement, each parent should prefer that the other do the lion's share of the work when it comes to taking care of the kids. During intrabrood conflict, young try to obtain resources that parents prefer to distribute to other members of the current brood, and during interbrood conflict, young try to obtain resources that parents prefer to save for future offspring. As an example of conflict among family members over parental investment, we consider intrabrood conflict.

It is easy to see how differences in the distribution of resources by parents can lead to sibling rivalry. Bluntly put, each youngster derives a greater fitness benefit from the parental care it personally receives than from the care its siblings receive. In some species, sibling rivalry involves overt, substantial aggression and results in the death of one or more siblings; such fatal sibling rivalry is called **siblicide**. In other species, the rivalry is somewhat subtler, with lower levels of fighting or siblings racing to outcompete each other for parental resources (the latter is called scramble competition; Mock and Parker 1997). We first consider an example of sibling rivalry with relatively low levels of fighting, and then we describe the conditions under which siblicide, the most extreme form of sibling rivalry, typically occurs.

In domestic piglets (*Sus scrofa*), sibling competition begins before birth when certain parts of the uterus are not spacious enough to support maximum growth of embryos. Some embryos in these areas die early in pregnancy while others survive but are characterized by unusually low birth weight, a condition that puts these piglets at a severe disadvantage for the intense competition that characterizes the early postnatal period (Figure 15.2*a*). Shortly after birth, piglets compete for teats at which to suckle; large piglets typically locate and retain possession of a particular teat or pair of teats, while smaller piglets fight with littermates to secure a location but may be continually displaced, especially if there are many piglets in the litter. Displaced piglets fail to obtain milk and colostrum (the fluid released from the teats shortly after parturition that boosts the newborn immune system). Most piglet deaths occur in the first few days after birth, and many are due to starvation because small piglets expend energy fighting for teat locations only to be routinely displaced before obtaining a meal (studies reviewed by Drake et al. 2008). Battles among piglets for teats involve frantic shoving and can include wounding because newborns are armed with slashing teeth (Figure 15.2*b*). The canines and third incisors of piglets are fully erupted at birth and angled in such a way that quick sideways movements of the head can lacerate the faces of adjacent siblings (farmers routinely clip these teeth to prevent piglet injuries). These slashing teeth seem to function solely

a

b

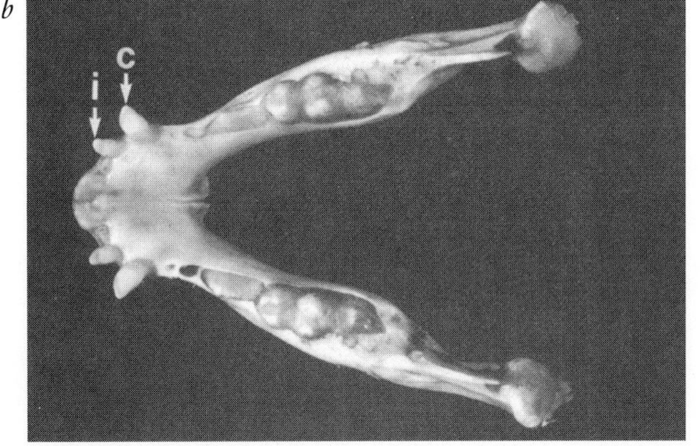

c

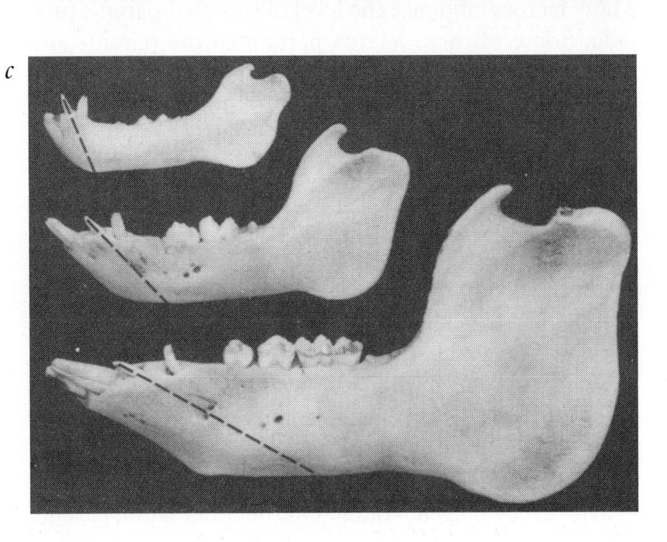

FIGURE 15.2 **Sibling rivalry occurs among piglets as they compete for access to the teats of the mother. (*a*) Rivalry is most intense in the first few days following birth when piglets compete for a particular teat or pair of teats to which they return at each nursing bout. (*b*) At birth, piglets have outwardly angled canines and third incisors, which they use as weapons during the early competition for teats. This jaw is from a newborn piglet; the canine and third incisor on the right side of the lower jaw have been labeled "c" and "i", respectively. (*c*) Canines and third incisors become increasingly insignificant as the piglet grows, other teeth erupt, and the orientation of the anterior part of the jaw changes. Shown here are jaws from a newborn piglet, a 21-day-old piglet, and an 84-day-old piglet (the typical age of weaning). The dashed line illustrates how the orientation of the anterior part of the jaw changes with age, so that the third incisor assumes a more forward orientation typical of incisors.**

in early sibling competition because they become much less significant as other teeth erupt and the jaws grow with age (Figure 15.2*c*). David Fraser has studied maternal care and sibling rivalry in domestic pigs for many years. Here, we include a poem that served as the abstract for one of his papers on sibling competition among piglets (From: D. Fraser, and B. K. Thompson. 1991. Armed sibling rivalry among suckling piglets. *Behavioral Ecology and Sociobiology* 29:9–15). Can you determine the experimental design and results of the study? For the record, the term *farrowing* refers to the time of parturition in pigs.

A piglet's most precious possession
Is the teat that he fattens his flesh on.
He fights for his teat with tenacity
Against any sibling's audacity.
The piglet, to arm for this mission,
Is born with a warlike dentition
Of eight tiny tusks, sharp as sabres,
Which help in impressing the neighbors;
But to render these weapons less harrowing,
Most farmers remove them at farrowing.
We studied pig sisters and brothers
When some had their teeth, but not others.
We found that when siblings aren't many,
The weapons help little if any,
But when there are many per litter,
The teeth help their owner grow fitter.
But how did selection begin
To make weapons to use against kin?

As mentioned above, in some species, sibling rivalry leads to one offspring attacking and killing its brother or sister. Siblicide is most common in species in which resources are limited and parents deposit eggs or young in a "nursery" with limited space (Mock and Parker 1998). The nursery can take various forms, including a uterus, a brood pouch, a parent's back, a nest, or a den. Although it seems odd at first, siblicide may be advantageous to the parents of some species. For example, when more young are produced than can be raised successfully, siblicide can save the parents time and energy by eliminating the young that are least likely to reach adulthood.

Why would a parent produce more young than it can raise successfully? Several answers to this question have been suggested. One idea is that overproduction of young is insurance in case some eggs or offspring fail to develop. In some species of eagles, for instance, the female typically lays two eggs, but only one chick reaches fledgling age. The eggs are usually laid a few days apart. If both eggs hatch, the older, stronger chick generally kills its younger nestmate (Mock et al. 1990). Another possible explanation is that the overproduction of young is an adaptation to a variable food supply. In years when food is plentiful, all the young may reach adulthood. However, in years of food scarcity, when sibling competition for resources is severe, the weaker siblings will be killed (Clutton-Brock and Godfray 1991). In other cases, extra offspring may be produced to benefit the stronger siblings, either by helping them raise offspring when they become adults or by serving as critical meals to provide nourishment when conditions are particularly harsh (Mock and Parker 1998).

In view of the many conflicting and competing interests associated with parental care, we can ask to what extent do parents respond to other family members when deciding how much care to provide? The data available suggest great variation in the responsiveness of parents to the actions and needs of family members. For example, some parents appear oblivious to the behavior of their mate (e.g., house sparrows; Schwagmeyer et al. 2002), while others are extremely responsive to the level of parental effort put forth by their partner and adjust their own effort accordingly (e.g., burying beetles; Smiseth and Moore 2004). Similarly, some parents seem remarkably insensitive to the needs of their offspring (e.g., guinea pigs; Laurien-Kehnen and Trillmich 2003), while others respond to signals of offspring need by adjusting the amount of care provided (e.g., canary; Kilner 1995). In addition, the responsiveness of parents to other family members varies not only across species, but also within species (Hinde and Kilner 2007). Let's examine some of the specific factors that influence parental allocation of resources.

SOME FACTORS THAT INFLUENCE THE ALLOCATION OF PARENTAL RESOURCES

Many factors influence the level of care that parents provide. Some of these factors pertain to the parents and some to the young. Here we consider how life history, certainty of paternity, and gender of offspring influence the amount of care provided by parents.

Life History

An obvious factor influencing how much effort parents invest in current offspring is the likelihood that the parents will have future opportunities to breed. This, in turn, will be affected by the parents' age (Trivers 1974) and the life span of individuals of that species (Linden and Møller 1989). On one hand, it might be expected that in short-lived species with little hope of producing additional young in the future, parents would invest more heavily in the present young. On the other hand, parents of long-lived species might spend more of their resources on their continued growth and survival because they might have the opportunity to breed again.

This hypothesis has been tested experimentally with several bird species by handicapping the parents so that more parental effort was required to raise the young and then determining whether the parents would bear the increased costs of reproduction themselves or pass the costs on to their young. Leach's storm-petrel (*Oceanodroma leucorhoa*) is a relatively long-lived seabird. Adult petrels make long journeys to ephemeral food patches to gather planktonic crustaceans, drops of oil, and small fish to feed their chicks. A foraging trip usually lasts two to three days. About 30% of that time is spent airborne, so the cost of flight for a parent that is provisioning chicks is significant. Shortening the wing span by clipping feathers increases the energetic cost of flight, raising the cost of reproduction. When parent petrels were handicapped in this way, they passed the increased reproductive costs to their offspring and maintained their own nutritional condition. Feather growth, which is a measure of a bird's nutritional state, did not differ between adult birds whose wings were clipped and untreated control birds. However, the chicks whose parents' wings had been clipped grew more slowly and spent more nights without food than chicks with untreated parents (Mauck and Grubb 1995). In contrast, when parents of short-lived species, such as starlings (Wright and Cuthill 1990), flycatchers (Slagsvold and Lifjeld 1988), and tits (Slagsvold and Lifjeld 1990), were handicapped in a way that increased their reproductive costs, they bore at least part of the increased costs themselves and continued to allocate nearly the same amount of resources to their chicks. Whereas the single chick raised each year by long-lived petrels represents only a small part of the parent's lifetime reproductive success, in these short-lived species, each clutch represents a large proportion of the parent's lifetime reproductive success. Thus, in these studies, we do find that expected life span influences a parent's allocation of resources in a way that maximizes lifetime reproductive success.

Certainty of Paternity

Trivers (1972) was one of the first to suggest that parental solicitude toward young should be correlated with the likelihood of genetic relatedness. Females can usually be fairly certain that they are related to their offspring. Certainty of maternity guarantees that 50% of a mother's genes are present in each of her progeny. Males, especially of species with internal fertilization, cannot be so confident. Recall from Chapter 14 that even though a male copulates with a female, he has no guarantee that his sperm, rather than the sperm of a competitor, will fertilize her eggs. In short, because males of internally fertilizing species run the risk of investing time and energy in raising another male's offspring, the odds run against the evolution of paternal behavior. Reliability of paternity is assumed to be greater when eggs are fertilized externally instead of inside the female. Nevertheless, certainty of paternity can still be an issue for externally fertilizing species (see below).

Although appealing to our sense of intrigue, the idea that certainty of paternity should influence paternal care has a mixed history of support from the scientific community. Early models, such as those developed by Maynard Smith (1977) and Werren et al. (1980) raised questions about the usefulness of the paternity hypothesis as a general explanation for the evolution of patterns of parental care. Later models, using different assumptions, showed that paternity could influence paternal care, under certain conditions (Houston 1995; Westneat and Sherman 1993). Finally, in a more recent evaluation of this issue, Sheldon (2002) concludes that testing models relating paternal care to paternity is extremely difficult (for example, how does one measure or manipulate "certainty"?) and that we are a long way from understanding whether males of most species adjust parental care in relation to certainty of paternity. Keeping this history in mind, let's look at one species for which there is strong evidence that certainty of paternity influences level of paternal care.

As described in Chapter 14, male bluegill sunfish (*Lepomis macrochirus*) display a discrete life history polymorphism in that they are either parentals or cuckholders (Gross 1982; Neff et al. 2003; Stoltz and Neff 2006). Parental males compete for sites at which to build their nest, guard females that enter their nest to spawn, and then provide sole care to the developing young (care entails guarding and fanning the eggs for two to three days until they hatch and then defending the fry from predators for another week or so). Cuckholders, on the other hand, steal fertilizations from parental males. When small, cuckholders are termed "sneakers" because they hide in vegetation at the edge of a parental male's nest and then dart into the nest to release sperm at the time the parental male is spawning with a female. Once cuckholders reach a large body size, they become female mimics that gain entry into the nests of parental males by their physical and behavioral resemblance to females. Once inside the nest, a female mimic releases sperm along with the parental male at the time of spawning. Sneakers seem to be particularly effective in fertilizing eggs.

Given the interesting and well-studied reproductive lives of bluegill sunfish, Bryan Neff (2003) decided to manipulate the perceived paternity of parental males by using two cues. The first cue was the presence of sneaker males during spawning; this is an indirect cue of reduced paternity. The second cue was a water-borne chemical released by newly hatched fry, possibly in their urine; parental males use this direct cue to distinguish their offspring from those of sneakers and female mimics

(Neff and Sherman 2003). The chemical cue is not released by eggs, only by fry.

Neff performed two experiments using a population of bluegills in Lake Opinicon Canada; this particular population has been studied for more than 30 years. In the first experiment, parental males in the midst of spawning were exposed to four sneaker males held in clear containers to prevent them from fertilizing any eggs. Parental males exposed to sneaker males were "treatment males." Parental males in the control group ("control males") were exposed to empty containers near their nests. The day after spawning, treatment males and control males were tested for their willingness to defend their brood against an egg predator: a pumpkinseed sunfish held in a clear bag at the outskirts of each parental male's nest. Treatment males displayed lower levels of egg defense than control males (Figure 15.3*a*; compare results for the egg phase). The eggs were allowed to hatch, and then males were tested again, this time for their willingness to defend their fry. Note that for this test, parental males had available to them chemical cues from fry; such cues should allow males to reassess their paternity. There was no difference between treatment and control males in the level with which they defended fry (Figure 15.3*a*; compare results for the fry phase). Viewing the overall results from the first experiment, we see that treatment males increased their level of defense from the egg to the fry stage more than control males. Experiment 1 thus provides evidence that male bluegills increase their care in response to increases in their certainty of paternity.

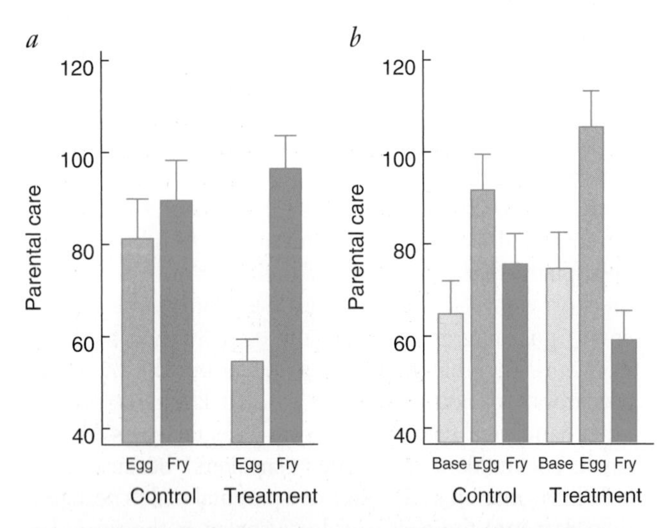

FIGURE 15.3 **Certainty of paternity influences level of parental care by male bluegill sunfish. (*a*) In experiment 1, males exposed to sneaker males during spawning (treatment group) display less parental care toward eggs than males not exposed to sneaker males (control group). Treatment and control groups do not differ during the fry phase, when males can reassess their paternity using chemical cues from the fry. (*b*) In experiment 2, males whose clutches had been manipulated (one-third of the eggs were exchanged with eggs from another male) do not differ from control males either before the manipulation (baseline) or the day after the manipulation (egg phase). However, treatment males do differ from control males after the eggs hatch and males can assess paternity using chemical cues from fry (fry phase). These data indicate that male bluegill sunfish increase (as in experiment 1) or decrease (as in experiment 2) their parental care in response to certainty of paternity. (From Neff 2003.)**

In the second experiment, Neff removed one-third of a clutch of eggs from parental males in the treatment group and replaced them with unrelated eggs from other parental males. Parental males in the control group were exposed to sham swaps of eggs (i.e., Neff went through the motions of swapping eggs, which entailed taking the eggs, swimming away from the nest, but then returning the very same eggs to the nest). He assessed the willingness of parental males to defend young before the egg swap (baseline), the day of the swap, and the day after the eggs hatched. Note that in this experiment, only the second cue (the chemical released by fry) was available to influence male behavior. There was no difference between treatment and control males in their willingness to defend young either before the swap (baseline) or the day after the swap (during the egg phase) (Figure 15.3*b*). However, after the eggs hatched (and the chemical cues of fry were available), treatment males decreased their level of defense more than controls. Experiment 2 provides evidence that male bluegills decrease their care in response to decreases in their certainty of paternity. Taken together, experiments 1 and 2 show that male bluegill sunfish adjust their level of care in response to changes in certainty of paternity.

Gender of Offspring

Sometimes patterns of parental investment are influenced by gender of the offspring. The manner in which parents distribute resources between the production of sons and daughters is called **sex allocation**. Parents can bias their allocation of resources in two main ways: they can either produce more offspring of one sex, or they can provide more (or better) resources to offspring of one sex. We focus on the second option concerning resources. Although in most animals resources are divided about equally between sons and daughters, there are some species in which parents distribute resources in a most biased fashion. Consider the case of the brown songlark (*Cinclorhamphus cruralis*) (Magrath et al. 2004, 2007). Songlarks are polygynous warblers endemic to Australia (polygyny is a mating system in which one male mates with more than one female during the breeding season; see later discussion of mating systems in this chapter). This species exhibits extreme sexual size dimorphism, with adult males being more than twice as heavy as adult females. Mother songlarks are largely responsible for feeding the young. At hatching, nestling males and females do not differ in body mass, but this changes dramatically over the next three weeks, when

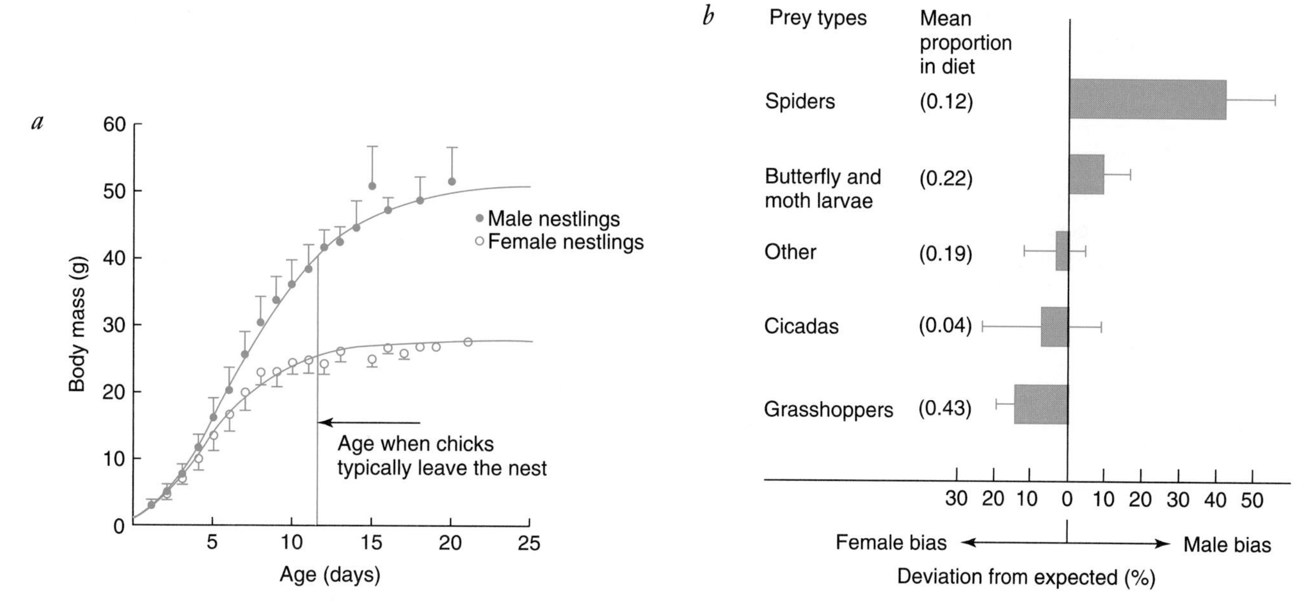

FIGURE 15.4 **Songlark mothers bias their parental investment toward their sons. (*a*) Although male and female nestlings have similar body masses at hatching, males become increasingly larger than females in the next few weeks. These data represent daily means (with standard deviations) for 180 male nestlings and 134 female nestlings. (*b*) Mothers not only deliver prey at higher rates to broods with more males, they feed male nestlings a higher quality diet than they feed female nestlings. For songlarks, a high-quality diet consists of more spiders and fewer grasshoppers. In the graph, the prey category "Other" represents prey that occurred very rarely in the diet or that could not be identified from the videotapes. (Modified from Magrath et al. 2004, 2007.)**

males become much heavier than females (Figure 15.4*a*). Indeed, by 10 days of age, male chicks are almost 50% heavier than their sisters! It turns out that male nestlings not only receive more prey than their smaller sisters, but they also receive higher quality prey. The diet fed to male nestlings contains more spiders and fewer grasshoppers than that fed to female nestlings (Figure 15.4*b*). This is important because spiders have certain amino acids that are important for early growth and development in birds. Also, compared with grasshoppers, spiders have less chitin (the indigestible carbohydrate of the exoskeleton). As you might imagine, raising male offspring is costly to mother songlarks; when experimental all-male and all-female broods were established, mothers raising all-male broods expended 27% more energy than those raising all-female broods. Given this cost in energy, what is the benefit of raising large male offspring? In a polygynous mating system, large body size is an important determinant of male reproductive success; it is typically less important for female reproductive success (Andersson 1994). Thus, by producing especially large sons who, when adult, will likely be successful in attracting and competing for mates, mothers ensure that their genes are well-represented in future generations. Producing large females would not yield as large a fitness benefit. We now turn our attention to the basic patterns of parental care displayed by animals.

STOP AND THINK

Given the above explanation for male-biased care in polygynous species, what patterns of parental investment would you expect in monogamous species (monogamy occurs when an individual male or female mates with only one partner per breeding season)? Would you predict that parental investment would be biased, and if so, in which direction? Explain your answer.

OVERALL PATTERNS OF PARENTAL CARE

Given the varied life histories of animals, we should not be surprised to find differences among taxa in the extent and pattern of parental care. Within vertebrates, for example, most teleost fishes (79% of families), frogs and toads (92% of genera), and lizards and snakes (97% of genera) show no parental care at all, yet all crocodilians and mammals display some form of parental care (Table 15.1) (Reynolds et al. 2002). Parental care is also typical of most birds. Indeed, only a small percentage of species (1%) dispenses with parental care by laying their eggs in the nests of others and relinquishing all care to the "host" parents (see discussion of brood parasitism later in this chapter) (Cockburn 2006).

TABLE 15.1 Patterns of Parental Care Exhibited by Some Groups of Vertebrates. Data Pertain to the Taxonomic Level Indicated in Parentheses After Each Group

Group	No care	Male-only care	Female-only care	Biparental care
Teleost fishes (422 families)	79%	10%	1%	3%
Frogs and toads (315 genera)[a]	92%	9%	9%	1%
Lizards and snakes (938 genera)	97%	0%	3%	0%
Crocodilians (21 species)	0%	0%	62%	38%
Birds (9456 species)[b]	1%	1%	8%	81%
Mammals (1117 genera)[c]	0%	0%	91%	9%

[a]Percentages add up to more than 100% because some genera show more than one form of care.

[b]Percentages add up to 91% because 9% of birds have three or more individuals providing care to young. This pattern, called cooperative breeding, is discussed in Chapter 19.

[c]Cooperative breeding also occurs in about 120 species of mammals.

Modified from Reynolds et al. (2002); data for birds from Cockburn (2006).

Who Provides the Care

In those taxa for which parental care is the norm, there are differences with regard to who provides the care (refer again to Table 15.1). For example, whereas female-only care is the most common form of care in mammals (91% of genera; Kleiman and Malcolm 1981), biparental care is the most common form of care in birds (81% of species; Cockburn 2006). Why the difference? In mammals, internal gestation and lactation necessitate a major parental role for the female and restrict the ability of the male to help in early offspring care. Male mammals cannot take over the duties of pregnancy and lactation for their mates, and rather than hang around during the period of early development, males, in most cases, seek mating opportunities elsewhere. In contrast to the extended period of internal development in mammals, birds develop outside the mother's body. Embryos, along with food in the form of yolk, are packed in eggs that develop largely in the external environment. Because male birds are just as capable as their mates at providing care, parental duties such as incubation, feeding, and guarding can be divided more equally between the sexes. Here we see that basic biological attributes of a lineage can constrain evolutionary possibilities.

A second example of taxon differences among vertebrates concerns whether male care occurs alone (male-only care) or in conjunction with female care (biparental care). When male care occurs in fishes and amphibians, it usually takes the form of solitary male care rather than shared male and female responsibilities. In contrast, paternal investment in birds almost always occurs in addition to maternal care. The higher frequency of biparental care in birds probably reflects the fact that parental care in this group usually involves both feeding and guarding the young. Two parents are better than one for feeding offspring, and thus biparental care evolves. Fish and amphibian parents, on the other hand, rarely feed their offspring, and parental duties consist largely of guarding, a task that may be performed almost as well by one parent as by two (Clutton-Brock 1991). Because male mammals cannot feed their offspring right after birth when the young are dependent on milk, early paternal care in mammals is always in conjunction with maternal care.

Having stated the broad generalization of the adequacy of solitary male care in fishes and amphibians, we should point out that there are exceptions to this rule, including some cases of biparental care in fishes. For example, cichlids, a lineage of freshwater teleosts in Africa, South and Central America, and India, display elaborate parental behavior that includes biparental care. Indeed, about 40% of cichlid genera display biparental care; for comparison, among teleosts overall, only about 3% of genera show biparental care (Goodwin et al. 1998; Reynolds et al. 2002). What is it about cichlids that has led to the evolution of biparental care? Biparental care may have evolved in some cichlids because their broods face intense predation pressure, and two parents may be necessary to successfully rear the young (McKaye 1977). Another explanation seems possible for the biparental discus (*Symphysodon discus*). Two days after fertilization, the fry of this cichlid hatch, aided by both parents who chew open the egg cases. The two parents then deposit their youngsters on aquatic vegetation, where the wrigglers dangle from threads as they live off their yolk supply. Within two or three more days, however, the brood can swim freely, and they attach themselves to their parents and feed off parental skin secretions (Figure 15.5). Biparental care may have evolved in this species because young that can feed off two parents may grow and develop more rapidly than those that have only a single

FIGURE 15.5 **Young cichlids of the biparental species *Symphysodon discus* graze on parental skin secretions.**

parent (Skipper and Skipper 1957). We see that in some fish—such as those exposed to extreme levels of predation or those whose care involves the feeding of young—biparental care may be necessary to ensure the survival and growth of the offspring.

Patterns of parental care among vertebrates are certainly diverse, and differences exist both among and within taxonomic groups. As we have seen, some of the differences among groups reflect basic biological differences, such as where the young develop and how they are fed. We have also seen that differences within groups can sometimes be understood in light of particular ecological conditions, such as intensity of predation. We continue our consideration of factors that influence the extent and type of parental care displayed by animals by looking at how mode of fertilization affects patterns of parental care.

Teleost fishes and anuran amphibians (frogs and toads) display four of the general categories of parental care in vertebrates: no care; male-only care; female-only care; and biparental care (Reynolds et al. 2002). (Note that in some animals young are cared for by more than two individuals, a system called cooperative breeding; see Chapter 19). Given their diverse forms of parental behavior, teleosts and anurans have often been the focus of studies on the evolution of parental care patterns. Both groups also have some members with external fertilization (fertilization in the external environment) and some with internal fertilization (fertilization within the female reproductive tract). This situation allows an examination of how mode of fertilization influences the evolution of patterns of parental care. We examine this relationship in ray-finned fishes, the larger group of bony fishes to which teleosts belong.

Several authors have suggested an association between external fertilization and male-only care and between internal fertilization and female-only care (Gross and Sargent 1985; Ridley 1978; Trivers 1972). To test this association in ray-finned fishes, Judith Mank and colleagues (2005) surveyed the literature and constructed a comprehensive phylogeny using phylogenies based largely on molecular data, though they also included some morphology-based phylogenies. The researchers then used the structure of the resulting phylogeny to examine the evolution of different modes of parental care (female-only, male-only, or biparental) in relation to several factors, including mode of fertilization. The results indicated that female-only, male-only, and biparental care have arisen repeatedly and independently in ray-finned fishes (Figure 15.6).

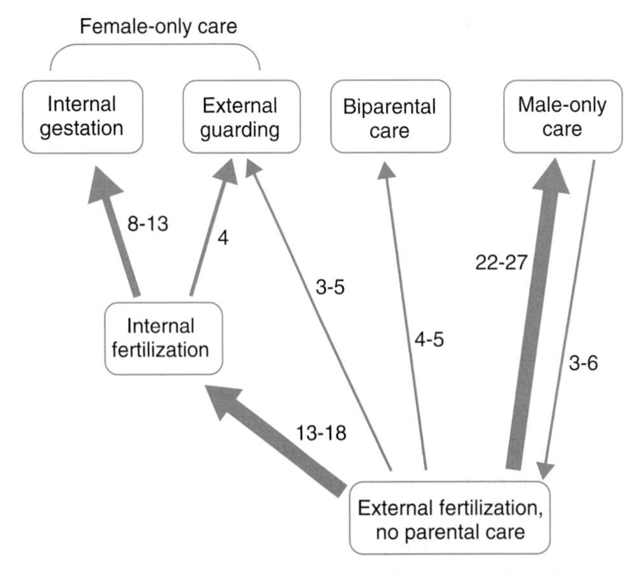

FIGURE 15.6 **Diagram showing the major independent evolutionary transitions among modes of parental care in ray-finned fishes. Size of arrows reflects the relative numbers of evolutionary transitions, and numbers next to arrows indicate the range of evolutionary transitions estimated. (Modified from Mank et al. 2005.)**

The most common pathway was from external fertilization and no care leading directly to male-only care. The second most common pathway was from external fertilization and no care to internal fertilization to female-only care. Occurring much less frequently were transitions from external fertilization and no care to either female-only care or biparental care.

But what is the source of the association between mode of fertilization and mode of parental care? One possibility is that it simply relates to the proximity of adults and offspring (Williams 1975). According to this hypothesis, external fertilization, particularly when it occurs in a territory defended by a male, would be associated with parental care by males. Because territorial males are almost always in the neighborhood of the eggs they fertilize, paternal behavior is likely to evolve. In contrast, with internal fertilization, it is usually the female that carries the embryos, and thus she is in the best position to care for the young when they enter the world. Paternal behavior is less feasible because fathers may not be in the vicinity when the eggs are laid or the young are born.

Sex-Role Reversals

In many animals, females provide more parental investment than males, and this is thought to explain sex differences in mating competition. Trivers (1972) suggested that the sex with greater parental investment (in this case females) becomes a limiting resource, essentially an object of competition among individuals of the sex investing less (here, males). The greater investment by females in their offspring also means that females are not available for other reproductive opportunities. In contrast, males, unencumbered by parental care, are able to take advantage of such opportunities. This results in two things. First, the **operational sex ratio**, defined as the ratio of potentially mating males to fertilizable females, becomes male-biased; from this we might also expect that males, as the more abundant sex, would be the more competitive sex (Emlen and Oring 1977). And second, the **potential reproductive rate**, defined as the maximum number of independent offspring that each parent can produce per unit of time, becomes greater for males than for females. Clutton-Brock and Vincent (1991) predicted that mating competition would be more intense in the sex with the higher potential reproductive rate. Taken together, the end result of greater investment by females in parental care is that males compete among themselves for access to females (a situation that favors large body size and aggressiveness in males), and females in turn are selective in their choice of mates. (Recall, however, from our discussion of sexual selection in Chapter 14, that patterns of parental investment are probably not the sole determinant of sexual selection.)

Sometimes, however, the sex roles are reversed, and the burden of parental care falls largely or entirely on the male. When parental investment by males is greater than that of females, we would expect males to be choosy and females to be competitive. We describe two species of birds known as jacanas that exhibit the phenomenon of sex-role reversal in parental care. These species exhibit full sex-role reversal (i.e., males provide almost all the parental care and are selective in their choice of mate, and females compete for access to males). There are other animal species that exhibit partial sex-role reversal whereby males are choosy and competitive or both sexes are choosy. With the notable exception of mammalian species, some level of sex-role reversal has been reported in insects and birds, as well as in crustaceans, fishes, and amphibians (see reviews by Eens and Pinxten 2000; Bonduriansky 2001). Overall, however, sex-role-reversed species make up a minority of animal species.

The northern jacana (*Jacana spinosa*) is a small bird of tropical marshes best identified by its spindly greenish legs and toes. Donald Jenni and Gerald Collier (1972) conducted an extensive study of the population dynamics, behavior, and social organization of individually marked jacanas in Costa Rica. They found that northern jacanas have a polyandrous mating system in which a given female is simultaneously paired with several males. At their study site, individuals breed year-round, and a female defends a large territory that encompasses from one to four male territories. In addition to helping her mates defend their individual territories, the female independently repels intruders from her entire territory. A female's critical role in territorial defense is matched by her dominant role in courtship. Because female northern jacanas keep harems of males, many females are excluded from breeding. The result is heightened competition among females for males. Morphological correlates appear to accompany the female jacana's male-like role in competition for males during courtship: breeding females weigh approximately 145 grams; adult males typically weigh in at a trim 89 grams.

Parental activities are the province of the male (Jenni and Betts 1978). Nest building and incubation are male duties exclusively, a division of labor underscored by the fact that only male jacanas have incubation patches (highly vascularized bare patches of skin on the belly). Once the precocial young have hatched, males provide most of the care to the chicks, including brooding and defending them. When necessary, females back up males in confronting potential predators; indeed, the female's larger body size seems to make her

FIGURE 15.7 Many jacana species exhibit sex-role reversal. Here a male African jacana cares for his chick and eggs.

more effective than the male at predator deterrence (Stephens 1984).

The pattern of sex-role reversal described for northern jacanas is also typical of several other jacana species (Figure 15.7). Observations of the wattled jacana (*Jacana jacana*) suggest that females no longer have incubation as part of their behavioral repertoire, but that they retain the ability to perform all of the behaviors involved in the care of chicks, though they rarely perform them. Indeed, during seven field seasons in Panama, Stephen Emlen and Peter Wrege (2004) never observed incubation by females, even when their mates disappeared when incubating eggs. Also, in 97% of the 242 observed broods, females did not associate at all with their dependent chicks.

DISPENSING WITH PARENTAL CARE—BROOD PARASITISM

Some birds exploit the parental behavior of other birds. These birds, known as brood parasites, lay their eggs in the nests of other birds and leave all subsequent parental care to the foster ("host") parents. Some brood parasites lay their eggs in the nests of conspecifics; these species are called **intraspecific brood parasites**. Included in this category are species such as cliff swallows, red-fronted coots, and wood ducks. In some cases of intraspecific brood parasitism, the brood parasite occasionally lays eggs in the nests of conspecifics, while still laying eggs in her own nest. In other instances, the brood parasite lays eggs in the nests of conspecifics and does not maintain a nest of her own; females in this group may resort to parasitizing nests because independent reproduction is impossible due to a shortage of nest sites or territories. Other brood parasites lay their eggs in the nests of other species and are called **interspecific brood parasites**; these species are also described as *obligate* brood parasites because they have no other reproductive option (i.e., they never build nests in which to lay eggs and raise their own young). The term *parasite* is used because of the apparent benefit experienced by the true parents and the harm that befalls the host parents, who typically experience a reduction in reproductive success (see below). We focus on interspecific (obligate) brood parasitism.

Approximately 1% of all species of birds lay their eggs in the nests of another species (Cockburn 2006). Obligate brood parasitism has arisen independently several times within birds, occurring in honeyguides, Old World cuckoos, New World cuckoos, viduine finches, cowbirds, and the black-headed duck (Sorenson and Payne 2002).

We mentioned that raising the young of brood parasites is costly to host parents. Damage to the host or its young may be directly inflicted by either the parasitic adult or its offspring (reviewed by Davies 2000). The adult, for example, must place its egg in the host's nest when the host is beginning to incubate its own eggs. Upon discovering a host's nest after incubation or hatching has begun, a female cuckoo may eat the eggs or kill the young of the host, causing the potential host to nest again. If the cuckoo is on time, however, she may throw out a host's egg before laying one of her own in the nest. To add insult to injury, cuckoos often lay their eggs from a perch above the nest, and their

FIGURE 15.8 A nestling parasitic cuckoo is evicting a host's egg from the nest.

thick-shelled eggs break the host's eggs when they strike them. Nestling cuckoos are renowned for methodically evicting eggs or young from the nest of their foster parents. By positioning itself under a nearby egg or nestling, a young cuckoo may lift a nestmate onto its back, slowly work its way to the edge of the nest, and nudge the host's egg or nestling to its death below (Figure 15.8). Nestling honeyguides employ an even more gruesome tactic to ensure full attention from their foster parents. At hatching, young honeyguides use their hooked bills to kill young with whom they share the nest.

An alternative strategy to the outright killing of foster siblings is simply to monopolize parental care. Brood parasites usually mature more rapidly than a host's young, thus gaining a critical head start in growth and development. Also, their huge mouths and persistent begging often elicit preferential and prolonged feeding by foster parents. The host's young may be no match for their larger, more aggressive foster sibling and may die from starvation, crowding, or trampling. Sometimes, however, parasitic young benefit by keeping a few of the host's young around. Parasitic brown-headed cowbird nestlings exhibit increased growth and survival when they share the nest with one or two host young as compared to when they are the sole occupant of the host's nest. Host parents apparently increase the rate at which they feed larger broods, and the young cowbird gets more than its fair share of the stepped-up feedings (Kilner 2003; Kilner et al. 2004). Also, killing host young might be a risky strategy for brood parasitic young if host parents, as part of their life history strategy, are more likely to desert single chick broods as compared to broods with two or more chicks, even when nests have not been parasitized (Broom et al. 2008).

In response to the devastating effects of brood parasitism, host species have developed ways to avoid being parasitized. The most common defenses are those used to reduce predation—the host species simply conceal their nests and defend them when they are discovered. Some hosts identify and remove the eggs (or young) of parasites from their nests. Recognition and rejection of parasitic eggs is a tricky business because the host could make a recognition mistake and erroneously throw out its own egg when parasitism has not occurred. Consistent with this view that egg recognition can be a challenging endeavor is the observation that hosts take longer to reject eggs that more closely resemble their own eggs (Antonov et al. 2008).

In response to the antiparasite devices of hosts, many aspects of the laying behavior of brood parasites are designed for better deception of hosts. For example, cuckoos time their laying for the late afternoon, when hosts are less attentive, and whereas some passerines spend at least 20 minutes laying an egg in their own nests, female cuckoos can deposit an egg in a host's nest in less than 10 seconds! Parasitic eggs or young often resemble those of the host species, a ploy that may reduce the chances of host parents recognizing them as different and rejecting them (Figure 15.9). Coevolution between the brood parasite and host may produce adaptations and counteradaptations of increasing complexity (Davies and Brooke 1988; Rothstein and Robinson 1998).

We turn now from the interactions between parents and offspring to those between mates.

a

b

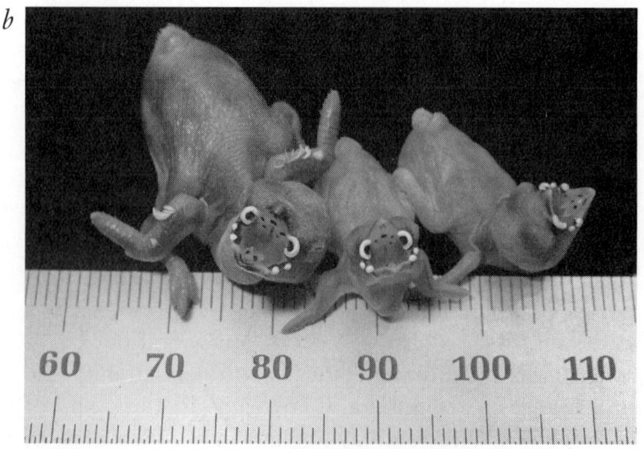

FIGURE 15.9 **Parasitic eggs or young often resemble those of host species. Presumably such resemblance reduces the chances that hosts will reject different-looking eggs or young from their nest. (*a*) The egg of the brood parasitic common cuckoo (upper left) closely resembles those of its host, the great reed-warbler, but is somewhat larger. (*b*) A nestling of the brood parasitic pin-tailed whydah (left) has elaborate mouth markings and begging behavior that resemble those of the young of its host, the common waxbill. The two smaller nestlings are common waxbills.**

STOP AND THINK

Viduine finches (whydahs and indigobirds; genus *Vidua*) are obligate brood parasites within the Family Estrildidae that typically parasitize a single host species. Hosts, such as waxbills and firefinches, are also members of Family Estrildidae, and all nestlings in this family have striking mouth markings. The mouth markings of parasitic nestlings precisely match those of host nestlings (look again at Figure 15.9*b*). Traditionally, the resemblance in mouth markings between parasite and host has been viewed as the outcome of a coevolutionary arms race, with host rejection of offspring that look different from their own as the driving force. Mark Hauber and Rebecca Kilner (2007) suggest a different scenario: perhaps the elaborate mouth markings have evolved through nestling competition for parental care. In a reversal of the traditional view that selection acts on parasitic young to look like their hosts, they further suggest that selection might be acting on host young to look like the more competitive parasitic young. How might you design an experiment to test whether the elaborate mouth markings of nestlings in Estrildidae function as signals of species identity (the traditional view) or as structures to attract parental provisioning?

MATING SYSTEMS

We have seen that parents and offspring often disagree over the details of their interactions. As we will see in the pages that follow, such conflict is not restricted to the parent–offspring relationship. In fact, adult males and females are often at odds over what constitutes an ideal mating relationship. Although the ultimate goal of reproduction for both sexes is to maximize fitness (the relative number of offspring that survive and reproduce), the reproductive success of males and females is constrained by different factors. In many species, a male's success is limited by access to females, while a female's is limited by access to resources (Davies 1991). Consequently, a male can often boost his reproductive success by mating with more than one female. In contrast, a female increases her reproductive success by gathering more resources, including male parental care and access to a high-quality territory. Generally speaking, then, males focus on mating effort, and females tend to emphasize parental effort. Therefore, we should not expect perfect parental harmony during the production of offspring. We would predict instead that each parent will attempt to maximize its own reproductive success, even if this is costly to the other (Trivers 1972).

CLASSIFYING MATING SYSTEMS

We begin our consideration of mating systems with some definitions. Mating systems can be defined based on the number of copulatory partners per individual per breeding season. Accordingly, monogamy occurs when a male and female have only a single mating partner per breeding season; polygyny when some males copulate with more than one female during the breeding season; and polyandry when some females mate with more than one male during the breeding season. Polygynandry, sometimes called promiscuity, occurs when both males and females mate with multiple individuals. Ecological factors, such as predation, resource quality and distribution, and availability of receptive mates, affect the need for parental care, the ability of males to monopolize females, and the ability of females to choose among potential suitors (Emlen and Oring 1977). Because such ecological conditions often vary within and between locations, considerable flexibility is usually associated with the mating patterns of a given species. For example, black howler monkeys, best known for their loud calls that can travel up to 3 miles through dense forest, were found to be primarily polygynous in a deciduous habitat and polygynandrous in a riparian (alongside a river) habitat, when studied in the Central American country of Belize (Jones et al. 2008). The lush riparian habitat, characterized as more productive and predictable than the deciduous habitat, also supported higher population densities of howler monkeys overall and larger groups of females, perhaps making the monopolization of females by a single male less likely. Our coverage of mating systems will focus on monogamy, polygyny, and polyandry.

At the start of our discussion we should say that regardless of mating system, sexual fidelity is hard to find among animals. Genetic analyses of parents and offspring often reveal that apparent social partners mate with other individuals: the social father is not always the genetic father of all the offspring. This has led researchers to distinguish between **social monogamy** (an exclusive living arrangement between one male and one female that makes no assumptions about mating exclusivity or biparental care) from **genetic monogamy** (an exclusive mating relationship between one male and one female). DNA studies of the offspring of 180 species of socially monogamous songbirds showed that only 10% of these species were genetically monogamous (Morell 1998). Although less studied than avian species in this regard, some species of socially monogamous fishes and mammals also engage in extra-pair fertilizations (Clutton-Brock and Isvaran 2006; Ophir et al. 2008; Sefc et al. 2008). Cuckoldry can be a problem for polygynous males as well. Indeed, extra-pair matings seem to be the rule rather than the exception.

Extra-pair matings have costs and benefits, especially if they result in fertilizations. A male's costs include the time and energy used in searching for receptive females other than his mate. While he's away, his primary mate may copulate with another male, which reduces his reproductive success. However, if he is successful in inseminating mates of other males, he can boost his reproductive success substantially.

Hypotheses for female benefits of extra-pair matings can be grouped into two categories: material or genetic. By mating with several males, a female may gain added assistance in raising her offspring. In some species, such as red-winged blackbirds, extra-pair males help defend the female's nest from predators and thereby improve her fledging success (Gray 1997). Females may also exchange copulations for a valuable resource, food, for instance. (In Chapter 14, the material benefits of nuptial gifts presented to the female at the time of copulation were discussed.) Extra-pair matings may provide sufficient sperm to fertilize all the eggs of a given female (Birkhead 1996).

There is also evidence that females seek extra-pair matings for genetic reasons, such as obtaining "good genes" for their offspring. Female reed warblers seek extra-pair matings with males with a larger song repertoire than their mate's. Although they don't get material benefits from the extra-pair mate, they apparently do get good genes for their offspring: postfledgling survivorship of the young is related to the genetic father (Hasselquist 1998; Hasselquist et al. 1996). Finally, females of some cooperatively breeding bird species engage in extra-pair matings to avoid mating with close relatives (recall from Chapter 14 that extreme inbreeding can be costly). Gray-crowned babblers live in social groups that consist of a dominant breeding pair and nonbreeding helpers (typically offspring from previous seasons). Sometimes, members of the dominant pair are related, and it is under these circumstances that extra-pair young are found in the nest. A recent field study in which DNA samples were taken from young and adult members of social groups showed that as relatedness of the dominant pair increased, so too did the proportion of extra-pair young in broods (Figure 15.10). In fact, most females that obtained extra-pair fertilizations did not obtain any fertilizations from their partner (Blackmore and Heinsohn 2008). Now we describe the major mating systems and the circumstances under which each is likely to evolve.

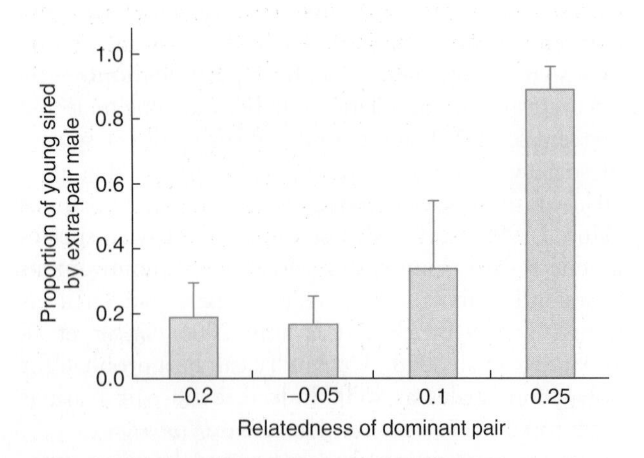

FIGURE 15.10 **In the cooperatively breeding gray-crowned babbler, the proportion of young sired by extra-pair males increases with relatedness of the dominant pair, suggesting that females of this species engage in extra-pair matings to avoid inbreeding. (From Blackmore and Heinsohn 2008.)**

MONOGAMY

Monogamy results when a male and female have only a single mating partner per breeding season. Because sperm from one male is often sufficient to fertilize a female's limited number of eggs, monogamy is often sufficient from the female perspective, but for males, confining copulation to a single female is a rather conservative means of ensuring genetic representation in the next generation. We might wonder, then, why a male would be monogamous and what ecological circumstances might favor monogamy over polygyny. Several hypotheses, which are not mutually exclusive, have been proposed (Kleiman 1977; Komers and Brotherton 1997; Whiteman and Côté 2004). Here, we consider a few of them.

Necessity of Biparental Care

When biparental care is necessary or at least important for offspring survival, monogamy may be favored. As we mentioned previously, biparental care is more common among birds than mammals. Nevertheless, males of some mammalian species do have important parental responsibilities, and the fitness of both mates depends on the male's parental investment. Monogamy is rare among mammals; it has been reported for only about 5% of species (Kleiman 1977). Some rodents are monogamous, and our examples come from this group.

Male California mice (*Peromyscus californicus*) provide extensive care to their young. Indeed, with the exception of nursing, fathers participate in all parental activities and to the same extent as mothers. The need for paternal care seems to have played a role in the evolution of monogamy in this species. Genetic analyses have shown that once paired, these mice never stray (Ribble 1991). Pups are typically born at the coldest time of the year, and because they cannot maintain their own body temperature until about 2 weeks of age, they need their parents' body heat to survive. Both parents take turns huddling over the pups to keep them warm. The importance of father presence to pup survival in California mice has been documented in the field and under challenging conditions in the laboratory (cold temperatures and when parents must wheel run for food); in both settings, experimental removal of fathers resulted in lower pup survival compared to control pairs (Table 15.2). Thus, in California mice, paternal assistance may improve offspring survival to such an extent that it is better for a

TABLE 15.2 Survival of Young California Mice When Fathers Are Present or Absent under Laboratory and Field Conditions.

Conditions	Pup survival (%) Father present	Father absent
Laboratory		
Cold temperatures[a]	90	55
Parents wheel run for food[b]	83	45
Field	81	26

[a]Mice were maintained in environmental chambers at 8.5–10.5°C; this temperature matched temperatures encountered in the field during the coldest months of the breeding season.

[b]Mice were maintained in foraging chambers equipped with a running wheel; adults were required to run 250 revolutions in a wheel for each pellet of food.

Laboratory data from Gubernick et al. (1993); field data from Gubernick and Teferi (2000).

male to remain with one female and invest in offspring than it is to seek additional mating opportunities (Gubernick et al. 1993; Gubernick and Teferi 2000).

Presence of the father can increase reproductive success in ways other than affecting survival of young—for example, it can shorten the interval between litters. A shorter interlitter interval can be critical to lifetime reproductive success, especially in short-lived species. Consider the case of the mound-building mouse (*Mus spicilegus*), a monogamous rodent found in Central and Eastern Europe. (By the way, the name reflects the mouse's habit of building enormous mud and vegetation mounds in which to spend the winter; once spring arrives, individuals disperse from the mounds to reproduce.) Christophe Féron and Patrick Gouat (2007) maintained pairs of mound-building mice in the laboratory and removed males from some pairs one day after delivery of their second litter (males and females mate shortly after the female gives birth, so even these pairs had time to mate). Males and females of other pairs were left together, and the researchers monitored production of third litters in the two groups. All females produced a third litter, but those whose mates were left with them produced their litter after a shorter interval than those whose mates were removed (Figure 15.11*a*). The extended pregnancies of females whose mates had been removed presumably were caused by delayed implantation of embryos, a response that may reflect the high energetic costs of concurrent pregnancy and lactation, along with the absence of help from their mate. Litter size at birth and survival of pups to weaning did not differ between the two groups.

In another experiment, Féron and Gouat (2007) continuously videotaped seven male–female pairs with young

a

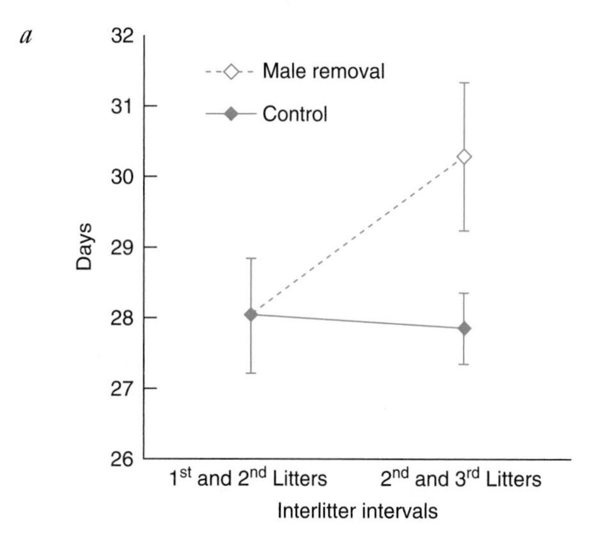

b

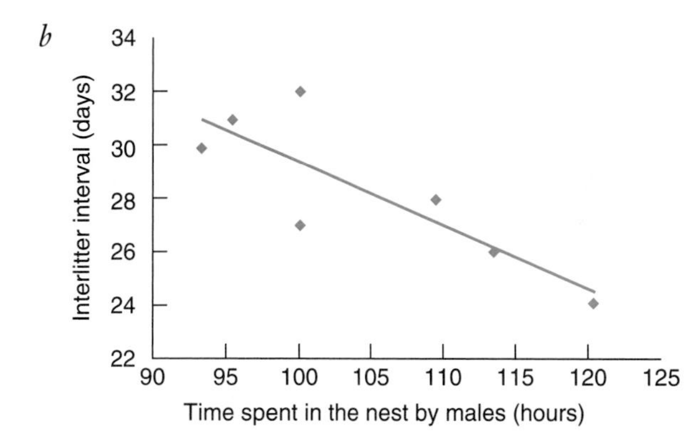

FIGURE 15.11 **Presence of fathers in mound-building mice leads to shorter intervals between litters. (*a*) Length of interval (mean ± SE, in days) between delivery of first and second litters and delivery of second and third litters for control pairs in which mates were left together and for pairs in which males were removed one day after the second litter was born. (*b*) Data for seven males videotaped over six days reveal a negative correlation between time spent by males in the nest and length of interval between litters: the more time that males spent in the nest with offspring, the shorter the interval until birth of their next litter. (From Féron and Gouat 2007.)**

for six days to determine the amount of time spent in the nest by individual males (taping began after postpartum sexual behavior had ended to ensure that time spent in the nest by males was devoted to care of young and not to mating activities). The data from this experiment show that male mound-building mice vary in the amount of time they spend in the nest with young: among the seven males, total time in the nest for six days ranged from about 94 to 120 hours. Especially interesting was the finding of a negative correlation between male time in nest and interlitter interval: the more time a male spent in the nest, the shorter the interval until the next litter (Figure 15.11*b*). Although the specific mechanism by which males shorten the interval between litters remains to be identified, this example shows that when male mound-building mice stay with their mate and provide paternal care to their offspring, both partners benefit by having shorter intervals between litters.

Distribution of Females

The distribution of females throughout the habitat is another factor that influences mating systems (Clutton-Brock and Isvaran 2006; Emlen and Oring 1977). Circumstances that make it difficult for a male to monopolize multiple mates will favor monogamy over polygyny. For example, when receptive females are uniformly distributed, perhaps because they defend exclusive territories, monogamy may evolve. Also, if females are widely dispersed, it might be more beneficial for a male to remain with a given female than to search endlessly for additional mates. With monogamy, a male is at least assured of access to one mate.

Consider the case of the symbiotic shrimp *Pontonia margarita*, which lives inside the mantle cavity of the pearl oyster. (The mantle cavity of oysters and other mollusks is a semi-internal cavity containing the gills and into which digestive, excretory, and reproductive systems discharge their products.) Pearl oysters are found at low densities in the Pacific Ocean from the Gulf of California to off the coast of northern Peru, and as permanent living quarters, they are quite small, even for shrimp. The maximum shell height of pearl oysters is about 14 cm, and the average carapace lengths of adult male and female shrimp are around 11 and 14 mm, respectively (the carapace is the hard covering of chitin on the back of the shrimp). Based on the relatively small size and scarcity of their hosts, J. Antonio Baeza (2008) predicted that *P. margarita* would be monogamous. He collected 68 oysters and recorded the number of shrimp per host, sex of each shrimp, and the reproductive condition of all females. His data revealed that the number of shrimp per host varied from zero to two (most oysters with one or no shrimp were small), with two being most common (Figure 15.12). All shrimp pairs consisted of a male and a female, and females were either close to spawning (eggs visible in ovaries) or brooding embryos of various stages. Body sizes of shrimp and host oysters were closely correlated, suggesting the particular associations are long lasting (frequent switching among hosts, as occurs in some symbiotic crustaceans, would result in a loose relationship between shrimp size and oyster size). These data are consistent with social monogamy for *P. margarita*. Although genetic data are not yet available, Baeza suggests that genetic monogamy may occur in *P. margarita* for at least two reasons. First, moving between hosts in search of extra-pair copulations would likely be rare in these shrimp because they live in predator-rich waters and are especially vulnerable when away from their hosts. Second, shrimp that temporarily leave their host may encounter difficulties trying to reenter their original oyster. Here, then, we have an example where several factors—scarcity of host, small size of host, and potentially high costs of searching for additional mates—seem to make monogamy the best option.

a

b

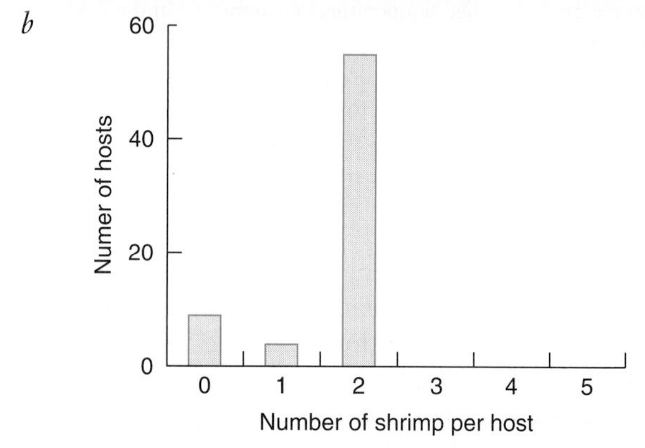

FIGURE 15.12 **Shrimp that live within oysters may be monogamous because their hosts are small and scarce. (*a*) A pair of symbiotic shrimp retrieved from their host, a pearl oyster. (*b*) A survey of 68 oysters revealed that most contained two shrimp, and only a few had one or no shrimp. All pairs were male–female pairs. (From Baeza 2008.)**

Mate Guarding

Sometimes monogamy evolves because a male can effectively guard only one female from other males. Kirk's dik-dik (*Madoqua kirkii*) is a small antelope (only about 15 inches tall) that lives in dry scrub in many parts of Africa. Dik-diks form monogamous bonds that last for several years, if not for an entire lifetime. Unlike many species, dik-diks seem to be faithful to their mates: genetic analyses revealed no evidence of extra-pair paternity (Brotherton et al. 1997). Given the rarity of monogamy among mammals, we might wonder why such devotion has evolved in dik-diks.

Several hypotheses can be ruled out. The need for biparental care can't be important in the evolution of monogamy in this species because paternal care is absent—the male does not defend resources or reduce predation risk, and there isn't a risk of infanticide (Brotherton and Rhodes 1996). Furthermore, at least some monogamous males defend territories that could support more than one female. In a rare population in which some males were polygynous, the territories of the polygynous males were not of higher quality than those of monogamous males. And it seems unlikely that monogamy evolved because constant male attention is necessary to keep females from wandering off the territory and mating with another male. Pair mates spend only about 64% of their time together, and so there would be ample opportunity to wander away.

Instead, it seems that monogamy evolved because the costs of guarding more than one female from intruding males would be too great. The male guards his female by preventing other males from knowing when she is in estrus. He does this by covering up the scent of his female's territorial markers—piles of dung—by scratching dirt over them and then defecating on top of them. It is also essential that a male advertise his territorial ownership because rivals bent on filling vacant territories are never far off. This is accomplished by marking his territorial borders with the scent from glands under his eyes (Komers 1997). If a male were to try to overmark the scent of two females on separate territories, he might fail to mark his territory sufficiently. As a result, he could lose ownership in the intense competition for vacant territories. In the few instances in which a male dik-dik succeeded in polygyny, the females shared a territory. The female accepts being guarded because an extra-pair mating might incite a fight between her male and the rival, which could prove harmful to her or her offspring (Brotherton and Manser 1997).

POLYGYNY

Recall that **polygyny** occurs when one male mates with more than one female during a breeding season. Natural selection's ledger sheet shows that polygyny has both costs and benefits for males and females. Because a male can generally fertilize more eggs than a female can produce, a male usually benefits from polygyny by producing more offspring (if paternal care is not required). As males maximize their reproductive output through multiple matings, polygyny results. Possible costs to a male include an increased chance of cuckoldry, because he is not guarding each female from other males, and the costs associated with achieving dominance or defending resources or territories.

Polygyny has several significant costs for females. For example, males in such cases usually do not help rear the young. In species in which the males do provide some parental care, it is divided among the offspring from more than one female or sometimes provided only to the first female, the so-called primary female. Females must also share essential resources, such as nest sites or territories. Also, the activity around these areas may attract predators, and other receptive females may increase the competition for commodities such as food.

Hypotheses for Females' Acceptance of Polygyny

Why would a female accept the costs of polygyny? There is no simple answer to this question. Indeed, the answers may vary among species and even within a species. Of the several hypotheses proposed to explain why females accept the costs of polygyny, we consider two—the polygyny threshold hypothesis and the "sexy son" hypothesis.

Polygyny Threshold Hypothesis Polygynous matings will be advantageous to females when the benefits achieved by mating with a high-quality male and gaining access to his resources more than compensate for costs. In other words, a female may reproduce more successfully as a secondary mate on a high-quality territory than as a monogamous mate on a low-quality territory. Jared Verner and Mary Willson (1966) coined the phrase *polygyny threshold* to describe the difference in a territory's quality needed to make secondary status a better reproductive option for females than primary status.

Gordon Orians (1969) elaborated on Verner and Willson's ideas and reasoned that if polygyny is always to a male's advantage and yet does not always occur, then

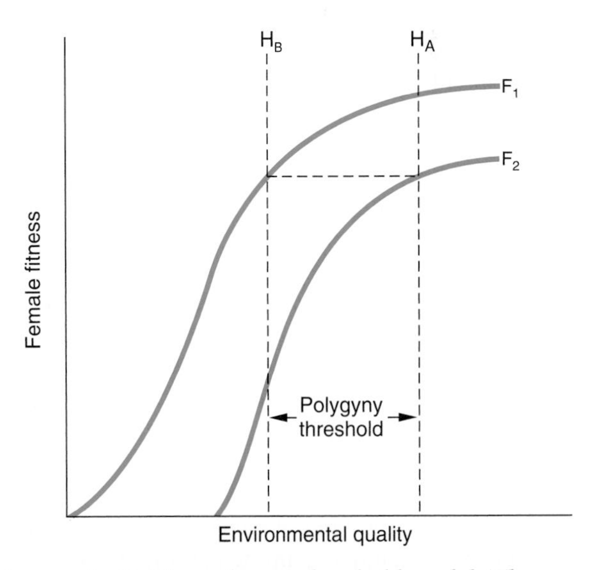

FIGURE 15.13 **The polygyny threshold model. The polygyny threshold is the difference in quality between H_A and H_B that would favor polygynous matings.**
F_1 = fitness curve for monogamous females
F_2 = fitness curve for secondary females
H_A = highest quality breeding habitat
H_B = marginal breeding habitat
(From Wittenberger 1979; modified from Orians 1969.)

the circumstances under which it does occur must be those in which there is some advantage to the female. He reasoned that females should join a harem when this decision confers greater reproductive success than monogamous alternatives. According to this argument, then, the average reproductive success of females should not decrease as the harem's size increases. Figure 15.13 illustrates this model's way of relating differences in a territory's quality to a female's choice of already mated versus unmated males.

Experimental field evidence in support of the polygyny threshold model has been obtained for a population of red-winged blackbirds in Ontario, Canada. First, one study established that all things being equal, females in this population prefer unmated males to already mated males when settling to breed in early spring (Pribil and Picman 1996). In this study, the researchers established 16 pairs of adjacent territories; the territories within pairs were similar in size, water depth (a critical determinant of territory quality; see below), and harem size the previous breeding season. Within each pair, one territory was defended by a monogamously mated male and the other by an unmated (bachelor) male. The researchers maintained these distinctions by removing females when necessary (e.g., females that had already settled in the bachelor territory before the start of the experiment were removed, as were any additional females that settled on the territory of the monogamously mated male). They then recorded where newly arriving females settled. Within about two weeks all territory pairs had been settled by new females. Female red-winged blackbirds were unanimous in their choice of where to settle: in all 16 pairs of territories, the first new female settled on the bachelor territory. Thus, females in this population prefer unmated males to mated males. A second study, using a similar experimental design, revealed that this basic preference could be reversed if the territories of mated males were made superior to those of unmated males. For female red-winged blackbirds, reproductive success is most closely correlated with water depth, with nests over deep water being more successful (presumably because such nests are harder for ground predators to detect and reach). Stanislaw Pribil and William Searcy (2001) manipulated pairs of adjacent territories. Within a pair of territories, one was designated to be that of a monogamously mated male (one female present) and the other to be that of an unmated male (no female present). As before, the pairing status of each male was maintained by trapping and removing any females that arrived on the territory before the start of the experiment. The researchers placed nesting platforms in open water in the territories of mated males and on dry land in the territories of unmated males; in this way, the territories of mated males were made superior and those of unmated males inferior. Each of the 19 territory pairs was then visited twice daily by the researchers to monitor presence of new females on the territories. Sixteen of the 19 territory pairs were settled by new females. Preference in 2 of the 16 settled pairs could not be ascertained because the territories of the mated and unmated males were settled simultaneously (i.e., an observer arrived to find both territories occupied). However, in 12 of the remaining 14 territory pairs, a female settled first on the territory of the mated male. Thus, in support of the polygyny threshold model, superior territory quality was shown to reverse the basic preference for pairing with an unmated male in this population of red-winged blackbirds.

"Sexy Son" Hypothesis According to this hypothesis, access to good genes for her offspring compensates a female for the costs of polygyny. A female may benefit from mating with an already mated male if her sons inherit the genes that made that male attractive. Her sexy sons will presumably provide her with many grandchildren, so the female's lifetime reproductive success may be enhanced by choosing to mate with a male that is attractive to many females (Weatherhead and Roberston 1979, 1981). According to this hypothesis, then, a female that chooses an already mated male may benefit *indirectly* if the good genes she acquires for her offspring boost their survival and reproductive success.

Thomas Huk and Wolfgang Winkel (2006) examined their 31 years of data on a German population of pied flycatchers (*Ficedula hypoleuca*) to see if there was evidence supporting the "sexy son" hypothesis. Their analyses showed that direct reproductive success (number of fledglings) was lower in females mated to polygynous males (here, males with two mates) than in females mated to monogamous males. This was especially true for secondary females without male assistance. Polygynous males of this species give priority to the young of primary females (females whose young hatch first) and show either reduced or no care to young of secondary females (females whose young hatch later than those of the primary female).

However, maybe it is not so bad being mated to a polygynous male: perhaps the sons of polygynous males are particularly attractive and produce many offspring of their own. In other words, might female pied flycatchers mated to polygynous males experience delayed compensation, if the sons they produced were especially attractive and produced many offspring? In order to test whether the "sexy son" hypothesis applied to their study population, the researchers compared the reproductive success of males hatched in monogamous broods with that of males hatched from polygynous broods; polygynous broods were further classified as primary or secondary. Males hatched in primary polygynous broods produced more fledglings over the course of their lifetimes than did males from monogamous broods. However, males hatched in secondary polygynous broods produced no more fledglings than those hatched

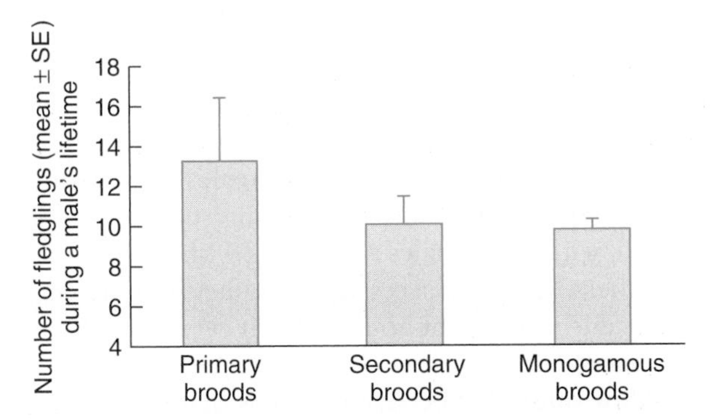

FIGURE 15.14 **Data from long-term monitoring of a population of pied flycatchers reveal that sons of secondary females do not differ from sons of monogamous females in number of fledglings produced over a lifetime. These data do not support the sexy son hypothesis. (From Huk and Winkel 2006.)**

in monogamous broods (Figure 15.14). These results indicate that good genes from fathers can increase the fitness of their sons, but only if fathers do not reduce paternal care (such as in the broods of primary females). The data for males hatched in secondary broods indicate that females do not benefit by pairing with an already mated male, even if he is sexy and his sexiness is heritable. For secondary females, then, choosing an already mated male with good genes does not compensate for the negative impact of his reduced paternal care. These data for secondary females do not support the "sexy son" hypothesis.

Types of Polygyny

There are three types of polygyny—female defense, resource defense, and lek—distinguished on the basis of what males are defending. We describe each in turn.

Female Defense Polygyny In female defense polygyny, a male defends a harem of females. This type of polygyny occurs when females live in groups that a male can easily defend. In some species, female gregariousness may be related to cooperative hunting or increased predator detection; in other species, clumping is more directly related to reproduction. For example, female elephant seals (*Mirounga angustirostris*) become sexually receptive less than one month after giving birth. Each year pregnant females haul themselves onto remote beaches of Año Nuevo Island, California. Female gregariousness, a shortage of suitable birth sites, and a tendency to return annually to traditional locations result in the formation of dense aggregations of receptive females. Under these crowded conditions, a single dominant male, weighing in the vicinity of 8000 pounds and sporting an enormous overhanging proboscis (Figure 15.15), can monopolize sexual access to 40 or more females

FIGURE 15.15 **A male elephant seal. Males of this species compete for access to, and defend, large numbers of females at birthing sites. This is an example of female defense polygyny.**

(LeBoeuf 1974). This male defends his harem against all other male intruders in bloody, and sometimes lethal, fighting.

Resource Defense Polygyny Resource defense polygyny occurs when males defend resources essential to female reproduction (e.g., nest sites or food) rather than defending females themselves. In other words, even though receptive females do not live together, a male can monopolize a number of mates by controlling critical resources. Conditions typical of resource defense polygyny include the following: (1) quality of the monopolized resource reflects male quality (i.e., higher quality resources should be defended by higher quality males); (2) females prefer males with resources to those without resources; and (3) males with resources have higher mating and reproductive success than those without resources (Emlen and Oring 1977; Thornhill 1981).

In some cases, there is good evidence that a female's choice is based on the quality of resources controlled by a male. An example is provided by several American species of scorpionflies (Mecoptera: Panorpidae), in which males fiercely defend the area around a dead arthropod (Thornhill 1981). Standing next to his nutritious find, a male will disperse a sex attractant and display with rapid wing movements and abdominal vibrations (Figure 15.16). If a female approaches, copulation is the fee that she must pay to gain access to this food. Thus, while males enjoy the benefits of mating with several females, females enjoy a hearty meal that may enhance their reproductive effort. In these American species, larger males obtain larger arthropods, while small males that are unable to capture large prey resort to less successful tactics such as stealing

FIGURE 15.16 **A male scorpionfly. American scorpionflies display resource defense polygyny in that a male will defend the area around a dead insect and females must mate with him to gain access to the meal.**

copulations or presenting salivary secretions (another option for a nuptial gift).

Based on these findings for several American species of scorpionfly, we might expect resource defense polygyny to characterize all species of *Panorpa*, but this is not the case. In experiments designed to be similar, if not identical, to those performed with American species, Merle Missoweit and Klaus Sauer (2007) obtained the following results for two European species: (1) males and females were equally adept at monopolizing food items, (2) males did not preferentially occupy large food items, (3) there was no relationship between male quality and resource quality, and (4) males that provided salivary secretions obtained more matings than those that provided dead arthropods. Thus, we see that even closely related species may differ dramatically in patterns of mating.

Lek Polygyny The third category of polygyny, lek polygyny, occurs when males defend "symbolic" territories that are often located at traditional display sites called leks (Figure 15.17). Males of lek species do not provide parental care and defend only their small territory on the lek, not groups of females that happen to be living together nor resources associated with specific areas. Females visit these display arenas, select a mate, copulate, and leave. This extreme form of polygyny occurs when environmental factors make it difficult for males to monopolize females either directly, as in female defense polygyny, or indirectly, as in resource defense polygyny.

In Chapter 14 we introduced you to sandflies of the *Lutzomyia longipalpis* species complex, and we return to them for an example of lekking behavior (reviewed in Jones and Quinnell 2002). Recall that at night, male sandflies gather on the back of a vertebrate host, often a chicken, where they defend small (radius of 2 cm), mobile territories by bumping and jostling one another with their abdomens. Females visit these nocturnal leks, evaluate several males, but copulate with just one. There is considerable agreement among females in their choice of mate, and as a result, male mating success is highly skewed, with some males copulating with many females and others being unsuccessful. Sandflies do not remain on leks during daylight hours but may return to leks over consecutive nights.

How might lek behavior such as that seen in sandflies have evolved? Several hypotheses have been suggested. These have been divided into two groups, based on potential benefits for males or females. From the male perspective, group, rather than individual, displays may increase the signal range or the amount of time that signals are emitted (e.g., Lack 1939; Snow 1963). Or males may aggregate because they require specific display habitats that are limited and patchily distributed (e.g., Snow 1974). Leks may provide protection from predators through increased vigilance—with more eyes watching, a predator would have a harder time sneaking up on them (e.g., Wiley 1974). Leks may also serve as information centers, where males exchange the latest news on good foraging sites (e.g., Vos 1979). Males may gather near "hot spots," areas in which females are most likely to be encountered (e.g., Bradbury et al. 1986; Ryder et al. 2006). Finally, leks may arise because less successful males generally have better mating chances in the vicinity of highly successful males. Because certain males are extremely successful at attracting females, other, less successful males gather around these "hotshots" and obtain more copulations than they would have had they displayed by themselves (Beehler and Foster 1988).

From the female perspective, large groups of males may facilitate mate choice (Alexander 1975; Bradbury 1981). After all, it might be easier to distinguish between superior and inferior males when comparison

FIGURE 15.17 **Two black grouse males displaying on a lek.**

shopping is possible. Mating within a group of males may reduce the vulnerability of females to predation since any predator might be distracted by so many displaying individuals (Wittenberger 1978); or if males aggregate in a less desirable habitat, lek mating may reduce competition between the sexes for resources (Wrangham 1980). Finally, females living in large, unstable, mixed-sex social groups may visit leks to avoid harassment from males in their group (Clutton-Brock et al. 1992). All of these hypotheses could be useful as explanations for the development of leks but for now are best regarded as only possible explanations.

POLYANDRY

Polyandry occurs when a female has more than one mate during the breeding season. Recall from Chapter 14 that Bateman's principle (1948) suggested that male, but not female, reproductive success exhibits substantial increases with number of mates. As the thinking went, because most males offer only sperm in return for copulation and because the sperm from one male is often sufficient to fertilize all the eggs of a given female, then females would have little to gain by mating with more than one male, especially given the costs associated with mating (exposure to predators, disease, and aggressive males). We now know, however, that mating with multiple males is widespread across taxonomic groups (Jennions and Petrie 2000; Rivas and Burghardt 2005; Zeh and Zeh 2001). We also have evidence from an increasing number of species that female reproductive success, like male reproductive success, can significantly increase with number of mates. For example, if copulation opens the door to critical resources or male parental assistance, then mating with several males may result in reproductive benefits for females. In insects, polyandry has been shown to increase the number of eggs laid and the hatching success of those eggs (studies reviewed by Arnqvist and Nilsson 2000). Explanations for clutch size effects include increased nutrients passed to females of species in which males provide nuptial gifts and increased receipt by females of hormonal stimulants in male ejaculates. Possible explanations for hatching success effects include avoidance of sperm depletion and genetic mechanisms such as increased genetic diversity among progeny and reduced risk of fertilization by genetically incompatible sperm (Simmons 2005; Yasui 2001; Zeh and Zeh 2001). Let's take a closer look at production of diverse offspring as a possible explanation for the evolution of polyandry.

Honeybees (genus *Apis*) are social insects that exhibit polyandry: the reproductive female of a colony (known as the queen) mates with multiple males. Reproductive males are called drones, and infertile females, known as workers, labor on behalf of the queen. Workers are closely related to the queen, so their helpful (altruistic) behavior can be understood from the standpoint of kin selection (see Chapter 19 for further discussion of social insects). This brief introduction to honeybee society raises an interesting question: if high levels of relatedness are the basis for altruistic behavior within honeybee colonies, then why would a queen mate with multiple males, a behavior that generates genetic diversity? Are there significant benefits associated with a genetically diverse workforce? Heather Mattila and Thomas Seeley (2007) set out to answer this question by establishing founding colonies of honeybees that were either genetically diverse (queens artificially inseminated, via an instrument, with sperm from 15 drones) or genetically uniform (queens inseminated

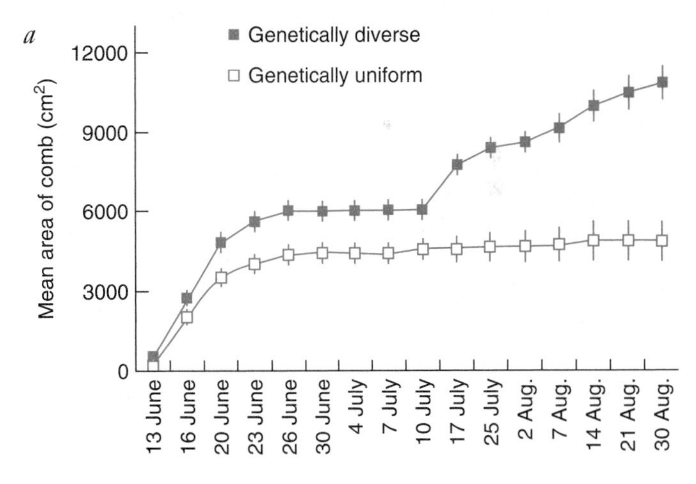

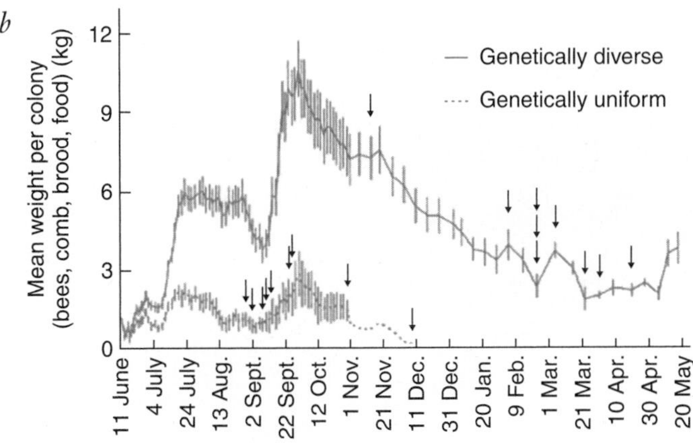

FIGURE 15.18 **Female honeybees mate with multiple males, and this results in genetic diversity within colonies. The benefits of such diversity have been demonstrated by experimentally creating founding colonies that were either genetically diverse or genetically uniform and by monitoring their progress. (*a*) Diverse colonies built more comb than did uniform colonies. (*b*) Diverse colonies also weighed more and survived longer than uniform colonies. Each arrow represents death of a colony. (From Mattila and Seeley 2007.)**

with the same volume of sperm but from one drone). They then monitored several features of colony development during the very challenging time of colony founding. Genetically diverse colonies were more efficient than genetically uniform colonies at building a comb (the latticework of cells that form the internal structure of the nest) (Figure 15.18*a*). Genetically diverse colonies also weighed more than genetically uniform colonies, and whereas all uniform colonies died by mid-December, several genetically diverse colonies survived the winter (Figure 15.18*b*). When compared with genetically uniform colonies, genetically diverse colonies also foraged at higher rates, produced more workers, and reared more drones. Here, we see that the genetic diversity generated by polyandry results in many benefits for honeybees.

SUMMARY

Parental care is one component of the overall life history of a species. Because animals have limited time, energy, and resources to devote to reproduction, evolutionary "decisions" must be made about the amount of care and who should assume parental responsibilities. Evolutionary conflicts over parental investment include sexual conflict, intrabrood conflict, and interbrood conflict. The issue of who provides the care appears to be related to factors such as mode of fertilization (male care being associated with external fertilization and female care with internal fertilization) and phylogenetic history (in mammals parental care usually rests with the female, whereas in birds parental duties are typically partitioned somewhat equally between the sexes). Parental behavior is not displayed by all species. For example, avian brood parasites dispense with all parental responsibilities by laying their eggs in the nests of other species.

Conflicts of interest are not restricted to parent–young interactions, and in fact such conflicts characterize most social behavior. In the case of mating relationships, males usually produce more offspring by seeking additional mates. In contrast, females tend to emphasize parental effort rather than mating effort and can usually produce more offspring by gaining male parental investment. The disparity between the sexes in parental investment interacts with ecological factors such as predation, resource quality and distribution, and the availability of receptive mates to shape the mating system of a species.

Mating systems can be defined on the basis of the number of individuals that a male or female copulates with during a breeding season. According to this system, monogamy occurs when a male and female have only a single mating partner per breeding season; polygyny when some males copulate with more than one female during the breeding season; and polyandry when some females mate with more than one male during the breeding season. Polygynandry occurs when both males and females mate with multiple partners.

Extra-pair matings are common regardless of the mating system. These have different costs and benefits for males and females and will affect the evolution of mating systems. Potential benefits to males include an increased number of offspring. Females gain material benefits, such as food or help from the extra-pair male in raising offspring, or genetic benefits, such as fertility insurance or high-quality genes.

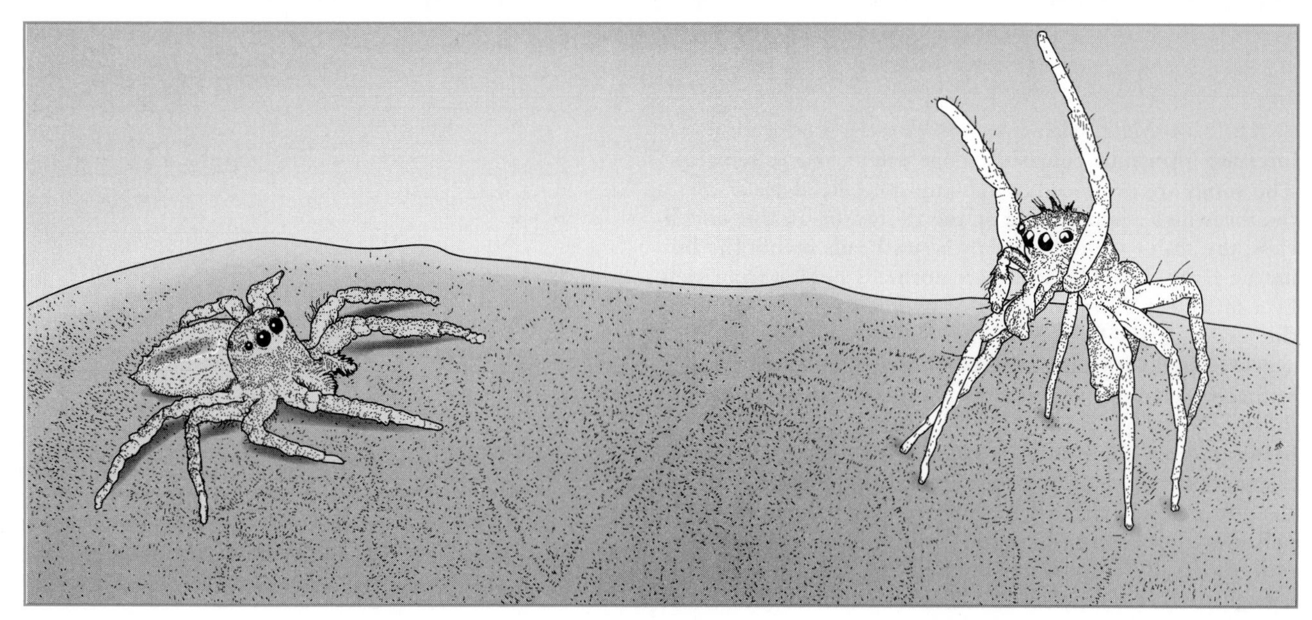

16

Communication: Channels and Functions

Sometimes important messages are whispered rather than shouted. This seems to be the case for Asian corn borer moths (*Ostrinia furnacalis*). Males of this species, when very close to a female (< 2.75 cm away), produce extremely low-intensity ultrasonic courtship songs (sounds with frequencies above the range of human hearing are described as ultrasonic; see section on audition). The songs are produced during wing strokes when males rub specialized scales on their forewings against those on their thorax (Figure 16.1). The scales are not found on females. These low-intensity courtship songs suppress escape behavior of the female, thus facilitating mating. The private nature of the courtship songs may reduce the risks of eavesdropping by rival males and predators (Nakano et al. 2008).

This brief foray into the confidential sexual communications of moths raises some general questions about animal communication. How do we define communication? What sensory channels do animals use when communicating? How are signals shaped by competitors, predators, and the environment? And what messages are conveyed?

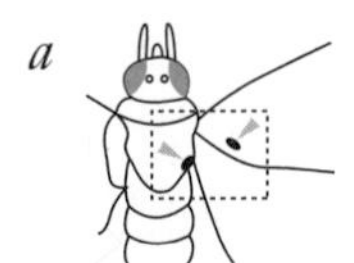

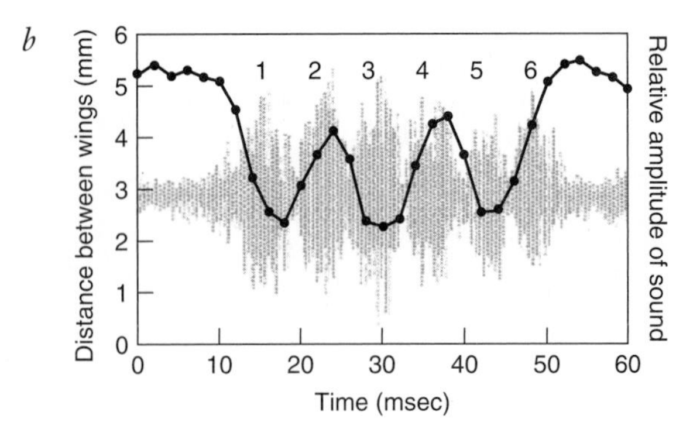

FIGURE 16.1 **Male Asian corn borer moths produce low-intensity ultrasonic courtship songs when close to females. The songs are produced by rubbing specialized scales on the forewings against those on the thorax. (*a*) In this dorsal view, the scales are indicated by arrow heads within the box on the right side. The scales are normally hidden from view by a small leaf-like structure that overlaps the base of the forewing. (*b*) The production of ultrasonic song is synchronized with wing strokes. Shown in light orange are six pulses of sound recorded with a special microphone during courtship. The black line traces the distances between distal edges of the right and left forewings; distances decrease during the upstroke and increase during the downstroke. These data show that the pattern in which pulses are generated corresponds with the up and down movements of the forewings (pulses 1, 3, and 5 coincide with upstrokes and pulses 2, 4, and 6 correspond with downstrokes). (Modified from Nakano et al. 2008.)**

THE DEFINITION OF COMMUNICATION

Defining *communication* is surprisingly difficult. One broad definition holds that communication occurs when a sender produces a signal that contains information (Batteau 1968). This signal is transmitted through the environment and detected by a receiver, who then interprets the signal and decides how to respond.

This definition covers most of what we would normally consider communication—the leaping courtship dance of a male crane, the growl and bared teeth of a wolf—but it also encompasses interactions that we might be more reluctant to include. For instance, the sound of a mouse rustling in a runway under dried leaves on its way to its food stores may be heard by an owl. The owl makes use of the information about the location of the mouse by swooping down to grab the sender. Should we call this communication? By the above definition it is, but this strikes many behaviorists as problematic. Clearly, the function of the noise, from the mouse's perspective, was not to alert the owl.

Many researchers add another element to the definition: the sender benefits from the transmission by altering the behavior of the receiver. This definition excludes examples such as the rustling mouse; instead, we would call the rustling a **cue**. In contrast, a courtship dance, a song, a brightly colored feather crest, or some other feature that does fit this more precise definition is called a **signal**. A stereotyped sequence of behaviors that has a signaling function is called a **display**.

Must the receiver also benefit from the information being transferred? Logically, *on average*, receivers must benefit from paying attention to a particular signal. If they did not, selection would favor receivers that ignored signals, and eventually senders would stop producing them. Therefore, some authors include in their definition of communication the idea that receivers also benefit from the transmission of information. The problem that arises with this definition is that receivers do not *always* benefit from receiving a signal. Sometimes senders manipulate receivers by sending dishonest signals, as we will explore in detail in the next chapter.

For our purposes, we'll use the definition that **communication** occurs when information is transferred from the sender to the receiver, and that it benefits the sender, on average. Keep in mind, however, that this definition is not ideal; as Bradbury and Vehrencamp (1998) point out, "Evolution hates definitions."

CHANNELS FOR COMMUNICATION

Communication can involve any of a variety of sensory channels—vision, audition, chemical, touch, and electrical fields. Each channel has both advantages and limitations (Table 16.1). As we will see, the channel used for a particular signal will depend on the biology and habitat of the species, as well as the function of the signal.

VISION

Visual signals have two obvious properties. The first is ease of localization. If the signal can be seen, the location of the sender is known. For example, when a male bird is engaging in a visual display to attract a mate, there is never any doubt about his precise location. The receiver can see him and, therefore, respond to him in terms of his exact location, as well as his general pres-

TABLE 16.1 Characteristics of Different Sensory Channels for Communication

	Type of signal				
Feature	Visual	Auditory	Chemical	Tactile	Electrical
Effective distance	Medium	Long	Long	Short	Short
Localization of sender	High	Medium	Variable	High	High
Ability to go around obstacles	Poor	Good	Good	Good	Good
Speed of transmission	Fast	Fast	Slow	Fast	Fast
Complexity	High	High	Low	Medium	Low
Durability	Variable	Low	High	Low	Low

ence. The second property is rapid transmission and rapid fade-out time. The message is sent literally at the speed of light, and as soon as the sender stops displaying the signal is gone. For instance, if the displaying bird suddenly spots a hawk and takes cover in a nearby hedge, its original position will not be revealed by any lingering images.

Visual systems can provide a rich variety of signals. This diversity is possible because of the number of stimulus variables that can be perceived by most animals. These include brightness and color, as well as spatial and temporal pattern that can be altered by the animal's movements and posturing. The presence and relative importance of one type of visual stimulus or another can be influenced by environmental conditions. For example, during agonistic displays aimed at conspecifics (or divers who invade their personal space), many shark species strongly depress both pectoral fins and hold them for prolonged periods in this symmetrical downward position (Martin 2007). Many species of sharks inhabiting clear water environments, such as near shore and reef habitats, have conspicuous markings on their pectoral fins, including black or white tips and margins that enhance the visibility of this postural display. In contrast, sharks living in habitats where light is scarcer rely simply on the posture (Figure 16.2).

FIGURE 16.2 **In many sharks, agonistic displays involve symmetrical depression of the pectoral fins. (*a*) A Galapagos shark using this visual signal of agitation. (*b*) Species that inhabit clear water habitats, such as the blacktip reef shark, often have markings on their pectoral fins that may enhance the agonistic display. (Drawn from photographs in Martin 2007.)**

Visual signals have disadvantages, of course. The most obvious is, quite simply, that if the sender cannot be seen, then its signals are useless, and vision is easily blocked by all sorts of obstructions, from vegetation to fog to sediment in water. Sometimes, however, the visual displays of animals can be timed to avoid these obstructions. Consider the case of the razorback sucker (*Xyrauchen texanus*), an endangered species of fish found in the Colorado River in the western United States. As the spawning season approaches, razorback suckers move from deep water to more shallow water where the males rest within territories along the river bottom. Should a roving male approach a territorial male, the territorial male quickly rolls his eyes downward, exposing the whites of his eyes to light coming from the water surface. These two quick flashes of reflected light signal the territorial male's presence to the interloper, who then retreats. (Females, by the way, move freely through male territories, without territorial males even batting an eye.) Spring runoff is a time when sediment increases dramatically in the Colorado River, making the transmission of visual signals very difficult. It is probably no mistake, then, that the spawning season of razorbacks is timed to occur *before* spring runoff, when eye flashes are still visible in the less turbid water. Thus, we see that two aspects of razorback life history—their move from deep to shallow water for spawning where light from the water's surface can penetrate to the river bottom, reflect off eyes, and be seen by conspecifics and the timing of spawning before spring runoff—make possible the use of visual signals in territorial communication (Flamarique et al. 2007).

At night or in dark places, light-producing species use visual signals. Most animals, however, cannot produce light, which leads us to our next question: do nocturnal species that cannot produce light ever use visual signals? The answer is yes, but as with the razorback sucker, timing is everything. It turns out that many nocturnal species are most active at dawn and dusk, when some light is available, so visual displays are still an option at these particular times of day. Nevertheless, colors are difficult to distinguish under conditions of low light, so we might expect visual signals at dawn and dusk to focus on contrast, rather than color, and to involve white. And that is exactly what we find for eagle owls (*Bubo bubo*). Males and females of this nocturnal species have a white badge of feathers on the throat (Figure 16.3). This badge is particularly visible when the throat is alternately inflated and deflated during vocal displays, which peak at dawn and dusk (Delgado and Penteriani 2007). Breeding eagle owls also appear to use white material to mark the area around their nest; they deposit white feces at obvious defecation posts and display the white feathers of prey species at conspicuous plucking sites (Penteriani and Delgado 2008). So we see that nocturnal species can use visual signals, providing that such displays focus on contrast rather than color, and occur when some light is available.

Finally, the apparent size of visual signals (their conspicuousness) diminishes with distance. Given this limitation, we might expect animals to adjust their visual signals with respect to receiver distance. Courting male fiddler crabs are well known for their conspicuous and energetically expensive claw-waving displays (see Chapter 14). During such displays, a male repeatedly waves his single, greatly enlarged claw. We will focus on *Uca perplexa*, a species of fiddler crab found in Australia, whose display is shown in Figure 16.4*a*. In the absence of a receiver (a female conspecific), males of this species broadcast their courtship displays. In the presence of a receiver, a courting male turns and faces the approaching female and directs his display toward her. Males, it turns out, do more than simply face the female at show time. Indeed, males adjust the temporal and structural elements of their claw-waving display in relation to female distance (How et al. 2008). As distance to the female decreases, the interval between claw waves decreases, as does the duration of each wave (Figure 16.4*b* and 16.4*c*, respectively). These changes in *timing* translate into a more intense display as the female nears, and may reflect a male's increasing willingness to invest energy in displaying as his chances of a successful mating increase. Changes in timing could also reflect a change in signal function. For example, rather than serving as a conspicuous beacon (as it does for broadcast displays), the claw-waving display may become a signal of male quality. The *structure* of the claw-waving display also changes as distance to the female decreases. For example, the horizontal distance swept by the tip of the enlarged claw decreases with decreasing distance to the female (Figure 16.4*d*, left side). This last result indicates that when claw-waving displays are used as long-range signals, the horizontal distance swept by the tip of the enlarged claw is at its greatest and displays are at their most conspicuous (see Figure 16.4*d*, right side). Thus, these long-range signals may serve as beacons to call in distant females. All of these adjustments in the timing and structure of the claw-waving display are quite remarkable because they require these small crustaceans, living on mudflats, to accurately measure the distance between themselves and an approaching female.

FIGURE 16.3 **Although eagle owls are nocturnal, they use visual signals at dawn and dusk when some light is available. The white throat patch used in these displays contrasts sharply with the surrounding darker plumage and is exposed during calls.**

AUDITION

Sound signals have several advantages. They can be transmitted over long distances, especially in water. Although sound is transmitted more slowly than light, it still can be a rapid means of sending a message, particularly at close range. The transitory nature of sound makes possible a rapid exchange and immediate modification, but it does not permit the signal to linger. After the message has been sent, the signal disappears without a trace. Sound signals have an additional advantage of being able to convey a message when there is limited visibility, such as at night or in deep water or dense vegetation. Finally, sound signals can be complex. Every music lover is aware of the tremendous variety of sound that can be produced by the temporal variation of just two of its parameters: frequency (pitch) and amplitude (loudness). Depending on the species, animals may vary either or both of these aspects of sound, or they may, like a drummer, simply alter the pattern of presentation.

The types of sounds used by a particular species in communication are determined by how the animal produces them. Sounds may be generated by respiratory

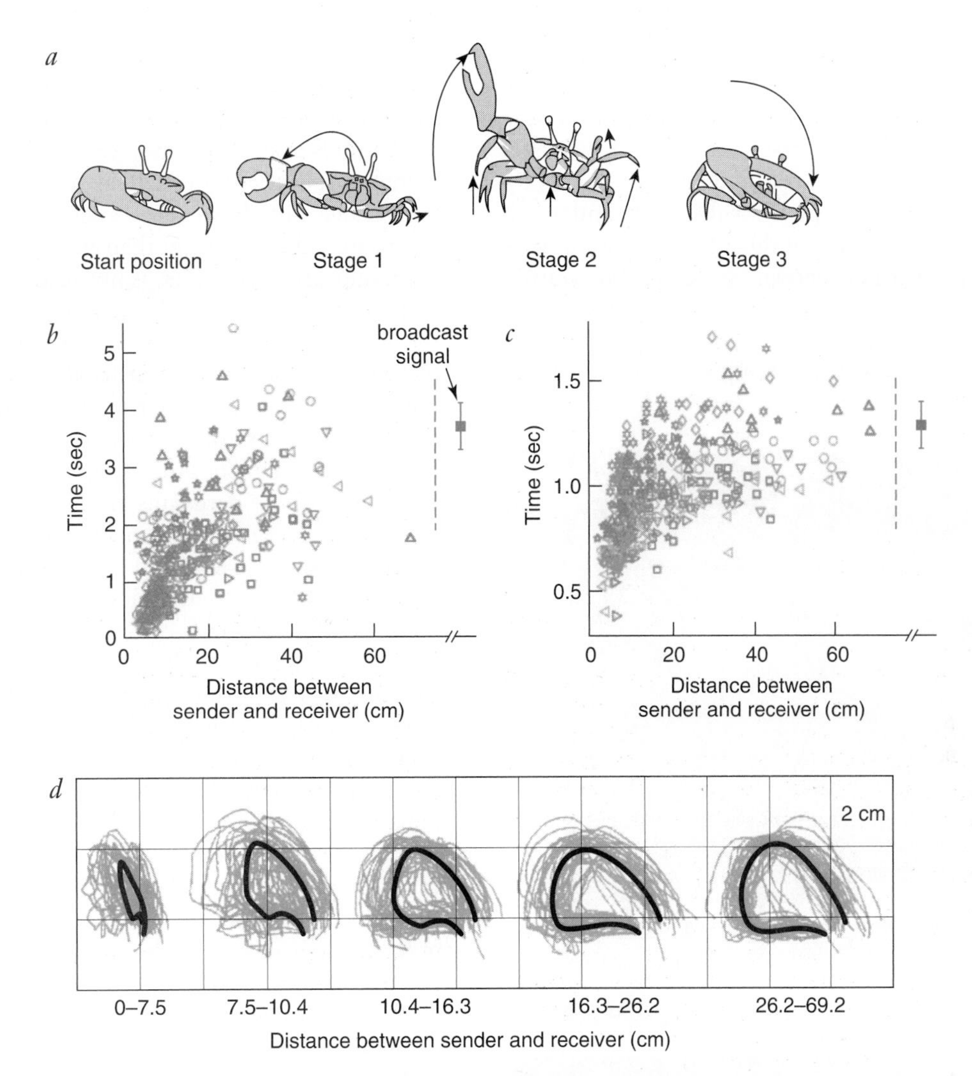

FIGURE 16.4 **Male fiddler crabs adjust the timing and structure of their claw-waving displays in relation to receiver distance. (*a*) Stages in the claw-waving display of male *Uca perplexa*. (*b*) The interval between claw waves, (*c*) the duration of claw waves, and (*d*) the horizontal sweep of the claw tip decrease with decreasing receiver distance. Thus, as distance to receiver decreases, male displays increase in intensity but become less conspicuous. Being conspicuous is more important when receivers are far away. (Modified from How et al. 2008.)**

structures, striking objects in the environment, or rubbing appendages together. Many structures specialized for sound production have evolved in association with respiratory structures. For example, mammals use their larynx and birds use their syrinx to produce sound. The anatomical structure and location of these organs are different, but both allow production of complex sounds. The environment may also be used to produce auditory signals. Humans often tap their toes when listening to music, but for some other animals, such as rabbits and deer, foot stamping itself is the signal. This musical theme has several variations—beavers slap the water with their tails and woodpeckers drum on trees with their bills. Insects most commonly produce auditory signals by rubbing together parts of their exoskeleton. Crickets, for example, produce sound by opening and closing their wings. Each wing has a thickened edge, called a scraper, that rubs against a row of ridges, the file, on the underside of the other wing cover. Whenever the scraper is moved across the file, a pulse of sound is produced. This method of generating sounds is called stridulation.

Stridulation is not confined to insects; it is also used by males of the club-winged manakin (*Machaeropterus deliciosus*), a small bird of the Neotropics (Bostwick and Prum 2005). During courtship displays, a perching male creates a tick-tick-ting sound, which is produced by specific movements of the wings and highly modified secondary feathers. (Secondary feathers arise from the ulna, a bone in the forearm, whereas primary feathers arise from bones of the hand). The "tick" sound is created when the male rotates his wings forward, rapidly flips the feathers above his back, and then forcefully brings the wings together, causing the secondary feathers to collide.

The "ting" sound is produced in much the same manner except that after the secondary feathers collide, the male shivers his wings, causing oscillations of the secondary feathers. The end result of these movements is that one highly modified feather (the fifth secondary feather, known as the "pick") rubs back and forth against the ribbed surface of the adjacent feather (the sixth secondary feather, known as the "file"), exciting resonance in the enlarged rachis (central shaft of the feather to which the vanes attach), and putting the finishing touches on this unique auditory signal (Figure 16.5).

FIGURE 16.5 Male club-winged manakins produce non-vocal courtship sounds, described as a tick-tick-ting sound. (*a*) The tick sound is created by hitting the secondary feathers of the wing together, above the back. (*b*) The ting sound is generated by highly modified secondary feathers of the wing. Following the collision of the feathers, the male shivers his wings, causing the fifth secondary feather (the pick) to scrape across ridges on the sixth secondary feather (the file).

Some animals make sounds that humans cannot hear. For example, several groups of mammals, including cetaceans, bats, and rodents, produce and detect ultrasounds as part of echolocation (see Chapter 10) or communication systems. **Ultrasounds** are sounds whose frequencies are above those audible to humans, which means frequencies greater than about 20 kHz. Ultrasonic communication, however, is not restricted to mammals, and it recently was described in an amphibian. Consider the fascinating case of the concave-eared torrent frog (*Odorrana tormota*), a nocturnal species that lives along noisy streams and waterfalls in Huangshan Hot Springs, central China. Its common name refers to the recessed ears of males. Unlike other frogs who typically lack ear canals and whose eardrums are located at the skin surface (and hence are visible), males in this species have ear canals and recessed eardrums, similar to mammals. Field experiments reveal that audible and ultrasonic components of male calls evoke vocal responses by male conspecifics (Feng et al. 2006). This ultrasonic hearing capacity was confirmed in the laboratory by electrophysiological recordings from an auditory processing center of the midbrain when frogs were exposed to bursts of ultrasound. More recent work showed that females, just before ovulation, also produce calls with ultrasonic components (Shen et al. 2008). When female calls were played in the field and in the laboratory, they evoked calls from nearby males, who then rapidly and precisely approached the loudspeaker, a behavior called positive phonotaxis. Researchers working with concave-eared torrent frogs suggest that ultrasonic communication allows this unusual species to avoid the masking effects of the constant low-frequency background noise of the local streams and waterfalls in their habitat.

We mentioned that humans cannot hear sounds whose frequencies are above 20 kHz (20,000 Hz) and that these sounds are called ultrasounds. Human hearing also has a lower limit: we typically cannot hear sounds whose frequencies are less than about 20 Hz, so sounds below this limit are called **infrasound**. When we think of elephants, we often think of the audible call known as the trumpet blast. In reality, however, most elephant calls are infrasonic, with fundamental frequencies of 15 to 25 Hz (Payne et al. 1986; Poole et al. 1988; note that some of these calls cross into the range of human hearing, and indeed humans can hear some of these low-frequency calls). The source of infrasonic calls is the same as that of audible calls: elephants drive air from the lungs to set the vocal folds of the larynx in motion (Garstang 2004).

Elephants are very social animals that live in matrilineal family groups in which daughters remain with mothers and sons leave to live in bachelor groups or as lone bulls. Family groups range over large areas in search of food and water, so long-distance communication between family members and between different family groups is critical, as is communication between sexually receptive females and males. (Estrus in elephants occurs every four years and lasts only five days, so males must be able to reliably, and quickly, find females over enormous distances.) When compared with high-frequency sounds, low-frequency sounds are much less degraded through refraction (the bending of sound waves as they pass from one medium to another of different density), reflection (the bouncing off of a new medium), and absorption (the conversion of sound energy to heat as it passes through a medium, resulting in decreased intensity). Therefore infrasound works especially well for long-distance communication (Garstang 2004; Langbauer 2000).

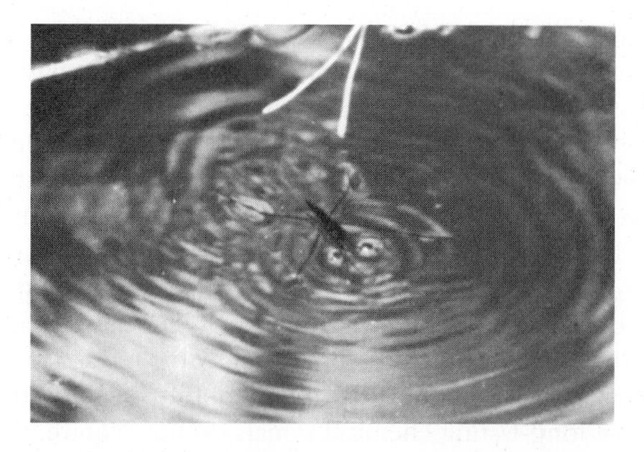

FIGURE 16.6 **Water striders communicate by tapping out messages on the surface of the water, thereby creating surface waves. The pattern of the wave determines the message. Here, a male taps out a repel signal.**

STOP AND THINK

The ultrasonic calls of laboratory rats have been studied in detail for many years, and three distinctly different ultrasonic vocalizations have been identified: (1) a 40- to 50-kHz call given by pups when separated from their mother, which can induce retrieval of the pup by the mother; (2) a 22-kHz call given by adolescent and adult rats during aversive situations (e.g., exposure to predators, handling, social isolation, or electric shock), which may serve to deter predators or signal alarm; and (3) a 50-kHz call given by adolescent and adult rats during encounters that are playful or sexual, which can stimulate social approach (Brudzynski 2005; Portfors 2007; Wøhr and Schwarting 2007, 2008). The first two calls reflect a negative affective (emotional) state, while the 50-kHz call of adolescents and adults indicates a positive state. Because the ultrasonic calls of rats reflect their affective state, some scientists suggest that these calls be routinely monitored (using ultrasonic microphones) by laboratory personnel to facilitate best care practices for rats housed in animal facilities. Do you think it would be worthwhile to use auditory signals to assess the well-being of laboratory rodents? Why or why not? If such a system were in place, how might you use the information from ultrasonic calls to organize laboratory practices? For example, what procedures do you think should not be performed in rooms housing other rats? What activities might be okay?

SUBSTRATE VIBRATIONS

Some animals communicate using seismic signals (Hill 2001). These signals are encoded in the pattern of vibrations of the environmental substrate, such as the ground or the water surface, and can be produced through percussion on the environmental substrate. For example, territory ownership is declared through foot drumming on the ground in kangaroo rats (Randall and Lewis 1997) and through head banging against the burrow ceiling in blind mole rats (Rado et al. 1987; Nevo et al. 1991). Tapping on the surface of water creates seismic signals of a slightly different sort—ripples, or surface waves. Ripple signals have been described in several species of water striders (Figure 16.6). These vibrational signals serve several functions, including sex identification, mate attraction, courtship, and territorial defense (Jablonski and Wilcox 1996).

Seismic signals also can be generated when an airborne vocalization couples with the ground. For example, the infrasonic calls of elephants travel not only through the air as auditory signals, but also through the ground as seismic signals (O'Connell-Rodwell 2007). Elephants detect and respond to seismic signals experimentally generated by researchers directing previously recorded vocalizations into the ground. Elephants also can distinguish subtle differences between seismic calls. For example, they can discriminate between the seismic alarm calls of familiar and unfamiliar conspecifics (O'Connell-Rodwell et al. 2007). How do elephants detect seismic signals? Two pathways have been suggested. The first is bone conduction. In this pathway, seismic signals travel from the ground through the feet, up the front legs to the shoulders, and on to the middle ear. The second potential pathway involves mechanoreceptors, such as Pacinian corpuscles, which are found in the skin of the trunk and feet. When compared with airborne signals, seismic signals generally travel farther and maintain their integrity longer, so vocalizations transmitted seismically may be detectable at greater distances, a feature that would be especially important to far-ranging species like elephants. It is also possible for elephants to simultaneously monitor groundborne signals and airborne signals, which travel at different speeds and therefore arrive at the receiving elephant at slightly different times, to determine the distance of the vocalizing individual (O'Connell-Rodwell 2007).

CHEMICAL SENSES

The chemical senses, smell and taste, are another channel for communication. These senses are based on the movement of odor molecules from signaler to receiver (Wyatt 2003). Information may be carried by chemicals over long distances, especially when assisted by currents of air or water. The rates of transmission and fade-out time are slower than for visual or auditory signals. Depending on the function of the signal, this may be an advantage. In the delineation of territorial boundaries, for example, a durable signal is more efficient because it remains after the signaler has gone (Figure 16.7). Some mammals increase the signal life of chemicals used to mark territories by secreting them with oily carrier substances, such as those from sebaceous glands, or by associating them with certain urinary proteins that slow the release of the signals (reviewed in Wyatt 2003). Another benefit is that these long-lasting chemical signals do not require continued energy expenditure by the sender. Furthermore, chemical signals can be used where visibility is limited. The ease with which the sender of a chemical signal can be located varies with the chemical emitted, but it is usually more difficult to locate a signaler that is using chemicals than one using visual or auditory signals.

a

b

FIGURE 16.7 **Muskrats mark their territories with secretions whose musky scent is long-lasting. (*a*) These relatively large, chunky rodents are found in wetlands throughout much of North America. (*b*) Paired glands near the genitalia emit an odorous secretion that is deposited with urine and feces on defecation posts. These posts are found on rocks along runways (shown here) and on their homes (lodges). The durable scent may serve to advertise territory boundaries and to attract mates.**

Some signals are complex blends of chemicals. In these situations, it may be the relative proportions of different chemicals that produce the effect on a sender rather than the presence of a particular chemical in the mixture. In other words, effects may be produced by the full chemical "image" of the signal, sometimes called the "odor mosaic" (Johnston 2000). Recall from Chapter 7 that **scent marking** is the act of strategically placing a chemical mark in the environment. Here we consider the chemical image left by common marmosets (*Callithrix jacchus*), small primates endemic to the forests of Brazil, and easily recognized by their white ear tufts and forehead blaze (Figure 16.8). Females of this species mark their environment by depositing a complex mixture of scent from several sources, including urine, feces, secretions from the reproductive tract and nearby oil and apocrine glands (apocrine glands are restricted to certain body regions, such as the anogenital area, where they secrete odorous substances onto hair follicles). In the laboratory, female common marmosets were trained (using positive reinforcement) to deposit a scent mark on a special collection device. Detailed analysis of the scent marks revealed the presence of 162 distinct chemicals! While the scent marks of most females contained all of the chemicals, the marks of individual females differed in the amounts of specific chemicals. Previous behavioral testing had shown that females could discriminate between the scent marks of familiar conspecifics and those of unfamiliar conspecifics. Thus, it appears that each female has a unique scent signature (or odor mosaic), which is based on the ratios of chemicals—particularly the volatile ones—in the scent mark (Smith 2006).

Sometimes the meaning of a particular signal may vary with the context in which it is given. This is the case with a chemical signal sent by a queen honeybee. The

FIGURE 16.8 **Common marmosets deposit scent marks that are complex combinations of many chemicals.**

chemical trans-9-keto-2-decenoic acid is picked up from the queen as the workers groom her and is distributed throughout the hive, along with the food that is shared by the workers. When attained in this manner, the chemical prevents the rearing of any additional queens. However, the queen also exudes the chemical as she soars skyward on her nuptial flight. In this context, the same chemical causes males to gather around her. In other words, it serves as a queen inhibitor or as a sex attractant, depending on the context (Robinson 1996).

The detection of chemical cues may occur at a distance, such as when volatile chemical cues become airborne and reach mammal noses or insect antennae; this is called remote chemoreception. Alternatively, chemical cues may be detected through direct contact with the chemical, which (not surprisingly) is called contact chemoreception. For example, you have probably seen one ant rapidly and repeatedly touch the body of another ant with its antennae. What you may not have appreciated at the time is that the touchy-feely individual is evaluating the mixture of chemicals on the outside of the other ant's body to determine whether the individual is a colony member or an intruder. Contact chemoreception is associated with relatively nonvolatile chemical cues.

Some species of amphibians, reptiles, and mammals have a **vomeronasal (or Jacobson's) organ** that is important in chemical communication between mates, parents and offspring, and rivals. It is anatomically separate from other chemosensory structures, and its neural wiring goes to brain regions different from those associated with the main olfactory system (Halpern and Martinez-Marcos 2003). The vomeronasal organ is located in the roof of the mouth or between the nasal cavity and the mouth, so communicative chemicals must reach it through the nose, mouth, or both. Because the chemicals that the vomeronasal organ is specialized to detect are primarily nonvolatile, they must be brought to the organ. In a snake, for example, the vomeronasal organ is on the roof of the mouth and the chemicals are delivered to it by the tongue. A mammal, however, must lick or touch its nose to the chemicals, which are usually found in urine or special body secretions left on surfaces by conspecifics. Following this contact, many mammals make a characteristic facial grimace, known as flehmen, which helps transfer the chemicals to the vomeronasal organ. In flehmen, the head is raised and the lips are curled back (Figure 16.9).

Chemicals produced to convey information to other members of the same species are called **pheromones**. Some of these, releaser pheromones, have an immediate effect on the recipient's behavior. A good example of a releaser pheromone is a sex attractant. The most famous sex attractant is probably that of the female silk moth, *Bombyx mori*. She emits a minuscule amount, only about 0.01 microgram, of her powerful sex attractant, bombykol, from a small sac at the tip of her abdomen (Figure 16.10). This pheromone, which is carried by the wind to

FIGURE 16.9 **Flehmen is a characteristic posture in which the head is raised and the upper lip is curled back. It serves to deliver nonvolatile communicatory chemicals, such as those found in urine or glandular secretions, to the vomeronasal organ.**

any males in the vicinity, binds to the receptor hairs on the male's antennae. As few as 200 molecules of bombykol have an immediate effect on the male's behavior—he turns and flies upwind in search of the emitting female (Schneider 1974). The highly specific olfactory receptors on the antennae of male silk moths have been identified and characterized (Sakurai et al. 2004). Other examples of releaser pheromones in insects are trail pheromones, which direct the foraging efforts of others, and alarm substances, which warn others of danger. Vertebrates also produce releaser pheromones. For example, lactating rabbits produce mammary pheromone, which stimulates their pups to search for and grasp onto a nipple (Hudson and Distel 1995; Moncomble et al. 2005; Schaal et al. 2003). Quick attachment by pups to a nipple is critical because mother rabbits return to their nest to nurse pups only once a day for about three to five minutes!

Primer pheromones exert their effect more slowly, by altering the physiology and subsequent behavior of the recipient. In insect societies, queens control the reproductive activities of nest mates largely through primer pheromones. For example, a queen honeybee produces several compounds from her mandibular gland that ensure that she will remain the only reproductive individual in the colony (Robinson 1996; Winston and Slessor 1992). This pheromone coats the queen's entire body surface but is most concentrated on her head and feet. Most of the pheromone spreads through the colony by the activities of the workers that are attending the queen, but some is spread through the wax of the comb (Naumann et al. 1991). The pheromone prevents the workers from feeding larvae the special diet that would cause them to develop into rival queens. When the queen dies, the inhibiting substance is no longer produced, and new queens can be reared (Wilson 1968).

a

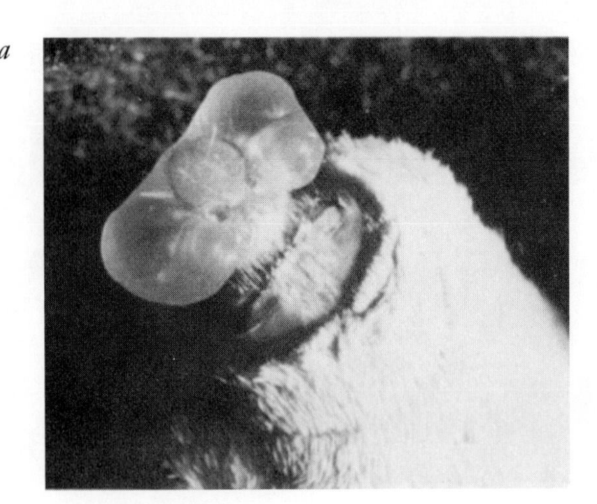

b

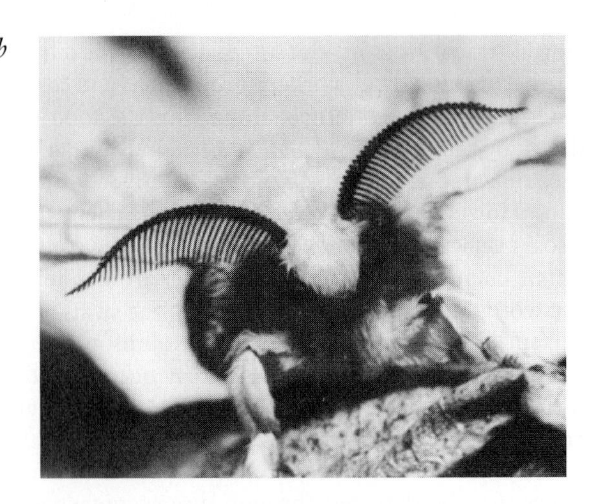

FIGURE 16.10 **Female silk moths produce the sex attractant bombykol. (*a*) The tip of the female's abdomen has a pair of glands, shown in an expanded active state here. Within these glands is a small quantity of bombykol. (*b*) The feathery appearance of the male's antennae is due to numerous branches, each containing many odor-receptor hairs. Half of the odor receptors on the male's antennae are tuned to bombykol, which is why the male is sensitive to such minute quantities of it. On detecting bombykol, he turns and flies upwind in search of the female.**

Vertebrates also produce primer pheromones that influence reproductive activity in various ways. They may help regulate reproductive activities so that reproduction occurs in the appropriate social or physical setting. Table 16.2 lists some of the effects of primer pheromones known in mice.

Before leaving the topic of chemical communication, we want to make sure that we have not left you with the impression that the vomeronasal organ is for pheromones and the main olfactory system (i.e., the olfactory epithelium, which is the layer of cells lining the nasal passages that contains olfactory receptors) is for smelling general odorants. In fact, there are no clear-cut functional differences between the vomeronasal organ and the main olfactory system (Baxi et al. 2006). The vomeronasal organ, it has been found, can be stimulated by substances other than pheromones. For example, the vomeronasal organ of a hunting snake responds to chemical cues of prey species brought to the organ by the flicking tongue. Chemicals from prey species clearly do not fit the definition of a pheromone (which requires communication to be *within* a species) and the behavior of interest is foraging, not communication. In addition, some chemical signals that fit the definition of pheromone, such as the rabbit mammary pheromone discussed earlier, appear to act via the main olfactory system and not through the vomeronasal organ. Consider, also, the scent marks left by female hamsters around their territories (Johnston 1998; Meredith 1998; Swann et al. 2001). These marks contain vaginal fluid pheromones that a male detects via his main olfactory system and that prompt him to locate the female. Once the male has found the female, another component of the vaginal secretion, this time perceived through his vomeronasal organ, prompts him to investigate and mount her. This is how it works for sexually inexperienced males. Sexually experienced males, however, have learned the odor cues of receptive females and no longer need the input via the vomeronasal organ to stimulate mounting. Thus, pheromones may be perceived by either the vomeronasal organ or the main olfactory system. In some situations, the two systems work together, although this relationship may be modified by experience.

TABLE 16.2 **Effects of Some Primer Pheromones in Mice.**

Origin	Recipient	Effect
Female urine	Female	Inhibits estrous cycling and ovulation
Male urine	Female	Induces estrous cycling and ovulation
Female urine	Male	Prompts release of testosterone and luteinizing hormone

TOUCH

Animals also communicate by touch (Figure 16.11). Tactile messages can be sent quickly, and it is easy to locate the sender, even in the dark. Honeybee scouts, for example, inform nest mates of the location of a food source by dancing. The recruits cannot see the choreography because the hive is so dark, but they follow the

FIGURE 16.11 **Social grooming, a form of tactile communication that builds and maintains social bonds, is displayed by many mammals, including horses.**

dancers' movements by touching them with their antennae (discussed shortly). Tactile signals are obviously only effective over short distances and are not effective around barriers. A message sent via touch may be varied in several ways, including how the recipient is touched (e.g., rubbing, patting, pinching), where the recipient is touched, the frequency and duration of touching, and the extent of surface area touched (Hertenstein et al. 2006a).

Humans also skillfully send and decode tactile signals. Consider one study in which two strangers were asked to sit at a table, where they were separated by a barrier (an opaque black curtain). One person was charged with sending a tactile signal that conveyed one of 12 different emotions and the other person was to receive the signal and decode it. The person to receive the tactile signal positioned his or her bare arm (from elbow to fingertips) on the sender's side of the curtain; this ensured that the receiver could not see any part of the tactile signal. The participants were asked not to speak to one another during the experiment. Twelve emotion words were presented serially to the sender on sheets of paper in random order; once presented with a sheet, the sender was asked to convey that particular emotion through touching the receiver's arm. The recipient of the touch was then asked to choose which of the 12 emotions the sender was trying to communicate. Study participants decoded fear, anger, love, gratitude, sympathy, and disgust at levels above chance, but performed less well with happiness, sadness, surprise, embarrassment, envy, and pride. In addition, specific types of touch were associated with specific emotions: for example, trembling with fear, hitting and squeezing with anger, stroking with love, shaking the hand with gratitude, patting with sympathy, and pushing with disgust (Hertenstein et al. 2006b).

ELECTRICAL FIELDS

Two distantly related groups of tropical freshwater fishes produce electrical signals used in both orientation (electrolocation, discussed in Chapter 10) and communication. These groups are the knifefishes (gymnotiforms) of South America and the elephant-nose fishes (mormyriforms) of Africa. Gymnotiforms and mormyriforms are described as "weakly electric" to contrast them with fishes, such as torpedo rays or electric eels, which generate very strong electric discharges to stun prey or predators.

The electrical signals are generated by electric organs derived from muscle. When a normal muscle cell contracts, a weak electrical current is generated. The modified muscle cells in an electric organ also generate a weak electrical current, but because they are arranged in stacks their currents are added, resulting in a stronger current. When an electric organ of a gymnotiform or mormyriform discharges, the tail end of the fish, where the electric organ is located, becomes momentarily negative with respect to the head. Thus, an electrical field is created around the fish (Figure 16.12*a*). This electrical field is the basis of the signal. Diverse signals can be created by varying the shape of the electrical field, the discharge frequency, and the timing patterns between signals from the sender and receiver, as well as by stopping the electrical discharge. Electric organ discharges are detected by special sensory receptors in the skin called electroreceptors (Bullock et al. 2005).

Two general patterns of electric organ discharge occur in weakly electric fish: wave-type and pulse (Hopkins 1974; Zakon et al. 2008). These two distinct patterns have independently evolved in both gymnotiforms and mormyriforms. Those species whose discharges are classified as the wave-type pattern produce signals continuously (whether active or inactive) and the waveform of the signal is monophasic, resembling a sine wave (Figure 16.12*b*). In contrast, species whose electric organ discharges are emitted as pulses produce discharges at higher rates when active and lower rates when resting, and the waveform of the signal has a complex multiphasic structure (Figure 16.12*c*).

What are the characteristics of electrical signals? When the electric organ discharges, an electrical field is created instantaneously. It also disappears at the instant the discharge stops. As a result, electrical signals are ideally suited for transmitting information that fluctuates quickly, such as aggressive tendencies (Hagedorn 1995). An electrical signal does not propagate away from the sender but instead exists as an electrical field around the sender. Because an electrical signal is not propagated, its waveform is not distorted during transmission. As a result, the waveform of the electrical signal may be a reliable indicator of the sender's identity (Hopkins 1986a). Although the waveform is generally constant for a particular individual, it is different in males and females and among different species (Stoddard et al. 2006).

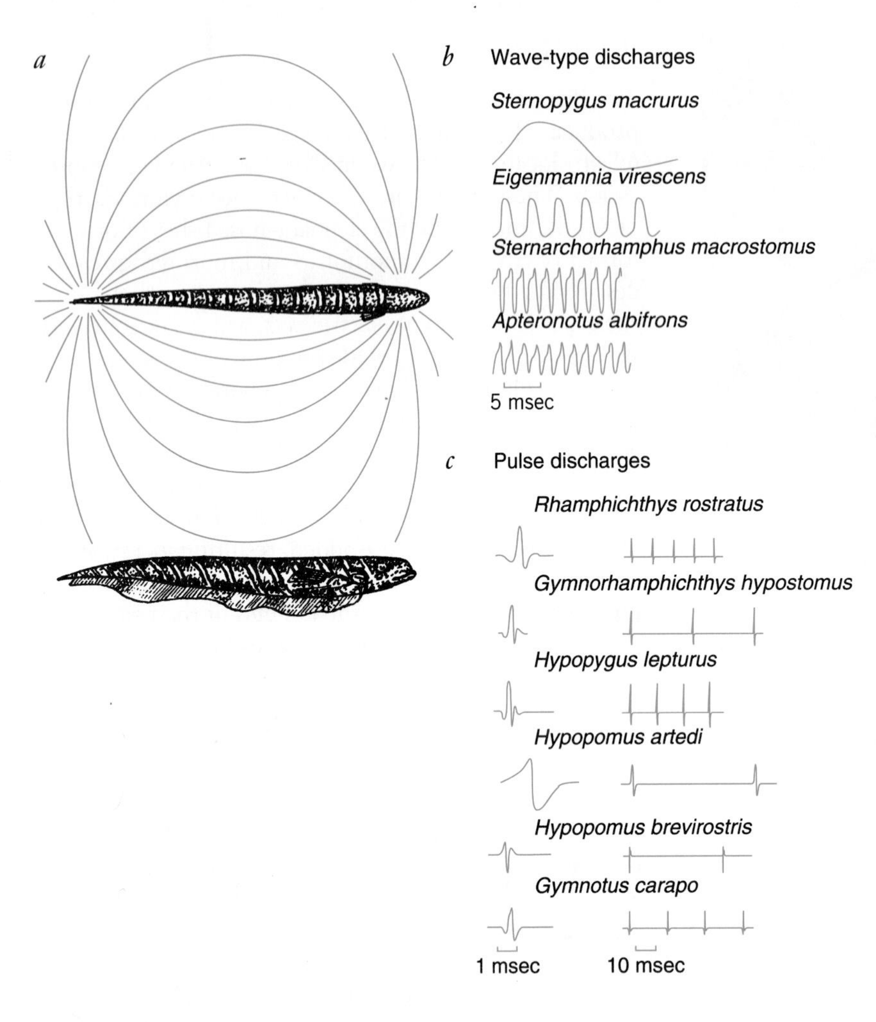

FIGURE 16.12 **Some tropical freshwater fishes communicate via weak electrical signals. (*a*) Discharges from an electric organ create an electrical field around a weakly electric fish that is used as a communication signal. The signal can be varied by altering the shape of the electrical field, the waveform of the electrical discharge, the discharge frequency, and the timing patterns between signals from the sender and receiver, as well as by stopping the electrical discharge. (*b*) Some species of weakly electric fish produce electrical signals continuously with monophasic waveforms. (*c*) Other species produce electrical signals in a pulse pattern, often with multiphasic waveforms. These so-called pulse-fish discharge at high rates when active and low rates when at rest. (From Hopkins 1974.)**

Electrical signals are well suited for communication in the environments in which the weakly electric fish live. Both gymnotiforms and mormyriforms are active at night and generally live in muddy tropical rivers and streams, or at depth in tropical rivers, where visibility is poor. Electrical signals can move around obstacles and are undisturbed by the suspended matter that creates murky water. However, they are effective only over short distances, about 1 to 2 m, depending on the depth of the water and the relative positions of the sender and receiver (Hopkins 1986b, 1999). The shortness of their effective distance may actually be an advantage. Different weakly electric species may coexist in an area, so the short effective distance of the signal may reduce electrical "noise" when many individuals are signaling at once.

Weakly electric fish use electrical signals to convey the same messages that other organisms send by other channels. For example, males of some species not only advertise their sex and species by electrical signals, but also court females by "singing" an electrical courtship song (Hagedorn 1986). Indeed, in some species, a courting male and female engage in electrical duetting, a coordinated pattern of communication in which the signals of one individual alternate with those of the other (Wong and Hopkins 2007). Electrical signals are also used during agonistic encounters, where certain patterns of discharge have been associated with aggression, dominance, and submission (Hupé and Lewis 2008; Triefenbach and Zakon 2008). Finally, there is some evidence that parents and offspring may communicate via electrical signals to maintain proximity to one another (Crampton and Hopkins 2005).

MULTIMODAL COMMUNICATION

Animals do not always use a single channel when communicating. In fact, the displays of many animals contain signals from two or more sensory modalities. This type of communication is called **multimodal communication**, and the signaling in different channels can occur either simultaneously or sequentially (Partan and Marler 2005). Here we focus on simultaneously produced multimodal signals. For example, the courtship display of a male bird may simultaneously contain visual and auditory signals. During allogrooming, a monkey may touch the skin of its companion while at the same time making a certain facial expression and vocalization. We know from experience that when humans communicate through speech, information is often simultaneously passed through the visual channel. Indeed, it is often much easier to understand a spoken message if we have both visual and auditory information from the sender (especially if the information from the two channels corresponds and does not conflict). Finally, we mentioned previously that an elephant vocalization could have seismic (ground borne) as well as auditory (airborne in this case) components.

During multimodal communication, the messages conveyed in different signaling channels can be redundant (convey the same thing) or nonredundant (convey different things). Whether the messages are redundant or nonredundant can be determined by presenting each component of the display individually to a recipient and seeing how the recipient responds (Partan and Marler 2005). In the case of redundant messages, the recipient should respond in the same way to each component presented individually. In the case of nonredundant messages, the recipient should respond differently to the different components.

As an example of a multimodal communication, consider the courtship display of male brush-legged wolf spiders (*Schizocosa ocreata*) (Gibson and Uetz 2008; Uetz and Roberts 2002; Uetz et al. 2002). Males of this species court females using a complex "jerky-tapping" display that simultaneously (and sequentially) contains both visual and seismic signals. The visual part of the display consists of the male rapidly raising and lowering his first pair of legs; each of these legs is adorned with a conspicuous tuft of bristles (Figure 16.13). The seismic components of the male courtship display include: (1) stridulation produced by organs on the pedipalps (appendages modified as copulatory organs in male spiders), (2) up and down bouncing of the entire body, and (3) striking the substrate with chelicerae (mouthparts). Experimental presentation of isolated visual and seismic components reveals that the signals in both channels elicit receptivity in females. In fact, the information content of visual and seismic signals appears quite similar; both allow a female to assess the quality (size and physical condition) of the displaying male. Thus, visual and seismic signals appear to contain largely redundant information in the male wolf spider's display. If females responded differently to individual presentations of visual and seismic signals, then nonredundant messages would be more likely.

Multimodal communication has benefits for senders and receivers (Hebets and Papaj 2005; Partan and Marler 2005). For nonredundant multimodal signals, more information can be sent per unit time. For redundant multimodal signals, a major benefit is insurance that the message will be received and accurately recognized, even if one sensory channel is especially noisy. In the case of the brush-legged wolf spiders, the leaf litter in which they live is a structurally complex habitat where communication is difficult at best. Under these challenging conditions, having redundant signals may be necessary to prevent miscommunication between males and females, which could lead to the male becoming the female's next meal rather than her mate. Miscommunication could also lead to mating with the wrong species, which would also be costly to all involved (Gibson and Uetz 2008).

FIGURE 16.13 **The courtship display of a male brush-legged wolf spider is multimodal, involving both visual and seismic components. The visual signal consists of rapidly raising and lowering the forelegs, which have conspicuous tufts.**

Multimodal communication also has costs for senders and receivers (Partan and Marler 2005). Signaling in multiple sensory modalities may require more of the sender's energy, and recipients may need more energy to receive and process multiple signals. Furthermore, signaling in multiple channels may make senders more susceptible to predation. Even receivers could be more susceptible to predation as a result of devoting more senses to detecting signals from conspecifics and fewer to detecting predators.

FUNCTIONS OF COMMUNICATION

Communication is integral to much of animal behavior. Some signals bring individuals together, and others help keep them apart. Some signals settle conflicts, whereas others incite them. Some cause alarm and others are pacifying. Some signals attract suitors and others repel them. Not surprisingly, descriptions of communication are woven throughout many chapters in this book. Here we will highlight just a few of the messages that signals convey.

SPECIES RECOGNITION

It's striking how attuned many species of animals are to the presence of conspecifics. For example, a male chickadee may give little or no observable response to the song of a gray-crowned rosy finch but immediately increases its call rate when a speaker broadcasts the song of another chickadee (Charrier and Sturdy 2005). It's no wonder. Conspecifics are likely to be competitors: they may want the same food, shelter, and mates, and thus should be dealt with accordingly. Other conspecifics might be friendly members of a social group, or potential mates that should be wooed. It's also adaptive not to mistake heterospecifics for conspecifics, as it is a waste to squander time and energy courting an animal with whom it is impossible to produce viable offspring, or to defend a territory from an individual that is not competing for resources or mates.

Virtually all sensory channels are used for species recognition in one species or another (reviewed in Bradbury and Vehrencamp 1998). Birds generally use song characteristics such as frequency (which notes are sung) and syntax (how the notes are strung together). Some animals, such as crickets that make sound by scraping together a file and rasper on their wings, don't have the ability to modulate the notes of their songs, so instead they rely on differences in their rhythmic patterns. Other species, particularly insects, use olfactory cues. For example, some species use species-specific pheromones to attract mates. Still others rely on visual cues, such as displays, color patterns, or other aspects of appearance.

Researchers can pinpoint exactly which part of a signal conveys information about species identification through experimental manipulation. Bird songs, for instance, can be easily digitized and manipulated so that components are omitted, added, or played in a different order or at a different speed. Researchers can then play back the modified calls to watch the response of free-living birds. By this technique, we know that the chickadees described earlier pay attention to a number of acoustic features when deciding whether a strange call is produced by a conspecific (Charrier and Sturdy 2005). In some cases, experimental manipulation leads to counterintuitive results. For example, let's take another look at fiddler crabs (genus *Uca*), which we mentioned earlier in the section on visual cues. In some places, many species can be found packed into close quarters on the beach; in one study, there were 12 species in an area slightly smaller than a quarter of a tennis court. Crabs signal by waving their large front claws, the chelipeds, and different *Uca* species wave in different patterns: *U. rhizophorae* waves up and down, *U. annulipes* in large circles, and *U. pugilator* in small circles (Crane 1941). These display patterns have long been thought to be important in species recognition. However, in a test of another pair of *Uca* species, *U. mjoebergi* and *U. capricornis*, species recognition depended not on display behavior but on the color of the cheliped. Female *U. mjoebergi* prefer a male with the bright yellow cheliped characteristic of the species, even if it is a male of another species with its claw painted yellow (Detto et al. 2006).

We expect that selection for species to have distinctly different signals will be strongest where species occur together in the same place, a prediction that is borne out in frogs. Male frogs and toads usually attract their mates by calling at night (Figure 16.14). Judging by the response of nearby females, these calls seem to be the amphibian equivalent of "I'm over here. Come find me." The calls of different species are impressive in their variety, ranging from high-pitched peeps to trills to bass-drum booms. Males of several species often serenade together in a chorus, and a female must choose one of her own kind from the variety of callers at the local pond. The female's ability to discriminate is important, as males generally are unselective and will grab any female (or male) of approximately the right size that happens by (Gerhardt 2001). As a result, selection has favored clear species differences in calls, and a reasonably skilled person can determine the species of many frogs by their calls alone. The ability of green tree frog females to distinguish conspecific calls from other calls has been shaped by selection. In parts of its range, *Hyla cinerea* shares its breeding ponds with the closely related *H. gratiosa*. In other regions, *H. gratiosa* is not present, and *H. cinerea* lives alone. Female *H. cinerea* that share the same area with *H. gratiosa* prefer calls with the acoustic properties that distinguish them from *H. gratiosa* calls (Höbel and Gerhardt 2003).

It is not always true that animals can unfailingly distinguish conspecifics from others. As you might predict from sexual selection theory (Chapter 14), males of many species indiscriminately court females. Australian *Julodimorpha bakewelli* beetles will even attempt to copulate with discarded beer bottles. Apparently, the fact that the bottles are, like females, shiny brown and bumpy is enough to overcome their noticeable lack of ability to

FIGURE 16.14 **A male American toad (*Bufo americanus*) is calling to attract a mate. The male amplifies the sound produced by his vocal cords by inflating the large vocal sac beneath his chin.**

give behavioral signals (Gwynne and Rentz 1983). Even females, which are generally the more selective sex, sometimes make erroneous choices (reviewed in Gröning and Hochkirch 2008). In some cases, such as recently introduced invasive species that share some traits with natives, natural selection has not had time to favor those individuals that can successfully make the distinctions.

MATE ATTRACTION

Animals that spend most of their time alone have the challenge of locating each other when it is time to breed. Besides being species-specific, the signals that attract a mate must be easy to locate and effective over long distances so that males and females can find each other even if the species members are widely distributed. For this reason, chemical and auditory signals are used commonly, but not exclusively, for attraction. For example, we have already described the sex attractant pheromone of the female silkmoth (*Bombyx mori*) in this chapter. Under the right wind conditions, the pheromone may attract males from perhaps 100 m away. The sex attractant pheromone of another species, *Actias selene*, is also potent. In one experiment, males that had been experimentally displaced 46 km away were still able to relocate newly hatched females at the original site (Immelmann 1980).

Auditory signals also carry well, especially when amplified by communal displaying or by special anatomical or environmental structures. We've already discussed the courtship songs of birds, frogs, and crickets, and how they can attract mates from long distances. As you can see in Figure 16.15, female crickets will approach a loudspeaker that broadcasts the courtship song of a male of its species.

FIGURE 16.15 **A female cricket will be attracted to a loudspeaker that is broadcasting the courtship song of a conspecific male.**

STOP AND THINK

When males signal to attract prospective mates, they usually give auditory or visual signals, whereas females that signal usually use the olfactory channel (Bradbury and Vehrencamp 1998). Why might this be so? Think in particular of the duration of receptivity, the costs of signals, and the dangers of signaling.

COURTSHIP AND MATING

Once the individuals are close enough to interact, they court before committing themselves to mating. Communication has several functions: identification of a partner of the appropriate species and sex, assessment of mate quality, coordination of the mates' behavior and physiology, and, in some species, maintenance of pair bonds after mating.

Identification of the Opposite Sex

We've already discussed how animals might signal their species identity. Animals might also communicate their sex. In many cases, the differences between males and females are readily apparent and are indicated by the development of antlers and other secondary sexual characteristics. Some species have displays that seem to showcase the aspects of their body that indicate their sex. When a female stickleback (*Gasterosteus aculeatus*) enters a male's territory, she reduces the probability of attack by assuming a head-up position that displays her egg-swollen abdomen and distinguishes her from an intruding male (Tinbergen 1952).

In other species, differences are more subtle. For example, male blue-ring octopuses (*Hapalochlaena lunulata*) apparently cannot distinguish males from females until quite late in the courtship sequence. Octopuses mate by inserting their modified third right arm into the mantle cavity of the female and releasing a spermatophore, or sperm packet. Male blue-ring octopuses insert their arms indiscriminately into both males and females, but only release spermatophores when inserted into females (Cheng and Caldwell 2000). Male–male interactions are brief and not aggressive; apparently the fitness costs of making an insertion into a male are low.

Mate Assessment

Courtship may allow a female to judge the qualities of her suitor so that she can choose the one most likely to enhance her own reproductive success (or, more rarely, it may allow the male to choose the characteristics of an appropriate female). This function of communication has been widely studied. We provide many examples and more details about the selection pressures on mate choice in both Chapters 14 and 17, and only briefly discuss some examples here.

Courtship displays may provide a means for evaluating a suitor's qualities, including his physical prowess, ability to provide food for the offspring, or even the extent of his commitment. For example, male common terns (*Sterna hirundo*) catch fish and offer them to the female. She compares the quantity of fish provided by her various suitors and usually chooses the best fisherman. The number of fish a male provides during courtship is correlated with the quantity he later provides to his chicks. The quality of the courtship offering, then, is a reliable indicator of the male's ability to provide for the pair's offspring (Wiggins and Morris 1986).

An odder example is the wheatear, a small bird from Spain and Morocco. Males demonstrate their devotion to their mates by collecting stones in their beaks, one by one, and carrying them to cavities that serve as potential nest sites. In one study, males carried an average of 277 stones weighing 1.8 kg over the course of a single week (Morena et al. 1994)—impressive for a 35- to 40-g bird! Females watch the males carry stones, and even occasionally heft stones that have been carried as if assessing their weight (Soler et al. 1996, 1999). Male wheatears that carried heavier stones scored better on a test of immunocompetence, an indicator of male health (Soler et al. 1999).

Coordination of Behavior and Physiology

Males and females have very different reproductive systems that may not always be synchronized. Courtship displays can function to coordinate the couple's behavior and physiology. Ring doves (*Streptopelia risoria*) are a well-studied example. Each step in the mating sequence, from initial courtship through nest construction, copulation, egglaying, and feeding nestlings, is choreographed by the behavior and hormonal state of each partner (Figure 16.16) (reviewed in Fusani 2008).

Because displays used for coordinating receptivity are generally between partners near each other many are visual or tactile. Some, however, rely on pheromones that are delivered at close range: a male mountain dusky salamander (*Desmognathus ochrophaeus*) applies a courtship pheromone to the female by pulling his lower jaw across the female's back and angling his snout so that his premaxillary teeth scrape the female's skin, thereby

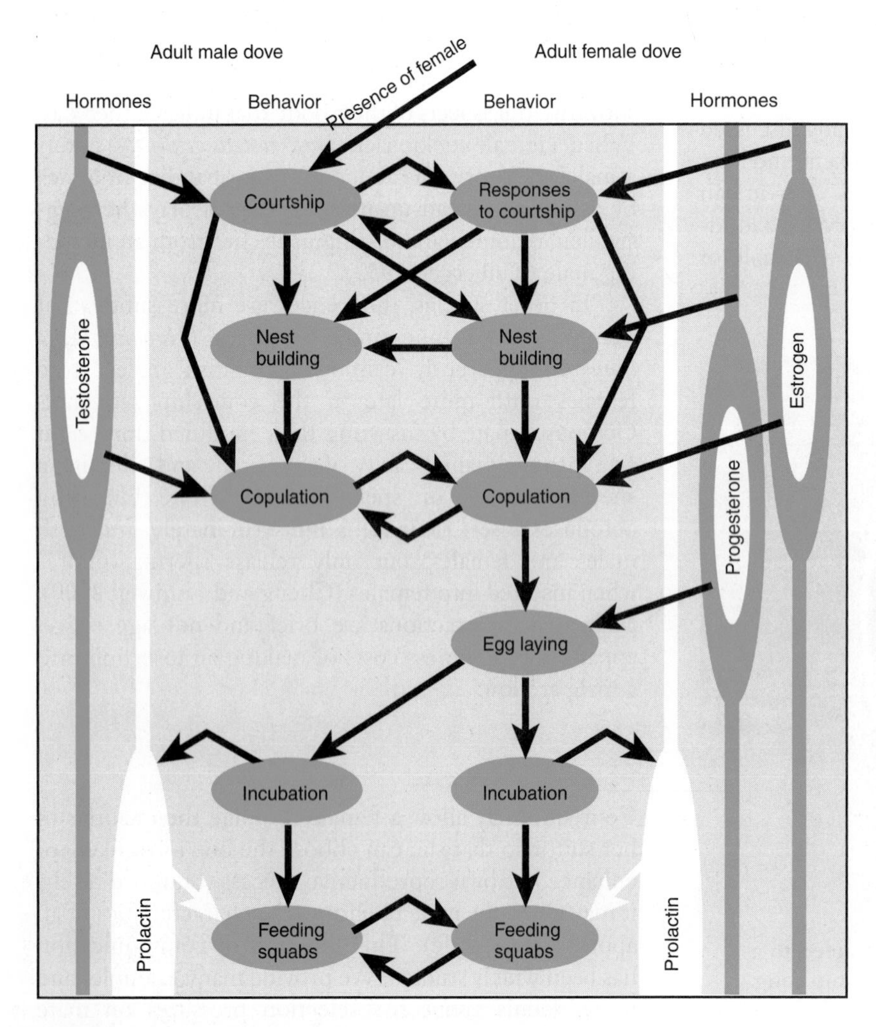

FIGURE 16.16 **The mating behavior of a pair of ring doves is choreographed by signals. The sight of the female causes the male to increase his testosterone production, and he begins to display. In response, the female coos, and her own vocalization stimulates estrogen production. Estrogen stimulates the development of the oviduct. Both begin nest building. In the female, estrogen declines and progesterone increases. The pair copulates and the female lays two eggs. The sight of the eggs induces incubation and suppresses courtship, testosterone, and progesterone. An increase in prolactin leads to incubation. (From Bradbury and Vehrencamp 1998, based on Nelson 1995.)**

"injecting" the pheromone into the female's circulatory system. The female then indicates her receptivity by assuming a tail-straddling position, and the male deposits a packet of sperm called a spermatophore (Figure 16.17*a*). Houck and Reagan (1990) demonstrated that the male's courtship pheromone makes the female more receptive. They staged a total of 200 courtship encounters between 50 pairs of salamanders on each of four nights. By removing the mental gland, which produces the pheromone, from each male, they prevented the males from delivering any pheromone during courtship. These glands were then used to create an elixir containing the courtship pheromone. Thirty minutes before some encounters, each female was treated with this pheromone-containing elixir. Before other encounters, the same female was treated with a saline solution. After a female received a pheromone treatment, she assumed a tail-straddle position, indicating receptivity, 43 minutes (26%) sooner and mated 59 minutes (28%) sooner than she did after receiving a saline injection (Figure 16.17*b*).

Maintenance of Pair Bonds

A final function of communication in the context of reproduction is the formation of bonds between (relatively) monogamous pairs of animals. As with displays that coordinate mating, pair-bond displays often occur at close range and are thus often visual or tactile, such as dusky titi monkeys that sit with their tails intertwined (Figure 16.18).

Let's examine a particular pair bond in detail (Sogabe and Yanagisawa 2007, 2008). Pipefish (*Corythoichthys haematopterus*) are long skinny fish related to seahorses. Like seahorses, males have a brood pouch into which females deposit eggs. The males care for the offspring for one to eight weeks. Pairs are monogamous and only form bonds with other individuals when their mate disappears. Male and female pipefish conduct a greeting ceremony every morning. They approach one another,

a

b

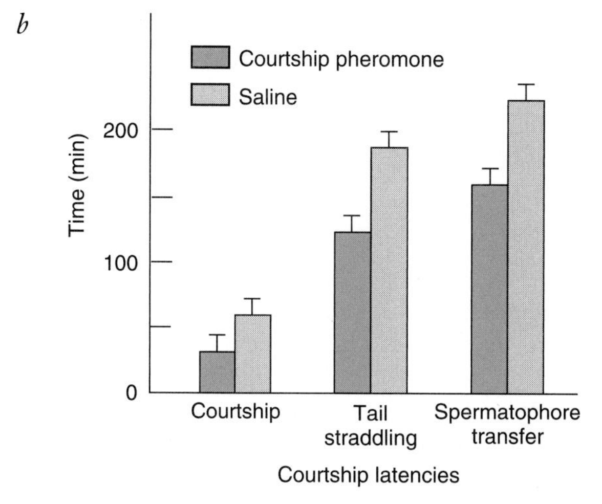

FIGURE 16.17 **During courtship, a male mountain dusky salamander injects a female with his courtship pheromone. (*a*) The male alternately scrapes the female's back with his teeth and swabs her with the pheromone, which is produced by a gland beneath his chin. The female signals her readiness to mate by placing her chin on the base of the male's tail and straddling his tail. (*b*) The courtship pheromone makes the female receptive. Tail straddling and mating occur more quickly when a female has been treated with a courtship pheromone compared to a control treatment with saline. (From Houck and Reagan 1990.)**

FIGURE 16.18 **Dusky titi monkeys maintain their pair bond with physical contact. (From Bradbury and Vehrencamp 1998, based on Moynihan 1966.)**

swim in parallel side-by-side, cross abruptly over each other's back, arch their bodies, or rise into a vertical position. Interestingly, although the home ranges of pairs overlap with those of other individuals, greeting ceremonies are only performed by members of a pair. Greetings are even carried out during the nonreproductive season. Members of a pair meet at a particular site every morning, exchange greetings for several minutes, and then have no further contact for the rest of the day. Because no other benefits seem obvious to this off-season greeting—the pairs do not cooperate in any other behavior, for example—it is thought that its function is solely to maintain the bond with the partner in preparation for the next breeding season.

MAINTAINING SOCIAL BONDS

In some species, social group members, and not just mated pairs, use communication to maintain their bonds. These communicatory signals are generally based on contact: resting together, nuzzling, and touching in general tend to firm social bonds (Eibl-Eibesfeldt 1975).

Many of the greeting signals exchanged by animals as they encounter one another serve as an assurance of nonaggression. Chimpanzees often greet by touching hands or sometimes by placing a hand on the companion's thigh (Goodall 1965). Sea lions rub noses, and lions rub cheeks. African wild dogs greet one another by pushing their muzzles into the corners of each other's mouths (Schaller 1972). You may have been head-bumped by your cat—you should take this as a compliment. (It is more appealing than engaging in the greeting ceremony of the domestic dog, which includes anal sniffing.)

In many mammals, social bonds appear to be built and maintained through social grooming, also called **allogrooming** (Figure 16.19). If we assume that self-grooming has skin care as its primary function, what evidence do we have to support a communicative function for social grooming? Maria Boccia compared several aspects of social grooming and self-grooming in rhesus monkeys: body site preference, duration (both overall and to specific areas of the body), and method (stroking or picking). She reasoned that if the primary function of both social grooming and self-grooming was hygiene, then these physical aspects of grooming would be the same in both. However, social grooming was found to be different from self-grooming in each of these respects. Therefore, she concluded that skin care is not the most important factor in molding the form of social grooming (Boccia 1983). Furthermore, she showed that the message of the tactile signal varies according to the body site being groomed (Boccia 1986). In other words, the recipient monkey's response depended on which part of its body was groomed. The animal being groomed was likely to move away from the groomer when the posterior

FIGURE 16.19 **Allogrooming between crab-eating macaques. Besides the benefits of skin and fur care and parasite removal, allogrooming is important in social bonding.**

part of the body was groomed. The responses to the five body regions typically groomed may reflect a continuum, from a tendency to maintain an affiliation at one extreme to a tendency to terminate the interaction at the other. Grooming can also facilitate a transition from one type of social interaction to another. For example, in primates, grooming can smooth over tension and restore relationships after conflicts (e.g., Watts 2006).

ALARM

Alarm signals warn another animal of danger. Usually, danger presents itself in the form of predators, but individuals may also have to guard against other members of their species bent on infanticide or some other form of aggression.

Alarms That Cause Animals to Flee

Alarm signals have different functions, and their characteristics depend on their function (Bradbury and Vehrencamp 1998). Many alarm signals cause those who hear them to flee or take cover. We can predict that "flee" signals should share several characteristics. These signals should be easy to make quickly, before it is too late. From the signaler's point of view, it's also best if the signal is difficult to locate. Alarm signals from several sensory channels fit the bill: examples include rapid visual signals such as the flash of a deer's tail; highly volatile pheromones that diffuse quickly; and high-pitched sounds that are hard to localize. As you might expect given the similar selective pressures, species often share similar alarm signals. For example, a number of species of passerine birds have the characteristics that would be expected to make them difficult to locate—they begin and end gradually and employ only a few wave-

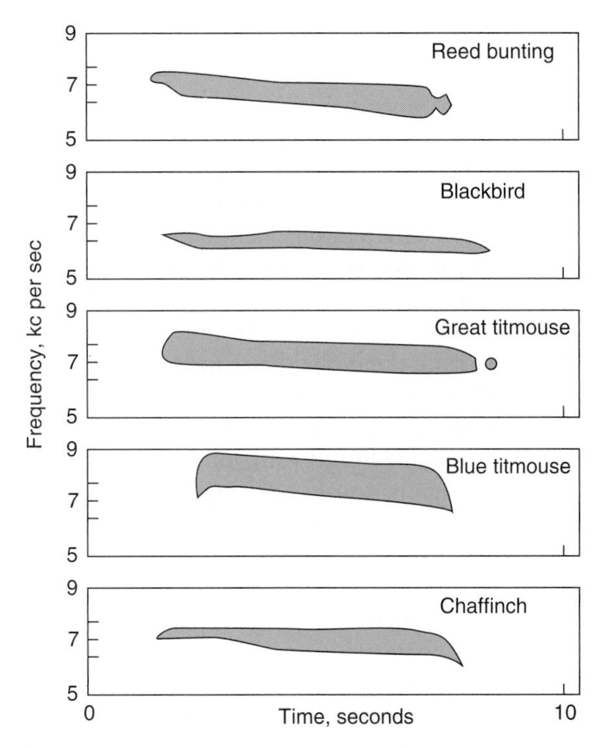

FIGURE 16.20 **Sonograms of alarm calls of several species of small song birds. Note the similarity of the calls. As a result, an alarm call will alert birds of many species. (Modified from Marler 1959.)**

lengths centered on 8 kHz (Figure 16.20). As mentioned in Chapter 13, some species even respond to the signals of other species: Eurasion red squirrels flee or increase their vigilance when they hear the alarm calls of Eurasian jays (Randler 2006).

Most species use the same signal to indicate any source of danger, although some use specific calls to designate the type of threat. For example, vervet monkeys (*Cercopithecus aethiops*) classify their most common predators into one of at least three groups—snakes (e.g., pythons), mammals (e.g., leopards), or birds (e.g., eagles). Considering the hunting strategy of each type of predator, the characteristics of the alarm call and the response of conspecifics within hearing distance seem to be adaptive. The low-amplitude alarm call emitted when a snake is encountered captures the attention of individuals near the caller that might be in danger from the slow-moving reptile without attracting other predators in the area. Other monkeys respond by looking at the ground, the most likely place to find a snake. However, when they see a large mammalian predator such as a leopard, monkeys emit very loud, low-pitched, and abrupt chirps. These properties make the call audible from a great distance and make it easy for conspecifics to locate the caller. The most common response to the chirp is to scatter and run for cover in the trees, a relatively safe haven from the ambush style of attack characteristic of a leopard. When monkeys spot an avian predator, they emit staccato grunts that are loud and low-pitched. These features allow the grunts to be easily located and to be transmitted over long distances, thereby broadcasting the position of the predator. Other monkeys run into thickets, where the dense brush makes it difficult for a swooping eagle to catch them (Struhsaker 1967). The responses just described are also typical of responses to playback tapes of these three types of call. The responses, then, are specific to the nature of the alarm call and not to the appearance of the particular type of predator (Seyfarth et al. 1980).

Other calls vary according to context rather than the species of the predator. For example, California ground squirrels whistle when a predator arrives suddenly and there is little time to escape, and give a chatter-chat call when predators are at a distance (Owings and Hennesy 1984; for other examples see Seyfarth and Cheney 2003). We'll talk more about the evolution of alarm calls in Chapter 19.

Alarms That Cause Animals to Assemble

Other types of alarm signals cause those who hear them to congregate into a group in order to defend a resource or drive off predators (Bradbury and Vehrencamp 1998). As you might expect, compared to signals that cause others to flee, assembly signals generally need to be easier to localize (so conspecifics can find the signaler) and need to be longer-lasting. Often they are repetitive. You may, for example, have heard a flock of crows cawing loudly and repeatedly as they mob a hawk.

Responses to assembly alarms can be behaviorally quite complex. Consider the general pattern of the response of ants to alarm pheromones (summarized in Yamagata et al. 2007). First they freeze, then they raise their heads and wave their antennae. They next move toward the source of the pheromone and perhaps release pheromone themselves. Finally, they begin biting the potential enemy. (If you have ever been bitten by fire ants, you can appreciate how effective a defense this is.) This complicated response is mediated by numerous neurons that are sensitive to alarm pheromone components.

AGGREGATION

Besides assembling in response to alarm signals, animals often aggregate for other reasons—for example, to hibernate, to share a resting place or a roost, or to prepare for migration. To select a particularly unpleasant example, consider bedbugs (*Cimex lectularis*). Bedbugs are household pests that come out from hiding places, such as in cracks in a bedframe, in the middle of the night and bite sleeping humans for a blood meal. They have recently been in the news because there has been a resurgence in bedbugs, even in expensive hotels. Bugs may benefit from aggregating for a number of reasons, including decreased sensitivity to desiccation, protection from predators, and ease of finding mates. To find one another, bedbugs release an aggregation pheromone (Siljander et al. 2008). Perhaps humans will be able to exploit this pheromone to control bedbugs.

AGONISTIC ENCOUNTERS

Animals can be in conflict with conspecifics over food, territory, mates, and their places in the dominance hierarchy. Agonistic behaviors are the actions involved in conflict, including both aggressive behaviors, such as threats and attacks, and submissive behaviors, such as appeasement or avoidance. Familiar examples include bighorn sheep butting heads, cats hissing, and dogs rolling on their backs to offer their vulnerable bellies to a stronger rival. We will talk about these behaviors extensively in Chapters 17 and 18.

COMMUNICATION ABOUT RESOURCES: A CASE STUDY

Group-living animals sometimes communicate the location of food and other resources to one another. For example, some animals are central-place foragers, leaving from a shared nest to collect food. In this section, we describe one remarkable and well-studied communication system, that of the honeybee (*Apis mellifera*).

Honeybees live in large colonies in carefully constructed hives. We explore honeybee sociality in more detail in Chapters 15 and 19; for now, what is most relevant is that all the labor required to maintain a colony is done by female worker bees. Workers divide up the tasks of caring for the hives and brood. Older bees are foragers and search for flowers that provide nectar and pollen. Depending on the neighboring habitat, flowers can be widely scattered, and their distribution changes rapidly over time as different species bloom and fade. It may take a few days for a newly opened flower to be discovered. However, once a bee finds the resource, many additional recruits soon appear. Scientists have known for a very long time that the initial discoverer does not simply lead the others to the food, because new bees are recruited even if the discoverer is captured as she leaves the hive on a return trip (Maeterlinck 1901). Thus, successful foragers must be communicating something about the new food source to other bees.

In the darkness of a hive, returning foragers do a very obvious and characteristic dance, which we describe in detail below. Other bees very attentively follow the dancers, touching them with their antennae and heads. Although many naturalists had noted these curious dances, their function was unknown. In the late 1910s, a scientist named Karl von Frisch was captivated by the mystery of the dances and began to suspect that it had some communicatory role (Munz 2005). Von Frisch was an ethologist who had worked on several different species. (Munz reports that one of von Frisch's earlier papers had the intriguing title of "A Catfish that Comes When Summoned by Whistling," in which he described training a blinded catfish to associate whistling with a food reward, thereby settling an argument over whether catfish can hear airborne sound.) Von Frisch's study of the dance language of honeybees spanned 50 years. The remarkable story that unfolded from his research eventually earned him the Nobel Prize in 1973 (see Chapter 2). Research on bee communication has been, at times, fraught with controversy, and there is still much to learn, but it is inextricably linked with von Frisch's name.

In his research, von Frisch used very simple techniques, many of which are still employed today. One technique was to put out feeders of sugar water that bees readily visit. Another was to individually mark bees by painstakingly applying unique patterns of paint (these days you can buy premade numbered tags, ready to glue on your bee). Glass-walled observation hives allowed him to observe bee behavior inside the hive. With these techniques, von Frisch could follow individuals as they foraged at the feeders, watch them dance in the hive, and then see whether marked bees who attended a dancing bee then visited the appropriate feeder.

Two Types of Dances

Von Frisch classified the dances he saw into two categories. The round dance is a circling dance, just as the name implies: the bee runs in a circle, then reverses direction and circles again (Figure 16.21*a*). The waggle dance is in the shape of a figure 8. The bee runs in a straight line through the center, circles to the right, runs straight through the center again, and then circles to the left (Figure 16.21*b*). During the central run, called the "waggle run," the bee waggles her abdomen about 15 times per second and buzzes her wings at about 250 beats per second.

Von Frisch originally suggested that these different types of dances indicate different food sources: he noticed that bees performed a round dance after visiting a feeder and a waggle dance after visiting a flower. However, he soon realized a potential confound in his experiment. He had initially placed all of his feeders close to the hive for convenience's sake, and flowers were further away. Instead of indicating the type of the food source, he wondered whether instead the two types of dances indicated distance information. Let's examine some of his experiments.

Here is one of the protocols that von Frisch used (and that is often used today). First, bees are trained to come to a feeding station by setting out a weak, unscented sugar solution. Because the solution is weak, the foragers do not dance upon their return to the hive.

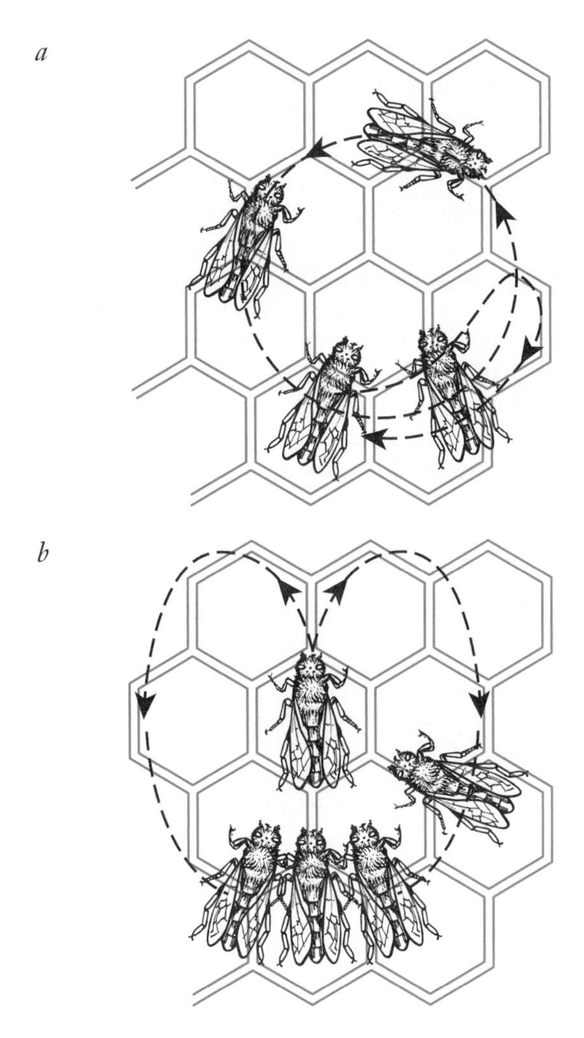

FIGURE 16.21 **(*a*) The round dance of a honeybee, performed after finding food near the hive. The bee circles alternately to the left and right, while recruits follow her. (*b*) The waggle dance, performed after finding food greater than about 50 m from the hive. During the waggle run through the center of the figure 8, the bee waggles her abdomen and buzzes her wings.**

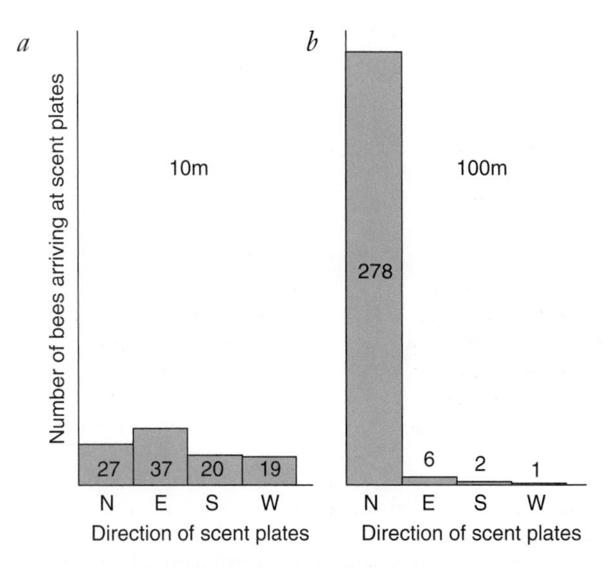

FIGURE 16.22 **Number of bees arriving at scent plates positioned to the north, south, east, and west of the hive when the training station was (*a*) 10 m to the east of the hive and (*b*) 100 m to the north of the hive. When the food is close to the hive, the recruits search randomly within a certain radius. When the food source is distant, they search primarily in the direction of the training station. (Modified from von Frisch 1971.)**

On the test day, a stronger sucrose solution, with a scent such as lavender oil, is set out at the feeding station. Foragers feeding on this solution return to the hive and dance. During the dancing, the hive is closed to prevent bees from exiting. Meanwhile, food is removed from the feeding station, and scent plates containing lavender fragrance but no food reward are set out. Different arrangements of these plates are used to test different hypotheses. When the hive is open, recruits leave and search for the food. The number of bees arriving at each scent plate indicates the area in which the bees are searching.

To test whether the round dance and the waggle dance are used to indicate different distances, bees were trained to a feeding station 10 m to the east of the hive. After sipping the strong scented sucrose on the test day, the scout returned to the hive and did a round dance. While the hive was closed, the feeding station was emptied and scent plates were positioned to the north, south, east, and west. As you can see in Figure 16.22*a*, the recruits arrived at each scent plate in almost equal numbers. Apparently, they did not know the direction to the nectar. This result suggests that the round dance does not convey direction information, but only tells recruits to "search for nearby food." It also suggests that bees may pick up the lavender scent as a cue. We return to these points with updated information below.

When von Frisch moved the same feeding station 100 m north of the hive, the returning discoverer did a waggle dance. This time almost all the bees arrived at the northern scent plate, indicating that they had direction information (Figure 16.22*b*). The accuracy of the direction information was further demonstrated in fan experiments. Following his usual procedures, von Frisch trained bees to a feeding station and then placed scent plates so that they were arrayed like a fan. Each plate was 550 m from the hive and separated from its neighbors by 150 m. The overwhelming majority of recruits appeared at the scent plate nearest to the original feeding station (Figure 16.23). Then von Frisch performed a step experiment to examine the accuracy of the distance information. After training bees to a feeding station, scent plates were placed in line with the empty feeding station at intervals closer or farther away. Again, the recruits appeared at the scent plates closest to the

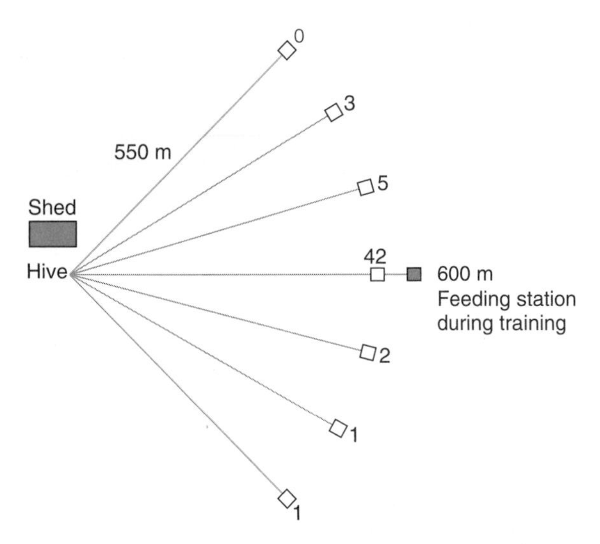

FIGURE 16.23 **The setup of a fan experiment to determine whether bees use directional information. The solid square shows the position of the feeding station during training, and squares show the position of the scent plates, which had no food. After following a waggle dance, most recruits arrive at the scent plate nearest the site of the feeding station. The number of bees arriving at each station is indicated.**

original feeding station (Figure 16.24). Thus, the waggle dance not only says "Food is far away," but also encodes information about direction and distance. As further experiments have shown, direction and distance are encoded by different aspects of the waggle dance.

How Dances Encode Information About Direction

Picture a forager poised at the exit of a hive, ready to set off for a particular flower. One piece of information that would help a bee determine its departure direction involves the sun. If a bee knows the angle formed by the sun's azimuth (the point on the horizon below it), the hive, and the flower, it would know which way to go (Figure 16.25). This angle is exactly the information that dancing bees convey. Honeybee dances are performed on the vertical comb of the hive, so it's not possible for a bee to simply waggle run in the direction of the food. Instead, the bees use the angle of the run with respect to gravity. The angle between the waggle run and "up" on the comb is the same as the angle formed between the flower, the hive and the azimuth. If a bee needs to fly toward the sun in order to reach the flower, the waggle run is oriented straight up; if the bee must fly directly away from the sun, the waggle run is oriented straight down. Likewise, a food source 20° to the right of the sun would be indicated by a waggle run directed 20° to the right of vertical.

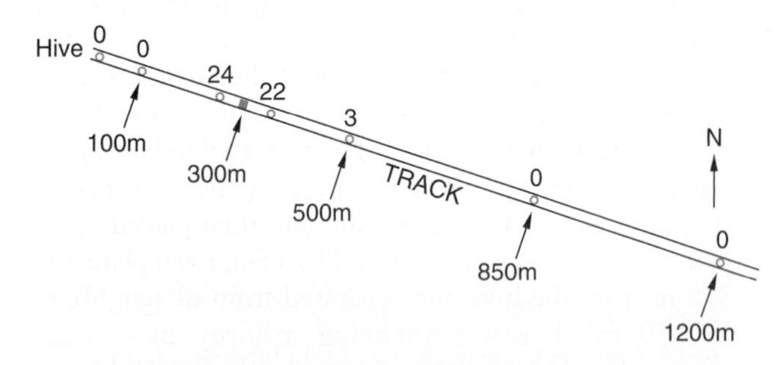

FIGURE 16.24 **The setup of an experiment to determine whether bees use distance information. The solid square shows the position of the feeding station during training, and circles show the position of the scent plates. After following a waggle dance, most recruits arrive at the scent plate nearest the site of the feeding station. The number of bees arriving at each station is indicated.**

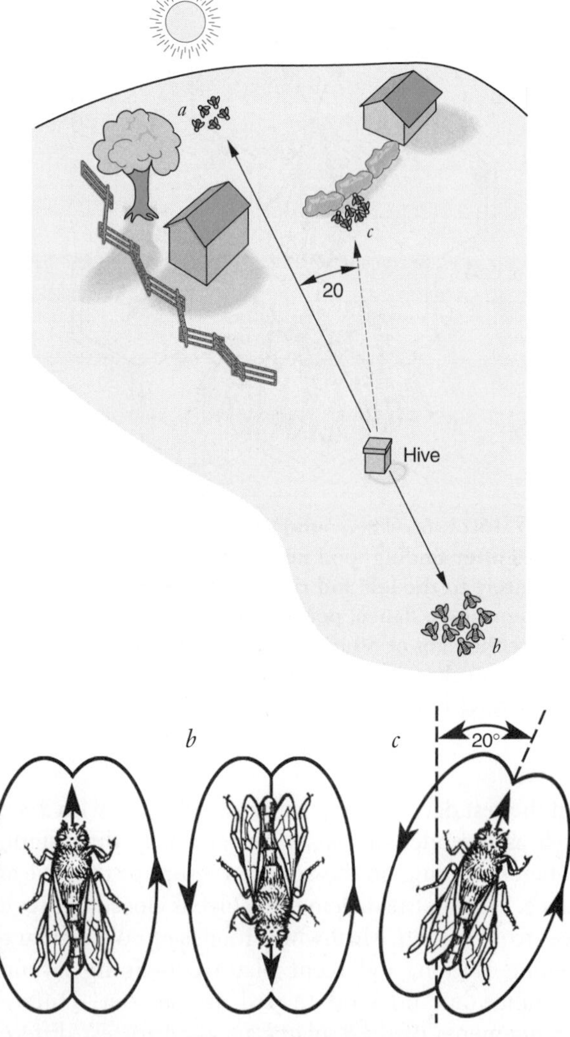

FIGURE 16.25 **Bees dance inside a dark hive on a vertical comb. During a waggle dance, the dancer indicates the direction to the food by the orientation of the waggle run relative to a vertical line. (*a*) If the food source is in the direction of the sun, the waggle run is oriented straight up. (*b*) If a bee should fly away from the sun to reach the food, the waggle run is oriented straight down. (*c*) If the food is 20° to the right of the sun, as in the diagram, the waggle run is oriented 20° to the right of vertical.**

One of the challenges with using the sun as a directional cue is that, as the earth rotates, the sun appears to move across the sky. Bees often dance for a long time in the darkness of the hive without going back outside to update their idea of where the sun lies. However, they don't need to, as the sun's apparent movement is very predictable. Dancing bees adjust the angle of their dances by approximately 15° an hour to compensate for the apparent movement of the sun.

How precise are these directions? Not that precise at all. Dancing bees repeat their dances over and over, and the direction of the waggle run is not exactly the same each time. It appears that observing bees take an average of the runs they've followed in order to select a direction in which to fly (Tanner and Visscher 2008). The directions encoded in the dance get bees to the general area of the resource, and from there they use scent to pinpoint the location of the flower.

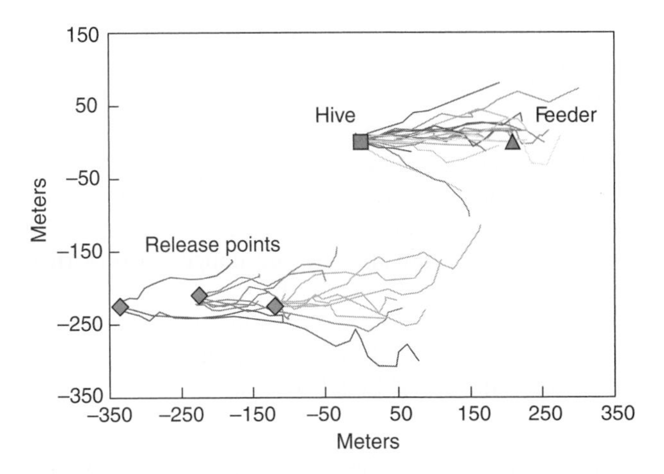

FIGURE 16.26 **The flight paths of bees that had attended a waggle dance, mapped by harmonic radar. The hive is at the center point of a large field. Bees recruited by a dance generally went straight to the feeder from the hive. Bees captured as they were leaving the hive and moved to a new release point 250 to 300 m away flew to the location where they expected the feeder to be. (From Riley et al. 2005.)**

How Dances Encode Information About Distance

Distance to the food source is correlated with two other features of the waggle run. The more waggles, the greater the distance to the food (von Frisch 1967). The distance to the food is also correlated with the duration of buzzing during the straight run. As the distance to the food increases, the sound trains of buzzes are longer (Wenner 1964).

A recent analysis of waggle and round dances suggests that perhaps these two types of dances are not as discrete as they first appeared to be (Gardner et al. 2008). Round dances often contain very brief waggle phases with acoustic signals that contain distance information. In addition, round dances also can contain directional information. However, there is a large amount of "noise" in round dances, thus making these dances less precise than waggle dances.

New Technology Offers Additional Insights

Most studies of bee navigation rely on simply seeing which bees show up at particular feeders. New technology has allowed researchers to map the paths of recruits. Riley et al. (2005) attached tiny harmonic transponders, weighing only a few milligrams, to bees. Unlike much heavier radio collars that are used in tracking larger animals, these passive transponders return signals to a radar. By positioning the radar at the edge of a large mowed field, researchers could accurately map the paths of individual bees. Bees that had followed a dance generally went straight to the feeder. Bees did not use odor cues to find the feeder, as no scented foods were used, and the prevailing breeze came from a different direction. If bees leaving the hive were captured and displaced, they did not fly in the true direction of the feeder, but rather searched where the dancer had led them to expect the feeder to be relative to their release point (Figure 16.26).

It has even been possible to "talk" to bees in their own language by using a mechanical model of a dancing bee (Figure 16.27). A computer controlled the model's movements and sound production. The dancing model bee fed recruits scented sugar water through a syringe, mimicking the manner in which a real bee would feed the recruits upon returning to the hive.

FIGURE 16.27 **A mechanical bee that can communicate to live bees is constructed of brass covered in beeswax. On its back is a single wing made from part of a razor blade. An electromagnet causes the wing to vibrate so that it mimics the sound patterns produced by real dancing bees. A tiny plastic tube connected to a syringe releases droplets of scented sugar water, simulating regurgitation by live bee scouts. Computer software choreographed the robot's dance so that it directed the bees to a food source in a location chosen by the experimenters.**

The model danced, indicating a food source 250 m to the south of the hive. Scent plates without food were placed 250 m to the north, south, southeast, and southwest of the hive. Although the model's dance was not as effective as a live bee's dance in recruiting others to look for food or in accurately directing them to the correct location, most of the recruits showed up at the feeding station indicated by the dance. Two critical components of the dance are the wagging movements and the buzzing. If either was missing from the model's dance, the recruits showed no preference for the direction indicated (Michelson et al. 1989).

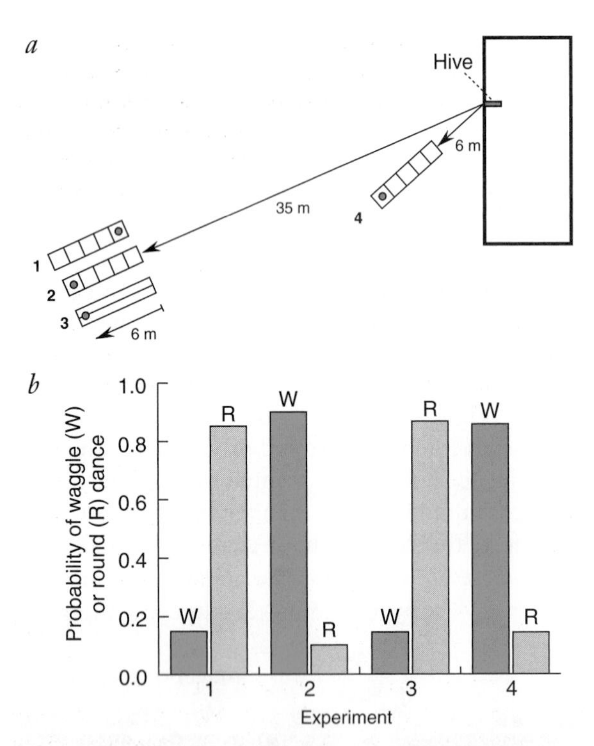

FIGURE 16.28 **(*a*) Layout for four experiments using tunnels with visual cues. Tunnels 1, 2, and 4 had vertical stripes that provide the bee with the perception of a great deal of image motion. Tunnel 3 had horizontal stripes that provided little perception of motion. The small circles inside the tunnels indicated the placement of the feeders. (*b*) The probability that bees did waggle (W) dances, indicating that they perceived the feeder to be far away, or round (R) dances, indicating that they perceived the feeder to be nearby, for each experiment. In experiment 1, the feeder was placed near the tunnel entrance, and bees perceived it to be close by. In experiment 2, the feeder was placed at the far end of a tunnel with vertical stripes, and bees perceived it to be far away. In experiment 3, the feeder was placed at the far end of a tunnel with horizontal stripes, and the bees perceived it to be close by. In experiment 4, the tunnel with vertical stripes was placed close to the hive, and bees perceived it to be far away. (From Srinivasan et al. 2000.)**

How Do Bees Measure Distance?

In order to communicate information about distance, bees must have some way of measuring it. There are a number of different possibilities, such as keeping track of the effort it takes to fly, or by keeping track of the landscape going by. In an elegant experiment, Srinivasan et al. (2000) discovered that foragers rely on visual cues to estimate how far they have traveled. These researchers trained bees to fly into tunnels (6.4 m long, 11 cm wide, and 20 cm high) in order to reach a feeder. The researchers manipulated the amount of image motion that the bees experienced, also called the optic flow, by manipulating the patterns on the inside walls of the tunnels (Figure 16.28*a*). Some tunnels had horizontal stripes lining the inside, parallel to the bees' movement. This design provided very little information to the bees about the distance they traveled. Other tunnels had vertical stripes, which provide the bees with the perception of a great deal of image motion, especially in the narrow confines of the tunnel. The tunnels were placed outdoors (Figure 16.28*a*), either 6 m or 35 m from the hive, so that part of the bees' journeys to the feeders was through the tunnel.

In order to determine how the bees perceived direction, the researchers took advantage of the two types of dances. The round dance is generally given when resources are within 50 m of the hive, and waggle dances for longer distances. All the tunnels were well within the 50-m mark. Researchers watched how bees danced when they returned to the hive after visiting a particular feeder. The results are shown in Figure 16.28*b*: When bees visited a feeder at the opening of a tunnel, or visited a feeder with horizontal stripes, they performed a round dance. However, when bees flew through a tunnel with vertical stripes, they acted as if they had flown a long way and performed a waggle dance, even when the tunnel was only 6 m from the hive. By analyzing the duration of the waggle runs, the researchers could determine by how much bees were overestimating the distance they flew. The tunnel increased the dancers' perception of the distance they had flown by a factor of 31! (Srinivasan 2000). The bees that were recruited by foragers confused by vertically striped tunnels searched for the food well beyond the tunnel, rather than looking inside the tunnel (Esch et al. 2001).

Thus, foraging bees learn about distance by using optic flow, and they convey that information to other bees. Normally this system works, as food sources are not hidden inside tunnels, and recruits are likely to perceive the same optic flow patterns as a dancing bee, given that the direction information is accurate. For example, a bee will perceive equally long flights over water or through a forest as very different distances, but if the recruits also take the same path, they will successfully find the food (reviewed in Dacke and Srinivasan 2007).

Other Cues Used in Foraging

As alluded to earlier, whether waggle dances truly convey information has been the subject of controversy (reviewed in Munz 2005). Adrian Wenner (e.g., Wenner 2002 and references therein) has been the major advocate of the idea that bees rely on odor cues and that the dance language is not used by recruits to gain information. Although most bee researchers accept the validity of the waggle dance as a communicatory system, Wenner's work and that of others helped focus attention on other cues used by bees, particularly odor cues.

The waggle dance directs bees to a particular area, but not that precisely, as we have discussed. The location of the flower is pinpointed by odor cues. Dance followers can detect food scents on the dancers and even contact dancers in the appropriate place: hivemates that foraged for pollen were more likely to contact the dancers' legs (where pollen is stored in "baskets"), whereas dancers that had collected sugar solutions were more likely to have head-to-head contacts (Díaz et al. 2007).

Besides the chemicals they pick up from food, dancing foragers also produce other chemicals from their abdomens. By comparing air samples taken from dancing and nondancing bees, and analyzing them with gas chromatography and mass spectrometry, Thom et al. (2007) identified four chemicals released by the dancers. When the researchers injected these chemicals into the air inside the hive, more foragers exited the hive. Thus, it seems that these chemicals cause the bees to become primed to look for food.

SUMMARY

Different definitions of *communication* have been used in the literature. Here, we will use the definition that communication occurs when information is transferred from the sender to the receiver and that the sender benefits, on average. A behavior, such as a call, or a feature, such as a colorful feathered crest, that transmits information is called a signal. Signals are thus distinct from cues, which may provide information to another animal but do not benefit the sender, such as the rustle of leaves that attracts an owl to a mouse.

Any sensory channel may be used for communication. Signals within each channel have characteristic properties that make them more or less useful, depending on the species, the environment, and the function of the signal.

Visual signals are easy to locate, are transmitted quickly, and disappear just as fast. However, visual signals must be seen and are, therefore, useful only when there is enough light and where there are few obstacles to obscure them. Visual signals include color and pattern, as well as movements and postures.

Auditory signals can be transmitted over long distances. The rate of transmission and fade-out is rapid. They do not require light and, in fact, work well under water. The sounds may be generated by respiratory structures, the rubbing of appendages, or beating on parts of the environment. Some animals produce sounds above (ultrasound) or below (infrasound) the limits of human hearing.

Seismic signals are those caused by vibration of the environment. They are well suited to communication over long distances. Seismic signals can be produced when an airborne vocalization couples with the ground.

Chemicals can convey messages over great distances, particularly when assisted by currents of air or water. They are transmitted more slowly, are more durable, and are usually more difficult to locate than visual or auditory signals. Furthermore, they are effective in environments with limited visibility. Chemicals used to convey information to conspecifics are called pheromones. When the chemical has an immediate effect on the behavior of the recipient, as occurs with sex attractants and trail and alarm substances, it is called a releaser pheromone. Primer pheromones act slowly, exerting their effect by altering the physiology of the recipient. Some species of amphibians, reptiles, and mammals have a vomeronasal organ, used (in addition to the olfactory epithelium) to sense chemical cues.

Tactile signals are effective only over short distances, and it is easy to locate the sender. These signals are rapidly transmitted. Allogrooming, a special form of tactile communication, is practiced by many species but is especially prominent among primates.

Electrical fields are used for communication among the mormyriform fishes of Africa and the gymnotiform fishes of South America. Transmission and fade-out are almost instantaneous, but the signals do not travel far. They are effective when visibility is limited.

Many animals communicate using displays that contain signals from more than one sensory channel; this is called multimodal communication. The messages sent by different channels can be redundant or nonredundant.

Communication has many functions, many of which we explore in more detail in other chapters. Many animals use communicatory signals to differentiate conspecifics from other animals. This distinction is clearly valuable, as conspecifics are likely to be competitors or potential mates. Virtually all sensory channels are used for species identification in one species or another. In regions where two closely related species overlap, we sometimes see greater differentiation between the signals of the two species than in regions where there is no overlap. This suggests that selection against making a mistake is particularly strong. In other examples, such as in invasive species that do not have a long evolutionary history with native species, we see many errors in species identification.

Communication is used throughout the reproductive process. First, many solitary animals have the challenge of locating potential mates and use communicatory signals to attract them. Because chemical and auditory channels work particularly well over long ranges, they are often (but not always) used for mate attraction. Second, animals might signal their identity, both their species and sex. Third, animals might advertise their qualities as a mate, as discussed in Chapters 14 and 17. Fourth, animals might use displays to coordinate their physiology, as illustrated by the well-studied case of the ring dove. Finally, some animals communicate in order to maintain their pair bond, even when it is not mating season.

Group-living animals maintain social bonds as well, generally through touch. Many animals lie or sit in contact with each other or have greeting ceremonies. Many species allogroom, or groom one another, which has the benefit of improving the health of skin, fur, or feathers, and also cements social bonds.

Alarm signals warn others of danger. Some alarm signals cause receivers to flee. These signals are generally quick to produce and difficult to locate, which benefits the signaler that is trying not to attract the attention of a predator. Alarm signals tend to be similar across species, and some animals respond to alarm calls of other species. Some alarm signals specify the nature of the danger, such as a predator approaching on the ground versus in the air.

Other alarm signals cause animals to assemble in order to defend a resource or drive off predators. This type of alarm signal tends to be easy to localize and longer-lasting than the signal that causes receivers to flee.

Animals must aggregate for other reasons besides alarm, such as to hibernate, roost, or prepare for migration. Many species use aggregation pheromones in order to attract conspecifics to a particular spot.

Animals are often in conflict with conspecifics over food, territory, mates, and their places in the dominance hierarchy. Many signals have evolved in the context of agonistic behavior. We will delay discussion of these to Chapters 17 and 18, where they are treated extensively.

Animals also communicate about the location of food resources. We end the chapter with a detailed look at a specific case study, the dance of the honeybee. Honeybee scouts return to the hive and communicate the direction of food with either a round dance (for nearby resources) or a waggle dance (for distant resources). The waggle dance encodes information about the direction of the resource (via the angle of the "waggle run" section of the dance) and the distance of the resource (via the duration of the "waggle run" and its accompanying buzzing). Bees measure information about distance via the amount of optic flow they perceive as they fly.

17

The Evolution of Communication

As we saw in the last chapter, we can often answer the who, what, where, and when of the communication process, but more difficult are questions of *how* and *why* have signals taken the various forms that they have. To answer these questions, we must consider the evolutionary costs and benefits of the signal, as well as the morphological and physiological features that influence the signal's production.

We will begin by examining in more detail something that we alluded to in the last chapter—the changing way that animal behaviorists have viewed communication. We'll consider when we expect communication to be honest, and when we expect it to be unreliable. Then we will discuss two hypotheses about the evolutionary origin of signals. Next we will look at some of the selective forces that influence signal design. We will examine language—what it is and whether it is uniquely human. Finally, we will discuss what the study of communication can tell us about animal cognition.

THE CHANGING VIEWS OF COMMUNICATION

SHARING INFORMATION

In the last chapter, we described how communication can be broadly defined as the transmission of information from one animal to others. Many researchers through the 1970s emphasized a cooperative view of communication, where both sender and receiver benefit from the accurate transfer of information (reviewed in Dawkins and Krebs 1978). Under this scenario, selection should act to make signals efficient, reliable, and unambiguous. Dishonest, or inaccurate, signaling was thought to be unlikely in animals; after all, even in humans, lying is socially risky and hard to do convincingly (Smith 1977).

The idea that communication benefits both partners does seem reasonable at first glance. Consider, for instance, the complex role of communication in coordinating mating activities. First, potential partners must locate each other. As described in Chapter 16, many species do this via species-specific pheromone signals. It is clear that females are adapted to broadcast news of their availability, while males are adapted to detect the particular chemical signal of prospective mates. After potential partners find each other, they may engage in a complex give-and-take of displays that culminate in mating. Surely, both partners receive fitness benefits by clearly communicating their intentions to each other. Thus, many familiar examples of communication—and indeed much of our own experience—seem to uphold the view that its function is to share information.

MANIPULATING OTHERS

This cooperative view of communication does not fit every situation, however. There are some situations in which the sender might gain by sending an inaccurate signal. In a territorial dispute, for example, the sender might bluff by sending signals that exaggerate its willingness to escalate the contest. Similarly, a male competing for a female's attention might increase his mating success by exaggerating his qualities. In these and many other cases, an animal that gave a dishonest signal would gain an advantage over animals that honestly communicated about their abilities or intentions. In fact, we expect deceptive signals will evolve on a regular basis. (Note that the terms *honesty*, *dishonesty*, and *deceptive* imply nothing about the cognitive abilities or the intentions of the sender, but are generally used in the literature as descriptive shorthand terms.)

How can we incorporate the possibility of deceptive signals into our understanding of communication? Richard Dawkins and John Krebs (1978) suggest that animals communicate not to convey information, but to manipulate the behavior of others to their own advantage. Thus, an animal will produce a signal when, *on average*, it results in the increase of its own reproductive success by influencing the behavior of others. We emphasize "on average" because not every signal an animal gives will necessarily be to its benefit. But for signals to evolve, they must benefit senders overall.

In this view, senders may not be trustworthy. So why would a receiver ever respond to a signal? If signals are potentially dishonest, why not just ignore them? *On average*, the receiver must benefit from responding to signals, even if it is sometimes deceived. After all, not all signals are dishonest. Often a signaler benefits from conveying accurate information, such as in the case of the female moth alerting males that she is ready to mate.

Thus, signals are normally reliable but occasionally deceptive. In the next section, we will examine the circumstances that favor honest signaling and those that allow for deception.

SIGNALS AND HONESTY

WHEN ARE HONEST SIGNALS LIKELY?

We will review four common circumstances under which we expect to see honest signals: (1) when senders and receivers share overlapping goals, (2) when signals indicate something about the sender that cannot be faked, (3) when signals are costly to produce, and (4) when dishonest signalers can be identified.

If Senders and Receivers Share Overlapping Goals

Sometimes both the sender and receiver share the same goal: the sender benefits by accurately transmitting information, and the receiver benefits by accurately interpreting it. The coordination of behaviors between animals that have chosen to mate provides an example. In cases like these, we expect that natural selection will favor unambiguous, honest signals.

When can we expect to see overlapping goals? One obvious place to look is the relationship between parent and offspring, although as described in Chapter 15, even this relationship can entail significant conflict. Nevertheless, if you've been around infants, you will have no trouble believing that they have been selected to convey their needs for food and comfort to their parents. Of course, human parents are not the only ones at the beck and call of their offspring. The human infant's need for near-constant care is rivaled by altricial baby birds. Naked and helpless, one of their few behavioral options is to beg their parents to feed them, and they do a lot of this. (Beware the undergraduate research assistant position that entails raising baby birds!) Nestlings beg with a hugely gaping mouth. As their begging intensifies, they add vocalizations. Begging calls are themselves graded in intensity, and can be emitted with great persistence and at a noise level that is hard to ignore. At the highest level of intensity, baby birds stand, gape, flap their wings, and call all at once (Figure 17.1). Experiments have shown that in many bird species, the vigor of begging increases with the degree of hunger (e.g., Redondo and Castro 1992). Begging chicks are quite obvious to predators as well as parents, so begging *too* much, when food is not needed, is a poor strategy (Haskell 1994). Parents usually respond to the signal by

FIGURE 17.1 **Levels of begging display in birds. (Modified from Redondo and Castro 1992.)**

providing food to those chicks that beg most vigorously (reviewed in Searcy and Nowicki 2005). Parents are under selection to respond because their own fitness depends on the survival of their offspring. Thus, in general, communication between baby birds and their parents appears to be honest.

Many other examples exist in which the goals of the sender and receiver of a signal are aligned. For example, as we will describe in more detail in Chapter 19, Belding's ground squirrels call to warn relatives of the presence of a predator. Because genes are shared between relatives, both the sender and receiver of the call reap a fitness benefit. In other species, members of a group share a common goal: every bee benefits when scouts convey accurately the location of a patch of flowers in full bloom. In such cases, honest communication is likely.

If Signals Cannot Be Faked

Sometimes signals are honest because they are tightly linked to a trait of the sender. Although the sender might benefit if it could lie, it simply is not possible to fake the signal. For instance, size is usually a good predictor of fighting success, and many displays allow opponents to judge one another's size. In some species, combatants can enhance their apparent size by puffing up their feathers, fluffing out their fur, or assuming an upright posture, but in other species, size is not so easily faked. In the threat display of male stalk-eyed flies (Diopsidae), flies face each other head-to-head, with their forelegs spread outward and parallel to the eyestalks (Figure 17.2). This pose allows each competitor to compare the length of its eyestalks to that of its rival. Eyestalk length increases with body size, and males with shorter eyestalks usually retreat without a fight (Burkhardt and de la Motte 1983; de la Motte and Burkhardt 1983). Thus, eyestalk length is an accurate index of size that cannot be faked.

FIGURE 17.2 **An aggressive display of male stalk-eyed flies. In this pose, each male can determine the distance between the eyes of its opponent and thereby the rival's size. Size is correlated with fighting success. Since there is little a combatant can do to alter its body size, this is an example of honest signaling.**

Other signals are honest because they are linked very strongly to an animal's health and physiological well-being. For example, the bright reds and yellows of feathers, scales, or fleshy necks or combs of some birds depend on chemicals called carotenoids. These chemicals cannot be synthesized by vertebrates but must be obtained in the diet. This makes these bright colors a good candidate as an honest signal of foraging ability and health. In a series of studies, Geoff Hill examined this signal in house finches, common backyard visitors throughout the United States (Hill 1990, 1991; Hill and Montgomerie 1994). Female house finches are a rather bland, streaky gray, but the head, throat, and breast of male house finches varies from pale yellow to bright red within a population. With controlled feeding experiments, Hill determined that the brightness of the color depends on carotenoids in the diet. Females prefer the brighter, redder males. The females benefit from their choice: brighter males are better parents, bringing more food to the nestlings.

If Signals Are Costly to Produce

Other signals seem to be more plastic or labile in their expression. For example, plumage colors may be bright or dull, and courtship dances may be given vigorously or lethargically. When signals are labile, and the sender and receiver do not share the same goals, the situation is ripe for deceit. For example, imagine a male attempting to convince a female that he, above all others, is the right one for her. It is to the male's advantage to convince the female to mate; however, it is to the female's advantage to accurately assess the quality of as many males as feasible in order to pick the best father for her offspring. What then is to prevent the male from exaggerating his qualities? How would selection favor honesty?

One hypothesis, proposed by Amotz Zahavi, is that reliable signals will be favored in a population when signals are costly to the sender (Zahavi 1975, 1977; Zahavi and Zahavi 1997). This is the **handicap principle** that you met in Chapter 14. Originally developed for signaling in the context of mate choice, it is also applicable to other types of communication. The idea is perhaps best described by an analogy. Imagine a person who buys expensive cars and $2000 wristwatches. These are hardly necessary to the owner: in fact, they divert resources from other things that might be more important. However, cars and watches might act as a signal to others, including prospective mates, that the owner has money to burn. Similarly, Zahavi suggested that some signals confer an advantage on their owner, not because they are useful themselves, but because their very extravagance indicates the owner's qualities. These signals are referred to as handicaps because their owners are perceived as doing well in spite of the handicap of investing in the signal. Zahavi's idea sparked lively debate and was followed by tests of its plausibility via mathematical models (e.g., Grafen 1990; Johnstone 1997). Eventually, several variations of the original formulation were introduced and are now generally regarded as plausible ways by which signals may evolve.

For a signal to evolve as a handicap, three criteria must be met: the signal must be costly, it must relate to the quality of the sender, and the receiver must be interested in the quality of the sender that is being signaled, and thus must benefit from attending to an honest signal. We've already seen a number of examples of how receivers benefit from correctly assessing the quality of a signaler, especially in the context of sexual selection (see Chapter 14), so let's look at some examples of signals that have been shown to meet the first two criteria.

Signals may be costly in several ways, but energetic costs are the most commonly measured. A way to monitor energy use is by measuring the amount of oxygen consumed during an activity. In insects and frogs, oxygen consumption increases 5 to 30 times when they call (Ryan 1988). Male red deer lose weight during the rut (mating season) because they constantly roar at one another as they compete for mates (Clutton-Brock et al. 1982). Another type of cost is a decrease in the ability to move about, escape predators, or forage. For instance, in several bird species, the long tail feathers that males sport as sexual ornaments impair their ability to fly (Jennions 1993).

Researchers have now accumulated many examples in which higher-quality senders are better able to pay the cost of producing higher-quality signals (brighter colors, more vigorous displays) than are lower-quality senders. Among red deer, for example, only the males in top physical condition can continue roaring long enough to win the vocal duel. Roaring uses many of the same muscles and behavior patterns involved in fighting, and so it serves as an honest signal of a male's fighting ability (Bradbury and Vehrencamp 1998). Sometimes, a particular quality of the signal rather than its repetition is condition dependent. For example, when a rival encroaches on his territory, a male common loon (*Gavia immer*) gives a "yodel" (a different call than the eerie wail that may be more familiar to you). Individuals in better condition (as measured by body mass adjusted for overall body size) produce lower-frequency (Hz) yodels. Over five years of study, the yodels of loons that gained mass dropped in frequency, and those that lost mass rose in frequency. Yodels were artificially changed in frequency and played back, and lower frequency yodels were greeted by livelier responses from other loons (Mager et al. 2007).

A particularly charming example of honest signaling is in the blue-footed booby (*Sula nebouxii*), seabirds with brightly colored feet. As part of their courtship ritual, boobies stand facing one another as the male lifts his feet in alternation to display them for the female's inspection (Figure 17.3*a*). Males in good condition have bright blue-green feet, but when held without food for 48 hours (which routinely happens to boobies in nature), their feet quickly fade to dull blue (Figure 17.3*b*, *c*). Females respond to this rapid change in signal, preferring males with brightly colored feet. In fact, in one research study females adjusted their own investment in offspring based on this signal: they laid smaller second eggs when researchers used makeup to color their mates' feet a dull blue (Velando et al. 2006).

a

b

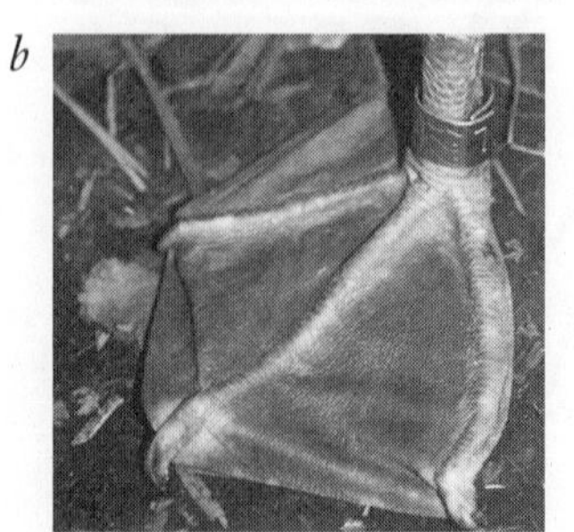

c

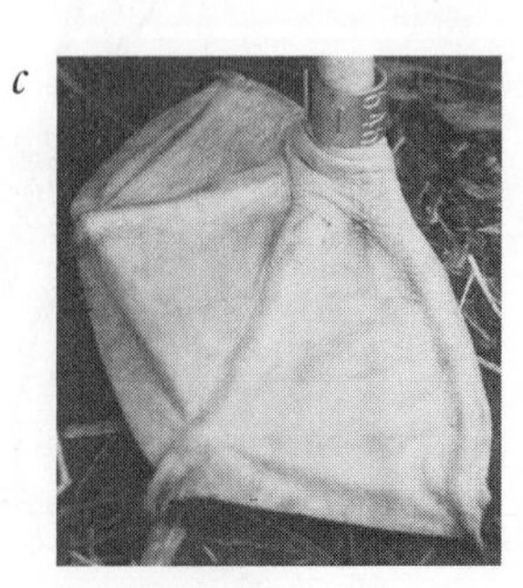

FIGURE 17.3 **(*a*) Blue-footed boobies lift their feet for examination by their partner during a courtship dance. Foot color changes with recent feeding, as shown by the feet of (*b*) food-deprived and (*c*) supplemented boobies.**

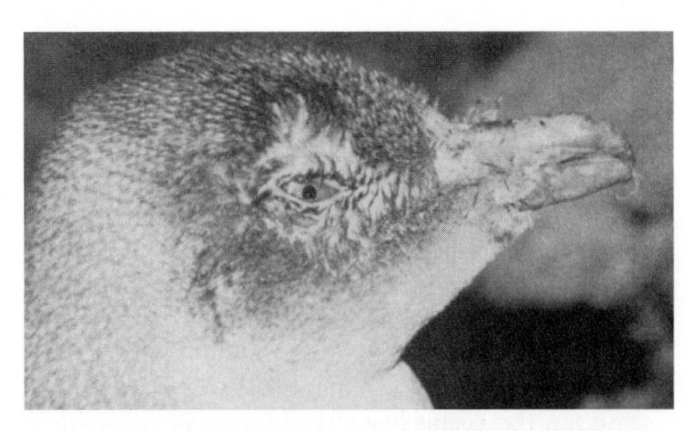

FIGURE 17.4 **These eye and head injuries in a little blue penguin resulted from fights. Little blue penguins have an extensive repertoire of aggressive displays, which vary in cost (the risk of injury). By its choice of display, a penguin indicates the extent of its willingness to fight. Signals remain honest because they are costly.**

Costs of signals may also come in the form of the receivers' response to the signal. Let's examine a situation that illustrates the dynamic relationship between the cost of a signal, the quality of the sender, and honesty. Little blue penguins (*Eudyptula minor*) have a repertoire of aggressive displays that differ in cost (risk of injury), effectiveness in deterring an opponent, and ability to predict an attack. Although two displays may involve the same ritualized posture, one that involves moving within the rival's striking distance is riskier to perform than one that is performed while stationary and out of the opponent's reach. By its choice of display, a penguin conveys information about both its willingness to sustain injury while performing the display and its willingness to fight. It chooses a display with costs that represent the value it places on the resource. Roughly 10% of penguin interactions are not settled by displays and end in fighting. During battles, losing penguins commonly suffer flesh wounds and sometimes eye loss (Figure 17.4). These injuries could make it more difficult to obtain sufficient food or to breed successfully, or the wounds might become infected and cause illness or death. Thus, attempting to intimidate an opponent into retreat by bluffing a strong motivation to attack could be quite costly, if the rival called the bluff and a fight ensued (Waas 1991).

The penguin encounters invariably begin with low-risk displays and escalate until one opponent retreats or a fight occurs. The process is somewhat analogous to human actions at an auction—the bids begin low and gradually increase until bidders unwilling to pay the price drop out of the process. The price that little blue penguins must pay is the risk of injury. As a territorial contest escalates and the price of the property increases, one "bidder" usually decides that the territory is not worth that great a risk. For these penguins, the signals remain honest because they are potentially costly, and the cost for a given level of signal is greater for weaker than for stronger individuals.

If Dishonest Signalers Can Be Identified

A stable social unit also favors honest communication. One reason to expect honesty is that individuals will both send and receive signals at different times. The advantages of sending dishonest signals will be reversed when the animal is the receiver. Therefore, the advantages of receiving honest signals might outweigh the advantages of sending dishonest ones, and honesty might come to predominate in the population.

Members of social units may be able to recognize one another and remember previous interactions, and thus learn whether a particular individual is honest. Working with vervet monkeys in Kenya's Amboseli National Park, Dorothy Cheney and Robert Seyfarth (1988) tested the idea that members of a social group would stop believing an individual that routinely gave unreliable signals. Vervet monkeys utter different vocalizations during different situations. They have two acoustically different calls that warn of the encroachment of another group of monkeys on their territory. One call, the "wrr," is given when another group is spotted in the distance, perhaps as far away as 200 m. The other, a "chutter," is given if the other group comes so close that there are threats, chases, or actual contact (Figure 17.5). Vervet monkeys also have a stable social group in which the animals recognize one another, including their calls. What would happen if one individual falsely signaled the approach of another group? Would other monkeys believe the "liar" again?

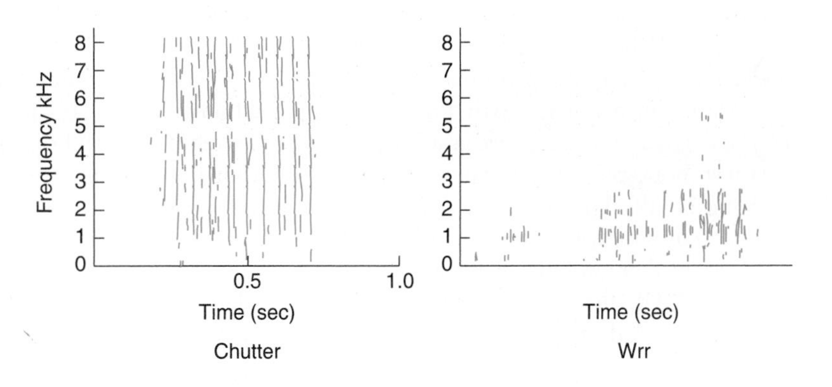

FIGURE 17.5 **Vervet monkey alarm calls that alert group members of the approach of a neighboring troop are of two types. The "chutter" warns that another group is nearby. The "wrr" is given when another group is spotted in the distance. The *x*-axis indicates time, and the horizontal lines show frequency in intervals of 1 kHz. (From Cheney and Seyfarth 1991.)**

To test this question, Cheney and Seyfarth broadcast tape recordings of the calls of a member of the group to create the illusion that it was giving inaccurate calls. The calls were broadcast from a speaker that was concealed where the group might expect that individual to be. First, an individual's "chutter" call was played to see what the baseline response of the others would be. The next day, when no other group was in sight, they played that individual's "wrr" call eight times, at approximately 20-minute intervals. They found that the other monkeys gradually stopped responding to that individual's warning that another group was in sight. The monkeys no longer believed the liar. However, if another monkey uttered the "wrr" call, the group still believed the warning and responded appropriately. So we see that in a stable social group in which individuals are recognized, others may soon learn not to believe a dishonest signaler.

WHEN ARE DISHONEST SIGNALS LIKELY?

Just as we can predict the conditions under which honest signals might be likely to occur, we can also predict when we expect to see dishonest signals.

If Senders and Receivers Have Different Goals

Just as overlapping goals between senders and receivers favor honesty in signaling, different goals set the stage for deception. Note, however, that this is not usually a black-and-white distinction. For example, a father bird and his chick share the general goal of the chick's survival. However, the chick may want to induce the father to give slightly more parental investment than he would like, perhaps at the cost of his other chicks or his future reproduction. Thus, we expect to see signals between parents and chicks that are essentially honest, but with some attempt by the chick at manipulation and deceit.

If Signals Are Costly to Assess or Challenge

Assessing signals can itself be costly. For example, a cricket assessing the quality of a singing male must invest time in listening and may risk attack by parasites or predators that are attracted to the male's song. Under conditions such as these, there may be more opportunity for senders to get away with deceit.

A receiver may pay dearly for challenging the honesty of a signal. If receivers are unwilling to accept the cost of a test, honest signals may be corrupted by occasional dishonest bluffs. For example, stomatopods (*Gonodactylus bredini*), commonly known as mantis shrimps, are marine crustaceans that ferociously defend the burrows and cavities in which they live. Stomatopods have two large forelimbs, called raptorial appendages, that can unfold and shoot forward in a manner similar to that of the praying mantis (Figure 17.6*a*). The raptorial appendages, which are adapted either for spearing or smashing, are used both in prey capture and territorial defense. Mantis shrimp are strong enough to easily break a human finger or the side of a glass aquarium. Combatants may be seriously injured or even killed during the battles over burrows. Readiness to attack is signaled by a threat display, called a meral spread, in which the raptorial appendages are splayed out (Figure 17.6*b*) (Caldwell and Dingle 1976; Dingle and Caldwell 1969).

So far, this seems like a typical case of signaling in aggressive contests. However, there is a twist: a newly molted stomatopod is virtually defenseless. Arthropods molt by pulling their bodies out of a break in their old exoskeleton, and in order to do so their new exoskeletons are very soft and pliable. A newly molted stomatopod gives the meral spread display even though it is unable to back up this signal with force. Its opponent is deceived and responds to the signal by retreating. A newly molted stomatopod can get away with the bluff because the receiver might pay dearly if it chose to test the honesty of the signal and the signaler was *not* newly molted (Adams and Caldwell 1990). Thus, the signal is stable because, on average, it is honest.

CAN HONEST AND DISHONEST SIGNALS COEXIST?

We see, then, that signals may be honest or dishonest, and often coexist in the same population. Exactly how coexistence occurs has been a bit of a puzzle. Consider the analysis of animal communication by John Maynard Smith and his colleagues (e.g., Maynard Smith 1974). In Chapter 4, we introduced the idea of an evolutionarily stable strategy (ESS). An ESS is a behavioral strategy that, when adopted by nearly all members of a population, cannot be beaten by a different strategy. Imagine, for example, a population in which every animal always gives an honest signal. Now imagine that a new mutant appears in the population that gives dishonest signals and that giving dishonest signals offers a fitness benefit. Over time, the strategy of being dishonest will spread through the population until every sender is dishonest. Receivers, in turn, will have greater fitness if they ignore the senders. Eventually "everyone would be lying and no one listening" (Rowell et al. 2006).

If this is the case, then why do we see populations with both honest and dishonest signals? There are several ideas. One idea, as described earlier, is that the costliness of signals can reinforce honesty. When signals are generally honest, a low level of deceit in a population can be stable and is, in fact, often expected to evolve. Another idea is that populations are in flux and that at any given point in time we may be seeing the

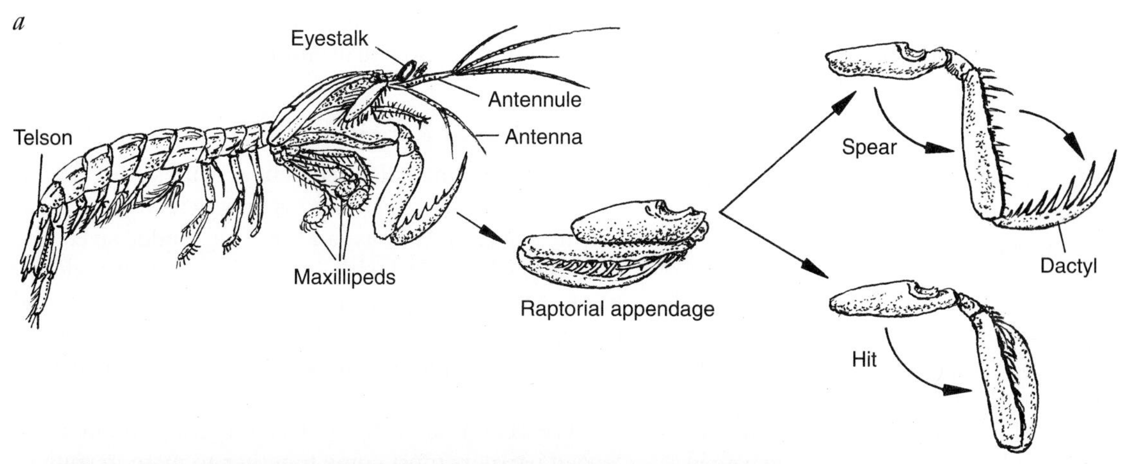

FIGURE 17.6 **(*a*) The mantid shrimp, a stomatopod, has two raptorial appendages that can be used in prey capture or combat. (*b*) The threat display of a stomatopod, called a meral spread, is a good predictor of attack. Here the display is seen from the front.**

spread of dishonesty—eventually it will take over the population, and the signal will fade from use. A third hypothesis for the coexistence of both honest and dishonest signals takes into account something we haven't explicitly considered before: that any given signal may be perceived by different receivers, some with the same goals as the sender, and some with different goals. For example, the same signal may be perceived simultaneously by a prospective mate and by a competitor. For a signaler under these conditions, a mixed signaling strategy, sometimes honest and sometimes deceptive, may be best (Rowell et al. 2006). These explanations are not mutually exclusive, nor are they the only scenarios possible.

THE EVOLUTIONARY ORIGINS OF SIGNALS

So far, we've been discussing established communication systems: a loon yodels, and another loon swims away. The signal apparently has meaning for both the sender and the receiver. However, these signals did not just spring fully formed into being. How did they originate? How does an incipient signal acquire meaning?

Behaviorists have taken two different approaches to thinking about the evolutionary origins of signals. The first focuses on identifying the behaviors of the senders that form the raw material for signals. This, the study of ritualization, was a focus of those meticulous observers of animal behavior, the ethologists. The second approach focuses on how signals exploit the receiver's sensory biases, or ability to detect some kinds of information better than others. These two evolutionary pathways are not mutually exclusive alternatives to one another: both may have played a role in a given system.

RITUALIZATION

The ethologists proposed that many signals get their start as part of another behavior or as a physiological response, and only later take on a signaling function. Then, evolution favors modification of the incipient signal so that it becomes more stereotyped and more unmistakable, thus facilitating the process of communication. Julian Huxley (1923) called this process **ritualization**. The study of ritualization was at its peak with the early ethologists, and we will illustrate our discussion with classic examples.

We will consider three sources of raw material for signals: intention movements, displacement activities, and autonomic responses.

Intention Movements

Animals often begin behavior patterns with some characteristic movements that prepare them for action. Because it is often possible to judge from these activities just what the animal intends to do, they have been named intention movements (Heinroth 1910). For example, wolves pull back their lips and bare their teeth before biting. It's likely to improve a wolf's fitness if he correctly interprets the bared teeth of an aggressive rival rather than waiting to feel jaws clamp down.

Numerous avian displays originated with intention movements for flight or walking (Daanje 1950). A bird about to take flight goes through a sequence of preparatory motions that are helpful in achieving takeoff. It will usually begin by crouching, pointing its beak upward, raising its tail, and spreading its wings slightly. Components of the takeoff leap have been ritualized into communicative signals in different species. The blue-footed booby, along with exhibiting its brightly colored feet to its prospective mate, also incorporates a posture called "sky pointing" into its courtship dance (Figure 17.7). Notice that the wings are spread and that the tip of the beak and the tail point upward. Although this display most likely had its origins in flight intention movements, it is difficult to see how a bird could ever take off from this ritualized pose. The movement has changed during the ritualization process.

FIGURE 17.7 **Sky pointing by a blue-footed booby. This display is part of courtship and probably evolved from the intention movements of flight.**

Displacement Activities

Ethologists defined displacement activities as irrelevant actions performed in situations in which an animal has conflicting motivations and is thus indecisive. For example, when faced with an aggressor, an animal may have conflicting motivations to fight and to flee, and may instead preen (groom) itself. Similar to intention movements, displacement activities are often incomplete actions.

Courtship is a time of conflicting tendencies. Sexual partners must come together to mate in spite of the aggressive tendencies that often tend to keep them apart. Lorenz (1972) suggested that the mock preening of the courtship displays of males of many duck species (Figure 17.8), including the familiar mallard, originated in displacement preening (Lorenz 1972). However, as Smith (1977) points out, it can be tricky to decide whether a display is really a result of displacement or whether it has a function that is unclear to the researcher. For example, duck courtship includes other vigorous displays, such as rearing up and splashing down into the water, which may well ruffle a drake's feathers. Preening during courtship may not be a display at all, but rather strictly functional.

Autonomic Responses

The autonomic nervous system regulates many of the basic body functions—digestion, circulatory activities such as heart rate and diameter of the blood vessels, and thermoregulation, to name a few. Many displays seem to have originated from autonomic functions (Morris 1956). For example, at times of stress or conflict, there is often a change in the distribution of blood through-

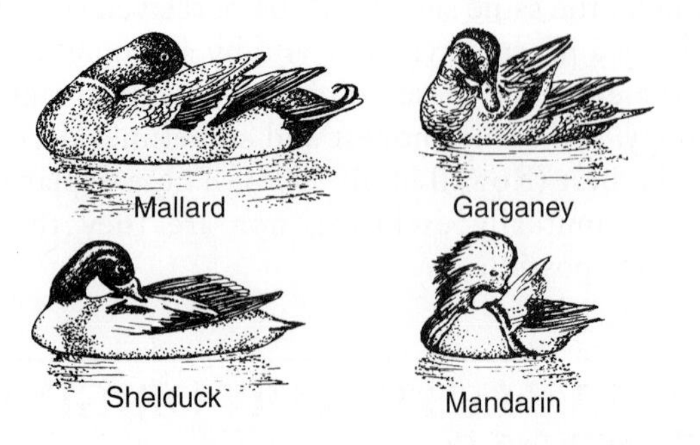

FIGURE 17.8 **Mock preening by courting male ducks, a display thought to have evolved from displacement preening. The movements emphasize the bright markings on the wings. (Modified from N. Tinbergen 1951.)**

FIGURE 17.9 **A courting male frigate bird inflates his huge, brilliant red throat pouch to capture the interest of females.**

out the body. In birds such as the turkey, jungle fowl, and bateleur eagle, the naked head or neck skin may flush and fleshy appendages may swell because of vasodilation. These changes now are part of signaling during courtship and aggressive encounters.

The respiratory system also has provided fodder for the evolution of signals. In air-breathing vertebrates, modifications of the respiratory system are used for the production of sound signals. Visual signals can also evolve from the respiratory system. Perhaps the best examples are the inflation displays of birds, during which the males fill pouches on their body with air to attract mates. For instance, the male frigate bird has a pouch on its throat that is inconspicuous when deflated, but when inflated its enormous size and brilliant scarlet color attract passing females (Figure 17.9).

Piloerection, or the erection of feathers and hair, has a thermoregulatory function: fluffed feathers or hair trap heat. Piloerection is frequently a part of aggressive and appeasement displays (Smith 1977). For example, a zebra finch that cannot escape from a dominant individual fluffs its feathers as an appeasement signal (Morris 1954). Feather position can also be part of courtship displays (Morris 1956). Sometimes only feathers in some regions are erected, such as the tail-raising courtship display of the peacock. In other species, the feathers used in signaling form a crest, as in the sulfur-crested cockatoo; an ear tuft, as in several species of owls; a throat plume as in some herons; or eye tufts, such as the shiny green patch between the eyes of the bird of paradise. In some mammals, such as the rufous-naped tamarin, a squirrel-sized South American monkey, the meaning of the message varies with the part of the body on which hair is erected. When all the hair is erect, the tamarin is likely to attack or behave indecisively. However, when only the tail hair is erected, it will probably flee (Moynihan 1970).

Other Behaviors as Raw Materials for Displays

Although these three classes of behaviors have traditionally been considered sources of displays, E. O. Wilson (1980) points out that "ritualization is a pervasive, highly opportunistic evolutionary process that can be launched from almost any convenient behavior pattern, anatomical structure or physiological change." For example, predatory behaviors have been ritualized in the male gray heron. During courtship, he erects his crest and certain other body feathers, points his head downward, as if to strike at an object, and snaps his mandibles closed, movements similar to those used during fishing (Verwey 1930). Food exchange has also been ritualized. Billing, the touching of bills together, is likely to be derived from the parental feeding of young and has taken on a variety of meanings in different species of birds. It is common in courtship and appeasement display, functioning to establish or maintain bonds. Mated pairs of masked lovebirds bill during greetings and after a spat (Figure 17.10).

FIGURE 17.10 **Lovebirds are billing. This display may be derived from parental feeding behavior. It now signals nonaggression during greetings and after conflicts.**

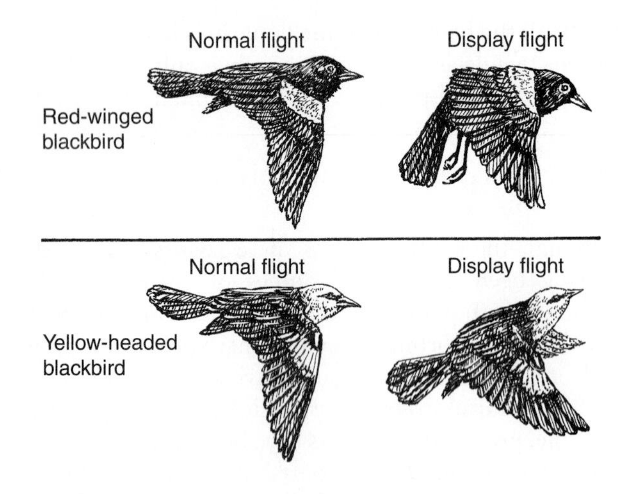

FIGURE 17.11 **Ritualized flight in blackbirds. Almost any behavior can be a starting point for ritualization. Notice the exaggeration of motion that developed during ritualization. (Modified from Orians and Christman 1968.)**

Locomotion can also be ritualized for communication. In a number of bird species, male courtship includes aerial displays of ritualized flight patterns. As you can see in Figure 17.11, ritualized flight is more conspicuous than normal flight because some movements are accentuated and because the movements reveal special plumage patterns (Orians and Christman 1968). Portions of the courtship display of the fiddler crab *Uca beebei* may have evolved from the movement of the male while entering his burrow. When a female approaches him and his burrow, he raises his body and flexed front claw to an almost vertical position, an exaggeration of his movements into his burrow during the final stages of courtship. This exposes his dark underside, which probably functions to guide a female to his burrow (Christy and Salmon 1991).

A final, very odd example of a ritualized display is pupil dilation in parrots. In birds, light regulation is performed by a layer of the retina rather than by dilation and constriction of the iris, as it is in mammals. Thus, pupil dilation is "freed up" for use in displays. Some species of parrots, when aggressive or excited, rapidly contract and expand their pupils (Bradbury and Vehrencamp 1998).

STOP AND THINK

Many of our examples are drawn from observations of behavior and reasonable guesses about the origins of signals. Is it possible to test whether a particular behavior has been ritualized from another behavior? How would you do so?

The Ritualization Process

Ethologists hypothesized that during the evolutionary process of ritualization, behaviors that take on a signaling function become more stereotyped. They may also change from the original movement, perhaps slowing down, speeding up, or becoming more exaggerated. Anatomical features, such as bright colors, large claws, and the like, might evolve in order to draw attention to the display.

In a highly ritualized signal, ethologists hypothesized that behaviors may become freed from the internal and external factors that originally caused them, in a process called **emancipation** (Tinbergen 1952b). In other words, the original proximate triggers no longer cause the behavior to occur, and the behavior has only a communicative function. Few experimental tests of emancipation exist (Sebbel et al. 1998). One of the displays explored most thoroughly from this perspective is the whistle-shake given by a species of duck (*Tadorna tadorna*). The hypothesized evolutionary precursor of the whistle-shake is the body-shake. Body-shakes, which are just what they sound like, have the function of drying and rearranging the feathers. Ducks give body-shakes when they are exposed to rain—or, in these experiments, sprayed with water. Whistle-shakes look much like body-shakes but are finished off by the duck tilting back its head and emitting a trill. Whistle-shakes are given in social situations, whereas body-shakes are not. However, like body-shakes, whistle-shakes are given in response to water sprays. Thus, they are not completely emancipated from their original proximate causes (Sebbel et al. 1998).

RECEIVER-BIAS MECHANISMS

Whereas ritualization focuses on characteristics of the sender that influence signal evolution, other hypotheses about the origin of signals focus on characteristics of the receiver. One hypothesis that specifically concerns signal origins is sensory exploitation. (We'll save the discussion of another receiver-bias hypothesis, sensory drive, for the next section.)

According to the **sensory exploitation** hypothesis, the receiver has a preexisting preference for a particular signal. In other words, features of the receiver's nervous system—either its peripheral sense organs or how its brain processes the signal—makes it more responsive to a particular form of stimulus. The sender takes advantage of the receiver's preexisting sensory biases when new signals are evolving (Ryan 1990; Ryan and Rand 1993a, b).

We already mentioned one example of sensory exploitation—that of the evolution of swordtails in fish (Chapter 4). Another example of sensory bias comes from túngara frogs (*Physalaemus pustulosus*). Like other

FIGURE 17.12 **A male túngara frog is calling to attract a mate. It is thought that "chuck" notes, added to the end of the "whine" part of the call, evolved to take advantage of the female's greater sensitivity to low-frequency sounds. Thus, it is an example of the sender's exploitation of a sensory bias in the receiver.**

frogs, males call to attract females (Figure 17.12). The calls consist of a descending whine, followed by a short "chuck" call. Females prefer males with lower frequency chucks. In fact, one of the membranes in the ears of females is tuned so that it is more sensitive to low-frequency chucks than to sounds of other frequencies. Females benefit from their preference: low-frequency chucks are produced by larger males, and mating with larger males results in more fertilized eggs (reviewed in Ryan 2005).

Comparative studies shed light on the evolution of the chuck call. Males of a closely related species of túngara frog, *P. coloradorum*, do not have chuck calls but simply give the whine portion of the signal. Strangely, female *P. coloradorum*, although they have never heard a chuck, prefer calls with chucks artificially added to the end over the normal calls of their own species (Ryan and Rand 1993a, b). The data suggest that the female preference for chuck calls existed prior to the evolution of chucks. (The alternate hypothesis, that *P. coloradorum* males once gave a chuck call but lost it over evolutionary time, is possible but less likely.)

There are also examples of male courtship signals that appear to have exploited the females' sensory adaptations for prey detection. An interesting example is the water mite (*Neumania papillator*). The eyes of water mites don't form images, and so they must find food and mates by other means. This species of water mite hunts by ambush. A mite waits on aquatic vegetation with its first four legs raised. This position allows it to detect and orient toward passing copepods by the characteristic water vibrations caused by the prey's swimming movements. When a courting male detects a female in the hunting posture, he performs a trembling display by moving his legs. The leg movements cause vibrations that mimic those of the prey, leading the female to grab him as she would a prey item. Food-deprived females are more likely than satiated ones to approach and clutch males. When the female detects that she has grabbed a male and not a meal, she releases the male. He then deposits his spermatophores (packets of sperm) in front of her and fans a pheromone contained in the spermatophore toward the female. The pheromone causes the female to pick up the packets and place them in her genital opening. It is to the male's advantage to elicit the predatory clutch because it allows him to orient to the female so that the spermatophores can be placed directly in front of her (Proctor 1991). Analysis of the evolutionary relationships among water mites suggests that males evolved courtship trembling *after* the sensitivity to water vibrations for prey detection. Thus, it seems that males exploit the prey-detecting mechanisms of the female to their own advantage (Proctor 1992). This exploitation works because females generally benefit from responding to this vibratory cue. Other examples of sensory bias are described in Chapter 14 and later in this chapter in the section on environmental influences on signal design.

SELECTIVE FORCES THAT SHAPE SIGNALS

As you saw in the last chapter, the variety of signals is astonishing: whiffs of chemicals, ear-splitting screeches, vigorous arm waves, electrical jolts. Even within a single channel of communication, the form of signals varies tremendously—think of the cacophony of different bird songs during a spring dawn. What causes this diversity of signal form? We'll organize our presentation by discussing in turn the influence of characteristics of the sender, the receiver, and the environment. However, these three influences do not work in isolation from one another, as illustrated by many of the examples presented here.

CHARACTERISTICS OF THE SENDER

It is obvious that anatomical structures form the foundation for producing signals, as illustrated by the many examples of signaling described in the last chapter. Some signals are produced by specialized structures, such as the human larynx or the sensory organs of electric fish, and it's clear that signal production is wedded to the evolution of these structures.

Even without highly specialized structure, body form influences signal design. For instance, consider four ways that the form of a display might be influenced by body size. (1) Small species might not be visible from great distances, and other sensory modalities might be favored for long-distance communication. (2) Body size may influence the evolution of physical characteristics that enhance visual signals. For example, in addition to whole-body movements and posturing, two forms of visual display are common among the primates of Central and South America: facial expression and the erection of hair on parts of the body. Which form is favored by evolution relates to body size, probably because it is easier to see hair raising on a small body than a change in expression on a diminutive face (Moynihan 1967). Squirrel-sized tamarins and marmosets have long, silky fur that makes their displays more effective. In contrast, larger species such as capuchins, spider monkeys, and woolly monkeys have a richer variety of facial expressions, although they tend to be poker-faced compared to their Old World cousins. (3) Body size influences agility, which in turn might influence the type of signal that can be given. Just as gymnasts are usually more petite than weight lifters, smaller animals are often better acrobats. For example, many bird species give impressive aerial displays to their mates. Small species of herons (Meyerriecs 1960) give more acrobatic displays than do larger species, as do smaller males within the same species of dunlin (Blomqvist et al. 1997). (4) Body size, at least in vertebrates, influences some characteristics of vocalizations. One such characteristic is the production of formants, particular frequencies produced by resonances of the vocal tract (reviewed in Searcy and Nowicki 2005). Vocal tract length influences formant production and also generally correlates with body size (Fitch 1997; Fitch and Hauser 2003). Formant production correlates with body size in several species, including rhesus monkeys (Fitch 1997). This relationship also holds for domestic dog breeds, with their 100-fold range in mass from chihuahua to St. Bernard (Riede and Fitch 1999).

These examples bring home a general point: the evolution of animal signals does not happen in a vacuum. Many signals are produced by structures that have other functions. Body size, hair, beaks, respiratory tracts—these traits are certain to be under selection for other reasons besides their role in producing signals. Thus, a complete understanding of how characteristics of the sender influences signal design must account for other selective pressures besides those directly on signals. This idea should be familiar. In Chapter 4, we described how beak shape in Darwin's finches is important in both feeding behavior and song production. Selection on feeding behavior drives the evolution of beak shape. Changes in beak shape then have secondary consequences for song production, which in turn influences female choice (Huber and Podos 2006; Podos and Nowicki 2004). Thus, a myriad of different selective forces act to shape a signal.

CHARACTERISTICS OF THE ENVIRONMENT

As described in Chapter 16, a species' habitat is key in determining which channel of communication (sound, chemical, visual, etc.) is likely to be used. For example, unless you can produce your own light like lantern fish do, visual signals are not much use in the ocean's depths. For this reason, whales rely on sound for long-distance communication (Payne and Webb 1971). Even within a sensory channel, the structure of the signal can be influenced by the environment. Let's look next at a few examples of signal evolution within sensory channels among closely related species.

The Structure of Bird Song

On a weekend night, walking near a college dormitory, you can hear for yourself some of the problems that face singing birds. As you pass very close by a dorm where a party is taking place, you might be able to hear the sounds of music, voices, and laughter. You might be able to make out words or even whole conversations, and you should be able to distinguish different elements of the music, perhaps the vocals, guitar, bass, and drums. As you walk further away, some of these elements will start to fade out and become less audible. If your path takes you through trees, the high-pitched sounds of laughter may become inaudible more quickly than the thumping of the drums; if you are walking between buildings, the sounds may bounce and echo and become distorted; and if you are walking across an open quad, all the sounds of the party may follow you for a long way. We describe characteristics of sound transmission as attenuation (weakening), or how far the sound will carry, and degradation, or how distorted the signal becomes during transmission. Both of these are affected by the characteristics of the habitat. If your goal was to make the noise of your party transmit as far as possible, you might consider the surrounding environment before making your choice of playing particular styles of music or whether to invite guests most apt to emit high-pitched squeals.

By playing recordings of bird songs and re-recording them at measured distances from the source, we can measure how songs attenuate and degrade in different environments. As you might expect, dense foliage means that songs are both more attenuated and degraded than if the habitat is open (e.g., Blumenrath and Dabelsteen 2004). However, not all songs are the same: for example, lower frequencies are less likely to become distorted in habitats with lots of foliage than are higher frequencies. The **acoustic adaptation hypothesis** proposes that the acoustic properties of bird song are shaped by habitat structure (Morton 1975). Specifically, the hypothesis predicts that in habitats with complex vegetational structure, songs should have low frequencies, narrow bandwidths (the bandwidth is the difference between the highest and lowest frequency), whistles, and long notes. In open habitat, like grasslands, songs should have high frequencies, broad bandwidths, trills, and short notes (Morton 1975; reviewed in Boncoraglio and Saino 2007).

How do these predictions bear out? In some studies, both across species and across populations of the same species, the predictions hold up quite well. For instance, the song of the great tit (*Parus major*) is ideal for this type of analysis because the species has such a large geographic distribution. Forest dwellers were found to have songs with a lower pitch, a narrower range of frequencies, and fewer notes per phrase than those of open woodland birds (Figure 17.13). In fact, birds from similar habitats in Oxfordshire, England, and Iran, separated by 5000 km, sing more similar melodies than two English populations occupying different habitats (Hunter and Krebs 1979). However, in other species, no differences in the predicted direction were found between birds in different habitat types (e.g., Hylton and Godard 2001).

So, here we have a challenge: some studies support the prediction of the hypothesis, whereas others do not. A mixture of different results is not unusual, not only in behavioral studies, but also in other areas of scientific research such as tests of the efficacy of different drugs or dietary supplements. So what can we say about the general validity of the acoustic adaptation hypothesis? One approach is to do a **meta-analysis**. In this technique, one performs a comprehensive literature review to find all the available studies that test the hypothesis. Then, the results of all the studies are combined and tested together using statistical techniques especially developed for such a purpose (e.g., Gurevitch and Hedges 1999). Boncoraglio and Saino (2007) carried out a meta-analysis for the acoustic adaptation hypothesis and found that habitat structure does significantly influence the acoustic properties of bird songs, though weakly.

Lizard Displays and Background Motion

Environmental characteristics influence not only the transmission of acoustic signals, but also that of other channels of communication. For instance, visual signals are easier to pick up when there is not so much "clutter" in the background, so one might predict that displays should differ in visually "noisy" environments versus environments with plain backgrounds. This hypothesis has been tested in lizards. Many species rely heavily on visual displays such as

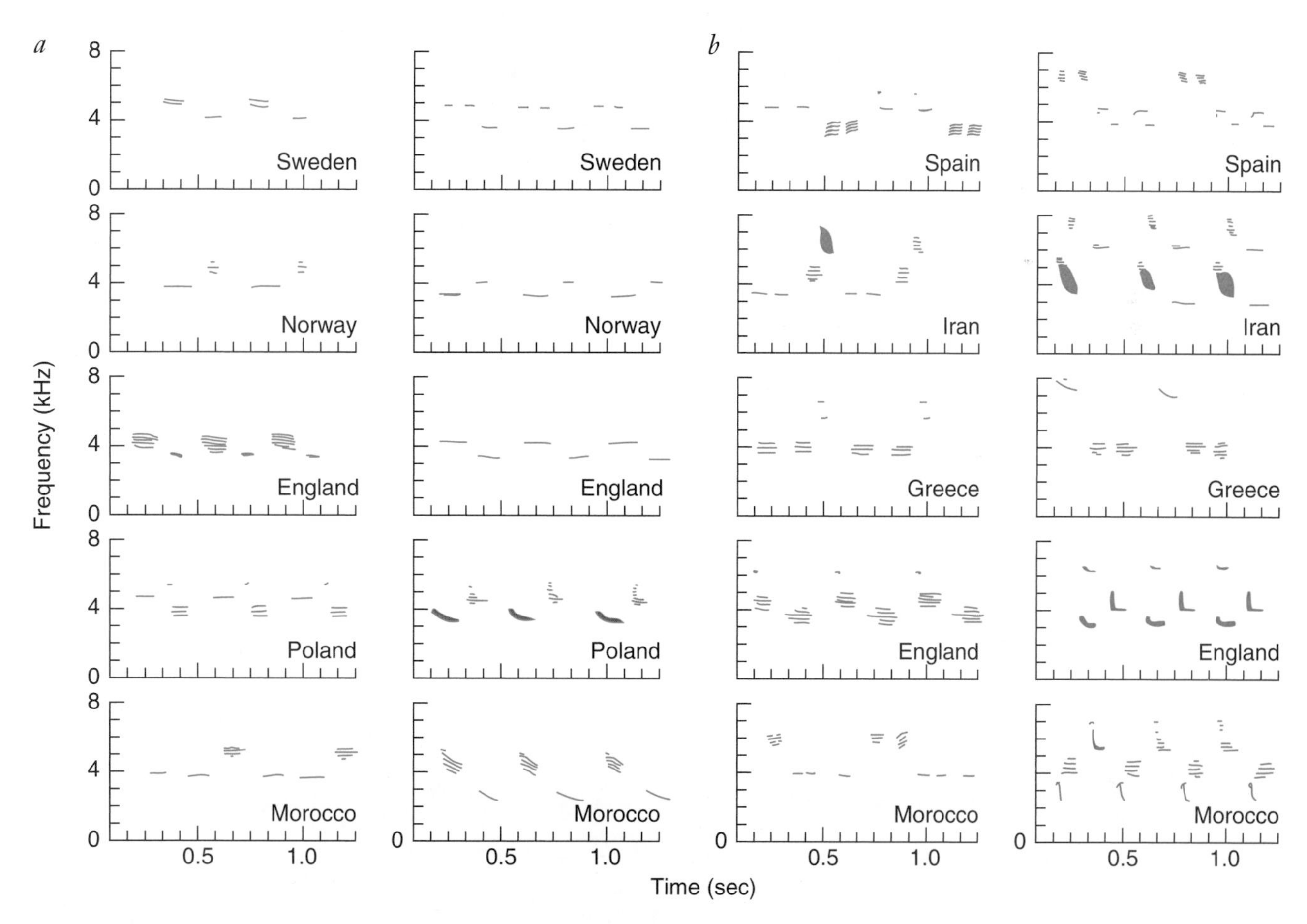

FIGURE 17.13 **Song structure of populations of great tits in many countries: (*a*) forest dwellers, (*b*) woodland inhabitants. The left column in each section shows songs that are typical of each area, and the column on the right shows songs that are most different from songs of the other habitat. Notice that the songs of the species that live in forests have a lower pitch, fewer frequencies, and fewer notes per phrase than those of the woodland species. (From Hunter and Krebs 1979.)**

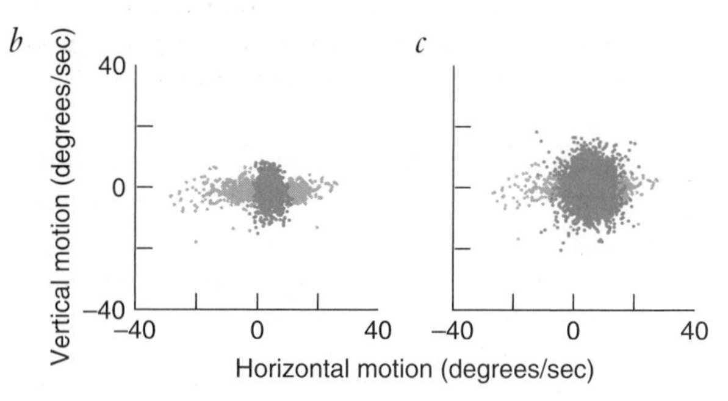

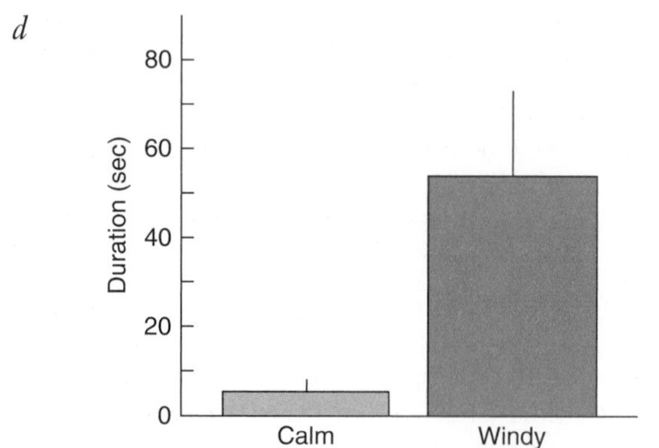

FIGURE 17.14 (*a*) A Jacky lizard from Australia in its complex habitat. (*b*) and (*c*) show the movement of the vegetation in the background under calm and windy conditions. (*d*) Signaling Jacky lizards flick their tail significantly longer under windy conditions.

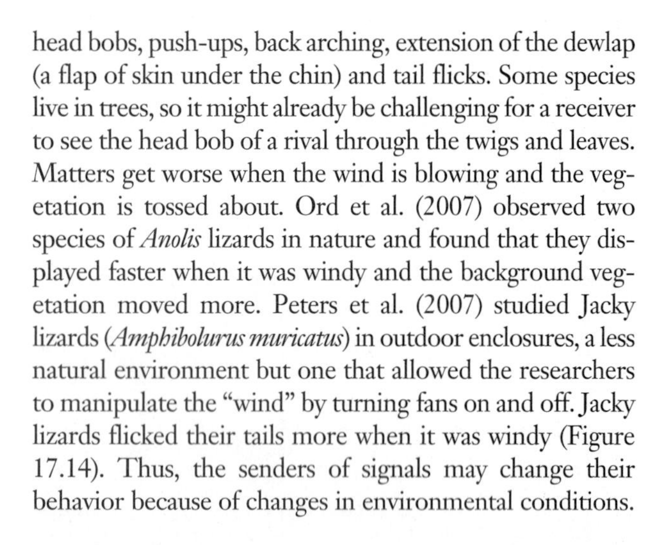

head bobs, push-ups, back arching, extension of the dewlap (a flap of skin under the chin) and tail flicks. Some species live in trees, so it might already be challenging for a receiver to see the head bob of a rival through the twigs and leaves. Matters get worse when the wind is blowing and the vegetation is tossed about. Ord et al. (2007) observed two species of *Anolis* lizards in nature and found that they displayed faster when it was windy and the background vegetation moved more. Peters et al. (2007) studied Jacky lizards (*Amphibolurus muricatus*) in outdoor enclosures, a less natural environment but one that allowed the researchers to manipulate the "wind" by turning fans on and off. Jacky lizards flicked their tails more when it was windy (Figure 17.14). Thus, the senders of signals may change their behavior because of changes in environmental conditions.

Habitat Changes Caused by Humans

Human activities can alter the environment in which animals communicate and alter the selection pressures on signal form. For instance, let's look at another study of the birds that so nicely fit the predictions of the acoustic adaptation hypothesis, the great tits. This species is quite willing to live in close contact with humans and does well in both urban and rural environments. Slabbekoorn and den Boer-Visser (2006) recorded great tit songs in ten big European cities, and compared them to ten forest sites near each city. In all ten urban/forest comparisons, songs sung by urban birds were shorter, sung faster, and had higher minimum frequencies (Figure 17.15), most likely due to competition with the cacophony of the urban background.

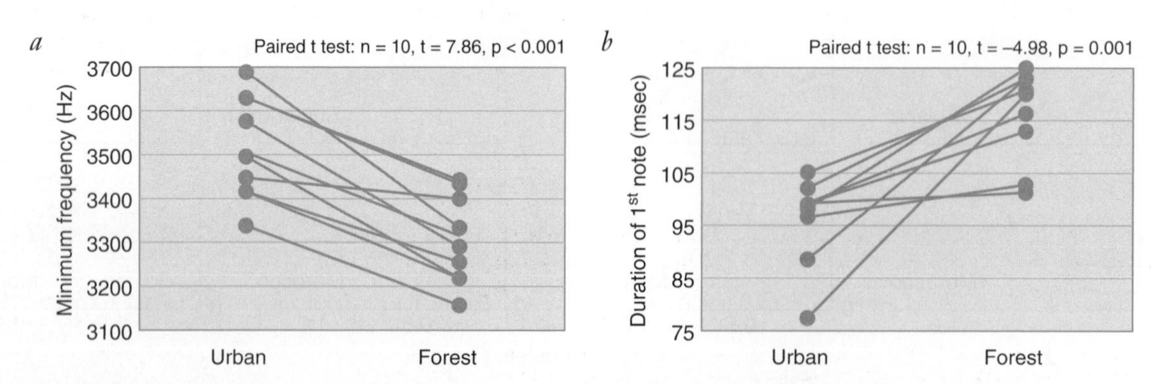

FIGURE 17.15 The song of great tits (*Parus major*) in urban habitats and in nearby forest habitats. Ten city-forest pairs were studied. Data from members of each pair are connected by a line. (*a*) The minimum frequency of songs was higher in the urban habitats in each pair. (*b*) The duration of the first note of the song was significantly shorter in urban habitats.

Visual signals can be impeded by human-induced environmental changes. For example, in the Baltic Sea, the water has grown turbid (cloudy) because of phytoplankton growth, caused by the introduction of excess nitrogen and phosphorus into the water. This is the habitat of stickleback fish (*Gasterosteus aculeatus*), a species in which males woo females with a courtship dance and bright coloration. Turbidity makes it harder for females to see males, and males in turbid water have to court much more vigorously in order to get the attention of the female (Engström-Öst and Candolin 2006). Males also can't see each other as well when the water is turbid. This is important from a female's perspective because male sticklebacks keep each other honest. Males in poor physical condition signal vigorously but make poor mates—for one thing, they are prone to cannibalize eggs, hardly the characteristic of an ideal father (Candolin 2000a, b). In clear water, males prevent other males in poor condition from dishonestly signaling their condition to females. In turbid conditions, males are not as likely to see each other. Thus, in turbid conditions, the honesty of male signals is reduced (Wong et al. 2007).

Research on the human impact on animal signaling systems is in its infancy. No doubt some species, like the great tits, will adapt successfully to changing conditions, whereas others will not.

CHARACTERISTICS OF THE RECEIVER

Sensory Drive

We delayed discussing characteristics of the receiver that affect signal evolution until after we discussed the role of the environment because receiver characteristics can themselves be shaped by the environment. This is called **sensory drive**. The sensory drive hypothesis predicts that if the constraints on a signal can be identified, then we can make testable predictions about the direction of signal evolution (Endler 1992).

Here is an example of sensory drive that is supported by phylogenetic, physiological, and behavioral evidence (Cummings 2004, 2007; Cummings and Partridge 2001). Surfperch (Family Embiotocidae) are marine fish that live in a variety of habitats differing in both overall light intensity and spectral qualities (color) (Figure 17.16). The retinas of surfperch species differ. For some, the visual pigments are best at detecting differences in color contrast. In others, the visual pigments are tuned to detect differences in brightness. Because of physiological limits associated with having only two classes of retinal cones, it is possible to be good at detecting differences in color contrast, or at differences in brightness, but it's not possible to be good at both. The first question Molly Cummings (2004) tested was whether surfperch species are adapted to detect prey in their particular light environment.

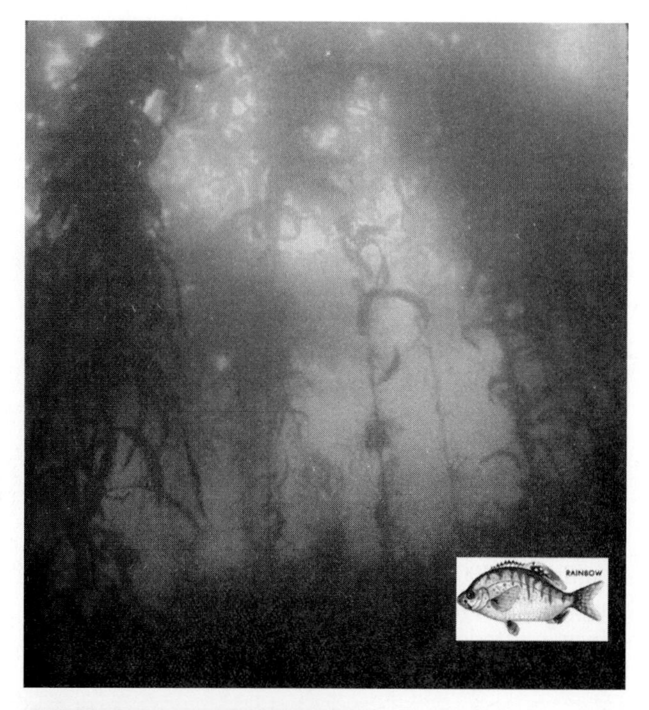

FIGURE 17.16 **Different optical habitats of related surfperch species. Rainbow surfperch (*Hypsurus caryi*) (top) live in an environment that is highly variable in light intensity. Striped surfperch (*Embiotoca lateralis*) (bottom) live in deeper water in the kelp forest where the background light is more even.**

Cummings made over 250 scuba dives in order to measure the light characteristics of each species' environment. By combining these data with what she knew about the performance of the photoreceptors under different conditions, and the visual characteristics of their food, she could predict the optimal "tuning" of the visual pigments for each environment. The species matched these predictions well. For example, deepwater surfperch are better at detecting brightness differences, whereas species living in environments with lots of brightness variation (e.g. shallow water with lots of light flickering by waves) are better able to see color differences (Figure 17.17*a*). Thus, it seems that natural selection to improve the ability to detect prey has produced fish with different biases in their visual systems.

Now, let's connect this with communicatory signals. Surfperch species vary in their color patterns, which males use in courting females. Male color patterns have evolved to take advantage of their conspecific female's biases: in species that are good at detecting brightness, male colors are more detectable in the brightness "channel," and in species that are good at distinguishing colors, males are more detectable in the color "channel" (Figure 17.17*b*) (Cummings 2007; Cummings and Partridge 2001). Thus, changes in habitat led to changes in the visual system, which in turn led to changes in color patterns.

Unintended Receivers

Unfortunately for signalers, it's not always just the intended target that senses a signal. Examples abound where predators, parasites, or conspecifics other than the intended receiver detect a signal, often to the detriment of the signaler. For example, recall the túngara frogs described earlier in this chapter, whose call is a whine followed by chucks. Fringe-lipped bats prey on frogs and are more likely to prey on frogs that give multiple chucks (Ryan et al. 1982). Thus, the presence of unintended receivers may also impose selection pressures on the signal. The effect of unintended receivers, both within and across species, on signal design is currently a lively area of research in communication.

LANGUAGE AND APES

Humans seem to be very interested in what separates us from other species. We mentioned in Chapter 5 that tool use was long considered to be confined to humans, but now we know that this is not the case—it's not even necessary to be a primate to be adept at using tools. Language, in contrast, seems to be a talent confined to humans. Or is it? A better question is a more specific one: which elements of language are unique to humans, and which are more broadly shared? Are differences between humans and other species a matter of kind or a matter of degree? We won't be able to answer all these questions fully, but we can at least begin to shed some light on them.

WHAT IS LANGUAGE?

There is no one accepted definition of **language**. Here, we will consider that true language contains the following four elements. First, words or signs must be used as

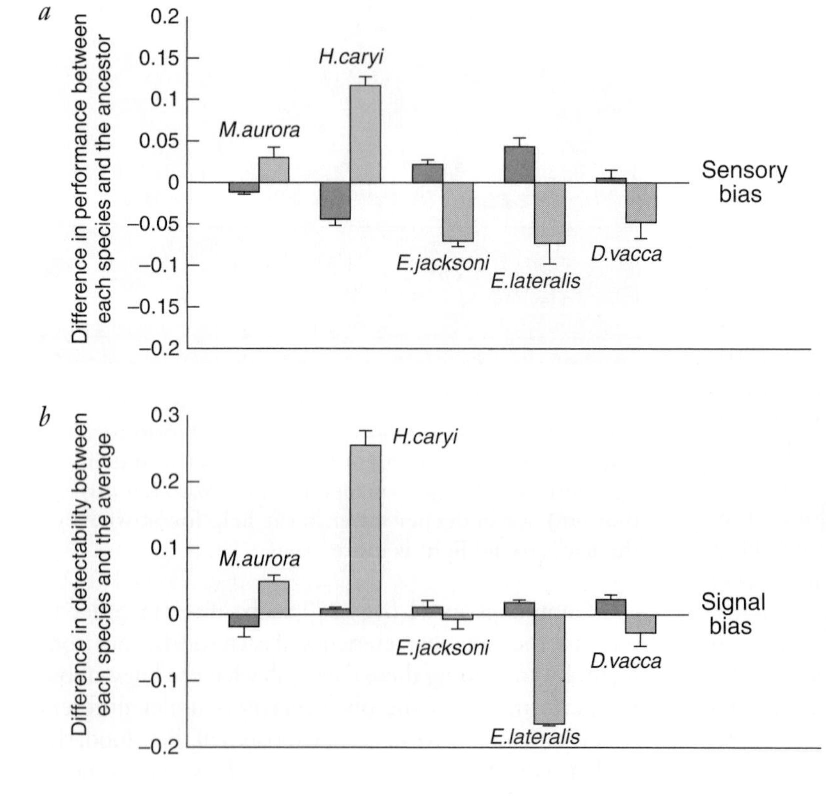

FIGURE 17.17 **(*a*) The sensory bias of the ability of five surfperch species to detect differences in brightness (dark bars) and color (light bars). The *y*-axis is the difference in performance between each modern-day species and that estimated for their shared ancestor. (*b*) The difference in detectability of male color patterns (signals) in the brightness (dark bars) and color (light bars) channels. The *y*-axis represents the difference in reflectance detectability between a species' measured reflectance and that of the average reflectance of species with similar colors. (From Cummings 2007.)**

true symbols that can stand for, or take the place of, a real object, event, person, action, or relationship. Second, symbols should permit reference to objects or events that are not present. Third, there should be some elements of grammar, or rules that determine the relationship between words. For example, true language necessitates knowledge of syntax because a change in the order of symbols can alter the meaning of the message. Finally, words or signs should be combined to form novel phrases or sentences that are understandable to others.

APE LANGUAGE STUDIES

Investigators have studied the ability of great apes—chimpanzees (*Pan troglodytes*); bonobos, a species within the same genus as chimpanzees (*P. paniscus*); western (lowland) gorillas (*Gorilla gorilla*); and orangutans (*Pongo pygmaeus*)—to learn language. As you will see, the approaches of different investigators have varied tremendously.

Early studies were designed to teach chimpanzees to talk. In the longest and most thorough of these attempts, Keith and Cathy Hayes (1951) managed to teach a chimp named Viki to say three words—"mama," "papa," and "cup"—in a voiceless aspiration. In retrospect, these attempts to teach chimps to speak were doomed to failure because chimpanzees lack the necessary vocal apparatus to make the range of sounds of human speech (Lenneberg 1967). Our precise motor control over our face and mouth is linked to a particular gene called *FOXP2* (Marcus and Fisher 2003), which became established in human populations at the relatively recent date of about 200,000 years ago (Enard et al. 2002).

A more fruitful approach to demonstrating the ability of apes to acquire language skills has been to use nonverbal languages. A well-known nonverbal language, and the first to be taught to a chimpanzee, is American Sign Language for the Deaf, ASL. In the 1960s, Allen and Beatrice Gardner trained Washoe, a young chimpanzee, to communicate using ASL (Figure 17.18). The Gardners believed that an interesting and intellectually stimulating environment would assist the development of language skills. For this reason, Washoe and other chimps were reared as much as possible like human children, even living inside the house with the Gardners. However, spoken English was not permitted around Washoe because it was feared that it might encourage her to ignore signs. After four years of training, Washoe had a reported vocabulary of 132 signs. Her signs were not restricted to requests. She used the signs to refer to more than just the original referent; she applied them correctly to a wide variety of referents. For example, Washoe extended the use of the sign for dog from the particular picture of a dog from which she learned it to all pictures of dogs, living dogs, and even the barking of

FIGURE 17.18 **The chimpanzee Washoe using American Sign Language.**

an unseen dog. She also invented combinations of signs to denote objects for which she had no name. Classic examples are her signing "water bird" for a swan on a lake and "rock berry" for a Brazil nut (Fouts 1974). By the time she knew eight to ten signs, Washoe had begun to string them together. Examples of typical early combinations are "please tickle," "gimme food," and "go in" (Gardner and Gardner 1969).

STOP AND THINK

Do you agree that the best way to test whether apes have linguistic skills is to incorporate them into human society as much as possible? What are the advantages and disadvantages of this approach?

Other ape-signing projects were begun by other teams. Roger Fouts (1973) and Herbert Terrace and his co-workers (Terrace et al. 1979) continued working with chimpanzees, Francine Patterson (1978, 1990) worked with a gorilla named Koko, and Lynn Miles (1990) extended the studies to an orangutan named Chantek. The techniques employed in these signing studies were similar to those originally used by the Gardners, except that spoken English was permitted in the presence of the apes. The emphasis of these studies was on the production of language (the use of signs) and not on the comprehension of language (understanding the meaning of signs) (Rumbaugh and Savage-Rumbaugh 1994).

Then a more critical voice was heard. Herbert Terrace and his co-workers (1979) also taught ASL signing to chimpanzees and initially felt they had some success. Most of their work was done with Nim Chimpsky, a young male chimp named in "honor" of

Noam Chomsky, the famous linguist who has argued that language is uniquely human. Terrace was writing a book on Nim's successes when, as he was reviewing slow-motion videotapes of the signing chimp, he came to the conclusion that Nim was not spontaneously signing after all. Their most important criticism of the work with "talking" chimps was that the animals were simply imitating their trainers. Terrace argued that the ape's signs were imitations of what the trainer had just signed and that the trainers were too liberal in their interpretation of the signs. Terrace's team then analyzed the tapes of the Gardners' work with Washoe and concluded that their analyses were plagued with similar problems.

Needless to say, Terrace's conclusions were not left unchallenged. Fouts, Patterson, and the Gardners argued that Nim's language abilities were stunted by the operant-conditioning procedures used in his training. Allen Gardner backed his claim "that you can turn it [imitation] on and off, depending on the type of training you give" with a videotape of a chimp who showed little or no imitation of his trainer's signs until the last third of the tape, when operant-conditioning techniques were begun. During this last section of the tape, 70% of the chimp's signs were imitative. The Rumbaughs argued that because of the way Nim was trained, he never understood the meaning of words and that is why he was unable to create a sentence. In addition, Nim's trainers changed so often that he may not have had the opportunity to form the relationships claimed to be essential for language development (Marx 1980).

Nonetheless, it became widely accepted that chimpanzees could not learn language, and the later successes of Nim and other chimps received little attention. Project Nim was discontinued, but other trainers began to work with him. His language skills improved impressively and no longer depended on imitation (O'Sullivan and Yeager 1989). The signs of the orangutan, Chantek, were more spontaneous than those of Nim and could not be attributed to imitation (Miles 1990). Washoe and the other language-trained chimps signed to other animals and objects (Gardner and Gardner 1989) and frequently to themselves (Bodamer et al. 1994). After Washoe's biological infant died, she adopted a ten-month-old infant named Loulis. For the next five years, humans avoided using any sign language in the presence of Loulis. Nonetheless, Loulis learned his first 55 signs during this time by observing other chimps (Gardner and Gardner 1989). Washoe and her family signed to one another during all their daily activities, including playing and eating and even family fights, until Washoe's death in 2007 (Fouts and Mills 1997).

At the same time that the Gardners were working with Washoe, David Premack (1976) was training another chimpanzee, Sarah, to use plastic chips of various shapes and colors as words (Figure 17.19). Most of her use of language consisted of using one word from a choice of several to complete a preformed statement or arranging four to five words into a sentence of a specific word order. Premack established certain criteria for accepting that Sarah was using a particular chip as a word. Sarah had to be able to use the plastic chip to request the object it stood for, to select the proper chip when asked the name of the referent, and to "describe" the referent of a particular chip by using other chips. Premack's strategy was to break down linguistic rules into simple units and to teach them to the chimp one at a time. In this way, Sarah was taught not only to name many objects but also to use more complicated relationships such as if-then and same-different.

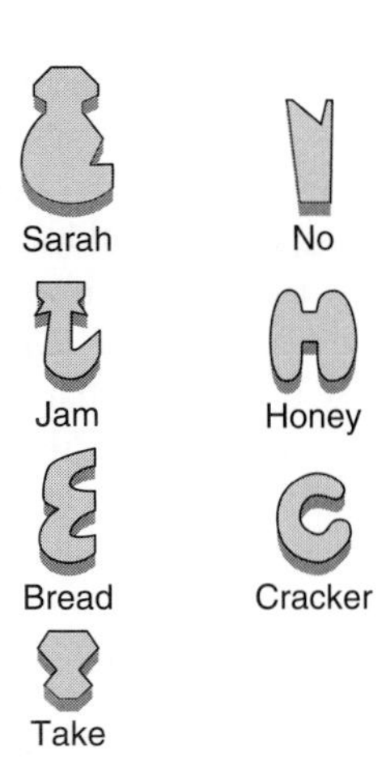

FIGURE 17.19 Symbols used as "words" by Sarah, a chimpanzee. Sarah learned to communicate using these plastic shapes.

The more successful aspects of the pioneering studies with Washoe and Sarah were combined in the LANA Project (Rumbaugh 1977; Rumbaugh et al. 1973). The chimpanzee Lana was trained to use a computer to communicate. This computerized language system eliminated social cueing and the difficulty of interpreting the symbols, problems that plagued the sign language studies. Lana communicated in a symbolic language, Yerkish, which was invented for the purpose. Yerkish words, called lexigrams, are geometric figures built from combinations of nine simple design elements such as lines, circles, and dots (Figure 17.20). Lana chose words by pressing a computer key labeled with the lexigram. When a key was depressed, it lit up and the lexigram simultaneously appeared on a projection screen. It is significant to note that Lana was required to use lexigrams in an appropriate order. That is, she had to learn syntax, the rules governing word order in a sentence. For example, she learned that pressing lexigram keys to say, "Please machine give . . ." might be rewarded but that pressing "Please give machine . . ." was not an acceptable way to make a request. Unlike Sarah, who was given

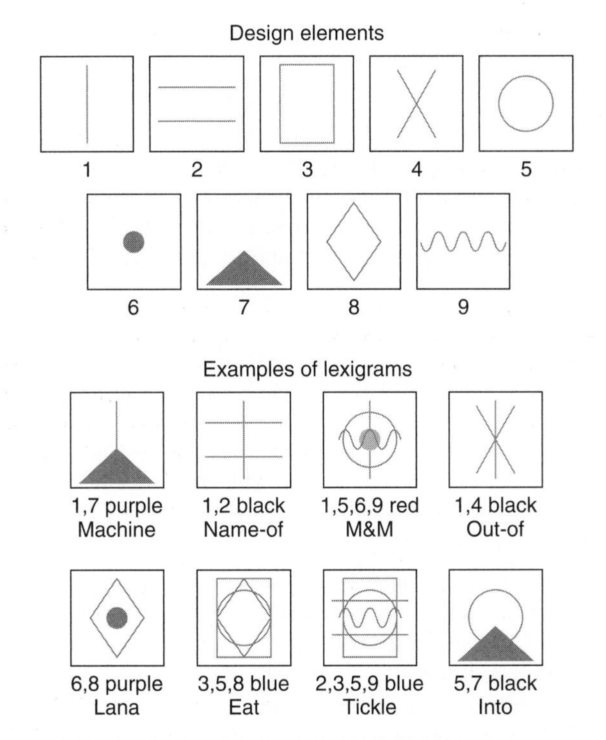

FIGURE 17.20 **Symbols of the Yerkish language. Each "word" is a combination of a few geometric shapes embossed on keys of a computer keyboard. Chimpanzees have learned to communicate by pressing the appropriate keys.**

a limited choice of words to use at one time, Lana always had her complete vocabulary available to her. Lana developed a large vocabulary and mastered Yerkish grammar. She also coined new words. For example, she called an orange soda a "Coke-which-is-orange" (Rumbaugh 1977).

The evidence presented up to this point suggests that apes do quite well on the checklist of language characteristics. The projects with Washoe, Sarah, and Lana demonstrated that apes are relatively adept at learning "words" in various forms, use them to refer to objects not present, readily string words together in short sequences so that the strings adhere to rules of grammar if that is required, and have the ability to coin new words. However, let's look more deeply.

Are these apes really using words as symbols? When a chimp uses a "word" (sign, plastic chip, or lexigram) to name an object, is it used as a symbol that stands for the object, one that can be used to refer to the object even when it is not present, or is the animal producing the "word" because it has been associated with the object through a reward system? When a pigeon pecks a red key for food and a green key for water, no one assumes that the animal is using the key to represent the item. However, when an animal as intelligent as an ape presses a key that is labeled, not with a color, but with a "word," the assumption is often made that it is using language and not simply performing a complicated behavior for a reward. This conclusion is especially tempting when the animal strings words together in a sequence for a reward. How could it be determined whether the animals were using the words as symbols or just mastering a complex conditioned response? It has been argued (Savage-Rumbaugh et al. 1980) that in the studies just described, the apes were not required to do anything that eliminates the possibility that they were simply using words as labels rather than as symbols.

Additional experiments were designed to test whether chimpanzees were really using "words" as true symbols (e.g., Savage-Rumbaugh et al. 1978a). Two chimpanzees, Sherman and Austin (Figure 17.21) communicated information to each other through the use of symbols, information that could not have been communicated without the symbols. They were trained, as Lana was, to communicate by pressing computer keys embossed with lexigrams. The emphasis in this program was on interanimal communication, so the animals were not taught to produce strings of lexigrams or to adhere to grammatical rules. In addition, they were raised in a social, preschool-style setting. The animals were first taught to name foods by pressing lexigram keys. It is important to note that they were taught to distinguish between the use of a food name as its name and the use of the food name as a request because they were never allowed to eat the same item that they named.

After this training, the animals' ability to communicate with each other symbolically was investigated. In one test, Sherman and Austin were able to specify foods to one another using lexigrams. The first chimpanzee was taken to a different room, where he watched the

FIGURE 17.21 **Sherman and Austin, two chimps, have shown that apes can cooperate with each other to solve problems by using the symbolic language Yerkish. They can communicate only when they have access to the computer keyboard.**

experimenter bait a container with one of 11 different foods or drinks. That animal was then led back to the keyboard and asked to name the food in the container. After observing the response of the first chimp, the second chimp was permitted to use the keyboard to request the food. If both animals were correct, they were given the food or drink. Sherman and Austin were able to communicate with one another in this manner whether or not they used the same keyboard and even if the experimenter was ignorant of the identity of the food in the container. Also, the animals communicated regardless of which of them was the observer. The animal that did the requesting based on the information provided by the knowledgeable chimp could demonstrate that he knew which item he was asking for by selecting its picture from a group. However, when the chimp who knew the identity of the food was prevented from using the keyboard to describe the contents of the container, he could not transmit the information to his buddy (Savage-Rumbaugh et al. 1978a).

The chimps also passed the next test—using symbols to inform each other of the appropriate tool to use to solve a problem. The animals were kept in separate rooms. One chimp had to decide which one of six tools he needed to obtain hidden food and then ask the other one for that tool via the keyboard. They could successfully cooperate in this manner only when the keyboard was turned on (Savage-Rumbaugh et al. 1978b). Clearly, Sherman and Austin were using words as symbols and not simply labeling objects.

The work with Sherman and Austin was important for reasons other than demonstrating that chimpanzees can use symbols in communication. It marked the beginning of a shift in emphasis from demonstrating that apes can *produce* language to showing that they can *understand* the symbols or words of language. In addition, it showed how important the learning environment is in the development of language comprehension.

The change in emphasis and learning environment led to great progress in the study of apes' language abilities. Consider the remarkable abilities of Kanzi and his half-sister, Panbanisha, bonobo or pygmy chimpanzees (*Pan paniscus*). Kanzi, born in 1980, was adopted and raised by Matata, who was part of a language study by Sue Savage-Rumbaugh. For two years, trainers futilely tried to teach Matata to use lexigrams to communicate. Kanzi, who was 6 months old at the beginning of the study, was always present during Matata's training sessions. But other than occasionally chasing the symbols projected over the keyboard, he showed little interest. When Kanzi was about 2.5 years old, he was separated from Matata so that she could be bred at another site. Much to the amazement of the experimenters, Kanzi began to use the symbols of the keyboard that they had tried to teach to Matata. Not only did he know the lexigrams, he also knew the English words that the lexigrams represented.

Kanzi had begun to learn communication skills simply by observing his mother's training. He was never trained to use lexigrams. Instead, he picked up the use of language in much the same way as a child would. Once the researchers recognized this, reward-based language training was stopped and replaced with conversation. Kanzi's constant human companions used lexigrams, gestures, and speech to communicate with one another and with him. In this way they served as communicative models. Once he learned to use lexigrams, he began to use them to refer to items like food or objects or to locations that were not in sight (Savage-Rumbaugh 1986; Savage-Rumbaugh and Lewin 1994). Kanzi became a language star able to communicate on a board with 256 lexigrams (Figure 17.22).

Besides being able to produce language, Kanzi and several other apes have demonstrated that they understand spoken English. In these tests, the words were presented through headphones or from behind a one-way mirror to avoid inadvertently cueing by gestures or facial expressions. The sentences were usually commands to perform some action with one or more objects or people. The person evaluating the response did not know what had been requested. Many of the requests were so unusual that it would be impossible to have carried them out without actually understanding the language. Consider for example, the directive, "Put the raisins in the shoe." Kanzi responded correctly to 72% of over 600 requests (Savage-

FIGURE 17.22 **Kanzi, a bonobo, has demonstrated the most advanced language skills so far. He communicates with a computer keyboard that has over 250 lexigrams. He was not trained by operant-conditioning techniques. Instead, he observed and interacted with humans who used gestures and lexigrams to communicate.**

Rumbaugh et al. 1998). Panbanisha, Kanzi's half-sister, was reared in the same type of learning environment as Kanzi had been. She, too, shows remarkable comprehension of spoken English, responding correctly to 77% of 145 sentences (Williams et al. 1997).

A different approach to understanding Kanzi's language skills comes from an analysis of his vocalizations. Kanzi was not trained to vocalize, but he spontaneously makes sounds many thousands of times per day. The question was whether the structure of these vocalizations was related to anything in the environment. Taglialatela et al. (2003) analyzed hundreds of hours of videotape and categorized vocalizations according to what Kanzi was doing at the time. For example, if Kanzi gestured to a grape or pointed to the lexigram representing a grape, his vocalizations were coded with "grape." Then, the acoustic structure of the vocalizations was analyzed with a computer. Vocalizations given within a particular context were statistically similar to each other, but different from those given in other contexts. Thus, Kanzi seemed to be spontaneously uttering different sounds that have different content (Savage-Rumbaugh et al. 2001).

Slocombe and Zuberbühler (2005) studied natural sounds produced by chimpanzees when they find food. Chimps produce acoustically distinct "rough grunts" when encountering different foods. In the chimps' enclosure, bread was always in one location and apples in another. When a chimp heard a recording of the "bread call," it looked in the bread location, and when it heard an "apple call," it looked in the apple location, suggesting that it correctly interpreted the calls to refer to each food type. This skill is found in other animals besides apes, including vervet monkeys (Seyfarth et al. 1980) and even chickens (Evans et al. 1993).

If we return to our definition, we see that there is some evidence that ape communication can fulfill each of the four requirements of language. However, their skills in some areas, especially using grammar and creating novel words or signs, are very limited compared to humans. Consider that from age 18 months to 4 years a human child's vocabulary grows from a few words to thousands, reaching an estimated rate of one new word per hour (Terrace 2005).

Many researchers have argued that it is more informative to study the natural communication systems of animals rather than artificially constructed systems in the laboratory. At this point, studies on the natural vocalizations of great apes in the wild are still few; monkeys are much better studied (Zuberbühler 2005). Still others have ethical objections to keeping primates captive in order to study their communication. Nonetheless, the experiments with Kanzi and other great apes have certainly provided food for thought and have gone a long way in defining both the capabilities and limits of language learning in apes.

COMMUNICATION AND ANIMAL COGNITION

Many people have wondered what it is like to be an animal—specifically, whether nonhuman animals have thoughts or subjective feelings and whether they are aware of other animals' feelings. Such musings have led some investigators to consider whether nonhuman animals are cognitive, conscious, aware beings. We examined some evidence for cognition in Chapter 5, including self-recognition and perspective-taking. Here, we will examine animal cognition and communication.

Donald Griffin (1978, 1981, 1984) has suggested that tapping animals' communication lines is a way to find out whether animals have conscious thoughts or feelings. After all, the only way we know about the thoughts or feelings of other people is when they *tell* us, through either verbal or nonverbal communication. So, if nonhuman animals also have thoughts and feelings, they probably communicate them to others through their communication signals. If we could learn to speak their language, we could eavesdrop and thereby get a glimpse into the animal mind. We might also learn about the animal mind through interspecies communication—teach the animal a language that we understand, such as in the great ape language studies, and then ask it how and what it thinks (Pepperberg 1993).

Most people agree that one sign of cognition is the ability to form mental representations of objects or events that are out of sight. So, one way to look for cognition is to ask whether animal signals are symbolic, that is, whether they refer to things that are not present (Smith 1991). We have seen that certain apes can learn a language that uses symbols, and they can use it to "talk" about things they don't currently see and things that occurred in the past (Savage-Rumbaugh 1986; Savage-Rumbaugh et al. 1998).

In Chapter 5, we discussed Alex, the African gray parrot who was able to vocally request more than 80 different items, even if they were out of sight. In addition, he could quantify and categorize these objects. He showed an understanding of the concepts of color, shape, and same versus different for both familiar and novel objects (Pepperberg 1991). Louis Herman and his colleagues have shown that bottlenosed dolphins (*Tursiops truncatus*) can also learn to understand symbolic languages. In one of these languages, the "words" are gestures, as in sign language. In the other, the words are sounds generated by a computer. The words can refer to objects, actions, and relationships among actions, and many other things. In tests of language comprehension, the dolphins show that they understand the experimenter's references to objects that are not present (Herman and Forestell 1985; Herman et al. 1993).

We can also learn about the way animals think from their natural communication systems. The use of language involves the ability to relate words to meanings. In some species, individuals classify signals by their meaning and not by some obvious physical property, such as the way the signal sounds. Consider, for example, the way rhesus monkeys respond to food calls. When a rhesus monkey finds food, it announces this to others with one or more of five different food calls. If the item is really good—of high quality or rare—the monkey will issue a warble, harmonic arch, or chirp. However, a low-quality, common food item is announced with a coo or a grunt. Thus, there are two categories of food calls, and the calls sound distinctively different (Hauser 1998).

You may recall from Chapter 5 that animals will habituate (gradually stop responding) to a stimulus that is repeated many times without consequence. Marc Hauser (1998) used habituation to determine how the monkeys classify food calls. The reasoning is that if the monkeys classify the call by its sound, after habituating to one type of call they will still respond to any other type of call because it sounds different. However, if they classify the call by its meaning, they will be unresponsive to a call in the same category but will remain responsive to a call in a different category. The calls were broadcast through a speaker, and a monkey was said to have responded if it turned its head toward the speaker and stared at it. As you can see in Figure 17.23, rhesus monkeys classify calls by their meaning. Habituation transferred from a harmonic arch to a warble or vice versa, two calls that advertise high-quality food items. But it did not transfer between calls that have different meanings, from a grunt to a warble or harmonic arch.

Additional evidence of cognition is the ability to understand the mental states of other individuals. Does a communicating animal really understand what potential receivers know? Vervet monkeys apparently do not. Mother vervet monkeys give alarm calls when they sense a predator, but they do not appear to take into account whether their offspring are ignorant or knowledgeable about the predator (Seyfarth and Cheney 2003).

Some intriguing data come from Thomas langurs (*Presbytis thomasi*) (Figure 17.24). Males give alarm calls, but only if there is an audience. In one study, researchers provoked alarm calls in wild langurs in the rainforest by having an observer carry a fake tiger skin over his shoulders. After spotting the "tiger," males called persistently, stopping only when every other member in the group also gave an alarm call (Wich and de Vries 2006). Thus, callers seemed to keep track of all the other members in their groups. Whether langurs understand each other's state of knowledge is not yet known, but this species is a good one in which to test such questions (Zuberbühler 2006).

The question of animal awareness is difficult to answer scientifically, and we only have space here to skim the surface of the relevant literature. The answer, however, has ethical ramifications, especially for researchers who study animal behavior. If the line between animals

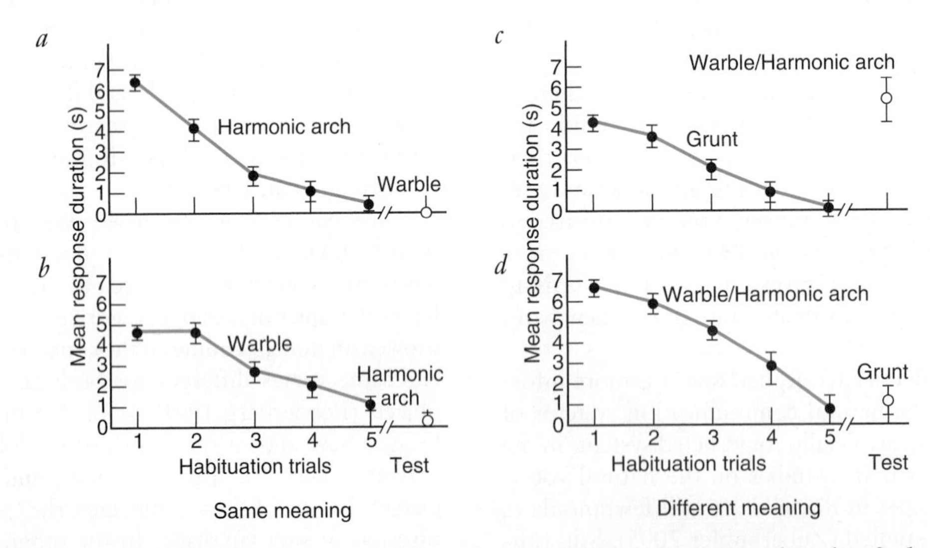

FIGURE 17.23 **Rhesus monkeys relate specific food calls to their meaning. A rhesus monkey that finds food announces this to troop members with a food call. Whereas warbles and harmonic arches announce high-quality, rare food items, grunts are used for low-quality, common items. (*a*, *b*) When a call is broadcast repeatedly through a loudspeaker, monkeys gradually stop responding; that is, they habituate. Although the calls sound different, habituation transfers from a warble to a harmonic arch or vice versa because the calls have the same meaning (only 1 of 12 monkeys responded to a test call with the same meaning). (*c*, *d*) However, habituation does not transfer from a grunt to a warble or harmonic arch because the calls have different meanings (13 of 14 monkeys responded to a test call with a different meaning). (Data from Hauser 1998.)**

FIGURE 17.24 **A Thomas langur. Males give alarm calls, but only when there is an audience.**

and humans is erased or even smudged a bit, should we rethink the way we treat animals? Should we keep them in zoos? Should great apes be used for language studies? What about dolphins?

So what's your opinion? Are animals aware, cognitive beings? All of them? Where do we draw the line?

SUMMARY

The view of communication that was most commonly held through the 1970s was that it is beneficial for animals to share information. Signals should thus be under selection to be honest, unambiguous, and informative. However, this view of communication is not always appropriate. Sometimes the sender of a signal benefits from being dishonest. In this view of communication, a sender should send a signal because, on average, it manipulates the behavior of others to its own advantage. Receivers should respond to signals when they benefit from them, on average, even if they are sometimes deceived.

We next examined four conditions under which we expect signals to be honest. (1) Honest signals are likely when senders and receivers share overlapping goals. For example, a baby bird and its parents both benefit when the baby honestly signals its need for food. (2) Some signals are honest because they cannot be faked. For example, the distance between the eyes of a stalk-eyed fly is an honest signal of its size that it cannot change. (3) Signals are also likely to be honest when they are costly to produce. This is the basis of the handicap hypothesis for the evolution of signaling. For example, an animal that is in good condition might be better able to pay the price of signal production than an animal in poor condition. Thus, signals will be honest indicators of condition. (4) Signals are also likely to be honest when dishonest signalers can be identified. Vervet monkeys, for instance, learn to ignore animals that give inappropriate signals.

We then examined two conditions that favor dishonest signals. (1) In some cases, senders and receivers have different goals. For instance, a male might only be interested in convincing a female to mate with him, but a female might be interested in assessing the quality of all males and choosing the best. (2) Signals might also be dishonest if they are costly to assess or to challenge. A stomatopod that challenges a conspecific that gives a threat display will be attacked with great force if the threat was not a bluff. Finally, we outlined conditions under which honest and dishonest signals can coexist in the same population.

How do signals originate? Earlier studies focused on identifying the behaviors of the sender that serve as raw material for the evolution of displays in a process called ritualization. The study of sensory exploitation focuses on how some senders take advantage of a preexisting sensory bias of the receiver.

In ritualization, three evolutionary sources for displays have been recognized: (1) intention movements, those behavior fragments that may precede a functional action, (2) displacement activities, or behaviors that occur out of context, and (3) autonomic responses, such as vasodilation and piloerection. Although these seem to be among the most common sources for signals, almost any behavior can be ritualized.

During the evolutionary process of ritualization, the form of the ancestral behavior is modified into a stereotyped signal. If one is reasonably sure of the precursor behavior, it can be compared to the current display to determine the changes in the original form. Typically, all or a part of the ancestral behavior pattern is exaggerated by changes in the duration or extent of movement, by alterations in the rate of performance, or by repetition. In addition, the original actions may be combined in a new order, or some parts may be deleted. As the signal becomes "emancipated" from the factors that originally caused it, the behavior may be shown in a new context and/or be motivated by different factors. Frequently, these changes in behavior are accentuated by anatomical modifications such as bright colors, antlers, and manes.

In sensory exploitation, the preexisting biases of the sense organs of the receivers drive the evolution of a signal. By using phylogenies to reconstruct the history of a group of species, it is sometimes possible to determine that the bias of the sensory system was in place before a particular signal evolved.

The design of the signal may be influenced by a variety of factors. The anatomy and physiology of the sender are important in directing the evolution of their signals. Characteristics of the habitat also play a key role in determining both which sensory channel is favored for communication, and exactly how signals within a sensory channel are structured. For example, the environment influences how well bird song is transmitted, and one can generate and test predictions about which sorts of songs will be found in different environments. Visual signals also differ in their detectability in different environments. Wind, for instance, can make visual signals harder to see, and several species of lizards modify their signals under windy conditions. Human activities are changing the conditions for signal transmission for many species. Finally, characteristics of the receiver also influence signal design. Receivers' sensory systems and the psychology of how they interpret signals have been modified by natural selection.

Most animal communication signals are not true language because animals do not use signals as symbols that can take the place of their referent and because they do not string signals together to form novel sentences. Researchers have used an array of approaches in their attempts to teach apes language. Their findings are fascinating but the interpretation is controversial.

Some people have suggested that knowledge about the communication systems of animals may provide an insight into the question of animal cognition. Some animals understand signals that represent items that are out of sight. The natural communication systems of some species, such as rhesus monkeys, reveal that they can relate meaning to signals. Researchers have also looked for evidence that senders of signals understand the mental states of those with whom they are communicating.

18

Conflict

If you've seen the *Trials of Life*, the PBS nature series with David Attenborough, you probably remember the ibex fight. Ibex are mountain goats with long, curved, ridged horns. At the start of the breeding season, males seek each other out in the barren mountains of the Middle East and fight for access to females on near-vertical slopes.

This particular confrontation begins with an assessment. The two males approach each other, lower their heads, and interlace their horns so that the horn of one individual is inside the horn of another. The horns make contact with a loud clack, and then the ibex begin to push one another, shoving each other's heads to the side.

The fight escalates. The ibex back away from each other. One rears up on its hind legs, walks a step or two toward its opponent, and crashes downward with its head turned to slam its horns against its opponent's horns. This sequence of shoving, then "rear clashing" (Alvarez 1990), repeats. Each time, the clash increases in ferocity. Finally, the fight ranges across a steep, rubble-strewn slope that provides only treacherous footing, and rocks and small boulders clatter and bounce down into the distant valley below. Sometimes similar fights end with the loser plunging down a cliff to a messy death.

You may see a very different battle in a British wood. Speckled wood butterflies, meeting each other in a sunny patch in the forest, spiral up and up, until finally one flies off and yields the spot to its rival. No blows are exchanged, and the loser flies off unharmed (Davies 1978).

Why are these fights so different? It's obvious, you may say—butterflies have no weapons, and ibex do.

But even when animals do have weapons, contestants often exhibit restraint. Many ibex fights—in fact, most—are low-risk shoving matches. Besides locking horns and pushing, ibex shove with their necks and shoulders in apparent tests of strength of their opponent. Many contests end at this point, with the weaker ibex galloping off.

In this chapter, we will address the nature of conflict. Why are some fights settled peaceably and quickly, whereas others are more dangerous? Why do some animals negotiate dominance status or a territorial boundary, and then honor the agreement?

AGGRESSION AND CONFLICT

Given the drama of many conflicts between animals, it's no wonder that behaviorists have been motivated to study them, both empirically, with field and laboratory experiments, and by developing theoretical models. Several terms have been used to describe conflict. **Aggression** has an everyday connotation for us but has been formally defined in the behavioral literature as a behavior that appears to be intended to inflict noxious stimulation or destruction on another organism (Moyer 1976). Interspecific interactions such as predation are often included under the umbrella of this term, but generally *not* included are behaviors that occur in response to aggression, such as fleeing. Another term that is meant to encompass the behavior of both the aggressor and the animal that is the focus of the aggression is **agonistic behavior** (note that this differs from agnostic behavior!). Agonistic behavior includes all conflict between conspecifics, including threats, submissive behavior, chasing, and physical combat. Agonistic behavior only includes interactions between conspecifics and excludes aggressive acts between species, such as predation.

WHY DO ANIMALS FIGHT?

Conflict is potentially dangerous. It is also metabolically costly: displays and attacks require energy. In many species, oxygen consumption increases and lactic acid accumulates in the blood or hemolymph (the fluid surrounding the organs in many invertebrates), requiring recovery time (Briffa and Sneddon 2007). If there were enough resources—such as food, shelter, and mates—for all animals, conflict would be far less common. Unfortunately, resources are frequently limited. We've already seen numerous examples of conflict over limited resources. In Chapter 12, you read examples of conflicts over food, and in Chapter 14, conflict over mates. Chapter 15 described conflict between parents and offspring over the amount of care that a parent should provide—here, the limiting resource is the amount of energy and time that a parent has available to lavish on a particular offspring versus investing in other offspring or its own survival. As we add more examples to our list in this chapter, keep in mind that we judge the value of a resource from an evolutionary perspective: that is, whether it increases the likelihood that an animal passes on its genes.

Conflict is not ubiquitous across species, but is extremely common indeed. It manifests itself in a variety of ways: sea anemones lash each other with their stinging tentacles, baboons charge at one another and slash with their alarmingly large canine teeth, and even limpets (shelled molluscs, resembling flattened snails) have slow-motion battles in which they try to pry each other off rocks into the swirling surf. The variation in the manner and intensity of conflict in different species is enormous. But can we find and apply broad principles that help us understand the evolution of conflict in a more general way? Let's first take a big-picture view of conflict. We'll then move to two special cases: conflict within groups, and conflict over territory. Finally, we'll briefly touch on proximate causes of conflict.

AN EVOLUTIONARY VIEW OF CONFLICT

THE EVOLUTION OF FIGHTING BEHAVIOR

At first glance, there seems to be an evolutionary puzzle in animal conflict. In the last chapter, we saw many examples of displays in which animals signal to one another rather than engaging in physical aggression. But if a resource is in limited supply, why don't animals always fight with their maximum effort? After all, an animal that only stands erect and fluffs its feathers to communicate its aggressive intentions will surely lose if its rival attacks violently.

Early ethologists thought that restraint in fighting evolved "for the good of the species," but we know now that this explanation is not plausible. Natural selection favors the individual that passes on more of its genes, not the individual that behaves for the sake of other individuals that are unrelated to it. Thus, we must seek other explanations for the variation we see in the intensity of animal fights. To do this, we turn to game theory.

USING GAME THEORY TO UNDERSTAND THE EVOLUTION OF CONFLICT

To construct hypotheses about the evolution of conflict behavior, researchers have used mathematical models. Models are logic-based tools to help us understand more precisely the costs and benefits of behavioral strategies. Recall that in Chapter 12 we used a mathematical model to predict which foods of a range of choices a forager should eat. For the problem of animal conflict, we need a different type of modeling approach than we used for foraging. Because the optimal strategy of one individual depends on what its opponent is doing, we need **game theory**. Game theory predicts an animal's optimal behavior while taking into account the behavior of other animals. Game theory was developed by economists to predict human behavior in economic markets, and was borrowed and modified by animal behaviorists (notably by John Maynard Smith, especially 1974, 1976).

Game theory shares some vocabulary with the other models we've discussed but adds some new terms. In a game-theory model, the combatants are called **players**. Players have available to them different decisions, or **strategies** (a term also used in foraging models). We measure the costs and benefits, or **payoff**, of each strategy using a **currency**. In economics, the currency of game-theory models is money. As in the other biological models we've examined, in behavior, the currency is something that relates to fitness, such as the number of offspring produced or the number of calories acquired. Strategies are assumed to be heritable, and thus successful strategies will increase in the population. A table called a **payoff matrix** is used to organize the values of the payoffs of each strategy against each of the other strategies.

The simplest game-theory model of aggression is called the hawk-dove game. Here is the scenario: two players are fighting over a resource. Each opponent has the option of playing one of two strategies, called **hawk** and **dove**. The hawk strategy is to immediately attack its opponent. The dove strategy is to flee immediately if confronted by a hawk and display if confronted by another dove. (Thus, "hawk" and "dove" do not refer to real hawks or doves, but are simply meant to evoke a picture of the "personalities" of these strategies.) If a hawk meets a hawk, or a dove meets a dove, each opponent has a 50% chance of winning.

It's very useful to look at the details of this simple game, as the predictions from this model drive the rationale for a great deal of empirical work described later in the chapter. First, let's work out the payoff matrix for this game. We must define three variables, each measured in a currency that relates to fitness:

V = the value of the resource being contested
W = the cost of being wounded in a fight
D = the cost of displaying to an opponent

What happens if an animal playing the hawk strategy meets another hawk? Both will attack immediately. One hawk wins the resource, so its payoff is V. The other hawk will be wounded, so its payoff will be $-W$. So, what is the *average* payoff for a hawk vs. hawk interaction? (Take a moment here to calculate this before reading ahead.)

The average payoff for a hawk vs. hawk interaction is the payoff for the winning hawk, plus the payoff for the losing hawk, divided by 2 to get the average.

$$\frac{\text{value of the resource} - \text{cost of being wounded}}{2} = \frac{V-W}{2}$$

We can add this value to the payoff matrix in Table 18.1.

Now, see if you can figure out, *without looking at the next paragraph*, the payoffs for the other three interactions: hawk against dove, dove against hawk, and dove against dove.

If a hawk meets a dove, the hawk immediately attacks, and the dove flees. The hawk wins the resource, so its payoff is V. If a dove meets a hawk, the dove immediately flees. The dove does not get injured, nor does it win anything. So, its payoff is 0.

If a dove meets a dove, both display. One eventually wins the resource, and the other walks away without getting injured. Both pay the cost of display. Thus, the payoff for the winning dove is $V - D$, and for the losing dove it is just $-D$. To get the average, sum these and divide by 2. We can rearrange these to make it a little easier on the eye:

$$\frac{V-D-D}{2} = \frac{V-2D}{2} = \frac{V}{2} - D$$

TABLE 18.1 The Payoff Matrix for the Hawk–Dove Game. Hawks always attack. Doves flee when attacked but otherwise display. The contested resource provides an increase in fitness of V, the cost of being wounded is a decrease in fitness of W, and the cost of displaying is a decrease in fitness of D

	When opponent is:	
Payoff to:	Hawk	Dove
Hawk	$\frac{V-W}{2}$	V
Dove	0	$\frac{V}{2} - D$

Check the payoff matrix in Table 18.1 to see that the values make sense to you. Now let's get more specific, and put some numbers into the payoff matrix. In this example, the cost of injury is high.

V = 30
W = 60
D = 5

Based on these numbers, fill in the payoff matrix. Check your answers in Table 18.2.

TABLE 18.2 Values of Payoff Matrix When $V = 30$, $W = 60$, and $D = 5$

	When opponent is:	
Payoff to:	Hawk	Dove
Hawk	$\frac{30 - 60}{2} = -15$	30
Dove	0	$\frac{30}{2} - 5 = 10$

Now what do we do with this information? Remember that the currency of these models is in units of fitness and that we assume that these strategies are heritable. Successful doves will have offspring that also play the dove strategy, and hawks will give rise to hawks. Thus, we can use game-theory models to predict whether strategies in a population will increase in frequency, remain stable over generations, or disappear.

In Chapter 4, we introduced the concept of an **evolutionarily stable strategy, or ESS**. An ESS is a strategy that, when played by most members of the population, cannot be invaded by another strategy. Let's look at our example with the numbers above, and ask if the dove strategy is an ESS. If the population were comprised entirely of animals playing the dove strategy, the average payoff would be 10. If an animal playing hawk entered the population, all of its opponents would be doves. The payoff to the hawk against a dove is 30, so the hawk strategy will do well. Because the units in the payoff matrix represent fitness, the hawk's genes increase in the population, and the hawk strategy increases in frequency.

So, will the population eventually become all hawks? In other words, is hawk an ESS? If the population is comprised of all hawks, the average payoff is –15. If a dove were to enter the population, it would not do too badly in comparison to the hawks—it won't win, but unlike the hawks, at least it won't be wounded during half its battles. So, the frequency of the dove strategy would increase.

Thus, with these values of W, V, and D, neither a "pure hawk" strategy nor a "pure dove" strategy is an ESS. However, there is a combination of hawk and dove strategies that *is* stable. This is called a **mixed ESS**. The stable proportion of hawks and doves occurs when the average payoff for the hawk strategy equals the average payoff for the dove strategy. A mixed ESS can come about either by a certain proportion of animals always playing hawk and another proportion always playing dove, or by all animals playing both hawk and dove with particular probabilities. See Box 18.1 for an example of how to calculate the proportions of hawks and doves at a stable equilibrium.

BOX 18.1: HOW MANY HAWKS AND DOVES?

Often, neither hawk nor dove is an evolutionarily stable strategy. Rather, the stable, equilibrium composition of the population is some combination of hawks and doves in a mixed ESS. The stable proportion of hawks and doves occurs when the average payoff for the hawk strategy equals the average payoff for the dove strategy. We can calculate these proportions with an equation with just a bit more math:

p = the proportion of hawks in a population
$1 - p$ = the proportion of doves (because the proportions of hawks and doves must add up to 1)

The average payoff for the hawk strategy can be put into words:

(chance it will meet another hawk) × (payoff when meeting a hawk)
+
(chance it will meet a dove) × (payoff when meeting a dove)

We assume that animals encounter each other randomly, so the chance an animal will meet a hawk is equal to the proportion of hawks in the population, or p. Similarly, the chance of meeting a dove is $(1 - p)$. We've already figured out the payoff matrix in Table 18.1. Substituting variables for the words, we find that the average payment for the hawk strategy is

$$p\frac{(V - W)}{2} + (1 - p)V$$

Now figure out what the equation would be for doves before you continue reading.

The average payoff for the dove strategy is:

(chance it will meet a hawk) × (payoff when meeting a hawk)
+
(chance it will meet another dove) × (payoff when meeting a dove)

Thus, the average payment for the dove strategy

$$= p(0) + (1 - p)\left(\frac{V}{2} - D\right)$$

At equilibrium, the average payment for hawks equals the average payment for doves. Thus, for any particular values of V, W, and D, we can solve for p and figure out the stable proportion of hawks and doves. Let's try it for $V = 30$, $W = 60$, and $D = 5$, the same values we used to figure out the payoff matrix in Table 18.2.

$$\text{Payoff for dove strategy} = p\frac{(V - W)}{2} + (1 - p)V$$

$$\text{Payoff for hawk strategy} = p(0) + (1 - p)\left(\frac{V}{2} - D\right)$$

At equilibrium,

$$p\frac{(V - W)}{2} + (1 - p)V = p(0) + (1 - p)\left(\frac{V}{2} - D\right)$$

$$p\frac{(30 - 60)}{2} + (1 - p)30 = p(0) + (1 + p)\left(\frac{30}{2} - 5\right)$$

Rearranging to solve for p gives $p = 0.57$. This is the proportion of hawks. The proportion of doves is $1 - p$, or 0.43.

A mixed ESS can occur in two ways. First, different individuals can play either hawk or dove. At equilibrium, there is a stable mixture of the different types of strategists in the population. Alternatively, every individual can play both strategies at the calculated frequency. In this example, an individual's optimal strategy would be to play hawk 57% of the time and dove 43% of the time.

FIGURE 18.1 **Male elephant seals are scarred and bloody from battle. Fights between bulls are brutal and often result in injury. Such battles are predicted by game theory when the value of the resource is very high, as in this example. Only the most dominant bull will be harem master and leave offspring.**

Are either hawk or dove stable strategies for *any* values of V and W? If the value of a resource, V, is greater than the cost of being wounded, W, a pure hawk strategy is an ESS. If $V < W$, then a mixed ESS will result. A pure dove strategy is never an ESS.

Testing the Predictions of Game Theory

Game-theory models generate a number of testable predictions. One prediction is that the ferocity of a contest depends on the value of the resource relative to the cost of injury. This makes intuitive sense—you would probably put up more of a struggle with a mugger if your life savings were at stake rather than a pocketful of change (in other words, if the value of the resource is relatively higher). You would probably even let a thief take everything you owned if you thought you would be killed if you put up a struggle (that is, if the cost of fighting were high). As you might expect, this prediction bears out in animal species as well. Let's take a look at a few examples.

In some species, the prize for winning a fight might be incredibly valuable, perhaps even a lifetime's worth of reproductive success. In these cases, we predict that animals should risk everything, even fighting to the death. A classic example of brutal and bloody fights are those among male elephant seals (*Mirounga angustirostris* and *M. leonina*) for the right to mate (Figure 18.1). Adult males bear many scars that serve as silent testimony to the intensity of combat. After a fight, when a male enters the sea, the water is often reddened with his blood. All matings are performed by a few dominant males who defend harems of females. The duels between males are so strenuous that a male can usually be harem master for only a year or two before he dies (LeBoeuf 1974; McCann 1981). The rivalry is intensified by the females, who vocalize loudly when a male tries to copulate (Cox and LeBoeuf 1977). The commotion attracts the attention of other males, who attempt to interfere. As a result, generally only the largest and strongest males are able to mate.

In other species, we see intensive hawk-like fighting, but for a different reason: not because the value of the resource is so enormously high as in elephant seals, but because fight costs are low. For example, toads have no real weapons to use against conspecifics, and fights rarely end in serious injury or death (Davies and Halliday 1978). Because fight costs are low, toads are especially willing to engage in lively battles over females (Figure 18.2). In contrast, in other species that have serious weapons, battles are often generally restricted to displays (Figure 18.3). Game theory reminds us that it is the cost

FIGURE 18.2 **A toad fight. Game theory predicts that fights are more likely to escalate when the costs (risk of injury) are low.**

FIGURE 18.3 **A rattlesnake fight. Each of these males could kill the other, but they do not bite. The males press against one another, belly to belly. Finally, the weaker individual yields, and his head is pushed to the ground by the stronger animal.**

of battle *relative* to the benefit of winning that drives fight intensity.

Compelling evidence of the influence of resource value on fight intensity comes from species in which the value of the resource changes over time or in different places, so that we can test within a single species whether fighting intensity correlates with resource value. For instance, the value of a female to a male is not always the same. In mammals, for example, the likelihood that a mated female conceives varies over time. Red deer stags fight most fiercely and, as a result, are wounded most frequently during the period when most calves are conceived (Figure 18.4) (Clutton-Brock et al. 1982).

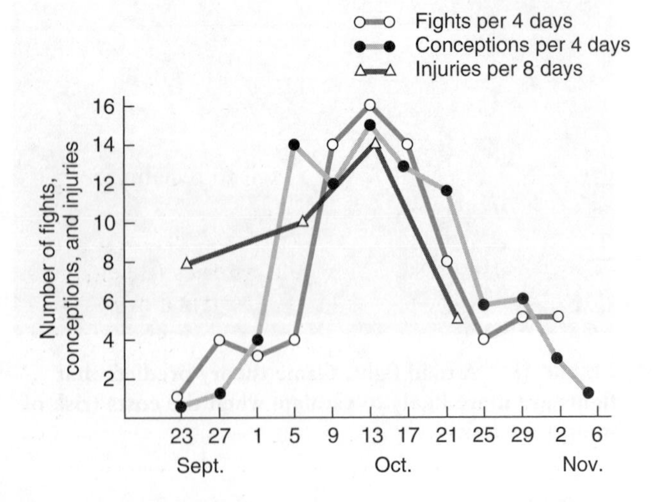

FIGURE 18.4 **The number of fights among male red deer (open circles) is compared to the number of conceptions during the rutting season. The number of fights and conceptions is indicated for four-day intervals. The number of injuries per eight-day interval during the rut is indicated by triangles. The number of fights and injuries peaks during the interval when conception is most likely. As predicted by game theory, male red deer fight harder when the value of the resource is greater. (From Clutton-Brock et al. 1982.)**

ASYMMETRIES IN CONTESTS

In the basic hawk–dove model, we assume that all animals value the contested resource to the same extent and that all individuals have the same ability to fight. But how often is this actually the case? In real life, rivals are rarely true equals, but generally differ in some quality. Thus, contests are usually asymmetric.

Inequalities, or asymmetries, among rivals can be grouped into categories (reviewed in Gherardi 2006). Here, we examine asymmetries in: (1) the ability of each contestant to defend the resource, (2) the experiences of each contestant in previous fights, (3) the value of the resource to each contestant, and (4) arbitrary asymmetries unrelated to either resource value or the ability to defend the resource (Maynard Smith and Parker 1976).

Asymmetry in Fighting Ability

One combatant may be larger or heavier, have bigger weapons, or be a more skilled fighter (reviewed in Hsu et al. 2006). Characteristics such as these that bear on an opponent's ability to defend a resource describe its **resource-holding potential** (RHP). It seems intuitive that contestants might increase their fitness by assessing their opponent's RHP relative to their own, and adjusting their fighting strategy accordingly (Archer 1988; Parker 1974). This is an example of a **conditional strategy**, adjusted based on the conditions of the particular fight.

How do animals assess RHP? We've already seen in Chapters 16 and 17 that many displays are designed to convey an impression of size and strength to an opponent. Thus, we expect to see animals evolve to assess one another accurately and to bluff convincingly when possible. Some traits are difficult to bluff, and such traits are often used in assessment. For example, a male mountain sheep with small horns will defer to a competitor with larger horns (Geist 1971). Similarly, male red deer judge each other's size by interlocking their antlers and pushing, and the one that is outclassed retreats (Clutton-Brock et al. 1982). The size of the claw that one male shore crab (*Carcinus maenus*) presents to his opponent during an agonistic contest is more important than his overall body size in determining the outcome of the conflict (Sneddon et al. 1997).

Some species may not be able to judge their opponent's RHP, but only their own. Combatants that assess themselves as being weak may give up more quickly than combatants that assess themselves as being strong. Careful experimental design and analysis is needed to determine whether a particular species is able to determine the RHP of its rivals, or only its own RHP (Taylor and Elwood 2003).

Asymmetry in Experience

After the initial contest, the experience of either winning or losing can influence the outcome of subsequent encounters. Fighting experience is important in many species, including insects, spiders, molluscs, fishes, birds, and mammals (Hsu et al. 2006). Depending on the species, winners become more likely to win subsequent fights, losers become more likely to lose, or both. In an example that illustrates a typical experimental design, Mary Whitehouse set *Argyrodes antipodiana* spiders against each other. These tiny spiders make their living primarily by stealing food from the webs of other spiders, and they will fight fiercely for packages of silk-wrapped prey. Because size is important in determining the outcome of a spider fight, spiders were paired by size so that neither had a natural advantage over the other. Each member of the pair was assigned randomly to fight against either larger spiders (so they generally lost their fights) or smaller opponents (so they generally won their fights). After these "training fights," the sized-matched spiders were pitted against each other. The results were very convincing: in 15 of 17 pairs tested, spiders that had experience as winners trounced their size-matched opponents (Whitehouse 1997).

Loser effects can be quite long-lasting, as Gordon Schuett (1997) demonstrated in male copperhead snakes (*Agkistrodon contortrix*) fighting for access to a female. First, for each trial, the experimenters placed two males with no fighting experience during the previous 6 to 12 months in an arena with a female snake (the prize). One of the males was 8% to 10% longer than the other, measured as the distance between the snout and the vent (the genital opening), and had greater body mass. In all cases, the larger male won the fight and the female. When the pairs were rematched 24 hours later, prior losers gave up without even challenging the competitor. The next day, prior losers were paired with unknown males that closely matched their own length but that had no recent fighting experience. This time the losers from the first day did give some challenge displays but significantly fewer than their opponent. No fighting occurred. Prior losers just gave up and were chased away.

Asymmetry in the Value of the Resource to Each Individual

It is not difficult to imagine conflicts in which the contested resource is more valuable to one contestant than to the other. Food is more precious to a starving animal than to a well-fed one, so we might expect the hungry contestant to fight harder for it. This prediction is borne out in northern harriers (*Circus cyaneus*). These hawks catch prey that requires some time to digest before another one can be consumed. Once the owners of a territory have eaten, the value of the remaining prey on the territory temporarily decreases. During this interval, harriers are not as aggressive toward territorial intruders that might steal food (Temeles 1989).

Asymmetry in resource value can even tip the outcome of fights in favor of a weaker opponent that has lower RHP. For instance, like many species that endure a long and rigorous migration, small birds called bluethroats (*Luscinia svecica*) prepare for departure by storing away as much body fat as they can. A fatter bird has a better chance of survival, and every bit of added fat is an extra safeguard. When experimenters offered food to groups of bluethroats, individuals with low body fat won access to the food more frequently than did those that were close to their maximum weight. Leaner birds were even able to chase away larger birds that were already positioned at the feeding bowl, presumably because the food was more important to the lean birds, and so they were more highly motivated to win (Lindstrom et al. 1990).

An animal that knows a great deal about a resource may value it more than an animal that knows less about it. For instance, a territory might be more important to the resident than to an intruder because a resident has learned the location of food sources, escape routes, and refuges (e.g., Stamps 1995). Thus, one might predict that an individual should grow to value a territory more as it becomes familiar with it. This expectation is supported in removal studies of male red-winged blackbirds (*Agelaius phoeniceus*). Territory owners were removed until replacement pairs moved in. When the original owners were released, they fought to reclaim their territory. When the original territory holders were released after being held in captivity for up to 49 hours, they nearly always won back their territories (Beletsky and Orians 1987). However, when the original owners were retained for up to a week, the new residents usually defeated the former owners. The released owners were just as persistent in their attempts to recover their territories after they had been removed for seven days as they were after two days. The difference in the outcome of the contests, therefore, seems to be due to a change in the behavior of the new residents. They were more willing to escalate contests as the territory became more valuable to them (Beletsky and Orians 1989).

We see other examples in which knowledge about a resource influences fighting behavior. Hermit crabs, for example, use empty snail shells as shelters, carrying them around on their abdomens. Much shell-swapping takes place as crabs attempt to steal shells from conspecifics. Researchers studying *Pagurus bernhardus* crabs made some shells less desirable by gluing sand deep inside the interior. Only the owner could tell that a particular shell had this unpleasantly scratchy surface; competitors for the shell could not. Owners of the poor shells changed their fighting strategies, fighting harder if they were in the role of attacker, but giving up more quickly if defending their shells (Arnott and Elwood 2007).

Arbitrary Asymmetry

The differences between contestants that we have discussed so far have a fairly straightforward relationship to the fighting strategy of each contestant. We have seen that a competitor might follow a conditional strategy such as "play hawk if larger; play dove if smaller" or "persist longer when the resource is more valuable." The association between a resource's value and resource holding potential is thus called a correlated asymmetry. A contestant maximizes its fitness by paying attention to correlated asymmetries because these are likely to predict the outcome of a battle.

Other differences between contestants are not logically connected to the fighting strategy of the opponents but do affect the outcome of the dispute. These arbitrary asymmetries (also called uncorrelated asymmetries) are, in essence, rules or conventions used to settle conflicts. Examples of arbitrary rules for settling differences among humans are flipping a coin or pulling straws. There is no reason that "heads" should win or that the short straw should lose. These are simply rules that are mutually agreed upon.

A potential example of an arbitrary asymmetry is that of prior ownership (or residency). Animals often appear to adhere to the principle that "possession is nine-tenths of the law." For example, a hamadryas baboon (*Papio hamadryas*) male permitted to associate with a female for as little as 20 minutes was perceived as the "owner" by a second, newly introduced male. That the second male was deferring ownership was revealed when, several weeks later, he was permitted to associate with the female. When the tables were turned, the first male did not challenge the second male's ownership (Kummer et al. 1974).

To add the possibility of prior ownership, a variation of the hawk–dove model includes a third strategy, bourgeois. The bourgeois strategy sets rules for dealing with prior ownership: play hawk if you had possession first; otherwise, play dove. If the bourgeois strategy is added to hawk and dove strategies in a population, it does better than either, so it is an ESS. If all animals are playing the bourgeois strategy, the owner always wins with barely a squabble, and the outcome of any dispute can be reversed by switching ownership.

Is the bourgeois strategy common in nature? This question is more difficult to answer than it first appears, as illustrated by the following example.

The Difficulty of Teasing Apart the Rules of Animal Fights: A Case Study

We return to the male speckled wood butterflies (*Pararge aegeria*), which defend spots of sunlight that serve as mating territories. Wood butterflies fight by flying upward together in a spiraling pattern. At the top of the spiral, one flies away. When Nick Davies netted resident butterflies, he found that the resident's sunspot was almost instantly claimed by another male. When the initial resident was released again, it always lost the fight to regain its spot. Only if two males were experimentally tricked into joint "ownership" did an escalated contest involving spiral flight occur. Wood butterflies thus seemed to be following the rule that "owner wins," and they became a classic example of an uncorrelated asymmetry (Davies 1978).

It is very difficult, however, to document conclusively that the outcome of a contest is truly uncorrelated with the resource-holding potential of the winner. Might there be some other reason that some butterflies are better fighters? One hypothesis is that a butterfly that is in a sunspot, even for a few minutes, gets warmer and is thus able to fly longer. This would produce the same result as Davies saw: the resident male would have a chance to stay warm, while the male waiting it out in Davies's net might cool off and then be at a disadvantage in the next contest.

To test the thermal hypothesis, Stutt and Willmer (1998) experimentally manipulated the body temperature of speckled wood butterflies. For each trial, two butterflies were caught and marked, and each was placed into one of two clear plastic boxes. One box was insulated with polystyrene, and the other was covered in black plastic. A 30-mm area at the top of each box was left transparent, allowing light to enter. The boxes were then placed next to each other and moved into a sunspot. The temperature in each box was monitored, and it quickly climbed higher in the black box than in the insulated one. Within five minutes the temperature in the black box rose to that previously measured in sunspots. However, the temperature in the insulated box still approximated temperatures in shady locations. Then the butterflies were released, each considering itself to be the resident owner of the sunspot. The duration of the escalated flight was recorded, and the winner was noted as the one who returned to the sunspot where the boxes had been placed.

When escalated flight occurred, the warmer male won significantly more often. This is consistent with the idea that the winner of the spiral contest will be the male with the higher RHP as measured by body temperature. Note that the result is the same as in Davies's experiments—the owner wins—but the underlying reason is different.

The matter is still not settled, however. Kemp and Wiklund (2004) argued that the experimental protocols used in previous studies were unnatural and probably traumatized the butterflies—and thus not all contestants were motivated to hold territories. Kemp and Wicklund used enclosures in which butterflies were allowed to discover each other naturally. The butterflies were allowed to have a contest over a sunspot. The initial loser was then granted sole ownership of the sunspot, while winners were temporarily stored in a cooler. Winners were then allowed to bask and initiate flights at their leisure. In another experiment, Kemp and Wiklund also manipulated temperature by chilling butterflies in a cooler. By tossing bark chips up into the sunspot, investigators could coax chilled butterflies out to investigate before they warmed up, and thus trigger a fight.

Contrary to what other researchers found, in Kemp and Wiklund's studies, the individual that won the first fight was highly likely to win the second fight when it was an intruder. Thus, there is no evidence of an uncorrelated asymmetry. In addition, they found no effect of temperature on winning. Instead, Kemp and Wiklund suggest that butterfly fights may be determined by intrinsic aggression or fighting ability, or by prior experience.

This example holds a few lessons. First, alternative hypotheses that produce the same results must be carefully identified and parsed out. Second, subtleties in experimental design can lead to dramatically different results—even very elegant experimental designs may have hidden problems. This species is sure to be the focus of additional research.

CONFLICT AMONG GROUP MEMBERS

Next let's consider a special case of conflict: conflict among animals in the same social group. In a stable social group, the same two individuals are likely to repeatedly encounter one another in competitive situations. In such cases, the animals don't usually fight each time they meet. Instead, relationships develop among them.

Dominance refers to the ability of one animal to assert itself over others in acquiring access to a limited resource, such as food, a mate, or a nesting site. A **submissive** animal predictably yields to a dominant one (Kaufmann 1983).

Dominance hierarchies vary among species, and within a species they can vary with conditions and over time. The simplest form of a dominance hierarchy is despotism, in which one individual rules over all others in the group and the subordinates are equal in rank. Hierarchies may also be linear. In this case, A is dominant over all other animals, B is dominant over all but A, and C is subordinate to A and B but dominant over the rest, as follows:

A —> B —> C —> D —> E

This is often called a pecking order because it was first described in chickens. Chickens commonly demonstrate their dominance by pecking lower-ranking animals.

Sometimes dominance hierarchies are even more complicated, with A dominant over B, B dominant over C, but C dominant over A, or with hierarchies shifting as circumstances change.

HOW DOMINANCE IS DETERMINED

All the determinants of fight outcome that are important in single encounters—size, strength, and experience as a winner or loser—are important in social groups as well. In addition, in social groups, dominance may also be attained through an association with a high-ranking individual. For example, when two flocks of dark-eyed juncos merge, all the birds of one flock tend to rank above those of the other. It is as if the subordinate birds ride the coattails of the highest ranking bird to achieve dominance in the combined flock. A possible explanation for the subordinates' rise in dominance is that the highest ranking individual behaves differently toward birds that are familiar because they were members of its original flock, than toward unfamiliar flockmates (Cristol 1995).

In other species, dominance may be a birthright based on the status of one's parents, just as it is in many human societies. This is the case among rhesus monkeys. Adult females have a linear dominance hierarchy, and offspring assume a dominance position just below their mother (de Waal 1991).

THE BENEFITS OF BEING DOMINANT

It seems entirely reasonable to expect that being dominant will translate into increased fitness. Two fitness-related benefits have been especially well studied: access to food and access to mates.

Dominant individuals get more food than subordinates across a wide range of species. For instance, food is an important benefit of dominance among brown hyenas (*Hyaena brunnea*) of the central Kalahari (Figure 18.5). Each sex has a clear linear dominance hierarchy, and the male and female at the top have equal rank.

FIGURE 18.5 **Top-ranking brown hyenas enjoy two important benefits of dominance—enhanced reproductive success and increased access to food. Dominant hyenas have more feeding time at carcasses than do subordinates.**

Although brown hyenas live in clans, they forage alone. During the rainy season, the primary component of their diet is the remains of kills, such as giraffe, gemsbok, and wildebeest, made by other predators. As many as six hyenas may arrive at a carcass together, but only one or two will feed together. The top-ranking animals of a clan have more feeding time at carcasses than subordinates. In addition, subordinate males and females are significantly more likely to leave the carcass without feeding if a dominant animal is present (Owens and Owens 1996).

In some social groups, all or nearly all the reproductive benefits in a group go to a single individual that is the best competitor. In some species, the dominant female (and in some cases the dominant male) suppresses reproduction by other members of the group. The most dramatic examples of reproductive suppression occur in the eusocial species, such as social insects and naked mole rats, in which only a single female—called the queen—reproduces. Eusociality is discussed further in Chapter 19.

More commonly, however, the dominant animals have a clear reproductive advantage, but they don't completely suppress reproduction by subordinates. For example, each pack of African wild dogs (*Lycaon pictus*) has a clear dominance hierarchy in each sex. In two study sites, top-ranking alpha females produced 76% and 81% of the litters. Although 82% of the dominant females gave birth each year, only 6% to 17% of the subordinate females did so (Creel et al. 1997).

THE BENEFITS OF BEING SUBORDINATE

With the benefits of a dominance hierarchy clearly stacked on the side of the high-ranking members, we may wonder what's in the relationship for a subordinate. If subordinates have no other choice but to stay in a group, repeatedly challenging other individuals to fights would lead to a risk of injury in conflicts they would most likely lose (e.g., Fournier and Festa-Bianchet 1995). In fact, in many groups, both subordinates and dominants suffer from a shake-up in the hierarchy. When the dominance hierarchy in chickens is stable, hens fight less and lay more eggs than when dominance relationships are still being established (Pusey and Packer 1997).

Sometimes dominance has its costs—and subordinates avoid incurring that cost (reviewed in Huntingford and Turner 1987). For instance, dominant great tits or pied flycatchers have higher resting metabolic rates than do subordinate birds and require more food (Røskaft et al. 1986). In some species (but not others), dominant individuals show increased glucocorticoid levels, a steroid hormone associated with stress (reviewed in Sands and Creel 2004). We'll discuss these hormonal data later in this chapter when we turn to a proximate view of dominance.

In some cases, subordinates have the option of leaving and joining another group. This, too, can be risky. For example, although subordinate red foxes have little hope of living long enough to become dominant in their natal group, the mortality rate of those who disperse is also very high (Baker et al. 1998). So, although the benefits of group living may not be as great for a subordinate animal as they are for a dominant one, they may still outweigh the costs of leaving the group. Besides, the situation in the group could get better for a subordinate. The dominant animal could die or become weak enough to be displaced. In the meantime, since many groups consist of family members, a subordinate animal may gain some fitness through kin selection by helping to raise its siblings. (A more detailed discussion of helping is found in Chapter 19.)

Sometimes an animal that is generally subordinate can occasionally win a fight to briefly gain access to a resource. Among Rocky Mountain bighorn sheep, for instance, subordinate rams occasionally manage to win a fight and get a few seconds with a female—all that is necessary to copulate. When lambs are born, the male most likely to be the father is identified by using a combination of genetic and behavioral data. The subordinate rams fathered 44% of the 142 lambs born in two natural populations (Hogg and Forbes 1997), so subordinates do quite well.

Sometimes subordinate members of a group employ more subversive techniques, such as alternative reproductive strategies. Sneaky males generally mimic female characteristics to get past a dominant male and copulate with his female. Satellite males generally position themselves so that they can intercept females who are attracted to a dominant male and copulate with them. We describe examples of these alternative reproductive strategies elsewhere in this book: sneaky male side-blotched lizards (Chapter 4), sneaky male plainfin midshipman fish (Chapter 7), and satellite male natterjack toads (Chapter 14). In all cases, the underlying strategy is to avoid the costs of achieving and maintaining dominance and still enjoy some reproductive success.

Subordinates may sometimes band together to challenge dominant individuals. For example, in a troop of savanna baboons with eight adult males, the three lowest-ranking males regularly formed alliances to oppose a single higher ranking male. The alliances gained reproductive access to the female on 18 of 28 attempts (Noë and Sluijter 1990). Alliances do not always overturn the current hierarchy, however. In some Old World monkeys, females band together and "gang up" on other females, and they generally target lower ranking monkeys (reviewed in Wittig et al. 2007). Alliances occur in other animals besides primates. For instance, in the ground-dwelling tropical bird called the white-winged trumpeter, which lives in social groups, subordinate males appear to collaborate in interrupting the copulation attempts of the dominant male (Eason and Sherman 1995).

CONFLICT OVER SPACE

A special category in the study of conflict is conflict over space. Whereas some animals peaceably coexist, even forming groups (Chapter 19), many species exclude conspecifics from particular areas.

HOME RANGES, CORE AREAS, AND TERRITORIES

We've already been talking about territories and space, but let's take a moment to formally define some terms. The **home range** of an individual animal is the area in which it carries out its normal activities. It includes space it defends from others, as well as space that is used by others. Within the home range there is often an area in which most activities are concentrated—the **core area**. In some cases, the core area may be the area immediately surrounding the nest site or perhaps a food or water source. Animals may have a home range and core area whether or not they share space with conspecifics.

Surprisingly, the definition of a **territory** generates little agreement. A survey of the literature on territoriality revealed no fewer than 48 different definitions (Maher and Lott 1995). The most common one, "a defended area" (Noble 1939), was used in only 50% of the papers. Although this definition of territory emphasizes active defense of an area, other definitions downplay defense and emphasize instead the exclusive use of space (Schoener 1968). This latter definition is often more practical for describing the space use patterns of animals when aggressive interactions over territory boundaries are difficult to observe in the field. Because of the secretive habits of many small mammals, for example, it is virtually impossible to state with any certainty that the exclusive use of an area is maintained by active defense (Ostfeld 1990). Since the definition affects the type of data collected, the lack of consistent terminology can make comparisons among studies difficult (Maher and Lott 1995). Here, we will define a territory as a defended space.

Territories may have different uses, depending on the resource being contested. They may be used solely for feeding, mating (recall our discussion of leks in Chapter 15), or raising young, or they may be used for a variety of purposes, in which case they are called multipurpose territories.

THE IDEAL FREE DISTRIBUTION AND SPACE USE

Before discussing how and why animals exclude each other from territories, it's useful for comparison purposes to think about how we might expect space to be divided if animals do not defend territories. You can see an example if you have two bags of bread, a friend, and a duck pond (or a flock of pigeons). Stand a short distance away from your friend. Both of you throw bread pieces into the water. You throw a piece of bread every 15 seconds, and your friend throws his bread three times faster than you do, at the rate of a piece every 5 seconds. What will the ducks do? The first duck to arrive at the feeding area should go to your friend, where the bread is coming at a faster pace. The second duck should also go to your friend—even if it only gets half the bread that your friend is throwing, it's still averaging one piece every 10 seconds, better than what you are providing. The third duck will get, on average, a piece of bread every 15 seconds no matter where it goes. If the third duck chooses your friend, where will the fourth duck go? See Fig. 18.6 for a general case.

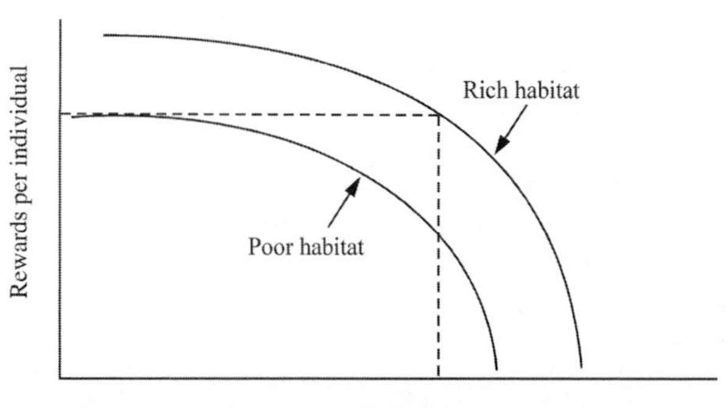

FIGURE 18.6 **The ideal free distribution. Illustrated is a case with two habitats, with habitat *A* of higher quality than habitat *B*. Thus, as animals arrive in the area, they should select habitat *A*. As the number of animals in habitat *A* increases, the number of resources available to each animal declines. Finally, the quality of *A* declines to the point that a new arrival will get the same benefits regardless of which habitat it selects.**

The pattern by which these ducks settle into different areas is an example of the ideal free distribution (Fretwell 1972): it's "ideal" because the animals know the value of each habitat and can instantly choose the best one, and "free" because every duck is free to choose its location without interference. Of course, animals are not always either "ideal" or "free." An "ideal" animal would have perfect knowledge of the quality of a site, but of course that's not always true—we already have seen in Chapter 12 that animals often need time to gather information before they can assess the quality of their environment. In addition, animals are not always "free"; instead, others constrain them from behaving optimally. For example, some ducks may be better competitors than others and grab more than their fair share of the food. However, in spite of the fact that deviations from the assumptions of the ideal free distribution are common, numerous species have been shown to behave in accord with its predictions (Giraldeau 2006).

THE ECONOMICS OF HOLDING A TERRITORY

What influences the decision of an animal to hold a territory versus simply share its space with other animals? Having a territory all to oneself has clear benefits, such as exclusive access to resources, be they shelter, food, areas of safety, mates, a high-quality site to raise offspring, or some combination of these things. However, holding a territory also has its costs. Energy is needed to patrol territory boundaries and display to or forcibly evict intruders, and sometimes boundary fights can be dangerous. Territory acquisition and defense also take time away from other essential activities such as foraging, an example of a "lost opportunity" cost. To take a single example, great tits, a bird we have met before, generally feed within 3 m of the ground, but most territorial defense, particularly singing, takes place high in trees, about 10 m above the ground. Obviously, then, at any given time these birds must choose between defense and foraging (Ydenberg and Krebs 1987).

We predict, then, that territoriality will occur only if the fitness benefits from enhanced access to the resource exceed the fitness costs of defending the resource—that is, when the territory is economically defendable (Brown 1964). What factors set the relative costs and benefits of holding a territory? Luckily for researchers, we can often answer this question very clearly because many species are flexible in their territorial behavior, defending territories under some conditions but not others (e.g., Lott 1991). By observing animals under different conditions, or even manipulating the conditions experimentally, we can identify exactly what leads them to hold territories or give them up. Let's look at a few variables that have been studied.

Resource Abundance Generally speaking, territoriality is favored when resources are moderately abundant. If resources are scarce, an individual may not gain enough benefits to pay the defense bill, and it is economically wiser just to let other animals enter the area. For instance, the golden-winged sunbird will abandon a territory when it no longer contains enough food to meet the energy costs of daily activities, including defense (Gill and Wolf 1975). Chickadees living in habitats disturbed by logging and thus low in quality were less likely to defend their territories against intruders than were birds in untouched habitats (Fort and Otter 2004). At the other extreme, if there are more than enough resources to go around, it is unnecessary to defend a territory. Water striders, insects that skate on top of ponds and streams, are among the species that will cease defending territories if supplied with abundant food (Wilcox and Ruckdeschel 1982). Similarly, female marine iguanas do not bother defending territories with nest sites on most of the Galápagos Islands. They defend territories only on Hood Island, the only Galápagos island where nest sites are in short supply (Eibl-Eibesfeldt 1966).

Resource Distribution All else being equal, we predict that animals are more likely to be territorial if resources are moderately clumped. A pile of food, for instance, is easier to defend than food that is spread thinly over a large expanse, as long as the number of competitors anxious to contest ownership is not too great (see discussion in Maher and Lott 2000).

Intruder Pressure The number of other individuals that are willing to compete for a territory is one of the factors determining territory cost. The more competitors, the greater the cost of defense. Male fruit flies (*Drosophila melanogaster*) defend small patches of food that are suitable for oviposition, particularly if there are females in the vicinity. As would be predicted, however, flies are less likely to hold territories when there is a higher density of males (Hoffmann and Cacoyianni 1990).

THE ECONOMICS OF TERRITORY SIZE

Costs and benefits influence not only whether a territory is held at all, but how large a territory should be. We can visualize what the optimal territory size might be by using a graph in which costs and benefits are plotted with separate lines. Examine Figure 18.7. Here, benefits initially increase as territory size increases, but the line begins to flatten out: there are just so many resources an animal can use, and any additional resources are not necessary. The exact shape and placement of the benefit curve depends on resource quality and distribution. For example, if resources are plentiful, the curve might be more like the top curve rather than the bottom. As you can see, costs of defense also increase with territory size. Larger territories mean a longer border to patrol and

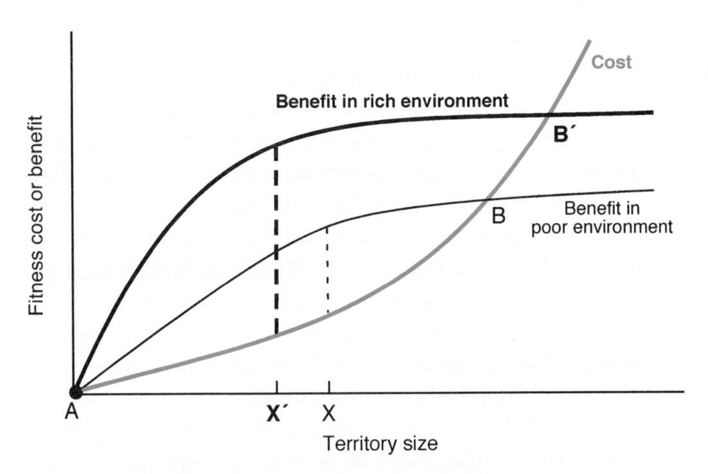

FIGURE 18.7 **The hypothetical relationship between the costs and benefits of territoriality. Both costs and benefits increase with territory size. The shape of the benefits curve varies with the quality of the territory. It is profitable to defend the territory as long as the benefits exceed the costs, between points *A* and *B*. If the animal is to maximize its net gain, the optimal territory size is at point *X* or *X′*, depending on the exact placement of the cost and benefit curves.**

more intruders to drive off. Territories will only be worth defending when benefits exceed costs. The optimal territory size is the size at which the benefits outweigh the costs by the greatest amount, marked by the dashed lines on the graph (Davies and Houston 1984).

STOP AND THINK

What factors might affect the shape and placement of the cost curve in Figure 18.7? In general, do you expect the cost and benefit curves will vary together, or independently?

We might predict, therefore, that in some species the size of an individual's territory would be adjusted to maximize energy gains. At least some individual rufous hummingbirds (*Selasphorus rufus*) appear to do this. During their southward migration, these birds pause for a few days in the mountain meadows of California to build the fat reserves needed to fuel the next leg of their journey. During this interval they feed on the nectar of the flowers of the Indian paintbrush. Each bird defends a group of flowers as a territory. The territory size and weight gain for a single individual are shown in Figure 18.8. As you can see, this bird adjusted the size of its territory so that it could gain weight as quickly as possible. This individual began with a small territory that contained few flowers, so its weight gain was minimal. It increased the territory size greatly on the third day. This territory had more flowers and the bird could obtain more energy, but it had to invest more energy in defense. Nonetheless, it gained somewhat more weight than possible on the smaller territory. It reduced the size of its territory slightly on the fourth and fifth days, thereby

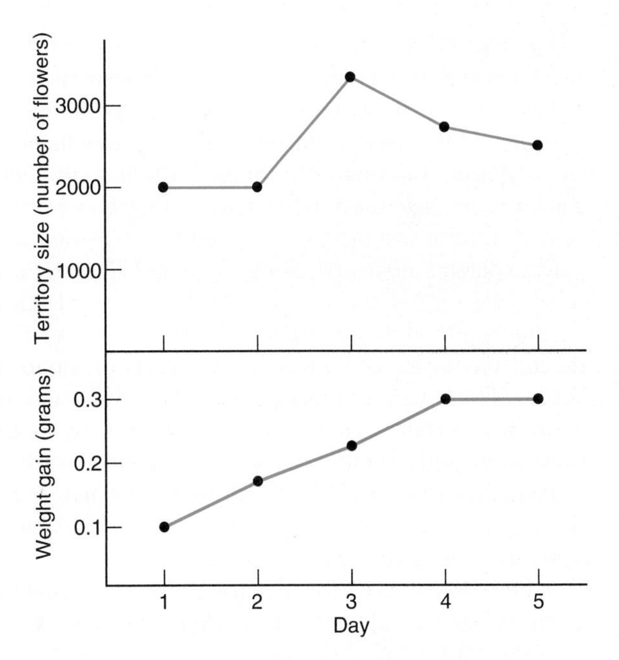

FIGURE 18.8 **The relationship between territory size and weight gain for one rufous hummingbird. It is important for these birds to gain weight maximally during their stopovers along their migratory route. These data indicate the weight gained by a single territorial bird on five successive days. This individual adjusted its territory size so that its weight gain was maximal. (From Carpenter et al. 1983.)**

cutting the energy costs of defense, and maximizing its weight gain (Carpenter et al.1983).

The exact shape of the benefit and cost curves in Figure 18.7 and their relative placement will vary among species and environments. The precise placements of these curves may alter or even reverse predictions of optimal territory size (Schoener 1983).

STRATEGIES FOR REDUCING THE COST OF TERRITORIAL DEFENSE

Territory holders can reduce their defense costs in a number of ways. One way is to band together and share a territory. We'll explore this issue a bit more in the next chapter.

Animals can also reduce their defense costs by selecting their territories wisely, as Perri Eason and her collaborators have shown in a variety of taxa. In general, the faster an intruder is detected, the easier it is for a resident to drive it away. Territories that include good vantage points, such as high trees, can improve the resident's chances of detecting an intruder quickly. Conversely, territories with obstacles can decrease the ability to detect intruders. Eason and Stamps (1992) tested the effect of visibility on territory settlement in juvenile *Anolis* lizards. Lizards were released into two types of habitats in the laboratory, identical except for the presence of a visual barrier. Lizards in habitats without a barrier defended compact territories similar to those in the field. Lizards in habitats with a barrier avoided including the barrier within their territories, adjusting the shape of their territories dramatically. Some birds have also been shown to pay attention to visibility as they choose their territories. For instance, red-capped cardinals (*Paroaria gularis*) defend territories along rivers and lakes in Peru. They feed on insects that they find in the vegetation. Strangely, although they do not feed over the water, they defend territories consisting of two parts on opposite shores. The benefit of this strategy is that they can more easily see intruders on the opposite shore than on the same shore and are more likely to detect and evict them immediately (Eason 1992). In addition, cardinals spend more time in areas of their territories that provide good visibility (Eason and Stamps 2001).

The costs of territorial defense can be reduced if the territory has natural boundary markers. It can be expensive, in both time and energy, to renegotiate a territorial boundary time and again, and selecting a territory with an obvious boundary can save trouble. (A look at maps of land ownership in humans also makes this point—features such as ponds, rivers, or roads often serve as property boundaries.) The use of boundary markers was nicely demonstrated in cicada killer wasps (*Sphecius speciosus*) (Figure 18.9*a*). The larvae of these wasps mature underground, and then adults emerge. Male wasps mature slightly before females and defend areas where females are likely to emerge. Eason et al. (1999) studied the behavior of males on a grassy, mowed, featureless lawn. The researchers began by mapping the territories of the males by watching them patrol their boundaries (Figure 18.9*b*). Then they added landmark cues: dowels (wooden sticks) laid flat on the grass so that they did not align with any territory boundary (Figure 18.9*c*). These dowels provided visual cues only, and not perching sites. The next day, all of the wasps had shifted

FIGURE 18.9 (*a*) An adult cicada killer wasp. (*b*) The original placement of the territories. (*c*) Wooden dowels were laid on the ground so that none of them aligned with territory boundaries. (*d*) The next day, the wasps shifted their territory boundaries so that they aligned with the dowels. (Eason et al. 1999.)

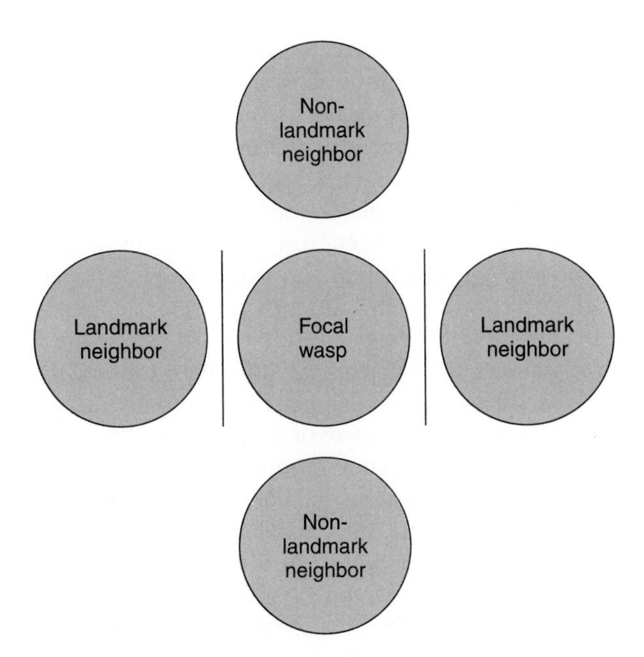

FIGURE 18.10 **In this experiment, pairs of dowels were placed parallel to another. Wasps established territories between the dowels. Thus, each experimental territory had two borders that were defined by a landmark, and two that were not. Defense costs were lower at the borders defined by a landmark.**

their territories so that the boundaries aligned with the dowels (Figure 18.9*d*). The benefits of this shift were documented in a second experiment. Two dowels were laid down parallel to one another, and wasps established territories between them. Thus, these more-or-less square-shaped territories had two boundaries defined by landmarks and two boundaries that had no landmarks (Figure 18.10). Focal wasps fought longer and more frequently with nonlandmark neighbors than with landmark neighbors.

Animals may also reduce the cost of territorial defense by paying attention to the early-warning system provided by neighboring territory holders. The Peruvian red-capped cardinals again provide an example. When a cardinal detects an intruder, it evicts it with a flurry of activity, including chasing and calling. This raucous behavior alerts neighbors that there is an intruder lurking nearby. Territory holders are more likely to immediately detect and evict an intruder if their neighbor has just evicted it (Eason and Stamps 1993).

A PROXIMATE VIEW OF CONFLICT

In previous chapters, we've looked in detail at the proximate mechanisms underlying many behaviors. We've already seen that aggressive tendencies can have a genetic basis (Chapter 4) and that hormones influence aggression (Chapter 7). Here we will revisit several concepts in the context of some interesting work on aggression. This is a large literature, and our hope is to entice you to explore it further (e.g., see texts by Adkins-Regan 2005; Nelson 2005) rather than to attempt a complete coverage.

AGGRESSION AND TESTOSTERONE

One of the best known examples of the influence of a hormone on aggressive behavior is that of testosterone. In Chapter 7, we discussed several examples from birds that illustrate that point. Let's begin with an example from another taxon, lizards, which illustrates the negative as well as the positive aspects of having high testosterone. In male spiny lizards (*Scleroporus jarrovi*), seasonal changes in testosterone concentrations are tightly correlated with the intensity of territorial aggression (Moore and Marler 1987). In winter, testosterone levels are low and territorial defense is lethargic at best. When males were implanted with testosterone-laden capsules in winter, they became very aggressive, spending so much time in territorial defense and vigorous displays that they failed to eat or rest enough, and died sooner than other males (Marler and Moore 1988, 1991; Marler et al. 1995).

Let's return to birds for another example that illustrates how testosterone affects several behaviors. Ellen Ketterson and her colleagues have many years of data on dark-eyed juncos (*Junco hyemalis*), a common visitor to bird feeders in North America. As in other species, testosterone fluctuates quickly in response to changing social situations. It is thought that it is advantageous to be able to raise and lower testosterone levels as needed to avoid carrying the costs of constantly high levels. To examine the ability of individual juncos to increase testosterone levels, researchers injected them with GnRH, or gonadotrophin-releasing hormone, which stimulates the production of transient (short term) increases in testosterone. The peak testosterone levels produced in response to this "GnRH challenge" predicted how strongly juncos responded to a simulated intruder on their territory. However, there is a trade-off: birds that had the highest response to the GnRH challenge were slackers when it came to delivering food to their nestlings (McGlothlin et al. 2007).

So, is testosterone's main function to make an individual more aggressive and thus better able to fend off an intruder? Interestingly, in nearly all species examined, testosterone increases *after* there is an aggressive response to an intruder (reviewed in Wingfield 2005). This suggests that testosterone allows a territory holder to maintain its high aggression levels, especially in the face of a persistent intruder (Wingfield et al. 1990).

STRESS, AGGRESSION, AND DOMINANCE

The experiments just described indicate that there is a link between aggression and testosterone. However, we are not yet done unraveling the proximate control of aggression. Aggressive behavior is regulated by both the endocrine system and the nervous system. That straightforward sentence glosses over a complicated truth: hormonal and nervous control are interconnected with many feedback loops. It doesn't help matters that the endocrine system and the nervous system are themselves sometimes hard to distinguish. For example, the brain not only responds to hormones, but also produces them. In addition, the presence of a hormone does not necessarily mean a particular behavior will follow: an animal that is primed to fight by testosterone will probably not initiate a battle if a predator is nearby. As Adkins-Regan (2005) says, hormones are not like light switches turning light bulbs on and off, but rather regulate or prime (increase the likelihood of) a behavior.

To illustrate this complexity, we'll briefly describe another area of intense research: the relationship among stress, aggression, and dominance. We are now beginning to understand the neurochemical and hormonal events that occur over the course of an aggressive interaction. Interestingly, in vertebrates, there is a tight relationship between the nervous and hormonal control of aggressive behavior, and the stress response. In particular, aggression and stress are linked by the neurotransmitter serotonin and by the hormones called glucocorticoids (corticosterone or cortisol, depending on the species), made by the adrenal cortex.

You doubtless know from experience that physiological changes accompany conflict and other stressful situations. Increased heart rate, rapid breathing, and sweating are all part of the rapidly induced fight-or-flight response that prepares the body to take immediate action. In addition, digestion, growth, and reproduction—all useful functions but not of utmost importance when you are about to be attacked—are shut down. These responses are under the control of short-term bursts of glucocorticoids. In addition, long-term, baseline patterns of glucocorticoid levels, present even before an animal fights, influence its predisposition toward aggression.

Summers and Winberg (2006) suggest a model to describe what generally happens to serotonin and glucocorticoids over the course of a fight, especially among fish and lizards, shown in Figure 18.11 (note that testosterone also primes aggression, but here we are omitting it for simplicity's sake). First, baseline levels of serotonin and glucocorticoids establish an animal's tendency to be aggressive: in dominant animals, glucocorticoids are high and serotonin is low, while the reverse is true in subordinate animals. At this stage, serotonin's function is to hold aggressive tendencies in check. However, the relationship between serotonin and aggression is not that simple. When a social interaction begins, and signals are first exchanged, dominant individuals show an increase in glucocorticoids, serotonin, and another neurotransmitter called dopamine. As the fight increases in intensity and becomes more stressful, both dominant and subordinate animals show an increase in serotonin and corticosterone. When dominance is established and the fight ends, serotonin remains high in subordinate animals, which are then less likely to initiate fights. This may remind you of the winner effect we discussed earlier in the chapter: winners are more likely to win subsequent fights, and losers are more likely to lose them. This is a mechanism by which this pattern may occur.

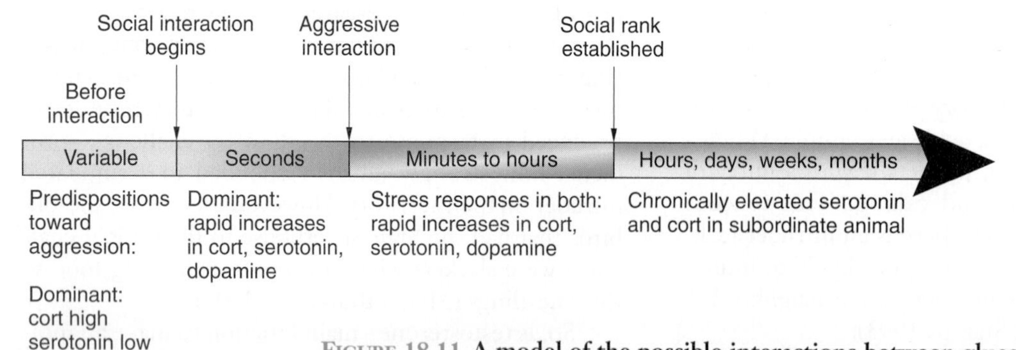

FIGURE 18.11 A model of the possible interactions between glucocorticoids, serotonin, and dopamine over the course of an interaction. Before the interaction begins, dominant individuals are more likely to have higher levels of corticosterone (abbreviated cort), and lower levels of serotonin. This predisposes dominant individuals to behave aggressively. During the interaction, both dominant and subordinate individuals show a stress response, and both cort and serotonin increase. After the interaction, chronic increases in serotonin and cort inhibit aggression. (From Summers and Winberg 2006.)

So is being subordinate always more stressful than being dominant? Not necessarily. In fact, the relationship between status and stress, as measured by glucocorticoid levels, varies across species. We now have data on glucocorticoid levels in several species of wild animals. These data can be quite tricky to collect! Here is Robert Sapolsky (1994) on the challenges of darting baboons in the wild:

> *You have to dart every animal at the same time of day to control for daily fluctuations in hormone levels. If you want to get a first blood sample in which hormone levels reflect basal, nonstressed conditions, you can't dart someone who is sick or injured or who has had a fight or intercourse that day. . . . If you are trying to measure resting hormone levels, you can't spend all morning making the same animal nervous as you repeatedly try to dart him; instead you get one shot, and you can't let him see it coming. Finally, once you dart him, you have to obtain the first blood sample rapidly, before hormone levels change in response to the dart. Quite a thing to do with your college education.*

(Luckily, we now have less invasive ways of measuring hormone levels using feces and hair, so you can become a behavioral endocrinologist even if your aim is poor.)

Sapolsky's (1992) long-term field studies show that subordinate baboons have higher levels of glucocorticoids, but that glucocorticoids in dominant animals also spike when dominance hierarchies are unstable. However, the pattern of glucorticoids across dominant and subordinate individuals varies across species. In recent reviews of field studies of several vertebrate taxa, including birds, primates, canids, and others, glucocorticoid levels were sometimes positively correlated, negatively correlated, or uncorrelated with rank (Creel 2005; Muller and Wrangham 2004; Sands and Creel 2004), or even varied in their correlation with rank throughout the day (Muller and Wrangham 2004). A comparative study of seven primate species showed that cortisol level varied according to the social situation: subordinates showed higher levels when they did not have close social support (Abbot et al. 2003). Currently, researchers are collecting data on dominance rank and hormone levels for an even wider variety of species, so perhaps we will better understand this relationship in the next few years.

Stress responses are adaptive in the short term, but persistent, long-term stress responses can lead to health problems (reviewed in Sapolsky et al. 2000; Sands and Creel 2004), such as increased susceptibility to disease and a shorter lifespan. Thus, understanding the relationship between rank and long-term stress will be necessary to evaluate the selection pressures on dominant and subordinate animals.

SUMMARY

Conflict is a part of the lives of many animals, from sea anemones to primates. Conflict occurs when resources, such as mates, shelter, or food, are limited. Terms used in the study of conflict include aggression, which is behavior intended to inflict noxious stimulation or destruction on another organism (which, according to some authors, includes predation), and agonistic behavior, which refers to the behavior of both the aggressor and the object of the aggression.

Animals generally do not engage in full-out fights to the limit of their abilities. Instead, as we have seen in previous chapters, fights are often limited to displays and other low-cost actions. Researchers have used a type of modeling called game theory in order to understand when conflicts escalate and when they do not. Like the foraging models we studied in Chapter 12, we use game-theory models to evaluate the relative fitness benefits that animals might acquire by pursuing different behavioral strategies. Game-theory models differ from foraging models because the payoff to playing a particular strategy depends on what other animals are doing.

One of the simplest game-theory models of conflict is the hawk–dove model. Here, animals can play one of two strategies. The hawk strategy is the strategy of escalation: immediately attack the opponent. The dove strategy is to display if one meets another dove, but to flee if encountering a hawk. The payoff of each strategy depends on the value of the resource, the cost of being wounded in a fight, and the cost of display. If the value of the resource is greater than the cost of being wounded, then a pure hawk strategy is an evolutionarily stable strategy: that is, a population of hawks cannot be invaded by a dove. This situation occurs in species where most or all of an animal's lifetime reproductive success depends on the outcome of a fight, such as in elephant seals fighting over access to a harem of females. More commonly, the value of the resource is less than the cost of being wounded. In that situation, the evolutionarily stable strategy is a mixture of hawk and dove strategies. It is possible to calculate what that mixture is for particular values.

In the basic hawk–dove game, it is assumed that all players are equal and can choose either strategy. However, this is not often the case. Instead, contests are often asymmetrical. Sometimes one animal is a stronger fighter than the other—larger, heavier, more aggressive. Experience can also create asymmetries: animals that have won a fight are more likely to win their next fight.

Resources may also differ in value to the contests: a hungry animal will value food more, and thus fight harder for it, than a well-fed animal. Finally, arbitrary asymmetries are not correlated with either the value of the resource or the ability of an animal to defend it, but are simply conventions that fighters follow. An example of an arbitrary asymmetry is "the current owner of the resource wins the fight." An example with speckled wood butterflies illustrates how difficult it can be to ascertain the rules of fights.

Next we turn to a special case of conflict: conflict among group members. Relationships develop among group members. Often some group members are dominant over others and routinely assert themselves in access to resources. Submissive animals are those that predictably yield to dominant ones. The structure of dominance relationships can be very simple straight-line hierarchies, with A dominant over B, who is dominant over C, and so on, or they can be more complex. Dominance may be determined by fights or, in some cases, as a birthright. Dominant animals often get more resources and access to mates than do subordinate animals. Subordinate animals often stay in a group because no other options are available. However, subordinates may occasionally win a fight, engage in sneaky alternative strategies, or band together with others.

The second special case that has received a great deal of attention is conflict over space. Territorial species defend space from one another (or, by another definition, have exclusive use of space). The ideal free distribution describes how animals should distribute themselves in space if they are "ideal" (with perfect knowledge of the location of resources) and "free" (not prevented from going where they wish by other animals). In contrast, territory holders prevent others from settling in their space. It is not always economically feasible to defend a territory: borders can be energetically costly and even dangerous to patrol. Territorial defense is favored when resources are moderately abundant (too scarce and the territory would need to be too large; too abundant and there is no need to defend it), when resources are moderately clumped (too clumped and too many intruders would flood in to try to wrest them away; too dispersed would mean boundaries are too long), and when the number of intruders is not too high. A similar logic can be applied to the optimal size of territories. Territory holders can reduce the cost of defense by selecting territories with good visibility so that intruders can be easily spotted, with landmarks that help to define boundaries, and by attending to the early-warning system provided by neighbors that alert them to the presence of intruders.

Proximate causes of aggression have been discussed in other chapters. Here, we describe the relationship between dominance, stress, and neurotransmitters and hormones, and we describe one proposed model for the hormonal pattern that occurs during an aggressive interaction. However, evidence is accumulating that not all species respond in the same way. In some species, dominant animals appear to exhibit the highest levels of stress, and in others, subordinate animals do. Finally, we look at an example of a species in which hormone levels and aggressiveness in territorial interactions are intertwined.

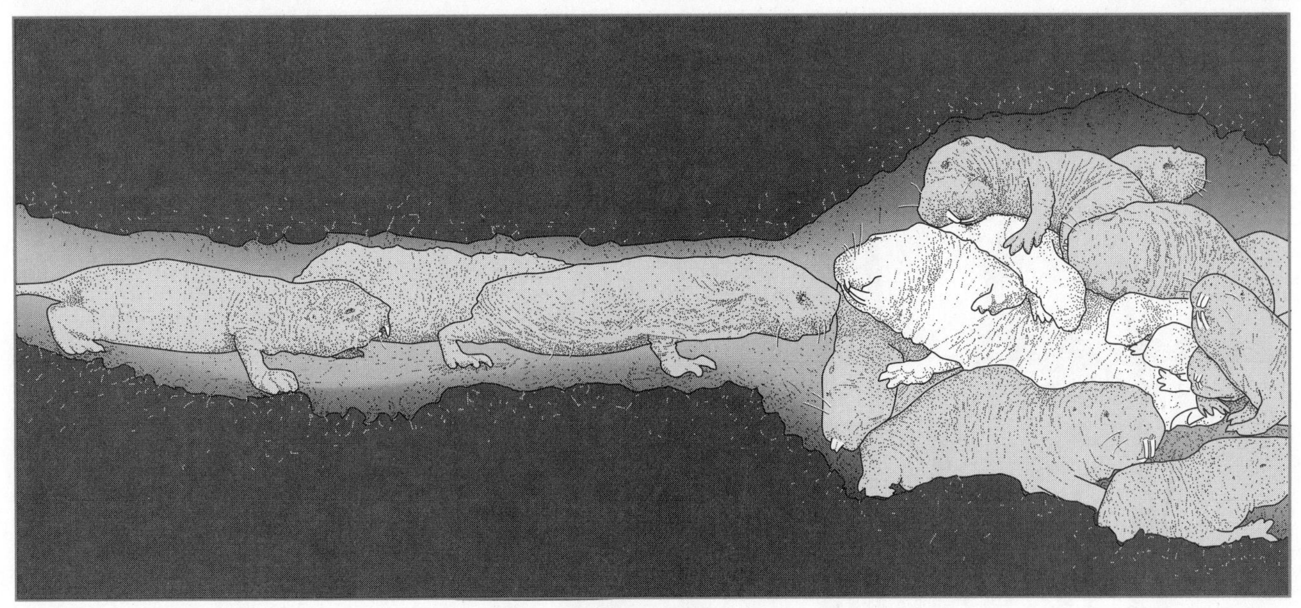

19

Group Living, Altruism, and Cooperation

LIVING IN GROUPS: FROM AGGREGATIONS TO STRUCTURED SOCIETIES

Brushing her teeth one night in her Massachusetts home in the dead of a snowy winter, one of the authors was smacked in the side of the head with . . . a ladybug? Sure enough. Some species of ladybug beetles overwinter in aggregations of thousands of individuals. In nature, they crawl inside rock crevices where they are sheltered from cold. When houses are available, they can find their way inside through vents and cracks. The species plaguing homes in eastern North America is a particularly unwelcome guest. Originally introduced from Asia in order to control invasive pests, *Harmonia axyridi* has become wildly successful and its numbers have soared. In some years, like this past one, *H. axyridi* find their way into houses in such numbers that they accumulate in drifts on the windowsills, flying around on warmer days and startling the unwary. They have more serious consequences too, as they outcompete and even eat native ladybug species, and sometimes cause allergic reactions in their human hosts (Goetz 2008; Snyder et al. 2004).

If we array animal groups from simple to complex, ladybug groups are at the "simpler" end of the spectrum. Their groups are seasonal rather than lifelong—the ladybugs will disperse in spring and forage and reproduce on their own. There's no evidence that ladybugs interact in complex ways, recognize each other as individuals, or cooperate, although they seem to be mutually attracted to one another via chemical cues. Some animal species that form groups do not even show mutual attraction, but simply respond to the same features of the physical environment and thus independently end up in the same spot.

Now consider this example. Two adult male chimpanzees have a noisy, aggressive conflict. The challenged male flees into a tree. The opponent slowly approaches,

FIGURE 19.1 **A chimpanzee offers an open hand to another chimp with whom he has just had a conflict.**

offering him an open hand (Figure 19.1). Seconds later, the chimpanzees touch, kiss, and climb to the ground to groom one another (de Waal 2005). Later, it is likely that these chimpanzees will reciprocally exchange favors, such as food for grooming. Chimpanzees, and other species, are members of structured groups called **societies**. Interaction with conspecifics is deeply ingrained in every part of life, so much so that Roger Fouts said, "One chimpanzee is no chimpanzee."

Not all animals live in groups, but group living occurs in nearly every major taxon. In this chapter, we'll first explore some selective advantages—and disadvantages—of living in groups. Next we'll discuss a particularly interesting aspect of many animal interactions called altruism, or helping behavior. Finally, we'll select some well-studied examples of animal cooperation to explore in more detail.

BENEFITS OF GROUP LIVING

What fitness benefits accrue to animals living in groups? Several themes recur in many taxa: improved foraging, decreased risk of predation, conservation of water and heat, and decreased energetic costs of movement (reviewed in Krause and Ruxton 2002, whose organization guides us here). Not every species profits in each way outlined below, but many do.

Improved Foraging

Foraging success in groups can be improved through several mechanisms. For example, some animals coordinate their foraging, sometimes with great precision, in **cooperative hunting**. In the southwestern United States, Harris's hawks (*Parabuteo unicinctus*) live and hunt in family groups (Bednarz 1988). In the early morning, family members typically gather at one perch site, from which the group then splits into smaller subgroups of one to three individuals. The subgroups take turns making short flights and "leapfrog" through their family's area. Upon discovering a rabbit, hawks employ one of three hunting tactics, or more typically a combination of the three. The most common, the surprise pounce, occurs when several hawks arrive from different directions and converge on a cottontail or jackrabbit unfortunate enough to be out in the open. Even if the rabbit escapes under vegetation, however, safety may only be temporary. At this point, the hawks employ their flush-and-ambush tactic, a strategy in which one or two hawks flush the rabbit from the cover, and then family members perching nearby pounce on it. Relay attack is the third, and least common, hunting tactic. Here, family members constantly chase the prey, with a new lead bird taking over each time there is a missed attempt to kill. Regardless of the tactic used, the dead prey is shared by all members of the hunting party.

In order for cooperative hunting to be favored by selection, individuals must average at least the same amount of food as they would get by hunting alone. Because the food is generally shared among all participants in the hunt, this must mean that hunting success must be increased a great deal in groups. In Harris's hawks, the chance of killing a prey correlates with group size—hunting parties of five to six individuals do better than smaller parties (Figure 19.2*a*). Even with more individuals sharing the prey, the average energy intake per individual from rabbit kills is higher in groups of five or six members than in smaller groups (Figure 19.2*b*).

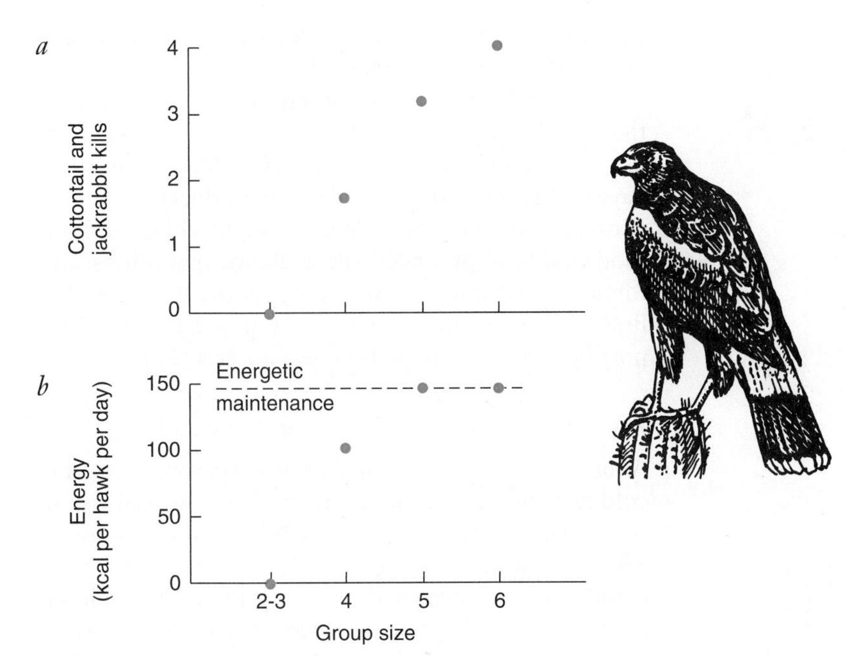

FIGURE 19.2 **In Harris's hawks, members of family groups engage in cooperative hunting. (*a*) Groups of five or six individuals are most successful at killing cottontails and jackrabbits, and (*b*) individuals in such groups have a higher average energy intake than individuals in smaller groups, and are more likely to meet their basic energetic needs. (From Bednarz 1988.)**

Another means by which animals improve their foraging while in groups is by **information sharing**. Many species pay attention when their conspecifics discover food, and use this information to guide their own foraging. For example, geese are more likely to land near artificial geese that have their heads down in a feeding position than to land near those standing erect (Drent and Swierstra 1977).

Communal roosts or colonies can act as **information centers**. If successful foragers return to the roost or colony, and then return to the food site, others might be able to identify them as successful foragers and follow them. Scientists debate how common this phenomenon is (Krause and Ruxton 2002), but it has been demonstrated in some species, including cliff swallows (Brown 1986) and the honeybees that we discussed in Chapter 16.

Decreasing Predation Risk

Recall from our discussion of antipredator behavior in Chapter 13 that group membership allows animals to employ several antipredator tactics that are not available to solitary individuals. To recap briefly: groups may be more likely than a single individual to detect a predator (the "many eyes" hypothesis). Animals in groups often give alarm signals when they detect a predator. (We'll examine alarm signals in more detail later in the chapter.) When a group is detected and attacked, an individual within a group has a smaller chance of becoming the next victim (the dilution effect), and animals in the center of the group often have a lower chance of being preyed upon than do animals at the edge (the selfish herd). By fleeing in different directions, group members can take advantage of the "confusion effect" and decrease a predator's ability to track and kill any one individual. And finally, group members can band together to drive a predator away by mobbing.

Conserving Heat and Water

By huddling together, animals can reduce the surface area exposed to the environment and thus reduce the loss of heat or water. For instance, many mammals sleep or overwinter together in communal burrows. Many birds perch snuggled up next to one another when they sleep. A male emperor penguin, balancing its egg on top of its feet and tucked under its belly, could not survive the long Antarctic winter without huddling with other males. The metabolic rates of penguins in small groups are reduced by 39% compared to isolated birds, and those of penguins in larger groups are reduced by another 21% (Gilbert et al. 2008). Even some not-very-cuddly animals huddle: some slugs rest in contact with one another in order to reduce water loss (Cook 1981).

Conserving Energy by Moving Together

Bicyclists know the value of drafting, or riding close behind in the slipstream of another bike: it reduces the amount of energy needed to pedal. Researchers have explored whether the same holds true for animal groups that travel together, such as schools of fish and flocks of birds flying in formation (reviewed in Krause and Ruxton 2002). It can be technically tricky to get these measurements and to tease apart the role of energetic savings from other potential benefits of traveling in a group, such as avoiding predators or using conspecifics as

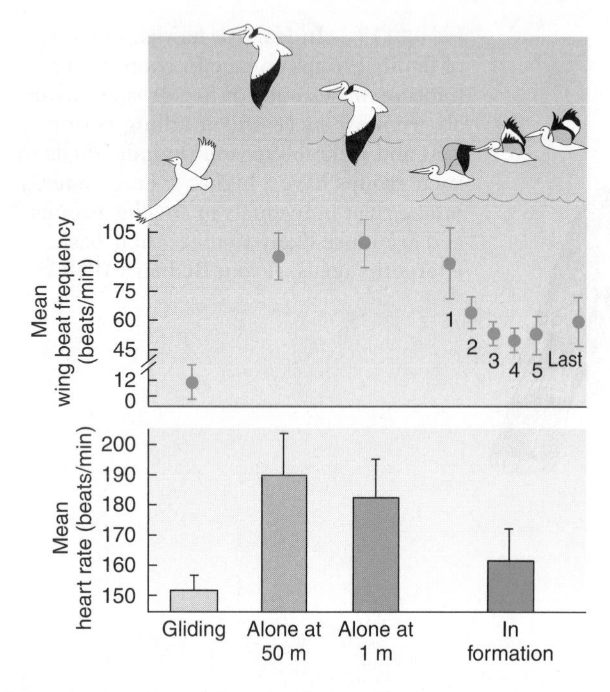

FIGURE 19.3 **Heart rates of pelicans flying in formation versus those flying solo. Compare pelicans that are gliding, flying solo at two different heights above the water, or in formation. The numbers next to the circles indicate the position of the birds in the group, with number 1 being the leader. (From Weimerskirch et al. 2001.)**

navigational cues. To illustrate what a daunting task this can be, consider Weimerskirch et al.'s (2001) study of why pelicans (*Pelecanus onocrotalus*) fly in a V formation. The effort expended by birds in flight can be measured by strapping heart monitors to their backs. But how to keep pelicans that are so equipped from simply flying away? The researchers trained eight pelicans to fly after a moving motor boat and an ultralight airplane. They could thus measure the heart rates of pelicans in a variety of group sizes. Birds flying in formation had lower heart rates than did birds flying solo (Figure 19.3).

COSTS OF LIVING IN GROUPS

Increased Competition

As you might expect, individuals that live in groups often compete with each other for mates, nest sites, or food. Consider the snail *Dendropoma maxima*, which forages in a slightly disgusting way: it secretes a sticky mucous net that floats in the water and traps plankton. The snail then draws the net back in, eating both the mucus and its catch. When snails are grouped together, their mucous nets frequently overlap and stick together, so that snails end up consuming the nets of their neighbors. The snail that is slow to retract its net loses food—and valuable slime!—to its neighbor, and thus grows more slowly. Snails seem to adjust for the presence of neighbors by retracting their nets more quickly when a conspecific is nearby (Gagern et al. 2008).

Group-living animals might often lose food to thieves—in fact, stealing can be considered a strategy. Giraldeau and his colleagues (e.g., Giraldeau et al. 1994) have extensively modeled and tested producer–scrounger situations in which an individual might either look for food (and be a "producer") or steal food that others have found (a "scrounger"). As you might predict, there is a limit on how many "scroungers" a population can support before the strategy becomes unrewarding.

Increased Risk of Disease and Parasites

As anyone who has been in a crowded classroom during cold and flu season can attest, another potential risk of social living is increased exposure to disease and parasites. For example, cliff swallows nest in large colonies, and colony size is correlated with the number of blood-sucking swallow bugs (*Oeciacus vicarius*). Swallow bugs are harmful to swallows in many ways: they reduce nestling mass, decrease survival of birds of all ages, increase white cell counts, and even cause asymmetrical growth of feathers (Brown and Brown 1986, 2002, 2004; Brown et al. 1995). Colony size is correlated with higher levels of glucocorticoid hormones, which we have already seen are released in response to stress (Chapter 18). Raouf et al. (2006) hypothesized that the stress response was related to the presence of the swallow bugs rather than to other aspects of group living, such as competition for food. The researchers tested this hypothesis by fumigating some colonies in order to kill the parasites. Fumigation lowered corticosteroid levels, indicating that exposure to the parasites caused the stress response.

Many group-living species have evolved behaviors that help fend off disease and parasites. Allogrooming animals lick each other and pick off ectoparasites, often focusing on those hard-to-reach places. Social insects, such as honeybees, ants, and termites, have several defenses against disease. They remove corpses and other waste from the colony, and they even wall off or remove infected individuals (Cremer et al. 2007). In fact, leaf-cutting ants are so effective in defending against a fatal fungus via grooming and antibiotic secretions that group living confers a net benefit in resisting disease (Hughes et al. 2002). For most species, however, proximity to conspecifics presents an increased risk of infection.

Interference With Reproduction

Animals in groups sometimes face interference with their own reproductive efforts. For instance, extra-pair copulations—copulations that occur outside of the pair bond—are extremely common, even among "monogamous" birds. Animals that live in groups have more opportunities for such dalliances than animals that do not. Thus, group-living males may be more likely than males of solitary species to invest resources in caring for young that are not theirs. Similarly, in the hustle and bustle of a large group, such as a colony of nesting birds, a female may fail to notice the arrival of another female at her nest and the quick deposit of an egg by the intruder. (For more extensive discussion of intraspecific brood parasitism, see Chapter 15.)

Mammalian mothers of 68 species have been documented to allonurse, or nurse offspring that are not their own (reviewed in Roulin 2002). Allonursing is common in animals that share roosts and reproduce communally, or where all the females reproduce in a small space (such as seals confined to suitable rocky ledges). Several hypotheses for the occurrence of allonursing have been proposed, some of which include fitness benefits to the nursing mother (e.g., feeding young that are related, or feeding the offspring of a mother that will later reciprocate). Another hypothesis is that nursing the young of others is costly misdirected parental care. For example, Mexican free-tailed bats spend their days in colonies that exceed several million individuals, and it is possible that allonursing occurs primarily because of mistaken identity (McCracken and Gustin 1991).

BALANCING COSTS AND BENEFITS

Of course, any particular species may face a combination of these costs and benefits. Although ideas on patterns of grouping in lions initially focused on hunting success (Figure 19.4), data suggest that this factor alone cannot explain the formation of prides in this species (Packer 1986; Packer et al. 1990). Scavenging may be a more efficient means of obtaining food than hunting, and groups may be necessary to defend carcasses against lions from other prides. Group living also appears to be important in protecting cubs from nomadic males that commit infanticide and in the defense of the pride's home area against intrusion by neighboring prides. Thus, for lions, the benefits of group living include the defense of food, young, and living areas against conspecifics.

FIGURE 19.4 **A group of lions shares a wildebeest kill. Studies of group living in lions have often focused on hunting success, but additional functions include the defense of cubs, space, and scavenged food.**

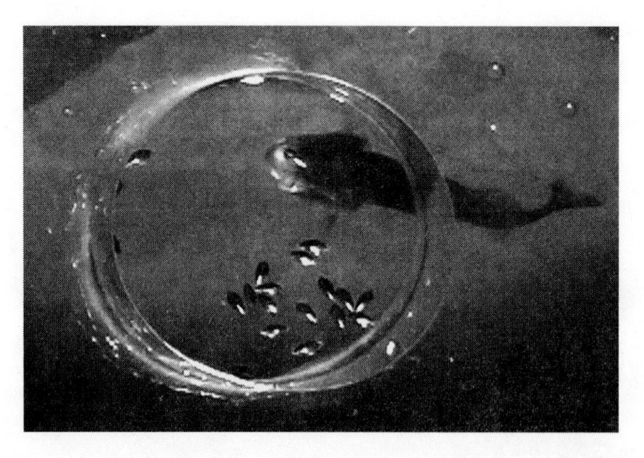

FIGURE 19.5 **Whirligig beetles aggregate on the surface of a pond. Those on the outside of the group get more food but are also more vulnerable to predators, such as the fish pictured here. In this experiment, beetles were confined to a ring.**

The costs and benefits of group living may differ across individuals. For instance, the aptly named whirligig beetles form aggregations on the surface of freshwater, each spinning in circles (Figure 19.5). Beetles at the outside edge of the group are more likely to get bits of food dropped into the water, but are also more likely to get attacked by fish (Romey 1995; Romey et al. 2008). How beetles weight the trade-off between predation risk and food availability depends on how hungry they are: hungry beetles move to the outside of the group, in spite of the risk (Romey 1995).

THE PUZZLE OF ALTRUISM

By now, you should be very comfortable with the idea that natural selection will cause a particular behavior to increase in a population if its fitness benefits outweigh its costs. What are we to make of the fact, then, that some animals seem to help other members of their species? For example, a Belding's ground squirrel (*Spermophilus beldingi*) increases its risk of being spotted by an approaching predator when it barks an alarm. Nevertheless, it barks at the sight of a badger, and all those in the area scurry to safety (Sherman 1977). In wild turkeys (*Meleagris gallopavo*), males form coalitions that court females and defend them against other groups (Figure 19.6). However, only dominant males father offspring: subordinate males expend effort in display, but to no benefit to themselves (Krakauer 2005; Watts and Stokes 1971). Members of still other species cooperate

FIGURE 19.6 **Wild turkey brothers strut together in order to attract mates. In general, only the dominant brother mates.**

in rearing offspring that do not belong to them. For example, dwarf mongooses (*Helogale parvula*) bring food to the young of others and guard the den from predators. Perhaps the most striking examples of helping behavior are found in eusocial insects (ants, termites, and some wasp and bee species). The workers toil tirelessly to care for their colony. They may even die in defense of the nest. However, the young they help rear are not their own: the workers are sterile.

Animal behaviorists call examples such as these **altruism**. An altruistic behavior appears to be costly to the altruist and beneficial to another member of its species. Although it is difficult to measure, biologists define altruism in terms of fitness: it is a behavior that raises the fitness (number of offspring produced that live to breed) of another individual *at the expense of the altruist's direct fitness*, as measured by the number of offspring it leaves (Hamilton 1964).

How could altruism possibly evolve? Shouldn't alleles that promote selfish behavior multiply more quickly in the population than alleles that promote altruism? Not surprisingly, many researchers, including Darwin, have puzzled over these questions.

INDIVIDUAL SELECTION AND "ALTRUISM"

In some cases, behavior *appears* to benefit others at the cost to the actor, but in fact, the actor might directly benefit from its behavior. For example, an animal that gives an alarm call may appear to be alerting others at its own expense, but it may actually improve its own survival by alerting the predator that it has been seen. Similarly, in some species of cichlid fish, adults adopt unrelated young into their own brood, caring for them and defending them from predators as if they were family. Although this may at first seem to be quite a generous act, on closer inspection it seems that the parents gain because the presence of the adopted young reduce the risk that their own young will be picked off by a predator (McKaye and McKaye 1977). Thus, before assuming that a behavior is truly altruistic, it is important to investigate carefully whether the actor benefits directly from the behavior.

KIN SELECTION

In more puzzling cases, the costs of a behavior to the actor's survival and reproductive success outweigh any benefit it might accrue. If an individual behaves so as to decrease its own reproductive success, how could natural selection possibly act to increase the frequency of its altruistic trait?

W. D. Hamilton's groundbreaking paper in 1964 revolutionized our thinking about altruism. Hamilton's key insight was that individuals can improve their fitness not only through their own offspring, also called their descendant kin, but also through the reproductive success of their other relatives, or nondescendant kin. Here is the logic: family members other than offspring also possess copies of some of the same alleles because they inherited the alleles from the same ancestor. Therefore, if family members are assisted in a way that increases their reproductive success, the alleles that the altruist has in common with them are also passed on, just as they would be if the altruist reproduced personally. Selection that works through relatives in this manner is called **kin selection**.

Of course, not all relatives have the same likelihood of sharing a particular allele. Intuitively, you can imagine that closer relatives, such as siblings, are more likely to share alleles than are more distant relatives, such as cousins. We can be even more precise: we can calculate the probability that particular pairs of relatives share the same allele through common descent. This probability is called the **coefficient of relatedness**, or ***r***. You may recall from our discussions in Chapter 3 that sexually reproducing animals have two alleles for each gene and that these separate during the formation of gametes (eggs or sperm). There is, therefore, a 50–50 chance (a probability of 0.5) that any particular allele will be found in an egg or a sperm produced by the parent. In other words, an individual shares 50% of its alleles with its parent, so a parent and offspring have a coefficient of relatedness of 0.5. The value of r ranges from 0 (nonrelatives) to 1 (identical twins or clones). (As an interesting aside, recent findings show that even identical twins might have small differences in DNA; Bruder et al. 2008.)

An easy way to calculate the coefficient of relatedness between more distant relatives is by using a family tree, as illustrated in Figure 19.7. By this method, we determine that, on average, an animal shares 50% (1/2) of its genes with a full sibling ($r = 0.5$); 25% (1/4) with a half sibling or grandparent ($r = 0.25$), and only 12.5% (1/8) with a first cousin ($r = 0.125$).

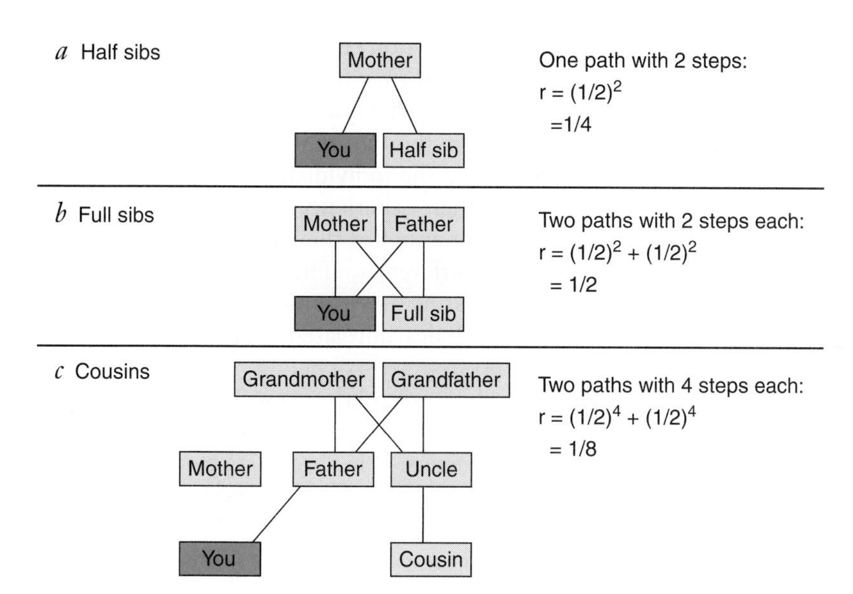

FIGURE 19.7 A method of calculating relatedness (*r*) with a family tree. First, draw in the family members that connect two individuals whose relationship you wish to know. Draw lines between all parents and each of their offspring. For each path between the two individuals of interest, count *n*, or the number of steps, and raise 1/2 to the power of *n*. Do this for each possible path, and sum the result to get *r*. (*a*) Only one path with two steps connects half sibs. (*b*) Two paths (one through the father and one through the mother), each with two steps, connect full sibs. (*c*) Two longer paths connect cousins.

Now let us see how the coefficient of relatedness affects altruism. When will the gene for an altruistic behavior spread in a population? To figure this out, we can employ **Hamilton's rule**:

$$\frac{B}{C} > \frac{1}{r}$$

where

B is the benefit to the recipient in terms of *extra* offspring that the relative produces because of the altruistic act;

C is the cost to the actor, measured as the number of offspring it *does not* produce because of the altruistic act; and

r is the coefficient of relatedness between the recipient and the actor. Note that $1/r$ has a value of 1 or greater, because r is a fraction ranging from 0–1.

To take an example: when should an individual forgo reproduction to help its sister reproduce? For siblings, $r = 0.5$, so $1/r = 2$. Therefore, in this example, the benefits of acting altruistically must outweigh the costs by 2:1 for an individual to help its sister reproduce.

Actually applying Hamilton's rule to real animals is quite tricky. We cannot just count the total number of offspring that a relative produces: this will inflate the value of helping. Instead, in order to calculate B in the equation above, we must count up the number of extra offspring that are produced *only because of the help of the altruist*. It is, of course, difficult to know what might have been. However, sometimes we can estimate this quantity by comparing the number of offspring produced by animals that followed one course of action with the number produced by those that did not. There are reasons that this estimate may be inaccurate (Grafen 1984), but it is a beginning. Later in the chapter, we'll see an example of how this might be accomplished.

STOP AND THINK

A bird will have two offspring if she raises them without help. However, if her altruistic sister helps her by bringing food to the nest and driving off predators, she will have five offspring. This behavior has a cost to the potential altruist: if she helps, she will not have any of her own offspring, whereas if she does not help, she will have one. Should she help?

We are now prepared to expand the definition of **fitness** that we have been using throughout the book. Thus far, we have primarily considered **direct fitness**, measured by the number of offspring that an individual has as a result of its own efforts. Hamilton proposed that we add to this another component of fitness, **indirect fitness**. To calculate indirect fitness, we count the number of extra offspring that an individual gains by helping a relative, devalued by the genetic distance between the individual and the relative who was helped (in other words, multiplied by r). **Inclusive fitness** is the sum of direct and indirect fitness.

Discriminating Between Kin and Nonkin

If animals have the ability to aid each other, we expect that it should be evolutionarily advantageous to discriminate kin from nonkin. Here, we'll discuss four ways in which animals may accomplish this.

Location As you may recall from Chapter 11, individual animals are often found in fairly predictable locations. In mammals, for instance, males generally disperse from home as they mature, whereas females tend to remain in their natal area. Thus, female mammals that help conspecifics located near their home are likely to be helping relatives, even if they do not have the ability to recognize them individually as kin. Simply following the rule of thumb of "If you are a female, help those near home" can lead to increased inclusive fitness. Thus, when relatives are distributed predictably in the habitat, kin selection can work if the altruistic deeds are directed toward those individuals in areas where relatives are most likely to be found.

Another example of how location is used as a mechanism for kin-biased behavior is when a parent identifies its offspring as those young in its nest or burrow. In many species of birds, parents will feed any young they find in their nest. We see this among bank swallows (*Riparia riparia*), which live in holes that they excavate in banks (Figure 19.8). Each pair of swallows lives in its own hole, but holes are clustered together. The parent bank swallow learns its nest-hole location and feeds any chicks inside, including any neighbor's chicks placed inside by experimenters. A parent will ignore its own chicks if they are moved to a nearby nest hole. After about two weeks, at the time the young begin to leave the nest and fledglings unrelated to the parent might enter the nest, the parent begins to recognize its young by its distinctive calls and also begins to reject foreign young experimentally placed in the nest (Beecher et al. 1981). At this point, location is no longer a reliable cue to relatedness. Interestingly, in barn swallows, where chick intermingling is rare, there is no evidence that parents can recognize the voices even of older offspring (Beecher et al. 1989).

The trouble, of course, is that this very simple system for discriminating kin from nonkin breaks down when animals are not where they are "supposed" to be. For kin discrimination to work in a variety of locations, we need a different mechanism.

FIGURE 19.8 Colonial bank swallows will feed any young they find in their nest.

Familiarity If kin don't always encounter each other in the same place, they must use some other method besides location to reliably identify one another. One frequently used method is familiarity. Here, the young learn to recognize the individuals with which they are raised through their experiences during early development, and then, later in life, they treat familiar and unfamiliar animals differently. The ideal setting for this learning is a rearing environment such as a nest or burrow that excludes unrelated individuals (reviewed in Mateo 2004).

Familiarity is apparently a mechanism used by a young spiny mouse when attempting to distinguish its siblings from strangers. When weanling pups are released into a test arena, they often huddle together in pairs, and the members of the pair are generally siblings. However, the spiny mice do not identify their siblings per se, but rather they prefer to huddle with familiar pups, their littermates. We know this because siblings separated soon after birth and raised apart treat one another as nonsiblings. However, if unrelated young are raised together, they respond to one another as siblings. So we see that kin-biased behavior among spiny mice seems to develop as a result of familiarity (Porter et al. 1981).

Like using location cues as a label of kinship, relying on learning about characteristics of conspecifics does not suit every occasion. For example, if nonkin are encountered during the learning process, they may be mistakenly classified as kin (reviewed in Mateo 2004).

Phenotype Matching Phenotype matching allows animals to identify kin even if they have never met them before (Alexander 1979; Holmes and Sherman 1982; Lacy and Sherman 1983). As a reminder, an animal's phenotype is its physical, behavioral, and physiological appearance (Chapter 3). As you know, family members often resemble one another, so phenotypic similarities can be a useful clue to relatedness.

Animals can learn the "kin phenotype" either by learning about the phenotype of familiar individuals (in informal shorthand, the equivalent to saying, "This is what my brothers look like") or by learning their own phenotype ("This is what I look like"). This template is then compared against strangers ("He looks like my brother" or "He looks like me"). Of course, we don't need to hypothesize any astonishing cognitive powers in order to invoke phenotype matching—animals may simply learn to respond favorably to a familiar cue that is also exhibited by a stranger.

Several species have been shown to learn about the phenotype of familiar individuals to develop a template, and then match new individuals to that template. Let's look at an example that uses both familiarity and phenotype matching. As we will see in more detail shortly, Belding's ground squirrels (*Spermophilus beldingi*) give alarm calls to warn their mothers, daughters, and sisters of an approaching predator. Pups apparently identify their siblings because they learn one another's odors while still in the same nest burrow. When pups from different nests are experimentally switched, unrelated pups that are raised together will treat one another as siblings later in life (Sherman et al. 1997). Experimental designs like these, where infants are reared by unrelated foster parents, are called "cross-fostering" experiments, and are very useful in teasing apart which mechanisms of kin recognition might be operating in a given species (Mateo and Holmes 2004). Later, juveniles and adults can use phenotype matching to discriminate relatedness among individuals that they have never encountered before (Holmes 1986a,b).

A slight variation on this sort of system comes from paper wasps (*Polistes fuscatus*), which discriminate between kin and nonkin by using the odor of hydrocarbons that become locked into the insect's cuticle before it hardens. The odor comes from the nest. Nest odor differs among colonies because it depends on the type of plant fiber used to build the nest, as well as on secretions produced by wasps that built the nest. Each colony uses a unique combination of plants to construct the nest, so each has a distinctive odor. The odoriferous hydrocarbons are transferred from the nest to the workers as they emerge from the pupal case. Even when a wasp meets a nestmate away from the nest, it can recognize it. Since a colony consists of a queen and her worker daughters, the nest odor is generally a reliable label of colony members as relatives (Breed 1998; Pfennig et al. 1983).

More challenging to demonstrate conclusively is the second form of phenotype matching, in which an animal uses its *own* phenotype as a template against which to compare strangers. This process is descriptively nicknamed the "armpit effect"—imagine an individual sniffing its own armpit and then that of a stranger. Several examples of self-referent phenotype matching have been proposed. For example, Mateo and Johnston (2000) cross-fostered golden hamster females (*Mesocricetus auratus*) on the day of birth. Because only one hamster was transferred into its new foster family, it had little opportunity to learn about the phenotypes of its relatives. Nevertheless, as adults, hamsters could discriminate between the odors of unfamiliar kin and unfamilar nonkin. This finding suggests that they were using their own odor to form a template. It is still possible that hamsters acquired some knowledge of kin in the brief period before they were transferred (Hare et al. 2003), although Mateo and Johnston (2003) argue that this seems unlikely given the developmental stage of their discrimination abilities. This work points out the challenge in controlling for every possible source of information available from kin.

Recognition Alleles Our fourth and final mechanism of kin discrimination is genetically based. In this mechanism, an individual inherits a "recognition allele" or group of alleles that enable it to recognize others with the same allele(s). The postulated allele would have three simultaneous effects: it would endow its bearer with a recognizable label, endow the bearer with the ability to perceive that label in others, and cause the bearer to behave preferentially toward others with the label. This recognition system has been named the "green beard effect" to indicate that the label could be any conspicuous trait, such as a green beard, as long as the allele responsible for it also causes its owner to behave appropriately to other labeled individuals (Dawkins 1976, 1982; Hamilton 1964). Demonstrating the existence of recognition alleles has proven quite difficult, primarily because it is so hard to eliminate all the possible opportunities for learning recognition cues during an animal's lifetime. In addition, theoretical models predict that the high degree of genetic polymorphism required for this mechanism to function may be unstable and disappear (Gardner and West 2007; Rousset and Roze 2007).

Laurent Keller and Kenneth Ross (1998) may have identified a green beard allele in the red fire ant (*Solenopsis invincta*). Originally from South America, the fire ant is a recently introduced pest in the southern United States. Its social organization is controlled by the protein-encoding gene, *Gp-9*, with two alleles, *B* and *b*. Workers that encounter individuals carrying the *b* allele form a template that they later use to determine whether to attack particular queens. Workers that came into contact with ants that bear *b* alleles when they were forming their template accept only *b*-bearing queens (*bb* and *Bb*), whereas workers that contacted only *BB* individuals during template formation accept only *BB* queens (Gotzek and Ross 2007). Thus, kin discrimination in fire ants is very closely tied to genetics, but memory (in the formation of a template) also plays a role.

Insects are not the only animals with potential for genetically based recognition mechanisms. For example, a particular region of DNA called the major histocompatibility complex (MHC) may be important in recognition. The MHC region codes for molecules on the surface of cells that allow the body to distinguish between "self" and "nonself." The MHC region can cause grafted tissue to be rejected and triggers protective immune responses when disease-causing organisms enter the body. These same genes may serve as direct cues of relatedness, allowing individuals to identify their kin (reviewed in Brown and Eklund 1994; Penn and Potts 1999).

The larvae of the sea squirt *Botryllus schlosseri* use MHC to discriminate kin. These larvae, which superficially resemble a frog tadpole, float in the water column for a short time and then settle, attach to the sea bottom, and develop to the adult form. When groups of siblings settle, they tend to clump together, but groups of unrelated larvae settle randomly. If the siblings do not share an allele in the MHC region, they do not settle together. Unrelated larvae that happen to share an allele are just as likely to settle together as are siblings that share an allele (Grosberg and Quinn 1986). MHC has been reported to play a role in kin recognition in a variety of other taxa, including amphibians (Villinger and Waldman 2008) and mammals, including humans (reviewed in Johnston 2003). Most effects of MHC-based recognition have been found in the context of mate choice and incest avoidance. To date there is no evidence that MHC genes influence the *perception* of odor, which is necessary for MHC to be a "recognition allele" under our strict definition above (Mateo 2004). Nonetheless, the MHC system remains an interesting case of a close link between genes and kin recognition.

RECIPROCAL ALTRUISM

As we have seen, helping relatives is favored by natural selection when the inclusive fitness of the helper is increased. However, in everyday speech, we typically use the word "altruist" not to describe a person helping a relative, but a person helping a nonrelative. Can altruism toward nonrelatives be favored by natural selection?

The answer is yes, but only under a narrow set of circumstances. Altruism between nonrelatives can evolve if there is an opportunity for payback in the future. We call this **reciprocal altruism**, evolution's version of "you scratch my back and I'll scratch yours" (Trivers 1971).

In order to think about the circumstances that might favor reciprocal altruism, it's helpful to follow the lead of many other behaviorists and think about a particular scenario: the Prisoner's Dilemma. In reciprocal altruism, the costs and benefits to the altruist depend on whether the recipient returns the favor. You may recall from Chapter 18 that evolutionary game theory is designed to handle situations such as this, in which the best course of action depends on what others are doing.

The name of the game comes from an imaginary story in which two suspects are arrested for a crime and kept in separate jail cells to prevent them from communicating. Certain that one of them is guilty, but lacking sufficient evidence for a conviction, the prosecutor offers each a deal. Each prisoner is told that there is enough incriminating evidence to guarantee a short jail term, but freedom can be obtained by providing enough evidence to send the other to jail for a long time. However, if each informs on the other, they both go to jail for an intermediate length of time. We can construct a payoff matrix (Table 19.1), just as we did for hawk–dove games of conflict in Chapter 18. The possible strategies available to each player are to cooperate (don't squeal on your partner) or to defect (squeal). The best that you can do is to defect while your partner cooperates (getting a payoff of T, which stands for the temptation to defect). If both you and your partner cooperate, the payoff is R (for reward for cooperation). If both partners defect, the payoff is P (for punishment for defection). If you cooperate and your partner defects, you get the lowest possible payoff, S (for sucker's payoff). For a game to be a Prisoner's Dilemma, the payoffs in the four cells illustrated in Table 19.1 must be in the order $T > R > P > S$.

In order to determine the best strategy, examine the payoff matrix. If you are Player A, what is your best strategy if your opponent cooperates? Because $T > R$, it is better to defect—you will go free. Similarly, if your opponent squeals, compare the payoffs of your two options. Because $P > S$, it is again better to defect, rather than taking the rap for both of you. Therefore, in a single round of playing the game, it is always better to

TABLE 19.1 Fitness Payoffs for Player A in the Prisoner's Dilemma. The payoffs are ordered $T > R > P > S$

Strategy of Player A		Strategy of Player B: Cooperate	Strategy of Player B: Defect
	Cooperate	*R* Reward for mutual cooperation	*S* Sucker's payoff
	Defect	*T* Temptation to defect	*P* Punishment for mutual defection

defect. If both players follow this logic, they will both defect and will each do worse than if they cooperated.

If we translate jail sentences into fitness payoffs, the solution to the Prisoner's Dilemma seems to imply that reciprocal altruism cannot evolve. Indeed, this may be true *if* the prisoners will never meet again. In real life, however, individuals often interact repeatedly, and when they do, reciprocal altruism may evolve.

One strategy, called "tit-for-tat," can be a winner in repeated games of Prisoner's Dilemma. In this strategy, an individual begins by being cooperative and in all subsequent interactions matches the other party's previous action. This strategy, then, has the characteristics of being "nice" (it begins with cooperation), it is retaliatory (if the partner defects, it defects in return), and forgiving (it immediately "forgets" a defection and cooperates if the partner later cooperates). If a population of individuals adopts this strategy, it cannot be overrun by a selfish mutant that always defects (Axelrod 1984). So, when the individuals have repeated encounters, reciprocal altruism can be an evolutionarily stable strategy (ESS), one that cannot be invaded by another strategy (see Chapter 4).

Given these results from game theory, let's step back and think more generally about when reciprocal altruism is likely to evolve. Trivers (1971) outlined several conditions that favor reciprocal altruism: (1) the benefit of the act to the recipient is greater than the cost to the actor, (2) the opportunity for repayment is likely to occur, and (3) the altruist and the recipient are able to recognize each other. These factors are most likely to occur in a highly social species with a good memory, long life span, and low dispersal rate (Trivers 1971).

One of the best-known examples of reciprocal altruism among nonhuman animals, partly because it is such a startling species in which to find this trait, is vampire bats (*Desmodus rotundus*). Vampire bats fly out at night to find their favorite prey, large mammals such as cattle or unsuspecting tropical biologists. The bats land near the sleeping victim and then skitter silently toward it along the ground. They then climb aboard and bite the prey with razor-sharp teeth specialized for slicing through skin, lapping up the blood. Anticoagulants in the bats' saliva keep the blood flowing. Successful bats can have giant meals—they can consume up to 50% of their body mass.

At the end of the night, vampire bats return to their communal roosts. Here is where altruistic acts take place: an unlucky bat that did not obtain a blood meal begs for food by licking one of its roostmates under the wings and on its lips. A receptive donor will then regurgitate blood (Figure 19.9). The regurgitated food is enough to sustain the hungry bat until the next night, when it may find its own meal. Donors may give blood to recipients that are not related to them (DeNault and McFarlane 1995; Wilkinson 1984, 1990).

Vampire bats meet the conditions that Trivers laid out that are favorable to the evolution of reciprocal altruism. First, the benefit to the recipient of the blood gift is greater than the cost to the donor. Since a bat's body weight decays exponentially after a meal, the recipient may gain 12 hours of life and, therefore, another chance to find food. However, the donor loses fewer than 12 hours of time until starvation and usually has about 36 hours, another two nights of hunting, before it would starve.

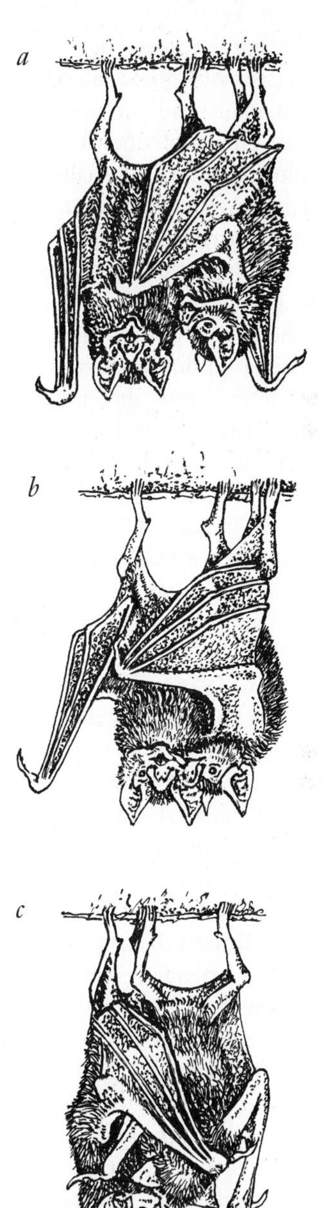

FIGURE 19.9 A vampire bat that was unsuccessful in obtaining a meal during a night's hunt begs for food from a roostmate. (*a*) First it grooms the roostmate by licking it under the wings, (*b*) then it licks it on the lips. (*c*) If receptive, a well-fed roostmate will respond by regurgitating blood to the hungry partner. (From Wilkinson 1990.)

Physiological studies of metabolism suggest that vampire bats are unusually susceptible to the effects of starvation (Freitas et al. 2003, 2005).

Second, bats are likely to have the opportunity to repay favors and to get favors repaid. Most bats are likely to be in the position of needing blood from roostmates, because on any given night, roughly 33% of the juveniles that are less than two years old and 7% of older bats fail to feed. Bats are also likely to encounter the same individuals time and again. Vampire bats roost in somewhat stable groups of both related and unrelated members. A typical group consists of 8 to 12 adult females and their pups, a dominant male, and perhaps a few subordinate males. Males leave their mothers when they are about 12 to 18 months old, but females usually remain well past reproductive maturity. New females occasionally join the group. Although the groups may change slightly over time, there are numerous opportunities to share food. Females may live as long as 18 years, and in one study, two tagged females shared the same roost for more than 12 years.

Trivers's third condition is that individuals that recognize each other can thus direct help appropriately. Generally, only bats that have had a prior association share food. In an experiment, a group of bats was formed in the laboratory. Aside from a grandmother and granddaughters, all the bats were unrelated. The bats were fed nightly from plastic measuring bottles so that the amount of blood consumed by each bat could be determined. Then, every night one bat was chosen at random, removed from the cage, and deprived of food. When it was reunited with its cagemates the following morning, the hungry bat would beg for food. In almost every instance, blood was shared by a bat that came from the starving bat's population in nature. Furthermore, there seemed to be pairs of unrelated bats that regurgitated almost exclusively to each other, suggesting a system of reciprocal exchange.

If Trivers's three characteristics of species likely to show reciprocal altruism bring our own species to mind, it is not surprising. Robert Trivers (1971) points out that reciprocal altruism is particularly common among humans. Not only do humans help the needy through social programs (with the expectation that they, too, might someday benefit from such a program), but they also help one another in times of danger and they share food, tools, and knowledge. Trivers even argues that our feelings of envy, guilt, gratitude, and sympathy have evolved to affect our ability to cheat, spot cheaters, or avoid being thought of as a cheater ourselves. We'll see more examples of reciprocal altruism among our primate kin later in this chapter.

MANIPULATION

In some cases, animals can coerce others to help them. This is especially likely to occur between parents and offspring (Alexander 1974, Trivers 1974; reviewed in Crespi and Ragsdale 2000). Recall that because parents and offspring are not genetically identical, their interests are not always perfectly aligned. Parents have an advantage in power struggles, as they are generally larger and more experienced than their offspring. Because offspring are related to their parents, their motivation to resist coercion is reduced—if, for example, they forgo their own breeding in order to help their parents reproduce, they at least gain indirect fitness.

EXAMPLES OF COOPERATION AMONG ANIMALS

As we consider various forms of cooperation among animals, we will note many similarities among distantly related groups. We will also notice that the selective forces leading to similar forms of cooperation may be quite different.

ALARM CALLS

Some of the classic long-term field studies of cooperation come from studies of alarm calls—warnings that animals give to alert others of danger. Let's examine a particularly well-studied taxon, ground squirrels and their relatives.

Belding's Ground Squirrels

High in the mountains of the Sierra Nevada in California, in Tioga Pass in Yosemite National Park, lives a very well-studied population of Belding's ground squirrels (*Spermophilus beldingi*). During most of the year, the alpine meadows are covered in deep snow, but in the short summer season, the meadows are awash in flowers and grasses. It is then that the ground squirrels become active, foraging in the fields during the day and returning to underground burrows to sleep and care for their offspring.

It's not only ground squirrels that call these lush meadows home. Squirrels are attacked both from the ground (badgers, snakes, weasels, coyotes) and the air (hawks and eagles). When squirrels spot a predator, they give an alarm call—a series of short sounds for a terrestrial predator, and a high-pitched whistle for an aerial predator (Figure 19.10).

Paul Sherman (1977; 1980a,b; 1985) studied the function of these calls: are they directed at the predator to let it know it has been detected? Or are they directed at kin? The answers seem to differ depending on the circumstances.

Individual Selection When a hawk is spotted overhead or when an alarm whistle is heard, near pandemonium breaks out in the colony. Following the first warning, others also whistle an alarm and all scurry to shelter. When a hawk is successful, the victim is most

FIGURE 19.10 **A female Belding's ground squirrel emits an alarm call.**

likely to be a noncaller. In one study, only 2% of the callers but 28% of the noncallers were caught (Table 19.2). The most frequent callers were those that were in exposed positions and close to the hawk, regardless of their sex or relationship to those around them. Thus, it seems that the alarm whistles given at the sight of a predatory bird directly benefit the caller by increasing its chances of escaping predation in the ensuing chaos (Sherman 1985).

Kin Selection In contrast, individual selection does not seem to be behind the evolution of the ground squirrels' alarm trills, given in response to terrestrial predators. In this case, the caller is truly assuming a risk; we know this because significantly more callers than noncallers are attacked. As can be seen in Table 19.2, 8% of the ground squirrels that called in response to terrestrial predators were captured, whereas only 4% of the noncallers were caught. The predators, even coyotes whose hunting success often relies on the element of surprise, did not give up when an alarm call was sounded. Furthermore, the caller was not manipulating its neighbors to its own advantage. Generally, the reaction of other ground squirrels was to sit up and look in the direction of the predator or to run to a rock. Their reaction did not create the chaos that might confuse a predator. Nor did the caller seek safety in the midst of aggregating conspecifics (Sherman 1977).

The evidence that kin selection is the basis for alarm trills to terrestrial predators by ground squirrels is strengthened by information on the structure of their society. Daughters tend to settle and breed near their birthplace, so the females within any small area usually are genetically related to one another. The sons, on the other hand, set off independently before the first winter hibernation, never to return to their natal burrow. Sherman knew the relationships and the identity of the squirrels (and could identify them from a distance because he marked them by painting their flanks with Lady Clairol hair dye).

TABLE 19.2 **Alarm Calling and Survival in Belding's Ground Squirrels at Tioga Pass, California. All Data Are from Observations Made During Attacks by Hawks (n = 58) and Predatory Mammals (n = 198) That Occurred Naturally During 1974–1982 (Data from Sherman 1985)**

	Number of Ground Squirrels			
	Captured	Escaped	% Captured	*P* value
Aerial predators				
Callers	1	41	2%	
Noncallers	11	28	28%	<0.01
Total	12	69	15%	
Terrestrial predators				
Callers	12	141	8%	
Noncallers	6	143	4%	<0.05
Total	18	284	6%	

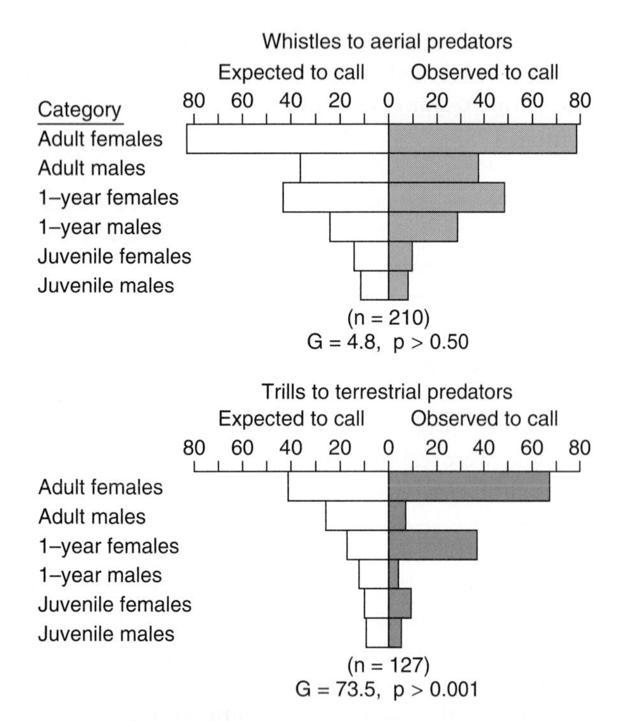

FIGURE 19.11 **Expected and observed frequencies of alarm calls by Belding's ground squirrels in response to aerial and terrestrial predators. Expected frequencies are those that would be predicted if the animals called randomly. The calls in response to aerial predators are close to the expected frequencies. However, the calls in response to terrestrial predators are more likely to be given by females with relatives nearby than would be predicted if the animals called randomly. (From Sherman 1985.)**

Sherman's data (Figure 19.11) suggest that the ground squirrels practice nepotism, or favoritism for family members. Notice in the figure that when a terrestrial predator appears, females are more likely than males to sound an alarm. This is consistent with kinship theory because it is females that are more likely to have nearby relatives that would benefit from the warning. In addition, reproductive females are more likely than nonreproductive females to call. An even finer distinction can be made: reproductive females with living relatives call more frequently than reproductive females with no living family members.

Alarm Calls in Other Rodents

Many rodent species, not only Belding's ground squirrels, give alarm calls, providing us with a powerful tool for comparative analysis. Species vary in whether alarm calling increases the caller's own chance of survival, primarily aids the caller's own offspring, or helps other animals. For example, yellow-bellied marmots are most likely to call when their own offspring are nearby, and (unlike Belding's ground squirrels) the presence of nonoffspring relatives does not significantly affect their likelihood to call (Blumstein et al. 1997). Other rodent species (e.g., mice of the genus *Peromyscus*) apparently do not call at all. Shelley and Blumstein (2005) carried out a comparative analysis (see Chapter 4 for more details on this research technique) in which they plotted traits onto a phylogenetic tree to see which traits were likely to have evolved together. Social species were more likely to call than nonsocial species. However, an even stronger relationship was that diurnal (day-active) rodent species were more likely to call than nocturnal (night-active) species and that the evolution of diurnality precedes the evolution of alarm calling. Shelley and Blumstein argue that, in general across rodent species, alarm calling serves primarily to communicate with predators and that benefits arising from kin selection are secondary. Belding's ground squirrels may benefit more from kin selection than some other rodent species because they live in high-density meadows where many relatives are likely to be within earshot (Blumstein 2007).

COOPERATION IN ACQUIRING A MATE

Males of some species cooperate in attracting a mate. Some even relinquish the opportunity to pass their alleles into the future generation personally, at least temporarily. Indeed, these males concentrate their efforts on making another male more attractive to females. This seems like a guaranteed route to lower fitness, but in some cases it can be advantageous. Let's look at four examples, two from birds and two from mammals. As you will see, sometimes the benefits to cooperators can arise via individual selection, kin selection, or reciprocal altruism, and sometimes a combination.

Wild Turkeys

Strangely, most male wild turkeys (*Meleagris gallopavo*) in some Texas populations never mate. Toward the end of a young male's first autumn, when he is about six to seven months old, he and his brothers forsake the others in their family and form a sibling group that will be an inseparable unit until death. This sibling group and all other juvenile male sibling units in the area flock together for the winter.

Only the dominant male in each group mates. During the first winter, each male's status within this fraternity is decided by the outcome of two contests. One competition is for dominance within the sibling group. Brothers battle by wrestling, spurring, striking with their wings, and pecking at each other's heads and necks. Endurance is the key to success: the turkeys fight until they are exhausted. When only one is able to do battle, however weakly, he is the winner. The second contest is between rival sibling groups. The groups challenge and fight one another until a dominance hierarchy is established. The sibling group with the most members is usually victorious. Renegotiation of rank is rare; the dominance hierarchy within and between sibling groups is stable.

When the breeding season begins, females interested in mating visit the open meadows, where the males congregate. Two to four siblings form display partnerships within larger aggregations of males. The brothers of each unit court the hens by strutting in unison, even though only the dominant male in the highest-ranking sibling group will mate. Of 170 tagged males displaying at four grounds, not more than 6 males accounted for all 59 observed matings. If a subordinate male is presumptuous enough to attempt a mating, the dominant male chases him away and then mates with the hen.

Watts and Stokes (1971) were the first to suggest that kin selection is the major driving force behind the evolution of this behavior, and this was confirmed recently by genetic analysis (Krakauer 2005). A subordinate male gains inclusive fitness by helping his brother to perpetuate his alleles. On the other hand, without his assistance, the brother could not be successful. The cooperative efforts of siblings are necessary for their unit to become dominant, and the synchronous strutting of siblings makes the dominant male more attractive to the hens. Thus, the subordinate brother reproduces by proxy.

Lions

Male lions (*Panthera leo*) also cooperate in acquiring mates. They form coalitions, or partnerships, that challenge the males of other prides. Coalitions may take over a pride by slowly driving out the resident males, or it may be a hostile takeover, involving serious fighting (Figure 19.12). In such contests, the larger coalition usually wins.

The reward for the victors is a harem of lionesses. When the females come into reproductive condition (which is sometimes hastened if the new males kill cubs that were sired by other males), they often do so simultaneously. During the two- to four-day period when a female is in reproductive condition, she mates about every 15 minutes around the clock. Any of the males in the coalition may be the first to find her, mate with her, and keep others away by his presence. A female may change mates during this period but generally not more than once a day (Bertram 1975, 1976).

In turkeys, we saw that males helped their brothers gain mates. Could kin selection also underlie the formation of coalitions in lions? Coalitions of more than three individuals usually consist of close relatives that left their natal pride as a group (Packer et al. 1991). They remain together, and after one to three years of traveling nomadically, they challenge the males of other prides. Thus, a male in a coalition with relatives has the chance to gain reproductive success indirectly (by helping his male relatives mate with the female).

Kin selection can't be the entire story behind cooperative groups of male lions, or even most of it. It is now known that roughly half of male coalitions contain at least one unrelated male, and coalitions of two or three usually consist of unrelated males (Packer 1986; Packer et al. 1991; Packer and Pusey 1982). When presented with a stuffed "intruder" lion accompanied by playback of recorded roars, lions attacked it (Figure 19.13),

FIGURE 19.12 **Two male lions fight for control of a pride.**

a

b

FIGURE 19.13 **A male lion attacks a stuffed lion after hearing the recorded roar of a male intruder.**

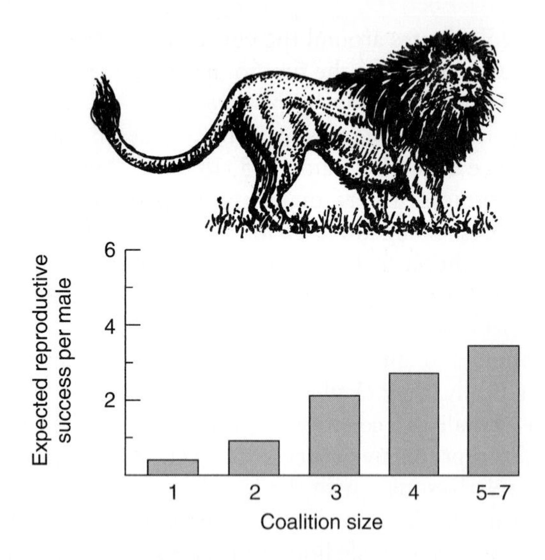

FIGURE 19.14 **Males in larger coalitions have greater reproductive success. (Data from Packer and Ruttan 1988.)**

regardless of whether they were related to other members of their coalition (Grinnell et al. 1995). Lions attacked the model regardless of the behavior of other coalition members and did not appear to monitor the actions of others, suggesting that reciprocal altruism is not playing an important role (Grinnell et al. 1995).

Why, then, do unrelated males gather into coalitions? The answer turns out to be quite simple—the larger the coalition, the greater a male's reproductive success (Figure 19.14). Larger coalitions have a better chance of ousting the current coalition in a pride, maintaining control of that pride, and perhaps even gaining residence in a succession of prides. A solitary male has little chance of reproducing and, therefore, much to gain by joining a coalition. A small coalition may also benefit by accepting an unrelated male because the extra member may help it take over prides. Indeed, coalitions accept unrelated companions only while coalition members are not yet resident in a pride (Packer and Pusey 1987).

Larger coalitions also remain in control of a pride longer than smaller ones. A coalition of three to six males may remain in control as long as two to three years. A coalition of two might be in possession for over a year. If a lone male manages to gain control of a pride, which happens infrequently, his tenure generally lasts only a few months (Bertram 1975). As a result, the lifetime success of a male lion increases by cooperating with other males in taking over a pride, even if all the males are not related. Packer et al. (1988) estimated that each additional member of a coalition increases individual reproductive success by 0.64 surviving cubs per male.

Long-Tailed Manakins

The Costa Rican rainforest is filled with fascinating animals, but a standout is a small bird called the long-tailed manakin (*Chiroxiphia linearis*). Walking through the forest near Monteverde, one can often hear in the distance their calls of "To-lay-do! To-lay-do!" A quiet approach yields a remarkable sight: in a small clearing, two or even three males work together to attract a mate. They begin with the call, emitted as many as 19 times a minute and 5000 times a day. Once a female arrives, they begin a visual display. In one of the most common variations of the display, called the up-down jump display, the birds perch side by side on a branch. One male jumps into the air, emitting a wheezy *buzzee* call, and hangs there momentarily; just as he lands, the other male jumps up. Alternately jumping up and down, the males are reminiscent of children on a see-saw. Even more spectacular is the cartwheel. Here, males again begin by perching next to each other. Then, one male jumps upward and backward over the second male. Meanwhile, the second male hops along the branch to take over the first male's spot. Now this male takes a turn at jumping up and moving to the rear. The dance looks as if the two birds were balls being juggled. The courtship sequence may be repeated only once or as many as a hundred times in succession. When the display bout is over, one male leaves the display branch and watches while the remaining male does a solo performance. If his gymnastics have impressed the female, she mates with him.

Although the males take turns jumping during their tandem courtship display, the same male always mates. The benefit of this elaborate dance to the mating male is obvious, but what about the other male? It is not indirect fitness because it seems unlikely that the two males are related. A typical brood consists of only one or two offspring, and there is no reason to assume that the siblings are necessarily the same sex. Furthermore, just before and after each breeding season, the young, particularly the subadult males, disperse. It seems unlikely, therefore, that male relatives would stay in proximity for the three to four years it takes them to acquire adult plumage. If the partners are not genetically related, the nonbreeder is not increasing his indirect fitness. So why does a subordinate bird stay?

A subordinate bird's chances of mating would not be increased by deserting his partner. Solitary males cannot mate. They cannot even perform the courtship display. If a male cannot dominate his current partner, his chances of becoming the dominant member of another pair will be low. However, if the subordinate male outlives his partner, it is likely that a younger male, one that can be dominated, will become his new associate. Then it will be his turn to mate and raise his direct fitness (Foster 1977; McDonald and Potts 1994). In addition, singing—in particular the ability to match a partner's

FIGURE 19.15 **An alliance between two male olive baboons. The two males on the right are cooperating to challenge the male on the left. At a later time, the male that was assisted will have to reciprocate to maintain the alliance.**

song, which impresses the females—improves with age and practice (Trainer and McDonald 1995; Trainer at al. 2002). Thus, the most likely reason for cooperative courtship in this species is that it increases the chances of obtaining direct fitness benefits in the future.

Olive Baboons

Male olive baboons (*Papio anubis*, Figure 19.15) also cooperate in attracting a mate, but here we see yet another evolutionary mechanism at work. A male who lacks a female consort sometimes enlists the help of a friend to win another male's mate. The following scenario is typical of what often occurs: Male A is associating (called consorting) with an estrous female. Male B, who has no female of his own, solicits the help of male C, and the two form an alliance and challenge male A. While the battle is in progress, male B gets away with the female. Male C has acted altruistically; he risked injury while assisting another to acquire a mate. However, at some time in the future, he will enlist the help of male B in winning a consort of his own (Packer 1977). You will recognize this as an example of reciprocal altruism.

COOPERATIVE BREEDING AND HELPING

When hiking in the California woodlands, keep alert for the unmistakable signs of the acorn woodpecker (*Melanerpes formicivorus*). If you don't recognize it by its harsh "waka-waka-waka" call, or by its black-and-white plumage topped off by a snazzy red cap, the dead giveaway that you are in its habitat is a dead tree or even a telephone pole peppered with a multitude of holes, each with an acorn inserted snugly in it. Acorn woodpeckers drill these holes to make their granaries, and they store their acorns for leaner times (Figure 19.16). (One group of woodpeckers got a bit carried away and stored 485 pounds (220 kg) of acorns in a water tank in Arizona!)

FIGURE 19.16 **An acorn woodpecker places an acorn in a hole drilled in a dead tree. Food stores must last the communal group through winter.**

What leads us to begin this section with acorn woodpeckers, however, is not their bizarre food-storing habits but their extensive social networks. These birds live in family groups of up to 15 or so, containing one to four breeding males, one or two breeding females, and from zero to ten nonbreeding helpers. It is these nonbreeding helpers that are of particular interest. They are adult birds, physiologically capable of breeding themselves, that instead stay to help their parents raise additional young. Acorn woodpeckers thus fit the definition of cooperative breeding: some individuals (helpers) assist in the care and rearing of another's young rather than producing offspring of their own. Cooperative breeding was first described in birds (Skutch 1935) but has since been documented in many taxa, including birds, mammals, insects, and spiders, although it is quite rare (e.g., it is found in only about 3% of birds and mammal species; Emlen 1997).

There are many variations on the theme of cooperative breeding. Instead of limiting ourselves to the woodpecker example, we will organize the remainder of this section by the questions that have intrigued researchers.

How Do Helpers Help?

In many species, helpers pitch in to feed offspring. The time-intensive task of rearing baby birds is easily shared because anybody, not just the parents, can collect and

a

b

FIGURE 19.17 **A jackal helper (*a*) prepares to regurgitate food to a pup and (*b*) chases away an intruder.**

carry food to the nest and pop it in a nestling's mouth. Acorn woodpeckers provide nestlings with food, both insects and acorn fragments (Koenig and Mumme 1987). Helpers in the well-studied Florida scrub jay (*Aphelocoma coeruslescens*) deliver about 30% of the food consumed by nestlings. The parents' job is thus reduced, and, as a result, they enjoy better health. In one study, 87% of the breeders with helpers survived to the next year, compared to 80% of breeders without helpers (Stallcup and Woolfenden 1978).

Although mammalian mothers are uniquely equipped to provide milk to their offspring, helpers can deliver other kinds of food by carrying it in their jaws or in their stomachs. Blackbacked jackals (*Canis mesomelas*) regurgitate food to eager pups (Figure 19.17). Helpers not only contribute 18 to 32% of all regurgitations to pups, but they sometimes also regurgitate to the lactating mother, allowing her more time to remain with the pups instead of hunting (Moehlman 1979).

Help can also come in the form of extra protection for the young. Florida scrub jays give alarm calls to predators such as snakes and even help to drive them away by mobbing them (Woolfenden 1975). Jackal families with helpers always have an adult on guard to drive away predators, whereas groups lacking helpers may have to leave the pups unattended while hunting. In fish, helpers mainly contribute by protecting offspring; they can contribute little to the nourishment of the young (Dugatkin 1997).

Provisioning of food and extra protection are the most common services offered by helpers, but other tasks also lend themselves to sharing. In some bird species, helpers may build and clean nests or incubate and brood the nestlings (Skutch 1987). In saddle-backed tamarins (*Saguinus fuscicollis*), a small primate, male helpers lug around the offspring. At birth a tamarin is almost 20% of its adult weight, and litters typically consist of twins. Thus, carrying these youngsters is burdensome, and if the duty were not shared the mother might not be able to obtain enough nourishment for herself and to ensure an adequate milk supply (Terborgh and Goldizen 1985).

Is "Helping" Really Helpful?

Given all the useful tasks helpers can perform, this may seem a silly question. However, not every study has found a relationship between the presence or number of helpers and the reproductive success of the breeding individuals (reviewed in Clutton-Brock 2002), so it is wise not to make assumptions.

In many species, researchers have shown that the number of helpers correlates with the survival of the young, the survival of the breeders, or both. For example, we have already mentioned the Florida scrub jay (Woolfenden 1975; Woolfenden and Fitzpatrick 1990). These birds have been studied for many years in the scrub habitat of Florida, currently threatened by development. The jays have proven to be easily tamed, allowing for close observation (Figure 19.18). The jays form territories that contain one breeding pair and a varying number of helpers—from none to as many as six. The breeding success of pairs with helpers clearly exceeds that of pairs without helpers (Woolfenden 1975). Figure 19.19

FIGURE 19.18 **Florida scrub jays are not difficult to observe.**

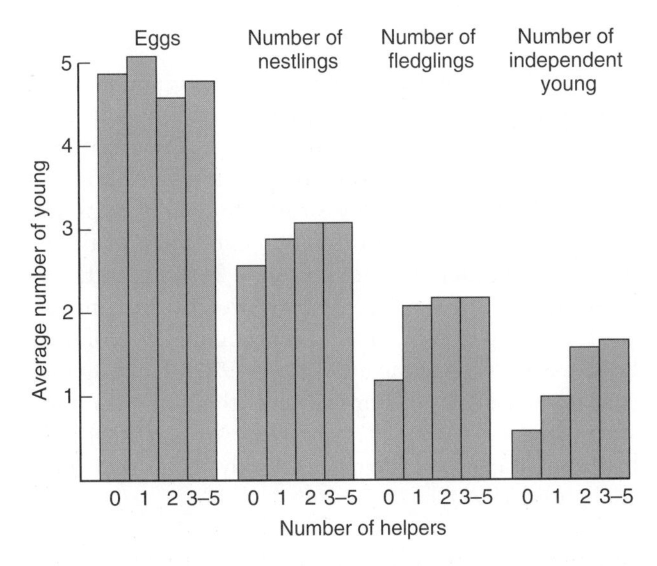

FIGURE 19.19 **The relationship between the number of Florida scrub jay helpers and the breeding success of the experienced parents. Helpers do not increase the number of eggs laid. They do, however, increase the chances that the eggs will hatch and that the young will survive to become independent. (Data from Woolfenden 1975.)**

shows the breeding success of experienced pairs with and without helpers during one five-season study. Notice that the presence of helpers has no effect on the number of eggs laid but does increase the chances that the young will hatch, leave the nest, and become independent birds.

Similarly, breeding success increases with the presence of helpers in some mammalian and fish species. For example, pairs of blackbacked jackals are joined by between one and three young from previous litters, who help them rear the next pups. The reproductive success of a pair of blackbacked jackals increases with the number of helpers (Moehlman 1979).

But we must be cautious: finding that there is a *correlation* between breeding success and the presence of helpers is not by itself sufficient to demonstrate that helping *causes* increased breeding success (Brown et al. 1982). It's quite possible that another factor, such as territory quality, causes both an increase in the number of helpers and increased breeding success, thus creating the correlation. In addition, helpers are often the offspring of the breeders they are assisting—perhaps there is a correlation between breeding success and the number of helpers simply because the breeding pair is of consistently high quality from year to year. How might we be sure that it is really the helpers that caused increased reproductive success of the breeders?

One way to increase our certainty that helping causes increased reproductive success in breeders is by removing helpers and measuring the consequences. For example, helpers were removed from the nests of gray-crowned babblers (*Pomatostomus temporalis*), a bird that lives in year-round territorial groups of 1 to 13 birds in the open woodland of Queensland, Australia. Parents are usually assisted by a variable number of their offspring from previous broods, but in this experiment, nine of the breeding groups were reduced to a single helper. These groups then raised an average of 0.8 young, less than half the number of fledglings produced by the 11 control groups, which had more assistance. Therefore, the positive relationship between breeding success and the number of helpers found in gray-crowned babblers does seem to be a result of the presence of helpers (Brown et al. 1982). However, a problem with this experimental approach is that removal or addition of helpers has other unintended effects: it can be very disruptive to the social structure of a group, and it also changes group size, which in turn can affect the group's success (reviewed in Wright and Russell 2008).

Another experimental approach to measuring the benefits of helping is to change the ratio of helpers to young in a different way, by temporarily removing or adding young while the number of helpers remains constant. When Clutton-Brock et al. (2001) performed this manipulation in meerkats, which are (quite adorable) mammals (Figure 19.20), pup weight gain was increased when the number of pups was reduced, and decreased when the number of pups was increased. Extra food helps the pups by reducing the age at which they first

FIGURE 19.20 **A meerkat helper with pups.**

reproduce and by increasing the chance that they can successfully compete for a high-ranking spot in the social hierarchy (Russell et al. 2007).

A third way to approach the problem of teasing apart correlation from causation, at least in long-term studies with plenty of data, is with statistical techniques by which different variables can be controlled (Cockburn et al. 2008; Wright and Russell 2008). Cockburn et al. (2008) used elegant statistics to analyze 19 years of data on the superb fairy-wren. They found that helpers did not increase the survival of offspring but instead increased the future survival of breeding females.

Do breeders always *want* help? Not always. After all, an extra helper around might mean competition for resources or increase the chance that your mate has an extra-pair copulation. In the pied kingfisher (*Ceryle rudis*), helpers are tolerated only when their services are needed. These birds usually have primary helpers, which are older offspring, but may also have secondary helpers, which are unrelated. Heinz-Ulrich Reyer (1980) compared two colonies of pied kingfishers in East Africa. Breeding pairs at Lake Naivasha typically have only one primary helper. When males apply for a job as secondary helpers, they are persistently chased away by the male territory holder. In contrast, at Lake Victoria, secondary helpers are eventually tolerated and permitted to stay and feed the young. Why? The answer is that the services of secondary helpers are needed to raise offspring at Lake Victoria but not at Lake Naivasha. These birds fish for a living, and Lake Victoria is a harder lake to fish. Victoria's rougher waters increase the time it takes to catch a fish, and the fish are smaller. Furthermore, the fishing grounds are farther from the colony. With the additional fish provided by secondary helpers, the breeding pair can raise more offspring (Table 19.3).

Is It Costly to Help?

Helping behavior becomes a more interesting evolutionary puzzle if it has a cost, and you won't be surprised to learn that it often does. For example, mongoose (*Suricata suricatta*) helpers forgo feeding and stay at the burrow to baby-sit for the young pups and guard them from predators for an entire day while the parents and others forage. During a 24-hour shift, the baby-sitter loses 1.3% of its body weight. In contrast, the foraging group members gain roughly 1.9% of their body weight (Heinsohn and Legge 1999). Helping may even reduce survival. For example, the helpers among stripe-backed wrens (*Campylorhynchus nuchalis*) that bring the most food die more quickly than other birds (Rabenoid 1990).

STOP AND THINK

Imagine you are spending your graduate school years studying helping behavior in a little-known bird species. You find that birds that remain behind on their natal territory to help their parents are far more likely to die than those that go off to breed on their own. In your dissertation defense, you would like to make the argument that helping is costly. How confident are you? What experiment(s) would you like to perform in order to increase your confidence?

If it is costly to help, why do it? We can break this question down into two parts: (1) Why would an offspring delay dispersal and stay near home? (2) Why would it help? It makes sense to address these as separate questions because not every offspring that delays dispersal offers assistance, and sometimes offspring that have dispersed to nearby areas return to their parents' area in order to provide care (reviewed in Pruett-Jones 2004; Ekman et al. 2004).

TABLE 19.3 The Effect of Helpers on the Reproductive Success of Pairs of Pied Kingfishers. Shown Are Mean, Standard Deviation, and Sample Size

	Lake Victoria			Lake Naivasha		
	Mean	SD	n	Mean	SD	n
Clutch size	4.9	0.6	22	5.0	0.6	8
Young hatched	4.6	0.5	14	4.5	0.7	2
Young fledged						
No helpers	1.8	0.6	14	3.7	0.9	9
1 helper	3.6	0.5	12	4.3	0.5	4
2 helpers	4.7	1.0	6	—	—	—

Why Should an Offspring Delay Dispersal?

In Chapter 11, we discussed the costs and benefits of dispersal and philopatry (staying near home). For example, animals may disperse in order to avoid inbreeding, reproductive suppression by their relatives, and competition. On the other hand, animals may exhibit philopatry because they are adapted to the local conditions, and familiar with the physical and social settings of home. Dispersal may be risky: for example, small individuals of a species of cichlid fish (with the evocative name of the Princess of Burundi) are very likely to be eaten by predators when they venture off their territory. Even when breeders are experimentally removed from a territory, potential helpers choose to stay at home instead of moving to the unoccupied territory (Taborsky 1985).

Animals might also delay dispersal because other options may be limited. Let's look at two major ways in which this might happen.

Habitat Saturation Florida scrub jays live only in a special scrub habitat, comprised of dry-adapted shrubs and understory plants in sandy soils, that is already limited in its distribution and growing ever scarcer because of human development. Habitat availability is so limited that virtually every territory is filled. Once a scrub jay is lucky enough to acquire a territory, it generally keeps it for life (Woolfenden 1975). The most common way for a male to acquire a territory is by inheriting a portion of his parents' property, either by replacing his father after his death or by subdivision of his father's territory. If there is more than one son helping, the dominant one is favored in the property settlement (Woolfenden and Fitzpatrick 1978). Otherwise, a scrub jay can only claim a territory of its own if it defeats a breeder or successfully competes for the territory of a breeder that has died. Thus, we believe that a major reason that scrub jays help their parents is that they are making the best of a bad set of circumstances.

Support for the idea that the availability of territories is important in determining helping has also been found in acorn woodpeckers. Study sites in California, New Mexico, and Arizona vary with respect to woodpecker density, territory turnover rate, and territory fidelity. In California, the habitat is extremely saturated: not a single territory became vacant during a three-year study (MacRoberts and MacRoberts 1976). Forty-nine percent of the juveniles remained at home, and 70% of the groups had helpers. Young acorn woodpeckers in the Magalena Mountains of New Mexico face somewhat better odds in their quest for suitable territories than do those on the West Coast: 19% of the territories in New Mexico became vacant over a three-year study (Stacey 1979). Here, 29% of the youngsters stayed at home, and 59% of the groups had helpers. In the Huachuca Mountains of southeastern Arizona, there is no shortage of territories (Stacey and Bock 1978). Only 16% of the breeding units had helpers. In short, the frequency of helping in populations of acorn woodpeckers varies directly with the scarcity of open territories.

Not all territories are equal in quality, and it might be better to stay home if the only other choice is a poor-quality territory. The interaction between habitat saturation and territory quality has been nicely demonstrated by Jan Komdeur (Komdeur 1992; Komdeur et al. 1995), who transplanted Seychelles warblers to unoccupied islands. At first, these small birds bred independently. When territories began to fill up, some birds stayed to help their parents rather than move to low-quality territories that had relatively few insects. When breeders on low-quality territories were removed, the territories were filled only by helpers from territories that were equivalent or worse in quality.

Lack of Mates In addition to habitat, mates can also be a limited resource (Emlen and Vehrencamp 1983). For example, splendid fairy-wren females (Figure 19.21) suffer much greater annual mortality than do males (57% and 29%, respectively; Rowley 1981), and thus females are frequently in short supply. Helpers tend to be males that are awaiting an available mate. For some individuals, the wait can be as long as five years. As patterns of mortality and the resultant sex ratios vary, so does the percentage of groups with helpers: when females are scarce, male helpers are plentiful (Russell and Rowley 1988; Rowley and Russell 1990).

FIGURE 19.21 **A male splendid wren carries an insect to feed its young. When adult females are in short supply in the population, this breeding male can count on his sons to help rear the next brood.**

Why Should a Helper Help?

Even if an animal does not have the opportunity to breed, that does not automatically mean it will become a helper. Many animals become "floaters" and wander around without a territory. Even if an individual remains on its own territory, it may not help (as we know from television sitcoms, if not our own experience), so there must be other reasons besides lack of other opportunities that underlie helping behavior.

As with other forms of cooperative behavior, we can categorize fitness benefits as direct benefits, in which a helper's own lifetime reproductive success is increased by its actions, and as indirect benefits, in which a helper benefits by increasing the production of relatives. The relative importance of direct and indirect fitness benefits depends on the species.

Helpers May Get Indirect Fitness Benefits As we have seen, in most species, helpers are older offspring who help their parents raise their younger siblings. Thus, in cases where helpers improved the survival of nestlings, helpers may gain indirect fitness benefits by increasing the number of their younger siblings.

Some long-term studies suggest that kin selection can be important. For example, researchers collected five years of data on white-fronted bee-eaters to test alternate hypotheses for helping behavior (Emlen 1991; Emlen and Wrege 1989). They found no evidence for direct benefits to the helpers (increased survival to the next breeding season, increased chance of mating in subsequent years, or increased success in rearing young), but concluded that helping led to increased production of related young, suggesting that indirect fitness benefits are important.

Another line of evidence supporting the hypothesis that indirect benefits can be important comes from cases in which animals have a choice of whom to help. In pied kingfishers, there are two kinds of helpers: as we defined previously, secondary helpers help unrelated offspring, and primary offspring help related individuals. Kingfishers become secondary helpers only when both their parents are dead; otherwise they are primary helpers (Reyer 1984, 1990). In long-tailed tits, helpers are failed breeders. When nests failed either naturally or because of the interference of researchers, the failed breeders helped at the nests of relatives rather than at equidistant nests of nonrelatives. In fact, if close relatives were not available, failed breeders didn't help at all (Russell and Hatchwell 2001).

Helpers May Get Direct Fitness Benefits Many social groups are comprised of relatives, so it is not surprising that indirect fitness benefits have drawn the attention of many researchers. However, many groups have unrelated members, and many studies find no relation between helping behavior and relatedness. As a result, the literature on cooperative breeding that once focused largely on studying indirect fitness benefits has shifted emphasis to a study of direct benefits (Cockburn 1998; Clutton-Brock 2002; Dickinson and Hatchwell 2004). To see how animals might improve their own reproductive success by helping, we'll look at several case studies.

FIGURE 19.22 **On the left is a nest built by a 4-year-old female Seychelles warbler with no experience. Note that it is built between a tree branch and a thin leaf stalk. On the right is a nest built by a 4-year-old female that had experience in being a helper. Note its placement in a sturdy tree fork.**

Our first example comes from the Seychelles warblers we met earlier. Although these birds normally remain on their natal territory to help their mothers raise additional offspring, they are not as related to the offspring they help as you might expect. In this species, extra-pair copulation is common, so often a helper will not have the same father as the offspring it helps. In addition, sometimes eggs are deposited in a nest by other birds. Indirect benefits are thus relatively lower in this species; instead, helpers gain direct benefits in several ways. Genetic analyses of offspring show that helpers may add their own eggs to a nest (Richardson et al. 2002). Second, subordinates may be able to take over a territory budded off from the main territory (Komdeur and Edelaar 2001). Finally, helpers have a chance to practice their parenting skills. When they finally acquire their own territory, they are superior breeders (Figure 19.22; Komdeur 1996).

When mates are scarce, a helper may increase its own reproductive success by mating with the original breeder in a later year or if the original mate dies. Remember that in pied kingfishers, the secondary helpers are unrelated to the breeders. More than half of these helpers return to the same area the following year. Of these, half succeed in mating with the female they had assisted (Reyer 1980, 1984, 1986).

STOP AND THINK

Helping behavior in humans has also been studied from an evolutionary perspective. Because human children are very dependent for years, and mothers can give birth to a number of children in rapid succession, most mothers require help from others to successfully raise their children. In a review of 45 studies done across many cultures, Sear and Mace (2008) found that the presence of maternal grandmothers (the mother's mother) and sibling helpers improves child survival, but surprisingly, fathers improved child survival in only one-third of studies. What is a hypothesis based on kin selection that might explain these data?

EUSOCIALITY

If you've spent any time watching a glass-walled observation hive of honeybees, you were probably impressed by the sense of purpose in the colony. The packed-in bees bustle around, busy at their tasks. Watch for longer, and you'll notice that younger workers stay inside the hive and feed the queen's helpless larval offspring, each tucked carefully away in its own cell. Other workers maintain the hive and carry out the dead. Older workers forage at flowers and, upon returning to the hive, communicate the location of these resources to their hive mates in an elaborate dance (Chapter 16). If you are foolish enough to threaten the hive, you will discover that some bees are guards and will quickly sting you, losing their lives in the process—as they pull away, their stingers rip out of their abdomens and remain in your flesh, into which venom continues to pump. There is a dearth of males in the hive—all those busy workers are female. Only one individual in the hive lays eggs, a bloated queen.

Bees are an example of a **eusocial** (literally "truly social") species. Eusocial species are defined by three characteristics: reproductive division of labor (some individuals have offspring, and others do not), cooperation in the care of young, and overlap of at least two generations capable of sharing in the colony's labor (Michener 1969; Wilson 1971). The astute reader may realize that these characteristics also apply to some of the cooperative breeders that we have just discussed. Species range along a eusociality continuum according to how evenly reproduction is shared among group members (Figure 19.23) (Lacey and Sherman 1997; Sherman et al. 1995). At one end, all or many of the group members breed. At the other end, breeding is restricted to one or several group members. One way that biologists have described the degree of eusociality is by measuring "reproductive skew," or the proportion of individuals that give up reproduction.

Eusociality is rare. For many years, the only species known to be eusocial were insects from only two taxonomic groups: either hymenopterans (ants, bees, and wasps) or isopterans (termites). More recently, the list of eusocial invertebrates has been expanded to include species of

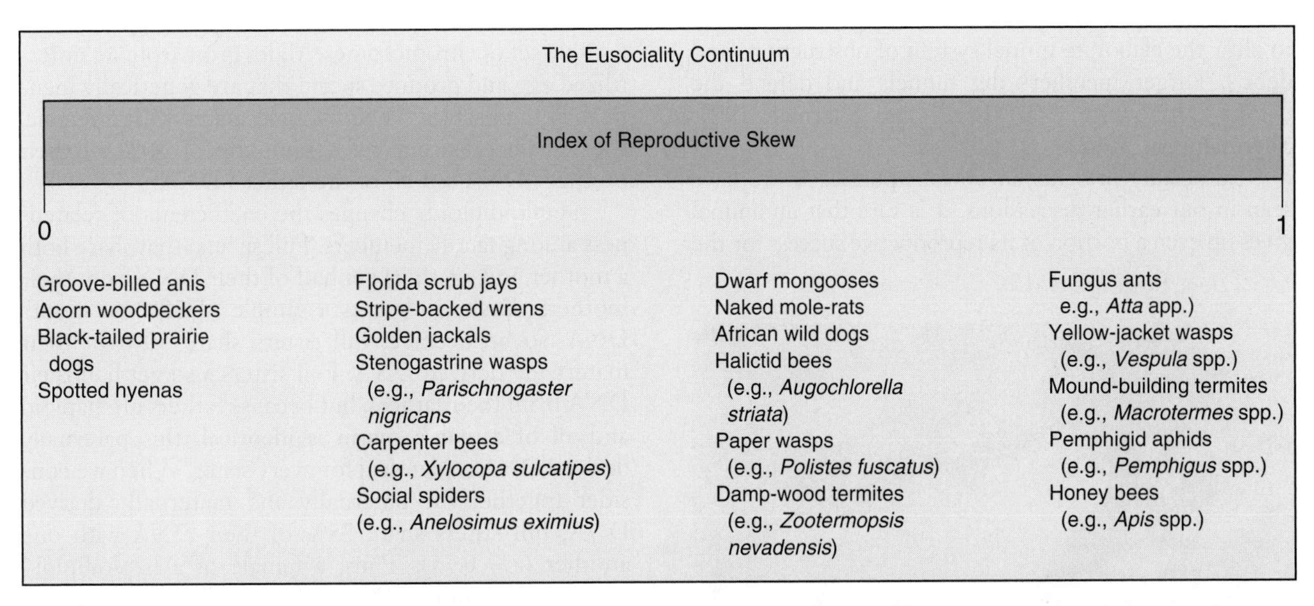

FIGURE 19.23 **The eusociality continuum blurs the distinction between cooperative breeders and eusocial species. The traditional definition of a eusocial society is one with reproductive division of labor, cooperation in caring for the young, and an overlap of adult generations. Cooperative breeding species share these characteristics with eusocial species, and so it has been suggested that they form a continuum of social systems. The primary differences among the social systems is the degree to which reproduction is shared among group members. When reproduction is restricted to a single individual, the reproductive skew is 1. A reproductive skew of 0 indicates that the lifetime reproductive success of all group members is equal. This diagram shows predicted locations of a variety of cooperatively breeding species along the eusociality continuum. (From Lacey and Sherman 1997.)**

aphids (e.g., Aoki 1972, 1979, 1982), an ambrosia beetle (Kent and Simpson 1992), tiny insects called thrips (Crespi 1992), spiders (Vollrath 1986), and snapping shrimp (Duffy 1996). In each case, the colonies meet the criteria described above but with variations across taxa. For example, some ant species are well known for their striking morphological castes. Whereas all worker honeybees look more or less the same, in some ant species individuals vary tremendously according to their job, such as intimidating soldier ants with their giant heads and strong jaws that deliver a memorable bite.

Eusociality is extremely rare in mammals, and in fact the only well-accepted examples are species in a group called the mole rats. Of mole rats, the most studied is the naked mole rat, *Heterocephalus glaber* (Jarvis 1981). This bizarre creature, furless and with long buck teeth, constructs massive burrows where it feeds on giant underground tubers. Naked mole rats fit the classical definition of eusociality originally applied to social insects. Breeding is restricted to a single female, the queen, even in groups with almost 300 members (Figure 19.24). Other adult females are smaller than the queen and neither ovulate nor breed. Only one to three males breed with the queen, although most adult males do produce sperm. Mole rat colonies contain overlapping generations of offspring, which are communally cared for, and there is division of labor among individuals within the colony. The duties assumed by the nonbreeding members seem to depend on their size and age. Smaller members generally gather food and transport nest material. As they grow, they begin to clear the elaborate tunnel system of obstructions and debris. Larger members dig tunnels and defend the colony (Honeycutt 1992; Lacey and Sherman 1991; Sherman et al. 1992).

Eusociality presents an obvious puzzle. As we have seen in our earlier discussions, it is rare that an animal gives up even a portion of its reproductive success for the good of another organism, but members of eusocial species may relinquish all chances of reproduction. It is no wonder that some researchers describe colonies such as these as "superorganisms"—much as cells and tissues function together for the survival and reproductive success of the body, members of a eusocial colony function efficiently together to ensure the survival and reproductive success of the colony.

FIGURE 19.24 **A queen naked mole rat, the only reproductive female of the colony, is resting on the workers that feed her and help care for the young.**

In the remainder of this section, we'll examine eusocial species from two perspectives: first, what favors the evolution of eusociality? Second, how do eusocial colonies function on a day-to-day basis?

The Evolutionary Origins of Eusociality

Haplodiploidy and the Hymenoptera Perhaps the most common feature of eusocial societies is that they are family groups (Strassman and Queller 2007). As we have seen, relatedness among individuals can favor the evolution of altruism, so the role of relatedness has long been a focus of research.

We'll begin by examining a special case of relatedness that has generated a great deal of ink. As we've already seen, eusociality is found only in a handful of species. What attracted the attention of researchers (especially before the recent expansion of the list of eusocial species) is that many of these species have an unusual genetic system called **haplodiploidy**. The females are diploid, with two sets of chromosomes, as are most familiar animals. The males, however, are haploid, with a single set of chromosomes. Males grow from an unfertilized egg and produce sperm that are genetically identical to themselves. When a male mates with a female, the female offspring get a sampling of 50% of their mother's DNA but all of the father's DNA.

Haploidiploidy changes the coefficients of relatedness among family members. Full sisters (that share both a mother and a father) get half of their DNA from their mothers. Each sister gets a sample of 50% of mom's DNA, so on average, full sisters share 50% of their maternally derived DNA. Full sisters also get half their DNA from their fathers, but because fathers are haploid and all of a male's sperm is identical, this paternally derived DNA is identical for every sister. When we consider together the paternally and maternally derived DNA, full sisters share 75% of their DNA with one another ($r = 0.75$). Thus, a female of a haplodiploid species shares more DNA with her full sister than with her mother or daughter ($r = 0.5$) (Figure 19.25).

Hamilton (1964), whose work we have already met earlier in this chapter (recall Hamilton's Rule), realized the potential implications of haplodiploidy for the evolution of eusociality. Because sisters are more related to one another than to their own daughters, they pass on more genes by rearing reproductively capable siblings than they would if they produced their own offspring.

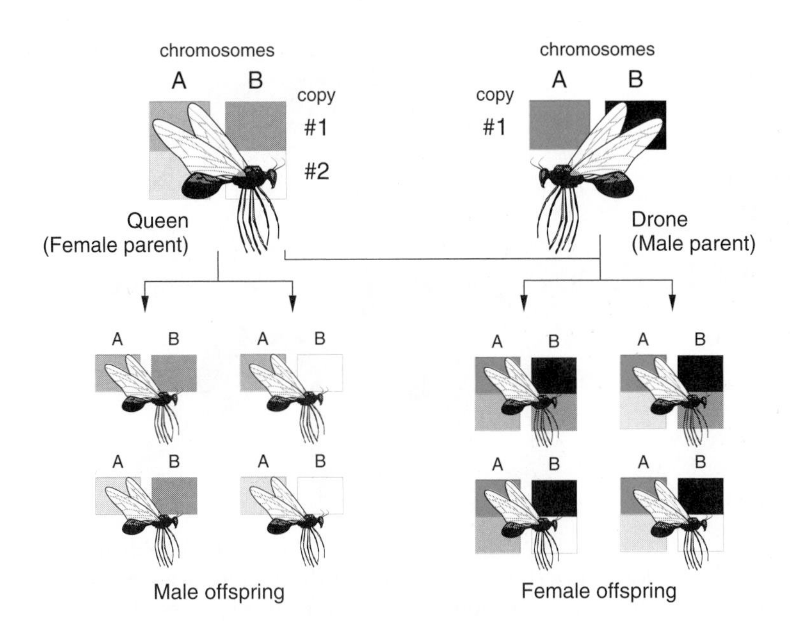

Degrees of relatedness in haplodiploid species

	Daughter	Son	Mother	Father	Sister	Brother
Female to her:	0.5	0.5	0.5	0.5	0.75	0.25
Male to his:	1	N/A	1	N/A	0.5	0.5

FIGURE 19.25 The genetic contributions of a male and female of a haplodiploid species to their offspring. Males are haploid and have only one set of chromosomes. They arise from unfertilized eggs of the female. All their sperm are identical. Females are diploid, and each of their haploid eggs contains half their genes. Females have half of their mother's genes and all of their father's genes. On average, two full sisters share all the genes they got from their father and half the genes they got from their mother.

This insight quickly captured the imagination of animal behaviorists. Could haplodiploidy be key to understanding eusociality?

Counterintuitively, some of the strongest evidence for the importance of relatedness in eusocial groups is not cooperation but conflict. The flip side to the argument that individuals should cooperate most with their closest relatives is that they should have the most conflict with those that are least related. Favoritism of close over distant relatives is demonstrated in a number of ways in eusocial colonies. For an example, let's look at whether individuals should favor rearing male or female larvae. As Figure 19.25 illustrates, a queen shares, on average, an equal percentage of alleles with her sons and her daughters. Thus, from a queen's perspective, she should invest equally in both sexes and should favor a 1:1 sex ratio. In contrast, as we have seen, worker bees (all females) share an average of only 25% of their genes with their brothers, but 75% with sisters. Thus, from a worker's point of view, the optimal sex ratio is 3:1 in favor of females.

So who gets to decide the sex ratio in a colony? Workers care for the brood and thus have an opportunity to manipulate the sex ratio. The investment in offspring is indeed nearly 3:1 in favor of females, in line with the idea that workers are manipulating the sex ratio for their own benefit (Trivers and Hare 1976). For comparison, consider the slave-making ants. Soldiers of slave-making species wage war against colonies of other ant species and drag back the pupae or larvae. When the captives grow into adults, they serve their colony as slaves, working diligently as nursemaids and performing other tasks. Because the nursemaids that care for the brood are unrelated to the colony, their fitness is not increased by altering the 1:1 sex ratio that is optimal for the queen. Although sex ratio data were available for only two slave-making species, Trivers and Hare (1976) found that the investment in females and males was about 1:1.

These results, as well as other evidence of the importance of relatedness in colony function, led researchers to hypothesize that there might be a tight relationship between haplodiploidy and eusociality. But how well has the haplodiploidy hypothesis stood the test of time? Unfortunately, not nearly as well as was initially hoped (reviewed in Linksvayer and Wade 2005, Wilson 2008). First, as noted earlier, we now know of a number of eusocial species, such as termites, that are not haplodiploid. Thus, haplodiploidy is not *necessary* for the evolution of eusociality. Second, there are many haplodiploid species that are not eusocial. Thus, haplodiploidy alone is not *sufficient* for the evolution of eusociality. Finally, even in haplodiploid species that are eusocial, the picture is complicated by the fact that females very often mate with more than one male (e.g., Laidlaw and Page 1984; Page and Metcalf 1982; Ross 1986; Strassmann et al. 1989). Multiple mating reduces the relatedness among sisters and also reduces the benefit of helping sisters compared to producing offspring. However, a recent comparative study shows that mating with a single female is ancestral in all eight eusocial lineages that were examined. The authors suggest that monogamy was critical to the origin of eusociality and that multiple mating evolves only after workers have lost the ability to breed (Hughes et al. 2008).

The extent to which haplodiploidy makes the evolution of eusociality more likely is still under debate. There may be an indirect route: haplodiploidy may make the evolution of maternal care more likely, which in turn is necessary for the evolution of eusociality (Linksvayer and Wade 2005). In any case, it is clear that we need to look beyond haplodiploidy for a complete explanation of eusociality.

Extended Parental Care and Long-Lasting Sibling Associations The existence of parental care may well be a prerequisite for the evolution of eusociality, as has long been noted (reviewed in Linksvayer and Wade 2005). Long-lasting associations between parents and offspring allow for kin selection and inclusive fitness, as we saw for other social species. In addition, long-lasting associations allow the opportunity for parents to manipulate their offspring by, for example, restricting food to make the offspring smaller, or by aggressively harassing offspring in order to make them cooperate (Crespi and Ragsdale 2000).

Sharing a Defendable Resource Most eusocial species rely on some sort of defendable resource, such as a nest, burrows, or tunnels bored in wood. For example, eusociality has arisen at least seven times in the 50,000–60,000 species of nest-building aculeate wasps, but never (to our knowledge) in the 70,000 species of wasps that do not build nests but instead lay their eggs on prey (Wilson and Hölldobler 2005). Our one example of a eusocial mammal, the mole rat, also vigorously defends its system of tunnels. There is strength in numbers: for example, Gamboa (1978) found that a paper wasp that founded a nest by herself more often lost it to a challenger than did wasps that founded nests with other females. Thus, defense of a resource can favor group living and cooperation. However, many animals share communal resources but do not have reproductive division of labor, so this alone cannot be the sole explanation for eusociality.

Day-to-Day Functioning of Eusocial Groups

Conflict Individuals within eusocial colonies are generally not genetically identical and thus may have conflicting goals. We described above how workers and queens in haplodiploid species might favor different sex ratios of the colony's offspring. Conflict occurs in other contexts as well: for example, some wasp colonies are initiated by multiple cofoundresses, often sisters. Given that the cofoundresses are not genetically identical, we predict—and generally see—conflict between them over who gets to produce more offspring. In fact, this conflict is often quite vicious and may even result in death (reviewed in Strassman and Queller 2007).

FIGURE 19.26 **In some species of ants, all individuals look very similar, whereas in others there are strikingly different morphological castes. In this species of *Acanthomyrmex*, there are minor workers (top) and major workers (bottom). The latter have giant heads, specialized for milling seeds and colony defense. (From Hölldobler and Wilson 1990.)**

Coordination Despite the existence of conflict, what is most impressive to the casual observer of eusocial groups is the level of cooperation. Although all eusocial colonies show coordinated behavior, perhaps the most impressive examples are found among the ants. In some species, different functions are carried out by morphologically distinct castes, as illustrated in Figure 19.26. Some castes have giant heads useful in milling grains and in defending the colony against intruders, whereas small castes specialize in foraging and nest maintenance.

In other species, the workers are all morphologically similar but divide up tasks according to the needs of the colony. If we saw such organization on a factory floor, we would assume that there was some sort of central command with a manager in charge of assigning tasks. But how can such complex behavior be carried out by individuals with brains smaller than pinheads? Is there some sort of a central organizational structure?

The short answer is no. Instead of a central command, ants make their decisions about which tasks to perform based on simple rules. These rules depend on the information they gain from their immediate environment and their interactions with others (reviewed in Bourke and Franks 1995). For example, red harvester ants (*Pogonomyrmex barbatus*) live in the deserts of the southwestern United States and Mexico. They rely on patchily distributed seeds that vary over time. After a heavy rain, many buried seeds are exposed, so it makes sense for foraging efforts to increase. When food is depleted, colony members should not bother going out to forage. Deborah Gordon and her colleagues (2008) tested whether ants inside a colony make decisions about whether to forage based on the success of other foragers. By the simple strategy of capturing and briefly detaining ants carrying seeds back to the nest, the researchers were able to slow down the apparent rate of foraging success. A decrease in the forager return rate of only three to five minutes led to an almost immediate reduction in the number of new foragers that left the colony. Instead, those ants could then turn their attention to other tasks. We see similar phenomena in other social insects: a bee's decision whether to collect nectar depends on the amount of nectar already stored, and a wasp's decision whether to collect more wood pulp for building depends on how long she had to wait since her last load was accepted at the nest (reviewed in Gordon 1996). The take-home message is that behaviors that appear to be very complex and highly coordinated can be generated by a handful of simple decision rules.

SUMMARY

A wide variety of animal species live in groups, from simple aggregations to highly coordinated societies. As with many other behaviors we discussed in this book, it's valuable to consider the costs and benefits that group living may provide. Benefits can include increased foraging success, decreased predation risk, conservation of heat and water by huddling together and conservation of energy by moving together. Costs include increased competition, risk of disease and parasites, and interference with reproduction. How costs and benefits are balanced vary according to the species, the ecological circumstances, and even across individuals within a group.

Altruism is the performance of a service that benefits a conspecific at a cost to the one that does the deed, all measured in terms of units of fitness. The occurrence of altruism is puzzling. If aiding a conspecific costs the altruist, the altruist should leave fewer offspring than do the beneficiaries of its services. As a result, the alleles for altruism would be expected to decrease in the population over generations. Hypotheses for the evolution of altruism can be classified into four overlapping classes.

Individual Selection The general thrust of these hypotheses is that when the interaction is examined closely enough, the altruist will be found to be gaining, rather than losing, by its actions. The benefit may not be immediate; sometimes the gain is in the individual's future reproductive potential. Another way of saying this is that an individual may receive direct benefits over the course of its life by being altruistic.

Kin Selection When the beneficiaries of the good deeds are genetically related to the altruist, the enhanced reproductive success they enjoy will also perpetuate the alleles that the altruist shares with them by virtue of their common descent. The altruist's relatives are more likely than nonrelatives to carry the alleles that lead to altruism. Another way of saying this is that an individual may raise its indirect fitness by helping relatives raise more offspring than they could without help. We make this more explicit with Hamilton's Rule. Hamilton's Rule states that an individual should help its relative raise offspring if $B/C > 1/r$, where B is the benefit to the recipient in terms of offspring produced because of the altruistic act, C is the cost to the actor in terms of offspring it does not produce because of the altruistic act, and r is the coefficient of relatedness between the recipient and the actor.

How might an animal identify its relatives? There are several possibilities. One way might be to use location as a cue: the individuals that share one's home are likely to be kin. Individuals might also be identified as kin because they are familiar. Animals that have never been encountered before may be recognized as kin through phenotype matching. In other words, an individual compares a stranger's traits to those of a known family member or even to itself. Finally, recognition may be genetically based. Perhaps there are alleles that in addition to labeling relatives with a noticeable characteristic cause the altruist to assist others that bear the label.

Reciprocal Altruism Altruism might also evolve in spite of the initial cost to the altruist if the service is repaid with interest. In other words, altruism will be favored if the final gain to the altruist exceeds its initial cost. However, for reciprocal altruism to work, individuals that fail to make restitution must be discriminated against. Because of this requirement, three factors make reciprocal altruism more likely: the benefit of the act to the recipient is greater than the cost to the actor; there should be a good chance that an opportunity for future repayment will arise; and the individuals involved must be able to recognize one another.

Manipulation Sometimes animals can coerce others into helping them by threatening even greater harm.

No single hypothesis applies to every example of altruistic behavior. In addition, these evolutionary mechanisms are not mutually exclusive; more than one may be responsible for a single example of altruism. To complicate matters even further, similar behaviors may evolve by different mechanisms in different species.

Members of some social species emit alarm calls to warn their neighbors of a predator's approach. Belding's ground squirrels emit two types of alarm calls—one in response to terrestrial predators and one in response to aerial predators. The calls seem to have been selected in different ways. Individual selection seems to be the best explanation for the evolution of alarm calls in response to aerial predators. However, kin selection seems to be the most likely mechanism for the evolution of ground squirrels' alarm calls in response to terrestrial predators. Belding's ground squirrels may be something of an exception: in a comparative study across all rodents, calls generally serve to communicate with predators rather than kin.

Many species cooperate in acquiring a mate. We reviewed four different examples. (1) In wild turkeys, males display in a group, but only one mates. Generally males display with their siblings, so they receive indirect benefits from helping. (2) The situation is more complicated in lions. Male lions form coalitions to challenge other prides. When members of a coalition are related, males can benefit from helping their brothers mate. Males also benefit directly from being in larger groups, even if they are composed of unrelated males. (3) Long-tailed manakins perform a carefully choreographed courtship dance in which displaying males fly around each other. Only one of the displaying males mates with the females that are attracted to the show. However, the subordinate male may become a dominant male if it outlives his partner. Males improve their dancing with practice, so a long apprenticeship may pay off. (4) Olive baboons form alliances to challenge another male in the troop. In return, they may receive help later, in an example of reciprocal altruism.

Another form of altruism is cooperative breeding. A helper is an individual that assists in the rearing of offspring that are not its own, usually by providing food or protecting or carrying the young. In most cases, helping appears to be beneficial, but it is a good idea to confirm this experimentally. For example, pairs on high-quality territories may attract additional helpers, and their increased reproductive output may be due to the help they have received or to the territory quality. An experiment is needed to tease these hypotheses apart.

Helping is generally costly to the helpers. We can ask about the conditions under which helping evolves by asking two questions. First, we can ask why an offspring might delay dispersal. As we saw in Chapter 11, dispersal may be dangerous or entail other costs. In addition, other options, such as empty habitat or available mates, may be unavailable. However, just because an offspring delays dispersal does not mean it will help. In a number of species, animals gain indirect fitness by helping to raise kin. In other species, help goes to nonrelatives, so direct benefits are important. These may include inheriting a territory or a mate.

Eusocial species present an extreme example of cooperative breeding. Eusocial species have reproductive division of labor (some animals have no offspring), cooperation in the care of young, and overlap of at least two generations that care for the colony. Eusociality was originally described in insects, but the list of eusocial species now includes representatives from other taxa, including mammals. Early researchers were intrigued by a correlation between eusociality and haplodiploidy (a genetic system in which males are haploid and females are diploid). This genetic system means that sisters have, on average, a higher level of relatedness to one another than to their own mothers or daughters. By Hamilton's Rule, a higher level of relatedness means that kin selection is more likely to evolve. However, haplodiploidy is neither necessary nor sufficient for eusociality to evolve. Other factors that favor the evolution of eusociality are extended parental care, long-lasting sibling associations, and sharing a central resource, such as a wasps' nest or the giant tubers fed upon by naked mole rats.

Recently, a great deal of work has focused on how eusocial groups function from day to day. It is common to see eusocial individuals favor colony-mates that are more closely related over those that are more distantly related. An impressive level of coordinated behavior among hundreds of individuals can arise from simple rules, and does not require any particular individual to have oversight over the colony's operations.

Glossary

Acoustic adaptation hypothesis The idea that the acoustic properties, such as pitch, of auditory signals are shaped by habitat structure. For example, songs in open habitat are predicted to have higher frequencies than those in heavily vegetated habitat.

Action potential A nerve impulse. An electrochemical signal conducted along an axon. A wave of depolarization caused by the inward flow of sodium ions followed by repolarization resulting from the outward flow of potassium ions.

Activational effects Effects of steroid hormones that typically occur in adulthood and tend to be transient, lasting only as long as the hormone is present at relatively high levels. For example, increases in circulating levels of sex steroids activate mating behavior. Activational effects may involve subtle changes in previously established connections rather than gross reorganization of neural pathways. Compare with organizational effects.

Adaptation (1) The process of natural selection in which evolutionary modification occurs in response to selection pressures. (2) The result of natural selection; that is, behavioral, morphological, developmental, or physiological changes that have been preserved because they have had a selective advantage. (3) The immediate physiological response of an organism to a change in conditions, such as an increase in temperature. In this book, we are generally concerned with one of the first two definitions.

Afferent neuron A sensory neuron. A nerve cell that carries information from the peripheral receptors toward the central nervous system.

Aggression A behavior apparently intended to inflict noxious effects or destruction on another organism.

Aggressive mimicry Phenomenon whereby a predator can get close to its prey because it imitates a signal that is not avoided by the prey or that even attracts it.

Agonistic behavior Behavior that encompasses all conflict between conspecifics, including threats, submissive behavior, chasing, and physical combat. It is not used to describe aggressive acts between species, such as predation.

Allee effect Phenomenon whereby having a few neighbors is beneficial, especially for gaining access to potential mates and defense against predators. According to this idea, individual fitness increases with number of conspecifics at low to moderate densities, and then declines from moderate to high densities. With regard to habitat selection, compare with ideal free distribution.

Allele Alternative form of a gene. One of two or more slightly different versions of a gene that code for different forms of the same trait.

Allogrooming Caring for the fur, skin, or feathers of a conspecific; it is distinguished from autogrooming (self-grooming). Allogrooming functions in hygiene and also builds and maintains social bonds.

Altricial Describes young that are virtually helpless and incapable of feeding on their own or following their parents for the first few weeks after birth or hatching. Compare with precocial.

Altruism One animal's apparent aid to another animal at its own expense.

Amplexus The reproductive embrace in which a male amphibian grasps a female with his front legs from a dorsal position.

Antihormones Drugs that can temporarily and reversibly suppress the actions of specific hormones.

Aposematism See warning coloration.

Apostatic selection A form of frequency-dependent selection that occurs when one morph in a polymorphic population is much more common than another morph. Predators may develop a search image for the more common, rather than the rare, morph, and take more of the common form relative to its frequency in the population. Predators need not hunt by search image to cause apostatic selection in prey; some predators may simply have an aversion to rare or unfamiliar prey.

Artificial selection The process by which humans change the frequency of traits within populations by intentional selective breeding. Dog breeds provide an example of artificial selection.

Associated reproductive pattern Pattern shown by some vertebrates in which there is a close temporal association between gonadal activity and mating. Gonadal growth and an increase in circulating levels of sex steroids activate mating behavior.

Associative learning The formation of some sort of mental connection between representations of two stimuli.

Audience effect The impact that passive onlookers have on the behavior and physiology of an individual performing a task.

Autotomy The ability to break off a body part when attacked.

Axon A long extension from the cell body of a neuron that carries an electrochemical message away from the cell body toward another neuron or effector (muscle or gland). The tips of the axon release a chemical called a neurotransmitter that can affect the activity of the receiving cell. Typically, there is one long axon on a neuron.

Batesian mimicry Phenomenon whereby a palatable species adopts the warning characteristics of a noxious or harmful species. The harmless species is called the mimic and the noxious one, the model. By resembling a noxious species, the mimic gains protection from predators.

Behavioral ecology A discipline that examines the evolutionary and ecological bases of the behavior of animals. It grew out of the ethological approach to studying animal behavior. Behavioral ecologists examine the costs and benefits of behaviors, such as foraging, mating, group living, and communicating.

Breeding dispersal The movement an animal makes between two successive breeding areas or social groups. It is sometimes called postbreeding dispersal.

Camouflage A collective term used to describe several antipredator devices, such as disruptive coloration, countershading, and transparency, whose general message is "I am not here."

Candidate gene A gene that is hypothesized to be involved in a particular behavior based on a search of the literature for genes known to be involved in producing a similar behavior in another organism or by comparing the sequence of the gene to the sequences of genes in other organisms using genome-sequencing data.

Carnivory Feeding on other animals.

Central pattern generator A neuron or network of neurons that is capable of generating a rhythmic pattern of activity in motor neurons, even when all sensory input has been removed from the system.

Chain of reactions A behavioral repertoire built from a sequence of fixed action patterns. Each fixed action pattern brings the animal into a situation that triggers the next fixed action pattern in the sequence.

Challenge hypothesis Hypothesis that levels of hormones involved in dominance and aggression rise during times of social challenge or instability, such as during the initial period of territory establishment.

Chase-away model An explanation for the evolution of extreme traits in males that exploit a sensory bias of females. Females receive no benefits from being selective in their choice of mate. The model is based on sexual conflict.

Circadian rhythm Biological rhythms that are about a day in length (*circa*, about; *diem*, a day). Used formally, it refers to a solar-day rhythm that persists in constant conditions of light and temperature with a period length that is slightly longer or shorter than 24 hours. Its period length is relatively independent of temperature, and its phase can be reset with pulses of light.

Circalunadian rhythm A basic lunar-day rhythm that persists in constant conditions with a period that is slightly longer or shorter than 24.8 hours. A circalunadian rhythm is usually bimodal.

Circamonthly rhythm A synodic monthly rhythm that persists in constant conditions with a period length that differs slightly from 29.5 days.

Circannual rhythm An annual rhythm that persists in a constant light-dark cycle with a period length that differs slightly from 365 days.

Circatidal rhythm A rhythm that persists in constant conditions with a period length that differs slightly from 12.4 hours. It is the same as a circalunadian rhythm.

Classical conditioning In the study of learning, the pairing of an innocuous stimulus with a rewarding stimulus, so that the animal learns to respond to the innocuous stimulus. This procedure was made famous by Pavlov and his study of dogs.

Coefficient of relatedness (*r*) The probability that a particular pair of animals share the same allele through common descent.

Coloration matching the visual background An antipredator device in which a prey animal's coloration resembles its background and thereby reduces risk of detection by visually hunting predators. It is sometimes called cryptic coloration.

Communication The process of transferring information from sender to receiver to the benefit of the sender, on average.

Comparative method Using the evolutionary relationships among species in order to gain insights into adaptation. For example, unrelated species that share similar environments may evolve similar traits, while related species that live in different environments may evolve different traits.

Comparative psychology A branch of psychology that studies the behavior of animals, with a focus on physiology, learning, and development. The experiments are quantitative and often laboratory-oriented.

Compass orientation The ability to find one's way without using landmarks. The animal keeps a certain angle toward an external reference system used as a compass. External reference systems include the sun, the stars, and the earth's magnetic field.

Conditional strategy In agonistic encounters, a strategy that is adjusted according to the conditions of a particular fight.

Conditioned response In the study of classical conditioning, the learned behavioral response elicited by the conditioned stimulus.

Conditioned stimulus In classical conditioning, an innocuous stimulus, such as a tone or a light, is paired with an unconditioned stimulus that is rewarding to an animal, such as food. Animals that have learned the pairing respond to the tone or light, which is then called the conditioned stimulus.

Confusion effect Situation in which predators are less successful in attacking grouped prey because they are unable to single out and attack an individual prey.

Conspecific attraction hypothesis Hypothesis proposing that individuals choose habitat patches based on the presence of established residents of their own species.

Constant reproductive pattern Pattern shown by some vertebrates inhabiting harsh environments where suitable breeding conditions occur suddenly and unpredictably. While waiting for suitable circumstances in which to breed, these species maintain large gonads, mature gametes, and high circulating levels of sex steroids for prolonged periods of time. Sometimes called the opportunistic reproductive pattern.

Constraint In mathematical modeling, a limitation that we place on a behavior. For example, we may create a model in which the amount of food an animal can eat is constrained by gut capacity.

Continuous reinforcement schedule In the study of learning, a reinforcement schedule in which each occurrence of a desired behavior is rewarded. This reinforcement schedule is effective when initially training an animal to perform a behavior.

Cooperative hunting Coordinated foraging efforts among predators.

Copulatory plug A plug made of thick, viscous secretions deposited by males in the reproductive tract of females at mating. Copulatory plugs occur in many vertebrates, including snakes, lizards, marsupials, rodents, bats, and primates. Several functions have been suggested for copulatory plugs, among them: (1) "enforcing chastity" in which the plug acts as a barrier to subsequent inseminations; (2) ensuring the retention of sperm in the female reproductive tract; (3) aiding the transport of sperm within the female reproductive tract; (4) providing for the gradual release of sperm as the plug disintegrates; and (5) providing a means by which a male can scent-mark a female's body and convey information regarding his identity and dominance status.

Core area The area of a home range in which the activities of an animal are concentrated.

Countershading Pattern of coloration characterized by dark backs and light bellies, which may achieve camouflage through either self-shadow concealment (obscuring the ventral shadow) or background matching.

Counting The process of assigning a tag such as "1, 2, 3" to individual numbers.

Cross-fostering A technique for detecting maternal influences by transferring shortly after birth the offspring from one strain (or species) to the mother of another strain (or species).

Cryptic female choice Selection by females who have mated with several males of the sperm that will fertilize their eggs. This ability is described as cryptic because it is a hidden, internal decision made after copulation.

Cue Information transmitted from one animal to another, but not necessarily to the benefit of the sender, such as the rustling noise made by a mouse and detected by an owl.

Currency In mathematical modeling, the common unit in which costs and benefits of different strategies are measured. Examples are the number of offspring produced or the number of calories gained.

Dead reckoning See path integration.

Dendrite A process of a neuron specialized to pick up messages and transmit them toward the cell body. There are typically many short branching dendrites on a neuron.

Deoxyribonucleic acid (DNA) The molecular basis of genetic inheritance. A category of nucleic acids that usually consists of a double helix of two nucleotide strands. The sequence of nucleotides carries the instructions for assembling proteins.

Depolarization A change in the difference in electrical charge across a membrane that moves it from a negative value toward 0 mV. During a nerve impulse (action potential), depolarization is caused by the inward flow of positively charged sodium ions.

Detouring The ability to identify an alternative route to a reward when the direct route is blocked.

Developmental homeostasis Ability of developmental processes to buffer themselves against potentially harmful influences to produce functional adults.

Dilution effect Phenomenon whereby individuals living in groups are safer from predators because each has a smaller chance of becoming the next victim.

Direct fitness Fitness gained by an individual through its own reproduction, generally measured as the number of surviving offspring it produces. Compare with indirect fitness.

Direct parental care Patterns of behavior performed by parents that have an immediate physical impact on offspring and their survival. In mammals, for example, direct parental care includes behaviors such as nursing (and feeding), grooming, transporting, and huddling with young.

Displacement activities Irrelevant actions performed in situations in which an animal has conflicting motivations.

Display A stereotyped sequence of behaviors that has a signaling function.

Disruptive coloration Coloration designed to prevent perception of a prey animal's form.

Dissociated reproductive pattern Pattern shown by some vertebrates in which mating behavior is completely uncoupled from gamete maturation and secretion of sex steroids. Gonadal activity occurs only after all breeding activity for the current season has ceased, and gametes are thus produced and stored for the next breeding season.

Dominance The ability of an animal to assert itself to others in acquiring access to a resource such as food, a mate, or a display or nesting site.

Dominant allele The allele that is fully expressed in a heterozygous individual.

Dove In game theory, the strategy of immediately fleeing if confronted by an opponent that is playing the hawk strategy, and displaying if confronted by an opponent that is also playing a dove strategy. See also hawk.

Ecological trap A low-quality habitat that animals prefer over a high-quality habitat.

Efferent neuron A motor neuron specialized to carry information away from the nervous system to an effector (muscle or a gland).

Emancipation Over evolutionary time, a behavior that has lost its original function and now serves only a communicative function.

Entrainment The process by which an environmental rhythm, such as a light-dark cycle, regulates the period and phase of a biological rhythm.

Environmental enrichment Efforts made to enrich the lives of captive animals. Such efforts usually involve social housing and the provision of larger cages, a more complex and variable physical environment with nesting material, foraging devices, toys, hiding places, and the opportunity for voluntary exercise.

Epigenetics A stable change in a gene that does not involve changes in nucleotide sequence. Genes become active or are shut down due to unwinding or winding of DNA.

Ethology The study of animal behavior that focuses largely on the evolution and function of behavior. Many early ethological studies were comparative and took place in the field.

Evolution A change in the frequencies of alleles in a population of organisms over generations.

Evolutionary stable strategy (ESS) A behavioral strategy that when adopted by all members of the population cannot be replaced by a different strategy. A pure ESS consists of a single strategy, while a mixed ESS consists of several strategies in a stable equilibrium.

Excitatory postsynaptic potential (EPSP) A temporary electrical change (depolarization) in the membrane of the postsynaptic neuron that is caused by the binding of an excitatory neurotransmitter. An EPSP makes it more likely that the postsynaptic neuron will generate an action potential.

Extinction In classical conditioning, presenting a conditioned, or learned, stimulus without the reinforcer, leading to a loss of the learned response over time.

Female defense polygyny Form of polygyny in which a male defends a harem of females. This type of polygyny occurs when females live in groups that a male can easily defend.

Filial imprinting Process by which a young bird learns, through exposure to its mother, her particular characteristics and then preferentially follows her. Filial imprinting may function to allow young birds to recognize close relatives and thereby distinguish their parents from other adults that might attack them.

Fitness The reproductive success of an allele or an individual compared to other alleles or individuals in the same population. See also direct fitness, indirect fitness, and inclusive fitness.

Fixed action pattern An innate stereotyped motor response that is initiated by a stimulus but that can continue to completion without ongoing stimulation.

Fixed ratio schedule In the study of learning, a reinforcement schedule in which the animal must respond a set number of times before a reward is given. This schedule generally results in very high response rates.

Foraging Finding, processing, and eating food.

Free-running period The period length of a biological clock when it is not being influenced by external time cues.

Frequency-dependent selection A type of natural selection in which an allele has a greater selective advantage when it is rare in a population and a smaller selective advantage when it is common. For example, one type of frequency-dependent selection occurs when females prefer to mate with males of a rare phenotype. Frequency-dependent selection results in the fluctuation of allele frequencies over time.

Full song In songbirds, the final adult song that in many species will remain virtually unchanged for the rest of the male's life. Sometimes called crystallized song.

Game theory A type of mathematical model in which the optimal strategy of one animal depends on the behavior of its opponent.

Gene flow The movement of genes from one population to another. Gene flow causes populations to become more similar to one another.

Genetic drift Evolutionary change in a population due to chance events. Genetic drift is more likely to be an important evolutionary force in small populations than in large populations.

Genetic monogamy An exclusive mating relationship between one male and one female. Compare with social monogamy.

Genomics The study of all the genes in an organism.

Genotype The genetic makeup of an individual. It refers to the precise alleles present.

Goal-directed emulation Behavior whereby an observer seems to learn from observation what goal is to be achieved but does not precisely copy what the demonstrator does. Compare with imitation.

Good genes models Models that assume that a particular trait in males indicates viability and that both the trait and viability have a genetic basis. If a female preference (also genetically based) for the male trait should arise, then these females mate with males carrying genes for the trait and enhanced viability. In this way, genes for the male trait, high viability, and the female preference become associated.

Habitat selection Process by which animals that disperse from their natal site or breeding site eventually select a new location in which to settle. Habitat selection has three phases: (1) search (animal searches for a new habitat); (2) settlement (animal arrives in a new habitat and begins to establish a home range or territory); and (3) residency (animal lives in the new habitat).

Habituation In learning, the waning of a behavioral response to a stimulus because during repeated presentations of that stimulus it was shown to be harmless.

Hamilton's rule A rule designed to determine whether an animal should behave altruistically toward another animal. It includes three variables: the coefficient of relatedness, the benefit to the receiver in terms of the number of additional offspring it produces because of the altruistic act, and the cost to the altruist in terms of the number of offspring it does not produce because of the altruistic act.

Handicap principle An example of a good genes model for mate choice. According to this principle, females prefer a male with a trait that reduces his chances of survival but announces his superior genetic quality precisely because he has managed to survive despite his "handicap." In short, male secondary sexual characteristics act as honest signals, indicating high fitness, and females choose males with the greatest handicaps because their superior genes may help produce viable offspring.

Handling time The time required to process a food item.

Hawk In game theory, the strategy of immediately attacking an opponent. See also dove.

Herbivory Feeding on plant material.

Heritability The portion of the variability in a specific trait in a population that is due to genetic factors; the ratio of the phenotypic variance due to genetic factors to total phenotypic variance.

Heterospecific attraction hypothesis Hypothesis proposing that individuals choose habitat patches based on the presence of established residents of another species.

Heterozygous The condition of having two different alleles for a particular gene.

Home range The area in which an animal carries out its normal activities.

Homing See true navigation.

Homozygous The condition of having two identical alleles for a particular gene.

Honest signals Signals that accurately convey information about the sender. Not all signals are honest, and honest signals are likely to evolve only under particular conditions.

Hot spots Areas in which males gather because they are most likely to encounter receptive females.

Hotshots Males that are extremely successful at attracting females. Less successful males may increase their chances of mating by staying near these highly successful males.

Hypothesis A logical, testable explanation for a specific set of observations that serves as the basis for experimentation.

Ideal free distribution A method to describe how animals distribute themselves in space if they know the value of each habitat and are free to go where they choose.

Imitation An observer's exact copying of what a demonstrator does. Compare with goal-directed emulation.

Inbred lines Domestic animals that have very little genetic variability. They are created by mating close family members with one another.

Inclusive fitness The sum of both direct fitness (fitness gained through offspring) and indirect fitness (fitness gained by helping relatives raise additional offspring they would not be able to raise without help).

Indirect fitness Fitness gained by helping relatives raise additional offspring that they would not have been able to raise on their own. Compare with direct fitness.

Indirect parental care Patterns of behavior performed by parents that may not involve direct physical contact with offspring, but still affect offspring survival. In mammals, for example, indirect forms of parental care include acquiring and defending critical resources, building and maintaining nests or dens, defending offspring against predators or infanticidal conspecifics, and caring for pregnant or lactating females.

Infanticide The killing of a conspecific infant to acquire copulations or resources. Can also occur in response to severe disturbance of the physical or social environment.

Information center Communal roosts or colonies from which animals can follow successful conspecifics to food sites.

Information sharing Improving foraging by paying attention to the behavior of group members.

Infrasound Sounds whose frequencies are below those audible to humans, which means frequencies lower than about 20 Hz. Some animals, such as elephants, use infrasound to communicate over long distances. Other animals may use infrasound during orientation.

Inhibitory postsynaptic potential (IPSP) A temporary electrical change (hyperpolarization) in the membrane of the postsynaptic neuron that is caused by the binding of an inhibitory neurotransmitter. An IPSP makes it less likely that the postsynaptic neuron will generate an action potential.

Insight An example of animal cognition in which understanding seems to occur suddenly and without practice.

Intention movements A behavior, such as spreading the wings for flight, that signals an animal's behavior to an observer. Intention movements are thought to be a route by which some signals evolve.

Intermediate-term memory Memory of new information that may last a day or so. Habituation and sensitization are examples of intermediate-term memory formation.

Interneuron An association neuron. Neurons located within the central nervous system between sensory and motor neurons that integrate information.

Intersexual selection Mechanism of sexual selection whereby members of the sex in demand (usually females) choose mates with certain preferred characteristics. Thus, males compete to attract females through the elaboration of structures or behavior patterns.

Interspecific brood parasites In birds, species that lay their eggs in the nests of other species. Interspecific brood parasites never build nests in which to lay eggs and raise their own young, so they are described as *obligate* brood parasites.

Intrasexual selection Mechanism of sexual selection whereby members of one sex (usually males) compete with one another for access to the other sex. Intense fighting and competition for mates can lead to selection for increased size and elaborate weapons.

Intraspecific brood parasites In birds, species that lay their eggs in the nests of conspecifics. In some cases of intraspecific brood parasitism, the brood parasite occasionally lays eggs in the nests of conspecifics, while still laying eggs in her own nest. In other instances, the brood parasite lays eggs in the nests of conspecifics and does not maintain a nest of her own.

Intrauterine position effects In rodents, effects on morphology, physiology, and behavior caused by exposure to hormones secreted by contiguous littermates during gestation.

Ion An atom or group of atoms that carries an electric charge resulting from the loss or gain of electrons.

Ion channel A protein-lined pore or channel in a plasma membrane through which one type or a few types of ions can pass. Ion channels in nerve cells are important in the generation and propagation of nerve impulses.

Kin selection A type of natural selection in which animals help their relatives reproduce, thereby helping pass the alleles that they share to the next generation. See also coefficient of relatedness and direct, indirect, and inclusive fitness.

Landmark An easily recognizable cue along a route that can be quickly stored in memory to guide a later journey.

Language A term that has four elements: first, words or signs must be used as true symbols that can stand for, or take the place of, a real object, event, person, action, or relationship. Second, symbols should permit reference to objects or events that are not present. Third, there should be some elements of grammar, or rules that determine the relationship between words. Fourth, words or signs should be combined to form novel phrases or sentences that are understandable to others.

Latent learning Learning that occurs without any obvious immediate reward. For instance, an animal can learn important characteristics of its environment during unrewarded explorations and then use this information later.

Learning A change in the capacity for behavior as a result of experience, excluding the effects of fatigue, sensory adaptation, or maturation of the nervous system.

Lek polygyny Form of polygyny in which males defend "symbolic" territories that are often located at traditional display sites called leks. Males of lek species do not provide parental care and defend only their small territory on the lek, not groups of females that happen to be living together nor resources associated with specific areas. Females visit these display arenas, select a mate, copulate, and leave.

Local enhancement A type of social learning in which an animal is attracted to a particular location because a conspecific is there.

Long-term depression (LTD) A lasting decrease in responsiveness of postsynaptic neurons after sensory neurons have received a slow train of stimuli. LTD is a mechanism that weakens the effectiveness of a synapse, decreasing the magnitude of a response by the postsynaptic cell. It may play a role in memory formation.

Long-term memory Memory that lasts weeks, months, or years. Changes in synapses underlie long-term memory.

Long-term potentiation (LTP) A stable and long-lasting responsiveness to an action potential by the receiving neuron caused by rapidly repeated strong stimulation. LTP strengthens the connections between the adjacent neurons. It is the molecular mechanism that underlies the acquisition and storage of memories.

Lordosis The copulatory posture that some female mammals assume when ready to mate. The posture typically involves ventral curvature of the vertebral column.

Macroevolution Large-scale evolutionary changes in lineages, generally observed over geological time. Compare with microevolution.

Major histocompatibility complex (MHC) A large chromosomal region that varies tremendously among individuals and is important in the immune responses that protect against disease-causing organisms. Some animals may identify kin and choose mates based on MHC genes.

Marginal value theorem A model that predicts when a foraging animal should leave a patch of food. It is generally depicted in a graph.

Mate guarding Strategies employed by males to increase the probability that their sperm, and not the sperm of a competitor, will fertilize the eggs of a particular female. Mate guarding can occur before or after copulation or both.

Meta-analysis A statistical approach in which one collects all the studies that test a particular hypothesis and then combine the results in order to look for significance in the overall pattern.

Microarray analysis An analysis that reveals which of hundreds or thousands of genes are active at any moment. Thousands of genes are stamped on a solid surface, and molecular tags are used to identify the messenger RNA (mRNA) produced by each gene. The more active a gene, the more mRNA is produced. Genes that are active only during a particular behavior may play a role in producing that behavior.

Microevolution Minor evolutionary changes within a species. Compare with macroevolution.

Migration Movement away from the home range that does not stop upon encountering the first suitable location. Migrating animals continue to move until they eventually become responsive to the presence of resources, such as nest sites and food, and then they stop. Within a particular species, migratory movements occur over greater distances than dispersal movements.

Mobbing An antipredator strategy whereby prey approach, gather around, and harass their predators. Mobbing is usually initiated by a single individual, and then conspecifics, or members of another species, join. The possible functions of mobbing include, but are not limited to, (1) confusing the predator; (2) discouraging the predator either through harassment or through the announcement that it has been spotted early in its hunting sequence; (3) alerting others, particularly relatives, of the danger; and (4) providing an opportunity for others, again particularly relatives, to learn to recognize and fear the object that is being mobbed.

Model organism A species chosen to study particular biological principles with the expectation that the information learned from the model organism can be applied to other species.

Monogamy Mating system in which a male and female have only a single mating partner per breeding season. Further classified as genetic or social monogamy.

Motor neuron A neuron specialized to carry information away from the nervous system to an effector (muscle or gland).

Müllerian mimicry Antipredator strategy in which two warningly colored species look alike. Two noxious species may benefit from a shared pattern because predators consume fewer of each species in the process of learning to avoid all animals of that general appearance.

Multimodal communication Communication that contains signals from two or more sensory modalities (e.g., vision, audition, and touch). Signaling in different channels can occur either simultaneously or sequentially and the messages can be redundant or nonredundant.

Mutation A change in the DNA sequence of an organism, such as an addition or a deletion of nucleotides.

Myelin sheath An insulating layer around axons of nerve cells that carry action potentials (nerve impulses) over relatively long distances. It is composed of multiple wrappings of the plasma membrane of certain glial cells. The myelin sheath greatly increases the speed at which impulses travel. The cells that form the myelin sheath are separated from one another by short regions of exposed axon. The nerve impulse "jumps" from one exposed region of axon to the next.

Natal dispersal The movement an animal makes from its natal area or social group to the area or social group where it first breeds.

Natal habitat preference induction (NHPI) Phenomenon whereby an animal's experience in its natal habitat induces a preference for a postdispersal habitat with similar qualities.

Natal philopatry Occurs when offspring remain at their natal area and share the home range or territory with their parents.

Natural selection Process by which organisms with favorable characteristics are more likely to survive and reproduce than those with unfavorable characteristics.

Negative-assortative mating A mating pattern in which animals preferentially mate with those with phenotypes different from their own.

Neurite A small-diameter process extending from a neuron's cell body. A neurite can be either an axon or a dendrite.

Neurogenesis The creation of new neurons.

Neuromodulators Chemicals that cause voltage changes that occur over seconds, minutes, hours, and perhaps even days. Neuromodulators alter neuronal activity slowly, by biochemical means. Different neuromodulators can act on the same neural elements to produce different behaviors.

Neurons Nerve cells involved in intercellular communication.

Neurosteroids Steroid hormones produced by the nervous system that act in the nervous system on either nearby cells or the same cell that produced the hormone. This mode of action contrasts with peripheral steroids produced by the gonads or adrenal glands that travel in the bloodstream and act on target cells some distance from their gland of origin.

Neurotransmitter A chemical released from the axon tip of a neuron that affects the activity of another cell (usually a nerve, muscle, or gland cell) by altering the electrical potential difference across the membrane of the receiving cell.

Nuptial gifts Food or other valuable substances offered by males to females during courtship or mating.

Observational conditioning A type of classical conditioning that takes place when an animal observes another animal rather than going through a conditioning experience directly.

Omnivory Feeding on both plants and animals.

Operant conditioning A form of associative learning in which the outcome (positive or negative) depends on the animal's behavior, such as an animal that learns to run a maze in order to receive a reward. Also called trial-and-error learning and instrumental conditioning.

Operational sex ratio The ratio of potentially mating males to fertilizable females.

Optimality modeling The use of mathematical procedures to weigh the costs and benefits of different strategies and determine which strategy provides the maximum award under a specified set of assumptions.

Organizational effects Effects of steroid hormones that occur early in life, usually just before or after birth or hatching, and tend to be permanent. This permanence implies structural changes in the brain or nonneural systems. For example, steroid hormones around hatching organize the brain of a male songbird to make him capable of singing in adulthood. Compare with activational effects.

Own-species bias In studies of bird song, the preference of young male songbirds to learn their songs from members of their own species.

Parental investment Any investment by parents in an offspring that increases the survival of that offspring while decreasing the ability of the parent to invest in other offspring.

Path integration Process by which an animal integrates information on the sequence of direction and distance traveled during each leg of the outward journey. Then, knowing its location relative to home, the animal can head directly there, using its compass(es). A compass may also be used to determine the direction traveled on each leg of the outward journey, or the direction may be estimated from the twists and turns taken, sounds, smells, or even the earth's magnetic field. Information from the outward journey is used to calculate the homeward direction (vector). Also called dead reckoning.

Payoff matrix In game theory, a table used to organize the values of each strategy when it is played against each of the other strategies.

Peptide hormones Water-soluble hormones that cannot pass through the plasma membrane of target cells on their own, so they influence cells indirectly through second messenger systems. These hormones activate existing enzymes. Examples of peptide hormones are luteinizing hormone (LH) and follicle-stimulating hormone (FSH) produced by the anterior pituitary gland.

Period The time interval of one complete cycle of a biological rhythm.

Peripheral-control hypothesis Hypothesis proposing that rhythmic behavior occurs because the first movement stimulates sensory receptors, which in turn trigger the next movement in the sequence. The second movement stimulates other sensory receptors that trigger the first component. Thus, sensory feedback is necessary for this hypothesis.

Phenotype The observable physical and physiological traits of an individual. Phenotype results from the inherited alleles and their interactions with the environment.

Phenotype matching A type of kin recognition accomplished by assessing the degree of similarity between one's phenotype and that of another individual, or by learning the phenotypes of relatives.

Pheromones Chemicals that convey information to other members of the same species. Some pheromones, called releaser pheromones, have an immediate effect on the recipient's behavior. Other pheromones, known as primer pheromones, exert their effect more slowly, by altering the physiology and subsequent behavior of the recipient.

Piloting The ability to find one's way using landmarks.

Play Behavior that borrows pieces of other behavior patterns, usually incomplete sequences and often in an exaggerated form. It consists of elements drawn from other, functionally different behavior patterns juxtaposed in new sequences. Play includes social play, locomotor play, and object play.

Players Individuals whose strategies are modeled in game theory.

Polyandry Mating system in which a female has more than one mate during a breeding season.

Polygyny Mating system in which one male mates with more than one female during a breeding season.

Polygyny threshold hypothesis Hypothesis proposing that polygynous matings will be advantageous to females when the benefits achieved by mating with a high-quality male and gaining access to his resources more than compensate for costs. Thus, a female may reproduce more successfully as a secondary mate on a high-quality territory than as a monogamous mate on a low-quality territory. The term *polygyny threshold* describes the difference in a territory's quality needed to make secondary status a better reproductive option for females than primary status.

Polymorphism Phenomenon whereby a species occurs in several different shapes and/or color forms. This may prevent predators from forming search images.

Potential reproductive rate The maximum number of independent offspring that each parent can produce per unit of time.

Precocial Describes young that are capable of moving about and feeding on their own just a short time after birth or hatching. Compare with altricial.

Quantitative trait A trait that is influenced by several genes acting together, rather than by any one gene acting alone.

Receiver-bias mechanism Hypotheses about the evolution of communication that focus on the characteristics of the receiver of signals rather than those of the sender.

Recessive allele The allele whose effects are usually masked in the heterozygous condition.

Reciprocal altruism One animal helps another, which in turn helps the first animal at a later time.

Recognition allele An allele or group of alleles that enable an animal to recognize others with the same genotype.

Recombination The production of a new combination of genes in offspring that differ from either of the parental genotypes, generally through the process of crossing over during meiosis.

Regulatory gene A gene that influences the activity of other genes. Regulatory genes often produce transcription factors.

Reinforcement schedule In the study of learning, the frequency and timing with which an experimenter rewards a particular behavior.

Reinforcer A stimulus that changes the probability that an animal will repeat its behavior. Reinforcers can be positive or negative.

Relative plasticity hypothesis The hypothesis relating alternative phenotypes to the different effects of steroid hormones. It states that developmentally fixed alternative phenotypes (i.e., individuals are one phenotype or the other and remain so for life) rely on organizational effects of steroid hormones, whereas developmentally plastic alternative phenotypes (i.e., individuals switch between phenotypes in response to environmental conditions) rely on activational effects of steroid hormones.

Releaser A sign stimulus that is emitted by a member of the same species.

Repolarization The return of the membrane potential to approximately its resting value. Repolarization of the nerve cell membrane during an action potential occurs because of the outflow of potassium ions.

Resource defense polygyny Form of polygyny in which males defend resources essential to female reproduction (e.g., nest sites or food) rather than defending females themselves.

Resource-holding potential The ability of an animal to defend a resource from competitors.

Resting potential The separation of charge across the plasma membrane of a neuron when the neuron is not transmitting an action potential. It is primarily caused by the unequal distribution of sodium ions, potassium ions, and large negatively charged proteins on either side of the plasma membrane. The resting potential of a neuron is about –70 mV.

Ribonucleic acid (RNA) A single-stranded nucleic acid that plays several roles in protein synthesis.

Risk sensitivity When an animal's decisions depend on the variance, or riskiness, of an option, it is said to be risk sensitive. An example of a risky option is a foraging patch that has a fluctuating amount of food. A risk-prone animal chooses a risky option because there is a chance it will produce a high payoff. A risk-averse animal chooses a safe option that has a low variance.

Ritualization Phenomenon whereby over evolutionary time, incipient signals become more stereotyped and unambiguous.

Round dance A circling dance performed by honeybees that indicates a food source is nearby.

Rule of thumb A term indicating that instead of following precisely the behavioral strategy that would be optimal, an animal may follow a rule of thumb that yields adequate results under most circumstances.

Runaway selection A form of sexual selection in which a positive feedback loop is created when genes for mate choice in the female become genetically linked to genes for the preferred traits in males. In this way, runaway selection can produce increasingly exaggerated male traits and a stronger female preference for them.

Satellite male A male that remains silent and associates closely with a signaling male, ready to intercept females that are attracted to the other male's signals or resources.

Scent-marking The act of strategically placing a chemical mark in the environment.

Schreckstoff An alarm chemical produced by some species of fish when physically attacked.

Scientific method A procedure underlying most scientific investigations that involves observation, formulating a hypothesis, making predictions, experimenting to test the predictions, and drawing conclusions. Experimentation usually includes a control group and an experimental group that differ in one or very few factors (variables). New hypotheses may be generated from the results of experimentation.

Search image The heightened ability to detect a particular target with experience.

Search time The time it takes a forager to find food.

Seismic signals Signals that are encoded in the pattern of vibrations of the environmental substrate. These signals can be produced through percussion on the environmental substrate, such as when rodents drum their feet on the ground or insects tap the surface of water to create ripples. Seismic signals can also be generated when an airborne vocalization couples with the ground, such as when elephants produce infrasonic calls.

Selfish herd An antipredator mechanism that considers the spatial arrangement of individuals within a group. In most groups, centrally located animals are safer than those at the edges. By obtaining a central position, animals can decrease their chances of being attacked and increase the probability that one of their more peripheral colleagues will be eaten instead. This mechanism emphasizes that although a given group appears to consist of members that coordinate their escape efforts, it is actually composed of selfish individuals, each trying to position as many others as possible between itself and the predator.

Self-shadow concealment The mechanism by which countershading can achieve camouflage by obscuring the ventral shadow. An alternative mechanism is background matching.

Sensitive period A time during development when certain experiences have a greater influence on the characteristics of an individual than at other stages.

Sensitization A simple type of learning that involves an enhanced response to repeated stimuli.

Sensorimotor stage In songbirds, phase when singing learned song actually begins. Birds retrieve a learned song from memory and rehearse it, constantly matching their sounds to those they memorized months earlier during the sensory phase.

Sensory bias model A model for mate choice which states that female preferences for certain traits in males could evolve because male traits stimulate an existing bias in the female's sensory system. The original bias might relate to feeding or avoiding predators. For example, females might have a sensory bias to help them find food of a particular color; males can then exploit this bias by using the same color to attract females during courtship.

Sensory drive In the study of communication, the hypothesis suggesting that receiver characteristics are shaped by the environment and will thus affect signal evolution.

Sensory exploitation In communication, when a receiver has a preexisting bias for a particular stimulus and a sender's signal evolves to exploit that bias.

Sensory neuron A nerve cell that carries information from the peripheral receptors toward the central nervous system.

Sensory phase In songbirds, phase during which songs are learned and stored in memory for months without rehearsal.

Sex allocation The manner in which parents distribute resources between the production of sons and daughters. Parents can bias their allocation of resources in two main ways: they can either produce more offspring of one sex or they can provide more (or better) resources to offspring of one sex.

Sex-role reversal Pattern shown by some species in which parental investment by males exceeds that of females. In these species, females often compete for access to males and males are selective in their choice of mates.

Sexual conflict A conflict between the evolutionary interests of males and females. It takes several forms, but the two main ones concern mating/fertilization and parental investment.

Sexual dimorphism A difference, for example in behavior or appearance, between the sexes.

Sexual imprinting Learning process through which experience with parents and siblings early in life influences sexual preferences in adulthood.

Sexual interference Any behavior that reduces a rival's fitness by decreasing his mating success.

Sexual selection A form of natural selection that occurs through male competition for access to mates and female choice of mates.

Sexy son hypothesis Hypothesis proposing that access to good genes for offspring compensates a female for the costs of polygyny. A female may benefit from mating with an already mated male if her sons inherit the genes that made that male attractive. Her sexy sons will presumably provide her with many grandchildren, so the female's lifetime reproductive success may be enhanced by choosing to mate with a male that is attractive to many females.

Shaping In learning, changing the behavior of an animal by rewarding increasingly closer approximations of the desired behavior. This procedure often is used in animal training.

Short-term memory Memory of new information that lasts for a few seconds or minutes.

Siblicide The killing of a sibling. Siblicide is most common in those species in which parents face limited resources and deposit eggs or young in a "nursery" with limited space; the nursery could be a uterus, a brood pouch, a parent's back, a nest, or a den. Siblicide can be advantageous to parents when more young are produced than can be raised successfully.

Sign stimulus A stimulus that triggers a fixed action pattern.

Social learning Learning in which animals acquire information from other animals.

Social monogamy An exclusive living arrangement between one male and one female that makes no assumptions about mating exclusivity or biparental care. Compare with genetic monogamy.

Societies Structured animal groups.

Sociobiology A discipline that applies the principles of evolution to social behavior.

Sodium-potassium pump A molecular mechanism in a plasma membrane that uses cellular energy in the form of adenosine triphosphate (ATP) to pump ions against their concentration gradients. Typically, each pump ejects three sodium ions from the cell while bringing in two potassium ions.

Sperm competition Competition that results when two or more males have deposited sperm in the reproductive tract of one female.

Sperm heteromorphism The simultaneous production by a single male of at least two types of sperm in the same ejaculate.

Spermatophore A packet used by many species to transfer sperm from the male to the female.

Star compass orientation The ability to use the star constellations in the northern part of the sky to determine compass directions.

Steroid hormones A group of closely related hormones chemically derived from cholesterol and secreted by the gonads and adrenal glands in vertebrates. The four major classes of steroids include progestogens, androgens, estrogens, and corticosteroids. Steroid hormones are fat-soluble, so they move easily through the plasma membranes of target cells into the cell interior, where they affect gene expression and protein synthesis. See neurosteroids for steroid hormones produced by the nervous system.

Stimulus enhancement A type of social learning in which an animal is attracted to a particular object because a conspecific is near it or is interacting with it.

Stotting A stiff-legged bounding display performed by many species of deer and antelope which appears to have several functions, including announcing to a predator that it has been detected.

Strategy A behavioral option available to an animal, often used in the context of mathematical models.

Submissive individual An animal that predictably yields to a dominant animal.

Subsong Vocalizations produced by young songbirds during the sensory phase that do not involve retrieval or rehearsal of previously learned material.

Sun compass orientation The ability to use the sun as a visual cue to determine compass directions. Because the sun appears to move across the sky throughout the day, an animal must constantly change the angle it maintains with the sun to remain headed in the same compass direction. Thus, sun compass orientation is usually time compensated.

Suspension feeding Removing small food particles suspended in the water by means of several techniques.

Synapse The site of communication between a neuron and another cell, such as another neuron or a muscle cell.

Synaptic remodeling A refinement of synaptic connections caused by the development of new synapses and the loss of others that often occurs during development.

Syntax In communication, a change in the order of symbols that alters the meaning of the message.

Territory As used here, a defended space. Many other definitions exist.

Tool use Use of an object in order to obtain a goal.

Traditions Learned behaviors that are stable in a group over time.

True navigation The ability to maintain or establish reference to a goal, regardless of its location, without the use of landmarks. Sometimes called homing.

Ultrasound Sounds whose frequencies are above those audible to humans, which means frequencies greater than about 20 kHz. Several groups of mammals, including cetaceans, bats, and rodents, produce and detect ultrasounds as part of echolocation or communication systems. Ultrasound is not restricted to mammals.

Unconditioned stimulus In classical conditioning, a rewarding stimulus (such as food) that is paired with an innocuous stimulus (such as a tone or light).

Variable ratio schedule In learning, a reinforcement schedule in which the number of responses required to receive a reward varies randomly. Because of the variability, the animal cannot easily detect when reinforcement has stopped, so the response tends to persist. This is the reinforcement schedule used in slot machines.

Vector navigation An inherited program that tells an animal the compass direction to head in and for how long. Certain species of migratory birds use this type of navigation on their first migration.

Vomeronasal organ An accessory olfactory sense organ found in some species of amphibians, reptiles, and mammals. Located in the roof of the mouth or between the nasal cavity and the mouth, the vomeronasal organ is anatomically separate from other chemosensory structures, and its neural wiring goes to brain regions other than those associated with the main olfactory system. Sometimes called Jacobson's organ, the vomeronasal organ can be stimulated by either pheromones or general odorants.

Waggle dance A figure-eight dance performed by honeybees that indicates the direction and distance of food resources.

Warning coloration The phenomenon by which conspicuous coloration advertises dangerous or unpleasant attributes.

References

Abbott, D. H., E. B. Keverne, F. B. Bercovitch, C. A. Shively, S. P. Mendoza, W. Saltzman, C. T. Snowdon, T. E. Ziegler, M. Banjevic, T. Garland, Jr., and R. M. Sapolsky. 2003. Are subordinates always stressed? A comparative analysis of rank differences in cortisol levels among primates. *Hormones and Behavior* 43:67–82.

Able, K. P. 1980. Mechanisms of orientation, navigation, and homing. In *Animal Migration Orientation and Navigation*, edited by S. A. Gauthreaux, Jr. New York: Academic Press.

Able, K. P. 1982. Field studies of avian nocturnal migratory orientation I. Interaction of sun, wind, and stars as directional cues. *Animal Behaviour* 30:761–767.

Able, K. P. 1993. Orientation cues used by migratory birds: A review of cue-conflict experiments. *Trends in Ecology and Evolution* 8:367–371.

Able, K. P. 1996. The debate over olfactory navigation by homing pigeons. *Journal of Experimental Biology* 199:121–124.

Able, K. P., and M. A. Able. 1996. The flexible migratory orientation system of the Savannah sparrow (*Passerculus sandwichensis*). *Journal of Comparative Neurology* 199:3–8.

Abramsky, M. V., M. L. Rosenzweig, and A. Subach. 2002. The costs of apprehensive foraging. *Ecology* 83:1330–1340.

Abu-Gideiri, Y. B. 1966. The behaviour and neuro-anatomy of some developing teleost fishes. *Journal of Zoology* 149:215–241.

Abzhanov, A., W. P. Kuo, C. Hartmann, B. R. Grant, P. R. Grant, and C. J. Tabin. 2006. The calmodulin pathway and evolution of elongated beak morphology in Darwin's finches. *Nature* 442:563–567.

Aceves-Pina, E. O., and W. G. Quinn. 1979. Learning in normal and mutant *Drosophila*. *Science* 206:93–96.

Adams, J., and R. L. Caldwell. 1990. Deceptive communication in asymmetric fights of the stomatopod crustacean *Gonodactylus bredini*. *Animal Behaviour* 39:706–717.

Adkins-Regan, E. 2005. *Hormones and Animal Social Behavior*. Princeton: Princeton University Press.

Adkins-Regan, E., and C. H. Leung. 2006. Sex steroids modulate changes in social and sexual preference during juvenile development in zebra finches. *Hormones and Behavior* 50:772–778.

Adkins-Regan, E., J.-P. Signoret, and P. Orgeur. 1989. Sexual differentiation of reproductive behavior in pigs: Defeminizing effects of prepubertal estradiol. *Hormones and Behavior* 23:290–303.

Adler, K. 1976. Extraocular photoreception in amphibians. *Photochemistry and Photobiology* 23:275–298.

Aidley, D. J. 1981. Questions about migration. In *Animal Migration*, edited by D. J. Aidley. New York: Cambridge University Press, pp. 1–8.

Åkesson, S., and A. Hedenström. 2007. How migrants get there: Migratory performance and orientation. *BioScience* 57:123–133.

Åkesson, S., and R. Wehner. 2002. Visual navigation in desert ants *Cataglyphis fortis*: are snapshots coupled to a celestial system of reference? *Journal of Experimental Biology* 205:1971–1978.

Alberts, S. C., J. C. Buchan, and J. Altmann. 2006. Sexual selection in wild baboons: from mating opportunities to paternity success. *Animal Behaviour* 72:1177–1196.

Alcock, J. 2001. *Animal Behavior*. Sunderland, MA: Sinauer.

Alerstam, T. 2003. Bird migration speed. In *Avian Migration*, edited by P. Berthold, E. Gwinner, and E. Sonnenschein. Berlin, Germany: Springer-Verlag, pp. 253–267.

Alexander, R. D. 1974. The evolution of social behavior. *Annual Review of Ecology and Systematics* 5:325–383.

Alexander, R. D. 1975. Natural selection and specialized chorusing behavior in acoustical insects. In *Insects, Science, and Society*, edited by D. Pimental. New York: Academic Press, pp. 35–77.

Alexander, R. D. 1979. *Darwinism and Human Affairs*. Seattle: University of Washington Press.

Allee, W. C. 1951. *The Social Life of Animals*. Boston: Beacon Press.

Allen, B. J., and J. S. Levinton. 2007. Costs of bearing a sexually selected ornamental weapon in a fiddler crab. *Functional Ecology* 21:154–161.

Allen, J. A. 1988. Frequency-dependent selection by predators. *Philosophical Transactions of the Royal Society of London B* 319:485–503.

Altmann, S. A., and J. Altmann. 2003. The transformation of behaviour field studies. *Animal Behaviour* 65:413–423.

Alvarez, F. 1990. Horns and fighting in male Spanish ibex, *Capra pyrenaica*. *Journal of Mammalogy* 71:608–616.

Anholt, R. R. H. 2004. Genetic modules and networks for behavior: lessons from *Drosophila*. *BioEssays* 26:1299–1306.

Anderson, J. B., and L. P. Brower. 1996. Freeze-protection of overwintering monarch butterflies in Mexico: Critical role of the forest as a blanket and an umbrella. *Ecological Entomology* 21:107–116.

Andersson, M. 1994. *Sexual Selection*. Princeton, NJ: Princeton University Press.

Andersson, M., and L. W. Simmons. 2006. Sexual selection and mate choice. *Trends in Ecology and Evolution* 21:296–302.

Andrade, M. C. B. 1996. Sexual selection for male sacrifice in the Australian redback spider. *Science* 271:70–72.

Antonov, A., B. G. Stokke, A. Moksnes, and E. Røskaft. 2008. Getting rid of the cuckoo *Cuculus canorus* egg: why do hosts delay rejection? *Behavioral Ecology* 19:100–107.

Anway, M. D., C. Leathers, and M. K. Skinner. 2006. Endocrine disruptor vinclozolin induced epigenetic transgenerational adult-onset disease. *Endocrinology* 147:5515–5523.

Aoki, S. 1972. *Colophina clematis* (Homoptera, Pemphigidae), an aphid species with "soldiers." *Kontyu* 45:276–282.

Aoki, S. 1979. Further observations on *Astegopteryx styracicola* (Homoptera, Pemphigidae), an aphid species with soldiers biting man. *Kontyu* 47:99–104.

Aoki, S. 1982. Soldiers and altruistic dispersal in aphids. In *The Biology of Social Insects*, edited by M. D. Breed, C. D. Michener, and H. E. Evans. Boulder, CO: Westview, pp. 154–158.

Arak, A. 1988. Callers and satellites in the natterjack toad: Evolutionarily stable decision rules. *Animal Behaviour* 36:416–432.

Archer, J. 1988. *The Behavioural Biology of Aggression*. Cambridge: Cambridge University Press.

Armstrong, D. P. 1991. Aggressiveness of breeding territorial honeyeaters corresponds to seasonal changes in nectar availability. *Behavioral Ecology and Sociobiology* 29:103–112.

Arnold, A. P., and S. M. Breedlove. 1985. Organizational and activational effects of sex steroids on brain and behavior: A reanalysis. *Hormones and Behavior* 19:469–498.

Arnold, E. N. 1988. Caudal autotomy as a defense. In *Biology of the Reptilia*, edited by C. Gans and R. B. Huey. New York: Alan Liss, pp. 235–273.

Arnold, S. J. 1976. Sexual behavior, sexual interference and sexual defense in the salamanders *Ambystoma maculatum*, *Ambystoma tigrinum* and *Plethodon jordani*. *Zeitschrift für Tierpsychologie* 42:247–300.

Arnott, G., and R. W. Elwood. 2007. Fighting for shells: How private information about resource value changes hermit crab pre-fight displays and escalated fight behavior. *Proceedings of the Royal Society of London B* 274:3011–3017.

Arnqvist, G., and T. Nilsson. 2000. The evolution of polyandry: multiple mating and female fitness in insects. *Animal Behaviour* 60:145–164.

Asdell, S. A. 1946. *Patterns of Mammalian Reproduction*. Ithaca, NY: Comstock.

Asher, G., D. Gatfield, M. Stratmann, H. Reinke, C. Dibner, F. Kreppel, R. Mostoslavsky, F. W. Alt, and U. Schibler. 2008. SIRT1 regulates circadian clock gene expression through per2 deacetylation. *Cell* 134:317–328.

Aubin-Horth, N., B. H. Letcher, and H. A. Hofmann. 2005a. Interaction of rearing environment and reproductive tactic on gene expression profiles in Atlantic salmon. *Journal of Heredity* 96:261–278.

Aubin-Horth, N., C. R. Landry, B. H. Letcher, and H. A. Hofmann. 2005b. Alternative life histories shape brain gene expression profiles in males of the same population. *Proceedings of the Royal Academy of Sciences* 272:1655–1662.

Avens, L., and K. J. Lohmann. 2004. Navigation and seasonal migratory orientation in juvenile sea turtles. *Journal of Experimental Biology* 207:1771–1778.

Axelrod, R. 1984. *The Evolution of Cooperation*. New York: Basic Books.

Ayala, F. J., and C. A. Campbell. 1974. Frequency-dependent selection. *Annual Review of Ecology and Systematics* 5:115–138.

Back, S. R., L. A. Beeler, R. L. Schaefer, and N. G. Solomon. 2002. Testing functional hypotheses for the behavior of resident pine voles, *Microtus pinetorum*, toward non-residents. *Ethology* 108:1023–1039.

Backwell, P. R. Y., J. H. Christy, S. R. Telford, M. D. Jennions, and N. I. Passmore. 2000. Dishonest signaling in a fiddler crab. *Proceedings of the Royal Society of London B* 267:719–724.

Baeza, J. A. 2008. Social monogamy in the shrimp *Pontonia margarita*, a symbiont of *Pinctada mazatlanica*, off the Pacific coast of Panama. *Marine Biology* 153:387–395.

Baker, R. R. 1980. *The Mystery of Migration*. London: Macdonald.

Baker, B. S., B. J. Taylor, and J. C. Hall. 2001. Are complex behaviors specified by dedicated regulatory genes? Reasoning from *Drosophila*. *Cell* 105:13–24.

Baker, P. J., C. P. J. Robertson, S. M. Funk, and S. Harris. 1998. Potential fitness benefits of group living in the red fox, *Vulpes vulpes*. *Animal Behaviour* 56:1411–1424.

Bakken, G. S., and A. R. Krochmal. 2006. The imaging properties and sensitivity of the facial pits of pit vipers as determined by optical and heat-transfer analysis. *Journal of Experimental Biology* 210:2801–2810.

Balda, R. P. 1980. Recovery of cached seeds by a captive *Nucifraga caryocatactes*. *Zeitschrift für Tierpsychologie* 52:331–346.

Balda, R. P., and A. C. Kamil. 1989. A comparative study of cache recovery by three corvid species. *Animal Behaviour* 38:486–495.

Balda, R. P., and A. C. Kamil. 1998. The ecology and evolution of spatial memory in corvids of the southwestern USA: the perplexing pinyon jay. In *Animal Cognition in Nature*, edited by R. P. Balda, I. M. Pepperberg, and A. C. Kamil. San Diego: Academic Press, pp. 29–64.

Balda, R. P., I. M. Pepperberg, and A. C. Kamil. 1998. Preface. In *Animal Cognition in Nature*, edited by R. P. Balda, I. M. Pepperberg and A. C. Kamil. San Diego: Academic Press.

Baldaccini, E. N., S. Benvenuti, V. Fiaschi, and F. Papi. 1975. Pigeon navigation: Effects of wind deflection at the home cage on homing behavior. *Journal of Comparative Physiology* 99:177–196.

Bandura, A. 1962. Social learning through imitation. In *Nebraska Symposium on Motivation*, edited by M. R. Jones. Lincoln: University of Nebraska Press.

Baptista, L. F., and L. Petrinovich. 1984. Social interaction, sensitive phases and the song template hypothesis in the white-crowned sparrow. *Animal Behaviour* 32:172–181.

Baptista, L. F., and L. Petrinovich. 1986. Song development in the white-crowned sparrow: Social factors and sex differences. *Animal Behaviour* 34:1359–1371.

Barbosa, M., and A. E. Magurran. 2006. Female mating decisions: maximizing fitness? *Journal of Fish Biology* 68: 1636–1661.

Barinaga, M. 1994. From fruit flies, rats, mice: Evidence of genetic influence. *Science* 264:1690–1693.

Barkley, C. L., and L. F. Jacobs. 2007. Sex and species differences in spatial memory in food-storing kangaroo rats. *Animal Behaviour* 73:321–329.

Barlow, G. W. 1968. Ethological units of behavior. In *The Central Nervous System and Fish Behavior*, edited by D. Ingle. Chicago: University of Chicago Press.

Barlow, G. W. 1989. Has sociobiology killed ethology or revitalized it? *Perspectives in Ethology* Volume 8, Whither ethology? pp. 1–45.

Barlow, G. W. 1991. Nature-nurture and the debates surrounding ethology and sociobiology. *American Zoologist* 31:286–296.

Barnea, A., and F. Nottebohm. 1994. Seasonal recruitment of hippocampal neurons in adult free-ranging black-capped chickadees. *Proceedings of the National Academy of Sciences* 91:11217–11221.

Bartz, J., and E. Hollander. 2006. The neuroscience of affiliation: Forging links between basic and clinical research on neuropeptides and social behavior. *Hormones and Behavior* 50:518–528.

Basil, J. A., A. C. Kamil, R. P. Balda, and K. V. Fite. 1996. Differences in hippocampal volume among food storing corvids. *Brain, Behavior and Evolution* 47:156–164.

Basolo, A. L. 1990. Female preference predates the evolution of the sword in swordtail fish. *Science* 250:808–810.

Basolo, A. L. 1995a. A future examination of a pre-existing bias favouring a sword in the genus *Xiphophorus*. *Animal Behaviour* 50:365–375.

Basolo, A. L. 1995b. Phylogenetic evidence for the role of a pre-existing bias in sexual selection. *Proceedings of the Royal Society of London B* 259:307–311.

Bass, A. H. 1996. Shaping brain sexuality. *American Scientist* 84:352–363.

Bass, A. H., D. A. Bodnar, and J. R. McKibben. 1997. From neurons to behavior: Vocal-acoustic communication in teleost fish. *Biological Bulletin* 192:158–160.

Bateman, A. J. 1948. Intra-sexual selection in *Drosophila*. *Heredity* 2:349–368.

Bateson, M. 2002. Recent advances in our understanding of risk-sensitive foraging preferences. *Proceedings of the Nutrition Society* 61:509–516.

Bateson, P. 1976. Specificity and the origins of behavior. In *Advances in the Study of Behavior*, edited by J. S. Rosenblatt, R. A. Hinde, and C. Beer. New York: Academic Press, pp. 1–20.

Bateson, P. 1979. How do sensitive periods arise and what are they for? *Animal Behaviour* 27:470–486.

Bateson, P. 1982. Preferences for cousins in Japanese quail. *Nature* 295:236–237.

Bateson, P. 1983. Optimal outbreeding. In *Mate Choice*, edited by P. Bateson. Cambridge: Cambridge University Press, pp. 257–277.

Bateson, P. 1990. Is imprinting such a special case? *Philosophical Transactions of the Royal Society of London B* 329:125–131.

Bateson, P. 2003. The promise of behavioural biology. *Animal Behaviour* 65:11–17.

Bateson, P. 2005. The return of the whole organism. *Journal of Biosciences* 30:31–39.

Bateson, P. P. G., and P. H. Klopfer, eds. 1989. Preface. In *Perspectives in Ethology*, Volume 8. Whither Ethology? New York: Plenum Press.

Bateson, P. P. G., and P. H. Klopfer, eds. 1991. *Perspectives in Ethology*, Volume 9. Human Understanding and Animal Awareness. New York: Plenum.

Batteau, D. W. 1968. The world as a source; the world as a sink. In *The Neuropsychology of Spatially Oriented Behavior*, edited by S. J. Freedman. Homewood, IL: Dorsey, pp. 197–203.

Battin, J. 2004. When good animals love bad habitats: Ecological traps and the conservation of animal populations. *Conservation Biology* 18:1482–1491.

Bauer, E. B., and B. B. Smuts. 2007. Cooperation and competition during dyadic play in domestic dogs, *Canis familiaris*. *Animal Behaviour* 73:489–499.

Baxi, K. N., K. M. Dorries, and H. L. Eisthen. 2006. Is the vomeronasal system really specialized for detecting pheromones? *Trends in Neurosciences* 29:1–7.

Beach, F. A. 1974. Effects of gonadal hormones on urinary behavior in dogs. *Physiology and Behavior* 12:1005–1013.

Beach, F. A. 1976. Sexual attractivity, proceptivity, and receptivity in female mammals. *Hormones and Behavior* 7:105–138.

Beauchamp, G. 1999. The evolution of communal roosting in birds: origins and secondary losses. *Behavioral Ecology* 10:675–687.

Beauchamp, G. 2007. Vigilance in a selfish herd. *Animal Behaviour* 73:445–451.

Beck, W., and W. Wiltschcko. 1988. Magnetic factors control the migratory direction of pied flycatchers (*Ficedula hypoleuca* Pallas). *Acta Congress International Ornithology* 19:1955–1962.

Beckerman, A. P., M. Uriarte, and O. J. Schmitz. 1997. Experimental evidence for a behavior-mediated trophic cascade in a terrestrial food chain. *Proceedings of the National Academy of Sciences* 94:10735–10738.

Bednarz, J. C. 1988. Cooperative hunting in Harris' hawks (*Parabuteo unicinctus*). *Science* 239:1525–1527.

Bee, M. A., and H. C. Gerhardt. 2001. Habituation as a mechanism of reduced aggression between neighboring territorial male bullfrogs (*Rana catesbeiana*). *Journal of Comparative Psychology* 115:68–82.

Beebe, T. 1995. Amphibian breeding and climate change. *Nature* 374:219–220.

Beecher, M. D., I. M. Beecher, and S. Hahn. 1981. Parent-offspring recognition in bank swallows (*Riparia riparia*): Development and acoustic basis. *Animal Behaviour* 29:95–101.

Beecher, M. D., and E. A. Brenowitz. 2005. Functional aspects of song learning in songbirds. *Trends in Ecology and Evolution* 20:143–149.

Beecher, M. D., J. M. Burt, A. L. O'Loghlen, C. N. Templeton, and S. E. Campbell. 2007. Bird song learning in an eavesdropping context. *Animal Behaviour* 73:929–935.

Beecher, M. D., S. E. Campbell, and J. C. Nordby. 2000. Territory tenure in song sparrows is related to song sharing with neighbours, but not to repertoire size. *Animal Behaviour* 59:29–37.

Beecher, M. D., P. Loesche, P. K. Stoddard, and M. B. Medvin. 1989. Individual recognition by voice in swallows: signal or perceptual adaptation? In *The Comparative Psychology of Audition*, edited by R. J. Dooling and S. H. Hulse. Mahwah, NJ: Erlbaum.

Beehler, M. N., and M. S. Foster. 1988. Hotshots, hotspots, and female preference in the organization of lek mating systems. *American Naturalist* 131:203–219.

Beer, M. F. 1999. Homosynaptic long-term depression: a mechanism for memory? *Proceedings of the National Academy of Sciences* 96:9457–9458.

Bekoff, A. 1992. Neuroethological approaches to the study of motor development in chicks: achievements and challenges. *Journal of Neurobiology* 23:1486–1505.

Bekoff, A., and J. A. Kauer. 1984. Neural control of hatching: Gate of the pattern generator for leg movements of hatching in post-hatching chicks. *Journal of Neuroscience* 4:2659–2666.

Beletsky, L. D., and G. H. Orians. 1987. Territoriality among male red-winged blackbirds. II. Removal experiments and site dominance. *Behavioral Ecology and Sociobiology* 20:339–349.

Beletsky, L. D., and G. H. Orians. 1989. Territoriality among male red-winged blackbirds. III. Testing hypotheses of territorial dominance. *Behavioral Ecology and Sociobiology* 24:333–339.

Beling, I. 1929. Über das Zeitgedächtnis der Bienen. *Zeitschrift für vergleichende Physiologie* 9:259–338.

Belovsky, G. E. 1978. Diet optimization in a generalist herbivore: the moose. *Theoretical Population Biology* 14:105–134.

Bemis, W. E., and B. Kynard. 1997. Sturgeon rivers: an introduction to acipenseriform biogeography and life history. *Environmental Biology of Fishes* 48:167–183.

Ben-Shahar, Y., A. Robichon, M. B. Sokolowski, and G. E. Robinson. 2002. Influence of gene action across different time scales on behavior. *Science* 296:741–744.

Berec, M., V. Krivan, and L. Berec. 2003. Are great tits (*Parus major*) really optimal foragers? *Canadian Journal of Zoology* 81:780–788.

Bereczkei, T., P. Gyuris, and G. E. Weisfeld. 2004. Sexual imprinting in human mate choice. *Proceedings of the Royal Society of London B* 271:1129–1134.

Berger, J. 2004. The last mile: How to sustain long-distance migration in mammals. *Conservation Biology* 18:320–331.

Berthold, P. 2001. *Bird Migration: A General Survey*. Edited by C. M. Perrins. 2nd ed. *Oxford Ornithological Series*. New York: Oxford University Press.

Bertram, B. C. R. 1975. Social factors influencing reproduction in wild lions. *Journal of Zoology* 177:463–482.

Bertram, B. C. R. 1976. Kin selection in lions and evolution. In *Growing Points in Ethology*, edited by P. P. G. Bateson and R. A. Hinde. New York: Cambridge University Press, pp. 281–301.

Bertram, S. M. 2002. Temporally fluctuating selection of sex-limited signaling traits in the Texas field cricket, *Gryllus texensis*. *Evolution* 56:1831–1839.

Biben, M. 1998. Squirrel monkey playfighting: Making the case for a cognitive training function for play. In *Animal Play: Evolutionary, Comparative, and Ecological Perspectives*, edited by M. Bekoff and J. A. Byers. Cambridge: Cambridge University Press, pp. 161–182.

Bingman, V. P. 1984. Night-sky orientation of migratory pied flycatchers raised in different magnetic fields. *Behavioral Ecology and Sociobiology* 15:77–80.

Bingman, V. P., and K. Cheng. 2005. Mechanisms of animal global navigation: Comparative perspectives and enduring challenges. *Ethology, Ecology and Evolution* 17:295–318.

Bingman, V., T. Jechura, and M. C. Kahn. 2006. Behavioral and neural mechanisms of homing and migration in birds. In *Animal Spatial Cognition: Comparative, Neural, and Computational Approaches. [On-line].*, edited by M. F. Brown and R. G. Cook: Available: *www.pigeon.psy.tufts.edu/asc/bingman/*.

Birkhead, T. R. 1996. Mechanisms of sperm competition in birds. *American Scientist* 84:254–262.

Birkhead, T. R., and G. A. Parker. 1997. Sperm competition and mating systems. In *Behavioural Ecology: An Evolutionary Approach*, edited by J. R. Krebs and N. B. Davies. Oxford: Blackwell Science, pp. 121–145.

Bischof, H.-J. 1994. Sexual imprinting as a two-stage process. In *Causal Mechanisms of Behavioural Development*, edited by J. A. Hogan and J. J. Bolhuis. Cambridge: Cambridge University Press, pp. 82–97.

Bischof, H.-J. 2003. Neural mechanisms of sexual imprinting. *Animal Biology* 53:89–112.

Bischof, H.-J. 2007. Behavioral and neuronal aspects of developmental sensitive periods. *NeuroReport* 18:461–465.

Blackmore, C. J., and R. Heinsohn. 2008. Variable mating strategies and incest avoidance in cooperatively breeding grey-crowned babblers. *Animal Behaviour* 75:63–70.

Blakemore, R. P., and R. B. Frankel. 1981. Magnetic navigation in bacteria. *Scientific American* 245:58–65.

Bloch Qazi, M. C., J. R. Aprille, and S. M. Lewis. 1998. Female role in sperm storage in the red flour beetle, *Tribolium castaneum*. *Comparative Biochemistry and Physiology A* 120:641–647.

Blomqvist, D., O. C. Johasson, U. Unger, M. Larsson, and L.-A. Flodin. 1997. Male aerial display and reversed sexual size dimorphism in the dunlin. *Animal Behaviour* 54:1291–1299.

Blumenrath, S. H., and T. Dabelsteen. 2004. Degradation of great tit (*Parus major*) song before and after foliation: implications for vocal communication in a deciduous forest. *Behaviour* 141:935–958.

Blumstein, D. T. 2006. The multipredator hypothesis and the evolutionary persistence of antipredator behavior. *Ethology* 112:209–217.

Blumstein, D. T. 2007. The evolution of alarm communication in rodents: structure, function, and the puzzle of apparently altruistic calling. In *Rodent Societies: An Ecological and Evolutionary Perspective*, edited by J. O. Wolff and P. W. Sherman. University of Chicago Press: Chicago, IL. pp. 317–327.

Blumstein, D. T., and K. B. Armitage. 1997. Does sociality drive the evolution of communicative complexity? A comparative test with ground-dwelling sciurid alarm calls. *American Naturalist* 150:179–200.

Blumstein, D. T., A. Bitton, and J. DaVeiga. 2006. How does the presence of predators influence the persistence of antipredator behavior? *Journal of Theoretical Biology* 239:460–468.

Blumstein, D. T., and J. C. Daniel. 2005. The loss of antipredator behaviour following isolation on islands. *Proceedings of the Royal Society of London B* 272:1663–1668.

Blumstein, D. T., and E. Fernández-Juricic. 2004. The emergence of conservation behavior. *Conservation Biology* 18:1175–1177.

Blumstein, D. T., J. Steinmetz, K. B. Armitage, and J. C. Daniel. 1997. Alarm calling in yellow-bellied marmots. 2. The importance of direct fitness. *Animal Behaviour* 53:173–184.

Boag, P. T., and P. R. Grant. 1981. Intense natural selection in a population of Darwin finches (Geospizinae) in the Galapagos. *Science* 214: 82–85.

Boal, C. W., and R. W. Mannan. 1999. Comparative breeding ecology of Cooper's hawks in urban and exurban areas of southeastern Arizona. *Journal of Wildlife Management* 63:77–84.

Boccia, M. L. 1983. A functional analysis of social grooming patterns through direct comparison with self grooming in rhesus monkeys. *International Journal of Primatology* 4:399–418.

Boccia, M. L. 1986. Grooming site preferences as a form of tactile communication and their role in the social relations of rhesus monkeys. In *Current Perspectives in Primate Social Dynamics*, edited by D. M. Taub and F. A. King. New York: Van Nostrand Reinhold, pp. 505–518.

Bodamer, M. D., D. H. Fouts, R. S. Fouts, and M. L. A. Jensvold. 1994. Functional analysis of chimpanzee (*Pan troglodytes*) private signing. *Human Evolution* 9:281–296.

Boinski, S., L. Kauffman, E. Ehmke, S. Schet, and A. Vreedzaam. 2005. Dispersal patterns among three species of squirrel monkeys (*Saimiri oerstedii*, *S. boliviensis* and *S. sciureus*): I. Divergent costs and benefits. *Behaviour* 142:525–632.

Boles, L. C., and K. J. Lohmann. 2003. True navigation and magnetic maps in spiny lobsters. *Nature* 421:60–63.

Bolhuis, J. J. 2008. Chasin' the trace: the neural substrate of birdsong memory. In *The Neuroscience of Birdsong*, edited by H. P. Zeigler and P. Marler. Cambridge: Cambridge University Press, pp. 271–281.

Bolhuis, J. J., and H. Eda-Fujiwara. 2003. Bird brains and songs: neural mechanisms of birdsong perception and memory. *Animal Biology* 53:129–145.

Bolhuis, J. J., and M. Gahr. 2006. Neural mechanisms of birdsong memory. *Nature Reviews Neuroscience* 7:347–357.

Bolhuis, J. J., and R. C. Honey. 1998. Imprinting, learning and development: from behaviour to brain and back. *Trends in Neurosciences* 21:306–311.

Bollenbacher, W. E., S. L. Smith, W. Goodman, and L. I. Gilbert. 1981. Ecdysone titer during larval-pupal–adult development in the tobacco hornworm, *Manduca sexta*. *General and Comparative Endocrinology* 44:302–306.

Bonadonna, F., S. Caro, P. Jouventin, and G. A. Nevitt. 2006. Evidence that blue petrel, *Halobaena caerulea*, fledglings can detect and orient to dimethyl sulfide. *Journal of Experimental Biology* 209:2165–2169.

Boncoraglio, G., and N. Saino. 2007. Habitat structure and the evolution of bird song: a meta-analysis of the evidence for the acoustic adaptation hypothesis. *Functional Ecology* 21:134–142.

Bond, A. B. 1983. Visual search and selection of natural stimuli in the pigeon: the attention threshold hypothesis. *Journal of Experimental Psychology—Animal Behavior Processes* 9:292–306.

Bond, A. B., and A. C. Kamil. 1998. Apostatic selection by blue jays produces balanced polymorphisms in virtual prey. *Nature* 395:594–596.

Bonduriansky, R. 2001. The evolution of male mate choice in insects: a synthesis of ideas and evidence. *Biological Reviews* 76:305–339.

Bonte, D., S. Van Belle, and J.-P. Maelfait. 2007. Maternal care and reproductive state-dependent mobility determine natal dispersal in a wolf spider. *Animal Behaviour* 74:63–69.

Borgia, G. 1986. Satin bowerbird parasites: A test of the bright male hypothesis. *Behavioral Ecology and Sociobiology* 19:355–358.

Borgia, G., I. M. Kaatz, and R. Condit. 1987. Flower choice and bower decoration in the satin bowerbird *Ptilonorhynchus violaceus*: A test of the hypothesis for the evolution of male display. *Animal Behaviour* 35:1129–1139.

Borries, C., K. Launhardt, C. Epplen, J. T. Epplen, and P. Winkler. 1999. DNA analyses support the hypothesis that infanticide is adaptive in langur monkeys. *Proceedings of the Royal Society of London B* 266:901–904.

Bostwick, K. S., and R. O. Prum. 2005. Courting bird sings with stridulating wing feathers. *Science* 309:736.

Bourke, A. F. G., and N. R. Franks. 1995. *Social Evolution in Ants*. Princeton: Princeton University Press.

Bowmaker, J. K., V. I. Govardovskii, S. A. Shukolykov, L. V. Zueva, D. M. Hunt, V. G. Sideleva, and O. G. Smirnova. 1994. Visual pigments and the photic environment: the cottoid fish of Lake Baikal. *Vision Research* 34:591–605.

Boyan, G. S., and J. H. Fullard. 1986. Interneurons responding to sound in the tobacco budworm moth *Heliothis virescens* (Noctuidae): morphological and physiological characteristics. *Journal of Comparative Physiology A* 158:391–404.

Boyan, G. S., and L. A. Miller. 1991. Parallel processing of afferent input by identified interneurones in the auditory pathway of the noctuid moth *Noctua pronumba*. *Journal of Comparative Physiology A* 168:727–738.

Boysen, S. T., and G. G. Berntson. 1989. Numerical competence in a chimpanzee (*Pan troglodytes*). *Journal of Comparative Psychology* 103:23–31.

Bradbury, J. W. 1981. The evolution of leks. In *Natural Selection and Social Behavior*, edited by R. D. Alexander and D. W. Tinkle. New York: Chiron, pp. 138–172.

Bradbury, J. W., R. M. Gibson, and M. B. Andersson. 1986. Hotspots and the evolution of leks. *Animal Behaviour* 34:1694–1709.

Bradbury, J. W., and S. L. Vehrencamp. 1998. *Principles of Animal Communication*. Sunderland, MA: Sinauer.

Brainard, M. S., and A. J. Doupe. 2000. Auditory feedback in learning and maintenance of vocal behaviour. *Nature Reviews* 1:31–40.

Brantley, R. K., and A. H. Bass. 1994. Alternative male spawning tactics and acoustic signals in the plainfin midshipman fish *Porichthys notatus* Girard (Teleostei, Batrachoididae). *Ethology* 96:213–232.

Braude, S., D. Ciszek, N. E. Berg, and N. Shefferly. 2001. The ontogeny and distribution of countershading in colonies of the naked mole-rat (*Heterocephalus glaber*). *Journal of Zoology* 253:351–357.

Breed, M. D. 1998. Recognition pheromones of the honeybee. *BioScience* 48:463–470.

Breedlove, S. M., and A. P. Arnold. 1983. Hormonal control of a developing neuromuscular system. II. Sensitive periods for the androgen-induced masculinization of the rat spinal nucleus of the bulbocavernosus. *Journal of Neuroscience* 3:424–432.

Breedlove, S. M., M. R. Rosenweig, and N. V. Watson. 2007. *Biological Psychology: An Introduction to Behavioral, Cognitive, and Clinical Neuroscience*. 5th ed. Sunderland MA: Sinauer.

Brenowitz, E. A. 2004. Plasticity of the adult avian song control system. *Annals of the New York Academy of Sciences* 1016:560–585.

Brenowitz, E. A., K. Lent, and E. W. Rubel. 2007. Auditory feedback and song production do not regulate seasonal growth of song control circuits in adult white-crowned sparrows. *The Journal of Neuroscience* 27:6810–6814.

Briffa, M., and L. U. Sneddon. 2007. Physiological constraints on contest behaviour. *Functional Ecology* 21:627–637.

Brockmann, J., A. Grafen, and R. Dawkins. 1979. Evolutionarily stable nesting strategy in a digger wasp. *Journal of Theoretical Biology* 77:473–496.

Brodin, A., and K. Lundborg. 2003. Is hippocampal volume affected by specialization for food hoarding in birds? *Proceedings of the Royal Society of London B* 270:1555–1563.

Broom, M., G. D. Ruxton, and R. M. Kilner. 2008. Host life-history strategies and the evolution of chick-killing by brood parasitic offspring. *Behavioral Ecology* 19:22–34.

Brotherton, P. N. M., and M. B. Manser. 1997. Female dispersion and the evolution of monogamy in the dik-dik. *Animal Behaviour* 54:1413–1424.

Brotherton, P. N. M., J. M. Pemberton, P. E. Komers, and G. Malarky. 1997. Genetic and behavioural evidence of monogamy in a mammal, Kirk's dik-dik (*Madoqua kirkii*). *Proceedings of the Royal Society of London B* 264:675–681.

Brotherton, P. N. M., and A. Rhodes. 1996. Monogamy without biparental care in a dwarf antelope. *Proceedings of the Royal Society of London B* 263:23–29.

Brower, L. P. 1996. Monarch butterfly orientation: missing pieces of a magnificent puzzle. *Journal of Experimental Biology* 199:93–103.

Brower, L. P., and W. H. Calvert. 1985. Foraging dynamics of bird predators on overwintering monarch butterflies. *Evolution* 39:852–868.

Brower, L. P., D. R. Kust, E. Rendon-Salinas, E. G. Serrano, K. R. Kust, J. Mller, C. Fernandez del Rey, and K. Pape. 2004. Catastrophic winter storm mortality of monarch butterflies in Mexico during January 2002. In *The Monarch Butterfly: Biology and Conservation*, edited by K. S. Oberhauser and M. J. Solensky. Ithaca: Cornell University Press, p. 155–198.

Brower, L. P., W. N. Ryerson, L. L. Coppinger, and S. C. Glazier. 1968. Ecological chemistry and palatability spectrum. *Science* 161:1349–1351.

Brown, C. R. 1986. Cliff swallow colonies as information centres. *Science* 234:83–85.

Brown, C. R., and M. B. Brown. 1986. Ectoparasitism as a cost of coloniality in cliff swallows (*Hirundo pyrrhonota*). *Ecology* 67:1206–1218.

Brown, C. R., and M. B. Brown. 2002. Ectoparasites cause increased bilateral asymmetry of naturally selected traits in a colonial bird. *Journal of Evolutionary Biology* 15:1067–1075.

Brown, C. R., and M. B. Brown. 2004. Group size and ectoparasitism affect daily survival probability in a colonial bird. *Behavioral Ecology and Sociobiology* 56:498–511.

Brown, C. R., M. B. Brown, and B. Rannala. 1995. Ectoparasites reduce long-term survival of their avian host. *Proceedings of the Royal Society of London B* 262:313–319.

Brown, F. A., Jr., J. W. Hastings, and J. D. Palmer. 1970. *The Biological Clock: Two Views*, New York: Academic Press.

Brown, J. L. 1964. The evolution of diversity in avian territorial systems. *Wilson Bulletin* 76:160–169.

Brown, J. L., and E. R. Brown. 1998. Are inbred offspring less fit? Survival in a natural population of Mexican jays. *Behavioral Ecology* 9:60–63.

Brown, J. L, E. R. Brown, and S. D. Brown. 1982. Morphological variation in a population of grey-crowned babblers—correlations with variables affecting social behavior. *Behavioral Ecology and Sociobiology* 10:281–287.

Brown, J. L., and A. Eklund. 1994. Kin recognition and the major histocompatibility complex: An integrative review. *American Naturalist* 143:435–461.

Brown, J. S., and B. P. Kotler. 2004. Hazardous duty pay and the foraging cost of predation. *Ecology Letters* 7:999–1014.

Brown, K. S. 1996. Infanticide as a way to get ahead. *BioScience* 46:174–176.

Brownell, P. H. 1984. Prey detection by the scorpion. *Scientific American* 251:86–98.

Brownell, P. H., and J. L. van Hemmen. 2001. Vibration sensitivity and a computational theory for prey-localizing behavior in sand scorpions. *American Zoologist* 41:1229–1240.

Bruder, C. E. G., A. Piotrowski, C. J. Gijsbers, R. Andersson, S. Erickson, T. D. de Ståhl, U. Menzel, J. Sandgren, D. von Tell, A. Poplawski, M. Crowley, C. Crasto, E. C. Partridge, H. Tiwari, D. B. Allison, J. Komorowski, G.-J. B. van Ommen, D. I. Boomsma, N. L. Pedersen, J. T. den Dunnen, K. Wirdefeldt, and J. P. Dumanski. 2008. Phenotypically concordant and discordant monozygotic twins display different DNA copy-number-variation profiles. *American Journal of Human Genetics* 82:763–781.

Brudzynski, S. M. 2005. Principles of rat communication: quantitative parameters of ultrasonic calls in rats. *Behavior Genetics* 35:85–92.

Bruel-Jungerman, E., S. Davis, and S. Laroche. 2007. Brain plasticity mechanisms and memory: A party of four. *Neuroscientist* 13:492–505.

Brunton, D. H. 1990. The effects of nesting stage, sex, and type of predator on parental defense by killdeer (*Charadrius vociferus*): Testing models of parental defense. *Behavioral Ecology and Sociobiology* 26:181–190.

Brusca, R. C., and G. J. Brusca.1990. *Invertebrates*. Sunderland MA: Sinauer.

Budzynski, C. A., R. Strasser, and V. P. Bingman. 1998. The effects of zinc sulfate anosmia on homing pigeons, *Columba livia*, in a homing and a non-homing experiment. *Ethology* 104:111–118.

Buler, J. J., F. R. Moore, and S. Woltmann. 2007. A multi-scale examination of stopover habitat use by birds. *Ecology* 88:1789–1802.

Bullock, T. H., C. D. Hopkins, A. N. Popper, and R. R. Fay (eds). 2005. *Electroreception*. New York: Springer Science + Business Media, Inc.

Bult, A., and C. B. Lynch. 1996. Multiple selection responses in house mice bidirectionally selected for thermoregulatory nest-building behavior: Crosses of replicate lines. *Behavior Genetics* 26:439–446.

Bult, A., and C. B.Lynch. 1997. Nesting and fitness: Lifetime reproductive success in house mice bidirectionally selected for thermoregulatory nest-building behavior. *Behavior Genetics* 27:231–240.

Burgess, K. S., J. Singfield, V. Melendez, and P. G. Kevan. 2004. Pollination biology of *Aristolochia grandiflora* (Aristolochiaceae) in Veracruz, Mexico. *Annals of the Missouri Botanical Garden* 91:346–356.

Burghardt, G. M. 1998. The evolutionary origins of play revisited. In *Animal Play: Evolutionary, Comparative, and Ecological Perspectives*, edited by M. Bekoff and J. A. Byers. Cambridge: Cambridge University Press, pp. 1–26.

Burghardt, G. M. 2005. *The Genesis of Animal Play*. Cambridge, MA: MIT press.

Burghardt, G. M., and H. W. Greene. 1988. Predator simulation and duration of death feigning in neonate hognose snakes. *Animal Behaviour* 36:1842–1844.

Burkhardt, D., and I. de la Motte. 1983. How stalk-eyed flies view stalk-eyed flies: Observations and measurements of the eyes of *Cyrtodiopsis whitei* (Diopsidea, Diptera). *Journal of Comparative Physiology A* 151:407–421.

Burmeister, S. S., E. D. Jarvis, and R. D. Fernald. 2005. Rapid behavioral and genomic responses to social opportunity. *PLoS Biology* 3:1996–2004.

Burt, J. M., A. L. O'Loghlen, C. N. Templeton, S. E. Campbell, and M. D. Beecher. 2007. Assessing the importance of social factors in bird song learning: A test using computer-simulated tutors. *Ethology* 113:917–925.

Butts, K. O., and J. C. Lewis. 1982. The importance of prairie dog towns to burrowing owls in Oklahoma. *Proceedings of the Oklahoma Academy of Science* 62:46–52.

Byers, D., R. L. Davis, and J. A. Kiger, Jr. 1981. Defect in cyclic AMP phosphodiesterase due to the *dunce* mutation of learning in *Drosophila melanogaster*. *Nature* 289:79–81.

Byers, J. A., and L. Waits. 2006. Good genes sexual selection in nature. *Proceedings of the National Academy of Sciences* 103:16343–16345.

Byers, J. A., and C. B. Walker. 1995. Refining the motor training hypothesis for the evolution of play. *American Naturalist* 146:25–40.

Byers, J. A., A. A. Byers, and S. J. Dunn. 2006. A dry summer diminishes mate search effort by pronghorn females: Evidence for a significant cost of mate search. *Ethology* 112:74–80.

Byers, J. A., J. D. Moodie, and N. Hall. 1994. Pronghorn females choose vigorous mates. *Animal Behaviour* 47:33–43.

Byrne, J. H. 1987. Cellular analysis of associative learning. *Physiological Reviews* 67:329–439.

Byrom, A. E., and C. J. Krebs. 1999. Natal dispersal of juvenile Arctic ground squirrels in the boreal forest. *Canadian Journal of Zoology* 77:1048–1059.

Cahalane, V. H. 1961. *Mammals of North America*. New York: Macmillan.

Caldwell, R. L., and H. Dingle. 1976. Stomatopods. *Scientific American* 234:80–89.

Calvert, W. H., and L. P. Brower. 1986. The location of the monarch butterfly (*Danaus plexippus* L.). *Journal of the Lepidopteran Society* 40:164–187.

Calvert, W. H., L. E. Hedrick, and L. P. Brower, 1979. Mortality of the monarch butterfly (*Danaus plexippus* L.): Avian predation at five overwintering sites in Mexico. *Science* 204:847–851.

Camhi, J. M. 1980. The escape system of the cockroach. *Scientific American* 248:158–172.

Camhi, J. M. 1984. *Neuroethology*. Sunderland, MA: Sinauer.

Camhi, J. M. 1988. Escape behavior in the cockroach: Distributed neural processing. *Experientia* 44:401–408.

Camhi, J. M. , W. Tom, and S. Volman. 1978. The escape behavior of the cockroach *Periplaneta americana*. II. Detection of natural predators by air displacement. *Journal of Comparative Physiology* 128:203–212.

Candolin, U. 2000a. Increased signalling effort when survival prospects decrease: male-male competition ensures honesty. *Animal Behaviour* 60:417–422.

Candolin, U. 2000b. Male-male competition ensures honest signaling of male parental ability in the three-spined stickleback (*Gasterosteus aculeatus*). *Behavioral Ecology and Sociobiology* 49:57–61.

Caplan, A. L., ed. 1978. *The Sociobiology Debate*. New York: Harper & Row Publishers.

Caputi, A. A., and R. Budelli. 2006. Peripheral electrosensory imaging by weakly electric fish. *Journal of Comparative Physiology A* 192:587–600.

Caputi, A. A., R. Budelli, K. Grant, and C. C. Bell. 1998. The electrical image in weakly electric fish: Physical images of resistive objects in *Gnathonemus petersii*. *Journal of Experimental Biology* 201:2115–2128.

Carballada, R., and P. Esponda. 1992. Role of fluid from seminal vesicles and coagulating glands in sperm transport into the uterus and fertility in rats. *Journal of Reproduction and Fertility* 95:639–648.

Carew, T. J. 2000. *Behavioral Neurobiology*. Sunderland MA: Sinauer.

Carew, T. J., H. M. Pinsker, and E. R. Kandel. 1972. Long-term habituation of a defensive withdrawal reflex in *Aplysia*. *Science* 175:451–454.

Carlier, P., and L. Lefebvre. 1997. Ecological differences in social learning between adjacent, mixing populations of Zenaida doves. *Ethology* 103:772–784.

Caro, T. M. 1986a. The functions of stotting: A review of hypotheses. *Animal Behaviour* 34:649–662.

Caro, T. M. 1986b. The functions of stotting in Thomson's gazelles: Some tests of the predictions. *Animal Behaviour* 34:663–684.

Caro, T. M. 1988. Adaptive significance of play: Are we getting closer? *Trends in Ecology and Evolution* 88:50–54.

Caro, T. M. 1994. *Cheetahs of the Serengeti Plains: Group Living in an Asocial Species*. Chicago: University of Chicago Press.

Caro, T. M. 1995. Short-term costs and correlates of play in cheetahs. *Animal Behaviour* 49:333–345.

Carpenter, F. L., D. C. Paton, and M. A. Hixon. 1983. Weight gain and adjustment of feeding territory size in migrant hummingbirds. *Proceedings of the National Academy of Sciences* 80:7259–7263.

Carroll, L. S., and W. K. Potts. 2007. Sexual selection: using social ecology to determine fitness differences. In *Rodent Societies: An Ecological and Evolutionary Perspective*, edited by J. O. Wolff and P. W. Sherman. Chicago: The University of Chicago Press, pp. 57–67.

Carroll, S. P., and L. Moore. 1993. Hummingbirds take their vitamins. *Animal Behaviour* 46:817–820.

Carruth, L. L., R. E. Jones, and D. O. Norris. 2002. Cortisol and Pacific salmon: a new look at the role of stress hormones in olfaction and home-stream migration. *Integrative and Comparative Biology* 42:574–581.

Castellucci, V., and E. R. Kandel. 1974. A quantal analysis of the synaptic depression underlying the gill-withdrawal reflex in *Aplysia*. *Proceedings of the National Academy of Sciences* 71:5004–5008.

Catania, K. C. 2006. Underwater 'sniffing' by semi-aquatic mammals. *Nature* 444:1024–1025.

Catania, K. C., and J. H. Kaas. 1996. The unusual nose and brain of the star-nosed mole. *BioScience* 46:578–586.

Catania, K. C., and F. E. Remple. 2004. Tactile foveation in the star-nosed mole. *Brain, Behavior and Evolution* 63:1–12.

Champagne, F. A., I. C. G. Weaver, J. Diorio, S. Dymov, M. Szyf, and M. J. Meaney. 2006. Maternal care associated with methylation of the estrogen receptor–alpha1b promoter and estrogen receptor alpha expression in the medial preoptic area of female offspring. *Endocrinology* 147:2909–2916.

Champalbert, A., and P.-P. Lachaud. 1990. Existence of a sensitive period during ontogenesis of social behavior in a primitive ant. *Animal Behaviour* 39:850–859.

Chang, H-S, M. S. Anway, S. S. Rekow, and M. K. Skinner. 2006. Transgenerational epigenetic imprinting of male germline by endocrine disruptor exposure during gonadal sex determination. *Endocrinology* 147:5524–5541.

Chapman, T., L. F. Liddle, J. M. Kalb, M. F. Wolfner, and L. Partridge. 1995. Cost of mating in *Drosophila melanogaster* females is mediated by male accessory gland products. *Nature* 373:241–244.

Chapman, T., D. M. Neubaum, M. F. Wolfner, and L. Partridge. 2000. The role of male accessory gland protein Acp36DE in sperm competition in *Drosophila melanogaster*. *Proceedings of the Royal Society of London B* 267:1097–1105.

Charnov, E. L. 1976. Optimal foraging: the marginal value theorem. *Theoretical Population Biology* 9:129–136.

Charrier, I., and C. B. Sturdy. 2005. Call-based recognition in black-capped chickadees. *Behavioural Processes* 70:271–281.

Cheetham, S. A., M. D. Thom, F. Jury, W. E. R. Ollier, R. J. Beynon, and J. L. Hurst. 2007. The genetic basis of individual-recognition signals in the mouse. *Current Biology* 17: 1771–1777.

Cheney, D. L., and R. M. Seyfarth. 1988. Assessment of meaning and detection of unreliable signals by vervet monkeys. *Animal Behaviour* 36:477–486.

Cheng, M. W., and R. L. Caldwell. 2000. Sex identification and mating in the blue-ringed octopus, *Hapalochlaena lunulata*. *Animal Behaviour* 60:27–33.

Chittka, L., T. C. Ings, and N. E. Raine. 2004. Chance and adaptation in the evolution of island bumblebee behaviour. *Population Ecology* 46:243–251.

Choleris, E., M. Kavaliers, and D. W. Pfaff. 2004. Functional genomics of social recognition. *Journal of Neuroendocrinology* 16:383–389.

Christy, J. H., and M. Salmon. 1991. Comparative studies of reproductive behavior in mantis shrimps and fiddler crabs. *American Zoologist* 31:329–337.

Clark, C., and M. Mangel. 2000. *Dynamic State Variable Models in Ecology: Methods and Applications*. Oxford: Oxford University Press.

Clark, R. B. 1960. Habituation of the polychaete *Nereis* to sudden stimuli. 1. General properties of the habituation process. *Animal Behaviour* 8:82–91.

Clark, R. W. 2005. Pursuit-deterrent communication between prey animals and timber rattlesnakes (*Crotalus horridus*): the response of snakes to harassment displays. *Behavioral Ecology and Sociobiology* 59:258–261.

Clarke, A. L., B.-R. Sæther, and E. Røskaft. 1997. Sex biases in avian dispersal: A reappraisal. *Oikos* 79:429–438.

Clarke, B. 1969. The evidence for apostatic selection. *Heredity* 24:347–352.

Clobert, J., J. O. Wolff, J. D. Nichols, E. Danchin, and A. A. Dhondt. 2001. Introduction. In *Dispersal*, edited by J. Clobert, E. Danhin, A. A. Dhondt, and J. D. Nichols. Oxford: Oxford University Press, pp. xvii–xxi.

Clucas, B., M. P. Rowe, D. H. Owings, and P. C. Arrowood. 2008. Snake scent application in ground squirrels, *Spermophilus* spp.: a novel form of antipredator behaviour? *Animal Behaviour* 75:298–307.

Clutton-Brock, T. H. 1989. Female transfer and inbreeding avoidance in social mammals. *Nature* 337:70–72.

Clutton-Brock, T. H. 1991. *The Evolution of Parental Care*. Princeton, NJ: Princeton University Press.

Clutton-Brock, T. H. 2002. Breeding together: kin selection and mutualism in cooperative vertebrates. *Science* 296:69–72.

Clutton-Brock, T. H., and C. Godfray. 1991. Parental investment. In *Behavioural Ecology: An Evolutionary Approach*, 3rd ed., edited by J. R. Krebs and N. B. Davies. Oxford: Blackwell Scientific, pp. 7–29.

Clutton-Brock, T. H., and P. H. Harvey. 1979. Comparison and adaptation. *Proceedings of the Royal Society of London B* 205:547–565.

Clutton-Brock, T. H., and P. H. Harvey. 1984. Comparative approaches to studying adaptation. In *Behavioural Ecology: An Evolutionary Approach*, 2nd ed., edited by J. R. Krebs and N. B. Davies. Sunderland, MA: Sinauer, pp. 7–29.

Clutton-Brock, T. H., and K. Isvaran. 2006. Paternity loss in contrasting mammalian societies. *Biology Letters* 2:513–516.

Clutton-Brock, T. H., and A. C. J. Vincent. 1991. Sexual selection and the potential reproductive rates of males and females. *Nature* 351:58–60.

Clutton-Brock, T. H., F. E. Guiness, and S. D. Albon. 1982. *Red Deer: The Behaviour and Ecology of Two Sexes*. Chicago: University of Chicago Press.

Clutton-Brock, T. H., O. F. Price, and A. D. C. MacColl. 1992. Mate retention, harassment, and the evolution of ungulate leks. *Behavioral Ecology* 3:234–242.

Clutton-Brock, T. H., S. J. Hodge, G. Spong, A. F. Russell, N. R. Jordan, N. C. Bennett, L. L. Sharpe, and M. B. Manser. 2006. Intrasexual competition and sexual selection in cooperative mammals. *Nature* 444:1065–1068.

Clutton-Brock, T. H., A. F. Russell, L. L. Sharpe, P. N. M. Brotherton, G. M. McIlrath, S. White, and E. Z. Cameron. 2001. Effects of helpers on juvenile development and survival in meerkats. *Science* 293:2446–2449.

Cockburn, A. 1998. Evolution of helping behavior in cooperatively breeding birds. *Annual Review of Ecology and Systematics* 29:141–472.

Cockburn, A. 2006. Prevalence of different modes of parental care in birds. *Proceedings of the Royal Society of London B* 273:1375–1383.

Cockburn, A., R. A. Sims, H. L. Osmond, D. J. Green, M. C. Double, and R. A. Mulder. 2008. Can we measure the benefits of help in cooperatively breeding birds: the case of superb fairy-wrens *Malurus cyaneus*? *Journal of Animal Ecology* 77:430–438.

Coleman, S. W., G. L. Patricelli, and G. Borgia. 2004. Variable female preferences drive complex male displays. *Nature* 428:742–746.

Collett, M., T. S. Collett, S. Bisch, and R. Wehner. 1998. Local and global vectors in desert ant navigation. *Nature* 394:269–272.

Collett, T. S., and M. F. Land. 1978. How hoverflies compute interception courses. *Journal of Comparative Physiology* 125:191–204.

Compagnone, N. A., and S. H. Mellon. 2000. Neurosteroids: biosynthesis and function of these neuromodulators. *Frontiers in Neuroendocrinology* 21:1–56.

Connor, V. M. 1986. The use of mucous trails by intertidal limpets to enhance food resources. *Biological Bulletin* 171:548–564.

Cook, A. 1981. Huddling and the control of water loss by the slug *Limax pseudoflavus* Evans. *Animal Behaviour* 29:289–298.

Cook, M., and S. Mineka. 1990. Selective associations in the observational conditioning of fear in rhesus monkeys. *Journal of Experimental Psychology* 16:372–389.

Cook, P. A., and N. Wedell. 1999. Non-fertile sperm delay female remating. *Nature* 397:486.

Cooper, J. B. 1985. Comparative psychology and ethology. In *Topics in the History of Psychology*, edited by G. A. Kimble and K. Schlesinger. Hillsdale, NJ: Erlbaum.

Cooper, J., A. T. Scholz, R. M. Horral, A. D. Hasler, and D. M. Madison. 1976. Experimental confirmation of the olfactory hypothesis with homing, artificially imprinted coho salmon (*Oncorhynchus kisutch*). *Journal of the Fisheries Research Board of Canada*. 28:703–710.

Cooper, W. E., V. Pérez-Mellado, and L. J. Vitt. 2004. Ease and effectiveness of costly autotomy vary with predation intensity among lizard populations. *Journal of Zoology* 262:243–255.

Cordoba-Aguilar, A. 1999. Male copulatory sensory stimulation induces female ejection of rival sperm in a damselfly. *Proceedings of the Royal Society of London B* 266:779–784.

Cott, H. B. 1940. *Adaptive Colouration in Animals*. London: Methuen.

Coulon, A., J.-F. Cosson, N. Morellet, J.-M. Angibault, B. Cargnelutti, M. Galan, S. Aulagnier, and A. J. M. Hewison. 2006. Dispersal is not female biased in a resource-defense mating ungulate, the European roe deer. *Proceedings of the Royal Society of London B* 273:341–348.

Coulson, J. C. 1966. The influence of the pair bond and age on the breeding biology of the kittiwake gull (*Risa tridactyla*). *Journal of Animal Biology* 35:269–279.

Cowlishaw, G. 1997. Trade-offs between foraging and predation risk determine habitat use in a desert baboon population. *Animal Behaviour* 53:667–686.

Cox, C. R., and B. J. LeBoeuf. 1977. Female incitation of male competition: A mechanism of mate selection. *American Naturalist* 111:317–335.

Crampton, W. G. R., and C. D. Hopkins. 2005. Nesting and paternal care in the weakly electric fish *Gymnotus* (Gymnotiformes: Gymnotidae) with descriptions of larval and adult electric organ discharges of two species. *Copeia* 2005:48–60.

Crane, J. 1941. Eastern Pacific expeditions of the New York Zoological Society XXVI: Crabs of the genus *Uca* from the West Coast of Central America. *Zoologica* 26:145–203.

Creel, S. 2005. Dominance, aggression, and glucocorticoid levels in social carnivores. *Journal of Mammalogy* 86:255–264.

Creel, S., and P. M. Waser. 1994. Inclusive fitness and reproductive strategies in dwarf mongooses. *Behavioral Ecology* 5:339–348.

Creel, S. R., and P. M. Waser. 1997. Variation in reproductive suppression among dwarf mongooses: Interplay between mechanisms and evolution. In *Cooperative Breeding in Mammals*, edited by N. G. Solomon and J. A. French. Cambridge: Cambridge University Press.

Creel, S., N. M. Creel, M. G. L. Mills, and S. L. Monfort. 1997. Rank and reproduction in cooperatively breeding African wild dogs—Behavioral and endocrine correlates. *Behavioral Ecology* 8:298–306.

Creel S., N. Creel, D. E. Wildt, and S. L. Monfort. 1992. Behavioral and endocrine mechanisms of reproductive suppression in Serengeti dwarf mongooses. *Animal Behaviour* 43:231–245.

Creel, S., S. L. Monfort, N. Marusha Creel, D. E. Wildt, and P. M. Waser. 1995. Pregnancy, oestrogens and future reproductive success in Serengeti dwarf mongooses. *Animal Behaviour* 50:1132–1135.

Cremer, S., S. Armitage, and P. Schmid-Hempel. 2007. Social immunity. *Current Biology* 17:R693–R702.

Crespi, B. J. 1992. Eusociality in Australian gall thrips. *Nature* 359:724–726.

Crespi, B. J., and J. E. Ragsdale. 2000. A skew model for the evolution of sociality via manipulation: why it is better to be feared than loved. *Proceedings of the Royal Society of London B* 267:821–828.

Crews, D. 1974. Effects of castration and subsequent androgen replacement therapy on male courtship behavior and environmentally induced ovarian recrudescence in the lizard *Anolis carolinensis*. *Journal of Comparative Physiology* 87:963–969.

Crews, D. 1979a. Neuroendocrinology of lizard reproduction. *Biology of Reproduction* 20:51–73.

Crews, D. 1979b. The hormonal control of behavior in a lizard. *Scientific American* 241:180–187.

Crews, D. 1983. Control of male sexual behavior in the Canadian red-sided garter snake. In *Hormones and Behavior in Higher Animals*, edited by J. Balthazart, E. Prove, and R. Gilles. New York: Springer-Verlag, pp. 398–406.

Crews, D. 1984. Gamete production, sex hormone secretion, and mating behavior uncoupled. *Hormones and Behavior* 18:22–28.

Crews, D. 1987. Diversity and evolution of behavioral controlling mechanisms. In *Psychobiology of Reproductive Behavior*, edited by D. Crews. Upper Saddle River, NJ: Prentice Hall, pp. 88–119.

Crews, D., and T. Groothuis. 2005. Tinbergen's fourth question, ontogeny: sexual and individual differentiation. *Animal Biology* 55:343–370.

Crews, D., and M. C. Moore. 2005. Historical contributions of research on reptiles to behavioral neuroendocrinology. *Hormones and Behavior* 48:384–394.

Crews, D., V. Hingorani, and R. J. Nelson. 1988. Role of the pineal gland in the control of annual reproductive behavioral and physiological cycles in the red-sided garter snake (*Thamnophis sirtalis parietalis*). *Journal of Biological Rhythms* 3:293–302.

Crews, D., V. Tranina, F. T. Wetzel, and C. Muller. 1978. Hormone control of male reproductive behavior in the lizard *Anolis carolinensis*: Role of testosterone, dihydrotestosterone, and estradiol. *Endocrinology* 103:1814–1821.

Crews, D., A. C. Gore, T. S. Hsu, N. L. Dangleben, M. Spinetta, T. Schaller, and M. D. Skinner Anway, M. L. 2007. Transgenerational epigenetic imprints on mate preference. *Proceedings of the National Academy of Sciences* 104:5942–5956.

Cristol, D. A. 1995. The coat-tail effect in merged flocks of dark-eyed juncos: Social status depends on familiarity. *Animal Behaviour* 50:151–159.

Cristol, D. A., and P. V. Switzer. 1999. Avian prey dropping behavior. II. American crows and walnuts. *Behavioral Ecology* 10:220–226.

Crook, J. H. 1964. The evolution of social organisation and visual communication in the weaver bird (Ploceinae). *Behaviour Supplement* 10:1–178.

Crook, J. H. 1970. The socio-ecology of primates. In *Social Behavior in Birds and Mammals*, edited by J. H. Crook. London: Academic Press, pp. 103–106.

Croze, H. 1970. Searching image in carrion crows. *Zeitschrift für Tierpsychologie Supplement* 5:1–86.

Cullen, E. 1957a. Adaptations in the kittiwake in the British Isles. *Bird Study* 10:147–179.

Cullen, E. 1957b. Adaptations in the kittiwake to cliff nesting. *Ibis* 99:275–302.

Cummings, M. E. 2004. Modelling divergence in luminance and chromatic detection performance across measured divergence in surfperch (Embiotocidae) habitats. *Vision Research* 44:1127–1145.

Cummings, M. E. 2007. Sensory trade-offs predict signal divergence in surfperch. *Evolution* 61:530–545.

Cummings, M. E., and J. C. Partridge. 2001. Visual pigments and optical habitats of surfperch (Embiotocidae) in the California kelp forest. *Journal of Comparative Physiology A* 187:875–889.

Curio, E. 1978. The adaptive significance of avian mobbing. I. Teleonomic hypotheses and predictions. *Ethology* 48:175–183.

Curio, E., and K. Regelmann. 1986. Predator harassment implies a deadly risk: A reply to Hennessy. *Ethology* 72:75–78.

Cuthill, I. C., M. Stevens, J. Sheppard, T. Maddocks, C. Alejandro Párraga, and T. S. Troscianko. 2005. Disruptive coloration and background pattern matching. *Nature* 434:72–74.

Daanje, A. 1950. On locomotor movements in birds and the intention movements derived from them. *Behaviour* 3:48–98.

Dacke, M., and M. V. Srinivasan. 2007. Honeybee navigation: distance estimation in the third dimension. *Journal of Experimental Biology* 210:845–853.

Daly, M., and M. Wilson. 1983. *Sex, Evolution and Behavior*. 2nd ed. Boston: PWS-Kent.

Daniels, S. J., and J. R. Walters. 2000. Between-year breeding dispersal in red-cockaded woodpeckers: multiple causes and estimated cost. *Ecology* 81:2473–2484.

Darst, C. R., M. E. Cummiings, and D. C. Cannatella. 2006. A mechanism for diversity in warning signals: conspicuousness versus toxicity in poison frogs. *Proceedings of the National Academy of Sciences* 103:5852–5857.

Darwin, C. 1859. *On the Origin of Species by Natural Selection or The Preservation of Favored Races in the Struggle for Life*. London: Murray.

Darwin, C. 1871. *The Descent of Man and Selection in Relation to Sex*. London: Murray.

Darwin, C. 1873. *Expression of the Emotions in Man and Animals*. London: Murray.

Davies, N. B. 1978. Territorial defense in the speckled wood butterfly, *Pararge aegeria*: the resident always wins. *Animal Behaviour* 26:148–147.

Davies, N. B. 1983. Polyandry, cloaca-pecking and sperm competition in dunnocks. *Nature* 302:334–336.

Davies, N. B. 1991. Mating systems. In *Behavioural Ecology: An Evolutionary Approach*, edited by J. R. Krebs and N. B. Davies. Oxford: Blackwell Scientific, pp. 263–294.

Davies, N. B. 2000. *Cuckoos, Cowbirds and Other Cheats*. London: T., and A. D. Poyser.

Davies, N. B., and M. de L. Brooke. 1988. Cuckoo versus reed warblers: Adaptations and counteradaptations. *Animal Behaviour* 36:262–284.

Davies, N. B., and T. R. Halliday. 1978. Deep croaks and fighting assessment in toads (*Bufo bufo*). *Nature* 274:683–685.

Davies, N. B., and A. I. Houston. 1984. Territory economics. In *Behavioural Ecology: An Evolutionary Approach*, 2nd edition, edited by J. R. Krebs and N. B. Davies. Sunderland, MA: Sinauer.

Davis, A. K., N. Cope, A. Smith, and M. J. Solensky. 2007. Wing color predicts future mating success in male monarch butterflies. *Annals of the Entomological Society of America* 100: 339–344.

Davis, J. M., and J. A. Stamps. 2004. The effect of natal experience on habitat preferences. *Trends in Ecology and Evolution* 19:411–416.

Davis, M. S. 1987. Acoustically mediated neighbor recognition in the North American bullfrog, *Rana catesbeiana*. *Behavioral Ecology and Sociobiology* 21:185–190.

Davis, R. L., J. Cherry, B. Dauwalder, P.-L. Han, and E. Skoulakis. 1995. The cyclic AMP system and *Drosophila* learning. *Molecular and Cellular Biochemistry* 149–150:271–178.

Dawkins, M. S. 1986. *Unravelling Animal Behavior*. Essex: Longman.

Dawkins, M. S. 1989. The future of ethology: How many legs are we standing on? In *Perspectives in Ethology*, edited by P. P. G. Bateson and P. H. Klopfer. New York: Plenum Press. pp. 47–54.

Dawkins, R. 1976. *The Selfish Gene*. New York: Oxford University Press.

Dawkins, R. 1980. Good strategy or evolutionarily stable strategy? In *Sociobiology: Beyond Nature/Nurture*, edited by G. Barlow and R. Silverberg. Boulder, CO: Westview, pp. 331–367.

Dawkins, R. 1982. *The Extended Phenotype*. Oxford:Freeman.

Dawkins, R., and J. R. Krebs. 1978. Animal signals: Information or manipulation? In *Behavioural Ecology: An Evolutionary Approach*, edited by J. R. Krebs and N. B. Davies. Sunderland, MA: Sinauer, pp. 282–309.

Dawson, B. V., and B. M. Foss. 1965. Observational learning in budgerigars. *Animal Behaviour* 13:470–474.

de Belle, J. S., A. J. Hilliker, and M. B. Sokolowski. 1989. Genetic localization of *foraging (for)*: A major gene for larval behavior in *Drosophila melanogaster. Genetics* 123:157–163.

de Belle, J. S., and M. B. Sokolowski. 1987. Heredity of *rover/sitter*: Alternative foraging strategies of *Drosophila melanogaster* larvae. *Heredity* 59:73–83.

de la Motte, I., and D. Burkhardt. 1983. Portrait of an Asian stalk-eyed fly. *Naturwissenschaften* 70:451–461.

de Vos, G. E. 1979. Adaptedness of arena behavior in black grouse (*Tetrao tetrix*) and other grouse species (Tetraonidae). *Behaviour* 68:277–314.

de Waal, F. B. M. 1991. Rank distance as a central feature of rhesus monkey social organization: A sociometric analysis. *Animal Behaviour* 41:383–395.

de Waal, F. B. M. 2001. *The Ape and the Sushi Master.* New York: Basic Books.

de Waal, F. B. M. 2005. A century of getting to know the chimpanzee. *Nature* 437:56–59.

Dean, J., D. J. Aneshansley, H. E. Edgerton, and T. Eisner. 1990. Defensive spray of the bombardier beetle: A biological pulse jet. *Science* 248:1219–1221.

DeCoursey, P. J. 1986. Light-sampling behavior in photoentrainment of a rodent circadian rhythm. *Journal of Comparative Physiology A* 159:161–169.

DeCoursey, P. J. 2004a. The behavioral ecology and evolution of biological timing systems. In *Chronobiology: Biological Timekeeping*, edited by J. C. Dunlap, J. J. Loros, and P. J. DeCoursey. Sunderland, MA: Sinauer, pp. 27–65.

DeCoursey, P. J. 2004b. Functional organization of circadian systems in multicellular organisms. In *Chronobiology: Biological Timekeeping*, edited by J. C. Dunlap, J. J. Loros, and P. J. DeCoursey. Sunderland, MA: Sinauer, pp. 145–178.

DeCoursey, P. J., J. K. Walker, and S. A. Smith. 2000. A circadian pacemaker in free-living chipmunks: essential for survival? *Journal of Comparative Physiology A* 186:169–180.

DeCoursey, P. J., J. R. Krulas, G. Mele, and D. C. Holley. 1997. Circadian performance of suprachiasmatic nuclei (SCN)-lesioned antelope ground squirrels in a desert enclosure. *Physiology and Behavior* 625:1099–1108.

Dehn, M. M. 1990. Vigilance for predators: Detection and dilution effects. *Behavioral Ecology and Sociobiology* 26:337–342.

Delaney, K. J., and L. G. Higley. 2006. An insect countermeasure impacts plant physiology: midrib vein cutting, defoliation, and leaf photosynthesis. *Plant Cell and Environment* 29:1245–1258.

Delcomyn, F. 1980. Neural basis of rhythmic behavior in animals. *Science* 210:492–498.

Delgado, M. M., and V. Penteriani. 2007. Vocal behaviour and neighbour spatial arrangement during vocal displays in eagle owls. *Journal of Zoology* 271:3–10.

DeNault, L. K., and D. A. McFarlane. 1995. Reciprocal altruism between male vampire bats, *Desmodus rotundus*. *Animal Behaviour* 49:855–856.

Dennis, T. E., M. J. Rayner, and M. M. Walker. 2007. Evidence that pigeons orient to geomagnetic intensity during homing. *Proceedings of the Royal Society of London B* 274:1153–1158.

Detto, T., P. R. Y. Backwell, J. M. Hemmi, and J. Zeil. 2006. Visually mediated species and neighbour recognition in fiddler crabs (*Uca mjoebergi* and *Uca capricornis*). *Proceedings of the Royal Society of London B* 273:1661–1666.

Devillard, S., D. Allaine, J.-M. Gaillard, and D. Pontier. 2004. Does social complexity lead to sex-biased dispersal in polygynous mammals? *Behavioral Ecology* 15:83–87.

Dewsbury, D. A. 1978. What is (was?) the 'fixed action pattern'? *Animal Behaviour* 26:310–311.

Dewsbury, D. A. 1982a. Ejaculate cost and male choice. *American Naturalist* 119:601–610.

Dewsbury, D. A. 1982b. Dominance rank, copulatory behavior, and differential reproduction. *Quarterly Review of Biology* 57:135–158.

Dewsbury, D. A. 1984. *Comparative Psychology in the Twentieth Century*. Stroudsburg, PA: Hutchinson Ross.

Dewsbury, D. A. 1988. A test of the role of copulatory plugs in sperm competition in deer mice (*Peromyscus maniculatus*). *Journal of Mammalogy* 69:854–857.

Dewsbury, D. A. 1989. A brief history in the study of animal behavior in North America. *Perspectives in Ethology* Volume 8, Whither ethology? pp. 85–122.

Dewsbury, D. A., and D. K. Sawrey. 1984. Male capacity as related to sperm production, pregnancy initiation, and sperm competition in deer mice (*Peromyscus maniculatus*). *Behavioral Ecology and Sociobiology* 16:37–47.

Dial, B. E., and L. C. Fitzpatrick. 1983. Lizard tail autotomy: Function and energetics of postautotomy tail movement in *Scincella lateralis*. *Science* 219:391–393.

Díaz, P. C., C. Grüter, and W. M. Farina. 2007. Floral scents affect the distribution of hive bees around dancers. *Behavioral Ecology and Sociobiology* 61:1589–1597.

Dickinson, J. L., and B. J. Hatchwell. 2004. Fitness consequences of helping. In *Ecology and Evolution of Cooperative Breeding in Birds*, edited by W. Koenig and J. Dickinson. Cambridge: Cambridge University Press, pp. 48–66.

Dickson, B. J. 2008. Wired for sex: The neurobiology of *Drosophila* mating decisions. *Science* 322:904–908.

Diesel, R. 1990. Sperm competition and reproductive success in the decapod *Inachus phalangium* (Majiidae): A male ghost spider crab that seals off rival's sperm. *Journal of Zoology* 220:213–223.

Dill, P. A. 1977. Development of behavior in alevins of Atlantic salmon, *Salmo salar*, and rainbow trout, *S. gairdneri*. *Animal Behaviour* 25:116–121.

Dingemanse, N. J., C. Both, A. J. van Noordwijk, A. L. Rutten, and P. J. Drent. 2003. Natal dispersal and personalities in great tits (*Parus major*). *Proceedings of the Royal Society of London B* 270:741–747.

Dingle, H. 1980. Ecology and evolution of migration. In *Animal Migration, Orientation, and Navigation*, edited by S. A. Gauthreaux. New York: Academic Press, pp. 1–101.

Dingle, H., and R. Caldwell. 1969. The aggressive and territorial behavior of the mantis shrimp *Gonodactylus bredini* Manning (Crustacea: Stomatopoda). *Behaviour* 33:115–136.

Dingle, H., and V. A. Drake. 2007. What is migration? *BioScience* 57:113–121.

Dittman, A. H., and T. P. Quinn. 1996. Homing in Pacific salmon: mechanisms and ecological basis. *Journal of Experimental Biology* 199:83–91.

Dittrich, G., F. Gilbert, P. Green, P. McGregor, and D. Grewcock. 1993. Imperfect mimicry—a pigeon's perspective. *Proceedings of the Royal Society of London B* 251:195.

Dixson, A. F. 1998. *Primate Sexuality: Comparative Studies of Prosimians, Monkeys, Apes, and Human Beings*. Oxford: Oxford University Press.

Dobson, F. S. 1982. Competition for mates and predominant male dispersal in mammals. *Animal Behaviour* 30:1183–1192.

Dobson, F. S., and W. T. Jones. 1985. Multiple causes of dispersal. *American Naturalist* 126:855–858.

Doligez, B., E. Danchin, and J. Clobert. 2002. Public information and breeding habitat selection in a wild bird population. *Science* 297:1168–1170.

Domenici, P., D. Booth, J. M. Blagburn, and J. P. Bacon. 2008. Cockroaches keep predators guessing using preferred escape trajectories. *Current Biology* 18:1792–1796.

Dong, W., X. Tang, Y. Yu, R. Nilsen, R. Kim, J. Griffith, J. Arnold, and H.-B. Schüttler. 2008. Systems biology of the clock in *Neurospora crassa*. *PLoS ONE* 3:e3105.

Dorries, K. M., E. Adkins-Regan, and B. P. Halpern. 1995. Olfactory sensitivity to the pheromone androsterone is sexually dimorphic in the pig. *Physiology and Behavior* 57:255–259.

Drake, A., D. Fraser, and D. M. Weary. 2008. Parent-offspring resource allocation in domestic pigs. *Behavioral Ecology and Sociobiology* 62:309–319.

Drent, R., and P. Swierstra. 1977. Goose flocks and food-finding: field experiments with barnacle geese in winter. *Wildfowl* 28:15–20.

Drickamer, L. C., P. A. Gowaty, and D. M. Wagner. 2003. Free mutual mate preferences in house mice affect reproductive success and offspring performance. *Animal Behaviour* 65:105–114.

Dudai, Y. 1979. Behavioral plasticity in a *Drosophila* mutant, dunce DB276. *Journal of Comparative Physiology* 130:271–275.

Dudai, Y., Y.-N. Jan, D. Byers, W. G. Quinn, and S. Benzer. 1976. *Dunce*, a mutant of *Drosophila* deficient in learning. *Proceedings of the National Academy of Sciences* 73:1684–1688.

Duffield, G. E. 2003. DNA microarray analyses of circadian timing: The genomic basis of biological time. *Journal of Neuroendocrinology* 15:991–1002.

Duffy, J. E. 1996. Eusociality in a coral-reef shrimp. *Nature* 381:512–514.

Dugatkin, L. A. 1997. *Cooperation Among Animals: An Evolutionary Perspective*. New York: Oxford University Press.

Dulac, C. 2005. Sex and the single splice. *Cell* 121:664–666.

Dunham, P. J. 1986. Mate guarding in amphipods: A role for brood patch stimuli. *Biological Bulletin* 170:526–531.

Dunham, P. J., and A. Hurshman. 1990. Precopulatory mate guarding in the amphipod, *Gammarus lawrencianus*: Effects of social stimulation during the post-copulation interval. *Animal Behaviour* 39:976–979.

Dupret, D., A. Fabre, M. D. Dobrossy, A. Panatier, J. J. Rodríguez, S. Lamarque, V. Lemaire, S. H. R. Oliet, P.-V. Pliazza, and D. N. Abrous. 2007. Spatial learning depends on both the addition and removal of new hippocampal neurons. *PLoS Biology* 5:1683–1694.

Dussourd, D. E. 1999. Behavioral sabotage of plant defense: Do vein cuts and trenches reduce insect exposure to exudate? *Journal of Insect Behavior* 12:501–515.

Dussutour, A., S. J. Simpson, E. Despland, and N. Colasurdo. 2007. When the group denies individual nutritional wisdom. *Animal Behaviour* 74:931–939.

Dyer, A. B., R. Lickliter, and G. Gottlieb. 1989. Maternal peer imprinting in mallard ducklings under experimentally simulated natural social conditions. *Developmental Psychobiology* 22:463–475.

Dzieweczynski, T. L., A. C. Eklund, and W. J. Rowland. 2006. Male 11-ketotestosterone levels change as a result of being watched in Siamese fighting fish, *Betta splendens*. *General and Comparative Endocrinology* 147:184–189.

Dzieweczynski, T. L., R. L. Earley, T. M. Green, and W. J. Rowland. 2005. Audience effect is context dependent in Siamese fighting fish, *Betta splendens*. *Behavioral Ecology* 16:1025–1030.

Earle, S. A. 1979. The gentle whales. *National Geographic* 155:2–17.

Eason, P. K. 1992. Optimization of territory shape in heterogeneous habitats—a field-study of the red-capped cardinal (*Paroaria gularis*). *Journal of Animal Ecology* 61:411–425.

Eason, P. K., and P. T. Sherman. 1995. Dominance status, mating strategies and copulation success in cooperatively polyandrous white-winged trumpeters, *Psophia leucoptera* (Aves: Psophiidae). *Animal Behaviour* 49:725–736.

Eason, P. K., and J. A. Stamps. 1992. The effect of visibility on territory size and shape. *Behavioral Ecology* 3:166–172.

Eason, P. K., and J. A. Stamps. 1993. An early warning system for detecting intruders in a territorial animal. *Animal Behaviour* 46:1005–1109.

Eason, P. K., and J. A. Stamps. 2001. The effect of visibility on space use by territorial red-capped cardinals. *Behaviour* 138:19–30.

Eason, P. K., G. A. Cobbs, and K. G. Trinca. 1999. The use of landmarks to define territorial boundaries. *Animal Behaviour* 58:85–91.

Ebensperger, L. A., and D. T. Blumstein. 2007. Nonparental infanticide. In *Rodent Societies: An Ecological and Evolutionary Perspective*, edited by J. O. Wolff and P. W. Sherman. Chicago: The University of Chicago Press, pp. 267–279.

Eberhard, W. G. 1996. *Female Control: Sexual Selection by Cryptic Female Choice*. Princeton, NJ: Princeton University Press.

Eckerle, K. P., and C. F. Thompson. 2006. Mate choice in house wrens: nest cavities trump male characteristics. *Behaviour* 143:253–271.

Edwards, J. S. 2006. The central nervous control of insect flight. *Journal of Experimental Biology* 209:4411–4413.

Eens, M., and R. Pinxten. 2000. Sex-role reversal in vertebrates: behavioral and endocrinological accounts. *Behavioural Processes* 51:135–147.

Eggert, A.-K., and S. K. Sakaluk. 1994. Sexual cannibalism and its relation to male mating success in sagebrush crickets, *Cyphoderris strepitans* (Haglidae: Orthoptera). *Animal Behaviour* 47:1171–1177.

Eibl-Eibesfeldt, I. 1966. Das Vertridigen der Eiablageplätze bei der Hood-Meerechse *(Amblyrhynchus cristatus venustissimus). Zeitschrift für Tierpsychologie* 23:627–631.

Eibl-Eibesfeldt, I. 1975. *Ethology: The Biology of Behavior*. 2nd ed. New York: Holt, Rinehart & Winston.

Eisenberg, J. F., N. A. Muckenhirn, and R. Rudran. 1972. The relation between ecology and social structure in primates. *Science* 176:963–874.

Eisner, T. 1958. The protective role of the spray mechanism of the bombardier beetle, *Brachynus ballistarius* Lec. *Journal of Insect Physiology* 2:215–220.

Eisner, T., and J. Meinwald. 1995. The chemistry of sexual selection. *Proceedings of the National Academy of Sciences* 92:50–55.

Eisner, T., D. F. Wiemer, L. W. Hayes, and J. Meinwald. 1978. Lucibufagins: defensive steroids from the fireflies *Photinus ignitus* and *P. marginellus* (Coleoptera: Lampyridae). *Proceedings of the National Academy of Sciences* 75:905–908.

Eisner, T., M. A. Goetz, D. E. Hill, S. R. Smedley, and J. Meinwald. 1997. Firefly 'femmes fatales' acquire defensive steroids (lucibufagins) from their firefly prey. *Proceedings of the National Academy of Sciences* 94:9723–9728.

Ekman, J., J. L. Dickinson, B. J. Hatchwell, and M. Griesser. 2004. Delayed dispersal. In *Ecology and Evolution of Cooperative Breeding in Birds*, edited by W. Koenig and J. Dickinson. Cambridge: Cambridge University Press, pp. 35–47.

Ellis, L. 1995. Dominance and reproductive success among nonhuman animals: a cross-species comparison. *Ethology and Sociobiology* 16:257–333.

Emery, N. J., and N. S. Clayton. 2005. Animal cognition. In *The Behavior of Animals: Mechanisms, Function, and Evolution*, edited by J. J. Bolhuis and L.-A. Giraldeau. Malden MA: Blackwell Publishing Ltd.

Emlen, S. T. 1967a. Migratory orientation in the indigo bunting (*Passerina cyanea*). Part I. Evidence for the use of celestial cues. *Auk* 84:309-342.

Emlen, S. T. 1967b. Migratory orientation in the indigo bunting (*Passerina cyanea*). Part II. Evidence for use of celestial cues. *Auk* 84:463-489.

Emlen, S. T. 1969. The development of migratory orientation in young Indigo Bunting. *Living Bird* 8:113–126.

Emlen, S. T. 1970. Celestial rotation: Its importance in the development of migratory orientation. *Science* 170:1198–1201.

Emlen, S. T. 1972. The ontogenetic development of orientation capabilities. *NASA Special Publication* NASA SP–262:191–210.

Emlen, S. T. 1975. Migration, orientation, navigation. In *Avian Biology*, edited by D. Farner. New York: Academic Press.

Emlen, S. T. 1991. Evolution of cooperative breeding in birds and mammals. In *Behavioural Ecology: An Evolutionary Approach*, 3rd ed., edited by J. R. Krebs and N. B. Davies. Oxford, Blackwell, pp. 301–337.

Emlen, S. T. 1997. Predicting family dynamics in social vertebrates. In *Behavioural Ecology: An Evolutionary Approach*, 4th ed., edited by J. R. Krebs and N. B. Davies. Oxford: Blackwell Science, pp. 228–253.

Emlen, S. T., and L. W. Oring. 1977. Ecology, sexual selection, and the evolution of mating systems. *Science* 197:215–223.

Emlen, S. T., and S. L. Vehrencamp. 1983. Cooperative breeding strategies among birds. In *Perspectives in Ornithology*, edited by A. H. Brush and G. A. Clark, Jr. Cambridge: Cambridge University Press, pp. 93–133.

Emlen, S. T., and P. H. Wrege. 1989. A test of alternative hypotheses for helping behavior in white-fronted bee-eaters of Kenya. *Behavioral Ecology and Sociobiology* 25:303–319.

Emlen, S. T., and P. H. Wrege. 2004. Division of labour in parental care behaviour of a sex-role-reversed shorebird, the wattled jacana. *Animal Behaviour* 68:847–855.

Enard, W., M. Przeworski, S. E. Fisher, C. S. L. Lai, V. Wiebe, T. Kitano, A. P. Monaco, and S. Pääbo. 2002. Molecular evolution of *FOXP2*, a gene involved in speech and language. *Nature* 418:869–872.

Endler, J. A. 1978. A predator's view of animal color patterns. *Evolutionary Biology* 11:319–364.

Endler, J. A. 1981. An overview of the relationships between mimicry and crypsis. *Biological Journal of the Linnean Society* 16:25–31.

Endler, J. A. 1992. Signals, signal conditions, and the direction of evolution. *American Naturalist* 139:S125–S153.

Endler, J. A. 2006. Disruptive and cryptic coloration. *Proceedings of the Royal Society of London B* 273:2425–2426.

Engel, J. E., and R. R. Hoy. 2000. A cGMP-dependent kinase gene, *foraging*, modifies habituation-like response decrement of the giant fiber escape circuit in *Drosophila*. *Learning and Memory* 7:341–352.

Engström-Öst, J., and U. Candolin. 2006. Human-induced water turbidity alters selection on sexual displays in sticklebacks. *Behavioral Ecology* 18:393–398.

Epstein, R., C. E. Kirshnit, R. P. Lanza, and L. C. Rubin. 1984. "Insight" in the pigeon: Antecedents and determinants of intelligent performance. *Nature* 308:61–62.

Esch, H. E., S. Zhang, M. V. Srinivasan, and J. Tautz. 2001. Honeybee dances communicate distances measured by optic flow. *Nature* 411:581–583.

Estep, D. Q., K. Nieuwenhuijsen, K. E. M. Bruce, K. J. De Neef, P. A. Walters, S. C. Baker, and A. K. Slob. 1988. Inhibition of sexual behavior among subordinate stumptail macaques, *Macaca arctoides*. *Animal Behaviour* 36:854–864.

Evans, C. S., L. Evans, and P. Marler. 1993. On the meaning of alarm calls: functional reference in an avian vocal system. *Animal Behaviour* 46:23–38.

Fagan, R. 1981. *Animal Play Behavior*. New York: Oxford University Press.

Falls, J. B. 1982. Individual recognition by sounds in birds. In *Acoustic Communication in Birds*, Volume 2, edited by D. E. Kroodsma and E. H. Miller. New York: Academic Press, pp. 237–278.

Farrell, W. J., and W. Wilczynski. 2006. Aggressive experience alters place preference in green anole lizards, *Anolis carolinensis*. *Animal Behaviour* 71:1155–1164.

Feder, M. E., and T. Mitchell-Olds. 2003. Evolutionary and ecological functional genetics. *Nature Reviews Genetics* 4:649–655.

Fedina, T. Y. 2007. Cryptic female choice during spermatophore transfer in *Tribolium castaneum* (Coleoptera: Tenebrionidae). *Journal of Insect Physiology* 53:93–98.

Fedina, T. Y., and S. M. Lewis. 2004. Female influence over offspring paternity in the red flour beetle *Tribolium castaneum*. *Proceedings of the Royal Society of London B* 271:1393–1399.

Fedina, T. Y., and S. M. Lewis. 2006. Proximal traits and mechanisms for biasing paternity in the red flour beetle *Tribolium castaneum* (Coleoptera: Tenebrionidae). *Behavioral Ecology and Sociobiology* 60:844–853.

Feltmate, B. W., and D. D. Williams. 1989. A test of crypsis and predator avoidance in the stonefly *Paragnetina media* (Plecoptera: Perlidae). *Animal Behaviour* 37:992–999.

Feng, A. S., P. M. Narins, C. H. Xu, W. Y. Lin, Z. L. Yu, Q. Qiu, Z. M. Xu, and J. X. Shen. 2006. Ultrasonic communication in frogs. *Nature* 440:333–336.

Fenton, M. B. 1983. *Just Bats*. Toronto: University of Toronto Press.

Fenton, M. B. 1992. *Bats*. New York: Facts on File.

Ferkau, C., and K. Fischer. 2006. Costs of reproduction in male *Bicyclus anynana* and *Pieris napi* butterflies: effects of mating history and food limitation. *Ethology* 112:1117–1127.

Ferkin, M. H., and I. Zucker. 1991. Seasonal control of odor preferences of meadow voles (*Microtus pennsylvanicus*) by photoperiod and ovarian hormones. *Journal of Reproduction and Fertility* 92:433–441.

Féron, C., and P. Gouat. 2007. Paternal care in the mound-building mouse reduces inter-litter intervals. *Reproduction, Fertility, and Development* 19:425–429.

Ferster, C. B., and B. F. Skinner. 1957. *Schedules of Reinforcement*. New York: Appleton-Century-Crofts.

Fields, R. D. 2007. The shark's electric sense. *Scientific American* 297:74–81.

Fink, L. S., and L. P. Brower. 1981. Birds can overcome the cardenolide defence of monarch butterflies in Mexico. *Nature* 291:67–70.

Finley, J., D. Ireton, W. M. Schleidt, and T. A. Thompson. 1983. A new look at the features of mallard courtship displays. *Animal Behaviour* 31:348–354.

Fisher, A. C., Jr. 1979. The mysteries of animal migration. *National Geographic* 156:154–193.

Fisher, J., and R. A. Hinde. 1949. The opening of milk bottles by birds. *British Birds* 42:347–357.

Fisher, R. A. 1930. *The Genetical Theory of Natural Selection*. Oxford: Oxford University Press.

Fitch, W. T. 1997. Vocal tract length and formant frequency dispersion correlate with body size in rhesus macaques. *Journal of the Acoustic Society of America* 102:1213–1222.

Fitch, W. T., and M. D. Hauser. 2003. Unpacking "honesty": vertebrate vocal production and the evolution of acoustic signals. In *Acoustic Communication*, edited by A. M. Simmons, R. R. Fay, and A. N. Popper. New York: Springer, pp. 65–137.

FitzGibbon, C. D., and J. H. Fanshawe. 1988. Stotting in Thomson's gazelles: An honest signal of condition. *Behavioral Ecology and Sociobiology* 23:69–74.

Fitzpatrick, M. J., and M. B. Sokolowski. 2004. In search of food: Exploring the evolutionary link between cGMP-dependent protein kinase (PKG) and behavior. *Integrative and Comparative Biology* 44:28–36.

Fitzpatrick, M. J., Y. Ben-Shahar, H. M. Smid, L. E. M. Vet, G. E. Robinson, and M. B. Sokolowski. 2005. Candidate genes for behavioral ecology. *Trends in Ecology and Evolution* 20:96–104.

Flamarique, N., G. A. Mueller, C. L. Cheng, and C. R. Figiel. 2007. Communication using eye roll reflexive signaling. *Proceedings of the Royal Society of London B* 274:877–882.

Foelix, R. F. 1996. *The Biology of Spiders*, 2nd ed. Oxford: Oxford University Press.

Foellmer, M. W., and D. J. Fairbairn. 2003. Spontaneous male death during copulation in an orb-weaving spider. *Proceedings of the Royal Society of London B* 270 (Suppl):183–185.

Forsman, J. T., J.-T. Seppänen, and M. Mönkkönen. 2002. Positive fitness consequences of interspecific interaction with a potential competitor. *Proceedings of the Royal Society of London B* 269:1619–1623

Fort, K. T., and K. A. Otter. 2004. Territorial breakdown of black-capped chickadees, *Poecile atricapillus*, in disturbed habitats. *Animal Behaviour* 68:407–415.

Foster, M. S. 1977. Odd couples in manakins: A study of social organization and cooperative breeding in *Chiroxiphia linearis*. *American Naturalist* 111:845–853.

Fournier, F., and M. Festa-Bianchet. 1995. Social dominance in adult female mountain goats. *Animal Behaviour* 49:1449–1459.

Fouts, R. S. 1973. Acquisition and testing of gestural signs in four young chimpanzees. *Science* 180:978–980.

Fouts, R. S. 1974. Language: Origins, definition and chimpanzees. *Journal of Human Evolution* 3:475–482.

Fouts, R., and S. T. Mills. 1997. *Next of Kin: What Chimpanzees Have Taught Me about Who We Are*. New York: Morrow.

Fox, S. F., and M. A. Rostker. 1982. Social cost of tail loss in *Uta stansburiana*. *Science* 218:692–693.

Francis, D., J. Diorio, D. Liu, and M. J. Meaney. 1999. Nongenomic transmission across generations of maternal and stress responses in the rat. *Science* 286:1155–1158.

Franks, N. R., J. W. Hooper, A. Dornhaus, P. J. Aukett, A. L. Hayward, and S. M Berghoff. 2007. Reconnaissance and latent learning in ants. *Proceedings of the Royal Society of London B* 274:1505–1509.

Fraser, D., and B. K. Thompson. 1991. Armed sibling rivalry among suckling piglets. *Behavioral Ecology and Sociobiology* 29:9–15.

Fraser, D., and D. M. Weary. 2005. Applied animal behavior and animal welfare. In *The Behavior of Animals: Mechanisms, Function, and Evolution*, edited by J. J. Bolhuis and L.-A. Giraldeau. Malden, MA: Blackwell Publishing. pp. 345–366.

Frei, U. 1982. Homing pigeons behavior in the irregular magnetic field of western Switzerland. In *Avian Navigation*, edited by F. Papi and H. G. Wallraff. Berlin: Springer-Verlag. pp. 129–139.

Frei, U., and G. Wagner. 1976. Die Anfangsorientierung von Brieftauben im erdmagnetisch gestörten Gebiet des Mont Jorat. *Revue Suisse De Zoologie* 83:891–897.

Freitas, M. B., C. B. C. Passos, R. B. Vasconcelos, and E. C. Pinheiro. 2005. Effects of short-term fasting on energy reserves of vampire bats (*Desmodus rotundus*). *Comparative Biochemistry and Physiology B* 140:59–62.

Freitas, M. B., A. F. Welker, S. F. Millan, and E. C. Pinheiro. 2003. Metabolic responses induced by fasting in the common vampire bat (*Desmodus rotundus*). *Journal of Comparative Physiology B* 173:703–707.

Fretwell, S. 1972. *Populations in a Seasonal Environment*. Princeton, NJ: Princeton University Press.

Frey-Roos, F., P. A. Brodmann, and H. U. Reyer. 1995. Relationships between food resources, foraging patterns, and reproductive success in the water pipit (*Anthus sp. spinoletta*). *Behavioral Ecology* 6:287–295.

Fromhage, L., and J. M. Schneider. 2006. Emasculation to plug up females: the significance of pedipalp damage in *Nephila fenestrata*. *Behavioral Ecology* 17:353–357.

Fryxell, J. M., J. Greever, and A. R. E. Sinclair. 1988. Why are migratory ungulates so abundant? *American Naturalist* 131:781–798.

Fujiwara, M., P. Sengupta, and S. L. McIntyre. 2002. Regulation of body size and behavioral state of *C. elegans* by sensory perception and the *elg-4* cGMP-dependent protein kinase. *Neuron* 36:1091–1102.

Fullard, J. H., J. W. Dawson, and D. S. Jacobs. 2003. Auditory encoding during the last moment of a moth's life. *Journal of Experimental Biology* 206:281–294.

Fusani, L. 2008. Testosterone control of male courtship in birds. *Hormones and Behavior* 54:227–233.

Gagern, A., T. Schurg, N. K. Michiels, G. Schult, D. Sprenger, and N. Anthes. 2008. Behavioural response to interference competition in a sessile suspension feeder. *Marine Ecology—Progress Series* 353:131–135.

Gagliardo, A., P. Ioalè, M. Savini, and J. M. Wild. 2006. Having the nerve to home: trigeminal magnetoreceptor *versus* olfactory mediation of homing in pigeons. *Journal of Experimental Biology* 209:2888–2892.

Gahr, M., E. Sonnenschein, and W. Wickler. 1998. Sex difference in the size of neural song control regions in a dueting songbird with similar song repertoire size of males and females. *Journal of Neuroscience* 18:1124–1131.

Galef, B. G., Jr. 1976. Social transmission of acquired behavior: A discussion of tradition and social learning in vertebrates. In *Advances in the Study of Behavior*, edited by J. S. Rosenblatt, R. A. Hinde, E. Shaw, and C. Beer. New York: Academic Press, pp. 77–100.

Galef, B. G., Jr. 1988. Imitation in animals: History, definition, and interpretation of data from the psychological laboratory. In *Social Learning: Psychological and Biological Perspectives*, edited by T. R. Zentall and B. G. Galef, Jr. Hillsdale, NJ: Erlbaum, pp. 3–28.

Galef, B. G., Jr. 1990a. An adaptationist perspective on social learning, social feeding, and social foraging in Norway rats. In *Contemporary Issues in Comparative Psychology*, edited by D. A. Dewsbury. Sunderland, MA: Sinauer, pp. 55–79.

Galef, B. G. 1990b. The question of animal culture. *Human Nature* 3:157–178.

Galef, B. G., Jr. 1993. Functions of social learning about food: A causal analysis of effects of diet novelty on preference transmission. *Animal Behaviour* 46:257–265.

Galef, B. G., Jr., and T. J. Wright. 1995. Groups of naive rats learn to select nutritionally adequate foods faster than do isolated naive rats. *Animal Behaviour* 49:403–409.

Gallup, G. G., Jr. 1970. Chimpanzees: self-recognition. *Science* 167:86–87.

Gamboa, G. J. 1978. Intraspecific defense: Advantage of social cooperation among paper wasp foundresses. *Science* 199:1463–1465.

Garamszegi, L. Z., and M. Eens. 2004. The evolution of hippocampus volume and brain size in relation to food hoarding in birds. *Ecology Letters* 7:1216–1224.

Gardner, A., and S. A. West. 2007. Social evolution: the decline and fall of genetic kin recognition. *Current Biology* 17:R810–R812.

Gardner, K. E., T. D. Seeley, and N. W. Calderone. 2008. Do honeybees have two discrete dances to advertise food sources? *Animal Behaviour* 75:1291–1300.

Gardner, R. A., and B. T. Gardner. 1969. Teaching sign language to a chimpanzee. *Science* 165:664–672.

Gardner, R. A., and B. T. Gardner. 1989. A cross-fostering laboratory. In *Teaching Sign Language to Chimpanzees*, edited by R. A. Gardner, B. T. Gardner, and T. E. Van Cantfort. Albany: State University of New York Press, pp. 1–28.

Garstang, M. 2004. Long-distance, low-frequency elephant communication. *Journal of Comparative Physiology A* 190:791–805.

Garthe, S., S. Benvenuti, and W. A. Montevecchi. 2000. Pursuit plunging by northern gannets (*Sula bassana*) feeding on capelin (*Mallotus villosus*). *Proceedings of the Royal Society of London B* 267:1717–1722.

Gazit, I., A. Goldblatt, and J. Terkel. 2005. Formation of an olfactory search image for explosives odours in sniffer dogs. *Ethology* 111:669–680.

Geist, V. 1971. *Mountain Sheep*. Chicago: University of Chicago Press.

Gerhardt, H. C. 2001. Acoustic communication in two groups of closely related treefrogs. In P. J. B. Slater, J. S. Rosenblatt, C. T. Snowdon, and T. J. Roper, eds. *Advances in Behavioral Biology*. New York, Academic Press, pp. 99–167.

Gherardi, F. 2006. Fighting behavior in hermit crabs: the combined effect of resource-holding potential and resource value in *Pagurus longicarpus*. *Behavioral Ecology and Sociobiology* 59:500–510.

Gibson, B. M., and A. C. Kamil. 2005. The fine-grained spatial abilities of three seed-caching corvids. *Learning and Behavior* 33:59–66.

Gibson, J. S., and G. W. Uetz. 2008. Seismic communication and mate choice in wolf spiders: components of male seismic signals and mating success. *Animal Behaviour* 75:1253–1262.

Gilbert, C., S. Blanc, Y. Le Maho, and A. Ancel. 2008. Energy saving processes in huddling emperor penguins: from experiments to theory. *Journal of Experimental Biology* 211:1–8.

Gilbert, S. F. 1988. *Developmental Biology*. Sunderland, MA: Sinauer.

Gill, F. B., and L. L. Wolf. 1975. Economics of feeding territoriality in the golden-winged sunbird. *Ecology* 56:33–45.

Gillette, R., M. U. Gillette, D. J. Green, and R. Huang. 1989. The neuromodulatory response: Integrating second messenger pathways. *American Zoologist* 29:1275–1286.

Gillis, E. A., and C. J. Krebs. 2000. Survival of dispersing versus philopatric snowshoe hares: Do dispersers die? *Oikos* 90:343–346,

Giraldeau, L.-A. 1997. The ecology of information use. In *Behavioural Ecology: An Evolutionary Approach*, edited by J. R. Krebs and N. B. Davies. Oxford: Blackwell Science, pp. 42–68.

Giraldeau, L.-A. 2006. The function of behavior. In *The Behavior of Animals: Mechanisms, Function and Evolution*, edited by J. Bolhuis and L.-A. Giraldeau. Malden, MA: Blackwell Publishing.

Giraldeau, L.-A., and T. Caraco. 2000. *Social Foraging Theory*. Princeton, NJ: Princeton University Press.

Giraldeau, L.-A., and D. L. Kramer. 1982. The marginal value theorem: A quantitative test using load size variation in a central place forager, the eastern chipmunk, *Tamias striatus*. *Animal Behaviour* 30:1036–1042.

Giraldeau, L.-A., C. Soos, G. Beauchamp. 1994. A test of the producer-scrounger foraging game in captive flocks of spice finches, *Lonchura puntulata*. *Behavioral Ecology and Sociobiology* 34:251–256.

Gittleman, J. L. 1989. The comparative approach in ethology: Aims and limitations. In *Perspectives in Ethology*, edited by P. P. G. Bateson and P. H. Klopfer. New York: Plenum, pp. 55–83.

Gobes, S. M. H., and J. J. Bolhuis. 2007. Birdsong memory: a neural dissociation between song recognition and production. *Current Biology* 17:789–793.

Goetz, D. W. 2008. *Harmonia axyridis* ladybug invasion and allergy. *Allergy and Asthma Proceedings* 29:123–129.

Goff, M., M. Salmon, and K. J. Lohmann. 1998. Hatchling sea turtles use surface waves to establish a magnetic compass direction. *Animal Behaviour* 55:69–77.

Goldman, S. A., and F. Nottebohm. 1983. Neuronal production, migration, and differentiation in a vocal control nucleus in the female canary brain. *Proceedings of the National Academy of Sciences* 80:2390–2394.

Gonzalez, A., C. Rossini, M. Eisner, and T. Eisner. 1999. Sexually transmitted chemical defense in a moth. *Proceedings of the National Academy of Sciences* 96:5570–5574.

Goodall, J. 1965. Chimpanzees of the Gombe Stream Reserve. In *Primate Behavior*, edited by I. DeVore. New York: Holt, Rinehart & Winston, pp. 425–473.

Goodson, J. L. 2005. The vertebrate social behavior network: Evolutionary themes and variation. *Hormones and Behavior* 48:11–22.

Goodson, J. L., A. K. Evans, L. Lindberg, and C. D. Allen. 2005. Neuro-evolutionary patterning of sociality. *Proceedings of the Royal Society of London B* 272:227–235.

Goodwin, N. B., S. Balshine-Earn, and J. D. Reynolds. 1998. Evolutionary transitions in parental care in cichlid fishes. *Proceedings of the Royal Society of London B* 265:2265–2272.

Gordon, D. M. 1996. The organization of work in social insect colonies. *Nature* 380:121–124.

Gordon, D. M., S. Holmes, and S. Nacu. 2008.The short-term regulation of foraging in harvester ants. *Behavioral Ecology* 19:217–22.

Gorzula, S. J. 1978. An ecological study of *Caiman crocodilus* inhabiting savanna lagoons in Venezuelan Guayana. *Oecologia* 35:21–34.

Goss-Custard, J. D., J. T. Cayford, and S. E. G. Lea. 1998. The changing trade-off between food finding and food stealing in juvenile oystercatchers. *Animal Behaviour* 55:745–760.

Gottlieb, G. 1965. Imprinting in relation to parental and species identification by avian neonates. *Journal of Comparative and Physiological Psychology* 59:345–356.

Gottlieb, G. 1968. Prenatal behavior of birds. *Quarterly Review of Biology* 43:148–174.

Gottlieb, G. 1978. Development of species identification in ducklings. IV. Change in species-specific perception caused by auditory deprivation. *Journal of Comparative and Physiological Psychology* 92:375–387.

Gottlieb, G. 1985. Development of species identification in ducklings. XI. Embryonic critical period for species-typical perception in the hatchling. *Animal Behaviour* 33:225–233.

Gotzek, D., and K. G. Ross. 2007. Genetic regulation of colony social organization in fire ants: an integrative overview. *Quarterly Review of Biology* 82:201–226.

Gould, J. L. 1980. The case for magnetic sensitivity in birds and bees (such as it is). *American Scientist* 68:256–267.

Gould, S. J., and R. C. Lewontin. 1979. The spandrels of San Marco and the Panglossian paradigm: A critique of the adaptationist programme. *Proceedings of the Royal Society of London B* 205:581–598.

Gowaty, P. A. 2003. Sexual natures: how feminism changed evolutionary biology. *Signs* 28:901–921.

Gowaty, P. A., R. Steinichen, and W. W. Anderson. 2003. Indiscriminate females and choosy males: within- and between-species variation in *Drosophila*. *Evolution* 57:2037–2045.

Grace, J. A., N. Amin, N. C. Singh, and F. E. Theunissen. 2003. Selectivity for conspecific song in the zebra finch auditory forebrain. *Journal of Neurophysiology* 89:472–487.

Grafen, A. 1984. Natural selection, kin selection, and group selection. In *Behavioural Ecology: An Evolutionary Approach*, edited by J. R. Krebs and N. B. Davies. Sunderland, MA: Sinauer, pp. 62–84.

Grafen, A. 1990. Biological signals as handicaps. *Journal of Theoretical Biology* 144:544–546.

Gray, E. M. 1997. Do female red-winged blackbirds benefit genetically from seeking extra-pair copulations? *Animal Behaviour* 54:605–623.

Greenwood, J. J. D. 1984. The functional basis of frequency-dependent food selection. *Biological Journal of the Linnean Society* 23:177–199.

Greenwood, J. J. D., P. A. Cotton, and D. M. Wilson. 1989. Frequency-dependent selection on aposematic prey: Some experiments. *Biological Journal of the Linnean Society* 36:213–226.

Greenwood, P. J. 1980. Mating systems, philopatry and dispersal in birds and mammals. *Animal Behaviour* 28:1140–1162.

Gregory, P. T., L. A. Isaac, and R. A. Griffiths. 2007. Death feigning by grass snakes (*Natrix natrix*) in response to handling by human "predators." *Journal of Comparative Psychology* 121:123–129.

Griffin, A. S., and S. A. West. 2002. Kin selection: fact and fiction. *Trends in Ecology and Evolution* 17:15–21.

Griffin, D. R. 1976. *The Question of Animal Awareness: Evolutionary Continuity of Mental Experience*. New York: Rockefeller University Press.

Griffin, D. R. 1978. Prospects for a cognitive ethology. *Behavior and Brain Science* 1:527–538.

Griffin, D. R. 1981. *The Question of Animal Awareness*. 2nd ed. New York: Rockefeller University Press.

Griffin, D. R. 1982. *Animal Mind—Human Mind*. Berlin: Springer-Verlag.

Griffin, D. R. 1984. *Animal Thinking*. Cambridge, MA: Harvard University Press.

Griffin, D. R. 1991. Progress toward a cognitive ethology. In *Cognitive Ethology, the Minds of Other Animals*, edited by C. A. Ristau. Hillsdale, NJ: Erlbaum, pp. 3–17.

Griffin, D. R. 2001. *Animal Minds: Beyond Cognition to Consciousness*. 2nd ed. Chicago: Chicago University Press.

Griffiths, M. 1988. The platypus. *Scientific American* 258:84–91.

Grinnell, J., C. Packer, and A. E. Pusey. 1995. Cooperation in male lions: kinship, reciprocity, or mutualism. *Animal Behaviour* 49:95–105.

Gröning, J., and A. Hochkirch. 2008. Reproductive interference between animal species. *Quarterly Review of Biology* 83:257–282.

Gronquist, M., F. C. Schroeder, H. Ghiradella, D. Hill, E. M. McCoy, J. Meinwald, and T. Eisner. 2006. Shunning the night to elude the hunter: diurnal fireflies and the 'femmes fatales.' *Chemoecology* 16:39–43.

Groothuis, T. G. G., W. Muller, N. von Engelhardt, C. Carere, and C. Eising. 2005. Maternal hormones as a tool to adjust offspring phenotype in avian species. *Neuroscience and Biobehavioral Reviews* 29:329–352.

Grosberg, R. K., and J. F. Quinn. 1986. The genetic control and consequences of kin recognition by the larvae of a colonial marine invertebrate. *Nature* 322:456–458.

Gross, M. R. 1982. Sneakers, satellites and parentals: Polymorphic mating strategies in North American sunfishes. *Zeitschrift für Tierpsychologie* 60:1–26.

Gross, M. R., and R. C. Sargent. 1985. The evolution of male and female parental care in fishes. *American Zoologist* 25:807–822.

Gubernick, D. J., and T. Teferi. 2000. Adaptive significance of male parental care in a monogamous mammal. *Proceedings of the Royal Society of London B* 267:147–150.

Gubernick, D. J., S. L. Wright, and R. E. Brown. 1993. The significance of father's presence for offspring survival in the monogamous California mouse, *Peromyscus californicus*. *Animal Behaviour* 46:539–546.

Guillette, L. M., K. L. Hollis, and A. Markarian. 2009. Learning in a sedentary insect predator: Antlions (Neuroptera: Myrmeleontidae) anticipate a long wait. *Behavioural Processes* 80:224–232.

Gunawan, R., and F. J. Doyle III. 2007. Phase sensitivity analysis of circadian rhythm entrainment. *Journal of Biological Rhythms* 22:180–194.

Gundersen, G., and H. P., Andreassen. 1998. Causes and consequences of natal dispersal in root voles *Microtus oeconomus*. *Animal Behaviour* 56:1355–1366.

Gurd, D. B. 2007. Predicting resource partitioning and community organization of filter-feeding dabbling ducks from functional morphology. *American Naturalist* 169:334–343.

Gurevitch, J., and L. V. Hedges 1999. Statistical issues in ecological meta-analyses. *Ecology* 80:1142–1149.

Gurney, M. E. 1981. Hormonal control of cell form and number in zebra finch song system. *Journal of Neuroscience* 1:658–673.

Gurney, M. E., and M. Konishi. 1980. Hormone induced sexual differentiation in brain and behavior in zebra finches. *Science* 208:1380–1382.

Gwinner, E. 1996. Circadian and circannual programmes in avian migration. *Journal of Experimental Biology* 199:39–48.

Gwinner, E., and W. Wiltschko. 1978. Endogenously controlled changes in migratory direction of the garden warbler, *Sylvia borin*. *Journal of Comparative Physiology* 125:267–273.

Gwynne, D. T., and W. D. Brown. 1994. Mate feeding, offspring investment, and sexual differences in katydids (Orthoptera: Tettigoniidae). *Behavioral Ecology* 5:267–272.

Gwynne, D. T., and D. C. F. Rentz. 1983. Beetles on the bottle: male buprestids mistake stubbies for females (Coleoptera). *Journal of the Australian Entomological Society* 22:79–80.

Gwynne, D. T., and L. W. Simmons. 1990. Experimental reversal of courtship roles in an insect. *Nature* 346:172–174.

Hagedorn, M. 1986. The ecology, courtship and mating of gymnotiform electric fish. In *Electroreception*, edited by T. H. Bullock and W. Heiligenberg. New York: John Wiley, pp. 497–525.

Hagedorn, M. 1995. The electric fish *Hypopomus occidentalis* can rapidly modulate the amplitude and duration of its electric organ discharges. *Animal Behaviour* 49:1409–1413.

Hailman, J. P. 1967. Cliff-nesting adaptations of the Galapagos swallow-tailed gull. *Wilson Bulletin* 77:346–362.

Hall, J. C. 1994. The mating of the fly. *Science* 264:1702–1714.

Halpern, M., and A. Martinez-Marcos. 2003. Structure and function of the vomeronasal system: an update. *Progress in Neurobiology* 70:245–318.

Halpin, Z. T. 1991. Introduction to the symposium: Animal behavior: Past, present, and future. *American Zoologist* 31:283–285.

Hamilton, W. D. 1964. The genetical evolution of social behaviour. *Journal of Theoretical Biology* 7:1–52.

Hamilton, W. D. 1971. Geometry for the selfish herd. *Journal of Theoretical Biology* 31:295–311.

Hamilton, W. D., and M. Zuk. 1982. Heritable true fitness and bright birds: A role for parasites? *Science* 218:384–387.

Hare, B., J. Call, and M. Tomassello. 2000. Chimpanzees know what conspecifics do and do not see. *Animal Behaviour* 59:771–785.

Hare, B., E. Addessi, J. Call, M. Tomasello, and E. Visalberghi. 2003. Do capuchin monkeys, *Cebus apella*, know what conspecifics do and do not see? *Animal Behaviour* 65:131–142.

Hare, J. F., S. G. Sealy, T. J. Underwood, K. S. Ellison, and R. L. M. Stewart. 2003. Evidence of self-referent phenotype matching revisited: airing out the armpit effect. *Animal Cognition* 6:65–68.

Hart, B. 1980. Neonatal spinal transection in male rats: Differential effects on penile reflexes and other reflexes. *Brain Research* 185:423–428.

Haskell, D. 1994. Experimental evidence that nestling begging behavior incurs a costs due to nest predation. *Proceedings of the Royal Society of London B* 257:161–164.

Hasler, A. D., and W. J. Wisby. 1951. Discrimination of stream odors by fishes in relation to parent stream behavior. *American Naturalist* 85:223–238.

Hasselquist, D. 1998. Polygyny in great reed warblers: A long term study of factors contributing to male fitness. *Ecology* 79:2376–2390.

Hasselquist, D., S. Bensch, and T. von Schantz. 1996. Correlation between male song repertoire, extra-pair paternity and offspring survival in the great reed warbler. *Nature* 381:229–232.

Hastings, M. H., A. B. Reddy, and E. S. Maywood. 2003. A clockwork web: Circadian timing in brain and periphery, in health and disease. *Nature Reviews Neuroscience* 4:649–651.

Hauber, M. E., and R. M. Kilner. 2007. Coevolution, communication, and host-chick mimicry in parasitic finches: who mimics whom? *Behavioral Ecology and Sociobiology* 61:497–503.

Hauber, M. E., and E. A. Lacey. 2005. Bateman's principle in cooperatively breeding vertebrates: the effects of non-breeding alloparents on variability in female and male reproductive success. *Integrative and Comparative Biology* 45:903–914.

Hauser, M. D. 1998. Functional referents and acoustic similarity: Field playback experiments with rhesus monkeys. *Animal Behaviour* 55:1647–1658.

Hausfater, G., and S. B. Hrdy. 1984. *Infanticide: Comparative and Evolutionary Perspectives*. New York: Aldine.

Hausfater, G., H. C. Gerhardt, and G. M. Klump. 1990. Parasites and mate choice in gray treefrogs, *Hyla versicolor*. *American Zoologist* 30:299–312.

Hausheer-Zarmakupi, Z., D. P. Wolfer, M.-C. Leisinger-Trigona, and H.-P. Lipp. 1996. Selective breeding for extremes in open-field activity of mice entails a differentiation of hippocampal mossy fibers. *Behavior Genetics* 26:167–176.

Hayakawa, Y. 2007. Parasperm: morphological and functional studies on nonfertile sperm. *Ichthyological Research* 54:111–130.

Hayden, D., A. Jennings, C. Müller, D. Pascoe, R. Bublitz, H. Webb, T. Breithaupt, L. Watkins, and J. Hardege. 2007. Sex-specific mediation of foraging in the shore crab, *Carcinus maenas*. *Hormones and Behavior* 52:162–168.

Hayes, K. J., and C. Hayes. 1951. The intellectual development of a home-raised chimpanzee. *Proceedings of the American Philosophical Society* 95:105.

Hazelett, D. J., and J. C. Weeks. 2005. Segment-specific muscle degeneration is triggered directly by a steroid hormone during insect metamorphosis. *Journal of Neurobiology* 62:164–177.

Hebets, E. A., and D. R. Papaj. 2005. Complex signal function: developing a framework of testable hypotheses. *Behavioral Ecology and Sociobiology* 57:197–214.

Hedenström, A., and T. Alerstam. 1997. Optimum fuel loads in migratory birds: Distinguishing between time and energy utilization. *Journal of Theoretical Biology* 189:227–234.

Hedrick, A. V., and S. E. Riechert. 1989. Population variation in the foraging behaviour of a spider: the role of genetics. *Oecologia* 80:533–539.

Hedwig, B., and R. Heinrich. 1997. Identified descending brain neurons control different stridulatory motor patterns in an acridid grasshopper. *Journal of Comparative Physiology A* 180:285–294.

Heinrich, B. 1995. An experimental investigation of insight in common ravens (*Corvus corax*). *Auk* 112:994–1003.

Heinrich, B., and T. Bugnyar. 2005. Testing problem solving in ravens: string-pulling to reach food. *Ethology* 111:962–976.

Heinroth, O. 1910. Beiträge zur Bilogie, insbiesonder Psychologie und Ethologie der Anatiden. In *Proceedings of the 5th International Ornithological Congress*, Berlin, pp. 589–702.

Heinsohn, R., and S. Legge. 1999. The cost of helping. *Trends in Ecology and Evolution* 14:53–57.

Helbig, A. J. 1991. Inheritance of migratory direction in a bird species: A cross-breeding experiment with SE- and SW-migrating blackcaps (*Sylvia atricapilla*). *Behavioral Ecology and Sociobiology* 28:9–12.

Helbig, A. J., P. Berthold, and W. Wiltschko. 1989. Migratory orientation of blackcaps (*Sylvia atricapilla*): Population specific shifts of direction during the autumn. *Ethology* 82:307–315.

Hennessy, D. F. 1986. On the deadly risk of predator harassment. *Ethology* 72:72–74.

Herman, L. M., and P. H. Forestell. 1985. Reporting presence or absence of named objects by a language-trained dolphin. *Neuroscience Biobehavioral Reviews* 9:667–681.

Herman, L. M., A. A. Pack, and P. Morrel-Samuels. 1993. Representational and conceptual skills of dolphins. In *Language and Communication: Comparative Perspectives*, edited by H. L. Roitblat, L. M. Herman, and P. E. Nachtigall. Hillsdale, NJ: Erlbaum, pp. 403–442.

Hertenstein, M. J., J. M. Verkamp, A. M. Kerestes, and R. M. Holmes. 2006a. The communicative functions of touch in humans, nonhuman primates, and rats: A review and synthesis of the empirical research. *Genetic, Social, and General Psychology Monographs* 132:5–94.

Hertenstein, M. J., D. Keltner, B. App, B. Bulleit, and A. Jaskolka. 2006b. Touch communicates distinct emotions. *Emotion* 6:528–533.

Herzog, E. D., and G. Tosini. 2001. The mammalian circadian clockshop. *Seminars in Cell and Developmental Biology* 12:295–303.

Hews, D. K. 1988. Alarm response in larval western toads, *Bufo boreas*: Release of larval chemical by a natural predator and its effect on predator capture efficiency. *Animal Behaviour* 36:125–133.

Heyers, D., M. Manns, H. Luksch, O. Güntürkün, and H. Moritsen. 2007. A visual pathway links brain structures active during magnetic compass orientation in migratory birds. *PLoS ONE* 2:e937.

Hill, G. E. 1990. Female house finches prefer colourful mates: sexual selection for a condition-dependent trait. *Animal Behaviour* 40:563–572.

Hill, G. E. 1991. Plumage coloration is a sexually selected indicator of male quality. *Nature* 350:337–339.

Hill, G. E., and R. Montgomerie. 1994. Plumage color signals nutritional condition in the house finch. *Proceedings of the Royal Society of London B* 258:47–52.

Hill, P. S. M. 2001. Vibration and animal communication: A review. *American Zoologist* 41:1135–1142.

Hinde, C. A., and R. M. Kilner. 2007. Negotiations within the family over the supply of parental care. *Proceedings of the Royal Society of London B* 274:53–60.

Hinde, R. A. 1970. *Animal Behavior: A Synthesis of Ethology and Comparative Psychology*. 2nd ed. New York: McGraw-Hill.

Hinde, R. A. 1982. *Ethology*. Oxford: Oxford University Press.

Hirschenhauser, K., and R. F. Oliveira. 2006. Social modulation of androgens in male vertebrates: meta-analyses of the challenge hypothesis. *Animal Behaviour* 71:265–277.

Hoage, R. J., and L. Goldman (eds.) 1986. *Animal Intelligence, Insights into the Animal Mind*. Washington, DC: Smithsonian Institution Press.

Hoar, W. S. 1988. The physiology of smolting salmonids. In: *Fish Physiology*, vol. XIB, edited by W. S. Hoar and D. Randall. New York: Academic Press, pp. 275–343.

Höbel, G., and H. C. Gerhardt. 2003. Reproductive character displacement in the acoustic communication system of green tree frogs (*Hyla cinerea*). *Evolution* 57:894–904.

Hochner, B., M. Klein, S. Schacher, and E. R. Kandel. 1986. Additional component in the cellular mechanism of presynaptic facilitation contributes to behavioral dishabituation in *Aplysia*. *Proceedings of the National Academy of Sciences* 83:8794–8798.

Hodos, W. C., and C. B. G. Campbell. 1969. Scala naturae: Why there is no theory in comparative psychology. *Psychological Review* 76:337–350.

Hoffmann, A. A., and Z. Cacoyianni. 1990. Territoriality in *Drosophila melanogaster* as a conditional strategy. *Animal Behaviour* 40:526–537.

Hofmann, H. A. 2003. Functional genomics of neural and behavioral plasticity. *Journal of Neurobiology* 54:272–282.

Hofmann, H. A. 2006. Gonadotropin-releasing hormone signaling in behavioral plasticity. *Current Opinion in Neurobiology* 16:343–350.

Hoffmann, K. 1954. Versuche zu der im Richtungsfinden der Vögel enthaltenen Zeitschätzung. *Zeitschrift für Tierpsychologie* 11:453–475.

Hoffmann, K. H., K. Dettner, and K.-H. Tomaschko. 2006. Chemical signals in insects and other arthropods: from molecular structure to physiological functions. *Physiological and Biochemical Zoology* 79:344–356.

Hogan, J. A. 2005. Causation: the study of behavioural mechanisms. *Animal Biology* 55:323–341.

Hogan, J. A., and J. J. Bolhuis. 2005. The development of behaviour: trends since Tinbergen (1963). *Animal Biology* 55:371–398.

Hogg, J. T. 1988. Copulatory tactics in relation to sperm competition in Rocky Mountain bighorn sheep. *Behavioral Ecology and Sociobiology* 22:49–59.

Hogg, J. T., and S. H. Forbes. 1997. Mating in bighorn sheep: frequent male reproduction via a high-risk "unconventional" tactic. *Behavioral Ecology and Sociobiology* 41:33–48.

Holder, C. F. 1901. A curious means of defense. *Scientific American* 2:186–187.

Holekamp, K. E., and P. W. Sherman. 1989. Why male ground squirrels disperse. *American Scientist* 77:232–239.

Holland, B., and W. R. Rice. 1998. Chase-away sexual selection: antagonistic seduction versus resistance. *Evolution* 52:1–7.

Holland, R. A., J. L. Kirschvink, T. G. Doak, and M. Wikelski. 2008. Bats use magnetite to detect the earth's magnetic field. *PLoS ONE* 3:e1676.

Hölldobler, B., and E. O. Wilson. 1990. *The Ants*. Cambridge: Belknap Press.

Hollis, K. L. 1984. The biological function of Pavlovian conditioning: The best defense is a good offense. *Journal of Experimental Psychology–Animal Behavior Processes* 10:413–425.

Hollis, K. L. 1990. The role of Pavlovian conditioning in territorial aggression and reproduction. In *Contemporary Issues in Comparative Psychology*, edited by D. A. Dewsbury. Sunderland, MA: Sinauer, pp. 197–219.

Hollis, K. L. 1999. The role of learning in the aggressive and reproductive behavior of blue gouramis, *Trichogaster trichopterus*. *Environmental Biology of Fishes* 54:355–369.

Hollis, K. L., M. J. Dumas, P. Singh, and P. Fackelman. 1995. Pavlovian conditioning of aggressive behavior in blue gourami fish (*Trichogaster trichopterus*)—Winners become winners and losers stay losers. *Journal of Comparative Psychology* 109:123–133.

Hollis, K. L., V. L. Pharr, M. J. Dumas, G. B. Britton, and J. Field. 1997. Classical conditioning provides paternity advantage for territorial male blue gouramis (*Trichogaster trichopterus*). *Journal of Comparative Psychology* 111:219–225.

Holman, L., and R. R. Snook. 2006. Spermicide, cryptic female choice and the evolution of sperm form and function. *Journal of Evolutionary Biology* 19:1660–1670.

Holmes, W. G. 1986a. Identification of paternal half-siblings by captive Belding's ground squirrels. *Animal Behaviour* 34:321–327.

Holmes, W. G. 1986b. Kin recognition by phenotype matching in female Belding's ground squirrels. *Animal Behaviour* 34:38–47.

Holmes, W. G., and P. W. Sherman. 1982. The ontogeny of kin recognition in two species of ground squirrels. *American Zoologist* 22:491–517.

Holmgren, N. A., and M. Enquist. 1999. Dynamics of mimicry evolution. *Biological Journal of the Linnean Society* 66:145–158.

Holveck, M-J., and K. Riebel. 2007. Preferred songs predict preferred males: consistency and repeatability of zebra finch females across three test contexts. *Animal Behaviour* 74:297–309.

Höner, O. P., B. Wachter, M. L. East, W. J. Streich, K. Wilhelm, T. Burke, and H. Hofer. 2007. Female mate-choice drives the evolution of male-biased dispersal in a social mammal. *Nature* 448:798–802.

Honess, P. E., and C. M. Marin. 2006. Enrichment and aggression in primates. *Neuroscience and Biobehavioral Reviews* 30:413–436.

Honeycutt, R. L. 1992. Naked mole-rats. *American Scientist* 80:43–53.

Honma, A., S. Oku, and T. Nishida. 2006. Adaptive significance of death feigning posture as a specialized inducible defence against gape-limited predators. *Proceedings of the Royal Society of London B*. 273:1631–1636.

Hopcraft, J. G., A. R. E. Sinclair, and C. Packer. 2005. Planning for success: Serengeti lions seek prey accessibility rather than abundance. *Journal of Animal Ecology* 74:559–566.

Hopkins, C. D. 1974. Electric communication in fish. *American Scientist* 62:426–437.

Hopkins, C. D. 1986a. Behavior of Mormyridae. In *Electroreception*, edited by T. H. Bullock and W. Heiligenberg. New York: John Wiley, pp. 527–576.

Hopkins, C. D. 1986b. Temporal structure of nonpropagated electric communication signals. *Brain, Behavior and Evolution* 28:43–59.

Hopkins, C. D. 1999. Design features for electric communication. *Journal of Experimental Biology* 202:1217–1228.

Hopkins, P. M. 1982. Growth and regeneration patterns in the fiddler crab, *Uca pugilator. Biological Bulletin* 163:301–319.

Horsmann, U., H. G. Heinzel, and G. Wendler. 1983. The phasic influence of self-generated air current modulations on the locust flight motor. *Journal of Comparative Physiology* 150:427–438.

Houck, L. D., and N. L. Reagan. 1990. Male courtship pheromones increase female receptivity in a plethodontid salamander. *Animal Behaviour* 39:729–734.

Houston, A. I. 1995. Parental effort and paternity. *Animal Behaviour* 50:1635–1644.

Houtman, A. M., and J. B. Falls. 1994. Negative assortative mating in the white-throated sparrow, *Zonotrichia albicollis*: The role of mate choice and intra-sexual competition. *Animal Behaviour* 48:377–383.

How, M. J., J. M. Hemmi, J. Zeil, and R. Peters. 2008. Claw waving display changes with receiver distance in fiddler crabs, *Uca perplexa*. *Animal Behaviour* 75:1015–1022.

Hsu, Y., R. L. Earley, and L. L. Wolf. 2006. Modulation of aggressive behaviour by fighting experience: mechanisms and contest outcomes. *Biological Reviews* 81:33–74.

Huang, Z. Y., and G. E. Robinson. 1992. Honeybee colony integration: Worker-worker interactions mediate hormonally regulated plasticity in division of labor. *Proceedings of the National Academy of Sciences* 89:11726–11729.

Huber, S. K., and J. Podos. 2006. Beak morphology and song features covary in a population of Darwin's finches (*Geospiza fortis*). *Biological Journal of the Linnean Society* 88:489–498.

Hudson, R., and H. Distel. 1995. On the nature and action of the nipple-search pheromone: a review. In *Chemical Signals in Vertebrates*, Volume 7, edited by R. L. Doty and D. Müller-Schwarze. New York: Plenum, pp. 223–232.

Hughes, K. A., L. Du, F. H. Rodd, and D. N. Reznick. 1999. Familiarity leads to female mate preference for novel males in the guppy, *Poecilia reticulata*. *Animal Behaviour* 58:907–916.

Hughes, W. O. H., J. Eilenberg, and J. J. Boomsma. 2002. Trade-offs in group living: transmission and disease resistance in leaf-cutting ants. *Proceedings of the Royal Society of London B* 269:1811–1819.

Hughes, W. O. H., B. P. Oldroyd, M. Beekman, and F. L. W. Ratnieks. 2008. Ancestral monogamy shows kin selection is key to the evolution of eusociality. *Science* 320:1213–1216.

Huk, T., and W. Winkel. 2006. Polygyny and its fitness consequences for primary and secondary female pied flycatchers. *Proceedings of the Royal Society of London B* 273:1681–1688.

Hunt, G. J., G. V. Amdam, D. Schlipalius, C. Emore, N. Sardesai, C. E. Williams, O. Rueppell, E. Guzmán-Novoa, M. Arechavaleta-Velasco, S. B. C. Chandra, M. K. Fondrk, M. Beye, and R. E. Page, Jr. 2007. Behavioral genomics of honeybee foraging and nest defense. *Naturwissenschaften* 94:247–267.

Hunt, G. R. 1996. Manufacture and use of hook-tools by New Caledonian crows. *Nature* 379:249–251.

Hunt, G. R., M. C. Corballis, and R. D. Gray. 2006. Design complexity and strength of laterality are correlated in New Caledonian crows' pandanus tool manufacture. *Proceedings of the Royal Society of London B* 273:1127–1133.

Hunt, G. R., and R. D. Gray. 2004. The crafting of hook tools by wild New Caledonian crows. *Proceedings of the Royal Society of London B* 271:S88–S90.

Hunter, M. L., and J. R. Krebs. 1979. Geographical variation in the song of the great tit *Parus major* in relation to ecological factors. *Journal of Animal Ecology* 48:759–785.

Huntingford, F. A. 1986. Development of behaviour in fish. In *The Behavior of Teleost Fishes*, edited by T. J. Pitcher. Baltimore, MD: Johns Hopkins University Press, pp. 47–68.

Huntingford, F. A., and A. K. Turner. 1987. *Animal Conflict*. London: Chapman and Hall.

Hupé, G. J., and J. E. Lewis. 2008. Electrocommunication signals in free swimming brown ghost knifefish, *Apteronotus leptorhynchus*. *Journal of Experimental Biology* 211:1657–1667.

Huxley, J. 1923. Courtship activities in the red-throated diver *Colymbus stellatus pontopp*; together with a discussion on the evolution of courtship in birds. *Journal of the Linnean Society of London* 2:491–562.

Hylton, R., and R. D. Godard. 2001. Song properties of Indigo Buntings in open and forested habitats. *Wilson Bulletin* 113:243–245.

Immelmann, K. 1963. Drought adaptations in Australian desert birds. In *Proceedings of the 13th International Ornithological Congress*, 1962, pp. 649–657.

Immelmann, K. 1969. Über den Einfluss frühkinlicher Erfahrungen auf die geschlechtliche Objektfixierung bei Estrildiden. *Zeitschrift für Tierpsychologie* 26:677–691.

Immelmann, K. 1972. The influence of early experience upon the development of social behavior in estrildine finches. In *Proceedings of the 15th International Ornithological Congress, The Hague*, 1970, pp. 316–338.

Immelmann, K. 1980. *Introduction to Ethology*. New York: Plenum.

Immelmann, K., and S. J. Suomi. 1981. Sensitive phases in development. In *Behavioral Development*, edited by K. Immelmann, W. Barlow, L. Petrinovich, and M. Main. Cambridge: Cambridge University Press, pp. 395–431.

Ingram, K. K., P. Oefner, and D. M. Gordon. 2005. Task-specific expression of the *foraging* gene in harvester ants. *Molecular Ecology* 14:813–818.

Inouye, S. T., and H. Kawamura. 1979. Persistence of circadian rhythmicity in mammalian hypothalamic "island" containing the suprachiasmatic nucleus. *Proceedings of the National Academy of Sciences* 76:5962–5966.

Insel, T. R. 2006. From species differences to individual differences. *Molecular Psychiatry* 11:424.

Ioalè, P., M. Nozzolini, and F. Papi. 1990. Homing pigeons do extract directional information from olfactory stimuli. *Behavioral Ecology and Sociobiology* 26:301–305.

Isaac, J. L. 2005. Potential causes and life-history consequences of sexual size dimorphism in mammals. *Mammal Review* 35:101–115.

Iwasaki, M., A. Delago, H. Nishino, and H. Aonuma. 2006. Effects of previous experience on the agonistic behaviour of male crickets, *Gryllus bimaculatus*. *Zoological Science* 23:863–872.

Jablonski, P. G., and R. S. Wilcox. 1996. Signaling asymmetry in the communication of the water strider *Aquarius remigis* in the context of dominance and spacing in the non-mating season. *Ethology* 102:353–359.

Jackson, R. R., and R. S. Wilcox. 1993a. Spider flexibly chooses aggressive mimicry signals for different prey by trial and error. *Behaviour* 127:21–36.

Jackson, R. R., and R. S. Wilcox. 1993b. Observations in nature of detouring behavior by *Portia fimbriata*, a web-invading aggressive mimic jumping spider from Queensland. *Journal of Zoology* 230:135–139.

Jacquot, J. J., and N. G. Solomon. 1997. Effects of site familiarity on movement patterns of male prairie voles *Microtus ochrogaster*. *American Midland Naturalist* 138:414–417.

Jan, Y., L. Jan, and M. Dennis. 1977. Two mutations of synaptic transmission in *Drosophila. Proceedings of the Royal Society of London B* 198:87–108.

Jarvis, E. D., H. Schwabl, S. Ribeiro, and C. V. Mello. 1997. Brain gene regulation by territorial singing behavior in freely ranging songbirds. *NeuroReport* 8:2073–2077.

Jarvis, J. U. M. 1981. Eusociality in a mammal: Cooperative breeding in naked mole-rat colonies. *Science* 212:571–573.

Jaynes, J. 1969. The historical origins of "ethology" and comparative psychology. *Animal Behaviour* 17:601–606.

Jenni, D. A., and B. J. Betts. 1978. Sex differences in nest construction, incubation, and parental behavior in the polyandrous American jacana (*Jacana spinosa*). *Animal Behaviour* 26:207–218.

Jenni, D. A., and G. Collier. 1972. Polyandry in the American jacana (*Jacana spinosa*). *Auk* 89:743–765.

Jennings, H. S. 1906. *Behavior of the Lower Organisms*. New York: Columbia University Press.

Jennions, M. D. 1993. Female choice in birds and the cost of long tails. *Trends in Ecology and Evolution* 8:230–232.

Jennions, M. D., and M. Petrie. 2000. Why do females mate multiply? A review of the genetic benefits. *Biological Reviews* 75:21–64.

Jeschke, J. M., and R. Tollrian. 2007. Prey swarming: which predators become confused and why? *Animal Behaviour* 74:387–393.

Jivoff, P. 1997. The relative roles of predation and sperm competition on the duration of the post-copulatory association between the sexes in the blue crab, *Callinectes sapidus*. *Behavioral Ecology and Sociobiology* 40:175–186.

Johnsen, S. 2001. Hidden in plain sight: the ecology and physiology of organismal transparency. *Biological Bulletin* 201:301–318.

Johnsgard, P. A. 1967. *Animal Behavior.* William C. Brown: Dubuque.

Johnson, C. H., J. A. Elliott, and R. Foster. 2003. Entrainment of circadian programs. *Chronobiology International* 20:741–774.

Johnson, M. H. 2005. Sensitive periods in functional brain development: problems and prospects. *Developmental Psychobiology* 46:287–292.

Johnson, M. L., and M. S. Gaines. 1990. Evolution of dispersal: Theoretical models and empirical tests using birds and mammals. *Annual Review of Ecology and Systematics* 21:449–480.

Johnston, R. E. 1998. Pheromones, the vomeronasal system, and communication—from hormonal responses to individual recognition. *Annals of the New York Academy of Sciences* 855:333–348.

Johnston, R. E. 2000. Chemical communication and pheromones: the types of chemical signals and the role of the vomeronasal system. In *The Neurobiology of Taste and Smell*, edited by T. E. Finger, W. L. Silver, and D. Restrepo. New York: Wiley-Liss, pp. 101–127.

Johnston, R. E. 2003. Chemical communication in rodents: from pheromones to individual recognition. *Journal of Mammalogy* 84:1141–1162.

Johnston, T. D. 1988. Developmental explanation and the ontogeny of birdsong: Nature/nurture redux. *Behavioral and Brain Sciences* 11:617–663.

Johnstone, R. A. 1997. The evolution of animal signals. In *Behavioural Ecology*, edited by J. R. Krebs and N. B. Davies. Oxford: Blackwell Science, pp. 155–178.

Jones, C. B., V. Milanov, and R. Hager. 2008. Predictors of male residence patterns in groups of black howler monkeys. *Journal of Zoology* 275:72–78.

Jones, T. M., and R. J. Quinnell. 2002. Testing predictions for the evolution of lekking in the sandfly, *Lutzomyia longipalpis*. *Animal Behaviour* 63:605–612.

Jones, T. M., R. J. Quinnell, and A. Balmford. 1998. Fisherian flies: benefits of female choice in a lekking sandfly. *Proceedings of the Royal Society of London B* 265:1651–1657.

Joron, M., and J. L. B. Mallet. 1998. Diversity in mimicry: Paradox or paradigm. *Trends in Ecology and Evolution* 13:461–466.

Kacelnik, A., and M. Bateson. 1996. Risky theories—the effect of variance on foraging decisions. *American Zoologist* 36:402–434.

Kalmijn, A. J. 1966. Electro-perception in sharks and rays. *Nature* 212:1232–1233.

Kalmijn, A. J. 1971. The electric sense of sharks and rays. *Journal of Experimental Biology* 55:371–383.

Kamil, A. C., and J. E. Mauldin. 1988. A comparative-ecological approach to the study of learning. In *Evolution and Learning*, edited by R. C. Bolles and M. D. Beecher. Hillsdale, NJ: Erlbaum, pp. 117–133.

Kamil, A. C., and S. I. Yoerg. 1982. Learning and foraging behavior. In *Perspectives on Ethology*, edited by P. P. G. Bateson. New York: Plenum, pp. 325–364.

Kamil, A. C., R. P. Balda, and D. J. Olson. 1994. Performance of four seed-caching corvid species in the radial-arm maze analog. *Journal of Comparative Psychology* 108:385–393.

Kamio, M., M. A. Reidenberg, and C. Derby. 2008. To paddle or not: context dependent courtship display by male blue crabs (*Callinectes sapidus*). *Journal of Experimental Biology* 211:1234–1248.

Kanda, L. L. 2005. Winter energetics of Virginia opossums *Didelphis virginiana* and implications for the species' northern distributional limit. *Ecography* 28:731–744.

Kanda, L. L., T. K. Fuller, and P. R. Sievert. 2006. Landscape associations of road-killed Virginia opossums (*Didelphis virginiana*) in central Massachusetts. *American Midland Naturalist* 156: 128–134.

Kandel, E. R. 1976. *Cellular Basis of Behavior: An Introduction to Behavioral Neurobiology*. San Francisco, CA: W. H. Freeman.

Kandel, E. R. 1979a. *The Behavioral Biology of Aplysia*. San Francisco: W. H. Freeman.

Kandel, E. R. 1979b. Small systems of neurons. *Scientific American* 241:66–76.

Kandel, E. R. 2001. The molecular biology of memory storage: A dialogue between genes and synapses. *Science* 294:1030–1038.

Kandel, E. R., and J. H. Schwartz. 1982. Molecular biology of learning: Modulation of transmitter release. *Science* 218:433–443.

Kaplan, A., and W. E. Trout III. 1969. The behavior of four neurological mutants of *Drosophila*. *Genetics* 61:399–409.

Kaufmann, J. H. 1983. On definitions and functions of dominance and territoriality. *Biological Reviews* 58:1–20.

Kawai, M. 1965. Newly acquired and pre-cultural behavior of the natural troop of Japanese monkeys on Koshima Islet. *Primates* 6:1–30.

Kawamura, S. 1959. Sub-culture propagation among Japanese macaques. *Primates* 2:43–60.

Keane, B., P. M. Waser, S. R. Creel, N. M. Creel, L. F. Elliott, and D. J. Minchella. 1994. Subordinate reproduction in dwarf mongooses. *Animal Behaviour* 47:65–75.

Keefer, M. L., C. C. Caudill, C.C. Peery, and C.T. Boggs. 2008. Non-direct homing behaviours by adult Chinook salmon in a large multi-stock river system. *Journal of Fish Biology* 72:27–44.

Keeton, W. T., T. S. Larkin, and D. M. Windsor. 1974. Normal fluctuations in the earth's magnetic field influence pigeon orientation. *Journal of Comparative Physiology* 95:95–103.

Keller, F. S. 1941. Light aversion in the white rat. *Psychological Record* 4:235–250.

Keller, L., and K. G. Ross. 1998. Selfish genes: A green beard in the red fire ant. *Nature* 394:573–575.

Kelley, D. B. 1996. Sexual differentiation in *Xenopus laevis*. In *The Biology of Xenopus*, edited by R. Tinsley and H. Kobel. Oxford: Oxford University Press, pp. 143–176.

Kelley, D. B., and D. I. Gorlick. 1990. Sexual selection and the nervous system. *BioScience* 40:275–283.

Kemp, D. J., and C. Wiklund. 2004. Residency effects in animal contests. *Proceedings of the Royal Society of London B* 271:1707–1711.

Kendrick, K. M., A. P. da Costa, A. E. Leigh, M. R. Hinton, and J. W. Peirce. 2007. Sheep don't forget a face. *Nature* 447:346.

Kendrick, K. M., M. A. Haupt, M. R. Hinton, K. D. Broad, and J. D. Skinner. 2001. Sex differences in the influence of mothers on the sociosexual preferences of their offspring. *Hormones and Behavior* 40:322–338.

Kendrick, K. M., M. R. Hinton, K. Atkins, M. A. Haupt, K. D. Broad, and J. D. Skinner. 1998. Mothers determine sexual preferences. *Nature* 395:229–230.

Kent, D. S., and J. A. Simpson. 1992. Eusociality in the beetle *Austroplatypus incompertus* (Coleoptera: Curculionidae). *Naturwissenschaften* 79:86–87.

Kessel, E. L. 1955. Mating activities of balloon flies. *Systematic Zoology* 4:97–104.

Kiepenheuer, J. 1978. A repetition of the deflector loft experiment. *Behavioral Ecology and Sociobiology* 3:393–395.

Kiepenheuer, J. 1979. Pigeon homing: Deprivation of olfactory information does not affect the deflector effect. *Behavioral Ecology and Sociobiology* 6:11–22.

Kilner, R. M. 1995. When do canary parents respond to nestling signals of need. *Proceedings of the Royal Society of London B* 260:343–348.

Kilner, R. M. 2003. How selfish is a cowbird nestling? *Animal Behaviour* 66:569–576.

Kilner, R. M., J. R. Madden, and M. E. Hauber. 2004. Brood parasitic cowbird nestlings use host young to procure resources. *Science* 305:877–879.

Kiltie, R. A. 1988. Countershading: Universally deceptive or deceptively universal? *Trends in Ecology and Evolution* 3:21–23.

Kiltie, R. A. 1989. Wildfire and the evolution of dorsal melanism in fox squirrels, *Sciurus niger*. *Journal of Mammalogy* 70:726–739.

Kimura, K.-I., M. Ote, T. Tazawa, and D. Yamamoto. 2005. *Fruitless* specifies sexually dimorphic neural circuitry in the *Drosophila* brain. *Nature* 438:229–233.

King, A. P., and M. J. West. 1983. Epigenesis of cowbird song—A joint endeavor of males and females. *Nature* 305:704–706.

Kirkpatrick, M. 1982. Sexual selection and the evolution of female choice. *Evolution* 36:1–12.

Kirschvink, J. L., S. Padmanabha, C. K. Boyce, and J. Oglesby. 1997. Measurement of the threshold of honeybees to weak, extremely low-frequency magnetic fields. *Journal of Experimental Biology* 200:1363–1368.

Klaassen, M. 1996. Metabolic constraints on long-distance migration in birds. *Journal of Experimental Biology* 199:57–64.

Kleiman, D. G. 1977. Monogamy in mammals. *Quarterly Review of Biology* 52:39–69.

Kleiman, D. G., and J. R. Malcolm. 1981. The evolution of male parental investment in mammals. In *Parental Care in Mammals*, edited by D. J. Gubernick and P. H. Klopfer. New York: Plenum, pp. 347–387.

Knaden, M., and R. Wehner. 2006. Ant navigation: resetting the path integrator. *Journal of Experimental Biology* 209:26–31.

Knapp, R., D. K. Hews, C. W. Thompson, L. Ray, and M. C. Moore. 2003. Environmental and endocrine correlates of tactic switching by non-territorial male tree lizards (*Urosaurus ornatus*). *Hormones and Behavior* 43:83–92.

Knudsen, E. I. 1981. The hearing of the barn owl. *Scientific American* 245:113–125.

Knudsen. E. I. 1982. Auditory and visual maps of space in the optic tectum of the owl. *Journal of Neuroscience* 2:1177–1194.

Knudsen, E. I. 2002. Instructed learning in the auditory localization pathway of the barn owl. *Nature* 417:322–328.

Knudsen, E. I., and M. Konishi. 1978. A neural map of auditory space in the owl. *Science* 200:795–797.

Koenig, W. D., and R. L. Mumme. 1987. *Population Ecology of the Cooperatively Breeding Acorn Woodpecker*. Princeton, NJ: Princeton University Press.

Köhler, W. 1927. *The Mentality of Apes*. New York: Harcourt Brace.

Kokko, H., R. Brooks, M. D. Jennions, and J. Morley. 2003. The evolution of mate choice and mating biases. *Proceedings of the Royal Society of London B* 270:653–664.

Koltermann, R. 1971. 24-std-Periodik in der Langzeiterinnerung an Duft-und Farbsignale ber der Honigbiene. *Zeitschrift für vergleichende Physiologie* 75:49–68.

Komdeur, J. 1992. Importance of habitat saturation and territory quality for evolution of cooperative breeding in the Seychelles warbler. *Nature* 358:493–495.

Komdeur, J. 1996. Influence of helping and breeding experience on reproductive performance in the Seychelles warbler: a translocation experiment. *Behavioral Ecology* 7:326–333.

Komdeur, J., and P. Edelaar. 2001. Male Seychelles warblers use territory budding to maximize lifetime fitness in a saturated environment. *Behavioral Ecology* 12:706–715.

Komdeur, J., T. Burke, and D. S. Richardson. 2007. Explicit experimental evidence for the effectiveness of proximity as mate-guarding behaviour in reducing extra-pair fertilization in the Seychelles warbler. *Molecular Ecology* 16:3679–3688.

Komdeur, J., A. Huffstadt, W. Prast, G. Castle, R. Mileto, and J. Wattel. 1995. Transfer experiments of Seychelles warblers to new islands—changes in dispersal and helping behavior. *Animal Behaviour* 49:695–708.

Komers, P. E. 1997. Property rites. *Natural History* 106:28–31.

Komers, P. E., and P. N. M. Brotherton. 1997. Female space use is the best predictor of monogamy in mammals. *Proceedings of the Royal Society of London B* 264:1261–1270.

Konishi, M. 1965. The role of auditory feedback in the control of vocalization in the white-crowned sparrow. *Zeitschrift für Tierpsychologie* 22:770–783.

Konishi, M. 1993a. Listening with two ears. *Scientific American* 286:66–73.

Konishi, M. 1993b. The neuroethology of sound localization in the owl. *Journal of Comparative Physiology A* 173:3–7.

Konishi, M. 2003. Coding of auditory space. *Annual Review of Neuroscience* 26:31–55.

Konishi, M. 2004. The role of auditory feedback in birdsong. *Annals of the New York Academy of Sciences* 1016:463–475.

Kowalski, U., R. Wiltschko, and E. Fuller. 1988. Normal fluctuations of the geomagnetic field may affect initial orientation in pigeons. *Journal of Comparative Physiology A* 163:593–600.

Krakauer, A. H. 2005. Kin selection and cooperative courtship in wild turkeys. *Nature* 434:69–72.

Krama, T., and I. Krams. 2005. Cost of mobbing call to breeding pied flycatcher, *Ficedula hypoleuca*. *Behavioral Ecology* 16:37–40.

Kramer, D. L., and D. M. Weary. 1991. Exploration versus exploitation: A field study of time allocation to environmental tracking by foraging chipmunks. *Animal Behaviour* 41:443–449.

Kramer, G. 1949. Über Richtungstendenzen bei der nächtichen Zugenruhe gekäfigter Vögel. In *Ornithologie als biologische Wissenschaft*, edited by E. Mayr and E. Schüz. Heidelberg: Carl Winter.

Kramer, G. 1950. Weitere Analyse der Faktoren, welche die Zugaktivität des gekäfigten Vogels orientieren. *Naturwissenschaften* 37:377–378.

Kramer, G. 1951. Eine neue Methode zur Erforschung der Zugorientierung und die bisher damit erzielten Ergebnisse. *Proceedings of the 10th International Ornithology Congress*: 269–280.

Krause, J. 1993. The effect of "Schreckstoff" on the shoaling behavior of the minnow: A test of Hamilton's selfish herd theory. *Animal Behaviour* 45:1019–1024.

Krause, J., and G. D. Ruxton. 2002. *Living in Groups*. Oxford: Oxford University Press.

Krebs, J. R., and M. I. Avery. 1984. Chick growth and prey quality in the European bee-eater (*Merops apiaster*). *Oecologia* 64:363–368.

Krebs, J. R., and N. B. Davies. 1997. *Behavioural Ecology: An Evolutionary Approach*. 4th ed. Oxford: Blackwell Science.

Krebs, J. R., and R. McCleery. 1984. Optimization in behavioural ecology. In *Behavioural Ecology, An Evolutionary Approach*, 2nd ed., edited by J. R. Krebs and N. B. Davies. Sunderland, MA: Sinauer.

Krebs, J. R., J. T. Erichsen, M. J. Webber, and E. L. Charnov. 1977. Optimal prey selection in the great tit (*Parus major*). *Animal Behaviour* 25:30–38.

Krebs, J. R., D. F. Sherry, S. D. Healy, V. H. Perry, and A. L. Vaccarino. 1989. Hippocampal specialization of food storing birds. *Proceedings of the National Academy of Sciences* 86:1388–1392.

Krebs, J. R., N. S. Clayton, S. D. Healy, D. A. Cristol, S. N. Patel, and A. R. Jolliffe. 1996. The ecology of the avian brain: Food-storing memory and the hippocampus. *Ibis* 138:34–46.

Krohmer, R. W., and D. Crews. 1987. Temperature activation of courtship behavior in the male red-sided garter snake (*Thamnophis sirtalis parietalis*): Role of the anterior hypothalamus-preoptic area. *Behavioral Neuroscience* 101:228–236.

Kroodsma, D. E. 2005. *The Singing Life of Birds: The Art and Science of Listening to Birdsong*. New York: Houghton Mifflin.

Kroodsma, D. E., and R. Pickert. 1980. Environmentally dependent sensitive periods for avian vocal learning. *Nature* 288:477–479.

Kroodsma, D. E., P. W. Houlihan, P. A. Fallon, and J. A. Wells. 1997. Song development by grey catbirds. *Animal Behaviour* 54:457–464.

Kuba, M. J., Byrne, R. A., D. V. Meisel, and J. A. Mather. 2006. When do octopuses play? Effects of repeated testing, object type, age, and food deprivation on object play in *Octopus vulgaris*. *Journal of Comparative Psychology* 120:184–190.

Kummer, H., W. Gotz, and W. Angst. 1974. Triadic differentiation: An inhibitory process protecting pair bonds in baboons. *Behaviour* 49:62–87.

Kunz, T. H., E. B. Arnett, B. M. Cooper, W. P. Erickson, R. P. Larkin, T. Mabee, M. L. Morrison, M. D. Strickland, and J. M. Szewczak. 2007. Assessing impacts of wind-energy development on nocturnally active birds and bats: a guidance document. *Journal of Wildlife Management* 71:2449–2486.

Kupfermann, I., V. Castellucci, H. Pinsker, and E. Kandel. 1970. Neuronal correlates of habituation and dishabituation of the gill withdrawal reflex in *Aplysia*. *Science* 167:1743–1745.

Kuvlesky, W. P., L. A. Brennan, M. L. Morrison, K. K. Boydston, B. M. Ballard, and F. C. Bryant. 2007. Wind energy development and wildlife conservation: challenges and opportunities. *Journal of Wildlife Management* 71:2487–2498.

LaBas, N. R., and L. R. Hockham. 2005. An invasion of cheats: The evolution of worthless nuptial gifts. *Current Biology* 15:64–67.

Lacey, E. A., and P. W. Sherman. 1991. Social organization of naked mole-rat colonies: Evidence of division of labor. In *The Biology of the Naked Mole-Rat*, edited by P. W. Sherman, J. U. M. Jarvis, and R. D. Alexander. Princeton, NJ: Princeton University Press, pp. 275–336.

Lacey, E. A., and P. W. Sherman. 1997. Cooperative breeding in naked mole-rats: implications for vertebrate and invertebrate sociality. In *Cooperative Breeding in Mammals*, edited by N. G. Solomon and J. A. French. Cambridge: Cambridge University Press, pp. 267–301.

Lack, D. 1939. The display of the black cock. *British Birds* 32:290–303.

Lack, D. 1943. *The Life of the Robin*. London: Penguin Books.

Lack, D. 1968. Bird migration and natural selection. *Oikos* 19:1–9.

Lacy, R. C., and P. W. Sherman. 1983. Kin recognition by phenotype matching. *American Naturalist* 121:489–512.

Laidlaw, H. H., Jr., and R. E. Page, Jr. 1984. Polyandry in honey bees (*Apis mellifera* L.) — Sperm utilization and intracolony genetic relationships. *Genetics* 108:985–997.

Laland, K. N., and K. Williams. 1997. Shoaling generates social learning of foraging information in guppies. *Animal Behaviour* 53:1149–1159.

Lande, R. 1981. Models of speciation by sexual selection on polygenic traits. *Proceedings of the National Academy of Sciences* 78:3721–3725.

Langbauer, W. R., Jr. 2000. Elephant communication. *Zoo Biology* 19:425–445.

Langkilde, T., R. A. Alford, and L. Schwarzkopf. 2005. No behavioural compensation for fitness costs of autotomy in a lizard. *Austral Ecology* 30:713–718.

Langley, C. M. 1996. Search images: selective attention to specific visual features of prey. *Journal of Experimental Psychology* 22:152–163.

Langley, C. M., D. A. Riley, A. B. Bond, and N. Goel. 1996. Visual search for natural grains in pigeons (*Columba livia*): Search images and selective attention. *Journal of Experimental Psychology—Animal Behavior Processes* 22:139–151.

Laposky, A. D., J. Bass, A. Kohsaka, and F. W. Turek. 2008. Sleep and circadian rhythms: Key components in the regulation of energy metabolism. *FEBS Letters* 582:142–151.

Lashley, K. 1950. In search of the engram. *Symposia of the Society for Experimental Biology* IV:454–482.

Laundre, J. W., Hernandez, L., and K. B. Altendorf. 2001. Wolves, elk, and bison: reestablishing the "landscape of fear" in Yellowstone National Park. *Canadian Journal of Zoology* 79:1401–1409.

Laurien-Kehnen, C., and F. Trillmich. 2003. Lactation performance of guinea pigs (*Cavia porcellus*) does not respond to experimental manipulation of pup demands. *Behavioral Ecology and Sociobiology* 53:145–152.

Lawson Handley, L. J., and N. Perrin. 2007. Advances in our understanding of mammalian sex-biased dispersal. *Molecular Ecology* 16:1559–1578.

Le Galliard, J.-F., R. Ferrière, and J. Clobert. 2003. Mother-offspring interactions affect natal dispersal in a lizard. *Proceedings of the Royal Society of London B* 270:1163–1169.

Le Galliard, J-F., G. Gundersen, H. P., Andreassen, and N. C. Stenseth. 2006. Natal dispersal, interactions among siblings, and intrasexual competition. *Behavioral Ecology* 17:733–740.

Leadbeater, E., and L. Chittka. 2007. Social learning in insects—from miniature brains to consensus building. *Current Biology* 17:R703–R713.

LeBas, N. R., L. R. Hockam, and M. G. Ritchie. 2004. Sexual selection in the gift-giving dance fly, *Rhamphomyia sulcata*, favors small males carrying small gifts. *Evolution* 58:1763–1772.

LeBoeuf, B. J. 1974. Male-male competition and reproductive success in elephant seals. *American Zoologist* 14:163–176.

Lee, J. S. F., and A. H. Bass. 2004. Does exaggerated morphology preclude plasticity to cuckoldry in the midshipman fish (*Porichthys notatus*)? *Naturwissenschaften* 91:338–341.

Lefebvre, L. 1995. Culturally-transmitted feeding behaviour in primates: Evidence for accelerating learning rates. *Primates* 36:227–239.

Lehrman, D. S. 1953. A critique of Konrad Lorenz's theory of instinctive behavior. *Quarterly Review of Biology* 28:337–363.

Leibrecht, B. C., and H. R. Askew. 1980. Habituation from a comparative perspective. In *Comparative Psychology: An Evolutionary Analysis of Animal Behavior*, edited by M. R. Denny. New York: John Wiley, pp. 208–229.

Leinders-Zufall, T., P. Brennan, P. Widmayer, P. Chandramani, A. Maul-Pavicic, M. Jäger, X. Li, H. Breer, F. Zufall, and T. Boehm. 2004. MHC class I peptides as chemosensory signals in the vomeronasal organ. *Science* 306:1033–1037.

Leitner, S., J. Nicholson, B. Leisler, T. J. DeVoogd, and C. K. Catchpole. 2002. Song and the song control pathway in the brain can develop independently of exposure to song in the sedge warbler. *Proceedings of the Royal Society of London B* 269:2519–2524.

Lenneberg, E. H. 1967. *Biological Foundations of Language.* New York: John Wiley.

Lent, C. M., M. H. Dickinson, and C. G. Marshall. 1989. Serotonin and leech feeding behavior: Obligatory neuromodulation. *American Zoologist* 29:1241–1254.

Leoncini, I., Y. Le Conte, G. Costagliola, E. Plettner, A. L. Toth, M. Wang, Z. Huang, M. R. Band, J.-M. Bécard, D. Crauser, K. N. Slessor, and G. E. Robinson. 2004. Regulation of behavioral maturation by a primer pheromone produced by adult worker honey bees. *Proceedings of the National Academy of Sciences* 101:17559–17564.

Levi, R., and J. M. Camhi. 1995. Distributing coordinated motor outputs to several body segments—Escape movements in the cockroach. *Journal of Comparative Physiology A* 177:427–437.

Levi, R., and J. M. Camhi. 2000. Population vector coding by the giant interneurons of the cockroach. *The Journal of Neuroscience* 20:3822–3829.

Levin, L. R., P.-L. Han, P. M. Hwaung, P. G. Feinstein, R. L. Davis, and R. R. Reed. 1992. The *Drosophila* learning and memory gene *rutabaga* encodes a Ca^{+2}/ calmodulin-adenylyl cyclase. *Cell* 68:479–489.

Levitan, D. R. 2005. The distribution of male and female reproductive success in a broadcast spawning marine invertebrate. *Integrative and Comparative Biology* 45:848–855.

Lewis, T. L., and D. Maurer. 2005. Multiple sensitive periods in human visual development: evidence from visually deprived children. *Developmental Psychobiology* 46:163–183.

Lewis, S. M., C. K. Cratsley, and J. A. Rooney. 2004. Nuptial gifts and sexual selection in *Photinus* fireflies. *Integrative and Comparative Biology* 44:234–237.

Lickliter, R., and G. Gottlieb. 1988. Social specificity: Interaction with own species is necessary to foster species-specific maternal preferences in ducklings. *Developmental Psychobiology* 21:311–321.

Lieberman, D. A. 1993. *Learning: Behavior and Cognition.* 2nd ed. Pacific Grove, CA: Brooks/Cole Publishing Company.

Light, P., M. Salmon, and K. J. Lohmann. 1993. Geomagnetic orientation of loggerhead sea turtles: evidence for an inclination compass. *Journal of Experimental Biology* 182:1–10.

Lim, M. M., and L. J. Young. 2006. Neuropeptidergic regulation of affiliative behavior and social bonding in animals. *Hormones and Behavior* 50:506–517.

Lim, M. M., I. F. Bielsky, and L. J. Young. 2005. Neuropeptides and the social brain: potential rodent models of autism. *International Journal of Developmental Neuroscience* 23:235–243.

Lim, M. M., Z. Wang, D. E. Olazábal, X. Ren, E. F. Terwilliger, and L. J. Young. 2004. Enhanced partner preference in a promiscuous species by manipulating the expression of a single gene. *Nature* 429:754–757.

Lima, S. L., and L. M. Dill. 1990. Behavioral decisions made under the risk of predation—a review and prospectus. *Canadian Journal of Zoology* 68:619–640.

Lima, S. L., and T. J. Valone. 1986. Influence of predation risk on diet selection: a simple example in the grey squirrel. *Animal Behaviour* 34:536–544.

Lima, S. L., T. J. Valone, and T. Caraco. 1985. Foraging-efficiency—predation-risk trade-off in the grey squirrel. *Animal Behaviour* 33:155–165.

Limongelli, L., S. T. Boysen, and E. Visalberghi. 1995. Comprehension of cause-effect relations in a tool-using task by chimpanzees (*Pan troglodytes*). *Journal of Comparative Psychology* 109:18–26.

Linden, M., and A. P. Møller. 1989. Cost of reproduction and covariation of life history traits in birds. *Trends in Ecology and Evolution* 4:367–371.

Lindström, Å., D. Hasselquist, S. Bensch, and M. Grahn. 1990. Asymmetric contests over resources for survival and migration: A field experiment with bluethroats. *Animal Behaviour* 40:453–461.

Lindström, L., R. V. Alatalo, and J. Mappes. 1997. Imperfect Batesian mimicry—The effects of the frequency and the distastefulness of the model. *Proceedings of the Royal Society of London B* 264:149–153.

Linksvayer, T. A., and M. J. Wade. 2005. The evolutionary origin and elaboration of sociality in the aculeate Hymenoptera: maternal effects, sib-social effects, and heterochrony. *Quarterly Review of Biology* 80:317–336.

Linn, C. D., Y. Molina, J. Difatta, and T. E. Christenson. 2007. The adaptive advantage of prolonged mating: a test of alternative hypotheses. *Animal Behaviour* 74:481–485.

Little, A. E. F., T. Murakami, U. G. Mueller, and C. R. Currie. 2006. Defending against parasites: fungus-growing ants combine specialized behaviours and microbial symbionts to protect their fungus gardens. *Biology Letters* 2:12–16.

Lloyd, J. E. 1975. Aggressive mimicry in *Photuris* fireflies: signal repertoires by femmes fatales. *Science* 187:452–453.

Lloyd, J. E. 1986. Firefly communication and deception: "oh, what a tangled web." In *Deception: Perspectives on Human and Nonhuman Deceit*, edited by R. W. Mitchell and N. S. Thompson. Albany, NY: SUNY Press, pp. 113–128.

Loeb, J. 1981. *The Foundations of Ethology*. New York: Springer Verlag.

Loeb, J. 1918. *Forced Movements, Tropisms, and Animal Conduct.* Philadelphia: J. B. Lippincott.

Lohmann, K. J. 1991. Magnetic orientation by hatchling loggerhead sea turtles (*Caretta caretta*). *Journal of Experimental Biology* 155:37–49.

Lohmann, K. J. 1992. How sea turtles navigate. *Scientific American* 266:100–106.

Lohmann, K. J., and C. M. F. Lohmann. 1992. Orientation to oceanic waves by green turtle hatchlings. *Journal of Experimental Biology* 171:1–13.

Lohmann, K. J., and C. M. F. Lohmann. 1996a. Detection of magnetic field intensity by sea turtles. *Nature* 380:59–61.

Lohmann, K. J., and C. M. F. Lohmann. 1996b. Orientation and open-sea navigation in sea turtles. *Journal of Experimental Biology* 199:73–81.

Lohmann, K. J., and C. M. F. Lohmann. 2006. Sea turtles, lobsters, and magnetic maps. *Marine Freshwater Behaviour and Physiology* 39:49–64.

Lohmann, K. J. , C. M. F. Lohmann, and C. S. Endres. 2008. The sensory ecology of ocean navigation. *Journal of Experimental Biology* 211:1719–1728.

Lohmann, K. J., C. M. F. Lohmann, and N. F. Putman. 2007. Magnetic maps in animals: nature's GPS. *Journal of Experimental Biology* 210:3697–3705.

Lohmann, K. J., N. F. Putnam, and C. M. F. Lohmann. 2008. Geomagnetic imprinting: A unifying hypothesis of long-distance natal homing in salmon and sea turtles. *Proceedings of the National Academy of Sciences* 105:19096–19101.

Lohmann, K. J., S. D. Cain, S. A. Dodge, and C. M. F. Lohmann. 2001. Regional magnetic fields as navigational markers for sea turtles. *Science* 294:364–366.

Lohmann, K. J., C. M. F. Lohmann, L. M. Ehrhart, D. A. Bagley, and T. Swing. 2004. Geomagnetic map used in sea-turtle navigation. *Nature* 428:909–910.

Lombardino, A. J., and F. Nottebohm. 2000. Age at deafening affects the stability of learned song in adult male zebra finches. *Journal of Neuroscience* 20:5054–5064.

London, S. E., and B. A. Schlinger. 2007. Steroidogenic enzymes along the ventricular proliferative zone in the developing songbird brain. *Journal of Comparative Neurology* 502:507–521.

Lorenz, K. 1935. Der Kumpan in der Umwelt des Vogels. *Journal für Ornithologie* 83:137–213.

Lorenz, K. 1952. *King Solomon's Ring*. New York: Crowell.

Lorenz, K. 1958. The evolution of behavior. *Scientific American* 199:67–78.

Lorenz, K. 1972. Comparative studies on the behavior of Anatinae. In *Function and Evolution of Behavior: An Historical Sample from the Pens of Ethologists*, edited by P. H. Klopfer and J. P. Hailman. Reading, MA: Addison-Wesley, pp. 231–258.

Lorenz, K., and N. Tinbergen. 1938. Taxis und Instinkhandlung in der Eirollbewegung der Graugans. *Zeitschrift für Tierpsychologie* 2:1–29.

Lott, D. F. 1991. *Intraspecific Variation in the Social Systems of Wild Vertebrates*. Cambridge: Cambridge University Press.

Low, B. S. 1990. Marriage systems and pathogen stress in human societies. *Amerian Zoologist* 30:325–339.

Lutz, C. K., and M. A. Novak. 2005. Environmental enrichment for nonhuman primates: theory and application. *ILAR Journal* 46:178–191.

Lynch, C. B. 1980. Response to divergent selection for nesting behavior in *Mus musculus*. *Genetics* 96:757–765.

Lynch, C. B. 1992. Clinal variation in cold adaptation in *Mus domesticus:* Verification of predictions from laboratory populations. *American Naturalist* 139:1219–1236.

Lyons, C., and C. J. Barnard. 2006. A learned response to sperm competition in the field cricket, *Gryllus bimaculatus* (de Geer). *Animal Behaviour* 72:673–680.

Lythgoe, J. N., W. R. A. Muntz, J. C. Partridge, J. Shand, and D. McB. Williams. 1994. The ecology of visual pigments of snappers (Lutjanidae) on the Great Barrier Reef. *Journal of Comparative Physiology A* 174:255–260.

Macdonald, D. W., and D. D. P. Johnson. 2001. Dispersal in theory and practice: consequences for conservation biology. In *Dispersal*, edited by J. Clobert, E. Danhin, A. A. Dhondt, and J. D. Nichols. Oxford: Oxford University Press, pp. 358–372.

Macphail, E. M., and J. J. Bolhuis. 2001. The evolution of intelligence: adaptive specializations versus general process. *Biological Reviews* 76:341–364.

MacRoberts, M. H., and B. R. MacRoberts. 1976. Social organization and behavior of the acorn woodpecker in central coastal California. *Ornithological Monographs* 21:1–115.

Madden, J. R., and K. Tanner. 2003. Preferences for coloured bower decorations can be explained in a nonsexual context. *Animal Behaviour* 65:1077–1083.

Madsen, T., and R. Shine. 1996. Seasonal migrations of predators and prey—A study of pythons and rats in tropical Australia. *Ecology* 77:149–156.

Maeterlinck, M. 1901. *The Life of the Bee*. New York: Dodd, Mead.

Mager, J. N., III, C. Walcott, and W. H. Piper. 2007. Male common loons, *Gavia immer*, communicate body mass and condition through dominant frequencies of territorial yodels. *Animal Behaviour* 73:683–690.

Magrath, M. J. L., E. van Lieshout, G. Henk Visser, and J. Komdeur. 2004. Nutritional bias as a new mode of adjusting sex allocation. *Proceedings of the Royal Society of London B* 271: S347–S349.

Magrath, M. J. L., E. van Lieshout, I. Pen, G. Henk Visser, and J. Komdeur. 2007. Estimating expenditure on male and female offspring in a sexually size-dimorphic bird: a comparison of different methods. *Journal of Animal Ecology* 76:1169–1180.

Maguire, E. A., D. G. Gadlian, I. S. Johnsrude, C. D. Good, J. Ashburner, R. S. J. Frackowiak, and C. D. Frith. 2000. Navigation-related structural change in the hippocampi of taxi drivers. *Proceedings of the National Academy of Sciences* 97:4398–4403.

Maher, C., and D. Lott. 1995. Definitions of territoriality used in the study of variation in vertebrate spacing systems. *Animal Behaviour* 49:1581–1597.

Maher, C., and D. Lott. 2000. A review of ecological determinants of territoriality within vertebrate species. *American Midland Naturalist* 143:1–29.

Malenka, R. C., and M. F. Bear. 2004. LTP and LTD: An embarrassment of riches. *Neuron* 44:5–21.

Mallard, S. T., and C. J. Barnard. 2003. Competition, fluctuating asymmetry and sperm transfer in male gryllid crickets (*Gryllus bimaculatus* and *Gyllodes sigillatus*). *Behavioral Ecology and Sociobiology* 53:190–197.

Mallot, R. W., and J. W. Siddall. 1972. Acquisition of the people concept in pigeons. *Psychological Reports* 31:3–13.

Manger, P. R., and J. D. Pettigrew. 1995. Electroreception and the feeding behavior of the platypus (*Ornithorhynchus anatinus*, Monotremata, Mammalia). *Philosophical Transactions of the Royal Society of London B* 347:359–381.

Mank, J. E., D. E. L. Promislow, and J. C. Avise. 2005. Phylogenetic perspectives in the evolution of parental care in ray-fined fishes. *Evolution* 59:1570–1578.

Manoli, D. S., M. Foss, A. Villelle, B. J. Taylor, J. C. Hall, and B. S. Baker. 2005. Male-specific *fruitless* specifies the neural substrates of *Drosophila* courtship. *Nature* 436:395–400.

Mappes, J., and R. V. Alatalo. 1997. Batesian mimicry and signal accuracy. *Evolution* 51:2050–2053.

Marcus, G. F., and S. E. Fisher. 2003. FOXP2 in focus: What can genes tell us about speech and language? *Trends in Cognitive Sciences* 7:257–262.

Marhold, S., W. Wiltschko, and H. Burda. 1997. A magnetic polarity compass for direction finding in a subterranean mammal. *Naturwissenschaften* 84:421–423.

Markham, J., and W. T. Greenough. 2004. Experience-driven brain plasticity. *Neuron Glia Biology* 1:351–363.

Marler, C. A., and M. C. Moore. 1988. Evolutionary costs of aggression revealed by testosterone manipulations in free-living male lizards. *Behavioral Ecology and Sociobiology* 23:21–26.

Marler, C. A., and M. C. Moore. 1991. Supplementary feeding compensates for testosterone-induced costs of aggression in male mountain spiny lizards, *Sceloporus jarrovi*. *Animal Behaviour* 42:209–219.

Marler, C. A., G. Walsberg, M. L. White, M. Moore. 1995. Increased energy expenditure due to increased territorial defense in male lizards after phenotypic manipulation. *Behavioral Ecology and Sociobiology* 37:225–231.

Marler, P. 1970. A comparative approach to vocal learning: Song development in white-crowned sparrows. *Journal of Comparative and Physiological Psychology* 71:1–25.

Marler, P. 1987. Sensitive periods and the roles of specific and general sensory stimulation in birdsong learning. In *Imprinting and Cortical Plasticity: Comparative Aspects of Sensitive Periods*, edited by J. R. Rauschecker and P. Marler. New York: John Wiley, pp. 99–135

Marler, P. 1996. Social cognition: Are primates smarter than birds? In *Current Ornithology, Vol. 13*, edited by V. J. Nolan and E. D. Ketterson. New York: Plenum, pp. 1–33.

Marler, P., and M. Tamura. 1964. Culturally transmitted patterns of vocal behavior in sparrows. *Science* 146:1483–1486.

Martan, J., and B. A. Shepherd. 1976. The role of the copulatory plug in reproduction of the guinea pig. *Journal of Experimental Zoology* 196:79–84.

Martín, J., J. J. Luque-Larena, and P. López. 2006. Collective detection in escape responses of temporary groups of Iberian green frogs. *Behavioral Ecology* 17:222–226.

Martin, R. A. 2007. A review of shark agonistic displays: comparison of display features and implications for shark-human interactions. *Marine and Freshwater Behaviour and Physiology* 40:3–34.

Martin-Gronert, M. S., and S. E. Ozanne. 2006. Maternal nutrition during pregnancy and health of the offspring. *Biochemical Society Transactions* 34:779–782.

Marx, J. L. 1980. Ape-language controversy flares up. *Science* 207:1330–1333.

Mason, G., R. Clubb, N. Latham, and S. Vickery. 2007. Why and how should we use environmental enrichment to tackle stereotypic behaviour? *Applied Animal Behaviour Science* 102:163–188.

Massey, A. 1988. Sexual interactions in red-spotted newt populations. *Animal Behaviour* 36:205–210.

Mateo, J. M. 2004. Recognition systems and biological organization: the perception component of social recognition. *Annales Zoologici Fennici* 41:729–745.

Mateo, J. M., and W. G. Holmes. 2004. Cross-fostering as a means to study kin recognition. *Animal Behaviour* 68:1451–1459.

Mateo, J. M., and R. E. Johnston. 2000. Kin recognition and the 'armpit effect': evidence of self-referent phenotype matching. *Proceedings of the Royal Society of London B* 267:695–700.

Mateo, J. M., and R. E. Johnston. 2003. Kin recognition by self-referent phenotype matching: weighing the evidence. *Animal Cognition* 6:73–76.

Mather, J. A., and R. C. Anderson. 1999. Exploration, play, and habituation in octopuses (*Octopus dofleini*). *Journal of Comparative Psychology* 113:333–338.

Mathis, A., D. P. Chivers, and R. J. F. Smith. 1996. Cultural transmission of predator recognition in fishes: Intraspecific and interspecific learning. *Animal Behaviour* 51:185–201.

Mattila, H. R., and T. D. Seeley. 2007. Genetic diversity in honey bee colonies enhances productivity and fitness. Science 317:362–364.

Mauck, R. A., and T. C. Grubb, Jr. 1995. Petrel parents shunt all experimentally increased reproductive costs to their offspring. *Animal Behaviour* 49:999–1008.

Maynard Smith, J. 1974. The theory of games and the evolution of animal conflicts. *Journal of Theoretical Biology* 47:209–221.

Maynard Smith, J. 1976. Evolution and the theory of games. *American Scientist* 64:41–45.

Maynard Smith, J. 1977. Parental investment: A prospective analysis. *Animal Behaviour* 25:1–9.

Maynard Smith, J. 1982. *Evolution and the Theory of Games*. Cambridge: Cambridge University Press.

Maynard Smith, J., and G. A. Parker. 1976. The logic of asymmetric contests. *Animal Behaviour* 24:159–175.

Maynard Smith, J., and S. E. Riechert. 1984. A conflicting tendency model of spider agonistic behaviour: hybrid-pure population line comparisons. *Animal Behaviour* 32:564–578.

Mayr, E. 1983. How to carry out the adaptationist program? *American Naturalist* 121:295–310.

Mays, H. L. Jr., and G. E. Hill. 2004. Choosing mates: good genes versus genes that are a good fit. *Trends in Ecology and Evolution* 19:554–559.

McBeath, M. K., D. M. Shaffer, and M. K. Kaiser. 1995. How baseball outfielders determine where to run to catch fly balls. *Science* 268:569–573.

McCann, T. S. 1981. Aggression and sexual activity of male southern elephant seals. *Journal of Zoology* 195:295–310.

McCormick, S. D. and D. Bradshaw. 2006. Hormonal control of salt and water balance in vertebrates. *General and Comparative Endocrinology* 147:3–8.

McCormick, S. D., L. P. Hansen, T. P. Quinn, and R. L. Saunders. 1998. Movement, migration, and smolting of Atlantic salmon (*Salmo salar*). *Canadian Journal of Fisheries and Aquatic Sciences* 55 (Supplement 1):77–92.

McCracken, G. F., and M. K. Gustin. 1991. Nursing behavior in Mexican free-tailed bat maternity colonies. *Ethology* 89:305–321.

McDonald, D. B., and W. K. Potts. 1994. Cooperative display and relatedness among males in a lek-mating bird. *Science* 266:1030–1032.

McEwen, B. S. 1976. Interactions between hormones and nerve tissue. *Scientific American* 235:48–58.

McFall-Ngai, M. J. 1990. Crypsis in the pelagic environment. *American Zoologist* 30:175–188.

McGlothlin, J. W., J. M. Jawor, and E. D. Ketterson. 2007. Natural variation in a testosterone-mediated trade-off between mating effort and parental effort. *American Naturalist* 170:864–875.

McGowan, K. J., and G. E. Woolfenden. 1989. A sentinel system in the Florida scrub jay. *Animal Behaviour* 37:1000–1006.

McGuire, B. 1988. The effects of cross-fostering on parental behavior of meadow voles (*Microtus pennsylvanicus*). *Journal of Mammalogy* 69:332–341.

McGuire, B., T. Pizzuto, W. E. Bemis, and L. L. Getz. 2006. General ecology of a rural population of Norway rats (*Rattus norvegicus*) based on intensive live trapping. *American Midland Naturalist* 155:221–236.

McGuire, B., L. L. Getz, J. Hofmann, T. Pizzuto, and B. Frase. 1993. Natal dispersal in prairie voles (*Microtus ochrogaster*) in relation to population density, season, and natal social environment. *Behavioral Ecology and Sociobiology* 32: 293–302.

McKaye, K. R. 1977. Competition for breeding sites between the cichlid fishes of Lake Jiloa, Nicaragua. *Ecology* 58:291–302.

McKaye, K. R., and N. M. McKaye. 1977. Communal care and kidnapping of young by parental cichlids. *Evolution* 31:674–681.

McLaren, A., and D. Michie. 1960. Control of prenatal growth in mammals. *Nature* 187:363–365.

McLean, I. G., C. Holzer, and B. J. S. Studholme. 1999. Teaching predator-recognition to a naive bird: implications for management. *Biological Conservation* 87:123–130.

Mech, L. D. 1970. *The Wolf: The Ecology and Behavior of an Endangered Species*. Minneapolis: University of Minnesota Press.

Melletti, M., V. Penteriani, and L. Boitani. 2007. Habitat preferences of the secretive forest buffalo (*Syncerus caffer nanus*) in Central Africa. *Journal of Zoology* 271:178–186.

Mellgren, R. L., ed. 1983. *Animal Cognition and Behavior*. Amsterdam: North Holland.

Mello, C., F. Nottebohm, and D. Clayton. 1995. Repeated exposure to one song leads to rapid and persistent decline in an immediate early gene's response to that song in zebra finch telencephalon. *Journal of Neuroscience* 15:6919–6925.

Mello, C., D. S. Vicario, and D. F. Clayton. 1992. Song presentation induces gene expression in the songbird forebrain. *Proceedings of the National Academy of Sciences* 89:6818–6822.

Meltzoff, A. N. 1988. The human infant as *Homo imitans*. In *Social Learning: Psychological and Biological Perspectives*, edited by T. R. Zentall and B. G. Galef, Jr. Hillsdale, NJ: Erlbaum, pp. 319–341.

Menzel, R. 2007. Searching for the memory trace in a mini-brain, the honeybee. *Learning and Memory* 8:53–62.

Meredith, M. 1998. Vomeronasal, olfactory, hormonal convergence in the brain: cooperation or coincidence? *Annals of the New York Academy of Sciences* 855:349–361.

Merilaita, S. 1998. Crypsis through disruptive coloration in an isopod. *Proceedings of the Royal Society of London B*. 265:1059–1064.

Merilaita, S., and J. Lind. 2005. Background-matching and disruptive coloration, and the evolution of cryptic coloration. *Proceedings of the Royal Society of London B* 272:665–670.

Mery, F., and T. J. Kawecki. 2002. Experimental evolution of learning ability in fruit flies. *Proceedings of the National Academy of Sciences* 99:14274–14279.

Mery, F., and T. J. Kawecki. 2004. An operating cost of learning in *Drosophila melanogaster*. *Animal Behaviour* 68:589–598.

Messenger, J. B. 2001. Cephalopod chromatophores: neurobiology and natural history. *Biological Reviews* 76:473–528.

Metzgar, L. H. 1967. An experimental comparison of screech owl predation on resident and transient white-footed mice (*Peromyscus leucopus*). *Journal of Mammalogy* 48:387–391.

Meyer, A. 1997. The evolution of sexually selected traits in male swordtail fishes (*Xiphophorus*: Poecilidae). *Heredity* 79:329–337.

Meyer, A., W. Salburger, and M. Schartl. 2006. Hybrid origin of a swordtail species (Teleostei: *Xiphophorus clemenciae*) driven by sexual selection. *Molecular Ecology* 15:721–730.

Meyerriecks, A. J. 1960. Comparative breeding behavior of four species of North American herons. *Publications of the Nuttall Ornithology Club* 2:1–158.

Michel, G. F., and A. N. Tyler. 2005. Critical period: a history of the transition from questions of when, to what, to how. *Developmental Psychobiology* 46:158–162.

Michelson, A., B. B. Andersen, W. H. Kirchner, and M. Lindauer. 1989. Honeybees can be recruited by a mechanical model of a dancing bee. *Naturwissenschaften* 76:277–280.

Michener, C. D. 1969. Comparative social behavior of bees. *Annual Review of Entomology* 14:299–342.

Miklósi, Á., V. Csányi, and R. Gerlai. 1997. Antipredator behavior in paradise fish (*Macropodus opercularis*) larvae: The role of genetic factors and paternal influence. *Behavior Genetics* 27:191–200.

Miles, H. L. W. 1990. The cognitive foundations for reference in a signing orangutan. In *"Language" and Intelligence in Monkeys and Apes: Comparative Developmental Perspectives*, edited by S. T. Parker and K. R. Gibson. New York: Cambridge University Press, pp. 511–539.

Miller, C. A., and J. D. Sweatt. 2007. Covalent modification of DNA regulates memory formation. *Neuron* 53:857–869.

Miller, D. B. 1980. Maternal vocal control of behavioral inhibition in mallard ducklings (*Anas platyrhynchos*). *Journal of Comparative and Physiological Psychology* 94:606–623.

Miller, D. B., and G. Gottlieb. 1978. Maternal vocalizations of mallard ducks (*Anas platyrhynchos*). *Animal Behaviour* 26:1178–1194.

Miller, R. C. 1922. The significance of the gregarious habit. *Ecology* 3:122–126.

Mineka, S., and M. Cook. 1988. Social learning and the acquisition of snake fear in monkeys. In *Social Learning: Psychological and Biological Perspectives*, edited by T. R. Zentall and B. G. J. Galef. Hillsdale, NJ: Erlbaum, pp. 51–73.

Missoweit, M., and K. P. Sauer. 2007. Not all *Panorpa* (Mecoptera: Panorpidae) scorpionfly mating systems are characterized by resource defence polygyny. *Animal Behaviour* 74:1207–1213.

Mistlberger, R. E., and B. Rusak. 2005. Biological Rhythms and Behavior. In *The Behavior of Animals: Mechanisms, Function, and Evolution*, edited by J. J. Bolhuis and L.-A. Giraldeau. Malden, MA: Blackwell Publishers. pp. 71–96.

Mock, D. W., H. Drummond, and C. H. Stinson. 1990. Avian siblicide. *American Scientist* 78:438–449.

Mock, D. W., and G. A. Parker. 1997. *The Evolution of Sibling Rivalry*. Oxford: Oxford University Press.

Mock, D. W., and G. A. Parker. 1998. Siblicide, family conflict and the evolutionary limits of selfishness. *Animal Behaviour* 56:1–10.

Moehlman, P. D. 1979. Jackal helpers and pup survival. *Nature* 277:382–383.

Molina, A., A. G. Castellano, and J. López-Barneo. 1997. Pore mutations in *Shaker* K^+ channels distinguish between the sites of tetraethylammonium blockade and C-type inactivation. *Journal of Physiology* 499:361–367.

Møller, A. P. 1988. Female choice selects for male sexual tail ornaments in the monogamous swallow. *Nature* 322:640–642.

Møller, A. P. 1990. Effects of a haemotaphagus mite on the barn swallow (*Hirundo rustica*): A test of the Hamilton and Zuk hypothesis. *Evolution* 44:771–784.

Møller, A. P. 1991. Parasite load reduces song output in a passerine bird. *Animal Behaviour* 41:723–730.

Møller, A. P. 1992. Parasites differentially increase the degree of fluctuating asymmetry in secondary sexual characters. *Journal of Evolutionary Biology* 5:691–699.

Møller, A. P., and A. Pomiankowski. 1993. Fluctuating asymmetry and sexual selection. *Genetica* 89:267–279.

Møller, A. P., and J. P. Swaddle. 1997. *Fluctuating Asymmetry, Developmental Stability and Evolution*. Oxford: Oxford University Press.

Møller, A. P., and T. R. Birkhead. 1989. Copulation behavior in mammals: Evidence that sperm competition is widespread. *Biological Journal of the Linnean Society* 38:119–131.

Møller, A. P., P. Christie, and E. Lux. 1999. Parasitism, host immune function, and sexual selection. *Quarterly Review of Biology* 74:3–20.

Moment, G. B. 1962. Reflexive selection: A possible answer to an old question. *Science* 136:262–263.

Moncomble, A.-S., G. Coureaud, B. Quennedey, D. Langlois, G. Perrier, and B. Schaal. 2005. The mammary pheromone of the rabbit: from where does it come? *Animal Behaviour* 69:29–38.

Mönkkönen, M., R. Hardling, J. T. Forsman, and J. Tuomi. 1999. Evolution of heterospecific attraction: using other species as cues in habitat selection. *Evolutionary Ecology* 13:91–104.

Moore, F. L., S. K. Boyd, and D. B. Kelley. 2005. Historical perspective: Hormonal regulation of behavior in amphibians. *Hormones and Behavior* 48:373–383.

Moore, F. L., and S. J. Evans. 1999. Steroid hormones use non-genomic mechanisms to control brain functions and behaviors: a review of evidence. *Brain, Behavior and Evolution* 54:41–50.

Moore, F. L., and L. J. Miller. 1984. Stress-induced inhibition of sexual behavior: corticosterone inhibits courtship behaviors of a male amphibian (*Taricha granulosa*). *Hormones and Behavior* 18:400–410.

Moore, F. R. 1986. Sunrise, skylight polarization, and the early morning orientation of night-migrating warblers. *Condor* 88:493–498.

Moore, J., and R. Ali. 1984. Are dispersal and inbreeding avoidance related? *Animal Behaviour* 32:94–112.

Moore, M. C. 1991. Application of organization-activation theory to alternative male reproductive strategies: a review. *Hormones and Behavior* 25:154–179.

Moore, M. C., and C. A. Marler. 1987. Effects of testosterone manipulations on nonbreeding season territorial aggression in free-living male lizards, *Sceloporus jarrovi*. *General and Comparative Endocrinology* 65:255–232.

Moore, M. C., D. K. Hews, and R. Knapp. 1998. Evolution and hormonal control of alternative male phenotypes. *American Zoologist* 38:133–151.

Moran, G. 1984. Vigilance behavior and alarm calls in a captive group of meerkats, *Suricata suricatta*. *Zeitschrift für Tierpsychologie* 65:228–240.

Moreira, P. L., P. López, and J. Martín. 2006. Femoral secretions and copulatory plugs convey chemical information about male identity and dominance status. *Behavioral Ecology and Sociobiology* 60:166–174.

Morell, V. 1998. A new look at monogamy. *Science* 281:1982–1983.

Morena, J., M. Soler, M. Lindén, and A. P. Møller. 1994. The function of stone carrying in the black wheatear, *Oenanthe leucura*. *Animal Behaviour* 47:1297–1309.

Morgan, C. L. 1894. *An Introduction to Comparative Psychology*. New York: Scribner.

Morris, D. 1954. The reproductive behavior of the zebra finch *Poephila guttata* with special reference to pseudofemale behavior and displacement activities. *Behaviour* 6:271–322.

Morris, D. 1956. The feather postures of birds and the problem of the origin of social signals. *Behaviour* 9:75–113.

Morris, D. 1958. The reproductive behavior of the ten-spined stickleback (*Pygosteus pungitius* L.). *Behaviour Supplement* 6:1–154.

Morris, R. J., and M. D. E. Fellowes. 2002. Learning and natal host influence host preference and sex allocation behaviour in a pupal parasitoid. *Behavioral Ecology and Sociobiology* 51:386–393.

Morrow, E. H., and G. Arnqvist. 2003. Costly traumatic insemination and a female counter-adaptation in bed bugs. *Proceedings of the Royal Society of London B* 270:2377–2381.

Morton, E. S. 1975. Ecological sources of selection in avian sounds. *American Naturalist* 109:17–34.

Mottley, K., and C. Heyes. 2003. Budgerigars (*Melopsittacus undulatus*) copy virtual demonstrators in a two-action test. *Journal of Comparative Psychology* 117:363–370.

Moum, S. E., and R. L. Baker. 1990. Colour change and substrate selection in larval *Ischnura verticalis*. *Canadian Journal of Zoology* 68:221–224.

Mouritsen, H., U. Janssen-Bienhold, M. Liedvogel, G. Feenders, J. Stalleicken, P. Dirks, and R. Weiler. 2004. Cryptochromes and neuronal-activity markers colocalize in the retina of migratory birds during magnetic orientation. *Proceedings of the National Academy of Sciences* 101:14294–14299.

Moyer, K. E. 1976. *Psychobiology of Aggression*. New York: Harper and Row.

Moynihan, M. 1966. Communication in *Callicebus*. *Journal of Zoology* 150:77–127.

Moynihan, M. 1967. Comparative aspects of communication in New World primates. In *Primate Ethology*, edited by D. Morris. London: Weidenfeld & Nicholson, pp. 236–266.

Moynihan, M. 1970. Some behavior patterns of platyrrhine monkeys. II. *Saguinus geoffroyi* and some other tamarins. *Smithsonian Contributions to Zoology* 28:1–77.

Mueller, U. G., N. M. Gerardo, D. K. Aanen, D. L. Six, and T. R. Schultz. 2005. The evolution of agriculture in insects. *Annual Review of Ecology, Evolution, and Systematics* 36:563–595.

Muller, M. N., and R. W. Wrangham. 2004. Dominance, cortisol and stress in wild chimpanzees (*Pan troglodytes schweinfurthii*). *Behavioral Ecology and Sociobiology* 55:332–340.

Müller, M., and R. Wehner. 2007. Wind and sky as compass cues in desert ant navigation. *Naturwissenschaften* 94:589–594.

Munz, T. 2005. The bee battles: Karl von Frisch, Adrian Wenner and the honey bee dance language controversy. *Journal of the History of Biology* 38:535–570.

Nagy, M., G. Heckel, C. C. Voigt, and F. Mayer. 2007. Female-biased dispersal and patrilocal kin groups in a mammal with resource-defense polygyny. *Proceedings of the Royal Society of London B* 274:3019–3025.

Nahallage, C., and M. A. Huffman. 2007. Age-specific functions of stone handling, a solitary-object play behavior, in Japanese macaques (*Macaca fuscata*). *American Journal of Primatology* 69:267–281.

Nakano, R., N. Skals, T. Takanashi, A. Surlykke, T. Koike, K. Yoshida, H. Maruyama, S. Tatsuki, and Y. Ishikawa. 2008. Moths produce extremely quiet ultrasonic courtship songs by rubbing specialized scales. *Proceedings of the National Academy of Sciences* 105:11812–11817.

Narendra, A., K. Cheng, D. Sulikowski, and R. Wehner. 2008. Search strategies of ants in landmark-rich environments. *Journal of Comparative Physiology A* 194:929–938.

Naumann, K., M. L. Winston, K. N. Slessor, G. D. Prestwich, and F. X. Webster. 1991. Production and transmission of honey bee queen (*Apis mellifera* L.) mandibular gland pheromone. *Behavioral Ecology and Sociobiology* 29:321–332.

Neff, B. D. 2003. Decisions about parental care in response to perceived paternity. *Nature* 422:716–719.

Neff, B. D., and P. W. Sherman. 2003. Nestling recognition via direct cues by parental male bluegill sunfish (*Lepomis macrochirus*). *Animal Cognition* 6:87–92.

Neff, B. D., P. Fu, and M. R. Gross. 2003. Sperm investment and alternative mating tactics in bluegill sunfish (*Lepomis macrochirus*). *Behavioral Ecology* 14:634–641.

Negro, J. J., J. Bustmante, J. Milward, and D. M. Bird. 1996. Captive fledgling American kestrels prefer to play with objects resembling natural prey. *Animal Behaviour* 52:707–714.

Neill, S. R. St. J., and J. M. Cullen. 1974. Experiments on whether schooling by their prey affects the hunting behavior of cephalopods and fish predators. *Journal of Zoology* 172:549–569.

Nelson, C. M., K. E. Ihle, M. K. Fondrk, and R. E. Page, Jr. 2007. The gene *vitellogenin* has multiple coordinating effects on social organization. *PLoS Biology* 5:0673–0677.

Nelson, D. A., and P. Marler. 1994. Selection-based learning in bird song development. *Proceedings of the National Academy of Sciences* 91:10498–10501.

Nelson, R. J. 2005. *An Introduction to Behavioral Endocrinology*. Sunderland, MA: Sinauer.

Neumann, D. 1976. Entrainment of a semi-lunar rhythm. In *Biological Rhythms in the Marine Environment*, edited by P. DeCoursey. Columbia: University of South Carolina Press.

Nevitt, G. A. 1999a. Foraging by seabirds on an olfactory landscape. *American Scientist* 87:46–53.

Nevitt, G. A. 1999b. Olfactory foraging in Antarctic seabirds: A species-specific attraction to krill odors. *Marine Ecology Progress Series* 177:235–241.

Nevitt, G. A., R. R. Veit, and P. Kareiva. 1995. Dimethyl sulfide as a foraging cue for Antarctic procellariiform seabirds. *Nature* 376:680–682.

Nevo, E., G. Heth, and H. Pratt. 1991. Seismic communication in a blind subterranean mammal: A major somatosensory mechanism in adaptive evolution underground. *Proceedings of the National Academy of Sciences* 88:1256–1260.

Newman, S. W. 1999. The medial extended amygdala in male reproductive behavior: a node in the mammalian social behavior network. *Annals of the New York Academy of Sciences* 877:242–257.

Newton, I. 2007. Weather-related mass-mortality events in migrants. *Ibis* 149:453–467.

Newton, P. N. 1986. Infanticide in an undisturbed population of Hanuman langurs, *Presbytis entellus*. *Animal Behaviour* 34:785–789.

Noble, G. K. 1939. The role of dominance in the social life of birds. *Auk* 56:263–273.

Noë, R., and A. A. Sluijter. 1990. Reproductive tactics of male savanna baboons. *Behaviour* 113:117–170.

Nonacs, P. 2001. State dependent behavior and the Marginal Value Theorem. *Behavioural Ecology* 12:71–83.

Nordby, J. C., S. E. Campbell, and M. D. Beecher. 2007. Selective attrition and individual song repertoire development in song sparrows. *Animal Behaviour* 74:1413–1418.

Nordeen, E. J., and K. W. Nordeen. 1992. Auditory feedback is necessary for the maintenance of stereotyped song in adult zebra finches. *Behavioral and Neural Biology* 57:58–66.

Norris, K. S., and C. H. Lowe. 1964. An analysis of background color-matching in amphibians and reptiles. *Ecology* 45:565–580.

Nottebohm, F. 1981. A brain for all seasons: cyclical anatomical changes in song control nuclei of the canary brain. *Science* 214:1368–1370.

Nottebohm, F. 2002. Why are some neurons replaced in the adult brain? *The Journal of Neuroscience* 22:624–628.

Nottebohm, F. 2005. The neural basis of birdsong. *PLoS Biology* 3:759–761.

Nottebohm, F., and J. P. Arnold. 1976. Sexual dimorphism in vocal control areas of the songbird brain. *Science* 194:211–213.

Nottebohm, F., T. M. Stokes, and C. M. Leonard. 1976. Central control of song in the canary (*Serinus canarius*). *Journal of Comparative Neurology* 165:457–468.

Nunes, S. 2007. Dispersal and philopatry. In *Rodent Societies: An Ecological and Evolutionary Perspective*, edited by J. O. Wolff and P. W. Sherman. Chicago: Chicago University Press, pp. 150–162.

Nunes, S., and K. E. Holekamp. 1996. Mass and fat influence the timing of natal dispersal in Belding's ground squirrels. *Journal of Mammalogy* 77:807–817.

O'Brien, E. L., A. E. Burger, and R. D. Dawson. 2005. Foraging decision rules and prey species preferences of northwestern crows (*Corvus caurinus*). *Ethology* 111:77–87.

O'Brien, S. J. 1994. A role for molecular genetics in biological conservation. *Proceedings of the National Academy of Sciences* 91:5748–5755.

O'Connell-Rodwell, C. E. 2007. Keeping an "ear" to the ground: Seismic communication in elephants. *Physiology* 22:287–294.

O'Connell-Rodwell, C. E., J. D. Wood, C. Kinzley, T. C. Rodwell, J. H. Poole, and S. Puria. 2007. Wild African elephants (*Loxodonta africana*) discriminate between familiar and unfamiliar conspecific seismic alarm calls. *Journal of the Acoustical Society of America* 122: 823–830.

O'Loghlen, A. L., and S. I. Rothstein. 1993. An extreme example of delayed vocal development: song learning in a population of wild brown-headed cowbirds. *Animal Behaviour* 46:293–304.

O'Sullivan, C., and C. Yeager. 1989. Communicative context and linguistic competence. In *Teaching Sign Language in Chimpanzees*, edited by R. G. Gardner, B. T. Gardner, and T. E. Van Cantfort. Albany: State University of New York Press, pp. 269–279.

Oetting, S., E. Pröve, and H.-J. Bischof. 1995. Sexual imprinting as a two-stage process: Mechanisms of information storage and stabilization. *Animal Behaviour* 50:393–403.

Olberg, R. M., A. H. Worthington, and K. R. Venator. 2000. Prey pursuit and interception in dragonflies. *Journal of Comparative Physiology A* 186:155–162.

Olson, D. J., A. C. Kamil, R. P. Balda, and P. J. Nims. 1995. Performance of four seed-caching corvid species in operant test of nonspatial and spatial memory. *Journal of Comparative Psychology* 109:173–181.

Olsson, M., T. Madsen, and R. Shine. 1997. Is sperm really so cheap? Costs of reproduction in male adders, *Vipera berus*. *Proceedings of the Royal Society of London B* 264:455–459.

Ophir, A. G., S. M. Phelps, A. B. Sorin, and J. O. Wolff. 2008. Social but not genetic monogamy is associated with greater breeding success in prairie voles. *Animal Behaviour* 75:1143–1154.

Orchinik, M., P. Licht, and D. Crews. 1988. Plasma steroid concentrations change in response to sexual behavior in *Bufo marinus*. *Hormones and Behavior* 22:338–350.

Orchinik, M., T. F. Murray, and F. L. Moore. 1991. A corticosteroid receptor in neuronal membranes. *Science* 252:1848–1851.

Ord, T. J., R. A. Peters, B. Clucas, and J. A. Stamps. 2007. Lizards speed up displays in noisy motion habitats. *Proceedings of the Royal Society B* 274:1057–1062.

Orians, G. H. 1969. On the evolution of mating systems of birds and mammals. *American Naturalist* 103:589–603.

Orians, G. H., and G. M. Christman. 1968. A comparative study of red-winged, tri-colored and yellow-headed blackbirds. *University of California Publications in Zoology* 84:1–81.

Osborne, K. A., A. Robichon, E. Burgess, S. Shaw Butland, R. A., A. Coulthard, H. S. Pereira, R. J. Greenspan, and M. B. Sokolowski. 1997. Natural behavior polymorphism due to a cGMP-dependent kinase of *Drosophila*. *Science* 277:834–836.

Ostfeld, R. S. 1990. The ecology of territoriality in small mammals. *Trends in Ecology and Evolution* 5:411–415.

Ostreiher, R. 2003. Is mobbing altruistic or selfish behaviour? *Animal Behaviour* 66:145–149.

Owen, D. 1980. *Camouflage and Mimicry*. Chicago: University of Chicago Press.

Owens, D., and M. Owens. 1996. Social dominance and reproductive patterns in brown hyaenas, *Hyaena brunnea*, of the central Kalahari desert. *Animal Behaviour* 51:535–551.

Owens, I. P. F. 2006. Where is behavioral ecology going? *Trends in Ecology and Evolution* 21:356–361.

Owings, D., and D. Hennessey. 1984. The importance of variation in sciurid visual and vocal communication. In *Biology of Ground-Dwelling Squirrels: Annual Cycles, Behavioral Ecology, and Sociality*, edited by J. O. Murie and G. R. Michener. Lincoln, NE: University of Nebraska Press, pp. 202–247.

Packer, C. 1977. Reciprocal altruism in *Papio anubis*. *Nature* 265:441–443.

Packer, C. 1986. The ecology of felid sociality. In *Ecological Aspects of Social Evolution*, edited by D. J. Rubenstein and R. W. Wrangham. Princeton, NJ: Princeton University Press, pp. 429–451.

Packer, C., and A. E. Pusey. 1982. Cooperation and competition within coalitions of male lions: Kin selection or game theory? *Nature* 296:740–742.

Packer, C., and A. E. Pusey. 1987. Intrasexual cooperation and the sex ratio in African lions. *American Naturalist* 130:636–642.

Packer, C., D. Scheel, and A. E. Pusey. 1990. Why lions form groups: Food is not enough. *American Naturalist* 119:263–281.

Packer, C., D. A. Gilbert, A. E. Pusey, and S. J. O'Brien. 1991. A molecular genetic analysis of kinship and cooperation in African lions. *Nature* 351:562–565.

Packer, C., L. Herbst, A. E. Pusey, J. D. Bygott, J. P. Hanby, S. J. Cairns, and M. B. Mulder. 1988. Reproductive success in lions. In *Reproductive Success, Studies of Individual Variation in Contrasting Breeding Systems*, edited by T. H. Clutton-Brock. Chicago: University of Chicago Press, pp. 363–383.

Page, R. E., Jr., and R. A. Metcalf. 1982. Multiple mating, sperm utilization, and social evolution. *American Naturalist* 119:263–281.

Palmer, J. D. 1990. The rhythmic lives of crabs: the same biological clock that controls circadian rhythms may also drive crustacean tidal rhythms. *BioScience* 40:352–358.

Palmer, J. D. 1995. *The Biological Rhythms and Clocks of Intertidal Animals*. New York: Oxford University Press.

Palmer, J. D. 2000. The clocks controlling the tide-associated rhythms of intertidal animals. *BioEssays* 22:32–37.

Papi, F., L. Fiore, V. Fiaschi, and S. Benvenuti. 1972. Olfaction and homing in pigeons. *Monitore Zoologico Italiano (N.S.)* 6:85–95.

Paranjpe, D. A., and V. K. Sharma. 2005. Evolution of temporal order in living organisms. *Journal of Circadian Rhythms* 3:7–19.

Parejo, D., J. White, and E. Danchin. 2007. Settlement decisions in blue tits: difference in the use of social cues according to age and individual success. *Naturwissenschaften* 94:749–757.

Parish, A. R., and F. B. M. de Waal. 2000. The other "closest living relative": How bonobos (*Pan paniscus*) challenge traditional assumptions about females, dominance, intra- and intersexual interactions, and hominid evolution. *Annals of the New York Academy of Sciences* 907:97–113.

Parker, G. A. 1970. Sperm competition and its evolutionary consequences in the insects. *Biological Reviews* 45:525–567.

Parker, G. A. 1974. Assessment strategy and the evolution of fighting behavior. *Journal of Theoretical Biology* 47:223–243.

Parker, G. A. 1979. Sexual selection and sexual conflict. In *Sexual Selection and Reproductive Competition in Insects*, edited by M. S. Blum and N. A. Blum. London: Academic Press, pp. 123–166.

Parker, G. A. 1984. Sperm competition and the evolution of mating strategies. In *Sperm Competition and the Evolution of Animal Mating Systems*, edited by R. L. Smith. Orlando, FL: Academic Press, pp. 2–60.

Parker, G. A. 2006. Sexual conflict over mating and fertilization: an overview. *Philosophical Transactions of the Royal Society of London B* 361:235–259.

Parrish, J. K. 1989. Re-examining the selfish herd: Are central fish safer? *Animal Behaviour* 38:1048–1053.

Partan, S. R., and P. Marler. 2005. Issues in the classification of multimodal communication signals. *American Naturalist* 166:231–245.

Patricelli, G., J. A. C. Uy, and G. Borgia. 2003. Multiple male traits interact: attractive bower decorations facilitate attractive behavioural displays in satin bowerbirds. *Proceedings of the Royal Society of London B* 270:2389–2395.

Patterson, F. 1978. The gestures of a gorilla: Sign language acquisition in another pongid species. *Brain and Language* 5:72–97.

Patterson, F. L. 1990. Language acquisition by a lowland gorilla: Koko's first ten years of vocabulary development. *Word* 41:97–143.

Pavlov, I. 1927. *Conditioned Reflexes*. Edited by G. V. Anrep. London: Oxford University Press.

Payne, R. B. 1962. How the barn owl locates its prey by hearing. *Living Bird* 1:151–159.

Payne, K., W. R. Langbauer Jr., and E. M. Thomas. 1986. Infrasonic calls of the Asian elephant (*Elephas maximus*). *Behavioral Ecology and Sociobiology* 18:297–301.

Payne, R. S., and D. Webb. 1971. Orientation by means of long range acoustic signaling in baleen whales. *Annals of the New York Academy of Sciences* 188:110–142.

Peeke, H. V. S. 1984. Habituation and the maintenance of territorial boundaries. In *Habituation, Sensitization, and Behavior*, edited by H. V. S. Peeke and L. Petrinovich. New York: Academic Press, pp. 393–421.

Penn, D. J., and W. K. Potts. 1998. How do major histocompatibility genes influence odor and mating preferences? *Advances in Immunology* 69:411–435.

Penn, D. J., and W. K. Potts. 1999. The evolution of mating preferences and the major histocompatibility complex genes. *American Naturalist* 153:145–164.

Penteriani, V., and M. M. Delgado. 2008. Owls may use faeces and prey feathers to signal current reproduction. *PloS 1* 3:e3014.

Pepperberg, I. M. 1987a. Evidence for conceptual quantitative abilities in the African grey parrot: Labeling of cardinal sets. *Ethology* 75:37–61.

Pepperberg, I. M. 1987b. Acquisition of same/different concept by an African grey parrot (*Psittacus erithacus*): Learning with respect to categories of color, shape and material. *Animal Learning and Behavior* 15:423–432.

Pepperberg, I. M. 1991. A communicative approach to animal cognition: A study of conceptual abilities of an African grey parrot. In *Cognitive Ethology, the Minds of Other Animals*, edited by C. A. Ristau. Hillsdale, NJ: Erlbaum, pp. 153–186.

Pepperberg, I. M. 1993. Cognition and communication in an African grey parrot (*Psittacus erithacus*). In *Language and Communication, Comparative Perspectives*, edited by H. L. Roitblat, L. M. Herman, and P. E. Nachtigall. Hillsdale, NJ: Lawrence Erlbaum, pp. 221–248.

Pepperberg, I. M. 1994. Numerical competence in an African grey parrot (*Psittacus erithacus*). *Journal of Comparative Psychology* 108:36–44.

Pepperberg, I. M. 2000. *The Alex Studies: Cognitive and Communicative Abilities of Grey Parrots*. Harvard University Press: Cambridge.

Pepperberg, I. M. 2006. Grey parrot (*Psittacus erithacus*) numerical abilities: addition and further experiments on a zero-like concept. *Journal of Comparative Psychology* 120:1–22.

Pepperberg, I. M., and J. D. Gordon. 2005. Number comprehension by a Grey parrot (*Psittacus erithacus*), including a zero-like concept. *Journal of Comparative Psychology* 119:197–209.

Perdeck, A. C. 1958. Two types of orientation in migrating starlings, *Sturnus vulgaris* L., and chaffinches, *Fringilla coelebs* L., as revealed by displacement experiments. *Ardea* 46:1–37.

Perdeck, A. C. 1967. Orientation of starlings after displacement to Spain. *Ardea* 51:91–104.

Pereira, H. S., and M. B. Sokolowski. 1993. Mutations in the larval foraging gene affect adult locomotory behavior. *Proceedings of the National Academy of Sciences* 90:5044–5046.

Pérez, M., and F. Coro. 1984. Physiological characteristics of the tympanic organ in noctuid moths. I. Responses to brief acoustic pulses. *Journal of Comparative Physiology A* 154:441–447.

Perfito, N., R. A. Zann, G. E. Bentley, and M. Hau. 2007. Opportunism at work: Habitat predictability affects reproductive readiness in free-living zebra finches. *Functional Ecology* 21:291–301.

Peroulakis, M. E., B. Goldman, and N. G. Forger. 2002. Perineal muscles and motoneurons are sexually monomorphic in the naked mole-rat (*Heterocephalus glaber*). *Journal of Neurobiology* 51:33–42.

Perrin, N., and J. Goudet. 2001. Inbreeding, kinship, and the evolution of natal dispersal. In *Dispersal*, edited by J. Clobert, E. Danchin, A. A. Dhondt, and J. D. Nichols. Oxford: Oxford University Press, pp. 123–142.

Perrin, N., and L. Lehmann. 2001. Is sociality driven by the costs of dispersal or the benefits of philopatry? A role for kin discrimination mechanisms. *American Naturalist* 158:471–483.

Peters, R. A., J. M. Hemmi, and J. Zeil. 2007. Signaling against the wind: modifying motion-signal structure in response to increased noise. *Current Biology* 17:1231–1234.

Petrie, M. 1992. Peacocks with low mating success are more likely to suffer predation. *Animal Behaviour* 44:585–586.

Petrie, M. 1994. Improved growth and survival of offspring of peacocks with more elaborate trains. *Nature* 371:598–599.

Petrie, M., and T. Halliday. 1994. Experimental and natural change in the peacock's (*Pavo cristatus*) train can affect mating success. *Behavioral Ecology Sociobiology* 35:213–217.

Petrie, M., T. Halliday, and C. Sanders. 1991. Peahens prefer peacocks with elaborate trains. *Animal Behaviour* 41:323–331.

Pettigrew, J. D., P. R. Manger, and S. L. B. Fine. 1998. The sensory world of the platypus. *Philosophical Transactions of the Royal Society of London B* 353:1199–1210.

Pfennig, D. W., G. J. Gamboa, H. K. Reeve, J. S. Reeve, and I. D. Ferguson. 1983. The mechanism of nestmate discrimination in social wasps (*Polistes*, Hymenoptera: Vespidae). *Behavioral Ecology and Sociobiology* 13:299–305.

Phillips, J. B. 1986. Two magnetoreception pathways in a migratory salamander. *Science* 233:765–766.

Phillips, J. B. 1987. Laboratory studies of homing orientation in the eastern red-spotted newt, *Notophthalmus viridescens*. *Journal of Experimental Biology* 131:215–229.

Phillips, J. B., and S. C. Borland. 1992. Wavelength specific effects of light on magnetic compass orientation in the eastern red-spotted newt *Notophthalmus viridescens*. *Ethology, Ecology and Evolution* 4:33–42.

Phillips, J. B., and S. C. Borland. 1994. Use of a specialized magnetoreception system for homing by the eastern red-spotted newt *Notophthalmus viridescens*. *Journal of Experimental Biology* 188:275–291.

Phillips, J. B., K. Schmidt-Koenig, and R. Muheim. 2006. True navigation: Sensory bases of gradient maps. In *Animal Spatial Cognition: Comparative, Neural, and Computational Approaches*, edited by M. F. Brown and R. G. Cook. [Online]. Available: *www.pigeon.psy.tufts.edu/asc/phillips/*.

Phoenix, C., R. Goy, A. Gerall, and W. Young. 1959. Organizing action of prenatally administered testosterone proprionate on the tissues mediating mating behavior in the female guinea pig. *Endocrinology* 65:369–382.

Pietrewicz, A. T., and A. C. Kamil. 1979. Search image formation in the blue jay (*Cyanocitta cristata*). *Science* 204:1332–1333.

Pietrewicz, A. T., and A. C. Kamil. 1981. Search images and the detection of cryptic prey: An operant approach. In *Foraging Behavior: Ecological, Ethological and Psychological Approaches*, edited by A. C. Kamil and T. D. Sargent. New York: Garland STPM Press, pp. 311–331.

Pinsker, H., I. Kupfermann, V. Castellucci, and E. R. Kandel. 1970. Habituation and dishabituation of the gill-withdrawal reflex in *Aplysia*. *Science* 167:1740–1742.

Plassart-Schiess, E., and E. E. Baulieu. 2001. Neurosteroids: recent findings. *Brain Research Reviews* 37:133–140.

Plautz, J. D., M. Kaneko, J. C. Hall, and S. A. Kay. 1997a. Independent photoreceptive circadian clocks throughout *Drosophila*. *Science* 278:1632–1635.

Plautz, J. D., M. Straume, R. Stanewsky, C. F. Jamieson, C. Brandes, H. B. Dowse, J. C. Hall, and S. A. Kay. 1997b. Quantitative analysis of *Drosophila period* gene transcription in living animals. *Journal of Biological Rhythms* 12:204–217.

Plomin, R. , J. C. DeFries, and G. E. McClearn. 1980. *Behavioral Genetics—A Primer*. San Francisco: W. H. Freeman.

Plomin, R., J. C. DeFries, I. W. Craig, and P. McGuffin. 2003. Behavioral genetics. In *Behavioral Genetics in the Postgenomic Era*, edited by R. Plomin, J. C. DeFries, I. W. Craig and P. McGuffin. Washington, DC: American Psychological Association.

Plotnik, J. J., F. de Waal, and D. Reiss. 2006. Self-recognition in an Asian elephant. *Proceedings of the National Academy of Sciences* 103:17053–17057.

Plummer, M. R., and J. M. Camhi. 1981. Discrimination of sensory signal from noise in the escape system of the cockroach: The role of wind acceleration. *Journal of Comparative Physiology A* 142:347–357.

Podos, J., and S. Nowicki. 2004. Beaks, adaptation, and vocal evolution in Darwin's finches. *BioScience* 54:501–510.

Poindron, P., F. Levy, and M. Keller. 2007. Maternal responsiveness and maternal selectivity in domestic sheep and goats: the two facets of maternal attachment. *Developmental Psychobiology* 49:54–70.

Pomiankowski, A., M. Denniff, K. Fowler, and T. Chapman. 2005. The costs and benefits of high early mating rates in male stalk-eyed flies (*Cyrtodiopsis dalmanni*). *Journal of Insect Physiology* 51:1165–1171.

Pongrácz, P., A. Miklósi, E. Kubinyi, K. Gurobi, J. Topál, and V. Csányi. 2001. Social learning in dogs: the effect of a human demonstrator on the performance of dogs in a detour task. *Animal Behaviour* 62:1109–1117.

Poole, J. H., K. Payne, W. R. Langbauer Jr., and C. Moss. 1988. The social contexts of some very low frequency calls of African elephants. *Behavioral Ecology and Sociobiology* 22:385–392.

Porter, R. H., V. J. Tepper, and D. M. White. 1981. Experimental influences on the development of huddling preferences and "sibling" recognition in spiny mice. *Developmental Psychobiology* 14:375–382.

Portfors, C. V. 2007. Types and functions of ultrasonic vocalizations in laboratory rats and mice. *Journal of the American Association for Laboratory Animal Science* 46:28–34.

Povinelli, D. J., G. G. Gallup, Jr., T. J. Eddy, D. T. Bierschwale, M. C. Engstron, H. K. Perilloux, and I. B. Toxopeus. 1997. Chimpanzees recognize themselves in mirrors. *Animal Behaviour* 53:1083–1088.

Pravosudov, V. V. 2006. On seasonality in food-storing behaviour in parids: Do we know the whole story? *Animal Behaviour* 71:1455–1460.

Pravosudov, V. V. 2007. The relationship between environment, food caching, spatial memory, and the hippocampus in chickadees. In *Ecology and Behavior of Chickadees and Titmice: an Integrated Approach*, edited by K. Otter. Oxford: Oxford University Press, pp. 25–41.

Pravosudov, V. V., and S. R de Kort. 2006. Is the western scrub-jay (*Aphelocoma californica*) really an underdog among food-caching corvids when it comes to hippocampal volume and food caching propensity? *Brain, Behavior and Evolution* 67:1–9.

Premack, D. 1976. *Intelligence in Ape and Man*. Hillsdale, NJ: Erlbaum.

Preston, B. T., I. R. Stevenson, J. M. Pemberton, and K. Wilson. 2001. Dominant rams lose out by sperm depletion. *Nature* 409:681–682.

Pribil, S., and J. Picman. 1996. Polygyny in the red-winged blackbird: Do females prefer monogamy or polygamy? *Behavioral Ecology and Sociobiology* 38:183–190.

Pribil, S., and W. A. Searcy. 2001. Experimental confirmation of the polygyny threshold model for red-winged blackbirds. *Proceedings of the Royal Society of London B* 268:1643–1646.

Price, T. D., P. R. Grant, H. L. Gibbs, and P. T. Boag. 1984. Recurrent patterns of natural selection in a population of Darwin's finches. *Nature* 309:787–789.

Proctor, H. C. 1991. Courtship in the water mite *Neumania papillator*: Males capitalize on female adaptations for predation. *Animal Behaviour* 42:589–598.

Proctor, H. C. 1992. Sensory exploitation and the evolution of male mating behavior: A cladistic test. *Animal Behaviour* 44:745–752.

Proctor, M., P. Yeo, and A. Lack. 1996. *The Natural History of Pollination*. Portland: Timber Press.

Pruett-Jones, S. 2004. Summary. In *Ecology and Evolution of Cooperative Breeding in Birds*, edited by W. Koenig and J. Dickinson. Cambridge: Cambridge University Press, pp. 228–238.

Pruetz, J. D., and P. Bertolani. 2007. Savanna chimpanzees, *Pan troglodytes verus*, hunt with tools. *Current Biology* 17:1–6.

Pusey, A. E. 1987. Sex-biased dispersal and inbreeding avoidance in birds and mammals. *Trends in Ecology and Evolution* 2:295–299.

Pusey, A. E., and C. Packer. 1997. The ecology of relationships. In *Behavioural Ecology: An Evolutionary Approach*, edited by J. R. Krebs and N. B. Davies. Oxford: Blackwell Science, pp. 254–283.

Putz, O., and D. Crews. 2006. Embryonic origin of mate choice in a lizard with temperature-dependent sex determination. *Developmental Psychobiology* 48:29–38.

Quinlan, R. J., and J. M. Cherrett. 1977. The role of substrate preparation in the symbiosis between the leaf-cutting ant *Acromyrmex octospinosus* (Reich) and its food fungus. *Ecological Entomology* 2:161–170.

Quinn, T. P., and A. H. Dittman. 1990. Pacific salmon migrations and homing: Mechanisms and adaptive significance. *Trends in Ecology and Evolution* 5:174–177.

Quinn, T. P., I. J. Stewart, and C. P. Boatright. 2006. Experimental evidence of homing to the site of incubation by mature sockeye salmon, *Oncorhynchus nerka*. *Animal Behaviour* 72:941–949.

Quinn, W. G., W. A. Harris, and S. Benzer. 1974. Conditioned behavior in *Drosophila melanogaster*. *Proceedings of the National Academy of Sciences* 71:708–712.

Rabenoid, K. N. 1990. *Campylorhynchus* wrens: The ecology of delayed dispersal and cooperation in the Venezuela savanna. In *Cooperative Breeding in Birds*, edited by P. B. Stacey and W. D. Koenig. Cambridge: Cambridge University Press, pp. 159–196.

Rado, R., N. Levi, H. Hauser, J. Witcher, N. Adler, N. Intrator, Z. Wollberg, and J. Terkel. 1987. Seismic signaling as a means of communication in a subterranean mammal. *Animal Behaviour* 35:1249–1266.

Ramirez, J. M., and K. G. Pearson. 1990. Chemical deafferentation of the locust flight system by phenotolamine. *Journal of Comparative Physiology A* 167:485–494.

Ramm, S. A., G. A. Parker, and P. Stockley. 2005. Sperm competition and the evolution of male reproductive anatomy in rodents. *Proceedings of the Royal Society of London B* 272:949–955.

Randall, J. A., and E. R. Lewis. 1997. Seismic communication between the burrows of kangaroo rats, *Dipodomys spectabilis*. *Journal of Comparative Physiology A* 181:525–531.

Randler, C. 2006. Red squirrels (*Sciurus vulgaris*) respond to alarm calls of Eurasian jays (*Garrulus glandarius*). *Ethology* 112:411–416.

Ranson, E., and F. A. Beach. 1985. Effects of testosterone on ontogeny of urinary behavior in male and female dogs. *Hormones and Behavior* 19:36–51.

Raouf, S. A., L. C. Smith, M. B. Brown, J. C. Wingfield, and C. R. Brown. 2006. Glucocorticoid hormone levels increase with group size and parasite load in cliff swallows. *Animal Behaviour* 71:39–48.

Rasa, O. A. E. 1986. Coordinated vigilance in dwarf mongoose family groups: The “Watchman’s Song” hypothesis and the costs of guarding. *Ethology* 71:340–344.

Rayor, L. S., and G. W. Uetz. 1990. Trade-offs in foraging success and predation risk with spatial position in colonial spiders. *Behavioral Ecology and Sociobiology* 27:77–85.

Redondo, T., and F. Castro. 1992. Signalling of nutritional need by magpie nestlings. *Ethology* 92:193–204.

Reeder, D. M. 2003. The potential for cryptic female choice in primates: behavioral, anatomical, and physiological considerations. In *Special Topics in Primatology, Volume 3. Sexual Selection and Reproductive Competition in Primates: New Perspectives and Directions*, edited by C. B. Jones. Norman: American Society of Primatologists, pp. 254–303.

Reimchen, T. E. 1989. Shell color ontogeny and tubeworm mimicry in a marine gastropod *Littorina mariae*. *Biological Journal of the Linnean Society* 36:97–109.

Reimers, M., F. Schwarzenberger, and S. Preuschoft. 2007. Rehabilitation of research chimpanzees: stress and coping after long-term social isolation. *Hormones and Behavior* 51:428–435.

Reiner, A., D. J. Perkel, C. V. Mello, and E. D. Jarvis. 2004. Songbirds and the revised avian brain nomenclature. *Annals of the New York Academy of Sciences* 1016:77–108.

Reiner, A., D. J. Perkel, L. L. Bruce, A. B. Butler, A. Csillag, W. Kuenzel, L. Medina, G. Paxinos, T. Shimizu, G. Strieddter, M. Wild, G. F. Ball, S. Durand, O. Gütürkün, D. W. Lee, C. V. Mello, A. Powers, S. A. White, G. Hough, L. Kubikova, T. V. Smulders, K. Wada, J. Dugas-Ford, S. Husband, K. Yamamoto, J. Yu, S. Siang, and E. D. Jarvis. 2004. Revised nomenclature for avian telencephalon and some related brainstem nuclei. *Journal of Comparative Neurology* 479:377–414.

Reiss, D., and L. Marino. 2001. Mirror self-recognition in the bottlenose dolphin: A case of cognitive convergence. *Proceedings of the National Academy of Sciences* 98:5937–5942.

Remage-Healey, L., and A. H. Bass. 2006. A rapid neuromodulatory role for steroid hormones in the control of reproductive behavior. *Brain Research* 1126:27–35.

Reppert, S. M., and D. R. Weaver. 2002. Coordination of circadian timing in mammals. *Nature* 418:935–941.

Rescorla, R. A. 1988a. Behavioral studies of Pavlovian conditioning. *Annual Review of Neuroscience* 11:329–352.

Rescorla, R. A. 1988b. Pavlovian conditioning: It’s not what you think it is. *American Psychologist* 43:151–160.

Reyer, H.-U. 1980. Flexible helper structure as an ecological adaptation in the red Kingfisher (*Ceryle rudis rudis* L.). *Behavioral Ecology and Sociobiology* 6:219–227.

Reyer, H.-U. 1984. Investment and relatedness: A cost/benefit analysis of breeding and helping in the pied kingfisher (*Ceryle rudis*). *Animal Behaviour* 32:1163–1178.

Reyer, H.-U. 1986. Breeder-helper interactions in the pied kingfisher reflect costs and benefits of cooperative breeding. *Behaviour* 96:277–303.

Reyer, H.-U. 1990. Pied kingfishers: ecological causes and reproductive consequences of cooperative breeding. In *Cooperative Breeding in Birds*, edited by P. B. Stacey and W. D. Koenig. Cambridge: Cambridge University Press, pp. 527–558.

Reynolds, J. D., N. B. Goodwin, and R. P. Freckleton. 2002. Evolutionary transitions in parental care and live bearing in vertebrates. *Philosophical Transactions of the Royal Society of London B* 357:269–281.

Ribble, D. O. 1991. The monogamous mating system of *Peromyscus californicus* as revealed by DNA fingerprinting. *Behavioral Ecology and Sociobiology* 29:161–166.

Richardson, D. S., T. Burke, and J. Komdeur. 2002. Direct benefits and the evolution of female-biased cooperative breeding in Seychelles warblers. *Evolution* 56:2313–2321.

Richardson, H., and N. A. M. Verbeek. 1986. Diet selection and optimization by Northwestern crows feeding on Japanese littleneck clams. *Ecology* 67:1219–1226.

Rickard, I. J., A. F. Russell, and V. Lummaa. 2007. Producing sons reduces lifetime reproductive success of subsequent offspring in pre-industrial Finns. *Proceedings of the Royal Society of London B* 274:2981–2988.

Ridley, M. 1978. Paternal care. *Animal Behaviour* 26:904–932.

Riechert, S. E. 1979. Games spiders play. II. Resource assessment strategies. *Behavioral Ecology and Sociobiology* 6:121–128.

Riechert, S. E. 1981. The consequences of being territorial: Spiders, a case study. *American Naturalist* 117:871–892.

Riechert, S. E. 1982. Spider interaction strategies: Communication vs. coercion. In *Spider Communication: Mechanisms and Ecological Significance*, edited by P. N. Witt and J. Rovner. Princeton, NJ: Princeton University Press, pp. 281–315.

Riechert, S. E. 1986. Between population variation in spider territorial behavior: Hybrid-pure population line comparisons. In *Evolutionary Genetics of Invertebrate Behaviour: Progress and Prospects*, edited by M. D. Huettel. New York: Plenum, pp. 33–42.

Riechert, S. E. 1993. Investigation of potential gene flow limitation of behavioral adaptation in an aridlands spider. *Behavioral Ecology and Sociobiology* 32:355–364.

Riechert, S. E., and R. F. Hall. 2000. Local population success in heterogenous habitats: reciprocal transplant experiments completed on a desert spider. *Journal of Evolutionary Biology* 13:541–550.

Riechert, S. E., and Hedrick, A. V. 1990. Levels of predation and genetically based anti-predatory behavior in the spider, *Agelenopsis aperta*. *Animal Behaviour* 40:679–687.

Riechert, S. E., and J. Maynard Smith. 1989. Genetic analyses of two behavioural traits linked to individual fitness in the desert spider, *Agelenopsis aperta*. *Animal Behaviour* 37:624–637.

Riechert, S. E., and T. R. Tracy. 1975. Thermal balance and prey availability: Bases for a model relating web-site characteristics to spider reproductive success. *Ecology* 56:265–285.

Riede, T., and T. Fitch. 1999. Vocal tract length and acoustics of vocalization in the domestic dog (*Canis familiaris*). *Journal of Experimental Biology* 202:2859–2867.

Riedl, C. A. L., S. J. Neal, A. Robichon, J. T. Westwood, and M. B. Sokolowski. 2005. *Drosophila* soluble guanylyl cyclase mutants exhibit increased foraging locomotion: behavioral and genomic investigations. *Behavior Genetics* 35:231–244.

Riley, J. R., U. Greggers, A. D. Smith, D. R. Reynolds and R. Menzel. 2005. The flight paths of honeybees recruited by the waggle dance. *Nature* 435:205–207.

Ristau, C.A., ed. 1991. *Cognitive Ethology, the Minds of Other Animals*. Hillsdale, NJ: Erlbaum.

Ritz, T., S. Adem, and K. Schulten. 2000. A model for photoreceptor-based magnetoreception in birds. *Biophysical Journal* 78:707–718.

Ritz, T., R. Wiltschko, P. J. Hore, C. T. Rodgers, K. Stapput, P. Thalau, C. R Timmel, and W. Wiltschko. 2009. Magnetic compass of birds is based on a molecule with optimal directional sensitivity. *Biophysical Journal* 96:3451-3451.

Ritzmann, R. E., and A. J. Pollack. 1986. Identification of thoracic interneurons that mediate giant interneuron-to-motor pathways. *Journal of Comparative Physiology A* 159:639–654.

Rivas, J. A., and G. M. Burghardt. 2005. Snake mating systems, behavior, and evolution: the revisionary implications of recent findings. *Journal of Comparative Psychology* 119:447–454.

Robbins, R. K. 1981. The "false head" hypothesis: Predation and wing pattern variation of Lycaenid butterflies. *American Naturalist* 118:770–775.

Roberts, T. S. 1907a. A Lapland longspur tragedy. *Auk* 24:369–377.

Roberts, T. S. 1907b. Supplemental note to "A Lapland longspur tragedy." *Auk* 24:449–450.

Robinson, G. E. 1996. Chemical communication in honeybees. *Science* 271:1824–1825.

Robinson, G. E. 1999. Integrative animal behaviour and sociogenomics. *Trends in Ecology and Evolution*. 14:202–205.

Robinson, G. E. 2002. Sociogenomics takes flight. *Science* 297:204–205.

Robinson, G. E. 2004. Beyond nature and nurture. *Science* 304:397–399.

Robinson, G. E., R. D. Fernald, and D. F. Clayton. 2008. Genes and social behavior. *Science* 322:896–900.

Robinson, G. E., C. M. Grozinger, and C. W. Whitfield. 2005. Sociogenomics: Social life in molecular terms. *Nature Reviews Genetics* 6:257–270.

Rodda, G. H. 1984. The orientation and navigation of juvenile alligators: evidence of magnetic sensitivity. *Journal of Comparative Physiology A* 154:649–658.

Rodl, T., S. Berger, and L. M. Romero. 2007. Tameness and stress physiology in a predator-naive island species confronted with novel predation threat. *Proceedings of the Royal Society of London B* 274:577–582.

Roeder, K. D. 1967. *Nerve Cells and Insect Behavior*. Cambridge, MA.: Harvard University Press.

Roeder, K. D., and R. S. Payne. 1966. Acoustic orientation of a moth in flight by means of two sense cells. *Symposia of the Society for Experimental Biology* 20:251–272.

Roeder, K. D., and A. E. Treat. 1957. Ultrasonic reception by the tympanic organ of noctuid moths. *Journal of Experimental Biology* 134:127–157.

Roeder, K. D., and A. E. Treat. 1961. The detection and evasion of bats by moths. *American Scientist* 49:135–148.

Rodgers, C. T., and P. J. Hore. 2009. Chemical magnetoreception in birds: The radical pair mechanism. *Proceedings of the National Academy of Sciences* 106:353-360.

Romanes, G. J. 1882. *Animal Intelligence*. New York: D. Appleton and Co.

Romanes, G. J. 1884. *Mental Evolution in Animals*. London: Keegan, Paul, Trench and Company.

Romanes, G. J. 1889. *Mental Evolution in Man*. New York: D. Appleton and Company.

Romey, W. L. 1995. Position preferences within groups—do whirligigs select positions which balance feeding opportunities with predator avoidance? *Behavioral Ecology and Sociobiology* 37:195–200.

Romey, W. L., A. R. Walston, and P. J. Watt. 2008. Do 3-D predators attack the margin of 2-D selfish herds? *Behavioral Ecology* 19:74–78.

Ronacher, B. 2008. Path integration as the basic navigation mechanism of the desert ant *Cataglyphis fortis* (Forel 1902) (Hymenoptera: Formicidae). *Myrmecological News* 11:53–62.

Rönn, J., M. Katvala, and G. Arnqvist. 2007. Coevolution between harmful male genitalia and female resistance in seed beetles. *Proceedings of the National Academy of Sciences* 104:10921–10925.

Rood, J. P. 1990. Group size, survival, reproduction, and routes to breeding in dwarf mongooses. *Animal Behaviour* 39:566–572.

Rooney, J., and S. M. Lewis. 2002. Fitness advantage from nuptial gifts in female fireflies. *Ecological Entomology* 27:373–377.

Rosa-Molinar, E., B. Fritzsch, and S. E. Hendricks. 1996. Organizational-activational concept revisited: sexual differentiation in an Atherinomorph teleost. *Hormones and Behavior* 30:563–575.

Rose, J. D., and F. L. Moore. 1999. A neurobehavioral model for rapid actions of corticosterone on sensory integration. *Steroids* 64:92–99.

Rose, J. D., and F. L. Moore. 2002. Behavioral neuroendocrinology of vasotocin and vasopressin and the sensorimotor processing hypothesis. *Frontiers in Neuroendocrinology* 23:317–341.

Rosen, R. A., and D. C. Hales. 1981. Feeding of paddlefish, *Polyodon spathula*. *Copeia* 2:441–455.

Rosenheim, J. A. 1993. Comparative and experimental approaches to understanding insect learning. In *Insect Learning: Ecological and Evolutionary Perspectives*, edited by D. R. Papaj and A. C. Lewis. New York: Chapman and Hall, pp. 273–307.

Røskaft, E., T. Järvi, T., M. Bakken, C. Beeh, and R. E. Reinersten. 1986. The relationship between social status and resting metabolic rate in great tits (*Parus major*) and pied flycatchers (*Ficedula hypoleuca*). *Animal Behaviour* 34:838–842.

Ross, K. G. 1986. Kin selection and the problem of sperm utilization in social insects. *Nature* 323:799–800.

Rothstein, S. I., and S. K. Robinson. (eds.) 1998. *Parasitic Birds and Their Hosts: Studies in Coevolution*. Oxford: Oxford University Press.

Roulin, A. 2002. Why do lactating females nurse alien offspring? A review of hypotheses and empirical evidence. *Animal Behaviour* 63:201–208.

Rousset, F., and D. Roze. 2007. Constraints on the origin and maintenance of genetic kin recognition. *Evolution* 61:2320–2330.

Rowell, J.T., S. P. Ellner, and H. K. Reeve. 2006. Why animals lie: How dishonesty and belief can coexist in a signaling system. *American Naturalist* 168:E180–E204.

Rowley, I. C. R. 1981. The communal way of life in the splendid wren, *Malurus splendens*. *Zeitschrift für Tierpsychologie* 55:228–267.

Rowley, I. C. R., and E. M. Russell. 1990. Splendid fairy-wrens: Demonstrating the importance of longevity. In *Cooperative Breeding in Birds*, edited by P. B. Stacey and W. D. Koenig. Cambridge: Cambridge University Press, pp. 3–30.

Rozhok, A. 2008. *Orientation and Navigation in Vertebrates*. Berlin: Springer.

Rubolini, D., P. Galeotti, F. Pupin, R. Sacchi, P. A. Nardi, and M. Fasola. 2007. Repeated matings and sperm depletion in the freshwater crayfish *Austropotamobius italicus*. *Freshwater Biology* 52:1898–1906.

Rumbaugh, D. M. 1977. The emergence and the state of ape language research. In *Progress in Ape Research*, edited by G. H. Bourne. New York: Academic Press, pp. 75–83.

Rumbaugh, D. M., and E. S. Savage-Rumbaugh. 1994. Language in comparative perspective. In *Animal Learning and Cognition*, edited by N. J. Macintosh. San Diego, CA: Academic Press, pp. 307–333.

Rumbaugh, D. M., T. V. Gill, and E. C. von Glaserfeld. 1973. Reading and sentence completion by a chimpanzee. *Science* 182:731–733.

Rüppell, O., T. Pankiw, and R. E. Page, Jr. 2004. Pleiotropy, epistasis, and new QTL: The genetic architecture of honey bee foraging behavior. *Journal of Heredity* 96:481–491.

Rüppell, O., S. B. C. Chandra, T. Pankiw, M. K. Fondrk, M. Beye, G. Hunt, and R. E. Page. 2006. The genetic architecture of sucrose responsiveness in the honeybee (*Apis mellifera* L.). *Genetics* 172:243–251.

Russell, A. F., and B. J. Hatchwell. 2001. Experimental evidence for kin-biased helping in a cooperatively breeding vertebrate. *Proceedings of the Royal Society of London B* 268:95–99.

Russell, E. M., and I. C. R. Rowley. 1988. Helper contributions to reproductive success in the splendid fairy-wren *Malurus splendens*. *Behavioral Ecology and Sociobiology* 22:131–140.

Russell, A. F., A. J. Young, G. Spong, N. R. Jordan, and T. H. Clutton-Brock. 2007. Helpers increase the reproductive potential of offspring in cooperative meerkats. *Proceedings of the Royal Society of London B* 274:513–520.

Rutz, C., L. A. Bluff, A. A. Weir, and A. Kacelnik. 2007. Video cameras on wild birds. *Science Express Brevia*, published online October 2007.

Ruxton, G. D., M. P. Speed, and D. J. Kelly. 2004. What, if anything, is the adaptive function of countershading? *Animal Behaviour* 68:445–451.

Ryan, M. J. 1988. Energy, calling, and selection. *American Zoologist* 28:885–898.

Ryan, M. J. 1990. Sexual selection, sensory systems, and sensory exploitation. *Oxford Survey of Evolutionary Biology* 7:157–195.

Ryan, M. J. 2005. The evolution of behaviour, and integrating it towards a complete and correct understanding of behavioural biology. *Animal Biology* 55:419–439.

Ryan, M. J. 2007. Sensory ecology: see me, hear me. *Current Biology* 17:R1019–R1021.

Ryan, B. C., and J. G. Vandenbergh. 2002. Intrauterine position effects. *Neuroscience and Biobehavioral Reviews* 26:665–678.

Ryan, M. J., and A. S. Rand. 1993a. Phylogenetic patterns of behavioral mate recognition systems in the *Physalaemus pustulosus* species group (Anura: Leptodactylidae): The role of ancestral and derived characters and sensory exploitation. In *Evolutionary Patterns and Processes*, edited by D. R. Lees and D. Edwards. London: Academic Press, pp. 251–267.

Ryan, M. J., and A. S. Rand. 1993b. Sexual selection and signal evolution: The ghost of biases past. *Philosophical Transactions of the Royal Society of London B* 340:187–195.

Ryan, M. J., Tuttle, M. D., and A. S. Rand. 1982. Bat predation and sexual advertisement in a neotropical anuran. *American Naturalist* 119:136–139.

Ryder, T. B., J. G. Blake, and B. A. Loiselle. 2006. A test of the environmental hotspot hypothesis for lek placement in three species of manakins (Pipridae) in Ecuador. *The Auk* 123:247–258.

Sack, R. L., R. W. Brandes, A. R. Kendall, and A. Lowy. 2000. Entrainment of free-running circadian rhythms by melatonin in blind people. *New England Journal of Medicine* 343:1070–1077.

Saether, S. A., P. Fiske, and J. A. Kalas. 2001. Male mate choice, sexual conflict and strategic allocation of copulations in a lekking bird. *Proceedings of the Royal Society of London B* 268:2097–2102.

Sahara, K., and Y. Takemura. 2003. Application of artificial insemination technique to eupyrene and/or apyrene sperm in *Bombyx mori*. *Journal of Experimental Zoology Part A – Comparative Experimental Biology* 297A:196–200.

Sakamoto, T., and S. D. McCormick. 2006. Prolactin and growth hormone in fish osmoregulation. *General and Comparative Endocrinology* 147:24–30.

Sakurai, T., T. Nakagawa, H. Mitsuno, H. Mori, Y. Endo, S. Tanoue, Y. Yasukochi, K. Touhara, and T. Nishioka. 2004. Identification and functional characterization of a sex pheromone receptor in the silkmoth *Bombyx mori*. *Proceedings of the National Academy of Sciences* 101:16653–16658.

Sands, J., and S. Creel. 2004. Social dominance, aggression and faecal glucocorticoid levels in a wild population of wolves, *Canis lupus*. *Animal Behaviour* 67:382–396.

Sapolsky, R. M. 1992. Cortisol concentrations and the social significance of rank instability among wild baboons. *Psychoneuroendocrinology* 17:701–709.

Sapolsky, R. M. 1994. *Why Zebras Don't Get Ulcers*. New York: Freeman and Company.

Sapolsky, R. M., L. M. Romero, and A.U. Munck. 2000. How do glucocorticoids influence stress responses? Integrating permissive, suppressive, stimulatory, and preparative actions. *Endocrine Reviews* 21:55–89.

Saranathan, V., D. Hamilton, G. V. N. Powell, D. E. Kroodsma, and R. O. Prum. 2007. Genetic evidence supports song learning in the three-wattled bellbird *Procnias tricarunculata* (Cotingidae). *Molecular Ecology* 16:3689–3702.

Sato, T., and S. Goshima. 2006. Impacts of male-only fishing and sperm limitation in manipulated populations of an unfished crab, *Hapalogaster dentata*. *Marine Ecology Progress Series* 313:193–204.

Sato, T., and S. Goshima. 2007. Female choice in response to risk of sperm limitation by the stone crab, *Hapalogaster dentata*. *Animal Behaviour* 73:331–338.

Sauer, E. G. F. 1957. Die Sternorientierung nächtlich ziehender Grasmücken (*Sylvia atriciapilla, borin and curruca*). *Zeitschrift für Tierpsychologie* 14:29–70.

Sauer, E. G. F. 1961. Further studies on the stellar orientation of nocturnally migrating birds. *Psychologische Forschung* 26:224–244.

Sauer, E. G. F., and E. M. Sauer. 1960. Star navigation in nocturnal migrating birds. The 1958 planetarium experiments. *Cold Spring Harbor Symposium of Quantitative Biology* 25:463–473.

Savage-Rumbaugh, E. S. 1986. *Ape Language from Conditioned Response to Symbol*. New York: Columbia University Press.

Savage-Rumbaugh, E. S., W. M. Fields, and J. P. Taglialatela. 2001. Language, speech, tools and writing: a cultural imperative. *Journal of Consciousness Studies* 8:273–292.

Savage-Rumbaugh, E. S., and R. Lewin. 1994. *Kanzi, the Ape at the Brink of the Human Mind*. New York: John Wiley.

Savage-Rumbaugh, E. S., D. M. Rumbaugh, and S. Boysen. 1978a. Symbolic communication between two chimpanzees (*Pan troglodytes*). *Science* 201:641–644.

Savage-Rumbaugh, E. S., D. M. Rumbaugh, and S. Boysen. 1978b. Linguistically mediated tool use and exchange by chimpanzees *Pan troglodytes*. *Behavioral and Brain Sciences* 1:539–554.

Savage-Rumbaugh, E. S., D. M. Rumbaugh, and S. Boysen. 1980. Do apes use language? *American Scientist* 68:49–61.

Savage-Rumbaugh, E. S., S. G. Shanker, and T. J. Taylor. 1998. *Apes, Language, and the Human Mind*. New York: Oxford University Press.

Sazima, I., L. N. Carvalho, F. P. Mendonca, and J. Zuanon. 2006. Fallen leaves on the water-bed: diurnal camouflage of three night active fish species in an Amazonian streamlet. *Neotropical Ichthyology* 4:119–122.

Schaal, B., G. Coureaud, D. Langlois, C. Ginies, E. Sémon, and G. Perrier. 2003. The mammary pheromone of the rabbit: chemical and behavioural characterisation. *Nature* 424:68–72.

Schaefer, P., G. V. Kondagunta, and R. E. Ritzman. 1994. Motion analysis of escape movements evoked by tactile stimulation in the cockroach *Periplaneta americana*. *Journal of Experimental Biology* 190:287–294.

Schaller, G. B. 1972. *The Serengeti Lion*. Chicago: University of Chicago Press.

Schanen, N. C. 2006. Epigenetics of autism spectrum disorders. *Human Molecular Genetics* 15:R138–R150.

Scheich, H., G. Langner, C. Tidemann, R. B. Coles, and A. Guppy. 1986. Electroreception and electrolocation in platypus. *Nature* 319:401–402.

Scheiner, R., M. B. Sokolowski, and J. Erber. 2004. Activity of cGMP-dependent protein kinase (PKG) affects sucrose responsiveness and habituation. *Learning and Memory* 11:303–311.

Schleidt, W. 1961a. Reaktionen von Truthühnern auf fliegende Raubvögel und Veruche zur Analyse ihrer AAM's. *Zeitschrift für Tierpsychologie* 18:534–560.

Schleidt, W. 1961b. Über die Auslösung des Kollern beim Truthahan (*Meleagris galopavo*). *Zeitschrift für Tierpsychologie* 11:417–435.

Schluter, A., J. Parzefall, and I. Schlupp. 1998. Female preference for symmetrical vertical bars in male sailfin mollies. *Animal Behaviour* 56:147–153.

Schmidt, V., H. M. Schaefer, and H. Winkler. 2004. Conspicuousness, not colour as foraging cue in plant animal signalling. *Oikos* 106:551–557.

Schmidt-Koenig, K., and H. J. Schlichte. 1972. Homing in pigeons with impaired vision. *Proceedings of the National Academy of Sciences* 69:2446–2447.

Schneider, D. 1974. The sex-attractant receptor of moths. *Scientific American* 231:28–35.

Schneider, J. S., M. K. Stone, K. E. Wynne-Edwards, T. H. Horton, J. Lydon, B. O'Malley, and J. E. Levine. 2003. Progesterone receptors mediate male aggression toward infants. *Proceedings of the National Academy of Sciences* 100:2951–2956.

Schoech, S. J. 1998. Physiology of helping in Florida scrub jays. *American Scientist* 86:70–77.

Schoech, S. J., R. Bowman, and S. J. Reynolds. 2004. Food supplementation and possible mechanisms underlying early breeding in the Florida scrub-jay (*Aphelocoma coerulescens*). *Hormones and Behavior* 46:565–573.

Schoech, S. J., R. L. Mumme, and J. C. Wingfield. 1996. Prolactin and helping behaviour in the cooperatively breeding Florida scrub-jay, *Aphelocoma coerulescens*. *Animal Behaviour* 52:445–456.

Schoener, T. W. 1968. Sizes of feeding territories among birds. *Ecology* 49:123–141.

Schoener, T. W. 1983. Simple models of optimal feeding-territory size: A reconciliation. *American Naturalist* 121:608–629.

Schuett, G. W. 1997. Body size and agnostic experience affect dominance and mating success in male copperheads. *Animal Behaviour* 54:213–224.

Schütz, D., and M. Taborsky. 2005. The influence of sexual selection and ecological constraints on an extreme sexual size dimorphism in a cichlid. *Animal Behaviour* 70:539–549.

Schutz, von F. 1965. Sexuelle Prägung bei Anatiden. *Zeitschrift für Tierpsychologie* 22:50–103.

Schwagmeyer, P., D. W. Mock, and G. A. Parker. 2002. Biparental care in house sparrows: negotiation or sealed bid? *Behavioral Ecology* 13:713–721.

Schwind, R. 1983. Zonation of the optical environment and zonation in the rhabdom structure within the eye of the backswimmer, *Notonecta glauca*. *Cell Tissue Research* 232:53–62.

Schwind, R. 1991. Polarization vision in water insects and insects living on a moist substrate. *Journal of Comparative Physiology A* 169:531–540.

Scott, J. P., and J. L. Fuller. 1965. *Genetics and the Social Behavior of the Dog*. Chicago: University of Chicago Press.

Sear, R., and R. Mace. 2008. Who keeps children alive? A review of the effects of kin on child survival. *Evolution and Human Behavior* 29:1–18.

Searcy, W. A. 1979. Female choice of mates: A general model for birds and its application to red-winged blackbirds (*Agelaius phoeniceus*). *American Naturalist* 114:77–100.

Searcy, W. A., and S. Nowicki. 2005. *The Evolution of Animal Communication*. Princeton, NJ: Princeton University Press.

Sebbel, P., H. Düttmann, and T. Groothius. 1998. Influence of comfort and social stimuli on a comfort movement and a display derived from it. *Animal Behaviour* 55:129–137.

Sefc, K. M., K. Mattersdorfer, C. Sturmbauer, and S. Koblmüller. 2008. High frequency of multiple paternity in broods of a socially monogamous cichlid fish with biparental care. *Molecular Ecology* 17:2531–2543.

Selonen, V., and I. K. Hanski. 2006. Habitat exploration and use in dispersing juvenile flying squirrels. *Journal of Animal Ecology* 75:1440–1449.

Semm, P., and R. C. Beason. 1990. Responses to small magnetic variations by the trigeminal system of the bobolink. *Brain Research Bulletin* 25:735–740.

Seney, B., D. Goldman, and N. G. Forger. 2006. Breeding status affects motoneuron number and muscle size in naked mole-rats: recruitment of perineal motoneurons? *Journal of Neurobiology* 66:1354–1364.

Serventy, D. L. 1971. Biology of desert birds. In *Avian Biology*, edited by D. S. Farner, J. R. King, and K. C. Parkes. New York: Academic Press, pp. 287–339.

Seyfarth, R. M., and D. L. Cheney. 2003. Signalers and receivers in animal communication. *Annual Review of Psychology* 54:145–173.

Seyfarth, R. M., D. L. Cheney, and P. Marler. 1980. Monkey responses to three different alarm calls: evidence of predator classification and semantic communication. *Science* 210:801–803.

Shaffer, D. M., S. M. Krauchunas, M. Eddy, and M. K. McBeath. 2004. How dogs navigate to catch frisbees. *Psychological Science* 15:437–441.

Sharma, V. K. 2003. Adaptive significance of circadian clocks. *Chronobiology International* 20:901–919.

Sharpe, F. A., and L. M. Dill. 1997. The behavior of Pacific herring schools in response to artificial humpback whale bubbles. *Canadian Journal of Zoology* 75:725–730.

Sheldon, B. C. 2002. Relating paternity to paternal care. *Philosophical Transactions of the Royal Society of London B* 357:341–350.

Shelley, E. L., and D. T. Blumstein. 2005. The evolution of vocal alarm communication in rodents. *Behavioral Ecology* 16:169–177.

Shen, J. X., A. S. Feng, Z. M. Xu, Z. L. Yu, V. S. Arch, X. J. Yu, and P. M. Narins. 2008. Ultrasonic frogs show hyperacute phonotaxis to female courtship calls. *Nature* 453:914–917.

Sherbrooke, W. C., and J. R. Mason. 2005. Sensory modality used by coyotes in responding to antipredator compounds in the blood of Texas horned lizards. *Southwest Naturalist* 50:216–222.

Sherbrooke, W. C., and G. A. Middendorf, III. 2004. Responses of kit foxes (*Vulpes macrotis*) to antipredator blood-squirting and blood of Texas horned lizards (*Phrynosoma cornutum*). *Copeia* 3:652–658.

Sheridan, M., and R. H. Tamarin. 1988. Space use, longevity, and reproductive success in meadow voles. *Behavioral Ecology and Sociobiology* 22:85–90.

Sherman, P. M. 1994. The orb-web: An energetic and behavioural estimator of a spider's dynamic foraging and reproductive strategies. *Animal Behaviour* 48:19–34.

Sherman, P. W. 1977. Nepotism and the evolution of alarm calls. *Science* 197:1246–1253.

Sherman, P. W. 1980a. The limits of ground squirrel nepotism. In *Sociobiology: Beyond Nature/Nurture?* edited by G. W. Barlow and J. Silverberg. Boulder, CO: Westview, pp. 505–544.

Sherman, P. W. 1980b. The meaning of nepotism. *American Naturalist* 116:604–606.

Sherman, P. W. 1985. Alarm calls of Belding's ground squirrels to aerial predators: Nepotism or self preservation? *Behavioral Ecology and Sociobiology* 17:313–323.

Sherman, P. W., J. U. M. Jarvis, and S. H. Braude. 1992. Naked mole-rats. *Scientific American* 267:72–78.

Sherman, P. W., H. K. Reeve, and D. W. Pfennig. 1997. Recognition systems. In *Behavioural Ecology*, edited by J. R. Krebs and N. B. Davies. Oxford: Blackwell Science, pp. 69–96.

Sherman, P. W., E. A. Lacey, H. K. Reeve, and L. Keller. 1995. The eusociality continuum. *Behavioral Ecology* 6:102–108.

Sherry, D. F. 2005a. Do ideas about function help in the study of causation? *Animal Biology* 55:441–456.

Sherry, D. F. 2005b. Brain and behavior. In *The Behavior of Animals: Mechanisms, Function, and Evolution*, edited by J. J. Bolhuis and L.-A. Giraldeau. Malden, MA: Blackwell Publishing.

Sherry, D. F., A. L. Vaccarino, K. Buckenham, and R. Herz. 1989. The hippocampal complex of food storing birds. *Brain, Behavior and Evolution* 34:308–317.

Shettleworth, S. J. 1995. Comparative studies of memory in food storing birds: From the field to the Skinner box. In *Behavioral Brain Research in Naturalistic and Semi-Naturalistic Settings*, NATO ASI Series, edited by E. Alleva, A. Fasolo, H. P. Lipp, L. Nadel, and L. Ricceri. Dordrecht: Kluwer, pp. 159–192.

Shettleworth, S. J. 1998. *Cognition, Evolution and Behavior.* Oxford: Oxford University Press.

Shields, W. M. 1982. *Philopatry, Inbreeding, and the Evolution of Sex*. Albany: State University of New York Press.

Shields, W. M. 1984. Barn swallow mobbing: Self-defence, collateral kin defence, group defence, or parental care? *Animal Behaviour* 32:132–148.

Shields, W. M. 1987. Dispersal and mating systems: Investigating their causal connections. In *Mammalian Dispersal Patterns*, edited by B. D. Chepko-Sade and Z. T. Halpin. Chicago: University of Chicago Press, pp. 3–24.

Shoji, T., Y. Yamamoto, D. Nishikawa, K. Kurihara, and H. Ueda. 2003. Amino acids in stream water are essential for salmon homing migration. *Fish Physiology and Biochemistry* 28:249–251.

Shoji, T., M. Ueda, T. Sakamoto, Y. Katsuragi, Y. Zohar, A. Urano, and K. Yamauchi. 2000. Amino acids dissolved in stream water as possible home stream odorant for masu salmon. *Chemical Senses* 25:533–540.

Shors, T. J., G. Miesegaes, A. Beylin, M. Zhao, T. Rydel, and E. Gould. 2001. Neurogenesis in the adult is involved in the formation of trace memories. *Nature* 410:372–376.

Siegel, R. G., and W. K. Honig. 1970. Pigeon concept formation: Successive and simultaneous acquisition. *Journal of the Experimental Analysis of Behavior* 13:385–390.

Sih, A. 2004. Behavioral syndromes: an integrated overview. *Quarterly Review of Biology* 79:241–277.

Sih, A., and B. Christensen. 2001. Optimal diet theory: when does it work, and when and why does it fail? *Animal Behaviour* 61:379–390.

Silberglied, R. E., J. G. Shepherd, and J. L. Dickinson. 1984. Eunuchs: The role of apyrene sperm in Lepidoptera. *American Naturalist* 123:255–265.

Siljander, E., R. Gries, G. Khaskin, and G. Gries. 2008. Identification of the airborne aggregation pheromone of the common bed bug, *Cimex lectularius*. *Journal of Chemical Ecology* 34:708–718.

Sillett, T. S., and R. T. Holmes. 2002. Variation in survivorship of a migratory songbird throughout its annual cycle. *Journal of Animal Ecology* 71:296–308.

Silver, R., J. LeSauter, P. A. Tresco, and M. N. Lehman. 1996. A diffusible coupling signal from the transplanted suprachiasmatic nucleus controlling circadian locomotor rhythms. *Nature* 382:810–813.

Simmons, L. W. 1988. Male size, mating potential and lifetime reproductive success in the field cricket, *Gryllus bimaculatus* (De Geer). *Animal Behaviour* 36:372–379.

Simmons, L. W. 2005. The evolution of polyandry: Sperm competition, sperm selection, and offspring viability. *Annual Review of Ecology and Systematics* 36:125–146.

Simpson, S. J., R. M. Sibly, K. P. Lee, S. T. Behmer, and D. Raubenheimer. 2004. Optimal foraging when regulating intake of multiple nutrients. *Animal Behaviour* 68:1299–1311.

Sinclair, A. R. E. 1983. The function of distance movements in vertebrates. In *The Ecology of Animal Movement*, edited by I. R. Swingland and P. J. Greenwood. Oxford: Clarendon, pp. 240–258.

Sinervo, B., and C. M. Lively. 1996. The rock-paper-scissors game and the evolution of alternative male strategies. *Nature* 380:240–243.

Singer, M. S., and E. A. Bernays. 2003. Understanding omnivory needs a behavioral perspective. *Ecology* 84:2532–2537.

Skinner, B. F. 1953. *Science and Human Behavior*. New York: Free Press, Macmillan.

Skipper, R., and G. Skipper. 1957. Those British-bred pompadours—The story completed. *Water Life Aquarium World* 12:63–64.

Skow, C. D., and E. M. Jakob. 2006. Jumping spiders attend to context during learned avoidance of aposematic prey. *Behavioral Ecology* 17:34–40.

Skutch, A. F. 1935. Helpers at the nest. *Auk* 52:257–273.

Skutch, A. F. 1987. *Helpers at Birds' Nests*. Iowa City: University of Iowa Press.

Skutelsky, O. 1996. Predation risk and state-dependent foraging in scorpions: effects of moonlight on foraging in the scorpion *Buthus occitanus*. *Animal Behaviour* 52:49–57.

Slabbekoorn, H., and A. den Boer-Visser. 2006. Cities change the songs of birds. *Current Biology* 16:2326–2331.

Slagsvold, T. 2004. Cross-fostering of pied flycatchers (*Ficedula hypoleuca*) to heterospecific hosts in the wild: a study of sexual imprinting. *Behaviour* 141:1079–1102.

Slagsvold, T., and J. T. Lifjeld. 1988. Ultimate adjustment of clutch size to parent feeding capacity in a passerine bird. *Ecology* 69:1918–1922.

Slagsvold, T., and J. T. Lifjeld. 1990. Influence of male and female quality on clutch size in tits (*Parus* spp.). *Ecology* 71:1258–1266.

Slagsvold, T., B. T. Hansen, L. E. Johannessen, and L. T. Lifjeld. 2002. Mate choice and imprinting in birds studied by cross-fostering in the wild. *Proceedings of the Royal Society of London B* 269:1449–1455.

Slocombe, K. E., and K. Zuberbühler. 2005. Functionally referential communication in a chimpanzee. *Current Biology* 15:1779–1784.

Smale, L., S. Nunes, and K. E. Holekamp. 1997. Sexually dimorphic dispersal in mammals: Patterns, causes, and consequences. *Advances in the Study of Behavior* 26:181–250.

Smedley, S. R., and T. Eisner. 1996. Sodium: A male moth's gift to its offspring. *Proceedings of the National Academy of Sciences* 93:809–813.

Smiseth, P. T., and A. J. Moore. 2004. Behavioural dynamics between caring males and females in a beetle with facultative biparental care. *Behavioral Ecology* 15:621–628.

Smith, M. D., and C. J. Conway. 2007. Use of mammal manure by nesting burrowing owls: a test of four hypotheses. *Animal Behaviour* 73:65–73.

Smith, T. 2006. Individual olfactory signatures in common marmosets (*Callithrix jacchus*). *American Journal of Primatology* 68:585–604.

Smith, V. A., A. P. King, and M. J. West. 2000. A role of her own: Female cowbirds, *Molothrus ater*, influence the development and outcome of song learning. *Animal Behaviour* 60:599–609.

Smith, W. J. 1977. *The Behavior of Communicating, An Ethological Approach*. Cambridge: Harvard University Press.

Smith, W. J. 1991. Animal communication and the study of cognition. In *Cognitive Ethology, the Minds of Other Animals*, edited by C. A. Ristau. Hillsdale, NJ: Erlbaum, pp. 209–230.

Sneddon, L. U., F. A. Huntingford, and A. C. Taylor. 1997. Weapon size versus body size as a predictor of winning in fights between shore crabs, *Carcinus maenus* (L.). *Behavioral Ecology and Sociobiology* 41:237–242.

Snow, B. K. 1974. Lek behavior and breeding of Guy's hermit hummingbird *Phaethornis guy*. *Ibis* 116:278–297.

Snow, D. W. 1963. The evolution of manakin displays. In *Proceedings of the 13th International Ornithological Congress* 1:553–561.

Snyder, W. E., G. M. Clevenger, and S. D. Eigenbrode. 2004. Intraguild predation and successful invasion by introduced ladybird beetles. *Oecologia* 140:559–565.

Sogabe, A., and Y. Yanagisawa. 2007. The function of daily greetings in a monogamous pipefish *Corythoichthys haematopterus*. *Journal of Fish Biology* 71:585–595.

Sogabe, A., and Y. Yanagisawa. 2008. Maintenance of pair bond during the non-reproductive season in a monogamous pipefish *Corythoichthys haematopterus*. *Journal of Ethology* 26:195–199.

Sokolowski, M. B. 1980. Foraging strategies of *Drosophila melanogaster*: a chromosomal analysis. *Behavior Genetics* 10:291–302.

Sokolowski, M. B., H. S. Pereira, and K. Hughes. 1997. Evolution of foraging behavior in *Drosophila* by density-dependent selection. *Proceedings of the National Academy of Sciences* 94:7373–7377.

Soler, M., M. Martín-Vivaldi, J. M. Marín, and A. P. Møller. 1999. Weight lifting and health status in the black wheatear. *Behavioral Ecology* 10:281–286.

Soler, M., J .J. Soler, A. P. Møller, J. Moreno, and M. Lindén. 1996. The functional significance of sexual display: stone carrying in the black wheatear. *Animal Behaviour* 51:247–254.

Solomon, N. G. 2003. A reexamination of factors influencing philopatry in rodents. *Journal of Mammalogy* 84:1182–1197.

Solomon, N. G., and B. Keane. 2007. Reproductive strategies in female rodents. In *Rodent Societies: An Ecological and Evolutionary Perspective*, edited by J. O. Wolff, and P. W. Sherman. Chicago: The University of Chicago Press, pp. 42–56.

Soma, K. K. 2006. Testosterone and aggression: Berthold, birds, and beyond. *Journal of Neuroendocrinology* 18:543–551.

Sommer, V. 1993. Infanticide among the langurs of Jodhpur: Testing the sexual selection hypothesis. In *Infanticide and Parental Care*, edited by S. Parmigiani and F. vom Saal. London: Harwood Academic Press, pp. 155–198.

Sorenson, M. D., and R. B. Payne. 2002. Molecular genetic perspectives on avian brood parasitism. *Integrative and Comparative Biology* 42:388–400.

Speed, M. P., and G. D. Ruxton. 2007. How bright and how nasty: explaining diversity in warning signal strength. *Evolution* 61:623–635.

Speed, M. P., and J. R. G. Turner. 1999. Learning and memory in mimicry: Do we understand the mimicry spectrum? *Biological Journal of the Linnean Society* 67:281–312.

Spencer, H. 1855. *Principles of Psychology*. New York: D. Appleton and Company.

Spencer, K. A., K. L. Buchanan, S. Leitner, A. R. Goldsmith, and C. K. Catchpole. 2005. Parasites affect song complexity and neural development in a songbird. *Proceedings of the Royal Society of London B* 272:2037–2043.

Spencer, K. A., S. Harris, P. J. Baker, and I. C. Cuthill. 2007. Song development in birds: the role of early experience and its potential effect on rehabilitation success. *Animal Welfare* 16:1–13.

Spritzer, M. D., N. G. Solomon, and D. B. Meikle. 2006. Social dominance among male meadow voles is inversely related to reproductive success. *Ethology* 112:1027–1037.

Srinivasan, M., and S. W. Zhang. 2003. Small brains, smart minds: Vision, perception, and "cognition" in honeybees. *IETE Journal of Research* 49:127–134.

Srinivasan, M. V., S. Zhang, M. Altwein, and J. Tautz. 2000. Honeybee navigation: Nature and calibration of the "odometer." *Science* 287:851–853.

Stacey, P. B. 1979. Habitat saturation and communal breeding in the acorn woodpecker. *Animal Behaviour* 27:1153–1166.

Stacey, P. B., and C. E. Bock. 1978. Social plasticity in the acorn woodpecker. *Science* 202:1298–1300.

Stallcup, J. A., and G. E. Woolfenden. 1978. Family status and contributions to breeding by Florida scrub jays. *Animal Behaviour* 26:1144–1156.

Stamps, J. A. 1987. Conspecifics as cues to territory quality: A preference of juvenile lizards (*Anolis aeneus*) for previously used territories. *American Naturalist* 129:629–642.

Stamps, J. A. 1988. Conspecific attraction and aggregation in territorial species. *American Naturalist* 131:329–347.

Stamps, J. A. 1991. Why evolutionary issues are reviving interest in proximate behavioral mechanisms. *American Zoologist* 31:338–348.

Stamps, J. A. 1994. Territorial behavior—testing the assumptions. *Advances in the Study of Behavior* 23:173–232.

Stamps, J. A. 1995. Motor learning and the value of familiar space. *American Naturalist* 146:41–58.

Stamps, J. A. 2001. Habitat selection by dispersers: integrating proximate and ultimate approaches. In *Dispersal*, edited by J. Clobert, E. Danchin, A. A. Dhondt, and J. D. Nichols. Oxford: Oxford University Press, pp. 230–242.

Stamps, J. A., and J. M. Davis. 2006. Adaptive effects of natal experience on habitat selection by dispersers. *Animal Behaviour* 72:1279–1289.

Stamps, J. A., and V. V. Krishnan. 2001. How territorial animals compete for divisible space: A learning based model with unequal competitors. *American Naturalist* 157:154–169.

Stamps, J. A., and R. R. Swaisgood. 2007. Someplace like home: Experience, habitat selection, and conservation biology. *Applied Animal Behaviour Science* 102:392–409.

Stankowich, T., and R. G. Coss. 2007a. Effects of risk assessment, predator behavior, and habitat on escape behavior in Columbian black-tailed deer. *Behavioral Ecology* 18:358–367.

Stankowich, T., and R. G. Coss. 2007b. The re-emergence of felid camouflage with the decay of predator recognition in deer under relaxed selection. *Proceedings of the Royal Society of London B* 274:175–182.

Stephens, D. W., and J. R. Krebs. 1986. *Foraging Theory*. Princeton, NJ: Princeton University Press.

Stephens, D. W., and E. L. Charnov. 1982. Optimal foraging: Some simple stochastic models. *Behavioral Ecology and Sociobiology* 10:251–263.

Stephens, M. L. 1984. Interspecific aggressive behavior of the polyandrous northern jacana (*Jacana spinosa*). *Auk* 101:508–518.

Stevens, M. 2007. Predator perception and the interrelation between different forms of protective coloration. *Proceedings of the Royal Society of London B* 274:1457–1464.

Stewart, I. J., S. M. Carlson, C. P. Boatright, G. B. Buck, and T.P. Quinn. 2004. Site fidelity of spawning sockeye salmon (*Oncorhynchus nerka* W.) in the presence and absence of olfactory cues. *Ecology of Freshwater Fish* 13:104–110.

Stoddard, P., H. Zakon, M. Markham, and M. McAnelly. 2006. Regulation and modulation of electric waveforms in gymnotiform electric fish. *Journal of Comparative Physiology A* 153:477–487.

Stoltz, J. A., and B. D. Neff. 2006. Male size and mating tactic influence proximity to females during sperm competition in bluegill sunfish. *Behavioral Ecology and Sociobiology* 59:811–818.

Storz, J. F., and H. E. Hoekstra. 2007. The study of adaptation and speciation in the genomic era. *Journal of Mammalogy* 88:1–4.

Stowe, M. K., J. H. Tumlinson, and R. R. Heath. 1987. Chemical mimicry: bolas spiders emit components of moth prey species sex pheromones. *Science* 236:964–966.

Strand, C. R., and P. Deviche. 2007. Hormonal and environmental control of song control region growth and new neuron addition in adult male house finches, *Carpodacus mexicanus*. *Developmental Neurobiology* 67:827–837.

Strandberg, R., and T. Alerstam. 2007. The strategy of fly-and-forage migration, illustrated for the osprey (*Pandion haliaetus*). *Behavioral Ecology and Sociobiology* 61:1865–1875.

Strassmann, J. E., and D. C. Queller. 2007. Insect societies as divided organisms: the complexities of purpose and cross-purpose. *Proceedings of the National Academy of Sciences* 104:8619–8628.

Strassmann, J. E., C. R. Hughes, D. C. Queller, S. Turlillazzi, R. Cervo, S. K. Davies, and K. F. Goodnight. 1989. Genetic relatedness in primitively eusocial wasps. *Nature* 342:268–269.

Struhsaker, T. T. 1967. Auditory communication among vervet monkeys (*Cercopithecus aethiops*). In *Social Communication among Primates*, edited by S. A. Altmann. Chicago: University of Chicago Press, pp. 281–324.

Stutt, A. D., and M. T. Siva-Jothy. 2001. Traumatic insemination and sexual conflict in the bed bug *Cimex lectularius*. *Proceedings of the National Academy of Sciences* 98:5683–5687.

Stutt, A. D., and P. Willmer. 1998. Territorial defence in speckled wood butterflies: Do the hottest males always win? *Animal Behaviour* 55:1341–1347.

Sugiyama, Y. 1965. Behavioral development and social structure in two troops of hanuman langurs (*Presbytis entellus*). *Primates* 6:213–247.

Sugiyama, Y. 1984. Proximate factors of infanticide among langurs at Dharwar: A reply to Bogess. In *Infanticide: Comparative and Evolutionary Perspectives*, edited by G. Hausfater and S. B. Hrdy. New York: Aldine, pp. 311–314.

Summers, C. H., and S. Winberg. 2006. Interactions between the neural regulation of stress and aggression. *Journal of Experimental Biology* 209:4581–4589.

Summers, K., and M. E. Clough. 2001. The evolution of coloration and toxicity in the poison frog family (Dendrobatidae). *Proceedings of the National Academy of Sciences* 98:6227–6232.

Surlykke, A. 1984. Hearing in notodontid moths: A tympanic organ with a single auditory neurone. *Journal of Experimental Biology* 113:323–335.

Sutherland, W. 1998. The importance of behavioural studies in conservation biology [Review]. *Animal Behaviour* 56:801–809.

Swaisgood, R. R. 2007. Current status and future directions of applied behavioral research for animal welfare and conservation. *Applied Animal Behaviour Science* 102:139–162.

Swann, J., F. Rahaman, T. Bijak, and J. Fiber. 2001. The main olfactory system mediates pheromone-induced *fos* expression in the extended amygdala and preoptic area of the male Syrian hamster. *Neuroscience* 105:695–706.

Taborsky, M. 1985. Breeder-helper conflict in a cichlid fish with broodcare helpers: An experimental analysis. *Behaviour* 95:45–75.

Taborsky, M. 2006. *Ethology* into a new era. *Ethology* 112:1–6.

Taglialatela, J. P., S. Savage-Rumbaugh, and L. A. Baker. 2003. Vocal production by a language-competent *Pan paniscus*. *International Journal of Primatology* 24:1–16.

Takahashi, D., and M. Kohda. 2004. Courtship in fast water currents by a male stream goby (*Rhinogobius brunneus*) communicates the parental quality honestly. *Behavioral Ecology and Sociobiology* 55:431–438.

Takeda, K. 1961. Classical conditioned response in the honey bee. *Journal of Insect Physiology* 6:168–179.

Tamura, N. 1989. Snake-directed mobbing by the Formosan squirrel *Callosciurus erythraeus thaiwanensis*. *Behavioral Ecology Sociobiology* 24:175–180.

Tan, E. J., and B. L. Tang. 2006. Looking for food: Molecular neuroethology of invertebrate feeding behavior. *Ethology* 112:826–832.

Tang-Martinez, Z., and T. B. Ryder. 2005. The problem with paradigms: Bateman's worldview as a case study. *Integrative and Comparative Biology* 45:821–830.

Tanner, D. A., and P. K. Visscher. 2008. Do honey bees average directions in the waggle dance to determine a flight direction? *Behavioral Ecology and Sociobiology* 62:1891–1898.

Tanouye, M. A., C. A. Ferrus, and S. C. Fujita. 1981. Abnormal action potentials associated with the *Shaker* complex locus in *Drosophila*. *Proceedings of the National Academy of Sciences* 78:6548–6552.

Tarsitano, M. S. 2006. Route selection by a jumping spider (*Portia labiata*) during the locomotory phase of a detour. *Animal Behaviour* 72:1437–1442.

Tarsitano, M. S., and A. R., Andrew. 1999. Scanning and route selection in the jumping spider *Portia labiata*. *Animal Behaviour* 58:255–265.

Tarsitano, M. S., and R. R. Jackson. 1997. Araneophagic jumping spiders discriminate between detour routes that do and do not lead to prey. *Animal Behaviour* 53:257–266.

Taylor, A. H., G. R. Hunt, J. C. Holzhaider, and R. D. Gray. 2007. Spontaneous metatool use by New Caledonian crows. *Current Biology* 17:1504–1507.

Taylor, P. W., and R. W. Elwood. 2003. The mismeasure of animal contests. *Animal Behaviour* 65:1195–1202.

Temeles, E. J. 1989. Effect of prey consumption on foraging activity of northern barriers. *Auk* 106:353–357.

Temeles, E. J., I. L. Pan, J. L. Brennan, and J. N. Horwitt. 2000. Evidence for ecological causation of sexual dimorphism in a hummingbird. *Science* 289:441–443.

Tennesen, M. 1999. Testing the depths of life. *National Wildlife* February/March, volume 2.

Tepperman, J. 1980. *Metabolic and Endocrine Physiology*. 4th ed. Chicago: Year Book Medical.

Terborgh, J., and A. W. Goldizen. 1985. On the mating system of the cooperatively breeding saddle-backed tamarin (*Saguinus fuscicollis*). *Behavioral Ecology and Sociobiology* 16:293–299.

Terrace, H. S. 2005. Metacognition and the evolution of language. In *The Missing Link in Cognition*, edited by H. S. Terrace and J. Metcalfe. Oxford: Oxford University Press, pp. 84–113.

Terrace, H. S., L. A. Petitto, R. J. Sanders, and T. G. Bever. 1979. Can an ape create a sentence? *Science* 206:891–900.

Terrick, T. D., R. L. Mumme, and G. M. Burghardt. 1995. Aposematic coloration enhances chemosensory recognition of noxious prey in the garter snake *Thamnophis radix*. *Animal Behaviour* 49:857–866.

Thayer, A. H. 1896. The law which underlies protective coloration. *Auk* 13:124–129.

Thom C., D. C. Gilley, J. Hooper, and H. E. Esch. 2007. The scent of the waggle dance. *PLoS Biology* 5:e228.

Thompson, C. K., G. E. Bentley, and E. A. Brenowitz. 2007. Rapid seasonal-like regression of the adult avian song control system. *Proceedings National Academy of Sciences* 104:15520–15525.

Thompson, K. V. 1996. Play-partner preferences and the function of social play in infant sable antelope, *Hippotragus niger*. *Animal Behaviour* 52:1143–1155.

Thorndike, E. L. 1898. Animal intelligence: An experimental study of the associative process in animals. *Psychological Monographs* 2:1–109.

Thorndike, E. L. 1911. *Animal Intelligence: Experimental Studies*. New York: Macmillan.

Thornhill, R. 1976. Sexual selection and nuptial feeding behavior in *Bittacus apicalis* (Insecta: Mecoptera). *American Naturalist* 110:529–548.

Thornhill, R. 1981. *Panorpa* (Mecoptera: Panorpidae) scorpionflies: Systems for understanding resource-defense polygyny. *Annual Review of Ecology and Systematics* 12:355–386.

Tian, L., B. Xiao, W. Lin, S. Zhang, R. Shu, and Y. Pan. 2007. Testing for the presence of magnetite in the upper-beak skin of homing pigeons. *BioMetals* 20:197–203.

Tinbergen, L. 1960. The natural control of insects in pine woods I. Factors influencing the intensity of predation by song birds. *Archives Néerlandaises de Zoologie* 13:265–343.

Tinbergen, N. 1951. *The Study of Instinct*. London: Oxford University Press.

Tinbergen, N. 1952a. The curious behavior of the stickleback. *Scientific American* 187:2–6.

Tinbergen, N. 1952b. Derived activities: their causation, biological significance, origin and emancipation during evolution. *Quarterly Review of Biology* 27:1–32.

Tinbergen, N. 1963. On aims and methods of ethology. *Zeitschrift für Tierpsycholgie* 20:410–433.

Tinbergen, N., and W. Kruyt. 1938. Über die Oreintierrung des Bienenwolfes (*Philanthus triangulum* Fabr.) III. Die Bevorzugung bestimmter Wegmarken. *Zeitschrift für vergleichende Physiologie* 25:292–334.

Tinbergen, N., G. J. Broekhuysen, F. Feekes, J. C. W. Houghton, H. Kruuk, and E. Szulc. 1962. Egg-shell removal by the black-headed gull, *Larus ridibundus* L., a behaviour component of camouflage. *Behaviour* 19:74–118.

Tobias, M. L., C. Barnard, R. O'Hagan, S. H. Horng, M. Rand, and D. B. Kelley. 2004. Vocal communication between male *Xenopus laevis*. *Animal Behaviour* 67:353–365.

Tobias, M. L., S. S. Viswanathan, and D. B. Kelley. 1998. Rapping, a female receptive call, initiates male-female duets in the South African clawed frog. *Proceedings of the National Academy of Sciences* 95:1870–1875.

Tomaschko, K.-H. 1994. Ecdysteroids from *Pycnogonum litorale* (Arthropoda, Pantopoda) act as a chemical defense against *Carcinus maenus* (Crustacea, Decapoda). *Journal of Chemical Ecology* 20:1445–1455.

Tomsic D., V. Massoni, H. Maldonado. 1993. Habituation to a danger stimulus in two semiterrestrial crabs—ontogenic, ecological and opioid modulation correlates. *Journal of Comparative Physiology A—Neuroethology* 173:621–633.

Tousson, E., and H. Meissl. 2004. Suprachiasmatic nuclei grafts restore circadian rhythm in the paraventricular nucleus of the hypothalamus. *The Journal of Neuroscience* 24:2983–2988.

Trainer, J. M., and D. B. McDonald. 1995. Singing performance, frequency matching and courtship success of long-tailed manakins (*Chiroxiphia linearis*). *Behavioral Ecology and Sociobiology* 37:249–254.

Trainer, J. M., D. B. McDonald, and W. A. Learn. 2002. The development of coordinated singing in cooperatively displaying long-tailed manakins. *Behavioral Ecology* 13:65–69.

Trainor, B. C., and H. A. Hofmann. 2006. Somatostatin regulates aggressive behavior in an African cichlid fish. *Endocrinology* 147:5119–5125.

Travers, S. E., M. D. Smith, J. Bai, S. H. Hulbert, J. E. Leach, P. S. Schnable, A. K. Knapp, G. A. Milliken, P. A. Fay, A. Saleh, and K. A. Garrett. 2007. Ecological genomics: Making the leap from model systems in the lab to native populations in the field. *Frontiers in Ecology and the Environment* 5:19–24.

Triefenbach, F. A., and H. H. Zakon. 2008. Changes in signaling during agonistic interactions between male weakly electric knifefish, *Apteronotus leptorhynchus*. *Animal Behaviour* 75: 1263–1272.

Trivers, R. L. 1971. The evolution of reciprocal altruism. *Quarterly Review of Biology* 46:35–57.

Trivers, R. L. 1972. Parental investment and sexual selection. In *Sexual Selection and the Descent of Man, 1871–1971*, edited by B. Campbell. Chicago: Aldine, pp. 136–179.

Trivers, R. L. 1974. Parent-offspring conflict. *American Zoologist* 14:249–264.

Trivers, R. L., and H. Hare. 1976. Haplodiploidy and the evolution of the social insects. *Science* 191:249–263.

Trumbo, S. T. 2006. Infanticide, sexual selection and task specialization in a biparental burying beetle. *Animal Behaviour* 72:1159–1167.

Trumbo, S. T. 2007. Can the "challenge hypothesis" be applied to insects? *Hormones and Behavior* 51:281–285.

Turchin, P., and P. Kareiva. 1989. Aggregation in *Aphis varians*: An effective strategy for reducing predation risk. *Ecology* 70:1008–1016.

Turek, F. W., C. Joshu, E. Lin, G. Ivanova, E. McDearmon, A. Laposky, S. Losee-Olson, A. Easton, D. R. Jensen, R. H. Eckel, J. S. Takahashi, and J. Bass. 2005. Obesity and metabolic syndrome in circadian *clock* mutant mice. *Science* 308:1043–1045.

Turner, J. R. G. 1977. Butterfly mimicry: The genetical evolution of an adaptation. *Evolutionary Biology*10:163–206.

Uesugi, K. 1996. The adaptive significance of Batesian mimicry in the swallowtail butterfly, *Papio polytes* (Insecta, Papilionidae): Associative learning in a predator. *Ethology* 102:762–775.

Uetz, G. W., and J. A. Roberts. 2002. Multisensory cues and multimodal communication in spiders: insights from video/audio playback studies. *Brain, Behavior, and Evolution* 59:222–230.

Uetz, G. W., R. Papke, and B. Kilinc. 2002. Influence of feeding regime on body size, body condition and a male secondary sexual character in *Schizocosa ocreata* wolf spiders (Araneae, Lycosidae): condition-dependence in a visual signaling trait. *Journal of Arachnology* 30:461–469.

Urquhart, F. A. 1987. *The Monarch Butterfly: International Traveler*. Chicago: Nelson-Hall.

Vahed, K. 1998. The function of nuptial feeding in insects: A review of empirical studies. *Biological Reviews* 73:43–78.

Vallin, A., S. Jakobsson, J. Lind, and C. Wiklund. 2005. Prey survival by predator intimidation: an experimental study of peacock butterfly defence against blue tits. *Proceedings of the Royal Society of London B* 272:1203–1207.

van Praag, H., G. Kempermann, and F. H. Gage. 2000. Neural consequences of environmental enrichment. *Nature Reviews Neuroscience* 1:191–198.

Van Shaik, C. P., and C. H. Janson. 2000. *Infanticide by Males and Its Implications*. Cambridge, MA: Cambridge University Press.

van Staaden, M. J. 1998. Ethology: at 50 and beyond. *Trends in Ecology and Evolution* 13:6–8.

Van Valen, L. 1973. A new evolutionary law. *Evolutionary Theory* 1:1–30.

Vander Wall, S. B. 1982. An experimental analysis of cache recovery in Clark's nutcracker. *Animal Behaviour* 30:84–94.

Vander Wall, S. B., and R. P. Balda. 1977. Coadaptations of the Clark's nutcracker and the pinyon pine for efficient seed harvest and dispersal. *Ecological Monographs* 47:89–111.

Velando, A., Beamonte-Barrientos, and R. Torres. 2006. Pigment-based skin colour in the blue-footed booby: an honest signal of current condition used by females to adjust reproductive investment. *Oecologia* 149:535–542.

Venner, S., and J. Casas. 2005. Spider webs designed for rare but life-saving catches. *Proceedings of the Royal Society of London B* 272:1587–1592.

Verner, J., and M. F. Willson. 1966. The influence of habitats on mating systems of North American passerine birds. *Ecology* 47:143–147.

Verwey, J. 1930. Die Paarungsbiologie des Fischreihers. *Zoölogische Jahrbucher-Abteilung für Allgemeine Zoölogie und Physiologie der Tiere* 48:1–120.

Villinger, J., and B. Waldman. 2008. Self-referent MHC type matching in frog tadpoles. *Proceedings of the Royal Society of London B* 275:1225–1230.

Visalberghi, E. and E. Alleva. 1979. Magnetic influences on pigeon homing. *Biological Bulletin* 125:246–256.

Visalberghi, E., and L. Limongelli. 1994. Lack of comprehension of cause-effect relations in tool-using capuchin monkeys (*Cebus apella*). *Journal of Comparative Psychology* 108:15–22.

Visalberghi, E., and L. Limongelli. 1996. Acting and understanding: tool use revisited through the minds of capuchin monkeys. In *Reaching into Thought: The Minds of the Great Apes*, edited by A. Russon, K. Bard, and S. Parker. Cambridge: Cambridge University Press, pp. 57–79.

Visalberghi, E., and L. Trinca. 1989. Tool use in capuchin monkeys: distinguishing between performing and understanding. *Primates* 30:511–521.

Vollrath, F. 1986. Eusociality and extraordinary sex ratios in the spider *Anelosimus eximius* (Araneae, Theridiidae). *Behavioral Ecology and Sociobiology* 4:283–287.

von der Emde, G. 1999. Active electrolocation of objects in weakly electric fish. *Journal of Experimental Biology* 202:1205–1215.

von Frisch, K. 1950. Die Sonne als Kompass im Leben der Bienen. *Experientia* 6:210–221.

von Frisch, K. 1967. *The Dance Language and Orientation of Bees*. Cambridge, MA: Harvard University Press.

von Frisch, K. 1971. *Bees: Their Vision, Chemical Senses and Language*. 2nd ed. Ithaca, NY: Cornell University Press.

Voss, R. S. 1979. Male accessory glands and the evolution of copulatory plugs in rodents. *Occasional Papers of the Museum of Zoology*. Ann Arbor, MI: University of Michigan 689:1–27.

Waage, J. K. 1979. Dual function of the damselfly penis: Sperm removal and transfer. *Science* 203:916–918.

Waas, J. R. 1991. The risks and benefits of signalling aggressive motivation: A study of cave-dwelling little blue penguins. *Behavioral Ecology and Sociobiology* 29:139–146.

Wade, J., and A. P. Arnold. 1996. Functional testicular tissue does not masculinize development of the zebra finch song system. *Proceedings of the National Academy of Sciences* 93:5264–5268.

Wade, J., and A. P. Arnold. 2004. Sexual differentiation of the zebra finch song system. *Annals of the New York Academy of Sciences* 1016: 540–559.

Wagner, G. 1976. Das Orientierungverhalten von Brieftauben im erdmagnetisch gestörten Gebiete des Chasseral. *Revue Suisse de Zoologie* 83:883–890.

Walcott, C. 1978. Anomalies in the earth's magnetic field increase scatter of pigeon vanishing bearings. In *Animal Migration, Navigation and Homing*, edited by K. Schmidt-Koenig and W. T. Keeton. Berlin: Springer-Verlag. pp. 143–151.

Walcott, C. 2005. Multi-modal orientational cues in homing pigeons. *Integrative and Comparative Biology* 45:574–581.

Walcott, C., and R. P. Green. 1974. Orientation of homing pigeons altered by a change in the direction of an applied magnetic field. *Science* 184:180–182.

Waldvogel, J. A., S. Benvenuti, W. T. Keeton, and F. Papi. 1978. Homing pigeon orientation influenced by deflected winds at home loft? *Journal of Comparative Physiology* 128:297–301.

Walker, M. M., and M. E. Bitterman. 1989. Honeybees can be trained to respond to very small changes in geomagnetic intensity. *Journal of Experimental Biology* 145:489–494.

Walker, M. M., C. E. Diebel, C. V. Haugh, P. M. Pankhurst, and J. C. Montgomery. 1997. Structure and function of the vertebrate magnetic sense. *Nature* 390:371–376.

Wallraff, H. G. 1980. Olfaction and homing in pigeons: Nerve section experiments, critique, hypotheses. *Journal of Comparative Physiology* 139:209–224.

Wallraff, H. G. 1981. The olfactory component of pigeon navigation: Steps of analysis. *Journal of Comparative Physiology* 143:411–422.

Wallraff, H. G. 2004. Avian olfactory navigation: its empirical foundation and conceptual state. *Animal Behaviour* 67:189–204.

Wallraff, H. G. 2005. *Avian Navigation: Pigeon Homing as a Paradigm*. Berlin: Springer-Verlag.

Waloff, Z. 1959. Notes on some aspects of the desert locust problem. *Report of the FAO Panels of aspects of the strategy of the desert locust plague control*. FAO Document 59-6-4737:23–26. (Cited in Alerstam, T., A. Hedenström, and S. Åkesson. 2003. Long-distance migration: evolution and determinants. *Oikos* 103:247–260. p. 249.)

Walter, H. 1979. *Eleanora's Falcon: Adaptations to Prey and Habitat in a Social Raptor*. Chicago: University of Chicago Press.

Walther, G. R., E. Post, P. Convey, A. Menzel, C. Parmesa, T. J. C. Beebee, J. M Fromentin, O. Hoegh-Guldberg, and F. Bairlein. 2002. Ecological responses to recent climate change. *Nature* 416:389–395.

Wang, Y., Y. Pan, S. Parsons, M. Walker, and S. Zhang. 2007. Bats respond to polarity of a magnetic field. *Proceedings of the Royal Society B* 274:2901–2905.

Waser, P. M., and W. T. Jones. 1983. Natal philopatry among solitary mammals. *Quarterly Review of Biology* 58:355–390.

Watanabe, S., J. Sakamoto, and M. Wakita. 1995. Pigeon's discrimination of paintings by Monet and Picasso. *Journal of the Experimental Analysis of Behavior* 63:165–174.

Waterhouse, J. M., and P. J. DeCoursey. 2004. The relevance of human circadian rhythms for human welfare. In *Chronobiology: Biological Timekeeping*, edited by J. C. Dunlap, J. J. Loros and P. J. DeCoursey. Sunderland, MA: Sinauer. pp. 325-356.

Waterman, T. H. 1989. *Animal Navigation*. New York: W. H. Freeman.

Watson, J. B. 1930. *Behaviorism*. New York: W.W. Norton & Company, Inc.

Watts , C. R., and A. W. Stokes. 1971. The social order of turkeys. *Scientific American* 224:112–118.

Watts, D. P. 1987. Effects of mountain gorilla foraging activities on the productivity of their food species. *African Journal of Ecology* 25:155–163.

Watts, D. P. 2006. Conflict resolution in chimpanzees and the valuable-relationships hypothesis. *International Journal of Primatology* 27:1337–1364.

Weatherhead, P. J., and R. J. Robertson. 1979. Offspring quality and the polygyny threshold: The "sexy son" hypothesis. *American Naturalist* 113:201–208.

Weatherhead, P. J., and R. J. Roberston. 1981. In defence of the "sexy son" hypothesis. *American Naturalist* 117:349–356.

Weaver, I. C. G., M. J. Meaney, and M. Szyf. 2006. Maternal care effects on the hippocampal transcriptome and anxiety-mediated behaviors in the offspring that are reversible in adulthood. *Proceedings of the National Academy of Sciences* 103:3480–3485.

Weber, N. A. 1972. The attines: The fungus-culturing ants. *American Scientist* 60:448–456.

Wedell, N., M. J. G. Gage, and G. A. Parker. 2002. Sperm competition, male prudence and sperm-limited females. *Trends in Ecology and Evolution* 17:313–320.

Wedell, N., C. Kvarnemo, C. M. Lessells, and T. Tregenza. 2006. Sexual conflict and life histories. *Animal Behaviour* 71:999–1011.

Weeks, J. C. 2003. Thinking globally, acting locally: steroid hormone regulation of the dendritic architecture, synaptic connectivity, and death of an individual neuron. *Progress in Neurobiology* 70:421–442.

Weeks, J. C., and J. W. Truman. 1984. Neural organization of peptide-activated ecdysis behaviors during metamorphosis of *Manduca sexta*. II. Retention of the proleg motor pattern despite the loss of prolegs at pupation. *Journal of Comparative Physiology A* 155:423–433.

Weeks, J. C., G. A. Jacobs, and C. I. Miles. 1989. Hormonally mediated modifications of neuronal structure, synaptic connectivity, and behavior during metamorphosis of the tobacco hornworm, *Manduca sexta*. *American Zoologist* 29:1331–1344.

Wehner, R. 1981. Spatial vision in arthropods. In *Handbook of Sensory Psychology*, edited by H. Autrum. Berlin: Springer-Verlag, pp. 287–616.

Wehner, R., and M. V. Srinivasan. 1981. Searching behavior of desert ants, genus *Cataglyphis* (Formicidae: Hymenoptera). *Journal of Comparative Physiology* 142:315–338.

Weimerskirch, H., J. Martin, Y. Clerquin, P. Alexandre, and S. Jarskova. 2001. Energy saving in flight formation. *Nature* 413:697–698.

Weiner, J. 1995. *The Beak of the Finch*. New York: Vintage Books.

Weir, A. S., J. Chappell, and A. Kacelnik. 2002. Shaping of hooks in New Caledonian crows. *Science* 297:981.

Weissburg, M. 1992. Functional analysis of fiddler crab foraging: sex specific mechanics and constraints. *Journal of Experimental Marine Biology and Ecology* 156:105–124.

Weissburg, M. 1993. Sex and the single forager: gender-specific foraging strategies in the fiddler crab *Uca pugnax*. *Ecology* 74:279–291.

Welsh, D. K., D. E. Logothetis, M. Meister, and S. M. Reppert. 1995. Individual neurons dissociated from rat suprachiasmatic nucleus express independently phased circadian firing patterns. *Neuron* 14:697–706.

Welsh, D. K., S.-H. Yoo, A. C. Liu, J. S. Takahashi, and S. A. Kay. 2004. Bioluminescence imaging of individual fibroblasts reveals persistent, independently phased circadian rhythms of clock gene expression. *Current Biology* 14:2289–2295.

Welty, J. C. 1962. *The Life of Birds*. 2nd ed. Philadelphia: Saunders.

Wenner, A. M. 1964. Sound communication in honeybees. *Scientific American* 210:116–125.

Wenner, A. M. 2002. The elusive honey bee dance "language" hypothesis. *Journal of Insect Behavior* 15:859–878.

Werren, J. H., M. R. Gross, and R. Shine. 1980. Paternity and the evolution of male parental care. *Journal of Theoretical Biology* 82:619–631.

West, M. J., and M. J. King. 1988. Female visual displays affect the development of male song in the cowbird. *Nature* 334:244–246.

West, M. J., A. P. King, and D. H. Eastzer. 1981. The cowbird: Reflections on development from an unlikely source. *American Scientist* 69:56–66.

Westneat, D. F., and P. W. Sherman. 1993. Parentage and the evolution of parental behavior. *Behavioral Ecology* 4:66–77.

White, D. J., A. P. King, and M. J. West. 2002. Facultative development of courtship and communication skills in juvenile male cowbirds, *Molothrus ater*. *Behavioral Ecology* 13:487–496.

Whitehouse, M. E. A. 1997. Experience influences male-male contests in the spider *Argyrodes antipodiana*. *Animal Behaviour* 53:913–923.

Whiteman, E. A., and I. M. Côté. 2004. Monogamy in marine fishes. *Biological Reviews* 79:351–375.

Whiten, A., D. M. Custance, J.-C. Gomez, P. Teixidor, and K. A. Bard. 1996. Imitative learning of artificial fruit processing in children (*Homo sapiens*) and chimpanzees (*Pan troglodytes*). *Journal of Comparative Psychology* 110:3–14.

Whiten, A., J. Goodall, W. C. McGrew, T Nishida, V. Reynolds, Y. Sugiyama, C. E. G. Tutin, R. W. Wrangham, and C. Boesch. 1999. Cultures in chimpanzees. *Nature* 399:682–685.

Whitfield, C. W., A. M. Cziko, and G. E. Robinson. 2003. Gene expression profiles in the brain predict behavior in individual honey bees. *Science* 302:296–299.

Whitlock, J. R., A. J. Heynen, M. G. Schler, and M. F. Bear. 2006. Learning induces long-term potentiation in the hippocampus. *Science* 313:1093–1097.

Whiten, A. 1996. When does smart behaviour-reading become mind-reading? In *Theories of Theories of Mind*, edited by P. Carruthers and P. K. Smith. Cambridge: Cambridge University Press, pp. 277–292.

Whitten, P. L., D. K. Brockman, and R. C. Stavisky. 1998. Recent advances in noninvasive techniques to monitor hormone-behavior interactions. *Yearbook of Physical Anthropology* 41:1–23.

Wich, S. A., and H. de Vries. 2006. Male monkeys remember which group members have given alarm calls. *Proceedings of the Royal Society of London B* 273:735–740.

Wickler, W. 1968. *Mimicry in Plants and Animals*. McGraw-Hill, New York. 253 pp.

Wickler, W. 1972. *Mimicry*. 2nd ed. New York: World University Library, McGraw-Hill.

Wiggins, D. A., and R. D. Morris. 1986. Criteria for female choice of mates: Courtship feeding and paternal care in the common tern. *American Naturalist* 128:126–129.

Wilbrecht, L., A. Crionas, and F. Nottebohm. 2002. Experience affects recruitment of new neurons but not adult neuron number. *The Journal of Neuroscience* 22:825–831.

Wilbrecht, L., H. Williams, N. Gangadhar, and F. Nottebohm. 2006. High levels of new neuron addition persist when the sensitive period for song learning is experimentally prolonged. *The Journal of Neuroscience* 26:9135–9141.

Wilcox, R. S., and T. Ruckdeschel. 1982. Food threshold territoriality in a water strider *(Gerris remigis)*. *Behavioral Ecology and Sociobiology* 11:85–90.

Wilcox, R. S., R. R. Jackson, and K. Gentile. 1996. Spiderweb smokescreens—Spider trickster uses background noise to mask stalking movements. *Animal Behaviour* 51:313–326.

Wiley, R. H. 1974. Evolution of social organization and life history patterns among grouse (Aves: Tetraonidae). *Quarterly Review of Biology* 49:209–227.

Wilkinson, G. S. 1984. Reciprocal food sharing in the vampire bat. *Nature* 308:181–184.

Wilkinson, G. S. 1990. Food sharing in vampire bats. *Scientific American* 262:76–82.

Williams, C. B. 1965. *Insect Migration*. London: Collins.

Williams, G. C. 1975. *Sex and Evolution*. Princeton, NJ: Princeton University Press.

Williams, S. L., K. E. Brakke, and E. S. Savage-Rumbaugh. 1997. Comprehension skills of language-competent and nonlanguage-competent apes. *Language and Communication* 17:301–317.

Williams, T. C., and J. M. Williams. 1978. An oceanic mass migration of land birds. *Scientific American* 239:166–176.

Willis, P. M., and L. M. Dill. 2007. Mate guarding in male Dall's porpoises (*Phocoenoides dalli*). *Ethology* 113:587–597.

Wilson, D. M. 1961. The central nervous control of flight in a locust. *Journal of Experimental Biology* 38:471–490.

Wilson, D. M., and T. Weis-Fogh. 1962. Patterned activity of coordinated motor units studied in flying locusts. *Journal of Experimental Biology* 40:643–674.

Wilson, E. O. 1968. Chemical systems. In *Animal Communication: Techniques of Study and Results of Research*, edited by T. A. Sebeok. Bloomington: Indiana University Press, pp. 75–102.

Wilson, E. O. 1971. *The Insect Societies*. Cambridge, MA: Belknap/Harvard University Press.

Wilson, E. O. 1975. *Sociobiology*. Cambridge, MA: Belknap Press of Harvard University Press.

Wilson, E. O. 1980. *Sociobiology, the Abridged Version*. Cambridge, MA: Belknap/Harvard University Press.

Wilson, E. O. 2008. One giant leap: How insects achieved altruism and colonial life. *BioScience* 58:17–25.

Wilson, E. O., and B. Hölldobler. 2005. Eusociality: origin and consequence. *Proceedings of the National Academy of Sciences* 102:13367–13371.

Wilson, P. L., M. C. Towner, and S. L. Vehrencamp. 2000. Survival and song-type sharing in a sedentary subspecies of the song sparrow. *Condor* 102:355–363.

Wiltschko, R., and W. Wiltschko. 1999. The orientation systems of birds I. Compass mechanisms. *Journal of Ornithology* 140:1–40.

Wiltschko, R., and W. Wiltschko. 2006. Magnetoreception. *BioEssays* 28:157–168.

Wiltschcko, R., M. Walker, and W. Wiltschko. 2000. Sun-compass orientation in homing pigeons: compensation for different rate of change in azimuth? *Journal of Experimental Biology* 203:889–894.

Wiltschko, R., T. Ritz, K. Stapput, P. Thalau, and W. Wiltschko. 2005. Two different types of light-dependent responses in magnetic fields in birds. *Current Biology* 15:1518–1523.

Wiltschko, W. 1978. Further analysis of the magnetic compass of migratory birds. In *Animal Migration, Navigation and Homing*, edited by K. Schmidt-Koenig and W. T. Keeton. Berlin: Springer-Verlag, pp. 302–310.

Wiltschko, W. 1982. The migratory orientation of Garden Warblers *Sylvia borin*. In *Avian Navigation*, edited by F. Papi and H. G. Wallraff. Berlin: Springer-Verlag, pp. 50–58.

Wiltschko, W., and R. Wiltschko. 1972. Magnetic compass of European robins. *Science* 176:62–64.

Wiltschko, W., and R. Wiltschko. 1996. Magnetic orientation in birds. *Journal of Experimental Biology* 199:29–38.

Wiltschko, W., and R. Wiltschko. 2005. Magnetic orientation and magnetoreception in birds and other animals. *Journal of Comparative Physiology A* 191:675–693.

Wiltschko, W., and R. Wiltschko. 2007. Magnetoreception in birds: two receptors for two different tasks. *Journal of Ornithology* 148:S61–S76.

Wiltschko, W., R. Wiltschko, and C. Walcott. 1987. Pigeon homing: different effects of olfactory deprivation in different countries. *Behavioral Ecology and Sociobiology* 21:333–342.

Wingfield, J. C. 1984. Environmental and endocrine control of reproduction in the song sparrow, *Melospiza melodia*. I. Temporal organization of the breeding cycle. *General and Comparative Endocrinology* 56:406–416.

Wingfield, J. C. 2005. A continuing saga: the role of testosterone in aggression. *Hormones and Behavior* 48:253–255.

Wingfield, J. C., and M. C. Moore. 1987. Hormonal, social, and environmental factors in the reproductive biology of free-living male birds. In *Psychobiology of Reproductive Behavior*, edited by D. Crews. Upper Saddle River, NJ: Prentice Hall, pp. 148–175.

Wingfield, J. C., and K. K. Soma. 2002. Spring and autumn territoriality in song sparrows: same behavior, different mechanisms? *Integrative and Comparative Biology* 42:11–20.

Wingfield, J. C., R. E. Hegner, A. M. Dufty, and G. F. Ball. 1990. The "Challenge Hypothesis": Theoretical implications for patterns of testosterone secretion, mating systems, and breeding strategies. *American Naturalist* 136:829–846.

Winston, M. L., and K. N. Slessor. 1992. The essence of royalty: Honey bee queen pheromone. *American Scientist* 80:374–385.

Wirant, S. C., and B. McGuire. 2004. Urinary behavior of female domestic dogs (*Canis familiaris*): Influence of reproductive condition, location and age. *Applied Animal Behaviour Science* 85:335–348.

Wirant, S. C., K. Halvorsen, and B. McGuire. 2007. Preliminary observations on the urinary behaviour of female Jack Russell Terriers in relation to stage of the oestrous cycle, location, and age. *Applied Animal Behaviour Science* 106:161–166.

Wisby, W. J., and A. D. Hasler. 1954. The effect of olfactory occlusion on migrating silver salmon (*O. kisutch*). *Journal of the Fisheries Research Board of Canada* 11:472–478.

Wittenberger, J. F. 1978. The evolution of mating systems in grouse. *Condor* 80:126–137.

Wittenberger, J. F. 1979. The evolution of mating systems in birds and mammals. In *Handbook of Behavioral Neurobiology*, edited by P. Marler and J. G. Vandenbergh. New York: Plenum Press, pp. 271–349.

Wittenberger, J. F. 1981. *Animal Social Behavior*. Boston: Duxbury.

Wittig, R. M., C. Crockford, R. M. Seyfarth, and D. L. Cheney. 2007. Vocal alliances in Chacma baboons (*Papio hamadryas ursinus*). *Behavioral Ecology and Sociobiology* 61:899–909.

Wittlinger, M., R. Wehner, and H. Wolf. 2007. The desert ant odometer: a stride integrator that accounts for stride length and walking speed. *Journal of Experimental Biology* 210:198–207.

Woelfle, M. A., and C. H. Johnson. 2006. No promoter left behind: Global circadian gene expression in cyanobacteria. *Journal of Biology Rhythms* 21:419–431.

Wöhr, M., and R. K. W. Schwarting. 2007. Ultrasonic communication in rats: Can playback of 50-kHz calls induce approach behavior? *PloS ONE* 2:e1365.

Wöhr, M., and R. K. W. Schwarting. 2008. Ultrasonic calling during fear conditioning in the rat: no evidence for an audience effect. *Animal Behaviour* 76:749–760.

Wojcieszek, J. M., J. A. Nicholls, and A. W. Goldizen. 2007. Stealing behavior and the maintenance of a visual display in the satin bowerbird. *Behavioral Ecology* 18:689–695.

Wolff, J. O. 1992. Parents suppress reproduction and stimulate dispersal in opposite-sex juvenile white-footed mice. *Nature* 359:409–410.

Wolff, J. O. 1994. More on juvenile dispersal in mammals. *Oikos* 71:349–352.

Wolff, J. O., and J. H. Plissner. 1998. Sex biases in avian natal dispersal: An extension of the mammalian model. *Oikos* 83:327–330.

Wolovich, C. K., S. Evans, and J. A. French. 2008. Dads do not pay for sex but do buy the milk: food sharing and reproduction in owl monkeys (*Aotus* spp.). *Animal Behaviour* 75:1155–1163.

Wong, B. B. M., U. Candolin, and K. Lindström. 2007. Environmental deterioration compromises socially enforced signals of male quality in three-spined sticklebacks. *American Naturalist* 170:184–189.

Wong, R. Y., and C. D. Hopkins. 2007. Electrical and behavioral courtship displays in the mormyrid fish *Brienomyrus brachyistius*. *Journal of Experimental Biology* 210:2244–2252.

Wood, D. E. 1995. Modulation of behavior by biogenic amines and peptides in the blue crab, *Callinectes sapidus*. *Journal of Comparative Physiology A* 177:331–333.

Wood, D. E., M. Nishikawa, and C. D. Derby. 1996. Proctolinlike immunoreactivity and identified neurosecretory cells as putative substrates for modulation of courtship display behavior in the blue crab, *Callinectes sapidus*. *Journal of Comparative Physiology A* 368:153–163.

Woods, W. A. Jr., H. Hendrickson, J. Mason, and S. M. Lewis. 2007. Energy and predation costs of firefly courtship signals. *American Naturalist* 170:702–708.

Woolfenden, G. E. 1975. Florida scrub jay helpers at the nest. *Auk* 92:1–15.

Woolfenden, G. E., and J. W. Fitzpatrick. 1978. The inheritance of territory in group-breeding birds. *BioScience* 28:104–108.

Woolfenden, G. E., and J. W. Fitzpatrick. 1990. Florida scrub jays: A synopsis after 18 years of study. In *Cooperative Breeding in Birds*, edited by P. B. Stacey and W. D. Koenig. Cambridge: Cambridge University Press, pp. 241–266.

Woolley, S. C., J. T. Sakata, and D. Crews. 2004. Evolutionary insights into the regulation of courtship behavior in male amphibians and reptiles. *Physiology and Behavior* 83:347–360.

Woolley, S. M., and E. W. Rubel. 1997. Bengalese finches *Lonchura striata domestica* depend upon auditory feedback for the maintenance of adult song. *Journal of Neuroscience* 17:6380–6390.

Worden, B. D., and D. R. Papaj. 2005. Flower choice copying in bumblebees. *Biology Letters* 1:504–507.

Wourms, M. K., and F. E. Wasserman. 1985. Butterfly wing markings are more advantageous during handling than during the initial strike of an avian predator. *Evolution* 39:845–851.

Wrangham, R. D. 1980. Female choice of least costly males; a possible factor in the evolution of leks. *Zeitschrift für Tierpsychologie* 54:357–367.

Wright, J., and I. Cuthill. 1990. Manipulation of sex difference in parental care: effect of brood size. *Animal Behaviour* 40:462–471.

Wright, J., and A. F. Russell. 2008. How helpers help: disentangling ecological confounds from the benefits of cooperative breeding. *Journal of Animal Ecology* 77:427–429.

Würbel, H. 2001. Ideal homes? Housing effects on rodent brain and behaviour. *Trends in Neuroscience* 24:207–211.

Wyatt, T. D. 2003. *Pheromones and Animal Behaviour: Communication by Smell and Taste*. Cambridge: Cambridge University Press.

Wyers, E. J., V. S. Peeke, and M. J. Herz. 1973. Behavioral habituation in invertebrates. In *Habituation*, edited by V. S. Peeke and M. J. Herz. New York: Academic Press, pp. 1–57.

Yamagata, N., H. Nishino, and M. Mizunami. 2007. Neural pathways for the processing of alarm pheromone in the ant brain. *Journal of Comparative Neurology* 505:424–442.

Yamamoto, Y., and H. Ueda. 2007. Physiological study on imprinting and homing related to olfactory functions in salmon. In *North Pacific A: North Pacific Anadromous Fish Commission Technical Report Number* 7, pp. 113–114.

Yamazaki, K., E. A. Boyse, V. Mike, H. T. Thaler, B. J. Mathieson, J. Abbott, J. Boyse, Z. A. Zayas, and L. Thomas. 1976. Control of mating preferences in mice by genes in the major histocompatability complex. *Journal of Experimental Medicine* 144:1324–1335.

Yamazaki, S., R. Numano, M. Abe, A. Hida, R.-I. Takahashi, M. Ueda, G. D. Block, Y. Sakai, M. Menaker, and H. Tei. 2000. Resetting central and peripheral circadian oscillators in transgenic rats. *Science* 288:288–290.

Yang, D., B. Kynard, Q. Wei, X. Chen, W. Zheng, and H. Du. 2006. Distribution and movement of Chinese sturgeon, *Acipenser sinensis*, on the spawning ground located below the Gezhouba Dam during spawning seasons. *Journal of Applied Ichthyology* 22 (Supplement 1):145–151.

Yasui, Y. 2001. Female multiple mating as a genetic bet-hedging strategy when mate choice criteria are unreliable. *Ecological Research* 16:605–16.

Ydenberg, R. C., R. W. Butler, and D. B. Lank. 2007. Effects of predator landscapes on the evolutionary ecology of routing, timing, and molt by long-distance migrants. *Journal of Avian Biology* 38:523–529.

Ydenberg, R. C., and J. R. Krebs. 1987. The trade-off between territorial defense and foraging in the great tit (*Parus major*). *American Zoologist* 27:337–346.

Yeargan, K. V. 1994. Biology of bolas spiders. *Annual Review of Ecology and Systematics* 31:81–99.

Yoo, S.-H., S. Yamazaki, P. L. Lowrey, K. Shimomura, C. H. Ko, E. D. Buhr, S. M. Siepka, H.-K. Hong, W. J. Oh, O. J. Yoo, M. Menaker, and J. S. Takahashi. 2004. Period2: Luciferase real-time reporting of circadian dynamics reveals persistent circadian oscillations in mouse peripheral tissues. *Proceedings of the National Academy of Sciences* 101:5339–5346.

Young, L. J., R. Nilsen, K. G. Waymire, G. R. MacGregor, and T. R. Insel. 1999. Increased affiliative response to vasopressin in mice expressing the vasopressin receptor from a monogamous vole. *Nature* 400:766–768.

Youthed, G. J., and R. C. Moran. 1969. The lunar day activity rhythm of myrmeleontid larvae. *Journal of Insect Physiology* 15:1259–1271.

Zach, R. 1978. Selection and dropping of whelks by Northwestern crows. *Behaviour* 67:134–148.

Zach, R. 1979. Shell-dropping: Decision making and optimal foraging in Northwestern crows. *Behaviour* 67:106–117.

Zagotta, W. N., S. Germeraad, S. S. Garber, T. Hoshi, and R. W. Aldrich. 1989. Properties of ShB A-type potassium channels expressed in *Shaker* mutant *Drosophila* by germline transformation. *Neuron* 3:773–782.

Zahavi, A. 1975. Mate selection—A selection for a handicap. *Journal of Theoretical Biology* 53:205–214.

Zahavi, A. 1977. Reliability in communication systems and the evolution of altruism. In *Evolutionary Ecology*, edited by B. Stonehouse. London: Macmillan, pp. 253–259.

Zahavi, A., and A. Zahavi. 1997. *The Handicap Principle: A Missing Piece of Darwin's Puzzle*. Oxford: Oxford University Press.

Zakon, H. H., D. J. Zwickl, Y. Lu, and D. M. Hillis. 2008. Molecular evolution of communication signals in electric fish. *Journal of Experimental Biology* 211:1814–1818.

Zedrosser, A., O.-G. Stoen, S. Saebo, and J. E. Swenson. 2007. Should I stay or should I go? Natal dispersal in the brown bear. *Animal Behaviour* 74:369–376.

Zeh, J. A., and D. W. Zeh. 1996. The evolution of polyandry I: Intragenomic conflict and genetic incompatability. *Proceedings of the Royal Society of London B* 263:1711–1717.

Zeh, J. A., and D. W. Zeh. 2001. Reproductive mode and the genetic benefits of polyandry. *Animal Behaviour* 61:1051–1063.

Zuberbühler, K. 2005. The phylogenetic roots of language: evidence from primate communication and cognition. *Current Directions in Psychological Science* 14:126–130.

Zuberbühler, K. 2006. Language evolution: the origin of meaning in primates. *Current Biology* 16:R123–R125.

Zucca, P., F. Antonelli, and G. Vallortigara. 2005. Detour behaviour in three species of birds: quails (*Coturnix* sp.), herring gulls (*Larus cachinnans*) and canaries (*Serinus canaria*). *Animal Cognition* 8:122–128.

Photo Credits

Chapter 1

Figure 1.1: Comstock, Inc.; Figure 1.2: Milton H. Tierney, Jr./Visuals Unlimited; Figure 1.3: S. Muller/Wildlife/Peter Arnold, Inc.

Chapter 2

Figure 2.1: American Museum of Natural History Library; Figure 2.3a: Culver Pictures; Figure 2.3b: Culver Pictures; Figure 2.3c: ©AP/Wide World Photos; Figure 2.8: Corbis–Bettmann; Figure 2.9: Nina Leen/©Time, Inc.

Chapter 3

Figure 3.1: Bruce Coleman, Inc./Alamy; Figure 3.4: from Lee Drickamer, *Animal Behavior*, Fifth Edition, 1997, McGraw–Hill Companies. Reproduced with the permission of Lee Drickamer and the McGraw–Hill Companies; Figure 3.10a: Photo from: Ben–Shahar, Y., A. Robichon, M. B. Sokolowski, and G. E. Robinson. 2002. Influence of gene action across different time scales on behavior. *Science* 296 (April 26):741–744; Figure 3.12a: Photo from: Whitfield, C. W., et al. 2003. Gene expression profiles in the brain predict behavior in individual honey bees. *Science* 302 (Oct. 10): 296–299; Figure 3.16a & b: Photo from Lim, M. M., Z. Wang, D. E. Olazábal, X. Ren, E. F. Terwilliger, and L. J. Young. 2004. Enhanced partner preference in a promiscuous species by manipulating the expression of a single gene. *Nature* 429:754–757; Figure 3.18a & b: Burmeister, S. S., et al. 2005. Rapid behavioral and genomic responses to social opportunity. *PLoS Biology* 3(11), 1996–2004; Figure 3.22a: Wegner, P./Peter Arnold, Inc.; Figure 3.22b: Courtesy Russell Fernald; Figure 3.22c: Image by Aaron1a12 from Wikipedia: http://commons.wikimedia.org/wiki/File:BeeCropped.jpg; Figure 3.22d: ©George McCarthy/Corbis Images; Figure 3.22e: Hans Reinhard/Photo Researchers, Inc.

Chapter 4

Figure 4.1: ©Biosphoto/Conchon Laurent/Peter Arnold, Inc.; Figure 4.2: Steve & Ann Toon/Photolibrary Group Limited; Figure 4.3: Sumio Harada/Minden Pictures, Inc.; Figure 4.4: Jorg & Petra Wegner/Minden Pictures, Inc.; Figure 4.6: Courtesy Susan Reichert, Dept. of Zoology, University of Tennessee; Figure 4.7a: ©Bob Gibbons/Alamy; Figure 4.7b: Courtesy Susan Reichert, Dept. of Zoology, University of Tennessee; Figure 4.8: Stan Tekiela/DRK Photo; Figure 4.9: Tui De Roy/Minden Pictures, Inc.; Figure 4.10: M. & C. Photography/Peter Arnold, Inc.; Figure 4.13: ©Eric and David Hosking/© Corbis.

Chapter 5

Figure 5.4: Nina Leen/Life Picture Service; Figure 5.6: Frank Lane Picture Agency; Figure 5.7: Courtesy Masao Kawai, Museum of Nature and Human Activities, Hyogo, Japan; Figure 5.8a: Michael Quinton/Minden Pictures, Inc.; Figure 5.8b: Helen Williams/Photo Researchers, Inc.; Figure 5.8c: Allan D. Cruickshank/Photo Researchers, Inc.; Figure 5.11: From "The Mentality of Apes" by W. Kohler, 1926 ©by Harcourt, Brace & Co., Routledge.; Figure 5.12: Courtesy Alex Weir, Department of Zoology, Oxford University, Oxford UK; Figure 5.14: Courtesy Robert R. Jackson, School of Biological Sciences, University of Canterbury, Christchurch, New Zealand; Figure 5.16: Courtesy Irene Pepperberg, Brandeis University, Psychology.

Chapter 6

Figure 6.11: G. Ronald Austing/Photo Researchers, Inc.; Figure 6.12: Michael Fairchild/Peter Arnold, Inc.; Figure 6.14: Anne Knudsen; Figure 6.21: Heather Cameron, NIMH/NIH; Figure 6.27: Stephen Dalton/Photo Researchers, Inc.

Chapter 7

Figure 7.5: Frank L. Moore, Department of Zoology, University of Oregon; Figure 7.11: Courtesy Stephen Goodenough; Figure 7.13: Courtesy Ronald Barfield, Dept. of Biological Sciences, Rutgers University; Figure 7.17: Courtesy David Crews, Departments of Zoology and Psychology, University of Texas; Figure 7.21a: Sharon Wirant; Figure 7.21b: Sharon Wirant; Figure 7.21c: Sharon Wirant.

Chapter 8

Figure 8.1: ©NHPA/Photoshot; Figure 8.3a: Scott Camazine/Photo Researchers, Inc.; Figure 8.3b: Steve Ross/Photo Researchers, Inc.; Figure 8.3c: Matt Meadows/Peter Arnold, Inc.; Figure 8.5: Courtesy W. E. Bemis, Department of Ecology and Evolutionary Biology, Cornell University; Figure 8.8: Stephen J. Krasemann/DRK Photo; Figure 8.9: Mary M. Thacher/Photo Researchers; Figure 8.10: Nina Leen/LIFE Magazine, Time Inc.; Figure 8.14: Courtesy Peter H. Klopfer, Dept. of Zoology, Duke University; Figure 8.16: ©Arco Images GmbH/Alamy; Figure 8.19: ©Robert Shantz/Alamy; Figure 8.21: E. R. Degginger/Photo Researchers, Inc.; Figure 8.22a: Arthur Morris/Visuals Unlimited.

Chapter 9

Figure 9.5a: Bert L. Dunne/Bruce Coleman, Inc.; Figure 9.8a: James L. Amos/Peter Arnold, Inc.; Figure 9.8b: Stephen Dalton/© Natural History Photographic Agency; Figure 9.9a: Courtesy Steve A. Kay, Jeffrey Plants/Scripps Research Institute; Figure 9.10a: Courtesy Steve A. Kay, Jeffrey Plants/Scripps Research Institute; Figure 9.14a: Courtesy Rae Silver; Figure 9.14b: Courtesy Rae Silver.

Chapter 10

Figure 10.5: Courtesy K. Schmidt–Koenig, Department of Zoology, Duke University, Durham, N.C.; Figure 10.6: James L. Amos/Peter Arnold, Inc.; Figure 10.8: Karl H. Maslonski/Photo Researchers; Figure 10.11a: Johathan Blair/Woodfin Camp & Associates; Figure 10.11b: Johathan Blair/Woodfin Camp & Associates; Figure 10.15a: Courtesy T. H. Waterman, Department of Molecular, Cellular and Developmental Biology, Yale University; Figure 10.15b: Courtesy T. H. Waterman, Department of Molecular, Cellular and Developmental Biology, Yale University; Figure 10.15c: Courtesy T. H. Waterman, Department of Molecular, Cellular and Developmental Biology, Yale University; Figure 10.16: David Dennis/Tom Stack & Associates; Figure 10.19a: Courtesy Charles Walcott, Laboratory of Ornithology, Cornell University; Figure 10.20: ©Mark Conlin/Alamy; Figure 10.21a: Kenneth Lohmann, Department of Biology, University of North Carolina.

Chapter 11

Figure 11.1: Photograph by Boyd Kynard, Department Natural Resources Conservation, University of Massachusetts Amherst; Figure 11.5a: Photo by Betty McGuire, Department of Ecology and Evolutionary Biology, Cornell University; Figure 11.5b: Photo by Betty McGuire, Department of Ecology and Evolutionary Biology, Cornell University; Figure 11.8: Norbert Rosing/Getty Images; Figure 11.9: Charles Ott/Photo Researchers, Inc.; Figure 11.11: Courtesy W. E. Bemis, Department of Ecology and Evolutionary Biology, Cornell University; Figure 11.12b: Gregory Dimijan/Photo Researchers, Inc.; Figure 11.14: Thomas D. Mangelsen www.mangelsen.com/ Images of Nature; Figure 11.15: Chris Newbert/Minden Pictures, Inc.

Chapter 12

Figure 12.3: Courtesy David Dussourd, Department of Biology, University of Central Arkansas; Figure 12.4: Ross Hutchins/Photo Researchers, Inc.; Figure 12.6a: Tom McHugh/Photo Researchers, Inc.; Figure 12.6b: Z. Leszczynski/Animals Animals/Earth Scenes; Figure 12.7: Francois Gohier/Ardea London; Figure 12.8: Jim Frazier/ Auscape International Pty. Ltd.; Figure 12.9a: Joe McDonald/ Animals Animals/Earth Scenes; Figure 12.9b: *Scientific American*, May, 1973, R. Igor Gamow and John F. Harris; Figure 12.12: Tom McHugh/Photo Researchers, Inc.; Figure 12.13a: Courtesy Barth Falkenberg, School of Biological Sciences, University of Nebraska, Lincoln; Figure 12.13b: Courtesy Al Kamil, School of Biological Sciences, University of Nebraska, Lincoln; Figure 12.13c: H. J. Vermes/Courtesy Theodore Sargent, Department of Biology, University of Massachusetts Amherst; Figure 12.15: J. B. & S. Bottomley/ Ardea London; Figure 12.17: Breck Kent/Animals Animals/ Earth Scenes; Figure 12.18: Z. Leszczynki/Animals Animals/ Earth Scenes; Figure 12.20a: David Hosking/Photo Researchers, Inc.E61; Figure 12.21: Courtesy Paul V. Switzer, Department of Biological Sciences, Eastern Illinois University.

Chapter 13

Figure 13.1: Courtesy Lincoln P. Brower, Department of Biology, Sweet Briar College; Figure 13.2: Hellio & Van Ingen/ NHPA/Photoshot; Figure 13.6a: Lydia M. Mäthger, Marine Biological Laboratory, Woods Hole; Figure 13.6b: Lydia M. Mäthger, Marine Biological Laboratory, Woods Hole; Figure 13.6c: Lydia M. Mäthger, Marine Biological Laboratory, Woods Hole; Figure 13.6d: Lydia M. Mäthger, Marine Biological Laboratory, Woods Hole; Figure 13.10: Braude, S., Biology Department, Washington University; Figure 13.11a: Robert and Linda Mitchell; Figure 13.11b: Ivan Sazima, Department Zoologia & Museu de História Natural, Universidade Estadual de Campinas; Figure 13.16: COLOR–PIC, INC./Animals Animals/Earth Scenes; Figure 13.18: Courtesy T. E. Reimchen, Department of Biology, University of Victoria, B.C.; Figure 13.20: Leonard Lee Rue/Photo Researchers, Inc.; Figure 13.23: Courtesy Edmund D. Brodie, Jr., Dept. of Biology, Utah State University; Figure 13.24: Adrian Vallin, Department of Zoology, Stockholm University; Figure 13.25: Courtesy Thomas Eisner & Daniel Aneshansley/Cornell University; Figure 13.26: Anup & Manoj Sham/Animals Animals/Earth Scenes.

Chapter 14

Figure 14.1: Piotr Naskrecki/Minden Pictures, Inc.; Figure 14.3: Courtesy Darryl Gwynne, Biology Department, University of Toronto; Figure 14.5: David Woodfall/Natural History Photographic Agency; Figure 14.8a: Hans Pfletschinger/Peter Arnold, Inc.; Figure 14.8b: Jonathan Waage, Division of Biology & Medicine, Brown University; Figure 14.9: Irene Vandermolen/ Photo Researchers, Inc.; Figure 14.10: Lutz Fromhage, School of Biological Sciences, University of Bristol; Figure 14.11a: S. J. Arnold, Department of Biology, Oregon State University; Figure 14.11b: S. J. Arnold, Department of Biology, Oregon State University; Figure 14.12: Richard Matthews/ Planet Earth Pictures; Figure 14.15: Eric Hosking; Figure 14.16a: Daisuke Takahashi, Department of Zoology, Kyoto University; Figure 14.17: John Cancalosi/DRK Photo; Figure 14.18: ©franzfoto.com/Alamy; Figure 14.20: Howard Earl Uible/Photo Researchers, Inc.; Figure 14.23a, b & c: Courtesy Göran Arnqvist. 2007. Coevolution between harmful male genitalia and female resistance in seed beetles. *Proceedings of the National Academy of Sciences* 104:10921–10925.

Chapter 15

Figure 15.2a: Betty McGuire, Department of Ecology and Evolutionary Biology, Cornell University; Figure 15.2b: David Fraser, Animal Welfare Program, University of British Columbia; Figure 15.2c: David Fraser, Animal Welfare Program, University of British Columbia; Figure 15.5: Aaron Norman; Figure 15.7: Tony Heald/Nature Picture Library; Figure 15.8: Eric & David Hosking; Figure 15.9a: George Reszeter/OSF/Photolibrary; Figure 15.9b: Justin G. Schuetz; Figure 15.12a: Antonio Baeza, Smithsonian Tropical Research Institute; Figure 15.15: Al Lowry/Photo Researchers, Inc.; Figure 15.16: C. Behnke/ Animals Animals/Earth Scenes; Figure 15.17: Konrad Wothe/ Minden Pictures, Inc.

Chapter 16

Figure 16.3: Photograph by Vincenzo Penteriani, Department of Conservation Biology, Estación Biológica de Doñana; Figure 16.5a: Tom Hince; Figure 16.5b: Photograph by Khoi Uong; Figure 16.6: Stimson Wilcox, Biological Sciences, Binghamton University; Figure 16.7a: ©tbkmedia.de/Alamy; Figure 16.7b: Kate Bemis; Figure 16.8: Pete Oxford/Minden Pictures, Inc.; Figure 16.9: Kate Bemis; Figure 16.10a: Courtesy Dietrich Schneider, Max–Planck Institute, Germany; Figure 16.10b: Courtesy Dietrich Schneider, Max–Planck Institute, Germany; Figure 16.11: Kate Bemis; Figure 16.13: Courtesy George W. Uetz, Department of Biological Sciences, University of Cincinnati; Figure 16.14: R. M. Meadows/Peter Arnold, Inc.; Figure 16.15: Time & Life Pictures/Getty Images, Inc.; Figure 16.17a: Courtesy Lynne D. Houck, Department of Zoology, Oregon State University, and S. J. Arnold, Department of Zoology, Oregon State University; Figure 16.19: Roy P. Fontaine/Photo Researchers; Figure 16.27: Mark Moffett/ Minden Pictures, Inc.

Chapter 17

Figure 17.3a: Tui De Roy/Minden Pictures, Inc.; Figure 17.3b: Alberto Velando; Figure 17.3c: Alberto Velando; Figure 17.4: Courtesy Joseph Waas, from Waas 1991; Figure 17.6b: Roy Caldwell, Department of Integrative Zoology, University of California at Berkeley; Figure 17.7: Courtesy John D. Palmer; Figure 17.9: Jen and Des Bartlett/Photo Researchers, Inc.; Figure 17.10: Toni Angermayer/Photo Researchers, Inc.; Figure 17.12: Phil Devries/Oxford Scientific Films Ltd.; Figure 17.14: Papilio/ Alamy; Figure 17.16: Molly Cummings, Division of Integrative Biology, University of Texas at Austin; Figure 17.18: Susan

Kuklin/Photo Researchers, Inc.; Figure 17.21: Courtesy Georgia State University & Yerkes/Emory; Figure 17.22: Courtesy Georgia State University & Yerkes/Emory; Figure 17.24: Jeffrey Oonk/Foto Natura/Minden Pictures, Inc.

Chapter 18

Figure 18.1: Hiroya Minakuchi/Minden Pictures, Inc.; Figure 18.2: Doug Weschler/Animals Animals/Earth Scenes; Figure 18.3: Frank Lane Picture Agency; Figure 18.5: Clem Haagner Photo Researchers, Inc.; Figure 18.6: Figure after Fretwell 1972, and Krebs and Davies 1993; Figure 18.9a: ©Brad Hamel/Pbase.

Chapter 19

Figure 19.1: Source: de Waal, F. B. M. 2005. A century of getting to know the chimpanzee. *Nature* 437:56–59, fig. 2; Figure 19.4: Courtesy W. E. Bemis, Department of Ecology and Evolutionary Biology, Cornell University; Figure 19.5: Photograph by Bill Romey, Department of Biology, SUNY Potsdam; Figure 19.6: Leonard Lee Rue, III/Animals Animals/Earth Scenes; Figure 19.8: terrygray; Figure 19.10: Richard R. Hansen/Photo Researchers; Figure 19.12: Ian Cleghorn/Photo Researchers, Inc.; Figure 19.13a: Image from: Grinnell, J., C. Packer, and A. E. Pusey. Cooperation in male lions: Kinship, reciprocity, or mutualism. *Animal Behaviour* 49:95–105; Figure 19.13b: Image from: Grinnell, J., C. Packer, and A. E. Pusey, 1995. Cooperation in male lions: kinship, reciprocity, or mutualism. *Animal Behaviour* 49:95–105; Figure 19.15: Leanne T. Nash; Figure 19.16: Steve & Dave Maslowski/Photo Researchers, Inc.; Figure 19.17a: ©Biosphoto/Fouquet Franck/Peter Arnold Inc.; Figure 19.17b: Werner Bollmann/Photolibrary Group Limited; Figure 19.18: ©Joe McDonald/©Corbis–Bettmann; Figure 19.20: ©Biosphoto/Dennis Nigel J. /Peter Arnold, Inc.; Figure 19.21: M. K. & I. M. Morcombe/© Natural History Photographic Agency; Figure 19.22: Komdeur, J. 1996. Influence of helping and breeding experience on reproductive performance in the Seychelles warbler: A translocation experiment. *Behavioral Ecology* 7:326–333; Figure 19.24: Raymond A. Mendez/Animals Animals/Earth Scenes.

Permission Credits

Chapter 2

Table 2.1 Fraser, D., and D. M. Weary. 2005. Applied animal behavior and animal welfare. In *The Behavior of Animals: Mechanisms, Function, and Evolution*, edited by J. J. Bolhuis and L.-A. Giraldeau. Wiley-Blackwell Publishing. Fig 2.2 Romanes, G. J. 1889. *Mental Evolution in Man: Origin of Human Faculty*. Appleton and Lange. Fig. 2.4 Kessel, E. L. 1955. *Systematic Zoology* 4. Society of Systematic Zoology. Fig. 2.5 Lorenz, K., and N. Tinbergen. 1938. *Zeitschrift für Tierpsychologie* 2, Paul Parey Scientific Publications. Fig. 2.6 Tinbergen, 1989. Courtship ritual of stickleback. *The Study of Instinct*. By permission of Oxford University Press. Fig. 2.7 Thorndike, E. L. 1911. *Animal Intelligence: Experimental Studies*. Macmillan Publishing Company.

Chapter 3

Table 3.1 Modified from Tan, E. J., and B. L. Tang. 2006. Looking for food: Molecular neuroethology of invertebrate feeding behavior. *Ethology* 112:826–832. Fig. 3.3 Miklósi, A., V. Csányi, and R. Gerlai. 1997. *Behavior Genetics* 27(3). Plenum Publishing. Fig. 3.5 Lynch, C. B. 1980. Response to divergent selection for nesting behavior in *Mus musculus*. *Genetics* 96:757–765. Fig. 3.6 de Belle, J. S., and M. B. Sokolowski. 1987. Heredity of *rover/sitter*: Alternative foraging strategies of *Drosophila melanogaster* larvae. *Heredity* 59:73–83. Fig. 3.7 de Belle, J. S., and M. B. Sokolowski. 1987. *Heredity* 59. Blackwell Science Ltd. Fig. 3.8 Fitzpatrick, M. J., Y. Ben-Shahar, H. M. Smid, L. E. M. Vet, G. E. Robinson, and M. B. Sokolowski. 2005. Candidate genes for behavioral ecology. *Trends in Ecology and Evolution* 20 (2):96–104. Fig. 3.9 Modified from Tan, E. J., and B. L. Tang. 2006. Looking for food: Molecular neuroethology of invertebrate feeding behavior. *Ethology* 112:826–832. Wiley-Blackwell Publishing. Fig. 3.10b–c From Ben-Shahar, Y., A. Robichon, M. B. Sokolowski, and G. E. Robinson. 2002. Influence of gene action across different time scales on behavior. *Science* 296:741–744. Reprinted with permission from AAAS. Fig. 3.11 Modified from Hunt, G. J., G. V. Amdam, D. Schlipalius, C. Emore, N. Sardesai, C. E. Williams, O. Rueppell, E. Guzmán-Novoa, M. Arechavaleta-Velasco, S. B. C. Chandra, M. K. Fondrk, M. Beye, and R. E. Page, Jr. 2007. Behavioral genomics of honeybee foraging and nest defense. *Naturwissenschaften* 94:247–267, Figure 1. Copyright 2007. With kind permission of Springer Science and Business Media. Fig. 3.12 From Whitfield, C. W., A. M. Cziko, and G. E. Robinson. 2003. Gene expression profiles in the brain predict behavior in individual honey bees. *Science* 302:296–299. Reprinted with permission from AAAS. Fig 3.13 Nelson, C. M., et al. 2007. The gene *vitellogenin* has multiple coordinating effects on social organization. *PLoS Biology* 5 (3):0673–0677. Fig. 3.14 Amholt, R. R. H. 2004. Interactions among genetic modules. Genetic modules and networks for behavior: Lessons from Drosophila. *BioEssay* 26 (12):1299–1306. Reprinted with permission of Wiley-Liss, Inc., a subsidiary of John Wiley & Sons, Inc. Fig. 3.15 Choleris, E., J. Å. Gustafson, K. S. Korach, J. J. Muglia, and S. Ogawa. An estrogen dependent micronet mediating social recognition. A study with oxytocin-and estrogen receptor aplha- and ß-knockout mice. *Proceedings of the National Academy of Sciences* USA 100:6192– 6197. Copyright (2003) National Academy of Sciences, U.S.A. Fig. 3.16c–d Lim, M. Zuoxin, W., Olazabal, D. E., Ren, X., Terwilliger, E., et al. 2004. Enhanced partner preference in a promiscuous species by manipulating the expression of a single gene. *Nature* 429:754–757. Copyright 2004. Reprinted by permission from Macmillan Publishers Ltd. Fig. 3.17 Based on Robinson, G. E. 2004. Beyond nature and nurture. *Science* 304:397–399. Fig. 3.18c–f Burmeister, S. S., E. D. Jarvis, and R. D. Fernald. 2005. Rapid behavioral and genomic responses to social opportunity. *PLoS Biology* 3:1996–2004. Fig. 3.19 Mello, C., D. S. Vicario, and D. F. Clayton. 1992. Song presentation induces gene expression in the songbird forebrain. *Proceedings of the National Academy of Sciences* 89:6818–6822. Fig. 3.20 *Journal of Neuroscience* 15, Mello, C., and F. Nottebohm, Society for Neuroscience, 1995. Fig. 3.21 Aubin-Horth, N., et al. 2005. Alternative life histories shape brain gene expression profiles in males of the same population. *Proceedings of the Royal Academy of Sciences* 272: 1655–1662. Figure 3, page 1659. The Royal Society. Fig. 3.22 From Robinson, G., et al. 2008. Genes and social behavior. *Science*. 322:896–900. Figure 1, page 897. Reprinted with permission from AAAS.

Chapter 4

Fig. 4.11 Brockmann J., A. Grafen and R. Dawkins, *Journal of Theoretical Biology* 77, Academic Press Ltd., 1979. Fig. 4.14 Modified from Rosenheim, J. A. 1993. Comparative and experimental approaches to understanding insect learning. In *Insect Learning: Ecological and Evolutionary Perspectives*, edited by D. R. Papaj and A. C. Lewis. New York: Chapman and Hall, with kind permission of Springer Science and Business Media.

Chapter 5

Fig. 5.1 Reprinted from Clark, R. B. 1960. Habituation of the polychaete Nereis to sudden stimuli. 1. General properties of the habituation process. *Animal Behaviour* 8 (1–2):87. Copyright 1960, by permission of the publisher Academic Press. Fig. 5.2 Hollis, K. L., V. L. Pharr, M. J. Dumas, G. B. Britton, and J. Field. 1997. Classical conditioning provides paternity advantage for territorial male blue gouramis (*Trichogaster trichopterus*). *Journal of Comparative Psychology* 111:219–225. Fig. 5.3 Reprinted from Lyons, C., and C. J. Barnard. 2006. A learned response to sperm competition in the field cricket, *Grullus bimaculatus* (de Geer). *Animal Behaviour* 72:673–680. With permission from Elsevier. Fig. 5.5 Franks, N. R., J. W. Hooper, A. Dornhaus, P. J. Aukett, A. L. Hayward, and S. M Berghoff. 2007. Reconnaissance and latent learning in ants. *Proceedings of the Royal Society of London B* 274:1505–1509. Fig. 5.9 Balda, R. P., and A. C. Kamil. 1989. A comparative study of cache recovery by three corvid species. *Animal Behaviour* 38:486–495. Fig. 5.10 Olson, D. J., A. C. Kamil, R. P. Balda, and P. J. Nims. Performance of four seed-caching corvid species in operant test of nonspatial and spatial memory. *Journal of Comparative Psychology* 109:173–181. Copyright 1995 by the American Psychological Association. Adapted with permission. Fig. 5.13 Shettleworth, S. J. 1998. *Cognition, Evolution and Behavior*. Oxford University Press. Fig. 5.15 Reprinted from Tarsitano, M. S., and A. R. Andrew. 1999. Scanning and route selection in the jumping spider *Portia labiata*. *Animal Behaviour* 58:255–265, Copyright 1999, with permission from Elsevier.

Chapter 6

Fig. 6.2 Camhi, J. M. 1980. The escape system of the cockroach. *Scientific American* 248:158–172. Fig. 6.2 Ritzmann, R. E., and A. J. Pollack. 1986. Identification of thoracic interneurons that mediate giant interneuron-to motor pathways. *Journal of Comparative Physiology A* 159:639–654. Fig 6.10 Alcock, J. 1975. *Animal Behavior*. Sinauer Associates. Fig. 6.13 From Knudsen, E. I., and M. Konishi, 1978. A neural map of auditory space in the owl. *Science* 200:795–797, Copyright 1978. Reprinted with permission from AAAS. Fig. 6.18 From Kandel, E. R. 2001. The molecular biology of memory storage: A dialogue between genes and synapses. *Science* 294:1030–1038. Reprinted with permission from AAAS. Fig. 6.20 Wilbrecht, L., A. Crionas, and F. Nottebohm. 2002. Experience affects recruitment of new neurons but not adult neuron number. *The Journal of Neuroscience* 22:825–831. Fig. 6.21 Cameron, H. A., and R. D. G. McKay. 2001. Adult neurogenesis produces a large pool of new granule cells in the dentate gyrus. *Journal of Comparative Neurology* 435:406–417. Figure 6, page 413. Reprinted with permission of Wiley-Liss, Inc., a subsidiary of John Wiley & Sons, Inc. Fig. 6.22 Reprinted by permission from Macmillan Publishers Ltd.: *Nature*. Shors, T. J., G. Miesegaes, A. Beylin, M. Zhao, T. Rydel, and E. Gould. Neurogenesis in the adult is involved in the formation of trace memories. *Nature* 410:372–376. Copyright 2001. Fig. 6.23 Dupret, D., A. Fabre, M. D. Dobrossy, A. Panatier, J. J. Rodríguez, S. Lamarque, V. Lemaire, S. H. R. Oliet, P.-V. Pliazza, and D. N. Abrous. 2007. Spatial learning depends on both the addition and removal of new hippocampal neurons. *PLoS Biology* 5:1683–1694. Fig. 6.24 Maguire, E. A., D. G. Gadlian, I. S. Johnsrude, C. D. Good, J. Ashburner, R. S. J. Frackowiak, and C. D. Frith. Navigation-related structural change in the hippocampi of taxi drivers. *Proceedings of the National Academy of Sciences USA* 97:4398–4403. Copyright (2000) National Academy of Sciences, U.S.A. Fig. 6.25 Newman, S. W. 1999. The medial extended amygdala in male reproductive behavior: a node in the mammalian social behavior network. *Annals of the New York Academy of Sciences* 877:742–257. Wiley-Blackwell Publishers. Fig. 6.26 Goodson, J. L., A. K. Evans, L. Lindberg, and C. D. Allen. 2005. Neuro-evolutionary patterning of sociality. *Proceedings of the Royal Society of London B* 272:227–235. Fig. 6.28a Weis-Fogh, T. 1956. *Philosophical Transactions of the Royal Society of London (Series B)*. The Royal Society. Fig. 6.28a Horsmann, U., H. G. Heinzel, and G. Wendler. 1983. The phasic influence of self-generated air current modulations on the locust flight motor. *Journal of Comparative Physiology* 150:427–438. Fig. 6.28b–c Wilson, D. M., and T. Weis-Fogh. 1962. *Journal of Experimental Biology* 39. Company of Biologists, Ltd.

Chapter 7

Fig. 7.1 Bass, A. H. 1996. Shaping brain sexuality. *American Scientist* 84:352–363. Fig. 7.6 Kelley, D. B., and D. I. Gorlick. 1990. Sexual selection and the nervous system. *BioScience* 40:275–283. Fig. 7.7 Adapted from M. Seney, B., D. Goldman, and N. G. Forger. 2006. Breeding status affects motoneuron number and muscle size in naked mole-rats: Recruitment of perineal motoneurons. *Journal of Neurobiology*, 66:1354–1364. (2006). Reprinted with permission of Wiley-Liss, Inc., a subsidiary of John Wiley & Sons, Inc. Fig. 7.8 Crews, D. 1979. The hormonal control of behavior in a lizard. *Scientific American* 241:180–187. Fig. 7.9 Crews, D. 1979. *Biology of Reproduction* 20. Society for the Study of Reproduction. Fig. 7.10a–b Schneider, J. S., M. K. Stone, K. E. Wynne-Edwards, T. H. Horton, J. Lydon, B. O'Malley, and J. E. Levine. Progesterone receptors mediate male aggression toward infants. *Proceedings of the National Academy of Sciences* 100:2951–2956. Copyright 2003 National Academy of Sciences, U.S.A. Fig. 7.12 Wingfield, J. C. 1984. *General and Comparative Endocrinology* 56. Academic Press, Inc. Fig. 7.15 Orchinik, M. P., Licht, and D. Crews. 1988. Plasma steroid concentrations change in response to sexual behavior in *Bufo marinus*. *Hormones and Behavior* 22:338–350. Academic Press, Inc. 1988. Fig. 7.16 Crews, D., ed. 1987. *Psychobiology of Reproductive Behavior: An Evolutionary Perspective*. Prentice-Hall, p. 92. Copyright 1987, Reprinted by permission of Prentice-Hall, Inc., Englewood Cliffs, NJ. Fig. 7.18 Dzieweczynski, T. L., R. L. Earley, T. M. Green, and W. J. Rowland. 2005. Audience effect is context dependent in Siamese fighting fish, *Betta splendens*. *Behavioral Ecology* 16:1025–1030, by permission of Oxford University Press. Fig. 7.19 Reprinted from Dzieweczynski, T. L., A. C. Eklund, and W. J. Rowland. 2006. Male 11-ketotestosterone levels change as a result of being watched in Siamese fighting fish, *Betta splendens*. *General and Comparative Endocrinology* 147:184–189. Copyright 2006, with permission from Elsevier. Fig. 7.20 Reprinted from Schoech, S. J., R. L. Mumme, and J. C. Wingfield. 1996. Prolactin and helping behaviour in the cooperatively breeding Florida scrub-jay, *Apheloma e. coerulesens*. *Animal Behaviour* 52 (3).

Chapter 8

Table 8.1 Gottlieb, G. *Journal of Comparative and Physiological Psychology*, Copyright 1978 by the American Psychological Association. Adapted with permission. Table 8.2 Gottlieb, G. 1985. *Animal Behavior* 33, Academic Press Ltd. Table 8.3 Rauschecker, J. R., and P. Marler. 1987. *Imprinting and Cortical Plasticity: Comparative Aspects of Sensitive Periods*. Reprinted by permission of John Wiley & Sons, Inc. Fig. 8.2 *Journal of Zoology London* 149, Abu-Gideiri, Y. B., Reprinted with the permission of Cambridge University Press, 1966, *Animal Behaviour*, Dill, 1977. Fig. 8.2 *Animal Behaviour*, Abu-Gideiri, Y. B., Academic Press Ltd., 1966., *Animal Behaviour*, Dill, 1977. Fig. 8.4 Weeks, J. C., and J. W. Truman. 1984. *Journal of Comparative Physiology* 155. Springer-Verlag. Fig. 8.6 McLaren, A., and D. Michie. 1960. Control of prenatal growth in mammals. *Nature* 187:363–365. Fig. 8.7 Crews, D., and T. Groothuis. Tinbergen's fourth question, ontogeny: Sexual and individual differentiation. *Animal Biology* 55:343–370. Copyright 2005 by Koninklijke Brill NV Leiden. Fig. 8.11 Copyright © 1965 by the American Psychological Association. Adapted with permission. Gottlieb, G. Imprinting in relation to parental and species identification by avian neonates. *Journal of Comparative and Physiological Psychology* 59: 345–356. Fig. 8.12 Bateson, P. 1982. *Nature* 295. Reprinted by permission from Macmillan Magazines Ltd. Fig. 8.13 Reprinted from Kendrick, et al. 2001. Sex differences in the influence of mothers on the sociosexual preferences of their offspring. *Hormones and Behavior* 40:322–338. Copyright 2001, with permission from

Elsevier. Fig. 8.15 Reprinted from Champalbert, A., and J. Lachaud. 1990. Existence of a sensitive period during the ontogenesis of social behavior in a primitive ant. *Animal Behaviour* 39:850–859. Copyright 1990, by permission of the publisher Academic Press Ltd. 8.17a Adapted from Bolhuis, J. J. 2008. Chasin' the trace: The neural substrate of birdsong memory. In *The Neuroscience of Birdsong*, edited by H. P. Zeigler and P. Marler. Cambridge University Press. Reprinted with the permission of Cambridge University Press. 8.17b From Nottebohm and Arnold. Sexual dimorphism in vocal control areas of the songbird brain. *Science* 194. Copyright 1976. Reprinted with permission from AAAS. Fig. 8.18 From Gurney, M. E., and M. Konishi. Hormone-induced sexual differentiation of brain and behavior in zebra finches. *Science* 208. Copyright 1980. Reprinted with permission from AAAS. Fig. 8.20 Konishi, M. 1965. *Zeitschrift für Tierpsychologie* 22. Paul Parey Scientific Publications. Fig. 8.22b West, M. J., A. P. King, and D. H. Eastzer. 1981. The cowbird: Reflections on development from an unlikely source. *American Scientist* 69:56–66. Scientific Research Society, 1981. Fig. 8.23 King, A. P., and M. J. West. 1983. *Nature* 305. Reprinted by permission from Macmillan Magazines Ltd. Fig. 8.24 Reprinted from Reimers, et al. 2007. Rehabilitation of research chimpanzees: Stress and coping after long-term isolation. *Hormones and Behavior* 51: 428–435. Copyright 2007, with permission from Elsevier.

Chapter 9

Fig. 9.2 This figure was published in *The Biological Clock-Two Views* by F., A. Brown, J. W. Hastings, and J. D. Palmer, Copyright Academic Press 1970. Fig. 9.3 Palmer, J. D. 1990. The rhythmic lives of crabs: the same biological clock that controls circadian rhythms may also drive crustacean tidal rhythms. *BioScience* 40:352–358. Copyright 1990 American Institute of Biological Sciences. Fig. 9.5b Youthed, G. J., and R. C. Moran. 1969. The lunar day activity rhythm of myrmeleontid larvae. *Journal of Insect Physiology* 15:1259–1271. Reprinted with permission from Pergamon Press, Inc. Fig. 9.6 Gwinner, E. 1996. Circadian and cirannual programmes in avian migration. *Journal of Experimental Biology* 199:39–48. The Company of Biologists Ltd. Fig. 9.7 Welsh, D. K., D. E. Logothetis, M. Meister, and S. M. Reppert. 1995. Individual neurons dissociated from rat suprachiasmatic nucleus express independently phased circadian firing rhythms. *Neuron* 14:697–706. Reprinted with permission from Elsevier. Fig. 9.8c Beling, I. 1929. Über das Zeitgedächtnis der Bienen. *Zeitschrift für vergleichende Physiologie* 9:259–338. Fig. 9.9b Plautz, J. D., M. Kaneko, J. C. Hall, and S. A. Kay. 1997. Independent photoreceptive circadian clocks throughout *Drosophila*. Reprinted with permission from AAAS. Fig. 9.10 Adapted Reppert, S. M., and D. R. Weaver. 2002. Coordination of circadian timing in mammals. *Nature* 418:935–941. By permission from Macmillan Publishers, Ltd. Fig. 9.11 Inouye, S. T., and H. Kawamura. 1979. Persistence of circadian rhythmicity in mammalian hypothalamic "island" containing the suprachiasmatic nucleus. *Proceedings of the National Academy of Sciences* 76:5962–5966. Fig. 9.12 Based on Breedlove, S. M., M. R. Rosenweig, and N. V. Watson. 2007. *Biological Psychology: An Introduction to Behavioral, Cognitive, and Clinical Neuroscience*. 5th ed. Sunderland MA: Sinauer. Fig. 9.12 Based on Reppert, S. M., and D. R. Weaver. 2002. Coordination of circadian timing in mammals. *Nature* 418:935–941. Fig. 9.13 Reprinted from Herzog, E. D., and G. Tosini. 2001. The mammalian circadian clockshop. *Seminars in Cell & Developmental Biology* 12:295–303. With permission from Elsevier. Fig. 9.14b Silver, R., J. LeSauter, P.A., Tresco, and M. N. Lehman. 1996. A diffusible coupling signal from the transplanted suprachiasmatic nucleus controlling circadian locomotor rhythms. *Nature* 382:810–813. Reprinted by permission of Macmillan Magazines Ltd. Fig. 9.15 Reprinted from Aschoff, J. 1967. *Life Sciences and Space Research*. North Holland Publishers; Elsevier Science Publishing Co., Inc., with permission from Elsevier Science.

Chapter 10

Tables 10.1–10.2 Wiltschko, R., and W. Wiltschko. 2006. Magnetoreception. *BioEssays* 28:157–168. Reprinted with permission of Wiley-Liss, Inc., a subsidiary of John Wiley & Sons, Inc. Fig. 10.2 Perdeck, A. C. 1958. Two types of orientation in migrating starlings, *Sturnus vulgaris* L., and chaffinches, *Fringilla coelebs* L., as revealed by displacement experiments. *Ardea* 46:1–37. Netherlands Ornithological Union, 1958. Fig. 10.10 Palmer, J. D., A. G. Gilbert, and The Bettmann Archive. 1966. *Natural History* 75. Fig. 10.13 *Quarterly Review of Biology* 62, Able and Bingman, The University of Chicago Press, 1987, modified from *Science* 170, Emlen, S. T., American Association for the Advancement of Science, 1970. Fig. 10.18 Wiltschko, W., and R. Wiltschko. 2005. Magnetic orientation and magnetoreception in birds and other animals. *Journal of Comparative Physiology A* 191:675–693, With kind permission from Springer Science+Business Media. Fig. 10.18 Data: S. Marhold. 1997. A magnetic polarity compass for direction finding in a subterranean mamma. *Naturwissenschaften* 84:421–423. With kind permission from Springer Science+Business Media. Fig 10.19 Walcott, C., and R. P. Green. 1974. *Science* 184. American Association for the Advancement of Science. Fig 10.21b Lohmann, K. J. 1991. *Journal of Experimental Biology* 155. The Company of Biologists Ltd. Fig. 10.22 Data from Beck, W., and W. Wiltschcko, 1988. Magnetic factors control the migratory direction of pied flycatchers (*Ficedula hypoleuca* Pallas) *Acta Congress International Ornithology* 19:1955–1962. In Lohmann, K. J., C.M.F. Lohmann, and N. F. Putman, 2007. Magnetic maps in animals: nature's GPS. *Journal of Experimental Biology*, 210:3697–3705. Fig. 10.23 From Lohmann, K. J., S. D. Cain, S. A. Dodge, and C. M. F. Lohmann. 2001. Regional magnetic fields as navigational markers for sea turtles. *Science* 294:364–366. Reprinted with permission from AAAS. Fig. 10.24 Gould, J. L. 1980. *American Science* 68. Sigma Xi Scientific Research Society. Fig 10.25 Lohmann, K. J., C. M. F. Lohmann, L. M. Ehrhart, D. A. Bagley, and T. Swing. 2004. Geomagnetic map used in sea-turtle navigation. *Nature* 428:909–910. Copyright 2004. Reprinted by permission from Macmillan Publishers Ltd: *Nature*. Fig 10.26 Ritz, T., S. Adem, and K. Schulten. 2000. A model for photoreceptor-based magnetoreception in birds. *Biophysical Journal* 78:707–718. Reprinted with permission from Elsevier. Fig. 10.27 Stewart, I. J., S. M. Carlson, C. P. Boatright, G. B. Buck, and T. P. Quinn. 2004. Site fidelity of spawning sockeye salmon (*Oncorhynchus nerka W.*) in the presence and absence of olfactory cues. *Ecology of Freshwater Fish* 13:104–110. Wiley-Blackwell Publishers. Fig 10.28 Baldaccini, E. N., S. Benvenuti, V. Fiaschi, and F. Papi. 1975. *Journal of Comparative Physiology* 99. Springer-Verlag. Fig 10.29 Ioale, P., M. Nozzolini, and F. Papi. 1990. *Behavioral Ecology and Sociobiology*. Springer-Verlag. Fig. 10.30 Adapted with permission. Gagliardo, A., P. Ioalè, M. Savini, and J. M. Wild. 2006. Having the nerve to home: trigeminal magnetoreceptor *versus* olfactory mediation of homing in pigeons. *Journal of Experimental Biology* 209:2888– 2892. Figs. 10.31–10.33 Von der Emde, G. 1999. *Journal of Experimental Biology*. The Company of Biologists Ltd.

Chapter 11

Table 11.1 Greenwood, P. J. 1980. Mating systems, philopatry, and dispersal in birds and mammals. *Animal Behavior* 28:1140–1162. Reprinted by permission of the publisher, Academic Press London. Table 11.2 Dobson, F. S. 1982. Competition for mates and predominant juvenile male dispersal in mammals. *Animal Behavior* 30:1183–1192. Reprinted by permission of the publisher, Academic Press London. Fig 11.2 Bonte, D., S. Van Belle, and J.-P. Maelfait. 2007. Maternal care and reproductive state-dependent mobility determine natal dispersal in a wolf spider. *Animal Behaviour* 74:63–69. Reprinted with permission from Elsevier. Fig 11.3 Höner, O. P., B. Wachter, M. L. East, W. J. Streich, K. Wilhelm, T. Burke, and H. Hofer. 2007. Female mate-choice drives the evolution of male-biased dispersal in a social mammal. *Nature* 448:798–802. Reprinted by permission from Macmillan Publishers Ltd: *Nature*. Fig. 11.4 Devillard, S., D. Allaine, J. M. Gaillard, and D. Pontier. 2004. Does social complexity lead to sex-biased dispersal in polygynous mammals? *Behavioral Ecology* 15:83–87, by permission of Oxford University Press. Fig. 11.6 Forsman, J. T., J.-T. Seppänen, and M. Mönkkönen. 2002. Positive fitness consequences of interspecific interaction with a potential competitor. *Proceedings of the Royal Society of London B* 269:1619–1623. The Royal Society. Fig 11.10 Strandberg, R., and T. Alerstam. 2007. The strategy of fly-and-forage migration, illustrated for the osprey (*Pandion haliaetus*). *Behavioral Ecology and Sociobiology* 61:1865–1875. With kind permission from Springer Science and Business Media. Fig 11.13 Anderson, J. B., and L. P. Brower. 1996. *Ecological Entomology* 21 (2). Blackwell Scientific Ltd.

Chapter 12

Table 12.1 Acknowledgement to the original publication. Zach, R. 1978, *Behaviour*. 67, E. J. Brill. Fig. 12.1 Brusca, R. C., and G. J. Brusca. 1990. *Invertebrates*. Sunderland, MA: Sinauer Associates. Fig. 12.2 Gurd, D. B. 2007. Predicting resource partitioning and community organization of filter-feeding dabbling ducks from functional morphology. *American Naturalist* 169:334–343. The University of Chicago Press. Fig. 12.5 Olberg, R. M., A. H. Worthington, and K. R. Venator. 2000. Prey pursuit and interception in dragonflies. *Journal of Comparative Physiology A* 186:155–162. With kind permission of Springer Science and Business Media. Fig. 12.11 Reprinted from Watt, M., C. S. Evas, and J. M. P. Joss. 1999. Use of electroreception during foraging by the Australian lungfish. *Animal Behaviour* 58:1039–1045. Reprinted by permission of the publisher, Academic Press Ltd. Fig. 12.14 Pietrewics, A. T., and A. C. Kamil. 1981. *Foraging Behavior: Ecological, Ethological, and Psychological Approaches*. Garland STPM Press. Fig. 12.19 *Behaviour* 68, Zach, R., E. J. Brill Leiden, 1979. Fig. 12.20b Cowlishaw, G. 1997. Trade-offs between foraging and predation risk determine habitat use in a desert baboon population. *Animal Behaviour* 53 (4):667–686. Reprinted by permission of the publisher, Academic Press Ltd.

Chapter 13

Table 13.1 Croze, H. 1970. *Zeitschrift für Tierpsychologie Supplement* 5, Paul Parey Scientific Publications. Table 13.2 Dial, B. E., and L. C. Fitzpatrick. 1983. *Science* 219. Reprinted with permission from American Association for the Advancement of Science. Table 13.3 Modifed from Honma, A., S. Oku, and T. Nishida. 2006. Adaptive significance of death feigning posture as a specialized inducible defence against gape-limited predators. *Proceedings of the Royal Society of London B* 273:1631–1636. Table 13.4 Modified from Vallin, A., S. Jakobsson, J. Lind, and C. Wiklund. 2005. Prey survival by predator intimidation: an experimental study of peacock butterfly defence against blue tits. *Proceedings of the Royal Society of London B* 272:1203–1207. Table 13.5 Caro, T. M. 1986a. The functions of stotting: A review of hypotheses. *Animal Behaviour* 34:649–662. Table 13.6 Caro, T. M. 1986b. The functions of stotting in Thomson's gazelles: Some tests of the predictions. *Animal Behaviour* 34:663–684. Reprinted from *Animal Behaviour*, Caro, 1986, by permission of the publisher Academic Press London. Figs. 13.3–13.4 S. Merilaita, J. Lind. 2005. Background-matching and disruptive coloration, and the evolution of cryptic coloration. *Proceedings of the Royal Society of London B* 272:665–670. The Royal Society. Fig. 13.5 Reprinted from Feltmate, B. W., and D. D. Williams. 1989. A test of crypsis and predator avoidance in the stonefly *Paragnetina media* (Plecoptera: Perlidae). *Animal Behaviour* 37:992–999. Figs. 13.7– 13.8 Cuthill, I. C., M. Stevens, J. Sheppard, T. Maddocks, C. Alejandro Párraga, and T. S. Troscianko. 2005. Disruptive coloration and background pattern matching. *Nature* 434:72–74. Copyright 2005. Reprinted by permission from Macmillan Publishers Ltd: *Nature*. Fig. 13.9 Cott, H. B. 1940. *Adaptive Colouration in Animals*. Methuen, London, London. Fig. 13.12 Endler, J. A. 1978. *Evolutionary Biology* 11:346–348, 354. With kind permission of Springer Science and Business Media. Figs. 13.13–13.14 Kiltie, R. A. 1989. *Journal of Mammalogy* 70. American Society of Mammalogists. Fig. 13.15 Moment, G. 1962. Reflexive selection: A possible answer to an old question. Science 136:262–263. Reprinted with permission from American Association for the Advancement of Science. Fig. 13.17 Terrick, T. D., R. L. Mumme, and G. M. Burghardt. 1995. Aposematic coloration enhances chemosensory recognition of noxious prey in the garter snake. *Animal Behaviour* 49 (4):857–866. By permission of the publisher, Academic Press Ltd. Fig. 13.19 Wickler, W. 1968. *Mimicry in Plants and Animals*. McGraw-Hill, New York. Fig. 13.21 Burghardt and Greene. 1988. Predator simulation and duration of death feigning in neonate hognose snakes. *Animal Behaviour* 36:1842–1844. By permission of the publisher, Academic Press London. Fig. 13.22 Johnsgard, P. A. 1967. *Animal Behavior*. William C. Brown Publishers. Fig. 13.24 Vallin, A., S. Jakobsson, J. Lind, and C. Wiklund. 2005. Prey survival by predator intimidation: an experimental study of peacock butterfly defence against blue tits. *Proceedings of the Royal Society of London B* 272:1203–1207. The Royal Society. Fig. 13.27 Reprinted from Hews, D. K. 1988. Alarm response in larval western toads, *Bufo boreas*: Release of larval chemical by a natural predator and its effect on predator capture efficiency. *Animal Behaviour* 36:125–133. By permission of the publisher, Academic Press London. Fig. 13.28 Turchin, P., and P. Kareiva. 1989. *Ecology* 70. Ecological Society of America. Fig. 13.29 Neill, S. R. St. J., and J. M. Cullen. 1974. *Journal of Zoology* 172. Reprinted with the permission of Cambridge University Press.

Chapter 14

Table 14.1 Modified from Schütz, D., and M. Taborsky. 2005. The influence of sexual selection and ecological constraints on an extreme sexual size dimorphism in a cichlid. *Animal Behaviour* 70:539–549. Table 14.2 Fisher, A. E. 1962. *Journal of*

Comparative and Physiological Psychology 55. American Psychological Association. Table 14.3 Modified from Holman, L., and R. R. Snook. 2006. Spermicide, cryptic female choice and the evolution of sperm form and function. *Journal of Evolutionary Biology* 19:1660–1670. Table 14.4 Modified from Ebensperger, L. A., and D. T. Blumstein. 2007. Nonparental infanticide. In *Rodent Societies: An Ecological and Evolutionary Perspective*, edited by J. O. Wolff, and P. W. Sherman. Chicago: The University of Chicago Press, 267–279. Fig. 14.2 Allen, B. J., and J. S. Levinton. 2007. Costs of bearing a sexually selected ornamental weapon in a fiddler crab. *Functional Ecology* 21:154–161, Blackwell Publishing. Fig. 14.4 Rooney, J., and S. M. Lewis. 2002. Fitness advantage from nuptial gifts in female fireflies. *Ecological Entomology* 27:373–377, Wiley-Blackwell Publishing. Fig. 14.7 Willis, P. M., and L. M. Dill. 2007. Mate guarding in male Dall's porpoises (*Phocoenoides dalli*). *Ethology* 113:587–597, Blackwell Publishing. Fig. 14.13 Sato, T., and S. Goshima. 2007. Female choice in response to risk of sperm limitation by the stone crab, *Hapalogaster dentata*. *Animal Behaviour* 73. Reprinted with permission from Elsevier. Fig. 14.14 Eggert, A.-K., and S. K. Sakaluk. 1994. Sexual cannibalism and its relation to male mating success in sagebrush crickets, *Cyphoderris strepitans* (Haglidae: Orthoptera). *Animal Behaviour* 47(5), Pages 1171–1177. Reprinted with permission of the publisher, Academic Press Ltd. Fig. 14.16b Takahashi, D., and M. Kohda. 2004. Courtship in fast water currents by a male stream goby (*Rhinogobius brunneus*) communicates the parental quality honestly. *Behavioral Ecology and Sociobiology* 55:431–438. Reprinted with kind permission from Springer Science and Business Media. Fig. 14.19 Byers, J. A., J. D. Moodie, and N. Hall. Pronghorn females choose vigorous mates. *Animal Behaviour* 47:33–43. Reprinted with permission from Elsevier. Fig. 14.21 Jones, T. M., R. J. Quinnell, and A. Balmford. 1998. Fisherian flies: Benefits of female choice in a lekking sandfly. *Proceedings of the Royal Society of London B* 265:1651–1657. The Royal Society. Fig. 14.22 Fedina, T. Y. 2007. Cryptic female choice during spermatophore transfer in *Tribolium castaneum* (Coleoptera: Tenebrionidae). *Journal of Insect Physiology* 53:93–98. Reprinted with permission from Elsevier.

Chapter 15

Table 15.1 Modified from Reynolds, J. D., N. B. Goodwin, and R. P. Freckleton. 2002. Evolutionary transitions in parental care and live bearing in vertebrates. *Philosophical Transactions of the Royal Society of London B* 357:269–281. Table 15.1. Cockburn, A. 2006. Prevalence of different modes of parental care in birds. *Proceedings of the Royal Society of London B* 273:1375–1383. Table 15.2 Gubernick, D. J., S. L. Wright, and R. E. Brown. 1993. The significance of father's presence for offspring survival in the monogamous California mouse, *Peromyscus californicus*. *Animal Behaviour* 46:539–546. Table 15.2 Gubernick, D. J., and T. Teferi. 2000. Adaptive significance of male parental care in a monogamous mammal. *Proceedings of the Royal Society of London B* 267:147–150. Fig. 15.1 Rickard, I. J., A. F. Russell, and V. Lummaa. 2007. Producing sons reduces lifetime reproductive success of subsequent offspring in pre-industrial Finns. *Proceedings of the Royal Society of London B* 274:2981–2988. The Royal Society. Fig. 15.3 Neff, B. D. 2003. Decisions about parental care in response to perceived paternity. *Nature* 422:716–719. Reprinted by permission from Macmillan Publishers Ltd: *Nature*. Copyright 2003. 15.4a Magrath, M. J. L., E. van Lieshout, I. Pen, G. Henk Visser, and J. Komdeur. 2007. Estimating expenditure on male and female offspring in a sexually size-dimorphic bird: A comparison of different methods. *Journal of Animal Ecology* 76:1169–1180, British Ecological Society, Wiley-Blackwell Publishing. Fig. 15.4b Magrath, M. J. L., E. van Lieshout, G. Henk Visser, and J. Komdeur. 2004. Nutritional bias as a new mode of adjusting sex allocation. *Proceedings of the Royal Society of London B* 271: S347–S349. The Royal Society. Fig. 15.6 Mank, J. E., D. E. L. Promislow, and J. C. Avise. 2005. Phylogenetic perspectives in the evolution of parental care in ray-fined fishes. *Evolution* 59:1570–1578, Wiley-Blackwell Publishers. Fig. 15.10 Blackmore, C. J. and R. Heinsohn. 2008. Variable mating strategies and incest avoidance in cooperatively breeding grey-crowned babblers. *Animal Behaviour* 75:63–70. Reprinted with permission from Elsevier. Fig. 15.11 Féron, C., and P. Gouat. 2007. Paternal care in the mound-building mouse reduces inter-litter intervals. *Reproduction, Fertility, and Development* 19:425–429. CSIRO Publishing, http://www.pulish.csiro.au/ nid/45/3365.htm. Fig. 15.12b Baeza, J. A. 2008. Social monogamy in the shrimp *Pontonia margarita*, a symbiont of *Pinctada mazatlanica*, off the Pacific coast of Panama. *Marine Biology* 153:387–395. With kind permission from Springer Science+Business Media. Fig. 15.13 Orians, G. H. 1969. On the evolution of mating systems in birds and mammals. *American Naturalist* 103. The University of Chicago Press. Fig. 15.14 Huk, T., and W. Winkel. 2006. Polygyny and its fitness consequences for primary and secondary female pied flycatchers. *Proceedings of the Royal Society of London B* 273:1681–1688. The Royal Society. Fig. 15.18 From Mattila, H. R., and T. D. Seeley. 2007. Genetic diversity in honey bee colonies enhances productivity and fitness. *Science* 317:362–364. Reprinted with permission from AAAS.

Chapter 16

Fig. 16.1 Nakano, R., N. Skals, T. Takanashi, A. Surlykke, T. Koike, K. Yoshida, H. Maruyama, S. Tatsuki, and Y. Ishikawa. 2008. Moths produce extremely quiet ultrasonic courtship songs by rubbing specialized scales. *Proceedings of the National Academy of Sciences* 105:11812–11817. Copyright 2008 National Academy of Sciences, U.S.A. Fig. 16.2 Aidan Martin, R. 2007. A review of shark agonistic displays: Comparison of display features and implications for shark-human interactions. *Marine and Freshwater Behaviour and Physiology* 40:3–34. Taylor & Francis, reprinted by permission of the publisher (Taylor & Francis Ltd., http://www.tandf.co.uk/journals). Fig. 16.4 How, M. J., J. M. Hemmi, J. Zeil, and R. Peters. 2008. Claw waving display changes with receiver distance in fiddler crabs, *Uca perplexa*. *Animal Behaviour* 75:1015–1022. Reprinted with permission from Elsevier. Fig. 16.12 Hopkins, C. 1974. *American Scientist* 62. Sigma Xi Scientific Research Society. Fig. 16.16 Bradbury, J. W., and S. L. Vehrencamp. 1998. *Principles of Animal Communication*. Sunderland, MA: Sinauer. Fig. 16.17 Houck and Reagan. 1990. Male courtship pheromones increase female receptivity in a plethodontid salamander. *Animal Behaviour* 39:729–734. Reprinted by permission of the publisher, Academic Press Ltd. Fig. 16.18 Bradbury, J. W., and S. L. Vehrencamp. 1998. *Principles of Animal Communication*. Sunderland MA: Sinauer. Based on Moynihan, M. 1966. Communication in *Callicebus*. *Journal of Zoology London* 150:77–127. Fig. 16.20 Marler, P., 1959. *Darwin's Biological Work*. Reprinted with the permission of Cambridge University Press. Fig. 16.22 Reprinted from von Frisch, K. 1950, 1971. *Bees: Their Vision, Chemical Senses, & Language*, revised edition. Copyright 1950, 1971, by Cornell University. Used by permission of the publisher, Cornell University Press. Figs.

16.23–16.24 Reprinted from von Frisch, K. 1950, 1971. *Bees: Their Vision, Chemical Senses, and Language*, revised edition. Copyright © 1950, 1971, by Cornell University. Used by permission of the publisher, Cornell University Press. Fig. 16.26 Riley, J. R., U. Greggers, A. D. Smith, D. R. Reynolds, and R. Menzel. 2005. The flight paths of honeybees recruited by the waggle dance. *Nature* 435:205–207. Reprinted with permission from Macmillan Publishers Ltd: *Nature*. Fig. 16.28 From Srinivasan, M. V., S. Zhang., M. Altwein, and J. Tautz. 2000. Honeybee navigation: Nature and calibration of the "odometer." *Science* 287:851–853. Reprinted with permission from AAAS.

Chapter 17

Fig. 17.1 Redondo, T., and F. Castro. 1992. Signaling of nutritional need by magpie nestlings. *Ethology* 92:193–204. Wiley-Blackwell Publishers. Fig. 17.5 Cheney D. L., and R. M. Seyfarth. 1988. Assessment of meaning and the detection of unreliable signals by vervet monkeys. Reprinted from Animal Behaviour 36:477–486. Reprinted by permission of the publisher, Academic Press London. Fig. 17.8 *The Study of Instinct*. Tinbergen, Oxford Clarendon, 1951. Willkinson, G. S. 1990. Food sharing in vampire bats. *Scientific American* 262:76–82. Fig. 17.11 Orians, G. H., and G. M. Christman. 1968. *Zoology* 84. University of California Press. Fig. 17.13 Hunter, M. L., and J. R. Krebs. 1979. *Journal of Animal Ecology* 48. Blackwell Scientific Publications Ltd. Fig. 17.14b–d Peters, R. A., J. M. Hemmi, and J. Zeil. 2007. Signaling against the wind: Modifying motion-signal structure in response to increased noise. *Current Biology* 17:1231–1234. Reprinted with permission from Elsevier. Fig. 17.15 Slabbekoorn, H., and A. den Boer-Visser. Cities change the songs of birds. *Current Biology* 16:2326–2331. Reprinted with permission from Elsevier. Fig. 17.17 Cummings, M. E. 2007. Sensory trade-offs predict signal divergence in surfperch. *Evolution* 61:530–545. Copyright 2007, Wiley-Blackwell Publishers. Fig. 17.23 Hauser, M. D. 1998. Functional referents and acoustic similarity: Field playback experiments with rhesus monkeys. *Animal Behaviour* 55 (6):1647–1658. Reprinted by permission of the publisher, American Press London.

Chapter 18

Fig. 18.4 Clutton-Brock, T. H., F. E. Guiness, and S. D. Albon, 1982. Red Deer: *The Behavior and Ecology of the Two Sexes*. The University of Chicago Press. Fig. 18.8 From Carpenter, F. L., D. C. Paton, and M. A. Hixon. 1983. Weight gain and adjustment of feeding territory size in migrant hummingbirds. *Proceedings of the National Academy of Sciences* 80:7259–7263. Fig. Figs. 18.9b–d and 18.10 Eason, P. K., G. A. Cobbs, and K. G. Trinca. 1999. The use of landmarks to define territorial boundaries. *Animal Behaviour* 58:85–91. Reprinted with permission from Elsevier. Fig. 18.11 Summers, C. H., and S. Winberg. 2006. Interactions between the neural regulation of stress and aggression. *Journal of Experimental Biology* 209:4581–4589. The Company of Biologists Ltd.

Chapter 19

Table 19.2 Sherman, P. W. 1985. *Behavioral Ecology and Sociobiology* 17. Springer-Verlag. Table 19.3 Reyer, H.-U. 1980. *Behavioral Ecology and Sociobiology* 6. Springer-Verlag. Fig. 19.1 de Waal, F. B. M. 2005. A century of getting to know the chimpanzee. *Nature* 437:56–59. Reprinted by permission from Macmillan Publishers Ltd: Nature. Fig. 19.2 From Bednarz, J. C. 1988. Cooperative Hunting Harris' Hawks (*Parabuteo unicinctus*). Science 239. Reprinted with permission from AAAS. Fig. 19.3 Weimerskirch, H., J. Martin, Y. Clerquin, P. Alexandre, and S. Jarskova. 2001. Energy saving in flight formation. Nature 413:697–698. Reprinted by permission from Macmillan Publishers Ltd: *Nature*. Fig. 19.9 From Wilkinson, G. S. 1990. Food sharing in vampire bats. *Scientific American* 262 (2):76–82. Scientific American Inc. Fig. 19.11 Sherman, P. W. 1985. Alarm callsof Belding's ground squirrels to aerial predators: nepotism or self-preservation? *Behavioral Ecology and Sociobiology* 17. With kind permission from Springer Science+Business Media. Springer-Verlag. Fig. 19.14 Adapted from Packer, C., L. Herbst, A. E. Pusey, J. D. Bygott, J. P. Hanby, S. J. Cairns, and M. B. Mulder. 1988. Reproductive success in lions. In *Reproductive Success: Studies of Individual Variation in Contrasting Breeding Systems*, edited by T. H. Clutton-Brock. Chicago: University of Chicago Press, pp. 363–383. Fig. 19.19 Woolfenden, G. E. 1975. *Auk* 92. The American Ornithologists Union. Fig. 19.23 Lacey, E. A., and P. W. Sherman. 1997. Cooperative breeding in naked mole-rats: implications for vertebrate and invertebrate sociality. In *Cooperative Breeding in Mammals*, edited by N. G. Solomon and J. A. French. Cambridge: Cambridge University Press, pp. 267–301. Fig. 19.26 Hölldobler-Forsyth, Turid. From Hölldobler, B. and E. O. Wilson. 1990. *The Ants*. Cambridge: Belknap Press.

Index